© Blue Lantern Studio/CORBIS

In Dodgson's era, mathematicians at times used the original Arabic term for algebra—"al-jabr w'al muquabala," or "restoration and opposition." This interesting bit of history is discussed in Chapter 13—the *Concepts and History of Calculus*.

Alice unsuccessfully tries to remember her multiplication tables, saying "Let me see: four times five is twelve, and four times six is thirteen, and four times seven is—oh dear!" Mathematicians know that $4 \times 5 = 12$, if we use base 18 rather than the familiar base 10, and that $4 \times 6 = 13$, in base 21. This is discussed in Chapter 7—Number Systems and Number Theory.

MATHEMATICS: A PRACTICAL ODYSSEY

is a brief survey of many different branches of mathematics. Paralleling *Alice's* journeys to Wonderland, the authors hope to take students on an odyssey throughout the amazing world of mathematics where they may encounter strange, wonderful, practical, and sometimes whimsical topics. The first chapter of the book is Logic, symbolized by the image of *Alice*. In fact, every image on this book's cover is a reference to one of the book's topics. We hope you have fun finding some of them.

A Mad Tea-party

© Michael Nicholson/CORBIS

MATHEMATICS

A PRACTICAL ODYSSEY

SEVENTH EDITION

DAVID B. JOHNSON

DIABLO VALLEY COLLEGE
PLEASANT HILL, CALIFORNIA

THOMAS A. MOWRY

DIABLO VALLEY COLLEGE
PLEASANT HILL, CALIFORNIA

BROOKS/COLE
CENGAGE Learning

Australia • Brazil • Japan • Korea • Mexico • Singapore • Spain • United Kingdom • United States

BROOKS/COLE
CENGAGE Learning™

Mathematics: A Practical Odyssey, 7e
David B. Johnson / Thomas A. Mowry

Publisher: Charles Van Wagner

Acquisitions Editor: Marc Bove

Developmental Editor: Stefanie Beeck

Assistant Editor: Shaun Williams

Media Editor: Guanglei Zhang

Marketing Manager: Ashley Pickering

Marketing Assistant: Angela Kim

Marketing Communications Manager: Mary Anne Payumo

Senior Content Project Manager: Carol Samet

Creative Director: Rob Hugel

Art Director: Vernon Boes

Print Buyer: Linda Hunt

Rights Acquisitions Specialist: Don Schlotman

Production Service: Anne Seitz/Hearthside Publishing Services

Text Designer: Terri Wright

Photo Researcher: Sarah Bonner, Bill Smith Group

Copy Editor: Barbara Willette/Hearthside Publishing Services

Illustrator: Jade Myers/Hearthside Publishing Services

Cover Designer: Terri Wright

Cover Image: All Royalty-free Photodisc/Getty Images unless specified below: Book illustrations from *Alice's Adventures in Wonderland* by Lewis Carroll: "Oh! The Duchess! by John Tenniel. © Lebrecht Music & Arts/Corbis; Tweedledee and Tweedledum, © The Granger Collection. Elections at Rockefeller Square, © AP Images/Kathy Willens Colored shapes, © Bard Sadowski/iStockphoto.comPassion flower © Jennifer Weinberg/Alamy Abstract, Royalty-free Dex Image Jackpot balls, © Wasabi/Alamy Gambling table, © Bill Bachman/Photoedit Seattle space needle © Brian Yarvin, The Image Works Hobbits from Middle Earth, Lord of the Rings, © Warner Brothers/Photofest Abacus, © Stock Connection/Superstock Background image: © Maciej Frowlow, Brand X Pictures/Getty Images

Compositor: MPS Limited, a Macmillan Company

For product information and technology assistance, contact us at
Cengage Learning Customer & Sales Support, 1-800-354-9706.

For permission to use material from this text or product, submit all requests online at **www.cengage.com/permissions.** Further permissions questions can be e-mailed to **permissionrequest@cengage.com.**

Library of Congress Control Number: 2010938098

Student Edition:

ISBN-13: 978-0-538-49505-9

ISBN-10: 0-538-49505-7

Brooks/Cole
20 Davis Drive
Belmont, CA 94002-3098
USA

Cengage Learning is a leading provider of customized learning solutions with office locations around the globe, including Singapore, the United Kingdom, Australia, Mexico, Brazil, and Japan. Locate your local office at **www.cengage.com/global.**

Cengage Learning products are represented in Canada by Nelson Education, Ltd.

To learn more about **Brooks/Cole,** visit **www.cengage.com Brooks/Cole.**

Purchase any of our products at your local college store or at our preferred online store **www.CengageBrain.com.**

Printed in the United States of America
3 4 5 6 7 14 13 12 11

CONTENTS

1 Logic 1

2 Sets and Counting 67

3 Probability 131

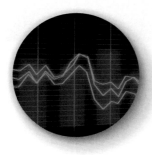

4 Statistics   223

5 Finance   329

6 Voting and Apportionment   407

7 Number Systems and Number Theory 473

8 Geometry 527

9 Graph Theory 663

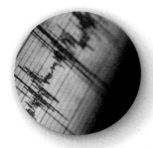

10 Exponential and Logarithmic Functions   731

11 Matrices and Markov Chains   807

12 Linear Programming   859

13 The Concepts and History of Calculus   13-1

Appendices

---

*Sections in color can be found at www.cengage.com/math/johnson

PREFACE

TO THE INSTRUCTOR ...

Course Prerequisite

Mathematics: A Practical Odyssey is written for the student who has successfully completed a course in intermediate algebra, not the student who excelled in it. It would be difficult for a student without background in intermediate algebra to succeed in a course using this book. However, some chapters are not algebra-based. These chapters require a level of critical thinking and mathematical maturity more commonly found in students who have taken intermediate algebra.

Topics

Algebra is the language of mathematics. A background in algebra allows you to learn mathematics that is usable and relevant to any educated person. In particular, it allows the student to do the following:

- Learn enough about *Logic* to analyze the validity of an argument.
- Learn enough about *Sets, Counting and Probability* to understand the risks of inherited diseases and to realize what an incredibly bad bet a lottery ticket is.
- Learn enough about *Statistics* to understand the accuracy and validity of a public opinion poll.
- Learn enough about *Finance* to calculate the monthly payment required by a car of home loan and to understand loans well enough to make an educated decision when selecting one.
- Learn enough about *Voting and Appointment* to understand that there is no perfect voting system or method of appointment; all methods have inherent flaws.
- Learn enough about *Number Systems and Number Theory* to understand why our commonly used base ten and base two number systems work, and to appreciate the prevalence of the Fibonacci numbers in nature and the use of the golden ratio in art.
- Learn enough about *Geometry* to understand its place in the history of Western civilization.
- Learn enough about *Graph Theory* to be able to use it in scheduling and to create networks.
- Learn enough about *Exponential and Logarithmic Functions* to understand how populations grow, how radiocarbon dating works, how the Richter scale measures earthquakes, and how sound is measured in decibels.
- Learn enough about *Matrices and Markov Chains* to understand how manufacturers can predict their products' success or failure in the marketplace.
- Learn enough about *Linear Programming* to understand how a small business can determine how to utilize its limited resources in order to maximize its profit.
- Learn enough about *Calculus* to understand just what the subject is and why it is so important.

New in the Seventh Edition

These special exercises are designed to help prepare the student for admissions examinations such as the GRE (required for graduate school) or the GMAT (required for graduate study in business).

---

q: I have an alibi.

express the following in words.

   a. $p \wedge q$ **b.** $p \rightarrow q$

   c. $\sim q \rightarrow \sim p$ **d.** $q \vee \sim p$

37. Using the symbolic representations

   p: I am an environmentalist.

   q: I recycle my aluminum cans.

   r: I recycle my newspapers.

   express the following in words.

   a. $(q \vee r) \rightarrow p$ **b.** $\sim p \rightarrow \sim (q \vee r)$

   c. $(q \wedge r) \vee \sim p$ **d.** $(r \wedge \sim q) \rightarrow \sim p$

38. Using the symbolic representations

   p: I am innocent.

   q: I have an alibi.

   r: I go to jail.

   express the following in words.

   a. $(p \vee q) \rightarrow \sim r$ **b.** $(p \wedge \sim q) \rightarrow r$

   c. $(\sim p \wedge q) \vee r$ **d.** $(p \wedge r) \rightarrow \sim q$

39. Which statement, #1 or #2, is more appropriate? Explain why.

   Statement #1: "Cold weather is necessary for it to snow."

   Statement #2: "Cold weather is sufficient for it to snow."

• **HISTORY QUESTIONS**

51. In what academic field did Gottfried Leibniz receive his degrees? Why is the study of logic important in this field?

52. Who developed a formal system of logic based on syllogistic arguments?

53. What is meant by *characteristica universalis?* Who proposed this theory?

🖱 **THE NEXT LEVEL**

If a person wants to pursue an advanced degree (something beyond a bachelor's or four-year degree), chances are the person must take a standardized exam to gain admission to a graduate school or to be admitted into a specific program. These exams are intended to measure verbal, quantitative, and analytical skills that have developed throughout a person's life. Many classes and study guides are available to help people prepare for the exams. The following questions are typical of those found in the study guides.

Exercises 54–58 refer to the following: A culinary institute has a small restaurant in which the students prepare various dishes.

FEATURED IN THE NEWS

CHURCH CARVING MAY BE ORIGINAL 'CHESHIRE CAT'

London—Devotees of writer Lewis Carroll believe they have found what inspired his grinning Cheshire Cat, made famous in his book "Alice's Adventures in Wonderland."

Members of the Lewis Carroll Society made the discovery over the weekend in a church at which the author's father was once rector in the Yorkshire village of Croft in northern England.

It is a rough-hewn carving of a cat's head smiling near an altar, probably dating to the 10th century. Seen from below and from the perspective of a small boy, all that can be seen is the grinning mouth.

Carroll's Alice watched the Cheshire Cat disappear "ending with the grin, which remained for some time after the rest of the head had gone."

Alice mused: "I have often seen a cat without a grin, but not a grin without a cat. It is the most curious thing I have seen in all my life."

Reprinted with permission from Reuters.

◀ **FEATURED IN THE NEWS**
Newspaper and magazine articles illustrate how the book's topics come up in the real world, in a way that might affect your students personally.

- **CHAPTER 1**, "Logic," now includes "necessary" and "sufficient" conditions as applies to conditional statements. Over 60 new exercises relating to necessary and sufficient conditions have been added.

- **CHAPTER 4**, "Statistics," now includes material on finding the minimum sample size needed to be approximately confident that the margin of error in a survey is at most a specified amount.

- **CHAPTER 5**, "Finance," now includes detailed descriptions of how to use a TI graphing calculator's "TVM" (time value of money) feature in a variety of financial situations.

- **CHAPTER 8**, "Geometry," now includes a new section on linear perspective. This application of geometry to art makes paintings seem more realistic by giving them a sense of depth.

- **THROUGHOUT THE BOOK:**
  - Chapter openers have been rewritten in a more engaging, student-oriented style.
  - Real world data has been updated.
  - Over 500 new exercises have been added.

Features

▶ **ALGEBRA REVIEW**

Many books review algebra. Usually, the algebra reviews are overly general and don't really help your students review the specific algebraic topics that arise in the book. Our algebra reviews occur at the very beginning of the chapters, and they only review the algebra that comes up in that chapter. there are many topics in algebra where students need to review and sharpen their skills.

10.0A Review of Exponentials and Logarithms

OBJECTIVES

- Define and graph an exponential function; define the natural exponential function
- Use a calculator to find values of the exponential function 10^x and the natural exponential function e^x
- Define and understand the meaning of a logarithm
- Rewrite a logarithm as an exponential equation and vice versa
- Use a calculator to find values of the common and natural logarithmic functions $\log x$ and $\ln x$

TOPIC X BLOOD TYPES: SET THEORY IN THE REAL WORLD

Human blood types are a classic example of set theory. As you may know, there are four categories (or sets) of blood types: A, B, AB, and O. Knowing someone's blood type is extremely important in case a blood transfusion is required; if blood of two different types is combined, the blood cells may begin to clump together, with potentially fatal consequences! (Do you know your blood type?)

What exactly are "blood types"? In the early 1900s, the Austrian scientist Karl Landsteiner observed the presence (or absence) of two distinct chemical molecules on the surface of all red blood cells in numerous samples of human blood. Consequently, he labeled one molecule "A" and the other "B." The presence or absence of these specific molecules is the basis of the universal classification of blood types. Specifically, blood samples containing only the A molecule are labeled type A, whereas those containing only the B molecule are labeled type B. If a blood sample contains both molecules (A and B) it is labeled type AB; and if neither is present, the blood is typed as O. The presence (or absence) of these molecules can be depicted in a standard Venn diagram as shown in Figure 2.27. In the notation of set operations, type A blood is denoted $A \cap B'$, type B is $B \cap A'$, type AB is $A \cap B$, and type O is $A' \cap B'$.

If a specific blood sample is mixed with blood containing a blood molecule (A or B) that it does not already have, the presence of the foreign molecule may cause the mixture of blood to clump. For example, type B blood cannot be mixed with any blood containing the B molecule (type B or type AB). Therefore, a person with type A blood can receive a transfusion only of type A or type O blood. Consequently, a person with type AB blood may receive a transfusion of any blood type; type AB is referred to as the "universal receiver." Because type O blood contains neither the A nor the B molecule, all blood types are compatible with type O blood; type O is referred to as the "universal donor."

It is not uncommon for scientists to study rhesus monkeys in an effort to learn more about human physiology. In so doing, a certain blood protein was discovered in rhesus monkeys. Subsequently, scientists found that the blood of some people contained this protein, whereas the blood of others did not. The presence, or absence, of this protein in human blood is referred to as the *Rh factor*; blood containing the protein is labeled "Rh+", whereas "Rh−" indicates the absence of the protein. The Rh factor of human blood is especially important for expectant mothers; a fetus can develop problems if its parents have opposite Rh factors.

When a person's blood is typed, the designation includes both the regular blood type and the Rh factor. For instance, type AB− indicates the presence of both the A and B molecules (type AB), along with the absence of the rhesus protein; type O+ indicates the absence of both the A and B molecules (type O), along with the presence of the rhesus protein. Utilizing the Rh factor, there are eight possible blood types as shown in Figure 2.28.

We will investigate the occurrence and compatibility of the various blood types in Example 5 and in Exercises 35–43.

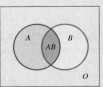

FIGURE 2.27 Blood types and the presence of the A and B molecules.

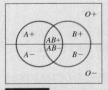

FIGURE 2.28 Blood types combined with the Rh factor.

◀ **REAL-WORLD APPLICATIONS**

"TOPIC X IN THE REAL WORLD"

Discussing current, powerful, real-world uses of the text's topics provides an opportunity to reinforce the practical emphasis of the text. To create the extended applications of Topic X in the Real World, we researched articles in professional journals, magazines, newspapers and the web, and distilled the information so that it is accessible to the liberal arts mathematics student. These highly practical applications are used as jumping-off points for both reality-based conceptual exercises. Below is a partial listing:

- Blood Type: *Set Theory in the Real World* (Section 2.2, page 88)
- The Business of Gambling: *Probabilities in the Real World* (Section 3.4, page 175)
- HIV/AIDS: *Probabilities in the Real World* (Section 3.6, page 198)
- NASA: *PERT Charts in the Real World* (Section 9.5, page 720)

For a complete listing and to access additional course materials and companion resources, please visit **www.cengagebrain.com**. At the CengageBrain.com home page, search for the ISBN of your title (from the back cover of this book) using the search box at the top of the page. This will take you to the product page where free companion resources can be found.

PROJECTS

Below is a partial listing:

CHAPTER 3: PROBABILITY

- Relative frequency of an odd phone number. (Section 3.2, Exercise 88, page 156)
- Compare relative frequency and theoretical probability with a coin toss. (Section 3.2, Exercise 89, page 156)
- Compare relative frequency and theoretical probability with a roll of a die. (Section 3.2, Exercise 90, page 157)
- Determine if the outcomes are equally likely with coin experiments. (Section 3.2, Exercises 91 and 92, page 157)
- Design a game of chance. Use probabilities and expected values to set house odds. (Section 3.5, Exercise 50, page 190)
- Investigate seemingly contradictory information regarding sex bias in graduate school and/or tuberculosis incidence in New York City and Richmond. (Section 3.6, Exercises 73 and 74, page 205)

 CHAPTER 13: CALCULUS

- Write a research paper on the contributions of Apollonius, Oresme, Descartes and Fermat to analytic geometry. (Section 13.1, Exercise 36, page 13-22)
- Compare and contrast the algebra of Apollonius, al-Khowarizmi, and Decartes. (Section 13.1, Exercise 37, page 13-22)
- Use dimensional analysis to convert the distance and speed falling object formulas from the English system to the metric system. (Section 13.7, Exercise 1, page 13-81)
- Use calculus to sketch the graph of a given polynomial function. (Section 13.7, Exercise 2, page 13-81)
- Use calculus to find the areas of some complicated shapes. (Section 13.7, Exercises 3–6, pages 13-81 and 13-82)

For a complete listing, please visit **www.cengagebrain.com.**

Organization

CHAPTER OPENERS

The chapter openers briefly discuss "What We Will Do in this Chapter," so that your students can get an idea of what to expect.

EXAMPLES

All math texts have examples. Many skip just enough steps to be frustrating. Ours don't skip steps, and ours include a verbal summary to aid in your students understanding.

BRIDGES BETWEEN CHAPTERS

As much as possible, this book has been written so that its chapters are independent of each other. The instructor therefore has wide latitude in selecting topics to cover and can teach a course that is responsive to the needs of his or her students and institution. Sometimes,

---

 Chapter is available online.

however, this independence is not desirable, because it does not allow connections to be made between seemingly unrelated topics. For this reason, the text features a number of bridges between chapters. These bridges feature both discussion and exercises. They are clearly labeled; that is, a bridge between an earlier Chapter X and a later Chapter Y is labeled "for those students who have completed Chapter X."

▶ **BUILDING BRIDGES**

There is a bridge between Chapter 1 (Logic) and Chapter 2 (Sets and Counting). At the end of Section 2.1, the similarities between the respective concepts and notation of logic and sets are discussed.

---

78   CHAPTER 2   Sets and Counting

Set Theory and Logic

If you have read Chapter 1, you have probably noticed that set theory and logic have many similarities. For instance, the union symbol $\cup$ and the disjunction symbol $\vee$ have the same meaning, but they are used in different circumstances; $\cup$ goes between sets, while $\vee$ goes between logical expressions. The $\cup$ and $\vee$ symbols are similar in appearance because their usages are similar. A comparison of the terms and symbols used in set theory and logic is given in Figure 2.13.

Set Theory		Logic		Common Wording
Term	**Symbol**	**Term**	**Symbol**	
union	$\cup$	disjunction	$\vee$	or
intersection	$\cap$	conjunction	$\wedge$	and
complement	$'$	negation	$\sim$	not
subset	$\subseteq$	conditional	$\rightarrow$	if . . . then . . .

FIGURE 2.13   Comparison of terms and symbols used in set theory and logic.

Applying the concepts and symbols of Chapter 1, we can define the basic operations of set theory in terms of logical biconditionals. The biconditionals in Figure 2.14 are tautologies (expressions that are always true); the first biconditional is read as "x is an element of the union of sets A and B if and only if x is an element of set A or x is an element of set B."

Basic Operations in Set Theory	Logical Biconditional
union	$[x \in (A \cup B)] \leftrightarrow [x \in A \vee x \in B]$
intersection	$[x \in (A \cap B)] \leftrightarrow [x \in A \wedge x \in B]$
complement	$(x \in A') \leftrightarrow \sim (x \in A)$
subset	$(A \subseteq B) \leftrightarrow (x \in A \rightarrow x \in B)$

FIGURE 2.14   Set theory operations as logical biconditionals.

- There is a bridge between Chapter 5 (Finance) and Chapter 10 (Exponential and Logarithmic Functions). That bridge discussed the relationship between the exponential growth model ($y = ae^{bt}$) and the common interest model [$FV = P(I + i)^n$], and the circumstances under which model can be used in place of the other.

- There is a bridge between Chapter 8 (Geometry) and Chapter 13 (Calculus). That bridge discussed the use of the trigonometric functions to determine the equation of a trajectory when the initial angle of elevation other that 30°, 45°, or 60°.

TOPIC SELECTION AND PREREQUISITE MAPS

Obviously, the book contains more material that you could ever cover in a one-quarter course. It is written in such a way that most chapters are independent of each other. The Instructor's Manual includes a "prerequisite map" so that you can easily tell which earlier topics must be covered. It also includes sample course outlines, suggesting specific chapters to be used in forming a typical course.

FLEXIBILITY AND COURSE LEVEL

The book contains a wide range of topics, varying in level of sophistication and difficulty. Chapters 1 through 9 cover topics that are not uncommon to beginning algebra-abased liberal arts mathematics texts. However, because of the intermediate algebra prerequisite, the topics are covered more thoroughly, and the acquisition of problem-solving skills is emphasized. Chapters 10 through 13 are more sophisticated, covering topics not commonly found in liberal arts texts. However, the treatment is such that the material is accessible to the students who enroll in this course.

A chapter need not be covered in its entirety; topics can easily be left out. In most cases, a chapter has a suggested core of key sections as well as a selection of optional sections, which tend to be more sophisticated. They are not labeled "optional" in the text; rather, this distinction is made in the prerequisite map in the Instructor's Resource Manual.

The text is designed so that the instructor determines the difficulty of the course by selecting the chapters and sections to be covered. The instructor can thus create his or her own course—one that will fit the students' needs.

USABILITY

This book is user-friendly:

- The examples don't skip steps.
- Key points are boxed for emphasis.
- Step-by-step procedures are given.
- There is an abundance of exposition.

ANSWER CHECKING

Throughout the text, we emphasize the importance of checking ones answers. It's important that students learn to evaluate the reasonableness of their answers, rather than accepting them on face value. To this end, some exercises do not have answers in the back of the book when students are instructed to check answers for themselves.

▶ HISTORY

The history of the subject matter is interwoven throughout most chapters. In addition, Historical Notes give in-depth biographies of the prominent people involved. It is our hope that students will see the human side of mathematics. After all, mathematics was invented by real people for real purposes and is a part of our culture. Interesting research topics are given, and writing assignments are suggested.

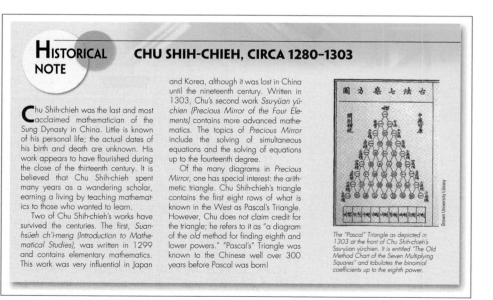

HISTORICAL NOTE   CHU SHIH-CHIEH, CIRCA 1280–1303

Chu Shih-chieh was the last and most acclaimed mathematician of the Sung Dynasty in China. Little is known of his personal life; the actual dates of his birth and death are unknown. His work appears to have flourished during the close of the thirteenth century. It is believed that Chu Shih-chieh spent many years as a wandering scholar, earning a living by teaching mathematics to those who wanted to learn.

Two of Chu Shih-chieh's works have survived the centuries. The first, *Suan-hsüeh ch'i-meng* (Introduction to Mathematical Studies), was written in 1299 and contains elementary mathematics. This work was very influential in Japan and Korea, although it was lost in China until the nineteenth century. Written in 1303, Chu's second work *Ssu-yüan yü-chien* (Precious Mirror of the Four Elements) contains more advanced mathematics. The topics of *Precious Mirror* include the solving of simultaneous equations and the solving of equations up to the fourteenth degree.

Of the many diagrams in *Precious Mirror*, one has special interest: the arithmetic triangle. Chu Shih-chieh's triangle contains the first eight rows of what is known in the West as Pascal's Triangle. However, Chu does not claim credit for the triangle; he refers to it as "a diagram of the old method for finding eighth and lower powers." "Pascal's" Triangle was known to the Chinese well over 300 years before Pascal was born!

The "Pascal" Triangle as depicted in 1303 at the front of Chu Shih-chieh's Ssu-yüan yü-chien. It is entitled "The Old Method Chart of the Seven Multiplying Squares" and tabulates the binomial coefficients up to the eighth power.

Brown University Library

Technology

WEB PROJECTS

These projects include links to web pages that the students can use as starting points in their research. Following is a partial listing:

CHAPTER 2: SETS AND COUNTING

- Determine the compatibility of the various blood types, including Rh factors. (Section 2.2, Exercise 53, Page 94)
- Write a research paper on an historic topic. (Section 2.4, Exercise 62, page 117; and section 2.5, Exercise 28, page 127)

CHAPTER 3: PROBABILITY

- Write a research paper on an historic topic. (Section 3.1, Exercise 46, page 139)
- Write a research paper on the successful screening of Jews for Tay-Sachs disease and the unsuccessful screening of blacks for sickle-cell anemia. (Section 3.2, Exercise 93, page 157)
- Determine the prevalence of the various blood types in the United States, the percentage of U.S. residents that can donate blood to people of each of each of the various blood types, and the percentage of U.S. residents that can receive blood from people of each of the various blood types. (Section 3.6, Exercise 71, page 204)
- Investigate automobile rollovers. Determine which type of vehicle—a sedan, and SUV or a van—is more prone to rollovers. (Section 3.6, Exercise 72, page 205)

CHAPTER 9: GRAPH THEORY

- Investigate some of the unsolved problems in graph theory. (Section 9.2, Exercise 33, page 683)
- Write an essay on the four-color map problem. (Section 9.2, Exercise 34, page 683)
- Investigate the traveling salesman problem on the web. Summarize their history, and describe specific problems that led to progress. (Section 9.3, Exercise 32, page 696)

> **WEB PROJECTS**
>
> 33. There are many interesting problems in graph theory. Some of these problems have been solved, and some remain unsolved. Many of these problems are discussed on the web. Visit several web sites and choose a specific problem. Describe the problem, its history, and its applications. If it has been solved, describe the method of solution if possible.

CHAPTER 13: CALCULUS

- Write a research paper on an historical topic, or on Fermat's last theorem. (Section 13.1, Exercise 38, page 13-22; and Section 13.2, Exercises 42 and 43, page 13-35)

For a complete listing, please visit **www.cengagebrain.com**.

GRAPHING AND SCIENTIFIC CALCULATORS

Calculator boxes give you all of the necessary keystrokes for both scientific calculators and graphing calculators. Calculator subsections help you learn how to use your calculator when a list of keystrokes is just not enough. The following graphing calculator topics are addressed:

- Section 3.3: Fractions on a Graphing Calculator (page 168)
- Section 5.2: Doubling Time with a TI's TVM Application (page 354)
- Section 5.5 Finding the APR with a TI's TVM Application (page 388)

- Section 11.0: Matrix Multiplication on a Graphing Calculator (page 819)
- Section 11.4: Solving Larger Systems of Linear Equations on a Graphing Calculator (page 847)

For a complete listing, please visit **www.cengagebrain.com**.

EXCEL AND COMPUTERS

Computers are ubiquitous at the workplace, and are becoming increasingly common in the classroom. However, many students have no mathematical experience using computer software such as Microsoft Excel. For this reason, we have included a number of subsections that give instruction on the use of Excel. We have also included a number of optional subsections that give instructions on the use of *Amortix*, custom text-specific software that is available on the text website (go to **http://www.brookscole.com/math_d/resources/ amortrix/**) The following topics are addressed:

- Section 4.1: Histograms and Pie Charts on a Computerized Spreadsheet (page 241)
- Section 4.3: Measures of Central Tendency and Dispersion on Excel (page 274)
- Section 5.4: Amortization Schedules on Amortix (page 383)
- Section 5.4: Amortization Schedules on Excel (page 384)
- Section 12.3: The Row Operations on Amortix (page 12-25)

These subsections allow instructors to incorporate the computer into their class if they so desire, but they are entirely optional, and the book is in no way computer dependent. The subsections do not assume any previous experience with Excel.

WEB SITE

To access additional course materials and companion resources, please visit **www .cengagebrain.com**. At the CengageBrain.com home page, search for the ISBN of your title (from the back cover of your book) using the search box at the top of the page. This will take you to the product page where free companion resources can be found.

The site offers book-specific student and instructor resources, as well as discipline-specific links. Student resources at the web site include a web-based version of Amortix (the software used in Chapters 5, 7, and 8), downloadable partially completed Excel spreadsheets that are keyed to examples in the text, graphing and calculation tools, and Internet links for further research.

EXERCISES

The exercises in this text are designed to solidify the students' understanding of the material and make them proficient in the calculations involved. It is assumed that most students who complete this course will not continue in their formal study of mathematics. Consequently, neither the exposition nor the exercises are designed to expose the students to all aspects of the topic.

The exercises vary in difficulty. Some are exactly like the examples, and others demand more of the students. The exercises are not explicitly graded into A, B and C categories, nor are any marked "optional;" students in this audience tend to react negatively if asked to do anything labeled in this manner. The more difficult the exercises are indicated in the Instructor's Resource Manual.

The short-answer historical questions are meant to focus and reinforce the students' understanding of the historical material. They also serve to warn them that history questions may appear on exams. The essay questions can be used as an integral part of the students' grades, as background for classroom discussion, or for extra credit work. Most are research topics and are kept as open-ended as possible.

Answers to the odd-numbered exercises are given in the back of the book, with two exceptions:

- Answers to historical questions and essay questions are not given
- Answers are not given when the exercises instruct the student to check the answers themselves.

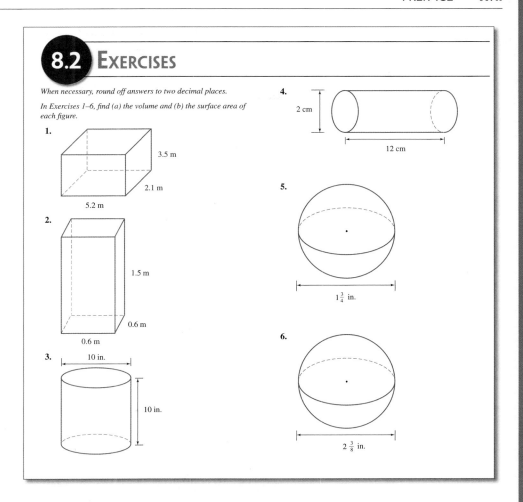

8.2 EXERCISES

When necessary, round off answers to two decimal places.

In Exercises 1–6, find (a) the volume and (b) the surface area of each figure.

1.   3.5 m   2.1 m   5.2 m

2.   1.5 m   0.6 m   0.6 m

3.   10 in.   10 in.

4.   2 cm   12 cm

5. $1\frac{3}{4}$ in.

6. $2\frac{3}{8}$ in.

Complete solutions to every other odd-numbered exercise are given in the Student Solutions Manual, with the above two exceptions.

Standards

This text is well-suited in addressing the AMATYC and NCTM standards.

Regarding *Standards for Intellectual Development*, the text contains a wide range of exercises designed to engage students in mathematical problem solving and modeling real-world situations. In addition, the exercises provide students an opportunity to expand their mathematical reasoning skills and to communicate their results effectively. Also, the text emphasizes the interrelationships between mathematics, human culture, and other disciplines. The use of appropriate technology is woven throughout the text, along with the material that encourages independent exploration and confidence building in mathematics.

Regarding *Standards for Content*, the text contains material covering all requisite topics including number sense, symbolism and algebra, geometry, function, combinatorics, probability and statistics, and deductive proof.

Regarding *Standards for Pedagogy*, the text provides ample opportunity for faculty to use technology in the teaching of mathematics, to foster interactive learning through collaborative activities and effective communication, to make connections between various branches of mathematics and between mathematics and the students' lives, and to use numerical, graphical, symbolic, and verbal approaches in the teaching of mathematics.

TO THE STUDENT, AS YOU EMBARK ON YOUR ODYSSEY ...

This textbook is designed for students in liberal arts programs and other fields that do not require a core of mathematics. The term *liberal arts* is a translation of a Latin phrase that means "studies befitting a free person." It was applied during the Middle Ages to seven branches of learning: arithmetic, geometry, logic, grammar, rhetoric, astronomy, and music. You might be surprised to learn that almost half of the original liberal arts are mathematics subjects.

In accordance with the tradition, handed down from the Middle Ages, that a broad-based education includes some mathematics, many institutions of higher education require their students to complete a college-level mathematics course. These schools award a bachelor's degree to a person who not only has acquired a detailed knowledge of his or her field but also has a broad background in the liberal arts.

The goal of this textbook is to expose you to topics in mathematics that are usable and relevant to any educated person. We hope that you will encounter topics that will be useful at some time during your life. In addition, you are encouraged to recognize the relevance of mathematics to a well-rounded education and to appreciate the creative, human aspect of mathematics.

This book is written for the student who has successfully completed a course in intermediate algebra, not the student who excelled in it. Your mathematical background doesn't have to be perfect, but algebra will come up all the time. It's also true that this book is written for a college-level math course, and that's a significant step up from your high school experience. You will have to work hard and put in a solid effort.

Your success in this course is important to us. To help you achieve that success, we have incorporated features in the textbook that promote learning and support various learning styles. Among these features are algebra review and instructions in using a calculator.

Our algebra reviews occur at the very beginning of the chapters, and they review only the algebra that comes up in that chapter. There are many topics in algebra in which students need to review and sharpen their skills. Calculator boxes give you all of the necessary keystrokes for scientific calculators and for graphing calculators. Calculator subsections help you learn how to use your calculator when a list of keystrokes is just not enough. We encourage you to examine these features and use them on your successful odyssey throughout this course.

SUPPLEMENTS

FOR THE STUDENT	FOR THE INSTRUCTOR
	Annotated Instructor's Edition (ISBN: 0840049137) The Annotated Instructor's Edition features an appendix containing the answers to all problems in the book as well as icons denoting which problems can be found in Enhanced WebAssign. (Print)
Student Solutions Manual for (ISBN: 0840053878) The Student Solutions Manual provides worked-out solutions to the odd-numbered problems in the text. Use of the solutions manual ensures that students learn the correct steps to arrive at an answer. (Print)	**Instructor's Resource Manual for** (ISBN: 0840053452) The Instructor's Resource Manual provides worked-out solutions to all of the problems in the text and includes suggestions for course syllabi and chapter summaries. (Print)
Text Specific DVDs (ISBN: 1111571570) Hosted by Dana, these professionally produced DVDs cover key topics of the text, offering a valuable alternative for classroom instruction or independent study and review. (Media)	**Text Specific DVDs** (ISBN: 1111571570) Hosted by Dana, these professionally produced DVDs cover key topics of the text, offering a valuable alternative for classroom instruction or independent study and review. (Media)
Enhanced WebAssign (ISBN: 0538738103) Enhanced WebAssign, used by over one million students at more than 1100 institutions, allows you to do homework assignments and get extra help and practice via the web. This proven and reliable homework system includes hundreds of algorithmically generated homework problems, and eBook, links to relevant textbook sections, video examples, problem specific tutorials, and more. (Online)	**Enhanced WebAssign** (ISBN: 0538738103) Enhanced WebAssign, used by over one million students at more than 1100 institutions, allows you to assign, collect, grade, and record homework assignments via the web. This proven and reliable homework system includes hundreds of algorithmically generated homework problems, and eBook, links to relevant textbook sections, video examples, and more. (Online) Note that the WebAssign problems for this text are highlighted by a ▶.
	PowerLecture with ExamView (ISBN: 0840054114) This CD-ROM provides the instructor with dynamic media tools for teaching. Create, deliver, and customize tests (both print and online) in minutes with *ExamView® Computerized Testing Featuring Algorithmic Equations*. Easily build solution sets for homework or exams using *Solution Builder's* online solutions manual. Microsoft® PowerPoint® lecture slides and figures from the book are also included on this CD-ROM. (CD)
	Solution Builder This online solutions manual allows instructors to create customizable solutions that they can print out to distribute or post as needed. This is a convenient and expedient way to deliver solutions to specific homework sets. Visit **www.cengage.com/ solutionbuilder**. (Online)

Acknowledgements

The authors would like to thank Marc Bove, Kyle O'Loughlin, Meaghan Banks, Stefanie Beeck, and all the fine people at Cengage Learning: Anne Seitz and Gretchen Miller at Hearthside Publishing Services; Ann Ostberg, and Rhoda Oden.

Special thanks go to the users of the text and reviewers who evaluated the manuscript for this edition, as well as those who offered comments on previous editions.

Reviewers

Dennis Airey, *Rancho Santiago College*
Francisco E. Alarcon, *Indiana University of Pennsylvania*
Judith Arms, *University of Washington*
Bruce Atkinson, *Palm Beach Atlantic College*
Wayne C. Bell, *Murray State University*
Wayne Bishop, *California State University—Los Angeles*
David Boliver, *Trenton State College*
Stephen Brick, *University of South Alabama*
Barry Bronson, *Western Kentucky University*
Frank Burk, *California State University—Chico*
Laura Cameron, *University of New Mexico*
Jack Carter, *California State University—Hayward*
Timothy D. Cavanaugh, *University of Northern Colorado*
Joseph Chavez, *California State University—San Bernadino*
Eric Clarkson, *Murray State University*
Rebecca Conti, *State University of New York at Fredonia*
S.G. Crossley, *University of Southern Alabama*
Ben Divers, Jr., *Ferrum College*
Al Dixon, *College of the Ozarks*
Joe S. Evans, *Middle Tennessee State University*
Hajrudin Fejzie, *California State University—San Bernardino*
Lloyd Gavin, *California State University—Sacramento*
William Greiner, *McLennan Community College*
Martin Haines, *Olympic College*
Ray Hamlett, *East Central University*
Virginia Hanks, *Western Kentucky University*
Anne Herbst, *Santa Rosa Junior College*
Linda Hinzman, *Pasadena City College*

Thomas Hull, *University of Rhode Island*
Robert W. Hunt, *Humboldt State University*
Robert Jajcay, *Indiana State University*
Irja Kalantari, *Western Illinois University*
Daniel Katz, *University of Kansas*
Katalin Kolossa, *Arizona State University*
Donnald H. Lander, *Brevard College*
Lee LaRue, *Paris Junior College*
Thomas McCready, *California State University—Chico*
Vicki McMillian, *Stockton State University*
Narendra L. Maria, *California State University—Stanislaus*
John Martin, *Santa Rosa Junior College*
Gael Mericle, *Mankato State University*
Robert Morgan, *Pima Community College*
Pamela G. Nelson, *Panhandle State University*
Carol Oelkers, *Fullerton College*
Michael Olinick, *Middlebury College*
Matthew Pickard, *University of Puget Sound*
Joan D. Putnam, *University of Northern Colorado*
J. Doug Richey, *Northeast Texas Community College*
Stewart Robinson, *Cleveland State University*
Eugene P. Schlereth, *University of Tennessee at Chattanooga*
Lawrence Somer, *Catholic University of America*
Michael Trapuzzano, *Arizona State University*
Pat Velicky, *Mid-Plains Community College*
Karen M. Walters, *University of Northern California*
Dennis W. Watson, *Clark College*
Denielle Williams, *Eastern Washington University*
Charles Ziegenfus, *James Madison University*

LOGIC

© Bart Sadowski/iStockPhoto

When writer Lewis Carroll took Alice on her journeys down the rabbit hole to Wonderland and through the looking glass, she had many fantastic encounters with the tea-sipping Mad Hatter, a hookah-smoking Caterpillar, the White Rabbit, the Cheshire Cat, the Red and White Queens, and Tweedledum and Tweedledee. On the surface, Carroll's writings seem to be delightful nonsense and mere children's entertainment. Many people are quite surprised to learn that *Alice's Adventures in Wonderland* is as much an exercise in logic as it is a fantasy and that Lewis Carroll was actually Charles Dodgson, an Oxford mathematician. Dodgson's many writings include the whimsical *The Game of Logic* and the brilliant *Symbolic Logic*, in addition to *Alice's Adventures in Wonderland* and *Through the Looking Glass.*

continued

WHAT WE WILL DO IN THIS CHAPTER

WE'LL EXPLORE DIFFERENT TYPES OF LOGIC OR REASONING:

- Deductive reasoning involves the application of a general statement to a specific case; this type of logic is typified in the classic arguments of the renowned Greek logician Aristotle.

- Inductive reasoning involves generalizing after a pattern has been recognized and established; this type of logic is used in the solving of puzzles.

WE'LL ANALYZE AND EXPLORE VARIOUS TYPES OF STATEMENTS AND THE CONDITIONS UNDER WHICH THEY ARE TRUE:

- A statement is a simple sentence that is either true or false. Simple statements can be connected to form compound, or more complicated, statements.

- Symbolic representations reduce a compound statement to its basic form; phrases that appear to be different may actually have the same basic structure and meaning.

continued

WE'LL ANALYZE AND EXPLORE CONDITIONAL, OR "IF . . . THEN . . .," STATEMENTS:

- In everyday conversation, we often connect phrases by saying "if *this*, then *that*." However, does "*this*" actually guarantee "*that*"? Is "*this*" in fact necessary for "*that*"?

- How does "if" compare with "only if"? What does "if and only if" really mean?

WE'LL DETERMINE THE VALIDITY OF AN ARGUMENT:

- What constitutes a valid argument? Can a valid argument yield a false conclusion?

- You may have used Venn diagrams to depict a solution set in an algebra class. We will use Venn diagrams to visualize and analyze an argument.

- Some of Lewis Carroll's whimsical arguments are valid, and some are not. How can you tell?

Webster's New World College Dictionary defines **logic** as "the science of correct reasoning; science which describes relationships among propositions in terms of implication, contradiction, contrariety, conversion, etc." In addition to being flaunted in Mr. Spock's claim that "your human emotions have drawn you to an illogical conclusion" and in Sherlock Holmes's immortal phrase "elementary, my dear Watson," logic is fundamental both to critical thinking and to problem solving. In today's world of misleading commercial claims, innuendo, and political rhetoric, the ability to distinguish between valid and invalid arguments is important.

In this chapter, we will study the basic components of logic and its application. Mischievous, wild-eyed residents of Wonderland, eccentric, violin-playing detectives, and cold, emotionless Vulcans are not the only ones who can benefit from logic. Armed with the fundamentals of logic, we can surely join Spock and "live long and prosper!"

Logic is the science of correct reasoning. Auguste Rodin captured this ideal in his bronze sculpture *The Thinker.*

Vanni/Art Resource, NY

In their quest for logical perfection, the Vulcans of *Star Trek* abandoned all emotion. Mr. Spock's frequent proclamation that "emotions are illogical" typified this attitude.

PARAMOUNT TELEVISION/THE KOBAL COLLECTION

1.1   Deductive versus Inductive Reasoning

OBJECTIVES

- Use Venn diagrams to determine the validity of deductive arguments
- Use inductive reasoning to predict patterns

Logic is the science of correct reasoning. *Webster*'s *New World College Dictionary* defines **reasoning** as "the drawing of inferences or conclusions from known or assumed facts." Reasoning is an integral part of our daily lives; we take appropriate actions based on our perceptions and experiences. For instance, if the sky is heavily overcast this morning, you might assume that it will rain today and take your umbrella when you leave the house.

Problem Solving

Logic and reasoning are associated with the phrases *problem solving* and *critical thinking.* If we are faced with a problem, puzzle, or dilemma, we attempt to reason through it in hopes of arriving at a solution.

The first step in solving any problem is to define the problem in a thorough and accurate manner. Although this might sound like an obvious step, it is often overlooked. Always ask yourself, "What am I being asked to do?" Before you can

Using his extraordinary powers of logical deduction, Sherlock Holmes solves another mystery. "Finding the villain was elementary, my dear Watson."

solve a problem, you must understand the question. Once the problem has been defined, all known information that is relevant to it must be gathered, organized, and analyzed. This analysis should include a comparison of the present problem to previous ones. How is it similar? How is it different? Does a previous method of solution apply? If it seems appropriate, draw a picture of the problem; visual representations often provide insight into the interpretation of clues.

Before using any specific formula or method of solution, determine whether its use is valid for the situation at hand. A common error is to use a formula or method of solution when it does not apply. If a past formula or method of solution is appropriate, use it; if not, explore standard options and develop creative alternatives. Do not be afraid to try something different or out of the ordinary. "What if I try this . . . ?" may lead to a unique solution.

Deductive Reasoning

Once a problem has been defined and analyzed, it might fall into a known category of problems, so a common method of solution may be applied. For instance, when one is asked to solve the equation $x^2 = 2x + 1$, realizing that it is a second-degree equation (that is, a quadratic equation) leads one to put it into the standard form ($x^2 - 2x - 1 = 0$) and apply the Quadratic Formula.

EXAMPLE **1**

USING DEDUCTIVE REASONING TO SOLVE AN EQUATION   Solve the equation $x^2 = 2x + 1$.

SOLUTION

The given equation is a second-degree equation in one variable. We know that all second-degree equations in one variable (in the form $ax^2 + bx + c = 0$) can be solved by applying the Quadratic Formula:

$$x = \frac{-b \pm \sqrt{b^2 - 4ac}}{2a}$$

Therefore, $x^2 = 2x + 1$ can be solved by applying the Quadratic Formula:

$$x^2 = 2x + 1$$

$$x^2 - 2x - 1 = 0$$

$$x = \frac{-(-2) \pm \sqrt{(-2)^2 - (4)1(-1)}}{2(1)}$$

$$x = \frac{2 \pm \sqrt{4 + 4}}{2}$$

$$x = \frac{2 \pm \sqrt{8}}{2}$$

$$x = \frac{2 \pm 2\sqrt{2}}{2}$$

$$x = \frac{2(1 \pm \sqrt{2})}{2}$$

$$x = 1 \pm \sqrt{2}$$

The solutions are $x = 1 + \sqrt{2}$ and $x = 1 - \sqrt{2}$.

In Example 1, we applied a general rule to a specific case; we reasoned that it was valid to apply the (general) Quadratic Formula to the (specific) equation $x^2 = 2x + 1$. This type of logic is known as **deductive reasoning**—that is, the application of a general statement to a specific instance.

Deductive reasoning and the formal structure of logic have been studied for thousands of years. One of the earliest logicians, and one of the most renowned, was Aristotle (384–322 B.C.). He was the student of the great philosopher Plato and the tutor of Alexander the Great, the conqueror of all the land from Greece to India. Aristotle's philosophy is pervasive; it influenced Roman Catholic theology through St. Thomas Aquinas and continues to influence modern philosophy. For centuries, Aristotelian logic was part of the education of lawyers and politicians and was used to distinguish valid arguments from invalid ones.

For Aristotle, logic was the necessary tool for any inquiry, and the syllogism was the sequence followed by all logical thought. A **syllogism** is an argument composed of two statements, or **premises** (the major and minor premises), followed by a **conclusion.** For any given set of premises, if the conclusion of an argument is guaranteed (that is, if it is inescapable in all instances), the argument is **valid.** If the conclusion is not guaranteed (that is, if there is at least one instance in which it does not follow), the argument is **invalid.**

Perhaps the best known of Aristotle's syllogisms is the following:

1. All men are mortal.   **major premise**
2. Socrates is a man.   **minor premise**
Therefore, Socrates is mortal.   **conclusion**

When the major premise is applied to the minor premise, the conclusion is inescapable; the argument is valid.

Notice that the deductive reasoning used in the analysis of Example 1 has exactly the same structure as Aristotle's syllogism concerning Socrates:

1. All second-degree equations in one variable can be   **major premise**
   solved by applying the Quadratic Formula.
2. $x^2 = 2x + 1$ is a second-degree equation in one variable.   **minor premise**
Therefore, $x^2 = 2x + 1$ can be solved by applying the   **conclusion**
Quadratic Formula.

Each of these syllogisms is of the following general form:

1. If *A*, then *B*.   **All *A* are *B*. (major premise)**

2. *x* is *A*.   **We have *A*. (minor premise)**

Therefore, *x* is *B*.   **Therefore, we have *B*. (conclusion)**

Historically, this valid pattern of deductive reasoning is known as *modus ponens*.

Deductive Reasoning and Venn Diagrams

The validity of a deductive argument can be shown by use of a Venn diagram. A **Venn diagram** is a diagram consisting of various overlapping figures contained within a rectangle (called the "universe"). To depict a statement of the form "All *A* are *B*" (or, equivalently, "If *A*, then *B*"), we draw two circles, one inside the other; the inner circle represents *A,* and the outer circle represents *B*. This relationship is shown in Figure 1.1.

Venn diagrams depicting "No *A* are *B*" and "Some *A* are *B*" are shown in Figures 1.2 and 1.3, respectively.

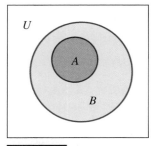

FIGURE 1.1

All *A* are *B*. (If *A*, then *B*.)

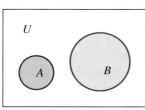

FIGURE 1.2 No *A* are *B*.

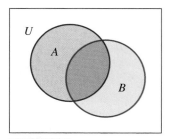

FIGURE 1.3 Some *A* are *B*. (At least one *A* is *B*.)

EXAMPLE **2**

ANALYZING A DEDUCTIVE ARGUMENT Construct a Venn diagram to verify the validity of the following argument:

1. All men are mortal.

2. Socrates is a man.

Therefore, Socrates is mortal.

SOLUTION

Premise 1 is of the form "All *A* are *B*" and can be represented by a diagram like that shown in Figure 1.4.

Premise 2 refers to a specific man, namely, Socrates. If we let *x* = Socrates, the statement "Socrates is a man" can then be represented by placing *x* within the circle labeled "men," as shown in Figure 1.5. Because we placed *x* within the "men" circle, and all of the "men" circle is inside the "mortal" circle, the conclusion "Socrates is mortal" is inescapable; the argument is valid.

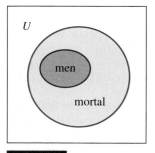

FIGURE 1.4 All men are mortal.

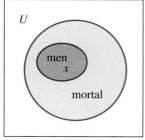

x = Socrates

FIGURE 1.5 Socrates is mortal.

EXAMPLE **3**

ANALYZING A DEDUCTIVE ARGUMENT   Construct a Venn diagram to determine the validity of the following argument:

1. All doctors are men.
2. My mother is a doctor.

Therefore, my mother is a man.

SOLUTION

Premise 1 is of the form "All *A* are *B*"; the argument is depicted in Figure 1.6.

No matter where *x* is placed within the "doctors" circle, the conclusion "My mother is a man" is inescapable; the argument is valid.

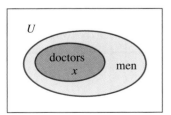

x = My mother

FIGURE 1.6

My mother is a man.

Saying that an argument is valid does not mean that the conclusion is true. The argument given in Example 3 *is* valid, but the conclusion is *false.* One's mother cannot be a man! Validity and truth do not mean the same thing. An argument is valid if the conclusion is inescapable, *given the premises.* Nothing is said about the truth of the premises. Thus, when examining the validity of an argument, we are not determining whether the conclusion is true or false. Saying that an argument is valid merely means that, *given the premises,* the reasoning used to obtain the conclusion is logical. However, if the premises of a valid argument are true, then the conclusion will also be true.

EXAMPLE **4**

ANALYZING A DEDUCTIVE ARGUMENT   Construct a Venn diagram to determine the validity of the following argument:

   1. All professional wrestlers are actors.
   2. The Rock is an actor.

Therefore, The Rock is a professional wrestler.

SOLUTION

Premise 1 is of the form "All *A* are *B*"; the "circle of professional wrestlers" is contained within the "circle of actors." If we let *x* represent The Rock, premise 2 simply requires that we place *x* somewhere within the actor circle; *x* could be placed in either of the two locations shown in Figures 1.7 and 1.8.

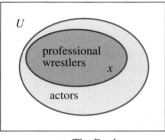

x = The Rock

FIGURE 1.7

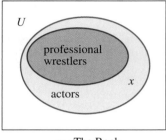

x = The Rock

FIGURE 1.8

If *x* is placed as in Figure 1.7, the argument would appear to be valid; the figure supports the conclusion "The Rock is a professional wrestler." However, the placement of *x* in Figure 1.8 does not support the conclusion; given the premises, we cannot *logically* deduce that "The Rock is a professional wrestler." Since the conclusion is *not* inescapable, the argument is invalid.

Saying that an argument is invalid does not mean that the conclusion is false. Example 4 demonstrates that an invalid argument can have a true conclusion; even though The Rock is a professional wrestler, the argument used to obtain the conclusion is invalid. In logic, validity and truth do not have the same meaning. *Validity* refers to the process of reasoning used to obtain a conclusion; *truth* refers to conformity with fact or experience.

Even though The Rock *is* a professional wrestler, the argument used to obtain the conclusion is invalid.

VENN DIAGRAMS AND INVALID ARGUMENTS

To show that an argument is invalid, you must construct a Venn diagram in which the premises are met yet the conclusion does not necessarily follow.

EXAMPLE **5**

ANALYZING A DEDUCTIVE ARGUMENT   Construct a Venn diagram to determine the validity of the following argument:

   1. Some plants are poisonous.
   2. Broccoli is a plant.

Therefore, broccoli is poisonous.

SOLUTION

Premise 1 is of the form "Some *A* are *B*"; it can be represented by two overlapping circles (as in Figure 1.3). If we let *x* represent broccoli, premise 2 requires that we place *x* somewhere within the plant circle. If *x* is placed as in Figure 1.9, the argument would appear to be valid. However, if *x* is placed as in Figure 1.10, the conclusion does not follow. Because we can construct a Venn diagram in which the premises are met yet the conclusion does not follow (Figure 1.10), the argument is invalid.

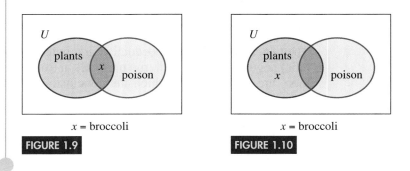

x = broccoli

FIGURE 1.9

x = broccoli

FIGURE 1.10

When analyzing an argument via a Venn diagram, you might have to draw three or more circles, as in the next example.

EXAMPLE **6**

ANALYZING A DEDUCTIVE ARGUMENT   Construct a Venn diagram to determine the validity of the following argument:

1. No snake is warm-blooded.
2. All mammals are warm-blooded.

Therefore, snakes are not mammals.

SOLUTION

Premise 1 is of the form "No *A* are *B*"; it is depicted in Figure 1.11. Premise 2 is of the form "All *A* are *B*"; the "mammal circle" must be drawn within the "warm-blooded circle." Both premises are depicted in Figure 1.12.

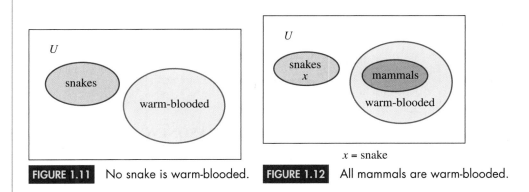

x = snake

FIGURE 1.11   No snake is warm-blooded.   **FIGURE 1.12**   All mammals are warm-blooded.

Because we placed *x* (= snake) within the "snake" circle, and the "snake" circle is outside the "warm-blooded" circle, *x* cannot be within the "mammal" circle (which is inside the "warm-blooded" circle). Given the premises, the conclusion "Snakes are not mammals" is inescapable; the argument is valid.

You might have encountered Venn diagrams when you studied sets in your algebra class. The academic fields of set theory and logic are historically intertwined; set theory was developed in the late nineteenth century as an aid in the study of logical arguments. Today, set theory and Venn diagrams are applied to areas other than the study of logical arguments; we will utilize Venn diagrams in our general study of set theory in Chapter 2.

Inductive Reasoning

The conclusion of a valid deductive argument (one that goes from general to specific) is guaranteed: Given true premises, a true conclusion must follow. However, there are arguments in which the conclusion is not guaranteed even though the premises are true. Consider the following:

1. Joe sneezed after petting Frako's cat.
2. Joe sneezed after petting Paulette's cat.

Therefore, Joe is allergic to cats.

Is the conclusion guaranteed? If the premises are true, they certainly *support* the conclusion, but we cannot say with 100% certainty that Joe is allergic to cats. The conclusion is *not* guaranteed. Maybe Joe is allergic to the flea powder that the cat owners used; maybe he is allergic to the dust that is trapped in the cats' fur; or maybe he has a cold!

Reasoning of this type is called inductive reasoning. **Inductive reasoning** involves going from a series of specific cases to a general statement (see Figure 1.13). Although it may seem to follow and may in fact be true, *the conclusion in an inductive argument is never guaranteed.*

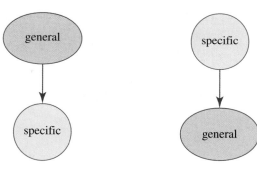

Deductive Reasoning
(Conclusion is guaranteed.)

Inductive Reasoning
(Conclusion may be probable but is not guaranteed.)

FIGURE 1.13

EXAMPLE **7**

INDUCTIVE REASONING AND PATTERN RECOGNITION   What is the next number in the sequence 1, 8, 15, 22, 29, . . . ?

SOLUTION

Noticing that the difference between consecutive numbers in the sequence is 7, we may be tempted to say that the next term is 29 + 7 = 36. Is this conclusion guaranteed? No! Another sequence in which numbers differ by 7 are dates of a given day of the week. For instance, the dates of the Saturdays in the year 2011 are (January) 1, 8, 15, 22, 29, (February) 5, 12, 19, 26, ... Therefore, the next number in the sequence 1, 8, 15, 22, 29, . . . might be 5. Without further information, we cannot determine the next number in the given sequence. We can only use inductive reasoning and give one or more *possible* answers.

TOPIC X   SUDOKU: *LOGIC IN THE REAL WORLD*

Throughout history, people have always been attracted to puzzles, mazes, and brainteasers. Who can deny the inherent satisfaction of solving a seemingly unsolvable or perplexing riddle? A popular new addition to the world of puzzle solving is **sudoku,** a numbers puzzle. Loosely translated from Japanese, *sudoku* means "single number"; a sudoku puzzle simply involves placing the digits 1 through 9 in a grid containing 9 rows and 9 columns. In addition, the 9 by 9 grid of squares is subdivided into nine 3 by 3 grids, or "boxes," as shown in Figure 1.14.

The rules of sudoku are quite simple: Each row, each column, and each box must contain the digits 1 through 9; and no row, column, or box can contain 2 squares with the same number. Consequently, sudoku does not require any arithmetic or mathematical skill; sudoku requires logic only. In solving a puzzle, a common thought is "What happens if I put this number here?"

Like crossword puzzles, sudoku puzzles are printed daily in many newspapers across the country and around the world. Web sites containing sudoku puzzles and strategies provide an endless source of new puzzles and help. See Exercise 62 to find links to popular sites.

FIGURE 1.14 A blank sudoku grid.

EXAMPLE **8** | **SOLVING A SUDOKU PUZZLE**   Solve the sudoku puzzle given in Figure 1.15.

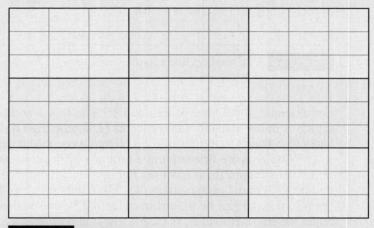

		2		6	4	8	5	
3	8						6	
		8	5				3	
		6		5	9			
6						1		5
7		8			1			
	6			7	9			
	4		1				9	8
	9	3	4	2		7		6

FIGURE 1.15 A sudoku puzzle.

SOLUTION

Recall that each 3 by 3 grid is referred to as a box. For convenience, the boxes are numbered 1 through 9, starting in the upper left-hand corner and moving from left to right, and each square can be assigned coordinates (x, y) based on its row number x and column number y as shown in Figure 1.16.

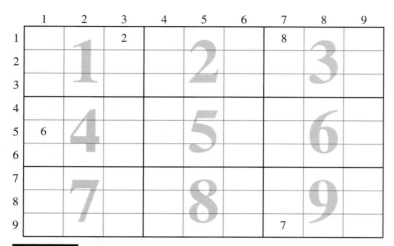

FIGURE 1.16 Box numbers and coordinate system in sudoku.

For example, the digit 2 in Figure 1.16 is in box 1 and has coordinates $(1, 3)$, the digit 8 is in box 3 and has coordinates $(1, 7)$, the digit 6 is in box 4 and has coordinates $(5, 1)$ and the digit 7 is in box 9 and has coordinates $(9, 7)$.

When you are first solving a sudoku puzzle, concentrate on only a few boxes rather than the puzzle as a whole. For instance, looking at boxes 1, 4, and 7, we see that boxes 4 and 7 each contain the digit 6, whereas box 1 does not. Consequently, the 6 in box 1 must be placed in column 3 because (shaded) columns 1 and 2 already have a 6. However, (shaded) row 2 already has a 6, so we can deduce that 6 must be placed in row 3, column 3, that is, in square $(3, 3)$ as shown in Figure 1.17.

		2		6	4	8	5	
3	8						**6**	
		6	8	5			3	
			6		5	9		
6						1		5
7		8			1			
	6			7	9			
	4		1				9	8
	9	3	4	2		7		6

FIGURE 1.17 The 6 in box 1 must be placed in square $(3, 3)$.

Examining boxes 1, 2, and 3, we see that boxes 2 and 3 each contain the digit 5, whereas box 1 does not. We deduce that 5 must be placed in square $(2, 3)$ because rows 1 and 3 already have a 5. In a similar fashion, square $(1, 4)$ must contain 3. See Figure 1.18.

		2	**3**	6	4	8	**5**	
3	8	**5**					6	
		6	8	**5**			**3**	
			6		5	9		
6						1		5
7		8			1			
	6			7	9			
	4		1				9	8
	9	3	4	2		7		6

FIGURE 1.18   Analyzing boxes 1, 2, and 3; placing the digits 3 and 5.

Because we have placed two new digits in box 1, we might wish to focus on the remainder of (shaded) box 1. Notice that the digit 4 can be placed only in square (3, 1), as row 1 and column 2 already have a 4 in each of them; likewise, the digit 9 can be placed only in square (1, 1) because column 2 already has a 9. Finally, either of the digits 1 or 7 can be placed in square (1, 2) or (3, 2) as shown in Figure 1.19. At some point later in the solution, we will be able to determine the exact values of squares (1, 2) and (3, 2), that is, which square receives a 1 and which receives a 7.

9	1,7	2	3	6	**4**	8	5	
3	8	5					6	
4	1,7	6	8	5			3	
			6		5	9		
6						1		5
7		8			1			
	6			7	9			
	4		1				9	8
	9	3	4	2		7		6

FIGURE 1.19   Focusing on box 1.

Using this strategy of analyzing the contents of three consecutive boxes, we deduce the following placement of digits: 1 must go in (2, 5), 6 must go in (6, 7), 6 must go in (8, 6), 7 must go in (8, 3), 3 must go in (8, 5), 8 must go in (9, 6), and 5 must go in (7, 4). At this point, box 8 is complete as shown in Figure 1.20. (Remember, each box must contain each of the digits 1 through 9.)

Once again, we use the three consecutive box strategy and deduce the following placement of digits: 5 must go in (8, 7), 5 must go in (9, 1), 8 must go in (7, 1), 2 must go in (8, 1), and 1 must go in (7, 3). At this point, box 7 is complete as shown in Figure 1.21.

9	1,7	2	3	6	4	8	5	
3	8	5		1			6	
4	1,7	6	8	5			3	
			6		5	9		
6						1		5
7		8			1	6		
	6		5	7	9			
	4	7	1	3	6		9	8
	9	3	4	2	8	7		6

FIGURE 1.20 Box 8 is complete.

9	1,7	2	3	6	4	8	5	
3	8	5		1			6	
4	1,7	6	8	5			3	
			6		5	9		
6						1		5
7		8			1	6		
8	6	1	5	7	9			
2	4	7	1	3	6	5	9	8
5	9	3	4	2	8	7		6

FIGURE 1.21 Box 7 is complete.

We now focus on box 4 and deduce the following placement of digits: 1 must go in (4, 1), 9 must go in (5, 3), 4 must go in (4, 3), 5 must go in (6, 2), 3 must go in (4, 2), and 2 must go in (5, 2). At this point, box 4 is complete. In addition, we deduce that 1 must go in (9, 8), and 3 must go in (5, 6) as shown in Figure 1.22.

9	1,7	2	3	6	4	8	5	
3	8	5		1			6	
4	1,7	6	8	5			3	
1	3	4	6		5	9		
6	2	9			3	1		5
7	5	8			1	6		
8	6	1	5	7	9			
2	4	7	1	3	6	5	9	8
5	9	3	4	2	8	7	1	6

FIGURE 1.22 Box 4 is complete.

Once again, we use the three consecutive box strategy and deduce the following placement of digits: 3 must go in (6, 9), 3 must go in (7, 7), 9 must go in (2, 4), and 9 must go in (6, 5). Now, to finish row 6, we place 4 in (6, 8) and 2 in (6, 4) as shown in Figure 1.23. (Remember, each row must contain each of the digits 1 through 9.)

9	1,7	2	3	6	4	8	5	
3	8	5	9	1			6	
4	1,7	6	8	5			3	
1	3	4	6		5	9		
6	2	9			3	1		5
7	5	8	2	9	1	6	4	3
8	6	1	5	7	9	3		
2	4	7	1	3	6	5	9	8
5	9	3	4	2	8	7	1	6

FIGURE 1.23   Row 6 is complete.

After we place 7 in (5, 4), column 4 is complete. (Remember, each column must contain each of the digits 1 through 9.) This leads to placing 4 in (5, 5) and 8 in (4, 5), thus completing box 5; row 5 is finalized by placing 8 in (5, 8) as shown in Figure 1.24.

9	1,7	2	3	6	4	8	5	
3	8	5	9	1			6	
4	1,7	6	8	5			3	
1	3	4	6	8	5	9		
6	2	9	7	4	3	1	8	5
7	5	8	2	9	1	6	4	3
8	6	1	5	7	9	3		
2	4	7	1	3	6	5	9	8
5	9	3	4	2	8	7	1	6

FIGURE 1.24   Column 4, box 5, and row 5 are complete.

Now column 7 is completed by placing 4 in (2, 7) and 2 in (3, 7); placing 2 in (2, 6) and 7 in (3, 6) completes column 6 as shown in Figure 1.25.

At this point, we deduce that the digit in (3, 2) must be 1 because row 3 cannot have two 7's. This in turn reveals that 7 must go in (1, 2), and box 1 is now complete. To complete row 7, we place 4 in (7, 9) and 2 in (7, 8); row 4 is finished with 2 in (4, 9) and 7 in (4, 8). See Figure 1.26.

9	1,7	2	3	6	4	8	5	
3	8	5	9	1	**2**	**4**	6	
4	1,7	6	8	5	7	2	3	
1	3	4	6	8	5	9		
6	2	9	7	4	3	1	8	5
7	5	8	2	9	1	6	4	3
8	6	1	5	7	9	3		
2	4	7	1	3	6	5	9	8
5	9	3	4	2	8	7	1	6

FIGURE 1.25   Columns 7 and 6 are complete.

9	7	2	3	6	4	8	5	
3	8	5	9	1	2	4	6	
4	**1**	6	8	5	7	2	3	
1	3	4	6	8	5	9	7	2
6	2	9	7	4	3	1	8	5
7	5	8	2	9	1	6	4	3
8	6	1	5	7	9	3	2	**4**
2	4	7	1	3	6	5	9	8
5	9	3	4	2	8	7	1	6

FIGURE 1.26   Box 1, row 7, and row 4 are complete.

To finish rows 1, 2, and 3, 1 must go in (1, 9), 7 must go in (2, 9), and 9 must go in (3, 9). The puzzle is now complete as shown in Figure 1.27.

9	7	2	3	6	4	8	5	**1**
3	8	5	9	1	2	4	6	**7**
4	1	6	8	5	7	2	3	9
1	3	4	6	8	5	9	7	2
6	2	9	7	4	3	1	8	5
7	5	8	2	9	1	6	4	3
8	6	1	5	7	9	3	2	4
2	4	7	1	3	6	5	9	8
5	9	3	4	2	8	7	1	6

FIGURE 1.27   A completed sudoku puzzle.

As a final check, we scrutinize each box, row, and column to verify that no box, row, or column contains the same digit twice. Congratulations, the puzzle has been solved!

In Exercises 1–20, construct a Venn diagram to determine the validity of the given argument.

1. **a.** 1. All master photographers are artists.
        2. Ansel Adams is a master photographer.

    Therefore, Ansel Adams is an artist.

    b. 1. All master photographers are artists.
        2. Ansel Adams is an artist.

    Therefore, Ansel Adams is a master photographer.

2. **a.** 1. All Olympic gold medal winners are role models.
        2. Michael Phelps is an Olympic gold medal winner.

    Therefore, Michael Phelps is a role model.

    b. 1. All Olympic gold medal winners are role models.
        2. Michael Phelps is a role model.

    Therefore, Michael Phelps is an Olympic gold medal winner.

3. **a.** 1. All homeless people are unemployed.
        2. Bill Gates is not a homeless person.

    Therefore, Bill Gates is not unemployed.

    b. 1. All homeless people are unemployed.
        2. Bill Gates is not unemployed.

    Therefore, Bill Gates is not a homeless person.

4. **a.** 1. All professional wrestlers are actors.
        2. Ralph Nader is not an actor.

    Therefore, Ralph Nader is not a professional wrestler.

    b. 1. All professional wrestlers are actors.
        2. Ralph Nader is not a professional wrestler.

    Therefore, Ralph Nader is not an actor.

5. 1. All pesticides are harmful to the environment.
    2. No fertilizer is a pesticide.

    Therefore, no fertilizer is harmful to the environment.

6. 1. No one who can afford health insurance is unemployed.
    2. All politicians can afford health insurance.

    Therefore, no politician is unemployed.

7. 1. No vegetarian owns a gun.
    2. All policemen own guns.

    Therefore, no policeman is a vegetarian.

8. 1. No professor is a millionaire.
    2. No millionaire is illiterate.

    Therefore, no professor is illiterate.

9. 1. All poets are loners.
    2. All loners are taxi drivers.

    Therefore, all poets are taxi drivers.

10. 1. All forest rangers are environmentalists.
    2. All forest rangers are storytellers.

    Therefore, all environmentalists are storytellers.

11. 1. Real men don't eat quiche.
    2. Clint Eastwood is a real man.

    Therefore, Clint Eastwood doesn't eat quiche.

12. 1. Real men don't eat quiche.
    2. Oscar Meyer eats quiche.

    Therefore, Oscar Meyer isn't a real man.

13. 1. All roads lead to Rome.
    2. Route 66 is a road.

    Therefore, Route 66 leads to Rome.

14. 1. All smiling cats talk.
    2. The Cheshire Cat smiles.

    Therefore, the Cheshire Cat talks.

15. 1. Some animals are dangerous.
    2. A tiger is an animal.

    Therefore, a tiger is dangerous.

16. 1. Some professors wear glasses.
    2. Mr. Einstein wears glasses.

    Therefore, Mr. Einstein is a professor.

17. 1. Some women are police officers.
    2. Some police officers ride motorcycles.

    Therefore, some women ride motorcycles.

18. 1. All poets are eloquent.
    2. Some poets are wine connoisseurs.

    Therefore, some wine connoisseurs are eloquent.

19. 1. All squares are rectangles.
    2. Some quadrilaterals are squares.

    Therefore, some quadrilaterals are rectangles.

20. 1. All squares are rectangles.
    2. Some quadrilaterals are rectangles.

    Therefore, some quadrilaterals are squares.

21. Classify each argument as deductive or inductive.

    a. 1. My television set did not work two nights ago.
        2. My television set did not work last night.

    Therefore, my television set is broken.

    b. 1. All electronic devices give their owners grief.
        2. My television set is an electronic device.

    Therefore, my television set gives me grief.

22. Classify each argument as deductive or inductive.

    a. 1. I ate a chili dog at Joe's and got indigestion.

       2. I ate a chili dog at Ruby's and got indigestion.

       Therefore, chili dogs give me indigestion.

    b. 1. All spicy foods give me indigestion.

       2. Chili dogs are spicy food.

       Therefore, chili dogs give me indigestion.

In Exercises 23–32, fill in the blank with what is most likely to be the next number. Explain (using complete sentences) the pattern generated by your answer.

23. 3, 8, 13, 18, _____

24. 10, 11, 13, 16, _____

25. 0, 2, 6, 12, _____

26. 1, 2, 5, 10, _____

27. 1, 4, 9, 16, _____

28. 1, 8, 27, 64, _____

29. 2, 3, 5, 7, 11, _____

30. 1, 1, 2, 3, 5, _____

31. 5, 8, 11, 2, _____

32. 12, 5, 10, 3, _____

In Exercises 33–36, fill in the blanks with what are most likely to be the next letters. Explain (using complete sentences) the pattern generated by your answers.

33. O, T, T, F, _____, _____

34. T, F, S, E, _____, _____

35. F, S, S, M, _____, _____

36. J, F, M, A, _____, _____

In Exercises 37–42, explain the general rule or pattern used to assign the given letter to the given word. Fill in the blank with the letter that fits the pattern.

37.

circle	square	trapezoid	octagon	rectangle
c	s	t	o	_____

38.

circle	square	trapezoid	octagon	rectangle
i	u	a	o	_____

39.

circle	square	trapezoid	octagon	rectangle
j	v	b	p	_____

40.

circle	square	trapezoid	octagon	rectangle
c	r	p	g	_____

41.

banana	strawberry	asparagus	eggplant	orange
b	z	t	u	_____

42.

banana	strawberry	asparagus	eggplant	orange
y	r	g	p	_____

43. Find two different numbers that could be used to fill in the blank.

    1, 4, 7, 10, _____

Explain the pattern generated by each of your answers.

44. Find five different numbers that could be used to fill in the blank.

    7, 14, 21, 28, _____

Explain the pattern generated by each of your answers.

45. Example 1 utilized the Quadratic Formula. Verify that

$$x = \frac{-b + \sqrt{b^2 - 4ac}}{2a}$$

is a solution of the equation $ax^2 + bx + c = 0$.

HINT: Substitute the fraction for x in $ax^2 + bx + c$ and simplify.

46. Example 1 utilized the Quadratic Formula. Verify that

$$x = \frac{-b - \sqrt{b^2 - 4ac}}{2a}$$

is a solution of the equation $ax^2 + bx + c = 0$.

HINT: Substitute the fraction for x in $ax^2 + bx + c$ and simplify.

47. As a review of algebra, use the Quadratic Formula to solve

$$x^2 - 6x + 7 = 0$$

48. As a review of algebra, use the Quadratic Formula to solve

$$x^2 - 2x - 4 = 0$$

Solve the sudoku puzzles in Exercises 49–54.

49.

	5							9
		7					3	4
						2	1	5	6
6			1	8	9			
	3							2
					3		6		5
5	9	2	1		4			
	6	3				8		
4							1	

50.

4		7					3	
	2		5		1		7	
				4	5			
	7		9					2
2		9		4		6		3
3				6		9		
		2	4					
	8		6		5		4	
	6					2		9

51.

	5		3			2	8	
		3			7			9
	7		9		2			
	8	7						
5	9			6			1	3
						4	6	
			8		9		2	
2			5			6		
	3	9			6		7	

52.

		2				1		
6	8			2			5	4
		3	8		6	2		
7			6		4			5
2			3		9			7
		6	7		5	4		
3	5			1			8	6
		7				5		

53.

7	6		5					
		2	7					8
		4	6				5	
	7	1			3		2	
5								9
	9		8			7	6	
	8				1	3		
3					6	1		
					7		9	6

54.

7	1	2					8	6
				7				
		4					1	
4		3	5	8				
1			2		4			8
			9	1	2			5
	8					4		
			3					
6	3					8	2	9

Answer the following questions using complete sentences and your own words.

• CONCEPT QUESTIONS

55. Explain the difference between deductive reasoning and inductive reasoning.

56. Explain the difference between truth and validity.

57. What is a syllogism? Give an example of a syllogism that relates to your life.

• HISTORY QUESTIONS

58. From the days of the ancient Greeks, the study of logic has been mandatory in what two professions? Why?

59. Who developed a formal system of deductive logic based on arguments?

60. What was the name of the school Aristotle founded? What does it mean?

61. How did Aristotle's school of thought differ from Plato's?

WEB PROJECT

62. Obtain a sudoku puzzle and its solution from a popular web site. Some useful links for this web project are listed on the text web site:

www.cengage.com/math/johnson

1.2 Symbolic Logic

OBJECTIVES

- Identify simple statements
- Express a compound statement in symbolic form
- Create the negation of a statement
- Express a conditional statement in terms of necessary and sufficient conditions

The syllogism ruled the study of logic for nearly 2,000 years and was not supplanted until the development of symbolic logic in the late seventeenth century. As its name implies, symbolic logic involves the use of symbols and algebraic manipulations in logic.

Statements

All logical reasoning is based on statements. A **statement** is a sentence that is either true or false.

EXAMPLE 1

IDENTIFYING STATEMENTS Which of the following are statements? Why or why not?

a. Apple manufactures computers.
b. Apple manufactures the world's best computers.
c. Did you buy a Dell?
d. A $2,000 computer that is discounted 25% will cost $1,000.
e. I am telling a lie.

SOLUTION

a. The sentence "Apple manufactures computers" is true; therefore, it is a statement.
b. The sentence "Apple manufactures the world's best computers" is an opinion, and as such, it is neither true nor false. It is true for some people and false for others. Therefore, it is not a statement.
c. The sentence "Did you buy a Dell?" is a question. As such, it is neither true nor false; it is not a statement.
d. The sentence "A $2,000 computer that is discounted 25% will cost $1,000" is false; therefore, it is a statement. (A $2,000 computer that is discounted 25% would cost $1,500.)
e. The sentence "I am telling a lie" is a self-contradiction, or paradox. If it were true, the speaker would be telling a lie, but in telling the truth, the speaker would be contradicting the statement that he or she was lying; if it were false, the speaker would not be telling a lie, but in not telling a lie, the speaker would be contradicting the statement that he or she was lying. The sentence is not a statement.

By tradition, symbolic logic uses lowercase letters as labels for statements. The most frequently used letters are *p, q, r, s,* and *t.* We can label the statement "It is snowing" as statement *p* in the following manner:

p: It is snowing.

If it *is* snowing, *p* is labeled true, whereas if it is *not* snowing, *p* is labeled false.

Compound Statements and Logical Connectives

It is easy to determine whether a statement such as "Charles donated blood" is true or false; either he did or he didn't. However, not all statements are so simple; some are more involved. For example, the truth of "Charles donated blood and did not wash his car, or he went to the library," depends on the truth of the individual pieces that make up the larger, compound statement. A **compound statement** is a statement that contains one or more simpler statements. A compound statement can be formed by inserting the word *not* into a simpler statement or by joining two or more statements with connective words such as *and, or, if . . . then . . . , only if,* and *if and only if.* The compound statement "Charles did *not* wash his car" is formed from the simpler statement "Charles did wash his car." The compound statement "Charles donated blood *and* did *not* wash his car, *or* he went to the library" consists of three statements, each of which may be true or false.

Figure 1.28 diagrams two equivalent compound statements.

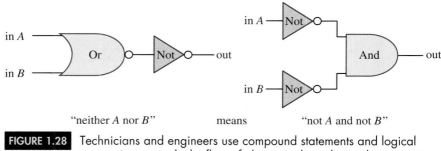

"neither *A* nor *B*"   means   "not *A* and not *B*"

FIGURE 1.28   Technicians and engineers use compound statements and logical connectives to study the flow of electricity through switching circuits.

When is a compound statement true? Before we can answer this question, we must first examine the various ways in which statements can be connected. Depending on how the statements are connected, the resulting compound statement can be a *negation,* a *conjunction,* a *disjunction,* a *conditional,* or any combination thereof.

The Negation ~*p*

The **negation** of a statement is the denial of the statement and is represented by the symbol ~. The negation is frequently formed by inserting the word *not.* For example, given the statement "*p:* It is snowing," the negation would be "~*p:* It is not snowing." If it *is* snowing, *p* is true and ~*p* is false. Similarly, if it is *not* snowing, *p* is false and ~*p* is true. A statement and its negation always have opposite truth values; when one is true, the other is false. Because the truth of the negation depends on the truth of the original statement, a negation is classified as a compound statement.

EXAMPLE **2**

WRITING A NEGATION   Write a sentence that represents the negation of each statement:

a. The senator is a Democrat.
b. The senator is not a Democrat.
c. Some senators are Republicans.
d. All senators are Republicans.
e. No senator is a Republican.

SOLUTION

a. The negation of "The senator is a Democrat" is "The senator is not a Democrat."
b. The negation of "The senator is not a Democrat" is "The senator is a Democrat."
c. A common error would be to say that the negation of "Some senators are Republicans" is "Some senators are not Republicans." However, "Some senators are Republicans" is not denied by "Some senators are not Republicans." The statement "Some senators *are* Republicans" implies that at least one senator is a Republican. The negation of this statement is "It is not the case that at least one senator is a Republican," or (more commonly phrased) the negation is "No senator is a Republican."
d. The negation of "All senators are Republicans" is "It is not the case that all senators are Republicans," or "There exists a senator who is not a Republican," or (more commonly phrased) "Some senators are not Republicans."
e. The negation of "No senator is a Republican" is "It is not the case that no senator is a Republican" or, in other words, "There exists at least one senator who *is* a Republican." If "some" is interpreted as meaning "at least one," the negation can be expressed as "Some senators are Republicans."

The words *some, all,* and *no* (or *none*) are referred to as **quantifiers.** Parts (c) through (e) of Example 2 contain quantifiers. The linked pairs of quantified statements shown in Figure 1.29 are negations of each other.

FIGURE 1.29   Negations of statements containing quantifiers.

The Conjunction *p* ∧ *q*

Consider the statement "Norma Rae is a union member and she is a Democrat." This is a compound statement, because it consists of two statements—"Norma Rae is a union member" and "she (Norma Rae) is a Democrat"—and the connective word *and.* Such a compound statement is referred to as a conjunction. A **conjunction** consists of two or more statements connected by the word *and.* We use the symbol ∧ to represent the word *and;* thus, the conjunction "*p* ∧ *q*" represents the compound statement "*p* and *q.*"

EXAMPLE **3**

TRANSLATING WORDS INTO SYMBOLS   Using the symbolic representations

> *p:* Norma Rae is a union member.
> *q:* Norma Rae is a Democrat.

express the following compound statements in symbolic form:

a. Norma Rae is a union member and she is a Democrat.
b. Norma Rae is a union member and she is not a Democrat.

SOLUTION

a. The compound statement "Norma Rae is a union member and she is a Democrat" can be represented as *p* ∧ *q*.
b. The compound statement "Norma Rae is a union member and she is not a Democrat" can be represented as *p* ∧ ~*q*.

The Disjunction $p \vee q$

When statements are connected by the word *or*, a **disjunction** is formed. We use the symbol $\vee$ to represent the word *or*. Thus, the disjunction "$p \vee q$" represents the compound statement "*p or q*." We can interpret the word *or* in two ways. Consider the statements

> *p:* Kaitlin is a registered Republican.
>
> *q:* Paki is a registered Republican.

The statement "Kaitlin is a registered Republican or Paki is a registered Republican" can be symbolized as $p \vee q$. Notice that it is possible that *both* Kaitlin and Paki are registered Republicans. In this example, *or* includes the possibility that both things may happen. In this case, we are working with the **inclusive *or*.**
   Now consider the statements

> *p:* Kaitlin is a registered Republican.
>
> *q:* Kaitlin is a registered Democrat.

The statement "Kaitlin is a registered Republican or Kaitlin is a registered Democrat" does *not* include the possibility that both may happen; one statement *excludes* the other. When this happens, we are working with the **exclusive *or*.** In our study of symbolic logic (as in most mathematics), we will always use the *inclusive or*. Therefore, "*p or q*" means "*p or q or both*."

EXAMPLE 4

TRANSLATING SYMBOLS INTO WORDS   Using the symbolic representations

> *p:* Juanita is a college graduate.
>
> *q:* Juanita is employed.

express the following compound statements in words:

a. $p \vee q$ **b.** $p \wedge q$
c. $p \vee \sim q$ **d.** $\sim p \wedge q$

SOLUTION

a. $p \vee q$ represents the statement "Juanita is a college graduate or Juanita is employed (or both)."
b. $p \wedge q$ represents the statement "Juanita is a college graduate and Juanita is employed."
c. $p \vee \sim q$ represents the statement "Juanita is a college graduate or Juanita is not employed."
d. $\sim p \wedge q$ represents the statement "Juanita is not a college graduate and Juanita is employed."

The Conditional $p \rightarrow q$

Consider the statement "If it is raining, then the streets are wet." This is a compound statement because it connects two statements, namely, "it is raining" and "the streets are wet." Notice that the statements are connected with "if . . . then . . ." phrasing. Any statement of the form "if p then q" is called a **conditional** (or an **implication**); p is called the **hypothesis** (or **premise**) of the conditional, and q is called the **conclusion** of the conditional. The conditional "if p then q" is represented by the symbols "$p \rightarrow q$" (p implies q). When people use conditionals in

HISTORICAL NOTE

GOTTFRIED WILHELM LEIBNIZ 1646–1716

In addition to cofounding calculus (see Chapter 13), the German-born Gottfried Wilhelm Leibniz contributed much to the development of symbolic logic. A precocious child, Leibniz was self-taught in many areas. He taught himself Latin at the age of eight and began the study of Greek when he was twelve. In the process, he was exposed to the writings of Aristotle and became intrigued by formalized logic.

At the age of fifteen, Leibniz entered the University of Leipzig to study law. He received his bachelor's degree two years later, earned his master's degree the following year, and then transferred to the University of Nuremberg.

Leibniz received his doctorate in law within a year and was immediately offered a professorship but refused it, saying that he had "other things in mind." Besides law, these "other things" included politics, religion, history, literature, metaphysics, philosophy, logic, and mathematics. Thereafter, Leibniz worked under the sponsorship of the courts of various nobles, serving as lawyer, historian, and librarian to the elite. At one point, Leibniz was offered the position of librarian at the Vatican but declined the offer.

Leibniz's affinity for logic was characterized by his search for a *characteristica universalis,* or "universal character." Leibniz believed that by combining logic and mathematics, a general symbolic language could be created in which all scientific problems could be solved with a minimum of effort. In this universal language, statements and the logical relationships between them would be represented by letters and symbols. In Leibniz's words, "All truths of reason would be reduced to a kind of calculus, and the errors would only be errors of computation." In essence, Leibniz believed that once a problem had been translated into this universal language of symbolic logic, it would be solved automatically by simply applying the mathematical rules that governed the manipulation of the symbols.

Leibniz's work in the field of symbolic logic did not arouse much academic curiosity; many say that it was too far ahead of its time. The study of symbolic logic was not systematically investigated again until the nineteenth century.

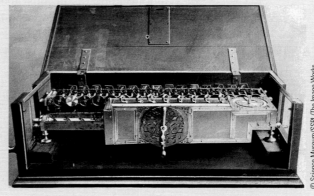

In the early 1670s, Leibniz invented one of the world's first mechanical calculating machines. Leibniz's machine could multiply and divide, whereas an earlier machine invented by Blaise Pascal (see Chapter 3) could only add and subtract.

everyday speech, they often omit the word *then,* as in "If it is raining, the streets are wet." Alternatively, the conditional "if *p* then *q*" may be phrased as "*q* if *p*" ("The streets are wet if it is raining").

EXAMPLE **5**

TRANSLATING WORDS INTO SYMBOLS   Using the symbolic representations

p: I am healthy.

q: I eat junk food.

r: I exercise regularly.

express the following compound statements in symbolic form:

a. I am healthy if I exercise regularly.

b. If I eat junk food and do not exercise, then I am not healthy.

SOLUTION

a. "I am healthy if I exercise regularly" is a conditional (*if . . . then . . .*) and can be rephrased as follows:

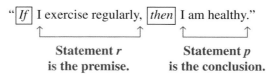

"$\boxed{If}$ I exercise regularly, $\boxed{then}$ I am healthy."

Statement r
is the premise.

Statement p
is the conclusion.

The given compound statement can be expressed as $r \to p$.

b. "If I eat junk food and do not exercise, then I am not healthy" is a conditional (*if . . . then . . .*) that contains a conjunction (*and*) and two negations (*not*):

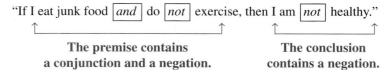

"If I eat junk food $\boxed{and}$ do $\boxed{not}$ exercise, then I am $\boxed{not}$ healthy."

The premise contains
a conjunction and a negation.

The conclusion
contains a negation.

The premise of the conditional can be represented by $q \wedge \sim r$, while the conclusion can be represented by $\sim p$. Thus, the given compound statement has the symbolic form $(q \wedge \sim r) \to \sim p$.

EXAMPLE **6**

TRANSLATING WORDS INTO SYMBOLS   Express the following statements in symbolic form:

a. All mammals are warm-blooded.
b. No snake is warm-blooded.

SOLUTION

a. The statement "All mammals are warm-blooded" can be rephrased as "If it is a mammal, then it is warm-blooded." Therefore, we define two simple statements p and q as

  p: It is a mammal.

  q: It is warm-blooded.

The statement now has the form

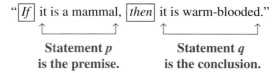

"$\boxed{If}$ it is a mammal, $\boxed{then}$ it is warm-blooded."

Statement p
is the premise.

Statement q
is the conclusion.

and can be expressed as $p \to q$. In general, any statement of the form "All p are q" can be symbolized as $p \to q$.

b. The statement "No snake is warm-blooded" can be rephrased as "If it is a snake, then it is not warm-blooded." Therefore, we define two simple statements p and q as

  p: It is a snake.

  q: It is warm-blooded.

The statement now has the form

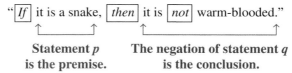

"$\boxed{If}$ it is a snake, $\boxed{then}$ it is $\boxed{not}$ warm-blooded."

Statement p
is the premise.

The negation of statement q
is the conclusion.

and can be expressed as $p \to \sim q$. In general, any statement of the form "No p is q" can be symbolized as $p \to \sim q$.

Necessary and Sufficient Conditions

As Example 6 shows, conditionals are not always expressed in the form "if *p* then *q*." In addition to "all *p* are *q*," other standard forms of a conditional include statements that contain the word *sufficient* or *necessary*.

Consider the statement "Being a mammal is sufficient for being warm-blooded." One definition of the word *sufficient* is "adequate." Therefore, "being a mammal" is an adequate condition for "being warm-blooded"; hence, "being a mammal" implies "being warm-blooded." Logically, the statement "Being a mammal is sufficient for being warm-blooded" is equivalent to saying "If it is a mammal, then it is warm-blooded." Consequently, the general statement "*p* is sufficient for *q*" is an alternative form of the conditional "if *p* then *q*" and can be symbolized as $p \rightarrow q$.

"Being a mammal" is a *sufficient* (adequate) condition for "being warm-blooded," but is it a *necessary* condition? Of course not: some animals are warm-blooded but are not mammals (chickens, for example). One definition of the word *necessary* is "required." Therefore, "being a mammal" is not required for "being warm-blooded." However, is "being warm-blooded" a necessary (required) condition for "being a mammal"? Of course it is: all mammals *are* warm-blooded (that is, there are no cold-blooded mammals). Logically, the statement "being warm-blooded is necessary for being a mammal" is equivalent to saying "If it is a mammal, then it is warm-blooded." Consequently, the general statement "*q* is necessary for *p*" is an alternative form of the conditional "if *p* then *q*" and can be symbolized as $p \rightarrow q$.

In summary, a sufficient condition is the hypothesis or premise of a conditional statement, whereas a necessary condition is the conclusion of a conditional statement.

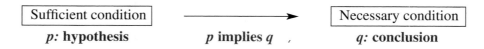

Sufficient condition		Necessary condition
p: **hypothesis**	*p* **implies** *q* ,	*q:* **conclusion**

EXAMPLE **7**

TRANSLATING WORDS INTO SYMBOLS   Using the symbolic representations

  p: A person obeys the law.
  q: A person is arrested.

express the following compound statements in symbolic form:

a. Being arrested is necessary for not obeying the law.
b. Obeying the law is sufficient for not being arrested.

SOLUTION

a. "Being arrested" is a necessary condition; hence, "a person is arrested" is the conclusion of a conditional statement.

A person does not obey the law.		A person is arrested.
		A necessary condition.

Therefore, the statement "Being arrested is necessary for not obeying the law" can be rephrased as follows:

"$\boxed{If}$ a person does $\boxed{not}$ obey the law, $\boxed{then}$ the person is arrested."

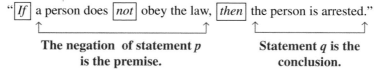

The negation of statement *p*   **Statement *q* is the**
   is the premise.   **conclusion.**

The given compound statement can be expressed as $\sim p \rightarrow q$.

b. "Obeying the law" is a sufficient condition; hence, "A person obeys the law" is the premise of a conditional statement.

| A person obeys the law. | ⟶ | A person is not arrested. |

A sufficient condition.

Therefore, the statement "Obeying the law is sufficient for not being arrested" can be rephrased as follows:

" $\boxed{If}$ a person obeys the law, $\boxed{then}$ the person is $\boxed{not}$ arrested."

Statement p   **The negation of statement q**
is the premise.   **is the conclusion.**

The given compound statement can be expressed as $p \rightarrow \sim q$.

We have seen that a statement is a sentence that is either true or false and that connecting two or more statements forms a compound statement. Figure 1.30 summarizes the logical connectives and symbols that were introduced in this section. The various connectives have been defined; we can now proceed in our analysis of the conditions under which a compound statement is true. This analysis is carried out in the next section.

Statement	Symbol	Read as . . .
negation	$\sim p$	not p
conjunction	$p \wedge q$	p and q
disjunction	$p \vee q$	p or q
conditional (implication)	$p \rightarrow q$	if p, then q p is sufficient for q q is necessary for p

FIGURE 1.30   Logical connectives.

1.2 EXERCISES

1. Which of the following are statements? Why or why not?

　a. George Washington was the first president of the United States.

　b. Abraham Lincoln was the second president of the United States.

　c. Who was the first vice president of the United States?

　d. Abraham Lincoln was the best president.

2. Which of the following are statements? Why or why not?

　a. $3 + 5 = 6$

　b. Solve the equation $2x + 5 = 3$.

　c. $x^2 + 1 = 0$ has no solution.

　d. $x^2 - 1 = (x + 1)(x - 1)$

　e. Is $\sqrt{2}$ a rational number?

3. Determine which pairs of statements are negations of each other.

　a. All of the fruits are red.

　b. None of the fruits is red.

　c. Some of the fruits are red.

　d. Some of the fruits are not red.

4. Determine which pairs of statements are negations of each other.

　a. Some of the beverages contain caffeine.

　b. Some of the beverages do not contain caffeine.

　c. None of the beverages contain caffeine.

　d. All of the beverages contain caffeine.

5. Write a sentence that represents the negation of each statement.

　a. Her dress is not red.

　b. Some computers are priced under $100.

c. All dogs are four-legged animals.

d. No sleeping bag is waterproof.

6. Write a sentence that represents the negation of each statement.

   a. She is not a vegetarian.

   b. Some elephants are pink.

   c. All candy promotes tooth decay.

   d. No lunch is free.

7. Using the symbolic representations

   p: The lyrics are controversial.

   q: The performance is banned.

   express the following compound statements in symbolic form.

   a. The lyrics are controversial, and the performance is banned.

   b. If the lyrics are not controversial, the performance is not banned.

   c. It is not the case that the lyrics are controversial or the performance is banned.

   d. The lyrics are controversial, and the performance is not banned.

   e. Having controversial lyrics is sufficient for banning a performance.

   f. Noncontroversial lyrics are necessary for not banning a performance.

8. Using the symbolic representations

   p: The food is spicy.

   q: The food is aromatic.

   express the following compound statements in symbolic form.

   a. The food is aromatic and spicy.

   b. If the food isn't spicy, it isn't aromatic.

   c. The food is spicy, and it isn't aromatic.

   d. The food isn't spicy or aromatic.

   e. Being nonaromatic is sufficient for food to be nonspicy.

   f. Being spicy is necessary for food to be aromatic.

9. Using the symbolic representations

   p: A person plays the guitar.

   q: A person rides a motorcycle.

   r: A person wears a leather jacket.

   express the following compound statements in symbolic form.

   a. If a person plays the guitar or rides a motorcycle, then the person wears a leather jacket.

   b. A person plays the guitar, rides a motorcycle, and wears a leather jacket.

   c. A person wears a leather jacket and doesn't play the guitar or ride a motorcycle.

   d. All motorcycle riders wear leather jackets.

   e. Not wearing a leather jacket is sufficient for not playing the guitar or riding a motorcycle.

   f. Riding a motorcycle or playing the guitar is necessary for wearing a leather jacket.

10. Using the symbolic representations

   p: The car costs $70,000.

   q: The car goes 140 mph.

   r: The car is red.

   express the following compound statements in symbolic form.

   a. All red cars go 140 mph.

   b. The car is red, goes 140 mph, and does not cost $70,000.

   c. If the car does not cost $70,000, it does not go 140 mph.

   d. The car is red and it does not go 140 mph or cost $70,000.

   e. Being able to go 140 mph is sufficient for a car to cost $70,000 or be red.

   f. Not being red is necessary for a car to cost $70,000 and not go 140 mph.

In Exercises 11–34, translate the sentence into symbolic form. Be sure to define each letter you use. (More than one answer is possible.)

11. All squares are rectangles.

12. All people born in the United States are American citizens.

13. No square is a triangle.

14. No convicted felon is eligible to vote.

15. All whole numbers are even or odd.

16. All muscle cars from the Sixties are polluters.

17. No whole number is greater than 3 and less than 4.

18. No electric-powered car is a polluter.

19. Being an orthodontist is sufficient for being a dentist.

20. Being an author is sufficient for being literate.

21. Knowing Morse code is necessary for operating a telegraph.

22. Knowing CPR is necessary for being a paramedic.

23. Being a monkey is sufficient for not being an ape.

24. Being a chimpanzee is sufficient for not being a monkey.

25. Not being a monkey is necessary for being an ape.

26. Not being a chimpanzee is necessary for being a monkey.

27. I do not sleep soundly if I drink coffee or eat chocolate.

28. I sleep soundly if I do not drink coffee or eat chocolate.

29. Your check is not accepted if you do not have a driver's license or a credit card.

30. Your check is accepted if you have a driver's license or a credit card.

31. If you drink and drive, you are fined or you go to jail.

32. If you are rich and famous, you have many friends and enemies.

33. You get a refund or a store credit if the product is defective.

34. The streets are slippery if it is raining or snowing.

35. Using the symbolic representations

p: I am an environmentalist.

q: I recycle my aluminum cans.

express the following in words.

a. $p \wedge q$ **b.** $p \rightarrow q$

c. $\sim q \rightarrow \sim p$ **d.** $q \vee \sim p$

36. Using the symbolic representations

p: I am innocent.

q: I have an alibi.

express the following in words.

a. $p \wedge q$ **b.** $p \rightarrow q$

c. $\sim q \rightarrow \sim p$ **d.** $q \vee \sim p$

37. Using the symbolic representations

p: I am an environmentalist.

q: I recycle my aluminum cans.

r: I recycle my newspapers.

express the following in words.

a. $(q \vee r) \rightarrow p$ **b.** $\sim p \rightarrow \sim (q \vee r)$

c. $(q \wedge r) \vee \sim p$ **d.** $(r \wedge \sim q) \rightarrow \sim p$

38. Using the symbolic representations

p: I am innocent.

q: I have an alibi.

r: I go to jail.

express the following in words.

a. $(p \vee q) \rightarrow \sim r$ **b.** $(p \wedge \sim q) \rightarrow r$

c. $(\sim p \wedge q) \vee r$ **d.** $(p \wedge r) \rightarrow \sim q$

39. Which statement, #1 or #2, is more appropriate? Explain why.

Statement #1: "Cold weather is necessary for it to snow."

Statement #2: "Cold weather is sufficient for it to snow."

40. Which statement, #1 or #2, is more appropriate? Explain why.

Statement #1: "Being cloudy is necessary for it to rain."

Statement #2: "Being cloudy is sufficient for it to rain."

41. Which statement, #1 or #2, is more appropriate? Explain why.

Statement #1: "Having 31 days in a month is necessary for it not to be February."

Statement #2: "Having 31 days in a month is sufficient for it not to be February."

42. Which statement, #1 or #2, is more appropriate? Explain why.

Statement #1: "Being the Fourth of July is necessary for the U.S. Post Office to be closed."

Statement #2: "Being the Fourth of July is sufficient for the U.S. Post Office to be closed."

 Answer the following questions using complete sentences and your own words.

● CONCEPT QUESTIONS

43. What is a negation?   **44.** What is a conjunction?

45. What is a disjunction?   **46.** What is a conditional?

47. What is a sufficient condition?

48. What is a necessary condition?

49. What is the difference between the inclusive *or* and the exclusive *or*?

50. Create a sentence that is a self-contradiction, or paradox, as in part (e) of Example 1.

● HISTORY QUESTIONS

51. In what academic field did Gottfried Leibniz receive his degrees? Why is the study of logic important in this field?

52. Who developed a formal system of logic based on syllogistic arguments?

53. What is meant by *characteristica universalis*? Who proposed this theory?

 ## THE NEXT LEVEL

If a person wants to pursue an advanced degree (something beyond a bachelor's or four-year degree), chances are the person must take a standardized exam to gain admission to a graduate school or to be admitted into a specific program. These exams are intended to measure verbal, quantitative, and analytical skills that have developed throughout a person's life. Many classes and study guides are available to help people prepare for the exams. The following questions are typical of those found in the study guides.

Exercises 54–58 refer to the following: A culinary institute has a small restaurant in which the students prepare various dishes. The menu changes daily, and during a specific week, the following dishes are to be prepared: moussaka, pilaf, quiche, ratatouille, stroganoff, and teriyaki. During the week, the restaurant does not prepare any other kind of dish. The selection of dishes the restaurant offers is consistent with the following conditions:

- *If the restaurant offers pilaf, then it does not offer ratatouille.*
- *If the restaurant does not offer stroganoff, then it offers pilaf.*
- *If the restaurant offers quiche, then it offers both ratatouille and teriyaki.*
- *If the restaurant offers teriyaki, then it offers moussaka or stroganoff or both.*

54. Which one of the following could be a complete and accurate list of the dishes the restaurant offers on a specific day?

a. pilaf, quiche, ratatouille, teriyaki

b. quiche, stroganoff, teriyaki

c. quiche, ratatouille, teriyaki

d. ratatouille, stroganoff

e. quiche, ratatouille

55. Which one of the following cannot be a complete and accurate list of the dishes the restaurant offers on a specific day?

   a. moussaka, pilaf, quiche, ratatouille, teriyaki

   b. quiche, ratatouille, stroganoff, teriyaki

   c. moussaka, pilaf, teriyaki

   d. stroganoff, teriyaki

   e. pilaf, stroganoff

56. Which one of the following could be the only kind of dish the restaurant offers on a specific day?

   a. teriyaki   b. stroganoff

   c. ratatouille   d. quiche

   e. moussaka

57. If the restaurant does not offer teriyaki, then which one of the following must be true?

   a. The restaurant offers pilaf.

   b. The restaurant offers at most three different dishes.

   c. The restaurant offers at least two different dishes.

   d. The restaurant offers neither quiche nor ratatouille.

   e. The restaurant offers neither quiche nor pilaf.

58. If the restaurant offers teriyaki, then which one of the following must be false?

   a. The restaurant does not offer moussaka.

   b. The restaurant does not offer ratatouille.

   c. The restaurant does not offer stroganoff.

   d. The restaurant offers ratatouille but not quiche.

   e. The restaurant offers ratatouille but not stroganoff.

1.3 Truth Tables

OBJECTIVES

- Construct a truth table for a compound statement
- Determine whether two statements are equivalent
- Apply De Morgan's Laws

Suppose your friend Maria is a doctor, and you know that she is a Democrat. If someone told you, "Maria is a doctor and a Republican," you would say that the statement was false. On the other hand, if you were told, "Maria is a doctor or a Republican," you would say that the statement was true. Each of these statements is a compound statement—the result of joining individual statements with connective words. When is a compound statement true, and when is it false? To answer these questions, we must examine whether the individual statements are true or false and the manner in which the statements are connected.

The **truth value** of a statement is the classification of the statement as true or false and is denoted by T or F. For example, the truth value of the statement "Santa Fe is the capital of New Mexico" is T. (The statement is true.) In contrast, the truth value of "Memphis is the capital of Tennessee" is F. (The statement is false.)

A convenient way of determining whether a compound statement is true or false is to construct a truth table. A **truth table** is a listing of all possible combinations of the individual statements as true or false, along with the resulting truth value of the compound statement. As we will see, truth tables also allow us to distinguish valid arguments from invalid arguments.

	p
1.	T
2.	F

FIGURE 1.31

Truth values for a statement p.

	p	**~p**
1.	T	F
2.	F	T

FIGURE 1.32

Truth table for a negation ~p.

	p	**q**
1.	T	T
2.	T	F
3.	F	T
4.	F	F

FIGURE 1.33

Truth values for two statements.

	p	**q**	**p ∧ q**
1.	T	T	T
2.	T	F	F
3.	F	T	F
4.	F	F	F

FIGURE 1.34

Truth table for a conjunction
p ∧ q.

	p	**q**	**p ∨ q**
1.	T	T	T
2.	T	F	T
3.	F	T	T
4.	F	F	F

FIGURE 1.35

Truth table for a disjunction
p ∨ q.

 # The Negation ~p

The **negation** of a statement is the denial, or opposite, of the statement. (As was stated in the previous section, because the truth value of the negation depends on the truth value of the original statement, a negation can be classified as a compound statement.) To construct the truth table for the negation of a statement, we must first examine the original statement. A statement p may be true or false, as shown in Figure 1.31. If the statement p is true, the negation $\sim p$ is false; if p is false, $\sim p$ is true. The truth table for the compound statement $\sim p$ is given in Figure 1.32. Row 1 of the table is read "$\sim p$ is false when p is true." Row 2 is read "$\sim p$ is true when p is false."

The Conjunction p ∧ q

A **conjunction** is the joining of two statements with the word *and*. The compound statement "Maria is a doctor and a Republican" is a conjunction with the following symbolic representation:

> *p:* Maria is a doctor.
>
> *q:* Maria is a Republican.
>
> *p ∧ q:* Maria is a doctor and a Republican.

The truth value of a compound statement depends on the truth values of the individual statements that make it up. How many rows will the truth table for the conjunction $p \wedge q$ contain? Because p has two possible truth values (T or F) and q has two possible truth values (T or F), we need four (2 · 2) rows in order to list all possible combinations of Ts and Fs, as shown in Figure 1.33.

For the conjunction $p \wedge q$ to be true, the components p and q must *both* be true; the conjunction is false otherwise. The completed truth table for the conjunction $p \wedge q$ is given in Figure 1.34. The symbols p and q can be replaced by any statements. The table gives the truth value of the statement "p and q," dependent upon the truth values of the individual statements "p" and "q." For instance, row 3 is read "The conjunction $p \wedge q$ is false when p is false and q is true." The other rows are read in a similar manner.

The Disjunction p ∨ q

A **disjunction** is the joining of two statements with the word *or*. The compound statement "Maria is a doctor or a Republican" is a disjunction (the *inclusive or*) with the following symbolic representation:

> *p:* Maria is a doctor.
>
> *q:* Maria is a Republican.
>
> *p ∨ q:* Maria is a doctor or a Republican.

Even though your friend Maria the doctor is not a Republican, the disjunction "Maria is a doctor or a Republican" is true. For a disjunction to be true, *at least one* of the components must be true. A disjunction is false only when *both* components are false. The truth table for the disjunction $p \vee q$ is given in Figure 1.35.

EXAMPLE **1**

CONSTRUCTING A TRUTH TABLE   Under what specific conditions is the following compound statement true? "I have a high school diploma, or I have a full-time job and no high school diploma."

SOLUTION

First, we translate the statement into symbolic form, and then we construct the truth table for the symbolic expression. Define p and q as

p: I have a high school diploma.

q: I have a full-time job.

The given statement has the symbolic representation $p \vee (q \wedge \sim p)$.

Because there are two letters, we need $2 \cdot 2 = 4$ rows. We need to insert a column for each connective in the symbolic expression $p \vee (q \wedge \sim p)$. As in algebra, we start inside any grouping symbols and work our way out. Therefore, we need a column for $\sim p$, a column for $q \wedge \sim p$, and a column for the entire expression $p \vee (q \wedge \sim p)$, as shown in Figure 1.36.

	p	q	$\sim p$	$q \wedge \sim p$	$p \vee (q \wedge \sim p)$
1.	T	T	F	F	T
2.	T	F	F	F	T
3.	F	T	F	T	T
4.	F	F	T	F	F

FIGURE 1.36   Required columns in the truth table. *Only x statement will be true*

In the $\sim p$ column, fill in truth values that are opposite those for p. Next, the conjunction $q \wedge \sim p$ is true only when both components are true; enter a T in row 3 and Fs elsewhere. Finally, the disjunction $p \vee (q \wedge \sim p)$ is false only when both components p and $(q \wedge \sim p)$ are false; enter an F in row 4 and Ts elsewhere. The completed truth table is shown in Figure 1.37.

	p	q	$\sim p$	$q \wedge \sim p$	$p \vee (q \wedge \sim p)$
1.	T	T	F	F	T
2.	T	F	F	F	T
3.	F	T	T	T	T
4.	F	F	T	F	F

FIGURE 1.37   Truth table for $p \vee (q \wedge \sim p)$.

As is indicated in the truth table, the symbolic expression $p \vee (q \wedge \sim p)$ is true under all conditions except one: row 4; the expression is false when both p and q are false. Therefore, the statement "I have a high school diploma, or I have a full-time job and no high school diploma" is true in every case except when the speaker has no high school diploma and no full-time job.

If the symbolic representation of a compound statement consists of two different letters, its truth table will have $2 \cdot 2 = 4$ rows. How many rows are required if a compound statement consists of three letters—say, p, q, and r? Because each statement has two possible truth values (T and F), the truth table must contain $2 \cdot 2 \cdot 2 = 8$ rows. In general, each time a new statement is added, the number of rows doubles.

> ## NUMBER OF ROWS
> *[handwritten: # of Rows $2^3 = 8$]*
>
> If a compound statement consists of n individual statements, each represented by a different letter, the number of rows required in its truth table is 2^n.

EXAMPLE 2

SOLUTION

CONSTRUCTING A TRUTH TABLE Under what specific conditions is the following compound statement true? "I own a handgun, and it is not the case that I am a criminal or police officer."

First, we translate the statement into symbolic form, and then we construct the truth table for the symbolic expression. Define the three simple statements as follows:

> *p:* I own a handgun.
> *q:* I am a criminal.
> *r:* I am a police officer.

The given statement has the symbolic representation $p \wedge \sim(q \vee r)$. Since there are three letters, we need $2^3 = 8$ rows. We start with three columns, one for each letter. To account for all possible combinations of *p, q,* and *r* as true or false, proceed as follows:

1. Fill the first half (four rows) of column 1 with Ts and the rest with Fs, as shown in Figure 1.38(a).
2. In the next column, split each half into halves, the first half receiving Ts and the second Fs. In other words, alternate two Ts and two Fs in column 2, as shown in Figure 1.38(b).
3. Again, split each half into halves; the first half receives Ts, and the second half receives Fs. Because we are dealing with the third (last) column, the Ts and Fs will alternate, as shown in Figure 1.38(c).

	p	*q*	*r*
1.	T		
2.	T		
3.	T		
4.	T		
5.	F		
6.	F		
7.	F		
8.	F		

(a)

	p	*q*	*r*
1.	T	T	
2.	T	T	
3.	T	F	
4.	T	F	
5.	F	T	
6.	F	T	
7.	F	F	
8.	F	F	

(b)

	p	*q*	*r*
1.	T	T	T
2.	T	T	F
3.	T	F	T
4.	T	F	F
5.	F	T	T
6.	F	T	F
7.	F	F	T
8.	F	F	F

(c)

FIGURE 1.38 Truth values for three statements.

(This process of filling the first half of the first column with Ts and the second half with Fs and then splitting each half into halves with blocks of Ts and Fs applies to all truth tables.)

   We need to insert a column for each connective in the symbolic expression $p \wedge \sim(q \vee r)$, as shown in Figure 1.39.

	p	*q*	*r*	*q* ∨ *r*	~(*q* ∨ *r*)	*p* ∧ ~(*q* ∨ *r*)
1.	T	T	T			
2.	T	T	F			
3.	T	F	T			
4.	T	F	F			
5.	F	T	T			
6.	F	T	F			
7.	F	F	T			
8.	F	F	F			

FIGURE 1.39 Required columns in the truth table.

Now fill in the appropriate symbol in the column under *q* ∨ *r*. Enter F if *both* *q* and *r* are false; enter T otherwise (that is, if at least one is true). In the ~(*q* ∨ *r*) column, fill in truth values that are opposite those for *q* ∨ *r*, as in Figure 1.40.

	p	*q*	*r*	*q* ∨ *r*	~(*q* ∨ *r*)	*p* ∧ ~(*q* ∨ *r*)
1.	T	T	T	T	F	
2.	T	T	F	T	F	
3.	T	F	T	T	F	
4.	T	F	F	F	T	
5.	F	T	T	T	F	
6.	F	T	F	T	F	
7.	F	F	T	T	F	
8.	F	F	F	F	T	

FIGURE 1.40 Truth values of the expressions *q* ∨ *r* and ~(*q* ∨ *r*).

The conjunction *p* ∧ ~(*q* ∨ *r*) is true only when *both* *p* and ~(*q* ∨ *r*) are true; enter a T in row 4 and Fs elsewhere. The truth table is shown in Figure 1.41.

	p	*q*	*r*	*q* ∨ *r*	~(*q* ∨ *r*)	*p* ∧ ~(*q* ∨ *r*)
1.	T	T	T	T	F	F
2.	T	T	F	T	F	F
3.	T	F	T	T	F	F
4.	T	F	F	F	T	T
5.	F	T	T	T	F	F
6.	F	T	F	T	F	F
7.	F	F	T	T	F	F
8.	F	F	F	F	T	F

FIGURE 1.41 Truth table for *p* ∧ ~(*q* ∨ *r*).

As indicated in the truth table, the expression $p \wedge \sim(q \vee r)$ is true only when p is true and both q and r are false. Therefore, the statement "I own a handgun, and it is not the case that I am a criminal or police officer" is true only when the speaker owns a handgun, is not a criminal, and is not a police officer—in other words, the speaker is a law-abiding citizen who owns a handgun.

The Conditional $p \rightarrow q$

A **conditional** is a compound statement of the form "If p, then q" and is symbolized $p \rightarrow q$. Under what circumstances is a conditional true, and when is it false? Consider the following (compound) statement: "If you give me $50, then I will give you a ticket to the ballet." This statement is a conditional and has the following representation:

p: You give me $50.

q: I give you a ticket to the ballet.

$p \rightarrow q$: If you give me $50, then I will give you a ticket to the ballet. T

The conditional can be viewed as a promise: *If* you give me $50, *then* I will give you a ticket to the ballet. Suppose you give me $50; that is, suppose p is true. I have two options: Either I give you a ticket to the ballet (q is true), or I do not (q is false). If I do give you the ticket, the conditional $p \rightarrow q$ is true (I have kept my promise); if I do not give you the ticket, the conditional $p \rightarrow q$ is false (I have not kept my promise). These situations are shown in rows 1 and 2 of the truth table in Figure 1.42. Rows 3 and 4 require further analysis.

Suppose you do not give me $50; that is, suppose p is false. Whether or not I give you a ticket, you cannot say that I broke my promise; that is, you cannot say that the conditional $p \rightarrow q$ is false. Consequently, since a statement is either true or false, the conditional is labeled true (by default). In other words, when the premise p of a conditional is false, it does not matter whether the conclusion q is true or false. In both cases, the conditional $p \rightarrow q$ is automatically labeled true, because it is not false.

The completed truth table for a conditional is given in Figure 1.43. Notice that the only circumstance under which a conditional is false is when the premise p is true and the conclusion q is false, as shown in row 2.

	p	q	$p \rightarrow q$
1.	T	T	T
2.	T	F	F
3.	F	T	?
4.	F	F	?

FIGURE 1.42

What if p is false?

	p	q	$p \rightarrow q$
1.	T	T	T
2.	T	F	F
3.	F	T	T
4.	F	F	T

FIGURE 1.43

Truth table for $p \rightarrow q$.

EXAMPLE **3**

CONSTRUCTING A TRUTH TABLE Under what conditions is the symbolic expression $q \rightarrow \sim p$ true?

SOLUTION

Our truth table has $2^2 = 4$ rows and contains a column for p, q, $\sim p$, and $q \rightarrow \sim p$, as shown in Figure 1.44.

	p	q	$\sim p$	$q \rightarrow \sim p$
1.	T	T	F	F
2.	T	F	F	T
3.	F	T	T	T
4.	F	F	T	F

FIGURE 1.44 Required columns in the truth table.

	p	q	$\sim p$	$q \rightarrow \sim p$
1.	T	T	F	F
2.	T	F	F	T
3.	F	T	T	T
4.	F	F	T	T

FIGURE 1.45 Truth table for $q \rightarrow \sim p$.

In the $\sim p$ column, fill in truth values that are opposite those for p. Now, a conditional is false only when its premise (in this case, q) is true and its conclusion (in this case, $\sim p$) is false. Therefore, $q \to \sim p$ is false only in row 1; the conditional $q \to \sim p$ is true under all conditions except the condition that both p and q are true. The completed truth table is shown in Figure 1.45.

EXAMPLE **4**

CONSTRUCTING A TRUTH TABLE Construct a truth table for the following compound statement: "I walk up the stairs if I want to exercise or if the elevator isn't working."

SOLUTION

Rewriting the statement so the word *if* is first, we have "If I want to exercise or (if) the elevator isn't working, then I walk up the stairs."

Now we must translate the statement into symbols and construct a truth table. Define the following:

p: I want to exercise.

q: The elevator is working.

r: I walk up the stairs.

The statement now has the symbolic representation $(p \vee \sim q) \to r$. Because we have three letters, our table must have $2^3 = 8$ rows. Inserting a column for each letter and a column for each connective, we have the initial setup shown in Figure 1.46.

	p	*q*	*r*	*~q*	*p ∨ ~q*	*(p ∨ ~q) → r*
1.	T	T	T			
2.	T	T	F			
3.	T	F	T			
4.	T	F	F			
5.	F	T	T			
6.	F	T	F			
7.	F	F	T			
8.	F	F	F			

FIGURE 1.46 Required columns in the truth table.

In the column labeled $\sim q$, enter truth values that are the opposite of those of q. Next, enter the truth values of the disjunction $p \vee \sim q$ in column 5. Recall that a disjunction is false only when both components are false and is true otherwise. Consequently, enter Fs in rows 5 and 6 (since both p and $\sim q$ are false) and Ts in the remaining rows, as shown in Figure 1.47.

The last column involves a conditional; it is false only when its premise is true and its conclusion is false. Therefore, enter Fs in rows 2, 4, and 8 (since $p \vee \sim q$ is true and r is false) and Ts in the remaining rows. The truth table is shown in Figure 1.48.

As Figure 1.48 shows, the statement "I walk up the stairs if I want to exercise or if the elevator isn't working" is true in all situations except those listed in rows 2, 4, and 8. For instance, the statement is false (row 8) when the speaker does not want to exercise, the elevator is not working, and the speaker does not walk up the stairs—in other words, the speaker stays on the ground floor of the building when the elevator is broken.

		p	q	r	$\sim q$	$p \vee \sim q$	$(p \vee \sim q) \to r$
1.		T	T	T	F	T	
2.		T	T	F	F	T	
3.		T	F	T	T	T	
4.		T	F	F	T	T	
5.		F	T	T	F	F	
6.		F	T	F	F	F	
7.		F	F	T	T	T	
8.		F	F	F	T	T	

FIGURE 1.47 Truth values of the expressions.

	p	q	r	$\sim q$	$p \vee \sim q$	$(p \vee \sim q) \to r$
1.	T	T	T	F	T	T
2.	T	T	F	F	T	F
3.	T	F	T	T	T	T
4.	T	F	F	T	T	F
5.	F	T	T	F	F	T
6.	F	T	F	F	F	T
7.	F	F	T	T	T	T
8.	F	F	F	T	T	F

FIGURE 1.48 Truth table for $(p \vee \sim q) \to r$.

Equivalent Expressions

When you purchase a car, the car is either new or used. If a salesperson told you, "It is not the case that the car is not new," what condition would the car be in? This compound statement consists of one individual statement ("*p:* The car is new") and two negations:

"It is not the case that the car is not new."

$\sim$ ~ ~ ~ ~ ~ ~ ~ ~ ~ $\sim p$

Does this mean that the car is new? To answer this question, we will construct a truth table for the symbolic expression $\sim(\sim p)$ and compare its truth values with those of the original p. Because there is only one letter, we need $2^1 = 2$ rows, as shown in Figure 1.49.

We must insert a column for $\sim p$ and a column for $\sim(\sim p)$. Now, $\sim p$ has truth values that are opposite those of p, and $\sim(\sim p)$ has truth values that are opposite those of $\sim p$, as shown in Figure 1.50.

	p
1.	T
2.	F

FIGURE 1.49
Truth values of p.

	p	$\sim p$	$\sim(\sim p)$
1.	T	F	T
2.	F	T	F

FIGURE 1.50
Truth table for $\sim(\sim p)$.

Notice that the values in the column labeled $\sim(\sim p)$ are identical to those in the column labeled p. Whenever this happens, the expressions are said to be equivalent and may be used interchangeably. Therefore, the statement "It is not the case that the car is not new" is equivalent in meaning to the statement "The car is new."

Equivalent expressions are symbolic expressions that have identical truth values in each corresponding entry. The expression $p \equiv q$ is read "p is equivalent to q" or "p and q are equivalent." As we can see in Figure 1.50, an expression and its double negation are logically equivalent. This relationship can be expressed as $p \equiv \sim(\sim p)$.

EXAMPLE 5

DETERMINING WHETHER STATEMENTS ARE EQUIVALENT Are the statements "If I am a homeowner, then I pay property taxes" and "I am a homeowner, and I do not pay property taxes" equivalent?

SOLUTION

We begin by defining the statements:

p: I am a homeowner.

q: I pay property taxes.

$p \to q$: If I am a homeowner, then I pay property taxes.

$p \wedge \sim q$: I am a homeowner, and I do not pay property taxes.

The truth table contains $2^2 = 4$ rows, and the initial setup is shown in Figure 1.51.

Now enter the appropriate truth values under $\sim q$ (the opposite of q). Because the conjunction $p \wedge \sim q$ is true only when both p and $\sim q$ are true, enter a T in row 2 and Fs elsewhere. The conditional $p \to q$ is false only when p is true and q is false; therefore, enter an F in row 2 and Ts elsewhere. The completed truth table is shown in Figure 1.52.

Because the entries in the columns labeled $p \wedge \sim q$ and $p \to q$ are not the same, the statements are not equivalent. "If I am a homeowner, then I pay property taxes" is *not* equivalent to "I am a homeowner and I do not pay property taxes."

p	q	$\sim q$	$p \wedge \sim q$	$p \to q$
T	T			
T	F			
F	T			
F	F			

FIGURE 1.51 Required columns in the truth table.

p	q	$\sim q$	$p \wedge \sim q$	$p \to q$
1. T	T	F	F	T
2. T	F	T	T	F
3. F	T	F	F	T
4. F	F	T	F	T

FIGURE 1.52 Truth table for $p \to q$.

Notice that the truth values in the columns under $p \wedge \sim q$ and $p \to q$ in Figure 1.52 are exact opposites; when one is T, the other is F. Whenever this happens, one statement is the negation of the other. Consequently, $p \wedge \sim q$ is the negation of $p \to q$ (and vice versa). This can be expressed as $p \wedge \sim q \equiv \sim(p \to q)$. The negation of a conditional is logically equivalent to the conjunction of the premise and the negation of the conclusion.

Statements that look or sound different may in fact have the same meaning. For example, "It is not the case that the car is not new" really means the same as "The car is new," and "It is not the case that if I am a homeowner, then I pay property taxes"

HISTORICAL NOTE

GEORGE BOOLE, 1815–1864

© CORBIS

George Boole is called "the father of symbolic logic." Computer science owes much to this self-educated mathematician. Born the son of a poor shopkeeper in Lincoln, England, Boole had very little formal education, and his prospects for rising above his family's lower-class status were dim. Like Leibniz, he taught himself Latin; at the age of twelve, he translated an ode of Horace into English, winning the attention of the local schoolmasters. (In his day, knowledge of Latin was a prerequisite to scholarly endeavors and to becoming a socially accepted gentleman.) After that, his academic desires were encouraged, and at the age of fifteen, he began his long teaching career. While teaching arithmetic, he studied advanced mathematics and physics.

In 1849, after nineteen years of teaching at elementary schools, Boole received his big break: He was appointed professor of mathematics at Queen's College in the city of Cork, Ireland. At last, he was able to research advanced mathematics, and he became recognized as a first-class mathematician. This was a remarkable feat, considering Boole's lack of formal training and degrees.

Boole's most influential work, *An Investigation of the Laws of Thought, on Which Are Founded the Mathematical Theories of Logic and Probabilities*, was published in 1854. In it, he wrote, "There exist certain general principles founded in the very nature of language and logic that exhibit laws as identical in form as with the laws of the general symbols of algebra." With this insight, Boole had taken a big step into the world of logical reasoning and abstract mathematical analysis.

Perhaps because of his lack of formal training, Boole challenged the status quo, including the Aristotelian assumption that *all* logical arguments could be reduced to syllogistic arguments. In doing so, he employed symbols to represent concepts, as did Leibniz, but he also developed systems of algebraic manipulation to accompany these symbols. Thus, Boole's creation is a marriage of logic and mathematics. However, as is the case with almost all new theories, Boole's symbolic logic was not met with total approbation. In particular, one staunch opponent of his work was Georg Cantor, whose work on the origins of set theory and the magnitude of infinity will be investigated in Chapter 2.

In the many years since Boole's original work was unveiled, various scholars have modified, improved, generalized, and extended its central concepts. Today, Boolean algebras are the essence of computer software and circuit design. After all, a computer merely manipulates predefined symbols and conforms to a set of preassigned algebraic commands.

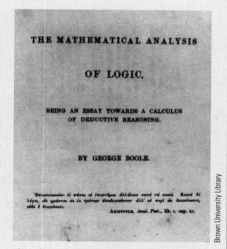

THE MATHEMATICAL ANALYSIS

OF LOGIC,

BEING AN ESSAY TOWARDS A CALCULUS OF DEDUCTIVE REASONING.

BY GEORGE BOOLE.

Brown University Library

Through an algebraic manipulation of logical symbols, Boole revolutionized the age-old study of logic. His essay The Mathematical Analysis of Logic *laid the foundation for his later book* An Investigation of the Laws of Thought.

actually means the same as "I am a homeowner, and I do not pay property taxes." When we are working with equivalent statements, we can substitute either statement for the other without changing the truth value.

De Morgan's Laws *algebra negatives*

Earlier in this section, we saw that the negation of a negation is equivalent to the original statement; that is, $\sim(\sim p) \equiv p$. Another negation "formula" that we discovered was $\sim(p \rightarrow q) \equiv p \wedge \sim q$, that is, the negation of a conditional. Can we find similar "formulas" for the negations of the other basic connectives, namely, the conjunction and the disjunction? The answer is yes, and the results are credited to the English mathematician and logician Augustus De Morgan.

[Handwritten margin note: Switch the inequality ≥ < Change the sign]

DE MORGAN'S LAWS

The negation of the conjunction $p \wedge q$ is given by $\sim(p \wedge q) \equiv \sim p \vee \sim q$.

"Not p and q" is equivalent to "not p or not q."

The negation of the disjunction $p \vee q$ is given by $\sim(p \vee q) \equiv \sim p \wedge \sim q$.

"Not p or q" is equivalent to "not p and not q."

De Morgan's Laws are easily verified through the use of truth tables and will be addressed in the exercises (see Exercises 55 and 56).

EXAMPLE 6

APPLYING DE MORGAN'S LAWS   Using De Morgan's Laws, find the negation of each of the following:

a. It is Friday and I receive a paycheck.
b. You are correct or I am crazy.

SOLUTION

a. The symbolic representation of "It is Friday and I receive a paycheck" is

  p: It is Friday.

  q: I receive a paycheck.

  $p \wedge q$: It is Friday and I receive a paycheck.

[Handwritten note: P = It is Friday, q = I receive a paycheck, p ∧ q = It is Friday and I receive a paycheck]

Therefore, the negation is $\sim(p \wedge q) \equiv \sim p \vee \sim q$, that is, "It is not Friday or I do not receive a paycheck."

b. The symbolic representation of "You are correct or I am crazy" is

  p: You are correct.

  q: I am crazy.

  $p \vee q$: You are correct or I am crazy.

Therefore, the negation is $\sim(p \vee q) \equiv \sim p \wedge \sim q$, that is, "You are not correct and I am not crazy."

As we have seen, the truth value of a compound statement depends on the truth values of the individual statements that make it up. The truth tables of the basic connectives are summarized in Figure 1.53.

Equivalent statements are statements that have the same meaning. Equivalent statements for the negations of the basic connectives are given in Figure 1.54.

	p	$\sim p$
1.	T	F
2.	F	T

Negation

	p	q	$p \wedge q$
1.	T	T	T
2.	T	F	F
3.	F	T	F
4.	F	F	F

Conjunction

	p	q	$p \vee q$
1.	T	T	T
2.	T	F	T
3.	F	T	T
4.	F	F	F

Disjunction

	p	q	$p \rightarrow q$
1.	T	T	T
2.	T	F	F
3.	F	T	T
4.	F	F	T

Conditional

FIGURE 1.53   Truth tables for the basic connectives.

1. $\sim(\sim p) \equiv p$	the negation of a negation
2. $\sim(p \wedge q) \equiv \sim p \vee \sim q$	the negation of a conjunction
3. $\sim(p \vee q) \equiv \sim p \wedge \sim q$	the negation of a disjunction
4. $\sim(p \rightarrow q) \equiv p \wedge \sim q$	the negation of a conditional

FIGURE 1.54   Negations of the basic connectives.

1.3 EXERCISES

In Exercises 1–20, construct a truth table for the symbolic expressions.

1. $p \vee \sim q$
2. $p \wedge \sim q$
3. $p \vee \sim p$
4. $p \wedge \sim p$
5. $p \rightarrow \sim q$
6. $\sim p \rightarrow q$
7. $\sim q \rightarrow \sim p$
8. $\sim p \rightarrow \sim q$
9. $(p \vee q) \rightarrow \sim p$
10. $(p \wedge q) \rightarrow \sim q$
11. $(p \vee q) \rightarrow (p \wedge q)$
12. $(p \wedge q) \rightarrow (p \vee q)$
13. $p \wedge \sim(q \vee r)$
14. $p \vee \sim(q \vee r)$
15. $p \vee (\sim q \wedge r)$
16. $\sim p \vee \sim(q \wedge r)$
17. $(\sim r \vee p) \rightarrow (q \wedge p)$
18. $(q \wedge p) \rightarrow (\sim r \vee p)$
19. $(p \vee r) \rightarrow (q \wedge \sim r)$
20. $(p \wedge r) \rightarrow (q \vee \sim r)$

In Exercises 21–40, translate the compound statement into symbolic form and then construct the truth table for the expression.

21. If it is raining, then the streets are wet.
22. If the lyrics are not controversial, the performance is not banned.
23. The water supply is rationed if it does not rain.
24. The country is in trouble if he is elected.
25. All squares are rectangles.
26. All muscle cars from the Sixties are polluters.
27. No square is a triangle.
28. No electric-powered car is a polluter.
29. Being a monkey is sufficient for not being an ape.

30. Being a chimpanzee is sufficient for not being a monkey.
31. Not being a monkey is necessary for being an ape.
32. Not being a chimpanzee is necessary for being a monkey.
33. Your check is accepted if you have a driver's license or a credit card.
34. You get a refund or a store credit if the product is defective.
35. If leaded gasoline is used, the catalytic converter is damaged and the air is polluted.
36. If he does not go to jail, he is innocent or has an alibi.
37. I have a college degree and I do not have a job or own a house.
38. I surf the Internet and I make purchases and do not pay sales tax.
39. If Proposition A passes and Proposition B does not, jobs are lost or new taxes are imposed.
40. If Proposition A does not pass and the legislature raises taxes, the quality of education is lowered and unemployment rises.

In Exercises 41–50, construct a truth table to determine whether the statements in each pair are equivalent.

41. The streets are wet or it is not raining.
    If it is raining, then the streets are wet.
42. The streets are wet or it is not raining.
    If the streets are not wet, then it is not raining.
43. He has a high school diploma or he is unemployed.
    If he does not have a high school diploma, then he is unemployed.
44. She is unemployed or she does not have a high school diploma.
    If she is employed, then she does not have a high school diploma.
45. If handguns are outlawed, then outlaws have handguns.
    If outlaws have handguns, then handguns are outlawed.
46. If interest rates continue to fall, then I can afford to buy a house.
    If interest rates do not continue to fall, then I cannot afford to buy a house.
47. If the spotted owl is on the endangered species list, then lumber jobs are lost.

If lumber jobs are not lost, then the spotted owl is not on the endangered species list.

48. If I drink decaffeinated coffee, then I do not stay awake.

If I do stay awake, then I do not drink decaffeinated coffee.

49. The plaintiff is innocent or the insurance company does not settle out of court.

The insurance company settles out of court and the plaintiff is not innocent.

50. The plaintiff is not innocent and the insurance company settles out of court.

It is not the case that the plaintiff is innocent or the insurance company does not settle out of court.

In Exercises 51–54, construct truth tables to determine which pairs of statements are equivalent.

51.
  i. Knowing Morse code is sufficient for operating a telegraph.
  ii. Knowing Morse code is necessary for operating a telegraph.
  iii. Not knowing Morse code is sufficient for not operating a telegraph.
  iv. Not knowing Morse code is necessary for not operating a telegraph.

52.
  i. Knowing CPR is necessary for being a paramedic.
  ii. Knowing CPR is sufficient for being a paramedic.
  iii. Not knowing CPR is necessary for not being a paramedic.
  iv. Not knowing CPR is sufficient for not being a paramedic.

53.
  i. The water being cold is necessary for not going swimming.
  ii. The water not being cold is necessary for going swimming.
  iii. The water being cold is sufficient for not going swimming.
  iv. The water not being cold is sufficient for going swimming.

54.
  i. The sky not being clear is sufficient for it to be raining.
  ii. The sky being clear is sufficient for it not to be raining.
  iii. The sky not being clear is necessary for it to be raining.
  iv. The sky being clear is necessary for it not to be raining.

55. Using truth tables, verify De Morgan's Law

$$\sim(p \wedge q) \equiv \sim p \vee \sim q.$$

56. Using truth tables, verify De Morgan's Law

$$\sim(p \vee q) \equiv \sim p \wedge \sim q.$$

In Exercises 57–68, write the statement in symbolic form, construct the negation of the expression (in simplified symbolic form), and express the negation in words.

57. I have a college degree and I am not employed.
58. It is snowing and classes are canceled.
59. The television set is broken or there is a power outage.
60. The freeway is under construction or I do not ride the bus.
61. If the building contains asbestos, the original contractor is responsible.
62. If the legislation is approved, the public is uninformed.
63. The First Amendment has been violated if the lyrics are censored.
64. Your driver's license is taken away if you do not obey the laws.
65. Rainy weather is sufficient for not washing my car.
66. Drinking caffeinated coffee is sufficient for not sleeping.
67. Not talking is necessary for listening.
68. Not eating dessert is necessary for being on a diet.

 Answer the following questions using complete sentences and your own words.

• CONCEPT QUESTIONS

69.
  a. Under what conditions is a disjunction true?
  b. Under what conditions is a disjunction false?
70.
  a. Under what conditions is a conjunction true?
  b. Under what conditions is a conjunction false?
71.
  a. Under what conditions is a conditional true?
  b. Under what conditions is a conditional false?
72.
  a. Under what conditions is a negation true?
  b. Under what conditions is a negation false?
73. What are equivalent expressions?
74. What is a truth table?
75. When constructing a truth table, how do you determine how many rows to create?

• HISTORY QUESTIONS

76. Who is considered "the father of symbolic logic"?
77. Boolean algebra is a combination of logic and mathematics. What is it used for?

1.4 More on Conditionals

OBJECTIVES

- Create the converse, inverse, and contrapositive of a conditional statement
- Determine equivalent variations of a conditional statement
- Interpret "only if" statements
- Interpret a biconditional statement

Conditionals differ from conjunctions and disjunctions with regard to the possibility of changing the order of the statements. In algebra, the sum $x + y$ is equal to the sum $y + x$; that is, addition is commutative. In everyday language, one realtor might say, "The house is perfect and the lot is priceless," while another says, "The lot is priceless and the house is perfect." Logically, their meanings are the same, since $(p \wedge q) \equiv (q \wedge p)$. The order of the components in a conjunction or disjunction makes no difference in regard to the truth value of the statement. This is not so with conditionals.

Variations of a Conditional

Given two statements p and q, various "if . . . then . . ." statements can be formed.

EXAMPLE 1

TRANSLATING SYMBOLS INTO WORDS Using the statements

> *p:* You are compassionate.
> *q:* You contribute to charities.

write an "if . . . then . . ." sentence represented by each of the following:

a. $p \rightarrow q$ **b.** $q \rightarrow p$ **c.** $\sim p \rightarrow \sim q$ **d.** $\sim q \rightarrow \sim p$

SOLUTION

a. $p \rightarrow q$: If you are compassionate, then you contribute to charities.
b. $q \rightarrow p$: If you contribute to charities, then you are compassionate.
c. $\sim p \rightarrow \sim q$: If you are not compassionate, then you do not contribute to charities.
d. $\sim q \rightarrow \sim p$: If you do not contribute to charities, then you are not compassionate.

Each part of Example 1 contains an "if . . . then . . ." statement and is called a conditional. Any given conditional has three variations: a converse, an inverse, and a contrapositive. The **converse** of the conditional "if p then q" is the compound statement "if q then p." That is, we form the converse of the conditional by interchanging the premise and the conclusion; $q \rightarrow p$ is the converse of $p \rightarrow q$. The statement in part (b) of Example 1 is the converse of the statement in part (a).

The **inverse** of the conditional "if p then q" is the compound statement "if not p then not q." We form the inverse of the conditional by negating both the premise and the conclusion; $\sim p \rightarrow \sim q$ is the inverse of $p \rightarrow q$. The statement in part (c) of Example 1 is the inverse of the statement in part (a).

The **contrapositive** of the conditional "if p then q" is the compound statement "if not q then not p." We form the contrapositive of the conditional by

43

negating *and* interchanging both the premise and the conclusion; $\sim q \to \sim p$ is the contrapositive of $p \to q$. The statement in part (d) of Example 1 is the contrapositive of the statement in part (a). The variations of a given conditional are summarized in Figure 1.55. As we will see, some of these variations are equivalent, and some are not. Unfortunately, many people incorrectly treat them all as equivalent.

Name	Symbolic Form	Read As . . .
a (given) conditional	$p \to q$	If p, then q.
the converse (of $p \to q$)	$q \to p$	If q, then p.
the inverse (of $p \to q$)	$\sim p \to \sim q$	If not p, then not q.
the contrapositive (of $p \to q$)	$\sim q \to \sim p$	If not q, then not p.

FIGURE 1.55 Variations of a conditional.

EXAMPLE **2**

CREATING VARIATIONS OF A CONDITIONAL STATEMENT Given the conditional "You did not receive the proper refund if you prepared your own income tax form," write the sentence that represents each of the following.

a. the converse of the conditional
b. the inverse of the conditional
c. the contrapositive of the conditional

SOLUTION

a. Rewriting the statement in the standard "if . . . then . . ." form, we have the conditional "If you prepared your own income tax form, then you did not receive the proper refund." The converse is formed by interchanging the premise and the conclusion. Thus, the converse is written as "If you did not receive the proper refund, then you prepared your own income tax form."
b. The inverse is formed by negating both the premise and the conclusion. Thus, the inverse is written as "If you did not prepare your own income tax form, then you received the proper refund."
c. The contrapositive is formed by negating *and* interchanging the premise and the conclusion. Thus, the contrapositive is written as "If you received the proper refund, then you did not prepare your own income tax form."

Equivalent Conditionals

We have seen that the conditional $p \to q$ has three variations: the converse $q \to p$, the inverse $\sim p \to \sim q$, and the contrapositive $\sim q \to \sim p$. Do any of these "if . . . then . . ." statements convey the same meaning? In other words, are any of these compound statements equivalent?

EXAMPLE **3**

DETERMINING EQUIVALENT STATEMENTS Determine which (if any) of the following are equivalent: a conditional $p \to q$, the converse $q \to p$, the inverse $\sim p \to \sim q$, and the contrapositive $\sim q \to \sim p$.

SOLUTION

To investigate the possible equivalencies, we must construct a truth table that contains all the statements. Because there are two letters, we need $2^2 = 4$ rows. The table must have a column for $\sim p$, one for $\sim q$, one for the conditional $p \to q$, and one for each variation of the conditional. The truth values of the negations $\sim p$ and $\sim q$ are readily entered, as shown in Figure 1.56.

given *converse* *inverse* *contrapositive*

	p	q	$\sim p$	$\sim q$	$p \to q$	$q \to p$	$\sim p \to \sim q$	$\sim q \to \sim p$
1.	T	T	F	F	T	T	T	T
2.	T	F	F	T	F	T	T	F
3.	F	T	T	F	T	F	F	T
4.	F	F	T	T	F	T	T	T

FIGURE 1.56 Required columns in the truth table.

equivalent = always

An "if . . . then . . ." statement is false only when the premise is true and the conclusion is false. Consequently, $p \to q$ is false only when p is T and q is F; enter an F in row 2 and Ts elsewhere in the column under $p \to q$.

Likewise, the converse $q \to p$ is false only when q is T and p is F; enter an F in row 3 and Ts elsewhere.

In a similar manner, the inverse $\sim p \to \sim q$ is false only when $\sim p$ is T and $\sim q$ is F; enter an F in row 3 and Ts elsewhere.

Finally, the contrapositive $\sim q \to \sim p$ is false only when $\sim q$ is T and $\sim p$ is F; enter an F in row 2 and Ts elsewhere.

The completed truth table is shown in Figure 1.57. Examining the entries in Figure 1.57, we can see that the columns under $p \to q$ and $\sim q \to \sim p$ are identical; each has an F in row 2 and Ts elsewhere. Consequently, a conditional and its contrapositive are equivalent: $p \to q \equiv \sim q \to \sim p$.

Likewise, we notice that $q \to p$ and $\sim p \to \sim q$ have identical truth values; each has an F in row 3 and Ts elsewhere. Thus, the converse and the inverse of a conditional are equivalent: $q \to p \equiv \sim p \to \sim q$.

	p	q	$\sim p$	$\sim q$	$p \to q$	$q \to p$	$\sim p \to \sim q$	$\sim q \to \sim p$
1.	T	T	F	F	T	T	T	T
2.	T	F	F	T	F	T	T	F
3.	F	T	T	F	T	F	F	T
4.	F	F	T	T	T	T	T	T

FIGURE 1.57 Truth table for a conditional and its variations.

We have seen that different "if . . . then . . ." statements can convey the same meaning—that is, that certain variations of a conditional are equivalent (see Figure 1.58). For example, the compound statements "If you are compassionate, then you contribute to charities" and "If you do not contribute to charities, then you are not compassionate" convey the same meaning. (The second conditional is the contrapositive of the first.) Regardless of its specific contents (p, q, $\sim p$, or $\sim q$), every "if . . . then . . ." statement has an equivalent variation formed by negating *and* interchanging the premise and the conclusion of the given conditional statement.

Equivalent Statements	Symbolic Representations
a conditional and its contrapositive	$(p \to q) \equiv (\sim q \to \sim p)$
the converse and the inverse (of the conditional $p \to q$)	$(q \to p) \equiv (\sim p \to \sim q)$

FIGURE 1.58 Equivalent "if . . . then . . ." statements.

EXAMPLE **4**

CREATING A CONTRAPOSITIVE Given the statement "Being a doctor is necessary for being a surgeon," express the contrapositive in terms of the following:

a. a sufficient condition
b. a necessary condition

SOLUTION

a. Recalling that a *necessary condition* is the *conclusion* of a conditional, we can rephrase the statement "Being a doctor is necessary for being a surgeon" as follows:

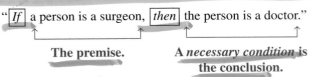

Therefore, by negating and interchanging the premise and conclusion, the contrapositive is

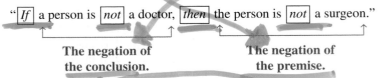

Recalling that a *sufficient condition* is the *premise* of a conditional, we can phrase the contrapositive of the original statement as "*Not being a doctor is sufficient for not being a surgeon.*"

b. From part (a), the contrapositive of the original statement is the conditional statement

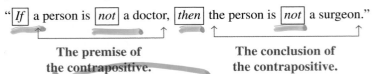

Because a necessary condition is the conclusion of a conditional, the contrapositive of the (original) statement "Being a doctor is necessary for being a surgeon" can be expressed as "*Not being a surgeon is necessary for not being a doctor.*"

The "Only If" Connective

(handwritten margin note: anything that follows if is the Premise)

Consider the statement "A prisoner is paroled only if the prisoner obeys the rules." What is the premise, and what is the conclusion? Rather than using p and q (which might bias our investigation), we define

(handwritten margin note: Premise)

r: A prisoner is paroled.
s: A prisoner obeys the rules.

The given statement is represented by "r only if s." Now, "r only if s" means that r can happen *only* if s happens. In other words, if s does not happen, then r does not happen, or $\sim s \rightarrow \sim r$. We have seen that $\sim s \rightarrow \sim r$ is equivalent to $r \rightarrow s$. Consequently, "r only if s" is equivalent to the conditional $r \rightarrow s$. The premise of the statement "A prisoner is paroled only if the prisoner obeys the rules" is "A prisoner is paroled," and the conclusion is "The prisoner obeys the rules."

The conditional $p \rightarrow q$ can be phrased "p only if q." Even though the word *if* precedes q, q is not the premise. *Whatever follows the connective "only if" is the conclusion of the conditional.*

EXAMPLE **5**

ANALYZING AN "ONLY IF" STATEMENT   For the compound statement "You receive a federal grant only if your artwork is not obscene," do the following:

a. Determine the premise and the conclusion.
b. Rewrite the compound statement in the standard "if . . . then . . ." form.
c. Interpret the conditions that make the statement false.

SOLUTION

a. Because the compound statement contains an "only if" connective, the statement that follows "only if" is the conclusion of the conditional. The premise is "You receive a federal grant." The conclusion is "Your artwork is not obscene."
b. The given compound statement can be rewritten as "If you receive a grant, then your artwork is not obscene."
c. First we define the symbols.

> *p:* You receive a federal grant.
> *q:* Your artwork is obscene.

	p	q	$\sim q$	$p \rightarrow \sim q$
1.	T	T	F	F
2.	T	F	T	T
3.	F	T	F	T
4.	F	F	T	T

FIGURE 1.59

Truth table for the conditional $p \rightarrow \sim q$.

Then the statement has the symbolic representation $p \rightarrow \sim q$. The truth table for $p \rightarrow \sim q$ is given in Figure 1.59.

  The expression $p \rightarrow q$ is false under the conditions listed in row 1 (when p and q are both true). Therefore, the statement "You receive a federal grant only if your artwork is not obscene" is false when an artist *does* receive a federal grant *and* the artist's artwork *is* obscene.

The Biconditional $p \leftrightarrow q$

What do the words *bicycle, binomial,* and *bilingual* have in common? Each word begins with the prefix *bi,* meaning "two." Just as the word *bilingual* means "two languages," the word *biconditional* means "two conditionals."

  In everyday speech, conditionals often get "hooked together" in a circular fashion. For instance, someone might say, "If I am rich, then I am happy, and if I am happy, then I am rich." Notice that this compound statement is actually the conjunction (*and*) of a conditional (if rich, then happy) and its converse (if happy, then rich). Such a statement is referred to as a biconditional. A **biconditional** is a statement of the form $(p \rightarrow q) \wedge (q \rightarrow p)$ and is symbolized as $p \leftrightarrow q$. The symbol $p \leftrightarrow q$ is read "*p* if and only if *q*" and is frequently abbreviated "*p* iff *q*." A biconditional is equivalent to the conjunction of two conversely related conditionals: $p \leftrightarrow q \equiv [(p \rightarrow q) \wedge (q \rightarrow p)]$.

  In addition to the phrase "if and only if," a biconditional can also be expressed by using "necessary" and "sufficient" terminology. The statement "*p* is sufficient for *q*" can be rephrased as "if *p* then *q*" (and symbolized as $p \rightarrow q$), whereas the statement "*p* is necessary for *q*" can be rephrased as "if *q* then *p*" (and symbolized as $q \rightarrow p$). Therefore, the biconditional "*p* if and only if *q*" can also be phrased as "*p* is necessary and sufficient for *q*."

EXAMPLE **6**

ANALYZING A BICONDITIONAL STATEMENT   Express the biconditional "A citizen is eligible to vote if and only if the citizen is at least eighteen years old" as the conjunction of two conditionals.

SOLUTION

The given biconditional is equivalent to "If a citizen is eligible to vote, then the citizen is at least eighteen years old, *and* if a citizen is at least eighteen years old, then the citizen is eligible to vote."

Under what circumstances is the biconditional $p \leftrightarrow q$ true, and when is it false? To find the answer, we must construct a truth table. Utilizing the equivalence $p \leftrightarrow q \equiv [(p \rightarrow q) \wedge (q \rightarrow p)]$, we get the completed table shown in Figure 1.60. (Recall that a conditional is false only when its premise is true and its conclusion is false and that a conjunction is true only when both components are true.) We can see that a biconditional is true only when the two components p and q have the same truth value—that is, when p and q are both true or when p and q are both false. On the other hand, a biconditional is false when the two components p and q have opposite truth value—that is, when p is true and q is false or vice versa.

	p	q	$p \rightarrow q$	$q \rightarrow p$	$(p \rightarrow q) \wedge (q \rightarrow p)$
1.	T	T	T	T	T
2.	T	F	F	T	F
3.	F	T	T	F	F
4.	F	F	T	T	T

FIGURE 1.60   Truth table for a biconditional $p \leftrightarrow q$.

Many theorems in mathematics can be expressed as biconditionals. For example, when solving a quadratic equation, we have the following: "The equation $ax^2 + bx + c = 0$ has exactly one solution if and only if the discriminant $b^2 - 4ac = 0$." Recall that the solutions of a quadratic equation are

$$x = \frac{-b \pm \sqrt{b^2 - 4ac}}{2a}$$

This biconditional is equivalent to "If the equation $ax^2 + bx + c = 0$ has exactly one solution, then the discriminant $b^2 - 4ac = 0$, and if the discriminant $b^2 - 4ac = 0$, then the equation $ax^2 + bx + c = 0$ has exactly one solution"—that is, one condition implies the other.

1.4   EXERCISES

In Exercises 1–2, using the given statements, write the sentence represented by each of the following.

a. $p \rightarrow q$
b. $q \rightarrow p$
c. $\sim p \rightarrow \sim q$
d. $\sim q \rightarrow \sim p$
e. Which of parts (a)–(d) are equivalent? Why?

1. *p:* She is a police officer.
    q: She carries a gun.

2. *p:* I am a multimillion-dollar lottery winner.
    q: I am a world traveler.

In Exercises 3–4, using the given statements, write the sentence represented by each of the following.

a. $p \rightarrow \sim q$
b. $\sim q \rightarrow p$
c. $\sim p \rightarrow q$
d. $q \rightarrow \sim p$
e. Which of parts (a)–(d) are equivalent? Why?

3. *p:* I watch television.
    q: I do my homework.

4. *p:* He is an artist.
    q: He is a conformist.

In Exercises 5–10, form (a) the inverse, (b) the converse, and (c) the contrapositive of the given conditional.

5. If you pass this mathematics course, then you fulfill a graduation requirement.

6. If you have the necessary tools, assembly time is less than thirty minutes.

7. The television set does not work if the electricity is turned off.

8. You do not win if you do not buy a lottery ticket.

9. You are a vegetarian if you do not eat meat.

10. If chemicals are properly disposed of, the environment is not damaged.

In Exercises 11–14, express the contrapositive of the given conditional in terms of (a) a sufficient condition and (b) a necessary condition.

11. Being an orthodontist is sufficient for being a dentist.

12. Being an author is sufficient for being literate.

13. Knowing Morse code is necessary for operating a telegraph.

14. Knowing CPR is necessary for being a paramedic.

In Exercises 15–20, (a) determine the premise and conclusion, (b) rewrite the compound statement in the standard "if . . . then . . ." form, and (c) interpret the conditions that make the statement false.

15. I take public transportation only if it is convenient.

16. I eat raw fish only if I am in a Japanese restaurant.

17. I buy foreign products only if domestic products are not available.

18. I ride my bicycle only if it is not raining.

19. You may become a U.S. senator only if you are at least thirty years old and have been a citizen for nine years.

20. You may become the president of the United States only if you are at least thirty-five years old and were born a citizen of the United States.

In Exercises 21–28, express the given biconditional as the conjunction of two conditionals.

21. You obtain a refund if and only if you have a receipt.

22. We eat at Burger World if and only if Ju Ju's Kitsch-Inn is closed.

23. The quadratic equation $ax^2 + bx + c = 0$ has two distinct real solutions if and only if $b^2 - 4ac > 0$.

24. The quadratic equation $ax^2 + bx + c = 0$ has complex solutions iff $b^2 - 4ac < 0$.

25. A polygon is a triangle iff the polygon has three sides.

26. A triangle is isosceles iff the triangle has two equal sides.

27. A triangle having a 90° angle is necessary and sufficient for $a^2 + b^2 = c^2$.

28. A triangle having three equal sides is necessary and sufficient for a triangle having three equal angles.

In Exercises 29–36, translate the two statements into symbolic form and use truth tables to determine whether the statements are equivalent.

29. I cannot have surgery if I do not have health insurance.
     If I can have surgery, then I do have health insurance.

30. If I am illiterate, I cannot fill out an application form.
     I can fill out an application form if I am not illiterate.

31. If you earn less than $12,000 per year, you are eligible for assistance.
     If you are not eligible for assistance, then you earn at least $12,000 per year.

32. If you earn less than $12,000 per year, you are eligible for assistance.
     If you earn at least $12,000 per year, you are not eligible for assistance.

33. I watch television only if the program is educational.
     I do not watch television if the program is not educational.

34. I buy seafood only if the seafood is fresh.
     If I do not buy seafood, the seafood is not fresh.

35. Being an automobile that is American-made is sufficient for an automobile having hardware that is not metric.
     Being an automobile that is not American-made is necessary for an automobile having hardware that is metric.

36. Being an automobile having metric hardware is sufficient for being an automobile that is not American-made.
     Being an automobile not having metric hardware is necessary for being an automobile that is American-made.

In Exercises 37–46, write an equivalent variation of the given conditional.

37. If it is not raining, I walk to work.

38. If it makes a buzzing noise, it is not working properly.

39. It is snowing only if it is cold.

40. You are a criminal only if you do not obey the law.

41. You are not a vegetarian if you eat meat.

42. You are not an artist if you are not creative.

43. All policemen own guns.

44. All college students are sleep deprived.

45. No convicted felon is eligible to vote.

46. No man asks for directions.

In Exercises 47–52, determine which pairs of statements are equivalent.

47. **i.** If Proposition 111 passes, freeways are improved.
     ii. If Proposition 111 is defeated, freeways are not improved.
     iii. If the freeways are improved, Proposition 111 passes.
     iv. If the freeways are not improved, Proposition 111 does not pass.

48. **i.** If the Giants win, then I am happy.
     ii. If I am happy, then the Giants win.

iii. If the Giants lose, then I am unhappy.

iv. If I am unhappy, then the Giants lose.

49. i. I go to church if it is Sunday.

ii. I go to church only if it is Sunday.

iii. If I do not go to church, it is not Sunday.

iv. If it is not Sunday, I do not go to church.

50. i. I am a rebel if I do not have a cause.

ii. I am a rebel only if I do not have a cause.

iii. I am not a rebel if I have a cause.

iv. If I am not a rebel, I have a cause.

51. i. If line 34 is greater than line 29, I use Schedule X.

ii. If I use Schedule X, then line 34 is greater than line 29.

iii. If I do not use Schedule X, then line 34 is not greater than line 29.

iv. If line 34 is not greater than line 29, then I do not use Schedule X.

52. i. If you answer yes to all of the above, then you complete Part II.

ii. If you answer no to any of the above, then you do not complete Part II.

iii. If you completed Part II, then you answered yes to all of the above.

iv. If you did not complete Part II, then you answered no to at least one of the above.

Answer the following questions using complete sentences and your own words.

• CONCEPT QUESTIONS

53. What is a contrapositive?
54. What is a converse?
55. What is an inverse?
56. What is a biconditional?
57. How is an "if . . . then . . ." statement related to an "only if" statement?

THE NEXT LEVEL

If a person wants to pursue an advanced degree (something beyond a bachelor's or four-year degree), chances are the person must take a standardized exam to gain admission to a graduate school or to be admitted into a specific program. These exams are intended to measure verbal, quantitative, and analytical skills that have developed throughout a person's life. Many classes and study guides are available to help people prepare for the exams. The following questions are typical of those found in the study guides.

Exercises 58–62 refer to the following: Assuming that a movie's popularity is measured by its gross box office receipts, six recently released movies—M, N, O, P, Q, and R—are ranked from most popular (first) to least popular (sixth). There are no ties. The ranking is consistent with the following conditions:

• *O is more popular than R.*

• *If N is more popular than O, then neither Q nor R is more popular than P.*

• *If O is more popular than N, then neither P nor R is more popular than Q.*

• *M is more popular than N, or else M is more popular than O, but not both.*

58. Which one of the following could be the ranking of the movies, from most popular to least popular?

a. N, M, O, R, P, Q
b. P, O, M, Q, N, R
c. Q, P, R, O, M, N
d. O, Q, M, P, N, R
e. P, Q, N, O, R, M

59. If N is the second most popular movie, then which one of the following could be true?

a. O is more popular than M.
b. Q is more popular than M.
c. R is more popular than M.
d. Q is more popular than P.
e. O is more popular than N.

60. Which one of the following cannot be the most popular movie?

a. M
b. N
c. O
d. P
e. Q

61. If R is more popular than M, then which one of the following could be true?

a. M is more popular than O.
b. M is more popular than Q.
c. N is more popular than P.
d. N is more popular than O.
e. N is more popular than R.

62. If O is more popular than P and less popular than Q, then which one of the following could be true?

a. M is more popular than O.
b. N is more popular than M.
c. N is more popular than O.
d. R is more popular than Q.
e. P is more popular than R.

- Identify a tautology
- Use a truth table to analyze an argument

Lewis Carroll's Cheshire Cat told Alice that he was mad (crazy). Alice then asked, "'And how do you know that you're mad?' 'To begin with,' said the cat, 'a dog's not mad. You grant that?' 'I suppose so,' said Alice. 'Well, then,' the cat went on, 'you see a dog growls when it's angry, and wags its tail when it's pleased. Now *I* growl when I'm pleased, and wag my tail when I'm angry. Therefore I'm mad!'"

Does the Cheshire Cat have a valid deductive argument? Does the conclusion follow logically from the hypotheses? To answer this question, and others like it, we will utilize symbolic logic and truth tables to account for all possible combinations of the individual statements as true or false.

Valid Arguments

When someone makes a sequence of statements and draws some conclusion from them, he or she is presenting an argument. An **argument** consists of two components: the initial statements, or hypotheses, and the final statement, or conclusion. When presented with an argument, a listener or reader may ask, "Does this person have a logical argument? Does his or her conclusion necessarily follow from the given statements?"

An argument is **valid** if the conclusion of the argument is guaranteed under its given set of hypotheses. (That is, the conclusion is inescapable in all instances.) For example, we used Venn diagrams in Section 1.1 to show the argument

> "All men are mortal.
> Socrates is a man. } the hypotheses
> Therefore, Socrates is mortal." } the conclusion

is a valid argument. Given the hypotheses, the conclusion is guaranteed. The term *valid* does not mean that all the statements are true but merely that the conclusion was reached via a proper deductive process. As shown in Example 3 of Section 1.1, the argument

> "All doctors are men.
> My mother is a doctor. } the hypotheses
> Therefore, my mother is a man." } the conclusion

is also a valid argument. Even though the conclusion is obviously false, the conclusion is guaranteed, *given the hypotheses.*

The hypotheses in a given logical argument may consist of several interrelated statements, each containing negations, conjunctions, disjunctions, and conditionals. By joining all the hypotheses in the form of a conjunction, we can form a single conditional that represents the entire argument. That is, if an argument has n hypotheses $(h_1, h_2, \ldots, h_n)$ and conclusion c, the argument will have the form "if $(h_1$ and $h_2 \ldots$ and $h_n)$, then c."

© Blue Lantern Studio/CORBIS

Using a logical argument, Lewis Carroll's Cheshire Cat tried to convince Alice that he was crazy. Was his argument valid?

CONDITIONAL REPRESENTATION OF AN ARGUMENT

An argument having n hypotheses $h_1, h_2, \cdots, h_n$ and conclusion c can be represented by the conditional $[h_1 \wedge h_2 \wedge \cdots \wedge h_n] \rightarrow c$.

If the conditional representation of an argument is always true (regardless of the actual truthfulness of the individual statements), the argument is valid. If there is at least one instance in which the conditional is false, the argument is invalid.

EXAMPLE 1

USING A TRUTH TABLE TO ANALYZE AN ARGUMENT   Determine whether the following argument is valid:

"If he is illiterate, he cannot fill out the application.

He can fill out the application.

Therefore, he is not illiterate."

SOLUTION   First, number the hypotheses and separate them from the conclusion with a line:

1. If he is illiterate, he cannot fill out the application.
2. He can fill out the application.

Therefore, he is not illiterate.

Now use symbols to represent each different component in the statements:

p: He is illiterate.

q: He can fill out the application.

We could have defined q as "He *cannot* fill out the application" (as stated in premise 1), but it is customary to define the symbols with a positive sense. Symbolically, the argument has the form

1. $p \rightarrow \sim q$ } **the hypotheses**
2. q
∴ $\sim p$ } **conclusion**

and is represented by the conditional $[(p \rightarrow \sim q) \wedge q] \rightarrow \sim p$. The symbol ∴ is read "therefore."

To construct a truth table for this conditional, we need $2^2 = 4$ rows. A column is required for the following: each negation, each hypothesis, the conjunction of the hypotheses, the conclusion, and the conditional representation of the argument. The initial setup is shown in Figure 1.61.

Fill in the truth table as follows:

$\sim q$: A negation has the opposite truth values; enter a T in rows 2 and 4 and an F in rows 1 and 3.

Hypothesis 1: A conditional is false only when its premise is true and its conclusion is false; enter an F in row 1 and Ts elsewhere.

	p	q	$\sim q$	Hypothesis 1 $p \rightarrow \sim q$	Hypothesis 2 q	Column Representing All the Hypotheses $1 \wedge 2$	Conclusion c $\sim p$	Conditional Representation of the Argument $(1 \wedge 2) \rightarrow c$
1.	T	T	F	F	T	F	F	T
2.	T	F	T	T	F	F	F	T
3.	F	T	F	T	T	T	T	T
4.	F	F	T	T	F	F	T	T

FIGURE 1.61   Required columns in the truth table.

Hypothesis 2: Recopy the q column.

1 $\wedge$ 2: A conjunction is true only when both components are true; enter a T in row 3 and Fs elsewhere.

Conclusion c: A negation has the opposite truth values; enter an F in rows 1 and 2 and a T in rows 3 and 4.

At this point, all that remains is the final column (see Figure 1.62).

	p	q	$\sim q$	1 $p \rightarrow \sim q$	2 q	$1 \wedge 2$	c $\sim p$	$(1 \wedge 2) \rightarrow c$
1.	T	T	F	F	T	F	F	
2.	T	F	T	T	F	F	F	
3.	F	T	F	T	T	T	T	
4.	F	F	T	T	F	F	T	

FIGURE 1.62 Truth values of the expressions.

The last column in the truth table is the conditional that represents the entire argument. A conditional is false only when its premise is true and its conclusion is false. The only instance in which the premise $(1 \wedge 2)$ is true is row 3. Corresponding to this entry, the conclusion $\sim p$ is also true. Consequently, the conditional $(1 \wedge 2) \rightarrow c$ is true in row 3. Because the premise $(1 \wedge 2)$ is false in rows 1, 2, and 4, the conditional $(1 \wedge 2) \rightarrow c$ is automatically true in those rows as well. The completed truth table is shown in Figure 1.63.

	p	q	$\sim q$	1 $p \rightarrow \sim q$	2 q	$1 \wedge 2$	c $\sim p$	$(1 \wedge 2) \rightarrow c$
1.	T	T	F	F	T	F	F	T
2.	T	F	T	T	F	F	F	T
3.	F	T	F	T	T	T	T	T
4.	F	F	T	T	F	F	T	T

FIGURE 1.63 Truth table for the argument $[(p \rightarrow \sim q) \wedge q] \rightarrow \sim p$.

The completed truth table shows that the conditional $[(p \rightarrow \sim q) \wedge q] \rightarrow \sim p$ is always true. The conditional represents the argument "If he is illiterate, he cannot fill out the application. He can fill out the application. Therefore, he is not illiterate." Thus, the argument is valid.

Tautologies

A **tautology** is a statement that is always true. For example, the statement

$$\text{``}(a + b)^2 = a^2 + 2ab + b^2\text{''}$$

is a tautology.

EXAMPLE **2**

SOLUTION

DETERMINING WHETHER A STATEMENT IS A TAUTOLOGY
Determine whether the statement $(p \wedge q) \rightarrow (p \vee q)$ is a tautology.

We need to construct a truth table for the statement. Because there are two letters, the table must have $2^2 = 4$ rows. We need a column for $(p \wedge q)$, one for $(p \vee q)$, and one for $(p \wedge q) \rightarrow (p \vee q)$. The completed truth table is shown in Figure 1.64.

	p	q	$p \wedge q$	$p \vee q$	$(p \wedge q) \rightarrow (p \vee q)$
1.	T	T	T	T	T
2.	T	F	F	T	T
3.	F	T	F	T	T
4.	F	F	F	F	T

FIGURE 1.64   Truth table for the statement $(p \wedge q) \rightarrow (p \vee q)$.

all true
are
Tautogy
is also
Valid
argument

Because $(p \wedge q) \rightarrow (p \vee q)$ is always true, it is a tautology.

As we have seen, an argument can be represented by a single conditional. If this conditional is always true, the argument is valid (and vice versa).

VALIDITY OF AN ARGUMENT

An argument having n hypotheses $h_1, h_2, \ldots, h_n$ and conclusion c is valid if and only if the conditional $[h_1 \wedge h_2 \wedge \ldots \wedge h_n] \rightarrow c$ is a tautology.

EXAMPLE **3**

SOLUTION

USING A TRUTH TABLE TO ANALYZE AN ARGUMENT   Determine whether the following argument is valid:
"If the defendant is innocent, the defendant does not go to jail. The defendant does not go to jail. Therefore, the defendant is innocent."

Separating the hypotheses from the conclusion, we have

1. If the defendant is innocent, the defendant does not go to jail.
2. The defendant does not go to jail.

Therefore, the defendant is innocent.

Now we define symbols to represent the various components of the statements:

p: The defendant is innocent.
q: The defendant goes to jail.

Symbolically, the argument has the form

1. $p \rightarrow \sim q$
2. $\sim q$

$\therefore p$

and is represented by the conditional $[(p \rightarrow \sim q) \wedge \sim q] \rightarrow p$.

Now we construct a truth table with four rows, along with the necessary columns. The completed table is shown in Figure 1.65.

		2	**1**		**c**	
p	**q**	**~q**	**p → ~q**	**1 ∧ 2**	**p**	**(1 ∧ 2) → c**
1. T	T	F	F	F	T	T
2. T	F	T	T	T	T	T
3. F	T	F	T	F	F	T
4. F	F	T	T	T	F	F

FIGURE 1.65 Truth table for the argument [(p → ~q) ∧ ~q] → p.

The column representing the argument has an F in row 4; therefore, the conditional representation of the argument is *not* a tautology. In particular, the conclusion does not logically follow the hypotheses when both *p* and *q* are false (row 4). The argument is not valid. Let us interpret the circumstances expressed in row 4, the row in which the argument breaks down. Both *p* and *q* are false—that is, the defendant is guilty and the defendant does *not* go to jail. Unfortunately, this situation can occur in the real world; guilty people do not *always* go to jail! As long as it is possible for a guilty person to avoid jail, the argument is invalid.

The following argument was presented as Example 6 in Section 1.1. In that section, we constructed a Venn diagram to show that the argument was in fact valid. We now show an alternative method; that is, we construct a truth table to determine whether the argument is valid.

EXAMPLE 4

USING A TRUTH TABLE TO ANALYZE AN ARGUMENT Determine whether the following argument is valid: "No snake is warm-blooded. All mammals are warm-blooded. Therefore, snakes are not mammals."

SOLUTION

Separating the hypotheses from the conclusion, we have

1. No snake is warm-blooded.
2. All mammals are warm-blooded.

Therefore, snakes are not mammals.

These statements can be rephrased as follows:

1. If it is a snake, then it is not warm-blooded.
2. If it is a mammal, then it is warm-blooded.

Therefore, if it is a snake, then it is not a mammal.

Now we define symbols to represent the various components of the statements:

p: It is a snake.
q: It is warm-blooded.
r: It is a mammal.

Symbolically, the argument has the form

1. $p \to \sim q$
2. $r \to q$
$\therefore p \to \sim r$

and is represented by the conditional $[(p \to \sim q) \wedge (r \to q)] \to (p \to \sim r)$.

Now we construct a truth table with eight rows ($2^3 = 8$), along with the necessary columns. The completed table is shown in Figure 1.66.

						1	**2**		**c**	
	p	q	r	$\sim q$	$\sim r$	$p \to \sim q$	$r \to q$	$1 \wedge 2$	$p \to \sim r$	$(1 \wedge 2) \to c$
1.	T	T	T	F	F	F	T	F	F	T
2.	T	T	F	F	T	F	T	F	T	T
3.	T	F	T	T	F	T	F	F	F	T
4.	T	F	F	T	T	T	T	T	T	T
5.	F	T	T	F	F	T	T	T	T	T
6.	F	T	F	F	T	T	T	T	T	T
7.	F	F	T	T	F	T	F	F	T	T
8.	F	F	F	T	T	T	T	T	T	T

FIGURE 1.66 Truth table for the argument $[(p \to \sim q) \wedge (r \to q)] \to (p \to \sim r)$.

The last column of the truth table represents the argument and contains all T's. Consequently, the conditional $[(p \to \sim q) \wedge (r \to q)] \to (p \to \sim r)$ is a tautology; the argument is valid.

The preceding examples contained relatively simple arguments, each consisting of only two hypotheses and two simple statements (letters). In such cases, many people try to employ "common sense" to confirm the validity of the argument. For instance, the argument "If it is raining, the streets are wet. It is raining. Therefore, the streets are wet" is obviously valid. However, it might not be so simple to determine the validity of an argument that contains several hypotheses and many simple statements. Indeed, in such cases, the argument's truth table might become quite lengthy, as in the next example.

EXAMPLE **5**

USING A TRUTH TABLE TO ANALYZE AN ARGUMENT   The following whimsical argument was written by Lewis Carroll and appeared in his 1896 book *Symbolic Logic:*

"No ducks waltz. No officers ever decline to waltz. All my poultry are ducks. Therefore, my poultry are not officers."

Construct a truth table to determine whether the argument is valid.

HISTORICAL NOTE

CHARLES LUTWIDGE DODGSON, 1832–1898

© Bettmann/CORBIS

To those who assume that it is impossible for a person to excel both in the creative worlds of art and literature and in the disciplined worlds of mathematics and logic, the life of Charles Lutwidge Dodgson is a wondrous counterexample. Known the world over as Lewis Carroll, Dodgson penned the nonsensical classics *Alice's Adventures in Wonderland* and *Through the Looking Glass*. However, many people are surprised to learn that Dodgson (from age eighteen to his death) was a permanent resident at the University at Oxford, teaching mathematics and logic. And as if that were not enough, Dodgson is now recognized as one of the leading portrait photographers of the Victorian era.

The eldest son in a family of eleven children, Charles amused his younger siblings with elaborate games, poems, stories, and humorous drawings. This attraction to entertaining children with fantastic stories manifested itself in much of his later work as Lewis Carroll. Besides his obvious interest in telling stories, the young Dodgson was also intrigued by mathematics. At the age of eight, Charles asked his father to explain a book on logarithms. When told that he was too young to understand, Charles persisted, "But please, explain!"

The Dodgson family had a strong ecclesiastical tradition; Charles's father, great-grandfather, and great-great-grandfather were all clergymen. Following in his father's footsteps, Charles attended Christ Church, the largest and most celebrated of all the Oxford colleges. After graduating in 1854, Charles remained at Oxford, accepting the position of mathematical lecturer in 1855. However, appointment to this position was conditional upon his taking Holy Orders in the Anglican church and upon his remaining celibate. Dodgson complied and was named a deacon in 1861.

The year 1856 was filled with events that had lasting effects on Dodgson. Charles Lutwidge created his pseudonym by translating his first and middle names into Latin (Carolus Ludovic), reversing their order (Ludovic Carolus), and translating them back into English (Lewis Carroll). In this same year, Dodgson began his "hobby" of photography. He is considered by many to have been an artistic pioneer in this new field (photography was invented in 1839). Most of Dodgson's work consists of portraits that chronicle the Victorian era, and over 700 photographs taken by Dodgson have been preserved. His favorite subjects were children, especially young girls.

Dodgson's affinity for children brought about a meeting in 1856 that would eventually establish his place in the history of literature. Early in the year, Dodgson met the four children of the dean of Christ Church: Harry, Lorina, Edith, and Alice Liddell. He began seeing the children on a regular basis, amusing them with stories and photographing them. Although he had a wonderful relationship with all four, Alice received his special attention.

On July 4, 1862, while rowing and picnicking with Alice and her sisters, Dodgson entertained the Liddell girls with a fantastic story of a little girl named Alice who fell into a rabbit hole. Captivated by

---

SOLUTION

Separating the hypotheses from the conclusion, we have

1. No ducks waltz.
2. No officers ever decline to waltz.
3. All my poultry are ducks.

Therefore, my poultry are not officers.

These statements can be rephrased as

1. If it is a duck, then it does not waltz.
2. If it is an officer, then it does not decline to waltz.
   (Equivalently, "If it is an officer, then it will waltz.")
3. If it is my poultry, then it is a duck.

Therefore, if it is my poultry, then it is not an officer.

Young Alice Liddell inspired Lewis Carroll to write Alice's Adventures in Wonderland. *This photo is one of the many Carroll took of Alice.*

episodes, Lewis Carroll gave the world *Alice's Adventures in Wonderland*. Although the book appeared to be a whimsical excursion into chaotic nonsense, Dodgson's masterpiece contained many exercises in logic and metaphor. The book was a success, and in 1871, a sequel, *Through the Looking Glass*, was printed. When asked to comment on the meaning of his writings, Dodgson replied, "I'm very much afraid I didn't mean anything but nonsense! Still, you know, words mean more than we mean to express when we use them; so a whole book ought to mean a great deal more than the writer means. So, whatever good meanings are in the book, I'm glad to accept as the meaning of the book."

In addition to writing "children's stories," Dodgson wrote numerous mathematics essays and texts, including *The Fifth Book of Euclid Proved Algebraically, Formulae of Plane Trigonometry, A Guide to the Mathematical Student,* and *Euclid and His Modern Rivals*. In the field of formal logic, Dodgson's books *The Game of Logic* (1887) and *Symbolic Logic* (1896) are still used as sources of inspiration in numerous schools worldwide.

the story, Alice Liddell insisted that Dodgson write it down for her. He complied, initially titling it *Alice's Adventure Underground*.

Dodgson's friends subsequently encouraged him to publish the manuscript, and in 1865, after editing and inserting new

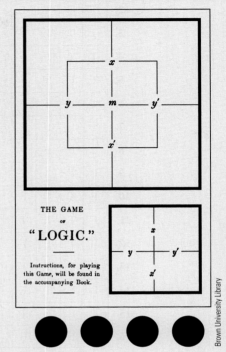

Carroll's book The Game of Logic *presents the study of formalized logic in a gamelike fashion. After listing the "rules of the game" (complete with gameboard and markers), Carroll captures the reader's interest with nonsensical syllogisms.*

Now we define symbols to represent the various components of the statements:

p: It is a duck.
q: It will waltz.
r: It is an officer.
s: It is my poultry.

Symbolically, the argument has the form

1. $p \rightarrow \sim q$
2. $r \rightarrow q$
3. $s \rightarrow p$
$$\therefore s \rightarrow \sim r$$

		p	q	r	s	~q	~r	1 p → ~q	2 r → q	3 s → p	1 ∧ 2 ∧ 3	c s → ~r	(1 ∧ 2 ∧ 3) → c
1.		T	T	T	T	F	F	F	T	T	F	F	T
2.		T	T	T	F	F	F	F	T	T	F	T	T
3.		T	T	F	T	F	T	F	T	T	F	T	T
4.		T	T	F	F	F	T	F	T	T	F	T	T
5.		T	F	T	T	T	F	T	F	T	F	F	T
6.		T	F	T	F	T	F	T	F	T	F	T	T
7.		T	F	F	T	T	T	T	T	T	T	T	T
8.		T	F	F	F	T	T	T	T	T	T	T	T
9.		F	T	T	T	F	F	T	T	F	F	F	T
10.		F	T	T	F	F	F	T	T	T	T	T	T
11.		F	T	F	T	F	T	T	T	F	F	T	T
12.		F	T	F	F	F	T	T	T	T	T	T	T
13.		F	F	T	T	T	F	T	F	F	F	F	T
14.		F	F	T	F	T	F	T	F	T	F	T	T
15.		F	F	F	T	T	T	T	T	F	F	T	T
16.		F	F	F	F	T	T	T	T	T	T	T	T

FIGURE 1.67 Truth table for the argument $[(p \to \sim q) \wedge (r \to q) \wedge (s \to p)] \to (s \to \sim r)$.

Now we construct a truth table with sixteen rows ($2^4 = 16$), along with the necessary columns. The completed table is shown in Figure 1.67.

The last column of the truth table represents the argument and contains all T's. Consequently, the conditional $[(p \to \sim q) \wedge (r \to q) \wedge (s \to p)] \to (s \to \sim r)$ is a tautology; the argument is valid.

1.5 EXERCISES

In Exercises 1–10, use the given symbols to rewrite the argument in symbolic form.

1. *p:* It is raining.
q: The streets are wet. } Use these symbols.

  1. If it is raining, then the streets are wet.
  2. It is raining.

Therefore, the streets are wet.

2. *p:* I have a college degree.
q: I am lazy. } Use these symbols.

  1. If I have a college degree, I am not lazy.
  2. I do not have a college degree.

Therefore, I am lazy.

3. *p:* It is Tuesday.
q: The tour group is in Belgium. } Use these symbols.

  1. If it is Tuesday, then the tour group is in Belgium.
  2. The tour group is not in Belgium.

Therefore, it is not Tuesday.

4. *p:* You are a gambler.
q: You have financial security. } Use these symbols.

  1. You do not have financial security if you are a gambler.
  2. You do not have financial security.

Therefore, you are a gambler.

5. *p:* You exercise regularly.
   q: You are healthy. } Use these symbols.

   1. You exercise regularly only if you are healthy.
   2. You do not exercise regularly.

   Therefore, you are not healthy.

6. *p:* The senator supports new taxes.
   q: The senator is reelected. } Use these symbols.

   1. The senator is not reelected if she supports new taxes.
   2. The senator does not support new taxes.

   Therefore, the senator is reelected.

7. *p:* A person knows Morse code.
   q: A person operates a telegraph. } Use these symbols.
   r: A person is Nikola Tesla.

   1. Knowing Morse code is necessary for operating a telegraph.
   2. Nikola Tesla knows Morse code.

   Therefore, Nikola Tesla operates a telegraph.

   HINT: Hypothesis 2 can be symbolized as $r \wedge p$.

8. *p:* A person knows CPR.
   q: A person is a paramedic. } Use these symbols.
   r: A person is David Lee Roth.

   1. Knowing CPR is necessary for being a paramedic.
   2. David Lee Roth is a paramedic.

   Therefore, David Lee Roth knows CPR.

   HINT: Hypothesis 2 can be symbolized as $r \wedge q$.

9. *p:* It is a monkey.
   q: It is an ape. } Use these symbols.
   r: It is King Kong.

   1. Being a monkey is sufficient for not being an ape.
   2. King Kong is an ape.

   Therefore, King Kong is not a monkey.

10. *p:* It is warm-blooded.
   q: It is a reptile. } Use these symbols.
   r: It is Godzilla.

   1. Being warm-blooded is sufficient for not being a reptile.
   2. Godzilla is not warm-blooded.

   Therefore, Godzilla is a reptile.

In Exercises 11–20, use a truth table to determine the validity of the argument specified. If the argument is invalid, interpret the specific circumstances that cause it to be invalid.

11. the argument in Exercise 1
12. the argument in Exercise 2
13. the argument in Exercise 3
14. the argument in Exercise 4
15. the argument in Exercise 5
16. the argument in Exercise 6
17. the argument in Exercise 7
18. the argument in Exercise 8
19. the argument in Exercise 9
20. the argument in Exercise 10

In Exercises 21–42, define the necessary symbols, rewrite the argument in symbolic form, and use a truth table to determine whether the argument is valid. If the argument is invalid, interpret the specific circumstances that cause the argument to be invalid.

21. 1. If the Democrats have a majority, Smith is appointed and student loans are funded.
   2. Smith is appointed or student loans are not funded.

   Therefore, the Democrats do not have a majority.

22. 1. If you watch television, you do not read books.
   2. If you read books, you are wise.

   Therefore, you are not wise if you watch television.

23. 1. If you argue with a police officer, you get a ticket.
   2. If you do not break the speed limit, you do not get a ticket.

   Therefore, if you break the speed limit, you argue with a police officer.

24. 1. If you do not recycle newspapers, you are not an environmentalist.
   2. If you recycle newspapers, you save trees.

   Therefore, you are an environmentalist only if you save trees.

25. 1. All pesticides are harmful to the environment.
   2. No fertilizer is a pesticide.

   Therefore, no fertilizer is harmful to the environment.

26. 1. No one who can afford health insurance is unemployed.
   2. All politicians can afford health insurance.

   Therefore, no politician is unemployed.

27. 1. All poets are loners.
   2. All loners are taxi drivers.

   Therefore, all poets are taxi drivers.

28. 1. All forest rangers are environmentalists.
   2. All forest rangers are storytellers.

   Therefore, all environmentalists are storytellers.

29. 1. No professor is a millionaire.
   2. No millionaire is illiterate.

   Therefore, no professor is illiterate.

30. 1. No artist is a lawyer.
   2. No lawyer is a musician.

   Therefore, no artist is a musician.

31. 1. All lawyers study logic.
   2. You study logic only if you are a scholar.
   3. You are not a scholar.

   Therefore, you are not a lawyer.

32. 1. All licensed drivers have insurance.
2. You obey the law if you have insurance.
3. You obey the law.

Therefore, you are a licensed driver.

33. 1. Drinking espresso is sufficient for not sleeping.
2. Not eating dessert is necessary for being on a diet.
3. Not eating dessert is sufficient for drinking espresso.

Therefore, not being on a diet is necessary for sleeping.

34. 1. Not being eligible to vote is sufficient for ignoring politics.
2. Not being a convicted felon is necessary for being eligible to vote.
3. Ignoring politics is sufficient for being naive.

Therefore, being naive is necessary being a convicted felon.

35. If the defendant is innocent, he does not go to jail. The defendant goes to jail. Therefore, the defendant is guilty.

36. If the defendant is innocent, he does not go to jail. The defendant is guilty. Therefore, the defendant goes to jail.

37. If you are not in a hurry, you eat at Lulu's Diner. If you are in a hurry, you do not eat good food. You eat at Lulu's. Therefore, you eat good food.

38. If you give me a hamburger today, I pay you tomorrow. If you are a sensitive person, you give me a hamburger today. You are not a sensitive person. Therefore, I do not pay you tomorrow.

39. If you listen to rock and roll, you do not go to heaven. If you are a moral person, you go to heaven. Therefore, you are not a moral person if you listen to rock and roll.

40. If you follow the rules, you have no trouble. If you are not clever, you have trouble. You are clever. Therefore, you do not follow the rules.

41. The water not being cold is sufficient for going swimming. Having goggles is necessary for going swimming. I have no goggles. Therefore, the water is cold.

42. I wash my car only if the sky is clear. The sky not being clear is necessary for it to rain. I do not wash my car. Therefore, it is raining.

The arguments given in Exercises 43–50 were written by Lewis Carroll and appeared in his 1896 book Symbolic Logic. *For each argument, define the necessary symbols, rewrite the argument in symbolic form, and use a truth table to determine whether the argument is valid.*

43. 1. All medicine is nasty.
2. Senna is a medicine.

Therefore, senna is nasty.

NOTE: Senna is a laxative extracted from the dried leaves of cassia plants.

44. 1. All pigs are fat.
2. Nothing that is fed on barley-water is fat.

Therefore, pigs are not fed on barley-water.

45. 1. Nothing intelligible ever puzzles me.
2. Logic puzzles me.

Therefore, logic is unintelligible.

46. 1. No misers are unselfish.
2. None but misers save eggshells.

Therefore, no unselfish people save eggshells.

47. 1. No Frenchmen like plum pudding.
2. All Englishmen like plum pudding.

Therefore, Englishmen are not Frenchmen.

48. 1. A prudent man shuns hyenas.
2. No banker is imprudent.

Therefore, no banker fails to shun hyenas.

49. 1. All wasps are unfriendly.
2. No puppies are unfriendly.

Therefore, puppies are not wasps.

50. 1. Improbable stories are not easily believed.
2. None of his stories are probable.

Therefore, none of his stories are easily believed.

 Answer the following questions using complete sentences and your own words.

• CONCEPT QUESTIONS

51. What is a tautology?

52. What is the conditional representation of an argument?

53. Find a "logical" argument in a newspaper article, an advertisement, or elsewhere in the media. Analyze that argument and discuss the implications.

• HISTORY QUESTIONS

54. What was Charles Dodgson's pseudonym? How did he get it? What classic "children's stories" did he write?

55. What did Charles Dodgson contribute to the study of formal logic?

56. Charles Dodgson was a pioneer in what artistic field?

57. Who was Alice Liddell?

WEB PROJECT

58. Write a research paper on any historical topic referred to in this chapter or a related topic. Below is a partial list of topics.
- Aristotle
- George Boole
- Augustus De Morgan
- Charles Dodgson/Lewis Carroll
- Gottfried Wilhelm Leibniz

Some useful links for this web project are listed on the text web site: **www.cengage.com/math/johnson**

TERMS	converse	invalid argument	sudoku
argument	deductive reasoning	inverse	sufficient
biconditional	disjunction	logic	syllogism
compound statement	equivalent expressions	necessary	tautology
conclusion	exclusive *or*	negation	truth table
conditional	hypothesis	premise	truth value
conjunction	implication	quantifier	valid argument
contrapositive	inclusive *or*	reasoning	Venn diagram
	inductive reasoning	statement	

REVIEW EXERCISES

1. Classify each argument as deductive or inductive.
   a. 1. Hitchcock's "Psycho" is a suspenseful movie.
   2. Hitchcock's "The Birds" is a suspenseful movie.
   Therefore, all Hitchcock movies are suspenseful.
   b. 1. All Hitchcock movies are suspenseful.
   2. "Psycho" is a Hitchcock movie.
   Therefore, "Psycho" is suspenseful.

2. Explain the general rule or pattern used to assign the given letter to the given word. Fill in the blank with the letter that fits the pattern.

day	morning	afternoon	dusk	night
y	r	f	s	_____

3. Fill in the blank with what is most likely to be the next number. Explain the pattern generated by your answer.

   1, 6, 11, 4, _____

In Exercises 4–9, construct a Venn diagram to determine the validity of the given argument.

4. 1. All truck drivers are union members.
   2. Rocky is a truck driver.
   Therefore, Rocky is a union member.

5. 1. All truck drivers are union members.
   2. Rocky is not a truck driver.
   Therefore, Rocky is not a union member.

6. 1. All mechanics are engineers.
   2. Casey Jones is an engineer.
   Therefore, Casey Jones is a mechanic.

7. 1. All mechanics are engineers.
   2. Casey Jones is not an engineer.
   Therefore, Casey Jones is not a mechanic.

8. 1. Some animals are dangerous.
   2. A gun is not an animal.
   Therefore, a gun is not dangerous.

9. 1. Some contractors are electricians.
   2. All contractors are carpenters.
   Therefore, some electricians are carpenters.

10. Solve the following sudoku puzzle.

	2			3			3	4	
			3		6				7
			7			9			8
		4			1	7			
5			4	7	8				3
		1	6			2			
2		6			4				
3			5		7				
	7	8						3	

11. Explain why each of the following is or is not a statement.
    a. The Golden Gate Bridge spans Chesapeake Bay.
    b. The capital of Delaware is Dover.
    c. Where are you spending your vacation?
    d. Hawaii is the best place to spend a vacation.

12. Determine which pairs of statements are negations of each other.
    a. All of the lawyers are ethical.
    b. Some of the lawyers are ethical.
    c. None of the lawyers is ethical.
    d. Some of the lawyers are not ethical.

13. Write a sentence that represents the negation of each statement.
    a. His car is not new.
    b. Some buildings are earthquake proof.
    c. All children eat candy.
    d. I never cry in a movie theater.

14. Using the symbolic representations

    p: The television program is educational.

    q: The television program is controversial.

    express the following compound statements in symbolic form.
    a. The television program is educational and controversial.
    b. If the television program isn't controversial, it isn't educational.
    c. The television program is educational and it isn't controversial.
    d. The television program isn't educational or controversial.
    e. Not being controversial is necessary for a television program to be educational.
    f. Being controversial is sufficient for a television program not to be educational.

15. Using the symbolic representations

    p: The advertisement is effective.

    q: The advertisement is misleading.

    r: The advertisement is outdated.

    express the following compound statements in symbolic form.
    a. All misleading advertisements are effective.
    b. It is a current, honest, effective advertisement.
    c. If an advertisement is outdated, it isn't effective.
    d. The advertisement is effective and it isn't misleading or outdated.
    e. Not being outdated or misleading is necessary for an advertisement to be effective.
    f. Being outdated and misleading is sufficient for an advertisement not to be effective.

16. Using the symbolic representations

    p: It is expensive.

    q: It is undesirable.

    express the following in words.
    a. $p \rightarrow \sim q$ b. $q \leftrightarrow \sim p$
    c. $\sim(p \vee q)$ d. $(p \wedge \sim q) \vee (\sim p \wedge q)$

17. Using the symbolic representations

    p: The movie is critically acclaimed.

    q: The movie is a box office hit.

    r: The movie is available on DVD.

    express the following in words.
    a. $(p \vee q) \rightarrow r$ b. $(p \wedge \sim q) \rightarrow \sim r$
    c. $\sim(p \vee q) \wedge r$ d. $\sim r \rightarrow (\sim p \wedge \sim q)$

In Exercises 18–25, construct a truth table for the compound statement.

18. $p \vee \sim q$ 19. $p \wedge \sim q$
20. $\sim p \rightarrow q$ 21. $(p \wedge q) \rightarrow \sim q$
22. $q \vee \sim(p \vee r)$ 23. $\sim p \rightarrow (q \vee r)$
24. $(q \wedge p) \rightarrow (\sim r \vee p)$ 25. $(p \vee r) \rightarrow (q \wedge \sim r)$

In Exercises 26–30, construct a truth table to determine whether the statements in each pair are equivalent.

26. The car is unreliable or expensive.
    If the car is reliable, then it is expensive.
27. If I get a raise, I will buy a new car.
    If I do not get a raise, I will not buy a new car.
28. She is a Democrat or she did not vote.
    She is not a Democrat and she did vote.
29. The raise is not unjustified and the management opposes it.
    It is not the case that the raise is unjustified or the management does not oppose it.
30. Walking on the beach is sufficient for not wearing shoes. Wearing shoes is necessary for not walking on the beach.

In Exercises 31–38, write a sentence that represents the negation of each statement.

31. Jesse had a party and nobody came.
32. You do not go to jail if you pay the fine.
33. I am the winner or you are blind.
34. He is unemployed and he did not apply for financial assistance.
35. The selection procedure has been violated if his application is ignored.
36. The jackpot is at least $1 million.
37. Drinking espresso is sufficient for not sleeping.
38. Not eating dessert is necessary for being on a diet.
39. Given the statements

    p: You are an avid jogger.

    q: You are healthy.

    write the sentence represented by each of the following.
    a. $p \rightarrow q$ b. $q \rightarrow p$
    c. $\sim p \rightarrow \sim q$ d. $\sim q \rightarrow \sim p$
    e. $p \leftrightarrow q$

40. Form (a) the inverse, (b) the converse, and (c) the contrapositive of the conditional "If he is elected, the country is in big trouble."

In Exercises 41 and 42, express the contrapositive of the given conditional in terms of (a) a sufficient condition, and (b) a necessary condition.

41. Having a map is sufficient for not being lost.
42. Having syrup is necessary for eating pancakes.

In Exercises 43–48, (a) determine the premise and conclusion and (b) rewrite the compound statement in the standard "if . . . then . . ." form.

43. The economy improves only if unemployment goes down.

44. The economy improves if unemployment goes down.

45. No computer is unrepairable.

46. All gemstones are valuable.

47. Being the fourth Thursday in November is sufficient for the U.S. Post Office to be closed.

48. Having diesel fuel is necessary for the vehicle to operate.

In Exercises 49 and 50, translate the two statements into symbolic form and use truth tables to determine whether the statements are equivalent.

49. If you are allergic to dairy products, you cannot eat cheese.

If you cannot eat cheese, then you are allergic to dairy products.

50. You are a fool if you listen to me.

You are not a fool only if you do not listen to me.

51. Which pairs of statements are equivalent?

  i. If it is not raining, I ride my bicycle to work.

  ii. If I ride my bicycle to work, it is not raining.

  iii. If I do not ride my bicycle to work, it is raining.

  iv. If it is raining, I do not ride my bicycle to work.

In Exercises 52–57, define the necessary symbols, rewrite the argument in symbolic form, and use a truth table to determine whether the argument is valid.

52. 1. If you do not make your loan payment, your car is repossessed.

  2. Your car is repossessed.

Therefore, you did not make your loan payment.

53. 1. If you do not pay attention, you do not learn the new method.

  2. You do learn the new method.

Therefore, you do pay attention.

54. 1. If you rent DVD, you will not go to the movie theater.

  2. If you go to the movie theater, you pay attention to the movie.

Therefore, you do not pay attention to the movie if you rent DVDs.

55. 1. If the Republicans have a majority, Farnsworth is appointed and no new taxes are imposed.

  2. New taxes are imposed.

Therefore, the Republicans do not have a majority or Farnsworth is not appointed.

56. 1. Practicing is sufficient for making no mistakes.

  2. Making a mistake is necessary for not receiving an award.

  3. You receive an award.

Therefore, you practice.

57. 1. Practicing is sufficient for making no mistakes.

  2. Making a mistake is necessary for not receiving an award.

  3. You do not receive an award.

Therefore, you do not practice.

In Exercises 58–66, define the necessary symbols, rewrite the argument in symbolic form, and use a truth table to determine whether the argument is valid.

58. If the defendant is guilty, he goes to jail. The defendant does not go to jail. Therefore, the defendant is not guilty.

59. I will go to the concert only if you buy me a ticket. You bought me a ticket. Therefore, I will go to the concert.

60. If tuition is raised, students take out loans or drop out. If students do not take out loans, they drop out. Students do drop out. Therefore, tuition is raised.

61. If our oil supply is cut off, our economy collapses. If we go to war, our economy doesn't collapse. Therefore, if our oil supply isn't cut off, we do not go to war.

62. No professor is uneducated. No monkey is educated. Therefore, no professor is a monkey.

63. No professor is uneducated. No monkey is a professor. Therefore, no monkey is educated.

64. Vehicles stop if the traffic light is red. There is no accident if vehicles stop. There is an accident. Therefore, the traffic light is not red.

65. Not investing money in the stock market is necessary for invested money to be guaranteed. Invested money not being guaranteed is sufficient for not retiring at an early age. Therefore, you retire at an early age only if your money is not invested in the stock market.

66. Not investing money in the stock market is necessary for invested money to be guaranteed. Invested money not being guaranteed is sufficient for not retiring at an early age. You do not invest in the stock market. Therefore, you retire at an early age.

Determine the validity of the arguments in Exercises 67 and 68 by constructing a

a. Venn diagram and a

b. truth table.

c. How do the answers to parts (a) and (b) compare? Why?

67. 1. If you own a hybrid vehicle, then you are an environmentalist.

  2. You are not an environmentalist.

Therefore, you do not own a hybrid vehicle.

68.
1. If you own a hybrid vehicle, then you are an environmentalist.
2. You are an environmentalist.

Therefore, you own a hybrid vehicle.

 Answer the following questions using complete sentences and your own words.

• CONCEPT QUESTIONS

69. What is a statement?

70.
 a. What is a disjunction? Under what conditions is a disjunction true?
 b. What is a conjunction? Under what conditions is a conjunction true?
 c. What is a conditional? Under what conditions is a conditional true?
 d. What is a negation? Under what conditions is a negation true?

71.
 a. What is a sufficient condition?
 b. What is a necessary condition?

72. What is a tautology?

73. When constructing a truth table, how do you determine how many rows to create?

• HISTORY QUESTIONS

74. What role did the following people play in the development of formalized logic?
- Aristotle
- George Boole
- Augustus De Morgan
- Charles Dodgson
- Gottfried Wilhelm Leibniz

SETS AND COUNTING

© Image copyright cristina popescu, 2009. Used under license from Shutterstock.com

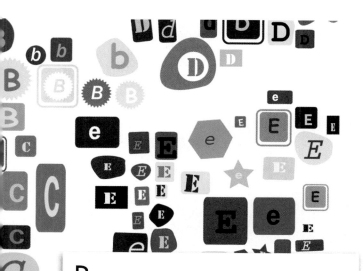

Recently, 1,000 college seniors were asked whether they favored increasing their state's gasoline tax to generate funds to improve highways and whether they favored increasing their state's alcohol tax to generate funds to improve the public education system. The responses were tallied, and the following results were printed in the campus newspaper: 746 students favored an increase in the gasoline tax, 602 favored an increase in the alcohol tax, and 449 favored increase in both taxes. How many of these 1,000 students favored an increase in at least one of the taxes? How many favored increasing only the gasoline tax? How many favored increasing only the alcohol tax? How many favored increasing neither tax?

The mathematical tool that was designed to answer questions like these is

continued

WHAT WE WILL DO IN THIS CHAPTER

WE'LL USE VENN DIAGRAMS TO DEPICT THE RELATIONSHIPS BETWEEN SETS:

- One set might be contained within another set.
- Two or more sets might, or might not, share elements in common.

WE'LL EXPLORE APPLICATIONS OF VENN DIAGRAMS:

- The results of consumer surveys, marketing analyses, and political polls can be analyzed by using Venn diagrams.
- Venn diagrams can be used to prove general formulas related to set theory.

WE'LL EXPLORE VARIOUS METHODS OF COUNTING:

- A fundamental principle of counting is used to determine the total number of possible ways of selecting specified items. For example, how many different student ID numbers are possible at your school?

continued

- In selecting items from a specified group, sometimes the order in which the items are selected matters (the awarding of prizes: first, second, and third), and sometimes it does not (selecting numbers in a lottery or people for a committee). How does this affect your method of counting?

WE'LL USE SETS IN VARIOUS CONTEXTS:

- In this text, we will use set theory extensively in Chapter 3 on probability.

- Many standardized admissions tests, such as the Graduate Record Exam (GRE) and the Law School Admissions Test (LSAT), ask questions that can be answered with set theory.

WE'LL EXPLORE SETS THAT HAVE AN INFINITE NUMBER OF ELEMENTS:

- One-to-one correspondences are used to "count" and compare the number of elements in infinite sets.

- Not all infinite sets have the same number of elements; some infinite sets are countable, and some are not.

the *set. Webster's New World College Dictionary* defines a **set** as "a prescribed collection of points, numbers, or other objects that satisfy a given condition." Although you might be able to answer the questions about taxes without any formal knowledge of sets, the mental reasoning involved in obtaining your answers uses some of the basic principles of sets. (Incidentally, the answers to the above questions are 899, 297, 153, and 101, respectively.)

The branch of mathematics that deals with sets is called **set theory.** Set theory can be helpful in solving both mathematical and nonmathematical problems. We will explore set theory in the first half of this chapter. As the above example shows, set theory often involves the analysis of the relationships between sets and counting the number of elements in a specific category. Consequently, various methods of counting, collectively known as **combinatorics,** will be developed and discussed in the second half of this chapter. Finally, what if a set has too many elements to count by using finite numbers? For example, how many integers are there? How many real numbers? The chapter concludes with an exploration of infinite sets and various "levels of infinity."

2.1 Sets and Set Operations

OBJECTIVES

- Learn the basic vocabulary and notation of set theory
- Learn and apply the union, intersection, and complement operations
- Draw Venn diagrams

A **set** is a collection of objects or things. The objects or things in the set are called **elements** (or *members*) of the set. In our example above, we could talk about the *set* of students who favor increasing only the gasoline tax or the *set* of students who do not favor increasing either tax. In geography, we can talk about the *set* of all state capitals or the *set* of all states west of the Mississippi. It is easy to determine whether something is in these sets; for example, Des Moines is an element of the set of state capitals, whereas Dallas is not. Such sets are called **well-defined** because there is a way of determining for sure whether a particular item is an element of the set.

EXAMPLE 1

DETERMINING WELL-DEFINED SETS Which of the following sets are well-defined?

a. the set of all movies directed by Alfred Hitchcock
b. the set of all great rock-and-roll bands
c. the set of all possible two-person committees selected from a group of five people

SOLUTION

a. This set is well-defined; either a movie was directed by Hitchcock, or it was not.
b. This set is *not* well-defined; membership is a matter of opinion. Some people would say that the Ramones (one of the pioneer punk bands of the late 1970s) are a member, while others might say they are not. (Note: The Ramones were inducted into the Rock and Roll Hall of Fame in 2002.)
c. This set is well-defined; either the two people are from the group of five, or they are not.

Notation

By tradition, a set is denoted by a capital letter, frequently one that will serve as a reminder of the contents of the set. **Roster notation** (also called *listing notation*) is a method of describing a set by listing each element of the set inside the symbols { and }, which are called *set braces*. In a listing of the elements of a set, each distinct element is listed only once, and the order of the elements doesn't matter.

The symbol $\in$ stands for the phrase *is an element of,* and $\notin$ stands for *is not an element of.* The **cardinal number** of a set A is the number of elements in the set and is denoted by $n(A)$. Thus, if R is the set of all letters in the name "Ramones," then $R = \{r, a, m, o, n, e, s\}$. Notice that m is an element of the set R, x is not an element of R, and R has 7 elements. In symbols, $m \in R$, $x \notin R$, and $n(R) = 7$.

The "Ramones" or The "Moaners"? The set R of all letters in the name "Ramones" is the same as the set M of all letters in the name "Moaners." Consequently, the sets are equal; $M = R = \{a, e, m, n, o, r, s\}$. (R.I.P. Joey Ramone 1951–2001, Dee Dee Ramone 1952–2002, Johnny Ramone 1948–2004.)

Two sets are **equal** if they contain exactly the same elements. *The order in which the elements are listed does not matter.* If M is the set of all letters in the name "Moaners," then $M = \{m, o, a, n, e, r, s\}$. This set contains exactly the same elements as the set R of letters in the name "Ramones." Therefore, $M = R = \{a, e, m, n, o, r, s\}$.

Often, it is not appropriate or not possible to describe a set in roster notation. For extremely large sets, such as the set V of all registered voters in Detroit, or for sets that contain an infinite number of elements, such as the set G of all negative real numbers, the roster method would be either too cumbersome or impossible to use. Although V could be expressed via the roster method (since each county compiles a list of all registered voters in its jurisdiction), it would take hundreds or even thousands of pages to list everyone who is registered to vote in Detroit! In the case of the set G of all negative real numbers, no list, no matter how long, is capable of listing all members of the set; there is an infinite number of negative numbers.

In such cases, it is often necessary, or at least more convenient, to use **set-builder notation,** which lists the rules that determine whether an object is an element of the set rather than the actual elements. A set-builder description of set G above is

$$G = \{x \mid x < 0 \quad \text{and} \quad x \in \Re\}$$

which is read as "the set of all x such that x is less than zero and x is a real number." A set-builder description of set V above is

$$V = \{\text{persons} \mid \text{the person is a registered voter in Detroit}\}$$

which is read as "the set of all persons such that the person is a registered voter in Detroit." In set-builder notation, the vertical line stands for the phrase "such that."

Whatever is on the left side of the line is the general type of thing in the set, while the rules about set membership are listed on the right.

EXAMPLE **2**

READING SET-BUILDER NOTATION Describe each of the following in words.

a. $\{x \mid x > 0 \text{ and } x \in \Re\}$ $\in (1,)^{\rightarrow}$
b. $\{\text{persons} \mid \text{the person is a living former U.S. president}\}$
c. $\{\text{women} \mid \text{the woman is a former U.S. president}\}$ *no elements*
empty set $\varnothing$ { }

SOLUTION

a. the set of all x such that x is a positive real number
b. the set of all people such that the person is a living former U.S. president
c. the set of all women such that the woman is a former U.S. president

The set listed in part (c) of Example 2 has no elements; there are no women who are former U.S. presidents. If we let W equal "the set of all women such that the woman is a former U.S. president," then $n(W) = 0$. A set that has no elements is called an **empty set** and is denoted by $\varnothing$ or by { }. Notice that since the empty set has no elements, $n(\varnothing) = 0$. In contrast, the set $\{0\}$ is not empty; it has one element, the number zero, so $n(\{0\}) = 1$.

Universal Set and Subsets

When we work with sets, we must define a universal set. For any given problem, the **universal set,** denoted by U, is the set of all possible elements of any set used in the problem. For example, when we spell words, U is the set of all letters in the alphabet. When every element of one set is also a member of another set, we say that the first set is a *subset* of the second; for instance, {p, i, n} is a subset of {p, i, n, e}. In general, we say that A is a **subset** of B, denoted by $A \subseteq B$, if for every $x \in A$ it follows that $x \in B$. Alternatively, $A \subseteq B$ if A contains no elements that are not in B. If A contains an element that is not in B, then A is not a subset of B (symbolized as $A \nsubseteq B$).

EXAMPLE **3**

DETERMINING SUBSETS Let $B = \{\text{countries} \mid \text{the country has a permanent seat on the U.N. Security Council}\}$. Determine whether A is a subset of B.

a. $A = \{\text{Russian Federation, United States}\}$
b. $A = \{\text{China, Japan}\}$
c. $A = \{\text{United States, France, China, United Kingdom, Russian Federation}\}$
d. $A = \{\ \}$

SOLUTION

We use the roster method to list the elements of set B.

$\qquad B = \{\text{China, France, Russian Federation, United Kingdom, United States}\}$

a. Since every element of A is also an element of B, A is a subset of B; $A \subseteq B$.
b. Since A contains an element (Japan) that is not in B, A is not a subset of B; $A \nsubseteq B$.
c. Since every element of A is also an element of B (note that $A = B$), A is a subset of B (and B is a subset of A); $A \subseteq B$ (and $B \subseteq A$). In general, every set is a subset of itself; $A \subseteq A$ for any set A.
d. Does A contain an element that is not in B? No! Therefore, A (an empty set) is a subset of B; $A \subseteq B$. In general, the empty set is a subset of all sets; $\varnothing \subseteq A$ for any set A.

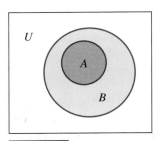

FIGURE 2.1

A is a subset of *B*. *A* ⊆ *B*.

We can express the relationship $A \subseteq B$ visually by drawing a Venn diagram, as shown in Figure 2.1. A **Venn diagram** consists of a rectangle, representing the universal set, and various closed figures within the rectangle, each representing a set. Recall that Venn diagrams were used in Section 1.1 to determine whether an argument was valid.

If two sets are equal, they contain exactly the same elements. It then follows that each is a subset of the other. For example, if $A = B$, then every element of A is an element of B (and vice versa). In this case, A is called an **improper subset** of B. (Likewise, B is an improper subset of A.) Every set is an improper subset of itself; for example, $A \subseteq A$. On the other hand, if A is a subset of B and B contains an element not in A (that is, $A \neq B$), then A is called a **proper subset** of B. To indicate a proper subset, the symbol $\subset$ is used. While it is acceptable to write $\{1, 2\} \subseteq \{1, 2, 3\}$, the relationship of a proper subset is stressed when it is written $\{1, 2\} \subset \{1, 2, 3\}$. Notice the similarities between the subset symbols, $\subset$ and $\subseteq$, and the inequality symbols, $<$ and $\leq$, used in algebra; it is acceptable to write $1 \leq 3$, but writing $1 < 3$ is more informative.

Intersection of Sets

Sometimes an element of one set is also an element of another set; that is, the sets may overlap. This overlap is called the **intersection** of the sets. If an element is in two sets *at the same time,* it is in the intersection of the sets.

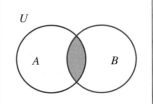

FIGURE 2.2

The intersection $A \cap B$ is represented by the (overlapping) shaded region.

INTERSECTION OF SETS

The **intersection** of set A and set B, denoted by $A \cap B$, is

$$A \cap B = \{x \mid x \in A \quad \text{and} \quad x \in B\}$$

The intersection of two sets consists of those elements that are common to both sets.

For example, given the sets $A = \{$Buffy, Spike, Willow, Xander$\}$ and $B = \{$Angel, Anya, Buffy, Giles, Spike$\}$, their intersection is $A \cap B = \{$Buffy, Spike$\}$.

Venn diagrams are useful in depicting the relationship between sets. The Venn diagram in Figure 2.2 illustrates the intersection of two sets; the shaded region represents $A \cap B$.

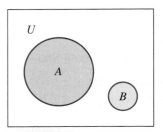

FIGURE 2.3

Mutually exclusive sets have no elements in common $(A \cap B = \varnothing)$.

Mutually Exclusive Sets

Sometimes a pair of sets has no overlap. Consider an ordinary deck of playing cards. Let $D = \{$cards $\mid$ the card is a diamond$\}$ and $S = \{$cards $\mid$ the card is a spade$\}$. Certainly, *no* cards are both diamonds and spades *at the same time;* that is, $S \cap D = \varnothing$.

Two sets A and B are **mutually exclusive** (or *disjoint*) if they have no elements in common, that is, if $A \cap B = \varnothing$. The Venn diagram in Figure 2.3 illustrates mutually exclusive sets.

Union of Sets   OR

What does it mean when we ask, "How many of the 500 college students in a transportation survey own an automobile or a motorcycle?" Does it mean "How many students own either an automobile or a motorcycle *or both*?" or does it mean "How many students own either an automobile or a motorcycle, *but not both*?" The former is called the *inclusive or,* because it includes the possibility of owning both; the latter is called the *exclusive or.* In logic and in mathematics, the word *or* refers to the *inclusive or,* unless you are told otherwise.

The meaning of the word *or* is important to the concept of union. The **union** of two sets is a new set formed by joining those two sets together, just as the union of the states is the joining together of fifty states to form one nation.

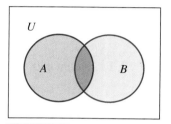

FIGURE 2.4

The union $A \cup B$ *is represented by the (entire) shaded region.*

UNION OF SETS

The **union** of set A and set B, denoted by $A \cup B$, is

$$A \cup B = \{x \mid x \in A \quad \text{or} \quad x \in B\}$$

The union of A and B consists of all elements that are in either A or B or both, that is, all elements that are in at least one of the sets.

For example, given the sets $A = \{\text{Conan, David}\}$ and $B = \{\text{Ellen, Katie, Oprah}\}$, their union is $A \cup B = \{\text{Conan, David, Ellen, Katie, Oprah}\}$, and their intersection is $A \cap B = \varnothing$. Note that because they have no elements in common, A and B are mutually exclusive sets. The Venn diagram in Figure 2.4 illustrates the union of two sets; the entire shaded region represents $A \cup B$.

EXAMPLE 4

FINDING THE INTERSECTION AND UNION OF SETS   Given the sets $A = \{1, 2, 3\}$ and $B = \{2, 4, 6\}$, find the following.

a. $A \cap B$ (the intersection of A and B)
b. $A \cup B$ (the union of A and B)

SOLUTION

a. The intersection of two sets consists of those elements that are common to both sets; therefore, we have

$$A \cap B = \{1, 2, 3\} \cap \{2, 4, 6\}$$
$$= \{2\}$$

b. The union of two sets consists of all elements that are in at least one of the sets; therefore, we have

$$A \cup B = \{1, 2, 3\} \cup \{2, 4, 6\}$$
$$= \{1, 2, 3, 4, 6\}$$

The Venn diagram in Figure 2.5 shows the composition of each set and illustrates the intersection and union of the two sets.

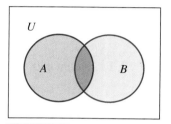

FIGURE 2.5

The composition of sets A and B in Example 4.

Because $A \cup B$ consists of all elements that are in A or B (or both), to find $n(A \cup B)$, we add $n(A)$ plus $n(B)$. However, doing so results in an answer that might be too big; that is, if A and B have elements in common, these elements will be counted twice (once as a part of A and once as a part of B). Therefore, to find the cardinal number of $A \cup B$, we add the cardinal number of A to the cardinal number of B and then *subtract* the cardinal number of $A \cap B$ (so that the overlap is not counted twice).

CARDINAL NUMBER FORMULA FOR THE UNION/INTERSECTION OF SETS

For any two sets A and B, the number of elements in their union is $n(A \cup B)$, where

$$n(A \cup B) = n(A) + n(B) - n(A \cap B)$$

and $n(A \cap B)$ is the number of elements in their intersection.

As long as any three of the four quantities in the general formula are known, the missing quantity can be found by algebraic manipulation.

EXAMPLE 5

ANALYZING THE COMPOSITION OF A UNIVERSAL SET   Given $n(U) = 169$, $n(A) = 81$, and $n(B) = 66$, find the following.

a. If $n(A \cap B) = 47$, find $n(A \cup B)$ and draw a Venn diagram depicting the composition of the universal set.
b. If $n(A \cup B) = 147$, find $n(A \cap B)$ and draw a Venn diagram depicting the composition of the universal set.

SOLUTION

a. We must use the Union/Intersection Formula. Substituting the three given quantities, we have

$$n(A \cup B) = n(A) + n(B) - n(A \cap B)$$
$$= 81 + 66 - 47$$
$$= 100$$

The Venn diagram in Figure 2.6 illustrates the composition of U.

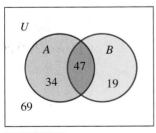

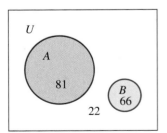

FIGURE 2.6 $n(A \cap B) = 47$.   **FIGURE 2.7** $n(A \cup B) = 147$.

b. We must use the Union/Intersection Formula. Substituting the three given quantities, we have

$$n(A \cup B) = n(A) + n(B) - n(A \cap B)$$
$$147 = 81 + 66 - n(A \cap B)$$
$$147 = 147 - n(A \cap B)$$
$$n(A \cap B) = 147 - 147$$
$$n(A \cap B) = 0$$

Therefore, A and B have no elements in common; they are mutually exclusive. The Venn diagram in Figure 2.7 illustrates the composition of U.

EXAMPLE **6**

ANALYZING THE RESULTS OF A SURVEY   A recent transportation survey of 500 college students (the universal set U) yielded the following information: 291 students own an automobile (A), 179 own a motorcycle (M), and 85 own both an automobile and a motorcycle ($A \cap M$). What percent of these students own an automobile or a motorcycle?

SOLUTION

Recall that "automobile or motorcycle" means "automobile or motorcycle or both" (the inclusive *or*) and that *or* implies union. Hence, we must find $n(A \cup M)$, the cardinal number of the union of sets A and M. We are given that $n(A) = 291$, $n(M) = 179$, and $n(A \cap M) = 85$. Substituting the given values into the Union/Intersection Formula, we have

$$n(A \cup M) = n(A) + n(M) - n(A \cap M)$$
$$= 291 + 179 - 85$$
$$= 385$$

Therefore, 385 of the 500 students surveyed own an automobile or a motorcycle. Expressed as a percent, $385/500 = 0.77$; therefore, 77% of the students own an automobile or a motorcycle (or both).

Complement of a Set

In certain situations, it might be important to know how many things are *not* in a given set. For instance, when playing cards, you might want to know how many cards are not ranked lower than a five; or when taking a survey, you might want to know how many people did not vote for a specific proposition. The set of all elements in the universal set that are *not* in a specific set is called the *complement* of the set.

COMPLEMENT OF A SET

The **complement** of set A, denoted by A' (read "A prime" or "the complement of A"), is

$$A' = \{x \mid x \in U \quad \text{and} \quad x \notin A\}$$

The complement of a set consists of all elements that are in the universal set but not in the given set.

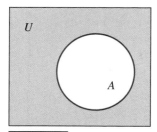

FIGURE 2.8

The complement A' is represented by the shaded region.

For example, given that $U = \{1, 2, 3, 4, 5, 6, 7, 8, 9\}$ and $A = \{1, 3, 5, 7, 9\}$, the complement of A is $A' = \{2, 4, 6, 8\}$. What is the complement of A'? Just as $-(-x) = x$ in algebra, $(A')' = A$ in set theory. The Venn diagram in Figure 2.8 illustrates the complement of set A; the shaded region represents A'.

Suppose A is a set of elements, drawn from a universal set U. If x is an element of the universal set ($x \in U$), then exactly one of the following must be true: (1) x is an element of A ($x \in A$), or (2) x is not an element of A ($x \notin A$). Since no element of the universal set can be in both A and A' at the same time, it follows that A and A' are mutually exclusive sets whose union equals the entire universal set. Therefore, the sum of the cardinal numbers of A and A' equals the cardinal number of U.

HISTORICAL NOTE

JOHN VENN, 1834–1923

John Venn is considered by many to be one of the originators of modern symbolic logic. Venn received his degree in mathematics from the University at Cambridge at the age of twenty-three. He was then elected a fellow of the college and held this fellowship until his death, some 66 years later. Two years after receiving his degree, Venn accepted a teaching position at Cambridge: college lecturer in moral sciences.

During the latter half of the nineteenth century, the study of logic experienced a rebirth in England. Mathematicians were attempting to symbolize and quantify the central concepts of logical thought. Consequently, Venn chose to focus on the study of logic during his tenure at Cambridge. In addition, he investigated the field of probability and published *The Logic of Chance*, his first major work, in 1866.

Venn was well read in the works of his predecessors, including the noted logicians Augustus De Morgan,

George Boole, and Charles Dodgson (a.k.a. Lewis Carroll). Boole's pioneering work on the marriage of logic and algebra proved to be a strong influence on Venn; in fact, Venn used the type of diagram that now bears his name in an 1876 paper in which he examined Boole's system of symbolic logic.

Venn was not the first scholar to use the diagrams that now bear his name. Gottfried Leibniz, Leonhard Euler, and others utilized similar diagrams years before Venn did. Examining each author's diagrams, Venn was critical of their lack of uniformity. He developed a consistent, systematic explanation of the general use of geometrical figures in the analysis of logical arguments. Today, these geometrical figures are known by his name and are used extensively in elementary set theory and logic.

Venn's writings were held in high esteem. His textbooks, *Symbolic Logic* (1881) and *The Principles of Empirical Logic* (1889), were used during the late

nineteenth and early twentieth centuries. In addition to his works on logic and probability, Venn conducted much research into historical records, especially those of his college and those of his family.

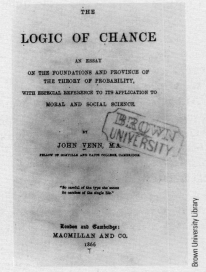

Courtesy The Masters and Fellows of Gonville and Caius College, Cambridge

Brown University Library

Set theory and the cardinal numbers of sets are used extensively in the study of probability. Although he was a professor of logic, Venn investigated the foundations and applications of theoretical probability. Venn's first major work, The Logic of Chance, exhibited the diversity of his academic interests.

It is often quicker to count the elements that are *not* in a set than to count those that are. Consequently, to find the cardinal number of a set, we can subtract the cardinal number of its complement from the cardinal number of the universal set; that is, $n(A) = n(U) - n(A')$.

CARDINAL NUMBER FORMULA FOR THE COMPLEMENT OF A SET

For any set A and its complement A',

$$n(A) + n(A') = n(U)$$

where U is the universal set.

Alternatively,

$$n(A) = n(U) - n(A') \qquad \text{and} \qquad n(A') = n(U) - n(A)$$

EXAMPLE **7**

USING THE COMPLEMENT FORMULA   How many letters in the alphabet precede the letter w?

SOLUTION

Rather than counting all the letters that precede w, we will take a shortcut by counting all the letters that do *not* precede w. Let $L = \{$letters $\mid$ the letter precedes w$\}$. Therefore, $L' = \{$letter $\mid$ the letter does not precede w$\}$. Now $L' = \{$w, x, y, z$\}$, and $n(L') = 4$; therefore, we have

$$n(L) = n(U) - n(L') \quad \textbf{Complement Formula}$$
$$= 26 - 4$$
$$= 22$$

There are twenty-two letters preceding the letter w.

Shading Venn Diagrams

In an effort to visualize the results of operations on sets, it may be necessary to shade specific regions of a Venn diagram. The following example shows a systematic method for shading the intersection or union of any two sets.

EXAMPLE **8**

SHADING VENN DIAGRAMS   On a Venn diagram, shade in the region corresponding to the indicated set.

a. $A \cap B'$ **b.** $A \cup B'$

SOLUTION

a. First, draw and label two overlapping circles as shown in Figure 2.9. The two "components" of the operation $A \cap B'$ are "A" and "B'." Shade each of these components in contrasting ways; shade one of them, say A, with horizontal lines, and the other with vertical lines as in Figure 2.10. Be sure to include a legend, or key, identifying each type of shading.

To be in the intersection of two sets, an element must be in *both* sets at the same time. Therefore, the intersection of A and B' is the region that is shaded in *both* directions (horizontal and vertical) at the same time. A final diagram depicting $A \cap B'$ is shown in Figure 2.11.

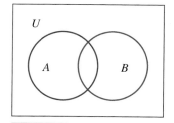

FIGURE 2.9

Two overlapping circles.

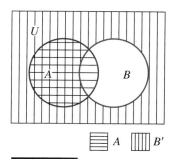

FIGURE 2.10

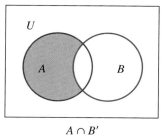

$A \cap B'$

FIGURE 2.11

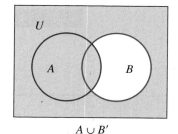

$A \cup B'$

FIGURE 2.12

b. Refer to Figure 2.10. To be in the union of two sets, an element must be in *at least one* of the sets. Therefore, the union of A and B' consists of all regions that are shaded in *any* direction whatsoever (horizontal or vertical or both). A final diagram depicting $A \cup B'$ is shown in Figure 2.12.

Set Theory and Logic

If you have read Chapter 1, you have probably noticed that set theory and logic have many similarities. For instance, the union symbol $\cup$ and the disjunction symbol $\vee$ have the same meaning, but they are used in different circumstances; $\cup$ goes between sets, while $\vee$ goes between logical expressions. The $\cup$ and $\vee$ symbols are similar in appearance because their usages are similar. A comparison of the terms and symbols used in set theory and logic is given in Figure 2.13.

Set Theory		Logic		Common Wording
Term	**Symbol**	**Term**	**Symbol**	
union	$\cup$	disjunction	$\vee$	or
intersection	$\cap$	conjunction	$\wedge$	and
complement	$'$	negation	$\sim$	not
subset	$\subseteq$	conditional	$\rightarrow$	if . . . then . . .

FIGURE 2.13 Comparison of terms and symbols used in set theory and logic.

Applying the concepts and symbols of Chapter 1, we can define the basic operations of set theory in terms of logical biconditionals. The biconditionals in Figure 2.14 are tautologies (expressions that are always true); the first biconditional is read as "x is an element of the union of sets A and B if and only if x is an element of set A or x is an element of set B."

Basic Operations in Set Theory	Logical Biconditional
union	$[x \in (A \cup B)] \leftrightarrow [x \in A \vee x \in B]$
intersection	$[x \in (A \cap B)] \leftrightarrow [x \in A \wedge x \in B]$
complement	$(x \in A') \leftrightarrow \sim (x \in A)$
subset	$(A \subseteq B) \leftrightarrow (x \in A \rightarrow x \in B)$

FIGURE 2.14 Set theory operations as logical biconditionals.

2.1 EXERCISES

1. State whether the given set is well defined.
   a. the set of all black automobiles
   b. the set of all inexpensive automobiles
   c. the set of all prime numbers
   d. the set of all large numbers

2. Suppose $A = \{2, 5, 7, 9, 13, 25, 26\}$.
   a. Find $n(A)$
   b. True or false: $7 \in A$
   c. True or false: $9 \notin A$
   d. True or false: $20 \notin A$

In Exercises 3–6, list all subsets of the given set. Identify which subsets are proper and which are improper.

3. B = {Lennon, McCartney}

4. N = {0}

5. S = {yes, no, undecided}

6. M = {classical, country, jazz, rock}

In Exercises 7–10, the universal set is U = {0, 1, 2, 3, 4, 5, 6, 7, 8, 9}.

7. If A = {1, 2, 3, 4, 5} and B = {4, 5, 6, 7, 8}, find the following.

    a. $A \cap B$ **b.** $A \cup B$

    c. A' **d.** B'

8. If A = {2, 3, 5, 7} and B = {2, 4, 6, 7}, find the following.

    a. $A \cap B$ **b.** $A \cup B$

    c. A' **d.** B'

9. If A = {1, 3, 5, 7, 9} and B = {0, 2, 4, 6, 8}, find the following.

    a. $A \cap B$ **b.** $A \cup B$

    c. A' **d.** B'

10. If A = {3, 6, 9} and B = {4, 8}, find the following.

    a. $A \cap B$ **b.** $A \cup B$

    c. A' **d.** B'

In Exercises 11–16, the universal set is U = {Monday, Tuesday, Wednesday, Thursday, Friday, Saturday, Sunday}. If A = {Monday, Tuesday, Wednesday, Thursday, Friday} and B = {Friday, Saturday, Sunday}, find the indicated set.

11. $A \cap B$ **12.** $A \cup B$

13. B' **14.** A'

15. $A' \cup B$ **16.** $A \cap B'$

In Exercises 17–26, use a Venn diagram like the one in Figure 2.15 to shade in the region corresponding to the indicated set.

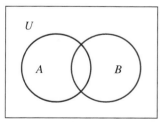

FIGURE 2.15 Two overlapping circles.

17. $A \cap B$ **18.** $A \cup B$

19. A' **20.** B'

21. $A \cup B'$ **22.** $A' \cup B$

23. $A' \cap B$ **24.** $A \cap B'$

25. $A' \cup B'$ **26.** $A' \cap B'$

27. Suppose n(U) = 150, n(A) = 37, and n(B) = 84.

    a. If $n(A \cup B) = 100$, find $n(A \cap B)$ and draw a Venn diagram illustrating the composition of U.

    b. If $n(A \cup B) = 121$, find $n(A \cap B)$ and draw a Venn diagram illustrating the composition of U.

28. Suppose n(U) = w, n(A) = x, n(B) = y, and $n(A \cup B) = z$.

    a. Why must x be less than or equal to z?

    b. If $A \neq U$ and $B \neq U$, fill in the blank with the most appropriate symbol: <, >, ≤, or ≥.

    w_____z, w_____y, y_____z, x_____w

    c. Find $n(A \cap B)$ and draw a Venn diagram illustrating the composition of U.

29. In a recent transportation survey, 500 high school seniors were asked to check the appropriate box or boxes on the following form:

> ☐ I own an automobile.
> ☐ I own a motorcycle.

The results were tabulated as follows: 102 students checked the automobile box, 147 checked the motorcycle box, and 21 checked both boxes.

    a. Draw a Venn diagram illustrating the results of the survey.

    b. What percent of these students own an automobile or a motorcycle?

30. In a recent market research survey, 500 married couples were asked to check the appropriate box or boxes on the following form:

> ☐ We own a DVD player.
> ☐ We own a microwave oven.

The results were tabulated as follows: 301 couples checked the DVD player box, 394 checked the microwave oven box, and 217 checked both boxes.

    a. Draw a Venn diagram illustrating the results of the survey.

    b. What percent of these couples own a DVD player or a microwave oven?

31. In a recent socioeconomic survey, 700 married women were asked to check the appropriate box or boxes on the following form:

> ☐ I have a career.
> ☐ I have a child.

The results were tabulated as follows: 285 women checked the child box, 316 checked the career box, and 196 were blank (no boxes were checked).

a. Draw a Venn diagram illustrating the results of the survey.

b. What percent of these women had both a child and a career?

32. In a recent health survey, 700 single men in their twenties were asked to check the appropriate box or boxes on the following form:

> ☐   I am a member of a private gym.
> ☐   I am a vegetarian.

The results were tabulated as follows: 349 men checked the gym box, 101 checked the vegetarian box, and 312 were blank (no boxes were checked).

a. Draw a Venn diagram illustrating the results of the survey.

b. What percent of these men were both members of a private gym and vegetarians?

For Exercises 33–36, let

$U = \{x \mid x$ is the name of one of the states in the United States$\}$

$A = \{x \mid x \in U$ and x begins with the letter A$\}$

$I = \{x \mid x \in U$ and x begins with the letter I$\}$

$M = \{x \mid x \in U$ and x begins with the letter M$\}$

$N = \{x \mid x \in U$ and x begins with the letter N$\}$

$O = \{x \mid x \in U$ and x begins with the letter O$\}$

33. Find $n(M')$.   **34.** Find $n(A \cup N)$.

35. Find $n(I' \cap O')$.   **36.** Find $n(M \cap I)$.

For Exercises 37–40, let

$U = \{x \mid x$ is the name of one of the months in a year$\}$

$J = \{x \mid x \in U$ and x begins with the letter J$\}$

$Y = \{x \mid x \in U$ and x ends with the letter Y$\}$

$V = \{x \mid x \in U$ and x begins with a vowel$\}$

$R = \{x \mid x \in U$ and x ends with the letter R$\}$

37. Find $n(R')$.   **38.** Find $n(J \cap V)$.

39. Find $n(J \cup Y)$.   **40.** Find $n(V \cap R)$.

In Exercises 41–50, determine how many cards, in an ordinary deck of fifty-two, fit the description. (If you are unfamiliar with playing cards, see the end of Section 3.1 for a description of a standard deck.)

41. spades or aces   **42.** clubs or 2's

43. face cards or black

44. face cards or diamonds

45. face cards and black

46. face cards and diamonds

47. aces or 8's   **48.** 3's or 6's

49. aces and 8's   **50.** 3's and 6's

51. Suppose $A = \{1, 2, 3\}$ and $B = \{1, 2, 3, 4, 5, 6\}$.

a. Find $A \cap B$.

b. Find $A \cup B$.

c. In general, if $E \cap F = E$, what must be true concerning sets E and F?

d. In general, if $E \cup F = F$, what must be true concerning sets E and F?

52. Fill in the blank, and give an example to support your answer.

a. If $A \subset B$, then $A \cap B =$ _____.

b. If $A \subset B$, then $A \cup B =$ _____.

53. a. List all subsets of $A = \{a\}$. How many subsets does A have?

b. List all subsets of $A = \{a, b\}$. How many subsets does A have?

c. List all subsets of $A = \{a, b, c\}$. How many subsets does A have?

d. List all subsets of $A = \{a, b, c, d\}$. How many subsets does A have?

e. Is there a relationship between the cardinal number of set A and the number of subsets of set A?

f. How many subsets does $A = \{a, b, c, d, e, f\}$ have?

HINT: Use your answer to part (e).

54. Prove the Cardinal Number Formula for the Complement of a Set.

HINT: Apply the Union/Intersection Formula to A and A'.

 Answer the following questions using complete sentences and your own words.

• CONCEPT QUESTIONS

55. If $A \cap B = \varnothing$, what is the relationship between sets A and B?

56. If $A \cup B = \varnothing$, what is the relationship between sets A and B?

57. Explain the difference between $\{0\}$ and $\varnothing$.

58. Explain the difference between 0 and $\{0\}$.

59. Is it possible to have $A \cap A = \varnothing$?

60. What is the difference between proper and improper subsets?

61. A set can be described by two methods: the roster method and set-builder notation. When is it advantageous to use the roster method? When is it advantageous to use set-builder notation?

62. Translate the following symbolic expressions into English sentences.

    a. $x \in (A \cap B) \leftrightarrow (x \in A \land x \in B)$

    b. $(x \in A') \leftrightarrow \sim (x \in A)$

    c. $(A \subseteq B) \leftrightarrow (x \in A \rightarrow x \in B)$

• HISTORY QUESTIONS

63. In what academic field was John Venn a professor? Where did he teach?

64. What was one of John Venn's main contributions to the field of logic? What new benefits did it offer?

 THE NEXT LEVEL

If a person wants to pursue an advanced degree (something beyond a bachelor's or four-year degree), chances are the person must take a standardized exam to gain admission to a school or to be admitted into a specific program. These exams are intended to measure verbal, Quantitative, and analytical skills that have developed throughout a person's life. Many classes and study guides are available to help people prepare for the exams. The following questions are typical of those found in the study guides.

Exercises 65–69 refer to the following: Two collectors, John and Juneko, are each selecting a group of three posters from a group of seven movie posters: J, K, L, M, N, O, *and* P. *No poster can be in both groups. The selections made by John and Juneko are subject to the following restrictions:*

- If K is in John's group, M must be in Juneko's group.
- If N is in John's group, P must be in Juneko's group.
- J and P cannot be in the same group.
- M and O cannot be in the same group.

65. Which of the following pairs of groups selected by John and Juneko conform to the restrictions?

John	Juneko
a. J, K, L	M, N, O
b. J, K, P	L, M, N
c. K, N, P	J, M, O
d. L, M, N	K, O, P
e. M, O, P	J, K, N

66. If N is in John's group, which of the following could not be in Juneko's group?

    a. J   **b.** K   **c.** L   **d.** M   **e.** P

67. If K and N are in John's group, Juneko's group must consist of which of the following?

    a. J, M, and O

    b. J, O, and P

    c. L, M, and P

    d. L, O, and P

    e. M, O, and P

68. If J is in Juneko's group, which of the following is true?

    a. K cannot be in John's group.

    b. N cannot be in John's group.

    c. O cannot be in Juneko's group.

    d. P must be in John's group.

    e. P must be in Juneko's group.

69. If K is in John's group, which of the following is true?

    a. J must be in John's group.

    b. O must be in John's group.

    c. L must be in Juneko's group.

    d. N cannot be in John's group.

    e. O cannot be in Juneko's group.

2.2 Applications of Venn Diagrams

OBJECTIVES

- Use Venn diagrams to analyze the results of surveys
- Develop and apply De Morgan's Laws of complements

As we have seen, Venn diagrams are very useful tools for visualizing the relationships between sets. They can be used to establish general formulas involving set operations and to determine the cardinal numbers of sets. Venn diagrams are particularly useful in survey analysis.

Surveys

Surveys are often used to divide people or objects into categories. Because the categories sometimes overlap, people can fall into more than one category. Venn diagrams and the formulas for cardinal numbers can help researchers organize the data.

EXAMPLE **1**

ANALYZING THE RESULTS OF A SURVEY: TWO SETS Has the advent of the DVD affected attendance at movie theaters? To study this question, Professor Redrum's film class conducted a survey of people's movie-watching habits. He had his students ask hundreds of people between the ages of sixteen and forty-five to check the appropriate box or boxes on the following form:

☐ I watched a movie in a theater during the past month.
☐ I watched a movie on a DVD during the past month.

After the professor had collected the forms and tabulated the results, he told the class that 388 people had checked the theater box, 495 had checked the DVD box, 281 had checked both boxes, and 98 of the forms were blank. Giving the class only this information, Professor Redrum posed the following three questions.

a. What percent of the people surveyed watched a movie in a theater or on a DVD during the past month?
b. What percent of the people surveyed watched a movie in a theater only?
c. What percent of the people surveyed watched a movie on a DVD only?

SOLUTION

a. To calculate the desired percentages, we must determine $n(U)$, the total number of people surveyed. This can be accomplished by drawing a Venn diagram. Because the survey divides people into two categories (those who watched a movie in a theater and those who watched a movie on a DVD), we need to define two sets. Let

$$T = \{\text{people} \mid \text{the person watched a movie in a theater}\}$$
$$D = \{\text{people} \mid \text{the person watched a movie on a DVD}\}$$

Now translate the given survey information into the symbols for the sets and attach their given cardinal numbers: $n(T) = 388$, $n(D) = 495$, and $n(T \cap D) = 281$.

Our first goal is to find $n(U)$. To do so, we will fill in the cardinal numbers of all regions of a Venn diagram consisting of two overlapping circles (because we are dealing with two sets). The intersection of T and D consists of 281 people, so we draw two overlapping circles and fill in 281 as the number of elements in common (see Figure 2.16).

Because we were given $n(T) = 388$ and know that $n(T \cap D) = 281$, the difference $388 - 281 = 107$ tells us that 107 people watched a movie in a theater but did not watch a movie on a DVD. We fill in 107 as the number of people who watched a movie only in a theater (see Figure 2.17).

Because $n(D) = 495$, the difference $495 - 281 = 214$ tells us that 214 people watched a movie on a DVD but not in a theater. We fill in 214 as the number of people who watched a movie only on a DVD (see Figure 2.18).

The only region remaining to be filled in is the region outside both circles. This region represents people who didn't watch a movie in a theater or on a DVD and is symbolized by $(T \cup D)'$. Because 98 people didn't check either box on the form, $n[(T \cup D)'] = 98$ (see Figure 2.19).

After we have filled in the Venn diagram with all the cardinal numbers, we readily see that $n(U) = 98 + 107 + 281 + 214 = 700$. Therefore, 700 people were in the survey.

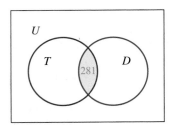

FIGURE 2.16
$n(T \cap D) = 281$.

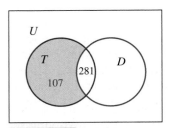
FIGURE 2.17
$n(T \cup D') = 107$.

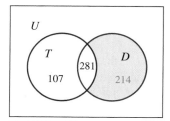

FIGURE 2.18

$n(D \cap T') = 214$.

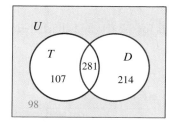

FIGURE 2.19

Completed Venn diagram.

To determine what *percent* of the people surveyed watched a movie in a theater *or* on a DVD during the past month, simply divide $n(T \cup D)$ by $n(U)$:

$$\frac{n(T \cup D)}{n(U)} = \frac{107 + 281 + 214}{700}$$

$$= \frac{602}{700}$$

$$= 0.86$$

Therefore, exactly 86% of the people surveyed watched a movie in a theater or on a DVD during the past month.

b. To find what *percent* of the people surveyed watched a movie in a theater only, divide 107 (the number of people who watched a movie in a theater only) by $n(U)$:

$$\frac{107}{700} = 0.152857142 \ldots$$

$$\approx 0.153 \text{ (rounding off to three decimal places)}$$

Approximately 15.3% of the people surveyed watched a movie in a theater only.

c. Because 214 people watched a movie on DVD only, $214/700 = 0.305714285 \ldots$, or approximately 30.6%, of the people surveyed watched a movie on DVD only.

When you solve a cardinal number problem (a problem that asks, "How many?" or "What percent?") involving a universal set that is divided into various categories (for instance, a survey), use the following general steps.

SOLVING A CARDINAL NUMBER PROBLEM

A cardinal number problem is a problem in which you are asked, "How many?" or "What percent?"

1. Define a set for each category in the universal set. If a category and its negation are both mentioned, define one set A and utilize its complement A'.
2. Draw a Venn diagram with as many overlapping circles as the number of sets you have defined.
3. Write down all the given cardinal numbers corresponding to the various given sets.
4. Starting with the innermost overlap, fill in each region of the Venn diagram with its cardinal number.
5. In answering a "what percent" problem, round off your answer to the nearest tenth of a percent.

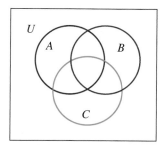

FIGURE 2.20

Three overlapping circles.

When we are working with three sets, we must account for all possible intersections of the sets. Hence, in such cases, we will use the Venn diagram shown in Figure 2.20

EXAMPLE **2**

ANALYZING THE RESULTS OF A SURVEY: THREE SETS   A consumer survey was conducted to examine patterns in ownership of notebook computers, cellular telephones, and DVD players. The following data were obtained: 213 people had notebook computers, 294 had cell phones, 337 had DVD players, 109 had all three, 64 had none, 198 had cell phones and DVD players, 382 had cell phones or notebook computers, and 61 had notebook computers and DVD players but no cell phones.

a. What percent of the people surveyed owned a notebook computer but no DVD player or cell phone?

b. What percent of the people surveyed owned a DVD player but no notebook computer or cell phone?

SOLUTION

a. To calculate the desired percentages, we must determine $n(U)$, the total number of people surveyed. This can be accomplished by drawing a Venn diagram. Because the survey divides people into three categories (those who own a notebook computer, those who own a cell phone, and those who own a DVD player), we need to define three sets. Let

C = {people | the person owns a notebook computer}

T = {people | the person owns a cellular telephone}

D = {people | the person owns a DVD player}

Now translate the given survey information into the symbols for the sets and attach their given cardinal numbers:

213 people had notebook computers ⟶	$n(C) = 213$
294 had cellular telephones ⟶	$n(T) = 294$
337 had DVD players ⟶	$n(D) = 337$
109 had all three ⟶ (C and T and D)	$n(C \cap T \cap D) = 109$
64 had none ⟶ (not C and not T and not D)	$n(C' \cap T' \cap D') = 64$
198 had cell phones and DVD players ⟶ (T and D)	$n(T \cap D) = 198$
382 had cell phones or notebook computers ⟶ (T or C)	$n(T \cup C) = 382$
61 had notebook computers and DVD players but no cell phones ⟶ (C and D and not T)	$n(C \cap D \cap T') = 61$

Our first goal is to find $n(U)$. To do so, we will fill in the cardinal numbers of all regions of a Venn diagram like that in Figure 2.20. We start by using information concerning membership in all three sets. Because the intersection of all three sets consists of 109 people, we fill in 109 in the region common to C and T and D (see Figure 2.21).

Next, we utilize any information concerning membership in two of the three sets. Because $n(T \cap D) = 198$, a total of 198 people are common to both T and D; some are in C, and some are not in C. Of these 198 people, 109 are in C (see Figure 2.21). Therefore, the difference $198 - 109 = 89$ gives the number not in C. Eighty-nine people are in T and D and *not* in C; that is, $n(T \cap D \cap C') = 89$. Concerning membership in the two sets C and D, we are given $n(C \cap D \cap T') = 61$. Therefore, we know that 61 people are in C and D and not in T (see Figure 2.22).

We are given $n(T \cup C) = 382$. From this number, we can calculate $n(T \cap C)$ by using the Union/Intersection Formula:

$$n(T \cup C) = n(T) + n(C) - n(T \cap C)$$
$$382 = 294 + 213 - n(T \cap C)$$
$$n(T \cap C) = 125$$

Therefore, a total of 125 people are in T and C; some are in D, and some are not in D. Of these 125 people, 109 are in D (see Figure 2.21). Therefore, the difference

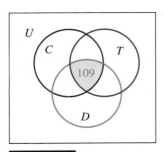

FIGURE 2.21

$n(C \cap T \cap D) = 109$.

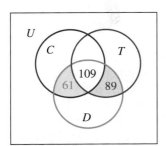

FIGURE 2.22

Determining cardinal numbers in a Venn diagram.

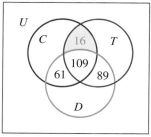

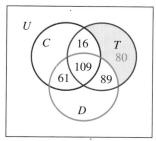

FIGURE 2.23 Determining cardinal numbers in a Venn diagram.

FIGURE 2.24 Determining cardinal numbers in a Venn diagram.

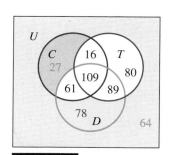

FIGURE 2.25

A completed Venn diagram.

$125 - 109 = 16$ gives the number not in D. Sixteen people are in T and C and *not* in D; that is, $n(C \cap T \cap D') = 16$ (see Figure 2.23).

Knowing that a total of 294 people are in T (given $n(T) = 294$), we are now able to fill in the last region of T. The missing region (people in T only) has $294 - 109 - 89 - 16 = 80$ members; $n(T \cap C' \cap D') = 80$ (see Figure 2.24).

In a similar manner, we subtract the known pieces of C from $n(C) = 213$, which is given, and obtain $213 - 61 - 109 - 16 = 27$; therefore, 27 people are in C only. Likewise, to find the last region of D, we use $n(D) = 337$ (given) and obtain $337 - 89 - 109 - 61 = 78$; therefore, 78 people are in D only. Finally, the 64 people who own none of the items are placed "outside" the three circles (see Figure 2.25).

By adding up the cardinal numbers of all the regions in Figure 2.25, we find that the total number of people in the survey is 524; that is, $n(U) = 524$.

Now, to determine what *percent* of the people surveyed owned only a notebook computer (no DVD player and no cell phone), we simply divide $n(C \cap D' \cap T')$ by $n(U)$:

$$\frac{n(C \cap D' \cap T')}{n(U)} = \frac{27}{524}$$
$$= 0.051526717\ldots$$

Approximately 5.2% of the people surveyed owned a notebook computer and did not own a DVD player or a cellular telephone.

b. To determine what *percent* of the people surveyed owned only a DVD player (no notebook computer and no cell phone), we divide $n(D \cap C' \cap T')$ by $n(U)$:

$$\frac{n(D \cap C' \cap T')}{n(U)} = \frac{78}{524}$$
$$= 0.148854961\ldots$$

Approximately 14.9% of the people surveyed owned a DVD player and did not own a notebook computer or a cell phone.

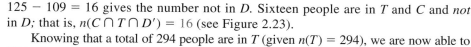

De Morgan's Laws

One of the basic properties of algebra is the distributive property:

$$a(b + c) = ab + ac$$

Given $a(b + c)$, the operation outside the parentheses can be distributed over the operation inside the parentheses. It makes no difference whether you add b and c first and then multiply the sum by a or first multiply each pair, a and b, a and c, and then add their products; the same result is obtained. Is there a similar property for the complement, union, and intersection of sets?

EXAMPLE **3**

INVESTIGATING THE COMPLEMENT OF A UNION Suppose $U = \{1, 2, 3, 4, 5\}$, $A = \{1, 2, 3\}$, and $B = \{2, 3, 4\}$.

a. For the given sets, does $(A \cup B)' = A' \cup B'$?
b. For the given sets, does $(A \cup B)' = A' \cap B'$?

SOLUTION

a. To find $(A \cup B)'$, we must first find $A \cup B$:

$$A \cup B = \{1, 2, 3\} \cup \{2, 3, 4\}$$
$$= \{1, 2, 3, 4\}$$

The complement of $A \cup B$ (relative to the given universal set U) is

$$(A \cup B)' = \{5\}$$

To find $A' \cup B'$, we must first find A' and B':

$$A' = \{4, 5\} \quad \text{and} \quad B' = \{1, 5\}$$

The union of A' and B' is

$$A' \cup B' = \{4, 5\} \cup \{1, 5\}$$
$$= \{1, 4, 5\}$$

Now, $\{5\} \neq \{1, 4, 5\}$; therefore, $(A \cup B)' \neq A' \cup B'$.

b. We find $(A \cup B)'$ as in part (a): $(A \cup B)' = \{5\}$. Now,

$$A' \cap B' = \{4, 5\} \cap \{1, 5\}$$
$$= \{5\}$$

For the given sets, $(A \cup B)' = A' \cap B'$.

Part (a) of Example 3 shows that the operation of complementation *cannot* be explicitly distributed over the operation of union; that is, $(A \cup B)' \neq A' \cup B'$. However, part (b) of the example implies that there *may* be some relationship between the complement, union, and intersection of sets. The fact that $(A \cup B)' = A' \cap B'$ *for the given sets* A and B does not mean that it is true *for all sets* A and B. We will use a general Venn diagram to examine the validity of the statement $(A \cup B)' = A' \cap B'$.

When we draw two overlapping circles within a universal set, four regions are formed. Every element of the universal set U is in exactly one of the following regions, as shown in Figure 2.26:

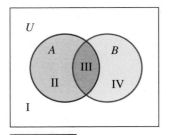

FIGURE 2.26

Four regions in a universal set U.

I   in neither A nor B

II   in A and not in B

III in both A and B

IV in B and not in A

The set $A \cup B$ consists of all elements in regions II, III, and IV. Therefore, the complement $(A \cup B)'$ consists of all elements in region I. A' consists of all elements in regions I and IV, and B' consists of the elements in regions I and II. Therefore, the elements common to both A' and B' are those in region I; that is, the set $A' \cap B'$ consists of all elements in region I. Since $(A \cup B)'$ and $A' \cap B'$ contain exactly the same elements (those in region I), the sets are equal; that is, $(A \cup B)' = A' \cap B'$ is true for all sets A and B.

The relationship $(A \cup B)' = A' \cap B'$ is known as one of **De Morgan's Laws.** Simply stated, "the complement of a union is the intersection of the complements." In a similar manner, it can be shown that $(A \cap B)' = A' \cup B'$ (see Exercise 33).

HISTORICAL NOTE

AUGUSTUS DE MORGAN, 1806–1871

© Bettmann/CORBIS

Being born blind in one eye did not stop Augustus De Morgan from becoming a well-read philosopher, historian, logician, and mathematician. De Morgan was born in Madras, India, where his father was working for the East India Company. On moving to England, De Morgan was educated at Cambridge, and at the age of twenty-two, he became the first professor of mathematics at the newly opened University of London (later renamed University College).

De Morgan viewed all of mathematics as an abstract study of symbols and of systems of operations applied to these symbols. While studying the ramifications of symbolic logic, De Morgan formulated the general properties of complementation that now bear his name. Not limited to symbolic logic, De Morgan's many works include books and papers on the foundations of algebra, differential calculus, and probability. He was known to be a jovial person who was fond of puzzles, and his witty and amusing book *A Budget of Paradoxes* still entertains readers today. Besides his accomplishments in the academic arena, De Morgan was an expert flutist, spoke five languages, and thoroughly enjoyed big-city life.

Knowing of his interest in probability, an actuary (someone who studies life expectancies and determines payments of premiums for insurance companies) once asked De Morgan a question concerning the probability that a certain group of people would be alive at a certain time. In his response, De Morgan employed a formula containing the number π. In amazement, the actuary responded, "That must surely be a delusion! What can a circle have to do with the number of people alive at a certain time?" De Morgan replied that π has numerous applications and occurrences in many diverse areas of mathematics. Because it was first defined and used in geometry, people are conditioned to accept the mysterious number only in reference to a circle. However, in the history of mathematics, if probability had been systematically studied before geometry and circles, our present-day interpretation of the number π would be entirely different. In addition to his accomplishments in logic and higher-level mathematics, De Morgan introduced a convention with which we are all familiar: In a paper written in 1845, he suggested the use of a slanted line to represent a fraction, such as 1/2 or 3/4.

De Morgan was a staunch defender of academic freedom and religious tolerance. While he was a student at Cambridge, his application for a fellowship was refused because he would not take and sign a theological oath. Later in life, he resigned his professorship as a protest against religious bias. (University College gave preferential treatment to members of the Church of England when textbooks were selected and did not have an open policy on religious philosophy.) Augustus De Morgan was a man who was unafraid to take a stand and make personal sacrifices when it came to principles he believed in.

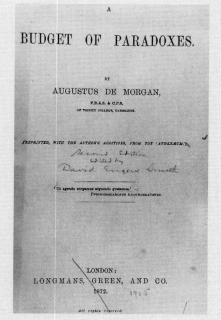

Gematria is a mystic pseudoscience in which numbers are substituted for the letters in a name. De Morgan's book A Budget of Paradoxes contains several gematria puzzles, such as, "Mr. Davis Thom found a young gentleman of the name of St. Claire busy at the Beast number: he forthwith added the letters in στκλαιρε (the Greek spelling of St. Claire) and found 666." (Verify this by using the Greek numeral system.)

DE MORGAN'S LAWS

For any sets A and B,

$$(A \cup B)' = A' \cap B'$$

That is, the complement of a union is the intersection of the complements. Also,

$$(A \cap B)' = A' \cup B'.$$

That is, the complement of an intersection is the union of the complements.

TOPIC X   BLOOD TYPES: SET THEORY IN THE REAL WORLD

Human blood types are a classic example of set theory. As you may know, there are four categories (or sets) of blood types: A, B, AB, and O. Knowing someone's blood type is extremely important in case a blood transfusion is required; if blood of two different types is combined, the blood cells may begin to clump together, with potentially fatal consequences! (Do you know your blood type?)

What exactly are "blood types"? In the early 1900s, the Austrian scientist Karl Landsteiner observed the presence (or absence) of two distinct chemical molecules on the surface of all red blood cells in numerous samples of human blood. Consequently, he labeled one molecule "A" and the other "B." The presence or absence of these specific molecules is the basis of the universal classification of blood types. Specifically, blood samples containing only the A molecule are labeled type A, whereas those containing only the B molecule are labeled type B. If a blood sample contains both molecules (A and B) it is labeled type AB; and if neither is present, the blood is typed as O. The presence (or absence) of these molecules can be depicted in a standard Venn diagram as shown in Figure 2.27. In the notation of set operations, type A blood is denoted $A \cap B'$, type B is $B \cap A'$, type AB is $A \cap B$, and type O is $A' \cap B'$.

If a specific blood sample is mixed with blood containing a blood molecule (A or B) that it does not already have,

the presence of the foreign molecule may cause the mixture of blood to clump. For example, type A blood cannot be mixed with any blood containing the B molecule (type B or type AB). Therefore, a person with type A blood can receive a transfusion only of type A or type O blood. Consequently, a person with type AB blood may receive a transfusion of any blood type; type AB is referred to as the "universal receiver." Because type O blood contains neither the A nor the B molecule, all blood types are compatible with type O blood; type O is referred to as the "universal donor."

It is not uncommon for scientists to study rhesus monkeys in an effort to learn more about human physiology. In so doing, a certain blood protein was discovered in rhesus monkeys. Subsequently, scientists found that the blood of some people contained this protein, whereas the blood of others did not. The presence, or absence, of this protein in

human blood is referred to as the *Rh factor;* blood containing the protein is labeled "Rh+", whereas "Rh−" indicates the absence of the protein. The Rh factor of human blood is especially important for expectant mothers; a fetus can develop problems if its parents have opposite Rh factors.

When a person's blood is typed, the designation includes both the regular blood type and the Rh factor. For instance, type AB− indicates the presence of both the A and B molecules (type AB), along with the absence of the rhesus protein; type O+ indicates the absence of both the A and B molecules (type O), along with the presence of the rhesus protein. Utilizing the Rh factor, there are eight possible blood types as shown in Figure 2.28.

We will investigate the occurrence and compatibility of the various blood types in Example 5 and in Exercises 35–43.

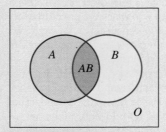

FIGURE 2.27 Blood types and the presence of the A and B molecules.

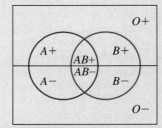

FIGURE 2.28 Blood types combined with the Rh factor.

EXAMPLE 4

APPLYING DE MORGAN'S LAW   Suppose $U = \{0, 1, 2, 3, 4, 5, 6, 7, 8, 9\}$, $A = \{2, 3, 7, 8\}$, and $B = \{0, 4, 5, 7, 8, 9\}$. Use De Morgan's Law to find $(A' \cup B)'$.

SOLUTION

The complement of a union is equal to the intersection of the complements; therefore, we have

$$(A' \cup B)' = (A')' \cap B' \qquad \text{**De Morgan's Law**}$$
$$= A \cap B' \qquad\qquad (A')' = A$$
$$= \{2, 3, 7, 8\} \cap \{1, 2, 3, 6\}$$
$$= \{2, 3\}$$

Notice that this problem could be done without using De Morgan's Law, but solving it would then involve finding first A', then $A' \cup B$, and finally $(A' \cup B)'$. This method would involve more work. (Try it!)

EXAMPLE 5

INVESTIGATING BLOOD TYPES IN THE UNITED STATES The American Red Cross has compiled a massive database of the occurrence of blood types in the United States. Their data indicate that on average, out of every 100 people in the United States, 44 have the A molecule, 15 have the B molecule, and 45 have neither the A nor the B molecule. What percent of the U.S. population have the following blood types?

a. Type O?　　**b.** Type AB?　　**c.** Type A?　　**d.** Type B?

SOLUTION

a. First, we define the appropriate sets. Let

$A = \{$Americans $|$ the person has the A molecule$\}$

$B = \{$Americans $|$ the person has the B molecule$\}$

We are given the following cardinal numbers: $n(U) = 100$, $n(A) = 44$, $n(B) = 15$, and $n(A' \cap B') = 45$. Referring to Figure 2.27, and given that 45 people (out of 100) have neither the A molecule nor the B molecule, we conclude that 45 of 100 people, or 45%, have type O blood as shown in Figure 2.29.

b. Applying De Morgan's Law to the Complement Formula, we have the following.

$$n(A \cup B) + n[(A \cup B)'] = n(U) \quad \textbf{Complement Formula}$$
$$n(A \cup B) + n(A' \cap B') = n(U) \quad \textbf{applying De Morgan's Law}$$
$$n(A \cup B) + 45 = 100 \quad \textbf{substituting known values}$$

Therefore, $n(A \cup B) = 55$.

Now, use the Union/Intersection Formula.

$$n(A \cup B) = n(A) + n(B) - n(A \cap B) \quad \textbf{Union/Intersection Formula}$$
$$55 = 44 + 15 - n(A \cap B) \quad \textbf{substituting known values}$$
$$n(A \cap B) = 44 + 15 - 55 \quad \textbf{adding } n(A \cap B) \textbf{ and subtracting 55}$$

Therefore, $n(A \cap B) = 4$. This means that 4 people (of 100) have both the A and the B molecules; that is, 4 of 100 people, or 4%, have type AB blood. See Figure 2.30.

c. Knowing that a total of 44 people have the A molecule, that is, $n(A) = 44$, we subtract $n(A \cap B) = 4$ and conclude that 40 have *only* the A molecule.

　　Therefore, $n(A \cap B') = 40$. This means that 40 people (of 100) have only the A molecule; that is, 40 of 100 people, or 40%, have type A blood. See Figure 2.31.

d. Knowing that a total of 15 people have the B molecule, that is, $n(B) = 15$, we subtract $n(A \cap B) = 4$ and conclude that 11 have *only* the B molecule.

　　Therefore, $n(B \cap A') = 11$. This means that 11 people (of 100) have only the B molecule; that is, 11 of 100 people, or 11%, have type B blood (see Figure 2.32).

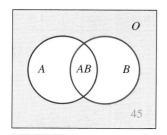

FIGURE 2.29

Forty-five of 100 (45%) have type O blood.

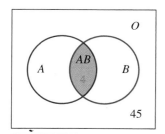

FIGURE 2.30

Four in 100 (4%) have type AB blood.

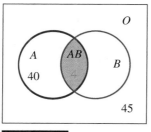

FIGURE 2.31　Forty in 100 (40%) have type A blood.

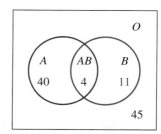

FIGURE 2.32　Eleven in 100 (11%) have type B blood.

The occurrence of blood types in the United States is summarized in Figure 2.33.

Blood Type	O	A	B	AB
Occurrence	45%	40%	11%	4%

FIGURE 2.33 Occurrence of blood types in the United States.

The occurrence of blood types given in Figure 2.33 can be further categorized by including the Rh factor. According to the American Red Cross, out of every 100 people in the United States, blood types and Rh factors occur at the rates shown in Figure 2.34.

Blood Type	O+	O–	A+	A–	B+	B–	AB+	AB–
Occurrence	38	7	34	6	9	2	3	1

FIGURE 2.34 Blood types per 100 people in the United States. (*Source:* American Red Cross.)

2.2 EXERCISES

1. A survey of 200 people yielded the following information: 94 people owned a DVD player, 127 owned a microwave oven, and 78 owned both. How many people owned the following?

   a. a DVD player or a microwave oven
   b. a DVD player but not a microwave oven
   c. a microwave oven but not a DVD player
   d. neither a DVD player nor a microwave oven

2. A survey of 300 workers yielded the following information: 231 workers belonged to a union, and 195 were Democrats. If 172 of the union members were Democrats, how many workers were in the following situations?

   a. belonged to a union or were Democrats
   b. belonged to a union but were not Democrats
   c. were Democrats but did not belong to a union
   d. neither belonged to a union nor were Democrats

3. The records of 1,492 high school graduates were examined, and the following information was obtained: 1,072 graduates took biology, and 679 took geometry. If 271 of those who took geometry did not take biology, how many graduates took the following?

   a. both classes
   b. at least one of the classes

   c. biology but not geometry
   d. neither class

4. A department store surveyed 428 shoppers, and the following information was obtained: 214 shoppers made a purchase, and 299 were satisfied with the service they received. If 52 of those who made a purchase were not satisfied with the service, how many shoppers did the following?

   a. made a purchase and were satisfied with the service
   b. made a purchase or were satisfied with the service
   c. were satisfied with the service but did not make a purchase
   d. were not satisfied and did not make a purchase

5. In a survey, 674 adults were asked what television programs they had recently watched. The following information was obtained: 226 adults watched neither the Big Game nor the New Movie, and 289 watched the New Movie. If 183 of those who watched the New Movie did not watch the Big Game, how many of the surveyed adults watched the following?

   a. both programs
   b. at least one program
   c. the Big Game
   d. the Big Game but not the New Movie

6. A survey asked 816 college freshmen whether they had been to a movie or eaten in a restaurant during the past week. The following information was obtained: 387 freshmen had been to neither a movie nor a restaurant, and 266 had been to a movie. If 92 of those who had been to a movie had not been to a restaurant, how many of the surveyed freshmen had been to the following?

   a. both a movie and a restaurant

   b. a movie or a restaurant

   c. a restaurant

   d. a restaurant but not a movie

7. A local 4-H club surveyed its members, and the following information was obtained: 13 members had rabbits, 10 had goats, 4 had both rabbits and goats, and 18 had neither rabbits nor goats.

   a. What percent of the club members had rabbits or goats?

   b. What percent of the club members had only rabbits?

   c. What percent of the club members had only goats?

8. A local anime fan club surveyed its members regarding their viewing habits last weekend, and the following information was obtained: 30 members had watched an episode of *Naruto,* 44 had watched an episode of *Death Note,* 21 had watched both an episode of *Naruto* and an episode of *Death Note,* and 14 had watched neither *Naruto* nor *Death Note.*

   a. What percent of the club members had watched *Naruto* or *Death Note?*

   b. What percent of the club members had watched only *Naruto?*

   c. What percent of the club members had watched only *Death Note?*

9. A recent survey of w shoppers (that is, $n(U) = w$) yielded the following information: x shoppers shopped at Sears, y shopped at JCPenney's, and z shopped at both. How many people shopped at the following?

   a. Sears or JCPenney's

   b. only Sears

   c. only JCPenney's

   d. neither Sears nor JCPenney's

10. A recent transportation survey of w urban commuters (that is, $n(U) = w$) yielded the following information: x commuters rode neither trains nor buses, y rode trains, and z rode only trains. How many people rode the following?

   a. trains and buses

   b. only buses

   c. buses

   d. trains or buses

11. A consumer survey was conducted to examine patterns in ownership of laptop computers, cellular telephones, and DVD players. The following data were obtained: 313 people had laptop computers, 232 had cell phones, 269 had DVD players, 69 had all three, 64 had none, 98 had cell phones and DVD players, 57 had cell phones but no computers or DVD players, and 104 had computers and DVD players but no cell phones.

   a. What percent of the people surveyed owned a cell phone?

   b. What percent of the people surveyed owned only a cell phone?

12. In a recent survey of monetary donations made by college graduates, the following information was obtained: 95 graduates had donated to a political campaign, 76 had donated to assist medical research, 133 had donated to help preserve the environment, 25 had donated to all three, 22 had donated to none of the three, 38 had donated to a political campaign and to medical research, 46 had donated to medical research and to preserve the environment, and 54 had donated to a political campaign and to preserve the environment.

   a. What percent of the college graduates donated to none of the three listed causes?

   b. What percent of the college graduates donated to exactly one of the three listed causes?

13. Recently, Green Day, the Kings of Leon, and the Black Eyed Peas had concert tours in the United States. A large group of college students was surveyed, and the following information was obtained: 381 students saw Black Eyed Peas, 624 saw the Kings of Leon, 712 saw Green Day, 111 saw all three, 513 saw none, 240 saw only Green Day, 377 saw Green Day and the Kings of Leon, and 117 saw the Kings of Leon and Black Eyed Peas but not Green Day.

   a. What percent of the college students saw at least one of the bands?

   b. What percent of the college students saw exactly one of the bands?

14. Dr. Hawk works in an allergy clinic, and his patients have the following allergies: 68 patients are allergic to dairy products, 93 are allergic to pollen, 91 are allergic to animal dander, 31 are allergic to all three, 29 are allergic only to pollen, 12 are allergic only to dairy products, and 40 are allergic to dairy products and pollen.

   a. What percent of Dr. Hawk's patients are allergic to animal dander?

   b. What percent of Dr. Hawk's patients are allergic only to animal dander?

15. When the members of the Eye and I Photo Club discussed what type of film they had used during the past month, the following information was obtained: 77 members used black and white, 24 used only black and white, 65 used color, 18 used only color, 101 used black and white or color, 27 used infrared, 9 used all three types, and 8 didn't use any film during the past month.

  a. What percent of the members used only infrared film?

  b. What percent of the members used at least two of the types of film?

16. After leaving the polls, many people are asked how they voted. (This is called an *exit poll.*) Concerning Propositions A, B, and C, the following information was obtained: 294 people voted yes on A, 90 voted yes only on A, 346 voted yes on B, 166 voted yes only on B, 517 voted yes on A or B, 339 voted yes on C, no one voted yes on all three, and 72 voted no on all three.

  a. What percent of the voters in the exit poll voted no on A?

  b. What percent of the voters voted yes on more than one proposition?

17. In a recent survey, consumers were asked where they did their gift shopping. The following results were obtained: 621 consumers shopped at Macy's, 513 shopped at Emporium, 367 shopped at Nordstrom, 723 shopped at Emporium or Nordstrom, 749 shopped at Macy's or Nordstrom, 776 shopped at Macy's or Emporium, 157 shopped at all three, 96 shopped at neither Macy's nor Emporium nor Nordstrom.

  a. What percent of the consumers shopped at more than one store?

  b. What percent of the consumers shopped exclusively at Nordstrom?

18. A company that specializes in language tutoring lists the following information concerning its English-speaking employees: 23 employees speak German; 25 speak French; 31 speak Spanish; 43 speak Spanish or French; 38 speak French or German; 46 speak German or Spanish; 8 speak Spanish, French, and German; and 7 speak English only.

  a. What percent of the employees speak at least one language other than English?

  b. What percent of the employees speak at least two languages other than English?

19. In a recent survey, people were asked which radio station they listened to on a regular basis. The following results were obtained: 140 people listened to WOLD (oldies), 95 listened to WJZZ (jazz), 134 listened to WTLK (talk show news), 235 listened to WOLD or WJZZ, 48 listened to WOLD and WTLK, 208 listened to WTLK or WJZZ, and 25 listened to none.

  a. What percent of people in the survey listened only to WTLK on a regular basis?

  b. What percent of people in the survey did not listen to WTLK on a regular basis?

20. In a recent health insurance survey, employees at a large corporation were asked, "Have you been a patient in a hospital during the past year, and if so, for what reason?" The following results were obtained: 494 employees had an injury, 774 had an illness, 1,254 had tests, 238 had an injury and an illness and tests, 700 had an illness and tests, 501 had tests and no injury or illness, 956 had an injury or illness, and 1,543 had not been a patient.

  a. What percent of the employees had been patients in a hospital?

  b. What percent of the employees had tests in a hospital?

In Exercises 21 and 22, use a Venn diagram like the one in Figure 2.35.

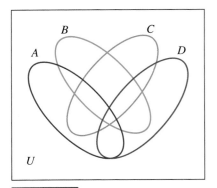

FIGURE 2.35 Four overlapping regions.

21. A survey of 136 pet owners yielded the following information: 49 pet owners own fish; 55 own a bird; 50 own a cat; 68 own a dog; 2 own all four; 11 own only fish; 14 own only a bird; 10 own fish and a bird; 21 own fish and a cat; 26 own a bird and a dog; 27 own a cat and a dog; 3 own fish, a bird, a cat, and no dog; 1 owns fish, a bird, a dog, and no cat; 9 own fish, a cat, a dog, and no bird; and 10 own a bird, a cat, a dog, and no fish. How many of the surveyed pet owners have no fish, no birds, no cats, and no dogs? (They own other types of pets.)

22. An exit poll of 300 voters yielded the following information regarding voting patterns on Propositions A, B, C, and D: 119 voters voted yes on A; 163 voted yes on B; 129 voted yes on C; 142 voted yes on D; 37 voted yes on all four; 15 voted yes on A only; 50 voted yes on B only; 59 voted yes on A and B; 70 voted yes on A and C; 82 voted yes on B and D; 93 voted yes on C and D; 10 voted yes on A, B, and C and no on D; 2 voted yes on A, B, and D and no on C; 16 voted yes on A, C, and D and no on B; and 30 voted yes on B, C, and D and no on A. How many of the surveyed voters voted no on all four propositions?

In Exercises 23–26, given the sets U = {0, 1, 2, 3, 4, 5, 6, 7, 8, 9}, A = {0, 2, 4, 5, 9}, and B = {1, 2, 7, 8, 9}, use De Morgan's Laws to find the indicated sets.

23. $(A' \cup B)'$

24. $(A' \cap B)'$

25. $(A \cap B')'$

26. $(A \cup B')'$

In Exercises 27–32, use a Venn diagram like the one in Figure 2.36 to shade in the region corresponding to the indicated set.

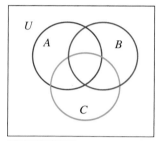

FIGURE 2.36 Three overlapping circles.

27. $A \cap B \cap C$

28. $A \cup B \cup C$

29. $(A \cup B)' \cap C$

30. $A \cap (B \cup C)'$

31. $B \cap (A \cup C')$

32. $(A' \cup B) \cap C'$

33. Using Venn diagrams, prove De Morgan's Law $(A \cap B)' = A' \cup B'$.

34. Using Venn diagrams, prove $A \cup (B \cap C) = (A \cup B) \cap (A \cup C)$.

Use the data in Figure 2.34 to complete Exercises 35–39. Round off your answers to a tenth of a percent.

35. What percent of all people in the United States have blood that is
   a. Rh positive?   **b.** Rh negative?

36. Of all people in the United States who have type O blood, what percent are
   a. Rh positive?   **b.** Rh negative?

37. Of all people in the United States who have type A blood, what percent are
   a. Rh positive?   **b.** Rh negative?

38. Of all people in the United States who have type B blood, what percent are
   a. Rh positive?   **b.** Rh negative?

39. Of all people in the United States who have type AB blood, what percent are
   a. Rh positive?   **b.** Rh negative?

40. If a person has type A blood, what blood types may the person receive in a transfusion?

41. If a person has type B blood, what blood types may the person receive in a transfusion?

42. If a person has type AB blood, what blood types may the person receive in a transfusion?

43. If a person has type O blood, what blood types may the person receive in a transfusion?

 Answer the following questions using complete sentences and your own words.

• HISTORY QUESTIONS

44. What notation did De Morgan introduce in regard to fractions?

45. Why did De Morgan resign his professorship at University College?

 THE NEXT LEVEL

If a person wants to pursue an advanced degree (something beyond a bachelor's or four-year degree), chances are the person must take a standardized exam to gain admission to a school or to be admitted into a specific program. These exams are intended to measure verbal, quantitative, and analytical skills that have developed throughout a person's life. Many classes and study guides are available to help people prepare for the exams. The following questions are typical of those found in the study guides.

Exercises 46–52 refer to the following: A nonprofit organization's board of directors, composed of four women (Angela, Betty, Carmen, and Delores) and three men (Ed, Frank, and Grant), holds frequent meetings. A meeting can be held at Betty's house, at Delores's house, or at Frank's house.

- Delores cannot attend any meetings at Betty's house.
- Carmen cannot attend any meetings on Tuesday or on Friday.
- Angela cannot attend any meetings at Delores's house.
- Ed can attend only those meetings that Grant also attends.
- Frank can attend only those meetings that both Angela and Carmen attend.

46. If all members of the board are to attend a particular meeting, under which of the following circumstances can it be held?
   a. Monday at Betty's
   b. Tuesday at Frank's
   c. Wednesday at Delores's
   d. Thursday at Frank's
   e. Friday at Betty's

47. Which of the following can be the group that attends a meeting on Wednesday at Betty's?

    a. Angela, Betty, Carmen, Ed, and Frank

    b. Angela, Betty, Ed, Frank, and Grant

    c. Angela, Betty, Carmen, Delores, and Ed

    d. Angela, Betty, Delores, Frank, and Grant

    e. Angela, Betty, Carmen, Frank, and Grant

48. If Carmen and Angela attend a meeting but Grant is unable to attend, which of the following could be true?

    a. The meeting is held on Tuesday.

    b. The meeting is held on Friday.

    c. The meeting is held at Delores's.

    d. The meeting is held at Frank's.

    e. The meeting is attended by six of the board members.

49. If the meeting is held on Tuesday at Betty's, which of the following pairs can be among the board members who attend?

    a. Angela and Frank

    b. Ed and Betty

    c. Carmen and Ed

    d. Frank and Delores

    e. Carmen and Angela

50. If Frank attends a meeting on Thursday that is not held at his house, which of the following must be true?

    a. The group can include, at most, two women.

    b. The meeting is at Betty's house.

    c. Ed is not at the meeting.

    d. Grant is not at the meeting.

    e. Delores is at the meeting.

51. If Grant is unable to attend a meeting on Tuesday at Delores's, what is the largest possible number of board members who can attend?

    a. 1   **b.** 2   **c.** 3

    d. 4   **e.** 5

52. If a meeting is held on Friday, which of the following board members *cannot* attend?

    a. Grant   **b.** Delores   **c.** Ed

    d. Betty   **e.** Frank

WEB PROJECT

53. A person's Rh factor will limit the person's options regarding the blood types he or she may receive during a transfusion. Fill in the following chart. How does a person's Rh factor limit that person's options regarding compatible blood?

If Your Blood Type Is:	You Can Receive:
O+	
O−	
A+	
A−	
B+	
B−	
AB+	
AB−	

Some useful links for this web project are listed on the text web site:

www.cengage.com/math/johnson

2.3 Introduction to Combinatorics

OBJECTIVES

● Develop and apply the Fundamental Principle of Counting

● Develop and evaluate factorials

If you went on a shopping spree and bought two pairs of jeans, three shirts, and two pairs of shoes, how many new outfits (consisting of a new pair of jeans, a new shirt, and a new pair of shoes) would you have? A compact disc buyers' club sends you a brochure saying that you can pick any five CDs from a group of 50 of today's hottest sounds for only $1.99. How many different combinations can you choose? Six local bands have volunteered to perform at a benefit concert, and there is some concern over the order in which the bands will perform. How many different line-ups are possible? The answers to questions like these can be obtained by listing all the possibilities or by using three shortcut counting methods: the **Fundamental Principle of Counting, combinations,** and **permutations.** Collectively, these methods are known as **combinatorics.** (Incidentally, the answers to the questions above are 12 outfits, 2,118,760 CD combinations, and 720 lineups.) In this section, we consider the first shortcut method.

The Fundamental Principle of Counting

Daily life requires that we make many decisions. For example, we must decide what food items to order from a menu, what items of clothing to put on in the morning, and what options to order when purchasing a new car. Often, we are asked to make a series of decisions: "Do you want soup or salad? What type of dressing? What type of vegetable? What entrée? What beverage? What dessert?" These individual components of a complete meal lead to the question "Given all the choices of soups, salads, dressings, vegetables, entrées, beverages, and desserts, what is the total number of possible dinner combinations?"

When making a series of decisions, how can you determine the total number of possible selections? One way is to list all the choices for each category and then match them up in all possible ways. To ensure that the choices are matched up in all possible ways, you can construct a **tree diagram.** A tree diagram consists of clusters of line segments, or *branches,* constructed as follows: A cluster of branches is drawn for each decision to be made such that the number of branches in each cluster equals the number of choices for the decision. For instance, if you must make two decisions and there are two choices for decision 1 and three choices for decision 2, the tree diagram would be similar to the one shown in Figure 2.37.

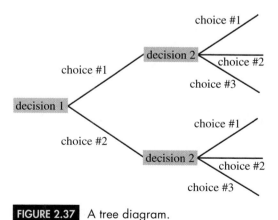

FIGURE 2.37   A tree diagram.

Although this method can be applied to all problems, it is very time consuming and impractical when you are dealing with a series of many decisions, each of which contains numerous choices. Instead of actually listing all possibilities via a tree diagram, using a shortcut method might be desirable. The following example gives a clue to finding such a shortcut.

EXAMPLE **1**

SOLUTION

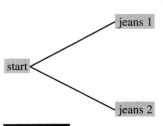

FIGURE 2.38

The first decision.

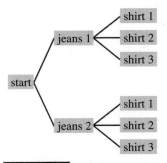

FIGURE 2.39

The second decision.

DETERMINING THE TOTAL NUMBER OF POSSIBLE CHOICES IN A SERIES OF DECISIONS   If you buy two pairs of jeans, three shirts, and two pairs of shoes, how many new outfits (consisting of a new pair of jeans, a new shirt, and a new pair of shoes) would you have?

Because there are three categories, selecting an outfit requires a series of three decisions: You must select one pair of jeans, one shirt, and one pair of shoes. We will make our three decisions in the following order: jeans, shirt, and shoes. (The order in which the decisions are made does not affect the overall outfit.)
    Our first decision (jeans) has two choices (jeans 1 or jeans 2); our tree starts with two branches, as in Figure 2.38.
    Our second decision is to select a shirt, for which there are three choices. At each pair of jeans on the tree, we draw a cluster of three branches, one for each shirt, as in Figure 2.39.
    Our third decision is to select a pair of shoes, for which there are two choices. At each shirt on the tree, we draw a cluster of two branches, one for each pair of shoes, as in Figure 2.40.

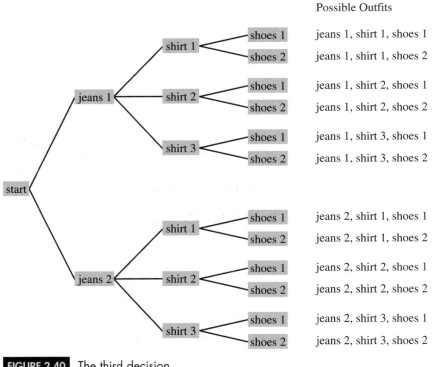

FIGURE 2.40   The third decision.

We have now listed all possible ways of putting together a new outfit; twelve outfits can be formed from two pairs of jeans, three shirts, and two pairs of shoes.

Referring to Example 1, note that each time a decision had to be made, the number of branches on the tree diagram was *multiplied* by a factor equal to the number of choices for the decision. Therefore, the total number of outfits could have been obtained by *multiplying* the number of choices for each decision:

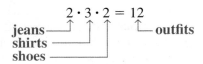

The generalization of this process of multiplication is called the Fundamental Principle of Counting.

THE FUNDAMENTAL PRINCIPLE OF COUNTING

The total number of possible outcomes of a series of decisions (making selections from various categories) is found by multiplying the number of choices for each decision (or category) as follows:

1. Draw a box for each decision.
2. Enter the number of choices for each decision in the appropriate box and multiply.

EXAMPLE **2**

APPLYING THE FUNDAMENTAL PRINCIPLE OF COUNTING   A serial number consists of two consonants followed by three nonzero digits followed by a vowel (A, E, I, O, U): for example, "ST423E" and "DD666E." Determine how many serial numbers are possible given the following conditions.

a. Letters and digits cannot be repeated in the same serial number.
b. Letters and digits can be repeated in the same serial number.

SOLUTION

a. Because the serial number has six symbols, we must make six decisions. Consequently, we must draw six boxes:

$$\square \quad \square \quad \square \quad \square \quad \square \quad \square$$

There are twenty-one different choices for the first consonant. Because the letters cannot be repeated, there are only twenty choices for the second consonant. Similarly, there are nine different choices for the first nonzero digit, eight choices for the second, and seven choices for the third. There are five different vowels, so the total number of possible serial numbers is

$$\boxed{21} \times \boxed{20} \times \boxed{9} \times \boxed{8} \times \boxed{7} \times \boxed{5} = 1{,}058{,}400$$

$$\underbrace{\qquad}_{\text{consonants}} \quad \underbrace{\qquad}_{\text{nonzero digits}} \quad \underbrace{\ }_{\text{vowel}}$$

There are 1,058,400 possible serial numbers when the letters and digits cannot be repeated within a serial number.

b. Because letters and digits can be repeated, the number of choices does not decrease by one each time as in part (a). Therefore, the total number of possibilities is

$$\boxed{21} \times \boxed{21} \times \boxed{9} \times \boxed{9} \times \boxed{9} \times \boxed{5} = 1{,}607{,}445$$

$$\underbrace{\qquad}_{\text{consonants}} \quad \underbrace{\qquad}_{\text{nonzero digits}} \quad \underbrace{\ }_{\text{vowel}}$$

There are 1,607,445 possible serial numbers when the letters and digits can be repeated within a serial number.

Factorials

EXAMPLE **3**

APPLYING THE FUNDAMENTAL PRINCIPLE OF COUNTING   Three students rent a three-bedroom house near campus. One of the bedrooms is very desirable (it has its own bath), one has a balcony, and one is undesirable (it is very small). In how many ways can the housemates choose the bedrooms?

SOLUTION

Three decisions must be made: who gets the room with the bath, who gets the room with the balcony, and who gets the small room. Using the Fundamental Principle of Counting, we draw three boxes and enter the number of choices for each decision. There are three choices for who gets the room with the bath. Once that decision has been made, there are two choices for who gets the room with the balcony, and finally, there is only one choice for the small room.

$$\boxed{3} \times \boxed{2} \times \boxed{1} = 6$$

There are six different ways in which the three housemates can choose the three bedrooms.

Combinatorics often involve products of the type $3 \cdot 2 \cdot 1 = 6$, as seen in Example 3. This type of product is called a **factorial,** and the product $3 \cdot 2 \cdot 1$ is written as 3!. In this manner, $4! = 4 \cdot 3 \cdot 2 \cdot 1 \, (= 24)$, and $5! = 5 \cdot 4 \cdot 3 \cdot 2 \cdot 1 \, (= 120)$.

FACTORIALS

If n is a positive integer, then n *factorial,* denoted by $n!$, is the product of all positive integers less than or equal to n.

$$n! = n \cdot (n - 1) \cdot (n - 2) \cdot \cdots \cdot 2 \cdot 1$$

As a special case, we define $0! = 1$.

Many scientific calculators have a button that will calculate a factorial. Depending on your calculator, the button will look like $\boxed{x!}$ or $\boxed{n!}$, and you might have to press a $\boxed{\text{shift}}$ or $\boxed{\text{2nd}}$ button first. For example, to calculate 6!, type the number 6, press the factorial button, and obtain 720. To calculate a factorial on most graphing calculators, do the following:

- Type the value of n. (For example, type the number 6.)
- Press the $\boxed{\text{MATH}}$ button.
- Press the right arrow button $\boxed{\rightarrow}$ as many times as necessary to highlight $\boxed{\text{PRB}}$.
- Press the down arrow $\boxed{\downarrow}$ as many times as necessary to highlight the "!" symbol, and press $\boxed{\text{ENTER}}$.
- Press $\boxed{\text{ENTER}}$ to execute the calculation.

To calculate a factorial on a Casio graphing calculator, do the following:

- Press the $\boxed{\text{MENU}}$ button; this gives you access to the main menu.
- Press 1 to select the RUN mode; this mode is used to perform arithmetic operations.
- Type the value of n. (For example, type the number 6.)
- Press the $\boxed{\text{OPTN}}$ button; this gives you access to various options displayed at the bottom of the screen.
- Press the $\boxed{\text{F6}}$ button to see more options (i.e., $\boxed{\rightarrow}$).
- Press the $\boxed{\text{F3}}$ button to select probability options (i.e., $\boxed{\text{PROB}}$).
- Press the $\boxed{\text{F1}}$ button to select factorial (i.e., $\boxed{x!}$).
- Press the $\boxed{\text{EXE}}$ button to execute the calculation.

The factorial symbol "$n!$" was first introduced by Christian Kramp (1760–1826) of Strasbourg in his *Élements d'Arithmétique Universelle* (1808). Before the introduction of this "modern" symbol, factorials were commonly denoted by ⎣n. However, printing presses of the day had difficulty printing this symbol; consequently, the symbol $n!$ came into prominence because it was relatively easy for a typesetter to use.

EXAMPLE **4**

EVALUATING FACTORIALS   Find the following values.

a. $6!$ **b.** $\dfrac{8!}{5!}$ **c.** $\dfrac{8!}{3! \cdot 5!}$

SOLUTION

a. $6! = 6 \cdot 5 \cdot 4 \cdot 3 \cdot 2 \cdot 1$
$= 720$

Therefore, $6! = 720$.

6 $\boxed{x!}$	
6 $\boxed{\text{MATH}}$ $\boxed{\text{PRB}}$ $\boxed{!}$ $\boxed{\text{ENTER}}$	
Casio 6 $\boxed{\text{OPTN}}$ $\boxed{\rightarrow}$ (i.e., $\boxed{\text{F6}}$) $\boxed{\text{PROB}}$ (i.e., $\boxed{\text{F3}}$) $\boxed{\text{x!}}$ (i.e., $\boxed{\text{F1}}$) $\boxed{\text{EXE}}$	

b. $\dfrac{8!}{5!} = \dfrac{8 \cdot 7 \cdot 6 \cdot 5 \cdot 4 \cdot 3 \cdot 2 \cdot 1}{5 \cdot 4 \cdot 3 \cdot 2 \cdot 1}$

$= \dfrac{8 \cdot 7 \cdot 6 \cdot \cancel{5} \cdot \cancel{4} \cdot \cancel{3} \cdot \cancel{2} \cdot \cancel{1}}{\cancel{5} \cdot \cancel{4} \cdot \cancel{3} \cdot \cancel{2} \cdot \cancel{1}}$

$= 8 \cdot 7 \cdot 6$
$= 336$

Therefore, $\frac{8!}{5!} = 336$.

Using a calculator, we obtain the same result.

8 $\boxed{x!}$ $\boxed{\div}$ 5 $\boxed{x!}$ $\boxed{=}$	
8 $\boxed{\text{MATH}}$ $\boxed{\text{PRB}}$ $\boxed{!}$ $\boxed{\div}$ 5 $\boxed{\text{MATH}}$ $\boxed{\text{PRB}}$ $\boxed{!}$ $\boxed{\text{ENTER}}$	

c. $\dfrac{8!}{3! \cdot 5!} = \dfrac{8 \cdot 7 \cdot 6 \cdot 5 \cdot 4 \cdot 3 \cdot 2 \cdot 1}{(3 \cdot 2 \cdot 1)(5 \cdot 4 \cdot 3 \cdot 2 \cdot 1)}$

$= \dfrac{8 \cdot 7 \cdot 6 \cdot \cancel{5} \cdot \cancel{4} \cdot \cancel{3} \cdot \cancel{2} \cdot \cancel{1}}{(3 \cdot 2 \cdot 1)(\cancel{5} \cdot \cancel{4} \cdot \cancel{3} \cdot \cancel{2} \cdot \cancel{1})}$

$= \dfrac{8 \cdot 7 \cdot 6}{3 \cdot 2 \cdot 1}$

$= 56$

Therefore, $\frac{8!}{3! \cdot 5!} = 56$.

Using a calculator, we obtain the same result.

8 $\boxed{x!}$ $\boxed{\div}$ $\boxed{(}$ 3 $\boxed{x!}$ $\boxed{\times}$ 5 $\boxed{x!}$ $\boxed{)}$ $\boxed{=}$
8 $\boxed{\text{MATH}}$ $\boxed{\text{PRB}}$ $\boxed{!}$ $\boxed{\div}$ $\boxed{(}$ 3 $\boxed{\text{MATH}}$ $\boxed{\text{PRB}}$ $\boxed{!}$ $\boxed{\times}$ 5 $\boxed{\text{MATH}}$ $\boxed{\text{PRB}}$ $\boxed{!}$ $\boxed{)}$ $\boxed{\text{ENTER}}$

1. A nickel, a dime, and a quarter are tossed.

   a. Use the Fundamental Principle of Counting to determine how many different outcomes are possible.

   b. Construct a tree diagram to list all possible outcomes.

2. A die is rolled, and a coin is tossed.

   a. Use the Fundamental Principle of Counting to determine how many different outcomes are possible.

   b. Construct a tree diagram to list all possible outcomes.

3. Jamie has decided to buy either a Mega or a Better Byte desktop computer. She also wants to purchase either Big Word, Word World, or Great Word word-processing software and either Big Number or Number World spreadsheet software.

   a. Use the Fundamental Principle of Counting to determine how many different packages of a computer and software Jamie has to choose from.

   b. Construct a tree diagram to list all possible packages of a computer and software.

4. Sammy's Sandwich Shop offers a soup, sandwich, and beverage combination at a special price. There are three sandwiches (turkey, tuna, and tofu), two soups (minestrone and split pea), and three beverages (coffee, milk, and mineral water) to choose from.

   a. Use the Fundamental Principle of Counting to determine how many different meal combinations are possible.

   b. Construct a tree diagram to list all possible soup, sandwich, and beverage combinations.

5. If you buy three pairs of jeans, four sweaters, and two pairs of boots, how many new outfits (consisting of a new pair of jeans, a new sweater, and a new pair of boots) will you have?

6. A certain model of automobile is available in six exterior colors, three interior colors, and three interior styles. In addition, the transmission can be either manual or automatic, and the engine can have either four or six cylinders. How many different versions of the automobile can be ordered?

7. To fulfill certain requirements for a degree, a student must take one course each from the following groups: health, civics, critical thinking, and elective. If there are four health, three civics, six critical thinking, and ten elective courses, how many different options for fulfilling the requirements does a student have?

8. To fulfill a requirement for a literature class, a student must read one short story by each of the following authors: Stephen King, Clive Barker, Edgar Allan Poe, and H. P. Lovecraft. If there are twelve King, six Barker, eight Poe, and eight Lovecraft stories to choose from, how many different combinations of reading assignments can a student choose from to fulfill the reading requirement?

9. A sporting goods store has fourteen lines of snow skis, seven types of bindings, nine types of boots, and three types of poles. Assuming that all items are compatible with each other, how many different complete ski equipment packages are available?

10. An audio equipment store has ten different amplifiers, four tuners, six turntables, eight tape decks, six compact disc players, and thirteen speakers. Assuming that all components are compatible with each other, how many different complete stereo systems are available?

11. A cafeteria offers a complete dinner that includes one serving each of appetizer, soup, entrée, and dessert for $6.99. If the menu has three appetizers, four soups, six entrées, and three desserts, how many different meals are possible?

12. A sandwich shop offers a "U-Chooz" special consisting of your choice of bread, meat, cheese, and special sauce (one each). If there are six different breads, eight meats, five cheeses, and four special sauces, how many different sandwiches are possible?

13. How many different Social Security numbers are possible? (A Social Security number consists of nine digits that can be repeated.)

14. To use an automated teller machine (ATM), a customer must enter his or her four-digit Personal Identification Number (PIN). How many different PINs are possible?

15. Every book published has an International Standard Book Number (ISBN). The number is a code used to identify the specific book and is of the form X-XXX-XXXXX-X, where X is one of digits 0, 1, 2, . . . , 9. How many different ISBNs are possible?

16. How many different Zip Codes are possible using (a) the old style (five digits) and (b) the new style (nine digits)? Why do you think the U.S. Postal Service introduced the new system?

17. Telephone area codes are three-digit numbers of the form XXX.

   a. Originally, the first and third digits were neither 0 nor 1 and the second digit was always a 0 or a 1. How many three-digit numbers of this type are possible?

   b. Over time, the restrictions listed in part (a) have been altered; currently, the only requirement is that the first digit is neither 0 nor 1. How many three-digit numbers of this type are possible?

c. Why were the original restrictions listed in part (a) altered?

18. Major credit cards such as VISA and MasterCard have a sixteen-digit account number of the form XXXX-XXXX-XXXX-XXXX. How many different numbers of this type are possible?

19. The serial number on a dollar bill consists of a letter followed by eight digits and then a letter. How many different serial numbers are possible, given the following conditions?
   a. Letters and digits cannot be repeated.
   b. Letters and digits can be repeated.
   c. The letters are nonrepeated consonants and the digits can be repeated.

20. The serial number on a new twenty-dollar bill consists of two letters followed by eight digits and then a letter. How many different serial numbers are possible, given the following conditions?
   a. Letters and digits cannot be repeated.
   b. Letters and digits can be repeated.
   c. The first and last letters are repeatable vowels, the second letter is a consonant, and the digits can be repeated.

21. Each student at State University has a student I.D. number consisting of four digits (the first digit is nonzero, and digits may be repeated) followed by three of the letters A, B, C, D, and E (letters may not be repeated). How many different student numbers are possible?

22. Each student at State College has a student I.D. number consisting of five digits (the first digit is nonzero, and digits may be repeated) followed by two of the letters A, B, C, D, and E (letters may not be repeated). How many different student numbers are possible?

In Exercises 23–38, find the indicated value.

23. 4! **24.** 5!
25. 10! **26.** 8!
27. 20! **28.** 25!
29. 6! · 4! **30.** 8! · 6!
31. a. $\frac{6!}{4!}$ **b.** $\frac{6!}{2!}$ **32. a.** $\frac{8!}{6!}$ **b.** $\frac{8!}{2!}$
33. $\frac{8!}{5! \cdot 3!}$ **34.** $\frac{9!}{5! \cdot 4!}$
35. $\frac{8!}{4! \cdot 4!}$ **36.** $\frac{6!}{3! \cdot 3!}$
37. $\frac{82!}{80! \cdot 2!}$ **38.** $\frac{77!}{74! \cdot 3!}$

39. Find the value of $\frac{n!}{(n-r)!}$ when $n = 16$ and $r = 14$.
40. Find the value of $\frac{n!}{(n-r)!}$ when $n = 19$ and $r = 16$.
41. Find the value of $\frac{n!}{(n-r)!}$ when $n = 5$ and $r = 5$.
42. Find the value of $\frac{n!}{(n-r)!}$ when $n = r$.
43. Find the value of $\frac{n!}{(n-r)!r!}$ when $n = 7$ and $r = 3$.
44. Find the value of $\frac{n!}{(n-r)!r!}$ when $n = 7$ and $r = 4$.
45. Find the value of $\frac{n!}{(n-r)!r!}$ when $n = 5$ and $r = 5$.
46. Find the value of $\frac{n!}{(n-r)!r!}$ when $n = r$.

 Answer the following questions using complete sentences and your own words.

• CONCEPT QUESTIONS

47. What is the Fundamental Principle of Counting? When is it used?
48. What is a factorial?

• HISTORY QUESTIONS

49. Who invented the modern symbol denoting a factorial? What symbol did it replace? Why?

 THE NEXT LEVEL

If a person wants to pursue an advanced degree (something beyond a bachelor's or four-year degree), chances are the person must take a standardized exam to gain admission to a school or to be admitted into a specific program. These exams are intended to measure verbal, quantitative, and analytical skills that have developed throughout a person's life. Many classes and study guides are available to help people prepare for the exams. The following questions are typical of those found in the study guides.

Exercises 50–54 refer to the following: In an executive parking lot, there are six parking spaces in a row, labeled 1 through 6. Exactly five cars of five different colors—black, gray, pink, white, and yellow—are to be parked in the spaces. The cars can park in any of the spaces as long as the following conditions are met:

• The pink car must be parked in space 3.
• The black car must be parked in a space next to the space in which the yellow car is parked.
• The gray car cannot be parked in a space next to the space in which the white car is parked.

50. If the yellow car is parked in space 1, how many acceptable parking arrangements are there for the five cars?
   a. 1 **b.** 2 **c.** 3 **d.** 4 **e.** 5

51. Which of the following must be true of any acceptable parking arrangement?
   a. One of the cars is parked in space 2.
   b. One of the cars is parked in space 6.

c. There is an empty space next to the space in which the gray car is parked.

d. There is an empty space next to the space in which the yellow car is parked.

e. Either the black car or the yellow car is parked in a space next to space 3.

52. If the gray car is parked in space 2, none of the cars can be parked in which space?

    a. 1   **b.** 3   **c.** 4   **d.** 5   **e.** 6

53. The white car could be parked in any of the spaces except which of the following?

    a. 1   **b.** 2   **c.** 4   **d.** 5   **e.** 6

54. If the yellow car is parked in space 2, which of the following must be true?

    a. None of the cars is parked in space 5.

    b. The gray car is parked in space 6.

    c. The black car is parked in a space next to the space in which the white car is parked.

    d. The white car is parked in a space next to the space in which the pink car is parked.

    e. The gray car is parked in a space next to the space in which the black car is parked.

2.4 Permutations and Combinations

OBJECTIVES

- Develop and apply the Permutation Formula
- Develop and apply the Combination Formula
- Determine the number of distinguishable permutations

The Fundamental Principle of Counting allows us to determine the total number of possible outcomes when a series of decisions (making selections from various categories) must be made. In Section 2.3, the examples and exercises involved selecting *one item each* from various categories; if you buy two pairs of jeans, three shirts, and two pairs of shoes, you will have twelve ($2 \cdot 3 \cdot 2 = 12$) new outfits (consisting of a new pair of jeans, a new shirt, and a new pair of shoes). In this section, we examine the situation when *more than one* item is selected from a category. If more than one item is selected, the selections can be made either *with* or *without* replacement.

With versus Without Replacement

Selecting items *with replacement* means that the same item *can* be selected more than once; after a specific item has been chosen, it is put back into the pool of future choices. Selecting items *without replacement* means that the same item *cannot* be selected more than once; after a specific item has been chosen, it is not replaced.

Suppose you must select a four-digit Personal Identification Number (PIN) for a bank account. In this case, the digits are selected with replacement; each time a specific digit is selected, the digit is put back into the pool of choices for the next selection. (Your PIN can be 3666; the same digit can be selected more than once.) When items are selected with replacement, we use the Fundamental Principle of Counting to determine the total number of possible outcomes; there are $10 \cdot 10 \cdot 10 \cdot 10 = 10{,}000$ possible four-digit PINs.

In many situations, items cannot be selected more than once. For instance, when selecting a committee of three people from a group of twenty, you cannot select the same person more than once. Once you have selected a specific person (say, Lauren), you do not put her back into the pool of choices. When selecting items without replacement, depending on whether the order of selection is important, *permutations* or *combinations* are used to determine the total number of possible outcomes.

Permutations

When more than one item is selected (without replacement) from a single category, and the order of selection *is* important, the various possible outcomes are called **permutations.** For example, when the rankings (first, second, and third place) in a talent contest are announced, the order of selection is important; Monte in first, Lynn in second, and Ginny in third place is different from Ginny in first, Monte in second, and Lynn in third. "Monte, Lynn, Ginny" and "Ginny, Monte, Lynn" are different permutations of the contestants. Naturally, these selections are made without replacement; we cannot select Monte for first place and reselect him for second place.

EXAMPLE 1

FINDING THE NUMBER OF PERMUTATIONS   Six local bands have volunteered to perform at a benefit concert, but there is enough time for only four bands to play. There is also some concern over the order in which the chosen bands will perform. How many different lineups are possible?

SOLUTION

We must select four of the six bands and put them in a specific order. The bands are selected without replacement; a band cannot be selected to play and then be reselected to play again. Because we must make four decisions, we draw four boxes and put the number of choices for each decision in each appropriate box. There are six choices for the opening band. Naturally, the opening band could not be the follow-up act, so there are only five choices for the next group. Similarly, there are four candidates for the third group and three choices for the closing band. The total number of different lineups possible is found by multiplying the number of choices for each decision:

$$\boxed{6} \times \boxed{5} \times \boxed{4} \times \boxed{3} = 360$$

$$\underset{\substack{\text{opening} \\ \text{band}}}{\uparrow} \qquad\qquad \underset{\substack{\text{closing} \\ \text{band}}}{\uparrow}$$

With four out of six bands playing in the performance, 360 lineups are possible. *Because the order of selecting the bands is important, the various possible outcomes, or lineups, are called permutations;* there are 360 permutations of six items when the items are selected four at a time.

The computation in Example 1 is similar to a factorial, but the factors do not go all the way down to 1; the product $6 \cdot 5 \cdot 4 \cdot 3$ is a "truncated" (cut-off) factorial. We can change this truncated factorial into a complete factorial in the following manner:

$$6 \cdot 5 \cdot 4 \cdot 3 = \frac{6 \cdot 5 \cdot 4 \cdot 3 \cdot (2 \cdot 1)}{(2 \cdot 1)} \quad \text{multiplying by } \frac{2}{2} \text{ and } \frac{1}{1}$$

$$= \frac{6!}{2!}$$

Notice that this last expression can be written as $\frac{6!}{2!} = \frac{6!}{(6-4)!}$. (Recall that we were selecting four out of six bands.) This result is generalized as follows.

PERMUTATION FORMULA

The number of **permutations,** or arrangements, of r items selected without replacement from a pool of n items ($r \leq n$), denoted by $_nP_r$, is

$$_nP_r = \frac{n!}{(n-r)!}$$

Permutations are used whenever more than one item is selected (without replacement) from a category and the order of selection is important.

Using the notation above and referring to Example 1, we note that 360 possible lineups of four bands selected from a pool of six can be denoted by $_6P_4 = \frac{6!}{(6-4)!} = 360$. Other notations can be used to represent the number of permutations of a group of items. In particular, the notations $_nP_r$, $P(n, r)$, P_r^n, and $P_{n,r}$ all represent the number of possible permutations (or arrangements) of r items selected (without replacement) from a pool of n items.

EXAMPLE **2**

FINDING THE NUMBER OF PERMUTATIONS Three door prizes (first, second, and third) are to be awarded at a ten-year high school reunion. Each of the 112 attendees puts his or her name in a hat. The first name drawn wins a two-night stay at the Chat 'n' Rest Motel, the second name wins dinner for two at Juju's Kitsch-Inn, and the third wins a pair of engraved mugs. In how many different ways can the prizes be awarded?

SOLUTION

We must select 3 out of 112 people (without replacement), and the order in which they are selected *is* important. (Winning dinner is different from winning the mugs.) Hence, we must find the number of permutations of 3 items selected from a pool of 112:

$$
\begin{aligned}
_{112}P_3 &= \frac{112!}{(112-3)!} \\
&= \frac{112!}{109!} \\
&= \frac{112 \cdot 111 \cdot 110 \cdot 109 \cdot 108 \cdots 2 \cdot 1}{109 \cdot 108 \cdots 2 \cdot 1} \\
&= 112 \cdot 111 \cdot 110 \\
&= 1,367,520
\end{aligned}
$$

There are 1,367,520 different ways in which the three prizes can be awarded to the 112 people.

In Example 2, if you try to use a calculator to find $\frac{112!}{109!}$ directly, you will not obtain an answer. Entering 112 and pressing $\boxed{x!}$ results in a calculator error. (Try it.) Because factorials get very large very quickly, most calculators are not able to find any factorial over 69!. ($69! = 1.711224524 \times 10^{98}$.)

EXAMPLE **3**

FINDING THE NUMBER OF PERMUTATIONS   A bowling league has ten teams. In how many different ways can the teams be ranked in the standings at the end of a tournament? (Ties are not allowed.)

SOLUTION

Because order is important, we find the number of permutations of ten items selected from a pool of ten items:

$$_{10}P_{10} = \frac{10!}{(10 - 10)!}$$

$$= \frac{10!}{0!} \qquad \textbf{Recall that } 0! = 1.$$

$$= \frac{10!}{1}$$

$$= 3,628,800$$

In a league containing ten teams, 3,628,800 different standings are possible at the end of a tournament.

Combinations

When items are selected from a group, the order of selection may or may not be important. If the order is important (as in Examples 1, 2, and 3), permutations are used to determine the total number of selections possible. What if the order of selection is *not* important? When more than one item is selected (without replacement) from a single category and the order of selection is not important, the various possible outcomes are called **combinations.**

EXAMPLE **4**

LISTING ALL POSSIBLE COMBINATIONS   Two adults are needed to chaperone a daycare center's field trip. Marcus, Vivian, Frank, and Keiko are the four managers of the center. How many different groups of chaperones are possible?

SOLUTION

In selecting the chaperones, the order of selection is *not* important; "Marcus and Vivian" is the same as "Vivian and Marcus." Hence, the permutation formula cannot be used. Because we do not yet have a shortcut for finding the total number of possibilities when the order of selection is not important, we must list all the possibilities:

Marcus and Vivian	Marcus and Frank	Marcus and Keiko
Vivian and Frank	Vivian and Keiko	Frank and Keiko

Therefore, six different groups of two chaperones are possible from the group of four managers. Because the order in which the people are selected is not important, the various possible outcomes, or groups of chaperones, are called *combinations;* there are six combinations when two items are selected from a pool of four.

Just as $_nP_r$ denotes the number of *permutations* of r elements selected from a pool of n elements, $_nC_r$ denotes the number of *combinations* of r elements selected from a pool of n elements. In Example 4, we found that there are six combinations of two people selected from a pool of four by listing all six of the combinations; that is, $_4C_2 = 6$. If we had a larger pool, listing each combination to find out how many there are would be extremely time consuming and tedious! Instead of listing, we take a different approach. We first find the number of permutations (with the permutation formula) and then alter that number to account for the distinction between permutations and combinations.

To find the number of combinations of two people selected from a pool of four, we first find the number of permutations:

$$_4P_2 = \frac{4!}{(4-2)!} = \frac{4!}{2!} = 12$$

This figure of 12 must be altered to account for the distinction between permutations and combinations.

In Example 4, we listed combinations; one such combination was "Marcus and Vivian." If we had listed permutations, we would have had to list both "Marcus and Vivian" and "Vivian and Marcus," because the *order* of selection matters with permutations. In fact, each combination of two chaperones listed in Example 4 generates two permutations; each pair of chaperones can be given in two different orders. Thus, there are twice as many permutations of two people selected from a pool of four as there are combinations. Alternatively, there are half as many combinations of two people selected from a pool of four as there are permutations. We used the permutation formula to find that $_4P_2 = 12$; thus,

$$_4C_2 = \frac{1}{2} \cdot {_4P_2} = \frac{1}{2}(12) = 6$$

This answer certainly fits with Example 4; we listed exactly six combinations.

What if three of the four managers were needed to chaperone the daycare center's field trip? Rather than finding the number of combinations by listing each possibility, we first find the number of permutations and then alter that number to account for the distinction between permutations and combinations.

The number of permutations of three people selected from a pool of four is

$$_4P_3 = \frac{4!}{(4-3)!} = \frac{4!}{1!} = 24$$

We know that some of these permutations represent the same combination. For example, the combination "Marcus and Vivian and Keiko" generates $3! = 6$ different permutations (using initials, they are: MVK, MKV, KMV, KVM, VMK, VKM). Because each combination of three people generates six different permutations, there are one-sixth as many combinations as permutations. Thus,

$$_4C_3 = \frac{1}{6} \cdot {_4P_3} = \frac{1}{6}(24) = 4$$

This means that if three of the four managers were needed to chaperone the daycare center's field trip, there would be $_4C_3 = 4$ possible combinations.

We just saw that when two items are selected from a pool of n items, each combination of two generates $2! = 2$ permutations, so

$$_nC_2 = \frac{1}{2!} \cdot {_nP_2}$$

We also saw that when three items are selected from a pool of n items, each combination of three generates $3! = 6$ permutations, so

$$_nC_3 = \frac{1}{3!} \cdot {_nP_3}$$

More generally, when r items are selected from a pool of n items, each combination of r items generates $r!$ permutations, so

$$_nC_r = \frac{1}{r!} \cdot {}_nP_r$$

$$= \frac{1}{r!} \cdot \frac{n!}{(n-r)!} \quad \text{using the Permutation Formula}$$

$$= \frac{n!}{r! \cdot (n-r)!} \quad \text{multiplying the fractions together}$$

COMBINATION FORMULA

The number of distinct **combinations** of r items selected without replacement from a pool of n items ($r \leq n$), denoted by $_nC_r$, is

$$_nC_r = \frac{n!}{(n-r)!\,r!}$$

Combinations are used whenever one or more items are selected (without replacement) from a category and the order of selection is not important.

EXAMPLE 5

FINDING THE NUMBER OF COMBINATIONS   A DVD club sends you a brochure that offers any five DVDs from a group of fifty of today's hottest releases. How many different selections can you make?

SOLUTION

Because the order of selection is *not* important, we find the number of combinations when five items are selected from a pool of fifty:

$$_{50}C_5 = \frac{50!}{(50-5)!\,5!}$$

$$= \frac{50!}{45!\,5!}$$

$$= \frac{50 \cdot 49 \cdot 48 \cdot 47 \cdot 46}{5 \cdot 4 \cdot 3 \cdot 2 \cdot 1}$$

$$= 2{,}118{,}760$$

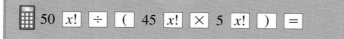

50 [x!] ÷ (45 [x!] × 5 [x!]) =

Graphing calculators have buttons that will calculate $_nP_r$ and $_nC_r$. To use them, do the following:

- Type the value of n. (For example, type the number 50.)
- Press the MATH button.
- Press the right arrow button → as many times as necessary to highlight PRB.
- Press the down arrow button ↓ as many times as necessary to highlight the appropriate symbol— $_nP_r$ for permutations, $_nC_r$ for combinations—and press ENTER.
- Type the value of r. (For example, type the number 5.)
- Press ENTER to execute the calculation.

On a Casio graphing calculator, do the following:

- Press the [MENU] button; this gives you access to the main menu.
- Press 1 to select the RUN mode; this mode is used to perform arithmetic operations.
- Type the value of *n*. (For example, type the number 50.)
- Press the [OPTN] button; this gives you access to various options displayed at the bottom of the screen.
- Press the [F6] button to see more options (i.e., [→]).
- Press the [F3] button to select probability options (i.e., [PROB]).
- Press the [F3] button to select combinations (i.e., [$_nC_r$]) or the [F2] button to select permutations (i.e., [$_nP_r$]).
- Type the value of *r*. (For example, type the number 5.)
- Press the [EXE] button to execute the calculation.

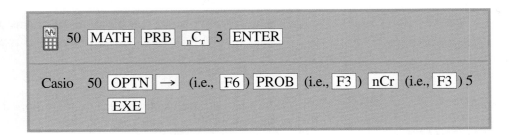

In choosing five out of fifty DVDs, 2,118,760 combinations are possible.

EXAMPLE **6**

FINDING THE NUMBER OF COMBINATIONS A group consisting of twelve women and nine men must select a five-person committee. How many different committees are possible if it must consist of the following?

a. three women and two men   **b.** any mixture of men and women

SOLUTION

a. Our problem involves two categories: women and men. The Fundamental Principle of Counting tells us to draw two boxes (one for each category), enter the number of choices for each, and multiply:

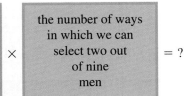

the number of ways in which we can select three out of twelve women × the number of ways in which we can select two out of nine men = ?

Because the order of selecting the members of a committee is not important, we will use combinations:

$$(_{12}C_3) \cdot (_9C_2) = \frac{12!}{(12-3)! \cdot 3!} \cdot \frac{9!}{(9-2)! \cdot 2!}$$

$$= \frac{12!}{9! \cdot 3!} \cdot \frac{9!}{7! \cdot 2!}$$

$$= \frac{12 \cdot 11 \cdot 10}{3 \cdot 2 \cdot 1} \cdot \frac{9 \cdot 8}{2 \cdot 1}$$

$$= 220 \cdot 36$$

$$= 7,920$$

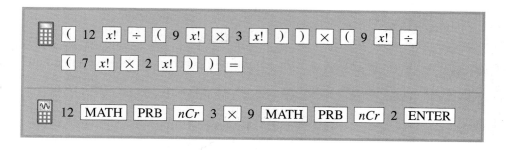

There are 7,920 different committees consisting of three women and two men.

b. Because the gender of the committee members doesn't matter, our problem involves only one category: people. We must choose five out of the twenty-one people, and the order of selection is not important:

$$_{21}C_5 = \frac{21!}{(21-5)! \cdot 5!}$$

$$= \frac{21!}{16! \cdot 5!}$$

$$= \frac{21 \cdot 20 \cdot 19 \cdot 18 \cdot 17}{5 \cdot 4 \cdot 3 \cdot 2 \cdot 1}$$

$$= 20,349$$

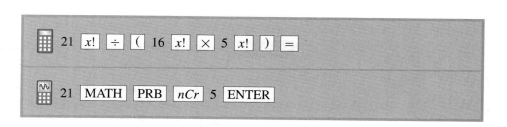

There are 20,349 different committees consisting of five people.

EXAMPLE **7**

EVALUATING THE COMBINATION FORMULA   Find the value of $_5C_r$ for the following values of r:

a. $r = 0$ **b.** $r = 1$ **c.** $r = 2$ **d.** $r = 3$ **e.** $r = 4$ **f.** $r = 5$

SOLUTION

a. $_5C_0 = \dfrac{5!}{(5-0)! \cdot 0!} = \dfrac{5!}{5! \cdot 0!} = 1$

b. $_5C_1 = \dfrac{5!}{(5-1)! \cdot 1!} = \dfrac{5!}{4! \cdot 1!} = 5$

c. $_5C_2 = \dfrac{5!}{(5-2)! \cdot 2!} = \dfrac{5!}{3! \cdot 2!} = 10$

d. $_5C_3 = \dfrac{5!}{(5-3)! \cdot 3!} = \dfrac{5!}{2! \cdot 3!} = 10$

e. $_5C_4 = \dfrac{5!}{(5-4)! \cdot 4!} = \dfrac{5!}{1! \cdot 4!} = 5$

f. $_5C_5 = \dfrac{5!}{(5-5)! \cdot 5!} = \dfrac{5!}{0! \cdot 5!} = 1$

The combinations generated in Example 7 exhibit a curious pattern. Notice that the values of $_5C_r$ are symmetric: $_5C_0 = {}_5C_5$, $_5C_1 = {}_5C_4$, and $_5C_2 = {}_5C_3$. Now examine the diagram in Figure 2.41. Each number in this "triangle" of numbers is the sum of two numbers in the row immediately above it. For example, $2 = 1 + 1$ and $10 = 4 + 6$, as shown by the inserted arrows. It is no coincidence that the values of $_5C_r$ found in Example 7 also appear as a row of numbers in this "magic" triangle. In fact, the sixth row contains all the values of $_5C_r$ for $r = 0, 1, 2, 3, 4$, and 5. In general, the $(n + 1)^{\text{th}}$ row of the triangle contains all the values of $_nC_r$ for $r = 0, 1, 2, \ldots, n$; alternatively, the nth row of the triangle contains all the values of $_{n-1}C_r$ for $r = 0, 1, 2, \ldots, n-1$. For example, the values of $_9C_r$, for $r = 0, 1, 2, \ldots, 9$, are in the tenth row, and vice versa, the entries in the tenth row are the values of $_9C_r$, for $r = 0, 1, 2, \ldots, 9$.

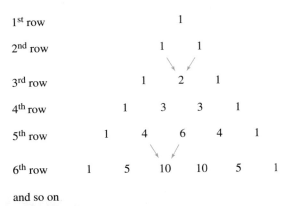

1st row   1

2nd row   1   1

3rd row   1   2   1

4th row   1   3   3   1

5th row   1   4   6   4   1

6th row   1   5   10   10   5   1

and so on

FIGURE 2.41 Pascal's triangle.

Historically, this triangular pattern of numbers is referred to as *Pascal's Triangle*, in honor of the French mathematician, scientist, and philosopher Blaise Pascal (1623–1662). Pascal is a cofounder of probability theory (see the Historical Note in Section 3.1). Although the triangle has Pascal's name attached to it, this "magic" arrangement of numbers was known to other cultures hundreds of years before Pascal's time.

The most important part of any problem involving combinatorics is deciding which counting technique (or techniques) to use. The following list of general steps and the flowchart in Figure 2.42 can help you to decide which method or methods to use in a specific problem.

WHICH COUNTING TECHNIQUE?

1. What is being selected?
2. If the selected items can be repeated, use the **Fundamental Principle of Counting** and multiply the number of choices for each category.
3. If there is only one category, use:
   combinations if the order of selection does not matter—that is, r items can be selected from a pool of n items in $_nC_r = \frac{n!}{(n-r)! \cdot r!}$ ways.
   permutations if the order of selection does matter—that is, r items can be selected from a pool of n items in $_nP_r = \frac{n!}{(n-r)!}$ ways.
4. If there is more than one category, use the **Fundamental Principle of Counting** with one box per category.
   a. If you are selecting one item per category, the number in the box for that category is the number of choices for that category.
   b. If you are selecting more than one item per category, the number in the box for that category is found by using step 3.

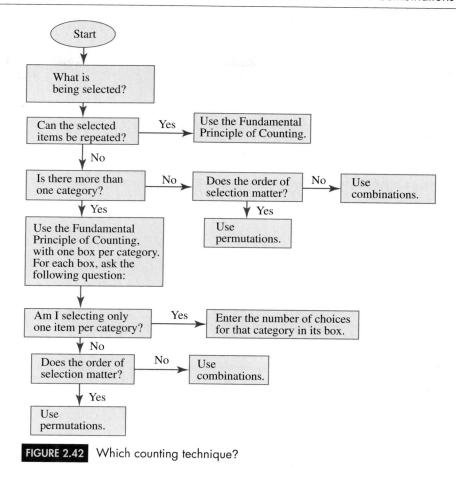

FIGURE 2.42   Which counting technique?

EXAMPLE **8**

USING THE "WHICH COUNTING TECHNIQUE?" FLOWCHART   A standard deck of playing cards contains fifty-two cards.

a. How many different five-card hands containing four kings are possible?
b. How many different five-card hands containing four queens are possible?
c. How many different five-card hands containing four kings or four queens are possible?
d. How many different five-card hands containing four of a kind are possible?

SOLUTION

a. We use the flowchart in Figure 2.42 and answer the following questions.

   Q. What is being selected?
   A. Playing cards.

   Q. Can the selected items be repeated?
   A. No.

   Q. Is there more than one category?
   A. Yes: Because we must have five cards, we need four kings and one non-king. Therefore, we need two boxes:

   $$\boxed{\text{kings}} \times \boxed{\text{non-kings}}$$

   Q. Am I selecting only one item per category?
   A. *Kings:* no. Does the order of selection matter? No: Use combinations. Because there are $n = 4$ kings in the deck and we want to select $r = 4$, we must compute $_4C_4$. *Non-kings:* yes. Enter the number of choices for that category: There are 48 non-kings.

HISTORICAL NOTE

CHU SHIH-CHIEH, CIRCA 1280–1303

Chu Shih-chieh was the last and most acclaimed mathematician of the Sung Dynasty in China. Little is known of his personal life; the actual dates of his birth and death are unknown. His work appears to have flourished during the close of the thirteenth century. It is believed that Chu Shih-chieh spent many years as a wandering scholar, earning a living by teaching mathematics to those who wanted to learn.

Two of Chu Shih-chieh's works have survived the centuries. The first, *Suan-hsüeh ch'i-meng (Introduction to Mathematical Studies)*, was written in 1299 and contains elementary mathematics. This work was very influential in Japan and Korea, although it was lost in China until the nineteenth century. Written in 1303, Chu's second work *Ssu-yüan yü-chien (Precious Mirror of the Four Elements)* contains more advanced mathematics. The topics of *Precious Mirror* include the solving of simultaneous equations and the solving of equations up to the fourteenth degree.

Of the many diagrams in *Precious Mirror*, one has special interest: the arithmetic triangle. Chu Shih-chieh's triangle contains the first eight rows of what is known in the West as Pascal's Triangle. However, Chu does not claim credit for the triangle; he refers to it as "a diagram of the *old* method for finding eighth and lower powers." "Pascal's" Triangle was known to the Chinese well over 300 years before Pascal was born!

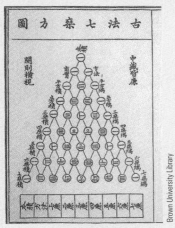

The "Pascal" Triangle as depicted in 1303 at the front of Chu Shih-chieh's *Ssu-yüan yü-chien. It is entitled "The Old Method Chart of the Seven Multiplying Squares" and tabulates the binomial coefficients up to the eighth power.*

Brown University Library

$$\boxed{\text{kings}} \times \boxed{\text{non-kings}} = \boxed{_4C_4} \times \boxed{48}$$

$$= \frac{4!}{(4-4)! \cdot 4!} \cdot 48$$

$$= \frac{4!}{0! \cdot 4!} \cdot 48$$

$$= 1 \cdot 48$$

$$= 48$$

There are forty-eight different five-card hands containing four kings.

b. Using the same method as in part (a), we would find that there are forty-eight different five-card hands containing four queens; the number of five-card hands containing four queens is the same as the number of five-card hands containing four kings.

c. To find the number of five-card hands containing four kings or four queens, we define the following sets:

$$A = \{\text{five-card hands} \mid \text{the hand contains four kings}\}$$
$$B = \{\text{five-card hands} \mid \text{the hand contains four queens}\}$$

Consequently,

$$A \cup B = \{\text{five-card hands} \mid \text{the hand contains four kings } or \text{ four queens}\}$$
$$A \cap B = \{\text{five-card hands} \mid \text{the hand contains four kings } and \text{ four queens}\}$$

(Recall that the union symbol, $\cup$, may be interpreted as the word "or," while the intersection symbol, $\cap$, may be interpreted as the word "and." See Figure 2.13 for a comparison of set theory and logic.)

Because there are no five-card hands that contain four kings *and* four queens, we note that $n(A \cap B) = 0$.

Using the Union/Intersection Formula for the Union of Sets, we obtain

$$n(A \cup B) = n(A) + n(B) - n(A \cap B)$$
$$= 48 + 48 - 0$$
$$= 96$$

There are ninety-six different five-card hands containing four kings or four queens.

d. *Four of a kind* means four cards of the same "denomination," that is, four 2's, or four 3's, or four 4's, or . . . , or four kings, or four aces. Now, regardless of the denomination of the card, there are forty-eight different five-card hands that contain four of any specific denomination; there are forty-eight different five-card hands that contain four 2's, there are forty-eight different five-card hands that contain four 3's, there are forty-eight different five-card hands that contain four 4's, and so on. As is shown in part (c), the word "or" implies that we *add* cardinal numbers. Consequently,

$$n(\text{four of a kind}) = n(\text{four 2's or four 3's or . . . or four kings or four aces})$$
$$= n(\text{four 2's}) + n(\text{four 3's} + \cdots + n(\text{four kings})$$
$$\quad + n(\text{four aces})$$
$$= 48 + 48 + \cdots + 48 + 48 \quad \text{(thirteen times)}$$
$$= 13 \times 48$$
$$= 624$$

There are 624 different five-card hands containing four of a kind.

As is shown in Example 8, there are 624 possible five-card hands that contain four of a kind. When you are dealt five cards, what is the likelihood (or probability) that you will receive one of these hands? This question, and its answer, will be explored in Section 3.4, "Combinatorics and Probability."

Permutations of Identical Items

In how many different ways can the three letters in the word "SAW" be arranged? As we know, arrangements are referred to as *permutations,* so we can apply the Permutation Formula, $_nP_r = \frac{n!}{(n-r)!}$.

Therefore,

$$_3P_3 = \frac{3!}{(3-3)!} = \frac{3!}{0!} = \frac{3 \cdot 2 \cdot 1}{1} = 6$$

The six permutations of the letters in SAW are

SAW   SWA   AWS   ASW   WAS   WSA

In general, if we have three *different* items (the letters in SAW), we can arrange them in $3! = 6$ ways. However, this method applies only if the items are all different (distinct).

What happens if some of the items are the same (identical)? For example, in how many different ways can the three letters in the word "SEE" be arranged? Because two of the letters are identical (E), we cannot use the Permutation Formula directly; we take a slightly different approach. Temporarily, let us assume that the E's are written in different colored inks, say, red and blue. Therefore, SEE could be expressed as SEE. These three symbols could be arranged in $3! = 6$ ways as follows:

SEE   SEE   ESE   ESE   EES   EES

If we now remove the color, the arrangements are

SEE   SEE   ESE   ESE   EES   EES

Some of these arrangements are duplicates of others; as we can see, there are only three different or **distinguishable permutations,** namely, SEE, ESE, and EES. Notice that when $n = 3$ (the total number of letters in SEE) and $x = 2$ (the number of identical letters), we can divide $n!$ by $x!$ to obtain the number of distinguishable permutations; that is,

$$\frac{n!}{x!} = \frac{3!}{2!} = \frac{3 \cdot 2 \cdot 1}{2 \cdot 1} = \frac{3 \cdot \cancel{2} \cdot \cancel{1}}{2 \cdot \cancel{1}} = 3$$

This method is applicable because dividing by the factorial of the repeated letter eliminates the duplicate arrangements; the method may by generalized as follows.

DISTINGUISHABLE PERMUTATIONS OF IDENTICAL ITEMS

The number of **distinguishable permutations** (or arrangements) of n items in which x items are identical, y items are identical, z items are identical, and so on, is $\frac{n!}{x!y!z!\cdots}$. That is, to find the number of distinguishable permutations, divide the total factorial by the factorial of each repeated item.

EXAMPLE **9**

FINDING THE NUMBER OF DISTINGUISHABLE PERMUTATIONS
Find the number of distinguishable permutations of the letters in the word "MISSISSIPPI."

SOLUTION

The word "MISSISSIPPI" has $n = 11$ letters; I is repeated $x = 4$ times, S is repeated $y = 4$ times, and P is repeated $z = 2$ times. Therefore, we divide the total factorial by the factorial of each repeated letter and obtain

$$\frac{n!}{x!y!z!} = \frac{11!}{4!4!2!} = 34{,}650$$

The letters in the word MISSISSIPPI can be arranged in 34,650 ways. (Note that if the 11 letters were all different, there would be $11! = 39{,}96{,}800$ permutations.)

2.4 EXERCISES

In Exercises 1–12, find the indicated value:

1. a. $_7P_3$ **b.** $_7C_3$
2. a. $_8P_4$ **b.** $_8C_4$

3. a. $_5P_5$ **b.** $_5C_5$
4. a. $_9P_0$ **b.** $_9C_0$

5. a. $_{14}P_1$ **b.** $_{14}C_1$
6. a. $_{13}C_3$ **b.** $_{13}C_{10}$

7. a. $_{100}P_3$ **b.** $_{100}C_3$
8. a. $_{80}P_4$ **b.** $_{80}C_4$

9. a. $_xP_{x-1}$ **b.** $_xC_{x-1}$
10. a. $_xP_1$ **b.** $_xC_1$

11. a. $_xP_2$ **b.** $_xC_2$
12. a. $_xP_{x-2}$ **b.** $_xC_{x-2}$

13. a. Find $_3P_2$.

  b. List all of the permutations of {a, b, c} when the elements are taken two at a time.

14. a. Find $_3C_2$.

  b. List all of the combinations of {a, b, c} when the elements are taken two at a time.

15. a. Find $_4C_2$.

  b. List all of the combinations of {a, b, c, d} when the elements are taken two at a time.

16. a. Find $_4P_2$.

  b. List all of the permutations of {a, b, c, d} when the elements are taken two at a time.

17. An art class consists of eleven students. All of them must present their portfolios and explain their work to the instructor and their classmates at the end of the semester.

  a. If their names are drawn from a hat to determine who goes first, second, and so on, how many presentation orders are possible?

b. If their names are put in alphabetical order to determine who goes first, second, and so on, how many presentation orders are possible?

18. An English class consists of twenty-three students, and three are to be chosen to give speeches in a school competition. In how many different ways can the teacher choose the team, given the following conditions?

    a. The order of the speakers is important.

    b. The order of the speakers is not important.

19. In how many ways can the letters in the word "school" be arranged? (See the photograph below.)

20. A committee of four is to be selected from a group of sixteen people. How many different committees are possible, given the following conditions?

    a. There is no distinction between the responsibilities of the members.

    b. One person is the chair, and the rest are general members.

    c. One person is the chair, one person is the secretary, one person is responsible for refreshments, and one person cleans up after meetings.

21. A softball league has thirteen teams. If every team must play every other team once in the first round of league play, how many games must be scheduled?

22. In a group of eighteen people, each person shakes hands once with each other person in the group. How many handshakes will occur?

23. A softball league has thirteen teams. How many different end-of-the-season rankings of first, second, and third place are possible (disregarding ties)?

24. Three hundred people buy raffle tickets. Three winning tickets will be drawn at random.

    a. If first prize is $100, second prize is $50, and third prize is $20, in how many different ways can the prizes be awarded?

    b. If each prize is $50, in how many different ways can the prizes be awarded?

25. A group of nine women and six men must select a four-person committee. How many committees are possible if it must consist of the following?

    a. two women and two men

    b. any mixture of men and women

    c. a majority of women

26. A group of ten seniors, eight juniors, six sophomores, and five freshmen must select a committee of four. How many committees are possible if the committee must contain the following:

    a. one person from each class

    b. any mixture of the classes

    c. exactly two seniors

Exercises 27–32 refer to a deck of fifty-two playing cards (jokers not allowed). If you are unfamiliar with playing cards, see the end of Section 3.1 for a description of a standard deck.

27. How many five-card poker hands are possible?

28. **a.** How many five-card poker hands consisting of all hearts are possible?

    b. How many five-card poker hands consisting of all cards of the same suit are possible?

AP Photo/News & Record, Joseph Rodriguez

Exercise 19: Right letters, wrong order. SHCOOL is painted along the newly paved road leading to Southern Guilford High School on Drake Road Monday, August 9, 2010, in Greensboro, N.C.

29. **a.** How many five-card poker hands containing exactly three aces are possible?
    b. How many five-card poker hands containing three of a kind are possible?
30. **a.** How many five-card poker hands consisting of three kings and two queens are possible?
    b. How many five-card poker hands consisting of three of a kind and a pair (a *full house*) are possible?
31. How many five-card poker hands containing two pair are possible?

    HINT: You must select two of the thirteen ranks, then select a pair of each, and then one of the remaining cards.
32. How many five-card poker hands containing exactly one pair are possible?

    HINT: After selecting a pair, you must select three of the remaining twelve ranks and then select one card of each.
33. A 6/53 lottery requires choosing six of the numbers 1 through 53. How many different lottery tickets can you choose? (Order is not important, and the numbers do not repeat.)
34. A 7/39 lottery requires choosing seven of the numbers 1 through 39. How many different lottery tickets can you choose? (Order is not important, and the numbers do not repeat.)
35. A 5/36 lottery requires choosing five of the numbers 1 through 36. How many different lottery tickets can you choose? (Order is not important, and the numbers do not repeat.)
36. A 6/49 lottery requires choosing six of the numbers 1 through 49. How many different lottery tickets can you choose? (Order is not important, and the numbers do not repeat.)
37. Which lottery would be easier to win, a 6/53 or a 5/36? Why?

    HINT: See Exercises 33 and 35.
38. Which lottery would be easier to win, a 6/49 or a 7/39? Why?

    HINT: See Exercises 34 and 36.
39. **a.** Find the sum of the entries in the first row of Pascal's Triangle.
    b. Find the sum of the entries in the second row of Pascal's Triangle.
    c. Find the sum of the entries in the third row of Pascal's Triangle.
    d. Find the sum of the entries in the fourth row of Pascal's Triangle.
    e. Find the sum of the entries in the fifth row of Pascal's Triangle.
    f. Is there a pattern to the answers to parts (a)–(e)? If so, describe the pattern you see.

g. Use the pattern described in part (f) to predict the sum of the entries in the sixth row of Pascal's triangle.
h. Find the sum of the entries in the sixth row of Pascal's Triangle. Was your prediction in part (g) correct?
i. Find the sum of the entries in the nth row of Pascal's Triangle.
40. **a.** Add adjacent entries of the sixth row of Pascal's Triangle to obtain the seventh row.
    b. Find $_6C_r$ for $r = 0, 1, 2, 3, 4, 5,$ and 6.
    c. How are the answers to parts (a) and (b) related?
41. Use Pascal's Triangle to answer the following.
    a. In which row would you find the value of $_4C_2$?
    b. In which row would you find the value of $_nC_r$?
    c. Is $_4C_2$ the second number in the fourth row?
    d. Is $_4C_2$ the third number in the fifth row?
    e. What is the location of $_nC_r$? Why?
42. Given the set $S = \{a, b, c, d\}$, answer the following.
    a. How many one-element subsets does S have?
    b. How many two-element subsets does S have?
    c. How many three-element subsets does S have?
    d. How many four-element subsets does S have?
    e. How many zero-element subsets does S have?
    f. How many subsets does S have?
    g. If $n(S) = k$, how many subsets will S have?

In Exercises 43–50, find the number of permutations of the letters in each word.

43. ALASKA   44. ALABAMA
45. ILLINOIS   46. HAWAII
47. INDIANA   48. TENNESSEE
49. TALLAHASSEE 50. PHILADELPHIA

The words in each of Exercises 51–54 are homonyms *(words that are pronounced the same but have different meanings). Find the number of permutations of the letters in each word.*

51. **a.** PIER   **b.** PEER
52. **a.** HEAR   **b.** HERE
53. **a.** STEAL   **b.** STEEL
54. **a.** SHEAR   **b.** SHEER

 Answer the following questions using complete sentences and your own words.

• CONCEPT QUESTIONS

55. Suppose you want to know how many ways r items can be selected from a group of n items. What determines whether you should calculate $_nP_r$ or $_nC_r$?
56. For any given values of n and r, which is larger, $_nP_r$ or $_nC_r$? Why?

THE NEXT LEVEL

If a person wants to pursue an advanced degree (something beyond a bachelor's or four-year degree), chances are the person must take a standardized exam to gain admission to a school or to be admitted into a specific program. These exams are intended to measure verbal, quantitative, and analytical skills that have developed throughout a person's life. Many classes and study guides are available to help people prepare for the exams. The following questions are typical of those found in the study guides.

Exercises 57–61 refer to the following: A baseball league has six teams: A, B, C, D, E, and F. All games are played at 7:30 P.M. on Fridays, and there are sufficient fields for each team to play a game every Friday night. Each team must play each other team exactly once, and the following conditions must be met:

- Team A plays team D first and team F second.
- Team B plays team E first and team C third.
- Team C plays team F first.

57. What is the total number of games that each team must play during the season?
   a. 3   **b.** 4   **c.** 5   **d.** 6   **e.** 7

58. On the first Friday, which of the following pairs of teams play each other?
   a. A and B; C and F; D and E
   b. A and B; C and E; D and F
   c. A and C; B and E; D and F

   d. A and D; B and C; E and F
   e. A and D; B and E; C and F

59. Which of the following teams must team B play second?
   a. A   **b.** C   **c.** D   **d.** E   **e.** F

60. The last set of games could be between which teams?
   a. A and B; C and F; D and E
   b. A and C; B and F; D and E
   c. A and D; B and C; E and F
   d. A and E; B and C; D and F
   e. A and F; B and E; C and D

61. If team D wins five games, which of the following must be true?
   a. Team A loses five games.
   b. Team A wins four games.
   c. Team A wins its first game.
   d. Team B wins five games.
   e. Team B loses at least one game.

WEB PROJECT

62. Write a research paper on any historical topic referred to in this section or in a previous section. Following is a partial list of topics:
   - John Venn
   - Augustus De Morgan
   - Chu Shih-chieh

   Some useful links for this web project are listed on the text web site: **www.cengage.com/math/johnson**

2.5 Infinite Sets

OBJECTIVES

- Determine whether two sets are equivalent
- Establish a one-to-one correspondence between the elements of two sets
- Determine the cardinality of various infinite sets.

> **WARNING:** Many leading nineteenth-century mathematicians and philosophers claim that the study of infinite sets may be dangerous to your mental health.

Consider the sets E and N, where $E = \{2, 4, 6, \ldots\}$ and $N = \{1, 2, 3, \ldots\}$. Both are examples of infinite sets (they "go on forever"). E is the set of all even counting numbers, and N is the set of all counting (or natural) numbers. Because every element of E is an element of N, E is a subset of N. In addition, N contains elements not in E; therefore, E is a *proper* subset of N. Which set is "bigger," E or N? Intuition might lead many people to think that N is twice as big as E because N contains all the even counting numbers *and* all the odd counting numbers. Not so! According to the work of Georg Cantor (considered by many to be the father of set theory), N and E have exactly the same number of elements! This seeming paradox, a *proper* subset that has the *same number* of elements as the set from which it came, caused a philosophic uproar in the late nineteenth century. (Hence, the warning at the beginning of this section.) To study Cantor's work (which is now accepted and considered a cornerstone in modern mathematics), we must first investigate the meaning of a one-to-one correspondence and equivalent sets.

One-to-One Correspondence

Is there any relationship between the sets $A = \{$one, two, three$\}$ and $B = \{$Pontiac, Chevrolet, Ford$\}$? Although the sets contain different types of things (numbers versus automobiles), each contains the same number of things; they are the same size. This relationship (being the same size) forms the basis of a one-to-one correspondence. A **one-to-one correspondence** between the sets A and B is a pairing up of the elements of A and B such that each element of A is paired up with exactly one element of B, and vice versa, with no element left out. For instance, the elements of A and B might be paired up as follows:

one	two	three
↕	↕	↕
Pontiac	Chevrolet	Ford

(Other correspondences, or matchups, are possible.) If two sets have the same cardinal number, their elements can be put into a one-to-one correspondence. Whenever a one-to-one correspondence exists between the elements of two sets A and B, the sets are **equivalent** (denoted by $A \sim B$). Hence, equivalent sets have the same number of elements.

If two sets have different cardinal numbers, it is not possible to construct a one-to-one correspondence between their elements. The sets $C = \{$one, two$\}$ and $B = \{$Pontiac, Chevrolet, Ford$\}$ do *not* have a one-to-one correspondence; no matter how their elements are paired up, one element of B will always be left over (B has more elements; it is "bigger"):

one	two	
↕	↕	
Pontiac	Chevrolet	Ford

The sets C and B are *not* equivalent.

Given two sets A and B, if any one of the following statements is true, then the other statements are also true:

1. There exists a one-to-one correspondence between the elements of A and B.
2. A and B are equivalent sets.
3. A and B have the same cardinal number; that is, $n(A) = n(B)$.

EXAMPLE 1

DETERMINING WHETHER TWO SETS ARE EQUIVALENT   Determine whether the sets in each of the following pairs are equivalent. If they are equivalent, list a one-to-one correspondence between their elements.

a. $A = \{$John, Paul, George, Ringo$\}$;
   $B = \{$Lennon, McCartney, Harrison, Starr$\}$
b. $C = \{\alpha, \beta, \chi, \delta\}$; $D = \{$I, O, $\Delta\}$
c. $A = \{1, 2, 3, \ldots, 48, 49, 50\}$; $B = \{1, 3, 5, \ldots, 95, 97, 99\}$

HISTORICAL NOTE

GEORG CANTOR, 1845–1918

© CORBIS

Georg Ferdinand Ludwig Philip Cantor was born in St. Petersburg, Russia. His father was a stockbroker and wanted his son to become an engineer; his mother was an artist and musician. Several of Cantor's maternal relatives were accomplished musicians; in his later years, Cantor often wondered how his life would have turned out if he had become a violinist instead of pursuing a controversial career in mathematics.

Following his father's wishes, Cantor began his engineering studies at the University of Zurich in 1862. However, after one semester, he decided to study philosophy and pure mathematics. He transferred to the prestigious University of Berlin, studied under the famed mathematicians Karl Weierstrass, Ernst Kummer, and Leopold Kronecker, and received his doctorate in 1867. Two years later, Cantor accepted a teaching position at the University of Halle and remained there until he retired in 1913.

Cantor's treatises on set theory and the nature of infinite sets were first published in 1874 in *Crelle's Journal*, which was influential in mathematical circles. On their publication, Cantor's theories generated much controversy among mathematicians and philosophers. Paradoxes concerning the cardinal numbers of infinite sets, the nature of infinity, and Cantor's form of logic were unsettling to many, including Cantor's former teacher Leopold Kronecker. In fact, some felt that Cantor's work was not just revolutionary but actually dangerous. Kronecker led the attack on Cantor's theories. He was an editor of *Crelle's Journal* and held up the publication of one of Cantor's subsequent articles for so long that Cantor refused to publish ever again in the *Journal*. In addition, Kronecker blocked Cantor's efforts to obtain a teaching position at the University of Berlin. Even though Cantor was attacked by Kronecker and his followers, others respected him. Realizing the importance of communication among scholars, Cantor founded the Association of German Mathematicians in 1890 and served as its president for many years. In addition, Cantor was instrumental in organizing the first International Congress of Mathematicians, held in Zurich in 1897.

As a result of the repeated attacks on him and his work, Cantor suffered many nervous breakdowns, the first when he was thirty-nine. He died in a mental hospital in Halle at the age of seventy-three, never having received proper recognition for the true value of his discoveries. Modern mathematicians believe that Cantor's form of logic and his concepts of infinity revolutionized all of mathematics, and his work is now considered a cornerstone in its development.

Written in 1874, Cantor's first major paper on the theory of sets, Über eine Eigenshaft des Inbegriffes aller reellen algebraischen Zahlen (On a Property of the System of All the Real Algebraic Numbers), sparked a major controversy concerning the nature of infinite sets. To gather international support for his theory, Cantor had his papers translated into French. This 1883 French version of Cantor's work was published in the newly formed journal Acta Mathematica. Cantor's works were translated into English during the early twentieth century.

Brown University Library

SOLUTION

a. If sets have the same cardinal number, they are equivalent. Now, $n(A) = 4$ and $n(B) = 4$; therefore, $A \sim B$.

Because A and B are equivalent, their elements can be put into a one-to-one correspondence. One such correspondence follows:

John	Paul	George	Ringo
$\updownarrow$	$\updownarrow$	$\updownarrow$	$\updownarrow$
Lennon	McCartney	Harrison	Starr

b. Because $n(C) = 4$ and $n(D) = 3$, C and D are not equivalent.

c. A consists of all natural numbers from 1 to 50, inclusive. Hence, $n(A) = 50$. B consists of all odd natural numbers from 1 to 99, inclusive. Since half of the natural numbers from 1 to 100 are odd (and half are even), there are fifty ($100 \div 2 = 50$) odd natural numbers less than 100; that is, $n(B) = 50$. Because A and B have the same cardinal number, $A \sim B$.

Many different one-to-one correspondences may be established between the elements of A and B. One such correspondence follows:

$$A = \{1, 2, 3, \ldots, \quad n, \quad \ldots, 48, 49, 50\}$$
$$\updownarrow \updownarrow \updownarrow \ldots \quad \updownarrow \quad \ldots \updownarrow \updownarrow \updownarrow$$
$$B = \{1, 3, 5, \ldots, (2n - 1), \ldots, 95, 97, 99\}$$

That is, each natural number $n \in A$ is paired up with the odd number $(2n - 1) \in B$. The $n \leftrightarrow (2n - 1)$ part is crucial because it shows *each* individual correspondence. For example, it shows that $13 \in A$ corresponds to $25 \in B$ ($n = 13$, so $2n = 26$ and $2n - 1 = 25$). Likewise, $69 \in B$ corresponds to $35 \in A$ ($2n - 1 = 69$, so $2n = 70$ and $n = 35$).

As we have seen, if two sets have the same cardinal number, they are equivalent, and their elements can be put into a one-to-one correspondence. Conversely, if the elements of two sets can be put into a one-to-one correspondence, the sets have the same cardinal number and are equivalent. Intuitively, this result appears to be quite obvious. However, when Georg Cantor applied this relationship to infinite sets, he sparked one of the greatest philosophical debates of the nineteenth century.

Countable Sets

Consider the set of all counting numbers $N = \{1, 2, 3, \ldots\}$, which consists of an infinite number of elements. Each of these numbers is either odd or even. Defining O and E as $O = \{1, 3, 5, \ldots\}$ and $E = \{2, 4, 6, \ldots\}$, we have $O \cap E = \varnothing$ and $O \cup E = N$; the sets O and E are mutually exclusive, and their union forms the entire set of all counting numbers. Obviously, N contains elements that E does not. As we mentioned earlier, the fact that E is a *proper* subset of N might lead people to think that N is "bigger" than E. In fact, N and E are the "same size"; N and E each contain the same number of elements.

Recall that two sets are equivalent and have the same cardinal number if the elements of the sets can be matched up via a one-to-one correspondence. To show the existence of a one-to-one correspondence between the elements of two sets of numbers, we must find an explicit correspondence between the general elements of the two sets. In Example 1(c), we expressed the general correspondence as $n \leftrightarrow (2n - 1)$.

EXAMPLE **2**

FINDING A ONE-TO-ONE CORRESPONDENCE BETWEEN TWO INFINITE SETS

a. Show that $E = \{2, 4, 6, 8, \ldots\}$ and $N = \{1, 2, 3, 4, \ldots\}$ are equivalent sets.
b. Find the element of N that corresponds to $1430 \in E$.
c. Find the element of N that corresponds to $x \in E$.

SOLUTION

a. To show that $E \sim N$, we must show that there exists a one-to-one correspondence between the elements of E and N. The elements of E and N can be paired up as follows:

$$N = \{1, 2, 3, 4, \ldots, n, \ldots\}$$
$$\updownarrow \updownarrow \updownarrow \updownarrow \ldots \quad \updownarrow$$
$$E = \{2, 4, 6, 8, \ldots, 2n, \ldots\}$$

Any natural number $n \in N$ corresponds with the even natural number $2n \in E$. Because there exists a one-to-one correspondence between the elements of E and N, the sets E and N are equivalent; that is, $E \sim N$.

b. $1430 = 2n \in E$, so $n = \frac{1430}{2} = 715 \in N$. Therefore, $715 \in N$ corresponds to $1430 \in E$.

c. $x = 2n \in E$, so $n = \frac{x}{2} \in N$. Therefore, $n = \frac{x}{2} \in N$ corresponds to $x = 2n \in E$.

We have just seen that the set of *even* natural numbers is equivalent to the set of *all* natural numbers. This equivalence implies that the two sets have the same number of elements! Although E is a proper subset of N, both sets have the same cardinal number; that is, $n(E) = n(N)$. Settling the controversy sparked by this seeming paradox, mathematicians today define a set to be an **infinite set** if it can be placed in a one-to-one correspondence with a proper subset of itself.

How many counting numbers are there? How many even counting numbers are there? We know that each set contains an infinite number of elements and that $n(N) = n(E)$, but how many is that? In the late nineteenth century, Georg Cantor defined the cardinal number of the set of counting numbers to be $\aleph_0$ (read "**aleph-null**"). Cantor utilized Hebrew letters, of which aleph, $\aleph$, is the first. Consequently, the proper response to "How many counting numbers are there?" is "There are aleph-null of them"; $n(N) = \aleph_0$. Any set that is equivalent to the set of counting numbers has cardinal number $\aleph_0$. A set is **countable** if it is finite or if it has cardinality $\aleph_0$.

Cantor was not the first to ponder the paradoxes of infinite sets. Hundreds of years before, Galileo had observed that part of an infinite set contained as many elements as the whole set. In his monumental *Dialogue Concerning the Two Chief World Systems* (1632), Galileo made a prophetic observation: "There are as many (perfect) squares as there are (natural) numbers because they are just as numerous as their roots." In other words, the elements of the sets $N = \{1, 2, 3, \ldots, n, \ldots\}$ and $S = \{1^2, 2^2, 3^2, \ldots, n^2, \ldots\}$ can be put into a one-to-one correspondence ($n \leftrightarrow n^2$). Galileo pondered which of the sets (perfect squares or natural numbers) was "larger" but abandoned the subject because he could find no practical application of this puzzle.

EXAMPLE 3

SHOWING THAT THE SET OF INTEGERS IS COUNTABLE   Consider the following one-to-one correspondence between the set I of all integers and the set N of all natural numbers:

$$N = \{1, \ 2, \ \ 3, \ \ 4, \ \ 5, \ldots\}$$
$$\updownarrow \ \updownarrow \ \ \updownarrow \ \ \updownarrow \ \ \updownarrow$$
$$I = \{0, \ 1, -1, \ 2, -2, \ldots\}$$

where an odd natural number n corresponds to a nonpositive integer $\frac{1-n}{2}$ and an even natural number n corresponds to a positive integer $\frac{n}{2}$.

a. Find the 613th integer; that is, find the element of I that corresponds to $613 \in N$.
b. Find the element of N that corresponds to $853 \in I$.
c. Find the element of N that corresponds to $-397 \in I$.
d. Find $n(I)$, the cardinal number of the set I of all integers.
e. Is the set of integers countable?

SOLUTION

a. $613 \in N$ is odd, so it corresponds to $\frac{1-613}{2} = \frac{-612}{2} = -306$. If you continued counting the integers as shown in the above correspondence, -306 would be the 613th integer in your count.

b. $853 \in I$ is positive, so

$$853 = \frac{n}{2} \qquad \textbf{multiplying by 2}$$

$$n = 1,706$$

$1,706 \in N$ corresponds to $853 \in I$.
This means that 853 is the 1,706th integer.

c. $-397 \in I$ is negative, so

$$-397 = \frac{1 - n}{2}$$

$$-794 = 1 - n \qquad \textbf{multiplying by 2}$$

$$-795 = -n \qquad \textbf{subtracting 1}$$

$$n = 795 \qquad \textbf{multiplying by } -1$$

$795 \in N$ corresponds to $-397 \in I$.
This means that -397 is the 795th integer.

d. The given one-to-one correspondence shows that I and N have the same (infinite) number of elements; $n(I) = n(N)$. Because $n(N) = \aleph_0$, the cardinal number of the set of all integers is $n(I) = \aleph_0$.

e. By definition, a set is called countable if it is finite or if it has cardinality $\aleph_0$. The set of integers has cardinality $\aleph_0$, so it is countable. This means that we can "count off" all of the integers, as we did in parts (a), (b), and (c).

We have seen that the sets N (all counting numbers), E (all even counting numbers), and I (all integers) contain the same number of elements, $\aleph_0$. What about a set containing fractions?

EXAMPLE **4**

SOLUTION

DETERMINING WHETHER THE SET OF POSITIVE RATIONAL NUMBERS IS COUNTABLE   Determine whether the set P of all positive rational numbers is countable.

The elements of P can be systematically listed in a table of rows and columns as follows: All positive rational numbers whose denominator is 1 are listed in the first row, all positive rational numbers whose denominator is 2 are listed in the second row, and so on, as shown in Figure 2.43.

Each positive rational number will appear somewhere in the table. For instance, $\frac{125}{66}$ will be in row 66 and column 125. Note that not all the entries in Figure 2.43 are in lowest terms; for instance, $\frac{2}{4}, \frac{3}{6}, \frac{4}{8}$, and so on are all equal to $\frac{1}{2}$.

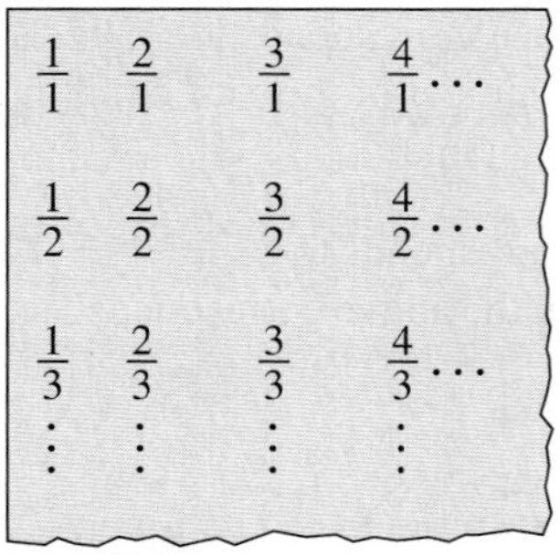

FIGURE 2.43

A list of all positive rational numbers.

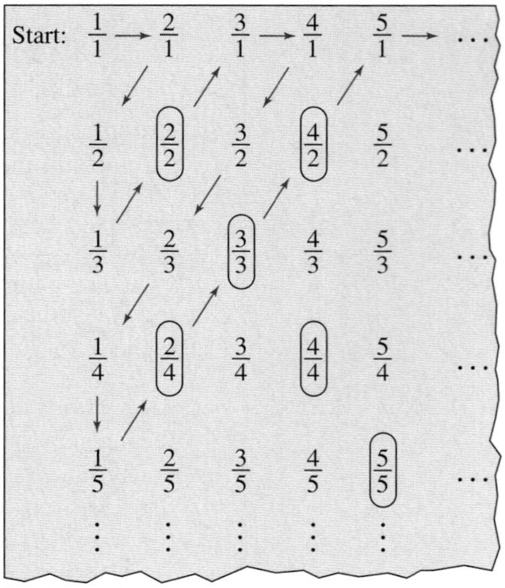

FIGURE 2.44   The circled rational numbers are not in lowest terms; they are omitted from the list.

Consequently, to avoid listing the same number more than once, an entry that is not in lowest terms must be eliminated from our list. To establish a one-to-one correspondence between P and N, we can create a zigzag diagonal pattern as shown by the arrows in Figure 2.44. Starting with $\frac{1}{1}$, we follow the arrows and omit any number that is not in lowest terms (the circled numbers in Figure 2.44). In this manner, a list of all positive rational numbers with no repetitions is created. Listing the elements of P in this order, we can put them in a one-to-one correspondence with N:

$$N = \{1, 2, 3, 4, 5, 6, 7, 8, 9, 10, 11, \ldots\}$$
$$\updownarrow \updownarrow \updownarrow \updownarrow \updownarrow \updownarrow \updownarrow \updownarrow \updownarrow \updownarrow \updownarrow$$
$$P = \left\{1, 2, \frac{1}{2}, \frac{1}{3}, 3, 4, \frac{3}{2}, \frac{2}{3}, \frac{1}{4}, \frac{1}{5}, 5, \ldots\right\}$$

Any natural number n is paired up with the positive rational number found by counting through the "list" given in Figure 2.44. Conversely, any positive rational number is located somewhere in the list and is paired up with the counting number corresponding to its place in the list.

Therefore, $P \sim N$, so the set of all positive rational numbers is countable.

Uncountable Sets

Every infinite set that we have examined so far is countable; each can be put into a one-to-one correspondence with the set of all counting numbers and consequently has cardinality $\aleph_0$. Do not be misled into thinking that all infinite sets are countable! By utilizing a "proof by contradiction," Georg Cantor showed that the infinite set $A = \{x \mid 0 \le x < 1\}$ is *not* countable. This proof involves logic that is different from what you are used to. Do not let that intimidate you.

Assume that the set $A = \{x \mid 0 \le x < 1\}$ is countable; that is, assume that $n(A) = \aleph_0$. This assumption implies that the elements of A and N can be put into a one-to-one correspondence; each $a \in A$ can be listed and counted. Because the elements of A are nonnegative real numbers less than 1, each $a_n = 0.\square\square\square\square\square\ldots$. Say, for instance, the numbers in our list are

$a_1 = 0.3750000\ldots$ **the first element of A**
$a_2 = 0.7071067\ldots$ **the second element of A**
$a_3 = 0.5000000\ldots$ **the third element of A**
$a_4 = 0.6666666\ldots$ **and so on.**

The *assumption* that A is countable implies that every element of A appears somewhere in the above list. However, we can create an element of A (call it b) that is *not* in the list. We build b according to the "diagonal digits" of the numbers in our list and the following rule: If the digit "on the diagonal" is not zero, put a 0 in the corresponding place in b; if the digit "on the diagonal" is zero, put a 1 in the corresponding place in b.

The "diagonal digits" of the numbers in our list are as follows:

$a_1 = 0.\boxed{3}750000\ldots$
$a_2 = 0.7\boxed{0}71067\ldots$
$a_3 = 0.50\boxed{0}0000\ldots$
$a_4 = 0.666\boxed{6}666\ldots$

Because the first digit on the diagonal is 3, the first digit of b is 0. Because the second digit on the diagonal is 0, the second digit of b is 1. Using all the "diagonal digits" of the numbers in our list, we obtain $b = 0.0110. \ldots$ Because $0 \le b < 1$, it follows that $b \in A$. However, the number b is not on our list of all elements of A. This is because

$b \ne a_1$ (*b* and a_1 differ in the first decimal place)

$b \ne a_2$ (*b* and a_2 differ in the second decimal place)

$b \ne a_3$ (*b* and a_3 differ in the third decimal place), **and so on**

This contradicts the assumption that the elements of A and N can be put into a one-to-one correspondence. Since the assumption leads to a contradiction, the assumption must be false; $A = \{x \mid 0 \le x < 1\}$ is not countable. Therefore, $n(A) \ne n(N)$. That is, A is an infinite set and $n(A) \ne \aleph_0$.

An infinite set that cannot be put into a one-to-one correspondence with N is said to be **uncountable.** Consequently, an uncountable set has *more* elements than the set of all counting numbers. This implies that there are different magnitudes of infinity! To distinguish the magnitude of A from that of N, Cantor denoted the cardinality of $A = \{x \mid 0 \le x < 1\}$ as $n(A) = c$ (*c* for **continuum**). Thus, Cantor showed that $\aleph_0 < c$. Cantor went on to show that A was equivalent to the entire set of all real numbers, that is, $A \sim \Re$. Therefore, $n(\Re) = c$.

Although he could not prove it, Cantor hypothesized that no set could have a cardinality between $\aleph_0$ and c. This famous unsolved problem, labeled the *Continuum Hypothesis,* baffled mathematicians throughout the first half of the twentieth century. It is said that Cantor suffered a devastating nervous breakdown in 1884 when he announced that he had a proof of the Continuum Hypothesis only to declare the next day that he could show the Continuum Hypothesis to be false!

The problem was finally "solved" in 1963. Paul J. Cohen demonstrated that the Continuum Hypothesis is independent of the entire framework of set theory; that is, it can be neither proved nor disproved by using the theorems of set theory. Thus, the Continuum Hypothesis is not provable.

Although no one has produced a set with cardinality between $\aleph_0$ and c, many sets with cardinality greater than c have been constructed. In fact, modern mathematicians have shown that there are *infinitely* many magnitudes of infinity! Using subscripts, these magnitudes, or cardinalities, are represented by $\aleph_0, \aleph_1, \aleph_2, \ldots$ and have the property that $\aleph_0 < \aleph_1 < \aleph_2 < \ldots$. In this sense, the set N of all natural numbers forms the "smallest" infinite set. Using this subscripted notation, the Continuum Hypothesis implies that $c = \aleph_1$; that is, given that N forms the smallest infinite set, the set $\Re$ of all real numbers forms the next "larger" infinite set.

Points on a Line

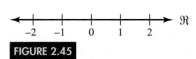

FIGURE 2.45

The real number line.

When students are first exposed to the concept of the real number system, a number line like the one in Figure 2.45 is inevitably introduced. The real number system, denoted by $\Re$, can be put into a one-to-one correspondence with all points on a line, such that every real number corresponds to exactly one point on a line and every point on a line corresponds to exactly one real number. Consequently, any (infinite) line contains c points. What about a line segment? For example, how many points does the segment $[0, 1]$ contain? Does the segment $[0, 2]$ contain twice as many points as the segment $[0, 1]$? Once again, intuition can lead to erroneous conclusions when people are dealing with infinite sets.

EXAMPLE **5**

SHOWING THAT LINE SEGMENTS OF DIFFERENT LENGTHS ARE EQUIVALENT SETS OF POINTS   Show that the line segments $[0, 1]$ and $[0, 2]$ are equivalent sets of points.

SOLUTION

Because the segment $[0, 2]$ is twice as long as the segment $[0, 1]$, intuition might tell us that it contains twice as many points. Not so! Recall that two sets are

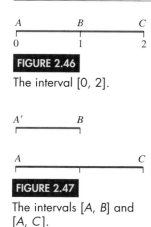

FIGURE 2.46

The interval [0, 2].

FIGURE 2.47

The intervals [A, B] and [A, C].

equivalent (and have the same cardinal number) if their elements can be put into a one-to-one correspondence.

On a number line, let A represent the point 0, let B represent 1, and let C represent 2, as shown in Figure 2.46. Our goal is to develop a one-to-one correspondence between the elements of the segments AB and AC. Now draw the segments separately, with AB above AC, as shown in Figure 2.47. (To distinguish the segments from each other, point A of segment AB has been relabeled as point A'.)

Extend segments AA' and CB so that they meet at point D, as shown in Figure 2.48. Any point E on A'B can be paired up with the unique point F on AC formed by the intersection of lines DE and AC, as shown in Figure 2.49. Conversely, any point F on segment AC can be paired up with the unique point E on A'B formed by the intersection of lines DF and A'B. Therefore, a one-to-one correspondence exists between the two segments, so [0, 1] ∼ [0, 2]. Consequently, the interval [0, 1] contains exactly the same number of points as the interval [0, 2]!

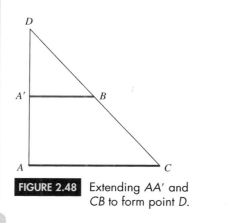

FIGURE 2.48   Extending AA' and CB to form point D.

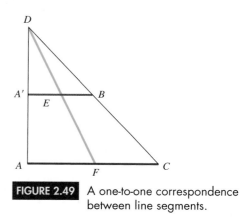

FIGURE 2.49   A one-to-one correspondence between line segments.

Even though the segment [0, 2] is twice as long as the segment [0, 1], each segment contains exactly the same number of points. The method used in Example 5 can be applied to any two line segments. Consequently, all line segments, regardless of their length, contain exactly the same number of points; a line segment 1 inch long has exactly the same number of points as a segment 1 mile long! Once again, it is easy to see why Cantor's work on the magnitude of infinity was so unsettling to many scholars.

Having concluded that all line segments contain the same number of points, we might ask how many points that is. What is the cardinal number? Given any line segment AB, it can be shown that $n(AB) = c$; the points of a line segment can be put into a one-to-one correspondence with the points of a line. Consequently, the interval [0, 1] contains the same number of elements as the entire real number system.

If things seem rather strange at this point, keep in mind that Cantor's pioneering work produced results that puzzled even Cantor himself. In a paper written in 1877, Cantor constructed a one-to-one correspondence between the points in a square (a two-dimensional figure) and the points on a line segment (a one-dimensional figure). Extending this concept, he concluded that a line segment and the entire two-dimensional plane contain exactly the same number of points, c. Communicating with his colleague Richard Dedekind, Cantor wrote, "I see it, but I do not believe it." Subsequent investigation has shown that the number of points contained in the interval [0, 1] is the same as the number of points contained in all of three-dimensional space! Needless to say, Cantor's work on the cardinality of infinity revolutionized the world of modern mathematics.

In Exercises 1–10, find the cardinal numbers of the sets in each given pair to determine whether the sets are equivalent. If they are equivalent, list a one-to-one correspondence between their elements.

1. $S =$ {Sacramento, Lansing, Richmond, Topeka}
   $C =$ {California, Michigan, Virginia, Kansas}
2. $T =$ {Wyoming, Ohio, Texas, Illinois, Colorado}
   $P =$ {Cheyenne, Columbus, Austin, Springfield, Denver}
3. $R =$ {a, b, c}; $G =$ {$\alpha, \beta, \chi, \delta$}
4. $W =$ {I, II, III}; $H =$ {one, two}
5. $C =$ {3, 6, 9, 12, . . . , 63, 66}
   $D =$ {4, 8, 12, 16, . . . , 84, 88}
6. $A =$ {2, 4, 6, 8, . . . , 108, 110}
   $B =$ {5, 10, 15, 20, . . . , 270, 275}
7. $G =$ {2, 4, 6, 8, . . . , 498, 500}
   $H =$ {1, 3, 5, 7, . . . , 499, 501}
8. $E =$ {2, 4, 6, 8, . . . , 498, 500}
   $F =$ {3, 6, 9, 12, . . . , 750, 753}
9. $A =$ {1, 3, 5, . . . , 121, 123}
   $B =$ {125, 127, 129, . . . , 245, 247}
10. $S =$ {4, 6, 8, . . . , 664, 666}
   $T =$ {5, 6, 7, . . . , 335, 336}

11. a. Show that the set O of all odd counting numbers, $O =$ {1, 3, 5, 7, . . .}, and $N =$ {1, 2, 3, 4, . . .} are equivalent sets.
   b. Find the element of N that corresponds to $1{,}835 \in O$.
   c. Find the element of N that corresponds to $x \in O$.
   d. Find the element of O that corresponds to $782 \in N$.
   e. Find the element of O that corresponds to $n \in N$.

12. a. Show that the set W of all whole numbers, $W =$ {0, 1, 2, 3, . . .}, and $N =$ {1, 2, 3, 4, . . .} are equivalent sets.
   b. Find the element of N that corresponds to $932 \in W$.
   c. Find the element of N that corresponds to $x \in W$.
   d. Find the element of W that corresponds to $932 \in N$.
   e. Find the element of W that corresponds to $n \in N$.

13. a. Show that the set T of all multiples of 3, $T =$ {3, 6, 9, 12, . . .}, and $N =$ {1, 2, 3, 4, . . .} are equivalent sets.
   b. Find the element of N that corresponds to $936 \in T$.
   c. Find the element of N that corresponds to $x \in T$.
   d. Find the element of T that corresponds to $936 \in N$.
   e. Find the element of T that corresponds to $n \in N$.

14. a. Show that the set F of all multiples of 5, $F =$ {5, 10, 15, 20, . . .}, and $N =$ {1, 2, 3, 4, . . .} are equivalent sets.
   b. Find the element of N that corresponds to $605 \in F$.
   c. Find the element of N that corresponds to $x \in F$.

d. Find the element of F that corresponds to $605 \in N$.
   e. Find the element of F that corresponds to $n \in N$.

15. Consider the following one-to-one correspondence between the set A of all even integers and the set N of all natural numbers:

$$N = \{1, 2, \ 3, \ 4, \ 5, \ldots\}$$
$$\updownarrow \updownarrow \ \updownarrow \ \updownarrow \ \updownarrow$$
$$A = \{0, 2, -2, 4, -4, \ldots\}$$

where an odd natural number n corresponds to the nonpositive integer $1 - n$ and an even natural number n corresponds to the positive even integer n.

   a. Find the 345th even integer; that is, find the element of A that corresponds to $345 \in N$.
   b. Find the element of N that corresponds to $248 \in A$.
   c. Find the element of N that corresponds to $-754 \in A$.
   d. Find $n(A)$.

16. Consider the following one-to-one correspondence between the set B of all odd integers and the set N of all natural numbers:

$$N = \{1, \ 2, \ 3, \ 4, \ 5, \ldots\}$$
$$\updownarrow \ \updownarrow \ \updownarrow \ \updownarrow \ \updownarrow$$
$$B = \{1, -1, 3, -3, 5, \ldots\}$$

where an even natural number n corresponds to the negative odd integer $1 - n$ and an odd natural number n corresponds to the odd integer n.

   a. Find the 345th odd integer; that is, find the element of B that corresponds to $345 \in N$.
   b. Find the element of N that corresponds to $241 \in B$.
   c. Find the element of N that corresponds to $-759 \in B$.
   d. Find $n(B)$.

In Exercises 17–22, show that the given sets of points are equivalent by establishing a one-to-one correspondence.

17. the line segments [0, 1] and [0, 3]
18. the line segments [1, 2] and [0, 3]
19. the circle and square shown in Figure 2.50

   HINT: Draw one figure inside the other.

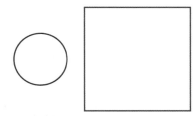

FIGURE 2.50

20. the rectangle and triangle shown in Figure 2.51

HINT: Draw one figure inside the other.

FIGURE 2.51

21. a circle of radius 1 cm and a circle of 5 cm

HINT: Draw one figure inside the other.

22. a square of side 1 cm and a square of side 5 cm

HINT: Draw one figure inside the other.

23. Show that the set of all real numbers between 0 and 1 has the same cardinality as the set of all real numbers.

HINT: Draw a semicircle to represent the set of real numbers between 0 and 1 and a line to represent the set of all real numbers, and use the method of Example 5.

Answer the following questions using complete sentences and your own words.

• CONCEPT QUESTIONS

24. What is the cardinal number of the "smallest" infinite set? What set or sets have this cardinal number?

• HISTORY QUESTIONS

25. What aspect of Georg Cantor's set theory caused controversy among mathematicians and philosophers?

26. What contributed to Cantor's breakdown in 1884?

27. Who demonstrated that the Continuum Hypothesis cannot be proven? When?

WEB PROJECT

28. Write a research paper on any historical topic referred to in this section. Following is a partial list of topics:
- Georg Cantor
- Richard Dedekind
- Paul J. Cohen
- Bernhard Bolzano
- Leopold Kronecker
- the Continuum Hypothesis

Some useful links for this web project are listed on the text web site: **www.cengage.com/math/johnson**

2 CHAPTER REVIEW

TERMS

aleph-null
cardinal number
combination
combinatorics
complement
continuum
countable set

De Morgan's Laws
distinguishable
  permutations
element
empty set
equal sets
equivalent sets
factorial
Fundamental Principle
  of Counting

improper subset
infinite set
intersection
mutually exclusive
one-to-one
  correspondence
permutation
proper subset
roster notation
set

set-builder notation
set theory
subset
tree diagram
uncountable set
union
universal set
Venn diagram
well-defined set

REVIEW EXERCISES

1. State whether the given set is well-defined.
   a. the set of all multiples of 5.
   b. the set of all difficult math problems
   c. the set of all great movies
   d. the set of all Oscar-winning movies

2. Given the sets
   $U = \{0, 1, 2, 3, 4, 5, 6, 7, 8, 9\}$
   $A = \{0, 2, 4, 6, 8\}$
   $B = \{1, 3, 5, 7, 9\}$
   find the following using the roster method.
   a. A' b. B'
   c. $A \cup B$ d. $A \cap B$

3. Given the sets $A = \{$Maria, Nobuko, Leroy, Mickey, Kelly$\}$ and $B = \{$Rachel, Leroy, Deanna, Mickey$\}$, find the following.
   a. $A \cup B$ b. $A \cap B$

4. List all subsets of $C = \{$Dallas, Chicago, Tampa$\}$. Identify which subsets are proper and which are improper.

5. Given $n(U) = 61$, $n(A) = 32$, $n(B) = 26$, and $n(A \cup B) = 40$, do the following.
   a. Find $n(A \cap B)$.
   b. Draw a Venn diagram illustrating the composition of U.

In Exercises 6 and 7, use a Venn diagram like the one in Figure 2.15 to shade in the region corresponding to the indicated set.

6. $(A' \cup B)'$

7. $(A \cap B')'$

In Exercises 8 and 9, use a Venn diagram like the one in Figure 2.36 to shade in the region corresponding to the indicated set.

8. $A' \cap (B \cup C')$

9. $(A' \cap B) \cup C'$

10. A survey of 1,000 college seniors yielded the following information: 396 seniors favored capital punishment, 531 favored stricter gun control, and 237 favored both.
    a. How many favored capital punishment or stricter gun control?
    b. How many favored capital punishment but not stricter gun control?
    c. How many favored stricter gun control but not capital punishment?
    d. How many favored neither capital punishment nor stricter gun control?

11. A survey of recent college graduates yielded the following information: 70 graduates earned a degree in mathematics, 115 earned a degree in education, 23 earned degrees in both mathematics and education, and 358 earned a degree in neither mathematics nor education.
    a. What percent of the college graduates earned a degree in mathematics or education?
    b. What percent of the college graduates earned a degree in mathematics only?
    c. What percent of the college graduates earned a degree in education only?

12. A local anime fan club surveyed 67 of its members regarding their viewing habits last weekend, and the following information was obtained: 30 members watched an episode of *Naruto*, 44 watched an episode of *Death Note*, 23 watched an episode of *Inuyasha*, 20 watched both *Naruto* and *Inuyasha*, 5 watched *Naruto* and *Inuyasha* but not *Death Note*, 15 watched both *Death Note* and *Inuyasha*, and 23 watched only *Death Note*.
    a. How many of the club members watched exactly one of the shows?
    b. How many of the club members watched all three shows?
    c. How many of the club members watched none of the three shows?

13. An exit poll yielded the following information concerning people's voting patterns on Propositions A, B, and C: 305 people voted yes on A, 95 voted yes only on A, 393 voted yes on B, 192 voted yes only on B, 510 voted yes on A or B, 163 voted yes on C, 87 voted yes on all three, and 249 voted no on all three. What percent of the voters voted yes on more than one proposition?

14. Given the sets $U = \{$a, b, c, d, e, f, g, h, i$\}$, $A = \{$b, d, f, g$\}$, and $B = \{$a, c, d, g, i$\}$, use De Morgan's Laws to find the following.
    a. $(A' \cup B)'$ b. $(A \cap B')'$

15. Refer to the Venn diagram depicted in Figure 2.32.
    a. In a group of 100 Americans, how many have type O or type A blood?
    b. In a group of 100 Americans, how many have type O and type A blood?
    c. In a group of 100 Americans, how many have neither type O nor type A blood?

16. Refer to the Venn diagram depicted in Figure 2.28.
    a. For a typical group of 100 Americans, fill in the cardinal number of each region in the diagram.
    b. In a group of 100 Americans, how many have type O blood or are Rh+?
    c. In a group of 100 Americans, how many have type O blood and are Rh+?
    d. In a group of 100 Americans, how many have neither type O blood nor are Rh+?

17. Sid and Nancy are planning their anniversary celebration, which will include viewing an art exhibit, having dinner, and going dancing. They will go either to the Museum of Modern Art or to the New Photo Gallery; dine either at Stars, at Johnny's, or at the Chelsea; and go dancing either at Le Club or at Lizards.

    a. In how many different ways can Sid and Nancy celebrate their anniversary?

    b. Construct a tree diagram to list all possible ways in which Sid and Nancy can celebrate their anniversary.

18. A certain model of pickup truck is available in five exterior colors, three interior colors, and three interior styles. In addition, the transmission can be either manual or automatic, and the truck can have either two-wheel or four-wheel drive. How many different versions of the pickup truck can be ordered?

19. Each student at State University has a student I.D. number consisting of five digits (the first digit is nonzero, and digits can be repeated) followed by two of the letters A, B, C, and D (letters cannot be repeated). How many different student numbers are possible?

20. Find the value of each of the following.

    a. $(17 - 7)!$ b. $(17 - 17)!$

    c. $\dfrac{82!}{79!}$ d. $\dfrac{27!}{20!7!}$

21. In how many ways can you select three out of eleven items under the following conditions?

    a. Order of selection is not important.

    b. Order of selection is important.

22. Find the value of each of the following.

    a. $_{15}P_4$ b. $_{15}C_4$ c. $_{15}P_{11}$

23. A group of ten women and twelve men must select a three-person committee. How many committees are possible if it must consist of the following?

    a. one woman and two men

    b. any mixture of men and women

    c. a majority of men

24. A volleyball league has ten teams. If every team must play every other team once in the first round of league play, how many games must be scheduled?

25. A volleyball league has ten teams. How many different end-of-the-season rankings of first, second, and third place are possible (disregarding ties)?

26. Using a standard deck of fifty-two cards (no jokers), how many seven-card poker hands are possible?

27. Using a standard deck of fifty-two cards and two jokers, how many seven-card poker hands are possible?

28. A 6/42 lottery requires choosing six of the numbers 1 through 42. How many different lottery tickets can you choose?

In Exercises 29 and 30, find the number of permutations of the letters in each word.

29. a. FLORIDA   b. ARIZONA   c. MONTANA

30. a. AFFECT   b. EFFECT

31. What is the major difference between permutations and combinations?

32. Use Pascal's Triangle to answer the following.

    a. In which entry in which row would you find the value of $_7C_3$?

    b. In which entry in which row would you find the value of $_7C_4$?

    c. How is the value of $_7C_3$ related to the value of $_7C_4$? Why?

    d. What is the location of $_nC_r$? Why?

33. Given the set $S = \{a, b, c\}$, answer the following.

    a. How many one-element subsets does S have?

    b. How many two-element subsets does S have?

    c. How many three-element subsets does S have?

    d. How many zero-element subsets does S have?

    e. How many subsets does S have?

    f. How is the answer to part (e) related to $n(S)$?

In Exercises 34–36, find the cardinal numbers of the sets in each given pair to determine whether the sets are equivalent. If they are equivalent, list a one-to-one correspondence between their elements.

34. $A = \{I, II, III, IV, V\}$ and $B = \{$one, two, three, four, five$\}$

35. $C = \{3, 5, 7, \ldots, 899, 901\}$ and $D = \{2, 4, 6, \ldots, 898, 900\}$

36. $E = \{$Ronald$\}$ and $F = \{$Reagan, McDonald$\}$

37. a. Show that the set S of perfect squares, $S = \{1, 4, 9, 16, \ldots\}$, and $N = \{1, 2, 3, 4, \ldots\}$ are equivalent sets.

    b. Find the element of N that corresponds to $841 \in S$.

    c. Find the element of N that corresponds to $x \in S$.

    d. Find the element of S that corresponds to $144 \in N$.

    e. Find the element of S that corresponds to $n \in N$.

38. Consider the following one-to-one correspondence between the set A of all integer multiples of 3 and the set N of all natural numbers:

$$N = \{1, 2, \ 3, \ 4, \ 5, \ldots\}$$
$$\updownarrow \updownarrow \ \updownarrow \ \updownarrow \ \updownarrow$$
$$A = \{0, 3, -3, 6, -6, \ldots\}$$

where an odd natural number n corresponds to the nonpositive integer $\frac{3}{2}(1 - n)$, and an even natural number n corresponds to the positive even integer $\frac{3}{2}n$.

    a. Find the element of A that corresponds to $396 \in N$.

    b. Find the element of N that corresponds to $396 \in A$.

    c. Find the element of N that corresponds to $-153 \in A$.

    d. Find $n(A)$.

39. Show that the line segments [0, 1] and [0, π] are equivalent sets of points by establishing a one-to-one correspondence.

 Answer the following questions using complete sentences and your own words.

• CONCEPT QUESTIONS

40. What is the difference between proper and improper subsets?

41. Explain the difference between {0} and ∅.

42. What is a factorial?

43. What is the difference between permutations and combinations?

• HISTORY QUESTIONS

44. What roles did the following people play in the development of set theory and combinatorics?
- Georg Cantor
- Augustus De Morgan
- Christian Kramp
- Chu Shih-chieh
- John Venn

 THE NEXT LEVEL

If a person wants to pursue an advanced degree (something beyond a bachelor's or four-year degree), chances are the person must take a standardized exam to gain admission to a school or to be admitted into a specific program. These exams are intended to measure verbal, quantitative, and analytical skills that have developed throughout a person's life. Many classes and study guides are available to help people prepare for the exams. The following questions are typical of those found in the study guides.

Exercises 45–48 refer to the following: Two doctors in a local clinic are determining which days of the week they will be on call. Each day, Monday through Sunday, is to be assigned to one of two doctors, A and B, such that the assignment is consistent with the following conditions:

- No day is assigned to both doctors.
- Neither doctor has more than four days.
- Monday and Thursday must be assigned to the same doctor.
- If Tuesday is assigned to doctor A, then so is Sunday.
- If Saturday is assigned to doctor B, then Friday is not assigned to doctor B.

45. Which one of the following could be a complete and accurate list of the days assigned to doctor A?
- **a.** Monday, Thursday
- **b.** Monday, Tuesday, Sunday
- **c.** Monday, Thursday, Sunday
- **d.** Monday, Tuesday, Thursday
- **e.** Monday, Thursday, Friday, Sunday

46. Which of the following cannot be true?
- **a.** Thursday and Sunday are assigned to doctor A.
- **b.** Friday and Saturday are assigned to doctor A.
- **c.** Monday and Tuesday are assigned to doctor B.
- **d.** Monday, Wednesday, and Sunday are assigned to doctor A.
- **e.** Tuesday, Wednesday, and Saturday are assigned to doctor B.

47. If Friday and Sunday are both assigned to doctor B, how many different ways are there to assign the other five days to the doctors?
- **a.** 1   **b.** 2   **c.** 3
- **d.** 5   **e.** 6

48. If doctor A has four days, none of which is Monday, which of the following days must be assigned to doctor A?
- **a.** Tuesday   **b.** Wednesday   **c.** Friday
- **d.** Saturday   **e.** Sunday

PROBABILITY

© Wasabi/Alamy

Uncertainty is a part of our lives. When we wake up, we check the weather report to see whether it might rain. We check the traffic report to see whether we might get stuck in a jam. We check to see whether the interest rate has changed for that car, boat, or home loan. **Probability theory** is the branch of mathematics that analyzes uncertainty.

Dennis Flaherty/photodisk/Jupiter Images

WHAT WE WILL DO IN THIS CHAPTER

GAMBLING IS ALL ABOUT PROBABILITIES:

- Gambling involves uncertainty. Which number will be the winning number?

- All casinos use probability theory to set house odds so that they will make a profit.

- We will use probability theory to determine which casino games of chance are better for you to play and which to avoid.

- Probability theory's roots in gambling are deep: Probability theory got its start when a French nobleman and gambler asked his mathematician friend why he had lost so much money at dice.

THE SCIENCE OF GENETICS COULDN'T EXIST WITHOUT PROBABILITIES:

- Humans inherit traits, such as hair color or an increased risk for a certain disease. A child will inherit some traits but not others, and geneticists use probability theory to analyze this uncertainty.

continued

- We will use probability theory to analyze the inheritance of hair color—the color of the parents' hair determines a range of possibilities for the color of their child's hair.

- We will also analyze the inheritance of sickle-cell anemia, cystic fibrosis, and other inherited diseases. Your future children might be at risk for these diseases because they can be inherited from disease-free parents.

- Genetics got its start when Gregor Mendel, a nineteenth-century monk, used probability theory to analyze the effect of randomness on heredity.

BELIEVE IT OR NOT, YOUR CELL PHONE USES PROBABILITIES:

- Your cell phone's predictive text feature ("T9" and "iTap" are two examples) allows you to type text messages with a single keypress for each letter, rather than the multitap approach required by older phones. When you type "4663," you could be wanting the word "good" or "home" or "gone." The software looks "4663" up in a dictionary and selects the word that you use with the highest probability.

BUSINESSES USE PROBABILITIES:

- All insurance companies use probability theory to set fees so that the companies will make a profit.

- Investment firms use probability theory to assess financial risk.

- Medical firms use probability theory in diagnosing ailments and determining appropriate treatments.

3.1 History of Probability

OBJECTIVES

- Understand the types of problems that motivated the invention of probability theory

- Become familiar with dice, cards, and the game of roulette

Origins in Gambling

Probability theory is generally considered to have began in 1654 when the French nobleman (and successful gambler) Antoine Gombauld, the Chevalier de Méré, asked his mathematician friend Blaise Pascal why he had typically made money with one dice bet but lost money with another bet that he thought was similar. Pascal wrote a letter about this to another prominent French mathematician, Pierre de Fermat. In answering the Chevalier's question, the two launched probability theory.

More than a hundred years earlier, in 1545, the Italian physician, mathematician, and gambler Gerolamo Cardano had used probabilities to help him win more often. His *Liber de ludo aleae* (*Book on Games of Chance*) was the first book on probabilities and gambling. It also contained a section on effective cheating methods. Cardano's work was ignored until interest in Pascal's and Fermat's work prompted its publication.

For a long time, probability theory was not viewed as a serious branch of mathematics because of its early association with gambling. This started to change when Jacob Bernoulli, a prominent Swiss mathematician, wrote *Ars Conjectandi* (*The Art of Conjecture*), published in 1713. In it, Bernoulli developed the mathematical theory of probabilities. Its focus was on gambling, but it also suggested applications of probability to government, economics, law, and genetics.

Other Early Uses: Insurance and Genetics

This **Bills of Mortality** (records of death) was published shortly after John Graunt's analysis.

In 1662, the Englishman John Graunt published his *Natural and Political Observations on the Bills of Mortality,* which used probabilities to analyze death records and was the first important incidence of the use of probabilities for a purpose other than gambling. (It preceded Bernoulli's book.) English firms were selling the first life insurance policies, and they used mortality tables to set fees that were appropriate to the risks involved but still allowed the companies to make a profit.

In the 1860s, Gregor Mendel, an Austrian monk, experimented with pea plants in the abbey garden. His experiments and theories allowed for randomness in the passing on of traits from parent to child, and he used probabilities to analyze the effect of that randomness. In combining biology and mathematics, Mendel founded the science of genetics.

Roulette

Probability theory was originally created to aid gamblers, and games of chance are the most easily and universally understood topic to which probability theory can be applied. Learning about probability and games of chance may convince you not to be a gambler. As we will see in Section 3.5, there is not a single good bet in a casino game of chance.

Roulette is the oldest casino game still being played. Its invention has variously been credited to Pascal, the ancient Chinese, a French monk, and the Italian mathematician Don Pasquale. It became popular when a French policeman introduced it to Paris in 1765 in an attempt to take the advantage away from dishonest gamblers. When the Monte Carlo casino opened in 1863, roulette was the most popular game, especially among the aristocracy. The American roulette wheel has thirty-eight numbered compartments around its circumference. Thirty-six of these compartments are numbered from 1 to 36 and are colored red or black. The remaining two are numbered 0 and 00 and are colored green. Players place their bets by putting their chips on an appropriate spot on the roulette table (see Figure 3.1) on page 134. The dealer spins the wheel and then drops a ball onto the spinning wheel. The ball eventually comes to rest in one of the compartments, and that

Roulette quickly became a favorite game of the French upper class, as shown in this 1910 photo of Monte Carlo.

compartment's number is the winning number. (See on page 183 for a photo of the roulette wheel and table.)

For example, if a player wanted to bet $10 that the ball lands in compartment number 7, she would place $10 worth of chips on the number 7 on the roulette table. This is a single-number bet, so house odds are 35 to 1 (see Figure 3.1 for house odds). This means that if the player wins, she wins $10 · 35 = $350, and if she loses, she loses her $10.

Similarly, if a player wanted to bet $5 that the ball lands either on 13 or 14, he would place $5 worth of chips on the line separating numbers 13 and 14 on the roulette table. This is a two-numbers bet, so house odds are 17 to 1. This means that if the player wins, he wins $5 · 17 = $85, and if he loses, he loses his $5.

Bet	House Odds
single number	35 to 1
two numbers ("split")	17 to 1
three numbers ("street")	11 to 1
four numbers ("square")	8 to 1
five numbers ("line")	6 to 1
six numbers ("line")	5 to 1
twelve numbers (column or section)	2 to 1
low or high (1 to 18 or 19 to 36, respectively)	1 to 1
even or odd (0 and 00 are neither even nor odd)	1 to 1
red or black	1 to 1

FIGURE 3.1   The roulette table and house odds for the various roulette bets.

HISTORICAL NOTE

BLAISE PASCAL, 1623–1662

As a child, Blaise Pascal showed an early aptitude for science and mathematics, even though he was discouraged from studying in order to protect his poor health. Acute digestive problems and chronic insomnia made his life miserable. Few of Pascal's days were without pain.

At age sixteen, Pascal wrote a paper on geometry that won the respect of the French mathematical community and the jealousy of the prominent French mathematician René Descartes. It has been suggested that the animosity between Descartes and Pascal was in part due to their religious differences. Descartes was a Jesuit, and Pascal was a Jansenist. Although Jesuits and Jansenists were both Roman Catholics, Jesuits believed in free will and supported the sciences, while Jansenists believed in predestination and mysticism and opposed the sciences and the Jesuits.

At age nineteen, to assist his father, a tax administrator, Pascal invented the first calculating machine. Besides co-founding probability theory with Pierre de Fermat, Pascal contributed to the advance of calculus, and his studies in physics culminated with a law on the effects of pressure on fluids that bears his name.

At age thirty-one, after a close escape from death in a carriage accident, Pascal turned his back on mathematics and the sciences and focused on defending Jansenism against the Jesuits. Pascal came to be the greatest Jansenist, and he aroused a storm with his anti-Jesuit *Provincial Letters*. This work is still famous for its polite irony.

Pascal turned so far from the sciences that he came to believe that reason is inadequate to solve humanity's difficulties or to satisfy our hopes, and he regarded the pursuit of science as a vanity. Even so, he still occasionally succumbed to its lure. Once, to distract himself from pain, he concentrated on a geometry problem. When the pain stopped, he decided that it was a signal from God that he had not sinned by thinking of mathematics instead of his soul. This incident resulted in the only scientific work of his last few years—and his last work. He died later that year, at age thirty-nine.

Pascal's calculator, the Pascaline.

HISTORICAL NOTE

GEROLAMO CARDANO, 1501–1576

Gerolamo Cardano is the subject of much disagreement among historians. Some see him as a man of tremendous accomplishments, while others see him as a plagiarist and a liar. All agree that he was a compulsive gambler.

Cardano was trained as a medical doctor, but he was initially denied admission to the College of Physicians of Milan. That denial was ostensibly due to his illegitimate birth, but some suggest that the denial was in fact due to his unsavory reputation as a gambler, since illegitimacy was neither a professional nor a social obstacle in sixteenth-century Italy. His lack of professional success left him with much free time, which he spent gambling and reading. It also resulted in a stay in the poorhouse.

Cardano's luck changed when he obtained a lectureship in mathematics, astronomy, and astrology at the University of Milan. He wrote a number of books on mathematics and became famous for publishing a method of solving third-degree equations. Some claim that Cardano's mathematical success was due not to his own abilities but rather to those of Ludovico Ferrari, a servant of his who went on to become a mathematics professor.

While continuing to teach mathematics, Cardano returned to the practice of medicine. (He was finally allowed to join the College of Physicians, perhaps owing to his success as a mathematician.) Cardano wrote books on medicine and the natural sciences that were well thought of. He became one of the most highly regarded physicians in Europe and counted many prominent people among his patients. He designed a tactile system, somewhat like braille, for the blind. He also designed an undercarriage suspension device that was later adapted as a universal joint for automobiles and that is still called a *cardan* in Europe.

Cardano's investment in gambling was enormous. He not only wagered (and lost) a great deal of money but also spent considerable time and effort calculating probabilities and devising strategies. His *Book on Games of Chance* contains the first correctly calculated theoretical probabilities.

Cardano's autobiography, *The Book of My Life*, reveals a unique personality. He admitted that he loved talking about himself, his accomplishments, and his illnesses and diseases. He frequently wrote of injuries done him by others and followed these complaints with gleeful accounts of his detractors' deaths. Chapter titles include "Concerning my friends and patrons," "Calumny, defamations, and treachery of my unjust accusers," "Gambling and dicing," "Religion and piety," "The disasters of my sons," "Successes in my practice," "Things absolutely supernatural," and "Things of worth which I have achieved in various studies."

Dice and Craps

A die. The other three faces have four, five, and six spots.

Dice have been cast since the beginning of time, for both divination and gambling purposes. The earliest **die** (singular of *dice*) was an animal bone, usually a knucklebone or foot bone. The Romans were avid dice players. The Roman emperor Claudius I wrote a book titled *How to Win at Dice*. During the Middle Ages, dicing schools and guilds of dicers were quite popular among the knights and ladies.

Hazard, an ancestor of the dice game craps, is an English game that was supposedly invented by the Crusaders in an attempt to ward off boredom during long, drawn-out sieges. It became quite popular in England and France in the nineteenth century. The English called a throw of 2, 3, or 12 *crabs,* and it is believed that *craps* is a French mispronunciation of that term. The game came to America with the French colonization of New Orleans and spread up the Mississippi.

Cards

The invention of playing cards has been credited to the Indians, the Arabs, the Egyptians, and the Chinese. During the Crusades, Arabs endured lengthy sieges by

A sixteenth-century 2 of swords.

A sixteenth-century king of clubs.

playing card games. Their European foes acquired the cards and introduced them to their homelands. Cards, like dice, were used for divination as well as gambling. In fact, the modern deck is derived from the Tarot deck, which is composed of four suits plus twenty-two *atouts* that are not part of any suit. Each suit represents a class of medieval society: swords represent the nobility; coins, the merchants; batons or clubs, the peasants; and cups or chalices, the church. These suits are still used in regular playing cards in southern Europe. The Tarot deck also includes a joker and, in each suit, a king, a queen, a knight, a knave, and ten numbered cards.

Around 1500, the French dropped the knights and *atouts* from the deck and changed the suits from swords, coins, clubs, and chalices to *piques* (soldiers' pikes), *carreaux* (diamond-shaped building tiles), *trèfles* (clover leaf-shaped trefoils), and *coeurs* (hearts). In sixteenth-century Spain, *piques* were called *espados,* from which we get our term *spades*. Our diamonds are so named because of the shape of the carreaux. Clubs was an original Tarot suit, and hearts is a translation of *coeurs*.

The pictures on the cards were portraits of actual people. In fourteenth-century Europe, the kings were Charlemagne (hearts), the biblical David (spades), Julius Caesar (diamonds), and Alexander the Great (clubs); the queens included Helen of Troy (hearts), Pallas Athena (spades), and the biblical Rachel (diamonds). Others honored as "queen for a day" included Joan of Arc, Elizabeth I, and Elizabeth of York, wife of Henry VII. Jacks were usually famous warriors, including Sir Lancelot (clubs) and Roland, Charlemagne's nephew (diamonds).

A modern deck of cards contains fifty-two cards (thirteen in each of four suits). The four suits are hearts, diamonds, clubs, and spades (♥, ♦, ♣, ♠). Hearts and diamonds are red, and clubs and spades are black. Each suit consists of cards labeled 2 through 10, followed by jack, queen, king, and ace. **Face cards** are the jack, queen, and king; and **picture cards** are the jack, queen, king, and ace.

Two of the most popular card games are poker and blackjack. Poker's ancestor was a Persian game called *dsands,* which became popular in eighteenth-century Paris. It was transformed into a game called *poque,* which spread to America via the French colony in New Orleans. *Poker* is an American mispronunciation of the word *poque*.

The origins of blackjack (also known as *vingt-et-un* or twenty-one) are unknown. The game is called twenty-one because high odds are paid if a player's first two cards total 21 points (an ace counts as either 1 or 11, and the ten, jack, queen, and king each count as 10). A special bonus used to be paid if those two cards were a black jack and a black ace, hence the name *blackjack*.

A modern deck of cards.

3.1 EXERCISES

1. The Chevalier de Méré generally made money betting that he could roll at least one 6 in four rolls of a single die. Roll a single die four times, and record the number of times a 6 comes up. Repeat this ten times. If you had made the Chevalier de Méré's bet (at $10 per game), would you have won or lost money? How much?

2. The Chevalier de Méré generally lost money betting that he could roll at least one pair of 6's in twenty-four rolls of a pair of dice. Roll a pair of dice twenty-four times, and record the number of times a 6 comes up. Repeat this five times. If you had made the Chevalier de Méré's bet (at $10 per game), would you have won or lost money? How much?

3. **a.** If you were to flip a pair of coins thirty times, approximately how many times do you think a pair of heads would come up? A pair of tails? One head and one tail?

   b. Flip a pair of coins thirty times, and record the number of times a pair of heads comes up, the number of times a pair of tails comes up, and the number of times one head and one tail come up. How closely do the results agree with your guess?

4. **a.** If you were to flip a single coin twenty times, approximately how many times do you think heads would come up? Tails?

   b. Flip a single coin twenty times, and record the number of times heads comes up and the number of times tails comes up. How closely do the results agree with your guesses?

5. **a.** If you were to roll a single die twenty times, approximately how many times do you think an even number would come up? An odd number?

   b. Roll a single die twenty times, and record the number of times an even number comes up and the number of times an odd number comes up. How closely do the results agree with your guess?

6. **a.** If you were to roll a pair of dice thirty times, approximately how many times do you think the total would be 7? Approximately how many times do you think the total would be 12?

   b. Roll a single die thirty times, and record the number of times the total is 7 and the number of times the total is 12. How closely do the results agree with your guess?

7. **a.** If you were to deal twenty-six cards from a complete deck (without jokers), approximately how many cards do you think would be red? Approximately how many do you think would be aces?

   b. Deal twenty-six cards from a complete deck (without jokers), and record the number of times a red card is dealt and the number of times an ace is dealt. How closely do the results agree with your guess?

8. **a.** If you were to deal twenty-six cards from a complete deck (without jokers), approximately how many cards do you think would be black? Approximately how many do you think would be jacks, queens, or kings?

   b. Deal twenty-six cards from a complete deck (without jokers), and record the number of times a black card is dealt and the number of times a jack, queen, or king is dealt. How closely do the results agree with your guess?

In Exercises 9–24, use Figure 3.1 to find the outcome of the bets in roulette, given the results listed.

9. You bet $10 on the 25.
   a. The ball lands on number 25.
   b. The ball lands on number 14.

10. You bet $15 on the 17.
    a. The ball lands on 18.
    b. The ball lands on 17.

11. You bet $5 on 17-20 split.
    a. The ball lands on number 17.
    b. The ball lands on number 20.
    c. The ball lands on number 32.

12. You bet $30 on the 22-23-24 street.
    a. The ball lands on number 19.
    b. The ball lands on number 22.
    c. The ball lands on number 0.

13. You bet $20 on the 8-9-11-12 square.
    a. The ball lands on number 15.
    b. The ball lands on number 9.
    c. The ball lands on number 00.

14. You bet $100 on the 0-00-1-2-3 line (the only five-number line on the table).
    a. The ball lands on number 29.
    b. The ball lands on number 2.

15. You bet $10 on the 31-32-33-34-35-36 line.
    a. The ball lands on number 5.
    b. The ball lands on number 33.

16. You bet $20 on the 13 through 24 section.
    a. The ball lands on number 00.
    b. The ball lands on number 15.

17. You bet $25 on the first column.
   a. The ball lands on number 13.
   b. The ball lands on number 14.

18. You bet $30 on the low numbers.
   a. The ball lands on number 8.
   b. The ball lands on number 30.

19. You bet $50 on the odd numbers.
   a. The ball lands on number 00.
   b. The ball lands on number 5.

20. You bet $20 on the black numbers.
   a. The ball lands on number 11.
   b. The ball lands on number 12.

21. You make a $20 single-number bet on number 14 and also a $25 single-number bet on number 15.
   a. The ball lands on number 16.
   b. The ball lands on number 15.
   c. The ball lands on number 14.

22. You bet $10 on the low numbers and also bet $20 on the 16-17-19-20 square.
   a. The ball lands on number 16.
   b. The ball lands on number 19.
   c. The ball lands on number 14.

23. You bet $30 on the 1-2 split and also bet $15 on the even numbers.
   a. The ball lands on number 1.
   b. The ball lands on number 2.
   c. The ball lands on number 3.
   d. The ball lands on number 4.

24. You bet $40 on the 1-12 section and also bet $10 on number 10.
   a. The ball lands on number 7.
   b. The ball lands on number 10.
   c. The ball lands on number 21.

25. How much must you bet on a single number to be able to win at least $100? (Bets must be in $1 increments.)

26. How much must you bet on a two-number split to be able to win at least $200? (Bets must be in $1 increments.)

27. How much must you bet on a twelve-number column to be able to win at least $1000? (Bets must be in $1 increments.)

28. How much must you bet on a four-number square to win at least $600? (Bets must be in $1 increments.)

29. a. How many hearts are there in a deck of cards?
   b. What fraction of a deck is hearts?

30. a. How many red cards are there in a deck of cards?
   b. What fraction of a deck is red?

31. a. How many face cards are there in a deck of cards?
   b. What fraction of a deck is face cards?

32. a. How many black cards are there in a deck of cards?
   b. What fraction of a deck is black?

33. a. How many kings are there in a deck of cards?
   b. What fraction of a deck is kings?

 Answer the following questions using complete sentences and your own words.

• HISTORY QUESTIONS

34. Who started probability theory? How?

35. Why was probability theory not considered a serious branch of mathematics?

36. Which authors established probability theory as a serious area of interest? What are some of the areas to which these authors applied probability theory?

37. What did Gregor Mendel do with probabilities?

38. Who was Antoine Gombauld, and what was his role in probability theory?

39. Who was Gerolamo Cardano, and what was his role in probability theory?

40. Which games of chance came to America from France via New Orleans?

41. Which implements of gambling were also used for divination?

42. What is the oldest casino game still being played?

43. How were cards introduced to Europe?

44. What is the modern deck of cards derived from?

45. What game was supposed to take the advantage away from dishonest gamblers?

WEB PROJECT

46. Write a research paper on any historical topic referred to in this section or a related topic. Following is a partial list of topics:
   - Jacob Bernoulli
   - Gerolamo Cardano
   - Pierre de Fermat
   - Blaise Pascal
   - The Marquis de Laplace
   - John Graunt
   - Gregor Mendel

   Some useful links for this web project are listed on the text web site: **www.cengage.com/math/johnson**

OBJECTIVES

- Learn the basic terminology of probability theory
- Be able to calculate simple probabilities
- Understand how probabilities are used in genetics

Much of the terminology and many of the computations of probability theory have their basis in set theory, because set theory contains the mathematical way of describing collections of objects and the size of those collections.

BASIC PROBABILITY TERMS

experiment: a process by which an observation, or **outcome,** is obtained
sample space: the set S of all possible outcomes of an experiment
event: any subset E of the sample space S

If a single die is rolled, the *experiment* is the rolling of the die. The possible *outcomes* are 1, 2, 3, 4, 5, and 6. The *sample space* (set of all possible outcomes) is $S = \{1, 2, 3, 4, 5, 6\}$. (The term *sample space* really means the same thing as *universal set;* the only distinction between the two ideas is that *sample space* is used only in probability theory, while *universal set* is used in any situation in which sets are used.) There are several possible *events* (subsets of the sample space), including the following:

$E_1 = \{3\}$ "a three comes up"
$E_2 = \{2, 4, 6\}$ "an even number comes up"
$E_3 = \{1, 2, 3, 4, 5, 6\}$ "a number between 1 and 6 inclusive comes up"

Notice that an event is not the same as an outcome. An *event* is a subset of the sample space; an *outcome* is an element of the sample space. "Rolling an odd number" is an event, not an outcome. It is the set $\{1, 3, 5\}$ that is composed of three separate outcomes. Some events are distinguished from outcomes only in that set brackets are used with events and not with outcomes. For example, $\{5\}$ is an event, and 5 is an outcome; either refers to "rolling a five."

The event E_3 ("a number between 1 and 6 inclusive comes up") is called a *certain event,* since $E_3 = S$. That is, E_3 is a sure thing. "Getting 17" is an *impossible event.* No outcome in the sample space $S = \{1, 2, 3, 4, 5, 6\}$ would result in a 17, so this event is actually the empty set.

MORE PROBABILITY TERMS

A **certain event** is an event that is equal to the sample space.
An **impossible event** is an event that is equal to the empty set.

THE FAR SIDE® BY GARY LARSON

Early shell games

An early certain event.

Finding Probabilities and Odds

The **probability** of an event is a measure of the likelihood that the event will occur. If a single die is rolled, the outcomes are equally likely; a 3 is just as likely to come up as any other number. There are six possible outcomes, so a 3 should come up about one out of every six rolls. That is, the probability of event E_1 ("a 3 comes up") is $\frac{1}{6}$. The 1 in the numerator is the number of elements in $E_1 = \{3\}$. The 6 in the denominator is the number of elements in $S = \{1, 2, 3, 4, 5, 6\}$.

If an experiment's outcomes are equally likely, then the probability of an event E is the number of outcomes in the event divided by the number of outcomes in the sample space, or $n(E)/n(S)$. (In this chapter, we discuss only experiments with equally likely outcomes.) Probability can be thought of as "success over a total."

PROBABILITY OF AN EVENT

The **probability** of an event E, denoted by $p(E)$, is

$$p(E) = \frac{n(E)}{n(S)}$$

if the experiment's outcomes are equally likely.
(*Think: Success over total.*)

Many people use the words *probability* and *odds* interchangeably. However, the words have different meanings. The **odds** in favor of an event are the number of ways the event can occur compared to the number of ways the event *can fail to occur,* or "success compared to *failure*" (if the experiment's outcomes are equally likely). The odds of event E_1 ("a 3 comes up") are 1 to 5 (or 1:5), since a three can come up in one way and can fail to come up in five ways. Similarly, the odds of event E_3 ("a number between 1 and 6 inclusive comes up") are 6 to 0 (or 6:0), since a number between 1 and 6 inclusive can come up in six ways and can fail to come up in zero ways.

ODDS OF AN EVENT

The **odds** of an event E with equally likely outcomes, denoted by $o(E)$, are given by

$$o(E) = n(E) : n(E')$$

(Think: Success compared with failure.)

$\dfrac{18}{38} \quad \dfrac{9}{18}$

In addition to the above meaning, the word *odds* can also refer to "house odds," which has to do with how much you will be paid if you win a bet at a casino. The odds of an event are sometimes called the **true odds** to distinguish them from the house odds.

EXAMPLE 1

FLIPPING A COIN   A coin is flipped. Find the following.

a. the sample space
b. the probability of event E_1, "getting heads"
c. the odds of event E_1, "getting heads"
d. the probability of event E_2, "getting heads or tails"
e. the odds of event E_2, "getting heads or tails"

SOLUTION

a. *Finding the sample space S:* The experiment is flipping a coin. The only possible outcomes are heads and tails. The sample space S is the set of all possible outcomes, so $S = \{h, t\}$.

b. *Finding the probability of heads:*

$$E_1 = \{h\} \ (\text{"getting heads"})$$
$$p(E_1) = \frac{n(E_1)}{n(S)} = \frac{1}{2}$$

This means that one out of every two possible outcomes is a success.

c. *Finding the odds of heads:*

$$E_1' = \{t\}$$
$$o(E_1) = n(E_1) : n(E_1') = 1:1$$

This means that for every one possible success, there is one possible failure.

d. *Finding the probability of heads or tails:*

$$E_2 = \{h, t\}$$
$$p(E_2) = \frac{n(E_2)}{n(S)} = \frac{2}{2} = \frac{1}{1}$$

This means that every outcome is a success. Notice that E_2 is a certain event.

e. *Finding the odds of heads or tails:*

$$E_2' = \varnothing$$
$$o(E_2) = n(E_2) : n(E_2') = 2 : 0 = 1:0$$

This means that there are no possible failures.

EXAMPLE **2**

ROLLING A DIE   A die is rolled. Find the following.

a. the sample space
b. the event "rolling a 5"
c. the probability of rolling a 5
d. the odds of rolling a 5
e. the probability of rolling a number below 5
f. the odds of rolling a number below 5

SOLUTION

a. The sample space is $S = \{1, 2, 3, 4, 5, 6\}$.
b. Event E_1 "rolling a 5" is $E_1 = \{5\}$.
c. The probability of E_1 is

$$p(E_1) = \frac{n(E_1)}{n(S)} = \frac{1}{6}$$

This means that one out of every six possible outcomes is a success (that is, a 5).
d. $E_1' = \{1, 2, 3, 4, 6\}$. The odds of E_1 are

$$o(E_1) = n(E_1):n(E_1') = 1:5$$

This means that there is one possible success for every five possible failures.
e. $E_2 = \{1, 2, 3, 4\}$ ("rolling a number below 5"). The probability of E_2 is

$$p(E_2) = \frac{n(E_2)}{n(S)} = \frac{4}{6} = \frac{2}{3}$$

This means that two out of every three possible outcomes are a success.
f. $E_2' = \{5, 6\}$. The odds of E_2 are

$$o(E_2) = n(E_2):n(E_2') = 4:2 = 2:1$$

This means that there are two possible successes for every one possible failure. Notice that odds are reduced in the same manner that a fraction is reduced.

Relative Frequency versus Probability

So far, we have discussed probabilities only in a theoretical way. When we found that the probability of heads was $\frac{1}{2}$, we never actually tossed a coin. It does not always make sense to calculate probabilities theoretically; sometimes they must be found empirically, the way a batting average is calculated. For example, in 8,389 times at bat, Babe Ruth had 2,875 hits. His batting average was $\frac{2,875}{8,389} \approx 0.343$. In other words, his probability of getting a hit was 0.343.

Sometimes a probability can be found either theoretically or empirically. We have already found that the theoretical probability of heads is $\frac{1}{2}$. We could also flip a coin a number of times and calculate (number of heads)/(total number of flips); this can be called the **relative frequency** of heads, to distinguish it from the theoretical probability of heads.

The Law of Large Numbers

Usually, the relative frequency of an outcome is not equal to its probability, but if the number of trials is large, the two tend to be close. If you tossed a coin a couple of times, anything could happen, and the fact that the probability of heads is

$\frac{1}{2}$ would have no impact on the results. However, if you tossed a coin 100 times, you would probably find that the relative frequency of heads was close to $\frac{1}{2}$. If your friend tossed a coin 1,000 times, she would probably find the relative frequency of heads to be even closer to $\frac{1}{2}$ than in your experiment. This relationship between probabilities and relative frequencies is called the **Law of Large Numbers.** Cardano stated this law in his *Book on Games of Chance,* and Bernoulli proved that it must be true.

LAW OF LARGE NUMBERS

If an experiment is repeated a large number of times, the relative frequency of an outcome will tend to be close to the probability of that outcome.

The graph in Figure 3.2 shows the result of a simulated coin toss, using a computer and a random number generator rather than an actual coin. Notice that when the number of tosses is small, the relative frequency achieves values such as 0, 0.67, and 0.71. These values are not that close to the theoretical probability of 0.5. However, when the number of tosses is large, the relative frequency achieves values such as 0.48. These values are very close to the theoretical probability.

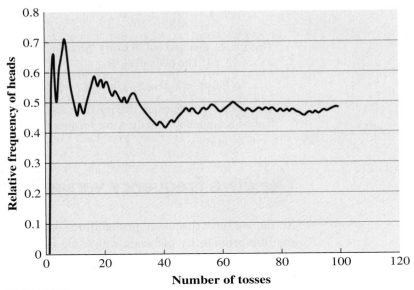

FIGURE 3.2   The relative frequency of heads after 100 simulated coin tosses.

What if we used a real coin, rather than a computer, and we tossed the coin a lot more? Three different mathematicians have performed such an experiment:

- In the eighteenth century, Count Buffon tossed a coin 4,040 times. He obtained 2,048 heads, for a relative frequency of $2,048/4,040 \approx 0.5069$.
- During World War II, the South African mathematician John Kerrich tossed a coin 10,000 times while he was imprisoned in a German concentration camp. He obtained 5,067 heads, for a relative frequency of $5,067/10,000 = 0.5067$.
- In the early twentieth century, the English mathematician Karl Pearson tossed a coin 24,000 times! He obtained 12,012 heads, for a relative frequency of $12,012/24,000 = 0.5005$.

At a casino, probabilities are much more useful to the casino (the "house") than to an individual gambler, because the house performs the experiment for a much larger number of trials (in other words, plays the game more often). In fact, the house plays the game so many times that the relative frequencies will be almost exactly the same as the probabilities, with the result that the house is not gambling at all—it knows what is going to happen. Similarly, a gambler with a "system" has to play the game for a long time for the system to be of use.

EXAMPLE **3**

FLIPPING A PAIR OF COINS   A pair of coins is flipped.

   a. Find the sample space.
   b. Find the event E "getting exactly one heads."
   c. Find the probability of event E.
   d. Use the Law of Large Numbers to interpret the probability of event E.

SOLUTION

The outcomes (h, h), (h, t), (t, h), and (t, t).

   a. The experiment is the flipping of a pair of coins. One possible outcome is that one coin is heads and the other is tails. A second and *different* outcome is that one coin is tails and the other is heads. These two outcomes seem the same. However, if one coin were painted, it would be easy to tell them apart. Outcomes of the experiment can be described by using ordered pairs in which the first component refers to the first coin and the second component refers to the second coin. The two different ways of getting one heads and one tails are (h, t) and (t, h).

      The sample space, or set of all possible outcomes, is the set $S = \{(h, h), (t, t), (h, t), (t, h)\}$. These outcomes are equally likely.

   b. The event E "getting exactly one heads" is

$$E = \{(h, t), (t, h)\}$$

   c. The probability of event E is

$$P(E) = \frac{n(E)}{n(S)} = \frac{2}{4} = \frac{1}{2} = 50\%$$

   d. According to the Law of Large Numbers, if an experiment is repeated a large number of times, the relative frequency of that outcome will tend to be close to the probability of the outcome. Here, that means that if we were to toss a pair of coins many times, we should expect to get exactly one heads about half (or 50%) of the time. Realize that this is only a prediction, and we might never get exactly one heads.

Mendel's Use of Probabilities

In his experiments with plants, Gregor Mendel pollinated peas until he produced pure-red plants (that is, plants that would produce only red-flowered offspring) and pure-white plants. He then cross-fertilized these pure reds and pure whites and obtained offspring that had only red flowers. This amazed him, because the accepted theory of the day incorrectly predicted that these offspring would all have pink flowers.

     He explained this result by postulating that there is a "determiner" that is responsible for flower color. These determiners are now called **genes.** Each plant has two flower color genes, one from each parent. Mendel reasoned that these offspring had to have inherited a red gene from one parent and a white gene from the other. These plants had one red flower gene and one white flower gene, but they had red flowers. That is, the red flower gene is **dominant,** and the white flower gene is **recessive.**

     If we use R to stand for the red gene and w to stand for the white gene (with the capital letter indicating dominance), then the results can be described with a **Punnett square** in Figure 3.3 on page 146.

	R	R	
	R	**R**	← first parent's genes
w	(R, w)	(R, w)	← possible offspring
w	(R, w)	(R, w)	← possible offspring

↑
second parent's genes

FIGURE 3.3 A Punnett square for the first generation.

When the offspring of this experiment were cross-fertilized, Mendel found that approximately three-fourths of the offspring had red flowers and one-fourth had white flowers (see Figure 3.4). Mendel successfully used probability theory to analyze this result.

	R	**w**	
	R	w	← first parent's genes
R	(R, R)	(w, R)	← possible offspring
w	(R, w)	(w, w)	← possible offspring

↑
second parent's genes

FIGURE 3.4 A Punnett square for the second generation.

Only one of the four possible outcomes, (w, w), results in a white-flowered plant, so event E_1 that the plant has white flowers is $E_1 = \{(w, w)\}$; therefore,

$$p(E_1) = \frac{n(E_1)}{n(S)} = \frac{1}{4}$$

This means that we should expect the actual relative frequency of white-flowered plants to be close to $\frac{1}{4}$. That is, about one-fourth, or 25%, of the second-generation plants should have white flowers (if we have a lot of plants).

Each of the other three outcomes, (R, R), (R, w), and (w, R), results in a red-flowered plant, because red dominates white. The event E_2 that the plant has red flowers is $E_2 = \{(R, R), (R, w), (w, R)\}$; therefore,

$$p(E_2) = \frac{n(E_2)}{n(S)} = \frac{3}{4} = 75\%$$

Thus, we should expect the actual relative frequency of red-flowered plants to be close to $\frac{3}{4}$. That is, about three-fourths, or 75%, of the plants should have red flowers.

Outcomes (R, w) and (w, R) are genetically identical; it does not matter which gene is inherited from which parent. For this reason, geneticists do not use the ordered-pair notation but instead refer to each of these two outcomes as "Rw." The only difficulty with this convention is that it makes the sample space appear to be S = {RR, Rw, ww}, which consists of only three elements, when in fact it consists of four elements. This distinction is important; if the sample space consisted of three equally likely elements, then the probability of a red-flowered offspring would be $\frac{2}{3}$ rather than $\frac{3}{4}$. Mendel knew that the sample space had to have four elements, because his cross-fertilization experiments resulted in a relative frequency very close to $\frac{3}{4}$, not $\frac{2}{3}$.

Ronald Fisher, a noted British statistician, used statistics to deduce that Mendel fudged his data. Mendel's relative frequencies were unusually close to the theoretical probabilities, and Fisher found that there was only about a 0.00007 chance of such close agreement. Others have suggested that perhaps Mendel did not willfully change his results but rather continued collecting data until the numbers were in close agreement with his expectations.*

*R. A. Fisher, "Has Mendel's Work Been Rediscovered?" *Annals of Science* 1, 1936, pp. 115–137.

HISTORICAL NOTE

GREGOR JOHANN MENDEL, 1822–1884

© Bettmann/CORBIS

Johann Mendel was born to an Austrian peasant family. His interest in botany began on the family farm, where he helped his father graft fruit trees. He studied philosophy, physics, and mathematics at the University Philosophical Institute in Olmütz. He was unsuccessful in finding a job, so he quit school and returned to the farm. Depressed by the prospects of a bleak future, he became ill and stayed at home for a year.

Mendel later returned to Olmütz. After two years of study, he found the pressures of school and work to be too much, and his health again broke down. On the advice of his father and a professor, he entered the priesthood, even though he did not feel called to serve the church. His name was changed from Johann to Gregor.

Relieved of his financial difficulties, he was able to continue his studies. However, his nervous disposition interfered with his pastoral duties, and he was assigned to substitute teaching. He enjoyed this work and was popular with the staff and students, but he failed the examination for certification as a teacher. Ironically, his lowest grades were in biology. The Augustinians then sent him to the University of Vienna, where he became particularly interested in his plant physiology professor's unorthodox belief that new plant varieties can be caused by naturally arising variations. He was also fascinated by his classes in physics, where he was exposed to the physicists' experimental and mathematical approach to their subject.

After further breakdowns and failures, Mendel returned to the monastery and was assigned the low-stress job of keeping the abbey garden. There he combined the experimental and mathematical approach of a physicist with his background in biology and performed a series of experiments designed to determine whether his professor was correct in his beliefs regarding the role of naturally arising variants in plants.

Mendel studied the transmission of specific traits of the pea plant—such as flower color and stem length—from parent plant to offspring. He pollinated the plants by hand and separated them until he had isolated each trait. For example, in his studies of flower color, he pollinated the plants until he produced pure-red plants (plants that would produce only red-flowered offspring) and pure-white plants.

At the time, the accepted theory of heredity was that of blending. In this view, the characteristics of both parents blend together to form an individual. Mendel reasoned that if the blending theory was correct, the union of a pure-red pea plant and a pure-white pea plant would result in a pink-flowered offspring. However, his experiments showed that such a union consistently resulted in red-flowered offspring.

Mendel crossbred a large number of peas that had different characteristics. In many cases, an offspring would have a characteristic of one of its parents, undiluted by that of the other parent. Mendel concluded that the question of which parent's characteristics would be passed on was a matter of chance, and he successfully used probability theory to estimate the frequency with which characteristics would be passed on. In so doing, Mendel founded modern genetics. Mendel attempted similar experiments with bees, but these experiments were unsuccessful because he was unable to control the mating behavior of the queen bee.

Mendel was ignored when he published his paper "Experimentation in Plant Hybridization." Sixteen years after his death, his work was rediscovered by three European botanists who had reached similar conclusions in plant breeding, and the importance of Mendel's work was finally recognized.

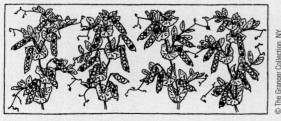

© The Granger Collection, NY

A nineteenth-century drawing illustrating Mendel's pea plants, showing the original cross, the first generation, and the second generation.

Probabilities in Genetics: Inherited Diseases

Cystic fibrosis is an inherited disease characterized by abnormally functioning exocrine glands that secrete a thick mucus, clogging the pancreatic ducts and lung passages. Most patients with cystic fibrosis die of chronic lung disease; until recently, most died in early childhood. This early death made it extremely unlikely that an afflicted person would ever parent a child. Only after the advent of Mendelian genetics did it become clear how a child could inherit the disease from two healthy parents.

In 1989, a team of Canadian and American doctors announced the discovery of the gene that is responsible for most cases of cystic fibrosis. As a result of that discovery, a new therapy for cystic fibrosis is being developed. Researchers splice a therapeutic gene into a cold virus and administer it through an affected person's nose. When the virus infects the lungs, the gene becomes active. It is hoped that this will result in normally functioning cells, without the damaging mucus.

In April 1993, a twenty-three-year-old man with advanced cystic fibrosis became the first patient to receive this therapy. In September 1996, a British team announced that eight volunteers with cystic fibrosis had received this therapy; six were temporarily cured of the disease's debilitating symptoms. In March 1999, another British team announced a new therapy that involves administering the therapeutic gene through an aerosol spray. Thanks to other advances in treatment, in 2006, infants born in the United States with cystic fibrosis had a life expectancy of twenty-seven years.

Cystic fibrosis occurs in about 1 out of every 2,000 births in the Caucasian population and only in about 1 in 250,000 births in the non-Caucasian population. It is one of the most common inherited diseases in North America. One in 25 Americans carries a single gene for cystic fibrosis. Children who inherit two such genes develop the disease; that is, cystic fibrosis is recessive.

There are tests that can be used to determine whether a person carries the gene. However, they are not accurate enough to use for the general population. They are much more accurate with people who have a family history of cystic fibrosis, so The American College of Obstetricians and Gynecologists recommends testing only for couples with a personal or close family history of cystic fibrosis.

EXAMPLE **4**

PROBABILITIES AND CYSTIC FIBROSIS   Each of two prospective parents carries one cystic fibrosis gene.

a. Find the probability that their child would have cystic fibrosis.
b. Find the probability that their child would be free of symptoms.
c. Find the probability that their child would be free of symptoms but could pass the cystic fibrosis gene on to his or her own child.

SOLUTION

We will denote the recessive cystic fibrosis gene with the letter c and the normal disease-free gene with an N. Each parent is Nc and therefore does not have the disease. Figure 3.5 shows the Punnett square for the child.

	N	c
N	(N, N)	(c, N)
c	(N, c)	(c, c)

FIGURE 3.5 A Punnett square for Example 4.

a. Cystic fibrosis is recessive, so only the (c, c) child will have the disease. The probability of such an event is 1/4.

b. The (N, N), (c, N), and (N, c) children will be free of symptoms. The probability of this event is

$$p(\text{healthy}) = p((N, N)) + p((c, N)) + p((N, c))$$
$$= \frac{1}{4} + \frac{1}{4} + \frac{1}{4} = \frac{3}{4}$$

c. The (c, N) and (N, c) children would never suffer from any symptoms but could pass the cystic fibrosis gene on to their own children. The probability of this event is

$$p((c, N)) + p((N, c)) = \frac{1}{4} + \frac{1}{4} = \frac{1}{2}$$

In Example 4, the Nc child is called a **carrier** because that child would never suffer from any symptoms but could pass the cystic fibrosis gene on to his or her own child. Both of the parents were carriers.

Sickle-cell anemia is an inherited disease characterized by a tendency of the red blood cells to become distorted and deprived of oxygen. Although it varies in severity, the disease can be fatal in early childhood. More often, patients have a shortened life span and chronic organ damage.

Newborns are now routinely screened for sickle-cell disease. The only true cure is a bone marrow transplant from a sibling without sickle-cell anemia; however, this can cause the patient's death, so it is done only under certain circumstances. Until recently, about 10% of the children with sickle-cell anemia had a stroke before they were twenty-one. But in 2009, it was announced that the rate of these strokes has been cut in half thanks to a new specialized ultrasound scan that identifies the individuals who have a high stroke risk. There are also medications that can decrease the episodes of pain.

Approximately 1 in every 500 black babies is born with sickle-cell anemia, but only 1 in 160,000 nonblack babies has the disease. This disease is **codominant:** A person with two sickle-cell genes will have the disease, while a person with one sickle-cell gene will have a mild, nonfatal anemia called **sickle-cell trait.** Approximately 8–10% of the black population has sickle-cell trait.

Huntington's disease, caused by a dominant gene, is characterized by nerve degeneration causing spasmodic movements and progressive mental deterioration. The symptoms do not usually appear until well after reproductive age has been reached; the disease usually hits people in their forties. Death typically follows 20 years after the onset of the symptoms. No effective treatment is available, but physicians can now assess with certainty whether someone will develop the disease, and they can estimate when the disease will strike. Many of those who are at risk choose not to undergo the test, especially if they have already had children. Folk singer Arlo Guthrie is in this situation; his father, Woody Guthrie, died of Huntington's disease. Woody's wife Marjorie formed the Committee to Combat Huntington's Disease, which has stimulated research, increased public awareness, and provided support for families in many countries.

In August 1999, researchers in Britain, Germany, and the United States discovered what causes brain cells to die in people with Huntington's disease. This discovery may eventually lead to a treatment. In 2008, a new drug that reduces the uncontrollable spasmodic movements was approved.

Tay-Sachs disease is a recessive disease characterized by an abnormal accumulation of certain fat compounds in the spinal cord and brain. Most typically,

HISTORICAL NOTE

NANCY WEXLER

Courtesy of Dr. Nancy Wexler

In 1993, scientists working together at six major research centers located most genes that cause Huntington's disease. This discovery will enable people to learn whether they carry a Huntington's gene, and it will allow pregnant women to determine whether their child carries the gene. The discovery could eventually lead to a treatment.

The collaboration of research centers was organized largely by Nancy Wexler, a Columbia University professor of neuropsychology, who is herself at risk for Huntington's disease—her mother died of it in 1978. Dr. Wexler, President of the Hereditary Disease Foundation, has made numerous trips to study and aid the people of the Venezuelan village of Lake Maracaibo, many of whom suffer from the disease or are at risk for it. All are related to one woman who died of the disease in the early 1800s. Wexler took blood and tissue samples and gave neurological and psychoneurological tests to the inhabitants of the village. The samples and test results enabled the researchers to find the single gene that causes Huntington's disease.

In October 1993, Wexler received an Albert Lasker Medical Research Award, a prestigious honor that is often a precursor to a Nobel Prize. The award was given in recognition for her contribution to the international effort that culminated in the discovery of the Huntington's disease gene. At the awards ceremony, she explained to then first lady Hillary Clinton that her genetic heritage has made her uninsurable—she would lose her health coverage if she switched jobs. She told Mrs. Clinton that more Americans will be in the same situation as more genetic discoveries are made, unless the health care system is reformed. The then first lady incorporated this information into her speech at the awards ceremony: "It is likely that in the next years, every one of us will have a pre-existing condition and will be uninsurable. . . . What will happen as we discover those genes for breast cancer, or prostate cancer, or osteoporosis, or any of the thousands of other conditions that affect us as human beings?"

AP Photo

Woody Guthrie's most famous song is "This Land is Your Land." This folksinger, guitarist, and composer was a friend of Leadbelly, Pete Seeger, and Ramblin' Jack Elliott and exerted a strong influence on Bob Dylan. Guthrie died at the age of fifty-five of Huntington's disease.

a child with Tay-Sachs disease starts to deteriorate mentally and physically at six months of age. After becoming blind, deaf, and unable to swallow, the child becomes paralyzed and dies before the age of four years. There is no effective treatment. The disease occurs once in 3,600 births among Ashkenazi Jews (Jews from central and eastern Europe), Cajuns, and French Canadians but only once in 600,000 births in other populations. Carrier-detection tests and fetal-monitoring tests are available. The successful use of these tests, combined with an aggressive counseling program, has resulted in a decrease of 90% of the incidence of this disease.

Genetic Screening

At this time, there are no conclusive tests that will tell a parent whether he or she is a cystic fibrosis carrier, nor are there conclusive tests that will tell whether a fetus has the disease. A new test resulted from the 1989 discovery of the location of most cystic fibrosis genes, but that test will detect only 85% to 95% of the cystic fibrosis genes, depending on the individual's ethnic background. The extent to which this test will be used has created quite a controversy.

Individuals who have relatives with cystic fibrosis are routinely informed about the availability of the new test. The controversial question is whether a massive genetic screening program should be instituted to identify cystic fibrosis carriers in the general population, regardless of family history. This is an important question, considering that four in five babies with cystic fibrosis are born to couples with no previous family history of the condition.

Opponents of routine screening cite a number of important concerns. The existing test is rather inaccurate; 5% to 15% of the cystic fibrosis carriers would be missed. It is not known how health insurers would use this genetic information—insurance firms could raise rates or refuse to carry people if a screening test indicated a presence of cystic fibrosis. Also, some experts question the adequacy of quality assurance for the diagnostic facilities and for the tests themselves.

Supporters of routine testing say that the individual should be allowed to decide whether to be screened. Failing to inform people denies them the opportunity to make a personal choice about their reproductive future. An individual who is found to be a carrier could choose to avoid conception, to adopt, to use artificial insemination by a donor, or to use prenatal testing to determine whether a fetus is affected—at which point the additional controversy regarding abortion could enter the picture.

The Failures of Genetic Screening

The history of genetic screening programs is not an impressive one. In the 1970s, mass screening of blacks for sickle-cell anemia was instituted. This program caused unwarranted panic; those who were told they had sickle-cell trait feared that they would develop symptoms of the disease and often did not understand the probability that their children would inherit the disease (see Exercises 73 and 74). Some people with sickle-cell trait were denied health insurance and life insurance. See the 1973 Newsweek article on "The Row over Sickle-Cell."

FEATURED IN THE NEWS   THE ROW OVER SICKLE-CELL

. . . Two years ago, President Nixon listed sickle-cell anemia along with cancer as diseases requiring special Federal attention. . . . Federal spending for sickle-cell anemia programs has risen from a scanty $1 million a year to $15 million for 1973. At the same time, in what can only be described as a head-long rush, at least a dozen states have passed laws requiring sickle-cell screening for blacks.

While all these efforts have been undertaken with the best intentions of both whites and blacks, in recent months the campaign has begun to stir widespread and bitter controversy. Some of the educational programs have been riddled with misinformation and have unduly

frightened the black community. To quite a few Negroes, the state laws are discriminatory—and to the extent that they might inhibit childbearing, even genocidal. . . .

Parents whose children have the trait often misunderstand and assume they have the disease. In some cases, airlines have allegedly refused to hire black stewardesses who have the trait, and some carriers have been turned down by life-insurance companies—or issued policies at high-risk rates.

Because of racial overtones and the stigma that attaches to persons found to have the sickle-cell trait, many experts seriously object to mandatory screening programs. They note, for example, that there are no laws requiring testing for

Cooley's anemia [or other disorders that have a hereditary basis]. . . . Moreover, there is little that a person who knows he has the disease or the trait can do about it. "I don't feel," says Dr. Robert L. Murray, a black geneticist at Washington's Howard University, "that people should be required by law to be tested for something that will provide information that is more negative than positive." . . . Fortunately, some of the mandatory laws are being repealed.

3.2   EXERCISES

In Exercises 1–14, use this information: A jar on your desk contains twelve black, eight red, ten yellow, and five green jellybeans. You pick a jellybean without looking.

1. What is the experiment?
2. What is the sample space?

In Exercises 3–14, find the following. Write each probability as a reduced fraction and as a percent, rounded to the nearest 1%.

3. the probability that it is black
4. the probability that it is green
5. the probability that it is red or yellow
6. the probability that it is red or black
7. the probability that it is not yellow
8. the probability that it is not red
9. the probability that it is white
10. the probability that it is not white
11. the odds of picking a black jellybean
12. the odds of picking a green jellybean
13. the odds of picking a red or yellow jellybean
14. the odds of picking a red or black jellybean

In Exercises 15–28, one card is drawn from a well-shuffled deck of fifty-two cards (no jokers).

15. What is the experiment?
16. What is the sample space?

In Exercises 17–28, (a) find the probability and (b) the odds of drawing the given cards. Also, (c) use the Law of Large Numbers to interpret both the probability and the odds. (You might want to review the makeup of a deck of cards in Section 3.1.)

17. a black card
18. a heart
19. a queen
20. a 2 of clubs
21. a queen of spades
22. a club
23. a card below a 5 (count an ace as high)
24. a card below a 9 (count an ace as high)
25. a card above a 4 (count an ace as high)
26. a card above an 8 (count an ace as high)
27. a face card
28. a picture card

Age (years)	0–4	5–19	20–44	45–64	65–84	85+	total
Male	10,748	31,549	53,060	38,103	14,601	1,864	149,925
Female	10,258	30,085	51,432	39,955	18,547	3,858	154,135

FIGURE 3.6   2008 U.S. population, in thousands, by age and gender. *Source:* 2008 Census, U.S. Bureau of the Census.

Age	White	Black	American Indian, Eskimo, Aleut	Asian, Pacific Islander
0–4	15,041	2,907	222	1,042
5–19	47,556	9,445	709	3,036
20–44	79,593	14,047	1,005	5,165
45–64	60,139	8,034	497	2,911
65–84	27,426	2,826	166	987
85+	4,467	362	29	110

FIGURE 3.7   2005 U.S. population (projected), in thousands, by age and race. *Source:* Annual Population Estimates by Age Group and Sex, U.S. Bureau of the Census

In Exercises 29–38, (a) find the probability and (b) the odds of winning the given bet in roulette. Also, (c) use the Law of Large Numbers to interpret both the probability and the odds. (You might want to review the description of the game in Section 3.1.)

29. the single-number bet
30. the two-number bet
31. the three-number bet
32. the four-number bet
33. the five-number bet
34. the six-number bet
35. the twelve-number bet
36. the low-number bet
37. the even-number bet
38. the red-number bet

39. Use the information in Figure 3.6 from the U.S. Census Bureau to answer the following questions.
  a. Find the probability that in the year 2008, a U.S. resident was female.
  b. Find the probability that in the year 2008, a U.S. resident was male and between 20 and 44 years of age, inclusive.

40. Use the information in Figure 3.7 from the U.S. Census Bureau to answer the following questions.
  a. Find the probability that in the year 2005, a U.S. resident was American Indian, Eskimo, or Aleut.
  b. Find the probability that in the year 2005, a U.S. resident was Asian or Pacific Islander and between 20 and 44 years of age.

41. The dartboard in Figure 3.8 is composed of circles with radii 1 inch, 3 inches, and 5 inches.
  a. What is the probability that a dart hits the red region if the dart hits the target randomly?
  b. What is the probability that a dart hits the yellow region if the dart hits the target randomly?
  c. What is the probability that a dart hits the green region if the dart hits the target randomly?

d. Use the Law of Large Numbers to interpret the probabilities in parts (a), (b), and (c).

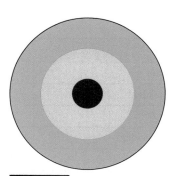

FIGURE 3.8   A dartboard for Exercise 41.

42. **a.** What is the probability of getting red on the spinner shown in Figure 3.9?
  b. What is the probability of getting blue?
  c. Use the Law of Large Numbers to interpret the probabilities in parts (a) and (b).

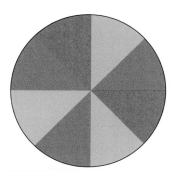

FIGURE 3.9   All sectors are equal in size and shape.

43. Amtrack's *Empire Service* train starts in Albany, New York, at 6:00 A.M., Mondays through Fridays. It arrives at New York City at 8:25 A.M. If the train breaks down, find the probability that it breaks down in the first half hour of its run. Assume that the train breaks down at random times.

44. Amtrack's *Downeaster* train leaves Biddeford, Maine, at 6:42 A.M., Mondays through Fridays. It arrives at Boston at 8:50 A.M. If the train breaks down, find the probability that it breaks down in the last 45 minutes of its run. Assume that the train breaks down at random times.

45. If $p(E) = \frac{1}{5}$, find $o(E)$.   46. If $p(E) = \frac{8}{9}$, find $o(E)$.

47. If $o(E) = 3:2$, find $p(E)$.   48. If $o(E) = 4:7$, find $p(E)$.

49. If $p(E) = \frac{a}{b}$, find $o(E)$.

In Exercises 50–52, find (a) the probability and (b) the odds of winning the following bets in roulette. In finding the odds, use the formula developed in Exercise 49. Also, (c) use the Law of Large Numbers to interpret both the probability and the odds.

50. the high-number bet   51. the odd-number bet

52. the black-number bet

53. In June 2009, sportsbook.com gave odds on who would win the 2009 World Series. They gave the New York Yankees 5:2 odds and the New York Mets 7:1 odds.

a. Are these house odds or true odds? Why?

b. Convert the odds to probabilities.

c. Who did they think was more likely to win the World Series: the Yankees or the Mets? Why?

54. In June 2009, linesmaker.com gave odds on who would win the 2009 World Series. They gave the New York Yankees 9:2 odds and the New York Mets 15:1 odds.

a. Are these house odds or true odds? Why?

b. Convert the odds to probabilities.

c. Who did they think was more likely to win the World Series: the Yankees or the Mets? Why?

d. Use the information in Exercise 53 to determine which source thought it was more probable that the Yankees would win the World Series: sportsbook.com or linesmaker.com.

In Exercises 55–60, use Figure 3.10.

55. a. Of the specific causes listed, which is the most likely cause of death in one year?

b. Which is the most likely cause of death in a lifetime? Justify your answers.

56. a. Of the causes listed, which is the least likely cause of death in one year?

Type of Accident or Injury	Odds of Dying in 1 Year	Lifetime Odds of Dying
All transportation accidents	1:6,121	1:79
Pedestrian transportation accident	1:48,816	1:627
Bicyclist transportation accident	1:319,817	1:4,111
Motorcyclist transportation accident	1:67,588	1:869
Car transportation accident	1:20,331	1:261
Airplane and space transportation accident	1:502,554	1:6,460
All nontransportation accidents	1:4,274	1:55
Falling	1:15,085	1:194
Drowning	1:82,777	1:1,064
Fire	1:92,745	1:1,192
Venomous animals and poisonous plants	1:2,823,877	1:36,297
Lightning	1:6,177,230	1:79,399
Earthquake	1:8,013,704	1:103,004
Storm	1:339,253	1:4,361
Intentional self-harm	1:9,085	1:117
Assault	1:16,360	1:210
Legal execution	1:5,490,872	1:70,577
War	1:10,981,743	1:141,154
Complications of medical care	1:111,763	1:1,437

FIGURE 3.10 The odds of dying due to an accident or injury in the United States in 2005. *Source: The odds of dying from . . . , 2009 edition, www.nsc.org.*

b. Which is the least likely cause of death in a lifetime? Justify your answers.

57. a. What is the probability that a person will die of a car transportation accident in one year?

b. What is the probability that a person will die of an airplane transportation accident in one year?

Write your answers as fractions.

58. a. What is the probability that a person will die from lightning in a lifetime?

b. What is the probability that a person will die from an earthquake in a lifetime?

Write your answers as fractions.

59. a. Of all of the specific forms of transportation accidents, which has the highest probability of causing death in one year? What is that probability?

b. Which has the lowest probability of causing death in one year? What is that probability?

Write your answers as fractions.

60. a. Of all of the specific forms of nontransportation accidents, which has the highest probability of causing death in a lifetime? What is that probability?

b. Which has the lowest probability of causing death in a lifetime? What is that probability?

Write your answers as fractions.

61. A family has two children. Using b to stand for boy and g for girl in ordered pairs, give each of the following.

a. the sample space

b. the event E that the family has exactly one daughter

c. the event F that the family has at least one daughter

d. the event G that the family has two daughters

| **e.** $p(E)$ | **f.** $p(F)$ | **g.** $p(G)$ |
| **h.** $o(E)$ | **i.** $o(F)$ | **j.** $o(G)$ |

(Assume that boys and girls are equally likely.)

62. Two coins are tossed. Using ordered pairs, give the following.

a. the sample space

b. the event E that exactly one is heads

c. the event F that at least one is heads

d. the event G that two are heads

| **e.** $p(E)$ | **f.** $p(F)$ | **g.** $p(G)$ |
| **h.** $o(E)$ | **i.** $o(F)$ | **j.** $o(G)$ |

63. A family has three children. Using b to stand for boy and g for girl and using ordered triples such as (b, b, g), give the following.

a. the sample space

b. the event E that the family has exactly two daughters

c. the event F that the family has at least two daughters

d. the event G that the family has three daughters

| **e.** $p(E)$ | **f.** $p(F)$ | **g.** $p(G)$ |
| **h.** $o(E)$ | **i.** $o(F)$ | **j.** $o(G)$ |

(Assume that boys and girls are equally likely.)

64. Three coins are tossed. Using ordered triples, give the following.

a. the sample space

b. the event E that exactly two are heads

c. the event F that at least two are heads

d. the event G that all three are heads

| **e.** $p(E)$ | **f.** $p(F)$ | **g.** $p(G)$ |
| **h.** $o(E)$ | **i.** $o(F)$ | **j.** $o(G)$ |

65. A couple plans on having two children.

a. Find the probability of having two girls.

b. Find the probability of having one girl and one boy.

c. Find the probability of having two boys.

d. Which is more likely: having two children of the same sex or two of different sexes? Why?

(Assume that boys and girls are equally likely.)

66. Two coins are tossed.

a. Find the probability that both are heads.

b. Find the probability that one is heads and one is tails.

c. Find the probability that both are tails.

d. Which is more likely: that the two coins match or that they don't match? Why?

67. A couple plans on having three children. Which is more likely: having three children of the same sex or of different sexes? Why? (Assume that boys and girls are equally likely.)

68. Three coins are tossed. Which is more likely: that the three coins match or that they don't match? Why?

69. A pair of dice is rolled. Using ordered pairs, give the following.

a. the sample space

HINT: S has 36 elements, one of which is (1, 1).

b. the event E that the sum is 7

c. the event F that the sum is 11

d. the event G that the roll produces doubles

| **e.** $p(E)$ | **f.** $p(F)$ | **g.** $p(G)$ |
| **h.** $o(E)$ | **i.** $o(F)$ | **j.** $o(G)$ |

70. Mendel found that snapdragons have no color dominance; a snapdragon with one red gene and one white gene will have pink flowers. If a pure-red snapdragon is crossed with a pure-white one, find the probability of the following.

a. a red offspring   **b.** a white offspring

c. a pink offspring

71. If two pink snapdragons are crossed (see Exercise 70), find the probability of the following.

a. a red offspring   **b.** a white offspring

c. a pink offspring

72. One parent is a cystic fibrosis carrier, and the other has no cystic fibrosis gene. Find the probability of each of the following.

a. The child would have cystic fibrosis.

b. The child would be a carrier.

c. The child would not have cystic fibrosis and not be a carrier.

d. The child would be healthy (i.e., free of symptoms).

73. If carrier-detection tests show that two prospective parents have sickle-cell trait (and are therefore carriers), find the probability of each of the following.

a. Their child would have sickle-cell anemia.

b. Their child would have sickle-cell trait.

c. Their child would be healthy (i.e., free of symptoms).

74. If carrier-detection tests show that one prospective parent is a carrier of sickle-cell anemia and the other has no sickle-cell gene, find the probability of each of the following.

a. The child would have sickle-cell anemia.

b. The child would have sickle-cell trait.

c. The child would be healthy (i.e., free of symptoms).

75. If carrier-detection tests show that one prospective parent is a carrier of Tay-Sachs and the other has no Tay-Sachs gene, find the probability of each of the following.

a. The child would have the disease.

b. The child would be a carrier.

c. The child would be healthy (i.e., free of symptoms).

76. If carrier-detection tests show that both prospective parents are carriers of Tay-Sachs, find the probability of each of the following.

a. Their child would have the disease.

b. Their child would be a carrier.

c. Their child would be healthy (i.e., free of symptoms).

77. If a parent started to exhibit the symptoms of Huntington's disease after the birth of his or her child, find the probability of each of the following. (Assume that one parent carries a single gene for Huntington's disease and the other carries no such gene.)

a. The child would have the disease.

b. The child would be a carrier.

c. The child would be healthy (i.e., free of symptoms).

 Answer the following questions using complete sentences and your own words.

• CONCEPT QUESTIONS

78. Explain how you would find the theoretical probability of rolling an even number on a single die. Explain how you would find the relative frequency of rolling an even number on a single die.

79. Give five examples of events whose probabilities must be found empirically rather than theoretically.

80. Does the theoretical probability of an event remain unchanged from experiment to experiment? Why or why not? Does the relative frequency of an event remain unchanged from experiment to experiment? Why or why not?

81. Consider a "weighted die"—one that has a small weight in its interior. Such a weight would cause the face closest to the weight to come up less frequently and the face farthest from the weight to come up more frequently. Would the probabilities computed in Example 1 still be correct? Why or why not? Would the definition $p(E) = \frac{n(E)}{n(S)}$ still be appropriate? Why or why not?

82. Some dice have spots that are small indentations; other dice have spots that are filled with the same material but of a different color. Which of these two types of dice is not fair? Why? What would be the most likely outcome of rolling this type of die? Why?

HINT: 1 and 6 are on opposite faces, as are 2 and 5, and 3 and 4.

83. In the United States, 52% of the babies are boys, and 48% are girls. Do these percentages contradict an assumption that boys and girls are equally likely? Why?

84. Compare and contrast theoretical probability and relative frequency. Be sure that you discuss both the similarities and differences between the two.

85. Compare and contrast probability and odds. Be sure that you discuss both the similarities and differences between the two.

• HISTORY QUESTIONS

86. What prompted Dr. Nancy Wexler's interest in Huntington's disease?

87. What resulted from Dr. Nancy Wexler's interest in Huntington's disease?

• PROJECTS

88. **a.** In your opinion, what is the probability that the last digit of a phone number is odd? Justify your answer.

b. Randomly choose one page from the residential section of your local phone book. Count how many phone numbers on that page have an odd last digit and how many have an even last digit. Then compute (number of phone numbers with odd last digits)/(total number of phone numbers).

c. Is your answer to part (b) a theoretical probability or a relative frequency? Justify your answer.

d. How do your answers to parts (a) and (b) compare? Are they exactly the same, approximately the same, or dissimilar? Discuss this comparison, taking into account the ideas of probability theory.

89. **a.** Flip a coin ten times, and compute the relative frequency of heads.

b. Repeat the experiment described in part (a) nine more times. Each time, compute the relative frequency of heads. After finishing parts (a) and (b), you will have flipped a coin 100 times, and you will have computed ten different relative frequencies.

c. Discuss how your ten relative frequencies compare with each other and with the theoretical probability heads. In your discussion, use the ideas discussed under the heading "Relative Frequency versus Probability" in this section. Do not plagiarize; use your own words.

d. Combine the results of parts (a) and (b), and find the relative frequency of heads for all 100 coin tosses. Discuss how this relative frequency compares with those found in parts (a) and (b) and with the theoretic probability of heads. Be certain to incorporate the Law of Large Numbers into your discussion.

90. a. Roll a single die twelve times, and compute the relative frequency with which you rolled a number below a 3.

b. Repeat the experiment described in part (a) seven more times. Each time, compute the relative frequency with which you rolled a number below a 3. After finishing parts (a) and (b), you will have rolled a die ninety-six times, and you will have computed eight different relative frequencies.

c. Discuss how your eight relative frequencies compare with each other and with the theoretical probability rolling a number below a 3. In your discussion, use the ideas discussed under the heading "Relative Frequency versus Probability" in this section. Do not plagiarize; use your own words.

d. Combine the results of parts (a) and (b), and find the relative frequency of rolling a number below a 3 for

all ninety-six die rolls. Discuss how this relative frequency compares with those found in parts (a) and (b) and with the theoretic probability of rolling a number less than a 3. Be certain to incorporate the Law of Large Numbers into your discussion.

91. Stand a penny upright on its edge on a smooth, hard, level surface. Then spin the penny on its edge. To do this, gently place a finger on the top of the penny. Then snap the penny with another finger (and immediately remove the holding finger) so that the penny spins rapidly before falling. Repeat this experiment fifty times, and compute the relative frequency of heads. Discuss whether or not the outcomes of this experiment are equally likely.

92. Stand a penny upright on its edge on a smooth, hard, level surface. Pound the surface with your hand so that the penny falls over. Repeat this experiment fifty times, and compute the relative frequency of heads. Discuss whether or not the outcomes of this experiment are equally likely.

WEB PROJECT

93. In the 1970s, there was a mass screening of blacks for sickle-cell anemia and a mass screening of Jews for Tay-Sachs disease. One of these was a successful program. One was not. Write a research paper on these two programs.

Some useful links for this web project are listed on the text web site:
www.cengage.com/math/johnson

3.3   Basic Rules of Probability

OBJECTIVES

- Understand what type of number a probability can be
- Learn about the relationships between probabilities, unions, and intersections

In this section, we will look at the basic rules of probability theory. The first three rules focus on why a probability can be a number like 5/7 or 32% but not a number like 9/5 or −27%. Knowing what type of number a probability can be is key to understanding what a probability means in real life. It also helps in checking your answers.

How Big or Small Can a Probability Be?

No event occurs less than 0% of the time. How could an event occur negative 15% of the time? Also, no event occurs more than 100% of the time. How could an event occur 125% of the time? Every event must occur between 0% and 100% of the time. For every event E,

$$0\% \leq p(E) \leq 100\%$$
$$0 \leq p(E) \leq 1 \qquad \textbf{converting from percents to decimals}$$

This means that *if you ever get a negative answer or an answer greater than 1 when you calculate a probability, go back and find your error.*

What type of event occurs 100% of the time? That is, what type of event has a probability of 1? Such an event must include *all possible* outcomes; if any possible outcome is left out, the event could not occur 100% of the time. An event that has a probability of 1 must be the sample space, because the sample space is the set of all possible outcomes. As we discussed earlier, such an event is called a *certain event.*

What type of event occurs 0% of the time? That is, what type of event has a probability of 0? Such an event must not include any possible outcome; if a possible outcome is included, the event would not have a probability of 0. An event with a probability of 0 must be the null set. As we discussed earlier, such an event is called an *impossible event.*

PROBABILITY RULES

Rule 1 $p(\varnothing) = 0$ The probability of the null set is 0.
Rule 2 $p(S) = 1$ The probability of the sample space is 1.
Rule 3 $0 \leq p(E) \leq 1$ Probabilities are between 0 and 1 (inclusive).

Probability Rules 1, 2, and 3 can be formally verified as follows:

Rule 1: $\quad p(\varnothing) = \dfrac{n(\varnothing)}{n(S)} = \dfrac{0}{n(S)} = 0$

Rule 2: $\quad p(S) = \dfrac{n(S)}{n(S)} = 1$

Rule 3: $\quad E$ is a subset of S; therefore,

$$0 \leq n(E) \leq n(S)$$
$$\frac{0}{n(S)} \leq \frac{n(E)}{n(S)} \leq \frac{n(S)}{n(S)} \qquad \textbf{dividing by } n(S)$$
$$0 \leq p(E) \leq 1$$

EXAMPLE 1

ILLUSTRATING RULES 1, 2, AND 3   A single die is rolled once. Find the probability of:

a. event E, "a 15 is rolled"
b. event F, "a number between 1 and 6 (inclusive) is rolled"
c. event G, "a 3 is rolled"

SOLUTION

The sample space is $S = \{1, 2, 3, 4, 5, 6\}$, and $n(S) = 6$.
a. It is *wrong* to say that $E = \{15\}$. Remember, we are talking about rolling a single die. Rolling a 15 is not one of the possible outcomes. That is, $15 \notin S$. There are no possible outcomes that result in rolling a 15, so the number of outcomes in event E is $n(E) = 0$. Thus,

$$p(E) = n(E)/n(S) = \frac{0}{6} = 0$$

The number 15 will be rolled 0% of the time. Event E is an impossible event.

Mathematically, we say that event E consists of no possible outcomes, so $E = \varnothing$. This agrees with rule 1, since

$$p(E) = p(\varnothing) = 0$$

b. $F = \{1, 2, 3, 4, 5, 6\}$, so $n(F) = 6$. Thus,

$$p(F) = n(F)/n(S) = \frac{6}{6} = 1$$

A number between 1 and 6 (inclusive) will be rolled 100% of the time. Event F is a certain event.

Mathematically, we say that event F consists of every possible outcome, so $F = S$. This agrees with rule 2, since

$$p(F) = p(S) = 1$$

c. $G = \{3\}$, so $n(G) = 1$. Thus

$$p(G) = n(G)/n(S) = 1/6$$

This agrees with rule 3, since

$$0 \le p(G) \le 1$$

Mutually Exclusive Events

Two events that cannot both occur at the same time are called **mutually exclusive.** In other words, E and F are mutually exclusive if and only if $E \cap F = \varnothing$.

EXAMPLE 2

DETERMINING WHETHER TWO EVENTS ARE MUTUALLY EXCLUSIVE A die is rolled. Let E be the event "an even number comes up," F the event "a number greater than 3 comes up," and G the event "an odd number comes up."

a. Are E and F mutually exclusive?
b. Are E and G mutually exclusive?

SOLUTION

a. $E = \{2, 4, 6\}$, $F = \{4, 5, 6\}$, and $E \cap F = \{4, 6\} \neq \varnothing$ (see Figure 3.11). Therefore, E and F are *not* mutually exclusive; the number that comes up could be *both* even *and* greater than 3. In particular, it could be 4 or 6.
b. $E = \{2, 4, 6\}$, $G = \{1, 3, 5\}$, and $E \cap G = \varnothing$. Therefore, E and G *are* mutually exclusive; the number that comes up could *not* be both even and odd.

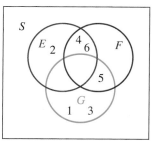

FIGURE 3.11 A Venn diagram for Example 2.

EXAMPLE 3

DETERMINING WHETHER TWO EVENTS ARE MUTUALLY EXCLUSIVE   Let M be the event "being a mother," F the event "being a father," and D the event "being a daughter."

a. Are events M and D mutually exclusive?
b. Are events M and F mutually exclusive?

SOLUTION

a. *M* and *D* are mutually exclusive if $M \cap D = \varnothing$. $M \cap D$ is the set of all people who are both mothers and daughters, and that set is not empty. A person can be a mother and a daughter at the same time. *M* and *D* are not mutually exclusive because being a mother does not exclude being a daughter.

b. *M* and *F* are mutually exclusive if $M \cap F = \varnothing$. $M \cap F$ is the set of all people who are both mothers and fathers, and that set is empty. A person cannot be a mother and a father at the same time. *M* and *F* are mutually exclusive because being a mother does exclude the possibility of being a father.

Pair-of-Dice Probabilities

To find probabilities involving the rolling of a pair of dice, we must first determine the sample space. The *sum* can be anything from 2 to 12, but we will not use {2, 3, 4, 5, 6, 7, 8, 9, 10, 11, 12} as the sample space because those outcomes are not equally likely. Compare a sum of 2 with a sum of 7. There is only one way in which the sum can be 2, and that is if each die shows a 1. There are many ways in which the sum can be a 7, including:

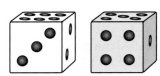

The event (3, 4).

- a 1 and a 6
- a 2 and a 5
- a 3 and a 4

What about a 4 and a 3? Is that the same as a 3 and a 4? They certainly *appear* to be the same. But if one die were blue and the other were white, it would be quite easy to tell them apart. So there are other ways in which the sum can be a 7:

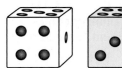

The event (4, 3).

- a 4 and a 3
- a 5 and a 2
- a 6 and a 1

Altogether, there are six ways in which the sum can be a 7. There is only one way in which the sum can be a 2. The outcomes "the sum is 2" and "the sum is 7" are not equally likely.

To have equally likely outcomes, we must use outcomes such as "rolling a 4 and a 3" and "rolling a 3 and a 4." We will use ordered pairs as a way of abbreviating these longer descriptions. So we will denote the outcome "rolling a 4 and a 3" with the ordered pair (4, 3), and we will denote the outcome "rolling a 3 and a 4" with the ordered pair (3, 4). Figure 3.12 lists all possible outcomes and the resulting sums. Notice that $n(S) = 6 \cdot 6 = 36$.

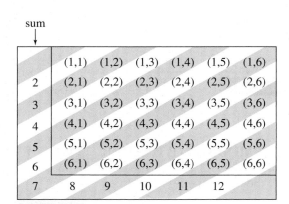

FIGURE 3.12 Outcomes of rolling two dice.

EXAMPLE **4**

SOLUTION

FINDING PAIR-OF-DICE PROBABILITIES   A pair of dice is rolled. Find the probability of each of the following events.

a. The sum is 7.   **b.** The sum is greater than 9.   **c.** The sum is even.

a. To find the probability that the sum is 7, let D be the event "the sum is 7." From Figure 3.12, $D = \{(1, 6), (2, 5), (3, 4), (4, 3), (5, 2), (6, 1)\}$, so $n(D) = 6$; therefore,

$$p(D) = \frac{n(D)}{n(S)} = \frac{6}{36} = \frac{1}{6}$$

This means that if we were to roll a pair of dice a large number of times, we should expect to get a sum of 7 approximately one-sixth of the time.

> Notice that $p(D) = \frac{1}{6}$ is between 0 and 1, as are all probabilities.

b. To find the probability that the sum is greater than 9, let E be the event "the sum is greater than 9."

$$E = \{(4, 6), (5, 5), (6, 4), (5, 6), (6, 5), (6, 6)\}, \text{ so } n(E) = 6; \text{ therefore,}$$

$$p(E) = \frac{n(E)}{n(S)} = \frac{6}{36} = \frac{1}{6}$$

This means that if we were to roll a pair of dice a large number of times, we should expect to get a sum greater than 9 approximately one-sixth of the time.

c. Let F be the event "the sum is even."

$$F = \{(1, 1), (1, 3), (2, 2), (3, 1), \dots, (6, 6)\}, \text{ so } n(F) = 18 \text{ (refer to Figure 3.12);}$$
therefore,

$$p(F) = \frac{n(F)}{n(S)} = \frac{18}{36} = \frac{1}{2}$$

This means that if we were to roll a pair of dice a large number of times, we should expect to get an even sum approximately half of the time.

A Roman painting on marble of the daughters of Niobe using knucklebones as dice. This painting was found in the ruins of Herculaneum, a city that was destroyed along with Pompeii by the eruption of Vesuvius.

© Scala/Art Resource, NY

EXAMPLE **5**

USING THE CARDINAL NUMBER FORMULAS TO FIND PROBABILITIES
A pair of dice is rolled. Use the Cardinal Number Formulas (where appropriate) and the results of Example 4 to find the probabilities of the following events.

a. the sum is not greater than 9.
b. the sum is greater than 9 and even.
c. the sum is greater than 9 or even.

SOLUTION

a. We could find the probability that the sum is not greater than 9 by counting, as in Example 4, but the counting would be rather excessive. It is easier to use one of the Cardinal Number Formulas from Chapter 2 on sets. The event "the sum is not greater than 9" is the complement of event E ("the sum is greater than 9") and can be expressed as E'.

$$n(E') = n(U) - n(E) \qquad \text{Complement Formula}$$
$$= n(S) - n(E) \qquad \text{"universal set" and "sample space" represent}$$
$$= 36 - 6 = 30 \qquad \text{the same idea.}$$
$$p(E') = \frac{n(E')}{n(S)} = \frac{30}{36} = \frac{5}{6}$$

This means that if we were to roll a pair of dice a large number of times, we should expect to get a sum that's not greater than 9 approximately five-sixths of the time.

b. The event "the sum is greater than 9 and even" can be expressed as the event $E \cap F$. $E \cap F = \{(4, 6), (5, 5), (6, 4), (6, 6)\}$, so $n(E \cap F) = 4$; therefore,

$$p(E \cap F) = \frac{n(E \cap F)}{n(S)} = \frac{4}{36} = \frac{1}{9}$$

This means that if we were to roll a pair of dice a large number of times, we should expect to get a sum that's both greater than 9 and even approximately one-ninth of the time.

c. Finding the probability that the sum is greater than 9 or even by counting would require an excessive amount of counting. It is easier to use one of the Cardinal Number Formulas from Chapter 2. The event "the sum is greater than 9 or even" can be expressed as the event $E \cup F$.

$$n(E \cup F) = n(E) + n(F) - n(E \cap F) \qquad \text{Union/Intersection Formula}$$
$$= 6 + 18 - 4 \qquad \text{from part (e), and Example 4}$$
$$= 20$$
$$p(E \cup F) = \frac{n(E \cup F)}{n(S)} = \frac{20}{36} = \frac{5}{9}$$

This means that if we were to roll a pair of dice a large number of times, we should expect to get a sum that's either greater than 9 or even approximately five-ninths of the time.

More Probability Rules

In part (b) of Example 4, we found that $p(E) = \frac{1}{6}$, and in part (a) of Example 5, we found that $p(E') = \frac{5}{6}$. Notice that $p(E) + p(E') = \frac{1}{6} + \frac{5}{6} = 1$. This should make sense to you. It just means that if E happens one-sixth of the time, then E' has to happen the other five-sixths of the time. This always happens—for any event E, $p(E) + p(E') = 1$.

As we will see in Exercise 79, the fact that $p(E) + p(E') = 1$ is closely related to the Cardinal Number Formula $n(E) + n(E') = n(S)$. The main difference is that one is expressed in the language of probability theory and the other is expressed in the language of set theory. In fact, the following three rules are all set theory rules (from Chapter 2) rephrased so that they use the language of probability theory rather than the language of set theory.

MORE PROBABILITY RULES

Rule 4 **The Union/Intersection Rule** $p(E \cup F)$
$\qquad\qquad = p(E) + p(F) - p(E \cap F)$

Rule 5 **The Mutually Exclusive Rule** $p(E \cup F) = p(E) + p(F)$, if E and F are mutually exclusive

Rule 6 **The Complement Rule** $p(E) + p(E') = 1$ or, equivalently, $p(E') = 1 - p(E)$

In Example 5, we used some Cardinal Number Formulas from Chapter 2 to avoid excessive counting. Some find it easier to use Probability Rules to calculate probabilities, rather than Cardinal Number Formulas.

EXAMPLE 6

USING RULES 4, 5, AND 6 Use a Probability Rule rather than a Cardinal Number Formula to find:

a. the probability that the sum is not greater than 9
b. the probability that the sum is greater than 9 or even

SOLUTION

We will use the results of Example 4.

a. $p(E') = 1 - p(E)$ **the Complement Rule**

$\qquad = 1 - \dfrac{1}{6}$ **from Example 4**

$\qquad = \dfrac{5}{6}$

b. $p(E \cup F) = p(E) + p(F) - p(E \cap F)$ **the Union/Intersection Rule**

$\qquad = \dfrac{1}{6} + \dfrac{1}{2} - \dfrac{1}{9}$ **from parts (b) and (c) of Example 4, and part (b) of Example 5**

$\qquad = \dfrac{3}{18} + \dfrac{9}{18} - \dfrac{2}{18} = \dfrac{5}{9}$ **getting a common denominator**

Probabilities and Venn Diagrams

Venn diagrams can be used to illustrate probabilities in the same way in which they are used in set theory. In this case, we label each region with its probability rather than its cardinal number.

EXAMPLE 7

USING VENN DIAGRAMS Zaptronics manufactures compact discs and their cases for several major labels. A recent sampling of Zaptronics' products has indicated that 5% have a defective case, 3% have a defective disc, and 7% have at least one of the two defects.

a. Find the probability that a Zaptronics product has both defects.
b. Draw a Venn diagram that shows the probabilities of each of the basic regions.

c. Use the Venn diagram to find the probability that a Zaptronics product has neither defect.

d. Find the probability that a Zaptronics product has neither defect, using probability rules rather than a Venn diagram.

SOLUTION

a. Let C be the event that the case is defective, and let D be the event that the disc is defective. We are given that $p(C) = 5\% = 0.05$, $p(D) = 3\% = 0.03$, and $p(C \cup D) = 7\% = 0.07$, and we are asked to find $p(C \cap D)$. To do this, substitute into Probability Rule 4.

$$p(C \cup D) = p(C) + p(D) - p(C \cap D) \quad \text{the Union/Intersection Rule}$$
$$0.07 = 0.05 + 0.03 - p(C \cap D) \quad \text{substituting}$$
$$0.07 = 0.08 - p(C \cap D) \quad \text{adding}$$
$$p(C \cap D) = 0.01 = 1\% \quad \text{solving}$$

This means that 1% of Zaptronics' products have a defective case *and* a defective disc.

b. The Venn diagram for this is shown in Figure 3.13. The 0.93 probability in the lower right corner is obtained by first finding the sum of the probabilities of the other three basic regions:

$$0.04 + 0.01 + 0.02 = 0.07$$

The rectangle itself corresponds to the sample space, and its probability is 1. Thus, the probability of the outer region, the region outside of the C and D circles, is

$$1 - 0.07 = 0.93$$

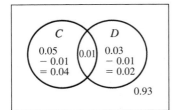

FIGURE 3.13

A Venn diagram for Example 6.

c. The only region that involves no defects is the outer region, the one whose probability is 0.93. This means that 93% of Zaptronics' products have neither defect.

d. If we are to not use the Venn diagram, we must first rephrase the event "a Zaptronics product has neither defect" as "a product does not have either defect." The "does not" part of this means *complement,* and the "either defect" means *either C or D*. This means that

- "has either defect" translates to $C \cup D$
- "does not have either defect" translates to $(C \cup D)'$

So we are asked to find $P((C \cup D)')$.

$$p((C \cup D)') = 1 - p(C \cup D) \quad \text{the Complement Rule}$$
$$= 1 - 0.07 \quad \text{substituting}$$
$$= 0.93 = 93\% \quad \text{subtracting}$$

This means that 93% of Zaptronics' products are defect-free.

3.3 EXERCISES

In Exercises 1–10, determine whether E and F are mutually exclusive. Write a sentence justifying your answer.

1. E is the event "being a doctor," and F is the event "being a woman."

2. E is the event "it is raining," and F is the event "it is sunny."

3. E is the event "being single," and F is the event "being married."

4. E is the event "having naturally blond hair," and F is the event "having naturally black hair."

5. E is the event "having brown hair," and F is the event "having gray hair."

6. E is the event "being a plumber," and F is the event "being a stamp collector."

7. E is the event "wearing boots," and F is the event "wearing sandals."

8. E is the event "wearing shoes," and F is the event "wearing socks."

9. If a die is rolled once, E is the event "getting a four," and F is the event "getting an odd number."

10. If a die is rolled once, E is the event "getting a four," and F is the event "getting an even number."

In Exercises 11–18, a card is dealt from a complete deck of fifty-two playing cards (no jokers). Use probability rules (when appropriate) to find the probability that the card is as stated. (Count an ace as high.)

11. a. a jack and red   **b.** a jack or red
    c. not a red jack

12. a. a jack and a heart   **b.** a jack or a heart
    c. not a jack of hearts

13. a. a 10 and a spade   **b.** a 10 or a spade
    c. not a 10 of spades

14. a. a 5 and black   **b.** a 5 or black
    c. not a black 5

15. a. under a 4
    b. above a 9
    c. both under a 4 and above a 9
    d. either under a 4 or above a 9

16. a. above a jack
    b. below a 3
    c. both above a jack and below a 3
    d. either above a jack or below a 3

17. a. above a 5
    b. below a 10
    c. both above a 5 and below a 10
    d. either above a 5 or below a 10

18. a. above a 7
    b. below a queen
    c. both above a 7 and below a queen
    d. either above a 7 or below a queen

In Exercises 19–26, use complements to find the probability that a card dealt from a full deck (no jokers) is as stated. (Count an ace as high.)

19. not a queen
20. not a 7
21. not a face card
22. not a heart
23. above a 3
24. below a queen
25. below a jack
26. above a 5

27. If $o(E) = 5:9$, find $o(E')$.
28. If $o(E) = 1:6$, find $o(E')$.
29. If $p(E) = \frac{2}{7}$, find $o(E)$ and $o(E')$.
30. If $p(E) = \frac{3}{8}$, find $o(E)$ and $o(E')$.
31. If $o(E) = a:b$, find $o(E')$.
32. If $p(E) = \frac{a}{b}$, find $o(E')$.

    HINT: Use Exercise 49 from Section 3.2 and Exercise 31 above.

In Exercises 33–38, use Exercise 32 to find the odds that a card dealt from a full deck (no jokers) is as stated.

33. not a king   **34.** not an 8
35. not a face card   **36.** not a club
37. above a 4   **38.** below a king

In Exercises 39–42, use the following information: To determine the effect their salespeople have on purchases, a department store polled 700 shoppers regarding whether or not they made a purchase and whether or not they were pleased with the service they received. Of those who made a purchase, 151 were happy with the service and 133 were not. Of those who made no purchase, 201 were happy with the service and 215 were not. Use probability rules (when appropriate) to find the probability of the event stated.

39. a. A shopper made a purchase.
    b. A shopper did not make a purchase.

40. a. A shopper was happy with the service received.
    b. A shopper was unhappy with the service received.

41. a. A shopper made a purchase and was happy with the service.
    b. A shopper made a purchase or was happy with the service.

42. a. A shopper made no purchase and was unhappy with the service.
    b. A shopper made no purchase or was unhappy with the service.

In Exercises 43–46, use the following information: A supermarket polled 1,000 customers regarding the size of their bill. The results are given in Figure 3.14.

Size of Bill	Number of Customers
below $20.00	208
$20.00–$39.99	112
$40.00–$59.99	183
$60.00–$79.99	177
$80.00–$99.99	198
$100.00 or above	122

FIGURE 3.14 Supermarket bills.

Use probability rules (when appropriate) to find the relative frequency with which a customer's bill is as stated.

43. a. less than $40.00
    b. $40.00 or more

44. a. less than $80.00
    b. $80.00 or more

45. a. between $40.00 and $79.99
    b. not between $40.00 and $79.99

46. a. between $20.00 and $79.99

b. not between $20.00 and $79.99

In Exercises 47–54, find the probability that the sum is as stated when a pair of dice is rolled.

47. a. 7   **b.** 9   **c.** 11

48. a. 2   **b.** 4   **c.** 6

49. a. 7 or 11   **b.** 7 or 11 or doubles

50. a. 8 or 10   **b.** 8 or 10 or doubles

51. a. odd and greater than 7

b. odd or greater than 7

52. a. even and less than 5

b. even or less than 5

53. a. even and doubles

b. even or doubles

54. a. odd and doubles

b. odd or doubles

In Exercises 55–58, use the following information: After examining their clients' records, Crashorama Auto Insurance calculated the probabilities in Figure 3.15.

	Miles/Year < 10,000	10,000 ≤ Miles/Year < 20,000	20,000 ≤ Miles/Year
Accident	0.05	0.10	0.3
No accident	0.20	0.15	0.2

FIGURE 3.15 Crashorama's Accident Incidence.

55. a. Find the probability that a client drives 20,000 miles/year or more or has an accident.

b. Find the probability that a client drives 20,000 miles/year or more and has an accident.

c. Find the probability that the client does not drive 20,000 miles/year or more and does not have an accident.

56. a. Find the probability that the client drives less than 10,000 miles/year and has an accident.

b. Find the probability that a client drives less than 10,000 miles/year or has an accident.

c. Find the probability that a client does not drive less than 10,000 miles per year and does not have an accident.

57. Are the probabilities in Figure 3.15 theoretical probabilities or relative frequencies? Why?

58. a. Find the probability that a client drives either less than 10,000 miles/year or 20,000 miles/year or more.

b. Find the probability that a client both has an accident and drives either less than 10,000 miles/year or 20,000 miles/year or more.

In Exercises 59–62, use the following information: The results of CNN's 2008 presidential election poll are given in Figure 3.16.

	Obama	McCain	Other/No Answer
Male Voters	23.0%	22.6%	1.4%
Female Voters	29.7%	22.8%	0.5%

FIGURE 3.16 Exit poll results, by gender. *Source:* CNN.

59. a. Find the probability that a polled voter voted for Obama or is male.

b. Find the probability that a polled voter voted for Obama and is male.

c. Find the probability that a polled voter didn't vote for Obama and is not male.

60. a. Find the probability that a polled voter voted for McCain or is female.

b. Find the probability that a polled voter voted for McCain and is female.

c. Find the probability that a polled voter didn't vote for McCain and is not female.

61. Are the probabilities in Figure 3.16 theoretical probabilities or relative frequencies? Why?

62. a. Find the probability that a polled voter didn't vote for Obama.

b. Find the probability that a polled voter is female.

63. Fried Foods hosted a group of twenty-five consumers at a tasting session. The consumers were asked to taste two new products and state whether or not they would be interested in buying them. Fifteen consumers said that they would be interested in buying Snackolas, twelve said that they would be interested in buying Chippers, and ten said that they would be interested in buying both.

a. What is the probability that one of the tasters is interested in buying at least one of the two new products?

b. What is the probability that one of the tasters is not interested in buying Snackolas but is interested in buying Chippers?

64. Ink Inc., a publishing firm, offers its 899 employees a cafeteria approach to benefits, in which employees can enroll in the benefit plan of their choice. Seven hundred thirteen employees have health insurance, 523 have dental insurance, and 489 have both health and dental insurance.

a. What is the probability that one of the employees has either health or dental insurance?

b. What is the probability that one of the employees has health insurance but not dental insurance?

65. ComCorp is considering offering city-wide Wi-Fi connections to the Web. They asked 3,000 customers about their use of the Web at home, and they received

1,451 responses. They found that 1,230 customers are currently connecting to the Web with a cable modem, 726 are interested in city-wide Wi-Fi, and 514 are interested in switching from a cable modem to city-wide Wi-Fi.

a. What is the probability that one of the respondents is interested in city-wide Wi-Fi and is not currently using a cable modem?

b. What is the probability that one of the respondents is interested in city-wide Wi-Fi and is currently using a cable modem?

66. The Video Emporium rents DVDs and blue-rays only. They surveyed their 1,167 rental receipts for the last two weeks. Eight hundred thirty-two customers rented DVDs, and 692 rented blue-rays.

a. What is the probability that a customer rents DVDs only?

b. What is the probability that a customer rents DVDs and blue-rays?

67. *Termiyak Magazine* conducted a poll of its readers, asking them about their telephones. Six hundred ninety-six readers responded. Five hundred seventy-two said they have a cell phone. Six hundred twelve said that they have a land line (a traditional, nonwireless telephone) at their home. Everyone has either a cell phone or a land line.

a. What is the probability that one of the readers has only a cell phone?

b. What is the probability that one of the readers has only a land line?

68. Which of the probabilities in Exercises 63–67 are theoretical probabilities and which are relative frequencies? Why?

69. Use probability rules to find the probability that a child will either have Tay-Sachs disease or be a carrier if (a) each parent is a Tay-Sachs carrier, (b) one parent is a Tay-Sachs carrier and the other parent has no Tay-Sachs gene, (c) one parent has Tay-Sachs and the other parent has no Tay-Sachs gene.

70. Use probability rules to find the probability that a child will have either sickle-cell anemia or sickle-cell trait if (a) each parent has sickle-cell trait, (b) one parent has sickle-cell trait and the other parent has no sickle-cell gene, (c) one parent has sickle-cell anemia and the other parent has no sickle-cell gene.

71. Use probability rules to find the probability that a child will neither have Tay-Sachs disease nor be a carrier if (a) each parent is a Tay-Sachs carrier, (b) one parent is a Tay-Sachs carrier and the other parent has no Tay-Sachs gene, (c) one parent has Tay-Sachs and the other parent has no Tay-Sachs gene.

72. Use probability rules to find the probability that a child will have neither sickle-cell anemia nor sickle-cell trait if (a) each parent has sickle-cell trait, (b) one parent has

sickle-cell trait and the other parent has no sickle-cell gene, (c) one parent has sickle-cell anemia and the other parent has no sickle-cell gene.

73. Mary is taking two courses, photography and economics. Student records indicate that the probability of passing photography is 0.75, that of failing economics is 0.65, and that of passing at least one of the two courses is 0.85. Find the probability of the following.

a. Mary will pass economics.

b. Mary will pass both courses.

c. Mary will fail both courses.

d. Mary will pass exactly one course.

74. Alex is taking two courses; algebra and U.S. history. Student records indicate that the probability of passing algebra is 0.35, that of failing U.S. history is 0.35, and that of passing at least one of the two courses is 0.80. Find the probability of the following.

a. Alex will pass history.

b. Alex will pass both courses.

c. Alex will fail both courses.

d. Alex will pass exactly one course.

75. Of all the flashlights in a large shipment, 15% have a defective bulb, 10% have a defective battery, and 5% have both defects. If you purchase one of the flashlights in this shipment, find the probability that it has the following.

a. a defective bulb or a defective battery

b. a good bulb or a good battery

c. a good bulb and a good battery

76. Of all the DVDs in a large shipment, 20% have a defective disc, 15% have a defective case, and 10% have both defects. If you purchase one of the DVDs in this shipment, find the probability that it has the following.

a. a defective disc or a defective case

b. a good disc or a good case

c. a good disc and a good case

77. Verify the Union/Intersection Rule.

HINT: Divide the Cardinal Number Formula for the Union of Sets from Section 2.1 by $n(S)$.

78. Verify the Mutually Exclusive Rule.

HINT: Start with the Union/Intersection Rule. Then use the fact that E and F are mutually exclusive.

79. Verify the Complement Rule.

HINT: Are E and E' mutually exclusive?

 Answer the following questions using complete sentences and your own words.

• CONCEPT QUESTIONS

80. What is the complement of a certain event? Justify your answer.

81. Compare and contrast mutually exclusive events and impossible events. Be sure to discuss both the similarities and the differences between these two concepts.

82. Explain why it is necessary to subtract $p(E \cap F)$ in Probability Rule 4. In other words, explain why Probability Rule 5 is not true for all events.

FRACTIONS ON A GRAPHING CALCULATOR

Some graphing calculators—including the TI-83, TI-84, and Casio—will add, subtract, multiply, and divide fractions and will give answers in reduced fractional form.

Reducing Fractions

The fraction 42/70 reduces to 3/5. To do this on your calculator, you must make your screen read "42/70 ▶ Frac" ("42 ⌐70" on a Casio). The way that you do this varies.

TI-83/84	• Type 42 ÷ 70, but do not press ENTER. This causes "42/70" to appear on the screen.
	• Press the MATH button.
	• Highlight option 1, "▶Frac." (Option 1 is automatically highlighted. If we were selecting a different option, we would use the ▲ and ▼ buttons to highlight it.)
	• Press ENTER. This causes "42/70▶Frac" to appear on the screen.
	• Press ENTER. This causes 3/5 to appear on the screen.
CASIO	• Type 42 a⅟ₓ 70.
	• Type EXE, and your display will read "3⌐5," which should be interpreted as "3/5."

EXAMPLE **8**

Use your calculator to compute

$$\frac{1}{6} + \frac{1}{2} - \frac{1}{9}$$

and give your answer in reduced fractional form.

SOLUTION

On a TI, make your screen read "1/6 + 1/2 − 1/9 ▶Frac" by typing

1 ÷ 6 + 1 ÷ 2 − 1 ÷ 9

and then inserting the "▶Frac" command, as described above. Once you press ENTER, the screen will read "5/9."

On a Casio, make your screen read "1⌄6 + 1⌄2 − 1⌄9" by typing

$$1 \;\boxed{a\tfrac{b}{c}}\; 6 + 1 \;\boxed{a\tfrac{b}{c}}\; 2 - 1 \;\boxed{a\tfrac{b}{c}}\; 9$$

Once you press EXE, the screen will read "5⌄9," which should be interpreted as "5/9."

EXERCISES

In Exercises 83–84, reduce the given fractions to lowest terms, both (a) by hand and (b) with a calculator. Check your work by comparing the two answers.

83. $\dfrac{18}{33}$ **84.** $-\dfrac{42}{72}$

In Exercises 85–90, perform the indicated operations, and reduce the answers to lowest terms, both (a) by hand and (b) with a calculator. Check your work by comparing the two answers.

85. $\dfrac{6}{15} \cdot \dfrac{10}{21}$

86. $\dfrac{6}{15} \div \dfrac{10}{21}$

87. $\dfrac{6}{15} + \dfrac{10}{21}$

88. $\dfrac{6}{15} - \dfrac{10}{21}$

89. $\dfrac{7}{6} - \dfrac{5}{7} + \dfrac{9}{14}$

90. $\dfrac{-8}{5} - \left(\dfrac{-3}{28} + \dfrac{5}{21}\right)$

91. How could you use your calculator to get decimal answers to the above exercises rather than fractional answers?

3.4 Combinatorics and Probability

OBJECTIVES

- Apply the concepts of permutations and combinations to probability calculations
- Use probabilities to analyze games of chance such as the lottery.

You would probably guess that it's pretty unlikely that two or more people in your math class share a birthday. It turns out, however, that it's surprisingly likely. In this section, we'll look at why that is.

Finding a probability involves finding the number of outcomes in an event and the number of outcomes in the sample space. So far, we have used probability rules as an alternative to excessive counting. Another alternative is combinatorics—that is, permutations, combinations, and the Fundamental Principle of Counting, as covered in Sections 2.3 and 2.4. The flowchart used in Chapter 2 is summarized below.

WHICH COMBINATORICS METHOD?

1. If the selection is done with replacement, use the Fundamental Principle of Counting.
2. If the selection is done without replacement and there is only one category, use:
   a. permutations if the order of selection does matter:

   $$_nP_r = \frac{n!}{(n-r)!}$$

   b. combinations if the order of selection does not matter:

   $$_nC_r = \frac{n!}{(n-r)!r!}$$

3. If there is more than one category, use the Fundamental Principle of Counting with one box per category. Use (2) above to determine the numbers that go in the boxes.

EXAMPLE 1

A BIRTHDAY PROBABILITY A group of three people is selected at random. What is the probability that at least two of them will have the same birthday?

SOLUTION

We will assume that all birthdays are equally likely, and for the sake of simplicity, we will ignore leap-year day (February 29).

- *The experiment* is to ask three people their birthdays. One possible outcome is (May 1, May 3, August 23).
- *The sample space S* is the set of all possible lists of three birthdays.
- *Finding $n(S)$* by counting the elements in S is impractical, so we will use combinatorics as described in "Which Combinatorics Method" above.
  - The selected items are birthdays. They are selected with replacement because people can share the same birthday. This tells us to use *the Fundamental Principle of Counting:*

  first person's birthday   second person's birthday   third person's birthday

  □   □   □

  - Each birthday can be any one of the 365 days in a year
  - $n(S) = \boxed{365} \cdot \boxed{365} \cdot \boxed{365} = 365^3$
- *The event E* is the set of all possible lists of three birthdays in which some birthdays are the same. It is difficult to compute $n(E)$ directly; instead, we will compute $n(E')$ and use the Complement Rule. E' is the set of all possible lists of three birthdays in which no birthdays are the same.
- *Finding $n(E')$* using "Which Combinatorics Method":
  - The birthdays are selected without replacement, because no birthdays are the same.
  - There is only one category: birthdays.

- The order of selection does matter—(May 1, May 3, August 23) is a different list from (May 3, August 23, May 1)—so use permutations
- $n(E') = {}_{365}P_3$
- *Finding $p(E')$ and $p(E)$:*

$$p(E') = \frac{n(E')}{n(S)} = \frac{{}_{365}P_3}{365^3} \qquad \text{using the above results}$$

$$p(E) = 1 - p(E') \qquad \text{the Complement Rule}$$

$$= 1 - \frac{{}_{365}P_3}{365^3} \qquad \text{using the above result}$$

$$= 0.008204 \dots$$

$$\approx 0.8\% \qquad \text{moving the decimal point left two places}$$

This result is not at all surprising. It means that two or more people in a group of three share a birthday slightly less than 1% of the time. In other words, this situation is extremely unlikely. However, we will see in Exercise 1 that it is quite likely that two or more people in a group of thirty share a birthday.

Lotteries

We're going to look at one of the most common forms of gambling in the United States: the lottery. Many people play the lottery, figuring that "someone's got to win—why not me?" Is this is a reasonable approach to lottery games?

EXAMPLE **2**

WINNING A LOTTERY Arizona, Connecticut, Missouri, and Tennessee operate 6/44 lotteries. In this game, a gambler selects any six of the numbers from 1 to 44. If his or her six selections match the six winning numbers, the player wins first prize. If his or her selections include five of the winning numbers, the player wins second prize. Find the probability of:

a. the event E, winning first prize
b. the event F, winning second prize

SOLUTION

a. *The sample space S is the set of all possible lottery tickets. That is, it is the set of all possible choices of six numbers selected from the numbers 1 through 44.*

- *Finding $n(S)$ by counting the elements in S is impractical, so we will use combinatorics as described in "Which Combinatorics Method" above.*
  - The selected is done without replacement, because you can't select the same lottery number twice.
  - Order does not matter, because the gambler can choose the six numbers in any order. Use combinations.
  - $n(S) = {}_{44}C_6$
- *Finding $n(E)$ is easy, because there is only one winning combination of numbers:*

$$n(E) = 1$$

- *Finding $p(E)$:*

$$p(E) = \frac{n(E)}{n(S)} = \frac{1}{{}_{44}C_6} = \frac{1}{7{,}059{,}052} \qquad \text{using the above results}$$

This means that only one out of approximately *seven million* combinations is the first-prize-winning combinations. This is an incredibly unlikely event. It is less probable than dying in one year by being hit by lightning (see Figure 3.10 on page 154.). You probably don't worry about dying from lightning, because it essentially doesn't happen. Sure, it happens to somebody sometime, but not to anyone you've ever known. Winning the lottery is similar. The next time that you get tempted to buy a lottery ticket, remember that you should feel less certain of winning first prize that you should of dying from lightning.

b. To find the probability of event F, winning second prize, we need to find $n(F)$. We already found $n(S)$ in part a:

$$n(S) = {}_{44}C_6$$

- *Finding $n(F)$:*
  - To win second prize, we must select five winning numbers and one losing number. This tells us to use *the Fundamental Principle of Counting:*

    winning numbers   losing numbers

    □   □

  - We will use combinations in each box, for the same reasons that we used combinations to find $n(S)$.
  - The state selects six winning numbers, and the second-prize-winner must select five of them, so there are ${}_6C_5$ ways of selecting the five winning numbers.
  - There are $44 - 6 = 38$ losing numbers, of which the player selects 1, so there are ${}_{38}C_{51}$ ways of selecting one losing number.
  - $n(F) = \boxed{{}_6C_5} \cdot \boxed{{}_{38}C_1}$ **using the above results**

- *Finding $p(F)$:*

  - $$p(F) = \frac{n(F)}{n(S)} = \frac{{}_6C_5 \cdot {}_{38}C_1}{{}_{44}C_6} = 0.000032\ldots$$

How do we make sense of a decimal with so many zeros in front of it? One way is to round it off at the first nonzero digit and convert the result to a fraction:

$$p(F) = 0.00003\ldots \quad \textbf{rounding to the first nonzero digit}$$
$$= 3/100{,}000 \quad \textbf{converting to a fraction}$$

This means that if you buy a lot of 6/44 lottery tickets, you will win second prize approximately three times out of every 100,000 times you play. It also means that in any given game, there are about three second-prize ticket for every 100,000 ticket purchases.

To check your work in Example 2 part (b), notice that in the event, there is a distinction between two categories (winning numbers and losing numbers); in the sample space, there is no such distinction. Thus, the numerator of

$$p(E) = \frac{{}_6C_5 \cdot {}_{38}C_1}{{}_{44}C_6}$$

has two parts (one for each category), and the denominator has one part. Also, the numbers in front of the Cs add correctly (6 winning numbers + 38 losing numbers = 44 total numbers to choose from), and the numbers after the Cs add correctly (5 winning numbers + 1 losing number = 6 total numbers to select).

HOW TO WRITE A PROBABILITY

A probability can be written as a fraction, a percentage, or a decimal. Our goal is to write a probability in a form that is intuitively understandable.

- If $n(E)$ and $n(S)$ are both small numbers, such as $n(E) = 3$ and $n(S) = 12$, then write $p(E)$ as a **reduced fraction:**

$$p(E) = \frac{n(E)}{n(S)} = \frac{3}{12} = \frac{1}{4}$$

It's intuitively understandable to say that something happens about one out of every four times.

- If $n(E)$ and $n(S)$ are not small numbers and $p(E)$ is a decimal with at most two zeros after the decimal point, then write $p(E)$ as a **percentage,** as in Example 1:

$$p(E) = 0.008204 \ldots \qquad \textbf{two zeros after the decimal point}$$
$$\approx 0.8\% \qquad \textbf{writing as a percentage}$$

A probability of $0.008204 \ldots$ is not easily understandable. But it is understandable to say that something happens slightly less than 1% of the time—that is, fewer than one out of every one hundred times.

- If $n(E)$ and $n(S)$ are not small numbers and $p(E)$ is a decimal with three or more zeros after the decimal point, then round off at the first nonzero digit and convert the result to a fraction, as in Example 2:

$$p(F) = \frac{n(F)}{n(S)} = \frac{6 \cdot 38}{7,059,059} = 0.000032298 \ldots$$

$$\approx 0.00003 \qquad \textbf{rounding at the first nonzero digit}$$
$$= 3/100,000 \qquad \textbf{converting to a fraction}$$

A probability of $\frac{6 \cdot 38}{7,059,059}$ or $0.000032298 \ldots$ or 0.003% is not so easily understood. It's more understandable to say that something happens about three times out of every 100,000 times.

EXAMPLE 3

WINNING POWERBALL Powerball is played in thirty states, the District of Columbia, and the U.S. Virgin Islands. It involves selecting any five of the numbers from 1 to 59, plus a "powerball number," which is any one of the numbers from 1 to 39. Find the probability of winning first prize.

SOLUTION

The sample space S is the set of all possible lottery tickets. That is, it is the set of all possible choices of five numbers from 1 to 59 plus a powerball number from 1 to 39.

- *Finding $n(S)$:*
  - The numbers are selected without replacement, because the gambler can't select the same number twice.

- There are two categories—regular numbers and the powerball number—so we use the Fundamental Principle of Counting with two boxes.

  regular   powerball

  □   □

- In each box, we will use combinations, because the player can choose the numbers in any order.
- The state selects five of fifty-nine regular numbers, so there are $_{59}C_5$ possible selections.
- The state selects one of thirty-nine powerball numbers, so there are $_{39}C_1$ possible powerball selections.
- The number of different lottery tickets is

$$n(S) = \boxed{_{59}C_5} \cdot \boxed{_{39}C_1} \quad \text{using the above results}$$

- *Finding $n(E)$*: Only one of these combinations is the winning lottery ticket, so $n(E) = 1$.
- *Finding $p(E)$*:

$$p(E) = \frac{n(E)}{n(S)} = \frac{1}{_{59}C_5 \cdot _{39}C_1} \quad \text{using the above results}$$

$= 0.00000000512...$	**more than three zeros after the decimal point**
≈ 0.000000005	**rounding to the first nonzero digit**
$= 5/1{,}000{,}000{,}000$	**converting to a fraction**
$= 1/200{,}000{,}000$	**reducing**

This means that if you play powerball a lot, you will win first prize about once every 200 million games. It also means that in any given game, there is about one first prize ticket for every 200 million ticket purchases.

The game of keno is a casino version of the lottery. In this game, the casino has a container filled with balls numbered from 1 to 80. The player buys a keno ticket, with which he or she selects anywhere from 1 to 15 (usually 6, 8, 9, or 10) of those 80 numbers; the player's selections are called "spots." The casino chooses 20 winning numbers, using a mechanical device to ensure a fair game. If a sufficient number of the player's spots are winning numbers, the player receives an appropriate payoff.

EXAMPLE **4**

WINNING AT KENO   In the game of keno, if eight spots are marked, the player wins if five or more of his or her spots are selected. Find the probability of having five winning spots.

SOLUTION

The sample space S is the set of all ways in which a player can select eight numbers from the eighty numbers in the game.

- *Finding $n(S)$*:
  - Selection is done without replacement.
  - Order doesn't matter, so use combinations.
  - $n(S) = {}_{80}C_8$
- *The event E is the set of all ways in which an eight-spot player can select five winning numbers and three losing numbers.*
- *Finding $n(E)$*:
  - There are two categories—winning numbers and losing numbers—so use the Fundamental Principle of Counting.

TOPIC X THE BUSINESS OF GAMBLING: *PROBABILITIES IN THE REAL WORLD*

It used to be that legal commercial gambling was not common. Nevada made casino gambling legal in 1931, and for more than thirty years, it was the only place in the United States that had legal commercial gambling. Then in 1964, New Hampshire instituted the first lottery in the United States since 1894. In 1978, New Jersey became the second state to legalize casino gambling. Now, state-sponsored lotteries are common, Native American tribes have casinos in more than half the states, and some states have casinos in selected cities or on riverboats.

It is very likely that you will be exposed to gambling if you have not been already. You should approach gambling with an educated perspective. If you are considering gambling, know what you are up against. The casinos all use probabilities and combinatorics in designing their games *to ensure that they make a consistent profit.* Learn this mathematics so that they do not take advantage of you.

Public lotteries have a long history in the United States. The settlement of Jamestown was financed in part by an English lottery. George Washington managed a lottery that paid for a road through the Cumberland Mountains. Benjamin Franklin used lotteries to finance cannons for the Revolutionary War, and John Hancock used lotteries to rebuild Faneuil Hall in Boston. Several universities, including Harvard, Dartmouth, Yale, and Columbia, were partly financed by lotteries. The U.S. Congress operated a lottery to help finance the Revolutionary War.

Today, public lotteries are a very big business. In 2008, Americans spent almost $61 billion on lottery tickets. Lotteries are quite lucrative for the forty-two states that offer them; on the average, 30% of the money went back into government budgets. Fewer than half of the states dedicate the proceeds to education. Frequently, this money goes into the general fund. The states' cuts vary quite a bit. In Oregon in 2008, 54% of the money went to the state, and the remaining 46% went to prizes and administrative costs. In Rhode Island, the state took only 15%.

Most lottery sales come from a relatively small number of people. In Pennsylvania, for example, 29% of the players accounted for 79% of the spending on the lottery in 2008. However, many people who don't normally play go berserk when the jackpots accumulate, partially because of the amazingly large winnings but also because of a lack of understanding of how unlikely it is that they will actually win. The largest cumulative jackpot was $390 million, which was spilt by winners in Georgia and New Jersey in 2007. The largest single-winner jackpot was $315 million in West Virginia in 2002. The winner opted to take a lump sum payment of $114 million, instead of receiving twenty years of regular payments that would have added up to $315 million.

In this section, we will discuss probabilities and gambling. Specifically, we will explore lotteries, keno, and card games. See Examples 2–7, and Exercises 5–22 and 35.

In Section 3.5, we will discuss how much money you can expect to win or lose when gambling. We will continue to discuss lotteries, keno, and card games, but we will also discuss roulette and raffles. See Exercises 1–12, 28–34, and 39–48 in that section.

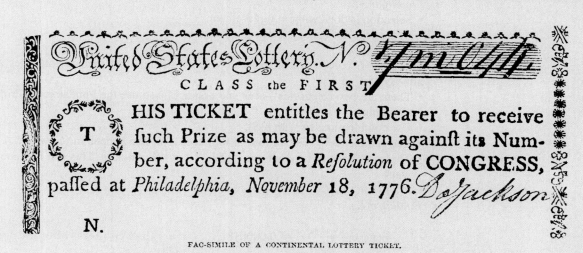

FAC-SIMILE OF A CONTINENTAL LOTTERY TICKET.

- Use combinations in each category, as with $n(S)$.
- The casino selects twenty winning numbers, from which the gambler is to select five winning spots. There are $_{20}C_5$ different ways of doing this.
- The casino selects $80 - 20 = 60$ losing numbers, from which the gambler is to select $8 - 5 = 3$ losing spots. There are $_{60}C_3$ different ways of doing this.

- $n(E) = {}_{20}C_5 \cdot {}_{60}C_3$ **using the above results**

- *Finding $p(E)$:*

$$p(E) = \frac{_{20}C_5 \cdot {}_{60}C_3}{_{80}C_8} = 0.0183 \ldots \qquad \text{one zero after the decimal point}$$

$$\approx 1.8\% \qquad\qquad\qquad \text{writing as a percentage}$$

This means that if you play eight-spot keno a lot, you will have five winning spots about 1.8% of the time. It also means that in any given game, about 1.8% of the players will have five winning spots.

Cards

One common form of poker is five-card draw, in which each player is dealt five cards. The order in which the cards are dealt is unimportant, so we compute probabilities with combinations rather than permutations.

EXAMPLE 5

SOLUTION

GETTING FOUR ACES   Find the probability of being dealt four aces.

The sample space consists of all possible five-card hands that can be dealt from a deck of fifty-two cards

- *Finding $n(S)$:*
  - Selection is done without replacement.
  - Order does not matter, so use combinations.
  - $n(S) = {}_{52}C_5$
- *The event E* consists of all possible five-card hands that include four aces and one non-ace.
  - There are two categories—aces and non-aces—so use the Fundamental Principle of Counting.
  - Use combinations as with $n(S)$.
  - The gambler is to be dealt four of four aces. This can happen in $_4C_4$ ways.
  - The gambler is to be dealt one of $52 - 4 = 48$ non-aces. This can be done in $_{48}C_1$ ways.
  - $n(E) = {}_4C_4 \cdot {}_{48}C_1$
- *Finding $p(E)$:*

$$p(E) = \frac{_4C_4 \cdot {}_{48}C_1}{_{52}C_5} = 0.000018 \ldots \qquad \text{more than two zeros after the decimal point}$$

$$\approx 0.00002 \qquad\qquad \text{rounding to the first nonzero digit}$$

$$= 2/100{,}000 \qquad\qquad \text{rewriting as a fraction}$$

$$= 1/50{,}000 \qquad\qquad \text{reducing}$$

This means that if you play cards a lot, you will be dealt four aces about once every 50,000 deals.

In the event, there is a distinction between two categories (aces and non-aces); in the sample space, there is no such distinction. Thus, the numerator of

$$p(E) = \frac{{}_4C_4 \cdot {}_{48}C_1}{{}_{52}C_5}$$

has two parts (one for each category), and the denominator has one part. Also, the numbers in front of the Cs add correctly (4 aces + 48 non-aces = 52 cards to choose from), and the numbers after the Cs add correctly (4 aces + 1 non-ace = 5 cards to select).

EXAMPLE **6**

GETTING FOUR OF A KIND   Find the probability of being dealt four of a kind.

SOLUTION

The sample space is the same as that in Example 5. The event "being dealt four of a kind" means "being dealt four 2's or being dealt four 3's or being dealt four 4's . . . or being dealt four kings or being dealt four aces." These latter events ("four 2's," "four 3's," etc.) are all mutually exclusive, so we can use the Mutually Exclusive Rule.

p (four of a kind)

$\quad = p(\text{four 2's} \cup \text{four 3's} \cup \cdots \cup \text{four kings} \cup \text{four aces})$

$\quad = p(\text{four 2's}) + p(\text{four 3's}) + \cdots + p(\text{four kings}) + p(\text{four aces})$

using the Mutually Exclusive Rule

Furthermore, these probabilities are all the same:

$$p(\text{four 2's}) = p(\text{four 3's}) = \cdots = p(\text{four aces}) = \frac{{}_4C_4 \cdot {}_{48}C_1}{{}_{52}C_5}$$

This means that the probability of being dealt four of a kind is

$p(\text{four of a kind}) = p(\text{four 2's}) + p(\text{four 3's}) + \cdots + p(\text{four kings})$
$\qquad\qquad\qquad\quad + p(\text{four aces})$

$$= \frac{{}_4C_4 \cdot {}_{48}C_1}{{}_{52}C_5} + \frac{{}_4C_4 \cdot {}_{48}C_1}{{}_{52}C_5} + \cdots + \frac{{}_4C_4 \cdot {}_{48}C_1}{{}_{52}C_5} + \frac{{}_4C_4 \cdot {}_{48}C_1}{{}_{52}C_5}$$

$$= 13 \cdot \frac{{}_4C_4 \cdot {}_{48}C_1}{{}_{52}C_5} \qquad \textbf{there are thirteen denominations}$$
$$\textbf{(2 through ace)}$$

$$= 13 \cdot \frac{1 \cdot 48}{2,598,960} \qquad \textbf{from Example 4}$$

$$= \frac{624}{2,598,960}$$

$$= 0.0002400 \ldots \qquad \textbf{three zeros after the decimal point}$$

$$\approx 0.0002 \qquad\qquad \textbf{rounding to the first nonzero digit}$$

$$= 2/10,000 \qquad\quad \textbf{converting to a fraction}$$

$$= 1/5,000 \qquad\qquad \textbf{reducing}$$

This means that if you play cards a lot, you will be dealt four of a kind about once every 5,000 deals.

See Example 8 in Section 2.4.

EXAMPLE **7**

SOLUTION

GETTING FIVE HEARTS   Find the probability of being dealt five hearts.

The sample space is the same as in Example 5. The event consists of all possible five-card hands that include five hearts and no non-hearts. This involves two categories (hearts and non-hearts), so we will use the Fundamental Counting Principle and multiply the number of ways of getting five hearts and the number of ways of getting no non-hearts. There are

$$_{13}C_5 = \frac{13!}{5! \cdot 8!} = 1{,}287$$

ways of getting five hearts, and there is

$$_{39}C_0 = \frac{39!}{0! \cdot 39!} = 1$$

ways of getting no non-hearts. Thus, the probability of being dealt five hearts is

$$p(E) = \frac{_{13}C_5 \cdot {}_{39}C_0}{_{52}C_5} = \frac{1287 \cdot 1}{2{,}598{,}960} \approx 0.000495198 \approx 0.0005 = 1/2000$$

> In the event, there is a distinction between two categories (hearts and non-hearts); in the sample space, there is no such distinction. Thus, the numerator of
>
> $$p(E) = \frac{_{13}C_5 \cdot {}_{39}C_0}{_{52}C_5}$$
>
> has two parts (one for each category), and the denominator has one part. Also, the numbers in front of the Cs add correctly (13 hearts + 39 non-hearts = 52 cards to choose from), and the numbers after the Cs add correctly (5 hearts + 0 non-hearts = 5 cards to select).

Notice that in Example 7, we could argue that since we're selecting only hearts, we can disregard the non-hearts. This would lead to the answer obtained in Example 7:

$$p(E) = \frac{_{13}C_5}{_{52}C_5} = \frac{1287}{2{,}598{,}960} \approx 0.000495198 \approx 0.0005 = 1/2000$$

However, this approach would not allow us to check our work in the manner described above; the numbers in front of the Cs don't add correctly, nor do the numbers after the Cs.

3.4 EXERCISES

1. A group of thirty people is selected at random. What is the probability that at least two of them will have the same birthday?

2. A group of sixty people is selected at random. What is the probability that at least two of them will have the same birthday?

3. How many people would you have to have in a group so that there is a probability of at least 0.5 that at least two of them will have the same birthday?

4. How many people would you have to have in a group so that there is a probability of at least 0.9 that at least two of them will have the same birthday?

5. In 1990, California switched from a 6/49 lottery to a 6/53 lottery. Later, the state switched again, to a 6/51 lottery.

    a. Find the probability of winning first prize in a 6/49 lottery.

    b. Find the probability of winning first prize in a 6/53 lottery.

    c. Find the probability of winning first prize in a 6/51 lottery.

    d. How much more probable is it that one will win the 6/49 lottery than the 6/53 lottery?

    e. Why do you think California switched from a 6/49 lottery to a 6/53 lottery? And why do you think the state then switched to a 6/51 lottery?

    (Answer using complete sentences.)

6. Find the probability of winning second prize—that is, picking five of the six winning numbers—with a 6/53 lottery.

7. Currently, the most common multinumber game is the 5/39 lottery. It is played in California, Georgia, Illinois, Michigan, New York, North Carolina, Ohio, Tennessee, and Washington.

    a. Find the probability of winning first prize.

    b. Find the probability of winning second prize.

8. Currently, the second most common multinumber game is the 6/49 lottery. It is played in Massachusetts, New Jersey, Ohio, Pennsylvania, Washington, and Wisconsin.

    a. Find the probability of winning first prize.

    b. Find the probability of winning second prize.

9. "Cash 5" is a 5/35 lottery. It is played in Arizona, Connecticut, Iowa, Massachusetts and South Dakota.

    a. Find the probability of winning first prize.

    b. Find the probability of winning second prize.

10. The 6/44 lottery is played in Arizona, Connecticut, and New Jersey.

    a. Find the probability of winning first prize.

    b. Find the probability of winning second prize.

11. Games like "Mega Millions" are played in thirty-nine states and the District of Columbia. It involves selecting any five of the numbers from 1 to 56, plus a number from 1 to 46.

Find the probability of winning first prize.

12. "Hot Lotto" is played in Delaware, Idaho, Iowa, Kansas, Minnesota, Montana, New Hampshire, New Mexico, North Dakota, Oklahoma, South Dakota, West Virginia, and the District of Columbia. It involves selecting any five of the numbers from 1 to 39 plus a number from 1 to 19. Find the probability of winning first prize.

13. "Wild Card 2" is played in Idaho, Montana, North Dakota, and South Dakota. It involves selecting any five of the numbers from 1 to 31, plus a card that's either a jack, queen, king or ace of any one of the four suits. Find the probability of winning first prize.

14. "2 by 2" is played in Kansas, Nebraska, and North Dakota. It involves selecting two red numbers from 1 to 26 and two white numbers from 1 to 26. Find the probability of winning first prize.

15. There is an amazing variety of multinumber lotteries played in the United States. Currently, the following lotteries are played: 4/26, 4/77, 5/30, 5/31, 5/32, 5/34, 5/35, 5/36, 5/37, 5/38, 5/39, 5/40, 5/43, 5/47, 5/50, 6/25, 6/35, 6/39, 6/40, 6/42, 6/43, 6/44, 6/46, 6/47, 6/48, 6/49, 6/52, 6/53, and 6/54. Which is the easiest to win? Which is the hardest to win? Explain your reasoning.

HINT: It isn't necessary to compute every single probability.

16. In the game of keno, if six spots are marked, the player wins if four or more of his or her spots are selected. Complete the chart in Figure 3.17.

Outcome	Probability
6 winning spots	
5 winning spots	
4 winning spots	
3 winning spots	
fewer than 3 winning spots	

FIGURE 3.17 Chart for Exercise 16.

17. In the game of keno, if eight spots are marked, the player wins if five or more of his or her spots are selected. Complete the chart in Figure 3.18.

Outcome	Probability
8 winning spots	
7 winning spots	
6 winning spots	
5 winning spots	
4 winning spots	
fewer than 4 winning spots	

FIGURE 3.18 Chart for Exercise 17.

18. In the game of keno, if nine spots are marked, the player wins if six or more of his or her spots are selected. Complete the chart in Figure 3.19.

Outcome	Probability
9 winning spots	
8 winning spots	
7 winning spots	
6 winning spots	
5 winning spots	
fewer than 5 winning spots	

FIGURE 3.19 Chart for Exercise 18.

19. "Pick three" games are played in thirty-six states. In this game, the player selects a three-digit number, such as 157. Also, the state selects a three-digit winning number.

a. How many different three-digit numbers are there? Explain your reasoning.

b. If the player opts for "straight play," she wins if her selecting matches the winning number *in exact order*. For example, if she selected the number 157, she wins only if the winning number is 157. How many different winning straight play numbers are there?

c. Find the probability of winning with straight play.

20. If a pick three player (see Exercise 19) opts for "box play," the player wins if his selection matches the winning number *in any order*. For example, if he selected the number 157, he wins if the winning number is 715 or any other reordering of 157.

a. How many different winning box play numbers are there if the three digits are different? That is, how many different ways are there to reorder a number such as 157?

b. Find the probability of winning with box play if the three digits are different.

c. How many different winning box play numbers are there if two digits are the same? That is, how many different ways are there to reorder a number such as 266?

d. Find the probability of winning with box play if two digits are the same.

21. a. Find the probability of being dealt five spades when playing five-card draw poker.

b. Find the probability of being dealt five cards of the same suit when playing five-card draw poker.

c. When you are dealt five cards of the same suit, you have either a *flush* (if the cards are not in sequence) or a *straight flush* (if the cards are in sequence). For each suit, there are ten possible straight flushes ("ace, two, three, four, five," through "ten, jack, queen, king, ace"). Find the probability of being dealt a straight flush.

d. Find the probability of being dealt a flush.

22. a. Find the probability of being dealt an "aces over kings" full house (three aces and two kings).

b. Why are there $13 \cdot 12$ different types of full houses?

c. Find the probability of being dealt a full house. (Round each answer off to six decimal places.)

You order twelve burritos to go from a Mexican restaurant, five with hot peppers and seven without. However, the restaurant forgot to label them. If you pick three burritos at random, find the probability of each event in Exercises 23–30.

23. All have hot peppers.

24. None has hot peppers.

25. Exactly one has hot peppers.

26. Exactly two have hot peppers.

27. At most one has hot peppers.

28. At least one has hot peppers.

29. At least two have hot peppers.

30. At most two have hot peppers.

31. Two hundred people apply for two jobs. Sixty of the applicants are women.

    a. If two people are selected at random, what is the probability that both are women?

    b. If two people are selected at random, what is the probability that only one is a woman?

    c. If two people are selected at random, what is the probability that both are men?

    d. If you were an applicant and the two selected people were not of your gender, do you think that the above probabilities would indicate the presence or absence of gender discrimination in the hiring process? Why or why not?

32. Two hundred people apply for three jobs. Sixty of the applicants are women.

    a. If three people are selected at random, what is the probability that all are women?

    b. If three people are selected at random, what is the probability that two are women?

    c. If three people are selected at random, what is the probability that one is a woman?

    d. If three people are selected at random, what is the probability that none is a woman?

    e. If you were an applicant and the three selected people were not of your gender, should the above probabilities have an impact on your situation? Why or why not?

33. In Example 2, $n(E) = 1$ because only one of the 7,059,052 possible lottery tickets is the first prize winner. Use combinations to show that $n(E) = 1$.

Answer the following questions using complete sentences and your own words.

● CONCEPT QUESTIONS

34. Explain why, in Example 6, $p(\text{four 2's}) = p(\text{four 3's}) = \cdots = p(\text{four kings}) = p(\text{four aces})$.

35. Do you think a state lottery is a good thing for the state's citizens? Why or why not? Be certain to include a discussion of both the advantages and disadvantages of a state lottery to its citizens.

36. Why are probabilities for most games of chance calculated with combinations rather than permutations?

37. Suppose a friend or relative of yours regularly spends (and loses) a good deal of money on lotteries. How would you explain to this person why he or she loses so frequently?

38. In Example 2, $n(E) = 1$ because only one of the 7,059,052 possible lottery tickets is the first prize winner. Does this mean that it is impossible for two people to each buy a first-prize-winning lottery ticket? Explain.

● HISTORY QUESTION

39. Are public lotteries relative newcomers to the American scene? Explain.

3.5   Expected Value

OBJECTIVES

● Understand how expected values take both probabilities and winnings into account

● Use expected values to analyze games of chance

● Use expected values to make decisions

Suppose you are playing roulette, concentrating on the $1 single-number bet. At one point, you were $10 ahead, but now you are $14 behind. How much should you expect to win or lose, on the average, if you place the bet many times?

The probability of winning a single-number bet is $\frac{1}{38}$, because there are thirty-eight numbers on the roulette wheel and only one of them is the subject of the bet. This means that if you place the bet a large number of times, it is most likely that you will win once for every thirty-eight times you place the bet (and lose the other thirty-seven times). When you win, you win $35, because the house odds are 35 to 1. When you lose, you lose $1. Your average profit would be

$$\frac{\$35 + 37 \cdot (-\$1)}{38} = \frac{-\$2}{38} \approx -\$0.053$$

per game. This is called the *expected value* of a $1 single-number bet, because you should expect to lose about a nickel for every dollar you bet if you play the game a long time. If you play a few times, anything could happen—you could win every single bet (though it is not likely). The house makes the bet so many times that it can be certain that its profit will be $0.053 per dollar bet.

The standard way to find the **expected value** of an experiment is to multiply the value of each outcome of the experiment by the probability of that outcome and add the results. Here the experiment is placing a $1 single-number bet in roulette. The outcomes are winning the bet and losing the bet. The values of the outcomes are +$35 (if you win) and −$1 (if you lose); the probabilities of the outcomes are $\frac{1}{38}$ and $\frac{37}{38}$, respectively (see Figure 3.20). The expected value would then be

$$35 \cdot \frac{1}{38} + (-1) \cdot \frac{37}{38} \approx -\$0.053$$

It is easy to see that this calculation is algebraically equivalent to the calculation done above.

Outcome	Value	Probability
winning	35	$\frac{1}{38}$
losing	−1	$\frac{37}{38}$

FIGURE 3.20 Finding the expected value.

Finding an expected value of a bet is very similar to finding your average test score in a class. Suppose that you are a student in a class in which you have taken four tests. If your scores were 80%, 76%, 90%, and 90%, your average test score would be

$$\frac{80 + 76 + 2 \cdot 90}{4} = 84\%$$

or, equivalently,

$$80 \cdot \frac{1}{4} + 76 \cdot \frac{1}{4} + 90 \cdot \frac{2}{4}$$

The difference between finding an average test score and finding the expected value of a bet is that with the average test score you are summarizing what *has* happened, whereas with a bet you are using probabilities to project what *will* happen.

EXPECTED VALUE

To find the **expected value** (or "long-term average") of an experiment, multiply the value of each outcome of the experiment by its probability and add the results.

Roulette, the oldest casino game played today, has been popular since it was introduced to Paris in 1765. Does this game have any good bets?

EXAMPLE **1**

COMPUTING AN EXPECTED VALUE By analyzing her sales records, a saleswoman has found that her weekly commissions have the probabilities in Figure 3.21. Find the saleswoman's expected commission.

Commission	0	$100	$200	$300	$400
Probability	0.05	0.15	0.25	0.45	0.1

FIGURE 3.21 Commission data for Example 1.

SOLUTION

To find the expected commission, we multiply each possible commission by its probability and add the results. Therefore,

$$\text{expected commission} = (0)(0.05) + (100)(0.15) + (200)(0.25)$$
$$+ (300)(0.45) + (400)(0.1)$$
$$= 240$$

On the basis of her history, the saleswoman should expect to average $240 per week in future commissions. Certainly, anything can happen in the future—she could receive a $700 commission (it's not likely, though, because it has never happened before).

Why the House Wins

Four of the "best" bets that can be made in a casino game of chance are the pass, don't pass, come, and don't come bets in craps. They all have the same expected value, −$0.014. In the long run, *there isn't a single bet in any game of chance with which you can expect to break even, let alone make a profit.* After all, the casinos are out to make money. The expected values for $1 bets in the more common games are shown in Figure 3.22.

Game	Expected Value of $1 Bet
baccarat	−$0.014
blackjack	−$0.06 to +$0.10 (varies with strategies)
craps	−$0.014 for pass, don't pass, come, don't come bets *only*
slot machines	−$0.13 to ? (varies)
keno (eight-spot ticket)	−$0.29
average state lottery	−$0.48

FIGURE 3.22 Expected values of common games of chance.

It is possible to achieve a positive expected value in blackjack. To do this, the player must keep a running count of the cards dealt, following a system that assigns a value to each card. Casinos use their pit bosses and video surveillance to watch for gamblers who use this tactic, and casinos will harass or kick such gamblers out when they find them. There is an application available for the iPhone and iPod Touch that will count cards. The casinos do everything in their power to eliminate its use. Some casinos use four decks at once and shuffle frequently to minimize the impact of counting.

Decision Theory

Which is the better bet: a $1 single-number bet in roulette or a lottery ticket? Each costs $1. The roulette bet pays $35, but the lottery ticket might pay several million dollars. Lotteries are successful in part because the possibility of winning a large amount of money distracts people from the fact that winning is extremely unlikely. In Example 2 of Section 3.4, we found that the probability of winning first prize in many state lotteries is $\frac{1}{7,059,052} \approx 0.00000014$. At the beginning of this section, we found that the probability of winning the roulette bet is $\frac{1}{38} \approx 0.03$.

A more informed decision would take into account not only the potential winnings and losses but also their probabilities. The expected value of a bet does just that, since its calculation involves both the value and the probability of each outcome. We found that the expected value of a $1 single-number bet in roulette is about −$0.053. The expected value of the average state lottery is −$0.48 (see Figure 3.22). The roulette bet is a much better bet than is the lottery. (Of course, there is a third option, which has an even better expected value of $0.00: not gambling!)

A decision always involves choosing between various alternatives. If you compare the expected values of the alternatives, then you are taking into account

FEATURED IN
THE NEWS

VIRGINIA LOTTERY HEDGES ON SYNDICATE'S BIG WIN

RICHMOND, VA.—Virginia lottery officials confirmed yesterday that an Australian gambling syndicate won last month's record $27 million jackpot after executing a massive block-buying operation that tried to cover all 7 million possible ticket combinations.

But lottery director Kenneth Thorson said the jackpot may not be awarded because the winning ticket may have been bought in violation of lottery rules.

The rules say tickets must be paid for at the same location where they are issued. The Australian syndicate, International Lotto Fund, paid for many of its tickets at the corporate offices of Farm Fresh Inc. grocery stores, rather than at the Farm Fresh store in Chesapeake

where the winning ticket was issued, Thorson said.

"We have to validate who bought the ticket, where the purchase was made and how the purchase was made," Thorson said. "It's just as likely that we will honor the ticket as we won't honor the ticket." He said he may not decide until the end of next week.

Two Australians representing the fund, Joseph Franck and Robert Hans Roos, appeared at lottery headquarters yesterday to claim the prize.

The group succeeded in buying about 5 million of the more than 7 million possible numerical combinations before the February 15 drawing. The tactic is not illegal, although lottery officials announced new rules earlier this week aimed at making such block purchases more difficult.

The Australian fund was started last year and raised about $13 million from an estimated 2,500 shareholders who each paid a minimum of $4,000, according to Tim Phillipps of the Australian Securities Commission.

Half the money went for management expenses, much of that to Pacific Financial Resources, a firm controlled by Stefan Mandel, who won fame when he covered all the numbers in a 1986 Sydney lottery. Roos owns 10 percent of Pacific Financial Resources.

Australian Securities Commission officials said last week that the fund is under investigation for possible violations of Australian financial laws.

the alternatives' potential advantages and disadvantages as well as their probabilities. This form of decision making is called **decision theory.**

EXAMPLE **2**   **USING EXPECTED VALUES TO MAKE A DECISION**   The saleswoman in Example 1 has been offered a new job that has a fixed weekly salary of $290. Financially, which is the better job?

SOLUTION   In Example 1, we found that her expected weekly commission was $240. The new job has a guaranteed weekly salary of $290. Financial considerations indicate that she should take the new job.

Betting Strategies

One very old betting strategy is to "cover all the numbers." In 1729, the French philosopher and writer François Voltaire organized a group that successfully implemented this strategy to win the Parisian city lottery by buying most if not all of the tickets. Their strategy was successful because, owing to a series of poor financial decisions by the city of Paris, the total value of the prizes was greater than the combined price of all of the tickets! Furthermore, there were not a great number of tickets to buy. This strategy is still being used. (See the above newspaper article on its use in 1992 in Virginia.)

A **martingale** is a gambling strategy in which the gambler doubles his or her bet after each loss. A person using this strategy in roulette, concentrating on the black numbers bet (which has 1-to-1 house odds), might lose three times before winning. This strategy would result in a net gain, as illustrated in Figure 3.23 below.

This seems to be a great strategy. Sooner or later, the player will win a bet, and because each bet is larger than the player's total losses, he or she has to come out ahead! We will examine this strategy further in Exercises 47 and 48.

Bet Number	Bet	Result	Total Winnings/Losses
1	$1	lose	−$1
2	$2	lose	−$3
3	$4	lose	−$7
4	$8	win	+$1

FIGURE 3.23 Analyzing the martingale strategy.

3.5 EXERCISES

In Exercises 1–10, (a) find the expected value of each $1 bet in roulette and (b) use the Law of Large Numbers to interpret it.

1. the two-number bet
2. the three-number bet
3. the four-number bet
4. the five-number bet
5. the six-number bet
6. the twelve-number bet
7. the low-number bet
8. the even-number bet
9. the red-number bet
10. the black-number bet

11. Using the expected values obtained in the text and in the preceding odd-numbered exercises, determine a casino's expected net income from a 24-hour period at a single roulette table if the casino's total overhead for the table is $50 per hour and if customers place a total of $7,000 on single-number bets, $4,000 on two-number bets, $4,000 on four-number bets, $3,000 on six-number bets, $7,000 on low-number bets, and $8,000 on red-number bets.

12. Using the expected values obtained in the text and in the preceding even-numbered exercises, determine a casino's expected net income from a 24-hour period at a single roulette table if the casino's total overhead for the table is $50 per hour and if customers place a total of $8,000 on single-number bets, $3,000 on three-number bets, $4,000 on five-number bets, $4,000 on twelve-number bets, $8,000 on even-number bets, and $9,000 on black-number bets.

13. On the basis of his previous experience, the public librarian at Smallville knows that the number of books checked out by a person visiting the library has the probabilities shown in Figure 3.24.

Number of Books	0	1	2	3	4	5
Probability	0.15	0.35	0.25	0.15	0.05	0.05

FIGURE 3.24 Probabilities for Exercise 13.

Find the expected number of books checked out by a person visiting this library.

14. On the basis of his sale records, a salesman knows that his weekly commissions have the probabilities shown in Figure 3.25.

Commission	0	$1,000	$2,000	$3,000	$4,000
Probability	0.15	0.2	0.45	0.1	0.1

FIGURE 3.25 Probabilities for Exercise 14.

Find the salesman's expected commission.

15. Of all workers at a certain factory, the proportions earning certain hourly wages are as shown in Figure 3.26.

Hourly Wage	$8.50	$9.00	$9.50	$10.00	$12.50	$15.00
Proportion	20%	15%	25%	20%	15%	5%

FIGURE 3.26 Data for Exercise 15.

Find the expected hourly wage that a worker at this factory makes.

16. Of all students at the University of Metropolis, the proportions taking certain numbers of units are as shown in Figure 3.27. Find the expected number of units that a student at U.M. takes.

Units	3	4	5	6	7	8
Proportion	3%	4%	5%	6%	5%	4%

Units	9	10	11	12	13	14
Proportion	8%	12%	13%	13%	15%	12%

FIGURE 3.27 Data for Exercise 16.

17. You have been asked to play a dice game. It works like this:
    * If you roll a 1 or 2, you win $50.
    * If you roll a 3, you lose $20.
    * If you roll a 4, 5, or 6, you lose $30.

    Should you play the game? Use expected values to justify your answer.

18. You have been asked to play a dice game. It works like this:
    * If you roll a 1, 2, 3, or 4, you win $50.
    * If you roll a 5 or 6, you lose $80.

    Should you play the game? Use expected values to justify your answer.

19. You are on a TV show. You have been asked to either play a dice game ten times or accept a $100 bill. The dice game works like this:
    * If you roll a 1 or 2, you win $50.
    * If you roll a 3, you win $20.
    * If you roll a 4, 5, or 6, you lose $30.

    Should you play the game? Use expected values and decision theory to justify your answer.

20. You are on a TV show. You have been asked to either play a dice game five times or accept a $50 bill. The dice game works like this:
    * If you roll a 1, 2, or 3, you win $50.
    * If you roll a 4 or 5, you lose $25.
    * If you roll a 6, you lose $90.

    Should you play the game? Use expected values and decision theory to justify your answer.

21. Show why the calculation at the top of page 182 is algebraically equivalent to the calculation in the middle of the same page.

22. In Example 1, the saleswoman's most likely weekly commission was $300. With her new job (in Example 2), she will always make $290 per week. This implies that she would be better off with the old job. Is this reasoning more or less valid than that used in Example 2? Why?

23. Maria just inherited $10,000. Her bank has a savings account that pays 4.1% interest per year. Some of her friends recommended a new mutual fund, which has been in business for three years. During its first year, the fund went up in value by 10%; during the second year, it went down by 19%; and during its third year, it went up by 14%. Maria is attracted by the mutual fund's potential for relatively high earnings but concerned by the possibility of actually losing some of her inheritance. The bank's rate is low, but it is insured by the federal government. Use decision theory to find the best investment. (Assume that the fund's past behavior predicts its future behavior.)

24. Trang has saved $8,000. It is currently in a bank savings account that pays 3.9% interest per year. He is considering putting the money into a speculative investment that would either earn 20% in one year if the investment succeeds or lose 18% in one year if it fails. At what probability of success would the speculative investment be the better choice?

25. Erica has her savings in a bank account that pays 4.5% interest per year. She is considering buying stock in a pharmaceuticals company that is developing a cure for cellulite. Her research indicates that she could earn 50% in one year if the cure is successful or lose 60% in one year if it is not. At what probability of success would the pharmaceuticals stock be the better choice?

26. Debra is buying prizes for a game at her school's fundraiser. The game has three levels of prizes, and she has already bought the second and third prizes. She wants the first prize to be nice enough to attract people to the game. The game's manufacturer has supplied her with the probabilities of winning first, second, and third prizes. Tickets cost $3 each, and she wants the school to profit an average of $1 per ticket. How much should she spend on each first prize?

Prize	Cost of Prize	Probability
1st	?	.15
2nd	$1.25	.30
3rd	$0.75	.45

FIGURE 3.28 Data for Exercise 26.

27. Few students manage to complete their schooling without taking a standardized admissions test such as

the Scholastic Achievement Test, or S.A.T. (used for admission to college); the Law School Admissions Test, or L.S.A.T.; and the Graduate Record Exam, or G.R.E. (used for admission to graduate school). Sometimes, these multiple-choice tests discourage guessing by subtracting points for wrong answers. In particular, a correct answer will be worth $+1$ point, and an incorrect answer on a question with five listed answers (a through e) will be worth $-\frac{1}{4}$ point.

a. Find the expected value of a random guess.

b. Find the expected value of eliminating one answer and guessing among the remaining four possible answers.

c. Find the expected value of eliminating three answers and guessing between the remaining two possible answers.

d. Use decision theory and your answers to parts (a), (b), and (c) to create a guessing strategy for standardized tests such as the S.A.T.

28. Find the expected value of a $1 bet in six-spot keno if three winning spots pays $1 (but you pay $1 to play, so you actually break even), four winning spots pays $3 (but you pay $1 to play, so your profit is $2), five pays $100, and six pays $2,600. (You might want to use the probabilities computed in Exercise 16 of Section 3.4.)

29. Find the expected value of a $1 bet in eight-spot keno if four winning spots pays $1 (but you pay $1 to play, so you actually break even), five winning spots pays $5 (but you pay $1 to play, so your profit is $4), six pays $100, seven pays $1,480, and eight pays $19,000. (You might want to use the probabilities computed in Exercise 17 of Section 3.4.)

30. Find the expected value of a $1 bet in nine-spot keno if five winning spots pays $1 (but you pay $1 to play, so you actually break even), six winning spots pays $50 (but you pay $1 to play, so your profit is $49), seven pays $390, eight pays $6,000, and nine pays $25,000. (You might want to use the probabilities computed in Exercise 18 of Section 3.4.)

31. Arizona's "Cash 4" is a 4/26 lottery. It differs from many other state lotteries in that its payouts are set; they do not vary with sales. If you match all four of the winning numbers, you win $10,000 (but you pay $1 to play, so your profit is $9,999). If you match three of the winning numbers, you win $25, and if you match two of the winning numbers, you win $2. Otherwise, you lose your $1.

a. Find the probabilities of winning first prize, second prize, and third prize.

b. Use the results of parts a and the Complement Rule to find the probability of losing.

c. Use the results of parts a and b to find the expected value of Cash 4.

32. New York's "Pick 10" is a 10/80 lottery. Its payouts are set; they do not vary with sales. If you match all ten winning numbers, you win $500,000 (but you pay $1 to play, so your profit is $499,999). If you match nine winning numbers, you win $6000. If you match eight, seven, or six you win $300, $40, or $10, respectively. If you match no winning numbers, you win $4. Otherwise, you lose your $1.

a. Find the probabilities of winning first prize, second prize, and third prize.

b. Use the results of part a and the Complement Rule to find the probability of losing.

c. Use the results of parts a and b to find the expected value of Pick 10.

33. Arizona and New York have Pick 3 games (New York's is called "Numbers"), as described in Exercises 19 and 20 of Section 3.4. If the player opts for "straight play" and wins, she wins $500 (but she pays $1 to play, so her profit is $499). If the player opts for "box play" with three different digits and wins, he wins $80. If the player opts for "box play" with two of the same digits and wins, he wins $160.

Courtesy of the New York Lottery

a. Use the probabilities from Exercises 19 and 20 of Section 3.4 to find the probability of losing.

b. Use the probabilities from part a and from Exercises 19 and 20 of Section 3.4 to find the expected value of the game.

34. Write a paragraph in which you compare the states' fiscal policies concerning their lotteries with the casinos' fiscal policies concerning their keno games. Assume that the expected value of a $1 lottery bet described in Exercises 31, 32, and/or 33 is representative of that of the other states' lotteries, and assume that the expected value of a $1 keno bet described in Exercises 28, 29, and/or 30 is representative of other keno bets.

35. Trustworthy Insurance Co. estimates that a certain home has a 1% chance of burning down in any one year. They calculate that it would cost $120,000 to rebuild that home. Use expected values to determine the annual insurance premium.

36. Mr. and Mrs. Trump have applied to the Trustworthy Insurance Co. for insurance on Mrs. Trump's diamond tiara. The tiara is valued at $97,500. Trustworthy estimates that the jewelry has a 2.3% chance of being stolen in any one year. Use expected values to determine the annual insurance premium.

37. The Black Gold Oil Co. is considering drilling either in Jed Clampett's back yard or his front yard. After thorough testing and analysis, they estimate that there is a 30% chance of striking oil in the back yard and a 40% chance in the front yard. They also estimate that the back yard site would either net $60 million (if oil is found) or lose $6 million (if oil is not found), and the front yard site would either net $40 million or lose $6 million. Use decision theory to determine where they should drill.

38. If in Exercise 37, Jed Clampett rejected the use of decision theory, where would he drill if he were an optimist? What would he do if he were a pessimist?

39. A community youth group is having a raffle to raise funds. Several community businesses have donated prizes. The prizes and their retail value are listed in Figure 3.29. Each prize will be given away, regardless of the number of raffle tickets sold. Tickets are sold for $15 each. Determine the expected value of a ticket, and discuss whether it would be to your financial advantage to buy a ticket under the given circumstances.

   a. 1000 tickets are sold.
   b. 2000 tickets are sold.
   c. 3000 tickets are sold.

Prize	Retail Value	Number of These Prizes to Be Given Away
new car	$21,580	1
a cell phone and a one-year subscription	$940	1
a one-year subscription to an Internet service provider	$500	2
dinner for two at Spiedini's restaurant	$100	2
a one-year subscription to the local newspaper	$180	20

FIGURE 3.29 Data for Exercise 39.

40. The Centerville High School PTA is having a raffle to raise funds. Several community businesses have donated prizes. The prizes and their retail value are listed in Figure 3.30. Each prize will be given away, regardless of the number of raffle tickets sold. Tickets are sold for $30 each. Determine the expected value of a ticket, and discuss whether it would be to your financial advantage to buy a ticket under the given circumstances.

   a. 100 tickets are sold.
   b. 200 tickets are sold.
   c. 300 tickets are sold.

Prize	Retail Value	Number of These Prizes to Be Given Away
one week in a condo in Hawaii, and airfare for two	$2,575	1
tennis lessons for two	$500	1
a one-year subscription to an Internet service provider	$500	2
dinner for two at Haute Stuff restaurant	$120	2
a one-year subscription at the Centerville Skate Park	$240	5
a copy of *Centerville Cooks,* a cookbook containing PTA members' favorite recipes	$10	20

FIGURE 3.30 Data for Exercise 40.

41. The Central State University Young Republicans Club is having a raffle to raise funds. Several community businesses have donated prizes. The prizes and their retail value are listed in Figure 3.31. Each prize will be given away, regardless of the number of raffle tickets sold. Tickets are sold for $5 each. Determine the expected value of a ticket, and discuss whether it would be to your financial advantage to buy a ticket under the given circumstances.

   a. 1000 tickets are sold.
   b. 2000 tickets are sold.
   c. 3000 tickets are sold.

Prize	Retail Value	Number of These Prizes to Be Given Away
laptop computer	$2,325	1
MP3 player	$425	2
20 CDs at Einstein Entertainment	$320	3
a giant pizza and your choice of beverage at Freddie's Pizza	$23	4
a one-year subscription to the *Young Republican Journal*	$20	15

FIGURE 3.31 Data for Exercise 41.

42. The Southern State University Ecology Club is having a raffle to raise funds. Several community businesses have donated prizes. The prizes and their retail value are listed in Figure 3.32. Each prize will be given away, regardless of the number of raffle tickets sold. Tickets are sold for $20 each. Determine the expected value of a ticket, and discuss whether it would be to your financial advantage to buy a ticket under the given circumstances.

   a. 100 tickets are sold.
   b. 200 tickets are sold.
   c. 300 tickets are sold.

Prize	Retail Value	Number of These Prizes to Be Given Away
one-week ecovacation to Costa Rica, including airfare for two	$2,990	1
Sierra Designs tent	$750	1
REI backpack	$355	2
Jansport daypack	$75	5
fleece jacket with Ecology Club logo	$50	10
Ecology Club T-shirt	$18	20

FIGURE 3.32 Data for Exercise 42.

43. Find the expected value of the International Lotto Fund's application of the "cover all of the numbers" strategy from the newspaper article on page 185. Assume that $5 million was spent on lottery tickets, that half of the $13 million raised went for management expenses, and that the balance was never spent. Also assume that Virginia honors the winning ticket.

44. One application of the "cover all the numbers" strategy would be to bet $1 on every single number in roulette.

   a. Find the results of this strategy.
   b. How could you use the expected value of the $1 single-number bet $\left(\frac{-\$2}{38}\right)$ to answer part (a)?

45. The application of the "cover all the numbers" strategy to a modern state lottery would involve the purchase of a large number of tickets.

   a. How many tickets would have to be purchased if you were in a state that has a 6/49 lottery (the player selects 6 out of 49 numbers)?
   b. How much would it cost to purchase these tickets if each costs $1?
   c. If you organized a group of 100 people to purchase these tickets and it takes one minute to purchase each ticket, how many days would it take to purchase the required number of tickets?

46. The application of the "cover all the numbers" strategy to Keno would involve the purchase of a large number of tickets.

   a. How many tickets would have to be purchased if you were playing eight-spot keno?
   b. How much would it cost to purchase these tickets if each costs $5?
   c. If it takes five seconds to purchase one ticket and the average keno game lasts twenty minutes, how many people would it take to purchase the required number of tickets?

47. If you had $100 and were applying the martingale strategy to the black-number bet in roulette and you started with a $1 bet, how many successive losses could you afford? How large would your net profit be if you lost each bet except for the last one?

48. If you had $10,000 and were applying the martingale strategy to the black-number bet in roulette and you started with a $1 bet, how many successive losses could you afford? How large would your net profit be if you lost each bet except for the last one?

 Answer the following questions using complete sentences and your own words.

• CONCEPT QUESTIONS

49. Discuss the meaningfulness of the concept of expected value to three different people: an occasional gambler, a regular gambler, and a casino owner. To whom is the concept most meaningful? To whom is it least meaningful? Why?

50. Discuss the advantages and disadvantages of decision theory. Consider the application of decision theory to a nonrecurring situation and to a recurring situation. Also consider Jed Clampett and the Black Gold Oil Co. in Exercises 37 and 38.

• PROJECTS

51. Design a game of chance. Use probabilities and expected values to set the house odds so that the house will make a profit. However, be certain that your game is not so obviously pro-house that no one would be willing to play. Your project should include the following:

   • a complete description of how the game is played
   • a detailed mathematical analysis of the expected value of the game (or of each separate bet in the game, whichever is appropriate)
   • a complete description of the bet(s) and the house odds

3.6 Conditional Probability

OBJECTIVES

- Use conditional probabilities to focus on one or two groups rather than the entire sample space
- Understand the relationship between intersections of events and products of probabilities
- Use tree diagrams to combine probabilities

Public opinion polls, such as those found in newspapers and magazines and on television, frequently categorize the respondents by such groups as sex, age, race, or level of education. This is done so that the reader or listener can make comparisons and observe trends, such as "people over 40 are more likely to support the Social Security system than are people under 40." The tool that enables us to observe such trends is conditional probability.

Probabilities and Polls

In a newspaper poll concerning violence on television, 600 people were asked, "What is your opinion of the amount of violence on prime-time television—is there too much violence on television?" Their responses are indicated in Figure 3.33.

	Yes	No	Don't Know	Total
Men	162	95	23	280
Women	256	45	19	320
Total	418	140	42	600

FIGURE 3.33 Results of "Violence on Television" poll.

Six hundred people were surveyed in this poll; that is, the sample space consists of 600 responses. Of these, 418 said they thought there was too much violence on television, so the probability of a "yes" response is $\frac{418}{600}$, or about 0.70 = 70%. The probability of a "no" response is $\frac{140}{600}$, or about 0.23 = 23%.

If we are asked to find the probability that a *woman* responded yes, we do not consider all 600 responses but instead limit the sample space to only the responses from women. See Figure 3.34.

	Yes	No	Don't Know	Total
Women	256	45	19	320

FIGURE 3.34 Women and violence on television.

The probability that a woman responded yes is $\frac{256}{320} = 0.80 = 80\%$.

Is there too much violence on TV?

Suppose we label the events in the following manner: W is the event that a response is from a woman, M is the event that a response is from a man, Y is the event that a response is yes, and N is the event that a response is no. Then the event that a woman responded yes would be written as

$Y \mid W$

The vertical bar stands for the phrase "given that"; the event $Y \mid W$ is read "a response is yes, given that the response is from a woman." The probability of this event is called a *conditional probability:*

$$p(Y \mid W) = \frac{256}{320} = \frac{4}{5} = 0.80 = 80\%$$

The numerator of this probability, 256, is the number of responses that are yes and are from women; that is, $n(Y \cap W) = 256$. The denominator, 320, is the number of responses that are from women; that is, $n(W) = 320$. A **conditional probability** is a probability whose sample space has been limited to only those outcomes that fulfill a certain condition. Because an event is a subset of the sample space, the event must also fulfill that condition. The numerator of $p(Y \mid W)$ is 256 rather than 418 even though there were 418 "yes" responses, because many of those 418 responses were made by men; we are interested only in the probability that a woman responded yes.

CONDITIONAL PROBABILITY DEFINITION

The **conditional probability** of event *A*, given event *B*, is

$$p(A \mid B) = \frac{n(A \cap B)}{n(B)}$$

EXAMPLE **1**

COMPUTING CONDITIONAL PROBABILITIES   Using the data in Figure 3.33, find the following.

a. the probability that a response is yes, given that the response is from a man.
b. the probability that a response is from a man, given that the response is yes.
c. the probability that a response is yes and is from a man.

SOLUTION

a. *Finding* $p(Y \mid M)$: We are told to consider only the male responses—that is, to limit our sample space to men. See Figure 3.35.

	Yes	No	Don't Know	Total
Men	162	95	23	280

FIGURE 3.35   Men and violence on television.

$$p(Y \mid M) = \frac{n(Y \cap M)}{n(M)} = \frac{162}{280} \approx 0.58 = 58\%$$

In other words, approximately 58% of the men responded yes. (Recall that 80% of the women responded yes. This poll indicates that men and women do not have the same opinion regarding violence on television and, in particular, that a woman is more likely to oppose the violence.)

b. *Finding* $p(M \mid Y)$: We are told to consider only the "yes" responses shown in Figure 3.36.

	Yes
Men	162
Women	256
Total	418

FIGURE 3.36

Limiting our sample space to "yes" responses.

$$p(M \mid Y) = \frac{n(M \cap Y)}{n(Y)} = \frac{162}{418} \approx 0.39 = 39\%$$

Therefore, of those who responded yes, approximately 39% were male.

c. *Finding* $p(Y \cap M)$: This is *not* a conditional probability (there is no vertical bar), so we do *not* limit our sample space.

$$p(Y \cap M) = \frac{n(Y \cap M)}{n(S)} = \frac{162}{600} = 0.27 = 27\%$$

Therefore, of all those polled, 27% were men who responded yes.

Each of the above three probabilities has the *same numerator*, $n(Y \cap M) = 162$. This is the number of responses that are yes and are from men.

But the three probabilities have *different denominators*. This means that we are comparing the group of men who said yes with three different larger groups. See Figure 3.37.

In the probability:	the denominator is:	so we're comparing the group of men who said yes with:
$p(Y \mid M)$	$n(M)$, the number of male responses	all of the men
$p(M \mid Y)$	$n(Y)$, the number of yes responses	all of the yes responses
$p(Y \cap M)$	$n(S)$, the number of responses	all of the people polled

FIGURE 3.37   The impact of the different denominators.

The Product Rule

If two cards are dealt from a full deck (no jokers), how would you find the probability that both are hearts? The probability that the first card is a heart is easy to find—it is $\frac{13}{52}$, because there are fifty-two cards in the deck and thirteen of them are hearts. The probability that the second card is a heart is more difficult to find. There are only fifty-one cards left in the deck (one was already dealt), but how many of these are hearts? The number of hearts left in the deck depends on the first card that was dealt. If it was a heart, then there are twelve hearts left in the deck; if it was not a heart, then there are thirteen hearts left. We could certainly say that the probability that the second card is a heart, *given that the first card was a heart, is $\frac{12}{51}$.*

Therefore, the probability that the first card is a heart is $\frac{13}{52}$, and the probability that the second card is a heart, given that the first was a heart, is $\frac{12}{51}$. How do we put these two probabilities together to find the probability that *both* the first and the second cards are hearts? Should we add them? Subtract them? Multiply them? Divide them?

The answer is obtained by algebraically rewriting the Conditional Probability Definition to obtain what is called the *Product Rule:*

$$p(A \mid B) = \frac{n(A \cap B)}{n(B)} \quad \textbf{Conditional Probability Definition}$$

$$p(A \mid B) \cdot n(B) = n(A \cap B) \quad \textbf{multiplying by } n(B)$$

$$\frac{p(A \mid B) \cdot n(B)}{n(S)} = \frac{n(A \cap B)}{n(S)} \quad \textbf{dividing by } n(S)$$

$$\frac{p(A \mid B)}{1} \cdot \frac{n(B)}{n(S)} = \frac{n(A \cap B)}{n(S)} \quad \textbf{since } 1 \cdot n(S) = n(S)$$

$$p(A \mid B) \cdot p(B) = p(A \cap B) \quad \textbf{definition of probability}$$

PRODUCT RULE

For any events A and B, the probability of A and B is
$$p(A \cap B) = p(A \mid B) \cdot p(B)$$

EXAMPLE **2**

USING THE PRODUCT RULE   If two cards are dealt from a full deck, find the probability that both are hearts.

SOLUTION

$$p(A \cap B) = \qquad p(A \mid B) \qquad \cdot \qquad p(B)$$

$$p(\text{2nd heart and 1st heart}) = p(\text{2nd heart} \mid \text{1st heart}) \cdot p(\text{1st heart})$$

$$= \qquad \frac{12}{51} \qquad \cdot \qquad \frac{13}{52}$$

$$= \qquad \frac{4}{17} \qquad \cdot \qquad \frac{1}{4}$$

$$= \qquad \frac{1}{17} \approx 0.06 = 6\%$$

Therefore, there is a 6% probability that both cards are hearts.

 **Tree Diagrams**

Many people find that a *tree diagram* helps them understand problems like the one in Example 2, in which an experiment is performed in stages over time. Figure 3.38 shows the tree diagram for Example 2. The first column gives a list of the possible outcomes of the first stage of the experiment; in Example 2, the first stage is dealing the first card, and its outcomes are "heart," and "not a heart." The branches leading to those outcomes represent their probabilities. The second column gives a list of the possible outcomes of the second stage of the experiment; in Example 2, the second stage is dealing the second card. A branch leading from a first-stage outcome to a second-stage outcome is the conditional probability $p(\text{2nd stage outcome} \mid \text{1st stage outcome})$.

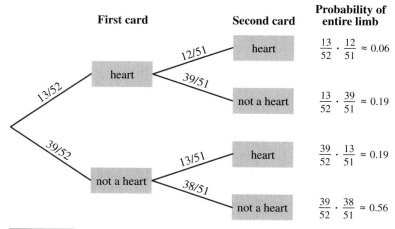

FIGURE 3.38   A tree diagram of dealing two hearts.

Looking at the top pair of branches, we see that the first branch stops at "first card is a heart" and the probability is $p(\text{1st heart}) = \frac{13}{52}$. The second branch starts at "first card is a heart" and stops at "second card is a heart" and gives the conditional probability $p(\text{2nd heart} \mid \text{1st heart}) = \frac{12}{51}$. The probability that we were asked to calculate in Example 2, p(1st heart and 2nd heart), is that of the top limb:

$$p(\text{1st heart and 2nd heart}) = p(\text{2nd heart} \mid \text{1st heart}) \cdot p(\text{1st heart})$$
$$= \frac{12}{51} \cdot \frac{13}{52}$$

(We use the word *limb* to refer to a sequence of branches that starts at the beginning of the tree.) Notice that the sum of the probabilities of the four limbs is 1.00. Because the four limbs are the only four possible outcomes of the experiment, they must add up to 1.

Conditional probabilities always start at their condition, never at the beginning of the tree. For example, p(2nd heart $\mid$ 1st heart) is a conditional probability; its condition is that the first card is a heart. Thus, its branch starts at the box "first card is a heart." However, p(1st heart) is not a conditional probability, so it starts at the beginning of the tree. Similarly, p(1st heart and 2nd heart) is not a conditional probability, so it too starts at the beginning of the tree. The product rule tells us that

$$p(\text{1st heart and 2nd heart}) = p(\text{2nd heart} \mid \text{1st heart}) \cdot p(\text{1st heart})$$

That is, the product rule tells us to multiply the branches that make up the top horizontal limb. In fact, "Multiply when moving horizontally across a limb" is a restatement of the product rule.

EXAMPLE **3**

USING TREE DIAGRAMS AND THE MUTUALLY EXCLUSIVE RULE
Two cards are drawn from a full deck. Use the tree diagram in Figure 3.38 on page 195 to find the probability that the second card is a heart.

SOLUTION

The second card can be a heart if the first card is a heart *or* if it is not. The event "the second card is a heart" is the union of the following two mutually exclusive events:

$$E = \text{1st heart and 2nd heart}$$
$$F = \text{1st not heart and 2nd heart}$$

We previously used the tree diagram to find that

$$p(E) = \frac{13}{52} \cdot \frac{12}{51}$$

Similarly,

$$p(F) = \frac{39}{52} \cdot \frac{13}{51}$$

Thus, we add the probabilities of limbs that result in the second card being a heart:

$$p(\text{2nd heart}) = p(E \cup F)$$
$$= p(E) + p(F) \qquad \text{the Mutually Exclusive Rule}$$
$$= \frac{13}{52} \cdot \frac{12}{51} + \frac{39}{52} \cdot \frac{13}{51} = 0.25$$

In Example 3, the first and third limbs represent the only two ways in which the second card can be a heart. These two limbs represent mutually exclusive events, so we used Probability Rule 5 [$p(E \cup F) = p(E) + p(F)$] to add their probabilities. In fact, "add when moving vertically from limb to limb" is a good restatement of Probability Rule 5.

TREE DIAGRAM SUMMARY

* Conditional probabilities start at their condition.
* Nonconditional probabilities start at the beginning of the tree.
* Multiply when moving horizontally across a limb.
* Add when moving vertically from limb to limb.

EXAMPLE 4

USING TREE DIAGRAMS, THE PRODUCT RULE, AND THE UNION/ INTERSECTION RULE Big Fun Bicycles manufactures its product at two plants, one in Korea and one in Peoria. The Korea plant manufactures 60% of the bicycles; 4% of the Korean bikes are defective; and 5% of the Peorian bikes are defective.

a. draw a tree diagram that shows this information.
b. use the tree diagram to find the probability that a bike is defective and came from Korea.
c. use the tree diagram to find the probability that a bike is defective.
d. use the tree diagram to find the probability that a bike is defect-free.

SOLUTION

a. First, we need to determine which probabilities have been given and find their complements, as shown in Figure 3.39.

Probabilities Given	Complements of These Probabilities
$p(\text{Korea}) = 60\% = 0.60$	$p(\text{Peoria}) = p(\text{not Korea}) = 1 - 0.60 = 0.40$
$p(\text{defective} \mid \text{Korea}) = 4\% = 0.04$	$p(\text{not defective} \mid \text{Korea}) = 1 - 0.04 = 0.96$
$p(\text{defective} \mid \text{Peoria}) = 5\% = 0.05$	$p(\text{not defective} \mid \text{Peoria}) = 1 - 0.05 = 0.95$

FIGURE 3.39 Probabilities for Example 4.

The first two of these probabilities [$p(\text{Korea})$ and $p(\text{Peoria})$] are not conditional, so they start at the beginning of the tree. The next two probabilities [$p(\text{defective} \mid \text{Korea})$ and $p(\text{not defective} \mid \text{Korea})$] are conditional, so they start at their condition (Korea). Similarly, the last two probabilities are conditional, so they start at their condition (Peoria). This placement of the probabilities yields the tree diagram in Figure 3.40.

b. The probability that a bike is defective and came from Korea is a nonconditional probability, so it starts at the beginning of the tree. Do not confuse it with the conditional probability that a bike is defective, *given that* it came from Korea, which starts at its condition (Korea). The former is the limb that goes through "Korea" and stops at "defective"; the latter is one branch of that limb. We use the product rule to multiply when moving horizontally across a limb:

$$p(\text{defective and Korea}) = p(\text{defective} \mid \text{Korea}) \cdot p(\text{Korea})$$
$$= 0.04 \cdot 0.60 = 0.024 \qquad \textbf{Product Rule}$$

This means that 2.4% of all of Big Fun's bikes are defective bikes manufactured in Korea.

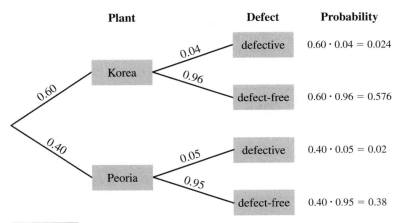

Plant	Defect	Probability

FIGURE 3.40 A tree diagram for Example 4.

c. The event that a bike is defective is the union of two mutually exclusive events:

> The bike is defective and came from Korea.
>
> The bike is defective and came from Peoria.

These two events are represented by the first and third limbs of the tree. We use the Union/Intersection Rule to add when moving vertically from limb to limb:

$$p(\text{defective}) = p(\text{defective and Korea} \cup \text{defective and Peoria})$$
$$= p(\text{defective and Korea}) + p(\text{defective and Peoria})$$
$$= 0.024 + 0.02 = 0.044$$

This means that 4.4% of Big Fun's bicycles are defective.

d. The probability that a bike is defect-free is the complement of part (c).

$$p(\text{defect-free}) = p(\text{not defective}) = 1 - 0.044 = 0.956$$

Alternatively, we can find the sum of all the limbs that stop at "defect-free":

$$p(\text{defect-free}) = 0.576 + 0.38 = 0.956$$

This means that 95.6% of Big Fun's bicycles are defect-free.

TOPIC X HIV/AIDS: PROBABILITIES IN THE REAL WORLD

The human immunodeficiency virus (or HIV) is the virus that causes AIDS. The Centers for Disease Control estimates that in 2005, 950,000 Americans had HIV/AIDS. They are of both genders, all ages and sexual orientations. Worldwide, nearly half of the 38 million people living with HIV/AIDS are women between 15 and 24 years old.

Most experts agree that the HIV/AIDS epidemic is in its early stages and that a vaccine is not on the immediate horizon. The only current hope of stemming the infection lies in education and prevention. Education is particularly important, because an infected person can be symptom-free for eight years or more. You may well know many people who are unaware that they are infected, because they have no symptoms.

Health organizations such as the Centers for Disease Control and Prevention routinely use probabilities to determine whether men or women are more likely to get HIV, which age groups are more likely to develop AIDS, and the different sources of exposure to HIV. See Exercises 49 and 50. In Section 3.7, we will investigate the accuracy of some HIV/AIDS tests. See Exercises 33 and 34 in Section 3.7.

3.6 EXERCISES

In Example 1, we wrote, "the probability that a response is yes, given that it is from a man" as p(Y | M), and we wrote, "the probability that a response is yes and is from a man" as p(Y ∩ M). In Exercises 1–4, write the given probabilities in a similar manner. Also, identify which are conditional and which are not conditional.

1. Let *H* be the event that a job candidate is hired, and let *Q* be the event that a job candidate is well qualified. Use the symbols *H*, *Q*, |, and ∩ to write the following probabilities.

   a. The probability that a job candidate is hired given that the candidate is well qualified.

   b. The probability that a job candidate is hired and the candidate is well qualified.

2. Let *W* be the event that a gambler wins a bet, and let *L* be the event that a gambler is feeling lucky. Use the symbols *W*, *L*, |, and ∩ to write the following probabilities.

   a. The probability that a gambler wins a bet and is feeling lucky.

   b. The probability that a gambler wins a bet given that the gambler is feeling lucky.

3. Let *S* be the event that a cell phone user switches carriers, and let *D* be the event that a cell phone user gets dropped a lot. Use the symbols *S*, *D*, |, and ∩ to write the following probabilities.

   a. The probability that a cell phone user switches carriers given that she gets dropped a lot.

   b. The probability that a cell phone user switches carriers and gets dropped a lot.

   c. The probability that a cell phone user gets dropped a lot given that she switches carriers.

4. Let *P* be the event that a student passes the course, and *S* be the event that a student studies hard. Use the symbols *P*, *S*, |, and ∩ to write the following probabilities.

   a. The probability that a student passes the course given that the student studies hard.

   b. The probability that a student studies hard given that the student passes the course.

   c. The probability that a student passes the course and studies hard.

In Exercises 5–8, use Figure 3.41.

5. **a.** Find $p(B | A)$
   b. Find $p(B \cap A)$

6. **a.** Find $p(B' | A)$
   b. Find $p(B' \cap A)$

7. **a.** Find $p(B | A')$
   b. Find $p(B \cap A')$

8. **a.** Find $p(B' | A')$
   b. Find $p(B' \cap A')$

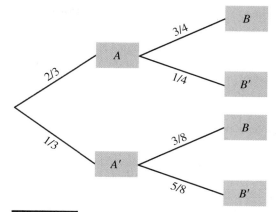

FIGURE 3.41 A tree diagram for Exercises 5–8.

9. Use the data in Figure 3.33 on page 191 to find the given probabilities. Also, write a sentence explaining what each means.

   a. $p(N)$ **b.** $p(W)$ **c.** $p(N | W)$
   d. $p(W | N)$ **e.** $p(N \cap W)$ **f.** $p(W \cap N)$

10. Use the data in Figure 3.33 on page 191 to find the given probabilities. Also, write a sentence explaining what each means.

    a. $p(N)$ **b.** $p(M)$ **c.** $p(N | M)$
    d. $p(M | N)$ **e.** $p(N \cap M)$ **f.** $p(M \cap N)$

Use the information in Figure 3.10 on page 154 to answer Exercises 11–14. Round your answers off to the nearest hundredth. Also interpret each of your answers using percentages and everyday English.

11. **a.** Find and interpret the probability that a U.S. resident dies of a pedestrian transportation accident, given that the person dies of a transportation accident in one year.

    b. Find and interpret the probability that a U.S. resident dies of a pedestrian transportation accident, given that the person dies of a transportation accident, in a lifetime.

    c. Find and interpret the probability that a U.S. resident dies of a pedestrian transportation accident, given that the person dies of a non-transportation accident, in a lifetime.

12. **a.** Find and interpret the probability that a U.S. resident dies from lightning, given that the person dies of a non-transportation accident, in one year.

    b. Find and interpret the probability that a U.S. resident dies from lightning, given that the person dies of a non-transportation accident, in a lifetime.

    c. Find and interpret the probability that a U.S. resident dies from lightning, given that the person dies of a transportation accident, in a lifetime.

13. **a.** Find and interpret the probability that a U.S. resident dies from an earthquake, given that the person dies from a non-transportation accident, in a lifetime.

   b. Find and interpret the probability that a U.S. resident dies from an earthquake, given that the person dies from a non-transportation accident, in one year.

   c. Find and interpret the probability that a U.S. resident dies from an earthquake, given that the person dies from an external cause, in one year.

14. **a.** Find and interpret the probability that a U.S. resident dies from a motorcyclist transportation accident, given that the person dies from a transportation accident, in a lifetime.

   b. Find and interpret the probability that a U.S. resident dies from a motorcyclist transportation accident, given that the person dies from a transportation accident, in one year.

   c. Find and interpret the probability that a U.S. resident dies from a motorcyclist transportation accident, given that the person dies from an external cause, in one year.

In Exercises 15–18, cards are dealt from a full deck of 52. Find the probabilities of the given events.

15. **a.** The first card is a club.

   b. The second card is a club, given that the first was a club.

   c. The first and second cards are both clubs.

   d. Draw a tree diagram illustrating this.

16. **a.** The first card is a king.

   b. The second card is a king, given that the first was a king.

   c. The first and second cards are both kings.

   d. Draw a tree diagram illustrating this. (Your diagram need not be a complete tree. It should have all the branches referred to in parts (a), (b), and (c), but it does not need other branches.)

17. **a.** The first card is a diamond.

   b. The second card is a spade, given that the first was a diamond.

   c. The first card is a diamond and the second is a spade.

   d. Draw a tree diagram illustrating this. (Your diagram need not be a complete tree. It should have all the branches referred to in parts (a), (b), and (c), but it does not need other branches.)

18. **a.** The first card is a jack.

   b. The second card is an ace, given that the first card was a jack.

   c. The first card is a jack and the second is an ace.

   d. Draw a tree diagram illustrating this.

*In Exercises 19 and 20, determine which probability the indicated branch in Figure 3.42 refers to. For example, the branch labeled * refers to the probability p(A).*

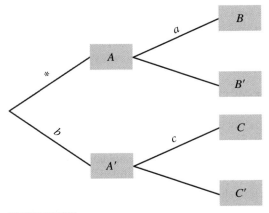

FIGURE 3.42 A tree diagram for Exercises 19 and 20.

19. **a.** the branch labeled *a*

   b. the branch labeled *b*

   c. the branch labeled *c*

20. **a.** Should probabilities (*) and (*a*) be added or multiplied? What rule tells us that? What is the result of combining them?

   b. Should probabilities (*b*) and (*c*) be added or multiplied? What rule tells us that? What is the result of combining them?

   c. Should the probabilities that result from parts (*a*) and (*b*) of this exercise be added or multiplied? What rule tells us that? What is the result of combining them?

In Exercises 21 and 22, a single die is rolled. Find the probabilities of the given events.

21. **a.** rolling a 6

   b. rolling a 6, given that the number rolled is even

   c. rolling a 6, given that the number rolled is odd

   d. rolling an even number, given that a 6 was rolled

22. **a.** rolling a 5

   b. rolling a 5, given that the number rolled is even

   c. rolling a 5, given that the number rolled is odd

   d. rolling an odd number, given that a 5 was rolled

In Exercises 23–26, a pair of dice is rolled. Find the probabilities of the given events.

23. **a.** The sum is 6.

   b. The sum is 6, given that the sum is even.

   c. The sum is 6, given that the sum is odd.

   d. The sum is even, given that the sum is 6.

24. **a.** The sum is 12.

   b. The sum is 12, given that the sum is even.

   c. The sum is 12, given that the sum is odd.

   d. The sum is even, given that the sum was 12.

25. **a.** The sum is 4.
   b. The sum is 4, given that the sum is less than 6.
   c. The sum is less than 6, given that the sum is 4.

26. **a.** The sum is 11.
   b. The sum is 11, given that the sum is greater than 10.
   c. The sum is greater than 10, given that the sum is 11.

27. A single die is rolled. Determine which of the following events is least likely and which is most likely. Do so without making any calculations. Explain your reasoning.

   E_1 is the event "rolling a 4."

   E_2 is the event "rolling a 4, given that the number rolled is even."

   E_3 is the event "rolling a 4, given that the number rolled is odd."

28. A pair of dice is rolled. Determine which of the following events is least likely and which is most likely. Do so without making any calculations. Explain your reasoning.

   E_1 is the event "rolling a 7."

   E_2 is the event "rolling a 7, given that the number rolled is even."

   E_3 is the event "rolling a 7, given that the number rolled is odd."

In Exercises 29 and 30, use the following information. To determine what effect the salespeople had on purchases, a department store polled 700 shoppers as to whether or not they had made a purchase and whether or not they were pleased with the service. Of those who had made a purchase, 125 were happy with the service and 111 were not. Of those who had made no purchase, 148 were happy with the service and 316 were not.

29. Find the probability that a shopper who was happy with the service had made a purchase (round off to the nearest hundredth). What can you conclude?

30. Find the probability that a shopper who was unhappy with the service had not made a purchase. (Round off to the nearest hundredth.) What can you conclude?

In Exercises 31–34, five cards are dealt from a full deck. Find the probabilities of the given events. (Round off to four decimal places.)

31. All are spades.

32. The fifth is a spade, given that the first four were spades.

33. The last four are spades, given that the first was a spade.

34. All are the same suit.

In Exercises 35–40, round off to the nearest hundredth.

35. If three cards are dealt from a full deck, use a tree diagram to find the probability that exactly two are spades.

36. If three cards are dealt from a full deck, use a tree diagram to find the probability that exactly one is a spade.

37. If three cards are dealt from a full deck, use a tree diagram to find the probability that exactly one is an ace.

38. If three cards are dealt from a full deck, use a tree diagram to find the probability that exactly two are aces.

39. If a pair of dice is rolled three times, use a tree diagram to find the probability that exactly two throws result in 7's.

40. If a pair of dice is rolled three times, use a tree diagram to find the probability that all three throws result in 7's.

In Exercises 41–44, use the following information: A personal computer manufacturer buys 38% of its chips from Japan and the rest from the United States. Of the Japanese chips, 1.7% are defective, and 1.1% of the American chips are defective.

41. Find the probability that a chip is defective and made in Japan.

42. Find the probability that a chip is defective and made in the United States.

43. Find the probability that a chip is defective.

44. Find the probability that a chip is defect-free.

45. The results of CNN's 2008 presidential election poll are given in Figure 3.43.

	Obama	McCain	Other/No Answer
Male Voters	23.0%	22.6%	1.4%
Female Voters	29.7%	22.8%	0.5%

FIGURE 3.43 Exit poll results, by gender. *Source:* CNN.

   a. Find the probability that a voter voted for Obama, given that the voter is male.
   b. Find the probability that a voter voted for Obama, given that the voter is female.
   c. What observations can you make?

46. Use the information in Figure 3.43 to answer the following questions.
   a. Find the probability that a voter is male, given that the voter voted for Obama.
   b. Find the probability that a voter is female, given that the voter voted for Obama.
   c. What observations can you make?

47. The results of CNN's 2008 presidential election poll are given in Figure 3.44 on page 202.
   a. Find that probability that a voter voted for McCain, given that the voter is under 45.
   b. Find the probability that a voter voted for McCain, given that the voter is 45 or over.
   c. What observations can you make?

Age of Voter	Obama	McCain	Other/No Answer
18–29	11.9%	5.8%	0.4%
30–44	15.1%	13.3%	0.6%
45–64	18.5%	18.1%	0.4%
65 and Older	7.2%	8.5%	0.3%

FIGURE 3.44 Exit poll results, by age. *Source:* CNN.

48. Use the information in Figure 3.44 to answer the following questions.
   a. Find the probability that a voter is under 45, given that the voter voted for McCain.
   b. Find the probability that a voter is 45 or over, given that the voter voted for McCain.
   c. What observations can you make?

49. Figure 3.45 gives the estimated number of diagnoses of AIDS among adults and adolescents in the United States by transmission category through the year 2007.

Transmission Category	Estimated Number of AIDS Cases, through 2007		
	Adult and Adolescent Male	Adult and Adolescent Female	Total
Male-to-male sexual contact	487,695	—	487,695
Injection drug use	175,704	80,155	255,859
Male-to-male sexual contact and injection drug use	71,242	—	71,242
High-risk heterosexual contact	63,927	112,230	176,157
Other*	12,108	6,158	18,266

FIGURE 3.45 AIDS sources. *Includes hemophilia, blood transfusion, perinatal exposure, and risk not reported or not identified. *Source:* Centers for Disease Control and Prevention.

   a. Find the probability that a U.S. adult or adolescent diagnosed with AIDS is male and the probability that a U.S. adult or adolescent diagnosed with AIDS is female.
   b. Find the probability that a male U.S. adult or adolescent diagnosed with AIDS was exposed by injection drug use (possibly combined with male-to-male sexual contact) and the probability that a female U.S. adult or adolescent diagnosed with AIDS was exposed by injection drug use.
   c. Find the probability that a U.S. adult or adolescent diagnosed with AIDS is male and was exposed by injection drug use (possibly combined with

male-to-male sexual contact) and the probability that a U.S. adult or adolescent diagnosed with AIDS is female and was exposed by injection drug use.
   d. Find the probability that a male U.S. adult or adolescent diagnosed with AIDS was exposed by heterosexual contact and the probability that a female U.S. adult or adolescent diagnosed with AIDS was exposed by heterosexual contact.
   e. Find the probability that a U.S. adult or adolescent diagnosed with AIDS is male and was exposed by heterosexual contact and the probability that a U.S. adult or adolescent diagnosed with AIDS is female and was exposed by heterosexual contact.
   f. Explain the difference between parts (b) and (c) and the difference between parts (d) and (e).

50. Figure 3.46 gives the estimated number of diagnoses of AIDS in the United States by age at the time of diagnosis, through 2007. Also note that in 2007, there were 301,621,157 residents of the United States, according to the U.S. Census Bureau.

Age (Years)	Cumulative Number of AIDS Cases
Under 13	9,209
13–14	1,169
15–24	44,264
25–34	322,370
35–44	396,851
45–54	176,304
55–64	52,409
65 or older	15,853

FIGURE 3.46 AIDS by age. *Source:* Centers for Disease Control and Prevention.

   a. Find the probability that a U.S. resident diagnosed with AIDS was 15 to 24 years old at the time of diagnosis.
   b. Find the probability that a U.S. resident was diagnosed with AIDS and was 15 to 24 years old at the time of diagnosis.
   c. Find the probability that a U.S. resident diagnosed with AIDS was 25 to 34 years old at the time of diagnosis.
   d. Find the probability that a U.S. resident was diagnosed with AIDS and was 25 to 34 years old at the time of diagnosis.
   e. Explain in words the difference between parts (c) and (d).

51. In November 2007, the National Center for Health Statistics published a document entitled "Obesity Among Adults in the United States." According to that

document, 32% of adult men and 35% of adult women in the United States were obese in 2006. At that time, there were 148 million adult men and 152 million adult women in the country.

Source: National Center for Health Statistics, C. Ogden et al: Obesity Among Adults in the United States.

a. Find the probability that an adult man was obese.

b. Find the probability that an adult was an obese man.

c. Find the probability that an adult woman was obese.

d. Find the probability that an adult was an obese woman.

e. Find the probability that an adult was obese.

f. Explain in words the difference between parts (c) and (d).

52. The document mentioned in Exercise 51 also says that 23.2% of adult men and 29.5% of adult women in the United States were overweight.

a. Find the probability that an adult man was overweight.

b. Find the probability that an adult was an overweight man.

c. Find the probability that an adult woman was overweight.

d. Find the probability that an adult was an overweight woman.

e. Find the probability that an adult was overweight.

53. In 1981 a study on race and the death penalty was released. The data in Figure 3.47 are from that study.

Death penalty imposed?	Victim's race	Defendant's race	Frequency
Yes	White	White	19
Yes	White	Black	11
Yes	Black	White	0
Yes	Black	Black	6
No	White	White	132
No	White	Black	152
No	Black	White	9
No	Black	Black	97

FIGURE 3.47 *Race and the death penalty. Source:* M. Radelet (1981) "Racial Characteristics and the Imposition of the Death Penalty." American Sociological Review, 46, 918–927.

a. Find p(death penalty imposed | victim white and defendant white).

b. Find p(death penalty imposed | victim white and defendant black).

c. Find p(death penalty imposed | victim black and defendant white).

d. Find p(death penalty imposed | victim black and defendant black).

e. What can you conclude from parts (a) through (d)?

f. Determine what other conditional probabilities would affect this issue, and calculate those probabilities.

g. Discuss your results.

54. The information in Exercise 53 is rather dated. Use the following information to determine whether things have improved since then. In April 2003, Amnesty International issued a study on race and the death penalty. The following quote is from that study:

"The population of the USA is approximately 75 percent white and 12 percent black. Since 1976, blacks have been six to seven times more likely to be murdered than whites, with the result that blacks and whites are the victims of murder in about equal numbers. Yet, 80 percent of the more than 840 people put to death in the USA since 1976 were convicted of crimes involving white victims, compared to the 13 percent who were convicted of killing blacks. Less than four percent of the executions carried out since 1977 in the USA were for crimes involving Hispanic victims. Hispanics represent about 12 percent of the US population. Between 1993 and 1999, the recorded murder rate for Hispanics was more than 40 percent higher than the national homicide rate."

(*Source:* Amnesty International. "United States of America: Death by discrimination—the continuing role of race in capital cases." AMR 51/046/2003)

a. In the above quote, what conditional probability is 80%? That is, find events A and B such that $p(A \mid B) = 80\%$.

b. In the above quote, what conditional probability is 13%? That is, find events C and D such that $p(C \mid D) = 13\%$.

c. Use the information in Exercise 53 to compute $p(A \mid B)$ for 1981.

d. Use the information in Exercise 53 to compute $p(C \mid D)$ for 1981.

e. Have things improved since 1981? Justify your answer.

55. A man and a woman have a child. Both parents have sickle-cell trait. They know that their child does not have sickle-cell anemia because he shows no symptoms, but they are concerned that he might be a carrier. Find the probability that he is a carrier.

56. A man and a woman have a child. Both parents are Tay-Sachs carriers. They know that their child does not have Tay-Sachs disease because she shows no symptoms, but they are concerned that she might be a carrier. Find the probability that she is a carrier.

In Exercises 57–60, use the following information. The University of Metropolis requires its students to pass an examination in college-level mathematics before they can graduate. The students are given three chances to pass the exam; 61% pass it on their first attempt, 63% of those that take it a second time pass it then, and 42% of those that take it a third time pass it then.

57. What percentage of the students pass the exam?

58. What percentage of the students are not allowed to graduate because of their performance on the exam?

59. What percentage of the students take the exam at least twice?

60. What percent of the students take the test three times?

61. In the game of blackjack, if the first two cards dealt to a player are an ace and either a ten, jack, queen, or king, then the player has a blackjack, and he or she wins. Find the probability that a player is dealt a blackjack out of a full deck (no jokers).

62. In the game of blackjack, the dealer's first card is dealt face up. If that card is an ace, then the player has the option of "taking insurance." "Insurance" is a side bet. If the dealer has blackjack, the player wins the insurance bet and is paid 2 to 1 odds. If the dealer does not have a blackjack, the player loses the insurance bet. Find the probability that the dealer is dealt a blackjack if his or her first card is an ace.

63. Use the data in Figure 3.33 to find the following probabilities, where N is the event "saying no," and W is the event "being a woman":
   a. $p(N' \mid W)$ **b.** $p(N \mid W')$ **c.** $p(N' \mid W')$
   d. Which event, $N' \mid W$, $N \mid W'$, or $N' \mid W'$, is the complement of the event $N \mid W$? Why?

64. Use the data in Figure 3.33 to find the following probabilities, where Y is the event "saying yes," and M is the event "being a man.":
   a. $p(Y' \mid M)$ **b.** $p(Y \mid M')$ **c.** $p(Y' \mid M')$
   d. Which event, $Y' \mid M$, $Y \mid M'$, or $Y' \mid M'$, is the complement of the event $Y \mid M$? Why?

65. If A and B are arbitrary events, what is the complement of the event $A \mid B$?

66. Show that $p(A \mid B) = \dfrac{p(A \cap B)}{p(B)}$.

   HINT: Divide the numerator and denominator of the Conditional Probability Definition by $n(S)$.

67. Use Exercise 66 and appropriate answers from Exercise 9 to find $P(N \mid W)$.

68. Use Exercise 66 and appropriate answers from Exercise 10 to find $P(Y \mid M)$.

Answer the following questions using complete sentences and your own words.

• CONCEPT QUESTIONS

69. Which must be true for any events A and B?
   • $P(A \mid B)$ is always greater than or equal to $P(A)$.
   • $P(A \mid B)$ is always less than or equal to $P(A)$.
   • Sometimes $P(A \mid B)$ is greater than or equal to $P(A)$, and sometimes $P(A \mid B)$ is less than or equal to $P(A)$, depending on the events A and B.

   Answer this without making any calculations. Explain your reasoning.

70. Compare and contrast the events A, $A \mid B$, $B \mid A$, and $A \cap B$. Be sure to discuss both the similarities and the differences between these events.

WEB PROJECTS

71. There are many different blood type systems, but the ABO and Rh systems are the most important systems for blood donation purposes. These two systems generate eight different blood types: A+, A−, B+, B−, AB+, AB−, O+, and O−.

   a. For each of these eight blood types, use the web to determine the percentage of U.S. residents that have that blood type. Interpret these percentages as probabilities. Use language like "the probability that a randomly-selected U.S. resident . . . ."

   b. You can always give blood to someone with the same blood type. In some cases, you can give blood to someone with a different blood type, or receive blood from someone with a different blood type. This depends on the donor's blood type and the receiver's blood type. Use the web to complete the chart in Figure 3.48.

   c. For each of the eight blood types, use the web to determine the percentage of U.S. residents that can donate blood to a person of that blood type. Interpret these percentages as conditional probabilities.

   d. For each of the eight blood types, use the web to determine the percentage of U.S. residents that can receive blood from a person of that blood type. Interpret these percentages as conditional probabilities.

Blood group	A+	A−	B+	B−	AB+	AB−	O+	O−
Can donate blood to								
Can receive blood from								

FIGURE 3.48 Blood types.

e. Are the probabilities in parts (a), (c), and (d) theoretical probabilities or relative frequencies? Why?

Some useful links for this web project are listed on the text web site:
www.cengage.com/math/johnson

72. According to the U.S. Department of Transportation's National Highway Traffic Safety Administration, "rollovers are dangerous incidents and have a higher fatality rate than other kinds of crashes. Of the nearly 11 million passenger car, SUV, pickup, and van crashes in 2002, only 3% involved a rollover. However, rollovers accounted for nearly 33% of all deaths from passenger vehicle crashes."
(*Source:* **http://www. safercar.gov/Rollover/pages/RolloCharFat.htm**)

Use probabilities and the web to investigate rollovers. How likely is a rollover if you are driving a sedan, an SUV, or a van? Which models have the highest probability of a rollover? Which models have the lowest probability? Wherever possible, give specific conditional probabilities.

Some useful links for this web project are listed on the text web site:
www.cengage.com/math/johnson

• PROJECTS

73. In 1973, the University of California at Berkeley admitted 1,494 of 4,321 female applicants for graduate study and 3,738 of 8,442 male applicants.
(*Source:* P.J. Bickel, E.A. Hammel, and J.W. O'Connell, "Sex Bias in Graduate Admissions: Data from Berkeley," *Science,* vol. 187, 7 February 1975.)

a. Find the probability that:
- an applicant was admitted
- an applicant was admitted, given that he was male
- an applicant was admitted, given that she was female

Discuss whether these data indicate a possible bias against women.

b. Berkeley's graduate students are admitted by the department to which they apply rather than by a campuswide admissions panel. When P(admission |

male) and P(admission | female) were computed for each of the school's more than 100 departments, it was found that in four departments, P(admission | male) was greater than P(admission | female) by a significant amount and that in six departments, P(admission | male) was less than P(admission | female) by a significant amount. Discuss whether this information indicates a possible bias against women and whether it is consistent with that in part (a).

c. The authors of "Sex Bias in Graduate Admissions: Data from Berkeley" attempt to explain the paradox by discussing an imaginary school with only two departments: "machismatics" and "social wafare." Machismatics admitted 200 of 400 male applicants for graduate study and 100 of 200 female applicants, while social warfare admitted 50 of 150 male applicants for graduate study and 150 of 450 female applicants. For *the school as a whole and for each of the two departments*, find the probability that:
- an applicant was admitted
- an applicant was admitted, given that he was male
- an applicant was admitted, given that she was female

Discuss whether these data indicate a possible bias against women.

d. Explain the paradox illustrated in parts (a)–(c).

e. What conclusions would you make, and what further information would you obtain, if you were an affirmative action officer at Berkeley?

74. a. Use the data in Figure 3.49 to find the probability that:
- a New York City resident died from tuberculosis
- a Caucasian New York City resident died from tuberculosis
- a non-Caucasian New York City resident died from tuberculosis
- a Richmond resident died from tuberculosis
- a Caucasian Richmond resident died from tuberculosis
- a non-Caucasian Richmond resident died from tuberculosis

b. What conclusions would you make, and what further information would you obtain, if you were a public health official?

	New York City		Richmond, Virginia	
	Population	**TB deaths**	**Population**	**TB deaths**
Caucasian	4,675,000	8400	81,000	130
Non-Caucasian	92,000	500	47,000	160

FIGURE 3.49 Tuberculosis deaths by race and location, 1910.
Source: Morris R. Cohen and Ernest Nagel, *An Introduction to Logic and Scientific Method* (New York: Harcourt Brace & Co, 1934.)

Independence; Trees in Genetics

O BJECTIVES

- Understand the difference between dependent and independent events
- Know the effect of independence on the Product Rule
- Be able to apply tree diagrams in medical and genetic situations

Dependent and Independent Events

Consider the dealing of two cards from a full deck. An observer who saw that the first card was a heart would be better able to predict whether the second card will be a heart than would another observer who did not see the first card. If the first card was a heart, there is one fewer heart in the deck, so it is slightly less likely that the second card will be a heart. In particular,

$$p(\text{2nd heart} \mid \text{1st heart}) = \frac{12}{51} \approx 0.24$$

whereas, as we saw in Example 3 of Section 3.6,

$$p(\text{2nd heart}) = 0.25$$

These two probabilities are different because of the effect the first card drawn has on the second. We say that the two events "first card is a heart" and "second card is a heart" are *dependent;* the result of dealing the second card depends, to some extent, on the result of dealing the first card. In general, two events E and F are **dependent** if $p(E \mid F) \neq p(E)$.

Consider two successive tosses of a single die. An observer who saw that the first toss resulted in a three would be *no better able* to predict whether the second toss will result in a three than another observer who did not observe the first toss. In particular,

$$p(\text{2nd toss is a three}) = \frac{1}{6}$$

and

$$p(\text{2nd toss is a three} \mid \text{1st toss was a three}) = \frac{1}{6}$$

These two probabilities are the same, because the first toss has no effect on the second toss. We say that the two events "first toss is a three" and "second toss is a three" are *independent;* the result of the second toss does *not* depend on the result of the first toss. In general, two events E and F are **independent** if $p(E \mid F) = p(E)$.

INDEPENDENCE/DEPENDENCE DEFINITIONS

Two events E and F are **independent** if $p(E \mid F) = p(E)$.
(Think: Knowing F does not affect E's probability.)
Two events E and F are **dependent** if $p(E \mid F) \neq p(E)$.
(Think: Knowing F does affect E's probability.)

Many people have difficulty distinguishing between *independent* and *mutually exclusive*. (Recall that two events E and F are mutually exclusive if $E \cap F = \varnothing$, that is, if one event excludes the other.) This is probably because the relationship between mutually exclusive events and the relationship between independent events both could be described, in a very loose sort of way, by saying that "the two events have nothing to do with each other." *Never think this way;* mentally replacing "mutually exclusive" or "independent" with "having nothing to do with each other" only obscures the distinction between these two concepts. E and F are independent if knowing that F has occurred *does not* affect the probability that E will occur. E and F are dependent if knowing that F has occurred *does* affect the probability that E will occur. E and F are mutually exclusive if E and F cannot occur simultaneously.

EXAMPLE 1

INDEPENDENT EVENTS AND MUTUALLY EXCLUSIVE EVENTS   Let F be the event "a person has freckles," and let R be the event "a person has red hair."

a. Are F and R independent?
b. Are F and R mutually exclusive?

SOLUTION

a. F and R are independent if $p(F \mid R) = p(F)$. With $p(F \mid R)$, we are given that a person has red hair; with $p(F)$, we are not given that information. Does knowing that a person has red hair affect the probability that the person has freckles? Yes, it does; $p(F \mid R) > p(F)$. Therefore, F and R are not independent; they are dependent.

b. F and R are mutually exclusive if $F \cap R = \varnothing$. Many people have both freckles and red hair, so $F \cap R \neq \varnothing$, and F and R are not mutually exclusive. In other words, having freckles does not exclude the possibility of having red hair; freckles and red hair can occur simultaneously.

EXAMPLE 2

INDEPENDENT EVENTS AND MUTUALLY EXCLUSIVE EVENTS   Let T be the event "a person is tall," and let R be the event "a person has red hair."

a. Are T and R independent?
b. Are T and R mutually exclusive?

SOLUTION

a. T and R are independent if $p(T \mid R) = p(T)$. With $p(T \mid R)$, we are given that a person has red hair; with $p(T)$, we are not given that information. Does knowing that a person has red hair affect the probability that the person is tall? No, it does not; $p(T \mid R) = p(T)$, so T and R are independent.

b. T and R are mutually exclusive if $T \cap R = \varnothing$. $T \cap R$ is the event "a person is tall and has red hair." There are tall people who have red hair, so $T \cap R \neq \varnothing$, and T and R are not mutually exclusive. In other words, being tall does not exclude the possibility of having red hair; being tall and having red hair can occur simultaneously.

In Examples 1 and 2, we had to rely on our personal experience in concluding that knowledge that a person has red hair does affect the probability that he or she has freckles and does not affect the probability that he or she is tall. It may be the case that you have seen only one red-haired person and she was short and without freckles. Independence is better determined by computing the appropriate probabilities than by relying on one's own personal experiences. This is especially crucial in determining the effectiveness of an experimental drug. *Double-blind* experiments, in which neither the patient nor the doctor knows whether the given medication is the experimental drug or an inert substance, are often done to ensure reliable, unbiased results.

Independence is an important tool in determining whether an experimental drug is an effective vaccine. Let D be the event that the experimental drug was administered to a patient, and let R be the event that the patient recovered. It is hoped that $p(R \mid D) > p(R)$, that is, that the rate of recovery is greater among those who were given the drug. In this case, R and D are dependent. Independence is also an important tool in determining whether an advertisement effectively promotes a product. An ad is effective if $p(\text{consumer purchases product} \mid \text{consumer saw ad}) > p(\text{consumer purchases product})$.

EXAMPLE **3**

DETERMINING INDEPENDENCE   Use probabilities to determine whether the events "thinking there is too much violence in television" and "being a man" in Example 1 of Section 3.6 are independent.

SOLUTION

Two events E and F are independent if $p(E \mid F) = p(E)$. The events "responding yes to the question on violence in television" and "being a man" are independent if $p(Y \mid M) = p(Y)$. We need to compute these two probabilities and compare them.

In Example 1 of Section 3.6, we found $p(Y \mid M) \approx 0.58$. We can use the data from the poll in Figure 3.33 to find $p(Y)$.

$$p(Y) = \frac{418}{600} \approx 0.70$$

$$p(Y \mid M) \neq p(Y)$$

The events "responding yes to the question on violence in television" and "being a man" are dependent. According to the poll, men are less likely to think that there is too much violence on television.

In Example 3, what should we conclude if we found that $p(Y) = 0.69$ and $p(Y \mid M) = 0.67$? Should we conclude that $p(Y \mid M) \neq p(Y)$ and that the events "thinking that there is too much violence on television" and "being a man" are dependent? Or should we conclude that $p(Y \mid M) \approx p(Y)$ and that the events are (probably) independent? In this particular case, the probabilities are relative frequencies rather than theoretical probabilities, and relative frequencies can vary. A group of 600 people was polled to determine the opinions of the entire viewing public; if the same question was asked of a different group, a somewhat different set of relative frequencies could result. While it would be reasonable to conclude that the events are (probably) independent, it would be more appropriate to include more people in the poll and make a new comparison.

Product Rule for Independent Events

The product rule says that $p(A \cap B) = p(A \mid B) \cdot p(B)$. If A and B are independent, then $p(A \mid B) = p(A)$. Combining these two equations, we get the following rule:

PRODUCT RULE FOR INDEPENDENT EVENTS

If A and B are independent events, then the probability of A and B is

$$p(A \cap B) = p(A) \cdot p(B)$$

A common error that is made in computing probabilities is using the formula $p(A \cap B) = p(A) \cdot p(B)$ without verifying that A and B are independent. In fact, the Federal Aviation Administration (FAA) has stated that this is the most frequently encountered error in probabilistic analysis of airplane component failures. If it is not known that A and B are independent, you must use the Product Rule $p(A \cap B) = p(A \mid B) \cdot p(B)$.

EXAMPLE **4**

USING THE PRODUCT RULE FOR INDEPENDENT EVENTS   If a pair of dice is tossed twice, find the probability that each toss results in a seven.

SOLUTION

In Example 4 of Section 3.3, we found that the probability of a seven is $\frac{1}{6}$. The two rolls are independent (one roll has no influence on the next), so the probability of a seven is $\frac{1}{6}$ regardless of what might have happened on an earlier roll; we can use the Product Rule for Independent Events:

$$p(A \cap B) = p(A) \cdot p(B)$$
$$p(\text{1st is 7 and 2nd is 7}) = p(\text{1st is 7}) \cdot p(\text{2nd is 7})$$
$$= \frac{1}{6} \cdot \frac{1}{6}$$
$$= \frac{1}{36}$$

See Figure 3.50. The thicker branch of the tree diagram starts at the event "1st roll is 7"and ends at the event "2nd roll is 7," so it is the conditional probability $p(\text{2nd is 7} \mid \text{1st is 7})$. However, the two rolls are independent, so $p(\text{2nd is 7} \mid \text{1st is 7}) = p(\text{2nd is 7})$. We are free to label this branch as either of these two equivalent probabilities.

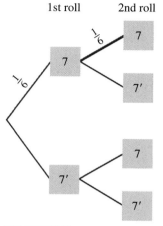

FIGURE 3.50
A tree diagram for Example 4.

Trees in Medicine and Genetics

Usually, medical diagnostic tests are not 100% accurate. A test might indicate the presence of a disease when the patient is in fact healthy (this is called a **false positive**), or it might indicate the absence of a disease when the patient does in fact have the disease (a **false negative**). Probability trees can be used to determine the probability that a person whose test results were positive actually has the disease.

EXAMPLE **5**

ANALYZING THE EFFECTIVENESS OF DIAGNOSTIC TESTS   Medical researchers have recently devised a diagnostic test for "white lung" (an imaginary disease caused by the inhalation of chalk dust). Teachers are particularly susceptible to this disease; studies have shown that half of all teachers are afflicted with it. The test correctly diagnoses the presence of white lung in 99% of the people who have it and correctly diagnoses its absence in 98% of the people who do not have it. Find the probability that a teacher whose test results are positive actually has white lung and the probability that a teacher whose test results are negative does not have white lung.

SOLUTION

First, we determine which probabilities have been given and find their complements, as shown in Figure 3.51. We use $+$ to denote the event that a person receives a positive diagnosis and $-$ to denote the event that a person receives a negative diagnosis.

Probabilities Given	Complements of Those Probabilities
$p(\text{ill}) = 0.50$	$p(\text{healthy}) = p(\text{not ill}) = 1 - 0.50 = 0.50$
$p(- \mid \text{healthy}) = 98\% = 0.98$	$p(+ \mid \text{healthy}) = 1 - 0.98 = 0.02$
$p(+ \mid \text{ill}) = 99\% = 0.99$	$p(- \mid \text{ill}) = 1 - 0.99 = 0.01$

FIGURE 3.51   Data from Example 5.

The first two of these probabilities [$p(\text{ill})$ and $p(\text{healthy})$] are not conditional, so they start at the beginning of the tree. The next two probabilities [$p(- \mid \text{healthy})$

and $p(+ \mid \text{healthy})$] are conditional, so they start at their condition (healthy). Similarly, the last two probabilities are conditional, so they start at their condition (ill). This placement of the probabilities yields the tree diagram in Figure 3.52.

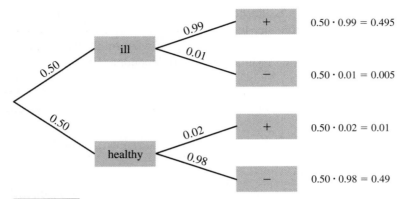

FIGURE 3.52 A tree diagram for Example 5.

The four probabilities to the right of the tree are

$p(\text{ill and } +) = 0.495$
$p(\text{ill and } -) = 0.005$
$p(\text{healthy and } +) = 0.01$
$p(\text{healthy and } -) = 0.49$

We are to find the probability that a teacher whose test results are positive actually has white lung; that is, we are to find $p(\text{ill} \mid +)$. Thus, we are given that the test results are positive, and we need only consider branches that involve positive test results: $p(\text{ill and } +) = 0.495$ and $p(\text{healthy and } +) = 0.01$. The probability that a teacher whose test results are positive actually has white lung is

$$p(\text{ill} \mid +) = \frac{0.495}{0.495 + 0.01} = 0.9801 \ldots \approx 98\%$$

The probability that a teacher whose test results are negative does not have white lung is

$$p(\text{healthy} \mid -) = \frac{0.49}{0.49 + 0.005} = 0.98989 \ldots \approx 99\%$$

The probabilities show that this diagnostic test works well in determining whether a teacher actually has white lung. In the exercises, we will see how well it would work with schoolchildren.

EXAMPLE **6**

FINDING THE PROBABILITY THAT YOUR CHILD INHERITS A DISEASE Mr. and Mrs. Smith each had a sibling who died of cystic fibrosis. The Smiths are considering having a child. They have not been tested to determine whether they are carriers. What is the probability that their child would have cystic fibrosis?

SOLUTION

In our previous examples involving trees, we were given probabilities; the conditional or nonconditional status of those probabilities helped us to determine the physical layout of the tree. In this example, we are not given any probabilities. To determine the physical layout of the tree, we have to separate what we know to be true from what is only possible or probable. The tree focuses on what is possible or probable. We know that both Mr. and Mrs. Smith had a sibling who died of cystic fibrosis, and it is possible that their child would inherit that disease. The tree's branches will represent the series of possible events that could result in the Smith child inheriting cystic fibrosis.

What events must take place if the Smith child is to inherit the disease? First, the grandparents would have to have had cystic fibrosis genes. Next, the Smiths themselves would have to have inherited those genes from their parents. Finally, the Smith child would have to inherit those genes from his or her parents.

Cystic fibrosis is recessive, which means that a person can inherit it only if he or she receives two cystic fibrosis genes, one from each parent. Mr. and Mrs. Smith each had a sibling who had cystic fibrosis, so each of the four grandparents must have been a carrier. They could not have actually had the disease because the Smiths would have known, and we would have been told.

We now know the physical layout of our tree. The four grandparents were definitely carriers. Mr. and Mrs. Smith were possibly carriers. The Smith child will possibly inherit the disease. The first set of branches will deal with Mr. and Mrs. Smith, and the second set will deal with the child.

The Smith child will not have cystic fibrosis unless Mr. and Mrs. Smith are both carriers. Figure 3.53 shows the Punnett square for Mr. and Mrs. Smith's possible genetic configuration.

	N	**c**	
N	NN	cN	← **possible offspring**
c	Nc	cc	← **possible offspring**

↑ one grandparent's genes

↑

other grandparent's genes

FIGURE 3.53 A Punnett square for Example 6.

Neither Mr. Smith nor Mrs. Smith has the disease, so we can eliminate the cc possibility. Thus, the probability that Mr. Smith is a carrier is $\frac{2}{3}$, as is the probability that Mrs. Smith is a carrier. Furthermore, these two events are independent, since the Smiths are (presumably) unrelated.

Using the Product Rule for Independent Events, we have

$p(\text{Mr. S is a carrier and Mrs. S is a carrier})$

$$= p(\text{Mr. S is a carrier}) \cdot p(\text{Mrs. S is a carrier}) = \frac{2}{3} \cdot \frac{2}{3} = \frac{4}{9}$$

The same Punnett square tells us that the probability that their child will have cystic fibrosis, given that the Smiths are both carriers, is $\frac{1}{4}$. Letting B be the event that both parents are carriers and letting F be the event that the child has cystic fibrosis, we obtain the tree diagram in Figure 3.54.

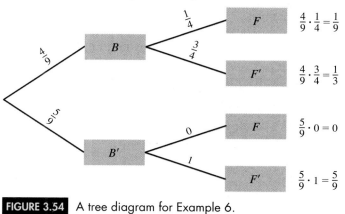

FIGURE 3.54 A tree diagram for Example 6.

The probability that the Smiths' child would have cystic fibrosis is $\frac{1}{9}$.

Notice that the tree in Figure 3.54 could have been drawn differently, as shown in Figure 3.55. It is not necessary to draw a branch going from the B' box to the F box, since the child cannot have cystic fibrosis if both parents are not carriers.

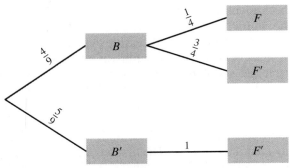

FIGURE 3.55 An alternative to the tree in Figure 3.54.

Hair Color

Like cystic fibrosis, hair color is inherited, but the method by which it is transmitted is more complicated than the method by which an inherited disease is transmitted. This more complicated method of transmission allows for the possibility that a child might have hair that is colored differently from that of other family members.

Hair color is determined by two pairs of genes: one pair that determines placement on a blond/brown/black spectrum and one pair that determines the presence or absence of red pigment. These two pairs of genes are independent.

Melanin is a brown pigment that affects the color of hair (as well as the colors of eyes and skin). The pair of genes that controls the brown hair colors does so by determining the amount of melanin in the hair. This gene has three forms, each traditionally labeled with an M (for melanin): M^{Bd}, or blond (a light melanin deposit); M^{Bw}, or brown (a medium melanin deposit); and M^{Bk}, or black (a heavy melanin deposit). Everyone has two of these genes, and their combination determines the brown aspect of hair color, as illustrated in Figure 3.56.

The hair colors in Figure 3.56 are altered by the presence of red pigment, which is determined by another pair of genes. This gene has two forms: R^-, or no red pigment, and R^+, or red pigment. Because everyone has two of these genes, there are three possibilities for the amount of red pigment: R^-R^-, R^+R^-, and R^+R^+. The amount of red pigment in a person's hair is independent of the brownness of his or her hair.

The actual color of a person's hair is determined by the interaction of these two pairs of genes, as shown in Figure 3.57.

Genes	Hair Color
$M^{Bd}M^{Bd}$	blond
$M^{Bd}M^{Bw}$	light brown
$M^{Bd}M^{Bk}$ $M^{Bw}M^{Bw}$	medium brown
$M^{Bw}M^{Bk}$	dark brown
$M^{Bk}M^{Bk}$	black

FIGURE 3.56 Melanin.

Genes	Blond ($M^{Bd}M^{Bd}$)	Light Brown ($M^{Bd}M^{Bw}$)	Medium Brown ($M^{Bd}M^{Bk}$, $M^{Bw}M^{Bw}$)	Dark Brown ($M^{Bw}M^{Bk}$)	Black ($M^{Bk}M^{Bk}$)
R^-R^-	blond	light brown	medium brown	dark brown	black
R^+R^-	strawberry blond	reddish brown	chestnut	shiny dark brown	shiny black
R^+R^+	bright red	dark red	auburn	glossy dark brown	glossy black

FIGURE 3.57 Melanin and red pigment.

EXAMPLE **7**

SOLUTION

PREDICTING A CHILD'S HAIR COLOR The Rosses are going to have a child. Mr. Ross has blond hair, and Mrs. Ross has reddish brown hair. Find their child's possible hair colors and the probabilities of each possibility.

The parent with blond hair has genes $M^{Bd}M^{Bd}$ and R^-R^-. The parent with reddish brown hair has genes $M^{Bd}M^{Bw}$ and R^+R^-. We need to use two Punnett squares, one for the brownness of the hair (Figure 3.58) and one for the presence of red pigment (Figure 3.59).

	M^{Bd}	M^{Bd}
M^{Bd}	$M^{Bd}M^{Bd}$	$M^{Bd}M^{Bd}$
M^{Bw}	$M^{Bd}M^{Bw}$	$M^{Bd}M^{Bw}$

FIGURE 3.58 Brownness.

	R^-	R^-
R^+	R^+R^-	R^+R^-
R^-	R^-R^-	R^-R^-

FIGURE 3.59 Red pigment.

$$p(M^{Bd}M^{Bd}) = \frac{2}{4} = \frac{1}{2}$$

$$p(M^{Bd}M^{Bw}) = \frac{2}{4} = \frac{1}{2}$$

$$p(R^+R^-) = \frac{2}{4} = \frac{1}{2}$$

$$p(R^-R^-) = \frac{2}{4} = \frac{1}{2}$$

We will use a tree diagram to determine the possible hair colors and their probabilities (see Figure 3.60).

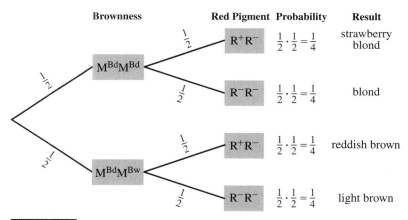

FIGURE 3.60 A tree diagram for Example 7.

The Rosses' child could have strawberry blond, blond, reddish brown, or light brown hair. The probability of each color is $\frac{1}{4}$. Notice that there is a 50% probability that the child will have hair that is colored differently from that of either parent.

3.7 EXERCISES

In Exercises 1–8, use your own personal experience with the events described to determine whether (a) E and F are independent and (b) E and F are mutually exclusive. (Where appropriate, these events are meant to be simultaneous; for example, in Exercise 2, "E and F" would mean "it's simultaneously raining and sunny.") Write a sentence justifying each of your answers.

1. *E* is the event "being a doctor," and *F* is the event "being a woman."

2. *E* is the event "it's raining," and *F* is the event "it's sunny."

3. *E* is the event "being single," and *F* is the event "being married."

4. *E* is the event "having naturally blond hair," and *F* is the event "having naturally black hair."

5. *E* is the event "having brown hair," and *F* is the event "having gray hair."

6. *E* is the event "being a plumber," and *F* is the event "being a stamp collector."

7. *E* is the event "wearing shoes," and *F* is the event "wearing sandals."

8. *E* is the event "wearing shoes," and *F* is the event "wearing socks."

In Exercises 9–10, use probabilities, rather than your own personal experience, to determine whether (a) E and F are independent and (b) E and F are mutually exclusive. (c) Interpret your answers to parts (a) and (b).

9. If a die is rolled once, *E* is the event "getting a 4," and *F* is the event "getting an odd number."

10. If a die is rolled once, *E* is the event "getting a 4," and *F* is the event "getting an even number."

11. Determine whether the events "responding yes to the question on violence in television" and "being a woman" in Example 1 of Section 3.6 are independent.

12. Determine whether the events "having a defect" and "being manufactured in Peoria" in Example 4 of Section 3.6 are independent.

13. A single die is rolled once.
    a. Find the probability of rolling a 5.
    b. Find the probability of rolling a 5 given that the number rolled is even.
    c. Are the events "rolling a 5" and "rolling an even number" independent? Why?
    d. Are the events "rolling a 5" and "rolling an even number" mutually exclusive? Why?
    e. Interpret the results of parts (c) and (d).

14. A pair of dice is rolled once.
    a. Find the probability of rolling a 6.
    b. Find the probability of rolling a 6 given that the number rolled is even.
    c. Are the events "rolling a 6" and "rolling an even number" independent? Why?
    d. Are the events "rolling a 6" and "rolling an even number" mutually exclusive? Why?
    e. Interpret the results of parts (c) and (d).

15. A card is dealt from a full deck (no jokers).
    a. Find the probability of being dealt a jack.
    b. Find the probability of being dealt a jack given that you were dealt a red card.
    c. Are the events "being dealt a jack" and "being dealt a red card" independent? Why?
    d. Are the events "being dealt a jack" and "being dealt a red card" mutually exclusive? Why?
    e. Interpret the results of parts (c) and (d).

16. A card is dealt from a full deck (no jokers).
    a. Find the probability of being dealt a jack.
    b. Find the probability of being dealt a jack given that you were dealt a card above a 7 (count aces high).
    c. Are the events "being dealt a jack" and "being dealt a card above a 7" independent? Why?
    d. Are the events "being dealt a jack" and "being dealt a card above a 7" mutually exclusive? Why?
    e. Interpret the results of parts (c) and (d).

In Exercises 17 and 18, use the following information: To determine what effect salespeople had on purchases, a department store polled 700 shoppers as to whether or not they made a purchase and whether or not they were pleased with the service. Of those who made a purchase, 125 were happy with the service and 111 were not. Of those who made no purchase, 148 were happy with the service and 316 were not.

17. Are the events "being happy with the service" and "making a purchase" independent? What conclusion can you make?

18. Are the events "being unhappy with the service" and "not making a purchase" independent? What conclusion can you make?

19. A personal computer manufacturer buys 38% of its chips from Japan and the rest from the United States. Of the Japanese chips, 1.7% are defective, whereas 1.1% of the U.S. chips are defective. Are the events "defective" and "Japanese-made" independent? What conclusion can you draw? (See Exercises 41–44 in Section 3.6.)

20. A skateboard manufacturer buys 23% of its ball bearings from a supplier in Akron, 38% from one in Atlanta, and the rest from one in Los Angeles. Of the ball bearings from Akron, 4% are defective; 6.5% of those from Atlanta are defective; and 8.1% of those from Los Angeles are defective.
    a. Find the probability that a ball bearing is defective.
    b. Are the events "defective" and "from the Los Angeles supplier" independent?
    c. Are the events "defective" and "from the Atlanta supplier" independent?
    d. What conclusion can you draw?

21. Over the years, a group of nutritionists have observed that their vegetarian clients tend to have fewer health problems. To determine whether their observation is accurate, they collected the following data:
    • They had 365 clients.
    • 281 clients are healthy.
    • Of the healthy clients, 189 are vegetarians.
    • Of the unhealthy clients, 36 are vegetarians.
    a. Use the data to determine whether the events "being a vegetarian" and "being healthy" are independent.
    b. Use the data to determine whether the events "being a vegetarian" and "being healthy" are mutually exclusive.
    c. Interpret the results of parts (a) and (b).

HINT: Start by organizing the data in a chart, similar to that used in the "Violence on Television" poll in Figure 3.33 in Section 3.6.

22. Over the years, a group of exercise physiologists have observed that their clients who exercise solely by running tend to have more ankle problems than do their clients who vary between running and other forms of exercise. To determine whether their observation is accurate, they collected the following data:
    * They had 422 clients.
    * 276 clients are have ankle problems.
    * Of the clients with ankle problems, 191 only run.
    * Of the clients without ankle problems, 22 only run.
    a. Use the data to determine whether the events "running only" and "having ankle problems" are independent.
    b. Use the data to determine whether the events "running only" and "having ankle problems" are mutually exclusive.
    c. Interpret the results of parts (a) and (b).

    HINT: Start by organizing the data in a chart, similar to that used in the "Violence on Television" poll in Figure 3.33 in Section 3.6.

23. The Venn diagram in Figure 3.61 contains the results of a survey. Event *A* is "supports proposition 3," and event *B* is "lives in Bishop."

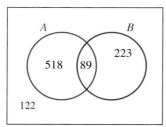

FIGURE 3.61   Venn diagram for Exercise 23.

    a. Are the events "supports proposition 3" and "lives in Bishop" independent?
    b. Are the events "supports proposition 3" and "lives in Bishop" mutually exclusive?
    c. Interpret the results of parts (a) and (b).

24. The Venn diagram in Figure 3.62 contains the results of a survey. Event *A* is "uses Ipana toothpaste," and event *B* is "has good dental checkups."

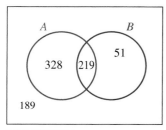

FIGURE 3.62   Venn diagram for Exercise 24.

a. Are the events "uses Ipana toothpaste" and "has good dental checkups" independent?
b. Are the events "uses Ipana toothpaste" and "has good dental checkups" mutually exclusive?
c. Interpret the results of parts (a) and (b).

In Exercises 25–28, you may wish to use Exercise 66 in Section 3.6.

25. Sixty percent of all computers sold last year were HAL computers, 4% of all computer users quit using computers, and 3% of all computer users used HAL computers and then quit using computers. Are the events "using a HAL computer" and "quit using computers" independent? What conclusion can you make?

26. Ten percent of all computers sold last year were Peach computers, 4% of all computer users quit using computers, and 0.3% of all computer users used Peach computers and then quit using computers. Are the events "using a Peach computer" and "quit using computers" independent? What conclusion can you make?

27. Forty percent of the country used SmellSoGood deodorant, 10% of the country quit using deodorant, and 4% of the country used SmellSoGood deodorant and then quit using deodorant. Are the events "used SmellSoGood" and "quit using deodorant" independent? What conclusion can you make?

28. Nationwide, 20% of all TV viewers use VideoLink cable, 8% of all TV viewers switched from cable to a satellite service, and 7% of all TV viewers used VideoLink cable and then switched from cable to a satellite service. Are the events "used VideoLink cable" and "switched from cable to satellite" independent? What conclusion can you make?

29. Suppose that the space shuttle has three separate computer control systems: the main system and two backup duplicates of it. The first backup would monitor the main system and kick in if the main system failed. Similarly, the second backup would monitor the first. We can assume that a failure of one system is independent of a failure of another system, since the systems are separate. The probability of failure for any one system on any one mission is known to be 0.01.
    a. Find the probability that the shuttle is left with no computer control system on a mission.
    b. How many backup systems does the space shuttle need if the probability that the shuttle is left with no computer control system on a mission must be $\frac{1}{1 \text{ billion}}$?

30. Use the information in Example 5 to find $p(\text{healthy} \mid +)$, the probability that a teacher receives a false positive.

31. Use the information in Example 5 to find $p(\text{ill} \mid -)$, the probability that a teacher receives a false negative.

32. Overwhelmed with their success in diagnosing white lung in teachers, public health officials decided to administer the test to all schoolchildren, even though only one child in 1,000 has contracted the disease. Recall from Example 5 that the test correctly diagnoses the presence of white lung in 99% of the people who have it and correctly diagnoses its absence in 98% of the people who do not have it.

  a. Find the probability that a schoolchild whose test results are positive actually has white lung.

  b. Find the probability that a schoolchild whose test results are negative does not have white lung.

  c. Find the probability that a schoolchild whose test results are positive does not have white lung.

  d. Find the probability that a schoolchild whose test results are negative actually has white lung.

  e. Which of these events is a false positive? Which is a false negative?

  f. Which of these probabilities would you be interested in if you or one of your family members tested positive?

  g. Discuss the usefulness of this diagnostic test, both for teachers (as in Example 5) and for schoolchildren.

33. In 2004, the Centers for Disease Control and Prevention estimated that 1,000,000 of the 287,000,000 residents of the United States are HIV-positive. (HIV is the virus that is believed to cause AIDS.) The SUDS diagnostic test correctly diagnoses the presence of AIDS/HIV 99.9% of the time and correctly diagnoses its absence 99.6% of the time.

  a. Find the probability that a person whose test results are positive actually has HIV.

  b. Find the probability that a person whose test results are negative does not have HIV.

  c. Find the probability that a person whose test results are positive does not have HIV.

  d. Find the probability that a person whose test results are negative actually has HIV.

  e. Which of the probabilities would you be interested in if you or someone close to you tested positive?

  f. Which of these events is a false positive? Which is a false negative?

  g. Discuss the usefulness of this diagnostic test.

  h. It has been proposed that all immigrants to the United States should be tested for HIV before being allowed into the country. Discuss this proposal.

34. Assuming that the SUDS test cannot be made more accurate, what changes in the circumstances described in Exercise 33 would increase the usefulness of the SUDS diagnostic test? Give a specific example of this change in circumstances, and demonstrate how it would increase the test's usefulness by computing appropriate probabilities.

35. Compare and contrast the circumstances and the probabilities in Example 5 and in Exercises 32, 33, and 34. Discuss the difficulties in using a diagnostic test when that test is not 100% accurate.

36. Use the information in Example 6 to find the following:

  a. the probability that the Smiths' child is a cystic fibrosis carrier

  b. the probability that the Smiths' child is healthy (i.e., has no symptoms)

  c. the probability that the Smiths' child is healthy and not a carrier

  HINT: You must consider three possibilities: Both of the Smiths are carriers, only one of the Smiths is a carrier, and neither of the Smiths is a carrier.

37. **a.** Two first cousins marry. Their mutual grandfather's sister died of cystic fibrosis. They have not been tested to determine whether they are carriers. Find the probability that their child will have cystic fibrosis. (Assume that no other grandparents are carriers.)

  b. Two unrelated people marry. Each had a grandparent whose sister died of cystic fibrosis. They have not been tested to determine whether they are carriers. Find the probability that their child will have cystic fibrosis. (Assume that no other grandparents are carriers.)

38. It is estimated that one in twenty-five Americans is a cystic fibrosis carrier. Find the probability that a randomly selected American couple's child will have cystic fibrosis (assuming that they are unrelated).

39. In 1989, researchers announced a new carrier-detection test for cystic fibrosis. However, it was discovered in 1990 that the test will detect the presence of the cystic fibrosis gene in only 85% of cystic fibrosis carriers. If two unrelated people are in fact carriers of cystic fibrosis, find the probability that they both test positive. Find the probability that they don't both test positive.

40. Ramon del Rosario's mother's father died of Huntington's disease. His mother died in childbirth, before symptoms of the disease would have appeared in her. Find the probability that Ramon will have the disease. (Assume that Ramon's grandfather had one bad gene, and that there was no other source of Huntington's in the family.) (Huntington's disease is discussed on page 149 in Section 3.2.)

41. Albinism is a recessive disorder that blocks the normal production of pigmentation. The typical albino has white hair, white skin, and pink eyes and is visually impaired. Mr. Jones is an albino, and although Ms. Jones is normally pigmented, her brother is an albino. Neither of Ms. Jones' parents is an albino. Find the probability that their child will be an albino.

42. Find the probability that the Joneses' child will not be an albino but will be a carrier. (See Exercise 41.)

43. If the Joneses' first child is an albino, find the probability that their second child will be an albino. (See Exercise 41.)

44. The Donohues are going to have a child. She has shiny black hair, and he has bright red hair. Find their child's possible hair colors and the probabilities of each possibility.

45. The Yorks are going to have a child. She has black hair, and he has dark red hair. Find their child's possible hair colors and the probabilities of each possibility.

46. The Eastwoods are going to have a child. She has chestnut hair ($M^{Bd}M^{Bk}$), and he has dark brown hair. Find their child's possible hair colors and the probabilities of each possibility.

47. The Wilsons are going to have a child. She has strawberry blond hair, and he has shiny dark brown hair. Find their child's possible hair colors and the probabilities of each possibility.

48. The Breuners are going to have a child. She has blond hair, and he has glossy dark brown hair. Find their child's possible hair colors and the probabilities of each possibility.

49. The Landres are going to have a child. She has chestnut hair ($M^{Bd}M^{Bk}$), and he has shiny dark brown hair. Find their child's possible hair colors and the probabilities of each possibility.

50. The Hills are going to have a child. She has reddish brown hair, and he has strawberry blond hair. Find their child's possible hair colors and the probabilities of each possibility.

51. Recall from Section 3.1 that Antoine Gombauld, the Chevalier de Méré, had made money over the years by betting with even odds that he could roll at least one 6 in four rolls of a single die. This problem finds the probability of winning that bet and the expected value of the bet.

  a. Find the probability of rolling a 6 in one roll of one die.

  b. Find the probability of not rolling a 6 in one roll of one die.

  c. Find the probability of never rolling a 6 in four rolls of a die.

  HINT: This would mean that the first roll is not a 6 *and* the second roll is not a 6 *and* the third is not a 6 *and* the fourth is not a 6.

  d. Find the probability of rolling at least one 6 in four rolls of a die.

  HINT: Use complements.

  e. Find the expected value of this bet if $1 is wagered.

52. Recall from Section 3.1 that Antoine Gombauld, the Chevalier de Méré, had lost money over the years by betting with even odds that he could roll at least one pair of 6's in twenty-four rolls of a pair of dice and that he could not understand why. This problem finds the probability of winning that bet and the expected value of the bet.

  a. Find the probability of rolling a double 6 in one roll of a pair of dice.

  b. Find the probability of not rolling a double 6 in one roll of a pair of dice.

  c. Find the probability of never rolling a double 6 in twenty-four rolls of a pair of dice.

  d. Find the probability of rolling at least one double 6 in twenty-four rolls of a pair of dice.

  e. Find the expected value of this bet if $1 is wagered.

53. Probability theory began when Antoine Gombauld, the Chevalier de Méré, asked his friend Blaise Pascal why he had made money over the years betting that he could roll at least one 6 in four rolls of a single die but he had lost money betting that he could roll at least one pair of 6's in twenty-four rolls of a pair of dice. Use decision theory and the results of Exercises 51 and 52 to answer Gombauld.

54. Use Exercise 66 of Section 3.6 to explain the calculations of $p(\text{ill} \mid +)$ and $p(\text{healthy} \mid -)$ in Example 5 of this section.

55. Dr. Wellby's patient exhibits symptoms associated with acute neural toxemia (an imaginary disease), but the symptoms can have other, innocuous causes. Studies show that only 25% of those who exhibit the symptoms actually have acute neural toxemia (ANT). A diagnostic test correctly diagnoses the presence of ANT in 88% of the persons who have it and correctly diagnoses its absence in 92% of the persons who do not have it. ANT can be successfully treated, but the treatment causes side effects in 2% of the patients. If left untreated, 90% of those with ANT die; the rest recover fully. Dr. Wellby is considering ordering a diagnostic test for her patient and treating the patient if test results are positive, but she is concerned about the treatment's side effects.

  a. Dr. Wellby could choose to test her patient and administer the treatment if the results are positive. Find the probability that her patient's good health will return under this plan.

  b. Dr. Wellby could choose to avoid the treatment's side effects by not administering the treatment. (This also implies not testing the patient.) Find the probability that her patient's good health will return under this plan.

  c. Find the probability that the patient's good health will return if he undergoes treatment regardless of the test's outcome.

  d. On the basis of the probabilities, should Dr. Wellby order the test and treat the patient if test results are positive?

Answer the following questions using complete sentences and your own words.

• CONCEPT QUESTIONS

56. In Example 5, we are given that $p(- \mid \text{healthy}) = 98\%$ and $p(+ \mid \text{ill}) = 99\%$, and we computed that $p(\text{ill} \mid +) \approx 98\%$ and that $p(\text{healthy} \mid -) \approx 99\%$. Which of these probabilities would be most important

to a teacher who was diagnosed as having white lung? Why?

57. Are the melanin hair color genes (M^{Bd}, M^{Bw}, and M^{Bk}) dominant, recessive, or codominant? Are the redness hair color genes (R^+ and R^-) dominant, or recessive, or codominant? Why?

58. Compare and contrast the concepts of independence and mutual exclusivity. Be sure to discuss both the similarities and the differences between these two concepts.

3 CHAPTER REVIEW

TERMS

carrier
certain event
codominant gene
conditional probability
decision theory

dependent events
dominant gene
event
expected value
experiment
false negative
false positive

genes
impossible event
independent events
Law of Large Numbers
mutually exclusive events
odds
outcome

probability
Punnett square
recessive gene
relative frequency
sample space

PROBABILITY RULES

1. The probability of the null set is 0: $p(\varnothing) = 0$
2. The probability of the sample space is 1: $p(S) = 1$
3. Probabilities are between 0 and 1 (inclusive): $0 \le p(S) \le 1$
4. $p(E \cup F) = p(E) + p(F) - p(E \cap F)$
5. $p(E \cup F) = p(E) + p(F)$ if E and F are mutually exclusive
6. $p(E) + p(E') = 1$

FORMULAS

If outcomes are equally likely, then:

- the **probability** of an event E is $p(E) = n(E)/n(S)$
- the **odds** of an event E are $o(E) = n(E){:}n(E')$

To find the **expected value** of an experiment, multiply the value of each outcome by its probability and add the results.

The **conditional probability** of A given B is

$p(A \mid B) = n(A \cap B)/n(B)$ if outcomes are equally likely

The **product rule:**

$p(A \cap B) = p(A \mid B) \cdot p(B)$ for any two events A and B

$p(A \cap B) = p(A) \cdot p(B)$ if A and B are independent

REVIEW EXERCISES

In Exercises 1–6, a card is dealt from a well-shuffled deck of fifty-two cards.

1. Describe the experiment and the sample space.

2. Find and interpret the probability and the odds of being dealt a queen.

3. Find and interpret the probability and the odds of being dealt a club.

4. Find and interpret the probability and the odds of being dealt the queen of clubs.

5. Find and interpret the probability and the odds of being dealt a queen or a club.

6. Find and interpret the probability and the odds of being dealt something other than a queen.

In Exercises 7–12, three coins are tossed.

7. Find the experiment and the sample space.

8. Find the event E that exactly two are tails.

9. Find the event F that two or more are tails.

10. Find and interpret the probability of E and the odds of E.

11. Find and interpret the probability of F and the odds of F.

12. Find and interpret the probability of E' and the odds of E'.

In Exercises 13–18, a pair of dice is tossed. Find and interpret the probability of rolling each of the following.

13. a 7

14. an 11

15. a 7, an 11, or doubles

16. a number that is both odd and greater than 8

17. a number that is either odd or greater than 8

18. a number that is neither odd nor greater than 8

In Exercises 19–24, three cards are dealt from a deck of fifty-two cards. Find the probability of each of the following.

19. All three are hearts.

20. Exactly two are hearts.

21. At least two are hearts.

22. The first is an ace of hearts, the second is a 2 of hearts, and the third is a 3 of hearts.

23. The second is a heart, given that the first is a heart.

24. The second is a heart, and the first is a heart.

In Exercises 25–30, a pair of dice is rolled three times. Find the probability of each of the following.

25. All three are 7's.

26. Exactly two are 7's.

27. At least two are 7's.

28. None are 7's.

29. The second roll is a 7, given that the first roll is a 7.

30. The second roll is a 7 and the first roll is a 7.

In Exercises 31–32, use the following information. A long-stemmed pea is dominant over a short-stemmed one. A pea with one long-stemmed gene and one short-stemmed gene is crossed with a pea with two short-stemmed genes.

31. Find and interpret the probability that the offspring will be long-stemmed.

32. Find and interpret the probability that the offspring will be short-stemmed.

In Exercises 33–35, use the following information: Cystic fibrosis is caused by a recessive gene. Two cystic fibrosis carriers produce a child.

33. Find the probability that that child will have the disease.

34. Find the probability that that child will be a carrier.

35. Find the probability that that child will neither have the disease nor be a carrier.

In Exercises 36–38, use the following information: Sickle-cell anemia is caused by a codominant gene. A couple, each of whom has sickle-cell trait, produce a child. (Sickle-cell trait involves having one bad gene. Sickle-cell anemia involves having two bad genes.)

36. Find the probability that that child will have the disease.

37. Find the probability that that child will have sickle-cell trait.

38. Find the probability that that child will have neither sickle-cell disease nor sickle-cell trait.

In Exercises 39–41, use the following information: Huntington's disease is caused by a dominant gene. One parent has Huntington's disease. This parent has a single gene for Huntington's disease.

39. Find the probability that that child will have the disease.

40. Find the probability that that child will be a carrier.

41. Find the probability that that child will neither have the disease nor be a carrier.

42. Find the probability of being dealt a pair of 10's (and no other 10's) when playing five-card draw poker.

43. Find the probability of being dealt a pair of tens and three jacks when playing five-card draw poker.

44. Find the probability of being dealt a pair of tens and a pair of jacks (and no other tens or jacks) when playing five-card draw poker.

45. In nine-spot keno, five winning spots breaks even, six winning spots pays $50, seven pays $390, eight pays $6000, and nine pays $25,000.

a. Find the probability of each of these events.

b. Find the expected value of a $1 bet.

46. Some $1 bets in craps (specifically, the pass, don't pass, come, and don't come bets) have expected values of −$0.014. Use decision theory to compare these bets with a $1 bet in nine-spot keno. (See Exercise 45.)

47. At a certain office, three people make $6.50 per hour, three make $7, four make $8.50, four make $10, four make $13.50, and two make $25. Find the expected hourly wage at that office.

In Exercises 48–55, use the following information: Jock O'Neill, a sportscaster, and Trudy Bell, a member of the state assembly, are both running for governor of the state of Erehwon. A recent telephone poll asked 800 randomly selected voters whom they planned on voting for. The results of this poll are shown in Figure 3.63.

	Jock O'Neill	**Trudy Bell**	**Undecided**
Urban residents	266	184	22
Rural residents	131	181	16

FIGURE 3.63 Data for Exercises 48–55.

48. Find the probability that an urban resident supports O'Neill and the probability that an urban resident supports Bell.

49. Find the probability that a rural resident supports O'Neill and the probability that a rural resident supports Bell.

50. Find the probability that an O'Neill supporter lives in an urban area and the probability that an O'Neill supporter lives in a rural area.

51. Find the probability that a Bell supporter lives in an urban area and the probability that a Bell supporter lives in a rural area.

52. Where are O'Neill supporters more likely to live? Where are Bell supporters more likely to live?

53. Which candidate do the urban residents tend to prefer? Which candidate do the rural residents tend to prefer?

54. Are the events "supporting O'Neill" and "living in an urban area" independent or dependent? What can you conclude?

55. On the basis of the poll, who is ahead in the gubernatorial race?

Are the following events independent or dependent? Are they mutually exclusive?

56. "It is summer" and "it is sunny." (Use your own personal experience.)

57. "It is summer" and "it is Monday." (Use your own personal experience.)

58. "It is summer" and "it is autumn." (Use your own personal experience.)

59. "The first card dealt is an ace" and "the second card dealt is an ace." (Do not use personal experience.)

60. "The first roll of the dice results in a 7" and "the second roll results in a 7" (Do not use personal experience.)

In Exercises 61–66, use the following information: Gregor's Garden Corner buys 40% of their plants from the Green Growery and the balance from Herb's Herbs. Twenty percent of the plants from the Green Growery must be returned, and 10% of those from Herb's Herbs must be returned.

61. What percent of all plants are returned?

62. What percent of all plants are not returned?

63. What percent of all plants are from the Green Growery and are returned?

64. What percent of the returned plants are from the Green Growery?

65. What percent of the returned plants are from Herb's Herbs?

66. Are the events "a plant must be returned" and "a plant was from Herb's Herbs" independent? What conclusion can you make?

In Exercises 67–73, use the following information: The Nissota Automobile Company buys emergency flashers from two different manufacturers, one in Arkansas and one in Nevada. Thirty-nine percent of its turn-signal indicators are purchased from the Arkansas manufacturer, and the rest are purchased from the Nevada manufacturer. Two percent of the Arkansas turn-signal indicators are defective, and 1.7% of the Nevada indicators are defective.

67. What percent of the indicators are defective and made in Arkansas?

68. What percentage of the indicators are defective and made in Nevada?

69. What percentage of the indicators are defective?

70. What percentage of the indicators are not defective?

71. What percentage of the defective indicators are made in Arkansas?

72. What percentage of the defective indicators are made in Nevada?

73. Are the events "made in Arkansas" and "a defective" independent? What conclusion can you make?

 Answer the following questions with complete sentences and your own words.

• CONCEPT QUESTIONS

74. What is a conditional probability?

75. Give two events that are independent, and explain why they are independent.

76. Give two events that are dependent, and explain why they are dependent.

77. Give two events that are mutually exclusive, and explain why they are mutually exclusive.

78. Give two events that are not mutually exclusive, and explain why they are not mutually exclusive.

79. Why are probabilities always between 0 and 1, inclusive?

80. Give an example of a permutation and a similar example of a combination.

81. Give an example of two events that are mutually exclusive, and explain why they are mutually exclusive.

82. Give an example of two events that are not mutually exclusive, and explain why they are not mutually exclusive.

83. How is set theory used in probability theory?

• HISTORY

84. What two mathematicians invented probability theory? Why?

85. Why was probability theory not considered to be a serious branch of mathematics? What changed that viewpoint? Give a specific example of something that helped change that viewpoint.

86. What role did the following people play in the development of probability theory?

Jacob Bernoulli

Gerolamo Cardano

Pierre de Fermat

Antoine Gombauld, the Chevalier de Méré

John Graunt

Pierre-Simon, the Marquis de Laplace

Gregor Mendel

Blaise Pascal

STATISTICS

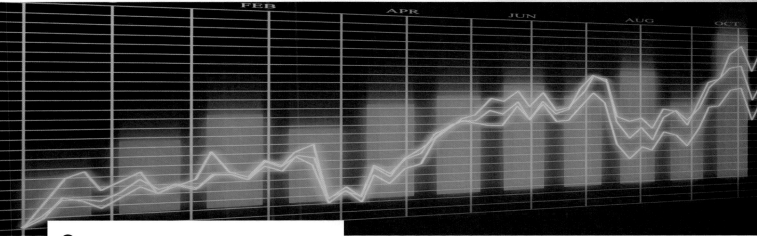

$\mathbf{4}$

Statistics are everywhere. The news, whether reported in a newspaper, on television, through the Internet, or over the radio, includes statistics of every kind. When shopping for a new car, you will certainly examine the statistics (average miles per gallon, acceleration times, braking distances, and so on) of the various makes and models you are considering. Statistics abound in government studies, and the interpretation of these statistics affects us all. Industry is driven by statistics; they are essential to the direction of quality control, marketing research, productivity, and many other factors. Sporting events are laden with statistics concerning the past performance of the teams and players.

A person who understands the nature of statistics is equipped to see beyond short-term and individual perspectives.

continued

$\mathbf{W}$HAT WE WILL DO IN THIS CHAPTER

WE'LL EXPLORE DIFFERENT TYPES OF STATISTICS, BOTH DESCRIPTIVE AND INFERENTIAL:

- Once data have been collected from a sample, how should they be organized and presented?

- Once they have been organized, how should data be summarized?

- Once data have been summarized, how are conclusions drawn and predictions made?

WE'LL ANALYZE AND EXPLORE DISTRIBUTIONS OF DATA THAT EXHIBIT SPECIFIC TRENDS OR PATTERNS:

- It is not uncommon for data to be clustered around a central value. In many instances, this type of pattern can be represented as a "bell-shaped" curve (high in the middle, low at each end).

- All bell curves share common features. What are these features and how can they be applied in specific situations?

continued

WE'LL ANALYZE AND INTERPRET SURVEYS AND OPINION POLLS:

- Once the opinions of a specific sample of people have been collected, how can they be used to predict the overall opinion of a large population?

- How accurate or reliable are opinion polls?

WE'LL EXAMINE RELATIONSHIPS BETWEEN TWO SETS OF DATA:

- Are the values of one variable related to the values of another variable?

- If two variables are related, how can the relationship be expressed so that predictions can be made?

He or she is also better prepared to deal with those who use statistics in misleading ways. To many people, the word *statistics* conjures up an image of an endless list of facts and figures. Where do statistics come from? What do they mean? In this chapter, you will learn to handle basic statistical problems and expand your knowledge of the meanings, uses, and misuses of statistics.

4.1 Population, Sample, and Data

OBJECTIVES

- Construct a frequency distribution
- Construct a histogram
- Construct a pie chart

The field of **statistics** can be defined as the science of collecting, organizing, and summarizing data in such a way that valid conclusions and meaningful predictions can be drawn from them. The first part of this definition, "collecting, organizing, and summarizing data," applies to **descriptive statistics.** The second part, "drawing valid conclusions and making meaningful predictions," describes **inferential statistics.**

Population versus Sample

population

sample

FIGURE 4.1

Population versus sample.

Who will become the next president of the United States? During election years, political analysts spend a lot of time and money trying to determine what percent of the vote each candidate will receive. However, because there are over 175 million registered voters in the United States, it would be virtually impossible to contact each and every one of them and ask, "Whom do you plan on voting for?" Consequently, analysts select a smaller group of people, determine their intended voting patterns, and project their results onto the entire body of all voters.

Because of time and money constraints, it is very common for researchers to study the characteristics of a small group in order to estimate the characteristics of a larger group. In this context, the set of all objects under study is called the **population,** and any subset of the population is called a **sample** (see Figure 4.1).

When we are studying a large population, we might not be able to collect data from every member of the population, so we collect data from a smaller, more manageable sample. Once we have collected these data, we can summarize by calculating various descriptive statistics, such as the average value. Inferential statistics, then, deals with drawing conclusions (hopefully, valid ones!) about the population, based on the descriptive statistics of the sample data.

Sample data are collected and summarized to help us draw conclusions about the population. A good sample is representative of the population from which it was taken. Obviously, if the sample is not representative, the conclusions concerning the population might not be valid. The most difficult aspect of inferential statistics is obtaining a representative sample. Remember that conclusions are only as reliable as the sampling process and that information will usually change from sample to sample.

What person who lived in the twentieth century do you admire most? The top ten responses in a Gallup poll taken on Dec. 20–21, 1999, were as follows: (1) Mother Teresa, (2) Martin Luther King, Jr., (3) John F. Kennedy, (4) Albert Einstein, (5) Helen Keller, (6) Franklin D. Roosevelt, (7) Billy Graham, (8) Pope John Paul II, (9) Eleanor Roosevelt, and (10) Winston Churchill.

© Tim Graham Photo Library/Getty Images

Frequency Distributions

The first phase of any statistical study is the collection of data. Each element in a set of data is referred to as a **data point.** When data are first collected, the data points might show no apparent patterns or trends. To summarize the data and detect any trends, we must organize the data. This is the second phase of descriptive statistics. The most common way to organize raw data is to create a **frequency distribution,** a table that lists each data point along with the number of times it occurs (its **frequency**).

The composition of a frequency distribution is often easier to see if the frequencies are converted to percents, especially if large amounts of data are being summarized. The **relative frequency** of a data point is the frequency of the data point expressed as a percent of the total number of data points (that is, made *relative* to the total). The relative frequency of a data point is found by dividing its frequency by the total number of data points in the data set. Besides listing the frequency of each data point, a frequency distribution should also contain a column that gives the relative frequencies.

EXAMPLE **1**

CREATING A FREQUENCY DISTRIBUTION: SINGLE VALUES While bargaining for their new contract, the employees of 2 Dye 4 Clothing asked their employers to provide daycare service as an employee benefit. Examining the personnel files of the company's fifty employees, the management recorded the number of children under six years of age that each employee was caring for. The following results were obtained:

0 2 1 0 3 2 0 1 1 0
0 1 1 2 4 1 0 1 1 0
2 1 0 0 3 0 0 1 2 1 $n = 50$
0 0 2 4 1 1 0 1 2 0
1 1 0 3 5 1 2 1 3 2 $= 50$

Organize the data by creating a frequency distribution.

SOLUTION

First, we list each different number in a column, putting them in order from smallest to largest (or vice versa). Then we use tally marks to count the number of times each data point occurs. The frequency of each data point is shown in the third column of Figure 4.2.

Number of Children under Six	Tally	Frequency	Relative Frequency
0	卌 卌 卌 I	16 / 50	$\frac{16}{50} = 0.32 = 32\%$
1	卌 卌 卌 III	18 / 50	$\frac{18}{50} = 0.36 = 36\%$
2	卌 IIII	9 / 50	$\frac{9}{50} = 0.18 = 18\%$
3	IIII	4 / 50	$\frac{4}{50} = 0.08 = 8\%$
4	II	2 / 50	$\frac{2}{50} = 0.04 = 4\%$
5	I	1 / 50	$\frac{1}{50} = 0.02 = 2\%$
		$n = 50$	total $= 100\%$

FIGURE 4.2 Frequency distribution of data. *No. of 6 yr olds*

To get the relative frequencies, we divide each frequency by 50 (the total number of data points) and change the resulting decimal to a percent, as shown in the fourth column of Figure 4.2.

Adding the frequencies, we see that there is a total of $n = 50$ data points in the distribution. This is a good way to monitor the tally process.

The raw data have now been organized and summarized. At this point, we can see that about one-third of the employees have no need for child care (32%), while the remaining two-thirds (68%) have at least one child under 6 years of age who would benefit from company-sponsored daycare. The most common trend (that is, the data point with the highest relative frequency for the fifty employees) is having one child (36%).

Grouped Data

When raw data consist of only a few distinct values (for instance, the data in Example 1, which consisted of only the numbers 0, 1, 2, 3, 4, and 5), we can easily organize the data and determine any trends by listing each data point along with its frequency and relative frequency. However, when the raw data consist of many nonrepeated data points, listing each one separately does not help us to see any trends the data set might contain. In such cases, it is useful to group the data into intervals or classes and then determine the frequency and relative frequency of each group rather than of each data point.

EXAMPLE **2**

CREATING A FREQUENCY DISTRIBUTION: GROUPED DATA   Keith Reed is an instructor for an acting class offered through a local arts academy. The class is open to anyone who is at least 16 years old. Forty-two people are enrolled; their ages are as follows:

26 16 21 34 45 18 41 38 22
48 27 22 30 39 62 25 25 38
29 31 28 20 56 60 24 61 28
32 33 18 23 27 46 30 34 62
49 59 19 20 23 24

Organize the data by creating a frequency distribution.

SOLUTION

This example is quite different from Example 1. Example 1 had only six different data values, whereas this example has many. Listing each distinct data point and its frequency might not summarize the data well enough for us to draw conclusions. Instead, we will work with grouped data.

First, we find the largest and smallest values (62 and 16). Subtracting, we find the range of ages to be $62 - 16 = 46$ years. In working with grouped data, it is customary to create between four and eight groups of data points. We arbitrarily choose six groups, the first group beginning at the smallest data point, 16. To find the beginning of the second group (and hence the end of the first group), divide the range by the number of groups, round off this answer to be consistent with the data, and then add the result to the smallest data point:

$46 \div 6 = 7.6666666 \ldots \approx 8$ **This is the width of each group.**

The beginning of the second group is $16 + 8 = 24$, so the first group consists of people from 16 up to (but not including) 24 years of age.

In a similar manner, the second group consists of people from 24 up to (but not including) 32 ($24 + 8 = 32$) years of age. The remaining groups are formed and the ages tallied in the same way. The frequency distribution is shown in Figure 4.3.

x = Age (years)	Tally	Frequency	Relative Frequency
$16 \leq x < 24$	‖‖‖ ‖‖‖ ‖	11	$\frac{11}{42} \approx 26\%$
$24 \leq x < 32$	‖‖‖ ‖‖‖ ‖‖‖	13	$\frac{13}{42} \approx 31\%$
$32 \leq x < 40$	‖‖‖ ‖‖	7	$\frac{7}{42} \approx 17\%$
$40 \leq x < 48$	‖‖‖	3	$\frac{3}{42} \approx 7\%$
$48 \leq x < 56$	‖‖	2	$\frac{2}{42} \approx 5\%$
$56 \leq x < 64$	‖‖‖ ‖	6	$\frac{6}{42} \approx 14\%$
		$n = 42$	total = 100%

FIGURE 4.3   Frequency distribution of grouped data.

Now that the data have been organized, we can observe various trends: Ages from 24 to 32 are most common (31% is the highest relative frequency), and ages from 48 to 56 are least common (5% is the lowest). Also, over half the people enrolled (57%) are from 16 to 32 years old.

When we are working with grouped data, we can choose the groups in any desired fashion. The method used in Example 2 might not be appropriate in all situations. For example, we used the smallest data point as the beginning of the first group, but we could have begun the first group at an even smaller number. The following box gives a general method for constructing a frequency distribution.

CONSTRUCTING A FREQUENCY DISTRIBUTION

1. If the raw data consist of many different values, create intervals and work with grouped data. If not, list each distinct data point. [When working with grouped data, choose from four to eight intervals. Divide the range (high minus low) by the desired number of intervals, round off this answer to be consistent with the data, and then add the result to the lowest data point to find the beginning of the second group.]
2. Tally the number of data points in each interval or the number of times each individual data point occurs.
3. List the frequency of each interval or each individual data point.
4. Find the relative frequency by dividing the frequency of each interval or each individual data point by the total number of data points in the distribution. The resulting decimal can be expressed as a percent.

Histograms

When data are grouped in intervals, they can be depicted by a **histogram,** a bar chart that shows how the data are distributed in each interval. To construct a histogram, mark off the class limits on a horizontal axis. If each interval has equal width, we draw two vertical axes; the axis on the left exhibits the frequency of an interval, and the axis on the right gives the corresponding relative frequency. We then draw a rectangle above each interval; the height of the rectangle corresponds to the number of data points contained in the interval. The vertical scale on the right gives the percentage of data contained in each interval. The histogram depicting the distribution of the ages of the people in Keith Reed's acting class (Example 2) is shown in Figure 4.4.

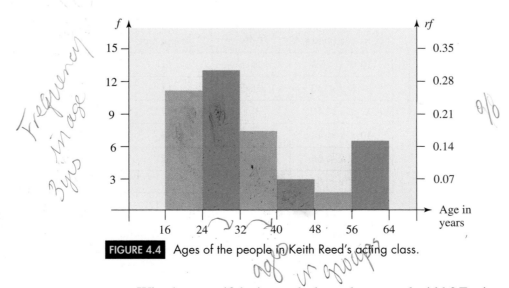

FIGURE 4.4 Ages of the people in Keith Reed's acting class.

What happens if the intervals do not have equal width? For instance, suppose the ages of the people in Keith Reed's acting class are those given in the frequency distribution shown in Figure 4.5.

$x =$ Age (years)	Frequency	Relative Frequency	Class Width
$15 \leq x < 20$	4	$\frac{4}{42} \approx 10\%$	5
$20 \leq x < 25$	11	$\frac{11}{42} \approx 26\%$	5
$25 \leq x < 30$	6	$\frac{6}{42} \approx 14\%$	5
$30 \leq x < 45$	11	$\frac{11}{42} \approx 26\%$	15
$45 \leq x < 65$	10	$\frac{10}{42} \approx 24\%$	20
	$n = 42$	Total $= 100\%$	

FIGURE 4.5 Frequency distribution of age.

With frequency and relative frequency as the vertical scales, the histogram depicting this new distribution is given in Figure 4.6.

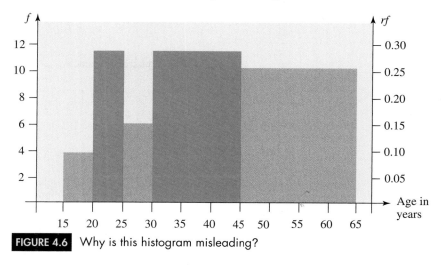

FIGURE 4.6 Why is this histogram misleading?

Does the histogram in Figure 4.6 give a truthful representation of the distribution? No; the rectangle over the interval from 45 to 65 appears to be larger than the rectangle over the interval from 30 to 45, yet the interval from 45 to 65 contains fewer data than the interval from 30 to 45. This is misleading; rather than comparing the heights of the rectangles, our eyes naturally compare the areas of the rectangles. Therefore, to make an accurate comparison, *the areas of the rectangles must correspond to the relative frequencies of the intervals*. This is accomplished by utilizing the **density** of each interval.

Histograms and Relative Frequency Density

Density is a ratio. In science, density is used to determine the concentration of weight in a given volume: Density = weight/volume. For example, the density of water is 62.4 pounds per cubic foot. In statistics, density is used to determine the concentration of data in a given interval: Density = (percent of total data)/(size of an interval). Because relative frequency is a measure of the percentage of data within an interval, we shall calculate the relative frequency density of an interval to determine the concentration of data within the interval.

For example, if the interval $20 \leq x < 25$ contains eleven out of forty-two data points, then the relative frequency density of the interval is

$$rfd = \frac{f/n}{\Delta x_i} = \frac{11/42}{5} = 0.052380952\ldots$$

DEFINITION OF RELATIVE FREQUENCY DENSITY

Given a set of n data points, if an interval contains f data points, then the **relative frequency density** (*rfd*) of the interval is

$$rfd = \frac{f/n}{\Delta x}$$

where Δx is the width of the interval.

If a histogram is constructed using relative frequency density as the vertical scale, the area of a rectangle will correspond to the relative frequency of the interval, as shown in Figure 4.7.

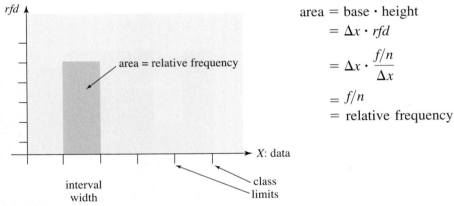

$$\begin{aligned}
\text{area} &= \text{base} \cdot \text{height} \\
&= \Delta x \cdot rfd \\
&= \Delta x \cdot \frac{f/n}{\Delta x} \\
&= f/n \\
&= \text{relative frequency}
\end{aligned}$$

FIGURE 4.7 Area equals relative frequency.

Adding a new column to the frequency distribution given in Figure 4.5, we obtain the relative frequency densities shown in Figure 4.8.

x = Age (years)	Relative Frequency	Class Frequency	Relative Width	Frequency Density
$15 \le x < 20$	4	$\frac{4}{42} \approx 10\%$	5	$\frac{4}{42} \div 5 \approx 0.020$
$20 \le x < 25$	11	$\frac{11}{42} \approx 26\%$	5	$\frac{11}{42} \div 5 \approx 0.052$
$25 \le x < 30$	6	$\frac{6}{42} \approx 14\%$	5	$\frac{6}{42} \div 5 \approx 0.028$
$30 \le x < 45$	11	$\frac{11}{42} \approx 26\%$	15	$\frac{11}{42} \div 15 \approx 0.017$
$45 \le x < 65$	10	$\frac{10}{42} \approx 24\%$	20	$\frac{10}{42} \div 20 \approx 0.012$
	$n = 42$	Total = 100%		

FIGURE 4.8 Calculating relative frequency density.

We now construct a histogram using relative frequency density as the vertical scale. The histogram depicting the distribution of the ages of the people in Keith Reed's acting class (using the frequency distribution in Figure 4.8) is shown in Figure 4.9.

Comparing the histograms in Figures 4.6 and 4.9, we see that using relative frequency density as the vertical scale (rather than frequency) gives a more truthful representation of a distribution when the interval widths are unequal.

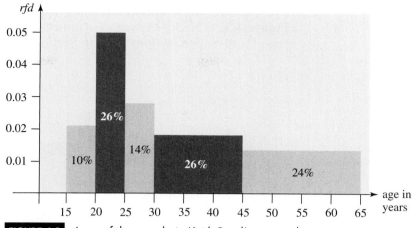

FIGURE 4.9   Ages of the people in Keith Reed's acting class.

EXAMPLE **3**

CONSTRUCTING A HISTOGRAM: GROUPED DATA   To study the output of a machine that fills bags with corn chips, a quality control engineer randomly selected and weighed a sample of 200 bags of chips. The frequency distribution in Figure 4.10 summarizes the data. Construct a histogram for the weights of the bags of corn chips.

x = Weight (ounces)	f = Number of Bags
$15.3 \leq x < 15.5$	10
$15.5 \leq x < 15.7$	24
$15.7 \leq x < 15.9$	36
$15.9 \leq x < 16.1$	58
$16.1 \leq x < 16.3$	40
$16.3 \leq x < 16.5$	20
$16.5 \leq x < 16.7$	12

FIGURE 4.10   Weights of bags of corn chips.

SOLUTION

Because each interval has the same width ($\Delta x = 0.2$), we construct a combined frequency and relative frequency histogram. The relative frequencies are given in Figure 4.11.

x	f	$rf = f/n$
$15.3 \leq x < 15.5$	10	0.05
$15.5 \leq x < 15.7$	24	0.12
$15.7 \leq x < 15.9$	36	0.18
$15.9 \leq x < 16.1$	58	0.29
$16.1 \leq x < 16.3$	40	0.20
$16.3 \leq x < 16.5$	20	0.10
$16.5 \leq x < 16.7$	12	0.06
	$n = 200$	Sum = 1.00

FIGURE 4.11   Relative frequencies.

We now draw coordinate axes with appropriate scales and rectangles (Figure 4.12). Notice the (near) symmetry of the histogram. We will study this type of distribution in more detail in Section 4.4.

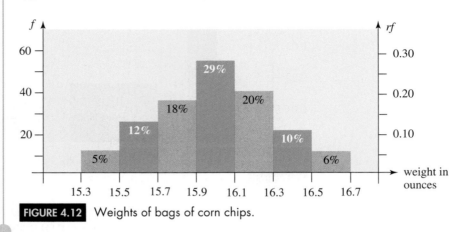

FIGURE 4.12   Weights of bags of corn chips.

Histograms and Single-Valued Classes

The histograms that we have constructed so far have all utilized intervals of grouped data. For instance, the first group of ages in Keith Reed's acting class was from 16 to 24 years old ($16 \leq x < 24$). However, if a set of data consists of only a few distinct values, it may be advantageous to consider each distinct value to be a "class" of data; that is, we utilize single-valued classes of data.

EXAMPLE **4**

CONSTRUCTING A HISTOGRAM: SINGLE VALUES   A sample of high school seniors was asked, "How many television sets are in your house?" The frequency distribution in Figure 4.13 summarizes the data. Construct a histogram using single-valued classes of data.

Number of Television Sets	Frequency
0	2
1	13
2	18
3	11
4	5
5	1

FIGURE 4.13   Frequency distribution.

SOLUTION

Rather than using intervals of grouped data, we use a single value to represent each class. Because each class has the same width ($\Delta x = 1$), we construct a combined frequency and relative frequency histogram. The relative frequencies are given in Figure 4.14.

We now draw coordinate axes with appropriate scales and rectangles. In working with single valued classes of data, it is common to write the single value at the midpoint of the base of each rectangle as shown in Figure 4.15.

Number of Television Sets	Frequency	Relative Frequency
0	2	2/50 = 4%
1	13	13/50 = 26%
2	18	18/50 = 36%
3	11	11/50 = 22%
4	5	5/50 = 10%
5	1	1/50 = 2%
	$n = 50$	Total = 100%

FIGURE 4.14 Calculating relative frequency.

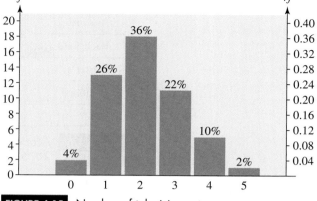

FIGURE 4.15 Number of television sets.

Pie Charts

Many statistical studies involve **categorical data**—that which is grouped according to some common feature or quality. One of the easiest ways to summarize categorical data is through the use of a **pie chart.** A pie chart shows how various categories of a set of data account for certain proportions of the whole. Financial incomes and expenditures are invariably shown as pie charts, as in Figure 4.16.

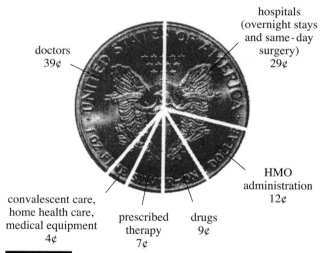

doctors
39¢

hospitals
(overnight stays
and same-day
surgery)
29¢

HMO
administration
12¢

convalescent care,
home health care,
medical equipment
4¢

prescribed
therapy
7¢

drugs
9¢

FIGURE 4.16 How a typical medical dollar is spent.

To draw the "slice" of the pie representing the relative frequency (percentage) of the category, the appropriate central angle must be calculated. Since a complete circle comprises 360 degrees, we obtain the required angle by multiplying 360° times the relative frequency of the category.

EXAMPLE **5** **CONSTRUCTING A PIE CHART** What type of academic degree do you hope to earn? The different types of degrees and the number of each type of degree conferred in the United States during the 2006–07 academic year is given in Figure 4.17. Construct a pie chart to summarize the data.

Type of Degree	Frequency (thousands)
Associate's	728
Bachelor's	1,524
Master's	605
Doctorate	61
Total	2,918

FIGURE 4.17 Academic degrees conferred, 2006–07. *Source:* National Center for Education Statistics.

SOLUTION

Find the relative frequency of each category and multiply it by 360° to determine the appropriate central angle. The necessary calculations are shown in Figure 4.18.

Type of Degree	Frequency (thousands)	Relative Frequency	Central Angle
Associate's	728	$\frac{728}{2,918} \approx 0.249$	$0.249 \times 360° = 89.64°$
Bachelor's	1,524	$\frac{1,524}{2,918} \approx 0.523$	$0.523 \times 360° = 188.28°$
Master's	605	$\frac{605}{2,918} \approx 0.207$	$0.207 \times 360° = 74.52°$
Doctorate	61	$\frac{61}{2,918} \approx 0.021$	$0.021 \times 360° = 7.56°$
	$n = 2,918$	Sum = 1.000	Total = 360°

FIGURE 4.18 Calculating relative frequency and central angles.

Now use a protractor to lay out the angles and draw the "slices." The name of each category can be written directly on the slice, or, if the names are too long, a legend consisting of various shadings may be used. Each slice of the pie should contain its relative frequency, expressed as a percent. Remember, the whole reason for constructing a pie chart is to convey information visually; pie charts should enable the reader to instantly compare the relative proportions of categorical data. See Figure 4.19.

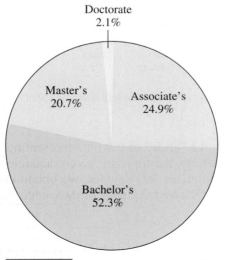

FIGURE 4.19 Academic degrees conferred, 2006–07.

1. To study the library habits of students at a local college, thirty randomly selected students were surveyed to determine the number of times they had been to the library during the last week. The following results were obtained:

```
1  5  2  1  1  4  2  1  5  4
5  2  5  1  2  3  4  1  1  2
3  5  4  1  2  2  4  5  1  2
```

   a. Organize the given data by creating a frequency distribution.
   b. Construct a pie chart to represent the data.
   c. Construct a histogram using single-valued classes of data.

2. To study the eating habits of students at a local college, thirty randomly selected students were surveyed to determine the number of times they had purchased food at the school cafeteria during the last week. The following results were obtained:

```
2  3  1  2  2  3  2  1  2  3
1  1  5  5  4  3  5  2  2  2
3  2  3  4  1  4  1  2  3  3
```

   a. Organize the given data by creating a frequency distribution.
   b. Construct a pie chart to represent the data.
   c. Construct a histogram using single-valued classes of data.

3. To study the composition of families in Manistee, Michigan, forty randomly selected married couples were surveyed to determine the number of children in each family. The following results were obtained:

```
2  1  3  0  1  0  2  5  1  2  0  2  2  1
4  3  1  1  3  4  1  1  0  2  0  0  2  2
1  0  3  1  1  3  4  2  1  3  0  1
```

   a. Organize the given data by creating a frequency distribution.
   b. Construct a pie chart to represent the data.
   c. Construct a histogram using single-valued classes of data.

4. To study the spending habits of shoppers in Orlando, Florida, fifty randomly selected shoppers at a mall were surveyed to determine the number of credit cards they carried. The following results were obtained:

```
2  5  0  4  2  1  0  6  3  5  4  3  4  0
5  2  5  2  0  2  5  0  2  5  2  5  4  3
5  6  1  0  6  3  5  3  4  0  5  2  2  5
2  0  2  0  4  2  1  0
```

   a. Organize the given data by creating a frequency distribution.
   b. Construct a pie chart to represent the data.
   c. Construct a histogram using single-valued classes of data.

5. The speeds, in miles per hour, of forty randomly monitored cars on Interstate 40 near Winona, Arizona, were as follows:

```
66  71  76  61  73  78  74  67  80  63  69
78  66  70  77  60  72  58  65  70  64  75
80  75  62  67  72  59  74  65  54  69  73
79  64  68  57  51  68  79
```

   a. Organize the given data by creating a frequency distribution. (Group the data into six intervals.)
   b. Construct a histogram to represent the data.

6. The weights, in pounds, of thirty-five packages of ground beef at the Cut Above Market were as follows:

```
1.0  1.9  2.5  1.2  2.0  0.7  1.3  2.4  1.1
3.3  2.4  0.8  2.3  1.7  1.0  2.8  1.4  3.0
0.9  1.1  1.4  2.2  1.5  3.2  2.1  2.7  1.8
1.6  2.3  2.6  1.3  2.9  1.9  1.2  0.5
```

   a. Organize the given data by creating a frequency distribution. (Group the data into six intervals.)
   b. Construct a histogram to represent the data.

7. To examine the effects of a new registration system, a campus newspaper asked freshmen how long they had to wait in a registration line. The frequency distribution in Figure 4.20 summarizes the responses. Construct a histogram to represent the data.

$x = $ Time (minutes)	Number of Freshmen
$0 \leq x < 10$	32
$10 \leq x < 20$	47
$20 \leq x < 30$	36
$30 \leq x < 40$	22
$40 \leq x < 50$	13
$50 \leq x < 60$	10

FIGURE 4.20 Time waiting in line.

8. The frequency distribution shown in Figure 4.21 lists the annual salaries of the managers at Universal Manufacturing of Melonville. Construct a histogram to represent the data.

x = Salary (thousands of dollars)	Number of Managers
$30 \leq x < 40$	6
$40 \leq x < 50$	12
$50 \leq x < 60$	10
$60 \leq x < 70$	5
$70 \leq x < 80$	7
$80 \leq x < 90$	3

FIGURE 4.21 Annual salaries.

9. The frequency distribution shown in Figure 4.22 lists the hourly wages of the workers at Universal Manufacturing of Melonville. Construct a histogram to represent the data.

x = Hourly Wage (dollars)	Number of Employees
$8.00 \leq x < 9.50$	21
$9.50 \leq x < 11.00$	35
$11.00 \leq x < 12.50$	42
$12.50 \leq x < 14.00$	27
$14.00 \leq x < 15.50$	18
$15.50 \leq x < 17.00$	9

FIGURE 4.22 Hourly wages.

10. To study the output of a machine that fills boxes with cereal, a quality control engineer weighed 150 boxes of Brand X cereal. The frequency distribution in Figure 4.23 summarizes her findings. Construct a histogram to represent the data.

x = Weight (ounces)	Number of Boxes
$15.3 \leq x < 15.6$	13
$15.6 \leq x < 15.9$	24
$15.9 \leq x < 16.2$	84
$16.2 \leq x < 16.5$	19
$16.5 \leq x < 16.8$	10

FIGURE 4.23 Weights of boxes of cereal.

11. The ages of the nearly 4 million women who gave birth in the United States in 1997 are given in Figure 4.24. Construct a histogram to represent the data.

Age (years)	Number of Women
$15 \leq x < 20$	486,000
$20 \leq x < 25$	948,000
$25 \leq x < 30$	1,075,000
$30 \leq x < 35$	891,000
$35 \leq x < 40$	410,000
$40 \leq x < 45$	77,000
$45 \leq x < 50$	4,000

FIGURE 4.24 Ages of women giving birth in 1997. *Source:* U.S. Bureau of the Census.

12. The age composition of the population of the United States in the year 2000 is given in Figure 4.25. Replace the interval "85 and over" with the interval $85 \leq x \leq 100$ and construct a histogram to represent the data.

Age (years)	Number of People (thousands)
$0 < x < 5$	19,176
$5 \leq x < 10$	20,550
$10 \leq x < 15$	20,528
$15 \leq x < 25$	39,184
$25 \leq x < 35$	39,892
$35 \leq x < 45$	44,149
$45 \leq x < 55$	37,678
$55 \leq x < 65$	24,275
$65 \leq x < 85$	30,752
85 and over	4,240

FIGURE 4.25 Age composition of the population of the United States in the year 2000. *Source:* U.S. Bureau of the Census.

In Exercises 13 and 14, use the age composition of the 14,980,000 students enrolled in institutions of higher education in the United States during 2000, as given in Figure 4.26.

13. Using the data in Figure 4.26, replace the interval "35 and over" with the interval $35 \leq x \leq 60$ and construct a histogram to represent the male data.

Age of Males (years)	Number of Students
$14 \le x < 18$	94,000
$18 \le x < 20$	1,551,000
$20 \le x < 22$	1,420,000
$22 \le x < 25$	1,091,000
$25 \le x < 30$	865,000
$30 \le x < 35$	521,000
35 and over	997,000
Total	6,539,000

Age of Females (years)	Number of Students
$14 \le x < 18$	78,000
$18 \le x < 20$	1,907,000
$20 \le x < 22$	1,597,000
$22 \le x < 25$	1,305,000
$25 \le x < 30$	1,002,000
$30 \le x < 35$	664,000
35 and over	1,888,000
Total	8,441,000

FIGURE 4.26 Age composition of students in higher education. *Source:* U.S. National Center for Education Statistics.

14. Using the data in Figure 4.26, replace the interval "35 and over" with the interval $35 \le x \le 60$ and construct a histogram to represent the female data.

15. The frequency distribution shown in Figure 4.27 lists the ages of 200 randomly selected students who received a bachelor's degree at State University last year. Where possible, determine what percent of the graduates had the following ages:

a. less than 23
b. at least 31
c. at most 20
d. not less than 19
e. at least 19 but less than 27
f. not between 23 and 35

Age (years)	Number of Students
$10 \le x < 15$	1
$15 \le x < 19$	4
$19 \le x < 23$	52
$23 \le x < 27$	48
$27 \le x < 31$	31
$31 \le x < 35$	16
$35 \le x < 39$	29
39 and over	19

FIGURE 4.27 Age of students.

16. The frequency distribution shown in Figure 4.28 lists the number of hours per day a randomly selected sample of teenagers spent watching television. Where possible, determine what percent of the teenagers spent the following number of hours watching television:

a. less than 4 hours
b. at least 5 hours

c. at least 1 hour
d. less than 2 hours
e. at least 2 hours but less than 4 hours
f. more than 3.5 hours

Hours per Day	Number of Teenagers
$0 \le x < 1$	18
$1 \le x < 2$	31
$2 \le x < 3$	24
$3 \le x < 4$	38
$4 \le x < 5$	27
$5 \le x < 6$	12
$6 \le x < 7$	15

FIGURE 4.28 Time watching television.

17. Figure 4.29 lists the top five reasons given by patients for emergency room visits in 2006. Construct a pie chart to represent the data.

Reason	Number of Patients (thousands)
Stomach pain	8,057
Chest pain	6,392
Fever	4,485
Headache	3,354
Shortness of breath	3,007

FIGURE 4.29 Reasons for emergency room visits, 2006. *Source:* National Center for Health Statistics.

18. Figure 4.30 lists the world's top six countries as tourist destinations in 2007. Construct a pie chart to represent the data.

Country	Number of Arrivals (millions)
France	81.9
Spain	59.2
United States	56.0
China	54.7
Italy	43.7
United Kingdom	30.7

FIGURE 4.30 World's top tourist destinations, 2007. *Source:* World Tourism Organization.

19. Figure 4.31 lists the race of new AIDS cases in the United States in 2005.

Race	Male	Female
White	10,027	1,747
Black	13,048	7,093
Hispanic	5,949	1,714
Asian	389	92
Native American	137	45

FIGURE 4.31 New AIDS cases in the United States, 2005. *Source:* National Center for Health Statistics.

a. Construct a pie chart to represent the male data.
b. Construct a pie chart to represent the female data.
c. Compare the results of parts (a) and (b). What conclusions can you make?
d. Construct a pie chart to represent the total data.

20. Figure 4.32 lists the types of accidental deaths in the United States in 2006. Construct a pie chart to represent the data.

21. Figure 4.33 lists some common specialties of physicians in the United States in 2005.

a. Construct a pie chart to represent the male data.
b. Construct a pie chart to represent the female data.
c. Compare the results of parts (a) and (b). What conclusions can you make?
d. Construct a pie chart to represent the total data.

Type of Accident	Number of Deaths
Motor vehicle	44,700
Poison	25,300
Falls	21,200
Suffocation	4,100
Drowning	3,800
Fire	2,800
Firearms	680

FIGURE 4.32 Types of accidental deaths in the United States, 2006. *Source:* National Safety Council.

Specialty	Male	Female
Family practice	54,022	26,305
General surgery	32,329	5,173
Internal medicine	104,688	46,245
Obstetric/gynecology	24,801	17,258
Pediatrics	33,515	36,636
Psychiatry	27,213	13,079

FIGURE 4.33 Physicians by Gender and Specialty, 2005. *Source:* American Medical Association.

22. Figure 4.34 lists the major metropolitan areas on intended residence for immigrants admitted to the United States in 2002. Construct a pie chart to represent the data.

Metropolitan Area	Number of Immigrants
Los Angeles/ Long Beach, CA	100,397
New York, NY	86,898
Chicago, IL	41,616
Miami, FL	39,712
Washington DC	36,371

FIGURE 4.34 Immigrants and areas of residence, 2002. *Source:* U.S. Immigration and Naturalization Service.

Answer the following questions using complete sentences and your own words.

• CONCEPT QUESTIONS

23. Explain the meanings of the terms *population* and *sample*.

24. The cholesterol levels of the 800 residents of Land-o-Lakes, Wisconsin, were recently collected and organized in a frequency distribution. Do these data represent a sample or a population? Explain your answer.

25. Explain the difference between frequency, relative frequency, and relative frequency density. What does each measure?

26. When is relative frequency density used as the vertical scale in constructing a histogram? Why?

27. In some frequency distributions, data are grouped in intervals; in others, they are not.

  a. When should data be grouped in intervals?

  b. What are the advantages and disadvantages of using grouped data?

HISTOGRAMS ON A GRAPHING CALCULATOR

In Example 2 of this section, we created a frequency distribution and a histogram for the ages of the students in an acting class. Much of this work can be done on a computer or a graphing calculator.

TECHNOLOGY AND STATISTICAL GRAPHS

Entering the Data

To enter the data from Example 2, do the following.

L₁	L₂	L₃
49		
59		
19		
20		
23		
24		

L₁(43)=

FIGURE 4.35

A TI-83/84's list screen.

ENTERING THE DATA ON A TI-83/84:

• *Put the calculator into statistics mode* by pressing [STAT].

• *Set the calculator up for entering the data* by selecting "Edit" from the "EDIT" menu, and the "list screen" appears, as shown in Figure 4.35. If data already appear in a list (as they do in list L_1 in Figure 4.35), use the arrow buttons to highlight the name of the list (i.e., "L_1" or "xStat") and press [CLEAR] [ENTER].

• *Enter the students' ages* in list L_1, in any order, using the arrow buttons and the [ENTER] button. When this has been completed, your screen should look similar to that in Figure 4.35. Notice the "$L_1(43)=$" at the bottom of the screen; this indicates that 42 entries have been made, and the calculator is ready to receive the 43rd. This allows you to check whether you have left any entries out.

Note: If some data points frequently recur, you can enter the data points in list L_1 and their frequencies in list L_2 rather than reentering a data point each time it recurs.

• Press [2nd] [QUIT].

ENTERING THE DATA ON A CASIO:

• *Put the calculator into statistics mode* by pressing MENU, highlighting STAT, and pressing EXE.

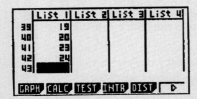

FIGURE 4.36

A Casio's list screen.

• *Enter the data* in List 1 in any order, using the arrow buttons and the EXE button. See Figure 4.36. If data already appear in a list and you want to clear the list, use the arrow keys to move to that list, press F6 and then F4 (i.e., DEL-A , which stands for "delete all"), and then press F1 (i.e., YES). Notice that the entries are numbered; this allows you to check whether you have left any entries out.

Drawing a Histogram

Once the data have been entered, you can draw a histogram.

DRAWING A HISTOGRAM ON A TI-83/84:

• Press Y= and clear any functions that may appear.
• *Enter the group boundaries* by pressing WINDOW , entering the left boundary of the first group as xmin (16 for this problem), the right boundary of the last group plus 1 as xmax (64 + 1 = 65 for this problem), and the group width as xscl (8 for this problem). (The calculator will create histograms only with equal group widths.) Enter 0 for ymin, and the largest frequency for ymax. (You may guess; it's easy to change it later if you guess wrong.)
• *Set the calculator up to draw a histogram* by pressing 2nd STAT PLOT and selecting "Plot 1." Turn the plot on and select the histogram icon.
• Tell the calculator to put the data entered in list L_1 on the *x*-axis by selecting "L_1" for "Xlist," and to consider each entered data point as having a frequency of 1 by selecting "1" for "Freq."

Note: If some data points frequently recur and you entered their frequencies in list L_2, then select "L_2" rather than "1" for "Freq" by typing 2nd L_2.

• *Draw a histogram* by pressing GRAPH . If some of the bars are too long or too short for the screen, alter ymin accordingly.
• Press TRACE to find out the left and right boundaries and the frequency of the bars, as shown in Figure 4.37. Use the arrow buttons to move from bar to bar.
• Press 2nd STAT PLOT , select "Plot 1" and turn the plot off, or else the histogram will appear on future graphs.

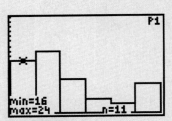

FIGURE 4.37

The first bar's boundaries are 16 and 24; its frequency is 11.

DRAWING A HISTOGRAM ON A CASIO:

• Press GRPH (i.e., F1).
• Press SET (i.e., F6).
• Make the resulting screen, which is labeled "StatGraph1," read as follows:

Graph Type   :Hist
Xlist   :List1
Frequency   : 1

To make the screen read as described above, do the following:

• Use the down arrow button to scroll down to "Graph Type."
• Press F6 .
• Press HIST (i.e., F1).
• In a similar manner, change "Xlist" and "Frequency" if necessary.

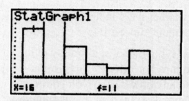

FIGURE 4.38

The first bar's left boundary is 16; its frequency is 11.

- Press $\boxed{\text{EXE}}$ and return to List1.
- Press $\boxed{\text{GPH1}}$ (i.e., $\boxed{\text{F1}}$) and make the resulting screen, which is labeled "Set Interval," read as follows:

  Start: 16   "**Start**" refers to the beginning of the first group
  Pitch: 8   "**Pitch**" refers to the width of each group

- Press $\boxed{\text{DRAW}}$ (i.e., $\boxed{\text{F6}}$), and the calculator will display the histogram.
- Press $\boxed{\text{SHIFT}}$ and then $\boxed{\text{TRCE}}$ (i.e., $\boxed{\text{F1}}$) to find out the left boundaries and the frequency of the bars. Use the arrow buttons to move from bar to bar. See Figure 4.38.

HISTOGRAMS AND PIE CHARTS ON A COMPUTERIZED SPREADSHEET

A **spreadsheet** is a large piece of paper marked off in rows and columns. Accountants use spreadsheets to organize numerical data and perform computations. A **computerized spreadsheet,** such as Microsoft Excel, is a computer program that mimics the appearance of a paper spreadsheet. It frees the user from performing any computations; instead, it allows the user merely to give instructions on how to perform those computations. The instructions in this subsection were specifically written for Microsoft Excel; however, all computerized spreadsheets work somewhat similarly.

When you start a computerized spreadsheet, you see something that looks like a table waiting to be filled in. The rows are labeled with numbers and the columns with letters, as shown in Figure 4.39.

◇	A	B	C	D
1				
2				
3				
4				
5				

FIGURE 4.39   A blank spreadsheet.

The individual boxes are called **cells.** The cell in column A row 1 is called cell A1; the cell below it is called cell A2, because it is in column A row 2.

A computerized spreadsheet is an ideal tool to use in creating a histogram or a pie chart. We will illustrate this process by preparing both a histogram and a pie chart for the ages of the students in Keith Reed's acting class, as discussed in Example 2.

Entering the Data

1. *Label the columns.* Use the mouse and/or the arrow buttons to move to cell A1, type in "age of student" and press "return" or "enter." (If there were other data, we could

enter it in other columns. For example, if the students' names were included, we could type "name of student" in cell A1, and "age of student" in cell B1.)

2. *Enter the students' ages in column A.* Move to cell A2, type in "26" and press "return" or "enter." Move to cell A3, type in "16" and press "return" or "enter." In a similar manner, enter all of the ages. You can enter the ages in any order. After you complete this step, your spreadsheet should look like that in Figure 4.40 (except that it should go down a lot further).

◇	A	B	C	D
1	age of student			
2	26			
3	16			
4	21			
5	34			

FIGURE 4.40 The spreadsheet after entering the students' ages.

3. *Save the spreadsheet.* Use your mouse to select "File" at the very top of the screen. Then pull your mouse down until "Save As" is highlighted, and let go. Your instructor may give you further instructions on where and how to save your spreadsheet.

Preparing the Data for a Chart

1. *Enter the group boundaries.* Excel uses "bin numbers" rather than group boundaries. A group's bin number is the highest number that should be included in that group. In Example 2, the first group was $16 \le x < 24$. The highest age that should be included in this group is 23, so the group's bin number is 23. Be careful; this group's bin number is not 24, since people who are 24 years old should be included in the second group.

- If necessary, use the up arrow in the upper-right corner of the spreadsheet to scroll to the top 6of the spreadsheet.
- Type "bin numbers" in cell B1.
- Determine each group's bin number, and enter them in column B.

After you complete this step, your spreadsheet should look like that in Figure 4.41. You might need to adjust the width of column A. Click on the right edge of the "A" label at the top of the first column, and move it.

◇	A	B
1	age of student	bin numbers
2	26	23
3	16	31
4	21	39
5	34	47
6	45	55
7	18	63
8	41	

FIGURE 4.41

The spreadsheet after entering the bin numbers.

2. *Have Excel determine the frequencies.*

- Use your mouse to select "Tools" at the very top of the screen. Then pull your mouse down until "Data Analysis" is highlighted, and let go.
- If "Data Analysis" is not listed under "Tools," then
  - Select "Tools" at the top of the screen, pull down until "Add-Ins" is highlighted, and let go.
  - Select "Analysis ToolPak-VBA."
  - Use your mouse to press the "OK" button.
  - Select "Tools" at the top of the screen, pull down until "Data Analysis" is highlighted, and let go.
- In the "Data Analysis" box that appears, use your mouse to highlight "Histogram" and press the "OK" button.
- In the "Histogram" box that appears, use your mouse to click on the white rectangle that follows "Input Range," and then use your mouse to draw a box around all of the ages. (To draw the box, move your mouse to cell A2, press the mouse button, and move the mouse down until all of the entered ages are enclosed in a box.) This should cause "$A$2:$A$43" to appear in the Input Range rectangle.

- Use your mouse to click on the white rectangle that follows "Bin Range," and then use your mouse to draw a box around all of the bin numbers. This should cause "$B$2:$B$7" to appear in the Bin Range rectangle.
- Be certain that there is a dot in the button to the left of "Output Range." Then click on the white rectangle that follows "Output Range," and use your mouse to click on cell C1. This should cause "$C$1" to appear in the Output Range rectangle.
- Use your mouse to press the "OK" button.

After you complete this step, your spreadsheet should look like that in Figure 4.42.

◇	A	B	C	D
1	age of student	bin numbers	Bin	Frequency
2	26	23	23	11
3	16	31	31	13
4	21	39	39	7
5	34	47	47	3
6	45	55	55	2
7	18	63	63	6
8	41		More	0
9	38			

FIGURE 4.42 The spreadsheet after Excel determines the frequencies.

3. *Prepare labels for the chart.*
   - In column E, list the group boundaries.
   - In column F, list the frequencies.

After you complete this step, your spreadsheet should look like that in Figure 4.43 (the first few columns are not shown).

C	D	E	F
Bin	Frequency	group boundaries	frequency
23	11	16≤x<24	11
31	13	24≤x<32	13
39	7	32≤x<40	7
47	3	40≤x<48	3
55	2	48≤x<56	2
63	6	56≤x<64	6
More	0		

FIGURE 4.43 The spreadsheet after preparing labels.

Drawing a Histogram

1. *Use the Chart Wizard to draw a bar chart.*
   - Use your mouse to press the "Chart Wizard" button at the top of the spreadsheet. (It looks like a histogram and it might have a magic wand.)
   - Select "Column" and then an appropriate style.

- Press the "Next" button.
- Use your mouse to click on the white rectangle that follows "Data range," and then use your mouse to draw a box around all of the group boundaries and frequencies from step 3 in "Preparing the data for a chart". This should cause "=Sheet1!\$E\$2: \$F\$7" to appear in the Data range rectangle.
- Press the "Next" button.
- Under "Chart Title" type an appropriate title, such as "Acting class ages."
- After "Category (X)" type an appropriate title for the x-axis, such as "Students' ages."
- After "Value (Y)" type an appropriate title for the y-axis, such as "frequencies."
- Press "Legend" at the top of the chart options box, and then remove the check mark next to "show legend."
- Press the "Finish" button and the bar chart will appear. See Figure 4.44.

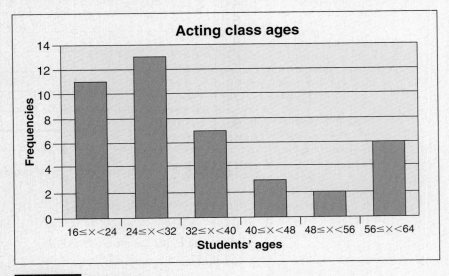

FIGURE 4.44 The bar chart.

2. *Save the spreadsheet.* Use your mouse to select "File" at the very top of the screen. Then pull your mouse down until "Save" is highlighted, and let go.

3. *Convert the bar graph to a histogram.* The graph is not a histogram because of the spaces between the bars.
   - Double click or right click on the bar graph and a "Format Data Series" box will appear.
   - Press the "Options" tab.
   - Remove the spaces between the bars by changing the "Gap width" to 0.
   - Press OK.
   - Save the spreadsheet. See Figure 4.45.

4. *Print the histogram.*
   - Click on the histogram, and it will be surrounded by a thicker border than before.
   - Use your mouse to select "File" at the very top of the screen. Then pull your mouse down until "Print" is highlighted, and let go.
   - Respond appropriately to the "Print" box that appears.

Drawing a Pie Chart

1. *Use the Chart Wizard to draw the pie chart.*
   - Use your mouse to press the "Chart Wizard" button at the top of the spreadsheet. (It looks like a histogram and it might have a magic wand.)

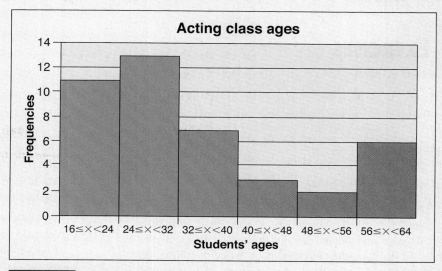

FIGURE 4.45 The histogram.

- Select "Pie" and then an appropriate style.
- Press the "Next" button.
- Use your mouse to click on the white rectangle that follows "Data range," and then use your mouse to draw a box around all of the group boundaries and frequencies from step 3 in "Preparing the data for a chart." This should cause "=Sheet1!\$E\$2: \$F\$7" to appear in the Data range rectangle.
- Press the "Next" button.
- Under "Chart Title" type an appropriate title.
- Press the "Finish" button and the pie chart will appear. See Figure 4.46.

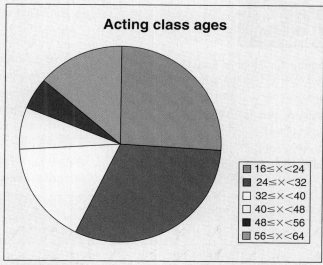

FIGURE 4.46 The pie chart.

2. *Save the spreadsheet.* Use your mouse to select "File" at the very top of the screen. Then pull your mouse down until "Save" is highlighted, and let go.

3. *Print the pie chart.*
   - Click on the chart, and it will be surrounded by a thicker border than before.
   - Use your mouse to select "File" at the very top of the screen. Then pull your mouse down until "Print" is highlighted, and let go.
   - Respond appropriately to the "Print" box that appears.

EXERCISES

Use a graphing calculator or Excel on the following exercises. Answers will vary depending on your choice of groups.

Graphing calculator instructions: *By hand, copy a graphing calculator's histogram onto paper for submission.*

Excel instructions: *Print out a histogram for submission.*

28. Do Double Stuf Oreos really contain twice as much "stuf"? In 1999, Marie Revak and Jihan William's "interest was piqued by the unusual spelling of the word stuff and the bold statement 'twice the filling' on the package." So they conducted an experiment in which they weighed the amount of filling in a sample of traditional Oreos and Double Stuf Oreos. The data from that experiment are in Figure 4.47 (weight in grams). An unformatted spreadsheet containing these data can be downloaded from the text web site: **www.cengage.com/math/johnson**

Forsling Research

  a. Create one histogram for Double Stuf Oreos and one for Traditional Oreos. Use the same categories for each histogram.

  b. Compare the results of part (a). What do they tell you about the two cookies?

29. Every year, CNNMoney.com lists "the best places to live." They compare the "best small towns" on several bases, one of which is the "quality of life."

See Figure 4.48. An unformatted spreadsheet containing these data can be downloaded from the text web site: **www.cengage.com/math/johnson**

  a. Find the mean of the air quality indices.

  b. Would the population standard deviation or the sample standard deviation be more appropriate here? Why?

  c. Find the standard deviation from part (b).

  d. What do your results say about the air quality index of all of the best places to live in the United States?

30. Every year, CNNMoney.com lists "the best places to live." They compare the "best small towns" on several bases, one of which is the "quality of life." See Figure 4.48. An unformatted spreadsheet containing these data can be downloaded from the text web site: **www.cengage.com/math/johnson**

  a. Find the mean of the commute times.

  b. Would the population standard deviation or the sample standard deviation be more appropriate here? Why?

  c. Find the standard deviation from part (b).

  d. What do your results say about commuting in all of the best places to live in the United States?

31. Every year, CNNMoney.com lists "the best places to live." They compare the "best small towns" on several bases, one of which is the "quality of life." See Figure 4.48. An unformatted spreadsheet containing these data can be downloaded from the text web site: **www.cengage.com/math/johnson**

  a. Find the mean of the property crime levels.

  b. Would the population standard deviation or the sample standard deviation be more appropriate here? Why?

  c. Find the standard deviation from part (b).

  d. What do your results say about property crime in all of the best places to live in the United States?

Double Stuf Oreos:																		
4.7	6.5	5.5	5.6	5.1	5.3	5.4	5.4	3.5	5.5	6.5	5.9	5.4	4.9	5.6	5.7	5.3	6.9	6.5
6.3	4.8	3.3	6.4	5.0	5.3	5.5	5.0	6.0	5.7	6.3	6.0	6.3	6.1	6.0	5.8	5.8	5.9	6.2
5.9	6.5	6.5	6.1	5.8	6.0	6.2	6.2	6.0	6.8	6.2	5.4	6.6	6.2					

Traditional Oreos:																		
2.9	2.8	2.6	3.5	3.0	2.4	2.7	2.4	2.5	2.2	2.6	2.6	2.9	2.6	2.6	3.1	2.9	2.4	2.8
3.8	3.1	2.9	3.0	2.1	3.8	3.0	3.0	2.8	2.9	2.7	3.2	2.8	3.1	2.7	2.8	2.6	2.6	3.0
2.8	3.5	3.3	3.3	2.8	3.1	2.6	3.5	3.5	3.1	3.1								

FIGURE 4.47 Amount of Oreo filling. *Source:* Revak, Marie A. and Jihan G. Williams, "Sharing Teaching Ideas: The Double Stuf Dilemma," *The Mathematics Teacher,* November 1999, Volume 92, Issue 8, page 674.

Rank	City	Air quality index (% days AQI ranked good)	Property crime— incidents per 1000	Median commute time (mins)
1	Louisville, CO	74.0%	15	19.5
2	Chanhassen, MN	N.A.	16	23.1
3	Papillion, NE	92.0%	16	19.3
4	Middleton, WI	76.0%	27	16.3
5	Milton, MA	81.0%	9	27.7
6	Warren, NJ	N.A.	21	28.7
7	Keller, TX	65.0%	14	29.4
8	Peachtree City, GA	84.0%	12	29.3
9	Lake St. Louis, MO	79.0%	16	26.8
10	Mukilteo, WA	89.0%	27	22.3

FIGURE 4.48   The best places to live.

32. Figure 4.49 gives EPA fuel efficiency ratings for 2010 family sedans with an automatic transmission and the smallest available engine. An unformatted spreadsheet containing these data can be downloaded from the text web site: **www.cengage.com/math/johnson**

   a. For each of the two manufacturing regions, find the mean and an appropriate standard deviation for city driving, and defend your choice of standard deviations.

   b. What do the results of part (a) tell you about the two regions?

33. Figure 4.49 gives EPA fuel efficiency ratings for 2010 family sedans with an automatic transmission and the smallest available engine. An unformatted spreadsheet containing these data can be downloaded from the text web site: **www.cengage.com/math/johnson**

   a. For each of the two manufacturing regions, find the mean and an appropriate standard deviation for highway driving, and defend your choice of standard deviations.

   b. What do the results of part (a) tell you about the two regions?

American cars	city mpg	highway mpg	Asian cars	city mpg	highway mpg
Ford Fusion hybrid	41	36	Toyota Prius	51	48
Chevrolet Malibu hybrid	26	34	Nissan Altima hybrid	35	33
Ford Fusion fwd	24	34	Toyota Camry Hybrid	33	34
Mercury Milan	23	34	Kia Forte	27	36
Chevrolet Malibu	22	33	Hyundai Elantra	26	35
Saturn Aura	22	33	Nissan Altima	23	32
Dodge Avenger	21	30	Subaru Legacy awd	23	31
Buick Lacrosse	17	27	Toyota Camry	22	33
			Hyundai Sonata	22	32
			Kia Optima	22	32
			Honda Accord	22	31
			Mazda 6	21	30
			Mitsubishi Galant	21	30
			Hyunda Azera	18	26

FIGURE 4.49   Fuel efficiency. Source: www.fueleconomy.gov

Measures of Central Tendency

BJECTIVES

- Find the mean
- Find the median
- Find the mode

Who is the best running back in professional football? How much does a typical house cost? What is the most popular television program? The answers to questions like these have one thing in common: They are based on averages. To compare the capabilities of athletes, we compute their average performances. This computation usually involves the ratio of two totals, such as (total yards gained)/(total number of carries) = average gain per carry. In real estate, the average-price house is found by listing the prices of all houses for sale (from lowest to highest) and selecting the price in the middle. Television programs are rated by the average number of households tuned in to each particular program.

Rather than listing every data point in a large distribution of numbers, people tend to summarize the data by selecting a representative number, calling it the average. Three figures—the *mean,* the *median,* and the *mode*—describe the "average" or "center" of a distribution of numbers. These averages are known collectively as the **measures of central tendency.**

The Mean

The **mean** is the average people are most familiar with; it can also be the most misleading. Given a collection of n data points, $x_1, x_2, \ldots, x_n$, the mean is found by adding up the data and dividing by the number of data points:

$$\text{mean of } n \text{ data points} = \frac{x_1 + x_2 + \cdots + x_n}{n}$$

If the data are collected from a sample, then the mean is denoted by $\bar{x}$ (read "x bar"); if the data are collected from an entire population, then the mean is denoted by μ (lowercase Greek letter "mu"). Unless stated otherwise, we will assume that the data represent a sample, so the mean will be symbolized by $\bar{x}$.

Mathematicians have developed a type of shorthand, called **summation notation,** to represent the sum of a collection of numbers. The Greek letter Σ ("sigma") corresponds to the letter S and represents the word *sum.* Given a group of data points $x_1, x_2, \ldots, x_n$, we use the symbol Σx to represent their sum; that is, $\Sigma x = x_1 + x_2 + \cdots + x_n$.

In 2001, Barry Bonds of the San Francisco Giants hit 73 home runs and set an all-time record in professional baseball. On average, how many home runs per season did Bonds hit? See Exercise 13.

© Reuters/Corbis

DEFINITION OF THE MEAN

Given a sample of n data points, $x_1, x_2, \ldots, x_n$, the **mean,** denoted by $\bar{x}$, is

$$\bar{x} = \frac{\Sigma x}{n} \qquad \text{or} \qquad \bar{x} = \frac{\text{the sum of the data points}}{\text{the number of data points}}$$

Many scientific calculators have statistical functions built into them. These functions allow you to enter the data points and press the "*x*-bar" button to obtain the mean. Consult your manual for specific instructions.

EXAMPLE 1

FINDING THE MEAN: SINGLE VALUES AND MIXED UNITS In 2005, Lance Armstrong won his seventh consecutive Tour de France bicycle race. No one in the 100-year history of the race has won so many times. (Miguel Indurain of Spain won five times from 1991 to 1995.) Lance's winning times are given in Figure 4.50. Find the mean winning time of Lance Armstrong's Tour de France victory rides.

Year	1999	2000	2001	2002	2003	2004	2005
Time (h:m:s)	91:32:16	92:33:08	86:17:28	82:05:12	83:41:12	83:36:02	86:15:02

FIGURE 4.50 Lance Armstrong's Tour de France winning times. *Source: San Francisco Chronicle.*

SOLUTION

To find the mean, we must add up the data and divide by the number of data points. However, before we can sum the data, they must be converted to a common unit, say, minutes. Therefore, we multiply the number of hours by 60 (minutes per hour), divide the number of seconds by 60 (seconds per minute), and add each result to the number of minutes. Lance Armstrong's 1999 winning time of 91 hours, 32 minutes, 16 seconds is converted to minutes as follows:

$$\left(91\, \text{hours} \times \frac{60\, \text{minutes}}{1\, \text{hour}}\right) + 32\, \text{minutes} + \left(16\, \text{seconds} \times \frac{1\, \text{minute}}{60\, \text{seconds}}\right)$$

$$= 5{,}460\, \text{minutes} + 32\, \text{minutes} + 0.266666 \ldots \text{minutes}$$

$$= 5{,}492.267\, \text{minutes} \quad \textbf{rounding off to three decimal places}$$

In a similar fashion, all of Lance Armstrong's winning times are converted to minutes, and the results are given in Figure 4.51.

Year	1999	2000	2001	2002	2003	2004	2005
Time (min)	5492.267	5553.133	5177.467	4925.200	5021.200	5016.033	5175.033

FIGURE 4.51 Winning time in minutes.

$$\bar{x} = \frac{\Sigma x}{n}$$

$$= \frac{5492.267 + 5553.133 + 5177.467 + 4925.200 + 5021.200 + 5016.033 + 5175.033}{7}$$

$$= \frac{36360.333}{7}$$

$$= 5194.333\, \text{minutes}$$

Converting back to hours, minutes, and seconds, we find that Lance Armstrong's mean winning time is 86 hours, 34 minutes, 20 seconds.

THE FAR SIDE® By GARY LARSON

© 1988 FarWorks, Inc. All Rights Reserved/Dist. by Creators Syndicate

**"Bob and Ruth! Come on in. … Have you met
Russell and Bill, our 1.5 children?"**

If a sample of 460 families have 690 children altogether, then the
mean number of children per family is $\bar{x} = 1.5$.

EXAMPLE 2

FINDING THE MEAN: GROUPED DATA In 2008, the U.S. Bureau of Labor
Statistics tabulated a survey of workers' ages and wages. The frequency
distribution in Figure 4.52 summarizes the age distribution of workers who
received minimum wage ($6.55 per hour). Find the mean age of a worker receiving
minimum wage.

y = Age (years)	Number of Workers
$16 \le y < 20$	108,000
$20 \le y < 25$	53,000
$25 \le y < 35$	41,000
$35 \le y < 45$	23,000
$45 \le y < 55$	29,000
$55 \le y < 65$	23,000
	$n = 277,000$

FIGURE 4.52 Age distribution of workers
receiving minimum wage.
Source: Bureau of Labor Statistics,
U.S. Department of Labor.

SOLUTION

To find the mean age of the workers, we must add up the ages of all the workers and divide by 277,000. However, because we are given grouped data, the ages of the individual workers are unknown to us. In this situation, we use the midpoint of each interval as the representative of the interval; consequently, our answer is an approximation.

To find the midpoint of an interval, add the endpoints and divide by 2. For instance, the midpoint of the interval $16 \leq y < 20$ is

$$\frac{16 + 20}{2} = 18$$

We can then say that each of the 108,000 people in the first interval is approximately eighteen years old. Adding up these workers' ages, we obtain

$$18 + 18 + 18 + \cdots + 18 \quad \text{(one hundred eight thousand times)}$$
$$= (108,000)(18)$$
$$= 1,944,000 \text{ years}$$

If we let f = frequency and x = the midpoint of an interval, the product $f \cdot x$ gives us the total age of the workers in an interval. The results of this procedure for all the workers are shown in Figure 4.53.

y = Age (years)	f = Frequency	x = Midpoint	$f \cdot x$
$16 \leq y < 20$	108,000	18	1,944,000
$20 \leq y < 25$	53,000	22.5	1,192,500
$25 \leq y < 35$	41,000	30	1,230,000
$35 \leq y < 45$	23,000	40	920,000
$45 \leq y < 55$	29,000	50	1,450,000
$55 \leq y < 65$	23,000	60	1,380,000
	n = 277,000		$\Sigma(f \cdot x) = 8,116,500$

FIGURE 4.53 Using midpoints of grouped data.

Because $f \cdot x$ gives the sum of the ages of the workers in an interval, the symbol $\Sigma(f \cdot x)$ represents the sum of all the $(f \cdot x)$; that is, $\Sigma(f \cdot x)$ represents the sum of the ages of *all* workers.

The mean age is found by dividing the sum of the ages of all the workers by the number of workers:

$$\bar{x} = \frac{\Sigma(f \cdot x)}{n}$$
$$= \frac{8,116,500}{277,000}$$
$$= 29.30144 \text{ years}$$

The mean age of the workers earning minimum wage is approximately 29.3 years.

One common mistake that is made in working with grouped data is to forget to multiply the midpoint of an interval by the frequency of the interval. Another common mistake is to divide by the number of intervals instead of by the total number of data points.

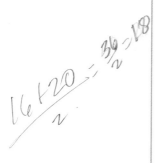

(handwritten in left margin: frequency · Total of frequencies · midpoint)

The procedure for calculating the mean when working with grouped data (as illustrated in Example 2) is summarized in the following box.

CALCULATING THE MEAN: GROUPED DATA

Given a frequency distribution containing several groups of data, the mean $\bar{x}$ can be found by using the following formula:

$$\bar{x} = \frac{\Sigma(f \cdot x)}{n}$$

where x = the midpoint of a group, f = the frequency of the group, and $n = \Sigma f$

(handwritten: frequency)

EXAMPLE 3

FINDING THE MEAN: REPEATED DATA Ten college students were comparing their wages earned at part-time jobs. Nine earned $10.00 per hour working at jobs ranging from waiting on tables to working in a bookstore. The tenth student earned $200.00 per hour modeling for a major fashion magazine. Find the mean wage of the ten students.

SOLUTION

The data point $10.00 is repeated nine times, so we multiply it by its frequency. Thus, the mean is as follows:

$$\bar{x} = \frac{\Sigma(f \cdot x)}{n}$$

$$= \frac{(9 \cdot 10) + (1 \cdot 200)}{10}$$

$$= \frac{290}{10}$$

$$= 29$$

The mean wage of the students is $29.00 per hour.

Example 3 seems to indicate that the average wage of the ten students is $29.00 per hour. Is that a reasonable figure? If nine out of ten students earn $10.00 per hour, can we justify saying that their average wage is $29.00? Of course not! Even though the mean wage *is* $29.00, it is not a convincing "average" for this specific group of data. The mean is inflated because one student made $200.00 per hour. This wage is called an **outlier** (or **extreme value**) because it is significantly different from the rest of the data. Whenever a collection of data has extreme values, the mean can be greatly affected and might not be an accurate measure of the average.

The Median

The **median** is the "middle value" of a distribution of numbers. To find it, we first put the data in numerical order. (If a number appears more than once, we include it as many times as it occurs.) If there is an odd number of data points, the median is the middle data point; if there is an even number of data points, the median is defined to be the mean of the two middle values. In either case, the median separates the distribution into two equal parts. Thus, the median can be viewed as an "average." (The word *median* is also used to describe the strip that runs down the middle of a freeway; half the freeway is on one side, and half is on the other. This common usage is in keeping with the statistical meaning.)

EXAMPLE **4**

FINDING THE MEDIAN: ODD VERSUS EVEN NUMBERS OF DATA
Find the median of the following sets of raw data.

a. 2   8   3   12   6   2   11
b. 2   8   3   12   6   2   11   8

SOLUTION

a. First, we put the data in order from smallest to largest. Because there is an odd number of data points, ($n = 7$), we pick the middle one:

$$2 \quad 2 \quad 3 \quad \text{⑥} \quad 8 \quad 11 \quad 12$$

middle value

The median is 6. (Notice that 6 is the fourth number in the list.)
b. We arrange the data first. Because there is an even number of data points ($n = 8$), we pick the two middle values and find their mean:

$$2 \quad 2 \quad 3 \quad 6 \quad 8 \quad 8 \quad 11 \quad 12$$

$$\frac{(6 + 8)}{2} = 7$$

Therefore, the median is 7. (Notice that 7 is halfway between the fourth and fifth numbers in the list.)

In Example 4, we saw that when $n = 7$, the median was the fourth number in the list, and that when $n = 8$, the median was halfway between the fourth and fifth numbers in the list. Consequently, the *location* of the median depends on *n,* the number of numbers in the set of data. The formula

$$L = \frac{n + 1}{2}$$

can be used to find the location, *L,* of the median; when $n = 7$,

$$L = \frac{7 + 1}{2} = 4$$

(the median is the fourth number in the list), and when $n = 8$,

$$L = \frac{8 + 1}{2} = 4.5$$

(the median is halfway between the fourth and fifth numbers in the list).

LOCATION OF THE MEDIAN

Given a sample of *n* data points, the location of the median can be found by using the following formula:

$$L = \frac{n + 1}{2}$$

Once the data have been arranged from smallest to largest, the median is the *L*th number in the list.

EXAMPLE **5**

FINDING THE LOCATION AND VALUE OF THE MEDIAN   Find the median wage for the ten students in Example 3.

SOLUTION

First, we put the ten wages in order:

10   10   10   10   10   10   10   10   10   200

Because there are $n = 10$ data points, the location of the median is

$$L = \frac{10 + 1}{2} = 5.5$$

That is, the median is halfway between the fifth and sixth numbers. To find the median, we add the fifth and sixth numbers and divide by 2. Now, the fifth number is 10 and the sixth number is 10, so the median is

$$\frac{10 + 10}{2} = 10$$

Therefore, the median wage is $10.00. This is a much more meaningful "average" than the mean of $29.00.

If a collection of data contains extreme values, the median, rather than the mean, is a better indicator of the "average" value. For instance, in discussions of real estate, the median is usually used to express the "average" price of a house. (Why?) In a similar manner, when the incomes of professionals are compared, the median is a more meaningful representation. The discrepancy between mean (average) income and median income is illustrated in the following news article. Although the article is dated (it's from 1992), it is very informative as to the differences between the mean and the median.

FEATURED IN THE NEWS

DOCTORS' AVERAGE INCOME IN U.S. IS NOW $177,000

Washington—The average income of the nation's physicians rose to $177,400 in 1992, up 4 percent from the year before, the American Medical Association said yesterday. . . .

The average income figures, compiled annually by the AMA and based on a telephone survey of more than 4,100 physicians, ranged from a low of $111,800 for general practitioners and family practice doctors to a high of $253,300 for radiologists.

Physicians' median income was $148,000 in 1992, or 6.5 percent more than a year earlier. Half the physicians earned more than that and half earned less.

The average is pulled higher than the median by the earnings of the highest paid surgeons, anesthesiologists and other specialists at the top end of the scale. . . .

Here are the AMA's average and median net income figures by specialty for 1992:

General/family practice:
   $111,800, $100,000.
Internal medicine:
   $159,300, $130,000.

Surgery: $244,600, $207,000.
Pediatrics: $121,700, $112,000.
Obstetrics/gynecology:
   $215,100, $190,000.
Radiology: $253,300, $240,000.
Psychiatry: $130,700, $120,000.
Anesthesiology:
   $228,500, $220,000.
Pathology: $189,800, $170,000.
Other: $165,400, $150,000.

EXAMPLE **6**

FINDING THE LOCATION AND VALUE OF THE MEDIAN   The students in Ms. Kahlo's art class were asked how many siblings they had. The frequency distribution in Figure 4.54 summarizes the responses. Find the median number of siblings.

Number of Siblings	Number of Responses
0	2
1	8
2	5
3	6

FIGURE 4.54 Frequency distribution of data.

SOLUTION

The frequency distribution indicates that two students had no (0) siblings, eight students had one (1) sibling, five students had two (2) siblings, and six students had three (3) siblings. Therefore, there were $n = 2 + 8 + 5 + 6 = 21$ students in the class; consequently, there are twenty-one data points. Listing the data in order, we have

$$0, 0, 1, \ldots 1, 2, \ldots 2, 3, \ldots 3$$

Because there are $n = 21$ data points, the location of the median is

$$L = \frac{21 + 1}{2} = 11$$

That is, the median is the eleventh number. Because the first ten numbers are 0s and 1s ($2 + 8 = 10$), the eleventh number is a 2. Consequently, the median number of siblings is 2.

 The Mode

The third measure of central tendency is the **mode.** The mode is the most frequent number in a collection of data; that is, it is the data point with the highest frequency. Because it represents the most common number, the mode can be viewed as an average. A distribution of data can have more than one mode or none at all.

EXAMPLE **7**

FINDING THE MODE   Find the mode(s) of the following sets of raw data:

a. 4   10   1   8   5   10   5   10
b. 4   9   1   10   1   10   4   9
c. 9   6   1   8   3   10   3   9

SOLUTION

a. The mode is 10, because it has the highest frequency (3).
b. There is no mode, because each number has the same frequency (2).
c. The distribution has two modes—namely, 3 and 9—each with a frequency of 2. A distribution that has two modes is called *bimodal.*

In summarizing a distribution of numbers, it is most informative to list all three measures of central tendency. It helps to avoid any confusion or misunderstanding in situations in which the word *average* is used. In the hands of someone with questionable intentions, numbers can be manipulated to mislead people. In his book *How to Lie with Statistics,* Darrell Huff states, "The secret language of statistics, so appealing in a fact-minded culture, is employed to sensationalize, inflate, confuse, and oversimplify. Statistical methods and statistical terms are necessary in reporting the mass data of social and economic trends, business conditions, 'opinion' polls, the census. But without writers who use the words with honesty and understanding, and readers who know what they mean, the result can only be semantic nonsense." An educated public should always be on the alert for oversimplification of data via statistics. The next time someone mentions an "average," ask "Which one?" Although people might not intentionally try to mislead you, their findings can be misinterpreted if you do not know the meaning of their statistics and the method by which the statistics were calculated.

4.2 EXERCISES

In Exercises 1–4, find the mean, median, and mode of the given set of raw data.

1. 9   12   8   10   9   11   12
    15   20   9   14   15   21   10

2. 20   25   18   30   21   25   32   27
    32   35   19   26   38   31   20   23

3. 1.2   1.8   0.7   1.5   1.0   0.7   1.9   1.7   1.2
    0.8   1.7   1.3   2.3   0.9   2.0   1.7   1.5   2.2

4. 0.07   0.02   0.09   0.04   0.10   0.08   0.07   0.13
    0.05   0.04   0.10   0.07   0.04   0.01   0.11   0.08

5. Find the mean, median, and mode of each set of data.
    a. 9   9   10   11   12   15
    b. 9   9   10   11   12   102
    c. How do your answers for parts (a) and (b) differ (or agree)? Why?

6. Find the mean, median, and mode for each set of data.
    a. 80   90   100   110   110   140
    b. 10   90   100   110   110   210
    c. How do your answers for parts (a) and (b) differ (or agree)? Why?

7. Find the mean, median, and mode of each set of data.
    a.   2   4   6   8   10   12
    b. 102   104   106   108   110   112
    c. How are the data in part (b) related to the data in part (a)?
    d. How do your answers for parts (a) and (b) compare?

8. Find the mean, median, and mode of each set of data.
    a. 12   16   20   24   28   32
    b. 600   800   1,000   1,200   1,400   1,600
    c. How are the data in part (b) related to the data in part (a)?
    d. How do your answers for parts (a) and (b) compare?

9. Kaitlin Mowry is a member of the local 4-H club and has six mini Rex rabbits that she enters in regional rabbit competition shows. The weights of the rabbits are given in Figure 4.55. Find the mean, median, and mode of the rabbits' weights.

| Weight (lb:oz) | 3:12 | 4:03 | 3:06 | 3:15 | 3:12 | 4:02 |

FIGURE 4.55 Weights of rabbits.

10. As was stated in Example 1, Lance Armstrong won the Tour de France bicycle race every year from 1999 to 2005. His margins of victory (time difference of the second place finisher) are given in Figure 4.56. Find the mean, median, and mode of Lance Armstrong's victory margins.

Year	1999	2000	2001	2002	2003	2004	2005
Margin (m:s)	7:37	6:02	6:44	7:17	1:01	6:19	4:40

FIGURE 4.56 Lance Armstrong's Tour de France victory margins. *Source: San Francisco Chronicle.*

11. Jerry Rice holds the all-time record in professional football for scoring touchdowns. The number of touchdown receptions (TDs) for each of his seasons is given in Figure 4.57. Find the mean, median, and mode of the number of touchdown receptions per year by Rice.

Year	TDs	Year	TDs
1985	3	1995	15
1986	15	1996	8
1987	22	1997	1
1988	9	1998	9
1989	17	1999	5
1990	13	2000	7
1991	14	2001	9
1992	10	2002	7
1993	15	2003	2
1994	13	2004	3

FIGURE 4.57 Touchdown receptions for Jerry Rice.
Source: http://sportsillustrated.cnn.com/football/nfl/players/.

12. Wayne Gretzky, known as "The Great One," holds the all-time record in professional hockey for scoring goals. The number of goals for each of his seasons is given in Figure 4.58. Find the mean, median, and mode of the number of goals per season by Gretzky.

Season	Goals	Season	Goals
1979–80	51	1989–90	40
1980–81	55	1990–91	41
1981–82	92	1991–92	31
1982–83	71	1992–93	16
1983–84	87	1993–94	38
1984–85	73	1994–95	11
1985–86	52	1995–96	23
1986–87	62	1996–97	25
1987–88	40	1997–98	23
1988–89	54	1998–99	9

FIGURE 4.58 Goals made by Wayne Gretzky.
Source: The World Almanac.

13. Barry Bonds of the San Francisco Giants set an all-time record in professional baseball by hitting 73 home runs in one season (2001). The number of home runs (HR) for each of his seasons in professional baseball is given in Figure 4.59. Find the mean, median, and mode of the number of home runs hit per year by Bonds.

Year	HR	Year	HR
1986	16	1997	40
1987	25	1998	37
1988	24	1999	34
1989	19	2000	49
1990	33	2001	73
1991	25	2002	46
1992	34	2003	45
1993	46	2004	45
1994	37	2005	5
1995	33	2006	26
1996	42	2007	28

FIGURE 4.59 Home runs hit by Barry Bonds.
Source: http://sportsillustrated.cnn.com/
baseball/mlb/players/.

14. Michael Jordan has been recognized as an extraordinary player in professional basketball, especially in terms of the number of points per game he has scored. Jordan's average number of points per game (PPG) for each of his seasons is given in Figure 4.60. Find the mean, median, and mode of the average points per game made per season by Jordan.

Season	PPG	Season	PPG
1984–85	28.2	1992–93	32.6
1985–86	22.7	1994–95	26.9
1986–87	37.1	1995–96	30.4
1987–88	35.0	1996–97	29.6
1988–89	32.5	1997–98	28.7
1989–90	33.6	2001–02	22.9
1990–91	31.5	2002–03	20.0
1991–92	30.1		

FIGURE 4.60 Points per game made by Michael Jordan.
Source: http://www.nba.com/playerfile/.

15. The frequency distribution in Figure 4.61 lists the results of a quiz given in Professor Gilbert's statistics class. Find the mean, median, and mode of the scores.

Score	Number of Students
10	3
9	10
8	9
7	8
6	10
5	2

FIGURE 4.61 Quiz scores in Professor Gilbert's statistics class.

16. Todd Booth, an avid jogger, kept detailed records of the number of miles he ran per week during the past year. The frequency distribution in Figure 4.62 summarizes his records. Find the mean, median, and mode of the number of miles per week that Todd ran.

Miles Run per Week	Number of Weeks
0	5
1	4
2	10
3	9
4	10
5	7
6	3
7	4

FIGURE 4.62 Miles run by Todd Booth.

17. To study the output of a machine that fills boxes with cereal, a quality control engineer weighed 150 boxes of Brand X cereal. The frequency distribution in Figure 4.63 summarizes his findings. Find the mean weight of the boxes of cereal.

x = Weight (ounces)	Number of Boxes
$15.3 \leq x < 15.6$	13
$15.6 \leq x < 15.9$	24
$15.9 \leq x < 16.2$	84
$16.2 \leq x < 16.5$	19
$16.5 \leq x \leq 16.8$	10

FIGURE 4.63 Amount of Brand X cereal per box.

18. To study the efficiency of its new price-scanning equipment, a local supermarket monitored the amount of time its customers had to wait in line. The frequency distribution in Figure 4.64 summarizes the findings. Find the mean amount of time spent in line.

x = Time (minutes)	Number of Customers
$0 \leq x < 1$	79
$1 \leq x < 2$	58
$2 \leq x < 3$	64
$3 \leq x < 4$	40
$4 \leq x \leq 5$	35

FIGURE 4.64 Time spent waiting in a supermarket checkout line.

19. Katrina must take five exams in a math class. If her scores on the first four exams are 71, 69, 85, and 83, what score does she need on the fifth exam for her overall mean to be
   a. at least 70?   b. at least 80?
   c. at least 90?

20. Eugene must take four exams in a geography class. If his scores on the first three exams are 91, 67, and 83, what score does he need on the fourth exam for his overall mean to be
   a. at least 70?   b. at least 80?
   c. at least 90?

21. The mean salary of twelve men is $58,000, and the mean salary of eight women is $42,000. Find the mean salary of all twenty people.

22. The mean salary of twelve men is $52,000, and the mean salary of four women is $84,000. Find the mean salary of all sixteen people.

23. Maria drove from Chicago, Illinois, to Milwaukee, Wisconsin, a distance of ninety miles, at a mean speed of 60 miles per hour. On her return trip, the traffic was much heavier, and her mean speed was 45 miles per hour. Find Maria's mean speed for the round trip.

   HINT: Divide the total distance by the total time.

24. Sully drove from Atlanta, Georgia, to Birmingham, Alabama, a distance of 150 miles, at a mean speed of 50 miles per hour. On his return trip, the traffic was much lighter, and his mean speed was 60 miles per hour. Find Sully's mean speed for the round trip.

   HINT: Divide the total distance by the total time.

25. The mean age of a class of twenty-five students is 23.4 years. How old would a twenty-sixth student have to be for the mean age of the class to be 24.0 years?

26. The mean age of a class of fifteen students is 18.2 years. How old would a sixteenth student have to be for the mean age of the class to be 21.0 years?

27. The mean salary of eight employees is $40,000, and the median is $42,000. The highest-paid employee gets a $6,000 raise.
   a. What is the new mean salary of the eight employees?
   b. What is the new median salary of the eight employees?

28. The mean salary of ten employees is $32,000, and the median is $30,000. The highest-paid employee gets a $5,000 raise.
   a. What is the new mean salary of the ten employees?
   b. What is the new median salary of the ten employees?

29. The number of civilians holding government jobs in various federal departments and their monthly payrolls for September 2007 are given in Figure 4.65.

Department	Number of Civilian Workers	Monthly Payroll (thousands of dollars)
Department of Education	4,201	44,497
Department of Health and Human Services	62,502	576,747
Environmental Protection Agency	18,119	195,320
Department of Homeland Security	159,447	1,251,754
Department of the Navy	175,722	712,172
Department of the Air Force	157,182	635,901
Department of the Army	245,599	678,044

FIGURE 4.65   Monthly earnings for civilian jobs (September 2007). *Source:* U.S. Office of Personnel Management.

a. Which department or agency has the highest mean monthly earnings? What is the mean monthly earnings for this department or agency?

b. Which department or agency has the lowest mean monthly earnings? What is the mean monthly earnings for this department or agency?

c. Find the mean monthly earnings of all civilians employed by the Department of Education, the Department of Health and Human Services, and the Environmental Protection Agency.

d. Find the mean monthly earnings of all civilians employed by the Department of Homeland Security, the Department of the Navy, the Department of the Air Force, and the Department of the Army.

30. The ages of the nearly 4 million women who gave birth in the United States in 2001 are given in Figure 4.66. Find the mean age of these women.

Age (years)	Number of Women
$15 \leq x < 20$	549,000
$20 \leq x < 25$	872,000
$25 \leq x < 30$	897,000
$30 \leq x < 35$	859,000
$35 \leq x < 40$	452,000
$40 \leq x < 45$	137,000

FIGURE 4.66   Ages of women giving birth in 2001. *Source:* U.S. Bureau of the Census.

31. The age composition of the population of the United States in the year 2000 is given in Figure 4.67.

a. Find the mean age of all people in the United States under the age of 85.

b. Replace the interval "85 and over" with the interval $85 \leq X \leq 100$ and find the mean age of all people in the United States.

Age (years)	Number of People (thousands)
$0 < x < 5$	19,176
$5 \leq x < 10$	20,550
$10 \leq x < 15$	20,528
$15 \leq x < 25$	39,184
$25 \leq x < 35$	39,892
$35 \leq x < 45$	44,149
$45 \leq x < 55$	37,678
$55 \leq x < 65$	24,275
$65 \leq x < 85$	30,752
85 and over	4,240

FIGURE 4.67   Age composition of the population of the United States in the year 2000. *Source:* U.S. Bureau of the Census.

In Exercises 32 and 33, use the age composition of the 14,980,000 students enrolled in institutions of higher education in the United States during 2000, as given in Figure 4.68.

Age of Males (years)	Number of Students
$14 \leq x < 18$	94,000
$18 \leq x < 20$	1,551,000
$20 \leq x < 22$	1,420,000
$22 \leq x < 25$	1,091,000
$25 \leq x < 30$	865,000
$30 \leq x < 35$	521,000
35 and over	997,000
total	6,539,000

FIGURE 4.68   Age composition of students in higher education. *Source:* U.S. National Center for Education Statistics.

Age of Females (years)	Number of Students
$14 \leq x < 18$	78,000
$18 \leq x < 20$	1,907,000
$20 \leq x < 22$	1,597,000
$22 \leq x < 25$	1,305,000
$25 \leq x < 30$	1,002,000
$30 \leq x < 35$	664,000
35 and over	1,888,000
total	8,441,000

FIGURE 4.68   *Continued.*

32. a. Find the mean age of all male students in higher education under 35.

   b. Replace the interval "35 and over" with the interval $35 \leq x \leq 60$ and find the mean age of all male students in higher education.

33. a. Find the mean age of all female students in higher education under 35.

   b. Replace the interval "35 and over" with the interval $35 \leq x \leq 60$ and find the mean age of all female students in higher education.

Answer the following questions using complete sentences and your own words.

• CONCEPT QUESTIONS

34. What are the three measures of central tendency? Briefly explain the meaning of each.

35. Suppose the mean of Group I is A and the mean of Group II is B. We combine Groups I and II to form Group III. Is the mean of Group III equal to $\frac{A+B}{2}$? Explain.

36. Why do we use the midpoint of an interval when calculating the mean of grouped data?

WEB PROJECT

37. What were last month's "average" high and low temperatures in your favorite city? Pick a city that interests you and obtain the high and low temperatures for each day last month. Print the data, and submit them as evidence in answering the following.

   a. Find the mean, median, and mode of the daily high temperatures.

   b. How does the mean daily high temperature last month compare to the mean seasonal high temperature for last month? That is, were the high temperatures last month above or below the normal high temperatures for the month?

   c. Find the mean, median, and mode of the daily low temperatures.

   d. How does the mean daily low temperature last month compare to the mean seasonal low temperature for last month? That is, were the low temperatures last month above or below the normal low temperatures for the month?

Some useful links for this web project are listed on the text web site:
www.cengage.com/math/johnson

4.3   Measures of Dispersion

OBJECTIVES

- Find the standard deviation of a data set
- Examine the dispersion of data relative to the mean

To settle an argument over who was the better bowler, George and Danny agreed to bowl six games, and whoever had the highest "average" would be considered best. Their scores were as shown in Figure 4.69.

George	185	135	200	185	250	155
Danny	182	185	188	185	180	190

FIGURE 4.69 Bowling scores.

Each bowler then arranged his scores from lowest to highest and computed the mean, median, and mode:

George $\begin{cases} \quad \\ \\ \\ \\ \\ \end{cases}$

$$135 \qquad 155 \qquad 185 \qquad 185 \qquad 200 \qquad 250$$

$$\text{mean} = \frac{\text{sum of scores}}{6} = \frac{1{,}110}{6} = 185$$

$$\text{median} = \text{middle score} = \frac{185 + 185}{2} = 185$$

$$\text{mode} = \text{most common score} = 185$$

Danny $\begin{cases} \quad \\ \\ \\ \\ \\ \end{cases}$

$$180 \qquad 182 \qquad 185 \qquad 185 \qquad 188 \qquad 190$$

$$\text{mean} = \frac{1{,}110}{6} = 185$$

$$\text{median} = \frac{185 + 185}{2} = 185$$

$$\text{mode} = 185$$

Much to their surprise, George's mean, median, and mode were exactly the same as Danny's! Using the measures of central tendency alone to summarize their performances, the bowlers appear identical. Even though their averages were identical, however, their performances were not; George was very erratic, while Danny was very consistent. Who is the better bowler? On the basis of high score, George is better. On the basis of consistency, Danny is better.

George and Danny's situation points out a fundamental weakness in using only the measures of central tendency to summarize data. In addition to finding the averages of a set of data, the consistency, or spread, of the data should also be taken into account. This is accomplished by using **measures of dispersion,** which determine how the data points differ from the average.

 Deviations

It is clear from George and Danny's bowling scores that it is sometimes desirable to measure the relative consistency of a set of data. Are the numbers consistently bunched up? Are they erratically spread out? To measure the dispersion of a set of data, we need to identify an average or typical distance between the data points and the mean. The difference between a single data point x and the mean $\bar{x}$ is called the **deviation from the mean** (or simply the **deviation**) and is given by $(x - \bar{x})$. A data point that is close to the mean will have a small deviation, whereas data points far from the mean will have large deviations, as shown in Figure 4.70.

To find the typical deviation of the data points, you might be tempted to add up all the deviations and divide by the total number of data points, thus finding the "average" deviation. Unfortunately, this process leads nowhere. To see why, we will find the mean of the deviations of George's bowling scores.

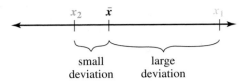

FIGURE 4.70 Large versus small deviations.

EXAMPLE **1**

SOLUTION

FINDING THE DEVIATIONS OF A DATA SET George bowled six games, and his scores were 185, 135, 200, 185, 250, and 155. Find the mean of the scores, the deviation of each score, and the mean of the deviations.

$$\bar{x} = \frac{\text{sum of scores}}{6} = \frac{1{,}110}{6} = 185$$

The mean score is 185.

To find the deviations, subtract the mean from each score, as shown in Figure 4.71.

$$\text{mean of the deviations} = \frac{\text{sum of deviations}}{6} = \frac{0}{6} = 0$$

The mean of the deviations is zero.

Score (x)	Deviation (x − 185)
135	−50
155	−30
185	0
185	0
200	15
250	65
	Sum = 0

FIGURE 4.71

Deviations of George's score.

In Example 1, the sum of the deviations of the data is zero. This *always* happens; that is, $\Sigma (x - \bar{x}) = 0$ for any set of data. The negative deviations are the "culprits"—they will always cancel out the positive deviations. Therefore, to use deviations to study the spread of the data, we must modify our approach and convert the negatives into positives. We do this by squaring each deviation.

Variance and Standard Deviation

Before proceeding, we must be reminded of the difference between a population and a sample. A population is the universal set of all possible items under study; a sample is any group or subset of items selected from the population. (Samples are used to study populations.) In this context, George's six bowling scores represent a sample, not a population; because we do not know the scores of *all* the games George has ever bowled, we are limited to a sample. Unless otherwise specified, we will consider any given set of data to represent a sample, not an entire population.

To measure the typical deviation contained within a set of data points, we must first find the **variance** of the data. Given a sample of n data points, the variance of the data is found by squaring each deviation, adding the squares, and then dividing the sum by the number $(n - 1)$.*

Because we are working with n data points, you might wonder why we divide by $n - 1$ rather than by n. The answer lies in the study of inferential statistics. Recall that inferential statistics deal with the drawing of conclusions concerning the nature of a population based on observations made within a sample. Hence, the variance of a sample can be viewed as an estimate of the variance of the population. However, because the population will vary more than the sample (a population has more data points), dividing the sum of the squares of the sample deviations by n would underestimate the true variance of the entire population.

*If the n data points represent the entire population, the population variance, denoted by σ^2, is found by squaring each deviation, adding the squares, and then dividing the sum by n.

SAMPLE VARIANCE DEVIATION

Given a sample of n data points, $x_1, x_2, \ldots, x_n$, the **variance** of the data, denoted by s^2, is

$$s^2 = \frac{\Sigma(x - \bar{x})^2}{n - 1}$$

The variance of a sample is found by dividing the sum of the squares of the deviations by $n - 1$. The symbol s^2 is a reminder that the deviations have been squared.

To compensate for this underestimation, statisticians have determined that dividing the sum of the squares of the deviations by $n - 1$ rather than by n produces the best estimate of the true population variance.

Variance is the tool with which we can obtain a measure of the typical deviation contained within a set of data. However, because the deviations have been squared, we must perform one more operation to obtain the desired result: We must take the square root. The square root of variance is called the **standard deviation** of the data.

STANDARD DEVIATION DEFINITION

Given a sample of n data points, $x_1, x_2, \ldots, x_n$, the **standard deviation** of the data, denoted by s, is

$$s = \sqrt{\text{variance}}$$

To find the standard deviation of a set of data, first find the variance and then take the square root of the variance.

EXAMPLE 2

FINDING THE VARIANCE AND STANDARD DEVIATION: SINGLE VALUES   George bowled six games, and his scores were 185, 135, 200, 185, 250, and 155. Find the standard deviation of his scores.

SOLUTION

To find the standard deviation, we must first find the variance. The mean of the six data points is 185. The necessary calculations for finding variance are shown in Figure 4.72.

Data (x)	Deviation (x − 185)	Deviation Squared (x − 185)²
135	−50	$(-50)^2 = 2{,}500$
155	−30	$(-30)^2 = 900$
185	0	$(0)^2 = 0$
185	0	$(0)^2 = 0$
200	15	$(15)^2 = 225$
250	65	$(65)^2 = 4{,}225$
		Sum = 7,850

FIGURE 4.72 Finding variance.

$$\text{variance} = \frac{\text{sum of the squares of the deviations}}{n-1}$$

$$s^2 = \frac{7{,}850}{6-1}$$

$$= \frac{7{,}850}{5}$$

$$= 1{,}570$$

The variance is $s^2 = 1{,}570$. Taking the square root, we have

$$s = \sqrt{1{,}570}$$

$$= 39.62322551\ldots$$

It is customary to round off s to one place more than the original data. Hence, the standard deviation of George's bowling scores is $s = 39.6$ points.

Because they give us information concerning the spread of data, variance and standard deviation are called **measures of dispersion.** Standard deviation (and variance) is a relative measure of the dispersion of a set of data; the larger the standard deviation, the more spread out the data. Consider George's standard deviation of 39.6. This appears to be high, but what exactly constitutes a "high" standard deviation? Unfortunately, because it is a relative measure, there is no hard-and-fast distinction between a "high" and a "low" standard deviation.

By itself, the standard deviation of a set of data might not be very informative, but standard deviations are very useful in comparing the relative consistencies of two sets of data. Given two groups of numbers of the same type (for example, two sets of bowling scores, two sets of heights, or two sets of prices), the set with the lower standard deviation contains data that are more consistent, whereas the data with the higher standard deviation are more spread out. Calculating the standard deviation of Danny's six bowling scores, we find $s = 3.7$. Since Danny's standard deviation is less than George's, we infer that Danny is more consistent. If George's standard deviation is less than 39.6 the next time he bowls six games, we would infer that his game has become more consistent (the scores would not be spread out as far).

Alternative Methods for Finding Variance

The procedure for calculating variance is very direct: First find the mean of the data, then find the deviation of each data point, and finally divide the sum of the squares of the deviations by $(n-1)$. However, using the definition of sample variance to find the variance can be rather tedious. Fortunately, many scientific calculators are programmed to find the variance (and standard deviation) if you just push a few buttons. Consult your manual to utilize the statistical capabilities of your calculator.

If your calculator does not have built-in statistical functions, you might still be able to take a shortcut in calculating variance. Instead of using the definition of variance (as in Example 2), we can use an alternative formula that contains the two sums Σx and Σx^2, where Σx represents the sum of the data and Σx^2 represents the sum of the squares of the data, as shown in the following box.

ALTERNATIVE FORMULA FOR SAMPLE VARIANCE

Given a sample of n data points, $x_1, x_2, \ldots, x_n$, the **variance** of the data, denoted by s^2, can be found by

$$s^2 = \frac{1}{(n-1)} \left[\Sigma x^2 - \frac{(\Sigma x)^2}{n} \right]$$

Note: Σx^2 means "square each data point, then add"; $(\Sigma x)^2$ means "add the data points, then square."

Although we will not prove it, this Alternative Formula for Sample Variance is algebraically equivalent to the sample variance definition; given any set of data, either method will produce the same answer. At first glance, the Alternative Formula might appear to be more difficult to use than the definition. Do not be fooled by its appearance! As we will see, the Alternative Formula is relatively quick and easy to apply.

EXAMPLE **3**

FINDING THE VARIANCE AND STANDARD DEVIATION: ALTERNA-TIVE FORMULA   Using the Alternative Formula for Sample Variance, find the standard deviation of George's bowling scores as given in Example 2.

SOLUTION

Recall that George's scores were 185, 135, 200, 185, 250, and 155. To find the standard deviation, we must first find the variance.

The Alternative Formula for Sample Variance requires that we find the sum of the data and the sum of the squares of the data. These calculations are shown in Figure 4.73. Applying the Alternative Formula for Sample Variance, we have

x	x^2
135	18,225
155	24,025
185	34,225
185	34,225
200	40,000
250	62,500
$\Sigma x = 1,110$	$\Sigma x^2 = 213,200$

FIGURE 4.73

Data and data squared.

$$s^2 = \frac{1}{(n-1)} \left[\Sigma x^2 - \frac{(\Sigma x)^2}{n} \right]$$

$$= \frac{1}{6-1} \left[213,200 - \frac{(1,110)^2}{6} \right]$$

$$= \frac{1}{5} \left[213,200 - 205,350 \right]$$

$$= \frac{7,850}{5} = 1,570$$

The variance is $s^2 = 1,570$. (Note that this is the same as the variance calculated in Example 2 using the definition of sample variance.)

Taking the square root, we have

$$s = \sqrt{1,570}$$

$$= 39.62322551 \ldots$$

Rounded off, the standard deviation of George's bowling scores is $s = 39.6$ points.

When we are working with grouped data, the individual data points are unknown. In such cases, the midpoint of each interval should be used as the representative value of the interval.

EXAMPLE **4**

FIND THE VARIANCE AND STANDARD DEVIATION: GROUPED DATA
In 2008, the U.S. Bureau of Labor Statistics tabulated a survey of workers' ages and wages. The frequency distribution in Figure 4.74 summarizes the age distribution of workers who received minimum wage ($6.55 per hour). Find the standard deviation of the ages these workers.

y = Age (years)	Number of Workers
$16 \leq y < 20$	108,000
$20 \leq y < 25$	53,000
$25 \leq y < 35$	41,000
$35 \leq y < 45$	23,000
$45 \leq y < 55$	29,000
$55 \leq y < 65$	23,000
	$n = 277,000$

FIGURE 4.74 Age distribution of workers receiving minimum wage.
Source: Bureau of Labor Statistics, U.S. Department of Labor.

SOLUTION

Because we are given grouped data, the first step is to determine the midpoint of each interval. We do this by adding the endpoints and dividing by 2.

To utilize the Alternative Formula for Sample Variance, we must find the sum of the data and the sum of the squares of the data. The sum of the data is found by multiplying each midpoint by the frequency of the interval and adding the results; that is, $\Sigma(f \cdot x)$. The sum of the squares of the data is found by squaring each midpoint, multiplying by the corresponding frequency, and adding; that is, $\Sigma(f \cdot x^2)$. The calculations are shown in Figure 4.75.

y = Age (years)	f = Frequency	x = Midpoint	$f \cdot x$	$f \cdot x^2$
$16 \leq y < 20$	108,000	18	1,944,000	34,992,000
$20 \leq y < 25$	53,000	22.5	1,192,500	26,831,250
$25 \leq y < 35$	41,000	30	1,230,000	36,900,000
$35 \leq y < 45$	23,000	40	920,000	36,800,000
$45 \leq y < 55$	29,000	50	1,450,000	72,500,000
$55 \leq y < 65$	23,000	60	1,380,000	82,800,000
	$n = 277,000$		$\Sigma(f \cdot x) = 8,116,500$	$\Sigma(f \cdot x^2) = 290,823,250$

FIGURE 4.75 Finding variance of grouped data.

Applying the Alternative Formula for Sample Variance, we have

$$s^2 = \frac{1}{n-1}\left[\Sigma(f\cdot x^2) - \frac{(\Sigma f\cdot x)^2}{n}\right]$$

$$= \frac{1}{277,000-1}\left[290,823,250 - \frac{(8,116,500)^2}{277,000}\right]$$

$$= \frac{1}{276,999}\left[290,823,250 - 237,825,170.6\right]$$

$$= \frac{52,998,079.4}{276,999}$$

The variance is $s^2 = 191.3294972$. Taking the square root, we have

$$s = \sqrt{191.3294972}$$

$$= 13.83219062\ldots$$

Rounded off, the standard deviation of the ages of the workers receiving minimum wage is $s = 13.8$ years.

The procedure for calculating variance when working with grouped data (as illustrated in Example 4) is summarized in the following box.

ALTERNATIVE FORMULA FOR SAMPLE VARIANCE: GROUPED DATA

Given a frequency distribution containing several groups of data, the variance s^2 can be found by

$$s^2 = \frac{1}{(n-1)}\left[\Sigma(f\cdot x^2) - \frac{(\Sigma f\cdot x)^2}{n}\right]$$

where x = the midpoint of a group, f = the frequency of the group, and $n = \Sigma f$

To obtain the best analysis of a collection of data, we should use the measures of central tendency and the measures of dispersion in conjunction with each other. The most common way to combine these measures is to determine what percent of the data lies within a specified number of standard deviations of the mean. The phrase "one standard deviation of the mean" refers to all numbers within the interval $[\bar{x} - s, \bar{x} + s]$, that is, all numbers that differ from $\bar{x}$ by at most s. Likewise, "two standard deviations of the mean" refers to all numbers within the interval $[\bar{x} - 2s, \bar{x} + 2s]$. One, two, and three standard deviations of the mean are shown in Figure 4.76.

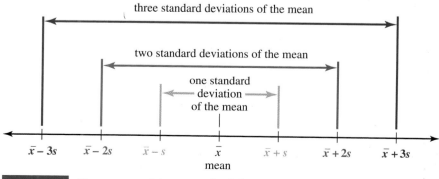

FIGURE 4.76   One, two, and three standard deviations of the mean.

EXAMPLE **5**

FINDING THE PERCENTAGE OF DATA WITHIN A SPECIFIED INTERVAL Paki Mowry is a rabbit enthusiast and has eleven mini Rex rabbits that she enters in regional rabbit competition shows. The weights of the rabbits are given in Figure 4.77. What percent of the rabbits' weights lie within one standard deviation of the mean?

Weight (lb:oz)	3:10	4:02	3:06	3:15	3:12	4:01
	3:11	4:00	4:03	3:13	3:15	

FIGURE 4.77 Weights of rabbits in pounds and ounces.

SOLUTION

First, the weights must be converted to a common unit, say, ounces. Therefore, we multiply the number of pounds by 16 (ounces per pound) and add the given number of ounces. For instance, 3 pounds, 10 ounces is converted to ounces as follows:

$$3 \text{ pounds, 10 ounces} = \left(3 \text{ pounds} \times \frac{16 \text{ ounces}}{\text{pound}}\right) + 10 \text{ ounces}$$
$$= 48 \text{ ounces} + 10 \text{ ounces}$$
$$= 58 \text{ ounces}$$

The converted weights are given in Figure 4.78.

Weight (ounces)	58	66	54	63	60	65
	59	64	67	61	63	

FIGURE 4.78 Weights of rabbits in ounces.

Now we find the mean and standard deviation.

Summing the eleven data points, we have $\Sigma x = 680$. Summing the squares of the data points, we have $\Sigma x^2 = 42,186$. The mean is

$$\bar{x} = \frac{680}{11} = 61.81818181\ldots = 61.8 \text{ ounces} \quad \text{(rounded to one decimal place)}$$

Using the Alternative Formula for Sample Variance, we have

$$s^2 = \frac{1}{(n-1)}\left[\Sigma x^2 - \frac{(\Sigma x)^2}{n}\right]$$

$$= \frac{1}{(11-1)}\left[42,186 - \frac{(680)^2}{11}\right]$$

$$= \frac{1}{10}\left[\frac{1646}{11}\right] \qquad\qquad \text{subtracting fractions with LCD = 11}$$

$$= \frac{823}{55} \qquad\qquad \text{reducing}$$

The variance is $\frac{823}{55} = 14.963636363\ldots$ Taking the square root, we have

$$s = \sqrt{823/55} = 3.868285972\ldots$$

The standard deviation is 3.9 ounces (rounded to one decimal place).

To find one standard deviation of the mean, we add and subtract the standard deviation to and from the mean:

$$[\bar{x} - s, \bar{x} + s] = [61.8 - 3.9, 61.8 + 3.9]$$
$$= [57.9, 65.7]$$

Arranging the data from smallest to largest, we see that eight of the eleven data points are between 57.9 and 65.7:

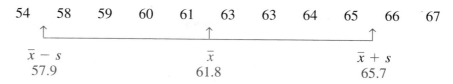

Therefore, $\dfrac{8}{11} = 0.7272727272\ldots$, or 72.7%, of the data lie within one standard deviation of the mean.

4.3 EXERCISES

1. Perform each task, given the following sample data:

    3 8 5 3 10 13

    a. Use the Sample Variance Definition to find the variance and standard deviation of the data.
    b. Use the Alternative Formula for Sample Variance to find the variance and standard deviation of the data.

2. Perform each task, given the following sample data:

    6 10 12 12 11 17 9

    a. Use the Sample Variance Definition to find the variance and standard deviation of the data.
    b. Use the Alternative Formula for Sample Variance to find the variance and standard deviation of the data.

3. Perform each task, given the following sample data:

    10 10 10 10 10 10

    a. Find the variance of the data.
    b. Find the standard deviation of the data.

4. Find the mean and standard deviation of each set of data.
    a. 2 4 6 8 10 12
    b. 102 104 106 108 110 112
    c. How are the data in (b) related to the data in (a)?
    d. How do your answers for (a) and (b) compare?

5. Find the mean and standard deviation of each set of data.
    a. 12 16 20 24 28 32
    b. 600 800 1,000 1,200 1,400 1,600
    c. How are the data in (b) related to the data in (a)?
    d. How do your answers for (a) and (b) compare?

6. Find the mean and standard deviation of each set of data.
    a. 50 50 50 50 50
    b. 46 50 50 50 54
    c. 5 50 50 50 95
    d. How do your answers for (a), (b), and (c) compare?

7. Joey and Dee Dee bowled five games at the Rock 'n' Bowl Lanes. Their scores are given in Figure 4.79.
    a. Find the mean score of each bowler. Who has the highest mean?
    b. Find the standard deviation of each bowler's scores.
    c. Who is the more consistent bowler? Why?

Joey	144	171	220	158	147
Dee Dee	182	165	187	142	159

FIGURE 4.79 Bowling scores.

8. Paki surveyed the price of unleaded gasoline (self-serve) at gas stations in Novato and Lafayette. The raw data, in dollars per gallon, are given in Figure 4.80.
    a. Find the mean price in each city. Which city has the lowest mean?
    b. Find the standard deviation of prices in each city.
    c. Which city has more consistently priced gasoline? Why?

Novato	2.899	3.089	3.429	2.959	2.999	3.099
Lafayette	3.595	3.389	3.199	3.199	3.549	3.349

FIGURE 4.80 Price (in dollars) of one gallon of unleaded gasoline.

9. Kaitlin Mowry is a member of the local 4-H club and has six mini Rex rabbits that she enters in regional rabbit competition shows. The weights of the rabbits are given in Figure 4.81. Find the standard deviation of the rabbits' weights.

| Weight (lb:oz) | 4:12 | 5:03 | 4:06 | 3:15 | 4:12 | 4:08 |

FIGURE 4.81 Weights of rabbits.

10. As was stated in Example 1 of Section 4.2, Lance Armstrong won the Tour de France bicycle race every year from 1999 to 2005. His margins of victory (time difference of the second place finisher) are given in Figure 4.82. Find the standard deviation of Lance Armstrong's victory margins.

| Year | 1999 | 2000 | 2001 | 2002 | 2003 | 2004 | 2005 |
| Margin | 7:37 | 6:02 | 6:44 | 7:17 | 1:01 | 6:19 | 4:40 |

FIGURE 4.82 Lance Armstrong's Tour de France victory margins. *Source: San Francisco Chronicle.*

11. Barry Bonds of the San Francisco Giants set an all-time record in professional baseball by hitting 73 home runs in one season (2001). The number of home runs (HR) for each of his seasons in professional baseball is given in Figure 4.83. Find the standard deviation of the number of home runs hit per year by Bonds.

Year	HR	Year	HR
1986	16	1997	40
1987	25	1998	37
1988	24	1999	34
1989	19	2000	49
1990	33	2001	73
1991	25	2002	46
1992	34	2003	45
1993	46	2004	45
1994	37	2005	5
1995	33	2006	26
1996	42	2007	28

FIGURE 4.83 Barry Bonds's home runs.

12. Michael Jordan has been recognized as an extraordinary player in professional basketball, especially in terms of the number of points per game he has scored. Jordan's average number of points per game (PPG) for each of his seasons is given in Figure 4.84. Find the standard deviation of the average points per game made per season by Jordan.

Season	PPG	Season	PPG
1984–85	28.2	1992–93	32.6
1985–86	22.7	1994–95	26.9
1986–87	37.1	1995–96	30.4
1987–88	35.0	1996–97	29.6
1988–89	32.5	1997–98	28.7
1989–90	33.6	2001–02	22.9
1990–91	31.5	2002–03	20.0
1991–92	30.1		

FIGURE 4.84 Michael Jordan's points per game.

13. The Truly Amazing Dudes are a group of comic acrobats. The heights (in inches) of the ten acrobats are as follows:

    68   50   70   67   72   78   69   68   66   67

Is your height or weight "average"? These characteristics can vary considerably within any specific group of people. The mean is used to represent the average, and the standard deviation is used to measure the "spread" of a collection of data.

a. Find the mean and standard deviation of the heights.

b. What percent of the data lies within one standard deviation of the mean?

c. What percent of the data lies within two standard deviations of the mean?

14. The weights (in pounds) of the ten Truly Amazing Dudes are as follows:

152   196   144   139   166   83   186   157   140   138

a. Find the mean and standard deviation of the weights.

b. What percent of the data lies within one standard deviation of the mean?

c. What percent of the data lies within two standard deviations of the mean?

15. The normal monthly rainfall in Seattle, Washington, is given in Figure 4.85.

Month	Jan.	Feb.	Mar.	Apr.	May	June
Inches	5.4	4.0	3.8	2.5	1.8	1.6

Month	July	Aug.	Sept.	Oct.	Nov.	Dec.
Inches	0.9	1.2	1.9	3.3	5.7	6.0

FIGURE 4.85   Monthly rainfall in Seattle, WA. *Source:* U.S. Department of Commerce.

a. Find the mean and standard deviation of the monthly rainfall in Seattle.

b. What percent of the year will the monthly rainfall be within one standard deviation of the mean?

c. What percent of the year will the monthly rainfall be within two standard deviations of the mean?

16. The normal monthly rainfall in Phoenix, Arizona, is given in Figure 4.86.

Month	Jan.	Feb.	Mar.	Apr.	May	June
Inches	0.7	0.7	0.9	0.2	0.1	0.1

Month	July	Aug.	Sept.	Oct.	Nov.	Dec.
Inches	0.8	1.0	0.9	0.7	0.7	1.0

FIGURE 4.86   Monthly rainfall in Phoeniz, AZ. *Source:* U.S. Department of Commerce.

a. Find the mean and standard deviation of the monthly rainfall in Phoenix.

b. What percent of the year will the monthly rainfall be within one standard deviation of the mean?

c. What percent of the year will the monthly rainfall be within two standard deviations of the mean?

17. The frequency distribution in Figure 4.87 lists the results of a quiz given in Professor Gilbert's statistics class.

Score	Number of Students	Score	Number of Students
10	5	7	8
9	10	6	3
8	6	5	2

FIGURE 4.87   Quiz scores in Professor Gilbert's statistics class.

a. Find the mean and standard deviation of the scores.

b. What percent of the data lies within one standard deviation of the mean?

c. What percent of the data lies within two standard deviations of the mean?

d. What percent of the data lies within three standard deviations of the mean?

18. Amy surveyed the prices for a quart of a certain brand of motor oil. The sample data, in dollars per quart, is summarized in Figure 4.88.

Price per Quart (dollars)	Number of Stores
1.99	2
2.09	5
2.19	10
2.29	13
2.39	9
2.49	3

FIGURE 4.88   Price for a quart of motor oil.

a. Find the mean and the standard deviation of the prices.

b. What percent of the data lies within one standard deviation of the mean?

c. What percent of the data lies within two standard deviations of the mean?

d. What percent of the data lies within three standard deviations of the mean?

19. To study the output of a machine that fills boxes with cereal, a quality control engineer weighed 150 boxes of Brand X cereal. The frequency distribution in Figure 4.89 summarizes his findings. Find the standard deviation of the weight of the boxes of cereal.

x = Weight (ounces)	Number of Boxes
$15.3 \leq x < 15.6$	13
$15.6 \leq x < 15.9$	24
$15.9 \leq x < 16.2$	84
$16.2 \leq x < 16.5$	19
$16.5 \leq x < 16.8$	10

FIGURE 4.89   Amount of Brand X cereal per box.

20. To study the efficiency of its new price-scanning equipment, a local supermarket monitored the amount of time its customers had to wait in line. The frequency distribution in Figure 4.90 summarizes the findings. Find the standard deviation of the amount of time spent in line.

x = Time (minutes)	Number of Customers
$0 \leq x < 1$	79
$1 \leq x < 2$	58
$2 \leq x < 3$	64
$3 \leq x < 4$	40
$4 \leq x \leq 5$	35

FIGURE 4.90   Time spent waiting in a supermarket checkout line.

21. The ages of the nearly 4 million women who gave birth in the United States in 2001 are given in Figure 4.91. Find the standard deviation of the ages of these women.

Age (years)	Number of Women
$15 \leq x < 20$	549,000
$20 \leq x < 25$	872,000
$25 \leq x < 30$	897,000
$30 \leq x < 35$	859,000
$35 \leq x < 40$	452,000
$40 \leq x < 45$	137,000

FIGURE 4.91   Ages of women giving birth in 2001. *Source:* U.S. Bureau of the Census.

22. The age composition of the population of the United States in the year 2000 is given in Figure 4.92. Replace the interval "85 and over" with the interval

$85 \leq x \leq 100$ and find the standard deviation of the ages of all people in the United States.

Age	Number of People (thousands)
$0 < x < 5$	19,176
$5 \leq x < 10$	20,550
$10 \leq x < 15$	20,528
$15 \leq x < 25$	39,184
$25 \leq x < 35$	39,892
$35 \leq x < 45$	44,149
$45 \leq x < 55$	37,678
$55 \leq x < 65$	24,275
$65 \leq x < 85$	30,752
85 and over	4,240

FIGURE 4.92   Age composition of the population of the United States in the year 2000. *Source:* U.S. Bureau of the Census.

 Answer the following questions using complete sentences and your own words.

• CONCEPT QUESTIONS

23. **a.** When studying the dispersion of a set of data, why are the deviations from the mean squared?

   b. What effect does squaring have on a deviation that is less than 1?

   c. What effect does squaring have on a deviation that is greater than 1?

   d. What effect does squaring have on the data's units?

   e. Why is it necessary to take a square root when calculating standard deviation?

24. Why do we use the midpoint of an interval when calculating the standard deviation of grouped data?

WEB PROJECT

25. This project is a continuation of Exercise 37 in Section 4.2. How did last month's daily high and low temperatures vary in your favorite city? Pick a city that interests you, and obtain the high and low temperatures for each day last month. Print the data, and submit them as evidence in answering the following.

   a. Find the standard deviation of the daily high temperatures.

b. Find the standard deviation of the daily low temperatures.

c. Comparing your answers to parts (a) and (b), what can you conclude?

Some useful links for this web project are listed on the text web site:

www.cengage.com/math/johnson

● PROJECTS

26. The purpose of this project is to explore the variation in the pricing of a common commodity. Go to several different stores that sell food (the more stores the better) and record the price of one gallon of whole milk.

   a. Compute the mean, median, and mode of the data.

   b. Compute the standard deviation of the data.

c. What percent of the data lie within one standard deviation of the mean?

d. What percent of the data lie within two standard deviations of the mean?

e. What percent of the data lie within three standard deviations of the mean?

27. The purpose of this project is to explore the variation in the pricing of a common commodity. Go to several different gas stations (the more the better) and record the price of one gallon of premium gasoline (91 octane).

   a. Compute the mean, median, and mode of the data.

   b. Compute the standard deviation of the data.

   c. What percent of the data lie within one standard deviation of the mean?

   d. What percent of the data lie within two standard deviations of the mean?

   e. What percent of the data lie within three standard deviations of the mean?

MEASURES OF CENTRAL TENDENCY AND DISPERSION ON A GRAPHING CALCULATOR

In Examples 1, 2, and 3 of this section, we found the mean, median, mode, variance, and standard deviation of George's bowling scores. This work can be done quickly and easily on either a graphing calculator or Excel.

TECHNOLOGY AND MEASURES OF CENTRAL TENDENCY AND DISPERSION

Calculating the Mean, the Variance, and the Standard Deviation

ON A TI-83/84:

- Enter the data from Example 1 of this section in list L_1 as discussed in Section 4.1.
- Press STAT.
- Select "1-Var Stats" from the "CALC" menu.
- When "1-Var" appears on the screen, press 2nd L_1 ENTER.

ON A CASIO:

- Enter the data from Example 1 of this section as described in Section 4.1.
- Press CALC (i.e., F2).
- Press 1VAR (i.e., F1).

```
1-Var Stats
x̄=185
Σx=1110
Σx²=213200
Sx=39.62322551
σx=36.17089069
↓n=6
```

```
1-Var Stats
↑n=6
  minX=135
  Q₁=155
  Med=185
  Q₃=200
  maxX=250
```

After computing the mean, standard deviation, and more on a TI graphing calculator.

The above steps will result in the first screen in Figure 4.93. This screen gives the mean, the sample standard deviation (S_x on a TI, $\chi\sigma n\text{-}1$ on a Casio), the population standard deviation (σ_x on a TI, $\chi\sigma n$ on a Casio), and the number of data points (n). The second screen can be obtained by pressing the down arrow. It gives the minimum and maximum data points (minX and maxX, respectively) as well as the median (Med).

Calculating the Sample Variance

The above work does not yield the sample variance. To find it, follow these steps:

- Quit the statistics mode.
- Get S_x on the screen.
- Press VARS, select "Statistics," and then select "S_x" from the "X/Y" menu.
- Once S_x is on the screen, square it by pressing x^2 ENTER. The variance is 1,570.

Calculating the Mean, the Variance, and the Standard Deviation with Grouped Data

To calculate the mean and standard deviation from the frequency distribution that utilizes grouped data, follow the same steps except:

ON A TI-83/84:

- Enter the midpoints of the classes in list L_1.
- Enter the frequencies of those classes in list L_2. Each frequency must be less than 100.
- After "1-Var Stats" appears on the screen, press 2nd L_1 , 2nd L_2 ENTER.

MEASURES OF CENTRAL TENDENCY AND DISPERSION ON EXCEL

1. *Enter the data* from Example 1 of this section as discussed in Section 4.1. See column A of Figure 4.94.
2. *Have Excel compute the mean, standard deviation, etc.*
   - Use your mouse to select "Tools" at the very top of the screen. Then pull your mouse down until "Data Analysis" is highlighted, and let go. (If "Data Analysis" is not listed under "Tools," then follow the instructions given in Section 4.1.)
   - In the "Data Analysis" box that appears, use your mouse to highlight "Descriptive Statistics" and press the "OK" button.
   - In the "Descriptive Statistics" box that appears, use your mouse to click on the white rectangle that follows "Input Range," and then use your mouse to draw a box around all of the bowling scores. (To draw the box, move your mouse to cell A2, press the mouse button, and move the mouse down until all of the entered scores are enclosed in a box.) This should cause "$A$2:$A$7" to appear in the Input Range rectangle.

◇	A	B	C
1	George's scores	Column 1	
2	185		
3	135	Mean	185
4	200	Standard Errc	16.1761141
5	185	Median	185
6	250	Mode	185
7	155	Standard Dev	39.6232255
8		Sample Varia	1570
9		Kurtosis	0.838675
10		Skewness	0.60763436
11		Range	115
12		Minimum	135
13		Maximum	250
14		Sum	1110
15		Count	6

FIGURE 4.94 After computing the mean, standard deviation, and more on Excel.

- Be certain that there is a dot in the button to the left of "Output Range." Then click on the white rectangle that follows "Output Range," and use your mouse to click on cell B1. This should cause "$B$1" to appear in the Output Range rectangle.
- Select "Summary Statistics."
- Use your mouse to press the "OK" button.

After you complete this step, your spreadsheet should look like that in Figure 4.94. To compute the population standard deviation, first type "pop std dev" in cell B16, and then click on cell C16. Press the *fx* button on the Excel ribbon at the top of the screen. In the "Paste Function" box that appears, click on "Statistical" under "Function category" and on "STDEVP" under "Function name." Then press the "OK" button. In the "STDEVP" box that appears, click on the white rectangle that follows "Number 1." Then use your mouse to draw a box around all of the bowling scores. Press the "OK" button and the population standard deviation will appear in cell C16.

EXERCISES

28. Do Double Stuf Oreos really contain twice as much "stuf"? In 1999, Marie Revak and Jihan William's "interest was piqued by the unusual spelling of the word stuff and the bold statement 'twice the filling' on the package." So they conducted an experiment in which they weighed the amount of filling in a sample of traditional Oreos and Double Stuf Oreos. The data from that experiment are in Figure 4.95 (weight in grams). An unformatted spreadsheet containing these data can be downloaded from the text web site: **www.cengage.com/math/johnson**

a. Find the mean of these data.

b. Find the standard deviation.

c. Did you choose the population or sample standard deviation? Why?

d. What do your results say about Double Stuf Oreos? Interpret the results of both parts (a) and (b).

Double Stuf Oreos:

4.7	6.5	5.5	5.6	5.1	5.3	5.4	5.4	3.5	5.5	6.5	5.9	5.4	4.9	5.6	5.7	5.3	6.9	6.5
6.3	4.8	3.3	6.4	5.0	5.3	5.5	5.0	6.0	5.7	6.3	6.0	6.3	6.1	6.0	5.8	5.8	5.9	6.2
5.9	6.5	6.5	6.1	5.8	6.0	6.2	6.2	6.0	6.8	6.2	5.4	6.6	6.2					

Traditional Oreos:

2.9	2.8	2.6	3.5	3.0	2.4	2.7	2.4	2.5	2.2	2.6	2.6	2.9	2.6	2.6	3.1	2.9	2.4	2.8
3.8	3.1	2.9	3.0	2.1	3.8	3.0	3.0	2.8	2.9	2.7	3.2	2.8	3.1	2.7	2.8	2.6	2.6	3.0
2.8	3.5	3.3	3.3	2.8	3.1	2.6	3.5	3.5	3.1	3.1								

FIGURE 4.95 Amount of Oreo filling. *Source:* Revak, Marie A. and Jihan G. Williams, "Sharing Teaching Ideas: The Double Stuf Dilemma," *The Mathematics Teacher,* November 1999, Volume 92, Issue 8, page 674.

Rank	City	Air Quality Index (% days AQI ranked good)	Property Crime (incidents per 1000)	Median Commute Time (in minutes)
1	Louisville, CO	74.0%	15	19.5
2	Chanhassen, MN	N.A.	16	23.1
3	Papillion, NE	92.0%	16	19.3
4	Middleton, WI	76.0%	27	16.3
5	Milton, MA	81.0%	9	27.7
6	Warren, NJ	N.A.	21	28.7
7	Keller, TX	65.0%	14	29.4
8	Peachtree City, GA	84.0%	12	29.3
9	Lake St. Louis, MO	79.0%	16	26.8
10	Mukilteo, WA	89.0%	27	22.3

FIGURE 4.96   The best places to live. *Source:* CNNMoney.com.

29. Every year, CNNMoney.com lists "the best places to live." They compare the "best small towns" on several bases, one of which is the "quality of life." See Figure 4.96. An unformatted spreadsheet containing these data can be downloaded from the text web site: **www.cengage.com/math/johnson**

   a. Find the mean of the air quality indices.

   b. Would the population standard deviation or the sample standard deviation be more appropriate here? Why?

   c. Find the standard deviation from part (b).

   d. What do your results say about the air quality index of all of the best places to live in the United States?

30. Every year, CNNMoney.com lists "the best places to live." They compare the "best small towns" on several bases, one of which is the "quality of life." See Figure 4.96. An unformatted spreadsheet containing these data can be downloaded from the text web site: **www.cengage.com/math/johnson**

   a. Find the mean of the commute times.

   b. Would the population standard deviation or the sample standard deviation be more appropriate here? Why or why not?

   c. Find the standard deviation from part (b).

   d. What do your results say about commuting in all of the best places to live in the United States?

31. Every year, CNNMoney.com lists "the best places to live." They compare the "best small towns" on several bases, one of which is the "quality of life." See Figure 4.96. An unformatted spreadsheet containing these data can be downloaded from the text web site: **www.cengage.com/math/johnson**

   a. Find the mean of the property crime levels.

   b. Would the population standard deviation or the sample standard deviation be more appropriate here? Why or why not?

   c. Find the standard deviation from part (b).

   d. What do your results say about property crime in all of the best places to live in the United States?

32. Figure 4.97 gives EPA fuel efficiency ratings for 2010 family sedans with an automatic transmission and the smallest available engine. An unformatted spreadsheet containing these data can be downloaded from the text web site: **www.cengage.com/math/johnson**

   a. For each of the two manufacturing regions, find the mean and an appropriate standard deviation for city driving, and defend your choice of standard deviations.

   b. What do the results of part (a) tell you about the two regions?

33. Figure 4.97 gives EPA fuel efficiency ratings for 2010 family sedans with an automatic transmission and the smallest available engine. An unformatted spreadsheet containing these data can be downloaded from the text web site: **www.cengage.com/math/johnson**

   a. For each of the two manufacturing regions, find the mean and an appropriate standard deviation for highway driving, and defend your choice of standard deviations.

   b. What do the results of part (a) tell you about the two regions?

American Cars	City mpg	Highway mpg	Asian Cars	City mpg	Highway mpg
Ford Fusion hybrid	41	36	Toyota Prius	51	48
Chevrolet Malibu hybrid	26	34	Nissan Altima hybrid	35	33
Ford Fusion fwd	24	34	Toyota Camry Hybrid	33	34
Mercury Milan	23	34	Kia Forte	27	36
Chevrolet Malibu	22	33	Hyundai Elantra	26	35
Saturn Aura	22	33	Nissan Altima	23	32
Dodge Avenger	21	30	Subaru Legacy awd	23	31
Buick Lacrosse	17	27	Toyota Camry	22	33
			Hyundai Sonata	22	32
			Kia Optima	22	32
			Honda Accord	22	31
			Mazda 6	21	30
			Mitsubishi Galant	21	30
			Hyunda Azera	18	26

FIGURE 4.97 Fuel efficiency. *Source:* www.fueleconomy.gov.

4.4 The Normal Distribution

OBJECTIVES

- Find probabilities of the standard normal distribution
- Find probabilities of a nonstandard normal distribution
- Find the value of a normally distributed variable that will produce a specific probability

Sets of data may exhibit various trends or patterns. Figure 4.98 shows a histogram of the weights of bags of corn chips. Notice that most of the data are near the "center" and that the data taper off at either end. Furthermore, the histogram is nearly symmetric; it is almost the same on both sides. This type of distribution (nearly symmetric, with most of the data in the middle) occurs quite often in many different situations. To study the composition of such distributions, statisticians have created an ideal **bell-shaped curve** describing a **normal distribution,** as shown in Figure 4.99.

Before we can study the characteristics and applications of a normal distribution, we must make a distinction between different types of variables.

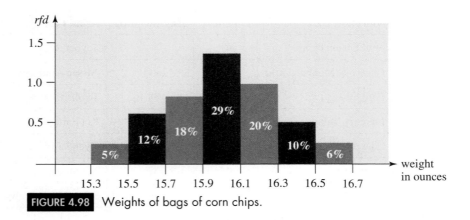

FIGURE 4.98 Weights of bags of corn chips.

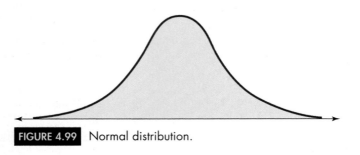

FIGURE 4.99 Normal distribution.

Discrete versus Continuous Variables

The number of children in a family is variable, because it varies from family to family. In listing the number of children, only whole numbers (0, 1, 2, and so on) can be used. In this respect, we are limited to a collection of discrete, or separate, values. A variable is **discrete** if there are "gaps" between the possible variable values. Consequently, any variable that involves counting is discrete.

On the other hand, a person's height or weight does not have such a restriction. When someone grows, he or she does not instantly go from 67 inches to 68 inches; a person grows continuously from 67 inches to 68 inches, attaining all possible values in between. For this reason, height is called a continuous variable. A variable is **continuous** if it can assume *any value* in an interval of real numbers (see Figure 4.100). Consequently, any variable that involves measurement is continuous; someone might claim to be 67 inches tall and to weigh 152 pounds, but the true values might be 67.13157 inches and 151.87352 pounds. Heights and weights are expressed (discretely) as whole numbers solely for convenience; most people do not have rulers or bathroom scales that allow them to obtain measurements that are accurate to ten or more decimal places!

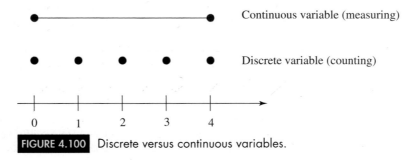

FIGURE 4.100 Discrete versus continuous variables.

Normal Distributions

The collection of all possible values that a discrete variable can assume forms a countable set. For instance, we can list all the possible numbers of children in a family. In contrast, a continuous variable will have an uncountable number of possibilities because it can assume any value in an interval. For instance, the weights (a continuous variable) of bags of corn chips could be *any* value x such that $15.3 \le x \le 16.7$.

When we sample a continuous variable, some values may occur more often than others. As we can see in Figure 4.98, the weights are "clustered" near the center of the histogram, with relatively few located at either end. If a continuous variable has a symmetric distribution such that the highest concentration of values is at the center and the lowest is at both extremes, the variable is said to have a **normal distribution** and is represented by a smooth, continuous, bell-shaped curve like that in Figure 4.99*.

The normal distribution, which is found in a wide variety of situations, has two main qualities: (1) the frequencies of the data points nearer the center or "average" are increasingly higher than the frequencies of data points far from the center, and (2) the distribution is symmetric (one side is a mirror image of the other). *Because of these two qualities, the mean, median, and mode of a normal distribution all coincide at the center of the distribution.*

Just like any other collection of numbers, the spread of normal distribution is measured by its standard deviation. It can be shown that for any normal distribution, slightly more than two-thirds of the data (68.26%) will lie within one standard deviation of the mean, 95.44% will lie within two standard deviations, and virtually all the data (99.74%) will lie within three standard deviations of the mean. Recall that μ (the Greek letter "mu") represents the mean of a population and σ (the Greek letter "sigma") represents the standard deviation of the population. The spread of a normal distribution, with μ and σ used to represent the mean and standard deviation, is shown in Figure 4.101.

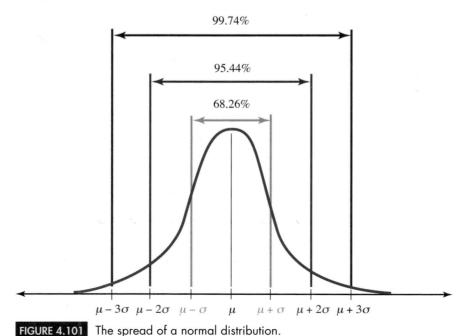

FIGURE 4.101   The spread of a normal distribution.

---

* This is an informal definition only. The formal definition of a normal distribution involves the number pi, the natural exponential e^x, and the mean and variance of the distribution.

EXAMPLE **1**

ANALYZING THE DISPERSION OF A NORMAL DISTRIBUTION   The heights of a large group of people are assumed to be normally distributed. Their mean height is 66.5 inches, and the standard deviation is 2.4 inches. Find and interpret the intervals representing one, two, and three standard deviations of the mean.

SOLUTION

The mean is $\mu = 66.5$, and the standard deviation is $\sigma = 2.4$.

1. *One standard deviation of the mean:*

$$\mu \pm 1\sigma = 66.5 \pm 1(2.4)$$
$$= 66.5 \pm 2.4$$
$$= [64.1, 68.9]$$

Therefore, approximately 68% of the people are between 64.1 and 68.9 inches tall.

2. *Two standard deviations of the mean:*

$$\mu \pm 2\sigma = 66.5 \pm 2(2.4)$$
$$= 66.5 \pm 4.8$$
$$= [61.7, 71.3]$$

Therefore, approximately 95% of the people are between 61.7 and 71.3 inches tall.

3. *Three standard deviations of the mean:*

$$\mu \pm 3\sigma = 66.5 \pm 3(2.4)$$
$$= 66.5 \pm 7.2$$
$$= [59.3, 73.7]$$

Nearly all of the people (99.74%) are between 59.3 and 73.7 inches tall.

In Example 1, we found that virtually all the people under study were between 59.3 and 73.7 inches tall. A clothing manufacturer might want to know what percent of these people are shorter than 66 inches or what percent are taller than 73 inches. Questions like these can be answered by using probability and a normal distribution. (We will do this in Example 6.)

Probability, Area, and Normal Distributions

In Chapter 3, we mentioned that relative frequency is really a type of probability. If 3 out of every 100 people have red hair, you could say that the relative frequency of red hair is $\frac{3}{100}$ (or 3%), or you could say that the probability of red hair $p(x = $ red hair) is 0.03. Therefore, to find out what percent of the people in a population are taller than 73 inches, we need to find $p(x > 73)$, the probability that x is greater than 73, where x represents the height of a randomly selected person.

Recall that a sample space is the set S of all possible outcomes of a random experiment. Consequently, the probability of a sample space must always equal 1; that is, $p(S) = 1$ (or 100%). If the sample space S has a normal distribution, its outcomes and their respective probabilities can be represented by a bell curve.

Recall that when constructing a histogram, relative frequency density (*rfd*) was used to measure the heights of the rectangles. Consequently, the *area* of a rectangle gave the relative frequency (percent) of data contained in an interval. In a similar manner, we can imagine a bell curve being a histogram composed of infinitely many "skinny" rectangles, as in Figure 4.102.

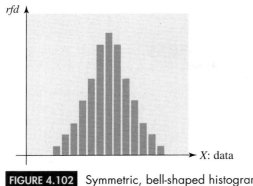

FIGURE 4.102 Symmetric, bell-shaped histogram.

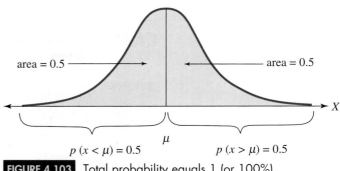

FIGURE 4.103 Total probability equals 1 (or 100%).

For a normal distribution, the outcomes nearer the center of the distribution occur more frequently than those at either end; the distribution is denser in the middle and sparser at the extremes. This difference in density is taken into account by consideration of the area under the bell curve; the center of the distribution is denser, contains more area, and has a higher probability of occurrence than the extremes. Consequently, we use the area under the bell curve to represent the probability of an outcome. Because $p(S) = 1$, we define the entire area under the bell curve to equal 1.

Because a normal distribution is symmetric, 50% of the data will be greater than the mean, and 50% will be less. (The mean and the median coincide in a symmetric distribution.) Therefore, the probability of randomly selecting a number x greater than the mean is $p(x > \mu) = 0.5$, and that of selecting a number x less than the mean is $p(x < \mu) = 0.5$, as shown in Figure 4.103.

To find the probability that a randomly selected number x is between two values (say a and b), we must determine the area under the curve from a to b; that is, $p(a < x < b) =$ area under the bell curve from $x = a$ to $x = b$, as shown in Figure 4.104(a). Likewise, the probability that x is greater than or less than any specific number is given by the area of the tail, as shown in Figure 4.104(b). To find probabilities involving data that are normally distributed, we must find the area of the appropriate region under the bell curve.

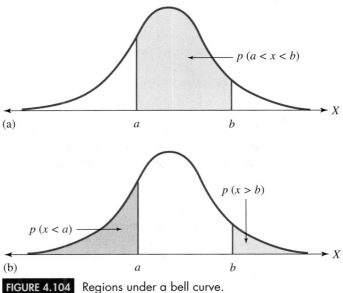

FIGURE 4.104 Regions under a bell curve.

CARL FRIEDRICH GAUSS, 1777–1855

© akg-images

Dubbed "the Prince of Mathematics," Carl Gauss is considered by many to be one of the greatest mathematicians of all time. At the age of three, Gauss is said to have discovered an arithmetic error in his father's bookkeeping. The child prodigy was encouraged by his teachers and excelled throughout his early schooling. When he was fourteen, Gauss was introduced to Ferdinand, the Duke of Brunswick. Impressed with the youth, the duke gave Gauss a yearly stipend and sponsored his education for many years.

In 1795, Gauss enrolled at Göttingen University, where he remained for three years. While at Göttingen, Gauss had complete academic freedom; he was not required to attend lectures, he had no required conferences with professors or tutors, and he did not take exams. Much of his time was spent studying independently in the library. For reasons unknown to us, Gauss left the university in 1798 without a diploma. Instead, he sent his dissertation to the University of Helmstedt and in 1799 was awarded his degree without the usual oral examination.

In 1796, Gauss began his famous mathematical diary. Discovered forty years after his death, the 146 sometimes cryptic entries exhibit the diverse range of topics that Gauss pondered and pioneered. The first entry was

Gauss's discovery (at the age of nineteen) of a method for constructing a seventeen-sided polygon with a compass and a straightedge. Other entries include important results in number theory, algebra, calculus, analysis, astronomy, electricity, magnetism, the foundations of geometry, and probability.

At the dawn of the nineteenth century, Gauss began his lifelong study of astronomy. On January 1, 1801, the Italian astronomer Giuseppe Piazzi discovered Ceres, the first of the known planetoids (minor planets or asteroids). Piazzi and others observed Ceres for forty-one days, until it was lost behind the sun. Because of his interest in the mathematics of astronomy, Gauss turned his attention to Ceres. Working with a minimal amount of data, he successfully calculated the orbit of Ceres. At the end of the year, the planetoid was rediscovered in exactly the spot that Gauss had predicted!

To obtain the orbit of Ceres, Gauss utilized his method of least squares, a technique for dealing with experimental error. Letting x represent the error between an experimentally obtained value and the true value it represents, Gauss's theory involved minimizing x^2—that is, obtaining the least square of the error. Theorizing that the probability of a small error was higher than that of a large

error, Gauss subsequently developed the normal distribution, or bell-shaped curve, to explain the probabilities of the random errors. Because of his pioneering efforts, some mathematicians refer to the normal distribution as the Gaussian distribution.

In 1807, Gauss became director of the newly constructed observatory at Göttingen. He held the position until his death some fifty years later.

THEORIA

MOTVS CORPORVM

COELESTIVM

IN

SECTIONIBVS CONICIS SOLEM AMBIENTIVM

AVCTORE

CAROLO FRIDERICO GAVSS

HAMBVRGI SVMTIBVS FRID. PERTHES ET I. H. BESSER
1809.

© Stock Montage

Published in 1809, Gauss's Theoria Motus Corporum Coelestium (Theory of the Motion of Heavenly Bodies) *contained rigorous methods of determining the orbits of planets and comets from observational data via the method of least squares. It is a landmark in the development of modern mathematical astronomy and statistics.*

The Standard Normal Distribution

All normal distributions share the following features: they are symmetric, bell-shaped curves, and virtually all the data (99.74%) lie within three standard deviations of the mean. Depending on whether the standard deviation is large or small, the bell curve will be either flat and spread out or peaked and narrow, as shown in Figure 4.105.

To find the area under any portion of any bell curve, mathematicians have devised a means of comparing the proportions of any curve with the proportions of a

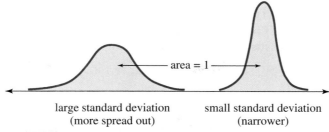

FIGURE 4.105 Large versus small standard deviation.

special curve defined as "standard." To find probabilities involving normally distributed data, we utilize the bell curve associated with the standard normal distribution.

The **standard normal distribution** is the normal distribution whose mean is 0 and standard deviation is 1, as shown in Figure 4.106. The standard normal distribution is also called the *z*-**distribution;** we will always use the letter *z* to refer to the standard normal. By convention, we will use the letter *x* to refer to any other normal distribution.

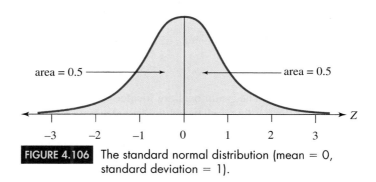

FIGURE 4.106 The standard normal distribution (mean = 0, standard deviation = 1).

Tables have been developed for finding areas under the standard normal curve using the techniques of calculus. Graphing calculators will also give these areas. We will use the table in Appendix F to find $p(0 < z < z^*)$, the probability that z is between 0 and a positive number z^*, as shown in Figure 4.107(a). The table in Appendix F is known as the **body table** because it gives the probability of an interval located in the middle, or body, of the bell curve.

The tapered end of a bell curve is known as a **tail.** To find the probability of a tail—that is, to find $p(z > z^*)$ or $p(z < z^*)$ where z^* is a positive real number—subtract the probability of the corresponding body from 0.5, as shown in Figure 4.107(b).

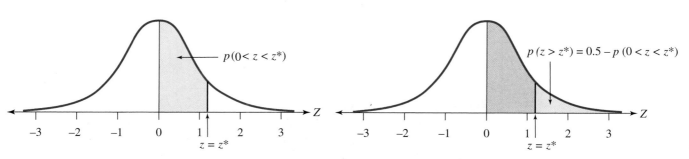

(a) Area found by using the body table (Appendix F)

(b) Area of a tail, found by subtracting the corresponding body area from 0.5

FIGURE 4.107

EXAMPLE **2**

FINDING PROBABILITIES OF THE STANDARD NORMAL DISTRIBUTION Find the following probabilities (that is, the areas), where z represents the standard normal distribution.

a. $p(0 < z < 1.25)$ **b.** $p(z > 1.87)$

SOLUTION

a. As a first step, it is always advisable to draw a picture of the z-curve and shade in the desired area. We will use the body table directly, because we are working with a central area (see Figure 4.108).

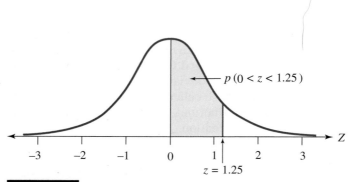

FIGURE 4.108 A central region, or body.

The z-numbers are located along the left edge and the top of the table. Locate the whole number and the first-decimal-place part of the number (1.2) along the left edge; then locate the second-decimal-place part of the number (0.05) along the top. The desired probability (area) is found at the intersection of the row and column of the two parts of the z-number. Thus, $p(0 < z < 1.25) = 0.3944$, as shown in Figure 4.109.

$z*$	0.00	0.01	0.02	0.03	0.04	0.05	0.06	0.07	0.08	0.09
○										
○										
○										
1.1	0.3643	0.3665	0.3686	0.3708	0.3729	0.3749	0.3770	0.3790	0.3810	0.3830
1.2	0.3849	0.3869	0.3888	0.3907	0.3925	0.3944	0.3962	0.3980	0.3997	0.4015
1.3	0.4032	0.4049	0.4066	0.4082	0.4099	0.4115	0.4131	0.4147	0.4162	0.4177
○										
○										

FIGURE 4.109 A portion of the body table.

Hence, we could say that about 39% of the z-distribution lies between $z = 0$ and $z = 1.25$.

b. To find the area of a tail, we subtract the corresponding body area from 0.5, as shown in Figure 4.110. Therefore,

$$p(z > 1.87) = 0.5 - p(0 < z < 1.87)$$
$$= 0.5 - 0.4692$$
$$= 0.0308$$

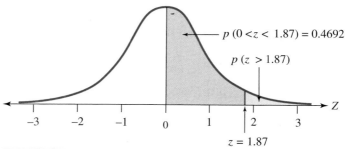

FIGURE 4.110 Finding the area of a tail.

The body table can also be used to find areas other than those given explicitly as $p(0 < z < z^*)$ and $p(z > z^*)$ where z^* is a positive number. By adding or subtracting two areas, we can find probabilities of the type $p(a < z < b)$, where a and b are positive or negative numbers, and probabilities of the type $p(z < c)$, where c is a positive or negative number.

EXAMPLE 3

FINDING PROBABILITIES OF THE STANDARD NORMAL DISTRIBUTION Find the following probabilities (the areas), where z represents the standard normal distribution.

a. $p(0.75 < z < 1.25)$ **b.** $p(-0.75 < z < 1.25)$

SOLUTION

a. Because the required region, shown in Figure 4.111, doesn't begin exactly at $z = 0$, we cannot look up the desired area directly in the body table. Whenever z is between two nonzero numbers, we will take an indirect approach to finding the required area.

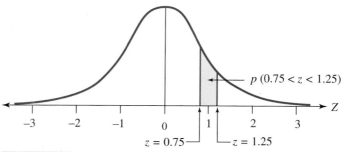

FIGURE 4.111 A strip.

The total area under the curve from $z = 0$ to $z = 1.25$ can be divided into two portions: the area under the curve from 0 to 0.75 and the area under the curve from 0.75 to 1.25.

To find the area of the "strip" between $z = 0.75$ and $z = 1.25$, we *subtract* the area of the smaller body (from $z = 0$ to $z = 0.75$) from that of the larger body (from $z = 0$ to $z = 1.25$), as shown in Figure 4.112.

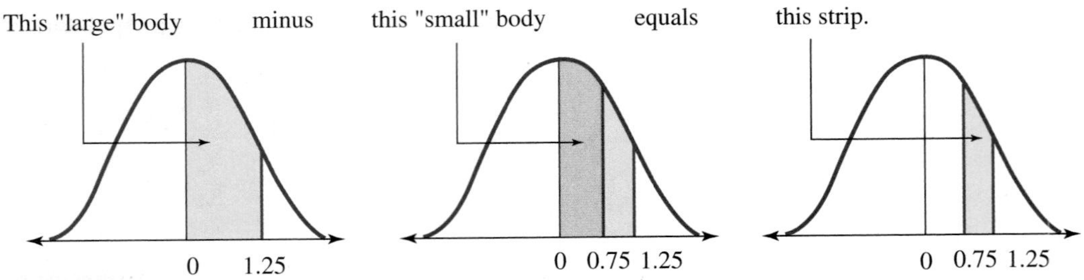

This "large" body   minus   this "small" body   equals   this strip.

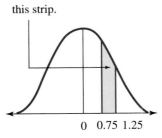

0   1.25   0 0.75 1.25   0 0.75 1.25

FIGURE 4.112   Area of a strip.

$$\text{area of strip} = \text{area of large body} - \text{area of small body}$$

$$p(0.75 < z < 1.25) = p(0 < z < 1.25) - p(0 < z < 0.75)$$

$$= 0.3944 - 0.2734$$

$$= 0.1210$$

Therefore, $p(0.75 < z < 1.25) = 0.1210$. Hence, we could say that about 12.1% of the z-distribution lies between $z = 0.75$ and $z = 1.25$.

b. The required region, shown in Figure 4.113, can be divided into two regions: the area from $z = -0.75$ to $z = 0$ and the area from $z = 0$ to $z = 1.25$.

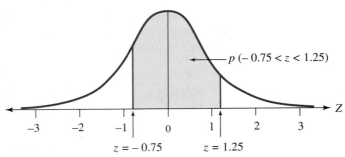

$p(-0.75 < z < 1.25)$

$$-3 \quad -2 \quad -1 \quad 0 \quad 1 \quad 2 \quad 3 \quad \rightarrow z$$

$z = -0.75$ $z = 1.25$

FIGURE 4.113   A central region.

To find the total area of the region between $z = -0.75$ and $z = 1.25$, we *add* the area of the "left" body (from $z = -0.75$ to $z = 0$) to the area of the "right" body (from $z = 0$ to $z = 1.25$), as shown in Figure 4.114.

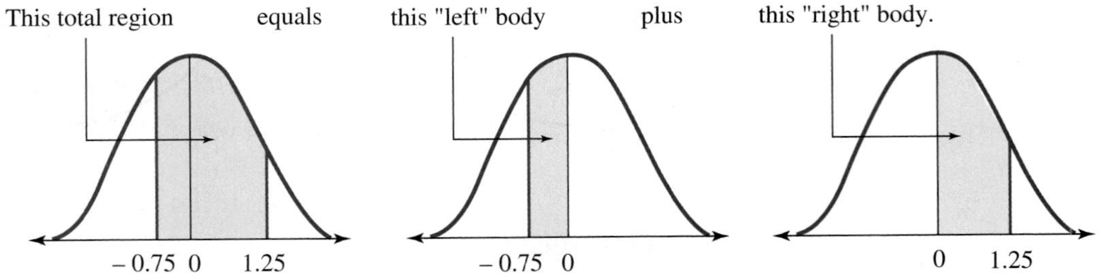

This total region   equals   this "left" body   plus   this "right" body.

$$-0.75\ 0 \quad 1.25 \qquad -0.75\ 0 \qquad\qquad 0 \quad 1.25$$

FIGURE 4.114   Left body plus right body.

This example is different from our previous examples in that it contains a negative z-number. A glance at the tables reveals that negative numbers are not included! However, recall that normal distributions are symmetric. Therefore, the area of the body from $z = -0.75$ to $z = 0$ is the same as that from $z = 0$ to $z = 0.75$; that is, $p(-0.75 < z < 0) = p(0 < z < 0.75)$. Therefore, total area of region = area of left body + area of right body

$$p(-0.75 < z < 1.25) = p(-0.75 < z < 0) + p(0 < z < 1.25)$$

$$= p(0 < z < 0.75) + p(0 < z < 1.25)$$

$$= 0.2734 + 0.3944$$

$$= 0.6678$$

Therefore, $p(-0.75 < z < 1.25) = 0.6678$. Hence, we could say that about 66.8% of the z-distribution lies between $z = -0.75$ and $z = 1.25$.

EXAMPLE 4

FINDING PROBABILITIES OF THE STANDARD NORMAL DISTRIBUTION   Find the following probabilities (the areas), where z represents the standard normal distribution.

a. $p(z < 1.25)$

b. $p(z < -1.25)$

SOLUTION

a. The required region is shown in Figure 4.115. Because 50% of the distribution lies to the left of 0, we can add 0.5 to the area of the body from $z = 0$ to $z = 1.25$:

$$p(z < 1.25) = p(z < 0) + p(0 < z < 1.25)$$

$$= 0.5 + 0.3944$$

$$= 0.8944$$

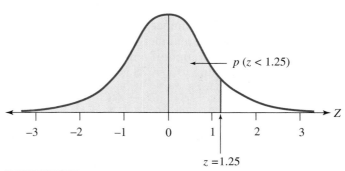

FIGURE 4.115   A body plus 50%.

Therefore, $p(z < 1.25) = 0.8944$. Hence, we could say that about 89.4% of the z-distribution lies to the left of $z = 1.25$.

b. The required region is shown in Figure 4.116. Because a normal distribution is symmetric, the area of the left tail ($z < -1.25$) is the same as the area of the corresponding right tail ($z > 1.25$).

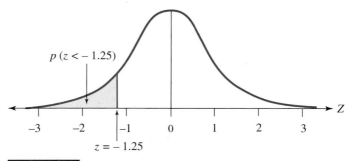

FIGURE 4.116   A left tail.

Therefore,

$$p(z < -1.25) = p(z > 1.25)$$
$$= 0.5 - p(0 < z < 1.25)$$
$$= 0.5 - 0.3944$$
$$= 0.1056$$

Hence, we could say that about 10.6% of the z-distribution lies to the left of $z = -1.25$.

Converting to the Standard Normal

Weather forecasters in the United States usually report temperatures in degrees Fahrenheit. Consequently, if a temperature is given in degrees Celsius, most people convert it to Fahrenheit in order to judge how hot or cold it is. A similar situation arises when we are working with a normal distribution. Suppose we know that a large set of data is normally distributed with a mean value of 68 and a standard deviation of 4. What percent of the data will lie between 65 and 73? We are asked to find $p(65 < x < 73)$. To find this probability, we must first convert the given normal distribution to the standard normal distribution and then look up the approximate z-numbers.

The body table (Appendix F) applies to the standard normal z-distribution. When we are working with any other normal distribution (denoted by X), we must first convert the x-distribution into the standard normal z-distribution. This conversion is done with the help of the following rule.

Given a number x, its corresponding z-number counts the number of standard deviations the number lies from the mean. For example, suppose the mean and standard deviation of a normal distribution are $\mu = 68$ and $\sigma = 4$. The z-number corresponding to $x = 78$ is

$$z = \frac{x - \mu}{\sigma} = \frac{78 - 68}{4} = 2.5$$

This implies that $x = 78$ lies two and one-half standard deviations above the mean, 68. Similarly, for $x = 65$,

$$z = \frac{65 - 68}{4} = -0.75$$

Therefore, $x = 65$ lies three-quarters of a standard deviation below the mean, 68.

CONVERTING A NORMAL DISTRIBUTION INTO THE STANDARD NORMAL Z

Every number x in a given normal distribution has a corresponding number z in the standard normal distribution. The z-**number** that corresponds to the number x is

$$z = \frac{x - \mu}{\sigma}$$

where μ is the mean and σ the standard deviation of the given normal distribution.

EXAMPLE **5**

**FINDING A PROBABILITY OF A NONSTANDARD NORMAL DISTRI-
BUTION**   Suppose a population is normally distributed with a mean of 24.6 and
a standard deviation of 1.3. What percent of the data will lie between 25.3
and 26.8?

SOLUTION

We are asked to find $p(25.3 < x < 26.8)$, the area of the region shown in Figure 4.117. Because we need to find the area of the strip between 25.3 and 26.8, we
must find the body of each and subtract, as in part (a) of Example 3.

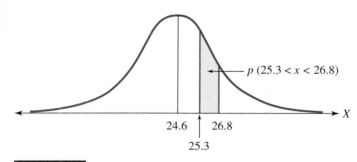

FIGURE 4.117   A strip.

   Using the Conversion Formula $z = (x - \mu)/\sigma$ with $\mu = 24.6$ and $\sigma = 1.3$,
we first convert $x = 25.3$ and $x = 26.8$ into their corresponding z-numbers.

Converting $x = 25.3$	Converting $x = 26.8$	
$z = \dfrac{x - \mu}{\sigma}$	$z = \dfrac{x - \mu}{\sigma}$	
$= \dfrac{25.3 - 24.6}{1.3}$	$= \dfrac{26.8 - 24.6}{1.3}$	
$= 0.5384615$	$= 1.6923077$	
$= 0.54$	$= 1.69$	**rounding off z-numbers to two decimal places**

Therefore,

$$p(25.3 < x < 26.8) = p(0.54 < z < 1.69)$$
$$= p(0 < z < 1.69) - p(0 < z < 0.54)$$
$$= 0.4545 - 0.2054 \qquad \textbf{using the body table}$$
$$= 0.2491$$

Assuming a normal distribution, approximately 24.9% of the data will lie between
25.3 and 26.8.

EXAMPLE **6**

**FINDING PROBABILITIES OF A NONSTANDARD NORMAL DISTRI-
BUTION**   The heights of a large group of people are assumed to be normally
distributed. Their mean height is 68 inches, and the standard deviation is 4 inches.
What percentage of these people are the following heights?

a. taller than 73 inches   **b.** between 60 and 75 inches

SOLUTION

a. Let x represent the height of a randomly selected person. We need to find
$p(x > 73)$, the area of a tail, as shown in Figure 4.118.

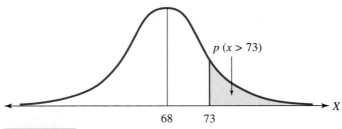

FIGURE 4.118 A right tail.

First, we must convert $x = 73$ to its corresponding z-number. Using the Conversion Formula with $x = 73$, $\mu = 68$, and $\sigma = 4$, we have

$$z = \frac{x - \mu}{\sigma}$$

$$= \frac{73 - 68}{4}$$

$$= 1.25$$

Therefore,

$$p(x > 73) = p(z > 1.25)$$

$$= 0.5 - p(0 < z < 1.25)$$

$$= 0.5 - 0.3944$$

$$= 0.1056$$

Approximately 10.6% of the people will be taller than 73 inches.

b. We need to find $p(60 < x < 75)$, the area of the central region shown in Figure 4.119. Notice that we will be adding the areas of the two bodies.

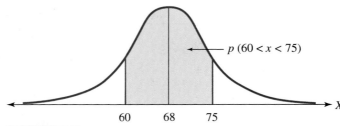

FIGURE 4.119 A central region.

First, we convert $x = 60$ and $x = 75$ to their corresponding z-numbers:

$$p(60 < x < 75) = p\left(\frac{60 - 68}{4} < z < \frac{75 - 68}{4} \right) \quad \text{using the Conversion Formula}$$

$$= p(-2.00 < z < 1.75) $$

$$= p(-2.00 < z < 0) + p(0 < z < 1.75) \quad \text{expressing the area as two bodies}$$

$$= p(0 < z < 2.00) + p(0 < z < 1.75) \quad \text{using symmetry}$$

$$= 0.4772 + 0.4599 \quad \text{using the body table}$$

$$= 0.9371$$

Approximately 93.7% of the people will be between 60 and 75 inches tall.

All the preceding examples involved finding probabilities that contained only the strict $<$ or $>$ inequalities, never $\leq$ or $\geq$ inequalities; the endpoints were never included. What if the endpoints are included? How does $p(a < x < b)$ compare with $p(a \leq x \leq b)$? Because probabilities for continuous data are found by determining *area* under a curve, including the endpoints does not affect the probability! The probability of a single point $p(x = a)$ is 0, because there is no "area" over a single point. (We obtain an area only when we are working with an interval of numbers.) Consequently, if x represents continuous data, then $p(a \leq x \leq b) = p(a < x < b)$; it makes no difference whether the endpoints are included.

EXAMPLE 7

SOLUTION

FINDING THE VALUE OF A VARIABLE THAT WILL PRODUCE A SPECIFIC PROBABILITY   Tall Dudes is a clothing store that specializes in fashions for tall men. Its informal motto is "Our customers are taller than 80% of the rest." Assuming the heights of men to be normally distributed with a mean of 67 inches and a standard deviation of 5.5 inches, find the heights of Tall Dudes' clientele.

Let c = the height of the shortest customer at Tall Dudes, and let x represent the height of a randomly selected man. We are given that the heights of all men are normally distributed with $\mu = 67$ and $\sigma = 5.5$.

Assuming Tall Dudes' clientele to be taller than 80% of all men implies that $x < c$ 80% of the time and $x > c$ 20% of the time. Hence, we can say that the probability of selecting someone shorter than the shortest tall dude is $p(x < c) = 0.80$ and that the probability of selecting a tall dude is $p(x > c) = 0.20$, as shown in Figure 4.120.

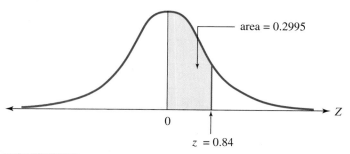

FIGURE 4.120 Finding a 20% right tail.

We are given that the area of the right tail is 0.20 (and that of the right body is 0.30), and we need to find the appropriate cutoff number c. This is exactly the reverse of all the previous examples, in which we were given the cutoff numbers and asked to find the area. Thus, our goal is to find the z-number that corresponds to a body of area 0.30 and convert it into its corresponding x-number.

When we scan through the *interior* of the body table, the number closest to the desired area of 0.30 is 0.2995, which is the area of the body when $z = 0.84$. This means that $p(0 < z < 0.84) = 0.2995$, as shown in Figure 4.121.

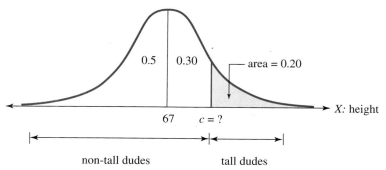

FIGURE 4.121 A body close to 30%.

Therefore, the number c that we are seeking lies 0.84 standard deviations above the mean. All that remains is to convert $z = 0.84$ into its corresponding x-number by substituting $x = c$, $z = 0.84$, $\mu = 67$, and $\sigma = 5.5$ into the Conversion Formula:

$$z = \frac{x - \mu}{\sigma}$$

$$0.84 = \frac{c - 67}{5.5}$$

$$(5.5)(0.84) = c - 67$$

$$4.62 = c - 67$$

$$c = 71.62 \quad (\approx 72 \text{ inches, or 6 feet})$$

Therefore, Tall Dudes caters to men who are at least 71.62 inches tall, or about 6 feet tall.

4.4 EXERCISES

1. The weights (in ounces) of several bags of corn chips are given in Figure 4.122. Construct a histogram for the data using the groups $15.40 \leq x < 15.60$, $15.60 \leq x < 15.80$, ..., $16.60 \leq x < 16.80$. Do the data appear to be approximately normally distributed? Explain.

16.08	16.49	15.61	16.66	15.80	15.87
16.02	15.82	16.48	16.08	15.63	16.02
16.00	16.25	15.41	16.22	16.04	15.68
16.45	16.41	16.01	15.82	16.08	15.82
16.29	16.26	16.05	16.25	15.86	

FIGURE 4.122   Weights (in ounces) of bags of corn chips.

2. The weights (in grams) of several bags of chocolate chip cookies are given in Figure 4.123. Construct a histogram for the data using the groups $420 \leq x < 430$, $430 \leq x < 440$, ..., $480 \leq x < 490$. Do the data appear to be approximately normally distributed? Explain.

451	435	482	449	454	451
479	448	432	423	461	475
467	453	448	459	454	444
475	461	450	466	446	458

FIGURE 4.123   Weights (in grams) of bags of chocolate chip cookies.

3. The time (in minutes) spent waiting in line for several students at the campus bookstore are given in Figure 4.124. Construct a histogram for the data using the groups $0 \leq x < 1$, $1 \leq x < 2$, ..., $4 \leq x < 5$. Do the data appear to be normally distributed? Explain.

1.75	2	0.5	1.5	0
1	2.5	0.75	2.25	3
1.5	3.5	2.5	1	0.5
1.75	0.25	4.5	0	1.5

FIGURE 4.124   Time (in minutes) spent waiting in line.

4. A die was rolled several times, and the results are given in Figure 4.125. Construct a histogram for the data using the single-values 1, 2, 3, 4, 5, and 6. Do the data appear to be normally distributed? Explain.

5	3	1	1	4	2
6	4	1	2	5	5
4	3	1	6	4	2
3	1	2	5	2	1
4	6	5	1	4	5

FIGURE 4.125   Rolling a die.

5. What percent of the standard normal z-distribution lies between the following values?

  a. $z = 0$ and $z = 1$

  b. $z = -1$ and $z = 0$

  c. $z = -1$ and $z = 1$

  (*Note:* This interval represents one standard deviation of the mean.)

6. What percent of the standard normal z-distribution lies between the following values?

  a. $z = 0$ and $z = 2$

  b. $z = -2$ and $z = 0$

  c. $z = -2$ and $z = 2$

  (*Note:* This interval represents two standard deviations of the mean.)

7. What percent of the standard normal z-distribution lies between the following values?

  a. $z = 0$ and $z = 3$

  b. $z = -3$ and $z = 0$

  c. $z = -3$ and $z = 3$

  (*Note:* This interval represents three standard deviations of the mean.)

8. What percent of the standard normal z-distribution lies between the following values?

  a. $z = 0$ and $z = 1.5$

  b. $z = -1.5$ and $z = 0$

  c. $z = -1.5$ and $z = 1.5$

  (*Note:* This interval represents one and one-half standard deviations of the mean.)

9. A population is normally distributed with mean 24.7 and standard deviation 2.3.

  a. Find the intervals representing one, two, and three standard deviations of the mean.

  b. What percentage of the data lies in each of the intervals in part (a)?

  c. Draw a sketch of the bell curve.

10. A population is normally distributed with mean 18.9 and standard deviation 1.8.

  a. Find the intervals representing one, two, and three standard deviations of the mean.

  b. What percent of the data lies in each of the intervals in part (a)?

  c. Draw a sketch of the bell curve.

11. Find the following probabilities.

  a. $p(0 < z < 1.62)$

  b. $p(1.30 < z < 1.84)$

  c. $p(-0.37 < z < 1.59)$

  d. $p(z < -1.91)$

  e. $p(-1.32 < z < -0.88)$

  f. $p(z < 1.25)$

12. Find the following probabilities.

  a. $p(0 < z < 1.42)$

  b. $p(1.03 < z < 1.66)$

  c. $p(-0.87 < z < 1.71)$

  d. $p(z < -2.06)$

  e. $p(-2.31 < z < -1.18)$

  f. $p(z < 1.52)$

13. Find c such that each of the following is true.

  a. $p(0 < z < c) = 0.1331$

  b. $p(c < z < 0) = 0.4812$

  c. $p(-c < z < c) = 0.4648$

  d. $p(z > c) = 0.6064$

  e. $p(z > c) = 0.0505$

  f. $p(z < c) = 0.1003$

14. Find c such that each of the following is true.

  a. $p(0 < z < c) = 0.3686$

  b. $p(c < z < 0) = 0.4706$

  c. $p(-c < z < c) = 0.2510$

  d. $p(z > c) = 0.7054$

  e. $p(z > c) = 0.0351$

  f. $p(z < c) = 0.2776$

15. A population X is normally distributed with mean 250 and standard deviation 24. For each of the following values of x, find the corresponding z-number. Round off your answers to two decimal places.

  a. $x = 260$ **b.** $x = 240$

  c. $x = 300$ **d.** $x = 215$

  e. $x = 321$ **f.** $x = 197$

16. A population X is normally distributed with mean 72.1 and standard deviation 9.3. For each of the following values of x, find the corresponding z-number. Round off your answers to two decimal places.

  a. $x = 90$ **b.** $x = 80$

  c. $x = 75$ **d.** $x = 70$

  e. $x = 60$ **f.** $x = 50$

17. A population is normally distributed with mean 36.8 and standard deviation 2.5. Find the following probabilities.

  a. $p(36.8 < x < 39.3)$ **b.** $p(34.2 < x < 38.7)$

  c. $p(x < 40.0)$ **d.** $p(32.3 < x < 41.3)$

  e. $p(x = 37.9)$ **f.** $p(x > 37.9)$

18. A population is normally distributed with mean 42.7 and standard deviation 4.7. Find the following probabilities.

  a. $p(42.7 < x < 47.4)$ **b.** $p(40.9 < x < 44.1)$

  c. $p(x < 50.0)$ **d.** $p(33.3 < x < 52.1)$

  e. $p(x = 45.3)$ **f.** $p(x > 45.3)$

19. The mean weight of a box of cereal filled by a machine is 16.0 ounces, with a standard deviation of 0.3 ounce. If the weights of all the boxes filled by the machine are normally distributed, what percent of the boxes will weigh the following amounts?

    a. less than 15.5 ounces

    b. between 15.8 and 16.2 ounces

20. The amount of time required to assemble a component on a factory assembly line is normally distributed with a mean of 3.1 minutes and a standard deviation of 0.6 minute. Find the probability that a randomly selected employee will take the given amount of time to assemble the component.

    a. more than 4.0 minutes

    b. between 2.0 and 2.5 minutes

21. The time it takes an acrylic paint to dry is normally distributed. If the mean is 2 hours 36 minutes with a standard deviation of 24 minutes, find the probability that the drying time will be as follows.

    a. less than 2 hours 15 minutes

    b. between 2 and 3 hours

    HINT: Convert everything to minutes (or to hours).

22. The shrinkage in length of a certain brand of blue jeans is normally distributed with a mean of 1.1 inches and a standard deviation of 0.2 inch. What percent of this brand of jeans will shrink the following amounts?

    a. more than 1.5 inches

    b. between 1.0 and 1.25 inches

23. The mean volume of a carton of milk filled by a machine is 1.0 quart, with a standard deviation of 0.06 quart. If the volumes of all the cartons are normally distributed, what percent of the cartons will contain the following amounts?

    a. at least 0.9 quart

    b. at most 1.05 quarts

24. The amount of time between taking a pain reliever and getting relief is normally distributed with a mean of 23 minutes and a standard deviation of 4 minutes. Find the probability that the time between taking the medication and getting relief is as follows.

    a. at least 30 minutes

    b. at most 20 minutes

25. The results of a statewide exam for assessing the mathematics skills of realtors were normally distributed with a mean score of 72 and a standard deviation of 12. The realtors who scored in the top 10% are to receive a special certificate, while those in the bottom 20% will be required to attend a remedial workshop.

    a. What score does a realtor need in order to receive a certificate?

    b. What score will dictate that the realtor attend the workshop?

26. Professor Harde assumes that exam scores are normally distributed and wants to grade "on the curve." The mean score was 58, with a standard deviation of 16.

    a. If she wants 14% of the students to receive an A, find the minimum score to receive an A.

    b. If she wants 19% of the students to receive a B, find the minimum score to receive a B.

27. The time it takes an employee to package the components of a certain product is normally distributed with $\mu = 8.5$ minutes and $\sigma = 1.5$ minutes. To boost productivity, management has decided to give special training to the 34% of employees who took the greatest amount of time to package the components. Find the amount of time taken to package the components that will indicate that an employee should get special training.

28. The time it takes an employee to package the components of a certain product is normally distributed with $\mu = 8.5$ and $\sigma = 1.5$ minutes. As an incentive, management has decided to give a bonus to the 20% of employees who took the shortest amount of time to package the components. Find the amount of time taken to package the components that will indicate that an employee should get a bonus.

 Answer the following questions using complete sentences and your own words.

• CONCEPT QUESTIONS

29. What are the characteristics of a normal distribution?

30. Are all distributions of data normally distributed? Support your answer with an example.

31. Why is the total area under a bell curve equal to 1?

32. Why are there no negative z-numbers in the body table?

33. When converting an x-number to a z-number, what does a negative z-number tell you about the location of the x-number?

34. Is it logical to assume that the heights of all high school students in the United States are normally distributed? Explain.

35. Is it reasonable to assume that the ages of all high school students in the United States are normally distributed? Explain.

• HISTORY QUESTIONS

36. Who is known as "the Prince of Mathematics"? Why?

37. What mathematician was instrumental in the creation of the normal distribution? What application prompted this person to create the normal distribution?

Polls and Margin of Error

OBJECTIVES

- Find the margin of error in a poll
- Determine the effect of sample size on the margin of error
- Find the level of confidence for a specific sample and margin of error
- Find the minimum sample size to obtain a specified margin of error

One of the most common applications of statistics is the evaluation of the results of surveys and public opinion polls. Most editions of the daily newspaper contain the results of at least one poll. Headlines announce the attitude of the nation toward a myriad of topics ranging from the actions of politicians to controversial current issues, such as abortion, as shown in the newspaper article on the next page. How are these conclusions reached? What do they mean? How valid are they? In this section, we investigate these questions and obtain results concerning the "margin of error" associated with the reporting of "public opinion."

Sampling and Inferential Statistics

The purpose of conducting a survey or poll is to obtain information about a population—for example, adult Americans. Because there are approximately 230 million Americans over the age of eighteen, it would be very difficult, time-consuming, and expensive to contact every one of them. The only realistic alternative is to poll a sample and use the science of inferential statistics to draw conclusions about the population as a whole. Different samples have different characteristics depending on, among other things, the age, sex, education, and locale of the people in the sample. Therefore, it is of the utmost importance that a sample be representative of the population. Obtaining a representative sample is the most difficult aspect of inferential statistics.

Another problem facing pollsters is determining *how many* people should be selected for the sample. Obviously, the larger the sample, the more likely that it will reflect the population. However, larger samples cost more money, so a limited budget will limit the sample size. Conducting surveys can be very costly, even for a small to moderate sample. For example, a survey conducted in 1989 by the Gallup Organization that contacted 1,005 adults and 500 teenagers would have cost $100,000 (the pollsters donated their services for this survey). The results of this poll indicated that Americans thought the "drug crisis" was the nation's top problem (stated by 27% of the adults and 32% of the teenagers).

After a sample has been selected and its data have been analyzed, information about the sample is generalized to the entire population. Because 27% of the 1,005 adults in a poll stated that the drug crisis was the nation's top problem, we would like to conclude that 27% of *all* adults have the same belief. Is this a valid generalization? That is, how confident is the pollster that the feelings of the people in the sample reflect those of the population?

MORE AMERICANS "PRO-LIFE" THAN "PRO-CHOICE" FOR THE FIRST TIME

Princeton, NJ—A new Gallup Poll, conducted May 7–10, 2009, finds 51% of Americans calling themselves "pro-life" on the issue of abortion and 42% "pro-choice." This is the first time a majority of U.S. adults have identified themselves as pro-life since Gallup began asking this question in 1995. The new results, obtained from Gallup's annual Values and Beliefs survey, represent a significant shift from a year ago, when 50% were pro-choice and 44% pro-life. Prior to now, the highest percentage identifying as pro-life was 46%, in both August 2001 and May 2002.

The source of the shift in abortion views is clear in the Gallup Values and Belief survey. The percentage of Republicans (including independents who lean Republican) calling themselves "pro-life" rose by 10 points over the past year, from 60% to 70%, while there has been essentially no change in the views of Democrats and Democratic leaners. Similarly, by ideology, all of the increase in pro-life sentiment is seen among self-identified conservatives and moderates; the abortion views of political liberals have not changed.

A year ago, Gallup found more women calling themselves pro-choice than pro-life, by 50% to 43%, while men were more closely divided: 49% pro-choice, 46% pro-life. Now, because of heightened pro-life sentiment among both groups, women as well as men are more likely to be pro-life. Men and women have been evenly divided on the issue in previous years; however, this is the first time in nine years of Gallup Values surveys that significantly more men and women are pro-life than pro-choice.

Results are based on telephone interviews with 1,015 national adults, aged 18 and older, conducted May 7–10, 2009. For results based on the total sample of national adults, one can say with 95% confidence that the maximum margin of sampling error is ± 3 percentage points.

By Lydia Saad
Gallup News Service

Sample Proportion versus Population Proportion

If x members (for example, people, automobiles, households) in a sample of size n have a certain characteristic, then the proportion of the sample, or **sample proportion,** having this characteristic is given by $\frac{x}{n}$. For instance, in a sample of $n = 70$ automobiles, if $x = 14$ cars have a defective fan switch, then the proportion of the sample having a defective switch is $\frac{14}{70} = 0.2$, or 20%. The true proportion of the entire population, or **population proportion,** having the characteristic is represented by the letter P. A sample proportion $\frac{x}{n}$ is an estimate of the population proportion P.

Sample proportions $\frac{x}{n}$ vary from sample to sample; some will be larger than P, and some will be smaller. Of the 1,005 adults in the Gallup Poll sample mentioned on page 295, 27% viewed the drug crisis as the nation's top problem. If a different sample of 1,005 had been chosen, 29% might have had this view. If still another 1,005 had been selected, this view might have been shared by only 25%. We will assume that the sample proportions $\frac{x}{n}$ are normally distributed around the population proportion P. The set of all sample proportions, along with their probabilities of occurring, can be represented by a bell curve like the one in Figure 4.126.

In general, a sample estimate is not 100% accurate; although a sample proportion might be close to the true population proportion, it will have an error term associated with it. The difference between a sample estimate and the true (population) value is called the **error of the estimate.** We can use a bell curve (like the one in Figure 4.126) to predict the probable error of a sample estimate.

Before developing this method of predicting the error term, we need to introduce some special notation. The symbol z_α (read "z alpha") will be used to represent the positive z-number that has a right body of area α. That is, z_α is the number such that $p(0 < z < z_\alpha) = \alpha$, as shown in Figure 4.127.

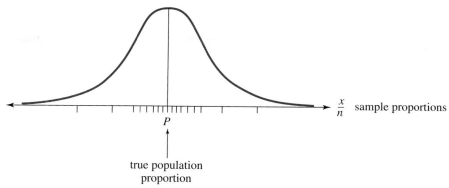

FIGURE 4.126 Sample proportions normally distributed around the true (population) proportion.

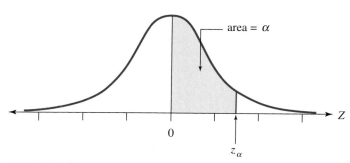

FIGURE 4.127 $p(0 < z < z_\alpha) = \alpha$.

EXAMPLE **1**

FINDING A VALUE OF THE STANDARD NORMAL DISTRIBUTION THAT WILL PRODUCE A SPECIFIC PROBABILITY   Use the body table in Appendix F to find the following values:

a. $z_{0.3925}$ **b.** $z_{0.475}$ **c.** $z_{0.45}$ **d.** $z_{0.49}$

SOLUTION

a. $z_{0.3925}$ represents the z-number that has a body of area 0.3925. Looking at the interior of the body table, we find 0.3925 and see that it corresponds to the z-number 1.24—that is, $p(0 < z < 1.24) = 0.3925$. Therefore, $z_{0.3925} = 1.24$.
b. In a similar manner, we find that a body of area 0.4750 corresponds to $z = 1.96$. Therefore, $z_{0.475} = 1.96$.
c. Looking through the interior of the table, we cannot find a body of area 0.45. However, we do find a body of 0.4495 (corresponding to $z = 1.64$) and a body of 0.4505 (corresponding to $z = 1.65$). Because the desired body is *exactly halfway* between the two listed bodies, we use a z-number that is exactly halfway between the two listed z-numbers, 1.64 and 1.65. Therefore, $z_{0.45} = 1.645$.
d. We cannot find a body of the desired area, 0.49, in the interior of the table. The closest areas are 0.4898 (corresponding to $z = 2.32$) and 0.4901 (corresponding to $z = 2.33$). Because the desired area (0.49) is *closer* to 0.4901, we use $z = 2.33$. Therefore, $z_{0.49} = 2.33$.

Margin of Error

Sample proportions $\frac{x}{n}$ vary from sample to sample; some will have a small error, and some will have a large error. Knowing that sample estimates have inherent errors, statisticians make predictions concerning the largest possible error associated with

HISTORICAL NOTE

GEORGE H. GALLUP, 1901–1984

© Corbis/Bettman

To many people, the name *Gallup* is synonymous with opinion polls. George Horace Gallup, the founder of the American Institute of Public Opinion, began his news career while attending the University of Iowa. During his junior year as a student of journalism, Gallup became the editor of his college newspaper, the *Daily Iowan*. After receiving his bachelor's degree in 1923, Gallup remained at the university nine years as an instructor of journalism.

In addition to teaching, Gallup continued his own studies of human nature and public opinion. Interest in how the public reacts to advertisements and perceives various issues of the day led Gallup to combine his study of journalism with the study of psychology. In 1925, he received his master's degree in psychology. Gallup's studies culminated in 1928 with his doctoral thesis, *A New Technique for Objective Methods for Measuring Reader Interest in Newspapers*. Gallup's new technique of polling the public was to utilize a stratified sample, that is, a sample that closely mirrors the composition of the entire population. Gallup contended that a stratified sample of 1,500 people was sufficient to obtain reliable estimates. For his pioneering work in this new field, Gallup was awarded his Ph.D. in journalism in 1928.

Gallup founded the American Institute of Public Opinion in 1935 with the stated purpose "to impartially measure and report public opinion on political and social issues of the day without regard to the rightness or wisdom of the views expressed." His first triumph was his prediction of the winner of the 1936 presidential election between Franklin D. Roosevelt and Alfred Landon. While many, including the prestigious *Literary Digest*, predicted that Landon would win, Gallup correctly predicted Roosevelt as the winner.

Gallup Polls have correctly predicted all presidential elections since, with the exception of the 1948 race between Thomas Dewey and Harry S Truman. Much to his embarrassment, Gallup predicted Dewey as the winner. Truman won the election with 49.9% of the vote, while Gallup had predicted that he would receive only 44.5%. Gallup's explanation was that he had ended his poll too far in advance of election day and had disregarded the votes of those who were undecided. Of his error, Gallup said, "We are continually experimenting and continually learning."

Although some people criticize the use of polls, citing their potential influence and misuse, Gallup considered the public opinion poll to be "one of the most useful instruments of democracy ever devised." Answering the charge that he and his polls influenced elections, Gallup retorted, "One might as well insist that a thermometer makes the weather!" In addition, Gallup confessed that he had not voted in a presidential election since 1928. Above all, Gallup wanted to ensure the impartiality of his polls.

Besides polling people regarding their choices in presidential campaigns, Gallup was the first pollster to ask the public to rate a president's performance and popularity. Today, these "presidential report cards" are so common we may take them for granted. In addition to presidential politics, Gallup also dealt with sociological issues, asking questions such as "What is the most important problem facing the country?"

Polling has become a multimillion-dollar business. In 2009, the Gallup Organization had revenues totaling $271.6 million, and it had over 2,000 employees. Today, Gallup Polls are syndicated in newspapers across the country and around the world.

AP Photo/Byron Rollins

The Gallup Organization predicted that Thomas Dewey would win the 1948 presidential election. Much to Gallup's embarrassment, Harry S. Truman won the election and triumphantly displayed a newspaper containing Gallup's false prediction. With the exception of this election, Gallup Polls have correctly predicted every presidential election since 1936.

a sample estimate. This error is called the **margin of error** of the estimate and is denoted by **MOE.** Because the margin of error is a prediction, we cannot guarantee that it is absolutely correct; that is, the probability that a prediction is correct might be 0.95, or it might be 0.75.

In the field of inferential statistics, the probability that a prediction is correct is referred to as the **level of confidence** of the prediction. For example, we might say that we are 95% confident that the maximum error of an opinion poll is plus or minus 3 percentage points; that is, if 100 samples were analyzed, 95 of them would have proportions that differ from the true population proportion by an amount less than or equal to 0.03, and 5 of the samples would have an error greater than 0.03.

Assuming that sample proportions are normally distributed around the population proportion (as in Figure 4.126), we can use the z-distribution to determine the margin of error associated with a sample proportion. In general, the margin of error depends on the sample size and the level of confidence of the estimate.

MARGIN OF ERROR FORMULA

Given a sample size n, the **margin of error,** denoted by **MOE,** for a poll involving sample proportions is

$$MOE = \frac{z_{\alpha/2}}{2\sqrt{n}}$$

where α represents the level of confidence of the poll. That is, the probability is α that the sample proportion has an error of at most MOE.

EXAMPLE 2

FINDING THE MARGIN OF ERROR   Assuming a 90% level of confidence, find the margin of error associated with each of the following:

a. sample size $n = 275$ **b.** sample size $n = 750$

SOLUTION

a. The margin of error depends on two things: the sample size and the level of confidence. For a 90% level of confidence, $\alpha = 0.90$. Hence, $\frac{\alpha}{2} = 0.45$, and $z_{\alpha/2} = z_{0.45} = 1.645$.
Substituting this value and $n = 275$ into the MOE formula, we have the following:

$$MOE = \frac{z_{\alpha/2}}{2\sqrt{n}}$$

$$= \frac{1.645}{2\sqrt{275}}$$

$$= 0.049598616\ldots$$

$$\approx 0.050 \qquad \textbf{rounding off to three decimal places}$$

$$= 5.0\%$$

When we are polling a sample of 275 people, we can say that we are 90% confident that the maximum possible error in the sample proportion will be plus or minus 5.0 percentage points.

b. For a 90% level of confidence, $\alpha = 0.90$, $\frac{\alpha}{2} = 0.45$, and $z_{\alpha/2} = z_{0.45} = 1.645$. Substituting this value and $n = 750$ into the MOE formula, we have the following:

$$MOE = \frac{z_{\alpha/2}}{2\sqrt{n}}$$

$$= \frac{1.645}{2\sqrt{750}}$$

$$= 0.030033453\ldots$$

$$\approx 0.030 \qquad \textbf{rounding off to three decimal places}$$

$$= 3.0\%$$

When we are polling a sample of 750 people, we can say that we are 90% confident that the maximum possible error in the sample proportion will be plus or minus 3.0 percentage points.

If we compare the margins of error in parts (a) and (b) of Example 2, we notice that by increasing the sample size (from 275 to 750), the margin of error was reduced (from 5.0 to 3.0 percentage points). Intuitively, this should make sense; a larger sample gives a better estimate (has a smaller margin of error).

EXAMPLE 3

FINDING A SAMPLE PROPORTION AND MARGIN OF ERROR To obtain an estimate of the proportion of all Americans who think the president is doing a good job, a random sample of 500 Americans is surveyed, and 345 respond, "The president is doing a good job."

a. Determine the sample proportion of Americans who think the president is doing a good job.
b. Assuming a 95% level of confidence, find the margin of error associated with the sample proportion.

SOLUTION

a. $n = 500$ and $x = 345$. The sample proportion is

$$\frac{x}{n} = \frac{345}{500} = 0.69$$

Sixty-nine percent of the sample think the president is doing a good job.
b. We must find MOE when $n = 500$ and $\alpha = 0.95$. Because $\alpha = 0.95$, $\frac{\alpha}{2} = 0.475$ and $z_{\alpha/2} = z_{0.475} = 1.96$.
   Therefore,

$$\text{MOE} = \frac{z_{\alpha/2}}{2\sqrt{n}}$$

$$= \frac{1.96}{2\sqrt{500}}$$

$$= 0.043826932\ldots$$

$$\approx 0.044 \qquad \textbf{rounding off to three decimal places}$$

$$= 4.4\%$$

The margin of error associated with the sample proportion is plus or minus 4.4 percentage points. We are 95% confident that 69% ($\pm 4.4\%$) of all Americans think the president is doing a good job. In other words, on the basis of our sample proportion $\left(\frac{x}{n}\right)$ of 69%, we predict (with 95% certainty) that the true population proportion (P) is somewhere between 64.6% and 73.4%.

Example 2 indicated that increasing the sample size will decrease the margin of error, that is, larger samples give better estimates. If larger samples give better estimates, how large should a sample be? This question can be answered by manipulating the margin of error formula. That is, in its original form, we plug values of $z_{\alpha/2}$ (based upon the level of confidence) and n (the sample size) into the formula, and we calculate the margin of error. However, if we first solve the margin of error formula for n, we will have a new version of the formula for determining how large a sample should be. We proceed as follows:

$$\text{MOE} = \frac{z_{\alpha/2}}{2\sqrt{n}} \qquad \text{the margin of error formula}$$

$$\sqrt{n} \cdot (\text{MOE}) = \frac{z_{\alpha/2}}{2} \qquad \text{multiplying each side by } \sqrt{n}$$

$$\sqrt{n} = \frac{z_{\alpha/2}}{2(\text{MOE})} \qquad \text{dividing each side by MOE}$$

$$n = \left(\frac{z_{\alpha/2}}{2(\text{MOE})} \right)^2 \qquad \text{squaring both sides}$$

SAMPLE SIZE FORMULA

The required sample size n, necessary to have a desired margin of error of at most MOE, is given

$$n = \left(\frac{z_{\alpha/2}}{2(\text{MOE})} \right)^2$$

where $z_{\alpha/2}$ is determined by the given level of confidence.

EXAMPLE **4**

SOLUTION

FINDING A SAMPLE SIZE   With a 98% level of confidence, how large should a sample be so that the margin of error is at most 4%?

For a 98% level of confidence, $\alpha = 0.98$, $\frac{\alpha}{2} = 0.49$, and $z_{\alpha/2} = z_{0.49} = 2.33$. Substituting this value and MOE = 0.04 into the sample size formula, we have

$$n = \left(\frac{z_{\alpha/2}}{2(\text{MOE})} \right)^2$$
$$= \left(\frac{2.33}{2(0.04)} \right)^2$$
$$= 848.265625\ldots$$

Therefore, we should have roughly 848 and "one-quarter" people ($0.265625 \approx 0.25 = \frac{1}{4}$) in the sample. However, we cannot have part of a person, so it is customary to round this number *up* to the next highest whole number (include the whole person!); thus, the required sample size is $n = 849$ people.

EXAMPLE **5**

FINDING THE MARGIN OF ERROR FOR DIFFERENT LEVELS OF CONFIDENCE   The article shown on the next page was released by the Gallup Organization in October 2009.

a. The poll states that 44% of the Americans questioned think that the laws covering firearm sales should be made more strict. Assuming a 95% level of confidence (the most commonly used level of confidence), find the margin of error associated with the survey.

b. Assuming a 98% level of confidence, find the margin of error associated with the survey.

IN U.S., RECORD-LOW SUPPORT FOR STRICTER GUN LAWS

Princeton, NJ—Gallup finds a new low of 44% of Americans saying the laws covering firearm sales should be made more strict. That is down 5 points in the last year and 34 points from the high of 78% recorded the first time the question was asked, in 1990.

Today, Americans are as likely to say the laws governing gun sales should be kept as they are now (43%) as to say they should be made more strict. Until this year, Gallup had always found a significantly higher percentage advocating stricter laws. At the same time, 12% of Americans believe the laws should be

less strict, which is low in an absolute sense but ties the highest Gallup has measured for this response.

These results are based on Gallup's annual Crime Poll, conducted October 1–4, 2009.

The Poll also shows a new low in the percentage of Americans favoring a ban on handgun possession except by the police and other authorized persons, a question that dates back to 1959. Only 28% now favor such a ban. The high point in support for a handgun-possession ban was 60% in the initial measurement in 1959. Since then, less than a majority has been in favor, and support has been below 40% since December 1993.

The trends on the questions about gun-sale laws and a handgun-possession ban indicate that Americans' attitudes have moved toward being more pro-gun rights. But this is not due to a growth in personal gun ownership, which has held steady around 30% this decade, or to an increase in household gun ownership, which has been steady in the low 40% range since 2000.

Results are based on telephone interviews with 1,013 national adults, aged 18 and older, conducted October 1–4, 2009.

By Jeffrey M. Jones
Gallup News Service

SOLUTION

a. We must find MOE when $n = 1{,}013$ and $\alpha = 0.95$. Because $\alpha = 0.95$, $\alpha/2 = 0.475$ and $z_{\alpha/2} = z_{0.475} = 1.96$.
Therefore,

$$
\begin{aligned}
\text{MOE} &= \frac{z_{\alpha/2}}{2\sqrt{n}} \\
&= \frac{1.96}{2\sqrt{1{,}013}} \\
&= 0.030790827\ldots \\
&\approx 0.031 \qquad \textbf{rounding off to three decimal places}
\end{aligned}
$$

The margin of error associated with the survey is plus or minus 3.1%. We are 95% confident that $44\% \pm 3.1\%$ of all Americans think that the laws covering firearm sales should be made more strict.

b. We must find MOE when $n = 1{,}013$ and $\alpha = 0.98$. Because $\alpha = 0.98$, $\alpha/2 = 0.49$ and $z_{\alpha/2} = z_{0.49} = 2.33$.
Therefore,

$$
\begin{aligned}
\text{MOE} &= \frac{z_{\alpha/2}}{2\sqrt{n}} \\
&= \frac{2.33}{2\sqrt{1{,}013}} \\
&= 0.036603381\ldots \\
&\approx 0.037 \qquad \textbf{rounding off to three decimal places}
\end{aligned}
$$

The margin of error associated with the survey is plus or minus 3.7%. We are 98% confident that $44\% \pm 3.7\%$ of all Americans think that the laws covering firearm sales should be made more strict.

WHO SUPPORTS MARIJUANA LEGALIZATION?
SUPPORT RISING; VARIES MOST BY AGE AND GENDER

Since the late 1960s, Gallup has periodically asked Americans whether the use of marijuana should be made legal in the United States. Although a majority of Americans have consistently opposed the idea of legalizing marijuana, public support has slowly increased over the years. In 1969, just 12% of Americans supported making marijuana legal, but by 1977, roughly one in four endorsed it. Support edged up to 31% in 2000, and now, about a third of Americans say marijuana should be legal.

Support for marijuana legalization varies greatest by gender and age. Overall, younger Americans (aged 18 to 29) are essentially divided, with 47% saying marijuana should be legal and 50% saying it should not be. Support for legalization is much lower among adults aged 30 to 64 (35%) and those aged 65 and older (22%). Men (39%) are somewhat more likely than women (30%) to support the legalization of marijuana in the country.

Americans residing in the western parts of the country are more likely than those living elsewhere to support the legalization of marijuana. These differences perhaps result from the fact that six Western states have, in various ways, already legalized marijuana for medicinal use. Overall, the data show that Westerners are divided about marijuana, with 47% saying it should be legal and 49% saying it should not be.

No more than a third of adults living in other parts of the country feel marijuana should be legal.

Support for legalizing marijuana is much lower among Republicans than it is among Democrats or independents. One in five Republicans (21%) say marijuana should be made legal in this country, while 37% of Democrats and 44% of independents share this view.

*Results are based on telephone interviews with 2,034 national adults, aged 18 and older, conducted Aug. 3–5, 2001, Nov. 10–12, 2003, and Oct. 21–23, 2005. For results based on the total sample of national adults, one can say with 95% confidence that the maximum margin of sampling error is ±2 percentage points.

By Joseph Carroll,
Gallup Poll Assistant Editor

If we compare the margins of error found in parts (a) and (b) of Example 4, we notice that as the level of confidence went up (from 95% to 98%) the margin of error increased (from 3.1% to 3.7%). Intuitively, if we want to be more confident in our predictions, we should give our prediction more leeway (a larger margin of error).

When the results of the polls are printed in a newspaper, the sample size, level of confidence, margin of error, date of survey, and location of survey may be given as a footnote, as shown in the above article.

EXAMPLE 6

VERIFYING A STATED MARGIN OF ERROR   Verify the margin of error stated in the article shown above.

SOLUTION

The footnote to the article states that for a sample size of 2,034 and a 95% level of confidence, the margin of error is ±2%, that is, MOE = 0.02 for $n = 2,034$ and $\alpha = 0.95$. Because $\alpha = 0.95$, $\frac{\alpha}{2} = 0.475$ and $z_{\alpha/2} = z_{0.475} = 1.96$. Therefore,

$$\text{MOE} = \frac{z_{\alpha/2}}{2\sqrt{n}}$$

$$= \frac{1.96}{2\sqrt{2,034}}$$

$$= 0.02195127\ldots$$

$$\approx 0.02 \qquad \textbf{rounding off to two decimal places}$$

Therefore, the stated margin of error of ±2% is correct.

TO MANY AMERICANS, UFOS ARE REAL AND HAVE VISITED EARTH IN SOME FORM

Most Americans appear comfortable with and even excited about the thought of the discovery of extraterrestrial life. More than half (56 percent) of the American public think that UFOs are something real and not just in people's imagination. Nearly as many (48 percent) believe that UFOs have visited earth in some form. Males are significantly more likely to believe in the reality of UFOs, as are those under the age of 65. A significant drop is witnessed in the percentage of believers among the 65+ age group.

Two-thirds (67 percent) of adults think there are other forms of intelligent life in the universe. This belief tends to be more prevalent among males, adults ages 64 or younger, and residents of the Northeast as opposed to North Central and South.

In the view of many adults (55 percent), the government does not share enough information with the public in general. An even greater proportion (roughly seven in ten) thinks that the government does not tell us everything it knows about extraterrestrial life and UFOs. The younger the age, the stronger the belief that the government is withholding information about these topics.

This study was conducted by RoperASW. The sample consists of 1,021 male and female adults (in approximately equal number), all 18 years of age and over. The telephone interviews were conducted from August 23 through August 25, 2002, using a Random Digit Dialing (RDD) probability sample of all telephone households in the continental United States. The margin of error for the total sample is ±3 percent.

The Roper Poll
UFOs & Extraterrestrial Life
Americans' Beliefs and Personal Experiences
(Prepared for the SCI FI Channel—September 2002)

News articles do not always mention the level of confidence of a survey. However, if the sample size and margin of error are given, the level of confidence can be determined, as shown in Example 7.

EXAMPLE 7

FINDING THE LEVEL OF CONFIDENCE OF AN OPINION POLL   Do you think that UFOs are real? The news article shown above presents the results of a Roper poll pertaining to this question. Find the level of confidence of this poll.

SOLUTION

We are given $n = 1,021$ and MOE $= 0.03$. To find α, the level of confidence of the poll, we must first find $z_{\alpha/2}$ and the area of the bodies, as shown in Figure 4.128.

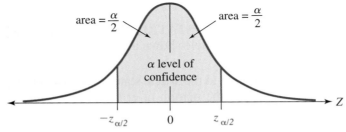

area $= \dfrac{\alpha}{2}$ area $= \dfrac{\alpha}{2}$

α level of confidence

$-z_{\alpha/2}$ 0 $z_{\alpha/2}$

FIGURE 4.128 Level of confidence equals central region.

Substituting the given values into the MOE formula, we have

$$\text{MOE} = \frac{z_{\alpha/2}}{2\sqrt{n}}$$

$$0.03 = \frac{z_{\alpha/2}}{2\sqrt{1,021}}$$

TOPIC X   RANDOM SAMPLING AND OPINION POLLS
STATISTICS IN THE REAL WORLD

National opinion polls are pervasive in today's world; hardly a day goes by without the release of yet another glimpse at the American psyche. Topics ranging from presidential performance and political controversy to popular culture and alien abduction vie for our attention in the media: "A recent survey indicates that a majority of Americans. . . ." Where do these results come from? Are they accurate? Can they be taken seriously, or are they mere entertainment and speculation?

Although many people are intrigued by the "findings" of opinion polls, others are skeptical or unsure of the fundamental premise of statistical sampling. How can the opinions of a diverse population of nearly 300 million Americans be measured by a survey of a mere 1,000 to 1,500 individuals? Wouldn't a survey of 10,000 people give better, or more accurate, results than those obtained from the typical sample of about a thousand people? The answer is "no, not necessarily." Although intuition might dictate that a survey's reliability is driven by the size of the sample (bigger is better), in reality, the most important factor in obtaining reliable results is the *method* by which the sample was selected. That is, depending on how it was selected, a sample of 1,000 people can yield far better results than a sample of 10,000, 20,000, or even 50,000 people.

The basic premise in statistical sampling is that the views of a small portion of a population *can* accurately represent the views of the entire population *if* the sample is selected properly, that is, if the sample is selected randomly. So what does it mean to say that a sample is selected "randomly"? The answer is simple: A sample is random if every member of the population has an *equal chance* of being selected. For example, suppose that a remote island has a population of 1,234 people and we wish to select a random sample of 30 inhabitants to interview. We could write each islander's name on a slip of paper, put all of the slips in a large box, shake up the box, close our eyes, and select 30 slips. Each person on the island would have an equal chance of being selected; consequently, we would be highly confident (say, 95%) that the views of the sample (plus or minus a margin or error) would accurately represent the views of the entire population.

The key to reliable sampling is the selection of a random sample; to select a random sample, each person in the population must have an equal chance of being selected. Realistically, how can this be accomplished with such a vast population of people in the United States or elsewhere? Years ago, the most accurate polls (especially Gallup polls) were based on data gathered from knocking on doors and conducting face-to-face interviews. However, in today's world, almost every American adult has a telephone, and telephone surveys have replaced the door-to-door surveys of the past. Although a poll might state that the target population is "all Americans aged 18 and over," it really means "all Americans aged 18 and over who have an accessible telephone number." Who will be excluded from this population? Typically, active members of the military forces and people in prisons, institutions, or hospitals are excluded from the sampling frame of today's opinion polls.

How can a polling organization obtain a list of the telephone numbers of all Americans aged 18 and over? Typically, no such list exists. Telephone directories are not that useful because of the large number of unlisted telephone numbers. However, using high-speed computers and a procedure known as random digit dialing, polling organizations are able to create a list of all possible phone numbers and thus are able to select a random sample. Finally, each telephone number in the sample is called, and "an American aged 18 and over" (or whatever group is being targeted) is interviewed. If no one answers the phone or the appropriate person is not at home, the polling organization makes every effort to establish contact at a later time. This ensures that the random sampling process is accurately applied and therefore that the results are true to the stated level of confidence with an acceptable margin of error.

$$z_{\alpha/2} = 0.03(2\sqrt{1,021})$$ **multiplying each side by** $2\sqrt{1,021}$

$$z_{\alpha/2} = 1.917185437\ldots$$

$$z_{\alpha/2} \approx 1.92$$ **rounding off to two decimal places**

Using the body table in Appendix F, we can find the area under the bell curve between $z = 0$ and $z = 1.92$; that is, $p(0 < z < 1.92) = 0.4726$.

Therefore, $\frac{\alpha}{2} = 0.4726$, and multiplying by 2, we have $\alpha = 0.9452$. Thus, the level of confidence is $\alpha = 0.9452$ (or 95%), as shown in Figure 4.129.

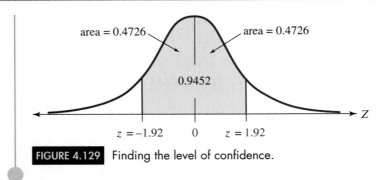

FIGURE 4.129 Finding the level of confidence.

4.5 EXERCISES

In Exercises 1–4, use the body table in Appendix F to find the specified z-number.

1. a. $z_{0.2517}$ **b.** $z_{0.1217}$ **c.** $z_{0.4177}$ **d.** $z_{0.4960}$

2. a. $z_{0.0199}$ **b.** $z_{0.2422}$ **c.** $z_{0.4474}$ **d.** $z_{0.4936}$

3. a. $z_{0.4250}$ **b.** $z_{0.4000}$ **c.** $z_{0.3750}$ **d.** $z_{0.4950}$

4. a. $z_{0.4350}$ **b.** $z_{0.4100}$ **c.** $z_{0.2750}$ **d.** $z_{0.4958}$

5. Find the z-number associated with a 92% level of confidence.

6. Find the z-number associated with a 97% level of confidence.

7. Find the z-number associated with a 75% level of confidence.

8. Find the z-number associated with an 85% level of confidence.

In Exercises 9–22, round off your answers (sample proportions and margins of error) to three decimal places (a tenth of a percent).

9. The Gallup Poll in Example 6 states that one-third (33%) of respondents support general legalization of marijuana. For each of the following levels of confidence, find the margin of error associated with the sample.

a. a 90% level of confidence

b. a 98% level of confidence

10. The Roper poll in Example 7 states that 56% of the Americans questioned think that UFOs are real. For each of the following levels of confidence, find the margin of error associated with the sample.

a. an 80% level of confidence

b. a 98% level of confidence

11. A survey asked, "How important is it to you to buy products that are made in America?"

Of the 600 Americans surveyed, 450 responded, "It is important." For each of the following levels of confidence, find the sample proportion and the margin of error associated with the poll.

a. a 90% level of confidence

b. a 95% level of confidence

12. In the survey in Exercise 11, 150 of the 600 Americans surveyed responded, "It is not important." For each of the following levels of confidence, find the sample proportion and the margin of error associated with the poll.

a. an 85% level of confidence

b. a 98% level of confidence

13. A survey asked, "Have you ever bought a lottery ticket?" Of the 2,710 Americans surveyed, 2,141 said yes, and 569 said no.*

a. Determine the sample proportion of Americans who have purchased a lottery ticket.

b. Determine the sample proportion of Americans who have not purchased a lottery ticket.

c. With a 90% level of confidence, find the margin of error associated with the sample proportions.

14. A survey asked, "Which leg do you put into your trousers first?" Of the 2,710 Americans surveyed, 1,138 said left, and 1,572 said right.*

a. Determine the sample proportion of Americans who put their left leg into their trousers first.

b. Determine the sample proportion of Americans who put their right leg into their trousers first.

c. With a 90% level of confidence, find the margin of error associated with the sample proportions.

---

* Data from Poretz and Sinrod, *The First Really Important Survey of American Habits* (Los Angeles: Price Stern Sloan Publishing, 1989).

15. A survey asked, "Do you prefer showering or bathing?" Of the 1,220 American men surveyed, 1,049 preferred showering, and 171 preferred bathing. In contrast, 1,043 of the 1,490 American women surveyed preferred showering, and 447 preferred bathing.*

    a. Determine the sample proportion of American men who prefer showering.

    b. Determine the sample proportion of American women who prefer showering.

    c. With a 95% level of confidence, find the margin of error associated with the sample proportions.

16. A survey asked, "Do you like the way you look in the nude?" Of the 1,220 American men surveyed, 830 said yes, and 390 said no. In contrast, 328 of the 1,490 American women surveyed said yes, and 1,162 said no.*

    a. Determine the sample proportion of American men who like the way they look in the nude.

    b. Determine the sample proportion of American women who like the way they look in the nude.

    c. With a 95% level of confidence, find the margin of error associated with the sample proportions.

Exercises 17–20 are based on a survey (published in April 2000) of 129,593 students in grades 6–12 conducted by USA WEEKEND *magazine.*

17. When asked, "Do you, personally, feel safe from violence in school?" 92,011 said yes, and 37,582 said no.

    a. Determine the sample proportion of students who said yes.

    b. Determine the sample proportion of students who said no.

    c. With a 95% level of confidence, find the margin of error associated with the sample proportions.

18. When asked, "Do kids regularly carry weapons in your school?" 14,255 said yes, and 115,338 said no.

    a. Determine the sample proportion of students who said yes.

    b. Determine the sample proportion of students who said no.

    c. With a 95% level of confidence, find the margin of error associated with the sample proportions.

19. When asked, "Is there a gun in your home?" 58,317 said yes, 60,908 said no, and 10,368 said they did not know.

    a. Determine the sample proportion of students who said yes.

    b. Determine the sample proportion of students who said no.

    c. Determine the sample proportion of students who said they did not know.

    d. With a 90% level of confidence, find the margin of error associated with the sample proportions.

20. When asked, "How likely do you think it is that a major violent incident could occur at your school?" 18,143 said very likely, 64,797 said somewhat likely, and 46,653 said not likely at all.

    a. Determine the sample proportion of students who said "very likely."

    b. Determine the sample proportion of students who said "somewhat likely."

    c. Determine the sample proportion of students who said "not likely at all."

    d. With a 90% level of confidence, find the margin of error associated with the sample proportions.

21. A survey asked, "Can you imagine a situation in which you might become homeless?" Of the 2,503 Americans surveyed, 902 said yes.*

    a. Determine the sample proportion of Americans who can imagine a situation in which they might become homeless.

    b. With a 90% level of confidence, find the margin of error associated with the sample proportion.

    c. With a 98% level of confidence, find the margin of error associated with the sample proportion.

    d. How does your answer to part (c) compare to your answer to part (b)? Why?

22. A survey asked, "Do you think that homeless people are responsible for the situation they are in?" Of the 2,503 Americans surveyed, 1,402 said no.*

    a. Determine the sample proportion of Americans who think that homeless people are not responsible for the situation they are in.

    b. With an 80% level of confidence, find the margin of error associated with the sample proportion.

    c. With a 95% level of confidence, find the margin of error associated with the sample proportion.

    d. How does your answer to part (c) compare to your answer to part (b)? Why?

23. A sample consisting of 430 men and 765 women was asked various questions pertaining to international affairs. With a 95% level of confidence, find the margin of error associated with the following samples.

    a. the male sample

    b. the female sample

    c. the combined sample

24. A sample consisting of 942 men and 503 women was asked various questions pertaining to the nation's economy. For a 95% level of confidence, find the margin of error associated with the following samples.

---

* Data from Poretz and Sinrod, *The First Really Important Survey of American Habits* (Los Angeles: Price Stern Sloan Publishing, 1989).

* Data from Mark Clements, "What Americans Say about the Homeless," *Parade Magazine,* Jan. 9, 1994: 4–6.

a. the male sample

b. the female sample

c. the combined sample

25. A poll pertaining to environmental concerns had the following footnote: "Based on a sample of 1,763 adults, the margin of error is plus or minus 2.5 percentage points." Find the level of confidence of the poll.

    HINT: See Example 7.

26. A poll pertaining to educational goals had the following footnote: "Based on a sample of 2,014 teenagers, the margin of error is plus or minus 2 percentage points." Find the level of confidence of the poll.

    HINT: See Example 7.

27. A recent poll pertaining to educational reforms involved 640 men and 820 women. The margin of error for the combined sample is 2.6%. Find the level of confidence for the entire poll.

    HINT: See Example 7.

28. A recent poll pertaining to educational reforms involved 640 men and 820 women. The margin of error is 3.9% for the male sample and 3.4% for the female sample. Find the level of confidence for the male portion of the poll and for the female portion of the poll.

    HINT: See Example 7.

In Exercises 29–32, you are planning a survey for which the findings are to have the specified level of confidence. How large should your sample be so that the margin of error is at most the specified amount?

29. 95% level of confidence

    a. margin of error = 3%

    b. margin of error = 2%

    c. margin of error = 1%

    d. Comparing your answers to parts (a)–(c), what conclusion can be made regarding the margin of error and the sample size?

30. 96% level of confidence

    a. margin of error = 3%

    b. margin of error = 2%

    c. margin of error = 1%

    d. Comparing your answers to parts (a)–(c), what conclusion can be made regarding the margin of error and the sample size?

31. 98% level of confidence

    a. margin of error = 3%

    b. margin of error = 2%

    c. margin of error = 1%

    d. Comparing your answers to parts (a)–(c), what conclusion can be made regarding the margin of error and the sample size?

32. 99% level of confidence

    a. margin of error = 3%

    b. margin of error = 2%

    c. margin of error = 1%

    d. Comparing your answers to parts (a)–(c), what conclusion can be made regarding the margin of error and the sample size?

 Answer the following questions using complete sentences and your own words.

● CONCEPT QUESTIONS

33. What is a sample proportion? How is it calculated?

34. What is a margin of error? How is it calculated?

35. If the sample size is increased in a survey, would you expect the margin of error to increase or decrease? Why?

36. What is a random sample?

37. What is random digit dailing (RDD)?

● HISTORY QUESTIONS

38. Who founded the American Institute of Public Opinion? When? In what two academic fields did this person receive degrees?

 WEB PROJECTS

39. What are the nation's current opinions concerning major issues in the headlines? Pick a current issue (such as abortion, same-sex marriage, gun control, war, or the president's approval rating) and find a recent survey regarding the issue. Summarize the results of the survey. Be sure to include the polling organization, date(s) of the survey, sample size, level of confidence, margin of error, and any pertinent information.

40. What is random digit dialing (RDD), and how is it used? Visit the web site of a national polling or news organization, and conduct a search of its FAQs regarding the organization's methods of sampling and the use of RDD. Write a report in which you summarize the use of RDD by the polling or news organization. Some useful links for this web project are listed on the text web site:

    www.cengage.com/math/johnson

● PROJECTS

41. The purpose of this project is to conduct an opinion poll. Select a topic that is relevant to you and/or your community. Create a multiple-choice question to gather people's opinions concerning this issue.

For example:

> If you could vote today, how would you vote on Proposition X?
> i. support   ii. oppose   iii. undecided

a. Ask fifty people your question, and record their responses. Calculate the sample proportion for each category of response. Use a 95% level of confidence, and calculate the margin of error for your survey.

b. Ask 100 people your question, and record their responses. Calculate the sample proportion for each category of response. Use a 95% level of confidence, and calculate the margin of error for your survey.

c. How does the margin of error in part (b) compare to the margin of error in part (a)? Explain.

4.6 Linear Regression

OBJECTIVES

- Find the line of best fit
- Use linear regression to make a prediction
- Calculate the coefficient of linear correlation

When x and y are variables and m and b are constants, the equation $y = mx + b$ has infinitely many solutions of the form (x, y). A specific ordered pair (x_1, y_1) is a solution of the equation if $y_1 = mx_1 + b$. Because every solution of the given equation lies on a straight line, we say that x and y are *linearly related*.

If we are given two ordered pairs (x_1, y_1) and (x_2, y_2), we should be able to "work backwards" and find the equation of the line passing through them; assuming that x and y are linearly related, we can easily find the equation of the line passing through the points (x_1, y_1) and (x_2, y_2). The process of finding the equation of a line passing through given points is known as **linear regression;** the equation thus found is called the **mathematical model** of the linear relationship. Once the model has been constructed, it can be used to make predictions concerning the values of x and y.

EXAMPLE 1

USING A LINEAR EQUATION TO MAKE A PREDICTION Charlie is planning a family reunion and wants to place an order for custom T-shirts from Prints Alive (the local silk-screen printer) to commemorate the occasion. He has ordered shirts from Prints Alive on two previous occasions; on one occasion, he paid $164 for twenty-four shirts; on another, he paid $449 for eighty-four. Assuming a linear relationship between the cost of T-shirts and the number ordered, predict the cost of ordering 100 shirts.

SOLUTION

Letting $x =$ the number of shirts ordered and $y =$ the total cost of the shirts, the given data can be expressed as two ordered pairs: $(x_1, y_1) = (24, 164)$ and $(x_2, y_2) = (84, 449)$. We must find $y = mx + b$, the equation of the line passing through the two points.

First, we find m, the slope:

$$m = \frac{y_2 - y_1}{x_2 - x_1}$$
$$= \frac{449 - 164}{84 - 24}$$
$$= \frac{285}{60}$$
$$= 4.75$$

Now we use one of the ordered pairs to find b, the y-intercept. Either point will work; we will use $(x_1, y_1) = (24, 164)$.

The slope-intercept form of a line is $y = mx + b$. Solving for b, we obtain

$$b = y - mx$$
$$= 164 - 4.75(24)$$
$$= 164 - 114$$
$$= 50$$

Therefore, the equation of the line is $y = 4.75x + 50$. We use this linear model to predict the cost of ordering $x = 100$ T-shirts.

$$y = 4.75x + 50$$
$$= 4.75(100) + 50$$
$$= 475 + 50$$
$$= 525$$

We predict that it will cost $525 to order 100 T-shirts.

Linear Trends and Line of Best Fit

Example 1 illustrates the fact that two points determine a unique line. To find the equation of the line, we must find the slope and the y-intercept. If we are given more than two points, the points might not be collinear. In collecting real-world data, this is usually the case. However, after the scatter of points is plotted on an x-y coordinate system, it may appear that the points "almost" fit on a line. If a sample of ordered pairs tend to "go in the same general direction," we say that they exhibit a **linear trend.** See Figure 4.130.

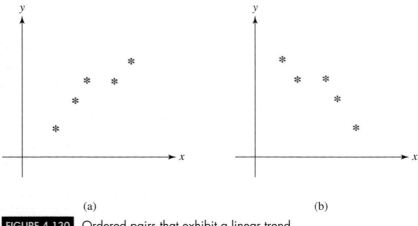

(a)   (b)

FIGURE 4.130 Ordered pairs that exhibit a linear trend.

When a scatter of points exhibits a linear trend, we construct the line that best approximates the trend. This line is called the **line of best fit** and is denoted by $\hat{y} = mx + b$. The "hat" over the y indicates that the calculated value of y is a prediction based on linear regression. See Figure 4.131.

To calculate the slope and y-intercept of the line of best fit, mathematicians have developed formulas based on the method of least squares. (See the Historical Note on Carl Gauss in Section 4.4.)

Recall that the symbol Σ means "sum." Therefore, Σx represents the sum of the x-coordinates of the points, and Σy represents the sum of the y-coordinates. To find Σxy, multiply the x- and y-coordinates of each point and sum the results.

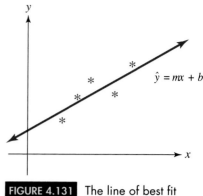

FIGURE 4.131   The line of best fit $\hat{y} = mx + b$.

LINE OF BEST FIT

Given a sample of n ordered pairs $(x_1, y_1), (x_2, y_2), \ldots, (x_n, y_n)$, the **line of best fit** (the line that best represents the data) is denoted by $\hat{y} = mx + b$, where the slope m and y-intercept b are given by

$$ m = \frac{n(\Sigma xy) - (\Sigma x)(\Sigma y)}{n(\Sigma x^2) - (\Sigma x)^2} \quad \text{and} \quad b = \bar{y} - m\bar{x} $$

$\bar{x}$ and $\bar{y}$ denote the means of the x- and y-coordinates, respectively.

EXAMPLE 2

FINDING AND GRAPHING THE LINE OF BEST FIT   We are given the ordered pairs (5, 14), (9, 17), (12, 16), (14, 18), and (17, 23).

a. Find the equation of the line of best fit.

b. Plot the given data and sketch the graph of the line of best fit on the same coordinate system.

SOLUTION

a. Organize the data in a table and compute the appropriate sums, as shown in Figure 4.132.

(x, y)	x	x²	y	xy
(5, 14)	5	25	14	5 · 14 = 70
(9, 17)	9	81	17	9 · 17 = 153
(12, 16)	12	144	16	12 · 16 = 192
(14, 18)	14	196	18	14 · 18 = 252
(17, 23)	17	289	23	17 · 23 = 391
$n = 5$ ordered pairs	$\Sigma x = 57$	$\Sigma x^2 = 735$	$\Sigma y = 88$	$\Sigma xy = 1,058$

FIGURE 4.132   Table of sums.

First, we find the slope:

$$m = \frac{n(\Sigma xy) - (\Sigma x)(\Sigma y)}{n(\Sigma x^2) - (\Sigma x)^2}$$

$$= \frac{5(1{,}058) - (57)(88)}{5(735) - (57)^2}$$

$$= 0.643192488\ldots$$

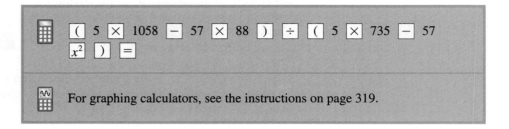

For graphing calculators, see the instructions on page 319.

Once m has been calculated, we store it in the memory of our calculator. We will need it to calculate b, the y-intercept.

$$b = \bar{y} - m\bar{x}$$

$$= \left(\frac{88}{5}\right) - 0.643192488\left(\frac{57}{5}\right)$$

$$= 10.26760564\ldots$$

Therefore, the line of best fit, $\hat{y} = mx + b$, is

$$\hat{y} = 0.643192488x + 10.26760564$$

Rounding off to one decimal place, we have

$$\hat{y} = 0.6x + 10.3$$

b. To graph the line, we need to plot two points. One point is the y-intercept $(0, b) = (0, 10.3)$. To find another point, we pick an appropriate value for x—say, $x = 18$—and calculate $\hat{y}$:

$$\hat{y} = 0.6x + 10.3$$

$$= 0.6(18) + 10.3$$

$$= 21.1$$

Therefore, the point $(x, \hat{y}) = (18, 21.1)$ is on the line of best fit.

Plotting $(0, 10.3)$ and $(18, 21.1)$, we construct the line of best fit. It is customary to use asterisks (*) to plot the given ordered pairs, as shown in Figure 4.133.

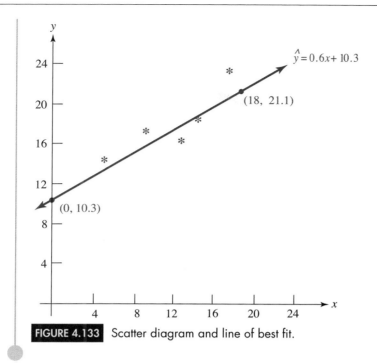

FIGURE 4.133   Scatter diagram and line of best fit.

Coefficient of Linear Correlation

Given a sample of n ordered pairs, we can always find the line of best fit. Does the line accurately portray the data? Will the line give accurate predictions? To answer these questions, we must consider the relative strength of the linear trend exhibited by the given data. If the given points are close to the line of best fit, there is a strong linear relation between x and y; the line will generate good predictions. If the given points are widely scattered about the line of best fit, there is a weak linear relation, and predictions based on it are probably not reliable.

One way to measure the strength of a linear trend is to calculate the **coefficient of linear correlation,** denoted by r. The formula for calculating r is shown in the following box.

COEFFICIENT OF LINEAR CORRELATION

Given a sample of n ordered pairs, $(x_1, y_1), (x_2, y_2), \ldots, (x_n, y_n)$
The **coefficient of linear correlation,** denoted by r, is given by

$$r = \frac{n(\Sigma xy) - (\Sigma x)(\Sigma y)}{\sqrt{n(\Sigma x^2) - (\Sigma x)^2}\sqrt{n(\Sigma y^2) - (\Sigma y)^2}}$$

The calculated value of r is always between -1 and 1, inclusive; that is, $-1 \leq r \leq 1$. If the given ordered pairs lie perfectly on a line whose slope is *positive,* then the calculated value of r will equal 1 (think 100% perfect with positive slope). In this case, both variables have the same behavior: As one increases (or decreases), so will the other. On the other hand, if the data points fall perfectly on a line whose slope is *negative,* the calculated value of r will equal -1 (think 100% perfect with negative slope). In this case, the variables have opposite behavior: As one increases, the other decreases, and vice versa. See Figure 4.134.

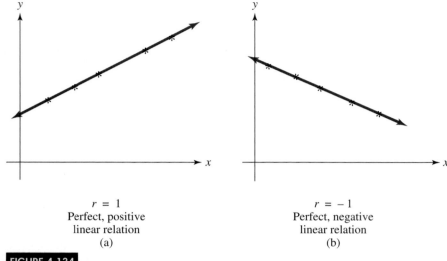

$r = 1$
Perfect, positive
linear relation
(a)

$r = -1$
Perfect, negative
linear relation
(b)

FIGURE 4.134

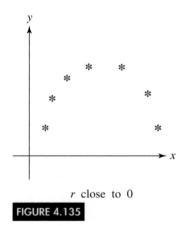

r close to 0

FIGURE 4.135

If the value of r is close to 0, there is little or no *linear* relation between the variables. This does not mean that the variables are not related. It merely means that no *linear* relation exists; the variables might be related in some nonlinear fashion, as shown in Figure 4.135.

In summary, the closer r is to 1 or -1, the stronger is the linear relation between x and y; the line of best fit will generate reliable predictions. The closer r is to 0, the weaker is the linear relation; the line of best fit will generate unreliable predictions. If r is positive, the variables have a direct relationship (as one increases, so does the other); if r is negative, the variables have an inverse relationship (as one increases, the other decreases).

EXAMPLE **3**

CALCULATING THE COEFFICIENT OF LINEAR CORRELATION
Calculate the coefficient of linear correlation for the ordered pairs given in Example 2.

SOLUTION

The ordered pairs are (5, 14), (9, 17), (12, 16), (14, 18), and (17, 23). We add a y^2 column to the table in Example 2, as shown in Figure 4.136.

(x, y)	x	x^2	y	y^2	xy
(5, 14)	5	25	14	196	$5 \cdot 14 = 70$
(9, 17)	9	81	17	289	$9 \cdot 17 = 153$
(12, 16)	12	144	16	256	$12 \cdot 16 = 192$
(14, 18)	14	196	18	324	$14 \cdot 18 = 252$
(17, 23)	17	289	23	529	$17 \cdot 23 = 391$
$n = 5$ ordered pairs	$\sum x = 57$	$\sum x^2 = 735$	$\sum y = 88$	$\sum y^2 = 1{,}594$	$\sum xy = 1{,}058$

FIGURE 4.136 Table of sums.

Now use the formula to calculate r:

$$r = \frac{n(\Sigma xy) - (\Sigma x)(\Sigma y)}{\sqrt{n(\Sigma x^2) - (\Sigma x)^2}\sqrt{n(\Sigma y^2) - (\Sigma y)^2}}$$

$$= \frac{5(1{,}058) - (57)(88)}{\sqrt{5(735) - (57)^2}\sqrt{5(1{,}594) - (88)^2}}$$

$$= 0.883062705\ldots$$

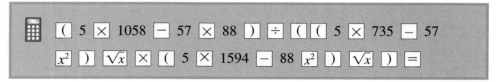

The coefficient of linear correlation is reasonably close to 1, so the line of best fit will generate reasonably reliable predictions. (Notice that the data points in Figure 4.133 are fairly close to the line of best fit.)

EXAMPLE **4**

USING LINEAR CORRELATION AND REGRESSION TO MAKE PREDICTIONS Unemployment and personal income are undoubtedly related; we would assume that as the national unemployment rate increases, total personal income would decrease. Figure 4.137 gives the unemployment rate and the total personal income for the United States for various years.

a. Use linear regression to predict the total personal income of the United States if the unemployment rate is 5.0%.
b. Use linear regression to predict the unemployment rate if the total personal income of the United States is $10 billion.
c. Are the predictions in parts (a) and (b) reliable? Why or why not?

Year	Unemployment Rate (percent)	Total Personal Income (billions)
1975	8.5	$1.3
1980	7.1	2.3
1985	7.2	3.4
1990	5.6	4.9
1995	5.6	6.1
2000	4.0	8.4
2006	4.6	11.0

FIGURE 4.137 *Sources:* Bureau of Labor Statistics, U.S. Department of Labor; Bureau of Economic Analysis, U.S. Department of Commerce.

SOLUTION

a. Letting x = unemployment rate and y = total personal income, we have $n = 7$ ordered pairs, as shown in Figure 4.138.
    Using a calculator, we find the following sums:

$$\Sigma x^2 = 274.38 \qquad \Sigma y^2 = 271.32 \qquad \Sigma xy = 197.66$$

x (percent)	y (billions)
8.5	$1.3
7.1	2.3
7.2	3.4
5.6	4.9
5.6	6.1
4.0	8.4
4.6	11.0
$\Sigma x = 42.6$	$\Sigma y = 37.4$

FIGURE 4.138

x = unemployment rate,
y = total personal income.

First we find the slope:

$$m = \frac{n(\Sigma xy) - (\Sigma x)(\Sigma y)}{n(\Sigma x^2) - (\Sigma x)^2}$$

$$= \frac{7(197.66) - (42.6)(37.4)}{7(274.38) - (42.6)^2}$$

$$= -1.979414542\ldots$$

Now we calculate b, the y-intercept:

$$b = \bar{y} - m\bar{x}$$

$$= \left(\frac{37.4}{7}\right) - (-1.979414542)\left(\frac{42.6}{7}\right)$$

$$= 17.3890085\ldots$$

Therefore, the line of best fit, $\hat{y} = mx + b$, is

$$\hat{y} = -1.979414542x + 17.3890085$$

Rounding off to two decimal places (one more than the data), we have

$$\hat{y} = -1.98x + 17.39$$

Now, substituting $x = 5.0$ (5% unemployment) into the equation of the line of best fit, we have

$$\hat{y} = -1.98x + 17.39$$

$$= -1.98(5.0) + 17.39$$

$$= -9.9 + 17.39$$

$$= 7.49$$

If the unemployment rate is 5.0%, we predict that the total personal income of the United States will be approximately $7.49 billion.

b. To predict the unemployment rate when the total personal income is $10 billion, we let $y = 10$, substitute into $\hat{y}$, and solve for x:

$$10 = -1.98x + 17.39$$

$1.98x + 10 = 17.39$	**adding 1.98x to both sides**
$1.98x = 17.39 - 10$	**subtracting 10 from both sides**
$1.98x = 7.39$	
$x = \dfrac{7.39}{1.98}$	**dividing by 1.98**
$\quad = 3.7323232\ldots$	

We predict that the unemployment rate will be approximately 3.7% when the total personal income is $10 billion.

c. To investigate the reliability of our predictions (the strength of the linear trend), we must calculate the coefficient of linear correlation:

$$r = \frac{n(\Sigma xy) - (\Sigma x)(\Sigma y)}{\sqrt{n(\Sigma x^2) - (\Sigma x)^2}\sqrt{n(\Sigma y^2) - (\Sigma y)^2}}$$

$$r = \frac{7(197.66) - (42.6)(37.4)}{\sqrt{7(274.38) - (42.6)^2}\sqrt{7(271.32) - (37.4)^2}}$$

$$= -0.9105239486\ldots$$

Because r is close to -1, we conclude that our predictions are very reliable; the linear relationship between x and y is high. Furthermore, since r is negative, we know that y (total personal income) decreases as x (unemployment rate) increases.

4.6 EXERCISES

1. A set of $n = 6$ ordered pairs has the following sums:

$$\Sigma x = 64 \qquad \Sigma x^2 = 814 \qquad \Sigma y = 85$$
$$\Sigma y^2 = 1{,}351 \qquad \Sigma xy = 1{,}039$$

   a. Find the line of best fit.
   b. Predict the value of y when $x = 11$.
   c. Predict the value of x when $y = 19$.
   d. Find the coefficient of linear correlation.
   e. Are the predictions in parts (b) and (c) reliable? Why or why not?

2. A set of $n = 8$ ordered pairs has the following sums:

$$\Sigma x = 111 \qquad \Sigma x^2 = 1{,}869 \qquad \Sigma y = 618$$
$$\Sigma y^2 = 49{,}374 \qquad \Sigma xy = 7{,}860$$

   a. Find the line of best fit.
   b. Predict the value of y when $x = 8$.
   c. Predict the value of x when $y = 70$.
   d. Find the coefficient of linear correlation.
   e. Are the predictions in parts (b) and (c) reliable? Why or why not?

3. A set of $n = 5$ ordered pairs has the following sums:

$$\Sigma x = 37 \qquad \Sigma x^2 = 299 \qquad \Sigma y = 38$$
$$\Sigma y^2 = 310 \qquad \Sigma xy = 279$$

   a. Find the line of best fit.
   b. Predict the value of y when $x = 5$.
   c. Predict the value of x when $y = 7$.
   d. Find the coefficient of linear correlation.
   e. Are the predictions in parts (b) and (c) reliable? Why or why not?

4. Given the ordered pairs $(4, 40)$, $(6, 37)$, $(8, 34)$, and $(10, 31)$:

   a. Find and interpret the coefficient of linear correlation.
   b. Find the line of best fit.
   c. Plot the given ordered pairs and sketch the graph of the line of best fit on the same coordinate system.

5. Given the ordered pairs $(5, 5)$, $(7, 10)$, $(8, 11)$, $(10, 15)$, and $(13, 16)$:

   a. Plot the ordered pairs. Do the ordered pairs exhibit a linear trend?
   b. Find the line of best fit.
   c. Predict the value of y when $x = 9$.
   d. Plot the given ordered pairs and sketch the graph of the line of best fit on the same coordinate system.
   e. Find the coefficient of linear correlation.
   f. Is the prediction in part (c) reliable? Why or why not?

6. Given the ordered pairs $(5, 20)$, $(6, 15)$, $(10, 14)$, $(12, 15)$, and $(13, 10)$:

   a. Plot the ordered pairs. Do the ordered pairs exhibit a linear trend?
   b. Find the line of best fit.
   c. Predict the value of y when $x = 8$.
   d. Plot the given ordered pairs and sketch the graph of the line of best fit on the same coordinate system.
   e. Find the coefficient of linear correlation.
   f. Is the prediction in part (c) reliable? Why or why not?

7. Given the ordered pairs $(2, 6)$, $(3, 12)$, $(6, 15)$, $(7, 4)$, $(10, 6)$, and $(11, 12)$:

   a. Plot the ordered pairs. Do the ordered pairs exhibit a linear trend?
   b. Find the line of best fit.
   c. Predict the value of y when $x = 8$.
   d. Plot the given ordered pairs and sketch the graph of the line of best fit on the same coordinate system.
   e. Find the coefficient of linear correlation.
   f. Is the prediction in part (c) reliable? Why or why not?

8. The unemployment rate and the amount of emergency food assistance (food made available to hunger relief organizations such as food banks and soup kitchens) provided by the federal government in the United States in certain years are given in Figure 4.139.

Year	Unemployment Rate	Emergency Food Assistance (in millions)
2000	4.0%	$225
2003	6.0	456
2004	5.5	420
2005	5.1	373
2006	4.6	300
2007	4.6	256

FIGURE 4.139 Unemployment rates and emergency food assistance. *Sources:* Bureau of Labor Statistics (U.S. Department of Labor); Food and Nutrition Service (U.S. Department of Agriculture).

   a. Letting $x =$ the unemployment rate and $y =$ the amount of emergency food assistance, plot the data. Do the data exhibit a linear trend?
   b. Find the line of best fit.

317

c. Predict the amount of emergency food assistance when the unemployment rate is 5.0%.

d. Predict the unemployment rate when the amount of emergency food assistance is $350 million.

e. Find the coefficient of correlation.

f. Are the predictions in parts (a) and (b) reliable? Why or why not?

9. The average hourly earnings and the average tuition at public four-year institutions of higher education in the United States for 1997–2002 are given in Figure 4.140.

Year	Average Hourly Earnings	Average Tuition at Four-Year Institutions
1997	$12.49	$3,110
1998	13.00	3,229
1999	13.47	3,349
2000	14.00	3,501
2001	14.53	3,735
2002	14.95	4,059

FIGURE 4.140 Average hourly earnings and average tuition, 1997–2002. *Source:* Bureau of Labor Statistics and National Center for Education Statistics.

a. Letting x = the average hourly earnings and y = the average tuition, plot the data. Do the data exhibit a linear trend?

b. Find the line of best fit.

c. Predict the average tuition when the average hourly earning is $14.75.

d. Predict the average hourly earning when the average tuition is $4,000.

e. Find the coefficient of correlation.

f. Are the predictions in parts (a) and (b) reliable? Why or why not?

10. The number of domestic and imported retail car sales (in hundreds of thousands) in the United States for 1999–2007 are given in Figure 4.141.

a. Letting x = the number of domestic car sales and y = the number of imported car sales, plot the data. Do the data exhibit a linear trend?

b. Find the line of best fit.

c. Predict the number of imported car sales when there are 5,600,000 domestic car sales.

d. Predict the number of domestic car sales when there are 2,200,000 imported car sales.

e. Find the coefficient of correlation.

f. Are the predictions in parts (a) and (b) reliable? Why or why not?

Year	Domestic Car Sales	Imported Car Sales
1999	69.8	17.2
2000	68.3	20.2
2001	63.2	21.0
2002	58.8	22.3
2003	55.3	20.8
2004	53.6	21.5
2005	54.8	21.9
2006	54.4	23.4
2007	52.5	23.7

FIGURE 4.141 Domestic and imported retail car sales (hundred thousands), 1999–2007. *Source:* Ward's Commission.

11. The numbers of marriages and divorces (in millions) in the United States are given in Figure 4.142.

Year	1965	1970	1975	1980	1985	1990	2000
Marriages	1.800	2.158	2.152	2.413	2.425	2.448	2.329
Divorces	0.479	0.708	1.036	1.182	1.187	1.175	1.135

FIGURE 4.142 Number of marriages and divorces (millions). *Source:* National Center for Health Statistics.

a. Letting x = the number of marriages and y = the number of divorces in a year, plot the data. Do the data exhibit a linear trend?

b. Find the line of best fit.

c. Predict the number of divorces in a year when there are 2,750,000 marriages.

d. Predict the number of marriages in a year when there are 1,500,000 divorces.

e. Find the coefficient of linear correlation.

f. Are the predictions in parts (c) and (d) reliable? Why or why not?

12. The median home price and average mortgage rate in the United States for 1999–2004 are given in Figure 4.143.

a. Letting x = the median price of a home and y = the average mortgage rate, plot the data. Do the data exhibit a linear trend?

b. Find the line of best fit.

c. Predict the average mortgage rate if the median price of a home is $165,000.

d. Predict the median home price if the average mortgage rate is 7.25%.

e. Find the coefficient of correlation.

f. Are the predictions in parts (a) and (b) reliable? Why or why not?

Year	Median Price (dollars)	Mortgage Rate (percent)
1999	133,300	7.33
2000	139,000	8.03
2001	147,800	7.03
2002	158,100	6.55
2003	180,200	5.74
2004	195,200	5.73

FIGURE 4.143 Median home price and average mortgage rate, 1999–2004. *Source:* National Association of Realtors.

Answer the following questions using complete sentences and your own words.

• CONCEPT QUESTIONS

13. What is a line of best fit? How do you find it?

14. How do you measure the strength of a linear trend?

15. What is a positive linear relation? Give an example.

16. What is a negative linear relation? Give an example.

• PROJECTS

17. Measure the heights and weights of ten people. Let x = height and y = weight.

a. Plot the ordered pairs. Do the ordered pairs exhibit a linear trend?

b. Use the data to find the line of best fit.

c. Find the coefficient of linear correlation.

d. Will your line of best fit produce reliable predictions? Why or why not?

TECHNOLOGY AND LINEAR REGRESSION

In Example 2 of this section, we computed the slope and y-intercept of the line of best fit for the five ordered pairs (5, 14), (9, 17), (12, 16), (14, 18), and (17, 23). These calculations can be tedious when done by hand, even with only five data points. In the real world, there are always a large number of data points, and the calculations are always done with the aid of technology.

LINEAR REGRESSION ON A GRAPHING CALCULATOR

Graphing calculators can draw a scatter diagram, compute the slope and y-intercept of the line of best fit, and graph the line.

ON A TI-83/84:

• *Put the calculator into statistics mode* by pressing STAT.

• *Set the calculator up for entering the data* from Example 2 by selecting "Edit" from the "EDIT" menu, and the "List Screen" appears, as shown in Figure 4.144. If data already appear in a list (as they do in Figure 4.144) and you want to clear the list, use the arrow buttons to highlight the name of the list and press CLEAR ENTER.

• Use the arrow buttons and the ENTER button to enter the x-coordinates in list L_1 and the corresponding y-coordinates in list L_2.

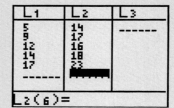

FIGURE 4.144 A TI-83/84's list screen.

ON A CASIO:

- *Put the calculator into statistics mode* by pressing MENU , highlighting STAT, and pressing EXE .

- Use the arrow buttons and the EXE button to enter the *x*-coordinates from Example 2 in List 1 and the corresponding *y*-coordinates in List 2. When you are done, your screen should look like Figure 4.145. If data already appears in a list and you want to erase it, use the arrow keys to move to that list, press F6 and then F4 (i.e., DEL-A , which stands for "delete all") and then F1 (i.e., YES).

FIGURE 4.145 A Casio's list screen.

Finding the Equation of the Line of Best Fit

Once the data have been entered, it is easy to find the equation. Recall from Example 2 that the equation is $y = 0.643192488x + 10.26760564$.

ON A TI-83/84:

- Press 2nd QUIT .

- The TI-83/84 does not display the correlation coefficient unless you tell it to. To do so, press 2nd CATALOG , scroll down, and select "DiagnosticOn." When "DiagnosticOn" appears on the screen, press ENTER.

- Press STAT , scroll to the right, and select the "CALC" menu, and select "LinReg(ax + b)" from the "CALC" menu.

- When "LinReg(ax + b)" appears on the screen, press ENTER .

- The slope, the *y*-intercept and the correlation coefficient will appear on the screen, as shown in Figure 4.146. They are not labeled *m, b,* and *r*. Their labels are different and are explained in Figure 4.147.

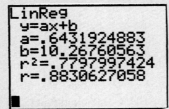

FIGURE 4.146 Finding the equation of the line of best fit on a TI-83/84.

	TI-83/84 Labels
slope ma	
y-intercept b	b
correlation coefficient r	r

FIGURE 4.147 Graphing calculator line of best fit labels.

ON A CASIO:

- Press ⌐CALC⌐ (i.e., ⌐F2⌐) to calculate the equation.
- Press ⌐REG⌐ (i.e., ⌐F3⌐); "REG" stands for "regression."
- Press ⌐X⌐ (i.e., ⌐F1⌐) for linear regression. (See Figure 4.148.)

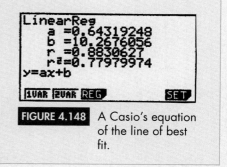

FIGURE 4.148 A Casio's equation of the line of best fit.

Drawing a Scatter Diagram and the Line of Best Fit

Once the equation has been found, the line can be graphed.

ON A TI-83/84:

- Press ⌐Y=⌐, and clear any functions that may appear.
- Set the calculator up to draw a scatter diagram by pressing ⌐2nd⌐ ⌐STAT PLOT⌐ and selecting "Plot 1." Turn the plot on and select the scatter icon.
- Tell the calculator to put the data entered in list L_1 on the x-axis by selecting "L_1" for "Xlist," and to put the data entered in list L_2 on the y-axis by selecting "L_2" for "Ylist."
- Automatically set the range of the window and obtain a scatter diagram by pressing ⌐ZOOM⌐ and selecting option 9: "ZoomStat."
- If you don't want the data points displayed on the line of best fit, press ⌐2nd⌐ ⌐STAT PLOT⌐ and turn off plot 1.
- Quit the statistics mode by pressing ⌐2nd⌐ ⌐QUIT⌐.
- Enter the equation of the line of best fit by pressing ⌐Y=⌐ ⌐VARS⌐, selecting "Statistics," scrolling to the right to select the "EQ" menu, and selecting "RegEQ" (for regression equation).
- Automatically set the range of the window and obtain a scatter diagram by pressing ⌐ZOOM⌐ and selecting option 9: "ZoomStat." (See Figure 4.149.)

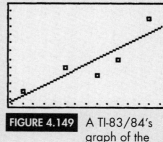

FIGURE 4.149 A TI-83/84's graph of the line of best fit and the scatter diagram.

- Press ⌐TRACE⌐ to read off the data points as well as points on the line of best fit. Use the up and down arrows to switch between data points and points on the line of best fit. Use the left and right arrows to move left and right on the graph.

ON A CASIO:

- Press ⌐SHIFT⌐ ⌐QUIT⌐ to return to List 1 and List 2.
- Press ⌐GRPH⌐ (i.e., ⌐F1⌐).
- Press ⌐SET⌐ (i.e., ⌐F6⌐).

- Make the resulting screen, which is labeled "StatGraph1," read as follows:

Graph Type   :Scatter

Xlist   :List1

Ylist   :List2

Frequency   :1

To make the screen read as described above,

- Use the down arrow button to scroll down to "Graph Type."
- Press $\boxed{\text{Scat}}$ (i.e., $\boxed{\text{F1}}$).
- In a similar manner, change "Xlist," "Ylist," and "Frequency" if necessary.
- Press $\boxed{\text{SHIFT}}$ $\boxed{\text{QUIT}}$ to return to List1 and List2.
- Press $\boxed{\text{GRPH}}$ (i.e., $\boxed{\text{F1}}$).
- Press $\boxed{\text{GPH1}}$ (i.e., $\boxed{\text{F1}}$) to obtain the scatter diagram.
- Press $\boxed{\text{X}}$ (i.e., $\boxed{\text{F1}}$).
- Press $\boxed{\text{DRAW}}$ (i.e., $\boxed{\text{F6}}$), and the calculator will display the scatter diagram as well as the line of best fit. (See Figure 4.150.)

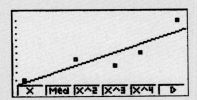

FIGURE 4.150 A Casio's graph of the line of best fit and the scatter diagram.

LINEAR REGRESSION ON EXCEL

Entering the Data

We will use the data from Example 2. Start by entering the information as shown in Figure 4.151. Then save your spreadsheet.

◇	A	B	C
1	x-coordinate	y-coordinate	
2	5	14	
3	9	17	
4	12	16	
5	14	18	
6	17	23	
7			

FIGURE 4.151 The given data.

Using the Chart Wizard to Draw the Scatter Diagram

- Press the "Chart Wizard" button at the top of the spreadsheet. (It looks like a bar chart, and it might have a magic wand.)

- Select "XY (Scatter)" and choose the Chart sub-type that displays points without connecting lines.
- Press the "Next" button and a "Chart Source Data" box will appear.
- Click on the triangle to the right of "Data range." This will cause the "Chart Source Data" box to shrink.
- Use your mouse to draw a box around all of the x-coordinates and y-coordinates. This will cause "= Sheet1!\$A\$2:\$B\$6" to appear in the "Chart Source Data" box.
- Press the triangle to the right of the "Chart Source Data" box and the box will expand.
- Press the "Next" button.
- Under "Chart Title" type an appropriate title. Our example lacks content, so we will use the title "Example 2."
- After "Category (X)" type an appropriate title for the x-axis. Our example lacks content, so we will use the title "x-axis."
- After "Category (Y)" type an appropriate title for the y-axis.
- Click on the "Gridlines" tab and check the box under "value (X) axis," next to "Major gridlines."
- Click on the "Legend" tab and remove the check from the box next to "show legend."
- Press the "Finish" button and the scatter diagram will appear.
- Save the spreadsheet.

After you have completed this step, your spreadsheet should include the scatter diagram shown in Figure 4.152.

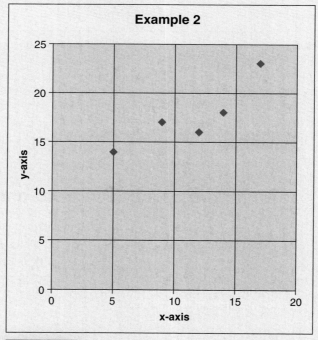

FIGURE 4.152 An Excel-generated scatter diagram.

Finding and Drawing the Line of Best Fit

- Click on the chart until it is surrounded by a thicker border.
- Use your mouse to select "Chart" at the very top of the screen. Then pull your mouse down until "Add Trendline . . ." is highlighted, and let go.
- Press the "Linear" button.

- Click on "Options".
- Select "Display Equation on Chart".
- Select "Display R-Squared value on Chart".
- Press the "OK" button, and the scatter diagram, the line of best fit, the equation of the line of best fit, and r^2 (the square of the correlation coefficient) appear. If the equation is in the way, you can click on it and move it to a better location.
- Use the square root button on your calculator to find r, the correlation coefficient.
- Save the spreadsheet.

After completing this step, your spreadsheet should include the scatter diagram, the line of best fit, and the equation of the line of best fit, as shown in Figure 4.153.

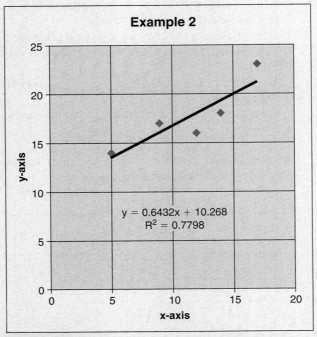

FIGURE 4.153 The scatter diagram with the line of best fit and its equation.

Printing the Scatter Diagram and Line of Best Fit

- Click on the diagram, until it is surrounded by a thicker border.
- Use your mouse to select "File" at the very top of the screen. Then pull your mouse down until "Print" is highlighted, and let go.
- Respond appropriately to the "Print" box that appears. Usually, it is sufficient to press a "Print" button or an "OK" button, but the actual way that you respond to the "Print" box depends on your printer.

EXERCISES

18. Throughout the twentieth century, the record time for the mile run steadily decreased, from 4 minutes 15.4 seconds in 1911 to 3 minutes 43.1 seconds in 1999. Some of the record times are given in Figure 4.154.

a. Convert the given data to ordered pairs (x, y), where x is the number of years since 1900 and y is the time in seconds.

b. Draw a scatter diagram, find the equation of the line of best fit, and find the correlation coefficient.

Year	1911	1923	1933	1942	1945	1954
Time (min:sec)	4:15.4	4:10.4	4:07.6	4:04.6	4:01.4	3:59.4

Year	1964	1967	1975	1980	1985	1999
Time (min:sec)	3:54.1	3:51.1	3:49.4	3:48.8	3:46.3	3:43.1

FIGURE 4.154 Record times for the mile run.

c. Predict the record time for the mile run in the year 2020.

d. Predict when the record time for the mile run will reach 3:30.

e. Predict when the record time for the mile run will reach 0:30.

f. On what assumption are these predictions based? If this assumption is correct, how accurate are these predictions?

19. Heart disease is much less of a problem in France than it is in the United States and other industrialized countries, despite a high consumption of saturated fats in France. One theory is that this lower heart disease rate is due to the fact that the French drink more wine. This so-called French paradox heightened an interest in moderate wine drinking. Figure 4.155 gives data on wine consumption and heart disease for nineteen industrialized countries, including France and the United States, in 1994.

a. Draw a scatter diagram, find the equation of line of best fit, and find the correlation for wine consumption versus annual deaths.

b. Do the data support the theory that wine consumption lowers the heart disease rate?

Country	Wine Consumption (liters per person)	Annual Deaths from Heart Disease per 100,000 People
Australia	2.5	211
Austria	3.9	167
Belgium	2.9	131
Canada	2.4	191
Denmark	2.9	220
Finland	0.8	297
France	9.1	71
Iceland	0.8	211
Ireland	0.6	300
Italy	7.9	107
Netherlands	1.8	266
New Zealand	1.9	266
Norway	0.8	227
Spain	6.5	86
Sweden	1.6	207
Switzerland	5.8	115
United Kingdom	1.3	285
United States	1.2	199
West Germany	2.7	172

FIGURE 4.155 Wine consumption and heart disease. *Source: New York Times,* December 28, 1994.

TERMS

bell-shaped curve
body table
categorical data
cells
coefficient of linear
  correlation
computerized
  spreadsheet
continuous variable
data point
density
descriptive statistics

deviation from the
  mean
discrete variable
error of the estimate
extreme value
frequency
frequency distribution
histogram
Inferential statistics
level of confidence
linear regression
linear trend
line of best fit
margin of error (MOE)

mathematical model
mean
measures of central
  tendency
measures of
  dispersion
median
mode
normal distribution
outlier
pie chart
population
population proportion
relative frequency

relative frequency
  density
sample
sample proportion
spreadsheet
standard deviation
standard normal
  distribution
statistics
summation notation
tail
variance
z-distribution
z-number

REVIEW EXERCISES

1. Find (a) the mean, (b) the median, (c) the mode, and (d) the standard deviation of the following set of raw data:

    5 8 10 4 8 10 6 8 7 5

2. To study the composition of families in Winslow, Arizona, forty randomly selected married couples were surveyed to determine the number of children in each family. The following results were obtained:

    3 1 0 4 1 3 2 2 0 2 0 2 2 1
    4 3 1 1 3 4 2 1 3 0 1 0 2 5
    1 2 3 0 0 1 2 3 1 2 0 2

  a. Organize the given data by creating a frequency distribution.

  b. Find the mean number of children per family.

  c. Find the median number of children per family.

  d. Find the mode number of children per family.

  e. Find the standard deviation of the number of children per family.

  f. Construct a histogram using single-valued classes of data.

3. The frequency distribution in Figure 4.156 lists the number of hours per day that a randomly selected sample of teenagers spent watching television. Where possible, determine what percent of the teenagers

$x =$ Hours per Day	Frequency
$0 \leq x < 2$	23
$2 \leq x < 4$	45
$4 \leq x < 6$	53
$6 \leq x < 8$	31
$8 \leq x \leq 10$	17

FIGURE 4.156 Time watching television.

spent the following number of hours watching television.

  a. less than 4 hours
  b. not less than 6 hours
  c. at least 2 hours
  d. less than 2 hours
  e. at least 4 hours but less than 8 hours
  f. more than 3.5 hours

4. To study the efficiency of its new oil-changing system, a local service station monitored the amount of time it took to change the oil in customers' cars. The frequency distribution in Figure 4.157 summarizes the findings.

x = Time (in minutes)	Number of Customers
$3 \leq x < 6$	18
$6 \leq x < 9$	42
$9 \leq x < 12$	64
$12 \leq x < 15$	35
$15 \leq x \leq 18$	12

FIGURE 4.157 Time to change oil.

a. Find the mean number of minutes to change the oil in a car.

b. Find the standard deviation of the amount of time to change the oil in a car.

c. Construct a histogram to represent the data.

5. If your scores on the first four exams (in this class) are 74, 65, 85, and 76, what score do you need on the next exam for your overall mean to be at least 80?

6. The mean salary of twelve men is $37,000, and the mean salary of eight women is $28,000. Find the mean salary of all twenty people.

7. Timo and Henke golfed five times during their vacation. Their scores are given in Figure 4.158.

Timo	103	99	107	93	92
Henke	101	92	83	96	111

FIGURE 4.158 Golf scores.

a. Find the mean score of each golfer. Who has the lowest mean?

b. Find the standard deviation of each golfer's scores.

c. Who is the more consistent golfer? Why?

8. Suzanne surveyed the prices for a quart of a certain brand of motor oil. The sample data, in dollars per quart, are summarized in Figure 4.159.

Price per Quart	Number of Stores
1.99	2
2.09	3
2.19	7
2.29	10
2.39	14
2.49	4

FIGURE 4.159 Price of motor oil.

a. Find the mean and standard deviation of the prices.

b. What percent of the data lies within one standard deviation of the mean?

c. What percent of the data lies within two standard deviations of the mean?

d. What percent of the data lies within three standard deviations of the mean?

9. Classify the following types of data as discrete, continuous, or neither.

a. weights of motorcycles

b. colors of motorcycles

c. number of motorcycles

d. ethnic background of students

e. number of students

f. amounts of time spent studying

10. What percent of the standard normal z-distribution lies in the following intervals?

a. between $z = 0$ and $z = 1.75$

b. between $z = -1.75$ and $z = 0$

c. between $z = -1.75$ and $z = 1.75$

11. A large group of data is normally distributed with mean 78 and standard deviation 7.

a. Find the intervals that represent one, two, and three standard deviations of the mean.

b. What percent of the data lies in each interval in part (a)?

c. Draw a sketch of the bell curve.

12. The time it takes a latex paint to dry is normally distributed. If the mean is $3\frac{1}{2}$ hours with a standard deviation of 45 minutes, find the probability that the drying time will be as follows.

a. less than 2 hours 15 minutes

b. between 3 and 4 hours

HINT: Convert everything to hours (or to minutes).

13. All incoming freshmen at a major university are given a diagnostic mathematics exam. The scores are normally distributed with a mean of 420 and a standard deviation of 45. If the student scores less than a certain score, he or she will have to take a review course. Find the cutoff score at which 34% of the students would have to take the review course.

14. Find the specified z-number.

a. $z_{0.4441}$ b. $z_{0.4500}$ c. $z_{0.1950}$ d. $z_{0.4975}$

15. A survey asked, "Do you think that the president is doing a good job?" Of the 1,200 Americans surveyed, 800 responded yes. For each of the following levels of confidence, find the sample proportion and the margin of error associated with the poll.

a. a 90% level of confidence

b. a 95% level of confidence

16. A survey asked, "Do you support capital punishment?" Of the 1,000 Americans surveyed, 750 responded no. For each of the following levels of confidence, find the sample proportion and the margin of error associated with the poll.
    a. an 80% level of confidence
    b. a 98% level of confidence

17. A sample consisting of 580 men and 970 women was asked various questions pertaining to international affairs. For a 95% level of confidence, find the margin of error associated with the following samples.
    a. the male sample
    b. the female sample
    c. the combined sample

18. A poll pertaining to environmental concerns had the following footnote: "Based on a sample of 1,098 adults, the margin of error is plus or minus 2 percentage points." Find the level of confidence of the poll.

19. You are planning a survey with a 94% level of confidence in your findings. How large should your sample be so that the margin of error is at most 2.5%?

20. A set of $n = 5$ ordered pairs has the following sums:

    $$\sum x = 66 \qquad \sum x^2 = 1{,}094 \qquad \sum y = 273$$
    $$\sum y^2 = 16{,}911 \qquad \sum xy = 4{,}272$$

    a. Find the line of best fit.
    b. Predict the value of y when $x = 16$.
    c. Predict the value of x when $y = 57$.
    d. Find the coefficient of linear correlation.
    e. Are the predictions in parts (b) and (c) reliable? Why or why not?

21. Given the ordered pairs (5, 38), (10, 30), (20, 33), (21, 25), (24, 18), and (30, 20),
    a. Plot the ordered pairs. Do the ordered pairs exhibit a linear trend?
    b. Find the line of best fit.
    c. Predict the value of y when $x = 15$.
    d. Plot the given ordered pairs and sketch the graph of the line of best fit on the same coordinate system.
    e. Find the coefficient of linear correlation.
    f. Is the prediction in part (c) reliable? Why or why not?

22. The value of agricultural exports and imports in the United States for 1999–2007 are given in Figure 4.160.
    a. Letting x = the value of agricultural exports and y = the value of agricultural imports, plot the data. Do the data exhibit a linear trend?
    b. Find the line of best fit.

Year	Agricultural Exports (billion dollars)	Agricultural Imports (billion dollars)
1999	48.4	37.7
2000	51.2	39.0
2001	53.7	39.4
2002	53.1	41.9
2003	59.5	47.3
2004	62.4	52.7
2005	62.5	57.7
2006	68.7	64.0
2007	82.2	70.1

FIGURE 4.160 Agricultural exports and imports (billion dollars), 1999–2007. *Source:* U.S. Department of Agriculture.

c. Predict the value of agricultural imports when the value of agricultural exports is 75.0 billion dollars.
d. Predict the value of agricultural exports when the value of agricultural imports is 60.0 billion dollars.
e. Find the coefficient of correlation.
f. Are the predictions in parts (a) and (b) reliable? Why or why not?

Answer the following questions using complete sentences and your own words.

• CONCEPT QUESTIONS

23. What are the three measures of central tendency? Briefly explain the meaning of each.
24. What does standard deviation measure?
25. What is a normal distribution? What is the standard normal distribution?
26. What is a margin of error? How is it related to a sample proportion?
27. How do you measure the strength of a linear trend?

• HISTORY QUESTIONS

28. What role did the following people play in the development of statistics?
    • George Gallup
    • Carl Friedrich Gauss

FINANCE

How you deal with money will have a big impact on the quality of your life. There's no question that you will borrow money. You probably already have a credit card loan or a student loan. Sooner or later, you'll borrow money for a car or a house. The question is whether or not you'll know enough that you can be wise about your borrowing.

Not everybody saves money, but your life will be less stressful and more successful if you do. You might now have a savings account at a bank or a money market account. Will you know enough about the mathematics of finance to be wise about your savings?

It's to your advantage to know the mathematics of finance. Right now, you probably know next to nothing about it. You might not even know what a money market account is or what an annuity is. You should know these things.

WHAT WE WILL DO IN THIS CHAPTER

BORROWING:

- Most loans require that you pay simple interest, so we will explore that.

- Many loans, including car loans and home loans, are amortized—that is, they require monthly payments—so we will explore amortized loans.

SAVING:

- Many investments, including savings accounts, certificates of deposit, and money market accounts, pay compound interest, so we will discuss how compound interest works.

- One of the best ways for an average person to save money is through an annuity, so we will explore that.

5.1 Simple Interest

OBJECTIVES

- Make simple interest calculations
- Determine a credit card finance charge
- Find the payment required by an add-on interest loan

Loans and investments are very similar financial transactions. Each involves:

- the flow of money from a source to another party
- the return of the money to its source
- the payment of a fee to the source for the use of the money

When you make a deposit in your savings account, you are making an investment, but the bank views it as a loan; you are lending the bank your money, which they will lend to another customer, perhaps to buy a house. When you borrow money to buy a car, you view the transaction as a loan, but the bank views it as an investment; the bank is investing its money in you in order to make a profit.

When an investor puts money into a savings account or buys a certificate of deposit (CD) or a Treasury bill, the investor expects to make a profit. The amount of money that is invested is called the **principal.** The profit is the **interest.** How much interest will be paid depends on the **interest rate** (usually expressed as a percent per year); the **term,** or length of time that the money is invested; and how the interest is calculated.

In this section, we'll explore simple interest in investments and short-term loans. **Simple interest** means that the amount of interest is calculated as a percent per year of the principal.

SIMPLE INTEREST FORMULA

The **simple interest** I on a principal P at an annual rate of interest r for a term of t years is

$$I = Prt$$

EXAMPLE **1**

USING THE SIMPLE INTEREST FORMULA   Tom and Betty buy a two-year CD that pays 5.1% simple interest from their bank for $150,000. (Many banks pay simple interest on larger CDs and compound interest on smaller CDs.) They invest $150,000, so the principal is $P = \$150,000$. The interest rate is $r = 5.1\% = 0.051$, and the term is $t = 2$ years.

a. Find the interest that the investment earns.
b. Find the value of the CD at the end of its term.

SOLUTION

a. Using the Simple Interest Formula, we have

$$I = Prt \qquad \text{the Simple Interest Formula}$$
$$= 150{,}000 \cdot 0.051 \cdot 2 \quad \text{substituting for } P, r, \text{ and } t$$
$$= \$15{,}300$$

b. Two years in the future, the bank will pay Tom and Betty

$150,000 principal + $15,300 interest = $165,300

This is called the *future value* of the CD, because it is what the CD will be worth in the future.

Future Value

In Example 1, we found the future value by adding the principal to the interest. The **future value** *FV* is always the sum of the principal *P* plus the interest *I*. If we combine this fact with the Simple Interest Formula, we can get a formula for the future value:

$FV = P + I$

$\qquad = P + Prt$ **from the Simple Interest Formula**

$\qquad = P(1 + rt)$ **factoring**

SIMPLE INTEREST FUTURE VALUE FORMULA

The **future value** *FV* of a principal *P* at an annual rate of interest *r* for *t* years is

$$FV = P(1 + rt)$$

Capital Letters for Money

There is an important distinction between the variables *FV*, *I*, *P*, *r*, and *t*: *FV*, *I*, and *P* measure amounts of money, whereas *r* and *t* do not. For example, consider the interest rate *r* and the interest *I*. Frequently, people confuse these two. However, the interest rate *r* is a percentage, and the interest *I* is an amount of money; Tom and Betty's interest rate is *r* = 5.1%, but their interest is *I* = $15,300. To emphasize this distinction, we will always use capital letters for variables that measure amounts of money and lowercase letters for other variables. We hope this notation will help you avoid substituting 5.1% = 0.051 for *I* when it should be substituted for *r*.

Finding the Number of Days: "Through" versus "To"

Clearly, there is only one day from January 1 to January 2. If the answer weren't so obvious, we could find it by subtracting:

$2 - 1 = 1$ day

This count of days includes January 1 but not January 2.

In contrast, there are *two* days from January 1 *through* January 2. The word "through" means to include both the beginning day (January 1) and the ending day (January 2). If the answer weren't so obvious, we could find it by doing the "January 1 to January 2" calculation above and then adding the one ending day:

$$\underbrace{2 - 1}_{\substack{\textbf{the 1st} \\ \textbf{to the 2nd}}} + \underbrace{1}_{\substack{\textbf{adding the one} \\ \textbf{ending day}}} = 2$$

The number of days from August 2 to August 30 is

$$30 - 2 = 28 \text{ days}$$

This count includes August 2 but not August 30.
The number of days from August 2 *through* August 30 is

$$30 - 2 + 1 = 29 \text{ days}$$

This count includes August 2 and August 30.

Short Term Loans

One of the more common uses of simple interest is a short-term (such as a year or less) loan that requires a single lump sum payment at the end of the term. Businesses routinely obtain these loans to purchase equipment or inventory, to pay operating expenses, or to pay taxes. A **lump sum** payment is a single payment that pays off an entire loan. Some loans require smaller monthly payments rather than a single lump sum payment. We will investigate that type of loan later in this section and in Section 5.4.

EXAMPLE **2**

SOLUTION

USING THE SIMPLE INTEREST FUTURE VALUE FORMULA Espree Clothing borrowed \$185,000 at $7\frac{1}{4}\%$ from January 1 to February 28.

a. Find the future value of the loan.
b. Interpret the future value.

a. • We are given:

$$P = 185{,}000$$
$$r = 7\tfrac{1}{4}\% = 0.0725$$

• Finding t:
  • January has 31 days.
  • February has 28 days.
  • There are $31 + 28 - 1 = 58$ days from January 1 to February 28.
  • $t = 58$ days

$$= 58 \text{ days} \cdot \frac{1 \text{ year}}{365 \text{ days}} = \frac{58}{365} \text{ years} \qquad \textbf{using dimensional analysis (see Appendix E)}$$

• Finding FV:

$$FV = P(1 + rt) \qquad \textbf{the Simple Interest Future Value}$$

$$= 185{,}000\left(1 + 0.0725 \cdot \frac{58}{365}\right) \qquad \textbf{Formula}$$

$$= 187{,}131.301\ldots \qquad \textbf{substituting for } P, r, \text{ and } t$$

$$\approx \$187{,}131.30 \qquad \textbf{rounding to the nearest penny}$$

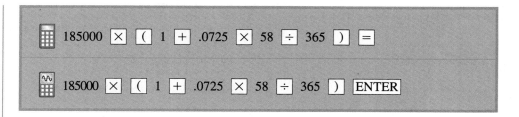

b. This means that Espree has agreed to make a lump sum payment to their lender $187,168.05 on February 28.

> As a rough check, notice that the answer is somewhat higher than the original $185,000. This is as it should be. The future value includes the original $185,000 plus interest.

Notice that in Example 2, we do not use $t = 2$ months $= \frac{2}{12}$ year. If we did, we would get $187,235.42, an inaccurate answer. Instead, we use the number of days, converted to years.

In Example 2, we naturally used 365 days per year, but some institutions traditionally count a year as 360 days and a month as thirty days (especially if that tradition works in their favor). This is a holdover from the days before calculators and computers—the numbers were simply easier to work with. Also, we used normal round-off rules to round $187,131.301 . . . to $187,131.30; some institutions round off some interest calculations in their favor. In this book, we will count a year as 365 days and use normal round-off rules (unless stated otherwise).

The issue of how a financial institution rounds off its interest calculations might seem unimportant. After all, we are talking about a difference of a fraction of a penny. However, consider one classic form of computer crime: the round-down fraud, performed on a computer system that processes a large number of accounts. Frequently, such systems use the normal round-off rules in their calculations and keep track of the difference between the theoretical account balance if no rounding off is done and the actual account balance with rounding off. Whenever that difference reaches or exceeds 1¢, the extra penny is deposited in (or withdrawn from) the account. A fraudulent computer programmer can write the program so that the extra penny is deposited in his or her own account. This fraud is difficult to detect because the accounts appear to be balanced. While each individual gain is small, the total gain can be quite large if a large number of accounts is processed regularly.

A written contract signed by the lender and the borrower is called a **loan agreement** or a **note**. The **maturity value** of the note (or just the **value** of the note) refers to the note's future value. Thus, the value of the note in Example 2 was $187,168.05. This is what the note is worth to the lender in the future—that is, when the note matures.

National Debt

In almost every year since 1931, the U.S. federal budget called for deficit spending, that is, spending more money than is received. In 2008, the total U.S. federal debt was $10,025 billion, and we paid about $214 billion for interest on that debt. In 2009, the debt grew to $11,343 billion, or about $37,000 per person in the United States.

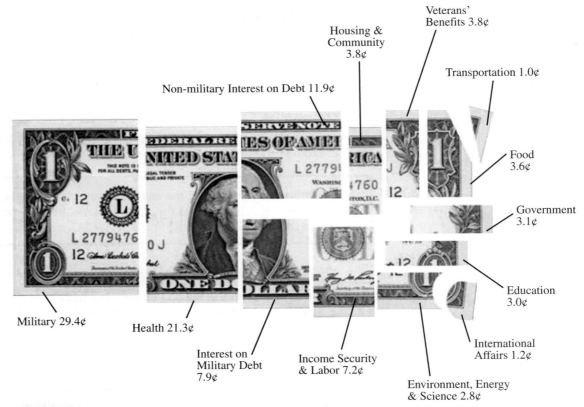

Non-military Interest on Debt 11.9¢

Housing &
Community
3.8¢

Veterans'
Benefits 3.8¢

Transportation 1.0¢

Food
3.6¢

Government
3.1¢

Education
3.0¢

International
Affairs 1.2¢

Military 29.4¢

Health 21.3¢

Interest on
Military Debt
7.9¢

Income Security
& Labor 7.2¢

Environment, Energy
& Science 2.8¢

FIGURE 5.1 Where your 2008 federal income tax dollar was spent. *Source:* National Priorities Project.

EXAMPLE 3

THE NATIONAL DEBT   Find the simple interest rate that was paid on the 2008 national debt.

SOLUTION

• We are given

$$I = \$214 \text{ billion}$$
$$P = \$10{,}025 \text{ billion}$$
$$t = 1 \text{ year}$$

• Finding *r*:

$$I = Prt \qquad\qquad \textbf{the Simple Interest Formula}$$
$$\$214 \text{ billion} = \$10{,}025 \text{ billion} \cdot r \cdot 1$$
$$r = \$21 \text{ billion}/\$10{,}025 \text{ billion} \cdot r \cdot 1 \quad \textbf{substituting solving for } r$$
$$= 0.0213 \ldots \approx 2.1\% \qquad\qquad \textbf{rounding}$$

Present Value

EXAMPLE 4

FINDING HOW MUCH TO INVEST NOW   Find the amount of money that must be invested now at a $5\frac{3}{4}\%$ simple interest so that it will be worth $1,000 in two years.

SOLUTION

We are asked to find the principal P that will generate a future value of $1,000.

- We are given

$$FV = \$1,000$$
$$r = 5\tfrac{3}{4}\% = 0.0575$$
$$t = 2 \text{ years}$$

- Finding P:

$$FV = P(1 + rt) \qquad \text{the Simple Interest Future Value Formula}$$
$$1,000 = P(1 + 0.0575 \cdot 2) \qquad \text{substituting}$$
$$P = \frac{1,000}{1 + 0.0575 \cdot 2} \qquad \text{solving for } P$$
$$= 896.86099\ldots \qquad \text{rounding}$$
$$\approx \$896.86$$

As a rough check, notice that the answer is somewhat smaller than $1,000. This is as it should be—the principal does not include interest, so it should be smaller than the future value of $1,000.

In Example 4, the investment is worth $1,000 two years *in the future;* that is, $1000 is the investment's *future value.* But the same investment is worth $896.86 *in the present.* For this reason, we say that $896.86 is the investment's **present value.** In this case, this is the same thing as the principal; it is just called the *present value* to emphasize that this is its value in the present.

Add-on Interest

An **add-on interest loan** is an older type of loan that was common before calculators and computers, because calculations of such loans can easily be done by hand. With this type of loan, the total amount to be repaid is computed with the Simple Interest Future Value Formula. The payment is found by dividing that total amount by the number of payments.

Add-on interest loans have generally been replaced with the more modern amortized loans (see Section 5.4). However, they are not uncommon at auto lots that appeal to buyers who have poor credit histories.

EXAMPLE **5**

AN ADD-ON INTEREST LOAN   Chip Douglas's car died, and he must replace it right away. Centerville Auto Sales has a nine-year-old Ford that's "like new" for $5,988. The sign in their window says, "Bad credit? No problem!" They offered Chip a 5% two-year add-on interest loan if he made a $600 down payment. Find the monthly payment.

SOLUTION

The loan amount is $P = 5,988 - 600 = 5,388$. The total amount due is

$$FV = P(1 + rt) \quad \text{the Simple Interest Future Value Formula}$$
$$= 5,388(1 + 0.05 \cdot 2) \quad \text{substituting}$$
$$= 5,926.80$$

The total amount due is spread out over twenty-four monthly payments, so the monthly payment is $5,926.8/24 = $246.95. This means that Chip has to pay $600 when he purchases the car and then $246.95 a month for twenty-four months.

Credit Card Finance Charge

Credit cards have become part of the American way of life. Purchases made with a credit card are subject to a finance charge, but there is frequently a grace period, and no finance charge is assessed if full payment is received by the payment due date. One of the most common methods of calculating credit card interest is the **average daily balance** method. To find the average daily balance, the balance owed on the account is found for each day in the billing period, and the results are averaged. The finance charge consists of simple interest, charged on the result.

EXAMPLE **6**

FINDING A CREDIT CARD FINANCE CHARGE The activity on Tom and Betty's Visa account for one billing is shown below. The billing period is October 15 through November 14, the previous balance was $346.57, and the annual in rate is 21%.

October 21	payment	$50.00
October 23	restaurant	$42.14
November 7	clothing	$18.55

a. Find the average daily balance.
b. Find the finance charge.

SOLUTION

a. To find the average daily balance, we have to know the balance for each day in the billing period and the number of days at that balance, as shown in Figure 5.2. The average daily balance is then the weighted average of each daily balance, with each balance weighted to reflect the number of days at that balance.

Time Interval	Days	Daily Balance
October 15–20	$20 - 14 = 6$	$346.57
October 21–22	$22 - 20 = 2$	$346.57 - $50 = $296.57
October 23–November 6 (October has 31 days)	$31 - 22 = 9$ $9 + 6 = 15$	$296.57 + $42.14 = $338.71
November 7–14	$14 - 6 = 8$	$338.71 + $18.55 = $357.26

FIGURE 5.2 Preparing to find the average daily balance.

$$\text{Average daily balance} = \frac{6 \cdot 346.57 + 2 \cdot 296.57 + 15 \cdot 338.71 + 8 \cdot 357.26}{6 + 2 + 15 + 8}$$
$$= 342.29967\ldots$$
$$\approx \$342.30$$

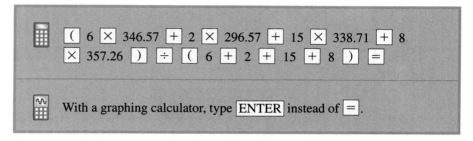

With a graphing calculator, type ENTER instead of =.

b. • We are given

$P = \$342.29967\ldots$

$r = 21\% = 0.21$

$t = 31$ days $= 31/365$ years

• Finding I:

$I = Prt$	**the Simple Interest Formula**
$= 342.29967\ldots \cdot 0.21 \cdot 31/365$	**substituting**
$= 6.1051258\ldots$	**multiplying**
$\approx \$6.11$	**rounding**

CREDIT CARD FINANCE CHARGE

To find the credit card finance charge with the average daily balance method, do the following:

1. Find the balance for each day in the billing period and the number of days at that balance.
2. The average daily balance is the weighted average of these daily balances, weighted to reflect the number of days at that balance.
3. The finance charge is simple interest applied to the average daily balance.

How Many Days?

Financial calculations usually involve computing the term t either in a whole number of years or in the number of days converted to years. This requires that you know the number of days in each month. As you can see from Figure 5.3, the months alternate between thirty-one days and thirty days, with two exceptions:

• February has twenty-eight days (twenty-nine in leap years).
• The alternation does not happen from July to August.

January	31 days	May	31 days	September	30 days
February	28 days	June	30 days	October	31 days
March	31 days	July	31 days	November	30 days
April	30 days	August	31 days	December	31 days

FIGURE 5.3 The number of days in a month.

HISTORICAL NOTE

THE HISTORY OF CREDIT CARDS

Library of Congress, Prints and Photographs Division

Credit cards and charge cards were first used in the United States in 1915, when Western Union issued a metal card to some of its regular customers. Holders of these cards were allowed to defer their payments and were assured of prompt and courteous service. Shortly thereafter, several gasoline companies, hotels, department stores, and railroads issued cards to their preferred customers.

The use of credit and charge cards virtually ceased during World War II, owing to government restrictions on credit. In 1950, a New York lawyer established the Diners' Club after being embarrassed when he lacked sufficient cash to pay a dinner bill. A year later, the club had billed more than $1 million. Carte Blanche and the American Express card soon followed. These "travel and entertainment" cards were and are attractive to the public because they provide a convenient means of paying restaurant, hotel, and airline bills. They eliminate both the possibility of an out-of-town check being refused and the need to carry a large amount of cash.

were accepted for use only in relatively small geographical areas. In 1965, the California-based Bank of America began licensing other banks (both in the United States and abroad) to issue BankAmericards. Other banks formed similar groups, which allowed customers to use the cards out of state.

In 1970, the Bank of America transferred administration of its bank card program to a new company owned by the banks that issued the card. In 1977, the card was renamed Visa. MasterCard was originally created by several California banks (United California Bank, Wells Fargo, Crocker Bank, and the Bank of California) to compete with BankAmericard. It was then licensed to the First

The first bank card was issued in 1951 by Franklin National Bank in New York. Within a few years, 100 other banks had followed suit. Because these bank cards were issued by individual banks, they

National Bank of Louisville, Kentucky and the Marine Midland Bank of New York.

Credit cards are now a part of the American way of life. In 2004, we started to use credit cards, debit cards and other forms of electronic bill paying more than we use paper checks. Credit card debt has soared, especially among college students who are already in debt with college loans. Credit card issuers have been accused of targeting college students, who tend to make minimum payments and thus incur much more interest. See Exercises 45–48.

Library of Congress, Prints & Photographs Division

Early Diners' Club, Carte Blanche and American Express cards

5.1 EXERCISES

In Exercises 1–4, find the number of days.

1. **a.** September 1 to October 31 of the same year
    b. September 1 through October 31 of the same year
2. **a.** July 1 to December 31 of the same year
    b. July 1 through December 31 of the same year
3. **a.** April 1 to July 10 of the same year
    b. April 1 through July 10 of the same year
4. **a.** March 10 to December 20 of the same year
    b. March 10 through December 20 of the same year

In Exercises 5–10, find the simple interest of the given loan amount.

5. $2,000 borrowed at 8% for three years
6. $35,037 borrowed at 6% for two years
7. $420 borrowed at $6\frac{3}{4}$% for 325 days
8. $8,950 borrowed at $9\frac{1}{2}$% for 278 days
9. $1,410 borrowed at $12\frac{1}{4}$% from September 1:
    a. to October 31 of the same year
    b. through October 31 of the same year

10. $5,682 borrowed at $11\frac{3}{4}\%$ from July 1:

    a. to December 31 of the same year

    b. through December 31 of the same year

In Exercises 11–14, find the future value of the given present value.

11. Present value of $3,670 at $2\frac{3}{4}\%$ for seven years

12. Present value of $4,719 at 14.1% for eleven years

13. Present value of $12,430 at $5\frac{7}{8}\%$ for 660 days

14. Present value of $172.39 at 6% for 700 days

In Exercises 15–20, find the maturity value of the given loan amount.

15. $1,400 borrowed at $7\frac{1}{8}\%$ for three years

16. $3,250 borrowed at $8\frac{1}{2}\%$ for four years

17. $5,900 borrowed at $14\frac{1}{2}\%$ for 112 days

18. $2,720 borrowed at $12\frac{3}{4}\%$ for 275 days

19. $16,500 borrowed at $11\frac{7}{8}\%$ from April 1 through July 10 of the same year

20. $2,234 borrowed at $12\frac{1}{8}\%$ from March 10 through December 20 of the same year

In Exercises 21–26 find the present value of the given future value.

21. Future value $8,600 at $9\frac{1}{2}\%$ simple interest for three years

22. Future value $420 at $5\frac{1}{2}\%$ simple interest for two years

23. Future value $1,112 at $3\frac{5}{8}\%$ simple interest for 512 days

24. Future value $5,750 at $4\frac{7}{8}\%$ simple interest for 630 days

25. Future value $1,311 at $6\frac{1}{2}\%$ simple interest from February 10 to October 15 of the same year

26. Future value $4,200 at $6\frac{3}{4}\%$ simple interest from April 12 to November 28 of the same year

27. If you borrow $1,000 at 8.5% interest and the loan requires a lump sum payment of $1,235.84, what is the term of the loan?

28. If you borrow $1,700 at 5.25% interest and the loan requires a lump sum payment of $1,863.12, what is the term of the loan?

29. The Square Wheel Bicycle store has found that they sell most of their bikes in the spring and early summer. On March 1, they borrowed $226,500 to buy bicycles. They are confident that they can sell most of these bikes by August 1. Their loan is at $6\frac{7}{8}\%$ interest. What size lump sum payment would they have to make on August 1 to pay off the loan?

30. Ernie Bilko has a business idea. He wants to rent an abandoned gas station for just the months of November and December. He will convert the gas station into a drive-through Christmas wrapping station. Customers will drive in, drop off their gifts, return the next day, and pick up their wrapped gifts. He needs $338,200 to rent the gas station, purchase wrapping paper, hire workers, and advertise. If he borrows this amount at $6\frac{1}{2}\%$ interest for those two months, what size lump sum payment will he have to make to pay off the loan?

31. Alice Cohen buys a two-year-old Honda from U-Pay-Less-Cars for $19,000. She puts $500 down and finances the rest through the dealer at 13% add-on interest. If she agrees to make thirty-six monthly payments, find the size of each payment.

32. Sven Lundgren buys a three-year-old Chevrolet from Skunk Motors for $14,600. He puts $300 down and finances the rest through the dealer at 12.5% add-on interest. If he agrees to make twenty-four monthly payments, find the size of each payment.

33. Ray and Teresa Martinez buy a used car from Fowler's Wholesale 2U for $6,700. They put $500 down and finance the rest through the car lot at 9.8% add-on interest. If they make thirty-six monthly payments, find the size of each payment.

34. Dick Davis buys a five-year-old used Toyota from Pioneer Auto Sales for $7,999. He puts $400 down and finances the rest through the car lot at 7.7% add-on interest. He agrees to make thirty-six monthly payments. Find the size of each payment.

In Exercises 35–36, use the following information. When "trading up," it is preferable to sell your old house before buying your new house because that allows you to use the proceeds from selling your old house to buy your new house. When circumstances do not allow this, the homeowner can take out a bridge loan.

35. Dale and Claudia have sold their house, but they will not get the proceeds from the sale for an estimated $2\frac{1}{2}$ months. The owner of the house they want to buy will not hold the house that long. Dale and Claudia have two choices: let their dream house go or take out a bridge loan. The bridge loan would be for $110,000, at 7.75% simple interest, due in ninety days.

    a. How big of a check would they have to write in 90 days?

    b. How much interest would they pay for this loan?

36. Tina and Mike have sold their house, but they will not get the proceeds from the sale for an estimated 3 months. The owner of the house they want to buy will not hold the house that long. Tina and Mike have two choices: let their dream house go or take out a bridge loan. The bridge loan would be for $85,000, at 8.5% simple interest, due in 120 days.

    a. How big of a check would they have to write in 120 days?

    b. How much interest would they pay for this loan?

37. The activity on Stuart Ratner's Visa account for one billing period is shown below. Find the average daily balance and the finance charge if the billing period is

April 11 through May 10, the previous balance was $126.38, and the annual interest rate is 18%.

April 15	payment	$15.00
April 22	DVD store	$25.52
May 1	clothing	$32.18

38. The activity on Marny Zell's MasterCard account for one billing period is shown below. Find the average daily balance and the finance charge if the billing period is June 26 through July 25, the previous balance was $396.68, and the annual interest rate is 19.5%.

June 30	payment	$100.00
July 2	gasoline	$36.19
July 10	restaurant	$53.00

39. The activity on Denise Hellings' Sears account for one billing period is shown below. Find the average daily balance and the finance charge if the billing period is March 1 through March 31, the previous balance was $157.14, and the annual interest rate is 21%.

| March 5 | payment | $25.00 |
| March 17 | tools | $36.12 |

40. The activity on Charlie Wilson's Visa account for one billing period is shown below. Find the average daily balance and the finance charge if the billing period is November 11 through December 10, the previous balance was $642.38, and the annual interest rate is 20%.

November 15	payment	$150
November 28	office supplies	$23.82
December 1	toy store	$312.58

41. Donovan and Pam Hamilton bought a house from Edward Gurney for $162,500. In lieu of a 10% down payment, Mr. Gurney accepted 5% down at the time of the sale and a promissory note from the Hamiltons for the remaining 5%, due in four years. The Hamiltons also agreed to make monthly interest payments to Mr. Gurney at 10% interest until the note expires. The Hamiltons obtained a loan from their bank for the remaining 90% of the purchase price. The bank in turn paid the sellers the remaining 90% of the purchase price, less a sales commission of 6% of the purchase price, paid to the sellers' and the buyers' real estate agents.

a. Find the Hamiltons' down payment.

b. Find the amount that the Hamiltons borrowed from their bank.

c. Find the amount that the Hamiltons borrowed from Mr. Gurney.

d. Find the Hamilton's monthly interest-only payment to Mr. Gurney.

e. Find Mr. Gurney's total income from all aspects of the down payment (including the down payment, the amount borrowed under the promissory note, and the monthly payments required by the promissory note).

f. Find Mr. Gurney's net income from the Hamiltons' bank.

g. Find Mr. Gurney's total income from all aspects of the sale.

42. George and Peggy Fulwider bought a house from Sally Sinclair for $233,500. In lieu of a 10% down payment, Ms. Sinclair accepted 5% down at the time of the sale and a promissory note from the Fulwiders for the remaining 5%, due in four years. The Fulwiders also agreed to make monthly interest payments to Ms. Sinclair at 10% interest until the note expires. The Fulwiders obtained a loan from their bank for the remaining 90% of the purchase price. The bank in turn paid the sellers the remaining 90% of the purchase price, less a sales commission of 6% of the purchase price, paid to the sellers' and the buyers' real estate agents.

a. Find the Fulwiders' down payment.

b. Find the amount that the Fulwiders borrowed from their bank.

c. Find the amount that the Fulwiders borrowed from Ms. Sinclair.

d. Find the Fulwiders' monthly interest-only payment to Ms. Sinclair.

e. Find Ms. Sinclair's total income from all aspects of the down payment (including the down payment, the amount borrowed under the promissory note, and the monthly payments required by the promissory note).

f. Find Ms. Sinclair's net income from the Fulwiders' bank.

g. Find Ms. Sinclair's total income from all aspects of the sale.

43. The Obamas bought a house from the Bushes for $389,400. In lieu of a 20% down payment, the Bushes accepted a 10% down payment at the time of the sale and a promissory note from the Obamas for the remaining 10%, due in 4 years. The Obamas also agreed to make monthly interest payments to the Bushes at 11% interest until the note expires. The Obamas obtained a loan for the remaining 80% of the purchase price from their bank. The bank in turn paid the sellers the remaining 80% of the purchase price, less a sales commission of 6% of the sales price paid to the sellers' and the buyers' real estate agents.

a. Find the Obamas' down payment.

b. Find the amount that the Obamas borrowed from their bank.

c. Find the amount that the Obamas borrowed from the Bushes.

d. Find the Obamas' monthly interest-only payment to the Bushes.

e. Find the Bushes' total income from all aspects of the down payment (including the down payment, the amount borrowed under the promissory note, and the monthly payments required by the note).

f. Find the Bushes' income from the Obamas' bank.

g. Find the Bushes' total income from all aspects of the sale.

44. Sam Needham bought a house from Sheri Silva for $238,300. In lieu of a 20% down payment, Ms. Silva accepted a 10% down payment at the time of the sale and a promissory note from Mr. Needham for the remaining 10%, due in four years. Mr. Needham also agreed to make monthly interest payments to Ms. Silva at 9% interest until the note expires. Mr. Needham obtained a loan for the remaining 80% of the purchase price from his bank. The bank in turn paid Ms. Silva the remaining 80% of the purchase price, less a sales commission (of 6% of the sales price) paid to the sellers' and the buyers' real estate agents.

a. Find Mr. Needham's down payment.

b. Find the amount that Mr. Needham borrowed from his bank.

c. Find the amount that Mr. Needham borrowed from the Ms. Silva.

d. Find Mr. Needham's monthly interest-only payment to Ms. Silva.

e. Find Ms. Silva's total income from all aspects of the down payment (including the down payment, the amount borrowed under the promissory note, and the monthly payments required by the note).

f. Find Ms. Silva's income from Mr. Needham's bank.

g. Find Ms. Silva's total income from all aspects of the sale.

45. Your credit card has a balance of $1,000. Its interest rate is 21%. You have stopped using the card, because you don't want to go any deeper into debt. Each month, you make the minimum required payment of $20.

a. During the January 10 through February 9 billing period, you pay the minimum required payment on January 25th. Find the average daily balance, the finance charge and the new balance. (The new balance includes the finance charge.)

b. During the February 10 through March 9 billing period, you pay the minimum required payment on February 25th. Find the average daily balance, the finance charge and the new balance. (The new balance includes the finance charge.)

c. During the March 10 through April 9 billing period, you pay the minimum required payment on March 25th. Find the average daily balance, the finance charge and the new balance. (The new balance includes the finance charge.)

d. Discuss the impact of making the minimum required payment, both on yourself and on the credit card issuer.

46. Your credit card has a balance of $1,200. Its interest rate is 20.5%. You have stopped using the card, because you don't want to go any deeper into debt. Each month, you make the minimum required payment of $24.

a. During the September 10 through October 9 billing period, you pay the minimum required payment on September 25th. Find the average daily balance, the finance charge and the new balance. (The new balance includes the finance charge.)

b. During the October 10 through November 9 billing period, you pay the minimum required payment on October 25th. Find the average daily balance, the finance charge and the new balance. (The new balance includes the finance charge.)

c. During the November 10 through December 9 billing period, you pay the minimum required payment on November 25th. Find the average daily balance, the finance charge and the new balance. (The new balance includes the finance charge.)

d. Discuss the impact of making the minimum required payment, both on yourself and on the credit card issuer.

47. Your credit card has a balance of $1,000. Its interest rate is 21%. You have stopped using the card, because you don't want to go any deeper into debt. Each month, you make the minimum required payment. Your credit card issuer recently changed their minimum required payment policy, in response to the Bankruptcy Abuse Prevention and Consumer Protection Act of 2005. As a result, your minimum required payment is $40.

a. During the January 10 through February 9 billing period, you pay the minimum required payment on January 25th. Find the average daily balance, the finance charge and the new balance. (The new balance includes the finance charge.)

b. During the February 10 through March 9 billing period, you pay the minimum required payment on February 25th. Find the average daily balance, the finance charge and the new balance. (The new balance includes the finance charge.)

c. During the March 10 through April 9 billing period, you pay the minimum required payment on March 25th. Find the average daily balance, the finance charge and the new balance. (The new balance includes the finance charge.)

d. Compare the results of parts (a) through (c) with those of Exercise 45. Discuss the impact of the credit card issuer's change in their minimum required payment policy.

48. Your credit card has a balance of $1,200. Its interest rate is 20.5%. You have stopped using the card, because you don't want to go any deeper into debt. Each month, you make the minimum required payment. Your credit card issuer recently changed their minimum required payment policy, in response to the Bankruptcy Abuse Prevention and Consumer Protection Act of 2005. As a result, your minimum required payment is $48.

a. During the September 10 through October 9 billing period, you pay the minimum required payment on September 25th. Find the average daily balance, the finance charge and the new balance. (The new balance includes the finance charge.)

b. During the October 10 through November 9 billing period, you pay the minimum required payment on October 25th. Find the average daily balance, the finance charge and the new balance. (The new balance includes the finance charge.)

c. During the November 10 through December 9 billing period, you pay the minimum required payment on November 25th. Find the average daily balance, the finance charge and the new balance. (The new balance includes the finance charge.)

d. Compare the results of parts (a) through (c) with those of Exercise 46. Discuss the impact of the credit card issuer's change in their minimum required payment policy.

Answer the following questions using complete sentences and your own words.

• **CONCEPT QUESTIONS**

49. Could Exercises 5–21 all be done with the Simple Interest Formula? If so, how? Could Exercises 5–21 all be done with the Simple Interest Future Value Formula? If so, how? Why do we have both formulas?

50. Which is always higher: future value or principal? Why?

• **HISTORY QUESTIONS**

51. Who offered the first credit card?
52. What was the first post–World War II credit card?
53. Who created the first post–World War II credit card?
54. What event prompted the creation of the first post–World War II credit card?
55. What was the first interstate bank card?

5.2 Compound Interest

OBJECTIVES

● Understand the difference between simple interest and compound interest

● Use the Compound Interest Formula

● Understand and compute the annual yield

Many forms of investment, including savings accounts, earn **compound interest,** in which interest is periodically paid on both the original principal and previous interest payments. This results in earnings that are significantly higher over a longer period of time. It is important that you understand this difference in order to make wise financial decisions. We'll explore compound interest in this section.

Compound Interest as Simple Interest, Repeated

EXAMPLE 1

UNDERSTANDING COMPOUND INTEREST Tom and Betty deposit $1,000 into their new bank account. The account pays 8% interest compounded quarterly. This means that interest is computed and deposited every quarter of a year. Find the account balance after six months, using the Simple Interest Future Value formula to compute the balance at the end of each quarter.

SOLUTION At the end of the first quarter, $P = \$1,000$, $r = 8\% = 0.08$, and $t =$ one quarter or $\frac{1}{4}$ year.

$$FV = P(1 + rt) \qquad \text{the Simple Interest Future Value Formula}$$
$$= 1{,}000(1 + 0.08 \cdot \tfrac{1}{4}) \quad \text{substituting}$$
$$= 1{,}000(1 + 0.02)$$
$$= \$1{,}020$$

This means that there is \$1,020 in Tom and Betty's account at the end of the first quarter. It also means that the second quarter's interest will be paid on this new principal. So at the end of the second quarter, $P = \$1{,}020$ and $r = 0.08$. Note that $t = \tfrac{1}{4}$, not $\tfrac{2}{4}$, because we are computing interest for one quarter.

$$FV = P(1 + rt) \qquad \text{the Simple Interest Future Value Formula}$$
$$= 1{,}020(1 + 0.08 \cdot \tfrac{1}{4}) \quad \text{substituting}$$
$$= 1{,}020(1 + 0.02)$$
$$= \$1{,}040.40$$

At the end of six months, the account balance is \$1,040.40.

> As a rough check, notice that the future value is slightly higher than the principal, as it should be.

The Compound Interest Formula

This process would become tedious if we were computing the balance after twenty years. Because of this, compound interest problems are usually solved with their own formula.

The **compounding period** is the time period over which any one interest payment is calculated. In Example 1, the compounding period was a quarter of a year. For each quarter's calculation, we multiplied the annual rate of 8% (0.08) by the time 1 quarter ($\tfrac{1}{4}$ year) and got 2% (0.02). This 2% is the **quarterly rate** (or more generally, the *periodic rate*). A **periodic rate** is any rate that is prorated in this manner from an annual rate.

If i is the periodic interest rate, then the future value at the end of the first period is

$$FV = P(1 + i)$$

Because this is the account balance at the beginning of the second period, it becomes the new principal. The account balance at the end of the second period is

$$FV = P(1 + i) \cdot (1 + i) \quad \text{substituting } P(1 + i) \text{ for } P$$
$$= P(1 + i)^2$$

This means that $P(1 + i)^2$ is the account balance at the beginning of the third period, and the future value at the end of the third period is

$$FV = \left[P(1 + i)^2\right] \cdot (1 + i) \quad \text{substituting } P(1 + i)^2 \text{ for } P$$
$$= P(1 + i)^3$$

If we generalize these results, we get the Compound Interest Formula.

COMPOUND INTEREST FORMULA

If initial principal P earns compound interest at a periodic interest rate i for n periods, the future value is
$$FV = P(1 + i)^n$$

Notice that we have maintained our variables tradition: i and n do not measure amounts of money, so they are not capital letters. We now have three interest-related variables:

* r, the annual interest rate (not an amount of money)
* i, the periodic interest rate (not an amount of money)
* I, the interest itself (an amount of money)

EXAMPLE **2**

USING THE COMPOUND INTEREST FORMULA Use the Compound Interest Formula to recompute Tom and Betty's account balance from Example 1.

SOLUTION

Their 8% interest is compounded quarterly, so each quarter they earn a quarter of $8\% = \frac{1}{4} \cdot 8\% = 2\% = 0.02$. Also, n counts the number of quarters, so $n = 2$.

$$FV = P(1 + i)^n \qquad \text{the Compound Interest Formula}$$
$$= 1{,}000(1 + 0.02)^2 = \$1{,}040.40 \quad \text{substituting}$$

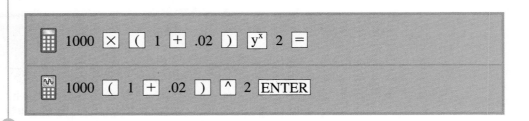

Compound Interest Compared with Simple Interest

As we saw at the beginning of this section, compound interest is just simple interest, repeated. However, there can be profound differences in their results.

EXAMPLE **3**

COMPOUND INTEREST OVER A LONG PERIOD OF TIME In 1777, it looked as though the Revolutionary War was about to be lost. George Washington's troops were camped at Valley Forge. They had minimal supplies, and the winter was brutal. According to a 1990 class action suit, Jacob DeHaven, a wealthy Pennsylvania merchant, saved Washington's troops and the revolutionary cause by loaning Washington $450,000. The suit, filed by DeHaven's descendants, asked the government to repay the still-outstanding loan plus compound interest at the then-prevailing rate of 6%. (*Source: New York Times*, May 27, 1990.) How much did the government owe on the 1990 anniversary of the loan if the interest is compounded monthly?

SOLUTION

The principal is $P = \$450{,}000$. If 6% interest is paid each year, then 1/12th of 6% is paid each month, and $i = \frac{1}{12} \cdot 6\% = 0.06/12$. The term is

$$n = 213 \text{ years}$$
$$= 213 \text{ years} \cdot \frac{12 \text{ months}}{1 \text{ year}} \qquad \text{using dimensional analysis}$$
$$= 2{,}556 \text{ months}$$

Using the Compound Interest Formula, we get

$$FV = P(1 + i)^n \qquad \text{the Compound Interest Formula}$$
$$= 450{,}000(1 + 0.06/12)^{2556} \quad \text{substituting}$$
$$= 1.547627234 \ldots \times 10^{11}$$
$$\approx \$154{,}762{,}723{,}400 \qquad \text{rounding}$$

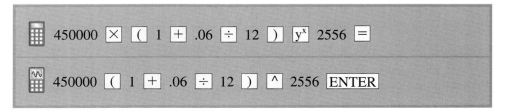

The DeHavens sued for this amount, but they also stated that they were willing to accept a more reasonable payment.

> As a rough check, notice that the future value is much higher than the principal, as it should be because of the long time period.

Some sources have questioned the DeHavens' claim: "There is also no evidence to support the claim of the DeHaven family that their ancestor Jacob DeHaven lent George Washington \$450,000 in cash and supplies while the army was encamped at Valley Forge. This tradition first appeared in print in a history of the DeHaven family penned by Howard DeHaven Ross. Periodically, the descendants of Jacob DeHaven make attempts to get the "loan" repaid with interest. . . . This remarkably persistent tradition has been thoroughly debunked by Judith A. Meier, of the Historical Society of Montgomery County, whose genealogical research revealed that there were no DeHavens living in the immediate area until after 1790 and that Jacob DeHaven had never been rich enough to make such a fabulous loan." (*Source:* Lorett Treese, *Valley Forge: Making and Remaking a National Symbol,* University Park, PA: Pennsylvania University Press, 1995.)

EXAMPLE 4

COMPARING SIMPLE INTEREST WITH COMPOUND INTEREST OVER A LONG PERIOD OF TIME   How much would the government have owed the DeHavens if the interest was simple interest?

SOLUTION

With simple interest, we use r and t, which are *annual* figures, rather than i and n, which are *periodic* figures. So $r = 6\% = 0.06$, and $t = 213$ years.

$$FV = P(1 + rt) \qquad \text{the Simple Interest Future Value Formula}$$
$$= 450000(1 + 0.06 \cdot 213) \quad \text{substituting}$$
$$= \$6{,}201{,}000$$

Simple interest would have required a payment of only \$6 million. This is a lot, but not in comparison with the \$155 billion payment required by compound interest.

In Examples 3 and 4, the future value with compound interest is almost 25,000 times the future value with simple interest. Over longer periods of time, compound interest is immensely more profitable to the investor than simple interest, because compound interest gives interest on interest. Similarly, compounding more frequently (daily rather than quarterly, for example) is more profitable to the investor.

The effects of the size of the time interval and the compounding period can be seen in Figure 5.4.

Type of Interest	Future Value of $1000 at 10% Interest		
	After 1 Year	After 10 Years	After 30 Years
Simple interest	$1100.00	$2000.00	$ 4,000.00
Compounded annually	$1100.00	$2593.74	$17,449.40
Compounded quarterly	$1103.81	$2685.06	$19,358.15
Compounded monthly	$1104.71	$2707.04	$19,837.40
Compounded daily	$1105.16	$2717.91	$20,077.29

FIGURE 5.4 Comparing simple interest and compound interest.

Finding the Interest and the Present Value

EXAMPLE 5

FINDING THE AMOUNT OF INTEREST EARNED Betty's boss paid her an unexpected bonus of $2,500. Betty and her husband decided to save the money for their daughter's education. They deposited it in account that pays 10.3% interest compounded daily. Find the amount of interest that they would earn in fifteen years by finding the future value and subtracting the principal.

SOLUTION

Compounding daily, we have $P = 2500, i = 1/365$th of $10.3\% = 0.103/365$, and $n = 15$ years $= 15$ years $\cdot$ 365 days/year $= 5,475$ days.

$$FV = P(1 + i)^n \qquad \textbf{the Compound Interest Formula}$$
$$= 2,500(1 + 0.103/365)^{5475} \quad \textbf{substituting}$$
$$= 11,717.374 \ldots$$
$$\approx \$11,717.37$$

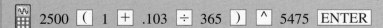

2500 ⊠ (1 ⊞ .103 ÷ 365) y^x 5475 =

2500 (1 ⊞ .103 ÷ 365) ^ 5475 ENTER

Warning: If you compute $\frac{0.103}{365}$ separately, you will get a long decimal. Do not round off that decimal, because the resulting answer will be inaccurate. Doing the calculation all at once (as shown above) avoids this difficulty.

The principal is $2,500, and the total of principal and interest is $11,717.37. Thus, the interest is $11,717.37 - \$2500 = \$9,217.37$.

EXAMPLE **6**

FINDING THE PRESENT VALUE Find the amount of money that must be invested now at $7\frac{3}{4}\%$ interest compounded annually so that it will be worth \$2,000 in 3 years.

SOLUTION

The question actually asks us to find the present value, or principal P, that will generate a future value of \$2,000. We have $FV = 2{,}000$, $i = 7\frac{3}{4}\% = 0.0775$, and $n = 3$.

$$FV = P(1 + i)^n \qquad \text{the Compound Interest Formula}$$
$$2{,}000 = P(1 + 0.0775)^3 \qquad \text{substituting}$$
$$P = \frac{2{,}000}{1.0775^3} \qquad \text{solving for } P$$
$$= 1{,}598.7412 \qquad \text{rounding}$$
$$\approx \$1{,}598.74$$

As a rough check, notice that the principal is somewhat lower than the future value, as it should be, because of the short time period.

Annual Yield

FIGURE 5.5 Most banks advertise their yields as well as their rates.

Which investment is more profitable: one that pays 5.8% compounded daily or one that pays 5.9% compounded quarterly? It is difficult to tell. Certainly, 5.9% is a better rate than 5.8%, but compounding daily is better than compounding quarterly. The two rates cannot be directly compared because of their different compounding frequencies. The way to tell which is the better investment is to find the annual yield of each.

The **annual yield** (also called the **annual percentage yield** or **APY**) of a compound interest deposit is the *simple interest rate* that has the same future value as the compound rate would have in one year. The annual yields of two different investments can be compared, because they are both simple interest rates. Annual yield provides the consumer with a uniform basis for comparison and banks display both their interest rates and their annual yields, as shown in Figure 5.5. The annual yield should be slightly higher than the compound rate, because compound interest is slightly more profitable than simple interest over a short period of time. The compound rate is sometimes called the **nominal rate** to distinguish it from the yield (here, *nominal* means "named" or "stated").

To find the **annual yield** r of a given compound interest rate, you find the simple interest rate that makes the future value under simple interest the same as the future value under compound interest in one year.

$$FV(\text{simple interest}) = FV(\text{compound interest})$$
$$P(1 + rt) = P(1 + i)^n$$

EXAMPLE **7**

SOLUTION

FINDING THE ANNUAL YIELD Find the annual yield of $2,500 deposited in an account in which it earns 10.3% interest compounded daily for 15 years.

Simple interest	Compound interest
$P = 2,500$	$P = 2,500$
$r = $ unknown annual yield	$i = \dfrac{1}{365}$ th of $10.3\% = \dfrac{0.103}{365}$
$t = 1$ year	$n = 1$ year $= 365$ days

(one year, not fifteen years—its *annual* yield)

$$FV\,(\text{simple interest}) = FV\,(\text{compounded monthly})$$

$$P(1 + rt) = P(1 + i)^n$$

$$2,500(1 + r \cdot 1) = 2,500(1 + 0.103/365)^{365} \qquad \text{substituting}$$

$$(1 + r \cdot 1) = (1 + 0.103/365)^{365} \qquad \text{dividing by 2,500}$$

$$r = (1 + 0.103/365)^{365} - 1 \qquad \text{solving for } r$$

$$r = 0.10847\ldots \approx 10.85\% \qquad \text{rounding}$$

As a rough check, notice that the annual yield is slightly higher than the compound rate, as it should be.

By tradition, we round this to the nearest hundredth of a percent, so the annual yield is 10.85%. This means that in one year's time, 10.3% compounded daily has the same effect as 10.85% simple interest. For any period of time longer than a year, 10.3% compounded daily will yield *more* interest than 10.85% simple interest would, because compound interest gives interest on interest.

Notice that in Example 7, the principal of $2,500 canceled out. If the principal were two dollars or two million dollars, it would still cancel out, and the annual yield of 10.3% compounded daily would still be 10.85%. The principal does not matter in computing the annual yield. Also notice that the fifteen years did not enter into the calculation—*annual* yield is always based on a *one-year* period.

EXAMPLE **8**

SOLUTION

FINDING THE ANNUAL YIELD Find the annual yield corresponding to a nominal rate of 8.4% compounded monthly.

We are told neither the principal nor the time, but (as discussed above) these variables do not affect the annual yield.

Compounding monthly, we have $i = 1/12$ of $8.4\% = \frac{0.084}{12}$, $n = 1$ year $= 12$ months, and $t = 1$ year.

$$FV\,(\text{simple interest}) = FV(\text{compounded monthly})$$

$$P(1 + rt) = P(1 + i)^n$$

$$(1 + rt) = (1 + i)^n \qquad \text{dividing by } P$$

$$(1 + r \cdot 1) = (1 + .084/12)^{12} \qquad \begin{array}{l}\text{annual, so } t = 1 \text{ year and}\\ n = 12 \text{ months}\end{array}$$

$$r = (1 + .084/12)^{12} - 1 \qquad \text{solving for } r$$

$$r = 0.08731\ldots \approx 8.73\% \qquad \text{rounding}$$

As a rough check, notice that the annual yield is slightly higher than the compound rate, as it should be.

TOPIC X   BENJAMIN FRANKLIN'S GIFT
COMPOUND INTEREST IN THE REAL WORLD

In 1789, when Benjamin Franklin was 83, he added a codicil to his will. That codicil was meant to aid young people, in Boston (where Franklin grew up), and Philadelphia (where he had been President of the state of Pennsylvania) over a time span of two hundred years.

In his codicil, Franklin wrote of his experience as an apprentice printer, of the friends who had loaned him the money to set up his own printing business, and of the importance of craftsmen to a city. He then went on to say, "To this End I devote Two thousand Pounds Sterling, which I give, one thousand thereof to the Inhabitants of the Town of Boston, in Massachusetts, and the other thousand to the Inhabitants of the City of Philadelphia, in Trust to and for the Uses, Interests and Purposes hereinafter mentioned and declared." At that time, £1,000 was the equivalent of about $4,500.

Franklin's plan called for each town to use the money as a loan fund for 100 years. The money would be loaned out to young tradesmen to help them start their own businesses. He calculated that at the end of 100 years, each fund would have grown to £131,000 (or about $582,000 in

Stock Montage/Stock Montage/Getty Images

1892 dollars). At that point, one-fourth of the money would continue to be used for a loan fund. The remainder would be used for public works.

He calculated that at the end of 200 years, each of the two city's loan funds would have grown to £4,061,000 (or about $7,000,000 in 1992 dollars). At that point, he called for the money to be given to the two cities and states.

Twenty-two tradesmen, including bricklayers, hairdressers, jewelers, and tanners, applied for loans from Franklin's fund in the first month after Franklin's death. The fund remained popular until the onset of the Industrial Revolution in the early 1800s, when young people stopped becoming tradesmen with their own shops. Instead, most went to work as mechanics in factories.

When Boston's fund reached its hundredth anniversary in 1891, its value was approximately $391,000. One-fourth of that money continued to be used for a loan fund, as Franklin

wished. The city of Boston decided to use the remainder to build a trade school, because it fit with Franklin's goal of helping young people.

Legal problems delayed the school's founding, but the Benjamin Franklin Institute of Technology was opened in Boston in 1908. At that point, the trade school part of the Boston fund had risen to $432,367, and the loan fund part had grown to $163,971.

The Institute continues to operate today. It has almost 400 students, 90% of whom receive financial aid. It awards Bachelor of Science degrees, Associate in Engineering degrees, and Associate in Science degrees. Ninety-eight percent of its graduates find work in their fields within six months of graduation.

Pennsylvania used its share of the money to fund the Franklin Institute of Philadelphia. Originally, the institute promoted the "mechanical arts." Now it houses a planetarium, an IMAX theater, and a science museum.

Exercises 57–62 explore some aspects of Franklin's bequest.

Source: The Benjamin Franklin Institute of Technology and the Franklin Institute of Philadelphia.

In Example 8, we found that 8.4% compounded monthly generates an annual yield of 8.73%. This means that 8.4% compounded monthly has the same effect as does 8.73% simple interest in one year's time. Furthermore, as Figure 5.6 indicates, 8.4% compounded monthly has the same effect as does 8.73% compounded annually *for any time period.*

For a Principal of $1,000	After 1 Year	After 10 Years
FV at 8.4% compounded monthly	$1,087.31	$2,309.60
FV at 8.73% simple interest	$1,087.30	$1,873.00
FV at 8.73% compounded annually	$1,087.30	$2,309.37

FIGURE 5.6 What annual yield means.

Notice that all three rates have the same future value after one year (the 1¢ difference is due to rounding off the annual yield to 8.73%). However, after ten years, the simple interest has fallen way behind, while the 8.4% compounded monthly and 8.73% compounded annually remain the same (except for the round-off error). This always happens. The annual yield is the simple interest rate that has the same future value that the compound rate would have in one year. It is also the annually compounded rate that has the same future value that the nominal rate would have after any amount of time.

An annual yield formula does exist, but the annual yield can be calculated efficiently without it, as was shown above. The formula is developed in the exercises. See Exercise 51.

5.2 EXERCISES

In Exercises 1–6, find the periodic rate that corresponds to the given compound rate, if the rate is compounded (a) quarterly, (b) monthly, (c) daily, (d) biweekly (every two weeks), and (e) semimonthly (twice a month). Do not round off the periodic rate.

1. 12%

2. 6%

3. 3.1%

4. 6.8%

5. 9.7%

6. 10.1%

In Exercises 7–10, find the number of periods that corresponds to the given time span, if a period is (a) a quarter of a year, (b) a month, and (c) a day. (Ignore leap years.)

7. $8\frac{1}{2}$ years

8. $9\frac{3}{4}$ years

9. 30 years

10. 45 years

In Exercises 11–16, (a) find and (b) interpret the future value of the given amount.

11. $3,000 at 6% compounded annually for 15 years

12. $7,300 at 7% compounded annually for 13 years

13. $5,200 at $6\frac{3}{4}$% compounded quarterly for $8\frac{1}{2}$ years

14. $36,820 at $7\frac{7}{8}$% compounded quarterly for 4 years

15. $1,960 at $4\frac{1}{8}$% compounded daily for 17 years (ignore leap years)

16. $12,350 at 6% compounded daily for 10 years and 182 days (ignore leap years)

In Exercises 17–20, (a) find and (b) interpret the annual yield corresponding to the given nominal rate.

17. 8% compounded monthly

18. $5\frac{1}{2}$% compounded quarterly

19. $4\frac{1}{4}$% compounded daily

20. $12\frac{5}{8}$% compounded daily

In Exercises 21 and 22, find and interpret the annual yield corresponding to the given nominal rate.

21. 10% compounded (a) quarterly, (b) monthly, and (c) daily

22. $12\frac{1}{2}$% compounded (a) quarterly, (b) monthly, and (c) daily

In Exercises 23–26, (a) find and (b) interpret the present value that will generate the given future value.

23. $1,000 at 8% compounded annually for 7 years

24. $9,280 at $9\frac{3}{4}$% compounded monthly for 2 years and 3 months

25. $3,758 at $11\frac{7}{8}$% compounded monthly for 17 years and 7 months

26. $4,459 at $10\frac{3}{4}$% compounded quarterly for 4 years

27. $10,000 is deposited in an account in which it earns 10% interest compounded monthly. No principal or interest is withdrawn from the account. Instead, both continue to earn interest over time. Find the account balance after six months:

　a. using the simple interest future value formula to compute the balance at the end of each compounding period.

　b. using the compound interest formula.

28. $20,000 is deposited in an account in which it earns 9.5% interest compounded quarterly. No principal or interest is withdrawn from the account. Instead, both continue to earn interest over time. Find the account balance after one year:

　a. using the simple interest future value formula to compute the balance at the end of each compounding period.

　b. using the compound interest formula.

29. $15,000 is deposited in an account in which it earns 6% interest compounded annually. No principal or interest is withdrawn from the account. Instead, both continue to earn interest over time. Find the account balance after three years:

　a. using the simple interest future value formula to compute the balance at the end of each compounding period.

　b. using the compound interest formula.

30. $30,000 is deposited in an account in which it earns 10% interest compounded annually. No principal or interest is withdrawn from the account. Instead, both continue to earn interest over time. Find the account balance after four years:

    a. using the simple interest future value formula to compute the balance at the end of each compounding period.

    b. using the compound interest formula.

31. Donald Trumptobe decided to build his own dynasty. He is considering specifying in his will that at his death, $10,000 would be deposited into a special account that would earn a guaranteed 6% interest compounded daily. This money could not be touched for 100 years, at which point it would be divided among his heirs. Find the future value.

32. How much would Donald Trumptobe in Exercise 31 have to have deposited so that his heirs would have $50,000,000 or more in 100 years if his money earns 7% compounded monthly?

33. Donald Trumptobe in Exercise 31 predicts that in 100 years, he will have four generations of offspring (that is, children, grandchildren, great-grandchildren, and great-great-grandchildren). He estimates that each person will have two children. How much will he have to have deposited so that each of his great-great-grandchildren would have $1,000,000 or more in 100 years if his money earns 7.5% compounded monthly?

34. Donald Trumptobe in Exercise 31 predicts that in 100 years, he will have four generations of offspring (that is, children, grandchildren, great-grandchildren, and great-great-grandchildren). He estimates that each person will have two children. How much will he have to have deposited so that each of his great-great-grandchildren would have $1,000,000 or more in 100 years if his money earns 9.25% compounded daily?

35. When Jason Levy was born, his grandparents deposited $3,000 into a special account for Jason's college education. The account earned $6\frac{1}{2}\%$ interest compounded daily.

    a. How much will be in the account when Jason is 18?

    b. If, on turning 18, Jason arranges for the monthly interest to be sent to him, how much will he receive each thirty-day month?

    c. How much would be in the account when Jason turns 18 if his grandparents started Jason's savings account on his tenth birthday?

36. When Alana Cooper was born, her grandparents deposited $5,000 into a special account for Alana's college education. The account earned $7\frac{1}{4}\%$ interest compounded daily,

    a. How much will be in the account when Alana is 18?

    b. If, on turning 18, Alana arranges for the monthly interest to be sent to her, how much will she receive each thirty-day month?

    c. How much would be in the account when Alana turns 18 if his grandparents started Alana's savings account on her tenth birthday?

For Exercises 37–40, note the following information: An **Individual Retirement Account (IRA)** *is an account in which the saver does not pay income tax on the amount deposited but is not allowed to withdraw the money until retirement. (The saver pays income tax at that point, but his or her tax bracket is much lower then.)*

37. At age 27, Lauren Johnson deposited $1,000 into an IRA, in which it earns $7\frac{7}{8}\%$ compounded monthly.

    a. What will it be worth when she retires at 65?

    b. How much would the IRA be worth if Lauren didn't set it up until she was 35?

38. At age 36, Dick Shoemaker deposited $2,000 into an IRA, in which it earns $8\frac{1}{8}\%$ compounded semiannually.

    a. What will it be worth when he retires at 65?

    b. How much would the IRA be worth if Dick didn't set it up until he was 48?

39. Marlene Silva wishes to have an IRA that will be worth $100,000 when she retires at age 65.

    a. How much must she deposit at age 35 at $8\frac{3}{8}\%$ compounded daily?

    b. If, at age 65, she arranges for the monthly interest to be sent to her, how much will she receive each thirty-day month?

40. David Murtha wishes to have an IRA that will be worth $150,000 when he retires at age 65.

    a. How much must he deposit at age 26 at $6\frac{1}{8}\%$ compounded daily?

    b. If, at age 65, he arranges for the monthly interest to be sent to him, how much will he receive each thirty-day month?

For Exercises 41–46, note the following information: A certificate of deposit (CD) is an agreement between a bank and a saver in which the bank guarantees an interest rate and the saver commits to leaving his or her deposit in the account for an agreed-upon period of time.

41. First National Bank offers two-year CDs at 9.12% compounded daily, and Citywide Savings offers two-year CDs at 9.13% compounded quarterly. Compute the annual yield for each institution and determine which is more advantageous for the consumer.

42. National Trust Savings offers five-year CDs at 8.25% compounded daily, and Bank of the Future offers five-year CDs at 8.28% compounded annually. Compute the annual yield for each institution, and determine which is more advantageous for the consumer.

43. Verify the annual yield for the five-year certificate quoted in the bank sign in Figure 5.7 on page 352, using interest that is compounded daily and:

    a. 365-day years

    b. 360-day years

    c. Various combinations of 360-day and 365-day years

CURRENT RATES *			
AVAILABLE THROUGH **MARCH**	Minimum	Interest Rate %	Annual Percentage Yield %
PASSBOOK SAVINGS		1.470	1.500
STATEMENT SAVINGS		1.470	1.500
INSURED Money Market Account		1.000	1.020
COMMERCIAL Money Market		0.750	0.760
3 MONTH Certificate		0.980	1.000
6 MONTH Certificate		1.030	1.050
12 MONTH Certificate		1.180	1.200
2 YEAR Certificate		1.320	1.350
30 MONTH Certificate		1.950	2.000
3 YEAR Certificate		1.570	1.600
4 YEAR Certificate		2.190	2.250
5 YEAR Certificate		3.390	3.500

*Ask us for further information about these accounts. Penalty for early withdrawal. Truth in savings disclosures available upon request.

© Robert Brenner / Photo Edit

FIGURE 5.7 A bank sign for Exercise 43.

44. Verify the yield for the one-year CDs quoted in Figure 5.8.

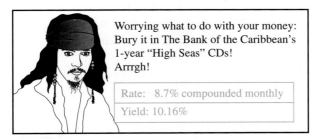

Worrying what to do with your money: Bury it in The Bank of the Caribbean's 1-year "High Seas" CDs! Arrrgh!

Rate: 8.7% compounded monthly
Yield: 10.16%

FIGURE 5.8 A savings bank advertisement for Exercise 44.

45. Verify the yield for the two-year CDs quoted in Figure 5.9.

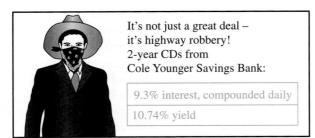

It's not just a great deal – it's highway robbery! 2-year CDs from Cole Younger Savings Bank:

9.3% interest, compounded daily
10.74% yield

FIGURE 5.9 A savings bank advertisement for Exercise 45.

46. Verify the yield for the six-month CDs quoted in the savings bank advertisement on page 347.

47. Recently, Bank of the West offered six-month CDs at 5.0% compounded monthly.
  a. Find the annual yield of one of these CDs.
  b. How much would a $1,000 CD be worth at maturity?
  c. How much interest would you earn?
  d. What percent of the original $1,000 is this interest?
  e. The answer to part (d) is not the same as that of part (a). Why?
  f. The answer to part (d) is close to, but not exactly half, that of part (a). Why?
  (*Source:* Bank of the West.)

48. Recently, Bank of the West offered six-month CDs at 5.0% interest compounded monthly and one-year CDs at 5.20% interest compounded monthly. Maria Ruiz bought a six-month $2,000 CD, even though she knew she would not need the money for at least a year, because it was predicted that interest rates would rise.
  a. Find the future value of Maria's CD.
  b. Six months later, Maria's CD has come to term, and in the intervening time, interest rates have risen. She reinvests the principal and interest from her first CD in a second six-month CD that pays 5.31% interest compounded monthly. Find the future value of Maria's second CD.
  c. Would Maria have been better off if she had bought a one-year CD instead of two six-month CDs?
  d. If Maria's second CD pays 5.46% interest compounded monthly, rather than 5.31%, would she be better off with the two six-month CDs or the one-year CD?
  (*Source:* Bank of the West.)

49. CD interest rates vary significantly with time. They hit a historical high in 1981, when the average rate was 16.7%. In 2000, it was 8.1%. In 2009, it was 2%. Find the interest earned by a $1000 two-year CD with interest compounded quarterly:
  a. in 1981   **b.** in 2000   **c.** in 2009

50. Use the data in Exercise 49 to find the yield of a $10,000 ten-year CD with interest compounded daily:
  a. in 1981   **b.** in 2000   **c.** in 2009

51. Develop a formula for the annual yield of a compound interest rate.
  HINT: Follow the procedure given in Example 8, but use the letters i and n in the place of numbers.)

In Exercises 52–56, use the formula found in Exercise 51 to compute the annual yield corresponding to the given nominal rate.

52. $9\frac{1}{2}\%$ compounded monthly

53. $7\frac{1}{4}\%$ compounded quarterly

54. $12\frac{3}{8}\%$ compounded daily

55. $5\frac{5}{8}\%$ compounded (a) semiannually, (b) quarterly, (c) monthly, (d) daily, (e) biweekly, and (f) semimonthly.

56. $10\frac{1}{2}\%$ compounded (a) semiannually, (b) quarterly, (c) monthly, (d) daily, (e) biweekly, and (f) semimonthly.

Exercises 57–62 refer to Benjamin Franklin's gift, discussed on page 349.

57. Would Benjamin Franklin have used simple or compound interest in projecting the 100-year future value of his bequest? Why? What interest rate did he use in calculating the 100-year future value? Use pounds, not dollars, in your calculation.

58. Would Benjamin Franklin have used simple or compound interest in projecting the 200-year future value of his bequest? Why? What interest rate did he use in calculating the 200-year future value? Use pounds, not dollars, in your calculation.

59. What interest rate did Franklin's Boston bequest actually earn in the first hundred years? Use dollars, not pounds, in your calculation.

60. What interest rate did Franklin's Boston bequest actually earn from 1891 to 1908? Use dollars, not pounds, in your calculation.

61. After 100 years, Franklin's total bequest to Boston was worth much more than his original bequest. The future value was what percentage of the original bequest? Use dollars, not pounds, in your calculation.

62. In 1908, Franklin's total bequest to Boston was worth much more than his original bequest. The future value was what percentage of the original bequest? Use dollars, not pounds, in your calculation.

 Answer the following questions using complete sentences and your own words.

● **CONCEPT QUESTIONS**

63. Explain how compound interest is based on simple interest.

64. Why is there no work involved in finding the annual yield of a given simple interest rate?

65. Why is there no work involved in finding the annual yield of a given compound interest rate when that rate is compounded annually?

66. Which should be higher: the annual yield of a given rate compounded quarterly or compounded monthly? Explain why, *without* performing any calculations or referring to any formulas.

67. Why should the annual yield of a given compound interest rate be higher than the compound rate? Why should it be only slightly higher? Explain why, *without* performing any calculations or referring to any formulas.

68. Explain the difference between simple interest and compound interest.

69. If money earns compound interest, why must the future value be *slightly* higher than the principal, after a short amount of time? Why must the future value be *much* higher than the principal after a long amount of time?

70. *Money* magazine and other financial publications regularly list the top-paying money-market funds, the top-paying bond funds, and the top-paying CDs and their yields. Why do they list yields rather than interest rates and compounding periods?

71. Equal amounts are invested in two different accounts. One account pays simple interest, and the other pays compound interest at the same rate. When will the future values of the two accounts be the same?

72. Suppose you invest some money in a new account that pays 5% interest compounded annually, and you do not make any further deposits into or withdrawals from that account. Which of the following must be true?

● The account grows by the same dollar amount in the second year as it did in the first year.

● The account grows by a larger dollar amount in the second year as it did in the first year.

● The account grows by a smaller dollar amount in the second year as it did in the first year. Why?

● **WEB PROJECT**

73. Go to the web sites of four different banks.
   a. For each bank, try to determine the following:
      ● The interest rate
      ● The compounding frequency
      ● The annual yield (also called the *annual percentage yield,* or A.P.Y.) for CDs for two different terms. (Use the same terms for each bank.) Some banks might not give all of the above information, but all will give the annual yield, as required by federal law.
   b. If a bank omits either the interest rate or the compounding frequency, calculate the omitted information.
   c. If a bank omits both the interest rate and the compounding frequency, assume that the compounding frequency is daily and calculate the interest rate.
   d. If a bank omits none of the information, verify the annual yield.
   e. Which bank offers the best deal? Why?

Some useful links for this web project are listed on the text web site: **www.cengage.com/math/johnson**

● **PROJECT**

74. Suppose you have $1,000 invested at 5% annual interest and you do not make any further deposits into that account. Let *x* measure years after you

invested the money, and let y measure the future value of the account. Draw a graph that shows the relationship between x and y, for $0 \le x \le 5$, if the interest rate is

a. simple interest

b. compounded annually

c. compounded daily

d. Discuss the differences between the three graphs. In your discussion, address the following:

- The difference in their shapes
- Where they coincide
- Which graph is above the others
- Which is below the others

DOUBLING TIME WITH A TI'S TVM APPLICATION

Simple interest is a very straightforward concept. If an account earns 5% simple interest, then 5% of the principal is paid for each year that principal is in the account. In one year, the account earns 5% interest; in two years, it earns 10% interest; in three years, it earns 15% interest, and so on.

It is not nearly so easy to get an intuitive grasp of compound interest. If an account earns 5% interest compounded daily, then it does not earn only 5% interest in one year, and it does not earn only 10% interest in two years.

Annual yield is one way of gaining an intuitive grasp of compound interest. If an account earns 5% interest compounded daily, then it will earn 5.13% interest in 1 year (because the annual yield is 5.13%), but it does not earn merely $2 \cdot 5.13\% = 10.26\%$ interest in two years.

Doubling time is another way of gaining an intuitive grasp of compound interest. **Doubling time** is the amount of time it takes for an account to double in value; that is, it's the amount of time it takes for the future value to become twice the principal. To find the doubling time for an account that earns 5% interest compounded daily, substitute $2P$ for the future value and solve the resulting equation.

$$FV = P(1 + i)^n \qquad \text{\textbf{Compound Interest Formula}}$$

$$2P = P\left(1 + \frac{0.05}{365}\right)^n \qquad \text{substituting}$$

$$2 = \left(1 + \frac{0.05}{365}\right)^n \qquad \text{dividing by } P$$

Solving this equation for n involves mathematics that will be covered in Section 10.0B. For now, we will use the TI-83/84's "Time Value of Money" (TVM) application.

EXAMPLE **9**

FINDING DOUBLING TIME Use a TI–83/84's TVM application to find the doubling time for an account that earns 5% interest compounded daily.

SOLUTION

1. Press APPS , select option 1: "Finance", and press ENTER .
2. Select option 1: "TVM Solver", and press ENTER.
3. Enter appropriate values for the variables:
   - N is the number of compounding periods. This is the number we're trying to find. We temporarily enter 0. Later, we'll solve for the actual value of N.
   - I% is the *annual* interest rate (not a periodic rate), as a percent, so enter 5 for I%. Note that we do not convert to a decimal or a periodic rate.
   - PV is the present value. The size of the present value doesn't matter, so we'll make it $1. However, it's an outgoing amount of money (since we give it to the bank), so we

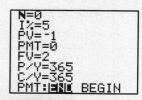

FIGURE 5.10

Preparing the TVM screen.

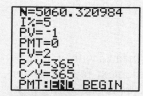

FIGURE 5.11

Solving for N.

enter –1 for PV. You must enter a negative number for any outgoing amount of money.

- PMT is the payment. There are no payments here, so we enter 0.
- FV is the future value. We're looking for the amount of time that it takes the present value to double, so enter 2. This is an incoming amount of money (since the bank gives it to us), so we use 2 rather than –2.
- P/Y is the number of periods per year. The interest is compounded daily, so there are 365 periods per year. Enter 365 for P/Y.
- C/Y is automatically made to be the same as P/Y. In this text, we will never encounter a situation in which C/Y is different from P/Y.

See Figure 5.10.

4. To solve for N, use the arrow buttons to highlight the 0 that we entered for N. Then press ALPHA SOLVE. (Pressing ALPHA makes the ENTER button becomes the SOLVE button.) As a result, we find that N is 5060.320984. See Figure 5.11.

This is the number of *days* for the money to double (a period is a day, because of our P/Y entry). This means that it takes about 5060.320984/365 = 13.86... ≈ 13.9 years for money invested at 5% interest compounded daily to double.

EXERCISES

75. If $1,000 is deposited into an account that earns 5% interest compounded daily, the doubling time is approximately 5,061 days.
   a. Find the amount in the account after 5,061 days.
   b. Find the amount in the account after 2 · 5,061 days.
   c. Find the amount in the account after 3 · 5,061 days.
   d. Find the amount in the account after 4 · 5,061 days.
   e. What conclusion can you make?

76. Do the following. (Give the number of periods and the number of years, rounded to the nearest hundredth.)
   a. Find the doubling time corresponding to 5% interest compounded annually.
   b. Find the doubling time corresponding to 5% interest compounded quarterly.
   c. Find the doubling time corresponding to 5% interest compounded monthly.
   d. Find the doubling time corresponding to 5% interest compounded daily.
   e. Discuss the effect of the compounding period on doubling time.

77. Do the following. (Give the number of periods and the number of years, rounded to the nearest hundredth.)

   a. Find the doubling time corresponding to 6% interest compounded annually.
   b. Find the doubling time corresponding to 7% interest compounded annually.
   c. Find the doubling time corresponding to 10% interest compounded annually.
   d. Discuss the effect of the interest rate on doubling time.

78. If you invest $10,000 at 8.125% interest compounded daily, how long will it take for you to accumulate $15,000? How long will it take for you to accumulate $100,000? (Give the number of periods and the number of years, rounded to the nearest hundredth.)

79. If you invest $15,000 at $9\frac{3}{8}\%$ interest compounded daily, how long will it take for you to accumulate $25,000? How long will it take for you to accumulate $100,000? (Give the number of periods and the number of years, rounded to the nearest hundredth.)

80. If you invest $20,000 at $6\frac{1}{4}\%$ interest compounded daily, how long will it take for you to accumulate $30,000? How long will it take for you to accumulate $100,000? (Give the number of periods and the number of years, rounded to the nearest hundredth.)

Annuities

OBJECTIVES

- Understand what an annuity is
- Use the Annuity Formulas
- Determine how to use an annuity to save for retirement.

Many people have long-term financial goals and limited means with which to accomplish them. Your goal might be to save $3,000 over the next four years for your college education, to save $10,000 over the next ten years for the down payment on a home, to save $30,000 over the next eighteen years to finance your new baby's college education, or to save $300,000 over the next forty years for your retirement. It seems incredible, but each of these goals can be achieved by saving only $50 a month (if interest rates are favorable)! All you need to do is start an annuity. We'll explore annuities in this section.

An **annuity** is simply a sequence of equal, regular payments into an account in which each payment receives compound interest. Because most annuities involve relatively small periodic payments, they are affordable for the average person. Over longer periods of time, the payments themselves start to amount to a significant sum, but it is really the power of compound interest that makes annuities so amazing. If you pay $50 a month into an annuity for the next forty years, then your total payment is

$$\frac{\$50}{\text{month}} \cdot \frac{12 \text{ months}}{\text{year}} \cdot 40 \text{ years} = \$24,000$$

A **Christmas Club** is an annuity that is set up to save for Christmas shopping. A Christmas Club participant makes regular equal deposits, and the deposits and the resulting interest are released to the participant in December when the money is needed. Christmas Clubs are different from other annuities in that they span a short amount of time—a year at most—and thus earn only a small amount of interest. (People set them up to be sure that they are putting money aside rather than to generate interest.) Our first few examples will deal with Christmas Clubs, because their short time span makes it possible to see how an annuity actually works.

Annuities as Compound Interest, Repeated

EXAMPLE **1**

UNDERSTANDING ANNUITIES On August 12, Patty Leitner joined a Christmas Club through her bank. For the next three months, she would deposit $200 at the beginning of each month. The money would earn $8\frac{3}{4}\%$ interest compounded monthly, and on December 1, she could withdraw her money for shopping. Use the compound interest formula to find the future value of the account.

SOLUTION

We are given $P = 200$ and $r = \frac{1}{12}$th of $8\frac{3}{4}\% = \frac{0.0875}{12}$.

First, calculate the future value of the first payment (made on September 1). Use $n = 3$ because it will receive interest during September, October, and November.

$$FV = P(1 + i)^n \qquad \text{the Compound Interest Formula}$$
$$= 200\left(1 + \frac{0.0875}{12}\right)^3 \qquad \text{substituting}$$
$$= 204.40698 \approx \$204.41 \qquad \text{rounding}$$

Next, calculate the future value of the second payment (made on October 1). Use $n = 2$ because it will receive interest during October and November.

$$FV = P(1 + i)^n \qquad \text{the Compound Interest Formula}$$
$$= 200\left(1 + \frac{0.0875}{12}\right)^2 \qquad \text{substituting}$$
$$= 202.9273 \approx \$202.93 \qquad \text{rounding}$$

Next, calculate the future value of the third payment (made on November 1). Use $n = 1$ because it will receive interest during November.

$$FV = P(1 + i)^n \qquad \text{the Compound Interest Formula}$$
$$= 200\left(1 + \frac{0.0875}{12}\right)^1 \qquad \text{substituting}$$
$$= 201.45833 \approx \$201.46 \qquad \text{rounding}$$

The payment schedule and interest earned are illustrated in Figure 5.12.

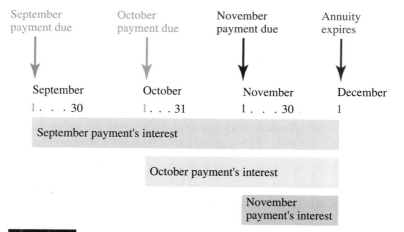

FIGURE 5.12 Patty's payment schedule and interest earning periods.

The future value of Patty's annuity is the sum of the future values of each payment:

$$FV \approx \$204.41 + \$202.93 + \$201.46$$
$$= \$608.80$$

Patty's deposits will total $600.00. She will earn $8.80 interest on her deposits.

The **payment period** of an annuity is the time between payments; in Example 1, the payment period was one month. The **term** is the time from the beginning of the first payment period to the end of the last; the term of Patty's Christmas Club was three months. When an annuity has **expired** (that is, when its term is over), the entire account or any portion of it may be withdrawn. Most annuities are **simple,** that is, their compounding period is the same as their payment period (for example, if payments are made monthly, then interest is compounded monthly). In this book, we will work only with simple annuities.

Ordinary Annuities and Annuities Due

An **annuity due** is one in which each payment is due at the beginning of its time period. Patty's annuity in Example 1 was an annuity due, because the payments were due at the *beginning* of each month. An **ordinary annuity** is an annuity for which each payment is due at the end of its time period. As the name implies, this form of annuity is more typical. As we will see in the next example, the difference is one of timing.

EXAMPLE **2**

UNDERSTANDING THE DIFFERENCE BETWEEN AN ORDINARY ANNUITY AND AN ANNUITY DUE Dan Bach also joined a Christmas Club through his bank. His was just like Patty's except that his payments were due at the end of each month, and his first payment was due September 30. Use the Compound Interest Formula to find the future value of the account.

SOLUTION

This is an *ordinary* annuity because payments are due at the *end* of each month. Interest is compounded monthly. From Example 1, we know that $P = 200$, $i = \frac{1}{12}$ of $8\frac{3}{4}\% = \frac{0.0875}{12}$.

To calculate the future value of the first payment (made on September 30), use $n = 2$. This payment will receive interest during October and November.

$$FV = P(1 + i)^n \qquad \text{the Compound Interest Formula}$$

$$= 200\left(1 + \frac{0.0875}{12}\right)^2 \qquad \text{substituting}$$

$$= 202.9273 \approx \$202.93 \qquad \text{rounding}$$

To calculate the future value of the second payment (made on October 31), use $n = 1$. This payment will receive interest during November.

$$FV = P(1 + i)^n \qquad \text{the Compound Interest Formula}$$

$$= 200\left(1 + \frac{0.0875}{12}\right)^1 \qquad \text{substituting}$$

$$= 201.45833 \approx \$201.46 \qquad \text{rounding}$$

To calculate the future value of the second payment (made on November 30), note that no interest is earned, because the payment is due November 30 and the annuity expires December 1. Therefore,

$$FV = \$200$$

Dan's payment schedule and interest payments are illustrated in Figure 5.13.

The future value of Dan's annuity is the sum of the future values of each payment:

$$FV \approx \$200 + \$201.46 + \$202.93$$

$$= \$604.39$$

Dan earned $4.39 interest on his deposits.

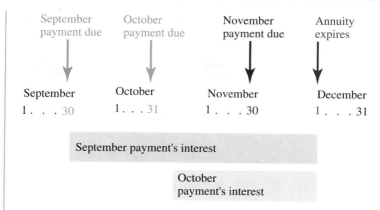

FIGURE 5.13 Dan's payment schedule and interest earning periods.

In Examples 1 and 2, why did Patty earn more interest than Dan? The reason is that each of her payments was made a month earlier and therefore received an extra month's interest. In fact, we could find the future value of Patty's account by giving Dan's future value one more month's interest.

$$\text{Patty's } FV = \text{Dan's } FV \cdot (1 + i)^1$$
$$\$608.80 = \$604.39 \cdot \left(1 + \frac{0.0875}{12}\right)^1$$

More generally, we can find the future value of an annuity due by giving an ordinary annuity's future value one more month's interest.

$$FV(\text{due}) = FV(\text{ordinary}) \cdot (1 + i)^1$$

The difference between an ordinary annuity and an annuity due is strictly a timing difference, because any ordinary annuity in effect will become an annuity due if you leave all funds in the account for one extra period.

The Annuity Formulas

The procedure followed in Examples 1 and 2 reflects what actually happens with annuities, and it works fine for a small number of payments. However, most annuities are long-term, and the procedure would become tedious if we were computing the future value after forty years. Because of this, long-term annuities are calculated with their own formula.

For an *ordinary* annuity with payment *pymt,* a periodic rate *i*, and a term of *n* payments, the first payment receives interest for $n - 1$ periods. The payment is made at the end of the first period, so it received no interest for that one period. Its future value is

$$FV(\text{first } pymt) = pymt(1 + i)^{n - 1}$$

The last payment receives no interest (under the annuity), because it is due at the end of the last period and it expires the next day. Its future value is

$$FV(\text{last } pymt) = pymt$$

The next-to-last payment receives one period's interest, so its future value is

$$FV(\text{next-to-last } pymt) = pymt(1 + i)^1$$

The future value of the annuity is the sum of all of these future values of individual payments:

$$FV = pymt + pymt(1 + i)^1 + pymt(1 + i)^2 + \cdots + pymt(1 + i)^{n-1}$$

To get a short-cut formula from all this, we will multiply each side of this equation by $(1 + i)$ and then subtract the original equation from the result. This leads to a lot of cancelling.

$$FV(1 + i) = \cancel{pymt(1 + i)} + \cancel{pymt(1 + i)^2} + \cdots + \cancel{pymt(1 + i)^{n-1}} + pymt(1 + i)^n$$

minus: $$FV = pymt + \cancel{pymt(1 + i)^1} + \cancel{pymt(1 + i)^2} + \cdots + \cancel{pymt(1 + i)^{n-1}}$$

equals: $FV(1 + i) - FV = pymt(1 + i)^n - pymt$ **subtracting**

$FV(1 + i - 1) = pymt[(1 + i)^n - 1]$ **factoring**

$FV(i) = pymt[(1 + i)^n - 1]$

$FV = pymt\dfrac{(1 + i)^n - 1}{i}$ **dividing**

This is the future value of the ordinary annuity.

ORDINARY ANNUITY FORMULA

The future value FV of an ordinary annuity with payment size $pymt$, a periodic rate i, and a term of n payments is

$$FV(\text{ord}) = pymt\frac{(1 + i)^n - 1}{i}$$

As we saw in Examples 1 and 2, the future value of an annuity due is the future value of an ordinary annuity plus one more period's interest.

$$FV(\text{due}) = FV(\text{ord}) \cdot (1 + i)$$

This gives us a formula for the future value of an annuity due.

ANNUITY DUE FORMULA

The future value FV of an annuity due with payment size $pymt$, a periodic rate i, and a term of n payments is

$$FV(\text{due}) = FV(\text{ord}) \cdot (1 + i)$$
$$= pymt\frac{(1 + i)^n - 1}{i}(1 + i)$$

Tax-Deferred Annuities

A **tax-deferred annuity (TDA)** is an annuity that is set up to save for retirement. Money is automatically deducted from the participant's paychecks until retirement, and the federal (and perhaps state) tax deduction is computed *after* the annuity payment has been deducted, resulting in significant tax savings. In some cases, the employer also makes a regular contribution to the annuity.

The following example involves a long-term annuity. Usually, the interest rate of a long-term annuity varies somewhat from year to year. In this case, calculations must be viewed as predictions, not guarantees.

EXAMPLE **3**

USING AN ANNUITY TO SAVE FOR RETIREMENT Tom and Betty decided that they should start saving for retirement, so they set up a tax-deferred annuity. They arranged to have $200 taken out of each of Tom's monthly checks, which will earn $8\frac{3}{4}\%$ interest. Because of the tax-deferring effect of the TDA, Tom's take-home pay went down by only $115. Tom just had his thirtieth birthday, and his ordinary annuity will come to term when he is 65.

a. Find the future value of the annuity.
b. Find Tom's contribution and the interest portion.

SOLUTION

a. This is an ordinary annuity, with $pymt = 200$, $i = \frac{1}{12}$th of $8\frac{3}{4}\% = 0.0875/12$, and $n = 35$ years = 35 years · 12 months/year = 420 monthly payments.

$$FV(\text{ord}) = pymt\frac{(1 + i)^n - 1}{i} \qquad \text{the Ordinary Annuity Formula}$$

$$= 200\frac{(1 + 0.0875/12)^{420} - 1}{0.0875/12} \qquad \text{substituting}$$

$$\approx \$552{,}539.96 \qquad \text{rounding}$$

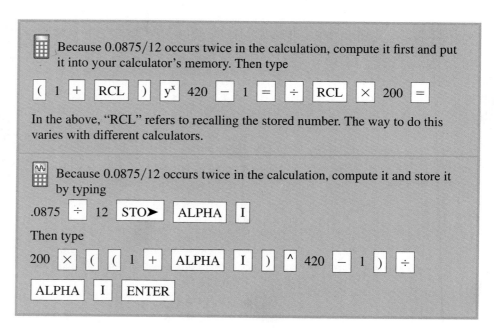

Because 0.0875/12 occurs twice in the calculation, compute it first and put it into your calculator's memory. Then type

(1 + RCL) y^x 420 − 1 = ÷ RCL × 200 =

In the above, "RCL" refers to recalling the stored number. The way to do this varies with different calculators.

Because 0.0875/12 occurs twice in the calculation, compute it and store it by typing

.0875 ÷ 12 STO▶ ALPHA I

Then type

200 × ((1 + ALPHA I) ^ 420 − 1) ÷

ALPHA I ENTER

b. The principal part of this $552,539.96 is Tom's contribution, and the rest is interest.

• *Tom's contribution* is 420 payments of $200 each = 420 · $200 = $84,000.
• The *interest portion* is then $552,539.96 − $84,000 = $468,539.96.

  In Example 3, the interest portion is almost six times as large as Tom's contribution! The magnitude of the earnings illustrates the amazing power of annuities and the effect of compound interest over a long period of time.

Sinking Funds

A **sinking fund** is an annuity in which the future value is a specific amount of money that will be used for a certain purpose, such as a child's education or the down payment on a home.

EXAMPLE **4**

USING AN ANNUITY TO SAVE A SPECIFIC AMOUNT   Tom and Betty have a new baby. They agreed that they would need $30,000 in eighteen years for the baby's college education. They decided to set up a sinking fund and have money deducted from each of Betty's biweekly paychecks. That money will earn $9\frac{1}{4}\%$ interest in Betty's ordinary annuity. Find their monthly payment.

SOLUTION

This is an ordinary annuity, with $i = \frac{1}{12}$th of $9\frac{1}{4}\% = 0.0925/26$, and $n = 18$ years $\cdot$ 26 periods/year $= 468$ periods, and $FV = \$30,000$.

$$FV(\text{ord}) = pymt\frac{(1 + i)^n - 1}{i} \qquad \text{the Ordinary Annuity Formula}$$

$$\$30,000 = pymt\frac{(1 + 0.0925/26)^{468} - 1}{0.0925/26} \qquad \text{substituting}$$

To find *pymt*, we must divide 30,000 by the fraction on the right side of the equation. Because the fraction is so complicated, it is best to first calculate the fraction and then multiply its reciprocal by 30,000.

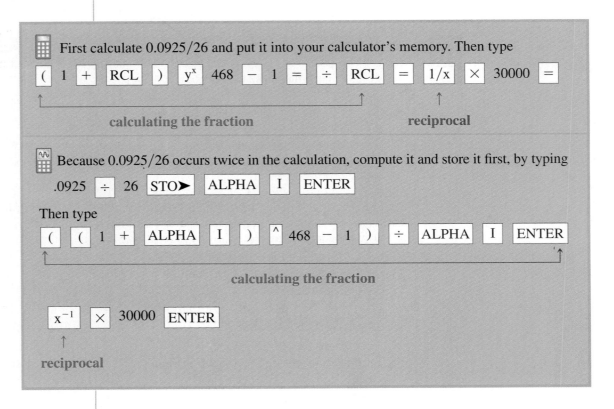

First calculate 0.0925/26 and put it into your calculator's memory. Then type

(1 + RCL) yˣ 468 − 1 = ÷ RCL = 1/x × 30000 =

calculating the fraction   *reciprocal*

Because 0.0925/26 occurs twice in the calculation, compute it and store it first, by typing

.0925 ÷ 26 STO▶ ALPHA I ENTER

Then type

((1 + ALPHA I) ^ 468 − 1) ÷ ALPHA I ENTER

calculating the fraction

x⁻¹ × 30000 ENTER

reciprocal

This gives *pymt* = 24.995038. . . . Betty would need to have only $25 taken out of each of her biweekly paychecks to save $30,000 in eighteen years. Notice that she will not have exactly $30,000 saved, because she cannot have exactly $24.995048 . . . deducted from each paycheck.

Present Value of an Annuity

The **present value of an annuity** is the lump sum that can be deposited at the beginning of the annuity's term, at the same interest rate and with the same compounding period, that would yield the same amount as the annuity. This value can help the saver to understand his or her options; it refers to an alternative way of saving the same amount of money in the same time. It is called the *present value* because it refers to the single action that the saver can take *in the present* (i.e., at the beginning of the annuity's term) that would have the same effect as would the annuity.

EXAMPLE **5**

FINDING THE PRESENT VALUE   Find the present value of Tom and Betty's annuity.

SOLUTION

$$FV = P(1 + i)^n \qquad \text{the Compound Interest}$$
$$30{,}005.95588 = P\left(1 + \frac{0.0925}{26}\right)^{468} \qquad \begin{array}{l}\text{Formula}\\ \text{substituting}\end{array}$$

$$P = \frac{30{,}005.95588}{\left(1 + \dfrac{0.0925}{26}\right)^{468}} \qquad \text{solving for } P$$

$$= 5{,}693.6451\ldots \approx \$5693.65 \qquad \text{rounding}$$

This means that Tom and Betty would have to deposit $5,693.55 as a lump sum to save as much money as the annuity would yield. They chose an annuity over a lump sum deposit because they could not afford to tie up $5,700 for eighteen years, but they could afford to deduct $25 out of each paycheck.

EXAMPLE **6**

FINDING THE PRESENT VALUE   Find the present value of an ordinary annuity that has $200 monthly payments for twenty-five years, where the account receives $10\frac{1}{2}\%$ interest.

SOLUTION

We could find the future value of the annuity and then find the lump sum deposit whose future value matches it, as we did in Example 5. However, it is simpler to do the calculation all at once. The key is to realize that the future value of the lump sum must equal the future value of the annuity:

Future value of lump sum = future value of annuity

$$P(1 + i)^n = pymt\frac{(1 + i)^n - 1}{i}$$

For both the lump sum and the annuity, $i = \frac{1}{12}$ of $10\frac{1}{2}\% = \frac{0.105}{12}$ and $n = 25$ years = 300 months. The annuity's payment is $pymt = \$200$.

$$P(1 + i)^n = pymt\frac{(1 + i)^n - 1}{i}$$

$$P(1 + 0.105/12)^{300} = 200\frac{(1 + 0.105/12)^{300} - 1}{0.105/12} \qquad \text{substituting}$$

First, calculate the right side, as with any annuity calculation. Then divide by $(1 + 0.105/12)^{300}$ to find P.

$$P = 21182.363\ldots$$
$$P \approx \$21{,}182.36$$

This means that one would have to make a lump sum deposit of more than $21,000 to have as much money after twenty-five years as with monthly $200 annuity payments.

PRESENT VALUE OF ANNUITY FORMULA

$$FV(\text{lump sum}) = FV(\text{annuity})$$

$$P(1 + i)^n = pymt\frac{(1 + i)^n - 1}{i}$$

The present value is the lump sum P.

5.3 EXERCISES

In Exercises 1–14, find the future value of the given annuity.

1. ordinary annuity, $120 monthly payment, $5\frac{3}{4}\%$ interest, one year

2. ordinary annuity, $175 monthly payment, $6\frac{1}{8}\%$ interest, eleven years

3. annuity due, $100 monthly payment, $5\frac{7}{8}\%$ interest, four years

4. annuity due, $150 monthly payment, $6\frac{1}{4}\%$ interest, thirteen years

5. On September 8, Bert Sarkis joined a Christmas Club. His bank will automatically deduct $75 from his checking account at the end of each month and deposit it into his Christmas Club account, where it will earn 7% interest. The account comes to term on December 1. Find the following:
   a. The future value of the account, using an annuity formula
   b. The future value of the account, using the compound interest formula
   c. Bert's total contribution to the account
   d. The total interest

6. On August 19, Rachael Westlake joined a Christmas Club. Her bank will automatically deduct $110 from her checking account at the end of each month and deposit it into her Christmas Club account, where it will earn $6\frac{7}{8}\%$ interest. The account comes to term on December 1. Find the following:
   a. The future value of the account, using an annuity formula
   b. The future value of the account, using the compound interest formula
   c. Rachael's total contribution to the account
   d. The total interest

7. On August 23, Ginny Deus joined a Christmas Club. Her bank will automatically deduct $150 from her checking account at the beginning of each month and deposit it into her Christmas Club account, where it will earn $7\frac{1}{4}\%$ interest. The account comes to term on December 1. Find the following:
   a. The future value of the account, using an annuity formula
   b. The future value of the account, using the compound interest formula
   c. Ginny's total contribution to the account
   d. The total interest

8. On September 19, Lynn Knight joined a Christmas Club. Her bank will automatically deduct $100 from her checking account at the beginning of each month and deposit it into her Christmas Club account, where it will earn 6% interest. The account comes to term on December 1. Find the following:
   a. The future value of the account, using an annuity formula
   b. The future value of the account, using the compound interest formula
   c. Lynn's total contribution to the account
   d. The total interest

9. Pat Gilbert recently set up a TDA to save for her retirement. She arranged to have $175 taken out of each of her monthly checks; it will earn $10\frac{1}{2}\%$ interest. She just had her thirty-ninth birthday, and her ordinary annuity comes to term when she is 65. Find the following:
   a. The future value of the account
   b. Pat's total contribution to the account
   c. The total interest

10. Dick Eckel recently set up a TDA to save for his retirement. He arranged to have $110 taken out of each of his biweekly checks; it will earn $9\frac{7}{8}\%$ interest. He just had his twenty-ninth birthday, and his ordinary annuity comes to term when he is 65. Find the following:
   a. The future value of the account
   b. Dick's total contribution to the account
   c. The total interest

11. Sam Whitney recently set up a TDA to save for his retirement. He arranged to have $290 taken out of each of his monthly checks; it will earn 11% interest. He just had his forty-fifth birthday, and his ordinary annuity comes to term when he is 65. Find the following:

    a. The future value of the account

    b. Sam's total contribution to the account

    c. The total interest

12. Art Dull recently set up a TDA to save for his retirement. He arranged to have $50 taken out of each of his biweekly checks; it will earn it will earn $9\frac{1}{8}$% interest. He just had his 30th birthday, and his ordinary annuity comes to term when he is 65. Find the following:

    a. The future value of the account

    b. Art's total contribution to the account

    c. The total interest

In Exercises 13–18, (a) find and (b) interpret the present value of the given annuity.

13. The annuity in Exercise 1
14. The annuity in Exercise 2
15. The annuity in Exercise 5
16. The annuity in Exercise 6
17. The annuity in Exercise 9
18. The annuity in Exercise 10

In Exercises 19–24, find the monthly payment that will yield the given future value.

19. $100,000 at $9\frac{1}{4}$% interest for thirty years; ordinary annuity
20. $45,000 at $8\frac{7}{8}$% interest for twenty years; ordinary annuity
21. $250,000 at $10\frac{1}{2}$% interest for forty years, ordinary annuity
22. $183,000 at $8\frac{1}{4}$% interest for twenty-five years, ordinary annuity
23. $250,000 at $10\frac{1}{2}$% interest for forty years, annuity due
24. $183,000 at $8\frac{1}{4}$% interest for twenty-five years, annuity due
25. Mr. and Mrs. Gonzales set up a TDA to save for their retirement. They agreed to have $100 deducted from each of Mrs. Gonzales's biweekly paychecks, which will earn $8\frac{1}{8}$% interest.

    a. Find the future value of their ordinary annuity if it comes to term after they retire in $35\frac{1}{2}$ years.

    b. After retiring, the Gonzales family convert their annuity to a savings account, which earns 6.1% interest compounded monthly. At the end of each month, they withdraw $650 for living expenses. Complete the chart in Figure 5.14 for their postretirement account.

Month Number	Account Balance at Beginning of the Month	Interest for the Month	With-drawal	Account Balance at End of the Month
1				
2				
3				
4				
5				

FIGURE 5.14 Chart for Exercise 25.

26. Mr. and Mrs. Jackson set up a TDA to save for their retirement. They agreed to have $125 deducted from each of Mrs. Jackson's biweekly paychecks, which will earn $7\frac{5}{8}$% interest.

    a. Find the future value of their ordinary annuity, if it comes to term after they retire in $32\frac{1}{2}$ years.

    b. After retiring, the Jacksons convert their annuity to a savings account, which earns 6.3% interest compounded monthly. At the end of each month, they withdraw $700 for living expenses. Complete the chart in Figure 5.15 for their postretirement account.

Month Number	Account Balance at Beginning of the Month	Interest for the Month	With-drawal	Account Balance at End of the Month
1				
2				
3				
4				
5				

FIGURE 5.15 Chart for Exercise 26.

27. Jeanne and Harold Kimura want to set up a TDA that will generate sufficient interest at maturity to meet their living expenses, which they project to be $950 per month.

    a. Find the amount needed at maturity to generate $950 per month interest if they can get $6\frac{1}{2}$% interest compounded monthly.

    b. Find the monthly payment that they would have to put into an ordinary annuity to obtain the future value found in part (a) if their money earns $8\frac{1}{4}$% and the term is thirty years.

28. Susan and Bill Stamp want to set up a TDA that will generate sufficient interest at maturity to meet their living expenses, which they project to be $1,200 per month.

a. Find the amount needed at maturity to generate $1,200 per month interest, if they can get $7\frac{1}{4}\%$ interest compounded monthly.

b. Find the monthly payment that they would have to make into an ordinary annuity to obtain the future value found in part (a) if their money earns $9\frac{3}{4}\%$ and the term is twenty-five years.

29. In June 2004, Susan set up a TDA to save for retirement. She agreed to have $200 deducted from each of her monthly paychecks. The annuity's interest rate was allowed to change once each year.

a. In 2004, the interest rate was 1%. Find the account balance in June 2005.

b. In 2005, the interest rate was 2.25%. Find the account balance in June 2006. To do this, think of the June 2005 account balance as a lump sum that earns compound interest.

c. In 2006, the interest rate was 4.5%. Find the account balance in June 2007.

Interest rate source: Mortgagex.com.

30. In June 2007, Manuel set up a TDA to save for retirement. He agreed to have $175 deducted from each of his monthly paychecks. The annuity's interest rate was allowed to change once each year.

a. In 2007, the interest rate was 5.25%. Find the account balance in June 2008.

b. In 2008, the interest rate was 3.8%. Find the account balance in June 2009. To do this, think of the June 2008 account balance as a lump sum that earns compound interest.

c. In 2009, the interest rate was 2.2%. Find the account balance in June 2010.

Interest rate source: Mortgagex.com.

In Exercises 31–34, use the following information. An Individual Retirement Account (IRA) is an annuity that is set up to save for retirement. IRAs differ from TDAs in that an IRA allows the participant to contribute money whenever he or she wants, whereas a TDA requires the participant to have a specific amount deducted from each of his or her paychecks.

31. When Shannon Pegnim was 14, she got an after-school job at a local pet shop. Her parents told her that if she put some of her earnings into an IRA, they would contribute an equal amount to her IRA. That year and every year thereafter, she deposited $1,000 into her IRA. When she became 25 years old, her parents stopped contributing, but Shannon increased her annual deposit to $2,000 and continued depositing that amount annually until she retired at age 65. Her IRA paid 8.5% interest. Find the following:

a. The future value of the account

b. Shannon's and her parents' total contributions to the account

c. The total interest

d. The future value of the account if Shannon waited until she was 19 before she started her IRA

e. The future value of the account if Shannon waited until she was 24 before she started her IRA

32. When Bo McSwine was 16, he got an after-school job at his parents' barbecue restaurant. His parents told him that if he put some of his earnings into an IRA, they would contribute an equal amount to his IRA. That year and every year thereafter, he deposited $900 into his IRA. When he became 21 years old, his parents stopped contributing, but Bo increased his annual deposit to $1,800 and continued depositing that amount annually until he retired at age 65. His IRA paid 7.75% interest. Find the following:

a. The future value of the account

b. Bo's and his parents' total contributions to the account

c. The total interest

d. The future value of the account if Bo waited until he was 18 before he started his IRA

e. The future value of the account if Bo waited until he was 25 before he started his IRA

33. If Shannon Pegnim from Exercise 31 started her IRA at age 35 rather than age 14, how big of an annual contribution would she have had to have made to have the same amount saved at age 65?

34. If Bo McSwine from Exercise 32 started his IRA at age 35 rather than age 16, how big of an annual contribution would he have had to have made to have the same amount saved at age 65?

35. Toni Torres wants to save $1,200 in the next two years to use as a down payment on a new car. If her bank offers her 9% interest, what monthly payment would she need to make into an ordinary annuity to reach her goal?

36. Fred and Melissa Furth's daughter Sally will be a freshman in college in six years. To help cover their extra expenses, the Furths decide to set up a sinking fund of $12,000. If the account pays 7.2% interest and they wish to make quarterly payments, find the size of each payment.

37. Anne Geyer buys some land in Utah. She agrees to pay the seller a lump sum of $65,000 in five years. Until then, she will make monthly simple interest payments to the seller at 11% interest.

a. Find the amount of each interest payment.

b. Anne sets up a sinking fund to save the $65,000. Find the size of her semiannual payments if her payments are due at the end of every six-month period and her money earns $8\frac{3}{8}\%$ interest.

c. Prepare a table showing the amount in the sinking fund after each deposit.

38. Chrissy Fields buys some land in Oregon. She agrees to pay the seller a lump sum of $120,000 in six years.

Until then, she will make monthly simple interest payments to the seller at 12% interest.

a. Find the amount of each interest payment.

b. Chrissy sets up a sinking fund to save the $120,000. Find the size of her semiannual payments if her money earns $10\frac{3}{4}\%$ interest.

c. Prepare a table showing the amount in the sinking fund after each deposit.

39. Develop a new formula for the present value of an ordinary annuity by solving the Present Value of Annuity Formula for P and simplifying.

40. Use the formula developed in Exercise 39 to find the present value of the annuity in Exercise 2.

41. Use the formula developed in Exercise 39 to find the present value of the annuity in Exercise 1.

42. Use the formula developed in Exercise 39 to find the present value of the annuity in Exercise 6.

43. Use the formula developed in Exercise 39 to find the present value of the annuity in Exercise 5.

 Answer the following questions using complete sentences and your own words.

• CONCEPT QUESTIONS

44. Explain the difference between compound interest and an annuity.

45. Explain how an annuity is based on compound interest.

46. Describe the difference between an ordinary annuity and an annuity due.

47. Compare and contrast an annuity with a lump sum investment that receives compound interest. Be sure to discuss both the similarity and the difference between these two concepts, as well as the advantages and disadvantages of each.

48. Which is always greater: the present value of an annuity or the future value? Why?

49. If you want to retire on $2,000 a month for twenty-five years, do you need to save $2,000 a month for twenty-five years before retiring? Why or why not?

50. *For those who have completed Section 1.1:* Is the logic used in deriving the Ordinary Annuity Formula inductive or deductive? Why? Is the logic used in deriving the relationship

$$FV(\text{due}) = FV(\text{ordinary}) \cdot (1 + i)$$

inductive or deductive? Why?

 **WEB PROJECTS**

51. Think of something that you would like to be able to afford but cannot—perhaps a car, boat, or motorcycle.

In this exercise, you will explore how to make that unaffordable dream a realistic goal.

a. Just what is it that you would like to be able to afford?

b. Determine an appropriate but realistic date for making the purchase. Justify your date.

c. Do some research and determine how much your goal would cost currently. Cite your sources.

d. Do some web research and determine the current rate of inflation. Inflation rate information is readily available on the web. Cite your sources.

e. Use the results of parts (b), (c), and (d), as well as either the simple or compound interest formula, to predict how much your goal will cost in the future.

f. Discuss why you chose to use the simple or compound interest formula in part (e).

g. Go to the web sites of four different banks and determine the current interest rate available for an annuity of an appropriate term.

h. Determine the necessary annuity payment that will allow you to meet your goal. If necessary, alter the date from part (b).

52. Suppose you had a baby in 2005 and you decided that it would be wise to start saving for his or her college education. In this exercise, you will explore how to go about doing that.

a. According to the College Board, four-year public schools cost an average of $5,491 per year in 2005–2006. Four-year private schools cost an average of $21,235 per year. These costs include tuition and fees, room and board, books and supplies, and personal expenses. Furthermore, four-year public schools increased 7.1% from 2004–2005, and four-year private schools increased 5.9% from 2004–2005. Use this information, as well as either the simple or compound interest formula, to predict the cost of your child's college education in his or her freshman, sophomore, junior, and senior years for both public and private institutions. Discuss your assumptions and justify your work.
(*Source:* **www.collegeboard.com/pay**.)

b. Will you save for a public or a private institution? Why?

c. Discuss why you chose to use the simple or compound interest formula in part (a).

d. Go to the web sites of four different banks and determine the current interest rate available for an annuity of an appropriate term.

e. Determine the necessary annuity payment that will allow you to save the total amount from part (a) in time to meet your goal.

Some useful links for these web projects are listed on the text web site:

www.cengage.com/math/johnson

ANNUITIES WITH A TI's TVM APPLICATION

EXAMPLE **7**

FINDING HOW LONG IT TAKES   Use a TI-83/84's TVM application to find the how long it takes for an ordinary annuity to have a balance of one million dollars if the monthly payments of $600 earn 6% interest.

SOLUTION

1. Press APPS , select option 1: "Finance", and press ENTER .
2. Select option 1: "TVM Solver", and press ENTER .
3. Enter appropriate values for the variables:

   • N is the number of compounding periods. This is the number we're trying to find, so we temporarily enter 0.
   • I% is the *annual* interest rate (not a periodic rate), as a percent, so enter 6 for I%. Note that we do not convert to a decimal or a periodic rate.
   • PV is the present value. There is no present value here, so we enter 0.
   • PMT is the payment. The payment is $600, but it's an outgoing amount of money (since we give it to the bank), so we enter −600 for PMT. You must enter a negative number for any outgoing amount of money.
   • FV is the future value. We're looking for the amount of time that it takes to have $1,000,000, so enter 1000000. This is an incoming amount of money (since the bank gives it to us), so we use 1000000 rather than −1000000.
   • P/Y is the number of periods per year. We make monthly payments, so there are twelve periods per year. Enter 12 for P/Y.
   • C/Y is automatically made to be the same as P/Y. In this text, we will never encounter a situation in which C/Y is different from P/Y. See Figure 5.16.
   • An ordinary annuity's payments are due at the end of each period so highlight "End"

```
N=█
I%=6
PV=0
PMT=-600
FV=1000000
P/Y=12
C/Y=12
PMT:END BEGIN
```

FIGURE 5.16

Preparing the TVM screen.

4. To solve for N, use the arrow buttons to highlight the 0 that we entered for N. Then press ALPHA SOLVE . (Pressing ALPHA makes the ENTER button becomes the SOLVE button.) As a result, we find that N is 447.8343121.

This is the number of *months* it takes (a period is a month, because of our P/Y entry). This means that it takes about 447.8343121/12 = 37.31 . . . ≈ 37.3 years to have a million dollars.

EXERCISES

53. Redo Example 7 for an annuity due.

54. Find how long it takes for an ordinary annuity to have a balance of two million dollars if the monthly payments of $1100 earn:

a. 2% interest

b. 4% interest

c. 6% interest

d. 10% interest

5.4 Amortized Loans

OBJECTIVES

- Understand what an amortized loan is
- Learn how to do home loan and car loan computations
- Learn how to make and use an amortization schedule

When you buy a house, a car, or a boat, chances are you will finance it with a simple interest amortized loan. A **simple interest amortized loan** is a loan in which each payment consists of principal and interest, and the interest is simple interest computed on the outstanding principal. We'll explore these loans in this section.

An **amortized loan** is a loan for which the loan amount, plus interest, is paid off with a series of regular equal payments. An add-on interest loan (discussed in Section 5.1) is an amortized loan. A simple interest amortized loan is also an amortized loan, but it is calculated differently from an add-on interest loan. You should be aware that the payments are smaller with a simple interest amortized loan than they are with an add-on interest loan (assuming, naturally, that the interest rates, loan amounts, and number of payments are the same).

A simple interest amortized loan is calculated as an ordinary annuity whose future value is the same as the loan amount's future value, under compound interest. Realize, though, that a simple interest amortized loan's payments are used to pay off a loan, whereas an annuity's payments are used to generate savings.

SIMPLE INTEREST AMORTIZED LOAN FORMULA

Future value of annuity = future value of loan amount

$$pymt \frac{(1 + i)^n - 1}{i} = P(1 + i)^n$$

where *pymt* is the loan payment, i is the periodic interest rate, n is the number of periods, and P is the present value or loan amount.

Algebraically, this formula could be used to determine any one of the four unknowns (*pymt*, i, n, and P) if the other three are known. We will use it to find the payment when the annual interest rate, number of periods, and loan amount are known.

EXAMPLE 1

BUYING A CAR   Tom and Betty decide that they need a more dependable car, now that they have a baby. They buy one for $13,518.77. They make a $1,000 down payment and finance the balance through a four-year simple interest amortized loan from their bank. They are charged 12% interest.

a. Find their monthly payment.

b. Find the total interest they will pay over the life of the loan.

SOLUTION

a. Use the simple interest amortized loan formula, with a loan amount of $P = $13,518.77 − $1,000 = $12,518.77$, a monthly interest rate of $12\%/12 = 1\% = 0.01$, and a term of $n = 4$ years $= 48$ months.

$$pymt\frac{(1 + i)^n - 1}{i} = P(1 + i)^n \qquad \text{the Simple Interest Amortized Loan Formula}$$

$$pymt\frac{(1 + 0.01)^{48} - 1}{0.01} = 12,518.77(1 + 0.01)^{48} \quad \text{substituting}$$

To find *pymt*, we need to divide the right side by the fraction on the left side or, equivalently, multiply by its reciprocal. First, find the fraction on the left side, as with any annuity calculation. Then multiply the right side by the fraction's reciprocal.

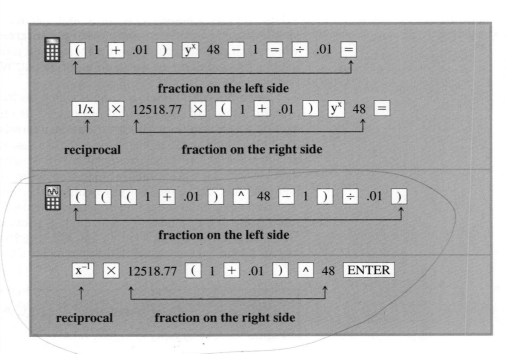

We get *pymt* = 329.667... ≈ $329.67.

As a rough check, note that Tom and Betty borrowed $12,518.77. If they paid no interest, each payment would be $12,518.77/48 ≈ $260.81. Since they must pay interest, each payment must be larger than $260.81. Our calculation checks, because $329.67 > $260.81.

b. The total of their payments is $48 \cdot \$329.67 = \$15,824.16$. This includes both principal and interest. Of this, $12,518.77 is principal, so $15,824.16 − $12,518.77 = $3,305.39 is interest. Over the life of the loan, Tom and Betty are paying $3,305.39 in interest.

Amortization Schedules

The simple interest amortized loan formula is an equation with the ordinary annuity formula on one side of the equal symbol and the compound interest formula on the other side. However, with such a loan, you do not pay compound interest. Instead, the interest portion of each payment is simple interest on the outstanding principal.

An **amortization schedule** is a list of several periods of payments, the principal and interest portions of those payments, and the **outstanding principal** (or **balance**) after each of those payments is made.

The data on the amortization schedule are important to the borrower for two reasons. The borrower needs to know the total interest paid, for tax purposes. Interest paid on a home loan is usually deductible from the borrower's income tax, and interest paid on a loan by a business is usually deductible. The borrower would also need the data on an amortization schedule if he or she is considering paying off the loan early. Such prepayment could save money because an advance payment would be all principal and would not include any interest; however, the lending institution may charge a **prepayment penalty** which would absorb some of the interest savings.

EXAMPLE 2

SOLUTION

PREPARING AN AMORTIZATION SCHEDULE   Prepare an amortization schedule for the first two months of Tom and Betty's loan.

For any simple interest loan, the interest portion of each payment is simple interest on the outstanding principal, so use the Simple Interest Formula, $I = Prt$, to compute the interest. Recall that r and t are annual figures. For each payment, $r = 12\% = 0.12$ and $t = 1$ month $= \frac{1}{12}$ year. For payment number 1, $P = \$12,418.77$ (the amount borrowed). The interest portion of payment number 1 is

$$I = Prt \quad \text{the Simple Interest Formula}$$
$$= 12,518.77 \cdot 0.12 \cdot \frac{1}{12} \quad \text{substituting}$$
$$\approx \$125.19 \quad \text{rounding}$$

The principal portion of payment number 1 is

Payment − interest portion
$$= \$329.67 - \$125.19$$
$$= \$204.48$$

The outstanding principal or balance is

$(P \times \%) \cdot 1 \div 12 =$

Previous principal − principal portion
$$= \$12,518.77 - \$204.48$$
$$= \$12,314.29$$

All of the above information goes into the amortization schedule. See Figure 5.17.

Payment number	Principal portion	Interest portion	Total payment	Balance
0	—	—	—	$12,518.77
1	$204.48	$125.19	$329.67	$12,314.29
2	$206.53	$123.14	$329.67	$12,107.76

FIGURE 5.17   An amortization schedule for the first two months of Tom and Betty's loan.

When it is time to make payment number 2, less money is owed. The outstanding principal is $12,314.29, so this is the new value of P. The interest portion of payment number 2 is

$$I = Prt \quad \text{the Simple Interest Formula}$$
$$= 12,314.29 \cdot 0.12 \cdot \frac{1}{12} \quad \text{substituting}$$
$$\approx \$123.14 \quad \text{rounding}$$

The principal portion is

Payment $-$ interest portion
$$= \$329.67 - \$123.14$$
$$= \$206.53$$

The outstanding principal or balance is

Previous principal $-$ principal portion
$$= \$12,314.29 - \$206.53$$
$$= \$12,107.76$$

All of the above information is shown in the amortization schedule in Figure 5.17.

Notice how in Example 2, the principal portion increases and the interest portion decreases. This continues throughout the life of the loan, and the final payment is mostly principal. This happens because after each payment, the amount due is somewhat smaller, so the interest on the amount due is somewhat smaller also.

EXAMPLE **3**

AMORTIZATION SCHEDULES AND WHY THE LAST PAYMENT IS DIFFERENT Comp-U-Rent needs to borrow $60,000 to increase their inventory of rental computers. The company is confident that their expanded inventory will generate sufficient extra income to allow them to pay off the loan in a short amount of time, so they wish to borrow the money for only three months. First National Bank offers them a simple interest amortized loan at $8\frac{3}{4}\%$ interest.

a. Find what their monthly payment would be with First National.
b. Prepare an amortization schedule for the entire term of the loan.

SOLUTION

a. $P = \$60,000$, $i = \frac{1}{12}$ of $8\frac{3}{4}\% = 0.0875/12$, and $n = 3$ months.

Future value of annuity $=$ future value of loan amount

$$pymt\frac{(1 + i)^n - 1}{i} = P(1 + i)^n$$

$$pymt\frac{\left(1 + \dfrac{0.0875}{12}\right)^3 - 1}{\dfrac{0.0875}{12}} = 60000\left(1 + \frac{0.0875}{12}\right)^3 \quad \text{substituting}$$

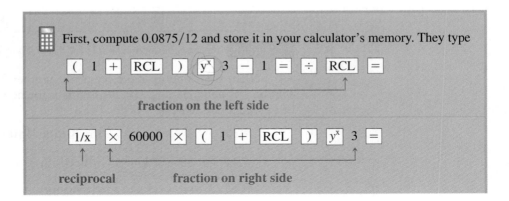

First, compute 0.0875/12 and store it in your calculator's memory. They type

(1 + RCL) y^x 3 $-$ 1 $=$ $\div$ RCL $=$

fraction on the left side

1/x $\times$ 60000 $\times$ (1 + RCL) y^x 3 $=$

reciprocal   **fraction on right side**

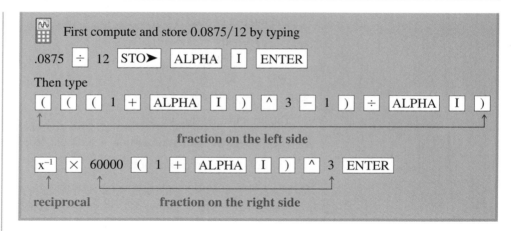

See Figure 5.18 on page 374.

We get

$$pymt = 20,292.375\ldots \quad \textbf{solving for } pymt$$
$$\approx \$20,292.38 \quad \textbf{rounding}$$

See the "Total Payment" column in Figure 5.18 on page 374.

> As a rough check, notice that Comp-U-Rent borrowed \$60,000, and each payment includes principal and interest, so the payment must be larger than $\frac{\$60,000}{3} = \$20,000$. Our calculation checks, because \$20,292.38 > \$20,000.

b. For each payment, $r = 8\frac{3}{4}\% = 0.0875$ and $t = 1$ month $= \frac{1}{12}$ year.

For payment number 1, $P = \$60,000.00$ (the amount borrowed). The interest portion of payment number 1 is

$$I = Prt \qquad\qquad \textbf{the Simple Interest Formula}$$
$$= 60,000.00 \cdot 0.0875 \cdot \frac{1}{12} \quad \textbf{substituting}$$
$$= \$437.50$$

See the "Interest portion" column in Figure 5.18.

The principal portion of payment number 1 is

Payment − interest portion
$$= \$20,292.38 - \$437.50 = \$19,854.88$$

See the "Principal portion" column in Figure 5.18.

The outstanding principal or balance is

Previous principal − principal portion
$$= \$60,000.00 - \$19,854.88 = \$40,145.12$$

For payment number 2, $P = \$40,145.12$ (the outstanding principal).

The interest portion of payment number 2 is

$$I = Prt \qquad\qquad \textbf{the Simple Interest Formula}$$
$$= 40,145.12 \cdot 0.0875 \cdot \frac{1}{12} \quad \textbf{substituting}$$
$$\approx \$292.72 \qquad\qquad \textbf{rounding}$$

The Principle portion of payment number 2 is

Payment − interest portion
$$= \$20,292.38 - \$292.72 = \$19,999.66$$

The outstanding principal or balance is

$$\$40,145.12 - \$19,999.66 = \$20,145.46$$

For payment number 3, $P = \$20,145.46$

The interest portion is

$$I = PRT$$

$$= \$20{,}145.46 \cdot 0.0875 \cdot \frac{1}{12}$$

$$= \$146.89$$

The principal portion is

$$\text{Payment} - \text{interest portion} = \$20{,}292.38 - \$146.89 = \$20{,}145.49$$

The outstanding principal or balance is

$$\text{Previous principal} - \text{principal portion}$$

$$= \$20{,}145.46 - \$20{,}145.49 = -\$0.03$$

Negative three cents cannot be correct. After the final payment is made, the amount due *must be* $0.00. The discrepancy arises from the fact that we rounded off the payment size from $20,292.375 . . . to $20,292.38. If there were some way in which the borrower could make a monthly payment that is not rounded off, then the calculation above would have yielded an amount due of $0.00. To repay the exact amount owed, we must compute the last payment differently.

The principal portion of payment number 3 *must be* $20,145.46, because that is the balance, and this payment is the only chance to pay it. The payment must also include $146.89 interest, as calculated above. The last payment is the sum of the principal due and the interest on that principal:

$$\$20{,}145.46 + \$146.89 = \$20{,}292.35$$

The closing balance is then $20,145.46 − $20,145.46 = $0.00, as it should be. The amortization schedule is given in Figure 5.18.

Payment number	Principal portion	Interest portion	Total payment	Balance
0	—	—	—	$60,000.00
1	$19,854.88	$437.50	$20,292.38	$40,145.12
2	$19,999.66	$292.72	$20,292.38	$20,145.46
3	$20,145.46	$146.89	$20,292.35	$0.00

FIGURE 5.18 The amortization schedule for Comp-U-Rent's loan.

AMORTIZATION SCHEDULE STEPS

For each payment, list the payment number, principal portion, interest portion, total payment, and balance.

For each payment:
1. Find the interest on the balance, using the simple interest formula.

For each payment except the last:
2. The principal portion is the payment minus the interest portion.
3. The new balance is the previous balance minus the principal portion.

For the last payment:
4. The principal portion is the previous balance.
5. The total payment is the sum of the principal portion and interest portion.

Figure 5.19 shows these steps in chart form.

Payment number	Principal portion	Interest portion	Total payment	Balance
0	—	—	—	Loan amount
First through next-to-last	Total payment minus interest portion	Simple interest on previous balance; use $I = Prt$	Use simple interest amortized loan formula	Previous balance minus this payment's principal portion
Last	Previous balance	Simple interest on previous balance; use $I = Prt$	Principal portion plus interest portion	$0.00

FIGURE 5.19   Amortization schedule steps.

In preparing an amortization schedule, the new balance is the previous balance minus the principal portion. This means that only the principal portion of a payment goes toward paying off the loan. The interest portion is the lender's profit.

Prepaying a Loan

Sometimes a borrower needs to pay his or her loan off early, before the loan's term is over. This is called **prepaying a loan.** This often happens when people sell their houses. It also happens when interest rates have fallen since the borrower obtained his or her loan and the borrower decides to refinance the loan at a lower rate.

When you prepay a loan, you pay off the unpaid balance. You can find that unpaid balance by looking at an amortization schedule. If an amortization schedule has not already been prepared, the borrower can easily approximate the **unpaid balance** by subtracting the current value of the annuity from the current value of the loan. This approximation is usually off by at most a few pennies.

UNPAID BALANCE FORMULA

Unpaid balance = current value of loan amount − current value of annuity

$$\approx P(1 + i)^n - pymt\,\frac{(1 + i)^n - 1}{i}$$

where *pymt* is the loan payment, *i* is the periodic interest rate, *n* is the number of periods *from the beginning of the loan to the present*, and *P* is the loan amount.

A common error in using this formula is to let *n* equal the number of periods in the entire life of the loan rather than the number of periods from the beginning of the loan until the time of prepayment. This results in an answer of 0. After all of the payments have been made, the unpaid balance should be 0.

EXAMPLE **4**

PREPAYING A LOAN Ten years ago, Tom and Betty bought a house for $140,000. They paid the sellers a 20% down payment and obtained a simple interest amortized loan for $112,000 from their bank at $10\frac{3}{4}\%$ for thirty years. Their monthly payment is $1,045.50.

Their old home was fine when there were just two of them. But with their new baby, they need more room. They are considering selling their existing home and buying a new home. This would involve paying off their home loan. They would use the income from the house's sale to do this. Find the cost of paying off the existing loan.

SOLUTION

The loan is a thirty-year loan, so if we were computing the payment, we would use $n = 30\text{ years} \cdot 12\text{ months/year} = 360$ months. However, we are computing the unpaid balance, not the payment, so n is not 360. Tom and Betty have made payments on their loan for 10 years = 120 months, so for this calculation, $n = 120$.

Unpaid balance
= current value of loan amount − current value of annuity

$$\approx P(1 + i)^n - pymt\,\frac{(1 + i)^n - 1}{i}$$

$$= 112{,}000\left(1 + \frac{0.1075}{12}\right)^{120} - 1{,}045.50\,\frac{\left(1 + \dfrac{0.1075}{12}\right)^{120} - 1}{\dfrac{0.1075}{12}}$$

$$= \$102{,}981.42$$

First, compute $0.1075/12$, and store it in the calculator's memory (as ALPHA I with a graphing calculator). Then type

⬛ 112000 ⊠ ⦅ 1 ⊞ ⟦RCL⟧ ⦆ ⟦yˣ⟧ 120 ⊟ 1045.50 ⊠ ⦅
⦅ 1 ⊞ ⟦RCL⟧ ⦆ ⟦yˣ⟧ 120 ⊟ 1 ⦆ ÷ ⟦RCL⟧ ⊟

〰 112000 ⦅ 1 ⊞ ⟦ALPHA⟧ ⟦I⟧ ⦆ ⟦^⟧ 120 ⊟ 1045.50
⦅ ⦅ 1 ⊞ ⟦ALPHA⟧ ⟦I⟧ ⦆ ⟦^⟧ 120 ⊟ 1 ⦆ ÷
⟦ALPHA⟧ ⟦I⟧ ⟦ENTER⟧

As a rough check, note that Tom and Betty borrowed $112,000, so their unpaid balance must be less. Our calculation checks, because $102,981.42 < $112,000.

The result of Example 4 means that after ten years of payments of over $1000 a month on a loan of $112,000, Tom and Betty still owe approximately $102,981.42! They were shocked. It is extremely depressing for first-time home purchasers to discover how little of their beginning payments actually goes toward paying off the loan. At the beginning of the loan, you owe a lot of money, so most of each payment is interest, and very little is principal. Later, you do not owe as much, so most of each payment is principal and very little is interest.

In the following exercises, all loans are simple interest amortized loans with monthly payments unless labeled otherwise.

In Exercises 1–6, find (a) the monthly payment and (b) the total interest for the given simple interest amortized loan.

1. $5,000, $9\frac{1}{2}\%$, 4 years
2. $8,200, $10\frac{1}{4}\%$, 6 years
3. $10,000, $6\frac{1}{8}\%$, 5 years
4. $20,000, $7\frac{3}{8}\%$, $5\frac{1}{2}$ years
5. $155,000, $9\frac{1}{2}\%$, 30 years
6. $289,000, $10\frac{3}{4}\%$, 35 years

7. Wade Ellis buys a new car for $16,113.82. He puts 10% down and obtains a simple interest amortized loan for the rest at $11\frac{1}{2}\%$ interest for four years.
   a. Find his monthly payment.
   b. Find the total interest.
   c. Prepare an amortization schedule for the first two months of the loan.

8. Guy dePrimo buys a new car for $9,837.91. He puts 10% down and obtains a simple interest amortized loan for the rest at $8\frac{7}{8}\%$ interest for four years.
   a. Find his monthly payment.
   b. Find the total interest.
   c. Prepare an amortization schedule for the first two months of the loan.

9. Chris Burditt bought a house for $212,500. He put 20% down and obtains a simple interest amortized loan for the rest at $10\frac{7}{8}\%$ for thirty years.
   a. Find his monthly payment.
   b. Find the total interest.
   c. Prepare an amortization schedule for the first two months of the loan.
   d. Most lenders will approve a home loan only if the total of all the borrower's monthly payments, including the home loan payment, is no more than 38% of the borrower's monthly income. How much must Chris make to qualify for the loan?

10. Shirley Trembley bought a house for $187,600. She put 20% down and obtains a simple interest amortized loan for the rest at $6\frac{3}{8}\%$ for thirty years.
    a. Find her monthly payment.
    b. Find the total interest.
    c. Prepare an amortization schedule for the first two months of the loan.
    d. Most lenders will approve a home loan only if the total of all the borrower's monthly payments, including the home loan payment, is no more than 38% of the borrower's monthly income. How much must Shirley make to qualify for the loan?

11. Dennis Lamenti wants to buy a new car that costs $15,829.32. He has two possible loans in mind. One loan is through the car dealer; it is a four-year add-on interest loan at $7\frac{3}{4}\%$ and requires a down payment of $1,000. The second is through his bank; it is a four-year simple interest amortized loan at at $7\frac{3}{4}\%$ and requires a down payment of $1,000.
    a. Find the monthly payment for each loan.
    b. Find the total interest paid for each loan.
    c. Which loan should Dennis choose? Why?

12. Barry Wood wants to buy a used car that costs $4,000. He has two possible loans in mind. One loan is through the car dealer; it is a three-year add-on interest loan at 6% and requires a down payment of $300. The second is through his credit union; it is a three-year simple interest amortized loan at 9.5% and requires a 10% down payment.
    a. Find the monthly payment for each loan.
    b. Find the total interest paid for each loan.
    c. Which loan should Barry choose? Why?

13. Investigate the effect of the term on simple interest amortized auto loans by finding the monthly payment and the total interest for a loan of $11,000 at $9\frac{7}{8}\%$ interest if the term is
    a. three years
    b. four years
    c. five years

14. Investigate the effect of the interest rate on simple interest amortized auto loans by finding the monthly payment and the total interest for a four-year loan of $12,000 at
    a. 8.5%   b. 8.75%
    c. 9%   d. 10%.

15. Investigate the effect of the interest rate on home loans by finding the monthly payment and the total interest for a thirty-year loan of $100,000 at
    a. 6%   b. 7%
    c. 8%   d. 9%
    e. 10%   f. 11%

16. Some lenders offer fifteen-year home loans. Investigate the effect of the term on home loans by finding the monthly payment and total interest for a loan of $100,000 at 10% if the term is
    a. thirty years   b. fifteen years

17. Some lenders offer loans with biweekly payments rather than monthly payments. Investigate the effect of this on home loans by finding the payment and total interest on a thirty-year loan of $100,000 at 10% interest if payments are made (a) monthly and (b) biweekly.

18. Some lenders are now offering loans with biweekly payments rather than monthly payments. Investigate the effect of this on home loans by finding the payment and total interest on a thirty-year loan of $150,000 at 8% interest if payments are made (a) monthly and (b) biweekly.

19. Verify (a) the monthly payments and (b) the interest savings in the savings and loan advertisement in Figure 5.20.

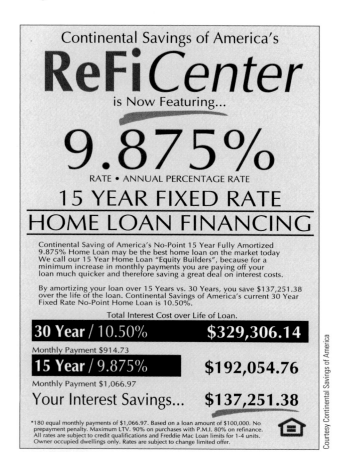

FIGURE 5.20 Loan terms for Exercise 19.

20. The home loan in Exercise 19 presented two options. The thirty-year option required a smaller monthly payment. A consumer who chooses the thirty-year option could take the savings in the monthly payment that this option generates and invest that savings in an annuity. At the end of fifteen years, the annuity might be large enough to pay off the thirty-year loan. Determine whether this is a wise plan if the annuity's interest rate is 7.875%. (Disregard the tax ramifications of this approach.)

21. Pool-N-Patio World needs to borrow $75,000 to increase its inventory for the upcoming summer season. The owner is confident that he will sell most, if not all, of the new inventory during the summer, so he wishes to borrow the money for only four months. His bank has offered him a simple interest amortized loan at $7\frac{3}{4}\%$ interest.

    a. Find the size of the monthly bank payment.
    b. Prepare an amortization schedule for all four months of the loan.

22. Slopes R Us needs to borrow $120,000 to increase its inventory of ski equipment for the upcoming season. The owner is confident that she will sell most, if not all, of the new inventory during the winter, so she wishes to borrow the money for only five months. Her bank has offered her a simple interest amortized loan at $8\frac{7}{8}\%$ interest.

    a. Find the size of the monthly bank payment.
    b. Prepare an amortization schedule for all five months of the loan.

23. The owner of Blue Bottle Coffee is opening a second store and needs to borrow $93,000. Her success with her first store has made her confident that she will be able to pay off her loan quickly, so she wishes to borrow the money for only four months. Her bank has offered her a simple interest amortized loan at $9\frac{1}{8}\%$ interest.

    a. Find the size of the monthly bank payment.
    b. Prepare an amortization schedule for all four months of the loan.

24. The Green Growery Nursery needs to borrow $48,000 to increase its inventory for the upcoming summer season. The owner is confident that he will sell most, if not all, of the new plants during the summer, so he wishes to borrow the money for only four months. His bank has offered him a simple interest amortized loan at $9\frac{1}{4}\%$ interest.

    a. Find the size of the monthly bank payment.
    b. Prepare an amortization schedule for all four months of the loan.

*For Exercises 25–28, note the following information. A **line of credit** is an agreement between a bank and a borrower by which the borrower can borrow any amount of money (up to a mutually agreed-upon maximum) at any time, simply by writing a check. Typically, monthly simple interest payments are required, and the borrower is free to make principal payments as frequently or infrequently as he or she wants. Usually, a line of credit is secured by the title to the borrower's house, and the interest paid to the bank by the borrower is deductible from the borrower's income taxes.*

25. Kevin and Roxanne Gahagan did not have sufficient cash to pay their income taxes. However, they had previously set up a line of credit with their bank. On April 15, they wrote a check to the Internal Revenue Service on their line of credit for $6,243. The line's interest rate is 5.75%.

    a. Find the size of the required monthly interest payment.
    b. The Gahagans decided that it would be in their best interests to get this loan paid off in eight months.

Find the size of the monthly principal-plus-interest payment that would accomplish this. (*HINT:* In effect, the Gahagans are converting the loan to an amortized loan.)

c. Prepare an amortization schedule for all eight months of the loan.

d. Find the amount of line of credit interest that the Gahagans could deduct from their taxes next year.

26. James and Danna Wright did not have sufficient cash to pay their income taxes. However, they had previously set up a line of credit with their bank. On April 15, they wrote a check to the Internal Revenue Service on their line of credit for $10,288. The line's interest rate is 4.125%.

a. Find the size of the required monthly interest payment.

b. The Wrights decided that it would be in their best interests to get this loan paid off in six months. Find the size of the monthly principal-plus-interest payment that would accomplish this. (*HINT:* In effect, the Wrights are converting the loan to an amortized loan.)

c. Prepare an amortization schedule for all six months of the loan.

d. Find the amount of line of credit interest that the Wrights could deduct from their taxes next year.

27. Homer Simpson had his bathroom remodeled. He did not have sufficient cash to pay for it. However, he had previously set up a line of credit with his bank. On July 12, he wrote a check to his contractor on his line of credit for $12,982. The line's interest rate is 8.25%.

a. Find the size of the required monthly interest payment.

b. Homer decided that it would be in his best interests to get this loan paid off in seven months. Find the size of the monthly principal-plus-interest payment that would accomplish this. (*HINT:* In effect, Simpson is converting the loan to an amortized loan.)

c. Prepare an amortization schedule for all seven months of the loan.

d. Find the amount of line of credit interest that Simpson could deduct from his taxes next year.

28. Harry Trask had his kitchen remodeled. He did not have sufficient cash to pay for it. However, he had previously set up a line of credit with his bank. On July 12, he wrote a check to his contractor on his line of credit for $33,519. The line's interest rate is 7.625%.

a. Find the size of the required monthly interest payment.

b. Harry decided that it would be in his best interests to get this loan paid off in seven months. Find the size of the monthly principal-plus-interest payment that would accomplish this. (*HINT:* In effect, Trask is converting the loan to an amortized loan.)

c. Prepare an amortization schedule for all seven months of the loan.

d. Find the amount of line of credit interest that Trask could deduct from his taxes next year.

29. Wade Ellis buys a car for $16,113.82. He puts 10% down and obtains a simple interest amortized loan for the balance at $11\frac{1}{2}\%$ interest for four years. Three years and two months later, he sells his car. Find the unpaid balance on his loan.

30. Guy de Primo buys a car for $9837.91. He puts 10% down and obtains a simple interest amortized loan for the balance at $10\frac{7}{8}\%$ interest for four years. Two years and six months later, he sells his car. Find the unpaid balance on his loan.

31. Gary Kersting buys a house for $212,500. He puts 20% down and obtains a simple interest amortized loan for the balance at $10\frac{7}{8}\%$ interest for thirty years. Eight years and two months later, he sells his house. Find the unpaid balance on his loan.

32. Shirley Trembley buys a house for $187,600. She puts 20% down and obtains a simple interest amortized loan for the balance at $11\frac{3}{8}\%$ interest for thirty years. Ten years and six months later, she sells her house. Find the unpaid balance on her loan.

33. Harry and Natalie Wolf have a three-year-old loan with which they purchased their house. Their interest rate is $13\frac{3}{8}\%$. Since they obtained this loan, interest rates have dropped, and they can now get a loan for $8\frac{7}{8}\%$ through their credit union. Because of this, the Wolfs are considering refinancing their home. Each loan is a thirty-year simple interest amortized loan, and neither has a prepayment penalty. The existing loan is for $152,850, and the new loan would be for the current amount due on the old loan.

a. Find their monthly payment with the existing loan.

b. Find the loan amount for their new loan.

c. Find the monthly payment with their new loan.

d. Find the total interest they will pay if they do *not* get a new loan.

e. Find the total interest they will pay if they *do* get a new loan.

f. Should the Wolfs refinance their home? Why or why not?

34. Russ and Roz Rosow have a ten-year-old loan with which they purchased their house. Their interest rate is $10\frac{5}{8}\%$. Since they obtained this loan, interest rates have dropped, and they can now get a loan for $9\frac{1}{4}\%$ through their credit union. Because of this, the Rosows are considering refinancing their home. Each loan is a 30-year simple interest amortized loan, and neither has a prepayment penalty. The existing loan is for $112,000, and the new loan would be for the current amount due on the old loan.

a. Find their monthly payment with the existing loan.

b. Find the loan amount for their new loan.

c. Find the monthly payment with their new loan.

d. Find the total interest they will pay if they do *not* get a new loan.

e. Find the total interest they will pay if they *do* get a new loan.

f. Should the Rosows refinance their home? Why or why not?

35. Michael and Lynn Sullivan have a ten-year-old loan for $187,900 with which they purchased their house. They just sold their highly profitable import-export business and are considering paying off their home loan. Their loan is a thirty-year simple interest amortized loan at 10.5% interest and has no prepayment penalty.

a. Find their monthly payment.

b. Find the unpaid balance of the loan.

c. Find the amount of interest they will save by prepaying.

d. The Sullivans decided that if they paid off their loan, they would deposit the equivalent of half their monthly payment into an annuity. If the ordinary annuity pays 9% interest, find its future value after 20 years.

e. The Sullivans decided that if they do not pay off their loan, they would deposit an amount equivalent to their unpaid balance into an account that pays $9\frac{3}{4}$% interest compounded monthly. Find the future value of this account after 20 years.

f. Should the Sullivans prepay their loan? Why or why not?

36. Charlie and Ellen Wilson have a twenty-five-year-old loan for $47,000 with which they purchased their house. The Wilsons are retired and are living on a fixed income, so they are contemplating paying off their home loan. Their loan is a thirty-year simple interest amortized loan at $4\frac{1}{2}$% and has no prepayment penalty. They also have savings of $73,000, which they have invested in a certificate of deposit currently paying $8\frac{1}{4}$% interest compounded monthly. Should they pay off their home loan? Why or why not?

37. Ray and Helen Lee bought a house for $189,500. They put 10% down, borrowed 80% from their bank for thirty years at $11\frac{1}{2}$%, and convinced the owner to take a second mortgage for the remaining 10%. That 10% is due in full in five years (this is called a *balloon payment*), and the Lees agree to make monthly interest-only payments to the seller at 12% simple interest in the interim.

a. Find the Lees' down payment

b. Find the amount that the Lees borrowed from their bank.

c. Find the amount that the Lees borrowed from the seller.

d. Find the Lees' monthly payment to the bank.

e. Find the Lees' monthly interest payment to the seller.

38. Jack and Laurie Worthington bought a house for $163,700. They put 10% down, borrowed 80% from their bank for thirty years at 12% interest, and convinced the owner to take a second mortgage for the remaining 10%. That 10% is due in full in five years, and the Worthingtons agree to make monthly interest-only payments to the seller at 12% simple interest in the interim.

a. Find the Worthingtons' down payment

b. Find the amount that the Worthingtons borrowed from their bank.

c. Find the amount that the Worthingtons borrowed from the seller.

d. Find the Worthingtons' monthly payment to the bank.

e. Find the Worthingtons' monthly interest payment to the seller.

39. **a.** If the Lees in Exercise 37 save for their balloon payment with a sinking fund, find the size of the necessary monthly payment into that fund if their money earns 6% interest.

b. Find the Lees' total monthly payment for the first five years.

c. Find the Lees' total monthly payment for the last twenty-five years.

40. **a.** If the Worthingtons in Exercise 38 save for their balloon payment with a sinking fund, find the size of the necessary monthly payment into that fund if their money earns 7% interest.

b. Find the Worthingtons' total monthly payment for the first five years.

c. Find the Worthingtons' total monthly payment for the last twenty-five years.

41. In July 2005, Tom and Betty bought a house for $275,400. They put 20% down and financed the rest with a thirty-year loan at the then-current rate of 6%. In 2007, the real estate market crashed. In August 2009, they had to sell their house. The best they could get was $142,000. Was this enough to pay off the loan? It so, how much did they profit? It not, how much did they have to pay out of pocket to pay off the loan?

42. In August 2004, Bonnie Martin bought a house for $395,000. She put 20% down and financed the rest with a thirty-year loan at the then-current rate of $5\frac{3}{4}$%. In 2007, the real estate market crashed. In June 2009, she had to sell her house. The best she could get was $238,000. Was this enough to pay off the loan? If so, how much did she profit? It not, how much did she have to pay out of pocket to pay off the loan?

43. Al-Noor Koorji bought a house for $189,000. He put 20% down and financed the rest with a thirty-year loan at $5\frac{3}{4}$%.

a. Find his monthly payment.

b. If he paid an extra $100 per month, how early would his loan be paid off?

(*HINT:* This involves some trial-and-error work.)

44. The Franklins bought a house for $265,000. They put 20% down and financed the rest with a thirty-year loan at $5\frac{3}{8}\%$.

a. Find their monthly payment.

b. If they paid an extra $85 per month, how early would their loan be paid off?

(*HINT:* This involves some trial-and-error work.)

 Answer the following questions using complete sentences and your own words.

• CONCEPT QUESTIONS

45. In Exercise 11, an add-on interest amortized loan required larger payments than did a simple interest amortized loan at the same interest rate. What is there about the structure of an add-on interest loan that makes its payments larger than those of a simple interest amortized loan?

(*HINT:* The interest portion is a percentage of what quantity?)

46. Why are the computations for the last period of an amortization schedule different from those for all preceding periods?

47. Give two different situations in which a borrower would need the information contained in an amortization schedule.

48. A borrower would need to know the unpaid balance of a loan if the borrower were to prepay the loan. Give three different situations in which it might be in a borrower's best interests to prepay a loan.

(*HINT:* See some of the preceding exercises.)

49. If you double the period of an amortized loan, does your monthly payment halve? Why or why not?

50. If you double the loan amount of an amortized loan, does your monthly payment double? Why or why not?

WEB PROJECTS

51. This is an exercise in buying a car. It involves choosing a car and selecting the car's financing. Write a paper describing all of the following points.

a. You might not be in a position to buy a car now. If that is the case, fantasize about your future. What job do you have? How long have you had that job? What is your salary? If you are married, does your spouse work? Do you have a family? What needs will your car fulfill? Make your fantasy realistic. Briefly describe what has happened between the present and your future fantasy. (If you are in a position to buy a car now, discuss these points on a more realistic level.)

b. Go shopping for a car. Look at new cars, used cars, or both. Read newspaper and magazine articles about your choices (see, for example, *Consumer Reports, Motor Trend,* and *Road and Track*). Discuss in detail the car you selected and why you did so. How will your selection fulfill your (projected) needs? What do newspapers and magazines say about your selection? Why did you select a new or a used car?

c. Go to the web sites of four different banks. Get all of the information you need about a car loan. Perform all appropriate computations yourself—do not have the lenders tell you the payment size, and do not use web calculators. Summarize the appropriate data (including the down payment, payment size, interest rate, duration of loan, type of loan, and loan fees) in your paper, and discuss which loan you would choose. Explain how you would be able to afford your purchase.

52. This is an exercise in buying a home. It involves choosing a home and selecting the home's financing. Write a paper describing all of the following points.

a. You might not be in a position to buy a home now. If this is the case, fantasize about your future. What job do you have? How long have you had that job? What is your salary? If you are married, does your spouse work? Do you have a family? What needs will your home fulfill? Make your future fantasy realistic. Briefly describe what has happened between the present and your future fantasy. (If you are in a position to buy a home now, discuss these points on a more realistic level.)

b. Go shopping for a home. Look at houses, condominiums, or both. Look at new homes, used homes, or both. Used homes can easily be visited by going to an "open house," where the owners are gone and the real estate agent allows interested parties to inspect the home. Open houses are probably listed in your local newspaper. Read appropriate newspaper and magazine articles (for example, in the real estate section of your local newspaper). Discuss in detail the home you selected and why you did so. How will your selection fulfill your (projected) needs? Why did you select a house or a condominium? Why did you select a new or a used home? Explain your choice of location, house size, and features of the home.

c. Go to the web sites of four different banks. Get all of the information you need about a home loan. Perform all appropriate computations yourself—do not have the lenders tell you the payment size, and do not use web calculators. Summarize the appropriate data in your paper and discuss which loan you would choose. Include in your discussion the down payment, the duration of the loan, the interest rate, the payment size, and other terms of the loan.

d. Also discuss the real estate taxes (your instructor will provide you with information on the local tax rate) and the effect of your purchase on your income taxes (interest paid on a home loan is deductible from the borrower's income taxes).

e. Most lenders will approve a home loan only if the total of all the borrower's monthly payments, including the home loan payment, real estate taxes, credit card payments, and car loan payments, is no more than 38% of the borrower's monthly income. Discuss your ability to qualify for the loan.

Some useful links for the above web projects are listed on the text web site: **www.cengage.com/math/johnson**

• PROJECTS

For Exercises 53–55, note the following information. An **adjustable-rate mortgage** *(or ARM) is, as the name implies, a mortgage in which the interest rate is allowed to change. As a result, the payment changes too. At first, an ARM costs less than a fixed-rate mortgage—its initial interest rate is usually two or three percentage points lower. As time goes by, the rate is adjusted. As a result, it might or might not continue to hold this advantage.*

53. Trustworthy Savings offers a thirty-year adjustable-rate mortgage with an initial rate of 5.375%. The rate and the required payment are adjusted annually. Future rates are set at 2.875 percentage points above the 11th District Federal Home Loan Bank's cost of funds. Currently, that cost of funds is 4.839%. The loan's rate is not allowed to rise more than two percentage points in any one adjustment, nor is it allowed to rise above 11.875%. Trustworthy Savings also offers a thirty-year fixed-rate mortgage with an interest rate of 7.5%.

a. Find the monthly payment for the fixed-rate mortgage on a loan amount of $100,000.

b. Find the monthly payment for the ARM's first year on a loan amount of $100,000.

c. How much would the borrower save in the mortgage's first year by choosing the adjustable rather than the fixed-rate mortgage?

d. Find the unpaid balance at the end of the ARM's first year.

e. Find the interest rate and the value of n for the ARM's second year if the 11th District Federal Home Loan Bank's cost of funds does not change during the loan's first year.

f. Find the monthly payment for the ARM's second year if the 11th District Federal Home Loan Bank's cost of funds does not change during the loan's first year.

g. How much would the borrower save in the mortgage's first two years by choosing the adjustable-rate mortgage rather than the fixed-rate mortgage if the cost of funds does not change?

h. Discuss the advantages and disadvantages of an adjustable-rate mortgage.

54. American Dream Savings Bank offers a thirty-year adjustable-rate mortgage with an initial rate of 4.25%. The rate and the required payment are adjusted annually. Future rates are set at three percentage points above the one-year Treasury bill rate, which is currently 5.42%. The loan's rate is not allowed to rise more than two percentage points in any one adjustment, nor is it allowed to rise above 10.25%. American Dream Savings Bank also offers a thirty-year fixed-rate mortgage with an interest rate of 7.5%.

a. Find the monthly payment for the fixed-rate mortgage on a loan amount of $100,000.

b. Find the monthly payment for the ARM's first year on a loan amount of $100,000.

c. How much would the borrower save in the mortgage's first year by choosing the adjustable rather than the fixed-rate mortgage?

d. Find the unpaid balance at the end of the ARM's first year.

e. Find the interest rate and the value of n for the ARM's second year if the one-year Treasury bill rate does not change during the loan's first year.

f. Find the monthly payment for the ARM's second year if the Treasury bill rate does not change during the loan's first year.

g. How much would the borrower save in the mortgage's first two years by choosing the adjustable-rate mortgage rather than the fixed-rate mortgage if the Treasury bill rate does not change?

h. Discuss the advantages and disadvantages of an adjustable-rate mortgage.

55. Bank Two offers a thirty-year adjustable-rate mortgage with an initial rate of 3.95%. This initial rate is in effect for the first six months of the loan, after which it is adjusted on a monthly basis. The monthly payment is adjusted annually. Future rates are set at 2.45 percentage points above the 11th District Federal Home Loan Bank's cost of funds. Currently, that cost of funds is 4.839%.

a. Find the monthly payment for the ARM's first year on a loan amount of $100,000.

b. Find the unpaid balance at the end of the ARM's first six months.

c. Find the interest portion of the seventh payment if the cost of funds does not change.

d. Usually, the interest portion is smaller than the monthly payment, and the difference is subtracted from the unpaid balance. However, the interest portion found in part (c) is larger than the monthly payment found in part (a), and the difference is added to the unpaid balance found in part (b). Why would this difference be added to the unpaid balance? What effect will this have on the loan?

e. The situation described in part (d) is called **negative amortization.** Why?

f. What is there about the structure of Bank Two's loan that allows negative amortization?

AMORTIZATION SCHEDULES ON A COMPUTER

Interest paid on a home loan is deductible from the borrower's income taxes, and interest paid on a loan by a business is usually deductible. A borrower with either of these types of loans needs to know the total interest paid on the loan during the final year. The way to determine the total interest paid during a given year is to prepare an amortization schedule for that year. Typically, the lender provides the borrower with an amortization schedule, but it is not uncommon for this schedule to arrive after taxes are due. In this case, the borrower must either do the calculation personally or pay taxes without the benefit of the mortgage deduction and then file an amended set of tax forms after the amortization schedule has arrived.

Computing a year's amortization schedule is rather tedious, and neither a scientific calculator nor a graphing calculator offers relief. The best tool for the job is a computer, combined either with the Amortrix computer program (available with this book) or with a computerized spreadsheet. Each is discussed below.

Amortization Schedules and Amortrix

Amortrix is one of the features of the text web site (**www.cengage.com/math/johnson**). This software will enable you to quickly and easily compute an amortization schedule for any time period. We'll illustrate this process by using Amortrix to prepare an amortization schedule for a fifteen-year $175,000 loan at 7.5% interest with monthly payments.

When you start Amortrix, a main menu appears. Click on the "Amortization Schedule" option. Once you're at the page labeled "Amortization Schedule," enter "175000" (without commas or dollar signs) for the Loan Amount (P), "7.5" (without a percent sign) for the Annual Interest Rate, and "180" for the Total Number of Payment Periods (n). See the top part of Figure 5.21. Click on "Calculate" and the software will create an amortization schedule. See the bottom part of Figure 5.21.

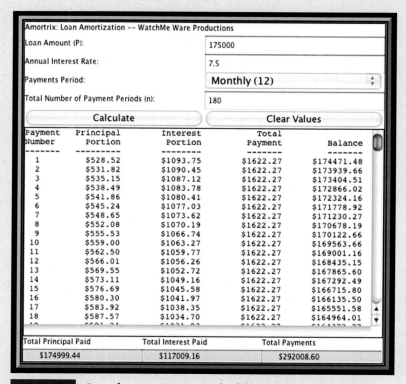

FIGURE 5.21 Part of an amortization schedule prepared with Amortrix.

The software will *not* correctly compute the last loan payment; it will compute the last payment in the same way it computes all other payments rather than in the way shown earlier in this section. You will have to correct this last payment if you use the computer to prepare an amortization schedule for a time period that includes the last payment.

Amortization Schedules and Excel

A **spreadsheet** is a large piece of paper marked off in rows and columns. Accountants use spreadsheets to organize numerical data and perform computations. A **computerized spreadsheet** such as Microsoft Excel is a computer program that mimics the appearance of a paper spreadsheet. It frees the user from performing any computations.

When you start a computerized spreadsheet, you see something that looks like a table waiting to be filled in. The rows are labeled with numbers and the columns are labeled with letters, as shown in Figure 5.22. The individual boxes are called **cells.** The cell in column A, row 1 is called cell A1.

◇	A	B	C	D	E	F	G
1	payment number	principal portion	interest portion	total payment	balance		
2	0				$175,000.00	rate	7.50%
3						years	15
4						payments/yr	12
5							

FIGURE 5.22 The spreadsheet after step 1.

Excel is an ideal tool to use in creating an amortization schedule. We will illustrate this process by preparing the amortization schedule for a fifteen-year $175,000 loan at 7.5% interest with monthly payments.

1. *Set up the spreadsheet.* Start by entering the information shown in Figure 5.22.

   * Adjust the columns' widths. To make column A wider, place your cursor on top of the line dividing the "A" and "B" labels. Then use your cursor to move that dividing line.
   * Format cell E2 as currency by highlighting that cell and pressing the "$" button on the Excel ribbon at the top of the screen.
   * Be certain that you include the % symbol in cell G2.
   * Save the spreadsheet.

   If you have difficulty creating a spreadsheet that looks like this, you may download it from our web site at **www.cengage.com/math/johnson**. See the file 5.4.xls and select the "step 1" sheet. Other sheets show later steps.

2. *Compute the monthly payment.* While you can do this with your calculator, as we discussed earlier in this section, you can also use the Excel "PMT" function to compute the monthly payment. This allows us to change the rate (in cell G2) or the loan amount (in cell E2) and Excel will automatically compute the new payment and change the entire spreadsheet accordingly. To use the "PMT" function, do the following:

   * Click on cell D3, where we will put the total payment.
   * Click on the *fx* button at the top of the screen.
   * In the resulting "Paste Function" box, select "Financial" under "Function Category", select "PMT" under "Function name", and press "OK." A "PMT" box will appear.

- Cell G2 holds 7.5%, the annual interest rate, and cell G4 holds 12, the number of periods/year. The periodic interest rate is $7.5\%/12 = G2/G4$. In the PMT box after "Rate", type "$G$2/$G$4".
- The number of periods is $15 \cdot 12 = G3 \cdot G4$. In the PMT box after "Nper", type "$G$3*$G$4."
- After "PV", type "$E$2". This is where we have stored the present value of the loan.
- Do not type anything after "Fv" or "type".
- Press "OK" at the bottom of the "PMT" box, and "($1,622.27)" should appear in cell D2. The parentheses mean that the number is negative.
- Save the spreadsheet.

Doing the above creates two problems that we must fix. The payment is negative. We can fix that with an extra minus sign. Also, the payment is not rounded to the nearest penny. We can fix that with the ROUND function.

In cell D2, Excel displays ($1622.27), but it stores $-1622.27163\ldots$, and all calculations involving cell D2 will be done by using $-1622.27163\ldots$. This will make the amortization schedule incorrect. To fix this, click on cell D3 and the long box at the top of the screen will have

=PMT($G$2/$G$4,$G$3*$G$4,$E$2)

in it. Change this to

=ROUND (−PMT($G$2/$G$4,$G$3*$G$4,$E$2),2)

   ↑   ↑ **a new minus sign**   ↑

the round fuction   **part of the round function**

The new parts here are "ROUND(−" at the beginning and ",2)" at the end. The "ROUND (,2)" part tells Excel to round to two decimal places, as a bank would. The minus sign after "ROUND(" makes the payment positive. Place your cursor where the new parts go, and add them.

3. *Fill in row 3.*

- We need to add a year each time we go down one row. Type "= A2 + 1" in cell A3.
- Cell C3 should contain instructions on computing simple interest on the previous balance.

$$I = P \cdot r \cdot t$$
$$= \text{the previous balance} \cdot 0.075 \cdot 1/12$$
$$= E2*\$G\$2/\$G\$4$$

Move to cell C3 and type in "=ROUND(E2*$G$2/$G$4,2)" and press "return". The "ROUND (,2)" part rounds the interest to two decimal places, as a bank would.

- Cell B3 should contain instructions on computing the payment's principal portion.

$$\text{Principal portion} = \text{payment} - \text{interest portion}$$
$$= D3 - C3$$

In cell B3, type "= D3 − C3".

- Cell E3 should contain instructions on computing the new balance.

$$\text{New balance} = \text{previous balance} - \text{principal portion}$$
$$= E2 - B3$$

In cell E3, type "= E2 − B3".

See Figure 5.23.

◇	A	B	C	D	E	F	G
1	payment number	principal portion	Interest portion	total payment	balance		
2	0				$175,000.00	rate	7.50%
3	1	$ 528.52	$1,093.75	$1,622.27	$174,471.48	years	15
4						payments/yr	12
5							

FIGURE 5.23 The spreadsheet after completing step 3.

4. *Fill in rows 4 and beyond.* All of the remaining payments' computations are just like payment 1 computations (except for the last payment), so all we have to do is copy the payment 1 instructions in row 3 and paste them in rows 4 and beyond. After completion of this step, the top and bottom of your spreadsheet should look like that in Figure 5.24.

◇	A	B	C	D	E	F	G
1	payment number	principal portion	interest portion	total payment	balance		
2	0				$175,000.00	rate	7.50%
3	1	$ 528.52	$1,093.75	$1,622.27	$174,471.48	years	15
4	2	$ 531.82	$1,090.45	$1,622.27	$173,939.66	payments/yr	12
5	3	$ 535.15	$1,087.12	$1,622.27	$173,404.51		
6	4	$ 538.49	$1,083.78	$1,622.27	$172,866.02		

The top of the spreadsheet, and ...

178	176	$ $1,572.51	$49.76	$1,622.27	$	6,389.50
179	177	$ $1,582.34	$39.93	$1,622.27	$	4,807.16
180	178	$ $1,592.23	$30.04	$1,622.27	$	3,214.93
181	179	$ $1,602.18	$20.09	$1,622.27	$	1,612.75
182	180	$ $1,612.19	$10.08	$1,622.27	$	0.56
183						

... the bottom of the spreadsheet, after completing step 4

FIGURE 5.24 The final spreadsheet.

5. *Fix the last payment.* Clearly, the ending balance of $0.56 in cell E182 is not correct—we cannot owe $0.56 or any other amount after making our last payment. See Exercise 56.

6. *Find the interest paid.* Typing "SUM(C3:C6)" in a cell will result in the sum of cells C4, C4, C5, and C6 appearing in that cell. Using this function makes it easy to find the total interest paid for any time period.

EXERCISES

In Exercises 56–62, first use Amortrix or Excel to create an amortization schedule for a fifteen-year $175,000 loan at 7.5% interest with monthly payments. (Figure 5.24 shows the first and last lines of this amortization schedule, created with Excel.) You will alter parts of this amortization schedule to answer Exercises 56–62.

56. **a.** The very last line of your schedule is incorrect, because the calculations for the last line of any amortization schedule are done differently from those of all other lines. What should the last line be?

   b. If you use Excel, state exactly what you should type in row 182 to fix the last payment.

57. Use Excel or Amortrix to find the total interest paid in
  a. the loan's first year and
  b. the loan's last year if the first month of the loan is January.
  c. If you use Excel, state exactly what you type in parts (a) and (b) to find the total interest paid.

58. Right before you signed your loan papers, the interest rate dropped from 7.5% to 7%.
  a. How does this affect the total payment?
  b. How does this affect the total interest paid in the loan's first year if the first month of the loan is January?
  c. How does this affect the total interest paid in the loan's last year?

59. Right before you signed your loan papers, you convinced the seller to accept a smaller price. As a result, your loan amount dropped from $175,000 to $165,000. (The interest rate remains at 7.5%)
  a. How does this affect the total payment?
  b. How does this affect the total interest paid in the loan's first year if the first month of the loan is January?
  c. How does this affect the total interest paid in the loan's last year?

60. You are considering a twenty-year loan. (The interest rate remains at 7.5% and the loan amount remains at $175,000.)
  a. How does this affect the total payment?
  b. How does this affect the total interest paid in the loan's first year if the first month of the loan is January?
  c. How does this affect the total interest paid in the loan's fifteenth year?
  d. Explain, without performing any calculations, why the total payment should go down and the total interest paid during the first year should go up.

61. You are considering a thirty-year loan. (The interest rate remains at 7.5%, and the loan amount remains at $175,000.)
  a. How does this affect the total payment?
  b. How does this affect the total interest paid in the loan's first year if the first month of the loan is January?
  c. How does this affect the total interest paid in the loan's fifteenth year?
  d. Explain, without performing any calculations, why the total payment should go down and the total interest paid during the first year should go up.

62. You are considering a loan with payments every two weeks. (The interest rate remains at 7.5%, the loan amount remains at $175,000, and the term remains at fifteen years.)
  a. How does this affect the total payment?
  b. How does this affect the total interest paid in the loan's first year if the first month of the loan is January?
  c. How does this affect the total interest paid in the loan's last year?

  d. Explain, without performing any calculations, why the total payment and the total interest paid during the first year should both go down.

In Exercises 63–66:
  a. *Use Amortrix or Excel to prepare an amortization schedule for the given loan.*
  b. *Find the amount that could be deducted from the borrower's taxable income (that is, find the total interest paid) in the loan's first year if the first payment was made in January. (Interest on a car loan is deductible only in some circumstances. For the purpose of these exercises, assume that this interest is deductible.)*
  c. *Find the amount that could be deducted from the borrower's taxable income (that is, find the total interest paid) in the loan's last year.*
  d. *If you sell your home or car after three years, how much will you still owe on the mortgage?*

63. A five-year simple interest amortized car loan for $32,600 at 12.25% interest

64. A four-year simple interest amortized car loan for $26,200 at 5.75% interest

65. A fifteen-year simple interest amortized home loan for $220,000 at 6.25% interest

66. A thirty-year simple interest amortized home loan for $350,000 at 14.5% interest

67. Use Amortrix or Excel to do Exercise 29 on page 379. Is your answer the same as that of Exercise 29? If not, why not? Which answer is more accurate? Why?

68. Use Amortrix or Excel to do Exercise 30 on page 379. Is your answer the same as that of Exercise 30? If not, why not? Which answer is more accurate? Why?

69. Use Amortrix or Excel to do Exercise 31 on page 379. Is your answer the same as that of Exercise 31? If not, why not? Which answer is more accurate? Why?

70. Use Amortrix or Excel to do Exercise 32 on page 379. Is your answer the same as that of Exercise 32? If not, why not? Which answer is more accurate? Why?

71. *For Excel users only:* When interest rates are high, some lenders offer the following type of home loan. You borrow $200,000 for thirty years. For the first five years, the interest rate is 5.75%, and the payments are calculated as if the interest rate were going to remain unchanged for the life of the loan. For the last twenty-five years, the interest rate is 9.25%.
  a. What are the monthly payments for the first five years?
  b. What are the monthly payments for the last twenty-five years?
  c. What is the total interest paid, during the loan's fifth year if the first month of the loan is January?
  d. What is the total interest paid during the loan's sixth year?
  e. Why would a lender offer such a loan?

Annual Percentage Rate with a TI's TVM Application

- Understand what an annual percentage rate (APR) is
- Find an APR
- Use APRs to compare loans

A **simple interest loan** is any loan for which the interest portion of each payment is simple interest on the outstanding principal. A *simple interest amortized loan* fulfills this requirement; in fact, we compute an amortization schedule for a simple interest amortized loan by finding the simple interest on the outstanding principal.

The APR of an Add-on Interest Loan

Whenever a loan is not a simple interest loan, the Truth in Lending Act requires the lender to disclose the annual percentage rate to the borrower. The **annual percentage rate (APR) of an add-on interest loan** is the simple interest rate that makes the dollar amounts the same if the loan is recomputed as a simple interest amortized loan.

We'll explore APRs in this section. They are important because they allow someone who is shopping for a loan to compare loans with different terms and to determine which loan is better.

EXAMPLE 1

FINDING THE APR In Example 5 in Section 5.1, Chip Douglas bought a nine-year-old Ford that's "like new" from Centerville Auto Sales for $5,988. He financed the purchase with the dealer's two-year, 5% add-on interest loan, which required a $600 down payment and monthly payments of $246.95 a month for twenty-four months. Use a TI-83/84's TVM application to find the APR of the loan.

SOLUTION

1. Press $\boxed{\text{APPS}}$, select option 1: "Finance", and press $\boxed{\text{ENTER}}$.
2. Select option 1: "TVM Solver", and press ENTER.
3. Enter appropriate values for the variables:

   - N is the number of payments. Enter 24 for N.
   - I% is the annual interest rate. This is the number we're trying to find, so temporarily enter 0. Later, we'll solve for the actual value of N.
   - PV is the present value, which is $5988 − $600 = $5388.
   - PMT is the payment. The payment is $246.95, but it's an outgoing amount of money (since we give it to the bank), so we enter −246.95 for PMT. You must enter a negative number for any outgoing amount of money.
   - FV is the future value. There is no future value here, so we enter 0.
   - P/Y is the number of periods per year. We make monthly payments, so there are twelve periods per year. Enter 12 for P/Y.
   - C/Y is automatically made to be the same as P/Y. In this text, we will never encounter a situation in which C/Y is different from P/Y. See Figure 5.25.

4. To solve for I%, use the arrow buttons to highlight the 0 that we entered for it earlier. Then press $\boxed{\text{ALPHA}}$ $\boxed{\text{SOLVE}}$. (Pressing $\boxed{\text{ALPHA}}$ makes the $\boxed{\text{ENTER}}$ button become the

```
N=24
I%=0
PV=5388
PMT=-246.95
FV=0
P/Y=12
C/Y=12
PMT:END BEGIN
```

FIGURE 5.25

Preparing the TVM screen.

$\boxed{\text{SOLVE}}$ button.) As a result, we find that I% is 9.323544% ≈ 9.32%, and the add-on interest loan has an APR of 9.32%. This means that the 5% add-on interest loan requires the same monthly payment that a 9.32% simple interest amortized loan would have!

The Truth in Lending Act, which requires a lender to divulge its loans' APR, allows a tolerance of one-eighth of 1% (0.125%) in the claimed APR. Thus, the dealership would be legally correct if it stated that the APR was between 9.323544% − 0.125% = 9.198544% and 9.323544% + 0.125% = 9.448544%.

The APR of a Simple Interest Loan

Sometimes **finance charges** other than the interest portion of the monthly payment are associated with a loan; these charges must be paid when the loan agreement is signed. For example, a **point** is a finance charge that is equal to 1% of the loan amount; a **credit report fee** pays for a report on the borrower's credit history, including any late or missing payments and the size of all outstanding debts; an **appraisal fee** pays for the determination of the current market value of the property (the auto, boat, or home) to be purchased with the loan. The Truth in Lending Act requires the lender to inform the borrower of the total finance charge, which includes the interest, the points, and some of the fees (see Figure 5.26). Most of the finance

Costs Included in the Prepaid Finance Charges		*Other Costs Not Included in the Finance Charge*	
2 points	$2,415.32	appraisal fee	$ 70.00
prorated interest	$1,090.10	credit report	$ 60.00
prepaid mortgage		closing fee	$ 670.00
insurance	$ 434.50	title insurance	$ 202.50
loan fee	$1,242.00	recording fee	$ 20.00
document		notary fee	$ 20.00
preparation fee	$ 80.00	tax and insurance	
tax service fee	$ 22.50	escrow	$ 631.30
processing fee	$ 42.75		
subtotal	$5,327.17	subtotal	$1,673.80

FIGURE 5.26 Sample portion of a federal truth-in-lending disclosure statement (loan amount = $120,765.90).

charges must be paid before the loan is awarded, so in essence the borrower must pay money now in order to get more money later. The law says that this means the lender is not really borrowing as much as he or she thinks. According to the law, the actual amount loaned is the loan amount minus all points and those fees included in the finance charge. The **APR of a simple interest amortized loan** is the rate that reconciles the payment and this actual loan amount.

EXAMPLE **2**

FINDING THE APR OF A SIMPLE INTEREST LOAN Glen and Tanya Hansen bought a home for $140,000. They paid the sellers a 20% down payment and obtained a simple interest amortized loan for the balance from their bank at $10\frac{3}{4}\%$ for thirty years. The bank in turn paid the sellers the remaining 80% of the purchase price, less a 6% sales commission paid to the sellers' and the buyers' real estate agents. (The transaction is illustrated in Figure 5.27.) The bank charged the Hansens 2 points plus fees totaling $3,247.60; of these fees, $1,012.00 were included in the finance charge.

a. Find the size of the Hansens' monthly payment.
b. Find the total interest paid.
c. Compute the total finance charge.
d. Find the APR

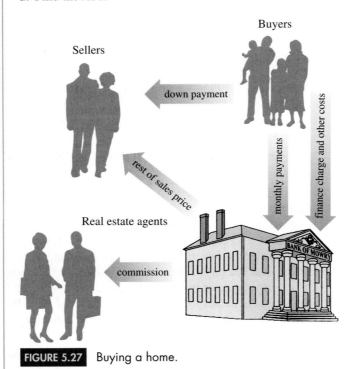

FIGURE 5.27 Buying a home.

SOLUTION

a. *Finding their monthly payment* We are given down payment = 20% of $140,000 = $28,000, P = loan amount = $140,000 - $28,000 = $112,000, $i = \frac{1}{12}$ of $10\frac{3}{4}\% = \frac{0.1075}{12}$, and n = 30 years = 30 years $\cdot$ (12 months)/(1 year) = 360 months.

Future value of annuity = future value of loan amount

$$pymt\frac{(1+i)^n - 1}{i} = P(1+i)^n$$

$$pymt\frac{\left(1 + \dfrac{0.1075}{12}\right)^{360} - 1}{\dfrac{0.1075}{12}} = 112,000\left(1 + \frac{0.1075}{12}\right)^{360}$$

Computing the fraction on the left side and multiplying its reciprocal by the right side, we get

$$pymt = 1,045.4991 \approx \$1,045.50$$

b. *Finding the total interest paid*   The total interest paid is the total amount paid minus the amount borrowed. The Hansens agreed to make 360 monthly payments of $1045.50 each, for a total of $360 \cdot \$1,045.50 = \$376,380.00$. Of this, $112,000 is principal; therefore, the total interest is

$$376,380 - 112,000 = \$264,380$$

c. *Computing their total finance charge*

$$
\begin{aligned}
\text{2 points} &= \text{2\% of }\$112,000 = \$\ \ 2,240 \\
\text{included fees} &= \$\ \ 1,012 \\
\text{total interest paid} &= \underline{\$264,380} \qquad \textbf{from part (b)} \\
\text{total finance charge} &= \$267,632
\end{aligned}
$$

d. *Finding the APR*   The APR is the simple interest rate that makes the dollar amounts the same if the loan is recomputed using the legal loan amount (loan amount less points and fees) in place of the actual loan amount. The legal loan amount is

$$P = \$112,000 - \$3252 = \$108,748$$

We'll find the APR with the TVM application.

1. Press $\boxed{\text{APPS}}$, select option 1: "Finance", and press $\boxed{\text{ENTER}}$.
2. Select option 1: "TVM Solver", and press ENTER.
3. Enter appropriate values for the variables.

- N is the number of payments. Enter 360.
- I% is the annual interest rate. This is the number we're trying to find, so temporarily enter 0. Later, we'll solve for the actual value of N.
- PV is the present value, which is $108,748.
- PMT is the payment. The payment is $1,045.50, but it's an outgoing amount of money (since we give it to the bank), so we enter −1045.5 for PMT. You must enter a negative number for any outgoing amount of money.
- FV is the future value. There is no future value here, so we enter 0.
- P/Y is the number of periods per year. We make monthly payments, so there are twelve periods per year. Enter 12 for P/Y.
- C/Y is automatically made to be the same as P/Y. In this text, we will never encounter a situation in which C/Y is different from P/Y. See Figure 5.28.

4. To solve for I%, use the arrow buttons to highlight the 0 that we entered for I%. Then press $\boxed{\text{ALPHA}}$ $\boxed{\text{SOLVE}}$. (Pressing $\boxed{\text{ALPHA}}$ makes the $\boxed{\text{ENTER}}$ button becomes the $\boxed{\text{SOLVE}}$ button.) As a result, we find that I% is $11.11993521\% \approx 11.12\%$, and the loan has an APR of 11.12%. This means the $10\frac{3}{4}\%$ loan requires the same monthly payment that a 11.12% loan with no points or fees would have. In other words, the points and fees in effect increase the interest rate from $10\frac{3}{4}\%$ to 11.12%.

FIGURE 5.28

Preparing the TVM screen.

If you need to obtain a loan, it is not necessarily true that the lender with the lowest interest rate will give you the least expensive loan. One lender may charge more points or higher fees than does another lender. One lender may offer an add-on interest loan, while another lender offers a simple interest amortized loan. These differences can have a significant impact on the cost of a loan. If two lenders offer loans at the same interest rate but differ in any of these ways, that difference will be reflected in the APR. "Lowest interest rate" does *not* mean "least expensive loan," but "lowest APR," *does* mean "least expensive loan."

Estimating Prepaid Finance Charges

As we have seen, a borrower usually must pay an assortment of fees when obtaining a home loan. These fees can be quite substantial, and they can vary significantly from lender to lender. For example, the total fees in Figure 5.26 were $7,000.97; the Hansens' total fees in Example 2 were $5,487.60. When shopping for a home loan, a borrower should take these fees into consideration.

All fees must be disclosed to the borrower when he or she signs the loan papers. Furthermore, the borrower must be given an estimate of the fees when he or she applies for the loan. Before an application is made, the borrower can use a loan's APR to obtain a reasonable approximation of those fees that are included in the finance charge (for a fixed rate loan). This would allow the borrower to make a more educated decision in selecting a lender.

EXAMPLE 3

ESTIMATING PREPAID FINANCE CHARGES   Felipe and Delores Lopez are thinking of buying a home for $152,000. A potential lender advertises an 80%, thirty-year simple interest amortized loan at $7\frac{1}{2}\%$ interest with an APR of 7.95%.

a. Find the size of the Lopezes' monthly payment.
b. Use the APR to approximate the fees included in the finance charge.

SOLUTION

a. *Finding the monthly payment*   The loan amount is 80% of $152,000 = $121,600, $i = \frac{1}{12}$ of 7.5% = $\frac{0.075}{12}$, and $n = 30$ years = 360 months.

Future value of annuity = future value of loan amount

$$pymt\frac{(1+i)^n - 1}{i} = P(1+i)^n$$

$$pymt\frac{\left(1+\frac{0.075}{12}\right)^{360} - 1}{\frac{0.075}{12}} = 121,600\left(1+\frac{0.075}{12}\right)^{360}$$

Computing the fraction on the left side and multiplying its reciprocal by the right side, we get

$pymt = 850.2448 \approx \$850.25$

b. *Approximating the fees included in the finance charge*   We approximate the fees by computing the legal loan amount (loan amount less points and fees), using the APR as the interest rate and the payment computed in part (a) as *pymt*. Therefore, $i = \frac{1}{12}$ of 7.95% = $\frac{0.0795}{12}$, *pymt* = 850.25.

Future value of annuity = future value of loan amount

$$pymt\frac{(1+i)^n - 1}{i} = P(1+i)^n$$

$$850.25\frac{\left(1+\frac{0.0795}{12}\right)^{360} - 1}{\frac{0.0795}{12}} = P\left(1+\frac{0.0795}{12}\right)^{360}$$

Computing the left side, and dividing by $(1+\frac{0.0795}{12})^{360}$, we get

$P = 116,427.6304 \approx \$116,427.63$

This is the legal loan amount, that is, the loan amount less points and fees. The Lopezes were borrowing $121,600, so that leaves $121,600 - $116,427.63 = $5,172.37 in points and fees. This is an estimate of the points and fees included in the finance charge, such as a loan fee, a document preparation fee, and a processing fee; it does not include fees such as an appraisal fee, a credit report fee, and a title insurance fee.

1. Wade Ellis buys a new car for $16,113.82. He puts 10% down and obtains a simple interest amortized loan for the balance at $11\frac{1}{2}\%$ interest for four years. If loan fees included in the finance charge total $814.14, find the APR.

2. Guy de Primo buys a new car for $9,837.91. He puts 10% down and obtains a simple interest amortized loan for the balance at $10\frac{7}{8}\%$ interest for four years. If loan fees included in the finance charge total $633.87, find the APR.

3. Chris Burditt bought a house for $212,500. He put 20% down and obtained a simple interest amortized loan for the balance at $10\frac{7}{8}\%$ interest for thirty years. If Chris paid 2 points and $4,728.60 in fees, $1,318.10 of which are included in the finance charge, find the APR.

4. Shirley Trembley bought a house for $187,600. She put 20% down and obtained a simple interest amortized loan for the balance at $11\frac{3}{8}\%$ for thirty years. If Shirley paid 2 points and $3,427.00 in fees, $1,102.70 of which are included in the finance charge, find the APR.

5. Jennifer Tonda wants to buy a used car that costs $4,600. The used car dealer has offered her a four-year add-on interest loan that requires no down payment at 8% annual interest, with an APR of $14\frac{1}{4}\%$.
   a. Find the monthly payment.
   b. Verify the APR.

6. Melody Shepherd wants to buy a used car that costs $5,300. The used car dealer has offered her a four-year add-on interest loan that requires a $200 down payment at 7% annual interest, with an APR of 10%.
   a. Find the monthly payment.
   b. Verify the APR.

7. Anne Scanlan is buying a used car that costs $10,340. The used car dealer has offered her a five-year add-on interest loan at 9.5% interest, with an APR of 9.9%. The loan requires a 10% down payment.
   a. Find the monthly payment.
   b. Verify the APR.

8. Stan Loll bought a used car for $9,800. The used car dealer offered him a four-year add-on interest loan at 7.8% interest, with an APR of 8.0%. The loan requires a 10% down payment.
   a. Find the monthly payment.
   b. Verify the APR.

9. Susan Chin is shopping for a car loan. Her savings and loan offers her a simple interest amortized loan for four years at 9% interest. Her bank offers her a simple interest amortized loan for four years at 9.1% interest. Which is the less expensive loan?

10. Stephen Tamchin is shopping for a car loan. His credit union offers him a simple interest amortized loan for four years at 7.1% interest. His bank offers him a simple interest amortized loan for four years at 7.3% interest. Which is the less expensive loan?

11. Ruben Lopez is shopping for a home loan. Really Friendly Savings and Loan offers him a thirty-year simple interest amortized loan at 9.2% interest, with an APR of 9.87%. The Solid and Dependable Bank offers him a thirty-year simple interest amortized loan at 9.3% interest, with an APR of 9.80%. Which loan would have the lower payments? Which loan would be the least expensive, taking into consideration monthly payments, points, and fees? Justify your answers.

12. Keith Moon is shopping for a home loan. Sincerity Savings offers him a thirty-year simple interest amortized loan at 8.7% interest, with an APR of 9.12%. Pinstripe National Bank offers him a thirty-year simple interest amortized loan at 8.9% interest, with an APR of 8.9%. Which loan would have the lower payments? Which loan would be the least expensive, taking into consideration monthly payments, points, and fees? Justify your answers.

13. The Nguyens are thinking of buying a home for $119,000. A potential lender advertises an 80%, thirty-year simple interest amortized loan at $8\frac{1}{4}\%$ interest, with an APR of 9.23%. Use the APR to approximate the fees included in the finance charge.

14. Ellen Taylor is thinking of buying a home for $126,000. A potential lender advertises an 80%, thirty-year simple interest amortized loan at $10\frac{3}{4}\%$ interest, with an APR of 11.57%. Use the APR to approximate the fees included in the finance charge.

15. James Magee is thinking of buying a home for $124,500. Bank of the Future advertises an 80%, thirty-year simple interest amortized loan at $9\frac{1}{4}\%$ interest, with an APR of 10.23%. R.T.C. Savings and Loan advertises an 80%, 30-year simple interest amortized loan at 9% interest with an APR of 10.16%.
   a. Find James's monthly payment if he borrows through Bank of the Future.
   b. Find James's monthly payment if he borrows through R.T.C. Savings and Loan.
   c. Use the APR to approximate the fees included in the finance charge by Bank of the Future.
   d. Use the APR to approximate the fees included in the finance charge by R.T.C. Savings and Loan.
   e. Discuss the advantages of each of the two loans. Who would be better off with the Bank of the Future loan? Who would be better off with the R.T.C. loan?

16. Holly Kresch is thinking of buying a home for $263,800. State Bank advertises an 80%, thirty-year simple interest amortized loan at $6\frac{1}{4}\%$ interest, with an APR of 7.13%.

Boonville Savings and Loan advertises an 80%, thirty-year simple interest amortized loan at $6\frac{1}{2}\%$ interest with an APR of 7.27%.

a. Find Holly's monthly payment if she borrows through State Bank.

b. Find Holly's monthly payment if she borrows through Boonville Savings and Loan.

c. Use the APR to approximate the fees included in the finance charge by State Bank.

d. Use the APR to approximate the fees included in the finance charge by Boonville Savings and Loan.

e. Discuss the advantages of the two loans. Who would be better off with the State Bank loan? Who would be better off with the Boonville loan?

Answer the following questions using complete sentences and your own words.

• CONCEPT QUESTIONS

17. If the APR of a simple interest amortized home loan is equal to the loan's interest rate, what conclusions could you make about the loan's required fees and points?

18. Compare and contrast the annual percentage rate of a loan with the annual yield of a compound interest rate. Be sure to discuss both the similarity and the difference between these two concepts.

19. Substitute the dollar amounts from Example 1 into the simple interest amortized loan formulas. Then discuss why the resulting equation can't be solved with algebra.

5.6 Payout Annuities

OBJECTIVES

- Understand the difference between a payout annuity and an ordinary annuity
- Use the payout annuity formulas
- Determine how to use a payout annuity to save for retirement

If you save for your retirement with an IRA or an annuity or both, you'll do a lot better if you start early. This allows you to take advantage of the "magic" effect of compound interest over a long period of time.

You'll also do better if your annuity turns into a *payout annuity* when it comes to term. This allows your money to earn a higher rate of interest *after* you retire and start collecting from it. We'll explore payout annuities in this section.

The annuities that we have discussed are all savings instruments. In Section 5.3, we defined an annuity as a sequence of equal, regular payments into an account in which each payment receives compound interest. A saver who utilizes such an annuity can accumulate a sizable sum.

Annuities can be payout instruments rather than savings instruments. After you retire, you might wish to have part of your savings sent to you each month for living expenses. You might wish to receive equal, regular payments from an account where each payment has earned compound interest. Such an annuity is called a **payout annuity.** Payout annuities are also used to pay for a child's college education.

Calculating Short-Term Payout Annuities

EXAMPLE **1** | UNDERSTANDING WHAT A PAYOUT ANNUITY IS On November 1, Debra Landre will make a deposit at her bank that will be used for a payout annuity. For the next three months, commencing on December 1, she will receive a payout of $1,000 per month. Use the Compound Interest Formula to find how much money she must deposit on November 1 if her money earns 10% compounded monthly.

SOLUTION

First, calculate the principal necessary to receive $1,000 on December 1. Use $FV = 1,000$ and $n = 1$ (interest is earned for one month).

$$FV = P(1 + i)^n$$ **the Compound Interest Formula**

$$1,000 = P\left(1 + \frac{0.10}{12}\right)^1$$ **substituting**

$$P = 1,000 \div \left(1 + \frac{0.10}{12}\right)^1$$ **solving for P**

$$= 991.7355\ldots \approx \$991.74$$ **rounding**

Next, calculate the principal necessary to receive $1,000 on January 1. Use $n = 2$ (interest is earned for two months).

$$FV = P(1 + i)^n$$ **the Compound Interest Formula**

$$1,000 = P\left(1 + \frac{0.10}{12}\right)^2$$ **substituting**

$$P = 1,000 \div \left(1 + \frac{0.10}{12}\right)^2$$ **solving for P**

$$= 983.5393\ldots \approx \$983.54$$ **rounding**

Now calculate the principal necessary to receive $1,000 on February 1. Use $n = 3$ (interest is earned for three months).

$$FV = P(1 + i)^n$$ **the Compound Interest Formula**

$$1,000 = P\left(1 + \frac{0.10}{12}\right)^3$$ **substituting**

$$P = 1,000 \div \left(1 + \frac{0.10}{12}\right)^3$$ **solving for P**

$$= 975.4109\ldots \approx \$975.41$$ **rounding**

Debra must deposit the sum of the above three amounts if she is to receive three monthly payouts of $1,000 each. Her total principal must be

$$\$991.74 + \$983.54 + \$975.41 = \$2,950.69$$

> If Debra's principal received no interest, then she would need $3 \cdot \$1,000 = \$3,000$. Since her principal does receive interest, she needs slightly less than $3,000.

Comparing Payout Annuities and Savings Annuities

Payout annuities and savings annuities are similar but not identical. It is important to understand their differences before we proceed. The following example is the "savings annuity" version of Example 1; that is, it is the savings annuity most similar to the payout annuity discussed in Example 1.

EXAMPLE **2**

UNDERSTANDING THE DIFFERENCE BETWEEN A PAYOUT ANNUITY AND AN ORDINARY ANNUITY On November 1, Debra Landre set up an ordinary annuity with her bank. For the next three months, she will make a payment of $1,000 per month. Each of those payments will receive compound interest. At the end of the three months (i.e., on February 1), she can withdraw her three $1,000 payments plus the interest that they will have earned. Use the Compound Interest Formula to find how much money she can withdraw.

SOLUTION

Her first payment of $1,000 would be due on November 30. It would earn interest for two months (December and January).

$$FV = P(1 + i)^n \qquad \text{the Compound Interest Formula}$$
$$FV = 1,000\left(1 + \frac{0.10}{12}\right)^2 \approx \$1,016.74 \quad \text{substituting}$$

Her second payment of $1,000 would be due on December 31. It would earn interest for one month (January).

$$FV = P(1 + i)^n \qquad \text{the Compound Interest Formula}$$
$$FV = 1,000\left(1 + \frac{0.10}{12}\right)^1 \approx \$1,008.33 \quad \text{substituting}$$

Her third payment would be due on January 31. It would earn no interest (since the annuity expires February 1), so its future value is $1,000. On February 1, Debra could withdraw

$$\$1,016.74 + \$1,008.33 + \$1,000 = \$3,025.07$$

Naturally, we could have found the future value of Debra's ordinary annuity more easily with the Ordinary Annuity Formula:

$$FV(\text{ordinary}) = pymt\frac{(1 + i)^n - 1}{i} \qquad \text{the Ordinary Annuity Formula}$$
$$= 1,000\frac{\left(1 + \frac{0.10}{12}\right)^3 - 1}{\frac{0.10}{12}} \qquad \text{substituting}$$
$$\approx \$3,025.07 \qquad \text{rounding}$$

The point of the method shown in Example 2 is to illustrate the differences between a savings annuity and a payout annuity.

Calculating Long-Term Payout Annuities

The procedure used in Example 1 reflects what actually happens with payout annuities, and it works fine for a small number of payments. However, most annuities are long-term. In the case of a savings annuity, we do not need to calculate the future value of each individual payment, as we did in Example 2; instead, we can use the Ordinary Annuity Formula. We need such a formula for payout annuities. We can find a formula if we look more closely at how the payout annuity from Example 1 compares with the savings annuity from Example 2.

In Example 2, we found that the future value of an ordinary annuity with three $1,000 payments is

$$FV = 1,000\left(1 + \frac{0.10}{12}\right)^2 + 1,000\left(1 + \frac{0.10}{12}\right)^1 + 1,000$$

In Example 1, we found that the total principal necessary to generate three $1,000 payouts is

$$P = 1,000 \div \left(1 + \frac{0.10}{12}\right)^1 + 1,000 \div \left(1 + \frac{0.10}{12}\right)^2 + 1,000 \div \left(1 + \frac{0.10}{12}\right)^3$$

This can be rewritten, using exponent laws, as

$$P = 1,000\left(1 + \frac{0.10}{12}\right)^{-1} + 1,000\left(1 + \frac{0.10}{12}\right)^{-2} + 1,000\left(1 + \frac{0.10}{12}\right)^{-3}$$

This is quite similar to the future value of the ordinary annuity—the only difference is the exponents. If we multiply each side by $(1 + \frac{0.10}{12})^3$, even the exponents will match.

$$P\left(1 + \frac{0.10}{12}\right)^3 = 1,000\left(1 + \frac{0.10}{12}\right)^{-1}\left(1 + \frac{0.10}{12}\right)^3$$
$$+ 1,000\left(1 + \frac{0.10}{12}\right)^{-2}\left(1 + \frac{0.10}{12}\right)^3$$
$$+ 1,000\left(1 + \frac{0.10}{12}\right)^{-3}\left(1 + \frac{0.10}{12}\right)^3$$
$$P\left(1 + \frac{0.10}{12}\right)^3 = 1,000\left(1 + \frac{0.10}{12}\right)^2 + 1,000\left(1 + \frac{0.10}{12}\right)^1 + 1,000$$

The right side of the above equation is the future value of an ordinary annuity, so we can use the Ordinary Annuity Formula to rewrite it.

$$P\left(1 + \frac{0.10}{12}\right)^3 = 1,000\frac{\left(1 + \frac{0.10}{12}\right)^3 - 1}{\frac{0.10}{12}}$$

If we generalize by replacing $\frac{0.10}{12}$ with i, 3 with n, and 1,000 with *pymt*, we have our Payout Annuity Formula.

PAYOUT ANNUITY FORMULA

$$P(1 + i)^n = pymt\frac{(1 + i)^n - 1}{i}$$

where P is the total principal necessary to generate n payouts, *pymt* is the size of the payout, and i is the periodic interest rate.

We have seen this formula before. In Section 5.3, we used it to find the present value P of a savings annuity. In Section 5.4, we used it to find the payment of a simple interest amortized loan. In this section, we use it to find the required principal for a payout annuity. This is a versatile formula.

The following example involves a long-term annuity. Usually, the interest rate of a long-term annuity varies somewhat from year to year. In this case, calculations must be viewed as predictions, not guarantees.

EXAMPLE **3**

FINDING THE NECESSARY DEPOSIT Fabiola Macias is about to retire, so she is setting up a payout annuity with her bank. She wishes to receive a payout of $1,000 per month for the next twenty-five years. Use the Payout Annuity Formula to find how much money she must deposit if her money earns 10% compounded monthly.

SOLUTION

We are given that $pymt = 1,000$, $i = \frac{1}{12}$ of $10\% = \frac{0.10}{12}$, and $n = 25 \cdot 12 = 300$.

$$P(1 + i)^n = pymt\frac{(1 + i)^n - 1}{i} \qquad \text{the Payout Annuity Formula}$$

$$P\left(1 + \frac{0.10}{12}\right)^{300} = 1,000\frac{\left(1 + \frac{0.10}{12}\right)^{300} - 1}{\frac{0.10}{12}} \qquad \text{substituting}$$

To find P, we need to calculate the right side and then divide by the $\left(1 + \frac{0.10}{12}\right)^{300}$ from the left side.

We get $110,047.23005 \approx \$110,047.23$. This means that if Fabiola deposits $110,047.23, she will receive monthly payouts of $1,000 each for twenty-five years, or a total of $25 \cdot 12 \cdot 1,000 = \$300,000$.

> If Fabiola's principal received no interest, then she would need $300 \cdot \$1,000 = \$300,000$. Since her principal does receive compound interest for a long time, she needs significantly less than $300,000.

If Fabiola Macias in Example 3 were like most people, she would not have $110,047.23 in savings when she retires, so she wouldn't be able to set up a payout annuity for herself. However, if she had set up a savings annuity thirty years before retirement, she could have saved that amount by making monthly payments of only $48.68. This is something that almost anyone can afford, and it's a wonderful deal. Thirty years of monthly payments of $48.68, while you are working, can generate twenty-five years of monthly payments of $1,000 when you are retired. In the exercises, we will explore this combination of a savings annuity and a payout annuity.

Payout Annuities with Inflation

The only trouble with Fabiola's retirement payout annuity in Example 3 is that she is ignoring inflation. In twenty-five years, she will still be receiving $1,000 a month, but her money won't buy as much as it does today. Fabiola would be better off if she allowed herself an annual **cost-of-living adjustment (COLA).**

EXAMPLE **4**

UNDERSTANDING WHAT A COLA IS After retiring, Fabiola Macias set up a payout annuity with her bank. For the next twenty-five years, she will receive payouts that start at $1,000 per month and then receive an annual COLA of 3%. Find the size of her monthly payout for

a. the first year
b. the second year
c. the third year
d. the 25th year

SOLUTION

a. During the first year, no adjustment is made, so she will receive $1,000 per month.

b. During the second year, her monthly payout of $1,000 will increase 3%, so her new monthly payout will be

$$1,000 \cdot (1 + .03) = \$1,030$$

c. During the third year, her monthly payout of $1,030 will increase 3%, so her new monthly payout will be

$$1,030 \cdot (1 + .03) = \$1,060.90$$

Since the 1,030 in the above calculation came from computing $1,000 \cdot (1 + .03)$, we could rewrite this calculation as

$$1000 \cdot (1 + .03)^2 = \$1,060.90$$

d. By the twenty-fifth year, she will have received twenty-four 3% increases, so her monthly payout will be

$$1,000 \cdot (1 + .03)^{24} = 2,032.7941 \ldots \approx \$2,032.79$$

If a payout annuity is to have automatic annual cost-of-living adjustments, the following formula should be used to find the principal that must be deposited. The COLA is an annual one, so all other figures must also be annual figures; in particular, r is the *annual* interest rate, t is the duration of the annuity in *years*, and we use an *annual* payout.

ANNUAL PAYOUT ANNUITY WITH COLA FORMULA

A payout annuity of t years, where the payouts receive an annual COLA, requires a principal of

$$P = (pymt)\frac{1 - \left(\dfrac{1+c}{1+r}\right)^t}{r - c}$$

where *pymt* is the annual payout for the first year, c is the annual COLA rate, and r is the annual rate at which interest is earned on the principal.

EXAMPLE 5

A PAYOUT ANNUITY WITH A COLA After retiring, Sam Needham set up a payout annuity with his bank. For the next twenty-five years, he will receive annual payouts that start at $12,000 and then receive an annual COLA of 3%. Use the Annual Payout Annuity with COLA Formula to find how much money he must deposit if his money earns 10% interest per year.

SOLUTION

We are given that the annual payout is $pymt = 12,000$, $r = 10\% = 0.10$, $c = 3\% = 0.03$, and $t = 25$.

$$P = (pymt)\frac{1 - \left(\dfrac{1+c}{1+r}\right)^t}{r - c} \qquad \textbf{the COLA Formula}$$

$$= (12,000)\frac{1 - \left(\dfrac{1 + 0.03}{1 + 0.10}\right)^{25}}{0.10 - 0.03} \qquad \textbf{substituting}$$

$$= (12,000)\frac{1 - \left(\dfrac{1.03}{1.10}\right)^{25}}{0.10 - 0.03} \qquad \textbf{simplifying}$$

$$= 138,300.4587 \approx \$138,300.46 \qquad \textbf{rounding}$$

1 − (1.03 ÷ 1.10) y^x 25 = ÷ (.10 − .03

) × 12000 =

With a graphing calculator, press ^ rather than y^x and ENTER rather than = .

This means that if Sam deposits $138,300.46 now, he will receive

$12,000 in one year

$12,000 · (1 + .03)¹ = $12,360 in two years

$12,000 · (1 + .03)² = $12,730.80 in three years

$12,000 · (1 + .03)²⁴ = $24,393.53 in twenty-five years

> If Sam's principal received no interest, he would need 25 · $12,000 = $300,000. Since his principal does receive compound interest for a long time, he needs significantly less than $300,000.

Compare Sam's payout annuity in Example 5 with Fabiola's in Example 3. Sam receives annual payouts that start at $12,000 and slowly increase to $24,393. Fabiola receives exactly $1,000 per month (or $12,000 per year) for the same amount of time. Sam was required to deposit $138,300, and Fabiola was required to deposit $110,047.23.

5.6 EXERCISES

1. Cheryl Wilcox is planning for her retirement, so she is setting up a payout annuity with her bank. She wishes to receive a payout of $1,200 per month for twenty years.

   a. How much money must she deposit if her money earns 8% interest compounded monthly?

   b. Find the total amount that Cheryl will receive from her payout annuity.

2. James Magee is planning for his retirement, so he is setting up a payout annuity with his bank. He wishes to receive a payout of $1,100 per month for twenty-five years.

   a. How much money must he deposit if his money earns 9% interest compounded monthly?

   b. Find the total amount that James will receive from his payout annuity.

3. Dean Gooch is planning for his retirement, so he is setting up a payout annuity with his bank. He wishes to receive a payout of $1,300 per month for twenty-five years.

   a. How much money must he deposit if his money earns 7.3% interest compounded monthly?

   b. Find the total amount that Dean will receive from his payout annuity.

4. Holly Krech is planning for her retirement, so she is setting up a payout annuity with her bank. She wishes to receive a payout of $1,800 per month for twenty years.

   a. How much money must she deposit if her money earns 7.8% interest compounded monthly?

   b. Find the total amount that Holly will receive from her payout annuity.

5. **a.** How large a monthly payment must Cheryl Wilcox (from Exercise 1) make if she saves for her payout annuity with an ordinary annuity, which she sets up thirty years before her retirement? (The two annuities pay the same interest rate.)

   b. Find the total amount that Cheryl will pay into her ordinary annuity, and compare it with the total amount that she will receive from her payout annuity.

6. **a.** How large a monthly payment must James Magee (from Exercise 2) make if he saves for his payout annuity with an ordinary annuity, which he sets up twenty-five years before his retirement? (The two annuities pay the same interest rate.)

   b. Find the total amount that James will pay into his ordinary annuity, and compare it with the total amount that he will receive from his payout annuity.

7. **a.** How large a monthly payment must Dean Gooch (from Exercise 3) make if he saves for his payout annuity with an ordinary annuity, which he sets up thirty years before his retirement? (The two annuities pay the same interest rate.)

   b. How large a monthly payment must he make if he sets the ordinary annuity up twenty years before his retirement?

8. **a.** How large a monthly payment must Holly Krech (from Exercise 4) make if she saves for her payout annuity with an ordinary annuity, which she sets up thirty years before her retirement? (The two annuities pay the same interest rate.)

   b. How large a monthly payment must she make if she sets the ordinary annuity up twenty years before her retirement?

9. Lily Chang is planning for her retirement, so she is setting up a payout annuity with her bank. For twenty years, she wishes to receive annual payouts that start at $14,000 and then receive an annual COLA of 4%.

   a. How much money must she deposit if her money earns 8% interest per year?

   b. How large will Lily's first annual payout be?

   c. How large will Lily's second annual payout be?

   d. How large will Lily's last annual payout be?

10. Wally Brown is planning for his retirement, so he is setting up a payout annuity with his bank. For twenty-five years, he wishes to receive annual payouts that start at $16,000 and then receive an annual COLA of 3.5%.

    a. How much money must he deposit if his money earns 8.3% interest per year?

    b. How large will Wally's first annual payout be?

    c. How large will Wally's second annual payout be?

    d. How large will Wally's last annual payout be?

11. Oshri Karmon is planning for his retirement, so he is setting up a payout annuity with his bank. He is now 30 years old, and he will retire when he is 60. He wants to receive annual payouts for twenty-five years, and he wants those payouts to receive an annual COLA of 3.5%.

    a. He wants his first payout to have the same purchasing power as does $13,000 today. How big should that payout be if he assumes inflation of 3.5% per year?

    b. How much money must he deposit when he is 60 if his money earns 7.2% interest per year?

    c. How large a monthly payment must he make if he saves for his payout annuity with an ordinary annuity? (The two annuities pay the same interest rate.)

    d. How large a monthly payment would he make if he waits until he is 40 before starting his ordinary annuity?

12. Shelly Franks is planning for her retirement, so she is setting up a payout annuity with her bank. She is now 35 years old, and she will retire when she is 65. She wants to receive annual payouts for twenty years, and she wants those payouts to receive an annual COLA of 4%.

    a. She wants her first payout to have the same purchasing power as does $15,000 today. How big should that payout be if she assumes inflation of 4% per year?

    b. How much money must she deposit when she is 65 if her money earns 8.3% interest per year?

    c. How large a monthly payment must she make if she saves for her payout annuity with an ordinary annuity? (The two annuities pay the same interest rate.)

    d. How large a monthly payment would she make if she waits until she is 40 before starting her ordinary annuity?

In Exercises 13–16, use the Annual Payout Annuity with COLA Formula to find the deposit necessary to receive monthly payouts with an annual cost-of-living adjustment. To use the formula, all figures must be annual figures, including the payout and the annual rate. You can adapt the formula for monthly payouts by using

- the future value of a one-year ordinary annuity in place of the annual payout, where *pymt* is the monthly payout, and
- the annual yield of the given compound interest rate in place of the annual rate *r*.

13. Fabiola Macias is about to retire, so she is setting up a payout annuity with her bank. She wishes to receive a monthly payout for the next twenty-five years, where the payout starts at $1,000 per month and receives an annual COLA of 3%. Her money will earn 10% compounded monthly.

    a. The annual payout is the future value of a one-year ordinary annuity. Find this future value.

    b. The annual rate *r* is the annual yield of 10% interest compounded monthly. Find this annual yield. (*Do not* round it off.)

    c. Use the Annual Payout Annuity with COLA Formula to find how much money she must deposit.

    d. Fabiola could have saved for her payout annuity with an ordinary annuity. If she had started doing so thirty years ago, what would the required monthly payments have been? (The two annuities pay the same interest rate.)

14. Gary Kersting is about to retire, so he is setting up a payout annuity with his bank. He wishes to receive a monthly payout for the next twenty years, where the payout starts at $1,300 per month and receives an annual COLA of 4%. His money will earn 8.7% compounded monthly.

    a. The annual payout is the future value of a one-year ordinary annuity. Find this future value.

    b. The annual rate r is the annual yield of 8.7% interest compounded monthly. Find this annual yield. (*Do not* round it off.)

    c. Use the Annual Payout Annuity with COLA Formula to find how much money he must deposit.

    d. Gary could have saved for his payout annuity with an ordinary annuity. If he had started doing so twenty-five years ago, what would the required monthly payments have been? (The two annuities pay the same interest rate.)

15. Conrad von Schtup is about to retire, so he is setting up a payout annuity with his bank. He wishes to receive a monthly payout for the next twenty-three years, where the payout starts at $1,400 per month and receives an annual COLA of 5%. His money will earn 8.9% compounded monthly.

    a. The annual payout is the future value of a one-year ordinary annuity. Find this future value.

    b. The annual rate r is the annual yield of 8.9% interest compounded monthly. Find this annual yield. (*Do not* round it off.)

    c. Use the Annual Payout Annuity with COLA Formula to find how much money he must deposit.

    d. Conrad could have saved for his payout annuity with an ordinary annuity. If he had started doing so twenty years ago, what would the required monthly payments have been? (The two annuities pay the same interest rate.)

16. Mitch Martinez is about to retire, so he is setting up a payout annuity with his bank. He wishes to receive a monthly payout for the next thirty years, where the payout starts at $1,250 per month and receives an annual COLA of 4%. His money will earn 7.8% compounded monthly.

    a. The annual payout is the future value of a one-year ordinary annuity. Find this future value.

    b. The annual rate r is the annual yield of 7.8% interest compounded monthly. Find this annual yield.

    c. Use the Annual Payout Annuity with COLA Formula to find how much money he must deposit.

    d. Mitch could have saved for his payout annuity with an ordinary annuity. If he had started doing so twenty years ago, what would the required monthly payments have been? (The two annuities pay the same interest rate.)

17. Bob Pirtle won $1 million in a state lottery. He was surprised to learn that he will not receive a check for $1 million. Rather, for twenty years, he will receive an annual check from the state for $50,000. The state finances this series of checks by buying Bob a payout annuity. Find what the state pays for Bob's payout annuity if the interest rate is 8%.

18. John-Paul Ramin won $2.3 million in a state lottery. He was surprised to learn that he will not receive a check for $2.3 million. Rather, for twenty years, he will receive an annual check from the state for $\frac{1}{20}$ of his winnings. The state finances this series of checks by buying John-Paul a payout annuity. Find what the state pays for John-Paul's payout annuity if the interest rate is 7.2%.

 Answer the following questions using complete sentences and your own words.

• **CONCEPT QUESTIONS**

19. Compare and contrast a savings annuity with a payout annuity. How do they differ in purpose? How do they differ in structure? How do their definitions differ?

    HINT: Compare Examples 1 and 2.

20. Under what circumstances would a savings annuity and a payout annuity be combined?

WEB PROJECTS

21. This is an exercise in saving for your retirement. Write a paper describing all of the following points.

    a. Go on the web and find out what the annual rate of inflation has been for each of the last ten years. Use the average of these figures as a prediction of the future annual rate of inflation.

    b. Estimate the total monthly expenses you would have if you were retired now. Include housing, food, and utilities.

    c. Use parts (a) and (b) to predict your total monthly expenses when you retire, assuming that you retire at age 65.

    d. Plan on financing your monthly expenses with a payout annuity. How much money must you deposit when you are 65 if your money earns 7.5% interest per year?

    e. How large a monthly payment must you make if you save for your payout annuity with an ordinary annuity, starting now? (The two annuities pay the same interest rate.)

    f. How large a monthly payment must you make if you save for your payout annuity with an ordinary annuity, starting ten years from now?

    g. How large a monthly payment must you make if you save for your payout annuity with an ordinary annuity, starting twenty years from now?

    Some useful links for this web project are listed on the text web site:

    www.cengage.com/math/johnson

TERMS

add-on interest loan
adjustable rate mortgage
amortization schedule
amortized loan
annual percentage rate (APR) of an add-on interest loan
annual percentage rate (APR) of a simple interest amortized loan
annual yield
annuity

annuity due
average daily balance
balance
compound interest
compounding period
cost-of-living adjustment (COLA)
doubling time
expire
finance
finance charges
future value
Individual Retirement Account (IRA)
interest
interest rate

line of credit
loan agreement
lump sum
maturity value
negative amortization
nominal rate
note
ordinary annuity
outstanding principal
payment period
payout annuity
periodic rate
point
prepaying a loan
prepayment penalty

present value
present value of an annuity
principal
simple interest
simple interest amortized loan
simple interest loan
sinking fund
term
tax-deferred annuity (TDA)
Truth in Lending Act
unpaid balance
yield

FORMULAS

The **simple interest** I on a principal P at a interest rate r for t years is

$$I = Prt$$

and the **future value** is

$$FV = P(1 + rt)$$

After n compounding periods, the future value FV of an initial principal P earning **compound interest** at a periodic interest rate i for n periods is

$$FV = P(1 + i)^n$$

The **annual yield** of a given compound rate is the simple interest rate r that has the same future value as the compound rate in 1 year. To find it, solve

$$FV(\text{compound interest}) = FV(\text{simple interest})$$
$$P(1 + i)^n = P(1 + rt)$$

for the simple interest rate r after making appropriate substitutions.

The future value FV of an **ordinary annuity** is

$$FV(\text{ordinary}) = pymt\frac{(1 + i)^n - 1}{i}$$

and the future value of an **annuity due** is

$$FV(\text{due}) = FV(\text{ordinary}) \cdot (1 + i)$$
$$= pymt\frac{(1 + i)^n - 1}{i}(1 + i)$$

The **present value of an ordinary annuity** is the lump sum P such that

$$FV(\text{lump sum}) = FV(\text{annuity})$$
$$P(1 + i)^n = pymt\frac{(1 + i)^n - 1}{i}$$

Simple interest amortized loan:

$$FV(\text{annuity}) = FV(\text{loan amount})$$
$$pymt\frac{(1 + i)^n - 1}{i} = P(1 + i)^n$$

Unpaid balance:

Unpaid balance = current value of loan amount − current value of annuity

$$\approx P(1 + i)^n - pymt\frac{(1 + i)^n - 1}{i}$$

where n is the number of periods *from the beginning of the loan to the present.*

Payout annuity: The total principal necessary to generate n payouts of size *pymt* is P, where

$$P(1 + i)^n = pymt \frac{(1 + i)^n - 1}{i}$$

Annual payout annuity with COLA: A payout annuity with a term of t years, a first-year payout of *pymt,* and an annual COLA rate c, requires a principal of

$$P = (pymt)\frac{1 - \left(\dfrac{1 + c}{1 + r}\right)^t}{r - c}$$

STEPS

To find the **credit card finance charge** with the **average daily balance method:**

1. Find the balance for each day in the billing period and the number of days at that balance.

2. The average daily balance is the weighted average of these daily balances, weighted to reflect the number of days at that balance.

3. The finance charge is simple interest applied to the average daily balance.

To create an **amortization schedule:**
For each payment, list the payment number, principal portion, interest portion, total payment and balance.

For each payment:
1. Find the interest on the balance—use the simple interest formula.

For each payment except the last:
2. The principal portion is the payment minus the interest portion.
3. The new balance is the previous balance minus the principal portion.

For the last payment:
4. The principal portion is the previous balance.
5. The total payment is the sum of the principal portion and interest portion.

REVIEW EXERCISES

1. Find the interest earned by a deposit of $8,140 at $9\frac{3}{4}\%$ simple interest for eleven years.

2. Find the interest earned by a deposit of $10,620 at $8\frac{1}{2}\%$ simple interest for twenty-five years.

3. Find the future value of a deposit of $12,288 at $4\frac{1}{4}\%$ simple interest for fifteen years.

4. Find the future value of a deposit of $22,880 at $5\frac{3}{4}\%$ simple interest for thirty years.

5. Find the maturity value of a loan of $3,550 borrowed at $12\frac{1}{2}\%$ simple interest for one year and two months.

6. Find the maturity value of a loan of $12,250 borrowed at $5\frac{1}{2}\%$ simple interest for two years.

7. Find the present value of a future value of $84,120 at $7\frac{1}{4}\%$ simple interest for twenty-five years.

8. Find the present value of a future value of $10,250 at $5\frac{3}{4}\%$ simple interest for twenty years.

9. Find the future value of a deposit of $8,140 at $9\frac{3}{4}\%$ interest compounded monthly for eleven years.

10. Find the future value of a deposit of $7,250 at $5\frac{1}{4}\%$ interest compounded monthly for twenty years.

11. Find the interest earned by a deposit of $7,990 at $4\frac{3}{4}\%$ interest compounded monthly for eleven years.

12. Find the interest earned by a deposit of $22,250 at $9\frac{1}{4}\%$ interest compounded monthly for twenty years.

13. Find the present value of a future value of $33,120 at $6\frac{1}{4}\%$ interest compounded daily for twenty-five years.

14. Find the present value of a future value of $10,600 at $7\frac{7}{8}\%$ interest compounded daily for four years.

15. Find the annual yield corresponding to a nominal rate of 7% compounded daily.

16. Find the annual yield corresponding to a nominal rate of $8\frac{1}{2}\%$ compounded monthly.

17. Find the future value of a twenty-year ordinary annuity with monthly payments of $230 at 6.25% interest.

18. Find the future value of a thirty-year ordinary annuity with biweekly payments of $130 at 7.25% interest.

19. Find the future value of a twenty-year annuity due with monthly payments of $450 at 8.25% interest.

20. Find the future value of a thirty-year annuity due with biweekly payments of $240 at 6.75% interest.

21. Find (a) the monthly payment and (b) the total interest for a simple interest amortized loan of $25,000 for five years at $9\frac{1}{2}\%$ interest.

22. Find (a) the monthly payment and (b) the total interest for a simple interest amortized loan of $130,000 for twenty years at $8\frac{1}{4}\%$ interest.

23. The Square Wheel Bicycle store has found that they sell most of their bikes in the spring and early summer. On February 15, they borrowed $351,500 to buy bicycles. They are confident that they can sell most of these bikes by September 1. Their loan is at $5\frac{7}{8}\%$ simple interest. What size lump sum payment would they have to make on September 1 to pay off the loan?

24. Mike Taylor buys a four-year old Ford from a car dealer for $16,825. He puts 10% down and finances the rest through the dealer at 10.5% add-on interest. If he agrees to make thirty-six monthly payments, find the size of each payment.

25. The activity on Sue Washburn's MasterCard account for one billing period is shown below. Find the average daily balance and the finance charge if the billing period is August 26 through September 25, the previous balance was $3,472.38, and the annual interest rate is $19\frac{1}{2}\%$.

August 30	payment	$100.00
September 2	gasoline	$34.12
September 10	restaurant	$62.00

26. George and Martha Simpson bought a house from Sue Sanchez for $205,500. In lieu of a 20% down payment, Ms. Sanchez accepted 5% down at the time of the sale and a promissory note from the Simpsons for the remaining 15%, due in eight years. The Simpsons also agreed to make monthly interest payments to Ms. Sanchez at 12% simple interest until the note expires. The Simpsons borrowed the remaining 80% of the purchase price from their bank. The bank paid that amount, less a commission of 6% of the purchase price, to Ms. Sanchez.

    a. Find the Simpsons' monthly interest-only payment to Ms. Sanchez.

    b. Find Ms. Sanchez's total income from all aspects of the down payment.

    c. Find Ms. Sanchez's total income from all aspects of the sale of the house, including the down payment.

27. Tien Ren Chiang wants to have an IRA that will be worth $250,000 when he retires at age 65.

    a. How much must he deposit at age 25 at $8\frac{1}{8}\%$ compounded quarterly?

    b. If he arranges for the monthly interest to be sent to him starting at age 65, how much would he receive each month? (Assume that he will continue to receive $8\frac{1}{8}\%$ interest, compounded monthly.)

28. Extremely Trustworthy Savings offers five-year CDs at 7.63% compounded annually, and Bank of the South offers five-year CDs at 7.59% compounded daily. Compute the annual yield for each institution and determine which offering is more advantageous for the consumer.

29. You are 32, and you have just set up an ordinary annuity to save for retirement. You make monthly payments of $200 that earn $6\frac{1}{8}\%$ interest. Find the future value when you reach age 65.

30. Find and interpret the present value of the annuity in Exercise 29.

31. Find the future value of an annuity due with monthly payments of $200 that earns $6\frac{1}{8}\%$ interest, after eleven years.

32. Matt and Leslie Silver want to set up a TDA that will generate sufficient interest on maturity to meet their living expenses, which they project to be $1,300 per month.

    a. Find the amount needed at maturity to generate $1,300 per month interest if they can get $8\frac{1}{4}\%$ interest compounded monthly.

    b. Find the monthly payment they would have to make into an ordinary annuity to obtain the future value found in part (a) if their money earns $9\frac{3}{4}\%$ and the term is thirty years.

33. Mr. and Mrs. Liberatore set up a TDA to save for their retirement. They agreed to have $100 deducted from each of Mrs. Liberatore's monthly paychecks, which will earn $6\frac{1}{8}\%$ interest.

    a. Find the future value of their ordinary annuity, if it comes to term after they retire in thirty years.

    b. After retiring, the Liberatores convert their annuity to a savings account, which earns 5.75% interest compounded monthly. At the end of each month, they withdraw $1,000 for living expenses. Complete the chart in Figure 5.29 for their postretirement account.

Month Number	Account Balance at Beginning of the Month	Interest for the Month	With-drawal	Account Balance at End of the Month
1				
2				
3				
4				
5				

FIGURE 5.29 Chart for Exercise 33.

34. Delores Lopez buys some land in Nevada. She agrees to pay the seller a lump sum of $235,000 in five years. Until then, she will make monthly simple interest payments to the seller at 10% interest.

    a. Find the amount of each interest payment.

    b. Delores sets up a sinking fund to save the $235,000. Find the size of her monthly payments if her payments are due at the end of every month and her money earns $9\frac{3}{8}\%$ interest.

    c. Prepare a table showing the amount in the sinking fund after each of the first two deposits.

35. Maude Frickett bought a house for $225,600. She put 20% down and obtains a simple interest amortized loan for the rest at $7\frac{3}{8}\%$ for thirty years.

    a. Find her monthly payment.

    b. Find the total interest.

    c. Prepare an amortization schedule for the first two months of the loan.

36. Navlet's Nursery needs to borrow $228,000 to increase its inventory for the upcoming spring season. The owner is confident that he will sell most if not all of the new plants during the summer, so he wishes to borrow the money for only six months. His bank has offered him a simple interest amortized loan at $8\frac{1}{4}\%$ interest.

    a. Find the size of the monthly bank payment.

    b. Prepare an amortization schedule for all six months of the loan.

37. Harry Carry had his kitchen remodeled. He did not have sufficient cash to pay for it. However, he had previously set up a line of credit with his bank. On May 16, he wrote a check to his contractor on his line of credit for $41,519. The line's interest rate is $7\frac{3}{4}\%$.

    a. Find the size of the required monthly interest payment.

    b. Harry decided that it would be in his best interests to get this loan paid off in seven months. Find the size of the monthly principal-plus-interest payment that would accomplish this.

    (*HINT:* In effect, Carry is converting the loan to an amortized loan.)

    c. Prepare an amortization schedule for all seven months of the loan.

    d. Find the amount of line of credit interest that Carry could deduct from his taxes next year.

38. Ben Suico buys a car for $13,487.31. He puts 10% down and obtains a simple interest amortized loan for the rest at $10\frac{7}{8}\%$ interest for five years.

    a. Find his monthly payment.

    b. Find the total interest.

    c. Prepare an amortization schedule for the first two months of the loan.

    d. Mr. Suico decides to sell his car two years and six months after he bought it. Find the unpaid balance on his loan.

39. Scott Frei wants to buy a used car that costs $6,200. The used car dealer has offered him a four-year add-on interest loan that requires a $200 down payment at 9.9% annual interest with an APR of 10%.

    a. Find the monthly payment.

    b. Verify the APR.

40. Miles Archer bought a house for $112,660. He put 20% down and obtains a simple interest amortized loan for the rest at $9\frac{7}{8}\%$ for thirty years. If Miles paid two points and $5,738.22 in fees, $1,419.23 of which are included in the finance charge, find the APR.

41. Susan and Steven Tamchin are thinking of buying a home for $198,000. A potential lender advertises an 80%, thirty-year simple interest amortized loan at $8\frac{1}{2}\%$ interest, with and APR of 9.02%.

    a. Find the size of the Tamchin's monthly payment.

    b. Use the APR to approximate the fees included in the finance charge.

42. Fred Rodgers is planning for his retirement, so he is setting up a payout annuity with his bank. He wishes to receive a payout of $1,700 per month for twenty-five years.

    a. How much money must he deposit if his money earns 6.1% interest compounded monthly?

    b. How large a monthly payment would Fred have made if he had saved for his payout annuity with an ordinary annuity, set up thirty years before his retirement? (The two annuities pay the same interest rate.)

    c. Find the total amount that Fred will pay into his ordinary annuity and the total amount that he will receive from his payout annuity.

43. Sue West is planning for her retirement, so she is setting up a payout annuity with her bank. She is now 30 years old, and she will retire when she is 60. She wants to receive annual payouts for twenty-five years, and she wants those payouts to have an annual COLA of 4.2%.

    a. She wants her first payout to have the same purchasing power as does $17,000 today. How big should that payout be if she assumes inflation of 4.2% per year?

    b. How much money must she deposit when she is sixty if her money earns 8.3% interest per year?

    c. How large a monthly payment must she make if she saves for her payout annuity with an ordinary annuity? (The two annuities pay the same interest rate.)

 Answer the following questions using complete sentences and your own words.

• CONCEPT QUESTIONS

44. What is the difference between simple interest and compound interest?

45. Describe a situation in which simple interest, rather than compound interest, would be expected.

46. Describe a situation in which compound interest, rather than simple interest, would be expected.

47. What is the difference between an account that earns compound interest and an annuity that earns compound interest?

48. What is the difference between a simple interest amortized loan and an add-on interest loan?

• HISTORY QUESTIONS

49. What does the Truth in Lending Act do for borrowers?

50. Who offered the first credit card?

51. How did the first credit card differ from the first post–World War II credit card?

VOTING AND APPORTIONMENT

6

Comstock/Jupiter Images

The right to vote for one's governing officials is the cornerstone of any democracy. In the United States, individual citizens have a voice in the selection of local mayors and city councils, county officials, state legislators, governors, members of the House of Representatives and the Senate, and even the ultimate office: the President of the United States. Elections of all types are commonplace, and their outcomes range from overwhelming landslides to razor-thin victories, legal suits, and tedious recounts.

When we vote to fill a political office, the outcome obviously depends on the number of votes cast for each candidate; when we vote to create laws, the numbers for and against a proposition determine its fate. However, determining the outcome of an election might not be as simple as some may think. The premises of "one

continued

WHAT WE WILL DO IN THIS CHAPTER

In this chapter, we will study systems of voting and methods of apportionment.

WE'LL DEVELOP AND EXPLORE VARIOUS SYSTEMS OF VOTING:

- Different systems of voting may produce different outcomes; one system may conclude that candidate A is the winner, whereas another system may proclaim that B wins. In any election, large or small, voters should understand the specific system of voting that is being used *before* they cast their votes!

- Ideally, any system of voting should produce fair results. What exactly is meant by the word *fair*? Is any system of voting ultimately fair in all situations?

WE'LL DEVELOP AND EXPLORE VARIOUS SYSTEMS OF APPORTIONMENT:

- Different methods of apportionment can lead to significantly different allocations of seats in a governing body. Some methods favor regions that have small populations, whereas others favor those with large. Who should decide which method will be used?

continued

- Ideally, any method of apportionment should produce fair results. However, in the pursuit of fairness, all methods of apportionment create paradoxes that undermine the fairness they seek! Which method is "best" to use?

- When the United States first became an independent nation, several of the "founding fathers" proposed various methods of apportionment for the House of Representatives. Which methods have actually been used? Which method is used today?

person, one vote" and "the candidate with the most votes wins" are neither universal nor absolute. In fact, several different voting systems have been used in democratic societies today and throughout history.

Elected governmental bodies range in scope from local to regional to national levels. That is, elected officials represent their constituencies in matters that affect others in a broader sense. For example, precincts within a city elect officials to represent the precinct on citywide matters, districts within a state elect officials to represent the district on statewide matters, and states elect officials to represent the individual states on nationwide matters. How many officials will a region receive to represent it in these matters? In some cases, each region may receive the same number of representative seats; in others, different regions may receive different numbers of seats. The determination of the number of representatives a region receives is known as apportionment.

6.1 Voting Systems

OBJECTIVES

- Develop and apply the plurality method of voting

- Develop and apply the plurality with elimination method of voting

- Develop and apply the instant runoff method of voting

- Develop and apply the Borda count method of voting

- Develop and apply the pairwise comparison method of voting

- Define and investigate the four fairness criteria for voting systems

Voting is an essential element in any democratic form of government. Whether it is the selection of leaders, the creation of laws and regulations, or deciding the outcome of issues ranging from mundane to volatile, decisions are made on the basis of the results of elections.

Voting is an essential element in any democratic form of government.

Once votes have been cast, they must be counted, and a winner must be declared. Many people assume that "the candidate with the most votes wins"; however, that is not always the case. Deciding who wins an election is ultimately determined by the type of voting system that is used. "The person with the most votes wins" is a commonly used system and is called the plurality method. However, the plurality method is not the only method that is used. In this section, we will examine several common voting systems: (1) the plurality method, (2) the plurality with elimination method, (3) the ranked-choice or instant run-off method (sometimes called the Australian method), (4) the Borda count method, and (5) the pairwise comparison method.

The Plurality Method

A common way to determine the outcome of an election is to declare the candidate with the most votes the winner. This is the basis of the **plurality method of voting.**

PLURALITY METHOD OF VOTING

Each person votes for his or her favorite candidate (or choice). The candidate (or choice) that receives the most votes is declared the winner. (In case of a tie, a special runoff election might be held.)

EXAMPLE **1**

APPLYING THE PLURALITY METHOD OF VOTING: MAJORITY SCENARIO City Cab is a new taxi service that will soon begin operation in a major metropolitan area. The board of directors for City Cab must purchase a fleet of new vehicles, and the vehicles must all be the same color. Four colors are available: black, white, red, and green. The five directors vote for their choice of color and the results are given in Figure 6.1. Using the plurality method of voting, which color choice wins?

	Director 1	**Director 2**	**Director 3**	**Director 4**	**Director 5**
Choice	green	red	red	green	red

FIGURE 6.1 Results of voting for the color of City Cab vehicles.

SOLUTION

Tallying the results, we see that green received two votes and red received three. Because red received the most votes (3), red is declared the winner.

In Example 1, it should be noted that in addition to receiving the most votes, red also received a majority of the votes, that is, more than 50% of the votes were for red. As we can see, three out of five, or 60%, of the votes were for red, while 40% were not for red.

Although the plurality method is easy to apply, this method can produce an unusual winner, as we shall see in the next example.

EXAMPLE **2**

APPLYING THE PLURALITY METHOD OF VOTING: NONMAJORITY SCENARIO Referring to City Cab in Example 1, suppose that the results of the election are those given in Figure 6.2. Using the plurality method of voting, which color choice wins?

	Director 1	**Director 2**	**Director 3**	**Director 4**	**Director 5**
Choice	green	red	red	black	white

FIGURE 6.2 Results of voting for the color of City Cab vehicles.

SOLUTION

Once again, because red received the most votes (two), red would be declared the winner. Although red is the winner, it should be noted that red did *not* receive a *majority* of the votes. Fewer than half of the votes, that is 40%, were for red, while significantly more than half (60%) of the directors did not want the color red. This type of "winner" can lead to very difficult situations in political elections. Imagine yourself as the "winner" of a local election when more than half of the voters do *not* support you.

The Plurality with Elimination Method

As was shown in Example 2, the plurality method of voting may produce a "winner" even though a majority of voters did not vote for the "winner." This dilemma can be avoided in several ways. One common way is to eliminate the candidate with the fewest votes and then hold another election. This is the basis of the **plurality with elimination method.** Consequently, the voters who supported the eliminated candidate now vote for someone else; that is, they must vote for their second choice. Typically, this second election will produce a winner with a majority of support. However, if a majority is still not attained, the process is repeated until a candidate obtains a majority of the votes.

PLURALITY WITH ELIMINATION METHOD OF VOTING

Each person votes for his or her favorite candidate. If a candidate receives a majority of votes, that candidate is declared the winner. If no candidate receives a majority, then the candidate with the fewest votes is eliminated and a new election is held. This process continues until a candidate receives a majority of the votes.

EXAMPLE **3**

APPLYING THE PLURALITY WITH ELIMINATION METHOD OF VOTING   The managers of We Deliver (a local shipping business) are planning a party for their 60 employees. There are three possible locations for the party: the warehouse, the park, and the beach. The employees voted on which location they preferred, and the results are given in Figure 6.3.

	Warehouse	Park	Beach
Number of votes	23	19	18

FIGURE 6.3   Results of voting for party location.

a. Using the plurality method of voting, which location wins?
b. Using the plurality with elimination method of voting, which location wins?

SOLUTION

a. Because the warehouse received the most votes (23), the party will be held at the warehouse.
b. Although the warehouse received the most votes, it did not receive a majority; the warehouse received only 23/60, or approximately 38.3%, of the vote. Because the beach received the fewest number of votes (18), it is eliminated, and a new election is held. The results of the second election are given in Figure 6.4.

	Warehouse	Park	Beach
Number of votes	25	35	0

FIGURE 6.4   Results of the second election.

Because the park received a majority of the votes ($35/60 \approx 58.3\%$), the party will be held at the park.

As is shown in Example 3, the outcome of an election may change depending on the type of voting system used. Therefore, it is imperative that the voters know ahead of time which system is being used!

As we have seen, the elimination of one candidate forces some people to shift their vote to their second choice in the new election. In the realm of real-world political elections, this method of elimination can be very expensive both for the local office of elections and for the candidates themselves, and it is overly time-consuming. Rather than holding a new election after eliminating the candidate with the fewest votes, a popular alternative is to have each voter rank each candidate during the first election. This is the basis of the ranked-choice or instant runoff method.

The Ranked-Choice or Instant Runoff Method

In Example 3, the employees of We Deliver had three choices for the location of their party. If each of the 60 employees was asked to rank these choices in order of preference, in how many different ways could the locations be ranked? Because the choices are being put in an order (that is, a person must make a first choice, a second choice, and a third choice), we conclude that *permutations* can be used to determine the number of different rankings (see Section 2.4). Specifically, there are

$$_3P_3 = \frac{3!}{(3-3)!} = \frac{3!}{0!} = \frac{3 \cdot 2 \cdot 1}{1} = 6$$

different rankings (permutations) of the three locations. One such ranking is "warehouse, park, beach" or, more conveniently, WPB. Another such ranking is BPW (beach, park, warehouse). Figure 6.5 lists all possible permutations (or rankings) of the three locations.

1st choice	W	W	P	P	B	B
2nd choice	P	B	B	W	W	P
3rd choice	B	P	W	B	P	W

FIGURE 6.5 All permutations (or rankings) of "Warehouse, Park, Beach."

After each employee ranked the locations and the ballots were tallied, it was determined that nine employees chose the WPB ranking, fourteen chose WBP, fifteen chose PBW, four chose PWB, two chose BWP, and sixteen chose BPW. These results are portrayed in the table shown in Figure 6.6.

		Number of Ballots Cast					
		9	14	15	4	2	16
Ranked Ballot	**1st choice**	W	W	P	P	B	B
	2nd choice	P	B	B	W	W	P
	3rd choice	B	P	W	B	P	W

FIGURE 6.6 Voter preference table.

Tables like that shown in Figure 6.6 are often called **voter preference tables,** that is, tables that list different rankings of the candidates along with the number of voters who chose each specific ranking. Tables of this nature are often used to depict the voting patterns in a ranked election.

Referring to the voter preference table in Figure 6.6, notice that the sum of the row of numbers is 60, that is, the sum of the row equals the number of people voting. In addition, to determine the number of first-choice votes for each candidate, go to the row labeled "1st choice" and observe that $9 + 14 = 23$ people had W (the warehouse) as their first choice, whereas $15 + 4 = 19$ people had P (the park) as their first choice, and $2 + 16 = 18$ had B (the beach) as their first choice. These subtotals agree with the values given in Example 3.

We now proceed as we did when using the plurality with elimination method; that is, we eliminate the candidate with the fewest first-choice votes (the beach). However, rather than holding a new election, we simply modify the original table by shifting the votes in the voter preference table upwards to fill the voids created by eliminating the beach, as shown in Figure 6.7.

		Number of Ballots Cast					
		9	14	15	4	2	16
Ranked Ballot	1st choice	W	W	P	P	~~B~~	~~B~~
	2nd choice	P	~~B~~	~~B~~	W	W↑	P↑
	3rd choice	~~B~~	P↑	W↑	~~B~~	P↑	W↑

FIGURE 6.7 Shifting votes in the voter preference table.

The modified voter preference table is shown in Figure 6.8.

	Number of Ballots Cast					
	9	14	15	4	2	16
1st choice	W	W	P	P	W	P
2nd choice	P	P	W	W	P	W

FIGURE 6.8 Modified voter preference table.

Referring to Figure 6.8, we now see that the Warehouse received $9 + 14 + 2 = 25$ first choice votes, whereas the Park received $15 + 4 + 16 = 35$ votes. Of course, these values agree with those given in Example 3.

The process described above is called the **ranked-choice or instant runoff method.** This voting system is commonly used in Australia. In March 2002, the voters of San Francisco, California, adopted the instant runoff method for the election of most political posts, including mayor, sheriff, district attorney, and many other high-visibility positions. It has been estimated that this voting method will save the City of San Francisco millions of dollars per election, and voters will not have to wait several weeks for a second election, as evidenced by the article on page 414 published in 2005.

The instant runoff method of voting is summarized as follows.

INSTANT RUNOFF METHOD OF VOTING

Each voter ranks all of the candidates; that is, each voter selects his or her first choice, second choice, third choice, and so on. If a candidate receives a majority of first-choice votes, that candidate is declared the winner. If no candidate receives a majority, then the candidate with the fewest first-choice votes is eliminated, and those votes are given to the next preferred candidate. If a candidate now has a majority of first-choice votes, that candidate is declared the winner. If no candidate receives a majority, this process continues until a candidate receives a majority.

FEATURED IN THE NEWS

VOTERS RANK ALL CHOICES

SUCCESS FOR INSTANT RUNOFF VOTING IN SAN FRANCISCO

Last November, San Francisco proved to be a beacon in an otherwise tumultuous election season. In a time of polarized national politics and an alienated electorate, San Francisco embarked on an important innovation that points American democracy toward the future.

San Francisco elected seven seats on the city council (called the Board of Supervisors) using a method known as instant runoff voting (IRV). Several races were hotly contested, one race drawing a remarkable 22 candidates. Observers long used to the blood sport of San Francisco politics were amazed to see how candidates in several races engaged in more coalition building and less vicious negative attacks. Winners were all decided either on election night or within 72 hours after the polls had closed, and even skeptics were won over. Two exit polls showed that city voters generally liked IRV and found it easy to use, including voters across racial and ethnic lines. National media including the New York Times, Washington Post, Associated Press and National Public Radio covered the successful election.

San Francisco will use IRV in future years for citywide offices like mayor and district attorney, joining the ranks of Ireland, Australia and London that use IRV to elect their highest offices. IRV simulates a series of runoff elections but finishes the job in a single election. Voters rank candidates for each race in order of choice: first, second, third. If your first choice gets eliminated from the "instant runoff," your vote goes to your second-ranked candidate as your backup choice. The runoff rankings are used to determine which candidate has support from a popular majority, and accomplish this in a single election. Voters are liberated to vote for the candidates they really like, no more spoiler candidates and "lesser of two evils" dilemmas.

Previously San Francisco decided majority winners in a December runoff election. Runoffs were expensive, costing the City more than $3 million citywide, and voter turnout often plummeted in the December election by as much as 50 percent. So San Francisco taxpayers will save millions of dollars by using IRV, and winners now are determined in the November election when voter turnout tends to be highest. Also, candidates didn't need to raise more money for a second election and independent expenditures declined, significantly improving the campaign finance situation.

by Steven Hill and Rob Richie
Published on Wednesday, January 12, 2005 by CommonDreams.org

EXAMPLE 4

APPLYING AND COMPARING THE PLURALITY AND INSTANT RUNOFF METHODS OF VOTING Townsville is electing a new mayor. The candidates are Alturas (A), Bellum (B), Chan (C), and Dushay (D). The ranked ballots are tallied, and the results are summarized as shown in the voter preference table given in Figure 6.9.

		Number of Ballots Cast					
		15,527	15,287	10,023	9,105	7,978	5,660
Ranked Ballot	1st choice	C	B	C	B	A	D
	2nd choice	B	C	D	A	B	A
	3rd choice	A	D	A	D	C	C
	4th choice	D	A	B	C	D	B

FIGURE 6.9 Voter preference table for Mayor of Townsville.

a. How many voters participated in the election?
b. Use the plurality method to determine the winner.
c. Use the instant runoff method to determine the winner.

SOLUTION

a. To determine the number of voters, we find the sum of the row of numbers representing the number of ballots cast for each ranking:

$$15,527 + 15,287 + 10,023 + 9,105 + 7,978 + 5,660 = 63,580$$

Therefore, 63,580 voters participated in the election.

b. Looking at the row of first-choice votes, we see that

A received 7,978 votes,

B received $15,287 + 9,105 = 24,392$ votes,

C received $15,527 + 10,023 = 25,550$ votes, and

D received 5,660 votes.

Because Chan (C) received the most votes (25,550), Mr. Chan becomes the new mayor of Townsville. However, it should be noted that Mr. Chan did not receive a majority of the votes; he received only $25,550/63,580 \approx 40.2\%$.

c. First, we determine the number of votes needed to obtain a majority. Because 63,580 votes were cast, a candidate needs more than half of 63,580, that is, more than $63,580 \div 2 = 31,790$ votes; in other words, a candidate needs 31,791 votes to win. Because D received the fewest number of votes (5,660), Dushay is eliminated, and we obtain the modified voter preference table shown in Figure 6.10.

	Number of Ballots Cast					
	15,527	15,287	10,023	9,105	7,978	5,660
1st choice	C	B	C	B	A	A
2nd choice	B	C	A	A	B	C
3rd choice	A	A	B	C	C	B

FIGURE 6.10 Modified voter preference table for Mayor of Townsville.

Examining the new row of first-choice votes, we see that

A received $7,978 + 5,660 = 13,638$ votes,

B received $15,287 + 9,105 = 24,392$ votes, and

C received $15,527 + 10,023 = 25,550$ votes.

Because no candidate has a majority, we eliminate the candidate with the fewest votes, that is, we eliminate Alturas and obtain the modified voter preference table shown in Figure 6.11.

	Number of Ballots Cast					
	15,527	15,287	10,023	9,105	7,978	5,660
1st choice	C	B	C	B	B	C
2nd choice	B	C	B	C	C	B

FIGURE 6.11 Modified voter preference table for Mayor of Townsville.

Looking at the new row of first-choice votes, we see that

B received $15,287 + 9,105 + 7,978 = 32,370$ votes, and

C received $15,527 + 10,023 + 5,660 = 31,210$ votes.

Consequently, Ms. Bellum becomes the new mayor with $32,370/63,580 \approx 50.9\%$ of the vote.

Whereas the plurality method utilizes only a voter's first choice, the instant runoff method is designed to accommodate a voter's alternative choices. Another popular method that requires voters to rank their choices is the Borda count method.

The Borda Count Method

To win an election using the instant runoff method, a candidate must capture a majority of first-choice votes. Rather than tallying the first-choice votes only, we might want to tally the first-choice votes, the second-choice votes, the third-choice votes, and so on. This is the basis of the **Borda count method,** which is summarized in the following box.

BORDA COUNT METHOD OF VOTING

Each voter ranks all of the candidates; that is, each voter selects his or her first choice, second choice, third choice, and so on. If there are k candidates, each candidate receives k points for each first-choice vote, $(k - 1)$ points for each second-choice vote, $(k - 2)$ points for each third-choice vote, and so on. The candidate with the most total points is declared the winner.

EXAMPLE **5**

APPLYING THE BORDA COUNT METHOD OF VOTING Use the Borda count method to determine the location of the party for the We Deliver employees referenced in Example 3.

SOLUTION

Recall that the employees were asked to rank the three locations (warehouse, park, and beach); the results are summarized in the voter preference table given in Figure 6.12.

		Number of Ballots Cast					
		9	14	15	4	2	16
Ranked Ballot	1st choice	W	W	P	P	B	B
	2nd choice	P	B	B	W	W	P
	3rd choice	B	P	W	B	P	W

FIGURE 6.12 Voter preference table.

First, we tally the votes for each location as shown in Figure 6.13. Because there were $k = 3$ candidates (warehouse, park, and beach), each first-choice vote is worth 3 points, each second-choice vote is worth 2 points, and each third-choice vote is worth 1 point.

	Warehouse	Park	Beach
1st choice	9 + 14 = 23 votes	15 + 4 = 19 votes	2 + 16 = 18 votes
2nd choice	4 + 2 = 6 votes	9 + 16 = 25 votes	14 + 15 = 29 votes
3rd choice	15 + 16 = 31 votes	14 + 2 = 16 votes	9 + 4 = 13 votes

FIGURE 6.13 Tally of votes.

Now determine the score for each candidate by multiplying the number of votes times the appropriate number of points as shown in Figure 6.14.

	Warehouse	Park	Beach
1st choice votes (3 points each)	23 × 3 points = 69 points	19 × 3 points = 57 points	18 × 3 points = 54 points
2nd choice votes (2 points each)	6 × 2 points = 12 points	25 × 2 points = 50 points	29 × 2 points = 58 points
3rd choice votes (1 point each)	31 × 1 points = 31 points	16 × 1 points = 16 points	13 × 1 points = 13 points

FIGURE 6.14 Points.

Summing each column, we obtain the results of the election:

Warehouse = 69 + 12 + 31 = 112 points
Park = 57 + 50 + 16 = 123 points
Beach = 54 + 58 + 13 = 125 points

When the Borda count method is used, the beach is declared the winner because it has the most points (125).

As is illustrated by the examples referring to the We Deliver employees' party location, the outcome of the election depends on the voting system that is used. Specifically, the plurality method produced the warehouse as the winner, whereas the plurality with elimination and instant runoff methods declared the park to be the winner, and the Borda count method chose the beach. Obviously, it is very important to inform voters *before* the election as to which system will be used!

EXAMPLE **6**

APPLYING THE BORDA COUNT METHOD OF VOTING Referring to Example 4, use the Borda count method to determine the next Mayor of Townsville.

SOLUTION

Referring to the voter preference table given in Figure 6.9, we tally the votes as shown in Figure 6.15.

	Alturas	Bellum	Chan	Dushay
1st choice	7,978 votes	15,287 + 9,105 = 24,392 votes	15,527 + 10,023 = 25,550 votes	5,660 votes
2nd choice	9,105 + 5,660 = 14,765 votes	15,527 + 7,978 = 23,505 votes	15,287 votes	10,023 votes
3rd choice	15,527 + 10,023 = 25,550 votes	0 votes	7,978 + 5,660 = 13,638 votes	15,287 + 9,105 = 24,392 votes
4th choice	15,287 votes	10,023 + 5,660 = 15,683 votes	9,105 votes	15,527 + 7,978 = 23,505 votes

FIGURE 6.15 Tally of votes.

Because there were $k = 4$ candidates (Alturas, Bellum, Chan, and Dushay), each first-choice vote is worth 4 points. Consequently, the second-, third-, and fourth-choice votes are worth 3, 2, and 1 points each, respectively. Now determine the score for each candidate by multiplying the number of votes times the appropriate number of points as shown in Figure 6.16.

	Alturas	**Bellum**	**Chan**	**Dushay**
1st choice (4 points each)	$7,978 \times 4$ = 31,912 points	$24,392 \times 4$ = 97,568 points	$25,550 \times 4$ = 102,200 points	$5,660 \times 4$ = 22,640 points
2nd choice (3 points each)	$14,765 \times 3$ = 44,295 points	$23,505 \times 3$ = 70,515 points	$15,287 \times 3$ = 45,861 points	$10,023 \times 3$ = 30,069 points
3rd choice (2 points each)	$25,550 \times 2$ = 51,100 votes	0×2 = 0 points	$13,638 \times 2$ = 27,276 points	$24,392 \times 2$ = 48,784 points
4th choice (1 point each)	$15,287 \times 1$ = 15,287 votes	$15,683 \times 1$ = 15,683 points	$9,105 \times 1$ = 9,105 points	$23,505 \times 1$ = 23,505 points

FIGURE 6.16 Points.

Summing each column, we obtain the results of the election:

Alturas $= 31,912 + 44,295 + 51,100 + 15,287 = 142,594$ points
Bellum $= 97,568 + 70,515 + 15,683 = 183,766$ points
Chan $= 102,200 + 45,861 + 27,276 + 9,105 = 184,442$ points
Dushay $= 22,640 + 30,069 + 48,784 + 23,505 = 124,998$ points

When the Borda count method is used, Mr. Chan becomes the new mayor because he has the most points (184,442).

The Pairwise Comparison Method

An election that has more than two candidates can be viewed as several mini-elections in which each possible pair of candidates compete against each other. That is, if there are three candidates, X, Y, and Z, we may consider the mini-elections or pairwise comparisons of X versus Y, X versus Z, and Y versus Z. We then determine the winner of each of these pairwise comparisons, and the candidate who wins the most of these is declared the winner of the overall election. This process is the basis of the **pairwise comparison method,** which is summarized in the following box.

PAIRWISE COMPARISON METHOD OF VOTING

Each voter ranks all of the candidates; that is, each voter selects his or her first choice, second choice, third choice, and so on. For each possible pairing of candidates, the candidate with the most votes receives 1 point; if there is a tie, each candidate receives 1/2 point. The candidate who receives the most points is declared the winner.

How many possible pairwise comparisons are there? That is, if there are k candidates and we must select two for a mini-election, how many different mini-elections are possible? The answer lies in using *combinations,* as illustrated in Section 2.4; that is, the order in which we select two candidates to compete does not matter (X versus Y is the same as Y versus X). Consequently, if there are k candidates, we must examine

$$_kC_2 = \frac{k!}{(k-2)!2!} = \frac{k(k-1)}{2}$$

possible pairwise comparisons.

EXAMPLE 7

APPLYING THE PAIRWISE COMPARISON METHOD OF VOTING Use the pairwise comparison method to determine the location of the party for the We Deliver employees referenced in Example 3.

SOLUTION

Recall that the employees were asked to rank the three locations (warehouse, park, and beach) and the results are summarized in the voter preference table in Figure 6.17.

		Number of Ballots Cast					
		9	14	15	4	2	16
Ranked Ballot	1st choice	W	W	P	P	B	B
	2nd choice	P	B	B	W	W	P
	3rd choice	B	P	W	B	P	W

FIGURE 6.17 Voter preference table.

Because there are $k = 3$ "candidates," we must examine

$$_3C_2 = \frac{3!}{(3-2)!2!} = \frac{3!}{1!2!} = \frac{3 \cdot 2 \cdot 1}{1 \cdot 2 \cdot 1} = 3$$

pairwise comparisons; specifically, we investigate W versus P, W versus B, and P versus B. For each comparison, we examine each column of the voter preference table to determine which candidate is preferred.

W versus P In the first column of Figure 6.17, we see that W is preferred over P; therefore, W receives the nine votes listed in column 1. In the second column, W is again ranked over P and receives fourteen votes. In the third column, P is preferred over W, so P receives fifteen votes. In the fourth column, P is again preferred over W and receives four votes. In the fifth column, W is preferred over P and receives two votes. Finally, P is preferred over W in the sixth column and receives sixteen votes. Tallying these results, we obtain the following totals:

votes for W = 9 + 14 + 2 = 25
votes for P = 15 + 4 + 16 = 35

For the comparison of W versus P, voters preferred P over W by a vote of 35 to 25. Consequently, P receives 1 point.

W versus B Examining Figure 6.17, we see that the voters in the first, second, and fourth columns all preferred W over B. Consequently, W receives the nine, fourteen, and four votes listed in those columns. We also see that the voters in the third, fifth, and sixth columns all preferred B over W, and so B receives the fifteen, two, and sixteen votes listed in those columns. Tallying these results, we obtain the following totals:

$$\text{votes for W} = 9 + 14 + 4 = 27$$
$$\text{votes for B} = 15 + 2 + 16 = 33$$

For the comparison of W versus B, voters preferred B over W by a vote of 33 to 27. Consequently, B receives 1 point.

P versus B   Figure 6.17 indicates that the voters in the first, third, and fourth columns all ranked P over B. Therefore, P receives the nine, fifteen, and four votes listed in those columns. The voters in the remaining columns (second, fifth, and sixth) preferred B over P, and B receives fourteen, two, and sixteen votes accordingly. Tallying these results, we obtain the following totals:

$$\text{votes for P} = 9 + 15 + 4 = 28$$
$$\text{votes for B} = 14 + 2 + 16 = 32$$

For the comparison of P versus B, voters preferred B over P by a vote of 32 to 28. Consequently, B receives 1 point.

Tallying the points, we see that B receives 2 points (B was preferred in two of the comparisons), P receives 1 point (P was preferred in one comparison), and W receives 0 points (W was not preferred in any of the comparisons). Consequently, B is declared the winner; the party will be held at the beach.

EXAMPLE **8**

APPLYING THE PAIRWISE COMPARISON METHOD OF VOTING
Use the pairwise comparison method to determine the new mayor of Townsville referenced in Example 4.

SOLUTION

Recall that the candidates are Alturas (A), Bellum (B), Chan (C), and Dushay (D) and that 63,580 votes were cast. The ranked ballots are tallied, and the results are summarized as shown in the voter preference table in Figure 6.18.

		Number of Ballots Cast					
		15,527	15,287	10,023	9,105	7,978	5,660
Ranked Ballot	**1st choice**	C	B	C	B	A	D
	2nd choice	B	C	D	A	B	A
	3rd choice	A	D	A	D	C	C
	4th choice	D	A	B	C	D	B

FIGURE 6.18   Voter preference table for Mayor of Townsville.

Because there are $k = 4$ candidates, we must examine

$$_4C_2 = \frac{4!}{(4-2)!2!} = \frac{4!}{2! \cdot 2!} = \frac{4 \cdot 3 \cdot 2 \cdot 1}{2 \cdot 1 \cdot 2 \cdot 1} = 6$$

pairwise comparisons; specifically, we investigate A versus B, A versus C, A versus D, B versus C, B versus D, and C versus D. For each comparison, we examine each column of the voter preference table to determine which candidate is preferred.

A versus B Figure 6.18 indicates that the voters in the third, fifth, and sixth columns all chose A over B. Therefore, A receives the $10,023 + 7,978 + 5,660 = 23,661$ votes listed in those columns. To determine the number of voters who prefer B over A, we could sum the votes listed in the remaining columns, or using a simpler approach, we subtract the 23,661 votes for A from the total number of votes to obtain the number of votes for B. Therefore, B receives $63,580 - 23,661 = 39,919$ votes. Consequently, B is preferred and receives 1 point.

A versus C Figure 6.18 indicates that the voters in the fourth, fifth, and sixth columns all preferred A over C. Therefore, the votes are tallied as follows:

$$A = 9,105 + 7,978 + 5,660 = 22,743 \text{ votes}$$
$$C = 63,580 - 22,743 = 40,837 \text{ votes}$$

Consequently, C is preferred and receives 1 point.

A versus D Figure 6.18 indicates that the voters in the first, fourth, and fifth columns all preferred A over D. Therefore, the votes are tallied as follows:

$$A = 15,527 + 9,105 + 7,978 = 32,610 \text{ votes}$$
$$D = 63,580 - 32,610 = 30,970 \text{ votes}$$

Consequently, A is preferred and receives 1 point.

B versus C Figure 6.18 indicates that the voters in the second, fourth, and fifth columns all preferred B over C. Therefore, the votes are tallied as follows:

$$B = 15,287 + 9,105 + 7,978 = 32,370 \text{ votes}$$
$$C = 63,580 - 32,370 = 31,210 \text{ votes}$$

Consequently, B is preferred and receives 1 point.

B versus D Figure 6.18 indicates that the voters in the first, second, fourth, and fifth columns all preferred B over D. Therefore, the votes are tallied as follows:

$$B = 15,527 + 15,287 + 9,105 + 7,978 = 47,897 \text{ votes}$$
$$D = 63,580 - 47,897 = 15,683 \text{ votes}$$

Consequently, B is preferred and receives 1 point.

C versus D Figure 6.18 indicates that the voters in the first, second, third, and fifth columns all preferred C over D. Therefore, the votes are tallied as follows:

$$C = 15,527 + 15,287 + 10,023 + 7,978 = 48,815 \text{ votes}$$
$$D = 63,580 - 48,815 = 14,765 \text{ votes}$$

Consequently, C is preferred and receives 1 point.

Tallying the points, we see that A receives 1 point, B receives 3 points, C receives 2 points, and D receives 0 points. B has the most points (3) and so is declared the winner; Ms. Bellum is the new mayor of Townsville.

Flaws of Voting Systems

As is illustrated by the examples involving the We Deliver employees' party location and the election of the mayor of Townsville, the outcome of an election may depend on the voting system that is used. Some people (especially the losing candidates) might claim that this is not fair. But what exactly is meant by the word *fair*? Political scientists and mathematicians have created a list of criteria that any "fair" voting system should meet. Specifically, the four fairness criteria are known as the *majority criterion*, the *head-to-head criterion*, the *monotonicity criterion*, and the *irrelevant alternatives criterion*.

If you were a candidate in an election and you received a **majority** of the votes (that is, you received more than half of the votes), you would expect to be declared the winner. This seems to be a fair and logical conclusion; most people would agree that if a candidate received more than half of the votes, that candidate should be declared the winner. This conclusion is known as the **majority criterion of fairness.**

THE MAJORITY CRITERION

If candidate X receives a majority of the votes, then candidate X should be declared the winner.

Unfortunately, not all systems of voting satisfy this criterion; that is, in some circumstances, a candidate might receive a majority of the votes, yet another candidate might be declared the winner! Specifically, Exercise 23 will illustrate that the Borda count method can at times violate the majority criterion.

If you were a candidate in an election running against two other candidates, say, A and B, and you received more votes when compared to either A or B individually, you would expect to be declared the winner. In other words, suppose that you were running only against candidate A and you beat candidate A; then suppose that you were running only against candidate B and you beat candidate B. The fair and logical conclusion would be that you triumphed over each candidate, A and B, and therefore, you should be declared the winner. This conclusion is known as the **head-to-head criterion of fairness.**

THE HEAD-TO-HEAD CRITERION

If candidate X is favored when compared head-to-head (individually) with each of the other candidates, then candidate X should be declared the winner.

Unfortunately, not all systems of voting satisfy this criterion; that is, in some circumstances, a candidate might triumph in every possible head-to-head comparison, yet another candidate might be declared the overall winner! Specifically, Exercise 24 will illustrate that the plurality method can in fact violate the head-to-head criterion.

Many times, regular elections are preceded by preliminary, nonbinding elections called *straw votes*. Suppose you were the winning candidate in a straw vote. Then, the regular election was held, and the only changes in the votes were changes in your favor; that is, some people now voted for you instead of their original candidate. Under these conditions, a fair and logical conclusion would be that you should be declared the winner. This conclusion is known as the **monotonicity criterion of fairness.**

THE MONOTONICITY CRITERION

If candidate X wins an election and, in a subsequent election, the only changes are changes in favor of candidate X, then candidate X should be declared the winner.

Unfortunately, not all systems of voting satisfy this criterion; that is, in some circumstances, a candidate might win an election (or straw vote) and gather more votes in a second election but lose out to another candidate in the second election. Specifically, Exercise 25 will illustrate that the instant runoff method can in fact violate the monotonicity criterion.

Suppose you win an election. However, one or more of the (losing) candidates are subsequently determined to be ineligible to run in the election; consequently, they are removed from the ballot, and a recount is initiated. Under these conditions, a fair and logical conclusion would be that you should still be declared the winner. This conclusion is known as the **irrelevant alternatives criterion of fairness.**

THE IRRELEVANT ALTERNATIVES CRITERION

If candidate X wins an election and, in a recount, the only changes are that one or more of the losing candidates are removed from the ballot, then candidate X should still be declared the winner.

Unfortunately, not all systems of voting satisfy this criterion; that is, in some circumstances, a candidate might win an election, but on the removal of other candidates from the ballot, another candidate might be declared the winner! Specifically, Exercise 26 will illustrate that the pairwise comparison method can in fact violate the irrelevant alternatives criterion.

Each of the common systems of voting studied in this section can be shown to violate at least one of the four fairness criteria. Is it possible to create an ultimate system that satisfies all four criteria? The answer is "no"; it is mathematically impossible to create a system of voting that satisfies all four fairness criteria. This result was proven in 1951 by Kenneth Arrow and is known as **Arrow's Impossibility Theorem.**

ARROW'S IMPOSSIBILITY THEOREM

It is mathematically impossible to create any system of voting (involving three or more candidates) that satisfies all four fairness criteria.

The systems of voting studied in this section are summarized in Figure 6.19.

System of Voting	Description
Plurality	Each person votes for his or her favorite candidate. The candidate who receives the most votes is declared the winner.
Instant Run-Off	Each voter ranks all of the candidates. If a candidate receives a majority (more than half) of first-choice votes, that candidate is declared the winner. If no candidate receives a majority, then the candidate with the fewest first-choice votes is eliminated and those votes are given to the next preferred candidate. If a candidate now has a majority of first-choice votes, that candidate is declared the winner. If no candidate receives a majority, this process continues until a candidate receives a majority.
Borda count	Each voter ranks all of the candidates. If there are k candidates, each candidate receives k points for each first-choice vote, $(k-1)$ points for each second-choice vote, $(k-2)$ points for each third-choice vote, and so on. The candidate with the most total points is declared the winner.
Pairwise comparison	Each voter ranks all of the candidates. For each possible pairing of candidates, the candidate with the most votes receives 1 point; if there is a tie, each candidate receives $\frac{1}{2}$ point. The candidate who receives the most points is declared the winner.

FIGURE 6.19 Systems of voting.

The systems of voting studied in this section each satisfy (or violate) the fairness criteria as summarized in Figure 6.20.

Fairness Criteria	System of Voting			
	Plurality Method	**Instant Runoff Method**	**Borda Count Method**	**Pairwise Comparison Method**
Majority Criterion	always satisfies	always satisfies	may not satisfy	always satisfies
Head-to-Head Criterion	may not satisfy	may not satisfy	may not satisfy	always satisfies
Monotonicity Criterion	always satisfies	may not satisfy	always satisfies	always satisfies
Irrelevant Alternatives Criterion	may not satisfy	may not satisfy	may not satisfy	may not satisfy

FIGURE 6.20 Fairness and flaws of the systems of voting.

6.1 EXERCISES

1. Four candidates, Alliotti, Baker, Cruz, and Daud, are running for president of the student government. After the polls close, votes are tallied, and the results in Figure 6.21 are obtained.

Candidate	Alliotti	Baker	Cruz	Daud
Number of votes	314	155	1,052	479

FIGURE 6.21 Table for Exercise 1.

 a. How many votes were cast?
 b. Using the plurality method of voting, which candidate wins?
 c. Did the winner receive a majority of the votes?

2. Five candidates, Edwards, Fischer, Gelinas, Horner, and Inclan, are running for president of the faculty senate. After the polls close, votes are tallied, and the results in Figure 6.22 are obtained.

Candidate	Edwards	Fischer	Gelinas	Horner	Inclan
Number of votes	32	25	17	102	24

FIGURE 6.22 Table for Exercise 2.

 a. How many votes were cast?
 b. Using the plurality method of voting, which candidate wins?
 c. Did the winner receive a majority of the votes?

3. Three candidates, Arce, Edelstein, and Spence, are running for president of academic affairs at a local college. After the polls close, votes are tallied, and the results in Figure 6.23 are obtained.

Candidate	Arce	Edelstein	Spence
Number of votes	3,021	4,198	3,132

FIGURE 6.23 Table for Exercise 3.

 a. How many votes were cast?
 b. Using the plurality method of voting, which candidate wins?
 c. Did the winner receive a majority of the votes?

4. Four candidates, Brecha, Parks, Wilcox, and Willett, are running for district supervisor. After the polls close, votes are tallied, and the results in Figure 6.24 are obtained.

Candidate	Brecha	Parks	Wilcox	Willett
Number of votes	3,007	2,957	10,541	2,851

FIGURE 6.24 Table for Exercise 4.

 a. How many votes were cast?
 b. Using the plurality method of voting, which candidate wins?
 c. Did the winner receive a majority of the votes?

5. The managers of Prints Alive (a local silk-screening business) are planning a party for their 30 employees. There are three possible locations for the party: the warehouse (W), the park (P), or the beach (B). The employees are asked to rank these choices in order of preference, and the results are summarized in Figure 6.25.

Number of Ballots Cast			
6	8	11	5
1st choice P	P	B	W
2nd choice B	W	W	B
3rd choice W	B	P	P

FIGURE 6.25 Voter preference table for Exercise 5.

   a. How many votes were cast?

   b. Use the plurality method of voting to determine the winner.

   c. What percent of the votes did the winner in part (b) receive?

   d. Use the instant runoff method to determine the winner.

   e. What percent of the votes did the winner in part (d) receive?

   f. Use the Borda count method to determine the winner.

   g. How many points did the winner in part (f) receive?

   h. Use the pairwise comparison method to determine the winner.

   i. How many points did the winner in part (h) receive?

6. The members of a local service club are volunteering to clean up and modernize the playground at one of the elementary schools in town. There are three schools: Hidden Lakes (H), Strandwood (S), and Valhalla (V). The members are asked to rank these choices in order of preference, and the results are summarized in Figure 6.26.

Number of Ballots Cast			
13	8	10	19
1st choice S	V	V	H
2nd choice H	S	H	S
3rd choice V	H	S	V

FIGURE 6.26 Voter preference table for Exercise 6.

   a. How many votes were cast?

   b. Use the plurality method of voting to determine the winner.

   c. What percent of the votes did the winner in part (b) receive?

   d. Use the instant runoff method to determine the winner.

   e. What percent of the votes did the winner in part (d) receive?

   f. Use the Borda count method to determine the winner.

   g. How many points did the winner in part (f) receive?

   h. Use the pairwise comparison method to determine the winner.

   i. How many points did the winner in part (h) receive?

7. The members of a youth club are raising money so that they can attend a summer camp. There are three camps in the area: Coastline (C), Pinewood (P), and The Ranch (R). The members are asked to rank these choices in order of preference, and the results are summarized in Figure 6.27.

Number of Ballots Cast				
19	12	10	11	13
1st choice C	C	P	R	R
2nd choice P	R	C	P	C
3rd choice R	P	R	C	P

FIGURE 6.27 Voter preference table for Exercise 7.

   a. How many votes were cast?

   b. Use the plurality method of voting to determine the winner.

   c. What percent of the votes did the winner in part (b) receive?

   d. Use the instant runoff method to determine the winner.

   e. What percent of the votes did the winner in part (d) receive?

   f. Use the Borda count method to determine the winner.

   g. How many points did the winner in part (f) receive?

   h. Use the pairwise comparison method to determine the winner.

   i. How many points did the winner in part (h) receive?

8. The members of a local charitable group are raising money to send a group of neighborhood children to a special event. There are three events to choose from: the circus (C), the ice show (I), and the symphony (S). The members are asked to rank these choices in order of preference, and the results are summarized in Figure 6.28.

   a. How many votes were cast?

   b. Use the plurality method of voting to determine the winner.

Number of Ballots Cast				
6	8	7	10	15
1st choice C	C	I	I	S
2nd choice I	S	C	S	C
3rd choice S	I	S	C	I

FIGURE 6.28 Voter preference table for Exercise 8.

   c. What percent of the votes did the winner in part (b) receive?

   d. Use the instant runoff method to determine the winner.

   e. What percent of the votes did the winner in part (d) receive?

   f. Use the Borda count method to determine the winner.

   g. How many points did the winner in part (f) receive?

   h. Use the pairwise comparison method to determine the winner.

   i. How many points did the winner in part (h) receive?

9. Three candidates, Budd (B), Nirgiotis (N), and Shattuck (S), are running for union president. After the polls close, ranked ballots are tallied, and the results are summarized in Figure 6.29.

Number of Ballots Cast					
25	13	19	27	30	26
1st choice B	B	N	N	S	S
2nd choice N	S	S	B	B	N
3rd choice S	N	B	S	N	B

FIGURE 6.29 Voter preference table for Exercise 9.

   a. How many votes were cast?

   b. Use the plurality method of voting to determine the winner.

   c. What percent of the votes did the winner in part (b) receive?

   d. Use the instant runoff method to determine the winner.

   e. What percent of the votes did the winner in part (d) receive?

   f. Use the Borda count method to determine the winner.

   g. How many points did the winner in part (f) receive?

   h. Use the pairwise comparison method to determine the winner.

   i. How many points did the winner in part (h) receive?

10. Three candidates, Maruyama (M), Peters (P), and Vilas (V), are running for district representative. After the polls close, ranked ballots are tallied, and the results are summarized in Figure 6.30.

Number of Ballots Cast					
675	354	451	387	601	297
1st choice M	M	P	P	V	V
2nd choice P	V	V	M	M	P
3rd choice V	P	M	V	P	M

FIGURE 6.30 Voter preference table for Exercise 10.

   a. How many votes were cast?

   b. Use the plurality method of voting to determine the winner.

   c. What percent of the votes did the winner in part (b) receive?

   d. Use the instant runoff method to determine the winner.

   e. What percent of the votes did the winner in part (d) receive?

   f. Use the Borda count method to determine the winner.

   g. How many points did the winner in part (f) receive?

   h. Use the pairwise comparison method to determine the winner.

   i. How many points did the winner in part (h) receive?

11. Four candidates, Dolenz (D), Jones (J), Nesmith (N), and Tork (T), are running for director of public relations. After the polls close, ranked ballots are tallied, and the results are summarized in Figure 6.31.

Number of Ballots Cast						
225	134	382	214	81	197	109
1st choice D	D	J	J	N	T	T
2nd choice J	N	D	T	J	D	J
3rd choice T	T	T	D	T	N	D
4th choice N	J	N	N	D	J	N

FIGURE 6.31 Voter preference table for Exercise 11.

   a. How many votes were cast?

   b. Use the plurality method of voting to determine the winner.

   c. What percent of the votes did the winner in part (b) receive?

   d. Use the instant runoff method to determine the winner.

e. What percent of the votes did the winner in part (d) receive?

f. Use the Borda count method to determine the winner.

g. How many points did the winner in part (f) receive?

h. Use the pairwise comparison method to determine the winner.

i. How many points did the winner in part (h) receive?

12. Four candidates, Harrison (H), Lennon (L), McCartney (M), and Starr (S), are running for regional manager. After the polls close, ranked ballots are tallied, and the results are summarized in Figure 6.32.

Number of Ballots Cast						
	23	98	45	82	32	17	21
1st choice	H	L	L	M	M	S	S
2nd choice	L	M	S	L	L	H	M
3rd choice	S	H	M	H	S	L	H
4th choice	M	S	H	S	H	M	L

FIGURE 6.32   Voter preference table for Exercise 12.

a. How many votes were cast?

b. Use the plurality method of voting to determine the winner.

c. What percent of the votes did the winner in part (b) receive?

d. Use the instant runoff method to determine the winner.

e. What percent of the votes did the winner in part (d) receive?

f. Use the Borda count method to determine the winner.

g. How many points did the winner in part (f) receive?

h. Use the pairwise comparison method to determine the winner.

i. How many points did the winner in part (h) receive?

13. Five candidates, Addley (A), Burke (B), Ciento (C), Darter (D), and Epp (E), are running for mayor. After the polls close, ranked ballots are tallied, and results are summarized in Figure 6.33.

a. How many votes were cast?

b. Use the plurality method of voting to determine the winner.

c. What percent of the votes did the winner in part (b) receive?

d. Use the instant runoff method to determine the winner.

e. What percent of the votes did the winner in part (d) receive?

f. Use the Borda count method to determine the winner.

g. How many points did the winner in part (f) receive?

h. Use the pairwise comparison method to determine the winner.

i. How many points did the winner in part (h) receive?

Number of Ballots Cast						
	1,897	1,025	4,368	2,790	6,897	9,571	5,206
1st choice	A	A	B	C	D	D	E
2nd choice	C	D	D	E	B	E	D
3rd choice	D	C	C	D	C	B	C
4th choice	B	E	E	B	E	A	B
5th choice	E	B	A	A	A	C	A

FIGURE 6.33   Voter preference table for Exercise 13.

14. Five candidates, Fino (F), Gempler (G), Holloway (H), Isho (I), and James (J), are running for president of the Polar Bear Swim Club. After the polls close, ranked ballots are tallied, and the results are summarized in Figure 6.34.

Number of Ballots Cast						
	12	18	29	11	21	19	18
1st choice	F	F	G	H	I	J	J
2nd choice	G	I	F	G	J	H	F
3rd choice	I	G	H	F	G	I	G
4th choice	J	H	J	I	F	G	I
5th choice	H	J	I	J	H	F	H

FIGURE 6.34   Voter preference table for Exercise 14.

a. How many votes were cast?

b. Use the plurality method of voting to determine the winner.

c. What percent of the votes did the winner in part (b) receive?

d. Use the instant runoff method to determine the winner.

e. What percent of the votes did the winner in part (d) receive?

f. Use the Borda count method to determine the winner.

g. How many points did the winner in part (f) receive?

h. Use the pairwise comparison method to determine the winner.

i. How many points did the winner in part (h) receive?

15. If there are six candidates in an election and voters are asked to rank all of the candidates, how many different rankings are possible?

16. If there are eight candidates in an election and voters are asked to rank all of the candidates, how many different rankings are possible?

17. If there are six candidates in an election and voters are asked to rank all of the candidates, how many different pairwise comparisons are there?

18. If there are eight candidates in an election and voters are asked to rank all of the candidates, how many different pairwise comparisons are there?

19. In an election, there are three candidates and 25 voters.

a. What is the maximum number of points that a candidate can receive using the Borda count method?

b. What is the minimum number of points that a candidate can receive using the Borda count method?

20. In an election, there are four candidates and 75 voters.

a. What is the maximum number of points that a candidate can receive using the Borda count method?

b. What is the minimum number of points that a candidate can receive using the Borda count method?

21. In an election, there are seven candidates.

a. What is the maximum number of points that a candidate can receive using the pairwise comparison method?

b. What is the minimum number of points that a candidate can receive using the pairwise comparison method?

22. In an election, there are nine candidates.

a. What is the maximum number of points that a candidate can receive using the pairwise comparison method?

b. What is the minimum number of points that a candidate can receive using the pairwise comparison method?

23. Candidates A, B, and C are being considered as supervisor of a local school district. There are thirteen directors on the school board, and they have ranked their choices as shown in Figure 6.35.

a. Does any candidate have a majority of first-choice votes? Who should win?

b. Use the Borda count method to determine the winner.

c. Does the Borda count method violate the majority criterion of fairness? Explain why or why not.

Number of Ballots Cast		
7	4	2
1st choice	A	B	C
2nd choice	B	C	B
3rd choice	C	A	A

FIGURE 6.35 Voter preference table for Exercise 23.

24. Candidates A, B, and C are being considered as chancellor of a local college district. There are twenty-one directors on the college board, and they have ranked their choices as shown in Figure 6.36.

Number of Ballots Cast			
9	2	4	6
1st choice	A	C	B	B
2nd choice	B	A	A	C
3rd choice	C	B	C	A

FIGURE 6.36 Voter preference table for Exercise 24.

a. Use the plurality method to determine the winner.

b. Who wins when A is compared to B?

c. Who wins when A is compared to C?

d. Who wins when B is compared to C?

e. Does the plurality method violate the head-to-head criterion of fairness? Explain why or why not.

25. Candidates A, B, and C are being considered as director of a public service agency. There are twenty-seven trustees on the executive board, and after the initial discussion and casting of straw votes, the trustees have ranked their choices as shown in Figure 6.37.

Number of Ballots Cast			
6	8	4	9
1st choice	A	B	A	C
2nd choice	B	C	C	A
3rd choice	C	A	B	B

FIGURE 6.37 Original voter preference table for Exercise 25.

a. Use the instant runoff method to determine the winner of the straw vote.

b. After discussing the results of the straw vote, four trustees changed their votes to support candidate C as shown in Figure 6.38. Use the instant runoff method to determine the winner of the subsequent election.

Number of Ballots Cast		
6	8	13
1st choice A	B	C
2nd choice B	C	A
3rd choice C	A	B

FIGURE 6.38   Revised voter preference table for Exercise 25.

c. Does the instant runoff method violate the monotonicity criterion of fairness? Explain why or why not.

26. Candidates A, B, C, and D are being considered as director of an endowment for the arts trust fund. There are twenty-eight trustees on the executive board, and they have ranked their choices as shown in Figure 6.39.

Number of Ballots Cast			
10	8	6	4
1st choice A	C	D	D
2nd choice B	B	A	A
3rd choice C	D	C	B
4th choice D	A	B	C

FIGURE 6.39   Original voter preference table for Exercise 26.

a. Use the pairwise comparison method to determine the winner.

b. After the trustees had ranked the candidates, it was discovered that candidates B and C did not meet the minimum qualifications and hence were ineligible for the position. Consequently, candidates B and C were removed, and the voter preference table in Figure 6.40 was established. Use the pairwise comparison method to determine the new winner.

	10	8	6	4
1st	A	D	D	D
2nd	D	A	A	A

FIGURE 6.40   Revised voter preference table for Exercise 26.

c. Does the pairwise comparison method violate the irrelevant alternatives criterion of fairness? Explain why or why not.

Answer the following questions using complete sentences and your own words.

• CONCEPT QUESTIONS

27. What is a majority?
28. What is a ranked ballot?
29. What is a voter preference table?
30. What are the four fairness criteria? Explain the meaning of each.
31. What is Arrow's Impossibility Theorem?

6.2   Methods of Apportionment

OBJECTIVES

- Develop and apply Hamilton's Method of apportionment
- Develop and apply Jefferson's Method of apportionment
- Develop and apply Adams's Method of apportionment
- Develop and apply Webster's Method of apportionment
- Develop and apply the Hill-Huntington Method of apportionment

After the thirteen former British colonies won their independence, the "founding fathers" had to come up with a constitution to govern the newly formed nation. Needless to say, it was an enormous task to create the framework of a government

In the Constitution, the "founding fathers" created two legislative bodies: the Senate (two members per state) and the House of Representatives (apportioned to each state's population).

rooted in revolution and driven by the underlying principle of "power to the people." The framers of the Constitution believed that the former colonies, because of their diversity, should be relatively independent states; however, there should also be some sort of central or federal government to serve as a cohesive binding element to strengthen and secure the sovereignty of the nation. To that effect, the Preamble of the Constitution reads, "We, the People of the United States, in Order to form a more perfect Union, establish Justice, insure domestic Tranquility, provide for the common defense, promote the general Welfare, and secure the Blessings of Liberty to ourselves and our Posterity, do ordain and establish this Constitution for the United States of America."

An important feature of any government is the method for creating the nation's laws; this is commonly called the legislative branch of government. When delegates of the original thirteen states met in Philadelphia in 1787, a common debate was the manner in which the states would be represented in the federal legislature. Some states, in particular the smaller ones, advocated that each state receive the same number of representatives. On the other hand, many of the larger states wanted representation in proportion to the number of people residing in the individual states.

The result of this great debate was the creation of two distinct legislative bodies. Article I, Section 1 of the Constitution states, "All legislative Powers herein granted shall be vested in a Congress of the United States, which shall consist of a Senate and House of Representatives." Article I, Section 3 created the Senate and states "The Senate of the United States shall be composed of two Senators from each state." In contrast, Article I, Section 2 created the House of Representatives and states, "Representatives and direct taxes shall be apportioned among the several States which may be included within this Union, according to their respective Numbers, which shall be determined by adding to the whole Number of free Persons, including those bound to Service for a Term of Years, and excluding Indians not taxed, three-fifths of all other persons. The actual Enumeration shall be made within three Years after the first Meeting of the Congress of the United States, and within every subsequent Term of ten Years, in such Manner as they shall by Law direct. The Number of Representatives shall not exceed one for every thirty Thousand, but each State shall have at Least one Representative."

It is very easy to interpret the phrase "The Senate of the United States shall be composed of two Senators from each state." However, the phrase "Representatives shall be *apportioned* among the several states according to their respective numbers" is nebulous at best. When we consult a standard dictionary, we find that the word **apportion** is defined as "to divide and distribute in shares according to a plan." Although the Constitution specifies that Representatives be apportioned or distributed according to a plan, it does not specify the plan. Consequently, several methods of apportionment have been proposed, and several different methods have actually been used since the first apportionment in 1790.

In this section, we will study the following methods of apportionment:

- **Hamilton's Method** was proposed by Alexander Hamilton (1757–1804) and was the first plan to be approved by Congress in 1790. However, President George Washington vetoed the plan (the first use of a presidential veto in the history of the United States). Eventually, Hamilton's method was used following each federal government census (every ten years) from 1850 to 1900.
- **Jefferson's Method** was proposed by Thomas Jefferson (1743–1826) and was the first method to actually be used; it was used following each census from 1790 to 1830.
- **Adams's Method** was proposed by John Quincy Adams (1767–1848) and has never been used. The method merits investigation, as it is viewed as the exact opposite of Jefferson's Method.
- **Webster's Method** was proposed by Daniel Webster (1782–1852). It was used following the census of 1840 and then again in 1910 and 1930. (There was no apportionment in 1920.)
- **Hill-Huntington Method** was proposed by Joseph Hill (1860–1938) and Edward Huntington (1874–1952), and it has been used following every census from 1940 to the present. During the early twentieth century, Hill was the chief statistician at the U.S. Census Bureau, and Huntington was a professor of mathematics and mechanics at Harvard.

Basic Terminology

Before we discuss any specific method of apportionment, we must introduce some basic concepts and define some basic terms.

STANDARD DIVISOR

The **standard divisor,** denoted by d, is the ratio of the total population to the total number of seats to be allocated; d is found by dividing the total population by the total number of seats to be allocated.

$$d = \text{standard divisor} = \frac{\text{total population}}{\text{total number of seats}}$$

It is customary to round the standard divisor to two decimal places.

For example, if the total population is 201,000 people and ten seats are to be allocated, the standard divisor will be

$$\frac{201{,}000 \text{ people}}{10 \text{ seats}} = 20{,}100$$

Therefore, $d = 20,100$ people per seat, or alternatively, each seat will represent 20,100 people. Once the standard divisor has been calculated, each state's **standard quota** must be determined

---

STANDARD QUOTA

The **standard quota** (of a specific state), denoted by q, is the ratio of a state's population to the standard divisor; q is found by dividing the state's population by the standard divisor.

$$q = \text{standard quota} = \frac{\text{state's population}}{\text{standard divisor}} = \frac{\text{state's population}}{d}$$

It is customary to round the standard quota to two decimal places.

For example, if the population of state A was 94,700 and the standard divisor was $d = 20,100$, then state A's standard quota would be

$$\frac{94,700}{20,100} = 4.711442786\ldots$$

Therefore, $q = 4.71$ (rounded to two decimal places). Subscripts may be used to identify a specific state's standard quota, that is, we can use q_A to represent the standard quota of state A; hence, $q_A = 4.71$. Similarly, we use q_B to represent the standard quota of state B, q_C to represent the standard quota of state C, and so on. Rounding to two decimal places provides satisfactory results in most cases. However, if rounding to two decimal places creates an integer (when the standard quota is not an exact integer), then round to three (or more) decimal places so that an integer is not obtained. That is, if we round $q = \frac{94,700}{18,950} = 4.997361478\ldots$ to two decimal places, we obtain $q = 5.00$, an integer. Instead, we round to three decimal places and obtain $q = 4.997$.

A standard quota can be interpreted as the number of seats allocated to a specific state; therefore, if $q_A = 4.71$, state A should receive 4.71 seats. Because a state cannot receive part of a seat, we conclude that state A should receive at least four seats. The "whole number" part of a standard quota is called the **lower quota** of the state.

LOWER QUOTA

The **lower quota** (of a specific state) is the standard quota of a state truncated (rounded down) to a whole number. That is, the lower quota is the whole number part of a standard quota. (Use the complete, original standard quota, not the rounded version, when finding the lower quota.) If the standard quota is an exact integer, then the lower quota equals the standard quota.

Hamilton's Method

We begin our study of the methods of apportionment with Hamilton's Method. Hamilton's Method is the easiest and most direct method to implement; it was the first plan to be approved by Congress in 1790, although it was not officially used until 1850.

HAMILTON'S METHOD OF APPORTIONMENT

1. Using the standard divisor d, calculate the standard quotas and the lower quotas of each state. Initially, each state receives a number of seats equal to its lower quota.
2. If the sum of the lower quotas equals the total number of seats to be apportioned, the apportionment process is complete.
3. If the sum of the lower quotas is less than the total number of seats to be apportioned, then assign a seat to the state that has the highest decimal part in its standard quota.
4. Repeat step 3 (using the next highest decimal part) until the total number of seats has been apportioned.

Inhabitants of Middle Earth.

EXAMPLE 1

APPLYING HAMILTON'S METHOD OF APPORTIONMENT   Middle Earth is a fantastic world created by the literary genius of J. R. R. Tolkien; Middle Earth's principle inhabitants are hobbits, dwarves, elves, wizards, men, and many mutant forms of evil. Let us suppose that the good inhabitants of the realms of Gondor, Eriador, and Rohan have agreed to form a federation to foster mutual protection, trade, and cultural exchange. The (estimated) populations of these realms (or states) are given in Figure 6.41. The high council of the federation is to have ten seats. Use Hamilton's Method to apportion the seats among these three realms (states).

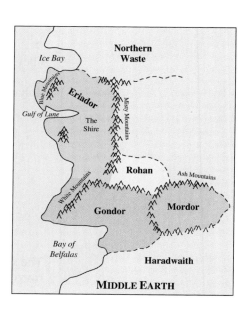

Realm (State)	Gondor	Eriador	Rohan	Total
Population (estimated)	94,700	72,600	33,700	201,000

FIGURE 6.41   Realms of Middle Earth.

SOLUTION

It is often advantageous to express large numbers that have several zeroes at the end in terms of simpler multiples. That is, rather than working with the numeral 94,700, we can express it as 94.7 thousand (or 94.7 × 1,000), and then use only the numeral 94.7. This will consistently simplify our calculations. Figure 6.42 shows the simplified population data in terms of thousands.

Realm (State)	Gondor	Eriador	Rohan	Total
Population (thousands)	94.7	72.6	33.7	201

FIGURE 6.42 Realms of Middle Earth.

First, we must find the standard divisor:

$$d = \text{standard divisor}$$
$$= \frac{\text{total population}}{\text{total number of seats}}$$
$$= \frac{201}{10} = 20.1$$

Therefore, $d = 20.1$.

Now calculate each realm's standard quota, starting with Gondor (G):

$$q_G = \text{Gondor's standard quota}$$
$$= \frac{\text{Gondor's population}}{d}$$
$$= \frac{94.7}{20.1}$$
$$= 4.711442786\ldots$$

Therefore $q_G = 4.71$.

In a similar fashion, we calculate the standard quotas for Eriador (E) and Rohan (R):

$$q_E = \text{Eriador's standard quota} \qquad q_R = \text{state C's standard quota}$$
$$= \frac{\text{Eriador's population}}{d} \qquad\qquad = \frac{\text{Rohan's population}}{d}$$
$$= \frac{72.6}{20.1} \qquad\qquad\qquad = \frac{33.7}{20.1}$$
$$q_E = 3.61 \quad \textbf{rounded to two} \qquad q_R = 1.68 \quad \textbf{rounded to two}$$
$$\textbf{decimal places} \qquad\qquad\qquad \textbf{decimal places}$$

The standard quotas and lower quotas for Gondor, Eriador, and Rohan are summarized in Figure 6.43.

Realm (State)	Gondor	Eriador	Rohan	Total
Population (thousands)	94.7	72.6	33.7	201
Standard Quota (using $d = 20.1$)	4.71	3.61	1.68	10
Lower Quota	4	3	1	8

FIGURE 6.43 Standard and lower quotas of the realms.

Because the sum of the lower quotas (8) is less than the total number of seats (10), we must decide who receives the two (10 − 8 = 2) "surplus" seats. According to Hamilton's method, Gondor will receive the first additional seat (Figure 6.44) because Gondor has the highest decimal part in its standard quota (0.71 is larger than either 0.61 or 0.68), and Rohan will receive the next seat (because 0.68 > 0.61).

Realm (State)	Gondor	Eriador	Rohan	Total
Population (thousands)	94.7	72.6	33.7	201
Standard Quota ($d = 20.1$)	4.71	3.61	1.68	10
Lower Quota	4	3	1	8
Additional Seats	1	0	1	2
Number of Seats	5	3	2	10

FIGURE 6.44   Hamilton's Method applied to the realms of Middle Earth.

By using Hamilton's Method, the ten seats will be apportioned as shown in Figure 6.45.

Realm (State)	Gondor	Eriador	Rohan	Total
Hamilton's Apportionment	5	3	2	10

FIGURE 6.45   Final apportionment of the seats of the high council.

Jefferson's Method

As we have seen, Hamilton's Method utilizes the standard divisor ($d = \frac{\text{state's population}}{\text{total number of seats}}$), and this method may result in surplus seats that must be distributed accordingly. Consequently, some states receive preferential treatment by obtaining extra seats. Thomas Jefferson proposed a method in which no surplus seats are created. The essence of Jefferson's Method is this: If the standard divisor creates surplus seats, then find a new divisor (called a modified divisor) that will result in all of the seats being allocated. As we shall see, Jefferson's Method requires a bit of trial and error.

A **modified divisor** is a number that is close to the standard divisor, and it is denoted by d_m. A modified divisor may be less than or greater than the standard divisor. Jefferson's method calls for a modified divisor that is *less than* the standard divisor, that is, $d_m < d$. Now, rather than calculating the standard quota, we calculate the **modified quota.**

MODIFIED QUOTA

The **modified quota** (of a specific state), denoted by q_m, is the ratio of a state's population to the modified divisor; q_m is found by dividing the state's population by the modified divisor.

$$q_m = \text{modified quota} = \frac{\text{state's population}}{\text{modified divisor}} = \frac{\text{state's population}}{d_m}$$

It is customary to round off the modified quota to two decimal places.

As a consequence, the modified quota (in Jefferson's Method) will always be larger than the standard quota (in Hamilton's Method) because we are dividing the state's population by a smaller number. (Dividing by a smaller number produces a larger result.) Because the modified quotas are larger than the standard quotas, our goal is to find a modified divisor that will automatically "use up" any surplus seats.

JEFFERSON'S METHOD OF APPORTIONMENT

1. Using the standard divisor d, calculate the standard quotas and the lower quotas of each state.
2. If the sum of the lower quotas equals the total number of seats to be apportioned, the apportionment process is complete, that is, each state receives a number of seats equal to its lower quota.
3. If the sum of the lower quotas does not equal the total number of seats to be apportioned, choose a modified divisor d_m less than the standard divisor, that is, $d_m < d$, and calculate the modified quotas and lower modified quotas.
4. Repeat step 3 until you find a modified divisor such that the sum of the lower modified quotas equals the total number of seats to be apportioned. Each state receives a number of seats equal to its lower modified quota, and the apportionment process is complete.

EXAMPLE 2

APPLYING JEFFERSON'S METHOD OF APPORTIONMENT As in Example 1, the populations of Gondor, Eriador, and Rohan are given in Figure 6.46. Use Jefferson's Method to apportion the ten seats of the high council (legislature) among these three realms.

Realm (State)	Gondor	Eriador	Rohan	Total
Population	94,700	72,600	33,700	201,000

FIGURE 6.46 Realms of Middle Earth.

SOLUTION

As we saw in Example 1, using Hamilton's Method and the standard divisor $d = 20.1$ results in two surplus seats, as shown in Figure 6.47.

Realm (State)	Gondor	Eriador	Rohan	Total
Population (thousands)	94.7	72.6	33.7	201
Standard Quota ($d = 20.1$)	4.71	3.61	1.68	10
Lower Quota	4	3	1	8

FIGURE 6.47 Standard and lower quotas of the realms.

Therefore, to apply Jefferson's Method, we must find a modified divisor, $d_m < d = 20.1$, such that no surplus is created. As a first guess, we try $d_m = 19$ and calculate each realm's modified quota and lower modified quota; that is, divide each population by $d_m = 19$, round to two decimal places, and then truncate (round down) to a whole number as shown in Figure 6.48.

Realm (State)	Gondor	Eriador	Rohan	Total
Population (thousands)	94.7	72.6	33.7	201
Standard Quota ($d = 20.1$)	4.71	3.61	1.68	10
Modified Quota ($d_m = 19$)	4.98	3.82	1.77	(not needed)
Lower Modified Quota	4	3	1	8

FIGURE 6.48 Modified and lower modified quotas using $d_m = 19$.

The modified divisor $d_m = 19$ still creates two surplus seats, so we must try an even smaller divisor, say, $d_m = 18$, and repeat the process; that is, divide each population by $d_m = 18$, round to two decimal places, and then truncate to a whole number as shown in Figure 6.49.

Realm (State)	Gondor	Eriador	Rohan	Total
Population (thousands)	94.7	72.6	33.7	201
Standard Quota ($d = 20.1$)	4.71	3.61	1.68	10
Modified Quota ($d_m = 19$)	4.98	3.82	1.77	(not needed)
Modified Quota ($d_m = 18$)	5.26	4.03	1.87	(not needed)
Lower Modified Quota	5	4	1	10

FIGURE 6.49 Jefferson's Method using $d_m = 18$.

The modified divisor $d_m = 18$ has properly apportioned the ten seats, and the final results are shown in Figure 6.50.

Realm (State)	Gondor	Eriador	Rohan	Total
Jefferson's Apportionment	5	4	1	10
Hamilton's Apportionment	5	3	2	10

FIGURE 6.50 Comparison of final apportionments.

Figure 6.50 shows that different methods of apportionment can lead to different allocations of legislative seats. Observe that Jefferson's Method favors larger states while Hamilton's Method favors smaller states. That is, under Jefferson's Method, Eriador received an extra seat (and the population of Eriador is greater than that of Rohan), while under Hamilton's Method, Rohan received an extra seat (and the population of Rohan is less than that of Eriador). We will explore these "flaws" in more detail in Section 6.3.

Adams's Method

As we have seen, Jefferson's Method utilizes a modified divisor that is less than the standard divisor, and the modified quotas are truncated (rounded down) to a whole number (lower modified quota). Like Jefferson, John Quincy Adams proposed using a modified divisor; however, his approach was the opposite of Jefferson's: Adams proposed using a modified divisor *greater than* the standard divisor, and he rounded his modified quotas *up* to the **upper quota.**

UPPER QUOTA

The **upper quota** (of a specific state) is the standard (or modified) quota of a state rounded up to the next whole number. (Use the complete, original standard quota, not the rounded version, when finding the upper quota.) If the standard quota is an exact integer, then the upper quota equals the standard quota plus 1.

ADAMS'S METHOD OF APPORTIONMENT

1. Using the standard divisor d, calculate the standard quotas and the upper quotas of each state.
2. If the sum of the upper quotas equals the total number of seats to be apportioned, the apportionment process is complete, that is, each state receives a number of seats equal to its upper quota.
3. If the sum of the upper quotas does not equal the total number of seats to be apportioned, choose a modified divisor d_m greater than the standard divisor, that is, $d_m > d$ and calculate the modified quotas and upper modified quotas.
4. Repeat step 3 until you find a modified divisor such that the sum of the upper modified quotas equals the total number of seats to be apportioned. Each state receives a number of seats equal to its upper modified quota, and the apportionment process is complete.

In Adams's Method, the modified quota will always be smaller than the standard quota (in Hamilton's Method) because we are dividing the state's population by a larger number. (Dividing by a larger number produces a smaller result.) Because the modified quotas are smaller than the standard quotas (and we will use upper modified quotas), our goal is to find a modified divisor that will automatically allocate the proper number of seats.

EXAMPLE **3**

APPLYING ADAMS'S METHOD OF APPORTIONMENT As in Example 1, the populations of Gondor, Eriador, and Rohan are given in Figure 6.51. Use Adams's Method to apportion the ten seats of the high council among these three realms.

Realm (State)	Gondor	Eriador	Rohan	Total
Population	94,700	72,600	33,700	201,000

FIGURE 6.51 Realms of Middle Earth.

SOLUTION

As we saw in Example 1, Hamilton's Method and the standard divisor $d = 20.1$ results in two unused seats as shown in Figure 6.52. Figure 6.52 also shows the upper quotas (standard quotas rounded up) for each realm.

Because the sum of the upper quotas (11) does not equal the total number of seats to be allocated (10), the apportionment is not complete (we have allocated

Realm (State)	Gondor	Eriador	Rohan	Total
Population (thousands)	94.7	72.6	33.7	201
Standard Quota ($d = 20.1$)	4.71	3.61	1.68	10
Lower Quota	4	3	1	8
Upper Quota	5	4	2	11

FIGURE 6.52 Standard, lower, and upper quotas of the realms.

one seat too many). Therefore, to apply Adams's Method, we must find a modified divisor, $d_m > d = 20.1$, such that the seats will be properly allocated. As a first guess, we try $d_m = 23$ and calculate each realm's modified quota and upper modified quota, that is, divide each population by $d_m = 23$, round to two decimal places, and then round up to a whole number as shown in Figure 6.53.

Realm (State)	Gondor	Eriador	Rohan	Total
Population (thousands)	94.7	72.6	33.7	201
Standard Quota ($d = 20.1$)	4.71	3.61	1.68	10
Modified Quota ($d_m = 23$)	4.12	3.16	1.47	(not needed)
Upper Modified Quota	5	4	2	11

FIGURE 6.53 Modified and upper modified quotas using $d_m = 23$.

The modified divisor $d_m = 23$ does not properly allocate the ten seats, so we must try an even larger divisor. Divisors need not equal whole numbers. For example, we can choose $d_m = 23.8$ and repeat the process; that is, we divide each population by $d_m = 23.8$, round to two decimal places, and find the upper quotas as shown in Figure 6.54.

Realm (State)	Gondor	Eriador	Rohan	Total
Population (thousands)	94.7	72.6	33.7	201
Standard Quota ($d = 20.1$)	4.71	3.61	1.68	10
Modified Quota ($d_m = 23$)	4.12	3.16	1.47	(not needed)
Modified Quota ($d_m = 23.8$)	3.98	3.05	1.42	(not needed)
Upper Modified Quota	4	4	2	10

FIGURE 6.54 Adams's Method using $d_m = 23.8$.

Because the sum of the upper quotas (10) equals the total number of seats to be allocated (10), the apportionment process is complete. By using Adams's Method, the ten seats will be apportioned as shown in Figure 6.55.

Realm (State)	Gondor	Eriador	Rohan	Total
Adams's Apportionment	4	4	2	10
Jefferson's Apportionment	5	4	1	10
Hamilton's Apportionment	5	3	2	10

FIGURE 6.55 Comparison of final apportionments.

Once again, we see (Figure 6.55) that different methods of apportionment can lead to significantly different allocations of seats in a legislature (high council). However, different methods of apportionment do not always lead to different allocations; different methods may lead to the same allocation of seats. Note that Adams's Method favors smaller states; that is, under Adams's Method, Eriador "took away" a seat from Gondor (and the population of Eriador is less than that of Gondor). Different methods of apportionment have "flaws," which will be examined in Section 6.3.

Webster's Method

As we have seen, both Hamilton's Method and Jefferson's Method consistently round *down* the quotas and thereby utilize *lower* quotas, whereas Adams's Method consistently rounds *up* the quotas and thereby utilizes *upper* quotas. Webster's Method differs from the previous methods in that this method utilizes *regular* rounding rules, that is, round down if a number is less than 5 and round up if it is 5 or more.

WEBSTER'S METHOD OF APPORTIONMENT

1. Using the standard divisor d, calculate the standard quotas of each state and use the regular rules of rounding to round each standard quota to a whole number.
2. If the sum of the rounded standard quotas equals the total number of seats to be apportioned, the apportionment process is complete, that is, each state receives a number of seats equal to its rounded standard quota.
3. If the sum of the rounded standard quotas does not equal the total number of seats to be apportioned, choose a modified divisor d_m that is different from the standard divisor (either less than or greater than d), and calculate the modified quotas and rounded modified quotas.
4. Repeat step 3 until you find a modified divisor such that the sum of the rounded modified quotas equals the total number of seats to be apportioned. Each state receives a number of seats equal to its rounded modified quota, and the apportionment process is complete.

EXAMPLE 4

APPLYING WEBSTER'S METHOD OF APPORTIONMENT As in Example 1, the populations of Gondor, Eriador, and Rohan are given in Figure 6.56. Use Webster's Method to apportion the ten seats of the high council among these three realms.

Realm (State)	Gondor	Eriador	Rohan	Total
Population	94,700	72,600	33,700	201,000

FIGURE 6.56 Realms of Middle Earth.

SOLUTION

As we saw in Example 1, the standard divisor $d = 20.1$ produces the standard quotas shown in Figure 6.57. Figure 6.57 also shows the rounded quotas (using the rules of regular rounding) for each realm.

Realm (State)	Gondor	Eriador	Rohan	Total
Population (thousands)	94.7	72.6	33.7	201
Standard Quota ($d = 20.1$)	4.71	3.61	1.68	10
Rounded Quota	5	4	2	11

FIGURE 6.57   Standard and (regular) rounded quotas of the realms.

The sum of the rounded quotas (11) is too large, so we must choose a modified divisor, $d_m > d = 20.1$, because dividing by a larger number creates a smaller result. As a first guess, we try $d_m = 21$ and calculate each realm's modified quota, that is, divide each population by $d_m = 21$, round to two decimal places, and then round to a whole number as shown in Figure 6.58.

Realm (State)	Gondor	Eriador	Rohan	Total
Population (thousands)	94.7	72.6	33.7	201
Standard Quota ($d = 20.1$)	4.71	3.61	1.68	10
Modified Quota ($d_m = 21$)	4.51	3.46	1.60	(not needed)
Rounded Quota	5	3	2	10

FIGURE 6.58   Webster's Method using $d_m = 21$.

Because the sum of the rounded quotas (10) equals the total number of seats to be allocated (10), the apportionment process is complete. By using Webster's Method, the ten seats will be apportioned as shown in Figure 6.59.

Realm (State)	Gondor	Eriador	Rohan	Total
Webster's Apportionment	5	3	2	10
Adams's Apportionment	4	4	2	10
Jefferson's Apportionment	5	4	1	10
Hamilton's Apportionment	5	3	2	10

FIGURE 6.59   Comparison of final apportionments.

In Example 4, the apportionment due to Webster's Method is the same as that due to Hamilton's Method. This does not always occur. Sometimes different methods lead to the same allocation of seats; other times, different methods may lead to different allocation of seats. It should also be noted that Webster's Method tends to favor smaller states.

Hill-Huntington Method

As we have seen, the methods of Hamilton and Jefferson consistently round *down* the quotas, the method advocated by Adams consistently rounds *up* the quotas, and quotas are subjected to the rules of *regular* rounding when Webster's Method is applied. The Hill-Huntington Method is similar to Webster's Method in that sometimes the quotas are rounded up and sometimes they are rounded down. Before this method is discussed, we must first explore the geometric mean.

When you study algebra, you undoubtedly encounter the calculation of an "average," that is, many people recall that the "average" of two numbers is the sum of the numbers divided by 2. (See Section 4.2.) Technically, this type of average is called the **arithmetic mean,** that is, the arithmetic mean of a group of numbers equals the sum of the numbers divided by the number of numbers. There are other types of means. In particular, the Hill-Huntington Method utilizes the geometric mean.

GEOMETRIC MEAN

Given two numbers x and y, the **geometric mean** of x and y, denoted by *gm*, is the square root of the product of x and y. That is,

$$gm = \text{geometric mean} = \sqrt{xy}$$

For example, the geometric mean of 2 and 18 is $gm = \sqrt{2 \times 18} = \sqrt{36} = 6$, whereas the arithmetic mean is $\frac{2+18}{2} = \frac{20}{2} = 10$. We now describe the Hill-Huntington Method of apportionment.

HILL-HUNTINGTON METHOD OF APPORTIONMENT

1. Using the standard divisor d, calculate the standard quotas, lower quotas, and upper quotas of each state.
2. For each state, calculate the geometric mean (rounded to two decimal places) of its lower quota and upper quota. If the standard quota is less than the geometric mean, round the quota down; if the standard quota is greater than or equal to the geometric mean, round the quota up.
3. If the sum of the rounded standard quotas equals the total number of seats to be apportioned, the apportionment process is complete, that is, each state receives a number of seats equal to its rounded standard quota.
4. If the sum of the rounded standard quotas does not equal the total number of seats to be apportioned, choose a modified divisor d_m different from the standard divisor (either less than or greater than d), and calculate the modified quotas and rounded modified quotas.
5. Repeat step 4 until you find a modified divisor such that the sum of the rounded modified quotas equals the total number of seats to be apportioned. Each state receives a number of seats equal to its rounded modified quota, and the apportionment process is complete.

EXAMPLE **5** | **APPLYING THE HILL-HUNTINGTON METHOD OF APPORTIONMENT**
As in Example 1, the populations of Gondor, Eriador, and Rohan are given in Figure 6.60. Use the Hill-Huntington Method to apportion the ten seats of the high council among these three realms.

Realm (State)	Gondor	Eriador	Rohan	Total
Population	94,700	72,600	33,700	201,000

FIGURE 6.60   Realms of Middle Earth.

SOLUTION

As we saw in Example 3, the standard divisor $d = 20.1$ produces the standard, lower, and upper quotas shown in Figure 6.61.

Beginning with Gondor, we now calculate the geometric mean of the lower and upper quotas for each state:

$$gm_G = \sqrt{4 \times 5} = \sqrt{20} = 4.472135955\ldots$$

Therefore, $gm_G = 4.47$.

Realm (State)	Gondor	Eriador	Rohan	Total
Population (thousands)	94.7	72.6	33.7	201
Standard Quota ($d = 20.1$)	4.71	3.61	1.68	10
Lower Quota	4	3	1	8
Upper Quota	5	4	2	11

FIGURE 6.61   Standard, lower, and upper quotas of the realms.

In a similar fashion, we calculate the geometric means for Eriador (E) and Rohan (R):

gm_E = Eriador's geometric mean gm_R = Rohan's geometric mean

$\quad = \sqrt{3 \times 4}$ $\qquad\qquad\qquad\qquad = \sqrt{1 \times 2}$

$\quad = \sqrt{12}$ $\qquad\qquad\qquad\qquad\quad = \sqrt{2}$

$gm_E = 3.46$ **rounded to two** $gm_R = 1.41$ **rounded to two**
$\qquad\qquad$ **decimal places** $\qquad\qquad\qquad\qquad$ **decimal places**

The standard quotas and geometric means for Gondor, Eriador, and Rohan are summarized in Figure 6.62.

Realm (State)	Gondor	Eriador	Rohan	Total
Population (thousands)	94.7	72.6	33.7	201
Standard Quota ($d = 20.1$)	4.71	3.61	1.68	10
Geometric Mean	4.47	3.46	1.41	(not needed)

FIGURE 6.62   Geometric means of the realms.

For each state, the standard quota is greater than the geometric mean; consequently, each standard quota is rounded up as shown in Figure 6.63.

Realm (State)	Gondor	Eriador	Rohan	Total
Population (thousands)	94.7	72.6	33.7	201
Standard Quota ($d = 20.1$)	4.71	3.61	1.68	10
Geometric Mean	4.47	3.46	1.41	(not needed)
Rounded Quota	5	4	2	11

FIGURE 6.63 Rounded quotas of the realms.

The sum of the rounded quotas (11) is too large, so we choose a modified divisor, $d_m > d = 20.1$, because dividing by a larger number creates a smaller result. As a first guess, we try $d_m = 21$ and calculate each realm's modified quota; that is, we divide each population by $d_m = 21$, round to two decimal places, and then round to a whole number (using the geometric mean) as shown in Figure 6.64.

Realm (State)	Gondor	Eriador	Rohan	Total
Population (thousands)	94.7	72.6	33.7	201
Standard Quota ($d = 20.1$)	4.71	3.61	1.68	10
Geometric Mean	4.47	3.46	1.41	(not needed)
Modified Quota ($d_m = 21$)	4.51	3.46	1.60	(not needed)
Rounded Quota	5	4	2	11

FIGURE 6.64 The Hill-Huntington Method using $d_m = 21$.

The sum of the rounded quotas (11) is still too large, so we must choose an even larger modified divisor, say, $d_m = 21.2$, and repeat the process. That is, we divide each population by $d_m = 21.2$, round to two decimal places, and then round to a whole number (using the geometric mean) as shown in Figure 6.65.

Realm (State)	Gondor	Eriador	Rohan	Total
Population (thousands)	94.7	72.6	33.7	201
Standard Quota ($d = 20.1$)	4.71	3.61	1.68	10
Geometric Mean	4.47	3.46	1.41	(not needed)
Modified Quota ($d = 21.2$)	4.47	3.42	1.59	(not needed)
Rounded Quota	5	3	2	10

FIGURE 6.65 The Hill-Huntington Method using $d_m = 21.2$.

Because the sum of the rounded quotas (10) equals the total number of seats to be allocated (10), the apportionment process is complete. By using the Hill-Huntington Method, the ten seats will be apportioned as shown in Figure 6.66.

Realm (State)	Gondor	Eriador	Rohan	Total
Hill-Huntington's Apportionment	5	3	2	10
Webster's Apportionment	5	3	2	10
Adams's Apportionment	4	4	2	10
Jefferson's Apportionment	5	4	1	10
Hamilton's Apportionment	5	3	2	10

FIGURE 6.66   Comparison of final apportionments.

Additional Seats

Once the seats of a legislature have been allocated, it might be decided that the size of the legislature should be increased; that is, new seats might be added to an existing apportionment. Who gets the new seats? The answer lies in the calculation of **Hill-Huntington numbers.**

HILL-HUNTINGTON NUMBER

The **Hill-Huntington number** for a state, denoted **HHN,** is the square of the state's population divided by the product of its current number of seats n and $n + 1$. That is,

$$HHN = \frac{(\text{state's population})^2}{n(n + 1)}$$

where n = state's current number of seats.

In allocating a new seat to an existing legislature, the state with the highest *HHN* should receive the seat.

EXAMPLE 6

CALCULATING HILL-HUNTINGTON NUMBERS   As we saw in Example 5, the populations of Gondor, Eriador, and Rohan and the apportionment of the ten seats via the Hill-Huntington Method are given in Figure 6.67. Suppose the high council decides to add an additional seat; that is, it is decided that the council should now consist of eleven seats. Use Hill-Huntington numbers to determine which realm should receive the new seat.

Realm (State)	Gondor	Eriador	Rohan	Total
Population (thousands)	94.7	72.6	33.7	201
Number of Seats	5	3	2	10

FIGURE 6.67   Apportionment of the high council.

SOLUTION

Beginning with Gondor, we calculate the Hill-Huntington number for each realm:

$$HHN_G = \frac{94.7^2}{5(5 + 1)} = \frac{94.7^2}{30} = 298.936333\ldots$$

Therefore, $HHN_G = 298.94$.

In a similar fashion, we calculate the Hill-Huntington numbers for Eriador (E) and Rohan (R):

HHN_E = Eriador's Hill-Huntington number

$$= \frac{72.6^2}{3(3 + 1)}$$

$$= \frac{72.6^2}{12}$$

$HHN_E = 439.23$ **rounded to two decimal places**

HHN_R = Rohan's Hill-Huntington number

$$= \frac{33.7^2}{2(2 + 1)}$$

$$= \frac{33.7^2}{6}$$

$HHN_R = 189.28$ **rounded to two decimal places**

The Hill-Huntington numbers for the three realms are summarized in Figure 6.68.

Realm (State)	Gondor	Eriador	Rohan
Hill-Huntington Number	298.94	439.23	189.28

FIGURE 6.68 Hill-Huntington numbers for each realm.

Because Eriador has the highest *HHN* (439.23 is greater than 298.23 or 189.28), Eriador should receive the new seat.

Why does this method work? Why does the state with the highest Hill-Huntington number deserve the additional seat? The answer lies in what is called *relative unfairness* and in the application of algebra. Consider the following scenario: A community college has a main campus and a satellite campus that is located several miles from the main campus. The main campus has an enrollment of 23,000 students with a faculty of 460; the satellite has an enrollment of 2,000 with a faculty of 40. If the college hires one new instructor, which campus is more deserving of receiving the new hire?

Using intuition, it might seem appropriate to calculate student-to-teacher ratios; that is, the campus that has more students per teacher would appear to be more deserving. However, in our scenario, both campuses have the same ratio, as shown in the following calculations:

Main Campus

number of students
─────────────────
number of teachers

$$= \frac{23,000}{460}$$

= 50 students per teacher

Satellite Campus

number of students
─────────────────
number of teachers

$$= \frac{2,000}{40}$$

= 50 students per teacher

To continue, we must investigate the meaning of *relative unfairness.*

Suppose two groups, *A* and *B*, are vying for an additional representative. Let p_A and p_B denote the populations of groups *A* and *B*, respectively, and let *a* and *b* denote the current number of representatives of groups *A* and *B*, respectively. The mean representations of *A* and *B* are given by $\frac{p_A}{a}$ and $\frac{p_B}{b}$, respectively (the student-to-teacher ratios shown above). Now, if *A* receives one new representative, its mean representation becomes $\frac{p_A}{a + 1}$, whereas if *B* receives the new representative, its mean representation becomes $\frac{p_B}{b + 1}$.

Our goal is to allocate the new representative in such a way as to minimize the relative unfairness of the allocation. Relative unfairness can be interpreted as the difference (subtraction) in mean representations compared to (divided by) the new mean representation. So if A receives one new representative, the difference in mean representations is $\frac{p_B}{b} - \frac{p_A}{a+1}$, and the relative unfairness to A may be expressed as

$$\frac{\dfrac{p_B}{b} - \dfrac{p_A}{a+1}}{\dfrac{p_A}{a+1}}$$

In a similar fashion, if B receives the new representative, the relative unfairness to B may be expressed as

$$\frac{\dfrac{p_A}{a} - \dfrac{p_B}{b+1}}{\dfrac{p_B}{b+1}}$$

For the sake of argument, let us now suppose that the relative unfairness to A is smaller than the relative unfairness to B; consequently, A deserves the new representative. We can express this analytically as follows:

Relative unfairness to A < relative unfairness to B

$$\left[\frac{\dfrac{p_B}{b} - \dfrac{p_A}{a+1}}{\dfrac{p_A}{a+1}}\right] < \left[\frac{\dfrac{p_A}{a} - \dfrac{p_B}{b+1}}{\dfrac{p_B}{b+1}}\right]$$

Because $\frac{p_A}{a+1}$ and $\frac{p_B}{b+1}$ are both positive, we multiply each side of the inequality by the expression $\left(\frac{p_A}{a+1}\right)\left(\frac{p_B}{b+1}\right)$ and obtain the following:

$$\left(\frac{p_A}{a+1}\right)\left(\frac{p_B}{b+1}\right)\left[\frac{\dfrac{p_B}{b} - \dfrac{p_A}{a+1}}{\dfrac{p_A}{a+1}}\right] < \left(\frac{p_A}{a+1}\right)\left(\frac{p_B}{b+1}\right)\left[\frac{\dfrac{p_A}{a} - \dfrac{p_B}{b+1}}{\dfrac{p_B}{b+1}}\right]$$

multiplying each side by the expression $\left(\dfrac{p_A}{a+1}\right)\left(\dfrac{p_B}{b+1}\right)$

$$\left(\frac{p_B}{b+1}\right)\left[\frac{p_B}{b} - \frac{p_A}{a+1}\right] < \left(\frac{p_A}{a+1}\right)\left[\frac{p_A}{a} - \frac{p_B}{b+1}\right]$$

canceling $\dfrac{p_A}{a+1}$ **on the left side and** $\dfrac{p_B}{b+1}$ **on the right side**

$$\frac{p_B^2}{b(b+1)} - \frac{p_A p_B}{(a+1)(b+1)} < \frac{p_A^2}{a(a+1)} - \frac{p_A p_B}{(a+1)(b+1)} \qquad \textbf{distributing}$$

$$\frac{p_B^2}{b(b+1)} < \frac{p_A^2}{a(a+1)} \qquad \textbf{adding } \frac{p_A p_B}{(a+1)(b+1)} \textbf{ to each side}$$

$$HHN_B < HHN_A \qquad \textbf{definition of Hill-Huntington numbers}$$

Therefore, if the relative unfairness to A is less than the relative unfairness to B, we see that $HHN_B < HHN_A$, and consequently, A deserves the new seat; that is, the group with the highest HHN should receive an additional seat. Regarding our scenario involving a community college and its satellite, we conclude that the main campus should receive the new instructor because $HHN_{main} > HHN_{satellite}$. (Verify this!)

We conclude this section with an example that recaps the five methods of apportionment.

EXAMPLE **7**

SUMMARIZING ALL FIVE METHODS OF APPORTIONMENT Suppose that the governors of six New England region states have agreed to form an interstate bureau to foster historical awareness, tourism, and commerce. The bureau will have sixteen seats, and the populations of the states are given in Figure 6.69.

State	Connecticut (CT)	Maine (ME)	Massachusetts (MA)	New Hampshire (NH)	Rhode Island (RI)	Vermont (VT)
Population	3,405,584	1,274,923	6,349,097	1,235,786	1,048,319	608,827

FIGURE 6.69 New England region states. *Source:* Bureau of the Census, 2000.

a. Use Hamilton's Method to apportion the seats.
b. Use Jefferson's Method to apportion the seats.
c. Use Adams's Method to apportion the seats.
d. Use Webster's Method to apportion the seats.
e. Use the Hill-Huntington Method to apportion the seats.

SOLUTION

To simplify calculations, we express the populations in millions, rounded to three decimal places, as shown in Figure 6.70.

State	CT	ME	MA	NH	RI	VT	Total
Population (millions)	3.406	1.275	6.349	1.236	1.048	0.609	13.923

FIGURE 6.70 New England region states.

a. The standard divisor is d = total population/total number of seats = 13.923/16 = 0.87, rounded to two decimal places. Now divide each state's population by 0.87 and round to two decimal places to obtain the standard quotas and lower quotas shown in Figure 6.71.

State	CT	ME	MA	NH	RI	VT	Total
Population (millions)	3.406	1.275	6.349	1.236	1.048	0.609	13.923
Standard Quota ($d = 0.87$)	3.91	1.47	7.30	1.42	1.20	0.70	16
Lower Quota	3	1	7	1	1	0	13

FIGURE 6.71 Standard and lower quotas of New England region states.

Finally, we allocate three additional seats to the three states (CT, VT, ME) that have the highest decimal parts (0.91, 0.70, 0.47) in their standard quotas and obtain the final apportionment shown in Figure 6.72.

State	CT	ME	MA	NH	RI	VT	Total
Lower Quota	3	1	7	1	1	0	13
Additional Seats	1	1	0	0	0	1	3
Hamilton's Apportionment	4	2	7	1	1	1	16

FIGURE 6.72 Final apportionment using Hamilton's Method.

b. Using the modified divisor $d_m = 0.7$ (less than d), we obtain the modified quotas, lower modified quotas, and final apportionment shown in Figure 6.73.

State	CT	ME	MA	NH	RI	VT	Total
Population (millions)	3.406	1.275	6.349	1.236	1.048	0.609	13.923
Modified Quota $(d_m = 0.7)$	4.87	1.82	9.07	1.77	1.50	0.87	(not needed)
Lower Modified Quota	4	1	9	1	1	0	16
Jefferson's Apportionment	4	1	9	1	1	0	16

FIGURE 6.73 Final apportionment using Jefferson's Method.

c. Using the modified divisor $d_m = 1.1$ (greater than d), we obtain the modified quotas, upper modified quotas, and final apportionment shown in Figure 6.74.

State	CT	ME	MA	NH	RI	VT	Total
Population (millions)	3.406	1.275	6.349	1.236	1.048	0.609	13.923
Modified Quota $(d_m = 1.1)$	3.10	1.16	5.77	1.12	0.95	0.55	(not needed)
Upper Modified Quota	4	2	6	2	1	1	16
Adams' Apportionment	4	2	6	2	1	1	16

FIGURE 6.74 Final apportionment using Adams's Method.

d. Using the modified divisor $d_m = 0.85$ (different from d), we obtain the modified quotas, rounded modified quotas, and final apportionment shown in Figure 6.75.

State	CT	ME	MA	NH	RI	VT	Total
Population (millions)	3.406	1.275	6.349	1.236	1.048	0.609	13.923
Modified Quota $(d_m = 0.85)$	4.01	1.50	7.47	1.45	1.23	0.72	(not needed)
Rounded Modified Quota	4	2	7	1	1	1	16
Webster's Apportionment	4	2	7	1	1	1	16

FIGURE 6.75 Final apportionment using Webster's Method.

e. Using the modified divisor $d_m = 0.9$ (different from d), we obtain the modified quotas, lower modified quotas, geometric means, rounded quotas, and final apportionment shown in Figure 6.76.

State	CT	ME	MA	NH	RI	VT	Total
Population (millions)	3.406	1.275	6.349	1.236	1.048	0.609	13.923
Modified Quota $(d_m = 0.9)$	3.78	1.42	7.05	1.37	1.16	0.68	16
Lower Modified Quota	3	1	7	1	1	0	(not needed)
Geometric Mean	$\sqrt{3 \times 4}$ $= 3.46$	$\sqrt{1 \times 2}$ $= 1.41$	$\sqrt{7 \times 8}$ $= 7.48$	$\sqrt{1 \times 2}$ $= 1.41$	$\sqrt{1 \times 2}$ $= 1.41$	$\sqrt{0 \times 1}$ $= 0$	(not needed)
Rounded Quota	4	2	7	1	1	1	16
Hill-Huntington Apportionment	4	2	7	1	1	1	16

FIGURE 6.76 Final apportionment using the Hill-Huntington Method.

TOPIC X THE U.S. HOUSE OF REPRESENTATIVES APPORTIONMENT IN THE REAL WORLD

After the adoption of the Constitution, the United States of America took its first census in the year 1790. This census, along with Jefferson's method of apportionment, was used to distribute the 105 seats of the first House of Representatives. The population (in descending magnitude) and apportionment of each of the states are listed in Figure 6.77.

It could be argued that in 1790, Virginia was the state with the most power or influence because it had the greatest number of representatives (nineteen). Notice that on the basis of the first apportionment, New York and North Carolina had equal power (each had ten representatives), while North Carolina had twice as many representatives as New Jersey (five).

Over time, the relative rankings of the "power" of these states has dramatically changed. For instance, on the basis of the 2000 census, New York now has more than twice as many representatives as North Carolina has (twenty-nine compared to thirteen), while North Carolina and New Jersey are now equal (thirteen each). Go to www.census.gov/population/www/censusdata/apportionment/index.html to see the current apportionment of the House of Representatives.

As we have seen, different methods of apportionment can lead to different allocations of seats. In the following example and in the exercises, we will explore the distribution of seats of the first House of Representatives, using several different methods of apportionment.

State	Population	Representatives
Virginia	630,560	19
Massachusetts	475,327	14
Pennsylvania	432,879	13
North Carolina	353,523	10
New York	331,589	10
Maryland	278,514	8
Connecticut	236,841	7
South Carolina	206,236	6
New Jersey	179,570	5
New Hampshire	141,822	4
Vermont	85,533	2
Georgia	70,835	2
Kentucky	68,705	2
Rhode Island	68,446	2
Delaware	55,540	1
Total	3,615,920	105

FIGURE 6.77 The first census and apportionment of the United States (1790).

For comparison, we have included all five apportionments in Figure 6.78.

State	CT	ME	MA	NH	RI	VT	Total
Hamilton's Apportionment	4	2	7	1	1	1	16
Jefferson's Apportionment	4	1	9	1	1	0	16
Adams's Apportionment	4	2	6	2	1	1	16
Webster's Apportionment	4	2	6	2	1	1	16
Hill-Huntington Apportionment	4	2	7	1	1	1	16

FIGURE 6.78 Comparison of final apportionments.

EXAMPLE **8**

VERIFYING THE APPORTIONMENT OF THE FIRST HOUSE OF REPRESENTATIVES OF THE UNITED STATES   Use Jefferson's Method of apportionment and the population data in Figure 6.77 to verify the apportionment of the 105 seats in the 1790 House of Representatives (as shown in Figure 6.77).

SOLUTION

To simplify calculations, we express the populations in hundred thousands, rounded to two decimal places, as shown in Figure 6.79.

State	Population (hundred thousands)
Virginia	6.31
Massachusetts	4.75
Pennsylvania	4.33
North Carolina	3.54
New York	3.32
Maryland	2.79
Connecticut	2.37
South Carolina	2.06
New Jersey	1.80
New Hampshire	1.42
Vermont	0.86
Georgia	0.71
Kentucky	0.69
Rhode Island	0.68
Delaware	0.56
Total	36.19

FIGURE 6.79   The census of the first United States.

First, we must find the standard divisor:

$$d = \frac{\text{total population}}{\text{total number of seats}} = \frac{36.19}{105} = 0.3446666667$$

Therefore, $d = 0.34$.

Now, calculate each state's standard quota, that is, divide each population by $d = 0.34$, round to two decimal places, and then truncate to a whole number to obtain the lower quotas, all as shown in Figure 6.80.

Because the sum of the lower quotas (100) is less than the number of seats to be allocated (105), the standard quota creates five surplus seats. Therefore, to apply Jefferson's Method, we must find a modified divisor, $d_m < d = 0.34$ such that no surplus is created. After trial and error, we find that the modified divisor $d_m = 0.33$ will properly apportion the 105 seats, and the final results are shown in Figure 6.81.

State	Population (hundred thousands)	Standard Quota (d = 0.34)	Lower Quota
Virginia	6.31	6.31/0.34 = 18.56	18
Massachusetts	4.75	4.75/0.34 = 13.97	13
Pennsylvania	4.33	4.33/0.34 = 12.74	12
North Carolina	3.54	3.54/0.34 = 10.41	10
New York	3.32	3.32/0.34 = 9.76	9
Maryland	2.79	2.79/0.34 = 8.21	8
Connecticut	2.37	2.37/0.34 = 6.97	6
South Carolina	2.06	2.06/0.34 = 6.06	6
New Jersey	1.80	1.80/0.34 = 5.29	5
New Hampshire	1.42	1.42/0.34 = 4.18	4
Vermont	0.86	0.86/0.34 = 2.53	2
Georgia	0.71	0.71/0.34 = 2.09	2
Kentucky	0.69	0.69/0.34 = 2.03	2
Rhode Island	0.68	0.68/0.34 = 2.00	2
Delaware	0.56	0.56/.34 = 1.65	1
Total	36.19		100

FIGURE 6.80 Standard and lower quotas of the first states.

State	Population (hundred thousands)	Modified Quota (d_m = 0.33)	Lower Modified Quota
Virginia	6.31	6.31/0.33 = 19.12	19
Massachusetts	4.75	4.75/0.33 = 14.39	14
Pennsylvania	4.33	4.33/0.33 = 13.12	13
North Carolina	3.54	3.54/0.33 = 10.73	10
New York	3.32	3.32/0.33 = 10.06	10
Maryland	2.79	2.79/0.33 = 8.45	8
Connecticut	2.37	2.37/0.33 = 7.18	7
South Carolina	2.06	2.06/0.33 = 6.24	6
New Jersey	1.80	1.80/0.33 = 5.45	5
New Hampshire	1.42	1.42/0.33 = 4.30	4
Vermont	0.86	0.86/0.33 = 2.61	2
Georgia	0.71	0.71/0.33 = 2.15	2
Kentucky	0.69	0.69/0.33 = 2.09	2
Rhode Island	0.68	0.68/0.33 = 2.06	2
Delaware	0.56	0.56/.33 = 1.70	1
Total	36.19		105

FIGURE 6.81 Jefferson's Method applied to the first states.

6.2 EXERCISES

1. A small country consists of three states, *A*, *B*, and *C*; the population of each state is given in Figure 6.82. The country's legislature is to have forty seats.

State	A	B	C
Population	900,000	700,000	400,000

FIGURE 6.82 Table for Exercise 1.

a. Express each state's population (and the total population) in terms of thousands.
b. Find the standard divisor.
c. Find each state's standard, lower, and upper quotas.

2. A small country consists of three states, *A*, *B*, and *C*; the population of each state is given in Figure 6.83. The country's legislature is to have fifty seats.

State	A	B	C
Population	810,000	720,000	510,000

FIGURE 6.83 Table for Exercise 2.

a. Express each state's population (and the total population) in terms of thousands.
b. Find the standard divisor.
c. Find each state's standard, lower and upper quotas.

3. Suppose that the governors of three Middle Atlantic region states have agreed to form an interstate bureau to foster historical awareness, tourism, and commerce. The bureau will have fifteen seats, and the populations of the states are given in Figure 6.84.

State	New York	Pennsylvania	New Jersey
Population	18,976,821	12,281,054	8,414,347

FIGURE 6.84 Table for Exercise 3. *Source:* Bureau of the Census, 2000.

a. Express each state's population (and the total population) in terms of millions, rounded to three decimal places.
b. Find the standard divisor.
c. Find each state's standard, lower, and upper quotas.

4. Suppose that the governors of three midwestern region states have agreed to form an interstate bureau to foster historical awareness, tourism, and commerce. The bureau will have ten seats, and the populations of the states are given in Figure 6.85.

State	Michigan	Wisconsin	Minnesota
Population	9,938,480	5,363,704	4,919,485

FIGURE 6.85 Table for Exercise 4. *Source:* Bureau of the Census, 2000.

a. Express each state's population (and the total population) in terms of millions, rounded to three decimal places.
b. Find the standard divisor.
c. Find each state's standard, lower, and upper quotas.

5. Use Hamilton's Method to apportion the legislative seats in Exercise 1.

6. Use Hamilton's Method to apportion the legislative seats in Exercise 2.

7. Use Hamilton's Method to apportion the bureau seats in Exercise 3.

8. Use Hamilton's Method to apportion the bureau seats in Exercise 4.

9. A local school district contains four middle schools: Applewood, Boatwright, Castlerock, and Dunsmuir. The number of students attending each school is given in Figure 6.86. The school district has received a generous donation of 200 graphing calculators for student use. Use Hamilton's Method to determine how the calculators should be divided among the schools.

School	Applewood	Boatwright	Castlerock	Dunsmuir
Students	1,768	1,357	1,091	893

FIGURE 6.86 Table for Exercise 9.

10. A local school district contains four elementary schools: Elmhurst, Fernwood, Greenbriar, and Hawthorne. The number of students attending each school is given in Figure 6.87. The school district has received a generous donation of sixty digital cameras for student use. Use Hamilton's Method to determine how the cameras should be divided among the schools.

School	Elmhurst	Fernwood	Greenbriar	Hawthorne
Students	1,214	1,008	901	766

FIGURE 6.87 Table for Exercise 10.

11. In J.R.R. Tolkien's Middle Earth, a region known as the Shire is homeland to numerous clans of hobbits. The

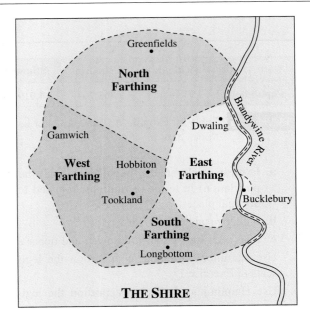

THE SHIRE

Shire is divided into four regions: North Farthing, South Farthing, East Farthing, and West Farthing. Suppose the hobbits decide to form an association to foster the preservation of their cultural and culinary traditions. If the association is to have twenty-four seats, use Hamilton's Method and the (estimated) population data given in Figure 6.88 to apportion the seats.

Region	North Farthing	South Farthing	East Farthing	West Farthing
Population (estimated)	2,680	6,550	2,995	8,475

FIGURE 6.88   Table for Exercise 11.

12. In J.R.R. Tolkien's Middle Earth, the regions in and around the Misty Mountains are inhabited by elves, dwarves, and Ents (tree shepherds). Elves reside in Rivendell and Lothlórien, dwarves dwell in Moria, and Ents roam the forests of Fangorn. Suppose that these inhabitants decide to form a federation to foster mutual protection, historical preservation, and cultural exchange. If the federation is to have twenty-one seats, use Hamilton's Method and the (estimated) population data given in Figure 6.89 to apportion the seats.

Region	Rivendell	Lothlórien	Moria	Fangorn
Population (estimated)	5,424	4,967	6,821	587

FIGURE 6.89   Table for Exercise 12.

13. Suppose that the governments of several Scandinavian countries have agreed to form an international bureau to

foster tourism, commerce, and cultural awareness. The bureau will have twenty seats, and the populations of the countries are given in Figure 6.90. Use Hamilton's Method to apportion the bureaucratic seats.

14. Suppose that the governments of several North African countries have agreed to form an international bureau to foster tourism, commerce, and education. The bureau will have twenty-five seats, and the populations of the countries are given in Figure 6.91. Use Hamilton's Method to apportion the bureau seats.

15. Suppose that the governments of several Central American countries have agreed to form an international bureau to foster tourism, commerce, and education. The bureau will have twenty-five seats, and the populations of the countries are given in Figure 6.92. Use Hamilton's Method to apportion the bureau seats.

16. Suppose that the governments of several Southeast Asian countries have agreed to form an international bureau to foster tourism, commerce, and education. The bureau will have thirty seats, and the populations of the countries are given in Figure 6.93. Use Hamilton's Method to apportion the bureau seats.

17. a. Use Jefferson's Method to apportion the legislative seats in Exercise 5.

   b. Use Adams's Method to apportion the legislative seats in Exercise 5.

   c. Use Webster's Method to apportion the legislative seats in Exercise 5.

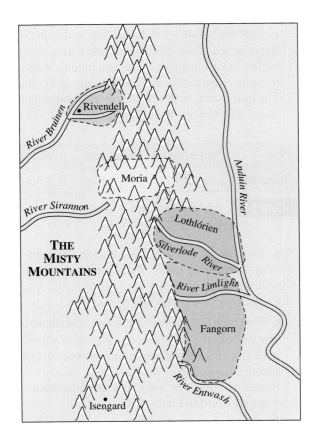

Country	Denmark	Finland	Iceland	Norway	Sweden
Population (thousands)	5,413	5,215	294	4,575	8,986

FIGURE 6.90 National populations of Scandinavia, 2004. *Source:* U.S. Department of Commerce.

Country	Algeria	Egypt	Libya	Morocco	Tunisia
Population (thousands)	32,129	76,117	5,632	32,209	9,975

FIGURE 6.91 National populations of North African countries, 2004. *Source:* U.S. Department of Commerce.

Country	Costa Rica	El Salvador	Guatemala	Honduras	Nicaragua	Panama
Population (thousands)	3,957	6,588	14,281	6,824	5,360	3,000

FIGURE 6.92 National populations of Central American countries, 2004. *Source:* U.S. Department of Commerce.

Country	Indonesia	Malaysia	Philippines	Taiwan	Thailand	Vietnam
Population (thousands)	238,453	23,522	86,242	22,750	64,866	82,690

FIGURE 6.93 National populations of Southeast Asian countries, 2004. *Source:* U.S. Department of Commerce.

18. a. Use Jefferson's Method to apportion the legislative seats in Exercise 6.
b. Use Adams's Method to apportion the legislative seats in Exercise 6.
c. Use Webster's Method to apportion the legislative seats in Exercise 6.

19. a. Use Jefferson's Method to apportion the bureau seats in Exercise 7.
b. Use Adams's Method to apportion the bureau seats in Exercise 7.
c. Use Webster's Method to apportion the bureau seats in Exercise 7.

20. a. Use Jefferson's Method to apportion the bureau seats in Exercise 8.
b. Use Adams's Method to apportion the bureau seats in Exercise 8.
c. Use Webster's Method to apportion the bureau seats in Exercise 8.

21. a. Use Jefferson's Method to determine how the calculators should be divided among the school in Exercise 9.
b. Use Adams's Method to determine how the calculators should be divided among the school in Exercise 9.
c. Use Webster's Method to determine how the calculators should be divided among the school in Exercise 9.

22. a. Use Jefferson's Method to determine how the cameras should be divided among the school in Exercise 10.
b. Use Adams's Method to determine how the cameras should be divided among the schools in Exercise 10.
c. Use Webster's Method to determine how the cameras should be divided among the schools in Exercise 10.

23. a. Use Jefferson's Method to apportion the associations seats in Exercise 11.
b. Use Adams's Method to apportion the association seats in Exercise 11.
c. Use Webster's Method to apportion the association seats in Exercise 11.

24. a. Use Jefferson's Method to apportion the federation seats in Exercise 12.
b. Use Adams's Method to apportion the federation seats in Exercise 12.
c. Use Webster's Method to apportion the federation seats in Exercise 12.

25. a. Use Jefferson's Method to apportion the bureau seats in Exercise 13.
b. Use Adams's Method to apportion the bureau seats in Exercise 13.
c. Use Webster's Method to apportion the bureau seats in Exercise 13.

26. **a.** Use Jefferson's Method to apportion the bureau seats in Exercise 14.

   b. Use Adams's Method to apportion the bureau seats in Exercise 14.

   c. Use Webster's Method to apportion the bureau seats in Exercise 14.

27. **a.** Use Jefferson's Method to apportion the bureau seats in Exercise 15.

   b. Use Adams's Method to apportion the bureau seats in Exercise 15.

   c. Use Webster's Method to apportion the bureau seats in Exercise 15.

28. **a.** Use Jefferson's Method to apportion the bureau seats in Exercise 16.

   b. Use Adams's Method to apportion the bureau seats in Exercise 16.

   c. Use Webster's Method to apportion the bureau seats in Exercise 16.

29. Use the Hill-Huntington Method to apportion the legislative seats in Exercise 5.

30. Use the Hill-Huntington Method to apportion the legislative seats in Exercise 6.

31. Use the Hill-Huntington Method to apportion the bureau seats in Exercise 7.

32. Use the Hill-Huntington Method to apportion the bureau seats in Exercise 8.

33. Use the Hill-Huntington Method to determine how the calculators should be divided among the school in Exercise 9.

34. Use the Hill-Huntington Method to determine how the cameras should be divided among the schools in Exercise 10.

35. Use the Hill-Huntington Method to apportion the association seats in Exercise 11.

36. Use the Hill-Huntington Method to apportion the federation seats in Exercise 12.

37. Use the Hill-Huntington Method to apportion the bureau seats in Exercise 13.

38. Use the Hill-Huntington Method to apportion the bureau seats in Exercise 14.

39. Use the Hill-Huntington Method to apportion the bureau seats in Exercise 15.

40. Use the Hill-Huntington Method to apportion the bureau seats in Exercise 16.

41. A community college has a main campus and a satellite campus that is located several miles away from the main campus. The main campus has an enrollment of 25,000 students with a faculty of 465; the satellite has an enrollment of 2,600 with a faculty of 47. The college decides to hire one new instructor. Use Hill-Huntington numbers to determine which location should receive the new instructor.

42. A waste management district operates two recycling locations: Alpha and Beta. Alpha services 20,000 households and has a staff of 23 workers; Beta's staff of 34 services 30,000 households. The district decides to hire one new worker. Use Hill-Huntington numbers to determine which location should receive the new staff member.

43. A school district has three elementary schools: Agnesi, Banach, and Cantor. If the district decides to hire one new teacher, use Hill-Huntington numbers to determine which school should receive the new instructor. The current enrollments and faculty of the schools are given in Figure 6.94.

	Agnesi	**Banach**	**Cantor**
Students	567	871	666
Teachers	21	32	25

FIGURE 6.94 Table for Exercise 43.

44. A school district has three middle schools: Descartes, Euclid, and Fermat. If the district decides to hire one new teacher, use Hill-Huntington numbers to determine which school should receive the new instructor. The current enrollments and faculty of the schools are given in Figure 6.95.

	Descartes	**Euclid**	**Fermat**
Students	752	984	883
Teachers	25	33	29

FIGURE 6.95 Table for Exercise 44.

45. Use Hamilton's Method of apportionment and the population data in Figure 6.77 to apportion the 105 seats in the 1790 House of Representatives. Does this hypothetical apportionment differ from the actual apportionment (shown in Figure 6.77)? If so, how?

46. Use Adams's Method of apportionment and the population data in Figure 6.77 to apportion the 105 seats in the 1790 House of Representatives. Does this hypothetical apportionment differ from the actual apportionment (shown in Figure 6.77)? If so, how?

47. Use Webster's Method of apportionment and the population data in Figure 6.77 to apportion the 105 seats in the 1790 House of Representatives. Does this hypothetical apportionment differ from the actual apportionment (shown in Figure 6.77)? If so, how?

48. Use the Hill-Huntington Method of apportionment and the population data in Figure 6.77 to apportion

the 105 seats in the 1790 House of Representatives. Does this hypothetical apportionment differ from the actual apportionment (shown in Figure 6.77)? If so, how?

 Answer the following questions using complete sentences and your own words.

• CONCEPT QUESTIONS

49. What is apportionment?

50. What is a standard divisor?

51. What is a modified divisor?

52. What is a standard quota?

53. What is a modified quota?

54. What are lower and upper quotas?

• HISTORY QUESTIONS

55. What was the first method of apportionment of the House of Representatives to be approved by Congress?

56. What was the first method of apportionment of the House of Representatives to actually be used?

57. Why are the answers to Exercises 55 and 56 different?

58. Since the founding of the United States, what methods of apportionment of the House of Representatives have been used? What is their chronology?

59. What method of apportionment of the House of Representatives is currently being used?

WEB PROJECT

60. In 1790, the United States had 3,615,920 people, and the House of Representatives had 105 seats. Therefore, on average, each seat represented approximately 34,437 people.

  a. What was the population of the United States in 1850? How many seats did the House have? On average, how many people did each seat represent?

  b. What was the population of the United States in 1900? How many seats did the House have? On average, how many people did each seat represent?

  c. What was the population of the United States in 1950? How many seats did the House have? On average, how many people did each seat represent?

  d. What was the population of the United States in 2000? How many seats did the House have? On average, how many people did each seat represent?

  e. When you compare the answers to parts (a) through (d), what conclusion(s) can you make about the average representation in the House of Representatives?

Some useful links for this web project are listed on the text website: **www.cengage.com/math/johnson**

6.3 Flaws of Apportionment

OBJECTIVES

● Define and investigate the Quota Rule

● Define and investigate the Alabama Paradox

● Define and investigate the New States Paradox

● Define and investigate the Population Paradox

In the previous section, we studied several different methods of apportionment. Quite often, different methods of apportionment produce different allotments of the items being apportioned. Are the methods "fair"? Is one method "better" than the rest? We will attempt to answer these questions in this section.

The Quota Rule

The first step in any method of apportionment is the calculation of the standard divisor and the standard quota. That is,

$$d = \text{standard divisor} = \frac{\text{total population}}{\text{total number of seats}}$$

$$q = \text{standard quota} = \frac{\text{state's population}}{\text{standard divisor}} = \frac{\text{state's population}}{d}$$

Most people would agree that in any fair method of apportionment, each state should receive either its lower quota (truncated standard quota) of seats or its upper quota (rounded up standard quota) of seats. After all, if a state received too few seats, the state would appear to have been penalized, whereas if it received too many, favoritism could be alleged. Consequently, this fundamental criterion of fairness is referred to as the **Quota Rule.**

THE QUOTA RULE

The apportionment of a group should equal either the lower quota of the group or the upper quota of the group.

EXAMPLE 1

APPLYING JEFFERSON'S METHOD AND INVESTIGATING THE QUOTA RULE   A small nation of 18 million people is composed of four states as shown in Figure 6.96. The national legislature has 120 seats.

State	A	B	C	D	Total
Population	1,548,000	2,776,000	3,929,000	9,747,000	18,000,000

FIGURE 6.96 National population figures.

a. Use Jefferson's Method to apportion the 120 legislative seats.
b. Compare the final apportionment with the lower and upper quotas of each state.

SOLUTION

a. After each state's population is expressed in terms of millions, we find the standard divisor:

$$d = \text{standard divisor} = \frac{\text{total population}}{\text{total number of seats}} = \frac{18.0}{120} = 0.15$$

However, the standard divisor $d = 0.15$ results in two surplus seats, as shown in Figure 6.97.

State	A	B	C	D	Total
Population (millions)	1.548	2.776	3.929	9.747	18.000
Standard Quota ($d = 0.15$)	10.32	18.51	26.19	64.98	120
Lower Quota	10	18	26	64	118

FIGURE 6.97 Standard and lower quotas of the states.

Therefore, to apply Jefferson's Method, we must find a modified divisor, $d_m < d = 0.15$, such that no surplus is created. As a first guess, we try $d_m = 0.147$ and calculate each state's modified quota and lower modified quota, that is, we divide each state's population by $d_m = 0.147$, round to two decimal places, and then truncate (round down) to a whole number, as shown in Figure 6.98.

State	A	B	C	D	Total
Population (millions)	1.548	2.776	3.929	9.747	18.000
Standard Quota ($d = 0.15$)	10.32	18.51	26.19	64.98	120
Modified Quota ($d_m = 0.147$)	10.53	18.88	26.73	66.31	122.45
Lower Modified Quota	10	18	26	66	120

FIGURE 6.98 Modified and lower modified quotas using $d_m = 0.147$.

The modified divisor $d_m = 0.147$ has properly apportioned the 120 seats, and the final results are shown in Figure 6.99.

State	A	B	C	D	Total
Jefferson's Apportionment	10	18	26	66	120

FIGURE 6.99 Final apportionment using Jefferson's Method.

b. The standard quota, lower quota, upper quota, and final apportionment for each state are shown in Figure 6.100.

State	A	B	C	D	Total
Standard Quota ($d = 0.15$)	10.32	18.51	26.19	64.98	120
Lower Quota	10	18	26	64	118
Upper Quota	11	19	27	65	122
Jefferson's Apportionment	10	18	26	66	120

FIGURE 6.100 Comparison of apportionment with lower and upper quotas.

Notice that the apportionments for A, B, and C are "fair" in that they equal either the lower quota or the upper quota of the state. However, the apportionment for D (66 seats) is "unfair" because it is greater than D's upper quota of 65.

Example 1 illustrates the fact that Jefferson's Method can violate the Quota Rule. In a similar fashion, it can be shown that Adams's Method, Webster's Method, and the Hill-Huntington Method can all produce apportionments that violate the Quota Rule. In contrast, Hamilton's Method always produces a fair apportionment with respect to the Quota Rule; it always allocates either the lower quota or the upper quota. However, even though it satisfies the Quota Rule, Hamilton's Method gives rise to several other problems. In particular, we shall investigate the Alabama Paradox, the New States Paradox, and the Population Paradox.

The Alabama Paradox

In 1870, the population of the United States was 38,558,371, and the House of Representatives had 292 seats. Ten years later, the 1880 census recorded a population of 50,189,209—a 30% increase. Subsequently, the number of seats in the House of Representatives was to be increased. A discussion on whether to increase to 299 or 300 seats ensued. At the time, there were thirty-eight states in the Union, and Hamilton's Method was the official means of apportionment. The U.S. Census Office calculated the apportionment for each state using both 299 and 300 seats. It was then discovered that with 299 seats, Alabama would receive eight seats, whereas with 300, Alabama's share would be only seven. That is, an increase of one seat overall would result in a reduction in a state's apportionment. Consequently, this scenario has become known as the **Alabama Paradox.**

THE ALABAMA PARADOX

Adding one new seat to the total number of seats being allocated causes one of the states to lose one of its seats (even though the population has not changed).

EXAMPLE **2**

APPLYING HAMILTON'S METHOD AND INVESTIGATING THE ALABAMA PARADOX   A small nation of 12 million people is composed of three states as shown in Figure 6.101. The national legislature has 150 seats.

State	A	B	C	Total
Population (thousands)	595	5,615	5,790	12,000

FIGURE 6.101   National population figures.

a. Use Hamilton's Method to apportion the 150 legislative seats.
b. Suppose the total number of seats increases by one. Reapportion the 151 legislative seats.
c. Compare the apportionments in parts (a) and (b).

SOLUTION

a. First, we must find the standard divisor:

$$d = \text{standard divisor} = \frac{\text{total population}}{\text{total number of seats}} = \frac{12,000}{150} = 80$$

Therefore, $d = 80$.

Now calculate each state's standard quota (divide each state's population by $d = 80$) and lower quota (truncate the standard quotas to whole numbers), as shown in

Figure 6.102. By using Hamilton's Method, state A receives the one additional seat because the decimal part of state A's standard quota is highest (0.44); the final apportionment is shown in the last row of Figure 6.102.

State	A	B	C	Total
Population (thousands)	595	5,615	5,790	12,000
Standard Quota ($d = 80$)	7.44	70.19	72.38	150.01
Lower Quota	7	70	72	149
Additional Seats	1	0	0	1
Number of Seats	8	70	72	150

FIGURE 6.102 Hamilton's apportionment of 150 seats.

b. The new standard divisor is

$$d = \frac{\text{total population}}{\text{total number of seats}} = \frac{12,000}{151} = 79.47019868\ldots$$

Therefore, $d = 79.47$ (rounded to two decimal places).

Now calculate each state's standard quota (divide each state's population by $d = 79.47$) and lower quota (truncate the standard quotas to whole numbers), as shown in Figure 6.103. By using Hamilton's Method, state C and state B each receive an additional seat because the decimal parts of their standard quotas are highest (0.86 and 0.66); the final apportionment is shown in the last row of Figure 6.103.

State	A	B	C	Total
Population (thousands)	595	5,615	5,790	12,000
Standard Quota ($d = 79.47$)	7.49	70.66	72.86	151.01
Lower Quota	7	70	72	149
Additional Seats	0	1	1	2
Number of Seats	7	71	73	151

FIGURE 6.103 Hamilton's apportionment of 151 seats.

c. Figure 6.104 summarizes the two apportionments found in parts (a) and (b). Although the states' populations did not change, state A lost one of its seats when one new seat was added to the legislature; this is an illustration of the Alabama Paradox. Notice that the state that lost a seat is also the smallest state. This is typical of the so-called Alabama Paradox; large states benefit at the expense of small states.

State	Apportionment of 150 seats	Apportionment of 151 seats
A	8	7
B	70	71
C	72	73

FIGURE 6.104 Comparison of apportionments.

The New States Paradox

Oklahoma was admitted to the United States in 1907 as the forty-sixth state. On the basis of its populations, it was determined that the new state should have five seats in the House of Representatives. Therefore, the House was increased from 386 seats to 391 with the intention of not altering the apportionment of the other forty-five states. However, when Hamilton's Method was applied to the new House of 391 seats, it was discovered that the apportionments for two other states would be affected; Maine would gain one seat at the expense of New York losing a seat! Therefore, after the 1910 census, Congress jettisoned Hamilton's Method, and Webster's Method was re-instated. Consequently, this scenario has become known as the **New States Paradox.**

THE NEW STATES PARADOX

Adding a new state, and the corresponding number of seats based on its population, alters the apportionments for some of the other states.

EXAMPLE 3

APPLYING HAMILTON'S METHOD AND INVESTIGATING THE NEW STATES PARADOX A small nation of 10 million people is composed of two states, as shown in Figure 6.105. The national legislature has 100 seats.

State	A	B	Total
Population	3,848,000	6,152,000	10,000,000

FIGURE 6.105 National population figures.

a. Use Hamilton's Method to apportion the 100 legislative seats.
b. Suppose state C has a population of 4,332,000 and joins the union. How many news seats should be added to the legislature?
c. Apportion the new total number of seats.
d. Compare the apportionments in parts (a) and (c).

SOLUTION

a. After expressing each state's population in terms of thousands (an arbitrary decision), we must find the standard divisor:

$$d = \text{standard divisor} = \frac{\text{total population}}{\text{total number of seats}} = \frac{10,000}{100} = 100$$

Therefore, $d = 100$.

Now calculate each state's standard quota (divide each state's population by $d = 100$) and lower quota (truncate the standard quotas to whole numbers), as shown in Figure 6.106. By using Hamilton's Method, state B receives the one additional seat

State	A	B	Total
Population (thousands)	3,848	6,152	10,000
Standard Quota ($d = 100$)	38.48	61.52	100
Lower Quota	38	61	99
Additional Seats	0	1	1
Number of Seats	38	62	100

FIGURE 6.106 Hamilton's apportionment of the original 100 seats.

because the decimal part of state B's standard quota is highest (0.52); the final apportionment is shown in the last row of Figure 6.106.

b. In terms of thousands, state C's population of 4,332,000 would be expressed as 4,332 thousand. Therefore, state C's quota of (new) seats is $\frac{4,332}{100} = 43.32$, or 43 seats.

That is, when state C is admitted to the union, forty-three seats should be added to the legislature; the total number of seats to be allocated is now $100 + 43 = 143$ seats.

c. We must now find the new standard divisor:

$$d = \text{standard divisor} = \frac{\text{total population}}{\text{total number of seats}} = \frac{10,000 + 4,332}{100 + 43}$$

$$= \frac{14,332}{143} = 100.2237762\ldots$$

Therefore, $d = 100.22$.

Now calculate each state's standard quota (divide each population by $d = 100.22$) and lower quota (truncate the standard quotas to whole numbers), as shown in Figure 6.107. By using Hamilton's Method, state A receives the one additional seat because the decimal part of state A's standard quota is highest (0.40); the final apportionment is shown in the last row of Figure 6.107.

State	A	B	C	Total
Population (thousands)	3,848	6,152	4,332	14,332
Standard Quota ($d = 100.22$)	38.40	61.38	43.22	143
Lower Quota	38	61	43	142
Additional Seat	1	0	0	1
Number of Seats	39	61	43	143

FIGURE 6.107 Hamilton's apportionment of the new 143 seats.

d. Figure 6.108 summarizes the apportionments from parts (a) and (c). On comparison, we see that when state C joined the union, state A gained 1 seat and state B lost one seat. That is, the addition of state C altered the apportionments for the other states; this is an illustration of the New States Paradox.

State	A	B	C	Total
(Original) Number of Seats	38	62	0	100
(New) Number of Seats	39	61	43	143

FIGURE 6.108 Comparison of apportionments.

The Population Paradox

Throughout history, the populations of the various states have been in flux; changes in population subsequently affect each state's apportionment of seats in the House of Representatives. Hypothetically, if one state grew at a fast rate while another grew at a slow rate, it might seem more appropriate that the state with the larger growth rate be entitled to an "extra" seat, if any became available. However, Hamilton's Method does not adhere to this premise. During the late 1800s and early 1900s, Virginia was growing at rates much faster than Maine; for example, from 1890 to 1910, Virginia's population increased by 24.5%, while Maine's growth was only 12.3%. To everyone's surprise, Virginia lost one of its seats to Maine in the early 1900s. Consequently, this scenario has become known as the **Population Paradox.**

> ## THE POPULATION PARADOX
> One state loses a seat to another state even though the first state is growing at a faster rate; state A loses a seat to state B even though state A is growing at a faster rate than state B.

EXAMPLE **4**

APPLYING HAMILTON'S METHOD AND INVESTIGATING THE POPULATION PARADOX A school district of slightly more than 6,000 students is composed of three campuses, as shown in Figure 6.109. The district has twelve educational specialists to assign to three locations.

Campus	A	B	C	Total
Enrollment (2009)	777	1,792	3,542	6,111
Enrollment (2010)	820	1,880	3,645	6,345

FIGURE 6.109 School district enrollments.

a. Use Hamilton's Method and the 2009 enrollments to apportion the twelve specialists.
b. Use Hamilton's Method and the 2010 enrollments to reapportion the twelve specialists.
c. Find the differences, if any, in the number of specialists apportioned to each campus in 2009 versus 2010.
d. Which campus had the highest growth rate (percentagewise) from 2009 to 2010? Do the growth rates of the campuses conform with the differences found in part (c)?

SOLUTION

a. First, we must find the standard divisor for 2009:

$$d = \text{standard divisor} = \frac{\text{total enrollment}}{\text{total number of specialists}} = \frac{6,111}{12} = 509.25$$

Therefore, $d = 509.25$.

Now calculate each campus's standard quota (divide each enrollment by $d = 509.25$) and lower quota (truncate the standard quotas to whole numbers), as shown in Figure 6.110. By using Hamilton's Method, campus C and campus A each receive one additional specialist because the decimal parts of their standard quotas are highest (0.96 and 0.53); the 2009 apportionment is shown in the last row of Figure 6.110.

Campus	A	B	C	Total
Enrollment	777	1,792	3,542	6,111
Standard Quota ($d = 509.25$)	1.53	3.52	6.96	12.01
Lower Quota	1	3	6	10
Additional Specialists	1	0	1	2
Number of Specialists (in 2005)	2	3	7	12

FIGURE 6.110 Hamilton's apportionment in 2009.

b. Now we find the standard divisor for 2010:

$$d = \text{standard divisor} = \frac{\text{total enrollment}}{\text{total number of specialists}} = \frac{6{,}345}{12} = 528.75$$

Therefore, $d = 528.75$.

Now calculate each campus's standard quota (divide each enrollment by $d = 528.75$) and lower quota (truncate the standard quotas to whole numbers), as shown in Figure 6.111. By using Hamilton's Method, campus C and campus B each receive one additional specialist because the decimal parts of their standard quotas are highest (0.89 and 0.56); the 2010 apportionment is shown in the last row of Figure 6.111.

Campus	A	B	C	Total
Enrollment	820	1,880	3,645	6,345
Standard Quota ($d = 509.25$)	1.55	3.56	6.89	12
Lower Quota	1	3	6	10
Additional Specialists	0	1	1	2
Number of Specialists (in 2010)	1	4	7	12

FIGURE 6.111   Hamilton's apportionment in 2010.

c. Comparing the number of specialists apportioned in 2009 and 2010 (Figure 6.112), we see that campus A lost one specialist, campus B gained one specialist, and the apportionment of campus C was unchanged.

Campus	A	B	C
Number of Specialists (in 2009)	2	3	7
Number of Specialists (in 2010)	1	4	7
Difference	−1	+1	0

FIGURE 6.112   Comparison of apportionment in 2009 and 2010.

d. To calculate the growth rate of each campus (expressed as a percentage), we must find the difference in the enrollment (2010 minus 2009), divide the difference by the original (2009) enrollment, and then multiply by 100. These calculations are summarized in Figure 6.113.

Campus	A	B	C
2009 Enrollment	777	1,792	3,542
2010 Enrollment	820	1,880	3,645
Difference (increase)	$820 - 777 = 43$	$1{,}880 - 1{,}792 = 88$	$3{,}645 - 3{,}542 = 103$
Proportion	$43/777$ $= 0.05534\ldots$	$88/1{,}792$ $= 0.04910\ldots$	$103/3{,}542$ $= 0.02907\ldots$
Percentage growth	5.5%	4.9%	2.9%

FIGURE 6.113   Growth rates from 2009 and 2010.

Consequently, campus A had the highest growth rate (5.5%) from 2009 to 2010. However, the growth rates do not conform with the differences in apportionment of specialists (Figure 6.114).

Campus	A	B	C
Enrollment growth rate	5.5%	4.9%	2.9%
Change in apportionment	lost 1 specialist	gained 1 specialist	no change

FIGURE 6.114 Comparison of growth rates and changes in apportionment.

Notice that campus A grew at a faster rate than campus B (5.5% > 4.9%), but campus A lost one of its specialists to campus B! This is an illustration of the Population Paradox; one group loses an apportioned position to another group even though the first group is growing at a faster rate.

The Balinski-Young Impossibility Theorem

We have studied several methods of apportionment, specifically the methods of Hamilton, Jefferson, Adams, Webster, and Hill-Huntington, We have also seen various problems associated with these methods of apportionment. Although Hamilton's Method does not violate the Quota Rule, it has several flaws in that it gives rise to the Alabama, New States, and Population Paradoxes. In contrast, it can be shown that the methods of Jefferson, Adams, Webster, and Hill-Huntington are all immune to these paradoxes; however, they may all violate the Quota Rule. In addition, some methods favor large states, whereas other methods favor small states. In other words, if a method of apportionment does not have one type of problem, it has another type of problem. In the quest for the ideal method of apportionment, can we ever hope to find a method that will satisfy the Quota Rule, be immune from paradoxes, and favor neither large nor small states? Unfortunately, the answer is "no." In 1980, mathematicians Michel Balinski and H. Payton Young proved that it is mathematically impossible for any method of apportionment to satisfy the Quota Rule and not produce any paradoxes. This is known as the **Balinski-Young Impossibility Theorem.**

THE BALINSKI-YOUNG IMPOSSIBILITY THEOREM

There is no perfect method of apportionment: If a method satisfies the Quota Rule, it must give rise to paradoxes; and if a method does not give rise to paradoxes, it must violate the Quota Rule.

Just as Arrow's Impossibility Theorem (Section 6.1) ruled out the possibility of the existence of a perfect system of voting, the Balinski-Young Impossibility Theorem rules out the possibility of creating a perfect method of apportionment.

Figure 6.115 lists several methods of apportionment and summarizes the flaws discussed in this section.

Flaw	Method of Apportionment				
	Hamilton	**Jefferson**	**Adams**	**Webster**	**Hill-Huntington**
May violate the Quota Rule.	No	Yes	Yes	Yes	Yes
May produce the Alabama Paradox.	Yes	No	No	No	No
May produce the New States Paradox.	Yes	No	No	No	No
May produce the Population Paradox.	Yes	No	No	No	No

FIGURE 6.115   Flaws of the methods of apportionment.

6.3 EXERCISES

1. A small country consists of three states: A, B, and C. The population of each state is given in Figure 6.116. The country's legislature is to have thirty-two seats.

State	A	B	C
Population	3,500,000	4,200,000	16,800,000

FIGURE 6.116   Table for Exercise 1.

a. Express each state's population (and the total population) in terms of millions.
b. Find the standard divisor, and round it off to two decimal places.
c. Find each state's standard, lower, and upper quotas.
d. Use Jefferson's Method to apportion the thirty-two seats.
e. Is the Quota Rule violated? Explain.

2. A small country consists of three states, A, B, C. The population of each state is given in Figure 6.117. The country's legislature is to have ninety-six seats.

State	A	B	C
Population	1,200,000	3,400,000	15,400,000

FIGURE 6.117   Table for Exercise 2.

a. Express each state's population (and the total population) in terms of millions.
b. Find the standard divisor, and round it off to two decimal places.

c. Find each state's standard, lower, and upper quotas.
d. Use Jefferson's Method to apportion the ninety-six seats.
e. Is the Quota Rule violated? Explain.

3. A small country consists of three states: A, B, and C. The population of each state is given in Figure 6.118. The country's legislature is to have seventy-nine seats.

State	A	B	C
Population	3,500,000	4,200,000	16,800,000

FIGURE 6.118   Table for Exercise 3.

a. Express each state's population (and the total population) in terms of millions.
b. Find the standard divisor, and round it off to two decimal places.
c. Find each state's standard, lower, and upper quotas.
d. Use Adams's Method to apportion the seventy-nine seats.
e. Is the Quota Rule violated? Explain.

4. A small country consists of three states: A, B, and C. The population of each state is given in Figure 6.119. The country's legislature is to have seventy-three seats.

State	A	B	C
Population	1,200,000	3,400,000	15,400,000

FIGURE 6.119   Table for Exercise 4.

a. Express each state's population (and the total population) in terms of millions.

b. Find the standard divisor, and round it off to two decimal places.

c. Find each state's standard, lower, and upper quotas.

d. Use Adams's Method to apportion the seventy-three seats.

e. Is the Quota Rule violated? Explain.

5. A small country consists of four states: A, B, C, and D. The population of each state is given in Figure 6.120. The country's legislature is to have 200 seats.

State	A	B	C	D
Population	1,200,000	3,400,000	17,500,000	19,400,000

FIGURE 6.120 Table for Exercise 5.

a. Express each state's population (and the total population) in terms of millions.

b. Find the standard divisor, and round it off to two decimal places.

c. Find each state's standard, lower, and upper quotas.

d. Use Webster's Method to apportion the 200 seats.

e. Is the Quota Rule violated? Explain.

6. A small country consists of four states: A, B, C, and D. The population of each state is given in Figure 6.121. The country's legislature is to have 177 seats.

State	A	B	C	D
Population	1,100,000	3,200,000	17,300,000	19,400,000

FIGURE 6.121 Table for Exercise 6.

a. Express each state's population (and the total population) in terms of millions.

b. Find the standard divisor, and round it off to two decimal places.

c. Find each state's standard, lower, and upper quotas.

d. Use Webster's Method to apportion the 177 seats.

e. Is the Quota Rule violated? Explain.

7. A small country consists of four states: A, B, C, and D. The population of each state is given in Figure 6.122. The country's legislature is to have 200 seats.

State	A	B	C	D
Population	1,200,000	3,400,000	17,500,000	19,400,000

FIGURE 6.122 Table for Exercise 7.

a. Express each state's population (and the total population) in terms of millions.

b. Find the standard divisor, and round it off to two decimal places.

c. Find each state's standard, lower, and upper quotas.

d. Use the Hill-Hunington Method to apportion the 200 seats.

e. Is the Quota Rule violated? Explain.

8. A small country consists of four states: A, B, C, and D. The population of each state is given in Figure 6.123. The country's legislature is to have 177 seats.

State	A	B	C	D
Population	1,100,000	3,200,000	17,300,000	19,400,000

FIGURE 6.123 Table for Exercise 8.

a. Express each state's population (and the total population) in terms of millions.

b. Find the standard divisor and round it off to two decimal places.

c. Find each state's standard, lower, and upper quotas.

d. Use the Hill-Huntington Method to apportion the 177 seats.

e. Is the Quota Rule violated? Explain.

9. A small country consists of three states: A, B, and C. The population of each state is given in Figure 6.124. The country's legislature is to have 120 seats.

State	A	B	C
Population	690,000	5,700,000	6,410,000

FIGURE 6.124 Table for Exercise 9.

a. Express each state's population (and the total population) in terms of thousands.

b. Find the standard divisor and round it off to two decimal places.

c. Use Hamilton's Method to apportion the 120 legislative seats.

d. Suppose the total number of seats increases by one. Reapportion the 121 legislative seats.

e. Is the Alabama Paradox exhibited? Explain.

10. A small country consists of three states: A, B, C. The population of each state is given in Figure 6.125. The country's legislature is to have 140 seats.

State	A	B	C
Population	700,000	5,790,000	6,510,000

FIGURE 6.125 Table for Exercise 10.

a. Express each state's population (and the total population) in terms of thousands.

b. Find the standard divisor and round it off to two decimal places.

c. Use Hamilton's Method to apportion the 140 legislative seats.

d. Suppose the total number of seats increases by one. Reapportion the 141 legislative seats.

e. Is the Alabama Paradox exhibited? Explain.

11. A small country consists of three states: A, B, and C. The population of each state is given in Figure 6.126. The country's legislature is to have 110 seats.

State	A	B	C
Population	1,056,000	1,844,000	2,100,000

FIGURE 6.126 Table for Exercise 11.

a. Express each state's population (and the total population) in terms of thousands.

b. Find the standard divisor and round it off to two decimal places.

c. Use Hamilton's Method to apportion the 110 legislative seats.

d. Suppose state D has a population of 2,440,000 and joins the union. How many new seats should be added to the legislature?

e. Apportion the new total number of seats.

f. Is the New States Paradox exhibited? Explain

12. A small country consists of three states: A, B, and C. The population of each state is given in Figure 6.127. The country's legislature is to have 150 seats.

State	A	B	C
Population	2,056,000	1,844,000	3,100,000

FIGURE 6.127 Table for Exercise 12.

a. Express each state's population (and the total population) in terms of thousands.

b. Find the standard divisor and round it off to two decimal places.

c. Use Hamilton's Method to apportion the 150 legislative seats.

d. Suppose state D has a population of 2,500,000 and joins the union. How many new seats should be added to the legislature?

e. Apportion the new total number of seats.

f. Is the New States Paradox exhibited? Explain.

13. A school district has three campuses with enrollments given in Figure 6.128. The district has eleven educational specialists to assign to the locations.

Campus	A	B	C
Enrollment (2005)	780	1,700	3,520
Enrollment (2006)	820	1,880	3,740

FIGURE 6.128 Table for Exercise 13.

a. Use Hamilton's Method and the 2005 enrollments to apportion the eleven specialists.

b. Use Hamilton's Method and the 2006 enrollments to reapportion the eleven specialists.

c. Find the growth rate (from 2005 to 2006) for each campus.

d. Is the Population Paradox exhibited? Explain.

14. A school district has three campuses with enrollments given in Figure 6.129. The district has twelve educational specialists to assign to the locations.

Campus	A	B	C
Enrollment (2005)	750	1,680	3,980
Enrollment (2006)	788	1,858	4,210

FIGURE 6.129 Table for Exercise 14.

a. Use Hamilton's Method and the 2005 enrollments to apportion the twelve specialists.

b. Use Hamilton's Method and the 2006 enrollments to reapportion the twelve specialists.

c. Find the growth rate (from 2005 to 2006) for each campus.

d. Is the Population Paradox exhibited? Explain.

 Answer the following questions using complete sentences and your own words.

• CONCEPT QUESTIONS

15. What is the Quota Rule? Explain its meaning.

16. What is the Alabama Paradox? Explain its meaning.

17. What is the New States Paradox? Explain its meaning.

18. What is the Population Paradox? Explain its meaning.

19. What is the Balinski-Young Impossibility Theorem? Explain its meaning.

REVIEW EXERCISES

1. The Metropolitan Symphonic Orchestra has volunteered to play a benefit concert, with proceeds to support music programs at local high schools. The organizers need to know what music will be featured at the concert. The conductor has asked each member of the orchestra to rank the following composers: Beethoven (B), Mozart (M), Tchaikovsky (T), and Vivaldi (V). The ranked ballots have been tallied, and the results are summarized in Figure 6.130.

   a. How many votes were cast?

   b. Use the plurality method of voting to determine the winner.

   c. What percent of the votes did the winner in part (b) receive?

   d. Use the instant runoff method to determine the winner.

   e. What percent of the votes did the winner in part (d) receive?

   f. Use the Borda count method to determine the winner.

   g. How many points did the winner in part (f) receive?

   h. Use the pairwise comparison method to determine the winner.

   i. How many points did the winner in part (h) receive?

2. A small country consists of three states: A, B, and C. The population of each state is given in Figure 6.131. The country's legislature is to have seventy-six seats.

State	A	B	C
Population	1,200,000	2,300,000	2,500,000

   FIGURE 6.131 Table for Exercise 2.

   a. Express each state's population (and the total population) in terms of thousands.

Number of Ballots Cast							
	5	7	6	12	13	17	14
1st choice	M	M	M	B	B	V	T
2nd choice	B	V	T	T	M	M	V
3rd choice	V	B	V	V	T	B	B
4th choice	T	T	B	M	V	T	M

FIGURE 6.130 Voter preference table for Exercise 1.

State	A	B	C
Population	3,500,000	4,100,000	16,800,000

FIGURE 6.134 Table for Exercise 14.

b. Find the standard divisor and round it off to two decimal places.

c. Find each state's standard, lower, and upper quotas.

3. Use Hamilton's Method to apportion the legislative seats in Exercise 2.

4. Use Jefferson's Method to apportion the legislative seats in Exercise 2.

5. Use Adams's Method to apportion the legislative seats in Exercise 2.

6. Use Webster's Method to apportion the legislative seats in Exercise 2.

7. Use the Hill-Huntington Method to apportion the legislative seats in Exercise 2.

8. Suppose that the governments of several South American countries have agreed to form an international bureau to foster tourism, commerce, and education. The bureau will have twenty-five seats, and the populations of the countries are given in Figure 6.132. Use Hamilton's Method to apportion the bureau seats.

Country	Argentina	Bolivia	Chile	Paraguay	Uruguay
Population (thousands)	39,145	8,724	15,824	6,191	3,399

FIGURE 6.132 National population figures, 2004. *Source:* U.S. Department of Commerce.

9. Use Jefferson's Method to apportion the bureau seats in Exercise 8.

10. Use Adams's Method to apportion the bureau seats in Exercise 8.

11. Use Webster's Method to apportion the bureau seats in Exercise 8.

12. Use the Hill-Huntington Method to apportion the bureau seats in Exercise 8.

13. A school district has three high schools: Leibniz, Maclaurin, and Napier. If the district decides to hire one new teacher, use Hill-Huntington numbers to determine which school gets the new instructor. The current enrollments and faculty of the schools are given in Figure 6.133.

	Leibniz	Maclaurin	Napier
Students	987	1,242	1,763
Teachers	25	31	44

FIGURE 6.133 Table for Exercise 13.

14. A small country consists of three states: A, B, and C. The population of each state is given in Figure 6.134. The country's legislature is to have thirty-one seats.

a. Express each state's population (and the total population) in terms of millions.

b. Find the standard divisor, and round it off to two decimal places.

c. Find each state's standard, lower, and upper quotas.

d. Use Jefferson's Method to apportion the thirty-one seats.

e. Is the Quota Rule violated? Explain.

15. A small country consists of three states: A, B, and C. The population of each state is given in Figure 6.135. The country's legislature is to have seventy-two seats.

State	A	B	C
Population	1,600,000	3,500,000	15,300,000

FIGURE 6.135 Table for Exercise 15.

a. Express each state's population (and the total population) in terms of millions.

b. Find the standard divisor, and round it off to two decimal places.

c. Find each state's standard, lower, and upper quotas.

d. Use Adams's Method to apportion the seventy-two seats.

e. Is the Quota Rule violated? Explain.

16. A small country consists of four states: A, B, C, and D. The population of each state is given in Figure 6.136. The country's legislature is to have 201 seats.

State	A	B	C	D
Population	1,350,000	3,430,000	17,600,000	19,650,000

FIGURE 6.136 Figure for Exercise 16.

a. Express each state's population (and the total population) in terms of millions.

b. Find the standard divisor, and round it off to two decimal places.

c. Find each state's standard, lower, and upper quotas.

d. Use Webster's Method to apportion the 201 seats.

e. Is the Quota Rule violated? Explain.

17. A small country consists of four states: A, B, C, and D. The population of each state is given in Figure 6.137. The country's legislature is to have 201 seats.

State	A	B	C	D
Population	1,100,000	3,500,000	17,600,000	19,400,000

FIGURE 6.137 Figure for Exercise 17.

a. Express each state's population (and the total population) in terms of millions.

b. Find the standard divisor, and round it off to two decimal places.

c. Find each state's standard, lower, and upper quotas.

d. Use the Hill-Huntington Method to apportion the 201 seats.

e. Is the Quota Rule violated? Explain.

18. A small country consists of three states: A, B, and C. The population of each state is given in Figure 6.138. The country's legislature is to have 140 seats.

State	A	B	C
Population	700,000	5,790,000	6,520,000

FIGURE 6.138 Figure for Exercise 18.

a. Express each state's population (and the total population) in terms of thousands.

b. Find the standard divisor and round it off to two decimal places.

c. Use Hamilton's Method to apportion the 140 legislative seats.

d. Suppose the total number of seats increases by one. Reapportion the 141 legislative seats.

e. Is the Alabama Paradox exhibited? Explain.

19. A small country consists of three states: A, B, and C. The population of each state is given in Figure 6.139. The country's legislature is to have 110 seats.

State	A	B	C
Population	1,057,000	1,942,000	2,001,000

FIGURE 6.139 Figure for Exercise 19.

a. Express each state's population (and the total population) in terms of thousands.

b. Find the standard divisor and round it off to two decimal places.

c. Use Hamilton's Method to apportion the 110 legislative seats.

d. Suppose state D has a population of 2,450,000 and joins the union. How many new seats should be added to the legislature?

e. Apportion the new total number of seats.

f. Is the New States Paradox exhibited? Explain.

20. A school district has three campuses with enrollments given in Figure 6.140. The district has twelve educational specialists to assign to the locations.

Campus	A	B	C
Enrollment (2005)	760	1,670	3,980
Enrollment (2006)	788	1,858	4,210

FIGURE 6.140 Figure for Exercise 20.

a. Use Hamilton's Method and the 2005 enrollments to apportion the twelve specialists.

b. Use Hamilton's Method and the 2006 enrollments to reapportion the twelve specialists.

c. Find the growth rate (from 2005 to 2006) for each campus.

d. Is the Population Paradox exhibited? Explain.

Answer the following questions using complete sentences and your own words.

• CONCEPT QUESTIONS

21. What is a majority?

22. What is a ranked ballot?

23. What is a voter preference table?

24. What is a standard divisor?

25. What is a standard quota?

26. What is apportionment?

27. What is the Quota Rule? Explain its meaning.

28. What is the Alabama Paradox? Explain its meaning.

29. What is the New States Paradox? Explain its meaning.

30. What is the Population Paradox? Explain its meaning.

31. What is the Balinski-Young Impossibility Theorem? Explain its meaning.

32. What are the four fairness criteria? Explain their meanings.

33. What is Arrow's Impossibility Theorem?

• HISTORY QUESTIONS

34. What methods of apportionment for the House of Representatives have actually been used? What is their chronology?

35. What method of apportionment for the House of Representatives is currently being used?

NUMBER SYSTEMS AND NUMBER THEORY

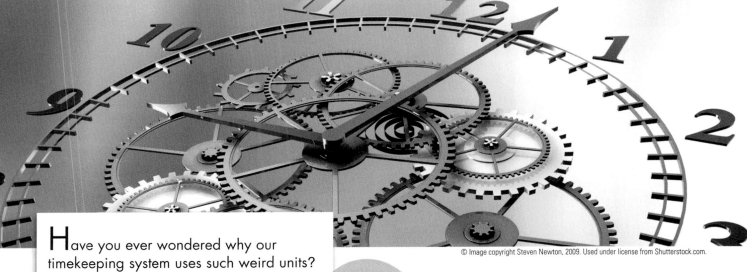

© Image copyright Steven Newton, 2009. Used under license from Shutterstock.com.

Have you ever wondered why our timekeeping system uses such weird units? Why are there *sixty* minutes in an hour and *sixty* seconds in a minute? Why not break hours and minutes down into tenths and hundredths, as we do in so many other cases? It turns out that we use sixty here because the ancient Babylonian number system is a base sixty system. That number system not only gave us our timekeeping system, but also is a forerunner of our own decimal system.

WHAT WE WILL DO IN THIS CHAPTER

WE LIVE IN A WORLD WITH *MANY DIFFERENT NUMBER SYSTEMS:*

- We use the *decimal system* (or *base ten system*) every day. This means that we count by tens.

- We use the *base twelve system* when we count things by twelves. Can you think of any examples of this?

- The French, Danish, Basque, Celtic, and Georgian languages all indicate that they might have once used a modified *base twenty system*—at times, they count by twenties.

- The Mayan and Nahuatl languages of South and Central America use *base twenty*.

- Other cultures use *base four* or *base eight* systems.

- Computers use *base two*, *base eight*, and *base sixteen* systems.

- Our modern navigational system uses the same *base sixty* system that we use for time.

continued

All of these different number systems work in the same way. We use one method to write a number, regardless of the base. The only difference is how many different symbols or digits we use—base ten uses ten digits and base eight uses eight digits. We also use one computational method, regardless of the base. Subtraction works the same in base two as it does in base sixteen or base ten. We will explore these methods in Sections 7.1, 7.2, and 7.3. Exploring them may give you some insight into why the subtraction and multiplication methods that you learned in grade school work the way they do.

NUMBER THEORY

This chapter's topics are quite varied. The second half of the chapter is about number theory. It might seem totally unrelated to the first half, which is about number systems, but there is a connection. Each topic in this chapter focuses on numbers. Of course, numbers are a part of every topic in mathematics. Usually, though, numbers are a tool rather than a principal focus. In this chapter, numbers are the principal focus.

NUMBER THEORY AND INTERNET SECURITY

Identity theft is one of the fastest-growing crimes in the United States. We all disclose personal information over the web: our names, addresses, credit card numbers, bank account numbers and Social Security numbers. If criminals could pry into your web transactions, they could use this information to steal your savings and even your name. Computer programmers have used number theory and *prime numbers* to keep criminals from doing this.

YOU MIGHT BE SURPRISED THAT ARTISTS USE THE *GOLDEN RATIO*, ANOTHER TOPIC FROM NUMBER THEORY:

- Leonardo da Vinci, Mondrian, Seurat, and other artists use the *golden ratio* in their compositions. Why?

- Some architects use the *golden ratio* in their designs.

- The *golden ratio* appears in the Great Pyramid of Egypt. Was this deliberate?

MOTHER NATURE USES *PRIME NUMBERS* AND *FIBONACCI NUMBERS*, ANOTHER TOPIC FROM NUMBER THEORY:

- The number of years in some insects' life cycles is a *prime number*. Why?

- The number of petals on many flowers is a *Fibonacci number*. The flower in the photograph on this page is loaded with *Fibonacci numbers*.

- *Fibonacci numbers* can also be found in pineapples, pinecones, cauliflowers, and sunflowers. They occur frequently in nature, for reasons that we are only beginning to understand.

We'll study number theory, including prime numbers, Fibonacci numbers, and the golden ratio, in Sections 7.4 and 7.5.

Jennifer Weinberg/Alamy.

7.1 Place Systems

OBJECTIVES

- Understand what a place system is
- Be able to convert a number from one base to another
- Understand how different base systems come up in your life

Counting Systems Are Based on Our Body's Configuration

We count by tens. That is, we use a **base ten system** (or **decimal system**). This is apparent if you consider our words for different numbers. "Thirteen" comes from the Old English for "ten more than three." "Thirty" means "three groups of ten." We probably count by tens because it's natural to count on our fingers.

We also count by twelves, when we count by dozens and grosses (a gross is a dozen dozen). We have a twelve-month year and a twelve-hour clock. There are twelve inches to a foot, and there are twelve pence to a shilling. We also count by sixties, when we measure time or angles. An hour and a degree both consist of sixty minutes, and each type of minute consists of sixty seconds. The ancient Sumerians and Babylonians counted by sixties and twelves, and apparently we got this from them.

Some cultures count by twenties, probably because we have twenty fingers and toes. The Mayan and Nahuatl languages of South and Central America use base twenty systems. The French and Danish peoples count by tens, but their languages indicate that at some point they might have counted by twenties. For example, in each language, "eighty" translates as "four twenties." The Basque, Celtic, and Georgian languages use a combination of base ten and base twenty systems as well.

Some cultures count by fours, because they use the spaces between the fingers to count, rather than the fingers themselves. Other cultures count by eights for the same reason. The Maori, Sulawesi, and Papua New Guinean languages use base four systems, and the Yuki language of California and the Pamean language of Mexico use base eight systems. In *Star Wars,* the Hutts have eight fingers and count in base eight.

Computers have switches rather than fingers. A computer contains thousands to trillions of electrical circuits, each of which can be turned off or on. When a circuit is on, it represents a 1, and when it is off, it represents a 0. As a result, computers use the **base two system** (or **binary system**), which uses only two numerals: 0 and 1.

Computers also use the **base eight system** (or **octal system**), which uses eight numerals (0 through 7) and the **base sixteen system** (or **hexidecimal system**), which uses sixteen numerals (0 through 9 followed by A through F). These systems are more compact than binary, and as we shall see, they're easily converted to binary.

The Abacus

The decimal system started with the **abacus,** one of the first counting devices. The earliest abacus was probably a flat surface with sand spread evenly across it. It was

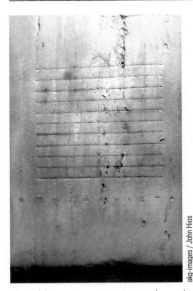

The oldest surviving counting board is made of marble and was used by the Babylonians in about 300 B.C.

used as a counting tool. In fact, the word *abacus* points to this early version—it is derived from either the Phoenician word *abak* ("sand") or the Hebrew word *avak* ("dust"). Merchants used abacuses to count how many items they bought or sold and to calculate the cost of those items. This type of abacus is also called a **counting board.**

Later, counting boards did not use sand. Instead, they had lines carved into them. Users placed stones in the areas between the lines. Every time an abacus user got ten stones in the first area, he would put one stone in the new second area and remove ten stones from the first area. And every time he got ten stones on the second area, he would put one stone in the new third area and remove ten stones from the second area.

How much is one stone in the third area worth? It is the same as ten stones in the second area, and each of those is itself worth ten stones in the first area. So one stone in the third area is worth $10 \cdot 10 = 100$ stones in the first area.

Later abacuses had beads and rods instead of stones and lines. The European abacus consists of several rods, each with ten beads, as shown in Figure 7.1. Some European abacuses had five beads per rod, with one or two extra beads for carrying. Many different cultures used abacuses, including the Chinese, Japanese, Phoenicians, and Egyptians. Each culture had either ten beads on each rod or five beads per rod and some carrying beads. It is thought that the abacus represents a refinement of counting on one's fingers and that each wire has ten beads because we have ten fingers.

Each bead on the rightmost wire counts as 1, each bead on the next wire counts as 10, and each bead on the third wire counts as $10^2 = 100$. The number 234 is indicated by raising four beads on the rightmost wire, three beads on the next wire, and two beads on the third wire. See Figure 7.1.

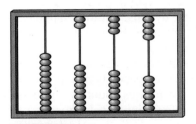

FIGURE 7.1

The number 234 on an abacus.

Hindu numerals (top two rows), Arabic numerals (middle row), and early European numerals (bottom two rows.)

The Hindu Number System

Hindu mathematicians in India developed the number system that we use today. They were the first people to write "234" to represent the number on the abacus in Figure 7.1.

The Hindu number system is a place system. A **place system** is a system in which the value of a digit is determined by its place. For example, 31 and 13 are different numbers because of the digits' places. Roman numerals do not comprise a place system; V means "five" regardless of where the V is placed. It is probable that the place system aspect of the Hindu number system came from the abacus.

The place system made arithmetic calculations much easier. Imagine what multiplication or long division would be like with a nonplace system such as Roman numerals! In fact, the Hindus invented the multiplication algorithm and the division algorithm that we use today. An **algorithm** is a logical procedure for solving a certain type of problem.

Arabic Mathematics

Mathematics and the sciences entered an extremely long period of stagnation when the Greek civilization fell and was replaced by the Roman Empire. This inactivity was uninterrupted until after the Islamic religion and the resulting Islamic culture were founded by the prophet Muhammad in A.D. 622. Within a century, the Islamic Empire stretched from Spain, Sicily, and Northern Africa to India.

Unlike the Roman Empire, Islamic culture encouraged the development of mathematics and the sciences as well as the arts. Arab scholars translated many Greek and Hindu works in mathematics and the sciences. In doing so, they came upon Hindu numerals.

The Arab mathematician Mohammed ibn Musa al-Khowarizmi wrote two important books around A.D. 830, which were translated into Latin and published in Europe in the twelfth century. Much of the mathematical knowledge of medieval Europe was derived from the Latin translations of al-Khowarizmi's two works.

Al-Khowarizmi's first book, on Hindu numerals, was titled *Algorithmi de numero Indorum* (or *al-Khowarizmi on Indian Numbers*). The Latin translation of this book introduced to Europe the

A calculation competition. One competitor uses an abacus, while the other uses Hindu numerals.

Hindu number system. Before this book came out, Europeans used Roman numerals, so a European wrote CCXXXIV instead of 234.

Al-Khowarizmi's book also introduced Europe to the simpler calculation techniques of the Hindu system, including the multiplication algorithm and the division algorithm that we use today. In fact, the title *Algorithmi de numero Indorum* is the origin of the word *algorithm*.

Al-Khowarizmi's second book, *Al-Jabr w'al Muqabalah,* is on algebra rather than Hindu numbers, multiplication, and division. In fact, this book marks the beginning of algebra, and the word *algebra* comes from *Al-Jabr* in its title. The book discusses linear and quadratic equations. Its Latin translation introduced Europe to algebra.

Place Values and Expanded Form

The Hindu number system is a *place* system. When we write the number 234, we put 2 in the hundreds *place,* 3 in the tens *place,* and 4 in the ones *place.* So 234 means 2 hundreds, 3 tens, and 4 ones. In symbols, it means

$$234 = 2 \cdot 100 + 3 \cdot 10 + 4 \cdot 1$$

The **place values** are $100 = 10^2$, $10 = 10^1$, and $1 = 10^0$. Using these powers of ten, we get

$$234 = 2 \cdot 10^2 + 3 \cdot 10^1 + 4 \cdot 10^0$$

When the number two hundred thirty-four is written as $2 \cdot 10^2 + 3 \cdot 10^1 + 4 \cdot 10^0$, it is said to be in **expanded form.** When it is written as 234, it is said to be in **standard form.**

The Decimal System

Recall that in the expression 10^2, 2 is called the exponent and 10 is called the base. The Hindu number system is called a **base ten** system because the system uses powers of 10; that is, it involves exponents with a base of 10.

Our base ten system uses the digits 1, 2, 3, 4, 5, 6, 7, 8, 9, and 0 and nothing else. We create bigger numbers by putting these digits in different places. Notice that there are ten digits in the above list. A base ten system uses ten digits.

The base ten system is also called the **decimal system.** The word *decimal* is derived from the Latin word *decim,* which means "ten."

BASE TEN FACTS

- It uses ten digits: 1, 2, 3, 4, 5, 6, 7, 8, 9, and 0.
- The place values are powers of ten: $10^0 = 1$, $10^1 = 10$, and $10^2 = 100$.
- It is called the decimal system.

 **Base Eight**

The base eight system is also called the **octal system.** The word *octal* is derived from the Greek word for "eight."

Base eight differs from base ten in several ways. It uses eight digits: 1, 2, 3, 4, 5, 6, 7, and 0. It does not use the digits 8 and 9. Also, the place values are not $10^0 = 1$, $10^1 = 10$, and $10^2 = 100$. Instead, they are $8^0 = 1$, $8^1 = 8$, and $8^2 = 64$. So the columns are ones, eights, and sixty-fours.

When we write 10_8, we mean something totally different from when we write 10. The subscript 8 indicates that the number is in base eight. (When a number is in base ten, no subscript is used.) There is a 1 in the eights column and a 0 in the ones column, so 10_8 means 1 eight and 0 ones—or, using expanded form,

$$10_8 = 1 \cdot 8^1 + 0 \cdot 8^0 = 8$$

We cannot write eight in base eight as "8" because that is not one of the digits that base eight uses.

Similarly, 11_8 has a 1 in the eights column and a 1 in the ones column. So 11_8 means 1 eight and 1 one.

$$11_8 = 1 \cdot 8^1 + 1 \cdot 8^0 = 9$$

We cannot write nine in base eight as "9" because that is not one of the digits that base eight uses.

BASE EIGHT FACTS

- It uses eight digits: 1, 2, 3, 4, 5, 6, 7, and 0.
- The place values are powers of eight: $8^0 = 1$, $8^1 = 8$, $8^2 = 64$, etc.
- It is called the octal system.

Reading Numbers in Different Bases

We do not read the number 234_8 as "two hundred thirty four, base 8" because "two hundred" means "$2 \cdot 10^2$" and "thirty" means "$3 \cdot 10^1$." Instead, we read it as "two three four, base eight."

Converting from Base Eight to Base Ten

EXAMPLE **1**

CONVERTING FROM BASE EIGHT TO BASE TEN

a. Rewrite 234_8 in expanded form.
b. Use part (a) to convert 234_8 to base ten.

SOLUTION

a.

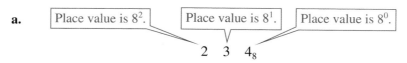

| Place value is 8^2. | Place value is 8^1. | Place value is 8^0. |

$$2 \quad 3 \quad 4_8$$

Expanded form for 234_8 is $2 \cdot 8^2 + 3 \cdot 8^1 + 4 \cdot 8^0$.

b. The computation described by the above expanded form gives 156, so $234_8 = 156$.

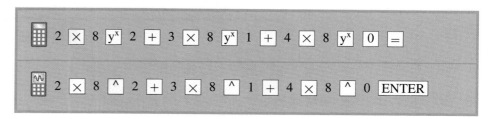

Converting from Base Ten to Base Eight

EXAMPLE **2**

CONVERTING FROM BASE TEN TO BASE EIGHT   Convert 234 to base eight.

SOLUTION

Some of the place values are $8^0 = 1$, $8^1 = 8$, and $8^2 = 64$. To convert 234 to base eight is to find how many sixty-fours, eights, and ones there are in 234. We will start by finding how many sixty-fours there are in 234.

$$\begin{array}{r} 3 \\ 64{\overline{\smash{\big)}\,234}} \\ -192 \\ \hline 42 \end{array}$$

Rewriting this result with multiplication, we get

$$234 = 3 \cdot 64 + 42$$

This tells us that there are three sixty-fours in 234.
    Next, we need to find out how many eights there are in 42.

$$\begin{array}{r} 5 \\ 8{\overline{\smash{\big)}\,42}} \\ -40 \\ \hline 2 \end{array}$$

Rewriting this result with multiplication, we get

$$42 = 5 \cdot 8 + 2$$

Substituting this into our previous result, we get

$$234 = 3 \cdot 64 + 42 \qquad \textbf{our previous result}$$
$$234 = 3 \cdot 64 + 5 \cdot 8 + 2 \quad \textbf{substituting for 42}$$
$$= 352_8$$

CONVERTING TO AND FROM BASE TEN

To convert a given number (in base b) to base ten:

Use base b's place values to rewrite the given number in expanded form. Then do the computation described by the expanded form.

To convert a given number from base ten to base b:

1. Find the highest power of the base b that is less than the given number.
2. Divide the given number by base b place values, starting with the number from step 1.
3. Use the result of step 2 to rewrite the given number in expanded form, using powers of base b.
4. Use the result of step 3 to rewrite the given number in standard form, using base b place values and base b digits.

Base Sixteen

Some computers use base sixteen internally, even though they display everything in base ten. The base sixteen system is also called the **hexadecimal system.** The current hexadecimal system was created by IBM in 1963. IBM decided that the term *sexidecimal,* from the Latin words for "six" and "ten," was too risqué. To avoid it, they welded together *hex,* the Greek word for "six," and *decim,* the Latin word for "ten."

Base sixteen uses sixteen numerals, which is more than we base ten users have at our disposal. So we use some letters as numerals. The sixteen numerals are shown in Figure 7.2.

Base Sixteen Numeral	1	2	3	4	5	6	7	8	9	A	B	C	D	E	F	0
Base Ten Equivalent	1	2	3	4	5	6	7	8	9	10	11	12	13	14	15	0

FIGURE 7.2 The base sixteen numerals.

BASE SIXTEEN FACTS

- It uses sixteen numerals: 1, 2, 3, 4, 5, 6, 7, 8, 9, A, B, C, D, E, F, and 0.
- The place values are powers of sixteen: $16^0 = 1$, $16^1 = 16$, $16^2 = 256$, etc.
- It is called the hexadecimal system.

Converting Between Base Sixteen and Base Ten

EXAMPLE **3**

CONVERTING FROM BASE SIXTEEN TO BASE TEN Convert 234_{16} to base ten.

SOLUTION

We will use base sixteen's place values to rewrite 234 in expanded form, as described in the box entitled "Converting to and from Base Ten" above.

| Place value is 16^2. | Place value is 16^1. | Place value is 16^0. |

$$2 \quad 3 \quad 4_{16}$$

$$234_{16} = 2 \cdot 16^2 + 3 \cdot 16^1 + 4 \cdot 16^0 = 564$$

We have found that the number 234_{16} has 2 two hundred fifty-sixes ($16^2 = 256$), 3 sixteens, and 4 ones. This amounts to the same as the base ten number 564, which has 5 hundreds, 6 tens, and 4 ones.

EXAMPLE **4**

CONVERTING FROM BASE TEN TO BASE SIXTEEN   Convert 234 to base sixteen.

SOLUTION

We will use the steps given in the box entitled "Converting to and from Base Ten" above.

Step 1   *Find the highest power of base sixteen that is less than the given number.*

$$16^2 = 256 \quad \text{and} \quad 256 \geqslant 234 \quad \text{\textbf{not the desired number}}$$
$$16^1 = 16 \quad \text{and} \quad 16 < 234 \quad \text{\textbf{the desired number}}$$

Step 2   *Divide the given number by base sixteen place values, starting with the number from step 1.*

$$\begin{array}{r} 14 \\ 16\overline{)234} \\ -16 \\ \hline 74 \\ -64 \\ \hline 10 \end{array}$$

$234 = 14 \cdot 16 + 10$

Step 3   *Use the result of step 2 to rewrite the given number in expanded form, using powers of base sixteen.*

$$234 = 14 \cdot 16 + 10 \qquad \text{\textbf{the result of step 2}}$$
$$= 14 \cdot 16^1 + 10 \cdot 16^0 \quad \text{\textbf{expanded form}}$$

Step 4   *Use the result of step 3 to rewrite the given number in standard form, using base sixteen place values and base sixteen digits.* The base sixteen digit for "14" is "E," and the base sixteen digit for "10" is "A," as shown in Figure 7.2.

$$234 = 14 \cdot 16^1 + 10 \cdot 16^0 \quad \text{\textbf{using base ten digits}}$$
$$= E \cdot 16^1 + A \cdot 16^0 \quad \text{\textbf{using base sixteen digits}}$$
$$= EA_{16} \qquad\qquad\quad \text{\textbf{standard form}}$$

The number 234 has E sixteens and A ones. Putting the E in the sixteens place and the A in the ones place, we get $234 = EA_{16}$.

The result of Example 4 is rather strange. Remember that in base sixteen, E and A are digits. They are just digits that are not used in base ten.

Base Two

All computers use *machine language,* which is a system of codes that are directly understandable by a computer's central processor. Machine language is composed only of the numerals 0 and 1. So at their deepest level, computers use base two. For certain applications, computers use base eight or base sixteen rather than base two. This is because programmers prefer to program in base eight or sixteen, because these systems are more compact. It takes a lot of space to write a big number in base two. It takes far less space to write the same number in base eight or base sixteen. Base ten is also more compact than base two. However, it is much easier to convert between base two and base eight or sixteen than it is to convert between base two and base ten.

The base two system is also called the **binary system.** The word *bi* is a Latin word that means "two."

BASE TWO FACTS

- It uses two digits: 1 and 0.
- The place values are powers of two: $2^0 = 1$, $2^1 = 2$, $2^2 = 4$, $2^3 = 8$, $2^4 = 16$, $2^5 = 32$, etc.
- It is called the binary system.

Converting Between Base Two and Base Ten

EXAMPLE 5

CONVERTING FROM BASE TWO TO BASE TEN Convert 234_2 to base ten.

SOLUTION

This cannot be done, because there is no such thing as 234_2. Base two uses only two numerals: 0 and 1. It does not use 2, 3, or 4.

EXAMPLE 6

CONVERTING FROM BASE TWO TO BASE TEN Convert 10010_2 to base ten.

SOLUTION

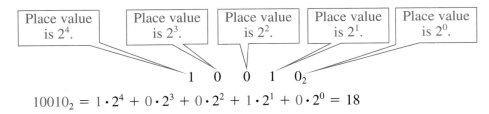

$$10010_2 = 1 \cdot 2^4 + 0 \cdot 2^3 + 0 \cdot 2^2 + 1 \cdot 2^1 + 0 \cdot 2^0 = 18$$

Example 6 shows that $18 = 10010_2$. It takes a lot of space to write 18 in base two. This bulkiness is one reason computer programmers prefer working in a base other than base two.

EXAMPLE 7

CONVERTING FROM BASE TEN TO BASE TWO Convert 25 to base two.

SOLUTION

Step 1 *Find the highest power of base two that is less than the given number.*

$2^5 = 32$ and $32 \geqslant 25$ **not the desired number**

$2^4 = 16$ and $16 < 25$ **the desired number**

Step 2

Divide the given number by base two place values, starting with the number from step 1.

$$16\overline{)25} \quad\quad 25 = 1 \cdot 16 + 9$$
$$\underline{-16}$$
$$9$$

$$8\overline{)9} \quad\quad 9 = 1 \cdot 8 + 1$$
$$\underline{-8}$$
$$1$$

Step 3

Use the result of step 2 to rewrite the given number in expanded form, using powers of base two.

$25 = 1 \cdot 16 + 9$	the first result
$= 1 \cdot 16 + 1 \cdot 8 + 1$	using the second result
$= 1 \cdot 16 + 1 \cdot 8 + 0 \cdot 4 + 0 \cdot 2 + 1$	inserting the missing place values
$= 1 \cdot 2^4 + 1 \cdot 2^3 + 0 \cdot 2^2 + 0 \cdot 2^1 + 1 \cdot 2^0$	expanded form

Step 4

Use the result of step 3 to rewrite the given number in standard form, using base two place values and base two digits.

$1 \cdot 2^4 + 1 \cdot 10^3 + 0 \cdot 2^2 + 0 \cdot 2^1 + 1 \cdot 2^0$	the result of step 3
$= 11001_2$	standard form

Base Sixty

The ancient Sumerians and Babylonians used base sixty, and no one really knows why. They used base sixty when they originated the system by which the day is divided into 24 hours, an hour is divided into 60 minutes, and a minute is divided into 60 seconds. They also originated the system by which a degree is divided into 60 minutes and a minute into 60 seconds.

This similarity of timekeeping and angle-measuring systems apparently started with Babylonian and Sumerian astronomers. Because of their perspective from the earth, they observed that the sun and planets seemed to move in relation to the fixed background of stars and constellations in a circular path, now called the *ecliptic*. (Of course, we now know that the earth moves, rather than the sun.) They

A small part of the ecliptic is easily seen in this photo from the 1994 Clementine spacecraft. We see, from right to left, the moon, the sun rising behind the moon, Saturn, Mars, and Mercury. Connect these dots, and you have drawn part of the ecliptic.

Courtesy of NASA

determined that if we could see the stars during the daytime, the sun would appear to pass through a different constellation about once every thirty days. These twelve constellations are called the *zodiac*. It takes the sun about $12 \cdot 30 = 360$ days to complete its path through the zodiac. They divided this circular path into 360 degrees to track each day's portion of the sun's journey. As a result, we now divide a circle into 360 degrees, and we now define a degree as one-three hundred sixtieth of a circle.

When we talk about 2 hours, 18 minutes, and 43 seconds, we are using the Babylonian base sixty system to measure time. In discussing base sixty, we will not use sixty different numerals. If we used all ten of our base ten numerals and all twenty-six letters in the alphabet, we would not have enough symbols. Instead, we will just use our base ten numerals and separate the places with commas. We will write 2 hours, 18 minutes, and 43 seconds in base sixty as

$$2,18,43_{60}$$

EXAMPLE 8

CONVERTING FROM BASE SIXTY TO BASE TEN   Convert $2,18,43_{60}$ to base ten.

SOLUTION

$$2,18,43_{60} = 2 \cdot 60^2 + 18 \cdot 60^1 + 43 \cdot 60^0 = 8323$$

This means that $2,18,43_{60} = 8323$. It also means that 2 hours, 18 minutes, and 43 seconds $= 8323$ seconds.

Converting between Computer Bases

Converting between computer bases (that is, base two, base eight, and base sixteen) is much easier than converting to or from base ten, because you don't have to work with the entire number all at once. Instead, you can work with individual digits separately.

CONVERTING BETWEEN BASE EIGHT AND BASE TWO

From base eight to base two:
1. Convert each base eight digit to base two.
2. Replace each base eight digit with a *three-digit version* of its base two equivalent.

From base two to base eight:
1. Break the base two number up into sets of *three* digits, adding zeros on the left as necessary.
2. Convert each of the base two numbers from step 1 to base eight.
3. Replace each base two three-digit number with its base eight equivalent.

EXAMPLE 9

CONVERTING FROM BASE EIGHT TO BASE TWO

a. Convert 43_8 to base two.
b. Check your work by converting both the problem and the answer to base ten.

SOLUTION

a. Converting to base two:

Step 1

Convert each base eight digit to base two.

43_8 has two digits: 4_8 and 3_8.

$$4_8 = 4 \qquad\qquad\qquad\quad \text{in base ten}$$
$$\quad = 1 \cdot 2^2 + 0 \cdot 2^1 + 0 \cdot 2^0 \quad \text{expanded form}$$
$$\quad = 100_2 \qquad\qquad\qquad \text{in base two}$$
$$3_8 = 3 \qquad\qquad\qquad\quad \text{in base ten}$$
$$\quad = 1 \cdot 2^1 + 1 \cdot 2^0 \qquad\quad \text{expanded form}$$
$$\quad = 11_2 \qquad\qquad\qquad\quad \text{in base two}$$
$$\quad = 011_2 \qquad\qquad\qquad \text{with three digits}$$

Step 2

Replace each base eight digit with a three-digit version of its base two equivalent.

In Base Eight:	4_8	3_8
In Base Two:	100_2	011_2

$43_8 = 100011_2$ **We've converted to base two.**

b. Checking:

$$43_8 = 4 \cdot 8^1 + 3 \cdot 8^0 = 35$$
$$100011_2 = 1 \cdot 2^5 + 0 \cdot 2^4 + 0 \cdot 2^3 + 0 \cdot 2^2 + 1 \cdot 2^1 + 1 \cdot 2^0 = 35 \quad \checkmark$$

Our answer checks because each number converts to 35 in base ten.

To convert from base two to base eight, reverse the above process.

EXAMPLE **10**

CONVERTING FROM BASE TWO TO BASE EIGHT

a. Convert 10101_2 to base eight.
b. Check your work by converting both the problem and the answer to base ten.

SOLUTION

a. Converting:

Step 1

Break the base two number up into sets of three digits, adding zeros on the left as necessary.

$$10101_2 = 010101_2 \quad \text{adding a zero on the left}$$

Step 2

Convert each of the base two numbers from step 1 to base eight.

In Base Two:	010_2	101_2
In Base Eight:	?	?

$$010_2 = 0 \cdot 2^2 + 1 \cdot 2^1 + 0 \cdot 2^0 = 2 = 2_8 \quad \text{in base eight}$$
$$101_2 = 1 \cdot 2^2 + 0 \cdot 2^1 + 1 \cdot 2^0 = 5 = 5_8 \quad \text{in base eight}$$

Step 3

Replace each base two three-digit number with its base eight equivalent.

In Base Two:	010_2	101_2
In Base Eight:	2_8	5_8

$010101_2 = 25_8$ **We've converted to base eight.**

b. Checking:

$$25_8 = 2 \cdot 8^1 + 5 \cdot 8^0 = 21$$
$$10101_2 = 1 \cdot 2^4 + 0 \cdot 2^3 + 1 \cdot 2^2 + 0 \cdot 2^1 + 1 \cdot 2^0 = 21 \quad \checkmark$$

CONVERTING BETWEEN BASE SIXTEEN AND BASE TWO

From base sixteen to base two:

1. Convert each base sixteen digit to base two.
2. Replace each base sixteen digit with a *four-digit version* of its base two equivalent.

From base two to base sixteen:

1. Break the base two number up into sets of *four* digits, adding zeros on the left as necessary.
2. Convert each of the base two numbers from step 1 to base sixteen.
3. Replace each base two three-digit number with its base sixteen equivalent.

EXAMPLE **11**

CONVERTING FROM BASE SIXTEEN TO BASE TWO

a. Convert $B3_{16}$ to base two.
b. Check your work by converting both the problem and the answer to base ten.

SOLUTION

a. Converting:

Step 1

Convert each base sixteen digit to base two.

$$
\begin{aligned}
B_{16} &= 11 && \text{in base ten} \\
&= 1 \cdot 2^3 + 0 \cdot 2^2 + 1 \cdot 2^1 + 1 \cdot 2^0 && \text{expanded form} \\
&= 1011_2 && \text{in base two} \\
3_{16} &= 3 && \text{in base ten} \\
&= 1 \cdot 2^1 + 1 \cdot 2^0 && \text{expanded form} \\
&= 11_2 && \text{in base two} \\
&= 0011_2 && \textit{with four digits}
\end{aligned}
$$

Step 2

Replace each base sixteen digit with a four-digit version of its base two equivalent.

In Base Sixteen:	B_{16}	3_{16}
In Base Two:	1011_2	0011_2

$$B3_{16} = 10110011_2 \quad \textbf{We've converted to base two.}$$

b. Checking:

$$10110011_2 = 1 \cdot 2^7 + 0 \cdot 2^6 + 1 \cdot 2^5 + 1 \cdot 2^4 + 0 \cdot 2^3 + 0 \cdot 2^2 + 1 \cdot 2^1$$
$$+ 1 \cdot 2^0 = 179$$

$$B3_{16} = 11 \cdot 16^1 + 3 \cdot 16^0 = 179 \quad \checkmark$$

To convert from base two to base sixteen, reverse the above process.

EXAMPLE **12**

CONVERTING FROM BASE TWO TO BASE SIXTEEN

a. Convert 110101_2 to base sixteen.
b. Check your work by converting both the problem and the answer to base ten.

TOPIC X COMPUTER IMAGING: THE BINARY SYSTEM IN THE REAL WORLD

The word *bit* is formed by combining the words *binary* and *digit*. A **bit** is a binary digit. That is, it is either of the digits 0 and 1.

A **bitmapped image** is a computer-generated image that is made up of **pixels,** which are tiny square dots. These pixels are lined up next to each other in a grid, forming a complete image. A computer forms a bitmapped image by associating a number, made up of bits, with each pixel. The number associated with one particular pixel determines the pixel's color.

The simplest of all bitmapped images is a **one-bit image,** in which each pixel is associated with one bit. If a pixel's bit is 0, then the pixel is colored white. If the pixel's bit is 1, then the pixel is black. A one-bit image is made up of blacks and whites, with no grays.

With a **grayscale image,** each pixel is typically associated with eight bits (also called one **byte**). That is, each pixel is associated with a number that has eight places when written in binary. The smallest eight-bit binary number is

$$00000000_2 = 0$$

A pixel with this number is colored white. The largest eight-bit binary number is

$$
\begin{aligned}
11111111_2 &= 1 \cdot 2^7 + 1 \cdot 2^6 \\
&\quad + 1 \cdot 2^5 + 1 \cdot 2^4 \\
&\quad + 1 \cdot 2^3 + 1 \cdot 2^2 \\
&\quad + 1 \cdot 2^1 + 1 \cdot 2^0 \\
&= 255
\end{aligned}
$$

A pixel with this number is colored black. A pixel with a number between 00000000_2 and 11111111_2 is colored a shade of gray. There are 256 numbers between 0 and 255, so each pixel in a grayscale image is colored one of 256 different shades of gray.

With an **RGB image,** each pixel is typically associated with three different one-byte numbers. One of these three numbers determines the amount of red (R) in the pixel, one determines the amount of green (G), and one determines the amount of blue (B). Each number has one byte (that is, eight bits), so an RGB image can contain 256 different shades of red, 256 different shades of green, and 256 different shades of blue. Computer monitors and TVs display RGB images. RGB color is also called *three-channel color.*

With a **CMYK image,** each pixel is typically associated with four different one-byte numbers. The four numbers determine the amount of cyan (C), magenta (M), yellow (Y), and black (K) in the pixel. Each number has one byte, so a CMYK image can produce 256 shades of the colors cyan, magenta, yellow, and black. Offset printing and color photography use the CMYK system. CMYK color is called *four-channel color,* and CMYK printing is called *four-color printing.*

Digital photographers and Adobe Photoshop users need to understand *bit*

Cyan, magenta, yellow, and black—the CMYK colors.

depth. **Bit depth** is the number of bits used to store color information about each pixel. The bit depth of a one-bit image is one. The bit depth of a grayscale image is 8, the bit depth of an RGB image is $3 \cdot 8 = 24$, and the bit depth of a CMYK image is $4 \cdot 8 = 32$.

See Exercise 95.

CMYK

Grayscale

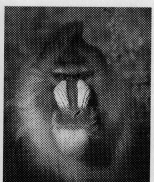

One-bit image

SOLUTION

a. Converting:

Step 1

Break the base two number up into sets of four digits, adding zeros on the left as necessary.

$$110101_2 = 00110101_2 \quad \textbf{adding two zeros on the left}$$

Step 2

Convert each of the base two numbers from step 1 to base sixteen.

In Base Two:	0011_2	0101_2
In Base Sixteen:	?	?

$$0011_2 = 1 \cdot 2^1 + 1 \cdot 2^0 = 3 = 3_{16} \qquad \textbf{in base sixteen}$$
$$0101_2 = 1 \cdot 2^2 + 0 \cdot 2^1 + 1 \cdot 2^0 = 5 = 5_{16} \quad \textbf{in base sixteen}$$

Step 3

Replace each base two three-digit number with its base sixteen equivalent.

In Base Two:	0011_2	0101_2
In Base Sixteen:	3_{16}	5_{16}

$$00110101_2 = 35_{16} \quad \textbf{We've converted to base sixteen.}$$

b. Checking:

$$35_{16} = 3 \cdot 16^1 + 5 \cdot 16^0 = 53$$
$$110101_2 = 1 \cdot 2^5 + 1 \cdot 2^4 + 0 \cdot 2^3 + 1 \cdot 2^2 + 0 \cdot 2^1 + 1 \cdot 2^0 = 53 \quad ✓$$

CONVERTING BETWEEN BASE EIGHT AND BASE SIXTEEN

From base sixteen to base eight:
1. Convert the base sixteen number to base two.
2. Convert the base two number from step one to base eight.

From base eight to base sixteen:
1. Convert the base eight number to base two.
2. Convert the base two number from step one to base sixteen.

7.1 EXERCISES

In Exercises 1–20, write the given number in expanded form, or explain why there is no such number.

1. 891
2. 458
3. 3,258
4. 86,420
5. 372_8
6. 543_8
7. 3592_{16}
8. 8403_{16}
9. $ABCDE0_{16}$
10. $FA3B02_{16}$
11. 1011001_2
12. 11001110_2
13. 1324_2
14. 2103_2
15. $5,32,85_{60}$
16. $27,19,01_{60}$
17. 4312_5
18. 643_7
19. 123_4
20. 8034_9

In Exercises 21–48, convert the given number to base ten.

21. 372_8 (See Exercise 5.)

22. 543_8 (See Exercise 6.)

23. 3592_{16} (See Exercise 7.)

24. 8403_{16} (See Exercise 8.)

25. $ABCDE0_{16}$ (See Exercise 9.)

26. $FA3B02_{16}$ (See Exercise 10.)

27. 1011001_2 (See Exercise 11.)

28. 11001110_2 (See Exercise 12.)

29. $5,32,85_{60}$ (See Exercise 15.)

30. $27,19,01_{60}$ (See Exercise 16.)

31. 4312_5 (See Exercise 17.)

32. 643_7 (See Exercise 18.)

33. 123_4 (See Exercise 19.)

34. 8034_9 (See Exercise 20.)

35. 1705_8

36. 2264_8

37. 11011011_2

38. 1011001_2

39. $55,28,33,59_{60}$

40. $43,56,19,20_{60}$

41. 5798_{16}

42. $AB839_{16}$

43. 7224_8

44. 11011101_2

45. $23,44,14_{50}$

46. $340GD1_{17}$

47. 253_7

48. 2412_5

49. Convert 452_8 to base two.

50. Convert 1502_8 to base two.

51. Convert 5260_8 to base two.

52. Convert 31570_8 to base two.

53. Convert 1010_2 to base eight.

54. Convert 11011_2 to base eight.

55. Convert 10110110_2 to base eight.

56. Convert 101001_2 to base eight.

57. Convert $53A2_{16}$ to base two.

58. Convert $FB0_{16}$ to base two.

59. Convert $3AB0_{16}$ to base two.

60. Convert $30AB_{16}$ to base two.

61. Convert 10100_2 to base sixteen.

62. Convert 110011_2 to base sixteen.

63. Convert 10111010_2 to base sixteen.

64. Convert 10100_2 to base sixteen.

65. **a.** Convert $2BA_{16}$ to base two.

   b. Convert the result of part (a) to base eight.

66. **a.** Convert 743_8 to base two.

   b. Convert the result of part (a) to base sixteen.

67. Convert 42_8 to base sixteen.

68. Convert 352_8 to base sixteen.

69. Convert 540_8 to base sixteen.

70. Convert 3670_8 to base sixteen.

71. Convert $AB8_{16}$ to base eight.

72. Convert $10C_{16}$ to base eight.

73. Convert $801D_{16}$ to base eight.

74. Convert $986B_{16}$ to base eight.

75. Convert $22,12_{60}$ to base ten.

76. Convert $57,02_{60}$ to base ten.

77. Convert $33,14,25_{60}$ to base ten.

78. Convert $52,58,12_{60}$ to base ten.

79. Convert 364 to base sixty.

80. Convert 9481 to base sixty.

81. Convert 129,845 to base sixty.

82. Convert 5,383,221 to base sixty.

In Exercises 83–86, you are to convert between bases other than two, eight, sixteen, and ten. To do this, first convert the given number to base ten. Then convert the result to the specified base.

83. Convert 534_9 to base seven.

84. Convert 512_6 to base four.

85. Convert 331_4 to base five.

86. Convert 58_9 to base eleven.

In Exercises 87–90, use a trial-and-error process along with what you know about the digits used in a given base.

87. **a.** If $121_x = 16$, find base x.

   b. If $121_x = 36$, find base x.

88. **a.** If $243_x = 129$, find base x.

   b. If $243_x = 73$, find base x.

89. **a.** If $370_x = 248$, find base x.

   b. If $370_x = 306$, find base x.

90. **a.** If $504_x = 409$, find base x.

   b. If $504_x = 184$, find base x.

91. **a.** What digits does base five use?

   b. What are the first four place values in base five?

   c. Write the first twenty-five whole numbers in base five.

92. **a.** What digits does base four use?

   b. What are the first four place values in base four?

   c. Write the first twenty-five whole numbers in base four.

93. a. What digits does base eleven use?

 b. What are the first four place values in base eleven?

 c. Write the first twenty-five whole numbers in base eleven.

94. a. What digits does base twelve use?

 b. What are the first four place values in base twelve?

 c. Write the first twenty-five whole numbers in base twelve.

95. With 8-bit color, each of the basic colors (red, green, and blue in the RGB system; cyan, magenta, yellow, and black in the CMYK system) has 256 different shades. How many different shades are there with 16-bit color? How many different shades are there with 24-bit color? Justify your answers.

 Answer the following using complete sentences and your own words.

• CONCEPT QUESTIONS

96. Why is there no base one?

97. Throughout history, most cultures have used either base ten or base twenty. Why are those bases the most popular ones?

98. Why did the digit 0 have to be invented before a written place system could be invented?

99. Describe how the number 531 would appear on a base seven abacus.

100. The words *calculate* and *calculator* are derived from the word *calculus*, which means "pebble" or "stone." Why?

101. What do place systems have in common with the abacus?

• HISTORY QUESTIONS

102. What did the Hindus do to further the decimal system?

103. What did the Arabs do to further the decimal system?

104. What was the subject matter of Mohammed ibn Musa al-Khowarizmi's two books?

105. What two English words come from the titles of Mohammed ibn Musa al-Khowarizmi's two books?

WEB PROJECT

106. Write an essay on one of the following topics:
- history of the abacus
- calculations using the abacus
- the abacus versus the calculator
- the Roman abacus
- the Chinese suan pan
- the Mesoamerican abacus
- the Mayan base 20

Some useful links for this web project are listed on the text web site: **www.cengage.com/math/johnson**

7.2 Addition and Subtraction in Different Bases

OBJECTIVES

- Be able to add and subtract in different bases
- Understand why the base ten addition and subtraction procedures you learned in grade school work

Why Study Arithmetic in Different Bases?

You've probably added, subtracted, multiplied, and divided (in base ten) so much that you just do it without thinking about it. This is a good thing, because this is something that you should be able to do.

Suppose you were trying to explain how to subtract or how to multiply to a child who didn't know how to do it. You could well get quite flustered and find that you are incapable of explaining something so basic. This would probably be because you've learned the processes without understanding what's really going on and why the processes work.

Everyone uses a calculator. You may well have one in your daypack or on your desk right now. Some people believe that calculators relieve them of any responsibility for being able to calculate by hand, let alone understanding why the process works. Some people are users and some are creators, and this is the attitude of a user. Which do you want to be?

When society loses knowledge, that knowledge is gone. The world loses a language every two weeks. The languages of the Apache and the Pomo are almost totally gone because a few generations learned English instead. "When we lose a language, we lose centuries of thinking about time, seasons, sea creatures, reindeer, edible flowers, mathematics, landscapes, myths, music, the unknown and the everyday," said K. David Harrison of the Living Tongues Institute for Endangered Languages. The ability to calculate by hand could go the same way. And if that goes, so does the ability to program the very computers we depend on.

Arithmetic in different bases works just like arithmetic in base ten. The algorithms that the Hindus invented are the same in base eight or base two as they are in base ten. What is different is how you write numbers. For example, we write "eight" as

- 8 in base ten
- 10_8 in base eight
- 1000_2 in base two

Addition

Base eight uses the digits 0 through 7. Because of this, whole numbers between zero and seven are written the same way in base eight as they are in base ten. For example, $7_8 = 7$. But numbers larger than 7_8 are written with multiple digits, using the place system. For example, $8 = 10_8$.

Compare the following two additions:

- $3 + 4 = 7$ **adding in base ten**
- $3_8 + 4_8 = 7_8$ **adding in base eight**

Everything is the same, except for the base. We are adding the same two numbers, we get the same answer, and we write the answers the same way (except for the subscript indicating base eight).

Compare another pair of additions:

- $4 + 4 = 8$ **adding in base ten**
- $4_8 + 4_8 = 10_8$ **adding in base eight**

The answers have the same meaning, but they are written differently. This happens whenever the sum is greater than or equal to eight, the base.

When we add numbers with multiple digits, such as $27 + 36$, we use the Hindu algorithm and add column by column. That is, we do not really add 27 and 36; instead, we add 7 and 6. Then we add 2 and 3 (and anything carried from the right column).

When we add in other bases, we also add column by column. But when a column's sum is greater than or equal to the base, we convert the sum from base ten to that base. This happened when we added $4_8 + 4_8$ above.

EXAMPLE **1**

BASE EIGHT ADDITION
a. Compute:

$$27_8$$
$$+\ 36_8$$

b. Check your answer by converting both the problem and the answer to base ten.

SOLUTION

a. As always, we start by adding the numbers in the right column. The column's sum is $7_8 + 6_8 = 13$, and $13 \geq 8$, the base. So we convert 13 from base ten to base eight.

$$7_8 + 6_8 = 13$$
$$= 1 \cdot 8^1 + 5 \cdot 8^0 = 15_8$$

As always, we put the 5 in the right column, and we carry the 1 to the second column.

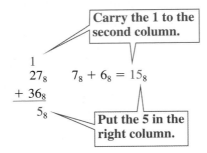

Now we add the digits in the second column.

$$1_8 + 2_8 + 3_8 = 6_8$$

The column's sum is 6, and $6 < 8$, the base. There is no need to convert 6 to base 8, because $6 = 6_8$. Instead, we put the 6 in the second column, and we are done.

$$1$$
$$27_8$$
$$+\ 36_8$$
$$\overline{65_8}$$

The sum is $27_8 + 36_8 = 65_8$.
b. We will check this answer by converting both the problem and the answer to base ten.

$$27_8 = 2 \cdot 8^1 + 7 \cdot 8^0 = 23$$
$$36_8 = 3 \cdot 8^1 + 6 \cdot 8^0 = 30$$
$$65_8 = 6 \cdot 8^1 + 5 \cdot 8^0 = 53$$

Our work checks because $23 + 30 = 53$. ✓

EXAMPLE **2**

BASE TWO ADDITION
a. Compute:

$$11110_2$$
$$+\ 1011_2$$

b. Check your answer by converting both the problem and the answer to base ten.

SOLUTION

a. Start by adding the numbers in the right column. We get $0_2 + 1_2 = 1_2$. The column's sum is 1, and $1 < 2$, the base. There is no need to convert 1 to base 2, because $1 = 1_2$. We put 1 in the right column and move on to the second column.

$$11110_2$$
$$+\ 1011_2$$
$$\overline{1_2}$$

In the second column, $1_2 + 1_2 = 2$. The column's sum is 2, and $2 \geq 2$, the base. So we convert 2 from base ten to base two.

$$1_2 + 1_2 = 2$$
$$= 1 \cdot 2^1 + 0 \cdot 2^0 = 10_2$$

As always, we put the 0 in the current column, and we carry the 1 to the next column.

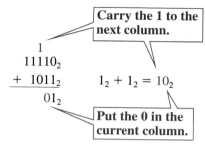

Now we add the digits in the third column.

$$1_2 + 1_2 + 0_2 = 2 = 10_2$$

Put the 0 in the current column, and carry the 1 to the next column.

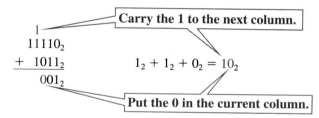

Now we add the digits in the fourth column.

$$1_2 + 1_2 + 1_2 = 3$$
$$= 1 \cdot 2^1 + 1 \cdot 2^0 = 11_2$$

Put the 1 in the current column, and carry the 1 to the next column.

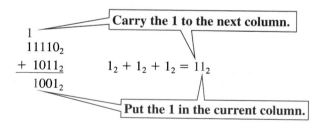

Now we add the digits in the fifth column, and we are done.

$$1_2 + 1_2 = 2$$
$$= 1 \cdot 2^1 + 0 \cdot 2^0 = 10_2$$

$$
\begin{array}{r}
1 \\
11110_2 \\
+\ \ 1011_2 \\
\hline
101001_2
\end{array}
$$
  The sum is 101001_2.

b.
$$11110_2 = 1 \cdot 2^4 + 1 \cdot 2^3 + 1 \cdot 2^2 + 1 \cdot 2^1 + 0 \cdot 2^0 = 30$$
$$1011_2 = 1 \cdot 2^3 + 0 \cdot 2^2 + 1 \cdot 2^1 + 1 \cdot 2^0 = 11$$
$$101001_2 = 1 \cdot 2^5 + 0 \cdot 2^4 + 1 \cdot 2^3 + 0 \cdot 2^2 + 0 \cdot 2^1 + 1 \cdot 2^0 = 41$$

Our work checks because $30 + 11 = 41$. ✓

Subtraction

EXAMPLE **3**

BASE EIGHT SUBTRACTION

a. Compute:

$$40_8$$
$$-26_8$$

b. Compute:

$$43_8$$
$$-26_8$$

c. Check your answers by converting each problem and its answer to base ten.

SOLUTION

a. In the right-hand column, we can't subtract $0_8 - 6_8$ without borrowing 1 from the second column. This column is the eights column, so we're borrowing 1 eight. When we subtract 6 from this 1 eight, we get 2.

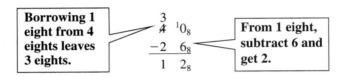

We conclude that $40_8 - 26_8 = 12_8$.

b. As in part (a), we borrow 1 eight from the second column, we subtract 6 from this 1 eight, and we get 2. What's different is that we now have a 3 where we once had a 0. When we combine the 3 with the 2 that came from subtracting, we get $3_8 + 2_8 = 5_8$.

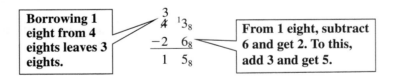

We conclude that $43_8 - 26_8 = 15_8$.

c. *Checking part (a):*

$$40_8 = 4 \cdot 8^1 + 0 \cdot 8^0 = 32$$
$$26_8 = 2 \cdot 8^1 + 6 \cdot 8^0 = 22$$
$$12_8 = 1 \cdot 8^1 + 2 \cdot 8^0 = 10$$

Part (a) checks because $32 - 22 = 10$. ✓
Checking part (b):

$$43_8 = 4 \cdot 8^1 + 3 \cdot 8^0 = 35$$
$$26_8 = 2 \cdot 8^1 + 6 \cdot 8^0 = 22$$
$$15_8 = 1 \cdot 8^1 + 5 \cdot 8^0 = 13$$

Part (b) checks because $35 - 22 = 13$. ✓

EXAMPLE **4**

BASE SIXTEEN SUBTRACTION

a. Compute:

$$683_{16}$$
$$-2AB_{16}$$

b. Check your answer by converting both the problem and the answer to base ten.

SOLUTION

a. In the right-hand column, we can't subtract $3_{16} - B_{16}$ without borrowing 1 from the second column. (Remember, $B_{16} = 11$.) This column is the sixteens column, so we're borrowing 1 sixteen. When we subtract $B_{16} = 11$ from this 1 sixteen, we get 5. To this 5, add 3 and get $5_{16} + 3_{16} = 8_{16}$.

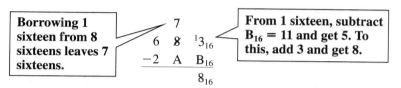

| Borrowing 1 sixteen from 8 sixteens leaves 7 sixteens. | | From 1 sixteen, subtract $B_{16} = 11$ and get 5. To this, add 3 and get 8. |

In the second column, we can't subtract $7_{16} - A_{16}$ without borrowing. (Remember, $A_{16} = 10$.)

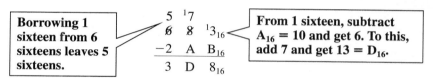

| Borrowing 1 sixteen from 6 sixteens leaves 5 sixteens. | | From 1 sixteen, subtract $A_{16} = 10$ and get 6. To this, add 7 and get $13 = D_{16}$. |

We conclude that $683_{16} - 2AB_{16} = 3D8_{16}$.

b.
$$683_{16} = 6 \cdot 16^2 + 8 \cdot 16^1 + 3 \cdot 16^0 = 1667$$
$$2AB_{16} = 2 \cdot 16^2 + 10 \cdot 16^1 + 11 \cdot 16^0 = 683$$
$$3D8_{16} = 3 \cdot 16^2 + 13 \cdot 16^1 + 8 \cdot 16^0 = 984$$

Our work checks because $1667 - 683 = 984$. ✓

7.2 EXERCISES

In Exercises 1–12, perform the given computation, working solely in the given base.

1. $3_8 + 7_8$

2. $5_8 + 6_8$

3. $4_8 + 5_8$

4. $7_8 + 7_8$

5. $9_{16} + 7_{16}$

6. $A_{16} + 2_{16}$

7. $B_{16} + 9_{16}$

8. $D_{16} + 6_{16}$

9. $1_2 + 1_2$

10. $10_2 + 11_2$

11. $11_2 + 11_2$

12. $101_2 + 10_2$

In Exercises 13–24, do the following.

a. *Perform the given computation, working solely in the given base.*

b. *Check your answer by converting both the problem and the answer to base ten. (Answers are not given in the back of the book.)*

13. $13_8 + 72_8$

14. $54_8 + 63_8$

15. $475_8 + 254_8$

16. $765_8 + 117_8$

17. $A9_{16} + 7B_{16}$

18. $A01_{16} + 236_{16}$

19. $BBC_{16} + CCD_{16}$

20. $1094_{16} + 2399_{16}$

21. $110_2 + 101_2$

22. $10110_2 + 11001_2$

23. $1100110_2 + 1101101_2$

24. $10110_2 + 11111_2$

In Exercises 25–30, perform the given computation, working solely in the given base.

25. $13_8 - 7_8$

26. $51_8 - 16_8$

27. $194_{16} - 52_{16}$

29. $1101_2 - 1010_2$

28. $A5_{16} - 61_{16}$

30. $10110_2 - 1010_2$

In Exercises 31–38, do the following.

a. Perform the given computation, working solely in the given base.

b. Check your answer by converting both the problem and the answer to base ten. (Answers are not given in the back of the book.)

31. $15_8 - 6_8$

33. $1B3_{16} - 105_{16}$

35. $11011_2 - 10110_2$

37. $341_5 - 243_5$

32. $55_8 - 26_8$

34. $A4_{16} - 31_{16}$

36. $101110_2 - 10010_2$

38. $4632_7 - 326_7$

Answer the following using complete sentences and your own words.

• CONCEPT QUESTIONS

39. Describe in detail how to add $38 + 99$ on a base ten abacus.

40. Describe in detail a base eight abacus and how to add $27_8 + 54_8$ on a base eight abacus.

41. Describe in detail how to subtract $251 - 89$ on a base ten abacus.

42. Describe in detail how to subtract $332_8 - 67_8$ on a base eight abacus.

7.3 Multiplication and Division in Different Bases

OBJECTIVES

- Be able to multiply and divide in different bases
- Understand why the base ten multiplication and division algorithms you learned in grade school work

In this section, we will discuss how to multiply and divide in different bases.

Multiplication

EXAMPLE 1

BASE EIGHT MULTIPLICATION

a. Compute:

$$\begin{array}{r} 32_8 \\ \times\ 5_8 \end{array}$$

b. Check your answer by converting both the problem and the answer to base ten.

SOLUTION

a. We start in the right column and multiply the 5 by the 2 above it.

$$5_8 \cdot 2_8 = 5_{10} \cdot 2_{10} = 10_{10}$$
$$= 1 \cdot 8^1 + 2 \cdot 8^0 = 12_8$$

Write the 2 in the right column, and carry the 1 to the next column.

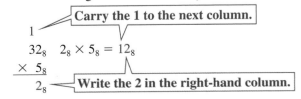

Next, we multiply the 5 by the 3 in the second column, and we add the carried 1.

$$(5_8 \cdot 3_8) + 1_8 = (5_{10} \cdot 3_{10}) + 1_{10} = 16_{10}$$
$$= 2 \cdot 8^1 + 0 \cdot 8^0 = 20_8$$

$$
\begin{array}{r}
1 \\
32_8 \\
\times\ 5_8 \\
\hline
202_8
\end{array}
$$

The answer is 202_8.

b. $32_8 = 3 \cdot 8^1 + 2 = 26$

$5_8 = 5$

$202_8 = 2 \cdot 8^2 + 0 \cdot 8^1 + 2 = 130$

Our work checks because $26 \times 5 = 130$. ✓

EXAMPLE 2

BASE EIGHT MULTIPLICATION
a. Compute:

$$
\begin{array}{r}
32_8 \\
\times\ 45_8
\end{array}
$$

b. Check your answer by converting both the problem and the answer to base ten.

SOLUTION

a. In Example 1, we did half of this problem—the half involving multiplication by 5. What remains is to do the half involving multiplication by 4.

$$
\left.
\begin{array}{r}
32_8 \\
\times\ 45_8 \\
\hline
202_8
\end{array}
\right\}
\text{ from Example 1}
$$

Applying the 4 to the right column, we have

$$4_8 \cdot 2_8 = 4_{10} \cdot 2_{10} = 8_{10}$$
$$= 1 \cdot 8^1 + 0 \cdot 8^0 = 10_8$$

We carry the 1, and we write the 0 under the 202_8 that we computed in Example 1, but one column to the left.

$$
\begin{array}{r}
1 \\
32_8 \\
\times\ 45_8 \\
\hline
202_8 \\
0_8
\end{array}
$$

Carry the 1.

$4_8 \times 2_8 = 10_8$

Put the 0 here.

In the next column, we have

$$(4_8 \cdot 3_8) + 1_8 = (4_{10} \cdot 3_{10}) + 1_{10} = 13_{10}$$
$$= 1 \cdot 8^1 + 5 \cdot 8^0 = 15_8$$

We write this next to the 0, at the bottom line. Then we add.

$$
\begin{array}{r}
32_8 \\
\times\ 45_8 \\
\hline
202_8 \\
150_8 \\
\hline
1702_8
\end{array}
$$

The answer is 1702_8

b.

$$32_8 = 3 \cdot 8^1 + 2 \cdot 8^0 = 26$$
$$45_8 = 4 \cdot 8^1 + 5 \cdot 8^0 = 37$$
$$1702_8 = 1 \cdot 8^3 + 7 \cdot 8^2 + 0 \cdot 8^1 + 2 \cdot 8^0 = 962$$

Our work checks because $26 \cdot 37 = 962$. ✓

Division

EXAMPLE 3

BASE EIGHT DIVISION

a. Compute:

$$3_8 \overline{)237_8}$$

b. Check your answer by converting both the problem and the answer to base ten.

SOLUTION

a. Start by dividing 3_8 into 23_8, just as we would if we were in base ten. Asking how many times 3_8 goes into 23_8 is the same as asking what we can multiply by 3_8 to get 23_8. To answer this question, we will create a base eight multiplication table for a multiplier of 3_8, as shown in Figure 7.3. The table's entries are found by using the multiplication process illustrated in Examples 1 and 2.

$3_8 \cdot$	1_8	2_8	3_8	4_8	5_8	6_8	7_8
$=$	3_8	6_8	11_8	14_8	17_8	22_8	25_8

FIGURE 7.3 A base eight multiplication table for a multiplier of 3_8.

Figure 7.3 shows that $3_8 \cdot 6_8 = 22_8$ (which is a bit too small) and $3_8 \cdot 7_8 = 25_8$ (which is a bit too big). We use the "bit too small" answer, as we would if we were in base ten.

$$3_8 \overline{)237_8} \quad \overset{6}{}$$

Use Figure 7.3 to multiply $6_8 \cdot 3_8 = 22_8$, and put the result underneath. Subtract, and bring down the 7_8.

$$\begin{array}{r} 6 \\ 3_8 \overline{)237_8} \\ -22 \\ \hline 17 \end{array}$$

$\boxed{23 - 22 = 1}$ ⟍ ⟋ $\boxed{\textbf{Bring down the 7.}}$

Now use Figure 7.3 to divide 3_8 into 17_8 and get 5_8.

$$\begin{array}{r} 65 \\ 3_8 \overline{)237_8} \\ -22 \\ \hline 17 \end{array}$$

Use Figure 7.3 to multiply $5_8 \cdot 3_8$, get 17_8, put the result underneath, and subtract.

$$\begin{array}{r} 65_8 \\ 3_8 \overline{)237_8} \\ -22 \\ \hline 17 \\ -17 \\ \hline 0 \end{array}$$

There is no remainder, so we are done. The quotient is 65_8.

b. $237_8 = 2 \cdot 8^2 + 3 \cdot 8^1 + 7 \cdot 8^0 = 159$

$3_8 = 3$

$65_8 = 6 \cdot 8^1 + 5 \cdot 8^0 = 53$

Our work checks because $159 \div 3 = 53$.   ✓

EXAMPLE 4

BASE EIGHT DIVISION
a. Compute: $3671_8 \div 12_8$, working solely in base eight.
b. Check your answer by multiplying, in base eight.

SOLUTION

a. We are dividing by 12_8, so we need a multiplication table for a multiplier of 12_8, as shown in Figure 7.4. The table's entries are found by using the multiplication process illustrated in Examples 1 and 2.

$12_8 \cdot$	1_8	2_8	3_8	4_8	5_8	6_8	7_8
$=$	12_8	24_8	36_8	50_8	62_8	74_8	106_8

FIGURE 7.4 A base eight multiplication table for a multiplier of 12_8.

Use Figure 7.4 to divide 12_8 into 36_8 and get 3_8.

$$12_8 \overline{)3671_8} \quad \overset{3}{}$$

Use Figure 7.4 to multiply $3_8 \cdot 12_8 = 36_8$, and put the result underneath. Subtract, and bring down the 7_8.

$$\begin{array}{r} 3 \\ 12_8 \overline{)3671_8} \\ -36 \\ \hline 07 \end{array}$$

$36 - 36 = 0$ **Bring down the 7.**

Now divide 12_8 into 7_8, get 0, and bring down the 1.

$$\begin{array}{r} 30 \\ 12_8 \overline{)3671_8} \\ -36 \\ \hline 071 \end{array}$$

Bring down the 1.

Use Figure 7.4 to divide 12_8 into 71_8 and get 5_8.

$$\begin{array}{r} 305 \\ 12_8 \overline{)3671_8} \end{array}$$

Use Figure 7.4 to multiply $5_8 \cdot 12_8 = 62_8$, put the result underneath, and subtract.

$$\begin{array}{r} 305_8 \\ 12_8 \overline{)3671_8} \\ -36 \\ \hline 071 \\ -62 \\ \hline 7_8 \end{array}$$

$71_8 - 62_8 = 7_8$

We are done, because 12_8 will not divide into 7_8. The quotient is 305_8, with a remainder of 7_8.

b. We will not check the answer by converting both the problem and the answer to base ten, as we have in the past, because the remainder means that we would have to work with fractions in base eight. It is less work to check by multiplying.

$$
\begin{array}{r}
305_8 \\
\times\ 12_8 \\
\hline
612 \\
305\ \ \\
\hline
3662_8
\end{array}
$$

Add the remainder of 7_8 to this and get $3662_8 + 7_8 = 3671_8$. Our answer to part (a) checks. ✓

7.3 EXERCISES

In Exercises 1–8, perform the given computation, working solely in the given base.

1. $12_8 \cdot 6_8$
2. $31_8 \cdot 4_8$
3. $C1_{16} \cdot 5_{16}$
4. $9D_{16} \cdot 11_{16}$
5. $101_2 \cdot 11_2$
6. $110_2 \cdot 10_2$
7. $32_4 \cdot 23_4$
8. $15_7 \cdot 21_7$

In Exercises 9–16, do the following.

a. *Perform the given computation, working solely in the given base.*

b. *Check your answer by converting both the problem and the answer to base ten. (Answers are not given in the back of the book.)*

9. $25_8 \cdot 5_8$
10. $53_8 \cdot 16_8$
11. $1BA_{16} \cdot 15_{16}$
12. $A41_{16} \cdot 11_{16}$
13. $1101_2 \cdot 10110_2$
14. $10110_2 \cdot 1010_2$
15. $31_5 \cdot 23_5$
16. $462_7 \cdot 26_7$

In Exercises 17 and 18, perform the given computation, working solely in the given base.

17. $46_8 \div 2_8$
18. $54_8 \div 4_8$

In Exercises 19–22, do the following.

a. *Perform the given computation, working solely in the given base.*

b. *Check your answer by converting both the problem and the answer to base ten. (Answers are not given in the back of the book.)*

19. $14D_{16} \div 3_{16}$
20. $A5_{16} \div 5_{16}$
21. $144_8 \div 31_8$
22. $234_8 \div 14_8$

In Exercises 23–34, do the following.

a. *Perform the given computation, working solely in the given base.*

b. *Check your answer by multiplying, in the given base. (Answers are not given in the back of the book.)*

23. $153_8 \div 15_8$
24. $227_8 \div 25_8$
25. $277_8 \div 152_8$
26. $7541_8 \div 370_8$
27. $79AB_{16} \div 38_{16}$
28. $A104_{16} \div 21_{16}$
29. $790B_{16} \div 218_{16}$
30. $A14B_{16} \div 2B1_{16}$
31. $1100111_2 \div 110_2$
32. $1010110_2 \div 1010_2$
33. $1101111_2 \div 1100_2$
34. $1010010_2 \div 10010_2$

35. Create a base five multiplication table, with entries ranging from $1_5 \cdot 1_5$ through $11_5 \cdot 11_5$.

36. Create a base six multiplication table, with entries ranging from $1_6 \cdot 1_6$ through $11_6 \cdot 11_6$.

37. Create a base fourteen multiplication table, with entries ranging from $1_{14} \cdot 1_{14}$ through $11_{14} \cdot 11_{14}$.

38. Create a base thirteen multiplication table, with entries ranging from $1_{13} \cdot 1_{13}$ through $11_{13} \cdot 11_{13}$.

39. Use the result of Exercise 35 to compute $4230_5 \div 23_5$.

40. Use the result of Exercise 36 to compute $5312_6 \div 35_6$.

41. Use the result of Exercise 37 to compute $BADC_{14} \div 102_{14}$.

42. Use the result of Exercise 38 to compute $C9AB_{13} \div B3_{13}$.

7.4 Prime Numbers and Perfect Numbers

OBJECTIVES

- Learn to determine whether a number is prime or composite
- Learn to determine if a number is abundant, perfect or deficient
- Know how to generate a Mersenne prime number
- Understand relationship between perfect numbers and Mersenne primes

Do you shop on the web? Do you pay your bills online? Do you surf the web using a wi-fi connection? If you're like most people, the answer is "yes."

Your credit cards are probably smart cards that track your purchases. Your computer might have a few tracking cookies that report the web sites you visit. Data miners make their living by obtaining and selling information about your purchases, your income, and the web sites you visit. There are plenty of buyers.

An incredible amount of your personal information, including your name, your mother's maiden name, your address, your Social Security number, your credit card account numbers and bank account numbers, is transmitted electronically and possibly available to prying eyes. As a result, you could be the next victim of identity theft. It happens all the time. In fact, it's one of the fastest-growing crimes in the United States. All it takes is for someone to get enough information to access your bank account, take over one of your credit cards, or open a new card in your name.

RSA encryption is one of the more successful defenses against this prying offense. RSA encryption is dependent on large prime numbers, and perfect numbers are used to find large prime numbers. In this section, we'll explore prime numbers and perfect numbers.

Factorizations and Primes

Counting numbers are the numbers that you count with: 1, 2, 3, 4, ...

To **factor** a counting number is to rewrite the number as a product of counting numbers. For example, the number 40 factors into $4 \cdot 10$. We say that 4 and 10 are **factors** of 40.

But $40 = 4 \cdot 10$ is not factored completely, because 4 and 10 can each be factored further.

$$40 = 4 \cdot 10$$
$$= 2 \cdot 2 \cdot 2 \cdot 5$$

This is **factored completely,** because it cannot be factored it further.

Numbers such as 2 and 5 that cannot be factored are called *primes*. Of course, $2 = 1 \cdot 2$, and $5 = 1 \cdot 5$. Any number can be written as the product of 1 and itself. We do not include these 1's when we factor completely.

A **prime number** is a counting number that has exactly two factors: 1 and itself. A complete factorization, such as $40 = 2 \cdot 2 \cdot 2 \cdot 5$, is also called a **prime factorization,** because it is a factorization into primes. A counting number that has factors other than 1 and itself is called a **composite number.** Every composite number has a prime factorization, as does the composite number 40.

501

In Figure 7.5, we list the numbers 2 through 20 and their prime factorizations. Prime numbers do not have prime factorizations, so when there is no prime factorization listed, the number is prime.

Number	Prime Factorization	Prime or Composite?
2		Prime
3		Prime
4	$2 \cdot 2$	Composite
5		Prime
6	$2 \cdot 3$	Composite
7		Prime
8	$2 \cdot 2 \cdot 2$	Composite
9	$3 \cdot 3$	Composite
10	$2 \cdot 5$	Composite
11		Prime
12	$2 \cdot 2 \cdot 3$	Composite
13		Prime
14	$2 \cdot 7$	Composite
15	$3 \cdot 5$	Composite
16	$2 \cdot 2 \cdot 2 \cdot 2$	Composite
17		Prime
18	$2 \cdot 3 \cdot 3$	Composite
19		Prime
20	$2 \cdot 2 \cdot 5$	Composite

FIGURE 7.5 Prime factorizations and primes.

Is 1 Prime or Composite?

Notice that the number 1 is not listed in Figure 7.5. Is 1 a prime number or a composite number? To see whether 1 is prime, we must determine whether 1 has exactly two factors: 1 and itself. The only way to factor 1 is to write $1 = 1 \cdot 1$. One does not have two factors; it has only one factor. So according to the definition, 1 is not a prime number.

Furthermore, 1 is not composite, because it does not have factors other than 1 and itself.

It can be hard to understand why 1 is not prime, because you cannot break 1 down into prime factors the way you can with $40 = 2 \cdot 2 \cdot 2 \cdot 5$. But the whole idea of factoring does not apply to 1. How do we factor 1? The best we can do is

$$1 = 1 \cdot 1 = 1 \cdot 1 \cdot 1 = 1 \cdot 1 \cdot 1 \cdot 1 = \cdots$$

The whole distinction of prime versus composite does not apply to 1.

TOPIC X   PERIODIC CICADAS: PRIME NUMBERS IN THE REAL WORLD

Cicadas live underground and suck sap from tree roots. They are common in the United States. Annual cicadas reach maturity after living about seven years underground. Each year, some annual cicadas in a particular area will reach maturity. These mature cicadas come out from under the ground to mate. If you are around when this happens, you might notice a few flying around.

Periodic cicadas do it differently. These insects reach maturity after seventeen years. Unlike annual cicadas, all periodic cicadas mature in the same year. So once every seventeen years, an onslaught of periodic cicadas emerges from under the ground. They cover sidewalks, cars, tree branches, and houses in Washington, D.C.; Baltimore; Long Island; Detroit; and other parts of the Northeast and Midwest. This last happened in 2004. If you are around in 2021 when this happens next, you will be astonished by how many there are. A related species of periodic cicada reaches maturity after thirteen years.

Glenn Webb, a mathematician at Vanderbilt University, argues that it is not a coincidence that the two types of periodic cicadas' lifespans are prime numbers. He has proposed that these prime numbers offer evolutionary benefits.

If cicadas emerged from the ground once every sixteen years, then predators with two-, four-, and eight-year life cycles would always be there to eat them. And eating them would be very easy, because when cicadas are present, they are present in big numbers. But they cycle at seventeen years, so a predator with a four-year life cycle would be around to eat them once every sixty-eight years. As a result, periodic cicadas do not have predators that specialize in eating them.

A cicada

Gary Meszaros/Photo Researchers, Inc.

Source: Webb, G. F. 2001. The prime number periodical cicada problem. *Discrete and Continuous Dynamical Systems,* Series B, Volume 1, Number 3, page 387–399.

How to Tell Whether a Number Is Prime or Composite

EXAMPLE 1

PRIME OR COMPOSITE?

a. Is 231 prime or composite?
b. Is 143 prime or composite?

SOLUTION

a. Every composite number has a prime factorization. If 231 isn't divisible by a prime number, then it does not have a prime factorization and isn't a composite number. If it is divisible by a prime number, then it has a prime factorization and is a composite number. We'll divide 231 by successive primes (2, 3, 5, 7, 11, . . .).

- Is 2 a factor? $231 \div 2 = 115.5 \rightarrow 231 = 2 \cdot 115.5$, but 115.5 is not a counting number, so 2 is not a factor of 231.
- Is 3 a factor? $231 \div 3 = 77 \rightarrow 231 = 3 \cdot 77 \rightarrow 3$ and 77 are factors, so 231 is a composite number

b. *Finding if 143 is prime or composite:*

Is 2 a factor? $143 \div 2 = 71.5$, so 2 is not a factor.
Is 3 a factor? $143 \div 3 = 47.6\ldots$, so 3 is not a factor.
Is 5, the next prime, a factor? $143 \div 5 = 28.6$, so 5 is not a factor.
Is 7 a factor? $143 \div 7 = 20.4\ldots$, so 7 is not a factor.
Is 11 a factor? $143 \div 11 = 13 \rightarrow 143 = 11 \cdot 13$, so 143 is a composite number.

Factors come in pairs. In part (a) of Example 1, we found a pair of factors: 3 and 77. In part (b), we also found a pair of factors: 11 and 13. Each pair has a smaller number and a bigger number. In part (b), we didn't have to divide 143 by 13 because 11 worked, and 11 is the smaller number paired with 13. We never have to divide by a bigger prime number because if it worked, the smaller number paired with it would have also worked. The smaller numbers and the bigger numbers separate at the original number's square root. In part (b), the original number's square root is $\sqrt{143} = 11.9\ldots$. The smaller prime numbers (2, 3, 5, 7, and 11) are at or below $11.9\ldots$, and the bigger prime numbers (13 and beyond) are above $11.9\ldots$.

FINDING WHETHER A NUMBER IS PRIME OR COMPOSITE

To find whether the counting number n is prime or composite:

1. Divide n by 2, the smallest prime. If there is no remainder, then 2 is a factor of n, and n is a composite number.
2. If there is a remainder, then the divisor is not a factor. Continue dividing n by each of the primes that follow 2. Stop if you find a factor or if you reach the largest prime that is less than or equal to $\sqrt{n}$.
3. If you find no factors in step 2, then n is a prime. If you find a factor, then n is a composite.

EXAMPLE **2**

SOLUTION

PRIME OR COMPOSITE?

We'll follow the steps given above.

1. Is 2 a factor? $257 \div 2 = 128.5$, so 2 is not a factor.
2. Is 3 a factor? $257 \div 3 = 85.6\ldots$, so 3 is not a factor.
   Is 5 a factor? $257 \div 5 = 51.4$, so 5 is not a factor.
   Is 7 a factor? $257 \div 7 = 36.7$, so 7 is not a factor.
   Is 11 a factor? $257 \div 11 = 23.3$, so 11 is not a factor.
   Is 13 a factor? $257 \div 13 = 19.7$, so 13 is not a factor.

   We can stop dividing, because the next prime is 17, and $17 > \sqrt{257} = 16.03\ldots$.
3. In step 2, we found no factors, so 257 is prime.

How Many Prime Numbers Are There?

If we were to continue the list in Figure 7.5, would we continue to find primes, or would we reach a point at which they just stopped occurring? It is easy to think that they would stop—a large number has so many more *possible* factors than does a small number. For example, 4, which is a small number, has only 1, 2, 3, and 4 as possible factors (1, 2, and 4 are the actual factors). But 1,000, which is a large number, has possible factors of 1, 2, 3, 4, . . . , 999, and, 1000. We might assume that as the list of factors gets larger, we would have to find a factor sooner or later.

But the ancient Greek mathematician Euclid proved in about 300 B.C. that this is not true. He proved that there is an infinite number of prime numbers. So no matter how many you list, they never stop.

Prime Numbers and Eratosthenes

In about 200 B.C., another ancient Greek mathematician named Eratosthenes developed a way to list prime numbers that used multiplication rather than division. Of course, calculators had not been invented, and division is much more difficult by hand than is multiplication. We will apply his method to the numbers 2 through 25.

First, go through and eliminate every multiple of 2 (except 2 itself, because 2 is prime). We will show eliminated numbers in black.

2	3	4	5	6	7	8	9	10	11	12	13
14	15	16	17	18	19	20	21	22	23	24	25

Next, go through and eliminate every multiple of 3 (except 3 itself, because 3 is prime).

2	3	4	5	6	7	8	9	10	11	12	13
14	15	16	17	18	19	20	21	22	23	24	25

There is no need to eliminate multiples of 4 because they are all multiples of 2 and were eliminated earlier. So eliminate every multiple of 5 (except for 5 itself). The only one that has not already been eliminated is 25.

2	3	4	5	6	7	8	9	10	11	12	13
14	15	16	17	18	19	20	21	22	23	24	25

Continue this process through the square root of the highest number on the list. The highest number on our list is 25, and $\sqrt{25} = 5$. So we are done, because we just eliminated multiples of 5. Of the numbers 2–25, the prime numbers are 2, 3, 5, 7, 11, 13, 17, 19, and 23.

This method is called the **Sieve of Eratosthenes,** because a *sieve* is a type of filter, and this method filters out the composites.

Why Search for Primes?

Eratosthenes was not the last person to search for primes. Many people are still actively searching for primes today. Why do people do this? Here are some reasons.

It is a tradition. Euclid started the search in approximately 300 B.C. Since then, many famous mathematicians have continued the search.

In the search for prime numbers, unforeseen benefits are discovered along the way. For example, prime numbers are used to encrypt credit card numbers and other personal information when they are transmitted over the web. Also, software that searches for prime numbers is used by Intel to test Pentium chips and by Cray Research, the manufacturer of some of the most powerful computers ever built, to test their computers. This finding of unforeseen benefits is a common occurrence in mathematical and the scientific research.

Some people search for primes just because they enjoy the glory of discovery.

The Search for Prime Numbers and Fermat

Many mathematicians tried to find a formula that could be used to generate primes. Because they are divisible by 2, it was clear that none of the even numbers are prime. Therefore, the focus should be on odd numbers. Some mathematicians studied the formula $2^n + 1$. The number 2^n is even, and adding 1 to an even number gives an odd number, so $2^n + 1$ must be odd. This formula fails to generate primes rather quickly. See Figure 7.6.

n	1	2	3
$2^n + 1$	$2^1 + 1 = 2 + 1$ $= 3$	$2^2 + 1 = 4 + 1$ $= 5$	$2^3 + 1 = 8 + 1$ $= 9$
Result	Prime	Prime	Composite

FIGURE 7.6 Is $2^n + 1$ always prime?

In the early seventeenth century, French mathematician Pierre de Fermat used the same formula but used only powers of 2 for n. See Figure 7.7. Fermat checked these numbers through $2^{16} + 1 = 65,537$, dividing by hand, and found that each outcome is prime. He conjectured that this formula always works, but he was unable to prove it. More than 100 years later, the Swiss mathematician Leonhard Euler showed that $2^{32} + 1 = 4,294,967,297$ is not prime, because it has 641 as a factor. (Euler comes up a lot in this book. He was instrumental in creating graph theory, studied in Chapter 9, and in exponential and logarithmic functions, studied in Chapter 10. There is a historical note about him in Section 10.0.A.)

n	$2^1 = 2$	$2^2 = 4$	$2^3 = 8$
$2^n + 1$	$2^2 + 1 = 4 + 1$ $= 5$	$2^4 + 1 = 16 + 1$ $= 17$	$2^8 + 1 = 256 + 1$ $= 257$
Result	Prime	Prime	Prime

n	$2^4 = 16$	$2^5 = 32$	$2^6 = 64$
$2^n + 1$	$2^{16} + 1$ $= 65,537$	$2^{32} + 1$ $= 4,294,967,297$	$2^{64} + 1 = ?$
Result	Prime	Composite	?

FIGURE 7.7 Is $2^n + 1$ always prime if n is a power of 2?

The Search for Prime Numbers and Mersenne

In the 1600s, the French monk Marin Mersenne studied the formula $2^n - 1$. The number 2^n is even, so $2^n - 1$ must be odd. It was proven that if n is composite, then $2^n - 1$ is also composite. So only prime numbers were used for n. See Figure 7.8.

n	2	3	5
$2^n - 1$	$2^2 - 1 = 4 - 1$ $= 3$	$2^3 - 1 = 8 - 1$ $= 7$	$2^5 - 1 = 32 - 1$ $= 31$
Result	Prime	Prime	Prime

n	7	11	13
$2^n - 1$	$2^7 - 1 = 128 - 1$ $= 127$	$2^{11} - 1 = 2047$	$2^{13} - 1 = 8191$
Result	Prime	?	?

FIGURE 7.8 Is $2^n - 1$ always prime if n is prime?

Unfortunately, this formula fails to generate *only* prime numbers, because $2^{11} - 1 = 2047 = 23 \cdot 89$ is not prime. However, it generates *many* prime numbers, and these Mersenne numbers continue to be one of the largest sources of prime numbers. In the late 1700s, Euler showed that $2^{31} - 1$ is a Mersenne prime. A **Mersenne number** is a number of the form $2^n - 1$, where n is prime. A **Mersenne prime** is a Mersenne number that is itself prime.

The search for primes changed in the 1950s, when mathematicians started to use computers to check Mersenne numbers to determine whether they are prime. There is now a collaborative group called GIMPS—the Great Internet Mersenne Prime Search. Using special computer software and unused computer power, its members have found thirteen Mersenne primes. The largest know prime number, $2^{243,112,609} - 1$, was found in 2008 by GIMPS member Edson Smith of the University of California at Los Angeles Mathematics Department. Its discovery qualified for an award of $100,000 from the Electronic Frontier Foundation and was described by *Time* magazine as one of the fifty best inventions of 2008.

Abundant, Deficient, and Perfect Numbers

In ancient times, numbers were thought to have mystical properties. One such property had to do with how many factors a number has. It turns out that this rather odd detail might help mathematicians with their search for prime numbers.

Look at the number 20. It has an abundance of factors, because there are many ways to factor it.

$20 = 1 \cdot 20$ **1 and 20 are factors.**
$20 = 2 \cdot 10$ **2 and 10 are factors.**
$20 = 4 \cdot 5$ **4 and 5 are factors.**

The factors of 20 are 1, 2, 4, 5, 10, and 20. But the *proper* factors of 20 are 1, 2, 4, 5, and 10. The **proper factors** of a number are that number's factors except for the number itself.

The sum of 20's proper factors is

$$1 + 2 + 4 + 5 + 10 = 22$$

Since 22 (the sum of the proper factors) is greater than 20 (the number itself), we say that 22 is an *abundant* number. A counting number is **abundant** if the sum of its proper factors is greater than the number itself.

Now consider the number 49:

$49 = 1 \cdot 49$ **1 and 49 are factors.**

$49 = 7 \cdot 7$ **7 is a factor.**

The proper factors of 49 are 1 and 7. The number 49 is rather deficient in its factors.

The sum of 49's proper factors is $1 + 7 = 8$. Since 8 (the sum of the proper factors) is less than 49 (the number itself), we say that 49 is a *deficient* number. A counting number is **deficient** if the sum of its proper factors is less than the number itself.

Finally, consider the number 6.

$6 = 1 \cdot 6$ **1 and 6 are factors.**

$6 = 2 \cdot 3$ **2 and 3 are factors.**

The proper factors of 6 are 1, 2 and 3. The sum of 6's proper factors is $1 + 2 + 3 = 6$. Since 6 (the sum of the proper factors) is equal to 6 (the number itself), we say that 6 is a *perfect* number. A counting number is **perfect** if the sum of its proper factors is equal to the number itself.

Euclid wrote about perfect numbers in about 300 B.C., but many ancient cultures studied perfect numbers before then. Pythagoras also studied them and their mystical properties. See the Historical Note for more on Pythagoras and his mystic view of numbers.

HISTORICAL NOTE

PYTHAGORAS (APPROXIMATELY 560–480 B.C.)

North Wind Picture Archives

Pythagoras founded a philosophical and religious school in Crotone, Italy, that had many followers. They lived permanently at the school, had no personal possessions, and were vegetarians. They worshiped Pythagoras as a demigod. All Pythagoreans were required to practice obedience, loyalty, secrecy, silence, ritualistic fasting, simplicity in dress and possessions, and the habit of frequent self-examination. In fact, the Pythagoreans were so loyal and secretive that the contributions of Pythagoras cannot be separated from those of his followers.

The famous Pythagorean Theorem was not discovered by Pythagoras; in fact, it was known by the Babylonians 1000 years earlier. However, Pythagoras was probably the first to prove it.

The Pythagoreans believed that geometry is the highest form of knowledge and that mathematics is the best, most direct approach to reality. In fact, they believed that mathematics constitutes the true nature of things.

The Pythagoreans observed that vibrating strings produce pleasant tones if one string's length divided by the other string's length is a rational number (that is, a fraction made of whole numbers, such as 1/2, 4/3, etc.). To them, this meant that even sound is reducible to mathematics.

The Pythagoreans saw a mystical significance in numbers.

- "One" is the generator of all numbers and therefore the number of reason.
- "Two" is the number of women.
- "Three" is the number of men.
- "Four" is the number of justice.
- "Five" is the number of marriage; it is the sum of woman and man.
- "Six" is the number of creation; it is the product of woman and man. Six is a perfect number, so creation is perfect.
- "Seven" is the number of opportunity.
- "Ten" is the number of the universe. Ten is the sum of the numbers one through four, so ten represents reason, women, men, and justice, combined.
- Numbers have special mystical properties if they are perfect, deficient, or abundant.

More Perfect Numbers

EXAMPLE **3**

DETERMINING WHETHER A NUMBER IS ABUNDANT, PERFECT, OR DEFICIENT

a. List all of 28's proper factors.
b. Is 28 an abundant, perfect, or deficient number?

SOLUTION

a. First, note that $\sqrt{28} = 5.2\ldots$, so in searching for 28's factors, it will be unnecessary to search above 5.

- $28/2 = 14$ $\Rightarrow$ $28 = 2 \cdot 14$ $\Rightarrow$ 2 and 14 are factors of 28
- $28/3 = 9.3\ldots$ $\Rightarrow$ 3 is not a factor of 28
- $28/4 = 7$ $\Rightarrow$ $28 = 4 \cdot 7$ $\Rightarrow$ 4 and 7 are factors of 28
- $28/5 = 5.6$ $\Rightarrow$ 5 is not a factor of 28
- It is unnecessary to search above 5, so we will not try 28/6.
- Clearly, $28 = 1 \cdot 28$, so 1 and 28 are factors of 28, but 28 is not a proper factor.

28's proper factors are 1, 2, 4, 7, and 14.
b. $1 + 2 + 4 + 7 + 14 = 28$, so 28 is a perfect number.

The first four perfect numbers are 6, 28, 496, and 8,128:

- 6's proper factors are 1, 2, and 3.
  Also, $6 = 1 + 2 + 3$.
- 28's proper factors are 1, 2, 4, 7, and 14.
  Also, $28 = 1 + 2 + 4 + 7 + 14$.
- 496's proper factors are 1, 2, 4, 8, 16, 31, 62, 124, and 248.
  Also, $496 = 1 + 2 + 4 + 8 + 16 + 31 + 62 + 124 + 248$.
- 8128's proper factors are 1, 2, 4, 8, 16, 32, 64, 127, 254, 508, 1016, 2032, and 4064.
  Also, $8128 = 1 + 2 + 4 + 8 + 16 + 32 + 64 + 127 + 254 + 508 + 1,016 + 2,032 + 4,064$.

These perfect numbers have been known since prehistoric times. They were discovered so long ago that there is no record of their discovery.

Perfect Numbers and Mersenne Primes

Perfect numbers contain an abundance of patterns. Let's look at some factorizations of some perfect numbers.

- $6 = 2 \cdot 3$
- $28 = 4 \cdot 7$
- $496 = 16 \cdot 31$
- $8,128 = 64 \cdot 127$

In each case, the first factor is a power of 2.

- 6's first factor is $2^1 = 2$
- 28's first factor is $2^2 = 4$
- 496's first factor is $2^4 = 16$
- 8,128's first factor is $2^6 = 64$

Furthermore, the powers are all 1 less than a prime number.

- 2^1's power is 1, 1 is one less than 2, and 2 is a prime number.
- 2^2's power is 2, 2 is one less than 3, and 3 is a prime number.

TOPIC X   RSA ENCRYPTION AND WEB PURCHASES: PRIME NUMBERS IN THE REAL WORLD

Most of us release personal information over the web. When you make an Internet purchase, you usually have to give your credit card number. There are people who eavesdrop electronically on transactions like this, hoping to steal your number. So for your protection, the credit card number is usually **encrypted**. That is, it is converted into another number that is useless to anyone who does not know how to **decrypt** it back to the actual credit card number. One common method is **RSA encryption**, which involves prime numbers and factoring.

The RSA algorithm was invented by Ron Rivest, Adi Shamir, and Leonard Adleman of the Massachusetts Institute of Technology. The algorithm is named after them.

Here's how RSA encryption works. Suppose you are buying something from Amazon.com. When you get to the web page at which you enter your credit card number, that page's address will begin with "https://www.amazon.com." The "s" in "https" indicates that you are now using a secure server and that your credit card number will be

encrypted. Amazon.com uses RSA encryption, so their secure server has stored a pair of two huge prime numbers, which we will call p and q. These two numbers are called the *private key*. They are kept private, because they are used to decrypt your credit card number.

Amazon.com's secure server transmits the product of p and q to your computer, which in turn uses it to encrypt your credit card number. This number, pq, cannot be used to decrypt. It is called the *public key*.

An electronic eavesdropper could easily find out the public key pq. But the eavesdropper would have to factor it into p and q to be able to decrypt. It is very easy to program a computer to factor a number, but it is also very time-consuming for the computer to succeed if the number is large. This is because the computer has to try dividing the number to be factored by almost every prime up to the number's square root.

Typically, the number pq has about 400 digits, so there are an incredibly large number of primes to try. According to an article published by the *Berkeley Math Circle*, "the computer would have

to run for 10^{176} times the life of the universe to factor" such a number. So it is essentially not factorable.

Of course, all of the encoding and decoding, as well as the creation of the public key and the private key, happen behind the scenes. The Internet user is never aware of it.

In summary:

- The public key is pq, which is the product of two huge prime numbers. It is used only to encrypt, so it can be made public without risk.
- The private key is p and q, the two prime factors of the public key. It is used only to decrypt, so it is never made public.
- Using a computer to try to factor pq would take far longer than all of human history.

Exercises 41–44 and 46–47 involve RSA encryption.

(See *Amazon.com Licenses RSA Security's RSA BSAFE® Crypto-C Software*. Press release from RSA Security, **http://www.rsasecurity.com/press_release.asp?doc_id=189** and *RSA Encryption*, by Tom Davis, **http://mathcircle.berkeley.edu/BMC3/rsa/rsa.html**.)

- 2^4's power is 4, 4 is one less than 5, and 5 is a prime number.
- 2^6's power is 6, 6 is one less than 7, and 7 is a prime number.

So the pattern, so far, is that the first factors are all of the form 2^{n-1}, where n is a prime number.

Now let's look at the second factors of the perfect numbers. Each is a Mersenne number. That is, each is of the form $2^n - 1$, where n is a prime number.

- 6's second factor is $3 = 2^2 - 1$.
- 28's second factor is $7 = 2^3 - 1$.
- 496's second factor is $31 = 2^5 - 1$.
- 8,128's second factor is $127 = 2^7 - 1$.

This relationship between perfect numbers and Mersenne primes is an example of why people—mystics and mathematicians alike—have a sense that there are some secrets waiting to be discovered here. Perhaps there is a formula that will generate all of the primes. Perhaps there is some theory that makes the existence, location, and importance of the primes clearer.

We have discussed $(2^{n-1})(2^n - 1)$, where n is one of the first prime numbers: 2, 3, 5, and 7. What about the next prime, $n = 11$?

$$(2^{n-1})(2^n - 1) = (2^{11-1})(2^{11} - 1) = 1{,}024 \cdot 2{,}047 = 2{,}096{,}128$$

It was not known for quite a while whether this is a perfect number, because of the difficulty in factoring a number this large without the aid of technology. In 1536, it was discovered that this is not a perfect number and that $2^{11} - 1 = 2{,}047$ is not a Mersenne prime, because $2{,}047 = 23 \cdot 89$.

The fifth Mersenne prime was not discovered by Europeans until 1461, but an Arab mathematician found it in about 1200. It comes from using the next prime number, 13, for n.

Every single number of the form $(2^{n-1})(2^n - 1)$, where n and $2^n - 1$ are prime numbers, is a perfect number. Euclid proved this in 300 B.C. See Figure 7.9.

n (must be prime)	2^{n-1}	$2^n - 1$	Is $2^n - 1$ Prime?	$(2^{n-1})(2^n - 1)$	Is $(2^{n-1})(2^n - 1)$ Perfect?
2	$2^1 = 2$	$2^2 - 1 = 3$	yes	$2 \cdot 3 = 6$	yes
3	$2^2 = 4$	$2^3 - 1 = 7$	yes	$4 \cdot 7 = 28$	yes
5	$2^4 = 16$	$2^5 - 1 = 31$	yes	$16 \cdot 31 = 496$	yes
7	$2^6 = 64$	$2^7 - 1 = 127$	yes	$64 \cdot 127 = 8128$	yes
11	$2^{10} = 1024$	$2^{11} - 1 = 2047$	no, $2047 = 23 \cdot 89$	$1024 \cdot 2047 = 2{,}096{,}128$	no
13	?	?	yes	?	yes
17	?	?	yes	?	yes
19	?	?	yes	?	yes

FIGURE 7.9 Mersenne Numbers and Perfect Numbers.

7.4 EXERCISES

In Exercises 1–6, do the following.

a. Give the prime factorization of the number, if it has one.
b. State whether the number is prime or composite.

1. 42   **2.** 66   **3.** 23
4. 52   **5.** 54   **6.** 29

In Exercises 7–12, determine whether the number is prime or composite.

7. 299   **8.** 301   **9.** 207
10. 501   **11.** 233   **12.** 283

13. Use the Sieve of Eratosthenes to determine which of the numbers from 26 to 49 are primes.

14. Use the Sieve of Eratosthenes to determine which of the numbers from 50 to 75 are primes.

15. Evaluate the formula $2^n + 1$ for $n = 4$. Is the result prime or composite? Why?

16. Evaluate the formula $2^n + 1$ for $n = 5$. Is the result prime or composite?

17. Evaluate the formula $2^n + 1$ for $n = 6$. Is the result prime or composite? Why?

18. Evaluate the formula $2^n + 1$ for $n = 7$. Is the result prime or composite?

19. Evaluate the formula $2^n - 1$ for $n = 4$. Is the result prime or composite? Why?

20. Evaluate the formula $2^n - 1$ for $n = 6$. Is the result prime or composite? Why?

21. Evaluate the formula $2^n - 1$ for $n = 9$. Is the result prime or composite? Why?

22. Evaluate the formula $2^n - 1$ for $n = 8$. Is the result prime or composite? Why?

23. List all of the proper factors of 42.

24. List all of the proper factors of 43.

25. List all of the proper factors of 54.

26. List all of the proper factors of 55.

27. Is 42 abundant, perfect, or deficient? Why? (See Exercise 23.)

28. Is 43 abundant, perfect, or deficient? Why? (See Exercise 24.)

29. Is 54 abundant, perfect, or deficient? Why? (See Exercise 25.)

30. Is 55 abundant, perfect, or deficient? Why? (See Exercise 26.)

31. Is 61 abundant, perfect, or deficient? Why?

32. Is 60 abundant, perfect, or deficient? Why?

33. Is 62 abundant, perfect, or deficient? Why?

34. Is 65 abundant, perfect, or deficient? Why?

35. In this exercise, you will fill in the first blank row in Figure 7.9. Justify each of your answers.
    a. What is the value of n?
    b. What is the corresponding value of 2^{n-1}?
    c. What is the corresponding value of $2^n - 1$?
    d. What is the corresponding value of $(2^{n-1})(2^n - 1)$?

36. In this exercise, you will fill in the second blank row in Figure 7.9. Justify each of your answers.
    a. What is the value of n?
    b. What is the corresponding value of 2^{n-1}?
    c. What is the corresponding value of $2^n - 1$?
    d. What is the corresponding value of $(2^{n-1})(2^n - 1)$?

37. In this exercise, you will fill in the third blank row in Figure 7.9. Justify each of your answers.
    a. What is the value of n?
    b. What is the corresponding value of 2^{n-1}?
    c. What is the corresponding value of $2^n - 1$?
    d. What is the corresponding value of $(2^{n-1})(2^n - 1)$?

38. You may use any of the information in Figure 7.9 or elsewhere in the text in answering the following.
    a. Which answers to Exercises 35, 36, and 37 are Mersenne numbers? Why?
    b. Which answers to Exercises 35, 36, and 37 are Mersenne primes? Why?

c. Which answers to Exercises 35, 36, and 37 are perfect numbers? Why?

39. $2^{127} - 1$ is a Mersenne prime. Find the perfect number associated with this number. Write your answer (a) using exponents with a base of 2 and (b) in scientific notation.

40. $2^{521} - 1$ is a Mersenne prime. Find the perfect number associated with this number. Write your answer using exponents with a base of 2.

41. In RSA encryption, if the public key is 35, what is the private key?

42. In RSA encryption, if the public key is 51, what is the private key?

43. In RSA encryption, if the public key is 143, what is the private key?

44. In RSA encryption, if the public key is 221, what is the private key?

45. a. Rewrite the first four perfect numbers with the binary system.
    b. Describe the pattern or patterns you see in the result of part (a).

 Answer the following, using complete sentences and your own words.

• CONCEPT QUESTIONS

46. What do encryption and decryption have to do with prime numbers?

47. What do encryption and decryption have to do with factoring?

48. In the article on periodic cicadas, we stated that "a predator with a four-year life cycle would be around to eat them once every sixty-eight years." Justify this statement.

• HISTORY QUESTIONS

49. What is odd about Pythagoras' most famous result?

50. What strange feature did the Pythagoreans see in numbers?

WEB PROJECT

51. Write an essay on one of the following topics:
    - Mersenne primes
    - GIMPS—the Great Internet Mersenne Prime Search
    - large prime numbers
    - prime numbers and perfect numbers
    - history of prime numbers
    - mysticism and numbers

Some useful links for this web project are listed on the text web site: **www.cengage.com/math/johnson**

Fibonacci Numbers and the Golden Ratio

- Be able to generate the Fibonacci sequence
- Know different occurrences of Fibonacci numbers in nature
- Know what the golden rectangle and golden ratio are
- Know different occurrences of the golden rectangle in art

Fibonacci Numbers

In 1202, the Italian mathematician Fibonacci tried to determine how quickly animals would breed in ideal circumstances. In discussing this issue, Fibonacci used rabbits as an example. However, the logic in Fibonacci's rabbit example was unnecessarily complicated. Henry Dudeny, an English puzzle book author in the early 1900s, came up with the following simplified version of Fibonacci's logic.

If a cow produces its first female calf at age two years and produces another female calf every year after that, how many female calves are there after five years if we start with one newborn female calf and no calves die?

- At the beginning, there is one female calf.
- After one year, there is still only one female calf.
- After two years, the one female has produced a calf, so there is now 1 new cow, and $1 + 1 = 2$ cows in total. This new cow will start to produce in two years.
- After three years, the original female has produced a second calf, so there is 1 new cow, and there are $1 + 2 = 3$ cows in total.
- After four years, the original female has produced yet another calf, and the calf born two years ago has also produced a calf, so there are now 2 new cows, and there are $2 + 3 = 5$ cows in total.

Notice that each year, the number of new cows is the same as the number of cows that were alive two years ago, because each of those cows has produced a calf.

- After five years, there are three new cows: one born to the original female, one to the second female, and one to the third female. There are now $3 + 5 = 8$ cows in total.

See Figure 7.10, in which the cows are color-coded. The original female is red, and the first-born calf is blue.

How many cows will there be in one more year? Two years earlier there were five cows, and each will produce a calf. So there will be $5 + 8 = 13$ cows.

The number of cows in any given year is the sum of those alive two years ago (because they will each produce a calf) and those alive last year (because they're still alive).

The pattern is as follows: $1, 1, 1 + 1 = 2, 1 + 2 = 3, 2 + 3 = 5, 3 + 5 = 8, 5 + 8 = 13, \ldots$. Each number in the sequence is the sum of the previous two numbers. This sequence of numbers is called the **Fibonacci sequence.** The numbers themselves are called **Fibonacci numbers.**

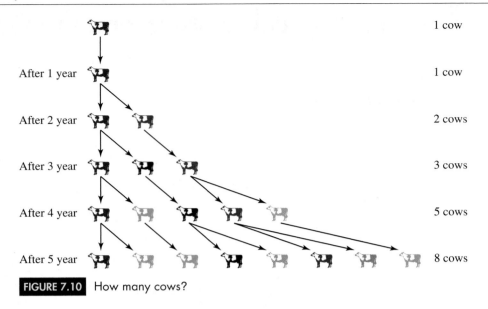

After 1 year ... 1 cow

After 2 year ... 2 cows

After 3 year ... 3 cows

After 4 year ... 5 cows

After 5 year ... 8 cows

1 cow

FIGURE 7.10 How many cows?

The Fibonacci Sequence and Honeybees

You might be thinking that the cow story is rather unrealistic. If so, you are right. But there are many realistic occurrences of Fibonacci numbers in nature, one of which has to do with honeybees.

There are two types of female honeybees: workers, which produce no eggs, and the queen, which produces eggs. Drone bees are males that are produced from the queen's unfertilized eggs. A drone's mother is the queen that laid the egg. But a drone *has no father,* because the egg was never fertilized.

Worker females and queen females are produced from the queen's fertilized eggs. So each type of female has a father as well as a mother, because the female eggs were fertilized.

Let's look at a drone's ancestors. A drone has:

- 1 parent—his mother
- 2 grandparents, who are his mother's parents—a male and a female
- 3 great-grandparents: his grandmother has two parents (his great-grandmother and great-grandfather), but his grandfather, like all males, has only one parent (his great-grandmother)
- 5 great-great-grandparents: each of his two great-grandmothers had two parents, and his one great-grandfather had one parent

The numbers of ancestors of a drone honeybee are the Fibonacci numbers.

Fibonacci Numbers and Plants

Frequently, the number of petals on a plant is a Fibonacci number. See Figures 7.11 and 7.12 on page 515.

Some plants have a spiral structure. For example, pinecones, pineapples, and cauliflowers all have a spiral structure. Frequently, the number of spirals is a Fibonacci number. And although you cannot see it in the photograph, the number of scales along any one spiral is usually a Fibonacci number. See Figures 7.13 and 7.14.

The number of petals on some plants is a Fibonacci number, and the number of spirals on some plants is a Fibonacci number. But these things do not always

One petal:
calla lily

Two petals:
euphorbia

Three petals:
trillium

Five petals:
hibiscus

Eight petals:
clematis

Thirteen petals:
black-eyed susan

Twenty-one petals:
shasta daisy

FIGURE 7.11 Fibonacci numbers and flowers.

FIGURE 7.12 This passionflower has three inner T-shaped petals, five white petals behind it, and then a set of five purple petals, behind which is another set of five light purple petals.

FIGURE 7.13 This pinecone has eight clockwise spirals. How many counterclockwise spirals does it have?

happen. You can find plants for which the number of petals is not a Fibonacci number. Nonetheless, it happens frequently. It is not a universal law but rather a tendency. Nature tends to use the Fibonacci numbers.

The Fibonacci Spiral

Here is another example of how nature tends to use Fibonacci numbers. The first two Fibonacci numbers are 1 and 1, so draw two adjacent squares, each with a side of length 1. See Figure 7.15.

FIGURE 7.14   This sunflower has fifty-five counterclockwise spirals. How many clockwise spirals does it have?

© John Foxx/Alamy

The left edge of those two squares is of length 2, and 2 is the next Fibonnaci number. Draw a square with a side of length 2 along this edge. See Figure 7.16.

The bottom edge is of length 3, and 3 is the next Fibonnaci number. Draw a square with a side of length 3 along this edge. See Figure 7.17.

The right edge is of length 5, and 5 is the next Fibonnaci number. Draw a square with a side of length 5 along this edge. See Figure 7.18.

The top edge is of length 8, and 8 is the next Fibonnaci number. Draw a square with a side of length 8 along this edge. See Figure 7.19.

FIGURE 7.15

Two squares with sides of length 1.

FIGURE 7.16

Squares with sides of length 1, 1, and 2.

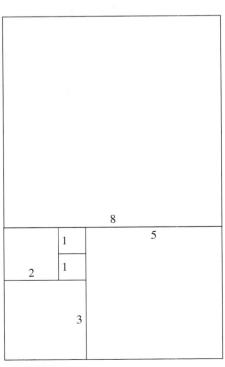

FIGURE 7.19   Squares with sides of length 1, 1, 2, 3, 5, and 8.

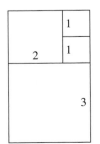

FIGURE 7.17

Squares with sides of length 1, 1, 2, and 3.

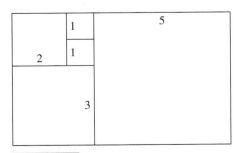

FIGURE 7.18   Squares with sides of length 1, 1, 2, 3, and 5.

We could continue this process indefinitely, and the next square would have an edge whose length is the next Fibonacci number. However, the result does not resemble anything found in nature. At least it doesn't until you insert quarter circles in the squares so that they form a spirallike curve. See Figure 7.20.

This **Fibonacci spiral** is a close approximation of the spirals found in nautilus shells, snail shells, and chameleon tails, but it does not perfectly match these natural spirals. See Figure 7.21.

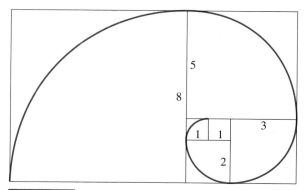

FIGURE 7.20 The Fibonacci spiral.

 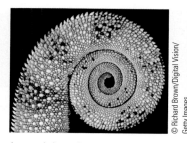 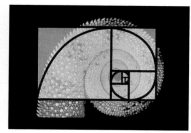

FIGURE 7.21 Nautilus shell, chameleon tails, and the Fibonacci spiral.

The Golden Rectangle and the Golden Ratio

Some artists and architects claim that rectangles shaped like the one that encloses the Fibonacci spiral in Figure 7.20 are more esthetically pleasing than others. What's your opinion? If the middle rectangle in Figure 7.22 has a more pleasant appearance than the others, then you agree, because it has the same shape as the one that encloses the Fibonacci spiral.

FIGURE 7.22 Three rectangles.

What do we mean when we say "the same shape"? The rectangle in Figure 7.20 has a short side of 8 and a long side of $8 + 2 + 3 = 13$. The ratio of its long side to its short side is $13/8 = 1.625$. The middle rectangle in Figure 7.22 is smaller than the rectangle in Figure 7.20, but the ratio of its sides is approximately the same. Without measuring, most people can't tell the difference between a rectangle whose ratio is 1.618 and one whose ratio is 1.625, so we can consider the rectangles to have the same shape even if their ratios are not exactly equal.

A rectangle with this ratio is called a **golden rectangle.** The ratio of a golden rectangle's long side to its short side is called the **golden ratio.**

The rectangle in Figure 7.23 consists of a square and another smaller rectangle. This smaller rectangle's ratio is

$$\frac{1}{0.618} = 1.618\ldots$$

and the larger rectangle's ratio is

$$\frac{1.618}{1} = 1.618$$

As a result, each rectangle is a golden rectangle. All golden rectangles can be broken up into a square and another smaller rectangle, where the smaller rectangle is another golden rectangle.

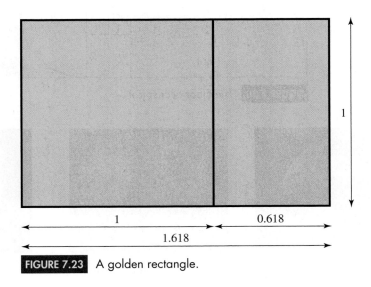

FIGURE 7.23 A golden rectangle.

In 1509, the Italian mathematician and artist Luca Pacioli wrote *De Divina Proportione* (*The Divine Proportion*), a book on the golden ratio and its use in art. The book was illustrated by Pacioli's close friend Leonardo da Vinci. Since then, it has been a major influence on many artists and architects.

Deliberate Users of the Golden Rectangle

Swiss architect Le Corbusier explicitly incorporated the golden rectangle (and the Fibonacci series) into his designs. He stated that the golden rectangle "resound(s) in man by an organic inevitability, the same fine inevitability which causes the tracing out of the Golden Section by children, old men, savages and the learned." See Figure 7.24 on page 519.

It is generally accepted that Salvadore Dali explicitly used the golden rectangle in *The Sacrament of the Last Supper* and other works. Juan Gris, Paul Serusier, and Giro Severini also deliberately made use of the golden rectangle in their works.

Leonardo da Vinci seldom discussed his art and never explicitly stated that he used the golden rectangle. Nevertheless, it occurs in many of his paintings including *Mona Lisa* and the unfinished St. Jerome in the *Wilderness*. His friendship with Pacioli and their collaboration on a book devoted to the golden ratio make it quite likely that Leonardo deliberately used the golden rectangle. See Figures 7.25 and 7.26.

FIGURE 7.24   Le Corbusier's Villa Stein.

FIGURE 7.25   Leonardo da Vinci's *Mona Lisa*.

FIGURE 7.26   Leonardo da Vinci's unfinished St. Jerome in the *Wilderness*.

Accidental Users

The golden rectangle occurs in the works of many artists and architects, apparently accidentally. This lack of deliberation only supports the thesis that the golden rectangle is esthetically pleasing.

Piet Mondrian's geometric paintings contain many golden rectangles, and Georges-Pierre Seurat's *Circus Sideshow* is easily seen as a collection of golden rectangles. See Figures 7.27 and 7.28. However, there is no evidence to support a claim that these artists knew about the golden ratio and used it in their art.

FIGURE 7.27   Piet Mondrian, *Composition with Red, Blue and Yellow*, 1930. Oil on canvas, 18 1/8 × 18 1/8 inches.

FIGURE 7.28   Georges-Pierre Seurat, *Circus Sideshow*, 1887–88. Oil on canvas, 39 1/4 × 59 inches.

The outline of the Parthenon in Athens, Greece, has several golden rectangles. See Figure 7.29. The Greeks knew about the golden ratio, but it's an open question as to whether they used the golden ratio in designing the Parthenon. The Great Pyramid of Giza, the Great Mosque of Kairouan, the General Assembly building of the United Nations, the Cathedral of Chartres, and the Notre Dame in Paris all exhibit the golden ratio. Some experts believe that this was deliberate, but many think that it was accidental.

FIGURE 7.29   The Parthenon—a golden rectangle.

FIGURE 7.30

Your own personal golden rectangle.

Businesses do many things to make their products more appealing to potential buyers. After all, that's the point of advertising. Do lenders deliberately make their credit cards in the shape of a golden rectangle to make the use of the cards more appealing? See Figure 7.30.

Deriving the Golden Ratio

Mathematically, where does the number 1.618 come from? It all starts with the fact that a golden rectangle can be broken up into a square and another smaller rectangle, where the smaller rectangle is another golden rectangle. The ratio for the bigger rectangle in Figure 7.31 is $\frac{a+b}{b}$, and the ratio for the smaller rectangle is $\frac{a}{b}$.

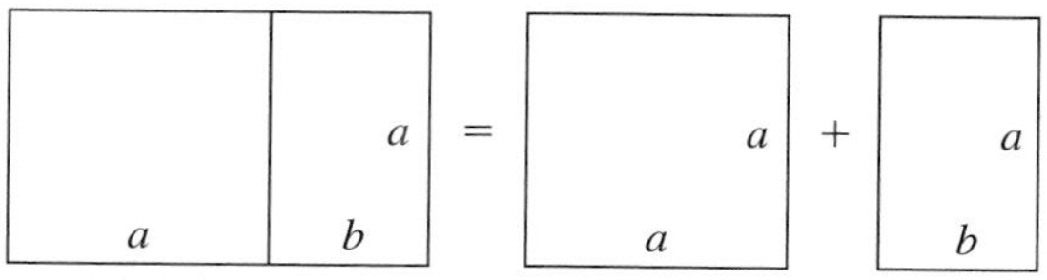

FIGURE 7.31 Finding the golden ratio.

If we make these ratios the same and solve the resulting equation for a/b, we get

$$a/b = \frac{1 + \sqrt{5}}{2} = 1.61803398\ldots$$

See Exercise 13. This number, which is called ϕ, is the golden ratio. (The symbol ϕ is the Greek letter phi.)

$$\phi = \frac{a}{b} = \frac{1 + \sqrt{5}}{2}$$

In its design, the golden rectangle is composed of two rectangles with the same ratio of length to width: a larger rectangle whose length is $a + b$ and whose width is a and a smaller rectangle whose length is a and whose width is b. See Figure 7.32.

Sometimes, artists use both golden rectangles. See Figure 7.33.

FIGURE 7.32

A golden rectangle within a golden rectangle.

© Musée du Louvre, Paris/Giraudon/The Bridgeman Art Library

FIGURE 7.33 A golden rectangle within a golden rectangle.

7.5 EXERCISES

Exercises 1 and 2 refer to Henry Dudeny's version of Fibonacci's puzzle.

1. How many female calves are there after six years, assuming that none die? Of these calves, how many are newly born?

2. How many female calves are there after seven years, assuming that none die? Of these calves, how many are newly born?

3. Give the ten Fibonacci numbers after 13.

4. Give the five Fibonacci numbers after those listed in Exercise 3.

5. How many great-great-great-grandparents does a drone bee have? How is each related to the drone?

6. How many great-great-great-great-grandparents does a drone bee have? How is each related to the drone?

7. How many counterclockwise spirals does the pinecone in Figure 7.13 have?

8. How many clockwise spirals does the sunflower in Figure 7.14 have?

9. Carefully draw the Fibonacci spiral composed of squares with sides of lengths 1, 1, 2, 3, 5, 8, and 13.

10. Carefully draw the Fibonacci spiral composed of squares with sides of lengths 1, 1, 2, 3, 5, 8, 13, and 21.

11. *Binet's Formula* states that the nth Fibonacci number is

$$\frac{1}{\sqrt{5}}\left[\left(\frac{1+\sqrt{5}}{2}\right)^n - \left(\frac{1-\sqrt{5}}{2}\right)^n\right]$$

 a. Use Binet's Formula to find the thirtieth and fortieth Fibonacci numbers.

 b. What advantage does Binet's Formula hold over the method of generating Fibonacci numbers discussed in this section?

 c. What disadvantage does Binet's Formula hold over the method of generating Fibonacci numbers discussed in this section?

12. a. Use Binet's Formula (see Exercise 11) to find the 50th and 60th Fibonacci numbers.

 b. What would you have to do to find the 50th and 60th Fibonacci numbers, without Binet's Formula?

13. Complete the derivation of the golden ratio that was started on page 521. Specifically, do the following:

 a. Make an equation by setting the larger rectangle's length to width ratio equal to that of the smaller rectangle.

 b. Find the lowest common denominator of the two fractions in part (a), and multiply each side of the equation in part (a) by it. This will eliminate denominators.

 c. Simplify the result of part (b) by distributing.

 d. Divide the result of part (c) by b^2.

 e. Simplify the result of part (d) by canceling.

 f. Substitute x for a/b in part (e). This should eliminate all a's and b's from the equation.

 g. Use the quadratic equation to solve the result of part (f). This generates two different answers. The positive one is the golden ratio.

 h. Why does a negative answer not make sense in this context?

14. The golden ratio is the only number whose square can be produced simply by adding 1.

 a. Use your calculator to approximate the square of the golden ratio.

 b. Use your calculator to approximate the sum of 1 and the golden ratio. What do you observe?

 c. Use the result of Exercise 13 part (f) to verify that the square of the golden ratio is exactly equal to the sum of 1 and the golden ratio?

15. Find a work of art that uses the golden ratio. Provide a sketch or photograph of the work of art, and discuss its use of the golden ratio. Do not use any work discussed in detail in the text.

16. Find a building that uses a golden rectangle. Provide a sketch or photograph of the building, and discuss its use of the golden rectangle. Do not use any work discussed in detail in the text.

17. Find something in nature that involves Fibonacci numbers. Provide a sketch, photograph, or sample of the thing itself. Describe its involvement with Fibonacci numbers. Do not use any work discussed in detail in the text.

18. Complete the chart in Figure 7.34. Do not round off the decimals in the third column. As you read down the third column, the numbers get closer to a certain number discussed in this section. What number is that?

Fibonacci number	Dividing consecutive Fibonacci numbers	Decimal approximation of the result
1		
1	1/1	1
2	2/1	2
3	3/2	1.5
5		
8		
13		
21		
34		
55		

FIGURE 7.34 Chart for Exercise 18.

522

19. The twenty-fourth and twenty-fifth Fibonacci numbers are 46,368 and 75,025, respectively. Divide the larger of these numbers by the smaller. What do you observe?

20. Exercises 18 and 19 indicate a relationship between Fibonacci numbers and the golden ratio. There is another relationship between them that has to do with Binet's Formula in Exercise 11 and the result of Exercise 13. Describe that relationship.

21. Carefully inspect the photograph of the Parthenon in Figure 7.29. Describe other occurrences of the golden rectangle or the golden ratio.

22. Carefully inspect *Composition in Red, Yellow and Blue* in Figure 7.27. Which rectangles are golden rectangles?

23. Carefully inspect *The Circus Sideshow* in Figure 7.28. Describe occurrences of golden rectangles and golden ratios.

24. Carefully inspect *Villa Stein* in Figure 7.24. Describe occurrences of golden rectangles.

• PROJECTS

25. It has been claimed that Fibonacci numbers are nature's numbers. Investigate this claim in some or all of the following situations.

- Get several pineapples. Pineapples have three sets of spirals, as shown in Figure 7.35. Count the number of spirals in each direction.

FIGURE 7.35

- Get several cauliflowers. Count the number of clockwise spirals and counterclockwise spirals on each. This can be difficult due to the wartiness of the flowerlets.
- Get several sunflowers. Count the number of clockwise spirals on the flower, the number of counterclockwise spirals on the flower, and the number of seeds on each spiral.
- Count the number of petals on many flowers.
- Usually, leaves are located on the stem of a plant in such a way that the leaves actually spiral around the stem (Figure 7.36). If you start at one leaf and follow one complete spiral, you might come to another leaf

that points in the same direction as does the first leaf. With some plants, one complete spiral is not enough, and you need to follow two complete spirals before you come to another leaf that points in the same direction as does the first leaf. For several different plants, count the number of spirals before you come to another leaf that points in the same direction as does the first leaf. Also, count the number of leaves from the first leaf to the next leaf that points in the same direction (not counting the first leaf).

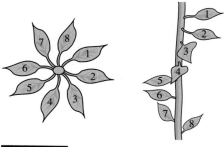

FIGURE 7.36

Write an essay in which you describe your research and discuss your results.

26. Obtain good copies of the following works of art:

- Leonardo da Vinci's *Mona Lisa* and *Vitruvian Man*
- Boticelli's *The Birth of Venus*

Search for golden rectangles in these works. Then determine whether or not those rectangles are actually golden by measuring their lengths and widths and dividing. You might also investigate works by Paul Serusier, Giro Severini, and Juan Gris as well as other works by Leonardo, Seurat, and Boticelli.

WEB PROJECT

27. Write an essay on one of the following topics:

- Fibonacci numbers in nature
- Fibonacci puzzles
- Fibonacci numbers and the golden rectangle
- the mathematics of Fibonacci numbers
- Fibonacci numbers and continued fractions
- Fibonacci's biography
- the golden rectangle and art
- Leonardo da Vinci's friendship and collaborative efforts with Fibonacci
- Fibonacci numbers and the stock market
- Fibonacci numbers and music
- the claim that the Egyptian pyramids were constructed using the golden ratio
- the claim that Debussy, Mozart, and Bartok used the golden ratio in their music and that some Gregorian chants are based on the golden ratio

Some useful links for this web project are listed on the text web site: **www.cengage.com/math/johnson**

REVIEW EXERCISES

In Exercises 1–18, write the given number in expanded form, or explain why there is no such number.

1. 5372
2. 1208
3. 325_8

4. 8642_8
5. 390_8
6. 2205_8

7. 905_{16}
8. 3265_{16}
9. ABC_{16}

10. $D013_{16}$
11. 11011001_2
12. 2103_2

13. 101112_2
14. 1011001_2
15. $39,22,54_{60}$

16. $12,14,75_{60}$
17. $25,44,34_{60}$
18. $12,18,05_{60}$

19. Convert 514_8 to base ten.

20. Convert $229A_{16}$ to base ten.

21. Convert 110101_2 to base ten.

22. Convert $32,55,49_{60}$ to base ten.

23. Convert $BC9A_{16}$ to base ten.

24. Convert $41,39,18_{60}$ to base ten.

25. Convert 111000_2 to base ten.

26. Convert 37292_8 to base ten.

27. Convert 1254_9 to base ten.

28. Convert 3241_5 to base ten.

29. Convert 514 to base eight.

30. Convert 3,922 to base two.

31. Convert 8,430 to base sixteen.

32. Convert 22,876 to base sixty.

33. Convert 39,287 to base sixty.

34. Convert 30,281 to base eight.

35. Convert 12,003 to base two.

36. Convert 6,438 to base sixteen.

37. Convert 59,375 to base seven.

38. Convert 44,886 to base three.

39. If $543_x = 207$, find base x.

40. If $123_x = 38$, find base x.

41. If $421_x = 343$, find base x.

42. If $111_x = 57$, find base x.

In Exercises 43–66, do the following.

a. *Perform the given computation, working solely in the given base.*
b. *Check your answer by converting it to base ten.*

43. $52_8 + 62_8$
44. $39_{16} + AB_{16}$

45. $101_2 + 110_2$

46. $22,12,53_{60} + 18,49,03_{60}$
47. $A9_{16} + 9B_{16}$

48. $1101_2 + 101_2$

49. $22,33,44_{60} + 55,44,33_{60}$

50. $32_8 + 157_8$
51. $201_8 - 112_8$

52. $50C_{16} - BA_{16}$
53. $1101_2 - 1010_2$

54. $9,22,32_{60} - 1,53,26_{60}$
55. $A08_{16} - 7B_{16}$

56. $1111_2 - 1001_2$

57. $22,33,44_{60} - 10,44,33_{60}$
58. $540_8 - 157_8$

59. $22_8 \cdot 72_8$
60. $3A_{16} \cdot A4_{16}$

61. $101_2 \cdot 110_2$
62. $42_7 \cdot 13_7$

63. $B3_{16} \cdot 92_{16}$
64. $1001_2 \cdot 101_2$

65. $34_5 \cdot 41_5$
66. $32_8 \cdot 57_8$

In Exercises 67–74, do the following:

a. *Perform the given computation, working solely in the given base.*
b. *Check your answer by multiplying in the given base.*

67. $301_8 \div 12_8$
68. $50C_{16} \div B1_{16}$

69. $1101_2 \div 101_2$
70. $147_8 \div 17_8$

71. $A08_{16} \div 1B_{16}$
72. $1111_2 \div 11_2$

73. $110110_2 \div 10_2$
74. $547_8 \div 17_8$

In Exercises 75–78, do the following:

a. *Give the prime factorization of the number, if it has one.*
b. *State whether the number is prime or composite.*

75. 33
76. 37
77. 41
78. 44

79. Use the Sieve of Eratosthenes to determine which of the numbers from 50 to 59 are primes.

80. Use the Sieve of Eratosthenes to determine which of the numbers from 60 to 69 are primes.

81. Evaluate the formula $2^n + 1$ for $n = 2$ and 3. Determine which of the results are prime and which are composites.

82. Evaluate the formula $2^n + 1$ for $n = 4$ and 5. Determine which of the results are prime and which are composites.

83. Evaluate the formula $2^n - 1$ for $n = 3$. Is the result prime or composite? Why?

84. Evaluate the formula $2^n - 1$ for $n = 4$. Is the result prime or composite? Why?

85. List all of the proper factors of 70.
86. List all of the proper factors of 66.
87. Is 20 abundant, perfect or deficient? Why?
88. Is 21 abundant, perfect or deficient? Why?
89. Is 49 abundant, perfect or deficient? Why?
90. Is 48 abundant, perfect or deficient? Why?

91. $2^{17} - 1$ is a Mersenne prime. Find the perfect number associated with this number. Write your answer (a) using exponents with a base of 2 and (b) in scientific notation.

92. $2^{19} - 1$ is a Mersenne prime. Find the perfect number associated with this number. Write your answer (a) using exponents with a base of 2 and (b) in scientific notation.

93. Give the first twenty Fibonacci numbers.

94. Referring to Henry Dudeny's version of Fibonacci's puzzle, how many female cows are there after five years, assuming that none die? Of these cows, how many are newly born?

95. How many great-great-grandparents does a drone bee have? How is each related to the drone?

96. How many great-grandparents does a drone bee have? How is each related to the drone?

97. Referring to Henry Dudeny's version of Fibonacci's puzzle, how many female calves are there after four years, assuming that none die? Of these calves, how many are newly born?

98. Carefully draw the Fibonacci spiral composed of squares with sides of lengths 1, 1, 2, 3, 5, and 8.

99. *Binet's Formula* states that the *n*th Fibonacci number is

$$\frac{1}{\sqrt{5}}\left[\left(\frac{1 + \sqrt{5}}{2}\right)^n - \left(\frac{1 - \sqrt{5}}{2}\right)^n\right]$$

a. Use Binet's Formula to find the twenty-fifth and twenty-sixth Fibonacci numbers.

b. Use the results of part (a) to find the twenty-seventh Fibonacci number.

100. Give six different examples of the presence of Fibonacci numbers in nature.

101. What is the golden rectangle?
102. What is the golden ratio?
103. Give an example of a work of art that uses a golden ratio.
104. Give an example of a building that uses a golden rectangle.
105. Describe two different relationships between the golden ratio and Fibonacci numbers.

Answer the following, using complete sentences and your own words.

• CONCEPT QUESTIONS

106. Why is there no base zero?
107. How does the place system use the number 0?
108. Do Roman numerals have a symbol for zero? Why or why not?
109. Would the Babylonian base sixty system have had some way of representing the number zero? Why or why not?
110. Describe in detail how to add $14_8 + 76_8$ on a base eight abacus.
111. Describe in detail how to subtract $321 - 74$ on a base ten abacus.

• HISTORY QUESTIONS

112. Describe the origins of Hindu-Arabic numerals.
113. What was the subject matter of Mohammed ibn Musa al-Khowarizmi's two books?
114. Describe the Pythagorean's beliefs regarding numbers.
115. Why were people originally interested in perfect numbers?
116. Is the interest in prime numbers a relatively new or an old part of the human pursuit of knowledge?
117. What was Fibonacci trying to do when he invented Fibonacci numbers?
118. Describe one of the earlier occurrences of the golden rectangle or golden ratio.

GEOMETRY

Geometry is one of the oldest branches of mathematics. Its roots include two very practical human endeavors: surveying and astronomy. From the beginning of recorded history, people have attempted to measure the land and the heavens. The word *geometry* literally means "earth measurement"; it is derived from the Greek words *geos*, meaning "earth," and *metros*, meaning "measure."

Over time, geometry evolved from the practical, empirical geometry typified by the Egyptians to the deductive, systematic, proof-laden geometry of the Greeks, typified by the works of Euclid. However, ancient geometry was devoid of symbolic algebra, which did not yet exist. With the development of algebra, the fundamentals of geometry were revisited; geometry and algebra were combined to form the classic analytic

continued

WHAT WE WILL DO IN THIS CHAPTER

WE WILL REVIEW THE STANDARD CONCEPTS AND CALCULATIONS OF LENGTH, AREA, AND VOLUME:

- How is the area of a standard geometric shape or the volume of a standard solid calculated?

- What exactly is π?

WE WILL PRESENT A BRIEF HISTORY OF SOME OF THE MAJOR ACCOMPLISHMENTS OF ANCIENT GEOMETRIES:

- What units of measurement were used by the ancient Egyptians?

- Did the Egyptians know the value of π?

- What is the Pythagorean Theorem, and why is it true?

WE WILL EXPLORE ANALYTIC GEOMETRY THROUGH THE STUDY OF CONIC SECTIONS:

- What is a parabola, and how can parabolas be applied to real-world situations?

continued

527

- What is an ellipse, and how are ellipses used in the study of astronomy?

- What is a hyperbola, and how can hyperbolas be used in navigation?

WE WILL PRESENT A BRIEF HISTORY OF NON-EUCLIDEAN GEOMETRIES:

- What is the Parallel Postulate, and why was it so controversial?

- Do parallel lines exist?

WE WILL DEVELOP AND APPLY THE FUNDAMENTAL RATIOS OF RIGHT TRIANGLE TRIGONOMETRY:

- Without using a tape measure, how can the height of a building or a tall tree be measured?

- How are angles and triangles used to measure astronomical distances, such as the distance between earth and the closest star (other than the sun)?

WE WILL EXPLORE THE USE OF LINEAR PERSPECTIVE IN THE WORLD OF ART:

- What is perspective?

- What does perspective give to a work of art?

- What does perspective have to do with geometry?

WE WILL EXPLORE THE FOUNDATIONS OF FRACTAL GEOMETRY:

- What is a fractal?

- How is mathematics used to create the fantastic images of fractal art?

- How are fractals used in movie production?

geometry of the seventeenth century, typified by the works of the Europeans Fermat and Descartes.

The evolution of geometry progressed to the point at which the fundamental beliefs of the Greeks were challenged, and consequently, in the nineteenth century, several non-Euclidean geometries were created. Just as the introduction of algebra had revolutionized the study of geometry centuries before, the introduction of computers in the late twentieth century gave rise to new geometries including the artistic fractal geometry of Benoit Mandelbrot.

8.1 Perimeter and Area

OBJECTIVES

- Find the perimeter of basic polygons, including squares, rectangles, and triangles

- Apply the Pythagorean Theorem to a right triangle

- Find the area of basic polygons, including squares, rectangles, triangles, parallelograms, and trapezoids

- Apply Heron's Formula to find the area of a triangle

- Use π in the calculation of circumference and area of a circle

- Understand and apply the units that are commonly used in astronomical measurement, including astronomical units, light-years, and parsecs

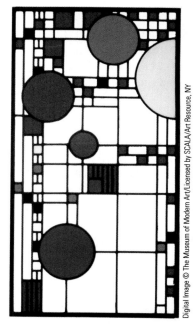

Architect Frank Lloyd Wright used simple geometric shapes to design this stained glass window in 1911.

Geometric shapes have intrigued people throughout history. Art, religion, science, engineering, architecture, psychology, and advertising are just a few of the many areas that make use of triangles, rectangles, squares, circles, cubes, pyramids, cones, spheres, and a host of other forms. Having seen many of these shapes in nature, geometers have constructed ideal representations of them and have developed various formulas for measuring their lengths (one-dimensional), areas (two-dimensional), and volumes (three-dimensional). In this section, we examine some features of the most commonly encountered two-dimensional figures.

Polygons

Two-dimensional figures can be classified by the number of sides they have. A **polygon** is a many-sided figure. A pentagon is a five-sided figure, a hexagon is a six-sided figure, and an octagon is an eight-sided figure. However, the names of polygons do not necessarily end with -*gon*. Although a three-sided figure could be called a *trigon,* we prefer *triangle*. Likewise, a four-sided figure is referred to as a quadrilateral rather than a quadragon.

Our study of polygons focuses on finding the distance around a figure and the amount of space enclosed within the figure. Some of the polygons we will examine are shown in Figure 8.1. The symbol ∟ represents an angle of 90° (90 degrees = a square corner).

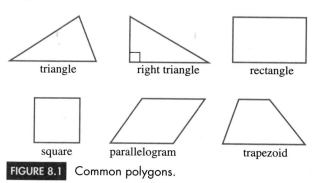

FIGURE 8.1 Common polygons.

The **perimeter** of (or distance around) a two-dimensional figure is the sum of the lengths of its sides. As shown in Figure 8.2, the perimeter of a rectangular scarf 18 inches wide and 2 feet long is 7 feet. (We must first convert 18 inches into 1.5 feet.)

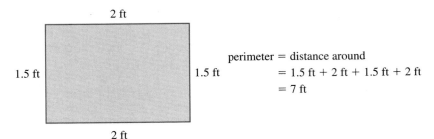

perimeter = distance around
= 1.5 ft + 2 ft + 1.5 ft + 2 ft
= 7 ft

FIGURE 8.2 Finding the perimeter of a rectangular scarf.

The **area** of a two-dimensional figure is the number of square units (for example, square inches, square miles) it takes to fill the interior of that figure. As shown in Figure 8.3, a rectangular rug 6 feet long and 3 feet 6 inches wide has an area of 21 square feet (or 21 ft^2).

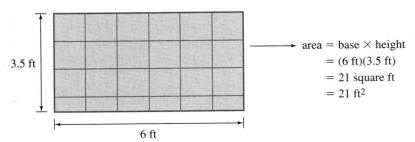

area = base × height
= (6 ft)(3.5 ft)
= 21 square ft
= 21 ft^2

FIGURE 8.3 Finding the area of a rectangular scarf.

It has been suggested that the concepts of square units and area have their origins in the weaving of fabric. Single strands of yarn have a linear, or one-dimensional, measurement, such as inches or feet. However, when an equal number of "horizontal" and "vertical" strands are woven together on a loom, a square figure is formed. Therefore, a natural way to measure the amount of cloth created from the strands is to employ units consisting of squares (square feet). (See Figure 8.4.)

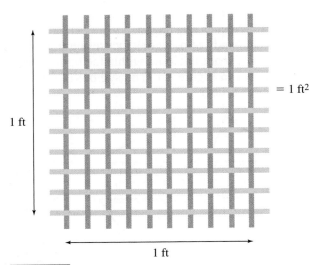

FIGURE 8.4 Woven cloth.

It is very easy to find the area of a rectangle or square. On the basis of these quadrilaterals, we can also find the areas of triangles, trapezoids, and parallelograms. A **trapezoid** is a quadrilateral with one pair of parallel sides; a **parallelogram** is a quadrilateral with two pairs of parallel sides. The area of a **triangle, rectangle,** parallelogram, or trapezoid can be found by use of the appropriate formula given in Figure 8.5, with A = area, b = base, h = height, and b_1 and b_2 = bases.

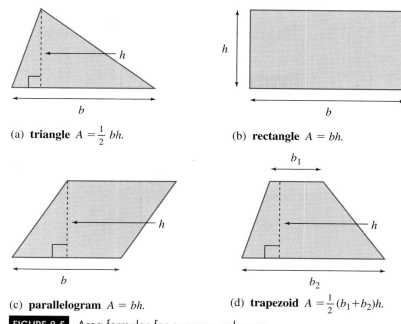

(a) **triangle** $A = \frac{1}{2} bh.$

(b) **rectangle** $A = bh.$

(c) **parallelogram** $A = bh.$

(d) **trapezoid** $A = \frac{1}{2}(b_1 + b_2)h.$

FIGURE 8.5   Area formulas for common polygons.

Why does the area of a triangle equal one-half the product of the base times the height? The answer lies in the formula for the area of a rectangle. A triangle can be divided into two smaller triangles. If copies of these smaller triangles are then "added on" to the original triangle, a rectangle of area $b \cdot h$ can be formed, as shown in Figure 8.6. Because the area of the original triangle is half that of the rectangle, we have the desired result.

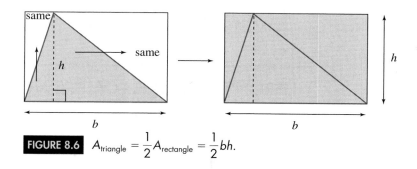

FIGURE 8.6 $A_{triangle} = \frac{1}{2} A_{rectangle} = \frac{1}{2} bh.$

Why does the area of a parallelogram equal the product of the base times the height? Once again, the answer lies in the formula for the area of a rectangle. A parallelogram can be rearranged to form a rectangle of area $b \cdot h$, as shown in Figure 8.7. That is, the area of the parallelogram is the same as the area of the rectangle, and we have the desired result.

Why does the area of a trapezoid equal one-half the product of the sum of the bases times the height? Yet again, the answer lies in the formula for the area of a rectangle! The two triangular "tips" of the trapezoid can be cut off and rearranged to form a rectangle, as shown in Figure 8.8. The base of the rectangle equals the average of the two bases of the trapezoid; that is, $b_{\text{rectangle}} = (b_1 + b_2)/2$. Thus, we have the desired result.

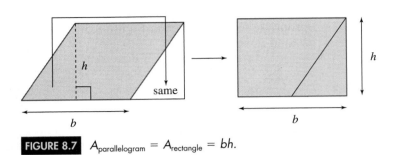

FIGURE 8.7 $A_{\text{parallelogram}} = A_{\text{rectangle}} = bh.$

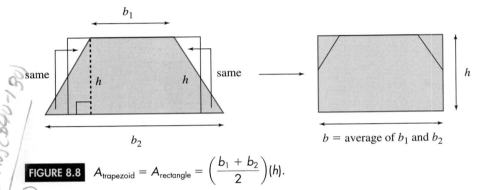

FIGURE 8.8 $A_{\text{trapezoid}} = A_{\text{rectangle}} = \left(\dfrac{b_1 + b_2}{2}\right)(h).$

Heron's Formula for the Area of a Triangle

If the height of triangle is not known, we cannot use the common formula $A = \frac{1}{2}bh$; an alternative method must be used. When the lengths of all three sides are known, the "semiperimeter" can be used to find the area of the triangle. As the name implies, the semiperimeter is half the perimeter. The formula is referred to as Heron's Formula, in honor of the ancient Greek mathematician Heron of Alexandria (circa A.D. 75).

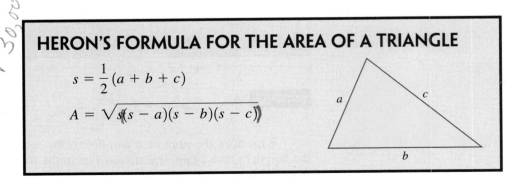

HERON'S FORMULA FOR THE AREA OF A TRIANGLE

$$s = \frac{1}{2}(a + b + c)$$

$$A = \sqrt{s(s - a)(s - b)(s - c)}$$

EXAMPLE **1** | **FINDING AREAS OF RECTANGLES AND TRIANGLES** Assuming the same growing conditions, which of the fields shown in Figures 8.9 and 8.10 would produce more grain?

a.

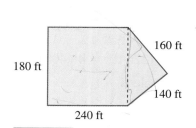

160 ft

180 ft

140 ft

240 ft

FIGURE 8.9

b.

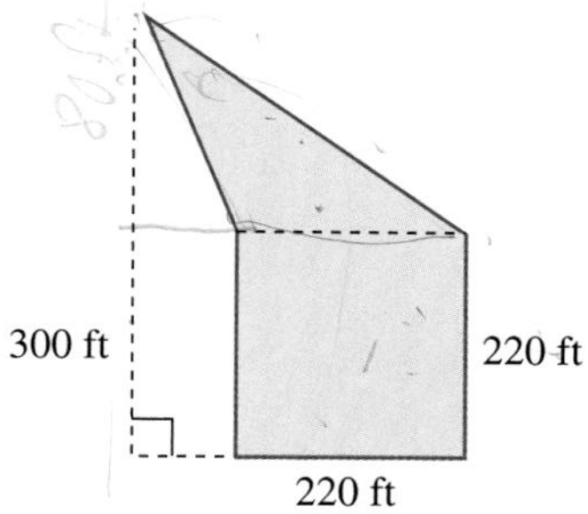

300 ft   220 ft

220 ft

FIGURE 8.10

SOLUTION

The field with the larger area would produce more grain.

Finding the Area of the Field in Part (a)   The field consists of a rectangle and a triangle, as shown in Figure 8.11. To find the total area, we must find the area of each separate shape and add the results together.

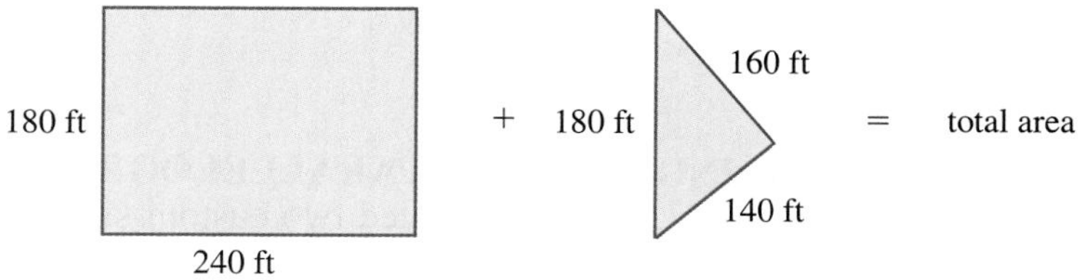

180 ft   240 ft $+$ 180 ft   160 ft   140 ft $=$ total area

FIGURE 8.11 Finding the area of a field.

$$A_{\text{rectangle}} + A_{\text{triangle}} = A_{\text{total}}$$

The area of the rectangle is

$$A_{\text{rectangle}} = bh = (240 \text{ ft})(180 \text{ ft}) = 43{,}200 \text{ ft}^2$$

To find the area of the triangle, we use the semiperimeter, because the lengths of all three sides are known:

$$s = \frac{1}{2}(a + b + c) = \frac{1}{2}(180 \text{ ft} + 160 \text{ ft} + 140 \text{ ft}) = 240 \text{ ft}$$

Using Heron's Formula, we have

$$
\begin{aligned}
A_{\text{triangle}} &= \sqrt{240(240 - 180)(240 - 160)(240 - 140)} \\
&= \sqrt{(240 \text{ ft})(60 \text{ ft})(80 \text{ ft})(100 \text{ ft})} \\
&= \sqrt{115{,}200{,}000 \text{ ft}^4} \\
&= 10{,}733.12629\ldots \text{ ft}^2 \\
&\approx 10{,}733 \text{ ft}^2 \\
A_{\text{total}} &= A_{\text{rectangle}} + A_{\text{triangle}} = 43{,}200 \text{ ft}^2 + 10{,}733 \text{ ft}^2 \\
&= 53{,}933 \text{ ft}^2
\end{aligned}
$$

The total area of the field in part (a) is 53,933 square feet.

Finding the Area of the Field in Part (b)   The field in part (b) is composed of a square and a triangle. The lengths of two of the sides of the triangle are not given, so the semiperimeter cannot be used. However, we do know that the base of the triangle is 220 feet, and we can find its height. The total height of the figure is 300 feet,

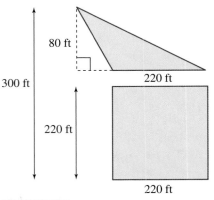

and the height of the square is 220 feet; we deduce that the height of the triangle is 300 ft − 220 ft = 80 ft. (See Figure 8.12.) Now the area of the square is

$$A_{square} = b^2 = (220 \text{ ft})^2 = 48,400 \text{ ft}^2$$

The area of the triangle is

$$A_{triangle} = \frac{1}{2}bh = \frac{1}{2}(220 \text{ ft})(80 \text{ ft})$$

$$= 8,800 \text{ ft}^2$$

$$A_{total} = A_{square} + A_{triangle}$$

$$= 48,400 \text{ ft}^2 + 8,800 \text{ ft}^2$$

$$= 57,200 \text{ ft}^2$$

FIGURE 8.12 Finding the area of a field.

The total area of the field in part (b) is 57,200 square feet.

Assuming the same growing conditions, the field in part (b) would produce more grain, because it has a larger area than the field in part (a).

EXAMPLE **2**

FINDING AREAS OF PARALLELOGRAMS AND TRAPEZOIDS A local art supply store has donated two custom-size canvases to be used for murals in a youth center. Before they can be used, they must be treated with a special primer. Of the two canvases shown in Figures 8.13 and 8.14, which requires more primer?

a.

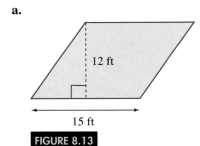

b.

FIGURE 8.13

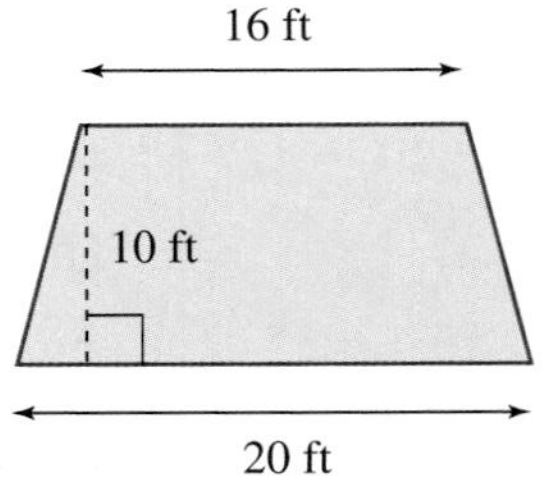

FIGURE 8.14

SOLUTION

The canvas with the larger area requires more primer.

Finding the Area of the Parallelogram in Part (a)

$$A_{parallelogram} = bh = (15 \text{ ft})(12 \text{ ft}) = 180 \text{ ft}^2$$

The canvas in part (a) contains 180 square feet of material to be primed.

Finding the Area of the Trapezoid in Part (b)

$$A_{trapezoid} = \frac{1}{2}(b_1 + b_2)h$$

$$= \frac{1}{2}(20 \text{ ft} + 16 \text{ ft})(10 \text{ ft})$$

$$= \frac{1}{2}(36 \text{ ft})(10 \text{ ft})$$

$$= 180 \text{ ft}^2$$

The canvas in part (b) also contains 180 square feet of material to be primed. Because the figures have the same area, each would require the same amount of primer.

c = hypotenuse

b = leg

a = leg

FIGURE 8.15

Right triangle.

Right Triangles

If one of the angles of a triangle is a right angle (a square corner, or 90°), the triangle is called a **right triangle.** The side opposite the right angle is called the **hypotenuse** and is labeled c, while the remaining two sides are called the **legs** and are labeled a and b, as shown in Figure 8.15.

An early military handbook utilized the Pythagorean Theorem. Because $18^2 + 24^2 = 30^2$, a 30-foot ladder is required to reach the top of a 24-foot wall from the opposite side of an 18-foot moat.

A special relationship exists between the hypotenuse and the legs of a right triangle. Over 2,000 years ago, early geometers observed that if the longest side of a right triangle (the hypotenuse) was squared, the number obtained was always the same as the sum of the squares of the two other sides (the legs). This observation was proved to be true for *all* right triangles. It is referred to as the **Pythagorean Theorem,** in honor of the ancient Greek mathematician Pythagoras of Samos (circa 580 B.C.), although the result was known to earlier peoples.

PYTHAGOREAN THEOREM

For any right triangle, the square of the hypotenuse equals the sum of the squares of the legs.

$$c^2 = a^2 + b^2$$

c = hypotenuse

b = leg

a = leg

---

c = 25 ft

a = 7 ft

b

FIGURE 8.16

A right triangle.

The converse of this theorem is also true; if the sides of a triangle satisfy the relationship $c^2 = a^2 + b^2$, then the triangle is a right triangle.

EXAMPLE **3**

APPLYING THE PYTHAGOREAN THEOREM AND FINDING THE AREA OF A RIGHT TRIANGLE.   The length of the hypotenuse of a right triangle is 25 feet, and the length of one of the legs is 7 feet. (See Figure 8.16.) Find the area of the triangle.

SOLUTION

To find the area of a triangle, we must know either a base and its corresponding (perpendicular) height or all three sides. In this case, we must find the missing side of the given triangle. Because we have a right triangle, we can apply the Pythagorean Theorem to find the missing leg, b:

$$a^2 + b^2 = c^2$$
$$b^2 = c^2 - a^2$$
$$b^2 = (25)^2 - (7)^2$$
$$b^2 = 625 - 49 = 576$$
$$b = \sqrt{576} = 24 \text{ ft}$$

Now we find the area of the triangle:

$$A = \frac{1}{2} bh = \frac{1}{2} (24 \text{ ft})(7 \text{ ft}) = 84 \text{ ft}^2$$

The area of the triangle is 84 square feet. (We obtain the same answer if we use Heron's Formula.)

Circles

Many people consider the circle a perfect geometric figure; it has no beginning or end, it has total symmetry, and it motivated the invention of the famous irrational (nonfraction) number π **(pi).** The early Greeks defined a **circle** as the set of all points in a plane equidistant from a fixed point. The fixed point is called the **center** of the circle; a line segment from the center to any point on the circle is called a **radius;** and any line segment connecting two points on the circle and passing through the center is called a **diameter.** (The length of a diameter is twice the length of a radius.) See Figure 8.17.

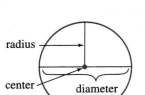

FIGURE 8.17

A circle.

Recall that the distance around a figure is called its *perimeter*; however, as a special case, the distance around a circle is called its **circumference.** Thousands of years ago, geometers observed a curious relationship: If they measured the circumference C of *any* circle (probably by using a string or a rope), they found that it was always a little bit longer than 3 times the diameter d of the circle, as shown in Figure 8.18. In other words, the ratio of circumference to diameter is constant: circumference/diameter = constant. This constant number, which is a little larger than 3, is represented by the Greek letter π (read "pi"; rhymes with *sly*). Hence,

$$\frac{\text{circumference}}{\text{diameter}} = \pi \qquad \text{or} \qquad \text{circumference} = \pi \cdot \text{diameter}$$

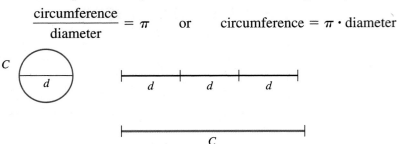

FIGURE 8.18   Circumference of a circle.

Mathematicians have shown that π is irrational; that is, it cannot be written exactly as a fraction, and its decimal expansion never repeats or terminates. Computers have calculated π to hundreds of thousands of decimal places, and we have

$$\pi \approx 3.14159265358979323846643383279\ldots$$

Various approximations of π have been used throughout history (some of them are investigated in Sections 8.3 and 8.4). Today, the most commonly used classroom estimates are 3.1416 and $\frac{22}{7}$. When π is needed in calculations, simply press the appropriate button on your calculator. If your calculator doesn't have a $\boxed{\pi}$ button, use $\pi = 3.1416$.

Pi is used to calculate the circumference and the area of a circle. We will investigate the origins of the area formula in Section 8.3.

CIRCUMFERENCE AND AREA OF A CIRCLE

The **circumference** C of a circle of radius r is

$$C = 2\pi r$$

The **area** A of a circle of radius r is

$$A = \pi r^2$$

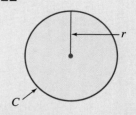

EXAMPLE 4

FINDING THE AREA AND PERIMETER OF A HYBRID SHAPE A Norman window consists of a rectangle with a semicircle mounted on top. For the window shown in Figure 8.19, find the following.

a. the area   **b.** the perimeter

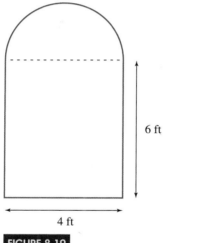

6 ft

4 ft

FIGURE 8.19

A Norman window.

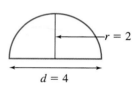

$r = 2$

$d = 4$

FIGURE 8.20

Half a circle.

SOLUTION

a. The area of the entire window is the sum of the areas of the rectangle and the semicircle. A semicircle is half a circle, so the semicircular region has area equal to one-half the area of a complete circle. The base of the rectangle is the same as the diameter of the semicircle. Hence, the diameter is 4 feet, and the radius is 2 feet (see Figure 8.20).

$$A_{\text{total}} = A_{\text{rectangle}} + A_{\text{semicircle}} = bh + \frac{1}{2}(\pi r^2)$$

$$= (4 \text{ ft})(6 \text{ ft}) + \frac{1}{2}\pi(2 \text{ ft})^2$$

$$= 24 \text{ ft}^2 + 2\pi \text{ ft}^2 = (24 + 2\pi) \text{ ft}^2 = 30.28318531 \ldots \text{ ft}^2$$

 24 $\boxed{+}$ 2 $\boxed{\times}$ $\boxed{\pi}$ $\boxed{=}$

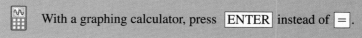 With a graphing calculator, press $\boxed{\text{ENTER}}$ instead of $\boxed{=}$.

Thus, the area of the Norman window is approximately 30.3 square feet.

b. The perimeter of the window consists of a semicircle (one-half the circumference of a circle), one horizontal line segment, and two vertical line segments.

$$P = \frac{1}{2}C + b + 2h$$

$$= \frac{1}{2}(\pi \cdot 4 \text{ ft}) + 4 \text{ ft} + 2(6 \text{ ft})$$

$$= 2\pi \text{ ft} + 16 \text{ ft}$$

$$= 22.28318531 \ldots \text{ ft}$$

Thus, the perimeter of the Norman window is approximately 22.3 feet.

EXAMPLE **5**

USING ASTRONOMICAL UNITS Like all the planets, Uranus revolves around the sun in an elliptical orbit. Consequently, the distance between Uranus and the sun varies from approximately 1,703 million to 1,866 million miles.

a. What is the (approximate) average distance from Uranus to the sun?
b. On average, how far from the sun is Uranus in astronomical units (AU)?
c. Interpret the answer to part (b) in terms of the earth's distance from the sun.

SOLUTION

a. To find the (approximate) average distance, we add the minimum and maximum distances and divide by 2 to obtain

$$\frac{1,703 + 1,866}{2} = 1,784.5 \text{ million miles}$$

b. To convert miles to astronomical units, we divide 1,784.5 million miles by 93 million miles (= 1 AU) to obtain

$$1,784.5 \text{ million miles} \times \frac{1 \text{ AU}}{93 \text{ million miles}} = 19.18817204 \ldots \text{ AU}$$

Therefore, the average distance from Uranus to the sun is approximately 19.2 AU.
c. Because 1 AU represents the earth's average distance from the sun, a distance of 19.2 AU indicates that Uranus is slightly more than 19 times farther from the sun than is the earth.

EXAMPLE **6**

CONVERTING VARIOUS UNITS USED IN ASTRONOMY The star closest to the earth (other than the sun) is Proxima Centauri. Scientists believe that this star is approximately 40 trillion kilometers from the earth.

a. How far, in astronomical units, is Proxima Centauri from the earth?
b. How far, in light-years, is Proxima Centauri from the earth?
c. How far, in parsecs, is Proxima Centauri from the earth?

SOLUTION

a. To convert kilometers to astronomical units, we divide 40 trillion km by 150 million km (= 1 AU) and obtain

$$40,000,000,000,000 \text{ km} \times \frac{1 \text{ AU}}{150,000,000 \text{ km}} = 266,667 \text{ AU}$$

Therefore, the distance from Proxima Centauri to the earth is approximately 266,667 AU; the distance between Proxima Centauri and the earth is 266,667 times the distance between our sun and the earth.

TOPIC X   ASTRONOMICAL MEASUREMENT
GEOMETRY IN THE REAL WORLD

What is the diameter of a tennis ball? How tall is the Statue of Liberty? What is the distance between London and Paris or between San Francisco and Singapore? When measuring objects or distances, most Americans use the familiar units of inches, feet, and miles; elsewhere in the world, the metric units of millimeters, centimeters, meters, and kilometers are prevalent. The answers to the previous questions can be found by browsing through an almanac; the diameter of a tennis ball is $2\frac{1}{2}$ inches (63.5 millimeters), Ms. Liberty is 151 feet (46 meters) tall (from base to torch), London and Paris are 210 miles (338 kilometers) apart, and there are 8,440 miles (13,583 kilometers) between San Francisco and Singapore.

How far is it from the earth is the sun? What is the distance between planets, say, Mercury and Pluto? How far must one travel to reach the next galaxy? Obviously, these distances are great; miles and kilometers pale in magnitude when astronomical distances are involved. Consequently, scientists have designed various standard units to express and calculate distances within our solar system and beyond. Three common units, from

smallest to largest, are the **astronomical unit (AU)**, **light-year (ly)**, and **parsec (pc)**.

The earth travels around the sun in an elliptical orbit. (See Section 8.7.) Consequently, the distance between the earth and the sun varies from approximately 91.4 million to 94.5 million miles. Scientists have made the following definition: 1 astronomical unit (AU) = the mean (or average) distance between the earth and the sun. Consequently, 1 AU ≈ (91.4 + 94.5)/2 = 92.95 million miles. It is common to round off this calculation to a whole number and say that 1 AU is about 93 million miles, or about 150 million kilometers. When distances within our solar system are measured, AU are the commonly used units.

Objects outside of our solar system are very, very far away, and AU are considered to be relatively small units. The next larger unit in measuring distance is the light-year. Scientists have made the following definition: 1 light-year (ly) = the distance that light can travel in one year through a vacuum (empty space). Because light travels at a speed of about 186,282.34 miles per second, 1 light-year equals approximately

$$\frac{186{,}282.34 \text{ miles}}{\text{second}} \times \frac{3{,}600 \text{ seconds}}{\text{hour}}$$
$$\times \frac{24 \text{ hours}}{\text{day}} \times \frac{365.25 \text{ days}}{\text{year}} \times 1 \text{ year}$$
$$= 5{,}878{,}623{,}573{,}000 \text{ miles}$$

Most references state that 1 light-year is about 5.88 trillion miles (mi), or about 9.46 trillion kilometers (km), or about 63,240 astronomical units (AU).

After the light-year, the next larger unit in measuring distance is the parsec. The exact definition of a parsec is based on a right triangle and an acute angle of measure: 1/3600th of a degree. This definition will be examined in Section 8.5. For our purposes here, we define 1 parsec (pc) = 3.26 light-years (ly).

Summarizing the commonly used units in astronomy, we have the following:

1 AU = 93 million mi
    = 150 million km
1 ly = 5.88 trillion mi
    = 9.46 trillion km
    = 63,240 AU
1 pc = 3.26 ly

b. To convert astronomical units to light-years, we divide 266,677 AU by 63,240 AU (= 1 ly) and obtain

$$266{,}667 \text{ AU} \times \frac{1 \text{ ly}}{63{,}240 \text{ AU}} = 4.216745731\ldots \text{ ly}$$

Therefore, the distance from Proxima Centauri to the earth is approximately 4.22 ly; light emitted from Proxima Centauri takes a little more than 4 years to reach the earth.

c. To convert light-years to parsecs, we divide 4.22 ly by 3.26 ly (= 1 pc) and obtain

$$4.22 \text{ ly} \times \frac{1 \text{ pc}}{3.26 \text{ ly}} = 1.294478528\ldots \text{ pc}$$

Therefore, the distance from Proxima Centauri to the earth is approximately 1.29 pc.

8.1 EXERCISES

When necessary, round off answers to one decimal place. In Exercises 1–8, find the area of each figure.

1.

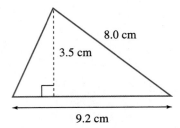

8.0 cm
3.5 cm
9.2 cm

2.

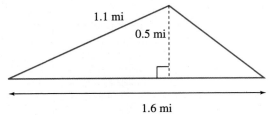

1.1 mi
0.5 mi
1.6 mi

3.

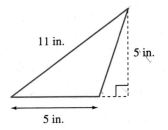

11 in.
5 in.
5 in.

4.

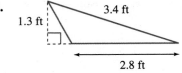

3.4 ft
1.3 ft
2.8 ft

5.

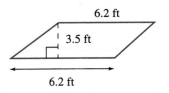

6.2 ft
3.5 ft
6.2 ft

6.

22 in.
5 in.
22 in.

7.

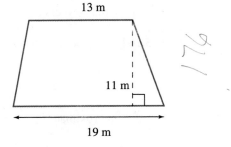

13 m
11 m
19 m

176

19-13

8.

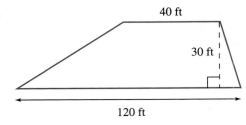

40 ft
30 ft
120 ft

In Exercises 9 and 10, find (a) the area and (b) the circumference of the circle.

9.

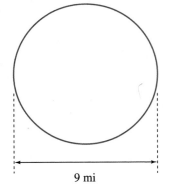

6.5 in.

r = 3.25

πr²

A = π 3.25²

10.

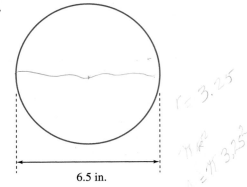

9 mi

In Exercises 11–20, find (a) the area and (b) the perimeter of each figure.

11.

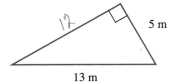

12
5 m
13 m

12.

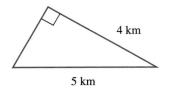

4 km
5 km

13.

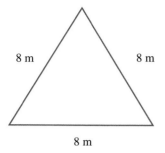

8 m   8 m

8 m

14.

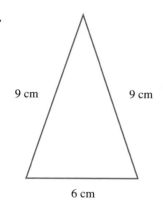

9 cm   9 cm

6 cm

15.

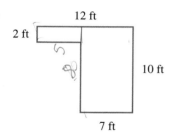

12 ft

2 ft

10 ft

7 ft

16.

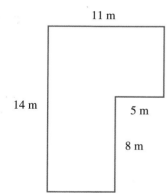

11 m

14 m

5 m

8 m

17.

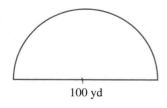

100 yd

18.

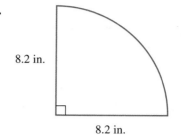

8.2 in.

8.2 in.

19.

$A = (6)(7) + \frac{1}{2}(6)(7) =$ 54 yd²

$P = 6 + 7 + 7 + 5 + 5$

$P = 30$

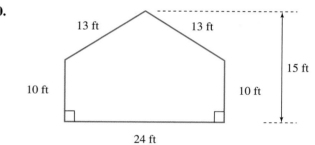

7 yd

5 yd

6 yd

½·3·4·4

5 yd

7 yd

11 yd

20.

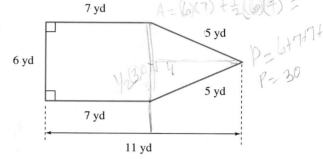

13 ft   13 ft

10 ft   10 ft   15 ft

24 ft

In Exercises 21 and 22, find (a) the area and (b) the perimeter of each Norman window.

HINT: See Example 4.

21.

a. 49.8 sq ft
b. 28.9 sq ft.

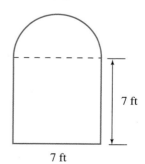

8 ft

5 ft

22.

7 ft

7 ft

23. A circular swimming pool has diameter 50 feet and is centered in a fenced-in square region measuring 80 feet by 80 feet. A concrete sidewalk 5 feet wide encircles the pool, and the rest of the region is grass, as shown in Figure 8.21.

   a. Find the surface area of the water.

   b. Find the area of the concrete sidewalk.

   c. Find the area of the grass.

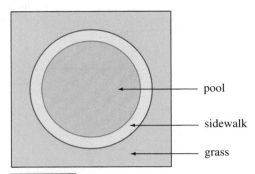

pool

sidewalk

grass

FIGURE 8.21   Diagram for Exercise 23.

24. A rectangular swimming pool 30 feet by 15 feet is surrounded by a uniform concrete sidewalk 5 feet wide. The pool is positioned inside a fenced-in circular grassy region of diameter 60 feet, as shown in Figure 8.22.

   a. Find the surface area of the water.

   b. Find the area of the concrete sidewalk.

   c. Find the area of the grass.

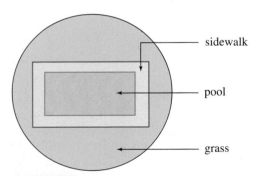

sidewalk

pool

grass

FIGURE 8.22   Diagram for Exercise 24.

25. The perimeter of a square window is 18 feet 8 inches. Find the area of the window (a) in square inches and (b) in square feet.

26. A square window has an area of 729 square inches. Find the perimeter of the window.

27. The circumference of a circular window is 10 feet. Find the area of the window.

28. A circular window has an area of 10 square feet. Find the circumference of the window.

29. You jog $\frac{3}{4}$ mile due north, then jog $1\frac{1}{2}$ miles due east, and then return to your starting point via a straight-line path. How many miles have you jogged?

30. You walk 100 yards due south, then 120 yards due west, and then 30 yards due north. How far are you from your starting point?

31. A 10-foot ladder leans against a wall. If the base of the ladder is 6 feet from the wall, how far up the wall does the ladder reach?

32. A ladder is leaning against a building. If the bottom of the ladder is $7\frac{1}{2}$ feet from the wall and the top of the ladder is 10 feet above the ground, how long is the ladder?

33. An oval athletic field is the union of a rectangle and semicircles at opposite ends, as shown in Figure 8.23.

   a. Find the total area of the field.

   b. Find the distance around the field.

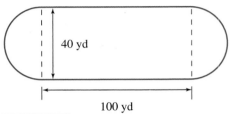

40 yd

100 yd

FIGURE 8.23   Diagram for Exercise 33.

34. An oval athletic field is the union of a square and semicircles at opposite ends, as shown in Figure 8.24. If the total area of the field is 1,300 square yards, find the dimensions of the square.

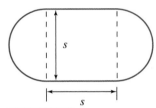

s

s

FIGURE 8.24   Diagram for Exercise 34.

35. The lengths of the edges of a triangular canvas are 100 feet, 140 feet, and 180 feet. You want to waterproof the tarp. If one can of waterproofing will treat 1,000 square feet, how many cans do you need?

36. Steve Loi wants to fertilize his backyard. The yard is a triangle whose sides are 110 feet, 120 feet, and 150 feet. If one bag of fertilizer will cover 1,200 square feet, how many bags does he need?

37. Reddie's Pizza Bash is famous for its Loads-o-Meat Deluxe Pizza. It comes in three sizes: small (13 inches) for $11.75, large (16 inches) for $14.75, and super (19 inches) for $22.75. Which size is the best deal?

   HINT: Find the price per square inch.

38. The best-selling pizza at Magic Mushroom Pizza Deli is the Vegetarian Surprise. It comes in three sizes: small (12 inches) for $11.25, large (14 inches) for $14.75, and super (18 inches) for $20.75. Which size is the best deal?

   HINT: Find the price per square inch.

In Exercises 39–44, use the following information: In the United States, land is commonly measured in terms of acres. One acre is equal to 43,560 square feet.

39. A rectangular parcel of land measures 99 feet by 110 feet. How many acres is the parcel?

40. A rectangular parcel of land measures 110 feet by 132 feet. How many acres is the parcel?

41. How many acres are in 1 square mile? (Note: 1 mile = 5,280 feet.)

42. In the "Old West," a common homestead parcel was 40 acres of land. How many 40-acre parcels are there in 1 square mile?

43. In 2009, California wildfires charred 336,020 acres of land. How many square miles of land were burned? Round off your answer to the nearest square mile.

44. In 2009, a major wildfire in Greece charred 21,000 hectares of land. How many square miles of land were burned? Round off your answer to the nearest square mile.

(*Note:* 1 hectare = 10,000 square meters = 107,639 square feet.)

45. Mercury (the closest planet to the sun) revolves around the sun in an elliptical orbit. Consequently, the distance between Mercury and the sun varies from approximately 28.6 million to 43.4 million miles.

  a. What is the (approximate) average distance from Mercury to the sun?

  b. On average, how far from the sun is Mercury in AU?

  c. Interpret the answer to part (b) in terms of the earth's distance from the sun.

46. Pluto (the farthest planet to the sun) revolves around the sun in an elliptical orbit. Consequently, the distance between Pluto and the sun varies from approximately 2,756 million to 4,539 million miles.

  a. What is the (approximate) average distance from Pluto to the sun?

  b. On average, how far from the sun is Pluto in AU?

  c. Interpret the answer to part (b) in terms of the earth's distance from the sun.

47. Scientists believe that the star Polaris (known as the North Star) is approximately 4,080 trillion kilometers from the earth.

  a. How far, in astronomical units, is Polaris from the earth?

  b. How far, in light-years, is Polaris from the earth?

  c. How far, in parsecs, is Polaris from the earth?

48. Scientists believe that the star Sirius (known as the Dog Star) is approximately 81.5 trillion kilometers from the earth.

  a. How far, in astronomical units, is Sirius from the earth?

  b. How far, in light-years, is Sirius from the earth?

  c. How far, in parsecs, is Sirius from the earth?

 THE NEXT LEVEL

If a person wants to pursue an advanced degree (something beyond a bachelor's or four-year degree), chances are the person must take a standardized exam to gain admission to a school or to be admitted into a specific program. These exams are intended to measure verbal, quantitative, and analytical skills that have developed throughout a person's life. Many classes and study guides are available to help people prepare for the exams. Exercises 49–56 are typical of those found in the study guides.

49. In Figure 8.25, polygon *RUST* is a square, and triangle *TRY* is an equilateral triangle.

  If $US = 4$, what is the area of the shaded region?

  a. $16 - 2\sqrt{3}$
  b. $16 - \sqrt{3}$
  c. $16 - 4\sqrt{3}$
  d. $4\sqrt{3}$
  e. 14
  f. None of these is correct.

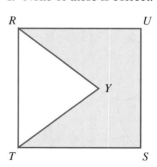

FIGURE 8.25 Diagram for Exercise 49.

50. In Figure 8.26, polygon *ADGJ* is a square whose sides have length 8. *BC*, *EF*, *HI*, and *KL* are each 6 and are the diameters of the four semicircles. What is the area of the shaded region?

  a. $64 - 36\pi$ **b.** $64 - 18\pi$
  c. $64 - 9\pi$ **d.** 46π
  e. 18π **f.** None of these is correct.

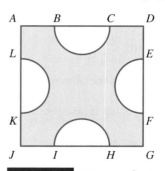

FIGURE 8.26 Diagram for Exercise 50.

51. In Figure 8.27, vertex *K* of square *CAKE* is on a circle with center *C*. If the area of the square is 12, what is the area of the circle?

   a. 36π **b.** 12π

   c. 24π **d.** $6\sqrt{6}\pi$

   e. $2\sqrt{6}\pi$ **f.** None of these is correct.

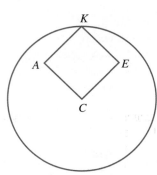

FIGURE 8.27 Diagram for Exercise 51.

52. In Figure 8.28, if the radius of the circle centered at *O* is 6, what is the length of diagonal *TP* of rectangle *OTEP*?

   a. $\sqrt{6}$ **b.** $2\sqrt{6}$

   c. $6\sqrt{2}$ **d.** 6

   e. $3\sqrt{2}$ **f.** None of these is correct.

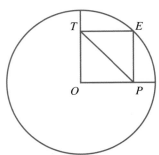

FIGURE 8.28 Diagram for Exercise 52.

53. What is the circumference of a circle whose area is 36π?

   a. 9π **b.** 6

   c. 6π **d.** 12

   e. 12π **f.** None of these is correct.

54. What is the area of a circle whose circumference is $\dfrac{\pi}{2}$?

   a. π **b.** $\dfrac{\pi}{2}$

   c. $\dfrac{\pi}{4}$ **d.** $\dfrac{\pi}{8}$

   e. $\dfrac{\pi}{16}$ **f.** None of these is correct.

55. What is the area of a circle that is inscribed in a square of area 8?

   a. π **b.** 2π

   c. 4π **d.** 8π

   e. $2\sqrt{2}\pi$ **f.** None of these is correct.

56. A square of area 8 is inscribed in a circle. What is the area of the circle?

   a. π **b.** 2π

   c. 4π **d.** 8π

   e. $2\sqrt{2}\pi$ **f.** None of these is correct.

Answer the following questions using complete sentences and your own words.

• PROJECTS

57. The purpose of this project is to verify the Pythagorean Theorem.

   a. On a sheet of paper that has a square corner, measure 3 inches from the corner along the edge of the paper and put a dot. Now measure 4 inches from the same corner along the other edge of the paper and put a dot. Use the Pythagorean Theorem to calculate the distance between the dots. Now, measure the distance between the dots. How does your measurement compare to your calculation?

   b. Repeat part (a) using 6 and 8 inches.

   c. Repeat part (a) using $7\frac{1}{2}$ and 10 inches.

   d. Repeat part (a) using 3 and 4 centimeters.

   e. Repeat part (a) using 5 and 12 centimeters.

   f. Repeat part (a) using 7 and 24 centimeters.

58. The purpose of this project is to calculate the cost of materials to resurface a floor.

   a. Pick a rectangular room in your apartment or house. Measure the dimensions of the floor, and calculate its area in square feet.

   b. Convert your answer from part (a) to square yards.

   (*HINT:* How many square feet are in 1 square yard?)

   c. Suppose you want to install a new floor covering in your selected room. Of the three materials, carpet, vinyl, and wood, which do you think would be most expensive? Least expensive?

   d. Go to a local carpet and floor covering store, and select a style of carpet for your room. What is the unit price of the product? Use the unit price to calculate the cost of the carpet for your room.

   e. Instead of carpet, select a vinyl product for the floor of your room. What is the unit price of the product? Use the unit price to calculate the cost of the vinyl for the floor of your room.

   f. Instead of carpet or vinyl, select a wood floor covering for your room. What is the unit price of the product? Use the unit price to calculate the cost of the wood floor for your room.

   g. How do your answers to parts (d)–(f) compare with your answer to part (c)?

8.2 Volume and Surface Area

OBJECTIVES

- Find the volume of basic three-dimensional objects, including rectangular solids, cylinders, spheres, cones, and pyramids
- Find the surface area of basic three-dimensional objects, including rectangular solids, cylinders, and spheres

Having examined the perimeter (one-dimensional measurement) and area (two-dimensional measurement) of common geometric figures in Section 8.1, we now explore the geometry of three-dimensional figures. In particular, we investigate the calculation of volume and surface area and apply the results to a variety of situations.

Problem Solving

Outdoors Unlimited, a camping supply store, is having a sale on its demonstration models, discontinued lines, and irregular items. Two water containers are on sale—one cylindrical and the other rectangular, as shown in Figure 8.29. You want to purchase the container that holds the most water, but unfortunately the original information on the capacity of each container has been lost. Which container would you choose? To compare the capacities of the two vessels, you must find their volumes.

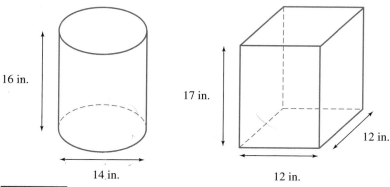

16 in.

14 in.

17 in.

12 in.

12 in.

FIGURE 8.29 Containers.

Volume is a measure of the amount of space occupied by a three-dimensional object. Volume was originally investigated in the course of bartering and commerce. When grain, wine, and oil were traded, various vessels or containers were used as standards. However, different cultures used different standard vessels, so elaborate conversion systems were necessary. Pints, quarts, gallons, hogsheads, cords, pecks, and bushels are the modern-day legacy of early vessel measurement. Today, the fundamental figure used in the calculation of volume is the cube; volume is expressed in terms of cubic units such as cubic feet (ft^3) and cubic centimeters (cc). (See Figure 8.30.)

The volume of a rectangular box is found by multiplying its length times its width times its height; that is, $V = L \cdot w \cdot h$. Notice that ($L \cdot w$) equals the area

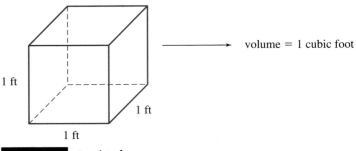

FIGURE 8.30 1 cubic foot.

of the (rectangular) base of the box. Consequently, the volume of the box can be found by multiplying the area of the base times the height of the box, as shown in Figure 8.31.

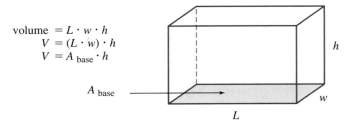

$$\text{volume} = L \cdot w \cdot h$$
$$V = (L \cdot w) \cdot h$$
$$V = A_{\text{base}} \cdot h$$

FIGURE 8.31 A rectangular box.

A rectangular box has the property that every cross section, or slice, taken parallel to the base produces an identical rectangle. It is precisely this property that allows us to calculate the volume of the box simply by multiplying the area of the base times the height. In general, the volume of any figure that has *identical* cross sections (same shape and size) from top to bottom is found by multiplying the area of the base times the height of the figure.

VOLUME OF A FIGURE HAVING IDENTICAL CROSS SECTIONS

If a three-dimensional figure has **height** h and identical cross sections from top to bottom, each of **area** A, the **volume** V of the figure is

$$V = A \cdot h$$

identical cross sections (top to bottom)

EXAMPLE **1**

FINDING VOLUMES OF RECTANGULAR AND CYLINDRICAL CONTAINERS Referring to the cylindrical and rectangular water containers shown in Figure 8.29, determine which holds more water.

SOLUTION

The holding capacity of each container is determined by its volume, so we must find the volume of each figure.

Finding the Volume of the Rectangular Container The rectangular box has identical cross sections from top to bottom; each cross section is a 12-inch-by-12-inch square.

$$V_{box} = A_{base} \cdot h$$
$$= A_{square} \cdot h$$
$$= (b^2) \cdot h$$
$$= (12 \text{ in.})^2 \cdot (17 \text{ in.})$$
$$= 2{,}448 \text{ in.}^3$$

The rectangular container holds 2,448 cubic inches of water.

Finding the Volume of the Cylindrical Container   The **cylinder** has identical cross sections from top to bottom; each is a circle of radius 7 inches (half the diameter).

$$V_{cylinder} = A_{base} \cdot h$$
$$= A_{circle} \cdot h$$
$$= (\pi r^2) \cdot h$$
$$= \pi \cdot (7 \text{ in.})^2 \cdot (16 \text{ in.})$$
$$= 2{,}463.00864 \ldots \text{ in.}^3$$

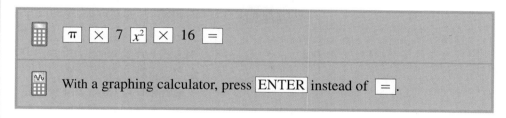

With a graphing calculator, press ENTER instead of =.

The cylindrical container holds approximately 2,463 cubic inches of water. Therefore, the cylindrical container holds about 15 cubic inches more than the rectangular container.

Surface Area

The **surface area** of a three-dimensional figure is the sum total of the areas of all the surfaces that compose the figure. Although a formula for the surface area of a rectangular box could be given, it is not necessary because we already know how to find the area of each rectangular face. Nonrectangular objects, however, deserve more attention. Cylinders and spheres have specific formulas requiring more analysis.

EXAMPLE 2

FINDING SURFACE AREAS OF RECTANGULAR AND CYLINDRICAL CONTAINERS   Referring to the cylindrical and rectangular water containers shown in Figure 8.29, determine which has the greater surface area.

SOLUTION

Finding the Surface Area of the Rectangular Container   The surface area of a box is composed of six pieces: the top, the bottom, and the four sides. In this case, the top and bottom are identical squares; each has area b^2. The sides are identical rectangles; each has area $b \cdot h$.

$$A_{box} = 2 \cdot A_{square} + 4 \cdot A_{rectangle}$$
$$= 2(b^2) + 4(b \cdot h)$$
$$= 2(12 \text{ in.})^2 + 4(12 \text{ in.})(17 \text{ in.})$$
$$= 288 \text{ in.}^2 + 816 \text{ in.}^2$$
$$= 1{,}104 \text{ in.}^2$$

The surface area of the rectangular container is 1,104 square inches.

Finding the Surface Area of the Cylindrical Container　The surface area of a cylinder is composed of three pieces: the top, the bottom, and the curved side. The top and bottom are circles, each with an area of πr^2. We find the area of the curved side by cutting it from top to bottom and rolling it out flat. The result is a rectangle with length equal to the circumference of the cylinder, as shown in Figure 8.32.

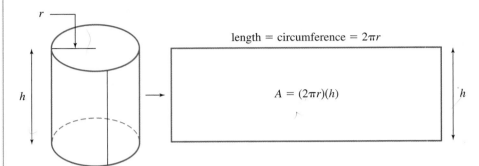

FIGURE 8.32　The curved side of a cylinder is a rectangle.

$$
\begin{aligned}
A_{\text{cylinder}} &= 2 \cdot A_{\text{circle}} + A_{\text{side}} \\
&= 2(\pi r^2) + (2\pi r) \cdot h \\
&= 2\pi \cdot (7 \text{ in.})^2 + 2\pi \cdot (7 \text{ in.})(16 \text{ in.}) \\
&= 98\pi \text{ in.}^2 + 224\pi \text{ in.}^2 \\
&= 322\pi \text{ in.}^2 \\
&= 1{,}011.592834 \ldots \text{ in.}^2
\end{aligned}
$$

The surface area of the cylindrical container is approximately 1,012 square inches. Therefore, the surface area of the rectangular container is about 92 square inches more than that of the cylindrical container.

Comparing Examples 1 and 2, we see that even though the volume of the cylinder is greater than that of the box, the surface area of the cylinder is less than that of the box. In essence, the cylinder is more "efficient" because of its curved surface.

Spheres

How much larger is a basketball than a soccer ball? Depending on which measurement you use for the comparison, different answers are possible. The "size" of a ball can be given in terms of its diameter (one-dimensional measurement), its surface area (two-dimensional measurement), or its volume (three-dimensional measurement).

A **sphere** is the three-dimensional counterpart of a circle; it is the set of all points *in space* equidistant from a fixed point. The fixed point is called the **center** of the sphere, a line segment connecting the center and any point on the sphere is a **radius,** and a line segment connecting any two points of the sphere *and* passing through the center is a **diameter.** The length of a diameter is twice that of a radius. (See Figure 8.33.)

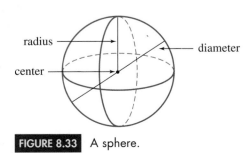

FIGURE 8.33　A sphere.

The ancient Greeks knew the following formulas for determining the volume and surface area of a sphere.

VOLUME AND SURFACE AREA OF A SPHERE

The **volume** V of a sphere of radius r is

$$V = \frac{4}{3}\pi r^3$$

The **surface area** A of a sphere of radius r is

$$A = 4\pi r^2$$

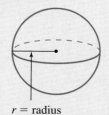

r = radius

EXAMPLE 3

FINDING VOLUMES AND SURFACE AREAS OF SPHERES   A regulation basketball has a circumference of 30 inches; the circumference of a standard soccer ball is 27 inches. (Each ball is a sphere.)

a. Find and compare the volumes of the two types of balls.
b. Find and compare the surface areas of the two types of balls.

SOLUTION

a. To find the volume of a sphere, we must know its radius. We are given the circumference, so we can find the radius as follows:

$$C = 2\pi r$$
$$r = \frac{C}{2\pi} \quad \text{dividing by } 2\pi$$

Finding the Volume of the Basketball   The circumference of a basketball is $C = 30$ inches, so we have

$$r_{\text{basketball}} = \frac{C}{2\pi} = \frac{30 \text{ in.}}{2\pi} = \frac{15}{\pi} \text{ in.}$$

Rather than using a calculator at this point, we substitute this exact value directly into the volume formula and then use a calculator:

$$V_{\text{basketball}} = \frac{4}{3}\pi r^3$$
$$= \frac{4}{3}\pi \left(\frac{15}{\pi} \text{ in.}\right)^3$$
$$= 455.9453264 \ldots \text{ in.}^3$$

| 4 | ÷ | 3 | × | π | × | (| 15 | ÷ | π |) | y^x | 3 | = |

| 4 | ÷ | 3 | × | π | × | (| 15 | ÷ | π |) | ^ | 3 | ENTER |

The volume of the basketball is approximately 456 cubic inches.

Finding the Volume of the Soccer Ball   The circumference of a soccer ball is $C = 27$ inches; thus,

$$r_{\text{soccer ball}} = \frac{C}{2\pi} = \frac{27 \text{ in.}}{2\pi} = \frac{13.5 \text{ in.}}{\pi}$$

Therefore,

$$V_{\text{soccer ball}} = \frac{4}{3}\pi r^3$$

$$= \frac{4}{3}\pi\left(\frac{13.5 \text{ in.}}{\pi}\right)^3$$

$$= 332.3841429\ldots \text{ in.}^3$$

The volume of the soccer ball is approximately 332 cubic inches.

To compare the volumes, consider the ratio of the larger ball to the smaller ball:

$$\frac{V_{\text{basketball}}}{V_{\text{soccer ball}}} = \frac{456 \text{ in.}^3}{332 \text{ in.}^3} \approx 1.37$$

$$V_{\text{basketball}} \approx 1.37 V_{\text{soccer ball}}$$

$$V_{\text{basketball}} \approx 137\% \text{ of } V_{\text{soccer ball}}$$

Therefore, the volume of the basketball is 37% more than the volume of the soccer ball.

b. To find the surface area of each ball, substitute the radii found in part (a) into the surface area formula.

Finding the Surface Area of the Basketball   Recall that $r_{\text{basketball}} = \frac{15}{\pi}$ in.

$$A = 4\pi r^2$$

$$= 4\pi\left(\frac{15}{\pi} \text{ in.}\right)^2$$

$$= 286.4788976\ldots \text{ in.}^2$$

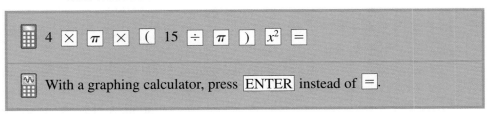

With a graphing calculator, press ENTER instead of =.

The surface area of the basketball is approximately 286 square inches.

Finding the Surface Area of the Soccer Ball   Recall that $r_{\text{soccer ball}} = \frac{13.5}{\pi}$ in.

$$A = 4\pi r^2$$

$$= 4\pi\left(\frac{13.5}{\pi} \text{ in.}\right)^2$$

$$= 232.047907\ldots \text{ in.}^2$$

The surface area of the soccer ball is approximately 232 square inches.

As with the volume comparison, use the ratio of the larger ball to the smaller ball to compare their surface areas:

$$\frac{A_{\text{basketball}}}{A_{\text{soccer ball}}} = \frac{286 \text{ in.}^2}{232 \text{ in.}^2} \approx 1.23$$

$$A_{\text{basketball}} \approx 1.23 A_{\text{soccer ball}}$$

$$A_{\text{basketball}} \approx 123\% \text{ of } A_{\text{soccer ball}}$$

Therefore, the surface area of the basketball is 23% more than the surface area of the soccer ball.

Even though the circumference of a basketball is 11% more than that of a soccer ball ($C_{\text{basketball}}/C_{\text{soccer ball}} = 30/27 = 1.11$), the surface area is 23% more, and the volume is 37% more. These comparisons differ because of the dimensions of the measurements; C is one-dimensional, A is two-dimensional, and V is three-dimensional. Each measurement involves the multiplication of an additional value of r.

Cones and Pyramids

Two other common three-dimensional objects are the cone and the pyramid. A **cone** has a circular base, whereas the base of a **pyramid** is a polygon. In either case, cross sections parallel to the base all have the same shape as the base, but they differ in size; starting at the base, the cross sections get progressively smaller until they reach a single point at the top of the figure. For both a cone and a pyramid, the volume is one-third the product of the area of the base times the height.

VOLUME OF A CONE OR A PYRAMID

If a cone or pyramid has height h and if the area of the base is A, the **volume** V of the figure is

$$v = \frac{1}{3} A \cdot h$$

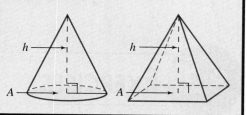

EXAMPLE 4

FINDING VOLUMES OF CONES AND PYRAMIDS   Find and compare the volumes of the cone and pyramid shown in Figure 8.34.

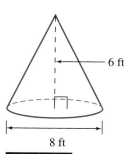 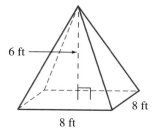

FIGURE 8.34   A cone and a pyramid.

SOLUTION

Finding the Volume of the Cone   We have $r = 8 \text{ ft}/2 = 4 \text{ ft}$ and $h = 6 \text{ ft}$.

$$V = \frac{1}{3} A \cdot h$$

$$= \frac{1}{3}(\pi r^2)h$$

$$= \frac{1}{3}\pi(4 \text{ ft})^2(6 \text{ ft})$$

$$= 32\pi \text{ ft}^3$$

$$= 100.5309649 \ldots \text{ft}^3$$

The volume of the cone is approximately 101 cubic feet.

Finding the Volume of the Pyramid

$$V = \frac{1}{3} A \cdot h = \frac{1}{3}(b^2)h = \frac{1}{3}(8 \text{ ft})^2(6 \text{ ft}) = 128 \text{ ft}^3$$

The volume of the pyramid is 128 cubic feet.

Comparing the Larger Volume to the Smaller Volume   We have

$$\frac{V_{\text{pyramid}}}{V_{\text{cone}}} = \frac{128 \text{ ft}^3}{101 \text{ ft}^3} \approx 1.27$$

$$V_{\text{pyramid}} \approx 1.27 \ V_{\text{cone}}$$

$$V_{\text{pyramid}} \approx 127\% \text{ of } V_{\text{cone}}$$

Therefore, the volume of the pyramid is 27% larger than the volume of the cone.

8.2 EXERCISES

When necessary, round off answers to two decimal places.

In Exercises 1–6, find (a) the volume and (b) the surface area of each figure.

1.

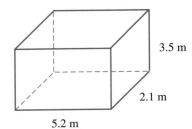

3.5 m

2.1 m

5.2 m

2.

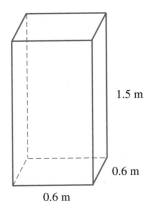

1.5 m

0.6 m

0.6 m

3.

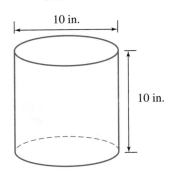

10 in.

10 in.

4.

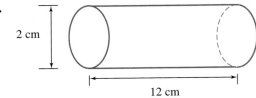

2 cm

12 cm

5.

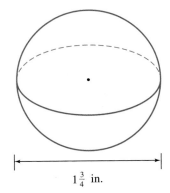

$1\frac{3}{4}$ in.

6.

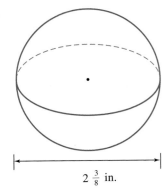

$2\frac{3}{8}$ in.

In Exercises 7–14, find the volume of each figure. All dimensions are given in feet.

7.

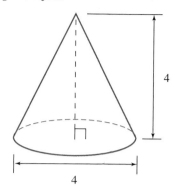

8.

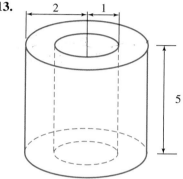

9.

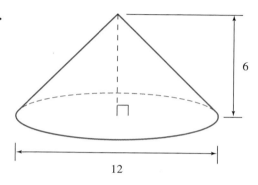

10.

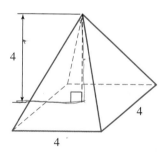

11.

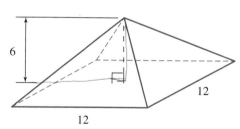

12.

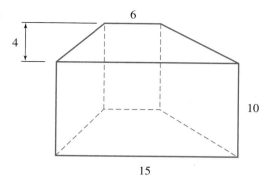

13.

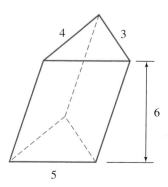

14.

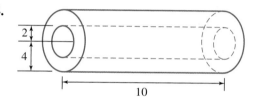

15. From a 10-inch-by-16-inch piece of cardboard, 1.5-inch-square corners are cut out, as shown in Figure 8.35, and the resulting flaps are folded up to form an open box. Find (a) the volume and (b) the external surface area of the box.

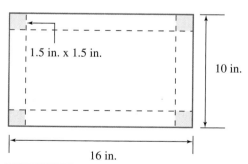

FIGURE 8.35 Diagram for Exercise 15.

16. From a 24-inch-square piece of cardboard, square corners are to be cut out as shown in Figure 8.36, and the resulting flaps folded up to form an open box. Find the volume of the resulting box if (a) 4-inch-square corners are cut out and (b) 8-inch-square corners are cut out.

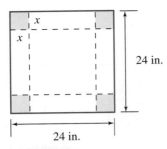

FIGURE 8.36 Diagram for Exercise 16.

17. A grain silo consists of a cylinder with a hemisphere on top. Find the volume of a silo in which the cylindrical part is 50 feet tall and has a diameter of 20 feet, as shown in Figure 8.37.

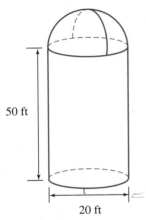

FIGURE 8.37 Diagram for Exercise 17.

18. A propane gas tank consists of a cylinder with a hemisphere at each end. Find the volume of the tank if the overall length is 10 feet and the diameter of the cylinder is 4 feet, as shown in Figure 8.38.

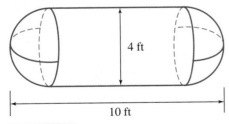

FIGURE 8.38 Diagram for Exercise 18.

19. A regulation baseball (hardball) has a circumference of 9 inches; a regulation softball has a circumference of 12 inches.
   a. Find and compare the volumes of the two types of balls.
   b. Find and compare the surface areas of the two types of balls.

20. A regulation tennis ball has a diameter of $2\frac{1}{2}$ inches; a Ping-Pong ball has a diameter of $1\frac{1}{2}$ inches.
   a. Find and compare the volumes of the two types of balls.
   b. Find and compare the surface areas of the two types of balls.

21. The diameter of the earth is approximately 7,920 miles; the diameter of the moon is approximately 2,160 miles. How many moons could fit inside the earth?

HINT: Compare their volumes.

22. The diameter of the earth is approximately 7,920 miles; the diameter of the planet Jupiter (the largest planet in our solar system) is approximately 88,640 miles. How many earths could fit inside Jupiter?

HINT: Compare their volumes.

23. The diameter of the earth is approximately 7,920 miles; the diameter of the planet Pluto (the smallest) is approximately 1,500 miles. How many Plutos could fit inside the earth?

HINT: Compare their volumes.

24. The diameter of the planet Jupiter (the largest planet in our solar system) is approximately 88,640 miles; the diameter of the planet Pluto (the smallest) is approximately 1,500 miles. How many Plutos could fit inside Jupiter?

HINT: Compare their volumes.

In Exercises 25 and 26, use the following information: Firewood is measured and sold by the cord. *A cord of wood is a rectangular pile of wood 4 feet wide, 4 feet high, and 8 feet long, as shown in Figure 8.39. Therefore, one cord = 128 cubic feet.*

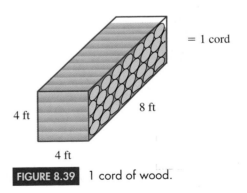

FIGURE 8.39 1 cord of wood.

25. A cord of seasoned almond wood costs $190. You paid $190 for a pile that was 4 feet wide, 2 feet high, and 10 feet long. Did you get an honest deal? If not, what should the cost have been?

26. A cord of seasoned oak wood costs $160. You paid $640 for a pile that was 4 feet wide, 6 feet high, and 20 feet long. Did you get an honest deal? If not, what should the cost have been?

27. Tennis balls are packaged three to a cylindrical can. If the diameter of a tennis ball is $2\frac{1}{2}$ inches, find the volume of the can.

28. Golf balls are packaged three to a rectangular box. If the diameter of a golf ball is $1\frac{3}{4}$ inches, find the volume of the box.

29. Ron Thiele bought an older house and wants to put in a new concrete driveway. The driveway will be 36 feet long, 9 feet wide, and 6 inches thick. Concrete (a mixture of sand, gravel, and cement) is measured by the cubic yard. One sack of dry cement mix costs $7.30, and it takes four sacks to mix up 1 cubic yard of concrete. How much will it cost Ron to buy the cement?

30. Marcus Robinson bought an older house and wants to put in a new concrete patio. The patio will be 18 feet long, 12 feet wide, and 3 inches thick. Concrete is measured by the cubic yard. One sack of dry cement mix costs $7.30, and it takes four sacks to mix up 1 cubic yard of concrete. How much will it cost Marcus to buy the cement?

31. The Great Pyramid of Cheops (erected about 2600 B.C.) originally had a square base measuring 756 feet by 756 feet and was 480 feet high. (It has since been eroded and damaged.) Find the volume of the original Great Pyramid.

32. The student union at the University of Utopia (U²) is a pyramid with a 150-foot square base and a height of 40 feet. Find the volume of the student union at U².

33. A water storage tank is an upside-down cone, as shown in Figure 8.40. If the diameter of the circular top is 12 feet and the length of the sloping side is 10 feet, how much water will the tank hold?

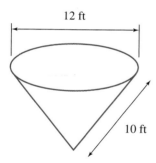

12 ft

10 ft

FIGURE 8.40   Diagram for Exercise 33.

34. The diameter of a conical paper cup is 3.5 inches, and the length of the sloping side is 4.55 inches, as shown in Figure 8.41. How much water will the cup hold?

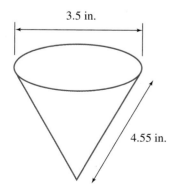

3.5 in.

4.55 in.

FIGURE 8.41   Diagram for Exercise 34.

 THE NEXT LEVEL

If a person wants to pursue an advanced degree (something beyond a bachelor's or four-year degree), chances are the person must take a standardized exam to gain admission to a school or to be admitted into a specific program. These exams are intended to measure verbal, quantitative, and analytical skills that have developed throughout a person's life. Many classes and study guides are available to help people prepare for the exams. Exercises 35–42 are typical of those found in the study guides.

35. How many cubic inches are in 1 cubic foot?
    a. 12   b. 36
    c. 144   d. 432
    e. 1,728   f. None of these is correct.

36. How many cubic feet are in 1 cubic yard?
    a. 3   b. 6
    c. 9   d. 12
    e. 27   f. None of these is correct.

37. The sum of the lengths of all the edges of a cube is 4 units. What is the volume, in cubic units, of the cube?
    a. $\dfrac{1}{27}$ b. $\dfrac{1}{8}$
    c. $\dfrac{1}{3}$ d. 1
    e. 8   f. None of these is correct.

38. The surface area of a cube is 24 square units. What is the volume, in cubic units, of the cube?
    a. 4   b. 8
    c. 16   d. 64
    e. 13,824   f. None of these is correct.

39. The volume of a cube is 216 cubic units. What is the surface area, in square units, of the cube?
    a. 36   b. 144
    c. 216   d. 288
    e. 864   f. None of these is correct.

40. A rectangular tank has a base that is 10 inches by 12 inches and a height of 18 inches. If the tank is one-third full of water, by how many inches will the water level rise if 420 cubic inches of water are poured into the tank?
    a. 3.5   b. 7
    c. 10.5   d. 11.5
    e. 18   f. None of these is correct.

41. If the height of a cylinder is 2 times its circumference, what is the volume of the cylinder in terms of its circumference, C?
    a. $\dfrac{C^3}{\pi}$ b. $\dfrac{C^3}{2\pi}$

c. $\dfrac{2C^2}{\pi}$ d. $\dfrac{C^2}{2\pi}$

e. $2C^3$ f. None of these is correct.

42. The height, h, of a cylinder is equal to the diameter, D, of a sphere. If the cylinder and sphere have the same volume, what is the radius of the cylinder?

a. $\dfrac{\sqrt{6}}{D}$ b. $\dfrac{D}{\sqrt{6}}$

c. $\dfrac{D^2}{6}$ d. $\dfrac{6}{D^2}$

e. $\sqrt{6}\,D$ f. None of these is correct.

Answer the following questions using complete sentences and your own words.

● **PROJECTS**

43. How do you convert fluid ounces to cubic inches or vice versa? The purpose of this project is to obtain appropriate conversion factors.

a. Select a canned food product or beverage such as a can of soup, soda, juice, or vegetable. According to the label, how many fluid ounces of product does the can contain?

b. Measure the height and diameter of the can in inches. Use these measurements to calculate the volume of the can in cubic inches.

c. Divide the volume calculated in part (b) by the amount given in part (a). This ratio will give the number of cubic inches per fluid ounce.

d. Your answer to part (c) should be close to 1.8; that is, a commonly used conversion factor is 1 fluid ounce = 1.8047 cubic inches. How accurate is your calculation?

e. If a can contains 32 fluid ounces of juice, how large is the can in terms of cubic inches?

f. Now measure the height and diameter of the can in centimeters. Use these measurements to calculate the volume of the can in cubic centimeters.

g. Divide the volume calculated in part (f) by the amount given in part (a). This ratio will give the number of cubic centimeters per fluid ounce.

h. Your answer to part (g) should be close to 29.6; that is, a commonly used conversion factor is 1 fluid ounce = 29.5735 cubic centimeters. How accurate is your calculation?

i. If a can contains 32 fluid ounces of juice, how large is the can in terms of cubic centimeters?

8.3   Egyptian Geometry

OBJECTIVES

● Explore ancient Egyptian methods of geometric calculation

● Explore and apply the ancient Egyptian approximation of π

● Use Egyptian units of measurement

The mathematics of the early Egyptians was very practical in its content and use. Mathematics was required in the surveying of land, the construction of buildings, the creation of calendars, and the recordkeeping necessary for commerce. All these activities contributed to the advancement of Egyptian civilization.

Repeated surveying was a necessity for the Egyptians because of the periodic flooding of the Nile. The floodbanks were very fertile, so this land was most valuable and was taxed accordingly. However, when the river flooded its banks, all landmarks and boundaries would be washed away, making it imperative to be able to lay out geometric figures of the correct size and shape.

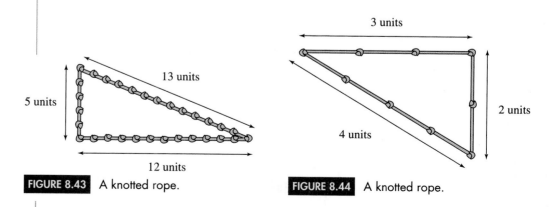

Found in the Tomb of Menna, this Egyptian wall painting depicts a harvest scene. Notice the use of knotted ropes to measure a field of grain.

The Egyptian surveyors used ropes with equally spaced knots to measure distance. In fact, surveyors were referred to as "rope stretchers." These knotted ropes were also used to construct right angles. A right triangle is formed when sides of lengths 3, 4, and 5 are used. A rope with twelve equally spaced knots can easily be stretched to form such a triangle, as shown in Figure 8.42.

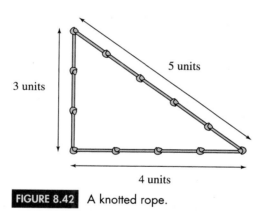

3 units

5 units

4 units

FIGURE 8.42 A knotted rope.

EXAMPLE **1** | **DETERMINING WHETHER A TRIANGLE IS A RIGHT TRIANGLE**
Which of the configurations of knotted ropes shown in Figures 8.43 and 8.44 would form a right triangle?

5 units

13 units

12 units

FIGURE 8.43 A knotted rope.

3 units

2 units

4 units

FIGURE 8.44 A knotted rope.

SOLUTION | **a.** Counting the number of segments on each side, we find the sides to be of lengths 5, 12, and 13 units. If the triangle is a right triangle, the sides will satisfy the Pythagorean theorem, $a^2 + b^2 = c^2$. Because 13 is the largest value, we let $c = 13$ and check the theorem:

$$a^2 + b^2 = 5^2 + 12^2$$
$$= 25 + 144$$
$$= 169$$
$$= 13^2 = c^2$$

The sides satisfy the Pythagorean Theorem; therefore, a right angle is formed at the vertex opposite the side of length 13.

b. The sides of the triangle have lengths 2, 3, and 4. Let $a = 2$ and $b = 3$ (the two shortest sides), and check the Pythagorean Theorem:

$$a^2 + b^2 = 2^2 + 3^2$$
$$= 4 + 9$$
$$= 13$$
$$\neq 4^2 = c^2$$

The lengths do not satisfy the Pythagorean Theorem; no right angle is formed.

Units of Measurement

The basic unit used by the ancient Egyptians for measuring length was the cubit. A *cubit* (*cubitum* is Latin for "elbow") is the distance from a person's elbow to the end of the middle finger. Just as a yard can be subdivided into smaller units of feet and inches, a cubit can be subdivided into smaller units of *palms* and *fingers*. One "royal" cubit (the unit used in official land measurement) equals seven palms, whereas one "common" cubit equals six palms; a royal cubit is longer than a common cubit. (In this book, we will take each cubit to be a royal cubit.) In either case, one palm equals four fingers. Since a cubit is a relatively small length, it is an inconvenient unit to use in measuring large distances. Consequently, the Egyptians defined a *khet* to equal 100 cubits; khets were used when land was surveyed.

The basic unit of area used by the Egyptians was the setat. A *setat* is equal to one square khet; a square whose sides each measure one khet (100 cubits) has an area of exactly one setat (or 10,000 square cubits). (See Figure 8.45.) One setat is approximately two-thirds of an acre.

If lengths are measured in terms of feet, volume is calculated in terms of cubic feet. Since the Egyptians measured lengths in terms of cubits, we would expect volume to be expressed in terms of cubic cubits. However, the basic unit of volume used by the Egyptians was the *khar;* one khar equals two-thirds of a cubic cubit (or 1 cubic cubit = $\frac{3}{2}$ khar). The Egyptian units of measurement are summarized in the following box.

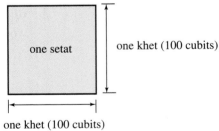

FIGURE 8.45

1 setat = 1 square khet
= 10,000 square cubits.

EGYPTIAN UNITS OF MEASUREMENT

1 cubit = 7 palms   1 setat = 1 square khet = 10,000 square cubits
1 palm = 4 fingers   1 khar = $\frac{2}{3}$ cubic cubit
1 khet = 100 cubits

Empirical Geometry (If It Works, Use It)

The body of mathematical knowledge possessed by the Egyptians was organized and presented quite differently from ours today. Whereas contemporary mathematics deals with general formulas, theorems, and proofs, most of the surviving written records of the Egyptians consist of single problems without mention of a general formula. The Egyptians lacked a formal system of variables and algebra; they were unable to write a formula such as $A = Lw$. Consequently, their work resembles a verbal narration on how to solve a specific problem or puzzle.

A specific example of this single-problem approach comes from what is now called the Moscow Papyrus, an Egyptian papyrus dating from about 1850 B.C. and containing 25 mathematical problems. It now resides in the Museum of Fine Arts in Moscow. Problem 14 of the Moscow Papyrus deals with finding the volume of a truncated pyramid (that is, a pyramid with its top cut off). The problem translates as follows: "Example of calculating a truncated pyramid. If you are told: a truncated pyramid of 6 for the vertical height by 4 on the base by 2 on the top: You are to square this 4; result 16. You are to double 4; result 8. You are to square this 2; result 4. You are to add the 16 and the 8 and the 4; result 28. You are to take $\frac{1}{3}$ of 6; result 2. You are to take 28 twice; the result 56. See, it is of 56. You will find it right."

The corresponding modern-day formula for finding the volume of a truncated pyramid like the one shown in Figure 8.46 is $V = \frac{h}{3}(a^2 + ab + b^2)$, where h is the height and a and b are the lengths of the sides of the square top and square base. The following line-by-line comparison shows the relationship between the Egyptians' work and the modern-day formula.

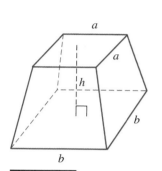

FIGURE 8.46

A truncated pyramid.

Problem from Moscow Papyrus	*Modern Counterparts*
Example of calculating a truncated pyramid.	$V = \frac{h}{3}(a^2 + ab + b^2)$
If you are told:	Given
a truncated pyramid of 6 for the vertical height	$h = 6,$
by 4 on the base	$b = 4,$ and
by 2 on the top:	$a = 2.$
You are to square this 4; result 16.	$b^2 = 16$
You are to double 4; result 8.	$ab = 8$
You are to square this 2; result 4.	$a^2 = 4$
You are to add the 16 and the 8 and the 4; result 28.	$a^2 + ab + b^2 = 28$
You are to take $\frac{1}{3}$ of 6; result 2.	$\frac{h}{3} = 2$
You are to take 28 twice;	$\frac{h}{3}(a^2 + ab + b^2)$
the result 56.	$V = 56$

EXAMPLE 2

FINDING VOLUMES USING EGYPTIAN UNITS OF MEASUREMENT
Which of the following structures has the greater storage capacity?

a. a truncated pyramid 15 cubits high with a 6-cubit-by-6-cubit top and an 18-cubit-by-18-cubit bottom (Figure 8.47)

b. a regular pyramid 20 cubits tall with an 18-cubit-by-18-cubit base (Figure 8.48)

SOLUTION

Because volume is a measure of the amount of space within a three-dimensional figure, it is used to compare the storage capacities of structures.

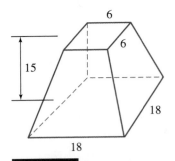

FIGURE 8.47

A truncated pyramid.

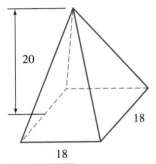

FIGURE 8.48

A regular pyramid.

a. To find the volume of the truncated pyramid, we use $h = 15$ cubits, $a = 6$ cubits, and $b = 18$ cubits:

$$V = \frac{h}{3}(a^2 + ab + b^2)$$

$$= \frac{15 \text{ cubits}}{3}[(6 \text{ cubits})^2 + (6 \text{ cubits})(18 \text{ cubits}) + (18 \text{ cubits})^2]$$

$$= (5 \text{ cubits})(36 \text{ cubits}^2 + 108 \text{ cubits}^2 + 324 \text{ cubits}^2)$$

$$= (5 \text{ cubits})(468 \text{ cubits}^2)$$

$$= 2{,}340 \text{ cubits}^3$$

$$= (2{,}340 \text{ cubits}^3)\left(\frac{\frac{3}{2} \text{ khar}}{1 \text{ cubit}^3}\right) \quad \textbf{converting cubic cubits to khar}$$

$$= 3{,}510 \text{ khar}$$

The volume of the truncated pyramid is 3,510 khar.

b. To find the volume of the pyramid, we use $h = 20$ cubits and $b = 18$ cubits. The pyramid is not truncated, so $a = 0$:

$$V = \frac{h}{3}(b^2)$$

$$= \frac{20 \text{ cubits}}{3}(18 \text{ cubits})^2$$

$$= 2{,}160 \text{ cubits}^3$$

$$= (2{,}160 \text{ cubits}^3)\left(\frac{\frac{3}{2} \text{ khar}}{1 \text{ cubit}^3}\right) \quad \textbf{converting cubic cubits to khar}$$

$$= 3{,}240 \text{ khar}$$

The volume of the regular pyramid is 3,240 khar.

Even though the two figures have the same base and the regular pyramid is taller, the truncated pyramid has a larger volume and thus a greater storage capacity. This result is due to the fact that the faces of the truncated pyramid have a steeper slope than those of the regular pyramid.

Because of its single-problem approach, Egyptian geometry consisted of a fragmented assortment of guesses, tricks, and rule-of-thumb procedures. It was an empirical subject in which approximate answers were usually good enough for practical purposes. This does not mean that the Egyptians weren't concerned with accuracy. The Great Pyramid of Cheops is a testament to the Egyptians' ability to calculate, measure, and construct with a high degree of precision.

The Great Pyramid of Cheops

Built around 2600 B.C. with an original height of 481.2 feet, the Great Pyramid is the most awesome of all ancient structures. The blocks of limestone composing the pyramid have a combined weight of over 5 million tons. The base is almost a perfect square; the average length of the four sides is 755.78 feet, with a maximum discrepancy of only 4.5 inches (a relative error of only 0.05%)! In addition, the sides of the pyramid are oriented to correspond to the four major compass headings of north, south, east, and west. This positioning has an error of less than 1 degree.

Because of its near perfection, some people believe that the pyramid contains mystic puzzles and the answers to the riddles of the universe. For instance, if you divide the semiperimeter of the base (the distance halfway around, or two times the length of a side) by the height of the pyramid, an interesting result is obtained:

$$\frac{\text{semiperimeter}}{\text{height}} = \frac{2(755.78)}{481.2} = 3.141230258\ldots$$

The precision of the Great Pyramid of Cheops is a testament to the ancient Egyptians' geometric ability. Some people believe that it also contains mystic puzzles.

The comparison of this number to the modern-day approximation of π is remarkable (π is approximately $3.141592654\ldots$). Was this planned, or is it merely coincidental? The debate continues.

As surveyors, astronomers, and architects, the early Egyptians were undoubtedly concerned with the mathematics of a circle. By taking careful measurements, they knew that the circumference of a circle was proportional to its diameter—that is, that circumference/diameter = constant. Today, we call this constant π. On the basis of a contemporary interpretation of an ancient document known as the Rhind Papyrus, it appears that the Egyptians would have concluded that $\pi = \frac{256}{81}$ (≈ 3.16).

The Rhind Papyrus

Most of our knowledge of early Egyptian mathematics was obtained from two famous papyri: the Moscow (or Golenischev) Papyrus and the Rhind Papyrus. In 1858, the Scottish lawyer and amateur archeologist Henry Rhind was vacationing in Luxor, Egypt. While investigating the buildings and tombs of Thebes, he came across an old rolled-up papyrus. The papyrus was written by the scribe Ahmes around 1650 B.C. and contained eighty-four mathematical problems and their solutions. We know that the papyrus is a copy of an older original, for it begins: "The entrance into the knowledge of all existing things and all obscure secrets. This book was copied in the year 33, the fourth month of the inundation season, under the King of Upper and Lower Egypt 'A-user-Re,' endowed with life, in likeness to writings made of old in the time of the King of Upper and Lower Egypt Ne-mat'et-Re. It is the scribe Ahmes who copies this writing." The reference to the ruler Ne-mat'et-Re places the original script around 2000 B.C. Of particular interest is Problem 48, in which the area of a circle (and hence π) is investigated.

Problem 48 of the Rhind Papyrus concludes that *"the area of a circle of diameter 9 is the same as the area of a square of side 8."* Today, we know that this is not exactly true; a circle of diameter 9 has area $A = \pi(4.5)^2 \approx 63.6$, whereas a square of side 8 has area $A = 8^2 = 64$. As we can see, the geometry of the early Egyptians was not perfect. But in many practical applications, the answers were close enough for their intended use.

The reasoning behind the conclusion of Problem 48 went like this: Construct a square whose sides are 9 units each, and inscribe a circle in the square. Thus, the diameter of the circle is 9 units. Now divide the sides of the square into three equal segments of 3 units each. Remove the triangular corners, as shown in Figure 8.49, and an irregular octagon is formed. The area of this octagon is then used as an approximation of the area of the circle: $A_{circle} \approx A_{octagon}$.

Notice that the area of the octagon is equal to the area of the large square minus the area of the four triangular corners. Rearranging these triangles, we can see that the area of two triangular corners is the same as the area of one square of side 3 units, as shown in Figure 8.50.

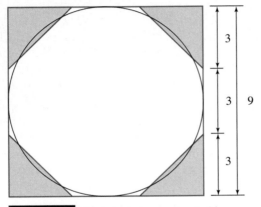

FIGURE 8.49 A circle is approximated by an irregular octagon.

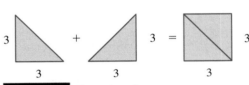

FIGURE 8.50 Two triangles = a square.

Therefore,

$$
\begin{aligned}
A_{circle} \approx A_{octagon} &= A_{square\ of\ side\ 9} - 4A_{triangle} \\
&= A_{square\ of\ side\ 9} - 2A_{square\ of\ side\ 3} \\
&= (9)^2 - 2(3)^2 \\
&= 81 - 18 \\
&= 63
\end{aligned}
$$

Therefore, $A_{circle} \approx 63$ square units. The Egyptians realized that the area of the circle was not 63 but close to it. They mistakenly thought that the area of the circle must be exactly 64, because 64 is close to 63 and 64 ($= 8^2$) is the area of a square with whole-number sides. (This 4,000-year-old "logic" doesn't seem reasonable today.) Thus, the ancient mathematicians reached their conclusion that

$$A_{circle\ of\ diameter\ 9} = A_{square\ of\ side\ 8}$$

Two approximations were used in Problem 48:

$$A_{circle\ of\ diameter\ 9} \approx A_{octagon}$$

$$A_{octagon} = 63 \approx 64 = A_{square\ of\ side\ 8}$$

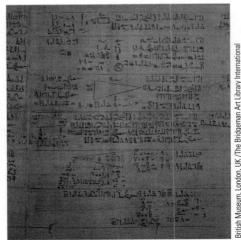

The Rhind Papyrus is one of the world's oldest known mathematics textbooks. Notice the triangles in this portion of the 18-foot-long papyrus.

Of these two estimates, the first (Figure 8.49) was an underestimate, and the second an overestimate; thus, the net "canceling" of errors contributed to the accuracy of their overall calculation. The Egyptians' geometry was not perfect, but it did give reasonable answers.

Pi and the Area of a Circle

The Egyptians obtained the "formula" $C = \pi d$ by observing that the ratio circumference/diameter is a constant. How did they obtain the "formula" for the area of a circle? (The Egyptians did not work with actual formulas; instead, they used verbal examples, as we saw with truncated pyramids.) The Egyptians might have used a method of rearrangement to find the area of any given circle. The method of rearrangement says that if a region is cut up into smaller pieces and rearranged to form a new shape, the new shape will have the same area as the original figure. We will apply this method to a circle, as the Egyptians might have done, and see what happens.

Suppose a circle of radius r is cut in half, and each semicircular region is then cut up into an equal number of uniform slices (like a pie). The circumference of the entire circle is known to be $2\pi r$, so the length of each semicircle is πr, as shown in Figure 8.51.

Now rearrange the slices as shown in Figure 8.52. The total area of these slices is still the same as the area of the circle. These pieces are then joined together as in Figure 8.53.

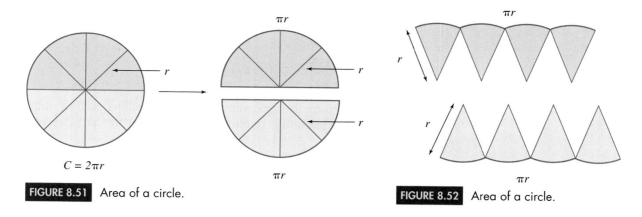

FIGURE 8.51 Area of a circle.

FIGURE 8.52 Area of a circle.

If we continue this process, cutting up the circle into more and more slices that get thinner and thinner, the rearranged figure becomes quite rectangular, as shown in Figure 8.54.

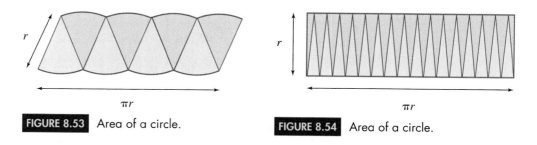

FIGURE 8.53 Area of a circle.

FIGURE 8.54 Area of a circle.

Rather than having a bumpy top and bottom, as in Figure 8.53, the top and bottom of this new figure are fairly straight, owing to the arcs of the circle becoming so short that they appear to be straight line segments.

Using the formula for the area of a rectangle (the rearranged figure), we have $A = bh = (\pi r)(r) = \pi r^2$. Because the area of the rearranged figure is πr^2, the area of the original circle is also πr^2.

EXAMPLE 3

FINDING THE EGYPTIAN APPROXIMATION OF π   Combining the result of Problem 48 of the Rhind Papyrus with the method of rearrangement of area, find the Egyptian approximation of π.

SOLUTION

Problem 48 of the Rhind Papyrus states that the area of a circle of diameter 9 (or radius $\frac{9}{2}$) is equal to the area of a square of side 8; that is,

$$A_{\text{circle of diameter 9}} = A_{\text{square of side 8}} = 64$$

Using rearrangement of area, we also have

$$A_{\text{circle of diameter 9}} = \pi \left(\frac{9}{2}\right)^2$$

Equating these two expressions, we have

$$\pi \left(\frac{9}{2}\right)^2 = 64$$

$$\frac{81}{4}\pi = 64$$

$$\pi = 64\left(\frac{4}{81}\right)$$

$$\pi = \frac{256}{81} \ (= 3.160493827 \ldots)$$

This is thought to be the procedure that the Egyptians used to obtain their value of π. This approximation is remarkably close to our modern-day value!

EXAMPLE 4

USING THE EGYPTIAN APPROXIMATION OF π AND COMPARING RESULTS   For a circle of radius 5 inches, do the following:

a. Use $\pi = \frac{256}{81}$ (the Egyptian approximation of pi) to find the area of the circle.
b. Use the value of π contained in a scientific calculator to find the area of the circle.
c. Find the error of the Egyptian calculation relative to the calculator value.

SOLUTION

a. $A = \pi r^2$

$$= \left(\frac{256}{81}\right)(5 \text{ in.})^2$$

$$= 79.01234568 \ldots \text{ in.}^2$$

Using the Egyptian value, we find that the area of the circle is approximately 79.0 square inches.

b. $A = \pi r^2$

$$= \pi (5 \text{ in.})^2$$

$$= 25\pi \text{ in.}^2$$

$$= 78.53981634 \ldots \text{ in.}^2$$

Using the calculator value, we find that the area of the circle is approximately 78.5 square inches. Note that if both decimals were rounded off to the nearest whole unit, they would yield the same answer.

c. To find the error of the Egyptian calculation relative to the calculator value, we must find the difference of the calculations and divide this difference by the calculator value:

$$A_{Egyptian} - A_{calculator} = 79.01234568 \text{ in.}^2 - 78.53981634 \text{ in.}^2$$

$$= 0.47252934 \text{ in.}^2$$

$$\text{Error relative to calculator} = \frac{0.47252934 \text{ in.}^2}{78.53981634 \text{ in.}^2}$$

$$= 0.00601643\ldots$$

$$= 0.6\%$$

The error of the Egyptian calculation relative to the calculator value is 0.6%.

8.3 EXERCISES

In Exercises 1–4, determine whether each configuration of knotted ropes would form a right triangle.

1.

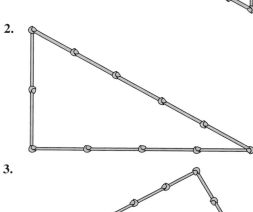

2.

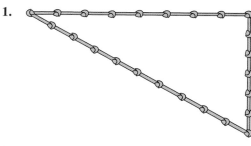

3.

4.

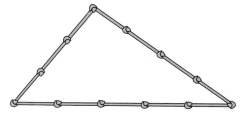

5. Which of the following structures has the larger storage capacity?

   a. a truncated pyramid that is 12 cubits high with a 2-cubit-by-2-cubit top and a 15-cubit-by-15-cubit bottom

   b. a regular pyramid that is 18 cubits tall with a 15-cubit-by-15-cubit base

6. Which of the following structures has the larger storage capacity?

   a. a truncated pyramid that is 33 cubits high with a 6-cubit-by-6-cubit top and a 30-cubit-by-30-cubit bottom

   b. a regular pyramid that is 50 cubits tall with a 30-cubit-by-30-cubit base

In Exercises 7 and 8, fill in the blanks as in the problem from the Moscow Papyrus (the Egyptian method for finding the volume of a truncated pyramid, immediately preceding Example 2).

7. "If you are told: a truncated pyramid of 9 for the vertical height by 6 on the base by 3 on the top: You are to square this 6; result _____. You are to triple 6; result _____. You are to square this 3; result _____. You are to add the _____ and the _____ and the _____; result _____. You are to take $\frac{1}{3}$ of _____; result _____. You are to take _____ three times; the result _____. See, it is of _____. You will find it right."

8. "If you are told: a truncated pyramid of 12 for the vertical height by 10 on the base by 4 on the top: You are to square this 10; result _____. You are to quadruple 10; result _____. You are to square this 4; result _____. You are to add the _____ and the _____ and the _____; result _____. You are to take $\frac{1}{3}$ of _____; result _____. You are to take _____ four times; the result _____. See, it is of _____. You will find it right."

9. Following the method of Problem 48 of the Rhind Papyrus, find a square, with a whole number as the length of its side, that has the same (approximate) area as a circle of diameter 24 units.

    HINT: Divide the sides of a 24-by-24 square into three equal segments.

10. Following the method of Problem 48 of the Rhind Papyrus, find a square, with a whole number as the length of its side, that has the same (approximate) area as a circle of diameter 42 units.

    HINT: Divide the sides of a 42-by-42 square into three equal segments.

11. Using the result of Exercise 9 and the method of Example 3, obtain an approximation of π.

12. Using the result of Exercise 10 and the method of Example 3, obtain an approximation of π.

13. For a circle of radius 3 palms, do the following.
    a. Use $\pi = \frac{256}{81}$ (the Egyptian approximation of pi) to find the area of the circle.
    b. Use the value of π contained in a scientific calculator to find the area of the circle.
    c. Find the error of the Egyptian calculation relative to the calculator value.

14. For a circle of radius 6 palms, do the following.
    a. Use $\pi = \frac{256}{81}$ (the Egyptian approximation of pi) to find the area of the circle.
    b. Use the value of π contained in a scientific calculator to find the area of the circle.
    c. Find the error of the Egyptian calculation relative to the calculator value.

15. For a sphere of radius 3 palms, do the following.
    a. Use $\pi = \frac{256}{81}$ (the Egyptian approximation of pi) to find the volume of the sphere.

b. Use the value of π contained in a scientific calculator to find the volume of the sphere.
    c. Find the error of the Egyptian calculation relative to the calculator value.

16. For a sphere of radius 6 palms, do the following.
    a. Use $\pi = \frac{256}{81}$ (the Egyptian approximation of pi) to find the volume of the sphere.
    b. Use the value of π contained in a scientific calculator to find the volume of the sphere.
    c. Find the error of the Egyptian calculation relative to the calculator value.

17. Find the perimeter of a triangle having sides with the following measurements: 2 cubits, 5 palms, 3 fingers; 3 cubits, 3 palms, 2 fingers; 4 cubits, 4 palms, 3 fingers.

18. It has been suggested that one palm equals 7.5 centimeters. Using a standard equivalence, one centimeter equals 0.3937 inch.
    a. How many inches are in one royal cubit?
    b. How many inches are in one common cubit?
    c. Measure your own personal cubit. How does it compare to the royal and common cubits?

19. Use part (a) of Exercise 18 to determine which is larger: one khar or one cubic foot.

20. A rectangular field measures 50 cubits by 200 cubits. Find the area of this field in setats.

21. A triangular field has sides that measure 150 cubits, 200 cubits, and 250 cubits. Find the area of this field in setats.

22. A rectangular room is 20 cubits long, 15 cubits wide, and 8 cubits high. Find the volume of this room in khar.

23. The area of a square is 4 setats. Find the length of a side of this square. Express your answer in the following units.
    a. khet   b. cubits

24. The volume of a cube is 12 khar. Find the length, in cubits, of an edge of this cube.

25. Problem 41 of the Rhind Papyrus pertains to finding the volume of a cylindrical granary. The diameter of the granary is 9 cubits, and its height is 10 cubits.
    a. Calculate the volume of this granary, in khar, using the Egyptian method (Problem 48 of the Rhind Papyrus) to calculate the area of a circle.
    b. Calculate the volume of this granary, in khar, using the conventional formula $A = \pi r^2$ to calculate the area of a circle.
    c. Find the error of the Egyptian calculation of volume relative to the conventional calculation.

Answer the following questions using complete sentences and your own words.

• HISTORY QUESTIONS

26. List four activities in which the ancient Egyptians used mathematics and, in so doing, contributed to the advancement of their civilization.

27. How did the Egyptians construct right angles?

28. Who wrote the Rhind Papyrus?

29. What does Problem 48 of the Rhind Papyrus state?

30. Write a research paper on a historical topic referred to in this section or on a related topic. Following is a list of possible topics:

   • Ancient agriculture and the Nile River (How did agriculture lead to mathematical and scientific advancement?)

   • Early astronomy (What motivated the early astronomers? What is the connection between early astronomy and mathematics?)

   • Early calendars (What purpose did calendars serve? Who created them?)

   • The Great Pyramid of Cheops (How was it constructed? What was its purpose?)

   • The history of π (Trace its history and applications.)

   • The Moscow Papyrus (What does it contain? When was it translated?)

   • The Rhind Papyrus (Is it intact? What does it contain? When was it translated?)

   • The Rosetta Stone (How was it used in mathematics? In politics? In science? In archeology?)

8.4 The Greeks

OBJECTIVES

● Understand and apply the concept that corresponding sides of similar triangles are proportional

● Prove that two triangles are congruent

● Understand and apply the concept that corresponding parts of congruent triangles are equal

● Explore the history and contributions of famous Greek mathematicians in the development of geometry

The primary question was not "What do we know," but "How do we know it."
—ARISTOTLE TO THALES

Many people accept common knowledge without questioning it, especially if they have firsthand experience of it. However, as Aristotle's comment to Thales points out, the ancient Greek scholars were not satisfied with mere facts; they were forever asking "why" in their search for absolute truth in the world around them. Whereas the Egyptians constructed a right angle by stretching a knotted rope to form a triangle with sides of three, four, and five units, the Greeks wanted to know *why* this method worked. The Greeks were not content to accept a claim simply because experience indicated that it worked. They wanted proof, and they obtained proof through the systematic application of logic and deductive reasoning.

Thales of Miletus

The empirical geometry of the Egyptians reached the ancient Greeks through commerce, travel, and warfare. One of the first Greek scholars to study this geometry was Thales of Miletus (625–547 B.C.). Thales is known as the first of the Seven Sages of Greece. (The Seven Sages of Greece were famous for their practical knowledge. In addition to Thales, the sages were Solon of Athens, Bias of Priene, Chilo of Sparta, Cleobulus of Rhodes, Periander of Corinth, and Pittacus of Mitylene.)

A successful businessman, statesman, and philosopher, Thales had occasion to travel to Egypt and consequently learned of Egyptian geometry. While in Egypt, he won the respect of the pharaoh by calculating the height of the Great Pyramid of Cheops. He did this by measuring the shadows of the pyramid and of his walking staff. Knowing the height of his staff, he used a proportion to calculate the height of the pyramid.

In the history of mathematics, Thales is credited with being the first to prove that corresponding sides of similar triangles are proportional. If the three angles of one triangle are equal to the three angles of another triangle, the triangles are **similar.** Consequently, similar triangles have the same shape but may differ in size.

THE SIDES OF SIMILAR TRIANGLES ARE PROPORTIONAL

If the three angles of one triangle equal the three angles of another triangle, then

$$\frac{a}{d} = \frac{b}{e} = \frac{c}{f}$$

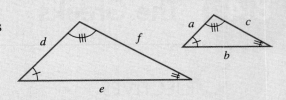

EXAMPLE **1**

FINDING THE LENGTHS OF SIDES OF SIMILAR TRIANGLES   For the two triangles in Figure 8.55, find the lengths of the unknown sides.

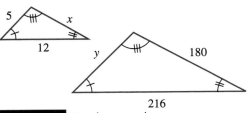

FIGURE 8.55 Similar triangles.

SOLUTION

Because the triangles are similar (they have equal angles), their sides are proportional. Comparing the small triangle to the large triangle, we have

$$\frac{x}{180} = \frac{12}{216}$$

$$x = \frac{12 \cdot 180}{216} \qquad \textbf{multiplying both sides by 180}$$

$$x = 10$$

Comparing the large triangle to the small triangle, we have

$$\frac{y}{5} = \frac{216}{12}$$

$$y = \frac{216 \cdot 5}{12} \quad \textbf{multiplying both sides by 5}$$

$$y = 90$$

Pythagoras of Samos

Pythagoras of Samos.

The name Pythagoras is synonymous with the famous formula that relates the lengths of the sides of a right triangle, $a^2 + b^2 = c^2$. However, the life and teachings of this famous Greek mathematician, philosopher, and mystic are shrouded in mystery, legend, and conjecture.

Born on the Aegean island of Samos about 500 B.C., Pythagoras traveled and studied extensively as a young man. His travels took him to Egypt, Phoenicia, and Babylonia. Returning to Samos when he was fifty, Pythagoras found his homeland under the rule of the tyrant Polycrates. For an unknown reason, Pythagoras was banned from Samos and migrated to Croton, in the south of present-day Italy, where he founded a school. His students and disciples, known as Pythagoreans, eventually came to hold considerable social power.

Pythagoras had his students concentrate on four subjects: the theory of numbers, music, geometry, and astronomy. To him, these subjects constituted the core of knowledge necessary for an educated person. This core of four subjects, later known as the *quadrivium*, persisted until the Middle Ages. At that time, three other subjects, known as the *trivium*, were added to the list: logic, grammar, and rhetoric. These seven areas of study came to be known as the seven liberal arts and formed the proper course of study for all educated people.

The Pythagoreans were a secretive sect. Their motto, "All is number," indicates their belief that numbers have a mystical quality and are the essence of the universe. Numbers had a mystic aura about them, and all things had a numerical representation. For instance, the number 1 stood for reason (the number of absolute truth), 2 stood for woman (the number of opinion), 3 stood for man (the number of harmony), 4 stood for justice (the product of equal terms, $4 = 2 \times 2 = 2 + 2$), 5 stood for marriage (the sum of man and woman, $5 = 2 + 3$), and 6 stood for creation (from which man and woman came, $6 = 2 \times 3$). In general, odd numbers were masculine, and even numbers were feminine. Indeed, the Pythagorean doctrine was a strange mixture of cosmic philosophy and number mysticism. The Pythagoreans' beliefs were augmented by many rites and taboos, which included refusing to eat beans (they were sacred), refusing to pick up a fallen object, and refusing to stir a fire with an iron.

In the realm of astronomy, Pythagoras had a curious theory. Being geocentric, he believed the earth to be the center of the universe. The sun, moon, and five known planets circled the earth, each traveling on its own crystal sphere. Because of the friction of these gigantic bodies whirling about, each body would produce a unique tone based on its distance from the earth. The combined effect of the seven bodies was the harmonious celestial music of the gods, which Pythagoras was the only mortal blessed to hear!

Near the end of his life, Pythagoras and his followers became more political. Their power and influence were felt in Croton and the other Greek cities of southern Italy. Legend has it that this power and the Pythagoreans' sense of autocratic supremacy initiated a conflict with the local peoples and governments that ultimately led to his death.

Historians do not agree on the exact circumstances of Pythagoras's death, but it is known that during a violent revolt against the Pythagoreans, the locals set fire to the school. Some say that Pythagoras died in the flames. Others say that he escaped the inferno but was chased to the edge of a bean field. Rather than trample the sacred beans, Pythagoras allowed the crowd to overtake and kill him. Some people believe that Pythagoras was murdered by a disgruntled disciple. In death, as in life, Pythagoras is shrouded in mystery.

Although Pythagoras is given credit for proving the theorem that bears his name, there is no hard evidence that he did so. All his teachings were verbal; there are no written records of his actual work. Furthermore, his disciples had to swear not to reveal any of his doctrines to outsiders.

Since the days of Pythagoras, many proofs of "his" theorem have been put forth. Most involve the method of rearrangement of area. A few years before being elected the twentieth president of the United States, James Garfield produced an original proof. His method, shown below, involved the creation of a trapezoid that consisted of three right triangles.

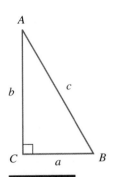

FIGURE 8.56

A right triangle.

James Garfield's Proof of the Pythagorean Theorem

Given: the right triangle *ABC*, as shown in Figure 8.56.

Extend segment *CA* a distance of *a*, and call this point *D*.

Construct a perpendicular segment of length *b* at *D*.

The endpoint of this segment is *E*. (See Figure 8.57.)

Triangles *ABC* and *EAD* are congruent, so $\angle EAB$ is a right angle. (Why?)

Figure *BCDE* is a trapezoid with parallel bases *CB* and *DE*.

The area of *BCDE* can be found by using the formula for the area of a trapezoid, or it can be found by adding the areas of the three right triangles. Apply both methods and set their results equal as follows:

$$\text{Area of trapezoid} = (\text{average base})(\text{height}) = \left(\frac{a+b}{2}\right)(a+b)$$

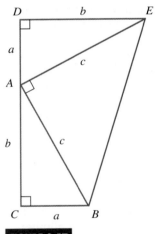

FIGURE 8.57

A trapezoid.

$$\text{Area of } \Delta\, ABC = \frac{1}{2}(\text{base})(\text{height}) \qquad = \frac{1}{2}ab$$

$$\text{Area of } \Delta\, EAD = \frac{1}{2}(\text{base})(\text{height}) \qquad = \frac{1}{2}ab$$

$$\text{Area of } \Delta\, EAB = \frac{1}{2}(\text{base})(\text{height}) \qquad = \frac{1}{2}c^2$$

Area of trapezoid = sum of the areas of the three triangles

$$\left(\frac{a+b}{2}\right)(a+b) = \left(\frac{1}{2}ab\right) + \left(\frac{1}{2}ab\right) + \left(\frac{1}{2}c^2\right)$$

$(a+b)(a+b) = ab + ab + c^2$ **multiplying each side by 2**

$a^2 + ab + ab + b^2 = ab + ab + c^2$ **multiplying** $(a+b)(a+b)$

$a^2 + b^2 = c^2$ **subtracting 2ab from each side**

Thus, given any right triangle with legs *a* and *b* and hypotenuse *c*, we have shown that $a^2 + b^2 = c^2$.

Euclid of Alexandria

The most widely published and read book in the history of the printed word is, of course, the Bible. And what do you suppose the second most widely published book is? Many people are quite surprised to discover that the number 2 all-time

Euclid's proof of the Pythagorean Theorem as it appeared in an Arabic translation of *Elements*.

best-seller is a mathematics textbook! Written by the Greek mathematician Euclid of Alexandria around 300 B.C., *Elements* was hand-copied by scribes for nearly 2,000 years. Since the coming of the printing press, over 1,000 editions have been printed, the first in 1482. *Elements* consists of 13 "books," or units. Although most people associate Euclid with geometry, *Elements* also contains number theory and elementary algebra, all the mathematics known to the scholars of the day.

Euclid's monumental contribution to the world of mathematics was his method of organization, not his discovery of new theorems. He took all the mathematical knowledge that had been compiled since the days of Thales (300 years earlier) and organized it in such a way that every result followed logically from its predecessors. To begin this chain of proof, Euclid had to start with a handful of assumptions, or things that cannot be proved. These assumptions, called *axioms* or *postulates*, are accepted without proof. By carefully choosing five geometric postulates, Euclid proceeded to prove 465 results, many of which were quite complicated and not at all intuitively obvious. The beauty of his work is that so much was proved from so few assumptions.

Euclid's Postulates of Geometry

1. A straight line segment can be drawn from any point to any other point.
2. A (finite) straight line segment can be extended continuously into an (infinite) straight line.
3. A circle may be described with any center and any radius.
4. All right angles are equal to one another.
5. Given a line and a point not on that line, there is one and only one line through the point parallel to the original line.

The only "controversial" item on this list is postulate 5, the so-called Parallel Postulate. This proved to be the unraveling of Euclidean geometry, and 2,000 years later non-Euclidean geometries were born. Non-Euclidean geometries will be discussed in Section 8.8.

The rigor of Euclid's work was unmatched in the ancient world. *Elements* was written for mature, sophisticated thinkers. Even today, the real significance of Euclid's work lies in the superb training it gives in logical thinking. Abraham Lincoln knew of the rigor of Euclid's *Elements*. At the age of forty, while still a struggling lawyer, Lincoln mastered the first six books in the *Elements* on his own, solely as training for his mind.

Teachers of mathematics invariably encounter resistance from students when abstract mathematics with little apparent application is presented. Even in the days of

Euclid by Joos van Ghent and Pedro Berruguete. Galleria Nazionale della Marche, Urbino, Italy.

Euclid, students were dismayed at the complexity of the proofs and asked the question so often echoed by today's students: "Why do we have to learn this stuff?" According to legend, one of Euclid's beginning students asked, "What shall I gain by learning these things?" Euclid quickly summoned his servant, responding, "Give him a coin, since he must make gain out of what he learns!" A second legend handed down through the years claims that King Ptolemy once

HISTORICAL
NOTE

ARCHIMEDES OF SYRACUSE (287–212 B.C.)

Archimedes demonstrated that a very heavy object could be raised with relatively little effort by using a series of pulleys (a compound pulley) or by using levers. He was so confident of his principles of levers and pulleys that he boasted that he could move anything. One of Archimedes' many famous quotes is "Give me a place to stand and I will move the earth!"

Aware of his genius, King Hieron turned to Archimedes on many occasions. One amusing legend concerns the problem of the king's gold crown. After giving a goldsmith a quantity of gold for the creation of an elaborate crown, King Hieron suspected the smith of keeping some of the gold and replacing it with an equal weight of silver. Since he had no proof, Hieron turned to Archimedes. While pondering the king's problem, Archimedes submerged himself in a full tub of water to bathe. As he climbed in, he noticed the water overflowing the tub. In a flash of brilliance, he discovered the solution to the king's problem. Overjoyed with his idea, Archimedes is rumored to have run down the street, totally naked, shouting, "Eureka! Eureka!" ("I have found it! I have found it!")

What Archimedes had found was that because silver is less dense than gold, an equal weight of silver would have a greater volume than an equal weight of gold and consequently would displace more water. When the crown and a piece of pure gold of the same weight were immersed in water, the smith's fate was sealed. Even though the crown and the pure gold had the same weight, the crown displaced more water, thus proving that it had been adulterated.

In 215 B.C., the Romans laid siege to Syracuse on the island of Sicily. The Roman commander Marcellus had no idea of the fierce resistance his troops were to face. The attack was from the sea. A native of Syracuse, Archimedes (then seventy-two years old) designed and helped to build a number of ingenious weapons to defend his home. He designed huge catapults that hurled immense

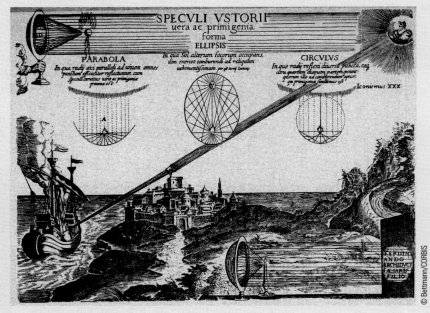

A seventeenth-century engraving depicting Archimedes' invention of a system of mirrors designed to focus the sun's rays on attacking ships to set them afire. The focusing properties of these shapes will be discussed in Section 8.7.

asked Euclid if there was a shorter way of learning geometry than the study of his *Elements,* to which Euclid replied, "There is no royal road to geometry." Sorry, no shortcuts today!

Deductive Proof

To investigate the Greeks' method of geometric deduction, we will examine congruent triangles. **Congruent triangles** are triangles that have exactly the same shape and size; their corresponding angles and sides are equal, as shown in Figure 8.58. "Triangle *ABC* is congruent to triangle *DEF*" (denoted by $\triangle ABC \cong \triangle DEF$) means that all six of the following statements are true:

$$\angle A = \angle D \qquad \angle B = \angle E \qquad \angle C = \angle F$$
$$AB = DE \qquad BC = EF \qquad CA = FD$$

boulders at the Roman ships. These catapults were set to throw the projectiles at different ranges so that no matter where the ships were, they were always under fire. When a ship managed to come in close, the defenders lowered large hooks over the city walls, grabbed the ship, lifted it into the air with pulleys, and tossed it back into the sea.

Needless to say, the Greek defenders put up a good fight. They fought so well that the siege lasted nearly three years.

However, one day, while the people of Syracuse were feasting and celebrating a religious festival, Roman sympathizers within the city informed the attackers of weaknesses in the city's defense, and a bloody sack began.

Marcellus had given strict orders to take Archimedes alive; no harm was to come to him or his house. At the time, Archimedes was studying the drawings of circles he had made in the sand. Preoccupied, he did not notice the Roman

soldier standing next to him until the soldier cast a shadow on his drawings in the sand. The agitated mathematician called out, "Don't disturb my circles!" At that, the insulted soldier used his sword and killed Archimedes of Syracuse.

The Roman commander Marcellus so grieved at the loss of Archimedes that when he learned of the mathematician's wish for the design of his tombstone, he fulfilled it.

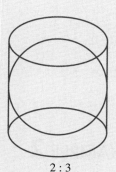

2 : 3

Archimedes' tombstone.

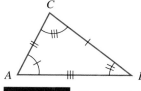 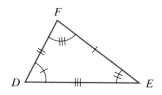

FIGURE 8.58 Congruent triangles.

Euclid proved that two triangles are congruent—that is, $\triangle ABC \cong \triangle DEF$—in any of the following three circumstances:

1. **SSS:** If the three sides of one triangle equal the three sides of the other triangle, then the triangles are congruent. (The corresponding angles will automatically be equal.) See Figure 8.59.

2. **SAS:** If two sides and the included angle of one triangle equal two sides and the included angle of the other triangle, then the triangles are congruent. (The other side and angles will automatically be equal.) See Figure 8.60.

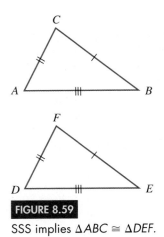

FIGURE 8.59

SSS implies △ABC ≅ △DEF.

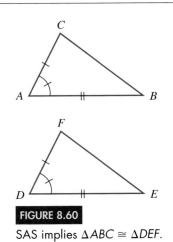

FIGURE 8.60

SAS implies △ABC ≅ △DEF.

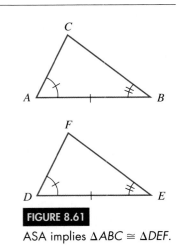

FIGURE 8.61

ASA implies △ABC ≅ △DEF.

3. **ASA:** If two angles and the included side of one triangle equal two angles and the included side of the other triangle, then the triangles are congruent. (The other angle and sides will automatically be equal.) See Figure 8.61.

When proving a result in geometry, we use a two-column method. The first column is a list of each valid statement used in the proof, starting with the given statements and progressing to the conclusion. The second column gives the reason for or justification of each statement in the first column.

EXAMPLE **2**

USING THE TWO-COLUMN STATEMENT-REASON METHOD TO PROVE A GEOMETRIC RESULT

Given: $CA = DB$
$\angle CAB = \angle DBA$

Prove: $CB = DA$

(See Figure 8.62.)

SOLUTION

Statements	Reasons
1. $CA = DB$	1. Given
2. $\angle CAB = \angle DBA$	2. Given
3. $AB = AB$	3. Anything equals itself.
4. $\triangle CAB \cong \triangle DBA$	4. SAS
5. $CB = DA$	5. Corresponding parts of congruent triangles are equal.

FIGURE 8.62

Hypotheses.

Archimedes of Syracuse

Many scholars consider Archimedes the father of physics; he was the first scientific engineer, for he combined mathematics and the deductive logic of Euclid with the scientific method of experimentation. Born around 287 B.C. in the Greek city of Syracuse on the island of Sicily, Archimedes gave the world many labor-saving devices, such as pulleys, levers, and a water pump that utilized a screw-shaped cylinder. In addition, he created formidable weapons that were used in the defense of Syracuse, developed the fundamental concept of buoyancy, and wrote many treatises on mathematics and geometry, including ones on the calculation of π and on the surface area and volume of spheres, cylinders, and cones.

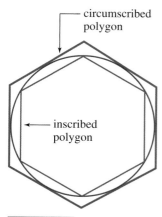

FIGURE 8.63

Inscribed and circumscribed polygons.

To calculate the value of π to any desired level of accuracy, Archimedes approximated a circle by using a regular polygon—a polygon whose sides all have the same length. His method was to inscribe a regular polygon of n sides inside a circle and determine the perimeter of the polygon. Because it was inside the circle, the perimeter of the polygon was smaller than the circumference of the circle. Then he would circumscribe an n-sided polygon around the circle and find its perimeter, which of course was larger than the circumference, as shown in Figure 8.63. By increasing n (the number of sides of the polygon) from 6 to 12 to 24 and so on, he got polygons whose perimeters were progressively closer and closer to the circumference of the circle. Using a polygon of 96 sides, he obtained

$$3\frac{10}{71} < \pi < 3\frac{1}{7}$$

This upper limit of $\frac{22}{7}$ is still used today as a common approximation for π.

EXAMPLE **3**

USING INSCRIBED AND CIRCUMSCRIBED HEXAGONS TO ESTIMATE THE VALUE OF π   A regular hexagon is inscribed in a circle of radius r, while another regular hexagon is circumscribed around the same circle.

a. Using the perimeter of the inscribed hexagon as an approximation of the circumference of the circle, obtain an estimate of π.

b. If the perimeter of the circumscribed hexagon has sides of length $s = (2\sqrt{3}/3)r$, obtain an estimate of π.

SOLUTION

a. Let s = length of a side of the inscribed hexagon; hence, $P_{\text{hexagon}} = 6s$. A regular hexagon is composed of six equilateral triangles; that is, triangles in which all sides are equal. Therefore, $s = r$, as shown in Figure 8.64.

$$C_{\text{circle}} \approx P_{\text{inscribed hexagon}}$$
$$2\pi r \approx 6s$$
$$2\pi r \approx 6r \quad \textbf{substituting } s \text{ for } r$$
$$\pi \approx \frac{6r}{2r} \quad \textbf{dividing by } 2r$$
$$\pi \approx 3$$

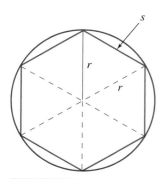

FIGURE 8.64

Inscribed hexagon.

Therefore, π is approximately 3. Note that an **inscribed polygon** will underestimate the circumference, so our approximation of π is too small.

b. Let s = length of a side of the circumscribed hexagon; hence, $P = 6s$, as shown in Figure 8.65. We are given that $s = (2\sqrt{3}/3)r$.

$$C_{\text{circle}} \approx P_{\text{circumscribed hexagon}}$$
$$2\pi r \approx 6s$$
$$2\pi r \approx 6\left(\frac{2\sqrt{3}}{3}r\right) \quad \textbf{substituting } s = (2\sqrt{3}/3)r$$
$$2\pi r \approx 4\sqrt{3}r$$
$$\pi \approx \frac{4\sqrt{3}r}{2r} \quad \textbf{dividing by } 2r$$
$$\pi \approx 2\sqrt{3}$$
$$\pi \approx 3.4641016$$

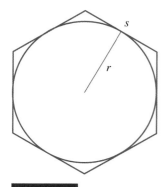

FIGURE 8.65

Circumscribed hexagon.

Therefore, π is approximately 3.4641016. Note that a **circumscribed polygon** will overestimate the circumference, so our approximation of π is too big.

Combining (b) with (a), we have

$$3 < \pi < 3.4641016$$

Of all the work Archimedes did in the field of solid geometry, he was proudest of one particular achievement. Archimedes found that if a sphere is inscribed in a cylinder, the volume of the sphere will be two-thirds the volume of the cylinder. In other words, the ratio of their volumes is 2 : 3. The result pleased him so much that he wanted a diagram of a sphere inscribed in a cylinder and the ratio 2 : 3 to be the only markings on his tombstone (see page 573).

8.4 EXERCISES

In Exercises 1–4, the given triangles are similar. Find the lengths of the missing sides.

1.

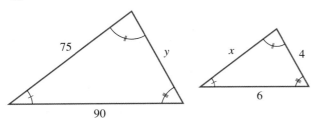

2.

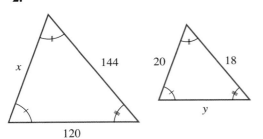

3.

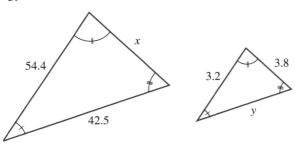

4.

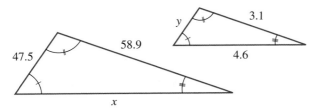

5. A 6-foot-tall man casts a shadow of 3.5 feet at the same instant that a tree casts a shadow of 21 feet. How tall is the tree?

6. A 5.4-foot-tall woman casts a shadow of 2 feet at the same instant that a telephone pole casts a shadow of 9 feet. How tall is the pole?

7. Use the diagrams in Figure 8.66 to show that $a^2 + b^2 = c^2$.

HINT: The area of square #1 = the area of square #2. (Why?)

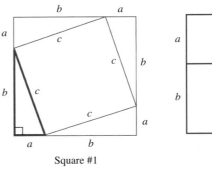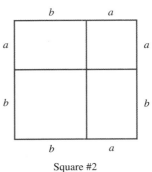

Square #1   Square #2

FIGURE 8.66 Diagrams for Exercise 7.

8. Use Figure 8.67 and the method of rearrangement to show that $a^2 + b^2 = c^2$.

HINT: Find the area of the figure in two ways, as was done in Garfield's proof.

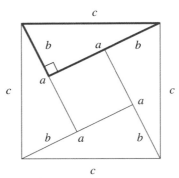

FIGURE 8.67 Diagram for Exercise 8.

9. Find the length of the longest object that will fit inside a rectangular box 4 feet long, 3 feet wide, and 2 feet high.

HINT: Draw a picture and use the Pythagorean Theorem.

10. Find the length of the longest object that will fit inside a cube 2 feet long, 2 feet wide, and 2 feet high.

   HINT: Draw a picture and use the Pythagorean Theorem.

11. Find the length of the longest object that will fit inside a cylinder that has a radius of $2\frac{1}{4}$ inches and is 6 inches high.

   HINT: Draw a picture and use the Pythagorean Theorem.

12. Find the length of the longest object that will fit inside a cylinder that has a radius of 1 inch and is 2 inches high.

   HINT: Draw a picture and use the Pythagorean Theorem.

In Exercises 13–18, use the two-column method to prove the indicated result.

13. Given: $AD = CD$

                  $AB = CB$

   Prove: $\angle DBA = \angle DBC$

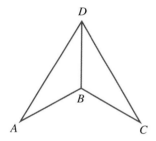

14. Given: $CA = DB$

                  $CB = DA$

   Prove: $\angle ACB = \angle ADB$

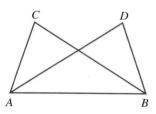

15. Given: $AD = BD$

                  $\angle ADC = \angle BDC$

   Prove: $AC = BC$

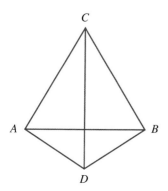

16. Given: $\angle ACD = \angle BCD$

                  $\angle ADC = \angle BDC$

   Prove: $AC = BC$

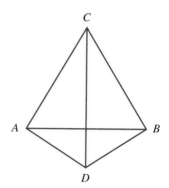

17. Given: $AE = CE$

                  $AB = CB$

   Prove: $\angle ADB = \angle CDB$

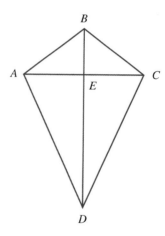

18. Given: $AE = CE$

                  $\angle AEB = \angle CEB$

   Prove: $AD = CD$

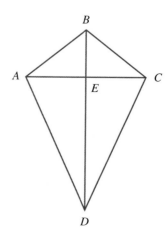

In Exercises 19–22, follow the method of Example 3.

19. A regular octagon (eight-sided polygon) is inscribed in a circle of radius r, while another regular octagon is circumscribed around the same circle.

   a. The length of a side of the inscribed octagon is $s = \sqrt{2 - \sqrt{2}}\,r$. Using the perimeter of this polygon as an approximation of the circumference of the circle, obtain an estimate of π.

   b. The length of a side of the circumscribed octagon is $s = 2(\sqrt{2} - 1)r$. Using the perimeter of this polygon as an approximation of the circumference of the circle, obtain an estimate of π.

20. A regular dodecagon (12-sided polygon) is inscribed in a circle of radius r, while another regular dodecagon is circumscribed around the same circle.

   a. The length of a side of the inscribed dodecagon is $s = [(\sqrt{6} - \sqrt{2})/2]r$. Using the perimeter of this polygon as an approximation of the circumference of the circle, obtain an estimate of π.

   b. The length of a side of the circumscribed dodecagon is $s = [(\sqrt{6} - \sqrt{2})/(\sqrt{2} + \sqrt{3})]r$. Using the perimeter of this polygon as an approximation of the circumference of the circle, obtain an estimate of π.

21. A regular 16-sided polygon is inscribed in a circle of radius r, while another regular 16-sided polygon is circumscribed around the same circle.

   a. The length of a side of the inscribed polygon is $s = \sqrt{2 - \sqrt{2 + \sqrt{2}}}\,r$. Using the perimeter of this polygon as an approximation of the circumference of the circle, obtain an estimate of π.

   b. The length of a side of the circumscribed polygon is
$$s = \left(\frac{2\sqrt{2 - \sqrt{2}}}{2 + \sqrt{2 + \sqrt{2}}} \right)r$$
   Using the perimeter of this polygon as an approximation of the circumference of the circle, obtain an estimate of π.

22. A regular 24-sided polygon is inscribed in a circle of radius r, while another 24-sided polygon is circumscribed around the same circle.

   a. The length of a side of the inscribed polygon is $s = [\sqrt{8 - \sqrt{24 - \sqrt{8}}}/2]r$. Using the perimeter of this polygon as an approximation of the circumference of the circle, obtain an estimate of π.

   b. The length of a side of the circumscribed polygon is
$$s = \left(\frac{2\sqrt{8 - \sqrt{24 - \sqrt{8}}}}{\sqrt{8} + \sqrt{24 + \sqrt{8}}} \right)r$$
   Using the perimeter of this polygon as an approximation of the circumference of the circle, obtain an estimate of π.

23. If a sphere is inscribed in a cylinder, the diameter of the sphere equals the height of the cylinder. Using this relationship, show Archimedes' favorite result, that is, that the volumes of the sphere and the cylinder are in the ratio $2:3$.

 Answer the following questions using complete sentences and your own words.

• HISTORY QUESTIONS

24. What is the connection between Thales and Egypt?

25. What was the motto of the Pythagoreans?

26. Which president of the United States developed an original proof of the Pythagorean Theorem?

27. What is the Parallel Postulate?

28. How did the geometry of the Greeks differ from that of the Egyptians?

29. Write a research paper on a historical topic referred to in this section or on a related topic. Following is a list of possible topics:

   • Alexandria as a center of knowledge (Why and when did Alexandria flourish as a center of knowledge in the ancient world?)
   • Ancient war machines (How were mathematics and physics used in the design and construction of ancient weapons of war?)
   • Archimedes
   • Classic construction problems (What do the problems "trisecting an angle," "squaring a circle," and "doubling a cube" have in common? Why are these problems special in the history of geometry?)
   • Eratosthenes of Cyrene (How did he calculate the earth's circumference? What did he contribute to the making of maps?)
   • Euclid
   • Eudoxus of Cnidos (What is the "Golden Section" or "Golden Ratio"?)
   • Heron of Alexandria (What is Heron's Formula?)
   • Hippocrates of Chios (What are the lunes of Hippocrates?)
   • Pappus of Alexandria (What did Pappus's *Mathematical Collection* contain?)
   • Pulleys and levers (How do pulleys and levers reduce the force required to lift an object?)
   • Pythagoras and the Pythagoreans
   • Thales of Miletus (How did he calculate the height of the Great Pyramid of Cheops?)
   • Zeno of Elea (What are Zeno's paradoxes?)

• PROJECT

30. The purpose of this project is to calculate the height of a tall building by using similar triangles. Choose a tall building, and measure the length of its shadow. Now have an assistant measure the length of your shadow. The top of the building, the base of the building, and the tip of the shadow of the building form a triangle, say, ΔABC. Likewise, the top of your head, your heels, and the tip of your shadow form a triangle, say,

ΔDEF. (See Figure 8.68.) Assuming that rays of sunlight are parallel, we conclude that ΔABC and ΔDEF are similar. Use the ratio of your height to the length of your shadow to calculate the height of the building.

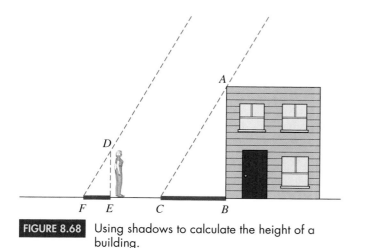

FIGURE 8.68   Using shadows to calculate the height of a building.

WEB PROJECTS

31. There are numerous web sites devoted to the works of Euclid; many are animated and interactive. Investigate some of these web sites and write an essay and/or make a presentation about your findings.

32. There are numerous web sites devoted to the works of Pythagoras; many are animated and interactive. Investigate some of these web sites, and write an essay and/or make a presentation about your findings.

Some useful links for this web project are listed on the text web site: **www.cengage.com/math/johnson**

8.5   Right Triangle Trigonometry

OBJECTIVES

- Use degrees to measure the angles of a triangle
- Define the basic trigonometric ratios: sine, cosine, and tangent
- Apply the basic trigonometric ratios to special triangles: 45°-45°-90° and 30°-60°-90°
- Use a calculator to find trigonometric ratios of arbitrary angles
- Use inverse trigonometric ratios and a calculator to find the measure of an angle
- Use trigonometric ratios to find unknown parts of a triangle

Plane geometry includes the study of many simple figures, such as circles, triangles, rectangles, squares, parallelograms, and trapezoids. Triangles are of special interest; they are the simplest figures that can be drawn with straight sides. **Trigonometry** is the branch of mathematics that details the study of triangles. The word *trigonometry* literally means "triangle measurement"; it is derived from the Greek words *trigon,* meaning "triangle," and *metros,* meaning "measure." Although modern trigonometry has numerous applications that do not involve triangles, trigonometry's roots lie in the study of right triangles; for example, we will use trigonometry to calculate the length of a cable that runs from the ground to the top of a warehouse, as shown in Figure 8.69.

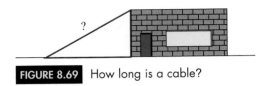

FIGURE 8.69   How long is a cable?

Angle Measurement

A triangle has three sides and three angles. The sides can be measured in terms of any linear unit, such as inches, feet, centimeters, or meters. How are angles measured? The two most common units are *degrees* and *radians*. We will consider only degrees in this textbook.

A **right angle** is an angle that makes a square corner. By definition, a right angle is said to have a measure of **90 degrees,** denoted 90°. See Figure 8.70(a). Consequently, a straight line can be viewed as an angle of measure 180°, as shown in Figure 8.70(b).

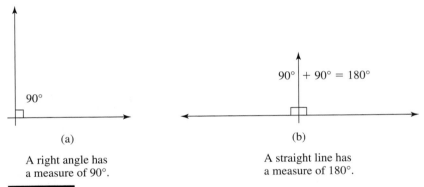

(a)

A right angle has
a measure of 90°.

(b)

A straight line has
a measure of 180°.

FIGURE 8.70

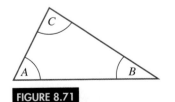

FIGURE 8.71

An arbitrary triangle.

If a right angle is "cut up" into 90 equal pieces, each "slice" would represent an angle of measure 1°. From a previous math course, you may recall that for any triangle, the sum of the measures of the three angles is 180°. Why is this true? Suppose the angles of a triangle are *A*, *B*, and *C*, as shown in Figure 8.71.

Now construct two copies of the triangle and place them next to the original, as shown in Figure 8.72.

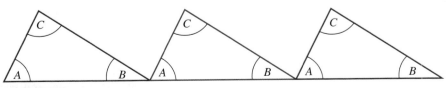

FIGURE 8.72   Multiple copies of a triangle.

If the triangle on the right is flipped backward and upside down, it will fit perfectly into the "notch" between the other two triangles, as shown in Figure 8.73.

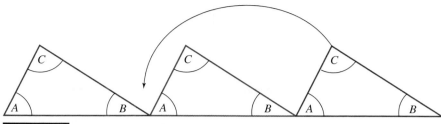

FIGURE 8.73 Moving a triangle.

Notice that the three angles A, B, and C now form a straight line, as shown on the bottom edge of Figure 8.74, consequently, the sum of the angles equals 180°.

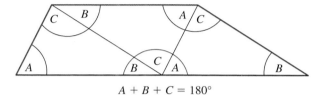

$$A + B + C = 180°$$

FIGURE 8.74 The sum of the angles in a triangle $= 180°$.

Special Triangles

There are two special triangles that frequently arise in trigonometry: the 30°-60°-90° triangle and the 45°-45°-90° triangle. The sides of each of these triangles have special relationships.

Each angle of a square has a measure of 90°. If a square with sides of length x is cut in two along one of its diagonals, each resulting right triangle will have two equal sides (of length x) and two 45° angles. See Figure 8.75.

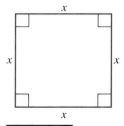
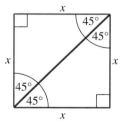

FIGURE 8.75 A square and its diagonal.

A triangle of this type is called an *isosceles right triangle*. (An isosceles triangle is a triangle that has two equal sides.) Applying the Pythagorean Theorem, we find the length of the hypotenuse as follows:

$$c^2 = a^2 + b^2 \quad \textbf{c = hypotenuse, a and b = legs}$$
$$c^2 = x^2 + x^2 \quad \textbf{substituting } a = x \textbf{ and } b = x$$
$$c^2 = 2x^2 \qquad \textbf{combining like terms}$$

Now take the (positive) square root of each side and simplify:

$$\sqrt{c^2} = \sqrt{2x^2}$$
$$c = \sqrt{2}x$$

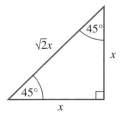

FIGURE 8.76

Isosceles right triangle.

Therefore, the hypotenuse has length $\sqrt{2}x$. See Figure 8.76.

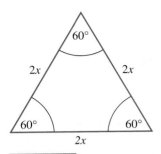

FIGURE 8.77

Equilateral triangle.

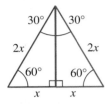

FIGURE 8.78

Bisected equilateral triangle.

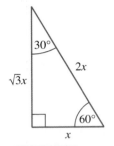

FIGURE 8.79

30°-60°-90° triangle.

An *equilateral triangle* is a triangle in which all three sides have the same length; consequently, all three angles have the same measure. Since the sum of the three angles must equal 180°, it follows that each angle in an equilateral triangle has a measure of 60°. For future convenience, suppose each side has length $2x$, as shown in Figure 8.77.

Suppose one of the angles of an equilateral triangle is bisected (cut into two equal pieces); the bisected 60° angle will generate two 30° angles. Consequently, two right triangles are created. In addition, the side opposite the bisected angle will be cut into two equal segments of length x. See Figure 8.78.

Applying the Pythagorean Theorem to one of these right triangles (called a *30°-60°-90° triangle*), we find the length of the missing (vertical) leg as follows:

$$a^2 + b^2 = c^2 \qquad \textbf{a and b = legs, c = hypotenuse}$$
$$x^2 + b^2 = (2x)^2 \qquad \textbf{substituting } a = x \textbf{ and } c = 2x$$
$$x^2 + b^2 = 4x^2 \qquad \textbf{multiplying}$$
$$b^2 = 3x^2 \qquad \textbf{subtracting } x^2 \textbf{ from both sides}$$

Now take the (positive) square root of each side and simplify:

$$\sqrt{b^2} = \sqrt{3x^2}$$
$$b = \sqrt{3}x$$

Therefore, the leg has length $\sqrt{3}x$. See Figure 8.79.

Trigonometric Ratios

An *acute angle* is an angle whose measure is between 0° and 90°. Let θ (the Greek letter theta) equal the measure of one of the acute angles of a right triangle. Using the abbreviations "opp," "adj," and "hyp" to represent, respectively, the side opposite angle θ, the side adjacent to angle θ, and the hypotenuse, a right triangle can be labeled, as in Figure 8.80.

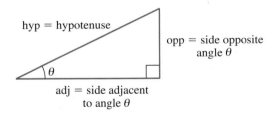

FIGURE 8.80 The sides of a right triangle.

In Section 8.4, we saw that for similar triangles (triangles that have the same shape but different size), corresponding sides are proportional; that is, ratios of corresponding sides are constant. Consequently, no matter how large or how small a right triangle is, the ratio of its sides is determined solely by the size of the acute angles. Three specific ratios of sides form the basis of trigonometry; these ratios are called the *sine, cosine,* and *tangent* and are defined as follows.

TRIGONOMETRIC RATIOS

Let θ equal the measure of one of the acute angles of a right triangle.

1. The **sine** of θ, denoted **sinθ**, is the ratio of
   the side opposite θ compared to the hypotenuse: $\sin\theta = \dfrac{\text{opp}}{\text{hyp}}$

2. The **cosine** of θ, denoted **cosθ**, is the ratio of
   the side adjacent θ compared to the hypotenuse: $\cos\theta = \dfrac{\text{adj}}{\text{hyp}}$

3. The **tangent** of θ, denoted **tanθ**, is the ratio of
   the side opposite θ compared to the side adjacent θ: $\tan\theta = \dfrac{\text{opp}}{\text{adj}}$

The following example examines the special isosceles right triangle.

EXAMPLE **1**

SOLUTION

FINDING TRIGONOMETRIC RATIOS OF A SPECIAL ANGLE: 45°   Find the sine, cosine, and tangent of 45°.

Recall the general isosceles right triangle shown in Figure 8.76 and reproduced in Figure 8.81. The length of the side opposite 45° is x, as is the length of the side adjacent 45°; the length of the hypotenuse is $\sqrt{2}x$. Applying the trigonometric ratios to the angle of measure 45°, we have the following:

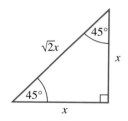

FIGURE 8.81

Isosceles right triangle.

$$\sin 45° = \frac{\text{opp}}{\text{hyp}} = \frac{x}{\sqrt{2}x} = \frac{1}{\sqrt{2}}$$ **To rationalize the denominator, multiply by** $\dfrac{\sqrt{2}}{\sqrt{2}}$.

Therefore, $\sin 45° = \dfrac{\sqrt{2}}{2}$.

$$\cos 45° = \frac{\text{adj}}{\text{hyp}} = \frac{x}{\sqrt{2}x} = \frac{1}{\sqrt{2}}$$ **To rationalize the denominator, multiply by** $\dfrac{\sqrt{2}}{\sqrt{2}}$.

Therefore, $\cos 45° = \dfrac{\sqrt{2}}{2}$.

$$\tan 45° = \frac{\text{opp}}{\text{adj}} = \frac{x}{x} = 1$$

Therefore, $\tan 45° = 1$.

TRIGONOMETRIC RATIOS FOR 45°

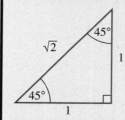

$$\sin 45° = \frac{1}{\sqrt{2}} = \frac{\sqrt{2}}{2}$$

$$\cos 45° = \frac{1}{\sqrt{2}} = \frac{\sqrt{2}}{2}$$

$$\tan 45° = \frac{1}{1} = 1$$

Notice that in Example 1, the value of x has no influence on the trigonometric ratios; x always "cancels" in both the numerator and denominator. In working with a 45° angle, an "easy" triangle to remember is the triangle in which $x = 1$. The results are summarized on the previous page.

The following example examines the special 30°-60°-90° triangle.

EXAMPLE **2**

FINDING TRIGONOMETRIC RATIOS OF TWO SPECIAL ANGLES: 30° AND 60°

a. Find the sine, cosine, and tangent of 30°.
b. Find the sine, cosine, and tangent of 60°.

SOLUTION

a. Recall the general 30°-60°-90° triangle shown in Figure 8.79 and reproduced in Figure 8.82. The length of the side opposite 30° is x, the length of the side adjacent 30° is $\sqrt{3}x$, and the length of the hypotenuse is $2x$. Applying the trigonometric ratios to the angle of measure 30°, we have the following:

FIGURE 8.82
30°-60°-90° triangle.

$$\sin 30° = \frac{\text{opp}}{\text{hyp}} = \frac{x}{2x} = \frac{1}{2}$$

Therefore, $\sin 30° = \dfrac{1}{2}$.

$$\cos 30° = \frac{\text{adj}}{\text{hyp}} = \frac{\sqrt{3}x}{2x} = \frac{\sqrt{3}}{2}$$

Therefore, $\cos 30° = \dfrac{\sqrt{3}}{2}$.

$$\tan 30° = \frac{\text{opp}}{\text{adj}} = \frac{x}{\sqrt{3}x} = \frac{1}{\sqrt{3}} \qquad \text{To rationalize the denominator, multiply by } \frac{\sqrt{3}}{\sqrt{3}}.$$

Therefore, $\tan 30° = \dfrac{\sqrt{3}}{3}$.

b. Applying the trigonometric ratios to the angle of measure 60°, we have the following:

$$\sin 60° = \frac{\text{opp}}{\text{hyp}} = \frac{\sqrt{3}x}{2x} = \frac{\sqrt{3}}{2}$$

Therefore, $\sin 60° = \dfrac{\sqrt{3}}{2}$.

$$\cos 60° = \frac{\text{adj}}{\text{hyp}} = \frac{x}{2x} = \frac{1}{2}$$

Therefore, $\cos 60° = \dfrac{1}{2}$.

$$\tan 60° = \frac{\text{opp}}{\text{adj}} = \frac{\sqrt{3}x}{x} = \frac{\sqrt{3}}{1}$$

Therefore, $\tan 60° = \sqrt{3}$.

Notice that in Example 2, the value of x has no influence on the trigonometric ratios; x always "cancels" in both the numerator and denominator. In working with a 30° or 60° angle, an "easy" triangle to remember is the triangle in which $x = 1$. The results are summarized on the following page.

TRIGONOMETRIC RATIOS FOR 30° AND 60°

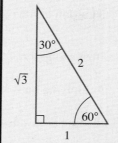

$$\sin 30° = \frac{1}{2}$$

$$\cos 30° = \frac{\sqrt{3}}{2}$$

$$\tan 30° = \frac{1}{\sqrt{3}} = \frac{\sqrt{3}}{3}$$

$$\sin 60° = \frac{\sqrt{3}}{2}$$

$$\cos 60° = \frac{1}{2}$$

$$\tan 60° = \sqrt{3}$$

Before we investigate acute angles other than the special ones (30°, 45°, and 60°), we first consider an application of trigonometric ratios.

EXAMPLE 3

USING TRIGONOMETRIC RATIOS TO CALCULATE HEIGHT AND LENGTH
A cable runs from the top of a warehouse to a point on the ground 54.0 feet from the base of the building. If the cable makes an angle of 30° with the ground (as shown in Figure 8.83), calculate

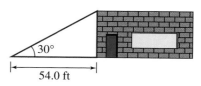

FIGURE 8.83 A cable attached to a warehouse.

a. the height of the warehouse
b. the length of the cable

SOLUTION

Let h = height of the warehouse and c = length of the cable. Assuming that the building makes a right angle with the ground, we have the right triangle shown in Figure 8.84.

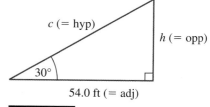

FIGURE 8.84 A right triangle.

a. The unknown height of the building is the side opposite the given angle of 30°; it is labeled *opp*. Similarly, the known side of 54.0 feet is adjacent to the angle; it is labeled *adj*. By comparing *opp* to *adj*, we use the tangent ratio to find the height of the building:

$$\tan \theta = \frac{\text{opp}}{\text{adj}} \qquad \textbf{definition of the tangent ratio}$$

$$\tan 30° = \frac{h}{54.0} \qquad \textbf{substituting } \boldsymbol{\theta = 30°, \text{opp} = h, \text{adj} = 54.0}$$

$$h = 54.0 \tan 30° \qquad \textbf{multiplying both sides by 54.0}$$

However, we know that $\tan 30° = \dfrac{\sqrt{3}}{3}$. Substituting, we obtain

$$h = 54.0\left(\frac{\sqrt{3}}{3}\right) = 18\sqrt{3} = 31.1769\ldots$$

The height of the warehouse is approximately 31.2 feet.

b. The unknown length of the cable is the hypotenuse; it is labeled *hyp*. By comparing *adj* (which is known) to *hyp*, we use the cosine ratio to find the length of the cable:

$$\cos \theta = \frac{\text{adj}}{\text{hyp}} \qquad \text{definition of the cosine ratio}$$

$$\cos 30° = \frac{54.0}{c} \qquad \text{substituting } \theta = 30°, \text{ adj} = 54.0, \text{ hyp} = c$$

$$c(\cos 30°) = 54.0 \qquad \text{multiplying both sides by } c$$

$$c = \frac{54.0}{\cos 30°} \qquad \text{dividing both sides by cos 30°}$$

However, we know that $\cos 30° = \frac{\sqrt{3}}{2}$. Substituting, we obtain

$$c = \frac{54.0}{\dfrac{\sqrt{3}}{2}}$$

$$c = \left(\frac{54.0}{1}\right)\left(\frac{2}{\sqrt{3}}\right) \qquad \text{dividing by a fraction: invert and multiply}$$

$$c = \frac{108}{\sqrt{3}} \cdot \frac{\sqrt{3}}{\sqrt{3}} \qquad \text{rationalizing the denominator}$$

$$c = \frac{108\sqrt{3}}{3} = 36\sqrt{3} = 62.3538\ldots$$

The length of the cable is approximately 62.4 feet.

Using a Calculator

So far, we have found trigonometric ratios of only three angles: 30°, 45°, and 60°. Historically, these were the "easiest," owing to their association with special triangles. How do we find the trigonometric ratios for *any* acute angle? In the past, mathematicians developed extensive tables that gave the desired values; today, we use calculators.

When using a calculator, the proper unit of angle measure must be selected. In this textbook, we measure angles in terms of degrees; the calculator must be in the "degree mode." If your calculator has a DRG button, press it as many times as necessary so that the display reads "D" or "DEG." (*DRG* signifies *Degrees-Radians-Gradients*.) Some scientific calculators have a reference chart printed near the display window; press MODE and the appropriate symbol to select "degrees." If you have a graphing calculator, press the MODE button and use the arrow buttons to select "degrees."

EXAMPLE **4**

USING A CALCULATOR TO FIND TRIGONOMETRIC RATIOS OF ARBITRARY ANGLES   Use a calculator to find the following values.

a. sin 17°   **b.** cos 25.3°   **c.** tan 83.45°

SOLUTION

a. Locate the [SIN] button. With most scientific calculators, you enter the measure of the angle first, then press [SIN]; with most graphing calculators, you press [SIN] first, type the measure of the angle, and then press [ENTER]. In any case, we find that sin 17° = 0.292371704723. . . .

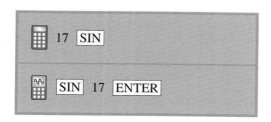

b. Using the [COS] button, we find that cos 25.3° = 0.9040825497. . . .

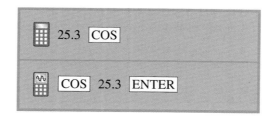

c. Using the [TAN] button, we find that tan 83.45° = 8.70930765678. . . .

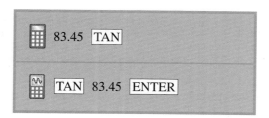

Finding an Acute Angle

If you know the measure of an acute angle, you can use a calculator to find the angle's trigonometric ratios. A calculator can also be used to work a problem in reverse; if you know a trigonometric ratio, a calculator can find the measure of the desired acute angle.

Specifically, let $a = opp$, $b = adj$, and $c = hyp$ be the lengths of the sides of a right triangle, where θ equals the measure of the acute angle opposite side a, as shown in Figure 8.85.

To find θ, note that $\sin \theta = opp/hyp = a/c$; alternatively, we can say that θ is *the angle whose sine is a/c*. This description of θ can be symbolized as $\theta = \sin^{-1}(a/c)$; that is, θ is the "inverse sine" of the ratio (a/c). Similar statements can be made concerning the cosine and tangent. These results are summarized on the next page.

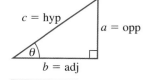

FIGURE 8.85

A right triangle.

INVERSE TRIGONOMETRIC RATIOS

If a, b, and c are the lengths of the sides of a right triangle and θ is the acute angle shown in the figure to the right, then

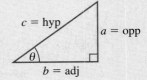

$$\theta = \sin^{-1}\left(\frac{a}{c}\right) \quad \text{means} \quad \theta \text{ is } \textit{the angle whose sine is } \frac{a}{c}; \quad \text{that is, } \sin\theta = \frac{a}{c}.$$

$$\theta = \cos^{-1}\left(\frac{b}{c}\right) \quad \text{means} \quad \theta \text{ is } \textit{the angle whose cosine is } \frac{b}{c}; \quad \text{that is, } \cos\theta = \frac{b}{c}.$$

$$\theta = \tan^{-1}\left(\frac{a}{b}\right) \quad \text{means} \quad \theta \text{ is } \textit{the angle whose tangent is } \frac{a}{b}, \quad \text{that is, } \tan\theta = \frac{a}{b}.$$

EXAMPLE 5

USING INVERSE TRIGONOMETRIC RATIOS AND A CALCULATOR TO FIND THE MEASURE OF AN ANGLE Use a calculator to find the measure of angle θ in each of the following right triangles:

a. 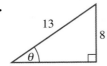 **b.** 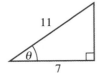 **c.**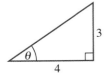

SOLUTION

a. Applying the sine ratio, we have $\sin\theta = \frac{8}{13}$ or, equivalently, $\theta = \sin^{-1}\left(\frac{8}{13}\right)$. On a calculator, pressing [INV] [SIN] or [2nd] [SIN] or [shift] [SIN] will summon the "inverse sine." Using a calculator, we obtain $\theta = 37.97987244\ldots°$, or $\theta \approx 37.98°$.

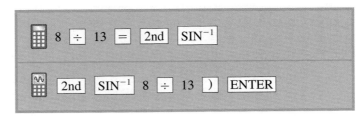

b. Applying the cosine ratio, we have $\cos\theta = \frac{7}{11}$ or, equivalently, $\theta = \cos^{-1}\left(\frac{7}{11}\right)$. Using a calculator, we obtain $\theta = 50.47880364\ldots°$, or $\theta \approx 50.48°$.

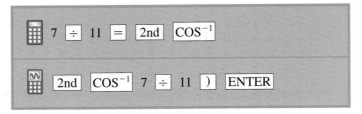

c. Applying the tangent ratio, we have $\tan\theta = \frac{3}{4}$ or, equivalently, $\theta = \tan^{-1}\left(\frac{3}{4}\right)$. Using a calculator, we obtain $\theta = 36.86989765\ldots°$, or $\theta \approx 36.87°$.

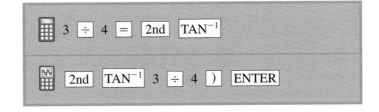

Angles of Elevation and Depression

If an object is above your eye level, you must raise your eyes to focus on the object. The angle measured from a horizontal line to the object is called the **angle of elevation.** If an object is below your eye level, you must lower your eyes to focus on the object. The angle measured from a horizontal line to the object is called the **angle of depression.** (See Figure 8.86.)

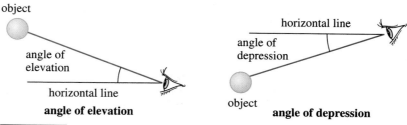

FIGURE 8.86

EXAMPLE 6

USING ANGLES OF ELEVATION AND TRIGONOMETRIC RATIOS TO CALCULATE HEIGHT   A communications antenna is on top of a building. You are 100.0 feet from the building. Using a transit (a surveyor's instrument), you measure the angle of elevation of the bottom of the antenna as 20.6° and the angle of elevation of the top of the antenna as 27.3°. How tall is the antenna?

SOLUTION

Let h = height of the antenna, and x = height of the building. The given information is depicted by the two (superimposed) right triangles, shown in Figure 8.87.

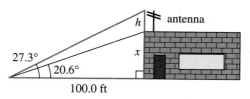

FIGURE 8.87   Finding the height of an antenna.

Referring to the large right triangle, the side opposite the 27.3° angle is $(x + h)$, and the side adjacent is 100.0 feet. Using the tangent ratio (*opp/adj*), we have

$$\tan 27.3° = \frac{x + h}{100.0}$$

Solving for h, we obtain

$$h = 100.0 \, (\tan 27.3°) - x \tag{1}$$

Using the tangent ratio for the small right triangle, we have

$$\tan 20.6° = \frac{x}{100.0}$$

Hence,

$$x = 100.0 \, (\tan 20.6°) \tag{2}$$

Substituting equation (2) into equation (1), we obtain

$$h = 100.0 \, (\tan 27.3°) - 100.0 \, (\tan 20.6°)$$

or

$$h = 100.0 \, (\tan 27.3° - \tan 20.6°) \quad \textbf{factoring 100.0 from each term}$$

Using a calculator, we have

$$h = 14.0263198\ldots$$

The antenna is approximately 14.0 feet tall.

Astronomical Measurement

In the feature Topic X: Astronomical Measurement: *Geometry in the Real World* presented in Section 8.1, a distance of 1 parsec was (arbitrarily) defined to equal 3.26 light-years (ly). What is the basis of this definition? The answer lies in right triangle trigonometry and in the measurement of very small angles. For most purposes, measuring an angle to the nearest degree is sufficient. However, if a finer measurement is required, 1 degree can be subdivided into smaller units called **minutes** (or arcminutes); by definition, 1 degree = 60 minutes. If still finer measurements are required, 1 minute can be subdivided into smaller units called **seconds** (or arcseconds); by definition, 1 minute = 60 seconds. Consequently, 1 degree = 3,600 seconds, or 1 second = 1/3600th of a degree.

Now, the unit of parsec can be defined precisely; 1 parsec (pc) is the length of the longer leg of a right triangle in which the shorter leg is exactly 1 AU (astronomical unit, or the mean distance between the earth and the sun) and the angle between the sun and the earth, as seen from an object in space (typically a star) is one arc-second, as shown in Figure 8.88.

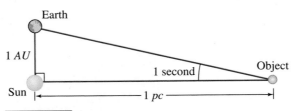

FIGURE 8.88 Definition of 1 parsec.

Given that 1 AU = 93 million miles and 1 second = 1/3600th of a degree, we can now calculate the length of 1 parsec by using the tangent ratio.

$$\tan(1 \text{ second}) = \frac{1 \text{ AU}}{1 \text{ pc}} \qquad \tan = \frac{opp}{adj}$$

$$1 \text{ pc} = \frac{1 \text{ AU}}{\tan(1 \text{ second})} \qquad \textbf{multiplying by 1 pc and dividing by tan (1 second)}$$

$$= \frac{93,000,000 \text{ mi}}{\tan(1/3600)°} \qquad \textbf{substituting}$$

$$= \frac{93,000,000 \text{ mi}}{\tan(1/3600)°} \times \frac{1 \text{ ly}}{5,880,000,000,000 \text{ mi}} \qquad \textbf{1 ly = 5.88 trillion mi}$$

$$= \frac{93}{\tan(1/3600)°} \times \frac{1 \text{ ly}}{5,880,000} \qquad \textbf{cancelling 1,000,000 mi}$$

$$= 3.262351527\ldots \text{ ly} \qquad \textbf{using a calculator}$$

Therefore, 1 parsec = 3.26 light-years, as was stated in Section 8.1.

In Exercises 1–8, use trigonometric ratios to find the unknown sides and angles in each right triangle. Do not use a calculator.

1.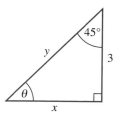

2.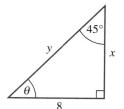

3.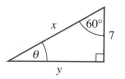

4.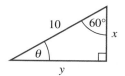

5.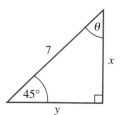

6.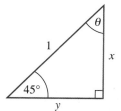

7.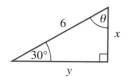

8.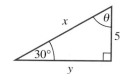

In Exercises 9–26, use the given information to draw a right triangle labeled like the one shown in Figure 8.89. Use trigonometric ratios and a calculator to find the unknown sides and angles. Round off sides and angles to the same number of decimal places as the given sides and angles; in Exercises 21–26, round off angles to one decimal place.

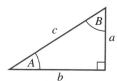

FIGURE 8.89 Standard triangle.

9. $a = 12.0$ and $A = 37°$

10. $a = 23.0$ and $A = 74°$

11. $b = 5.6$ and $A = 54.3°$

12. $b = 19.5$ and $A = 20.1°$

13. $c = 0.92$ and $B = 49.9°$

14. $c = 0.086$ and $B = 18.7°$

15. $a = 1,546$ and $B = 9.15°$

16. $a = 2,666$ and $B = 81.07°$

17. $c = 54.40$ and $A = 53.125°$

18. $c = 81.60$ and $A = 32.375°$

19. $b = 1.002$ and $B = 5.00°$

20. $b = 1.321$ and $B = 3.00°$

21. $a = 15.0$ and $c = 23.0$

22. $a = 6.0$ and $c = 17.0$

23. $a = 6.0$ and $b = 7.0$

24. $a = 52.1$ and $b = 29.6$

25. $b = 0.123$ and $c = 0.456$

26. $b = 0.024$ and $c = 0.067$

27. A cable runs from the top of a building to a point on the ground 66.5 feet from the base of the building. If the cable makes an angle of 43.9° with the ground (as shown in Figure 8.90), find

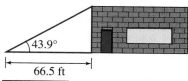

FIGURE 8.90 Cable and building.

a. the height of the building
b. the length of the cable
Round off answers to the nearest tenth of a foot.

28. A support cable runs from the top of a telephone pole to a point on the ground 42.7 feet from its base. If the cable makes an angle of 29.6° with the ground (as shown in Figure 8.91), find

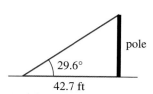

FIGURE 8.91 Cable and pole.

a. the height of the pole
b. the length of the cable
Round off answers to the nearest tenth of a foot.

29. A 20-foot-high diving tower is built on the edge of a large swimming pool. From the tower, you measure the angle of depression of the far edge of the pool. If the angle is 7.6°, how wide (to the nearest foot) is the pool?

30. You are at the top of a lighthouse and measure the angle of depression of an approaching ship. If the lighthouse is 90 feet tall and the angle of depression of the ship is 4.12°, how far from the lighthouse is the ship? Round your answer to the nearest multiple of ten.

31. The "pitch" of a roof refers to the vertical rise measured against a standard horizontal distance of 12 inches. If the pitch of a roof is 4 in 12 (the roof rises 4 inches for every 12 horizontal inches), find the acute angle the roof makes with a horizontal line. Round your answer to one decimal place.

32. If the pitch of a roof is 3 in 12, find the acute angle the roof makes with a horizontal line. Round your answer to one decimal place. See Exercise 31.

33. A bell tower is known to be 48.5 feet tall. You measure the angle of elevation of the top of the tower. How far from the tower (to the nearest tenth of a foot) are you if the angle is
  a. 15.4°?   **b.** 61.2°?

34. You are standing on your seat in the top row of an outdoor sports stadium, 200 feet above ground level. You see your car in the parking lot below and measure its angle of depression. How far (to the nearest foot) from the stadium is your car if the angle is
  a. 73.5°?   **b.** 22.9°?

35. A billboard is on top of a building. You are 125.0 feet from the building. You measure the angle of elevation of the bottom of the billboard as 33.4° and the angle of elevation of the top of the billboard as 41.0°. How tall is the billboard? Round your answer to one decimal place.

36. The distance between two buildings is 180 feet. You are standing on the roof of the taller building. You measure the angle of depression of the top and bottom of the shorter building. If the angles are 36.7° and 71.1°, respectively, find
  a. the height of the taller building
  b. the height of the shorter building
  Round your answers to the nearest foot.

37. You are hiking along a river and see a tall tree on the opposite bank. You measure the angle of elevation of the top of the tree and find it to be 61.0°. You then walk 50 feet directly away from the tree and measure the angle of elevation. If the second measurement is 49.5° (see Figure 8.92), how tall is the tree? Round your answer to the nearest foot.

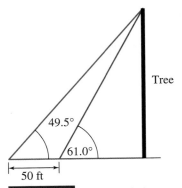

FIGURE 8.92 Angles of elevation.

38. While sightseeing in Washington, D.C., you visit the Washington Monument. From an unknown distance, you measure the angle of elevation of the top of the monument. You then move 100 feet backward (directly away from the monument) and measure the angle of elevation of the top of the monument. If the angles are 61.6° and 54.2°, respectively, how tall (to the nearest foot) is the Washington Monument?

39. While sightseeing in St. Louis, you visit the Gateway Arch. Standing near the base of the structure, you measure the angle of elevation of the top of the arch. You then move 120 feet backward and measure the angle of elevation of the top of the arch. If the angles are 73.72° and 64.24°, respectively, how tall (to the nearest foot) is the Gateway Arch?

©Scenics of America/PhotoLink/PhotoDisc/Getty Images

40. While sightseeing in Seattle, you visit the Space Needle. From a point 175 feet from the structure, you measure the angle of elevation of the top of the structure. If the angle is 73.87°, how tall (to the nearest foot) is the Space Needle?

a. opposite sides of the balloon. See Figure 8.93.

b. the same side of the balloon. See Figure 8.94.

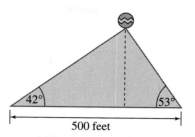

FIGURE 8.93 Observers on opposite sides.

41. The Willis Tower (formerly The Sears Tower), in Chicago, is the tallest building in the United States. From a point 900 feet from the building, you measure the angle of elevation of the top of the building. If the angle is 58.24°, how tall (to the nearest foot) is the Willis Tower?

42. A tree is growing on a hillside. From a point 100 feet downhill from the base of the tree, the angle of elevation to the base of the tree is 28.3° with an additional 12.5° to the top of the tree. How tall is the tree?

43. A tree is growing on a hillside. From a point 100 feet uphill from the base of the tree, the angle of depression to the top of the tree is 28.3° with an additional 12.5° to the base of the tree. How tall is the tree?

44. Two observers are 500 feet apart. Each measures the angle of elevation of a hot air balloon that lies in a vertical plane passing through their locations; the angles are 53° and 42°. Find the height of the balloon if the observers are on

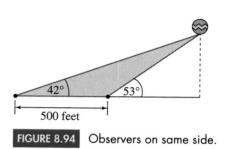

FIGURE 8.94 Observers on same side.

45. Calculate the distance (in parsecs) from the sun to an object in space if the angle between the sun and the earth (as seen from the object) is 2 seconds and the distance between the earth and the sun is 1 AU. (See Figure 8.88.)

46. Calculate the distance (in parsecs) from the sun to an object in space if the angle between the sun and the earth (as seen from the object) is 1/5 second and the distance between the earth and the sun is 1 AU. (See Figure 8.88.)

47. Suppose the distance from the sun to an object in space is 0.9 parsec and the distance between the earth and the sun is 1 AU (as in Figure 8.88).

a. How far from the sun is the object in terms of AU?

b. Calculate the angle between the sun and the earth, as seen from the object. Express your answer in terms of seconds, rounded off to the nearest hundredth of a second.

48. Suppose the distance from the sun to an object in space is 0.8 parsec and the distance between the earth and the sun is 1 AU (as in Figure 8.88).

a. How far from the sun is the object in terms of AU?

b. Calculate the angle between the sun and the earth, as seen from the object. Express your answer in terms of seconds, rounded off to the nearest hundredth of a second.

THE NEXT LEVEL

If a person wants to pursue an advanced degree (something beyond a bachelor's or four-year degree), chances are the person must take a standardized exam to gain admission to a school or to be admitted into a specific program. These exams are intended to measure verbal, quantitative, and analytical skills that have developed throughout a person's life. Many classes and study guides are available to help people prepare for the exams. Exercises 49–52 are typical of those found in the study guides.

49. Find the perimeter of $\triangle PQR$, given that $PS = 6$ and $PQ = 10$ as shown Figure 8.95.
  a. 32   **b.** 40
  c. $40 + 8\sqrt{3}$ **d.** $32 + 8\sqrt{2}$
  e. $32 + 8\sqrt{3}$ **f.** None of these is correct.

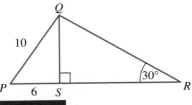

FIGURE 8.95
Diagram for Exercises 49 and 50.

50. Find the area of $\triangle PQR$ as shown Figure 8.95.
  a. 48   **b.** 96
  c. 80   **d.** $24 + 32\sqrt{2}$
  e. $24 + 32\sqrt{3}$ **f.** None of these is correct.

51. In Figure 8.96, what is the value of x?
  a. 12   **b.** 24
  c. 36   **d.** 48
  e. 60   **f.** None of these is correct.

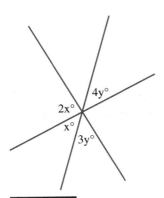

FIGURE 8.96 Diagram for Exercise 51.

52. In Figure 8.97, what is the value of $\frac{y + x}{y - x}$?
  a. $\frac{3}{5}$ **b.** $\frac{5}{3}$
  c. 2   **d.** 4
  e. 10   **f.** None of these is correct.

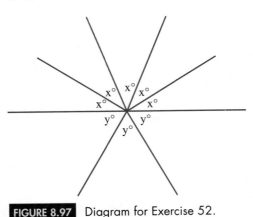

FIGURE 8.97 Diagram for Exercise 52.

• PROJECT

53. The purpose of this project is to calculate the height of a tall building by measuring an angle of elevation.
  a. First, we must construct a device to measure an angle of elevation. You will need a protractor, a plastic drinking straw, a bead (or button), and a piece of thread. Attach the straw to the straight edge of the protractor so that the straw lines up with the 0° marks on the protractor. Next, tie the thread to the bead and attach the thread to the center mark of the protractor as shown in Figure 8.98.

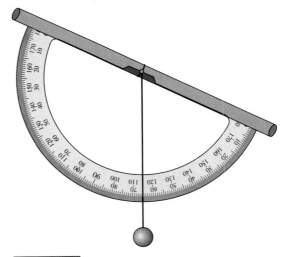

FIGURE 8.98 Home-made device.

Note that when the straw is horizontal, the thread is positioned over the 90° mark on the scale. To measure an angle of elevation, look through the straw and focus on the top of the object being measured. Allow the thread to hang vertically, and have an assistant read the larger of the two angles indicated by the thread. (One scale will give an angle greater than 90°, the other less than 90°.) Subtract 90° from the larger angle, and you will obtain the angle of elevation. The angle of elevation indicated by the device shown in Figure 8.98 is $115° - 90° = 25°$.

b. Choose a tall building. Pick an accessible point away from the building, and measure the angle of elevation of the top of the building. Now use a tape measure to determine the distance from your location to the base of the building. Set up the appropriate trigonometric ratio, and calculate the height of the building.

c. Pick a second location, and repeat part (b).

d. Your answers to parts (b) and (c) should be roughly the same. Are they? If not, give possible reasons for the discrepancy.

8.6 Linear Perspective

OBJECTIVES

- Understand how and why linear perspective makes works of art more realistic by giving them a sense of depth

- Learn to draw simple objects in one-point perspective and in two-point perspective

- Be able to draw an Albertian grid

- Learn to determine whether a work of art uses one-point perspective, two-point perspective, or no perspective

- In a work of art, be able to find:

  - the vanishing point(s)

  - the horizon

  - uses of foreshortening

  - uses of pavement

  - uses of an Albertian grid

Linear perspective is a technique artists use to impart a natural quality to drawings and paintings. It gives the viewer both a sense of depth and a correct impression of the painted objects' relative sizes and positions.

The painting by Giotto in Figure 8.99 on the next page does not use perspective. It lacks depth—everything seems to be two-dimensional, and the mountains seem too close to the people. Also, the mountains inappropriately appear to be about the same height as the people.

The painting by Perugino in Figure 8.100 uses perspective. It gives a strong sense of depth. In fact, the scene seems to be much deeper than it is wide. Also, the buildings appear to be appropriately sized relative to the people.

FIGURE 8.99   Giotto, *Joachim Among the Shepherds,* no perspective.

Cameraphoto Arte, Venice / Art Resource, NY

FIGURE 8.100   Perugino, *Christ Handing the Keys to St. Peter,* with perspective.

© The Art Gallery Collection / Alamy

The Theory of Linear Perspective

FIGURE 8.101   Railroad tracks.

© Image copyright Igumnova Irina, 2009. Used under license from Shutterstock.com

The theory of linear perspective is based on the fact that an object looks smaller when it's farther away than it does when it's close. In Figure 8.101, the width of the railroad tie that's farther away from the viewer is much smaller *in the photo* than the width of the closest tie, even though *in actuality,* the widths are the same. This makes the ties appear to go farther and farther away from the viewer, even though the photograph has no actual depth.

These same features occur in Perugino's painting. The people closer to the viewer are much bigger than the people farther away, just as the railroad ties closer to the viewer are much wider than the ties that are farther away. The tiles' borders form diagonal lines that meet at the horizon, as do the railroad track in the photograph. These two features give the painting a strong sense of depth and are characteristic features of linear perspective.

The point where the lines meet is called the **vanishing point.** Usually, it occurs at an important place in the painting, because the viewers' eyes are naturally drawn to that point. In the painting, the vanishing point is at the central building's door, as shown in Figure 8.102. This makes that building stand out more than the other two buildings.

FIGURE 8.102   The tiles' borders and the horizon meet at the vanishing point.

© The Art Gallery Collection / Alamy

Foreshortening is another feature of linear perspective. The artist used foreshortening in depicting the tiles. They are meant to be perceived as squares with equal sides. In the painting, however, the sides parallel to the viewer's line of sight are much shorter than the sides parallel to the bottom of the painting, as shown in Figure 8.103. This also gives the painting a sense of depth.

FIGURE 8.103   Foreshortening.

In the railroad track photograph (Figure 8.101), the viewer's line of sight is straight down the middle, between the two tracks. The tracks are *in actuality* parallel to the line of sight. *In the photograph,* though, they are diagonal lines that meet at the vanishing point. The railroad ties have a different orientation: They are perpendicular to the line of sight. They are parallel to each other in actuality as well as in the photograph.

The same two things occur in the painting. The tiles' borders that are parallel to the viewer's line of sight are depicted by diagonal lines that meet at the vanishing point. The tiles' borders that are perpendicular to the line of sight are depicted by parallel lines.

FEATURES OF LINEAR PERSPECTIVE

- Lines that that are parallel to the viewer's line of sight are depicted by diagonal lines that meet at a vanishing point.
- Vanishing points are on the horizon line.
- Lines that are perpendicular to the line of sight are depicted by parallel lines.
- Similar objects are drawn smaller if they are meant to be farther away from the viewer and bigger if they are meant to be closer.
- Equal lengths are drawn smaller if they are parallel to the viewer's line of sight and bigger if they are parallel to the bottom of the painting.

Linear Perspective and Mathematics

You might be wondering what perspective has to do with mathematics. If so, you question this because perspective uses little or no algebra, and you have studied mostly algebra. The theory of perspective employs geometry and optics.

The invention of linear perspective is generally attributed to Filippo Brunelleschi, a Renaissance engineer and architect. He designed and engineered the dome for the Duomo, the principal cathedral in Florence, Italy. (See Figure 8.104 on the next page.)

Brunelleschi was knowledgeable about geometry, owing to his background in engineering and architecture. He used that knowledge to make his paintings look more natural. In doing so, he invented linear perspective.

FIGURE 8.104 The dome of the Duomo.

FIGURE 8.105 The Baptistery.

The first painting that accurately used linear perspective was a Brunelleschi painting of the Baptistery (see Figure 8.105), a building adjacent to the Duomo. To demonstrate the effectiveness of his invention, Brunelleschi placed his painting so that it faced the Baptistery. He asked a viewer to stand behind the painting and look through a small hole in it. He put a mirror in front of it so that the viewer would see the painting reflected in the mirror. After removing the mirror, the viewer would see the real Baptistery through the hole. It was barely possible to tell which was the painting and which was the real thing. Brunellesco's demonstration was so successful that most Florentine artists started using linear perspective.

 ## Albertian Grids

If Brunelleschi wrote about the mathematics of perspective, the writings have been lost. His friend Leon Battista Alberti wrote two books on linear perspective and the mathematics behind it. Alberti also wrote about how to correctly draw a tile floor.

Tile floors, or **pavements,** are common in Renaissance art because of their ability to impart depth. Brunelleschi knew that to draw pavement borders parallel to the line of sight, you make them equally spaced in the foreground, and you make them meet at a vanishing point. He also knew that pavement borders perpendicular to the line of sight should be parallel to each other and progressively closer together. He did not know how to space these borders. Alberti found out how to do this.

Figure 8.106 contains two pavements, only one of which has horizontal lines that are correctly spaced. Can you tell which is correct?

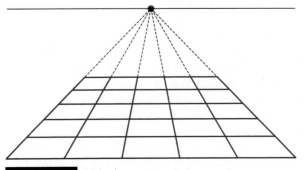

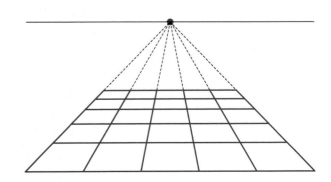

FIGURE 8.106 Which pavement is correct?

You may be able to see that the pavement on the right looks slightly more realistic, because its horizontal lines are located correctly. It's easy to tell why, if you insert a diagonal line in each pavement. In the pavement on the right in Figure 8.107, the diagonal line perfectly hits each of the corners. In the pavement on the left, the diagonal line misses them. A pavement that is correctly drawn in this manner is called an **Albertian grid.**

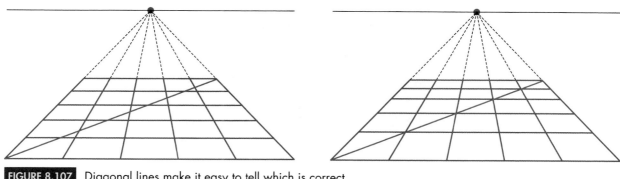

FIGURE 8.107 Diagonal lines make it easy to tell which is correct.

If a painting includes a fairly straight item, such as a stick, placed diagonally on a pavement, a painting with correctly spaced horizontal lines looks more realistic, and one with incorrectly spaced lines looks less realistic.

Perspective and Leonardo da Vinci

Brunelleschi invented perspective, and Alberti developed it further, but Leonardo da Vinci perfected it. Leonardo said that "perspective is the rein and rudder of painting." The study in Figure 8.108 shows many lines of perspective, a clear vanishing point, and an Albertian grid. Leonardo went so far as to incise lines of perspective on a panel before painting it.

FIGURE 8.108 Leonardo's study for *the Adoration of the Magi.*

Leonardo da Vinci/The Bridgeman Art Library/Getty Images

One-Point Perspective

Perugino's painting in Figure 8.109 has buildings that face forward. That is, their faces are parallel to the surface of the painting itself. This type of painting uses **one-point perspective,** in which there is only one vanishing point.

FIGURE 8.109 Perugino's use of one-point perspective.

DRAWING A BOX IN ONE-POINT PERSPECTIVE

1. Draw a horizon line with a vanishing point.
2. Draw two perspective lines coming out from the vanishing point.
3. Draw two vertical lines connecting the two perspective lines.

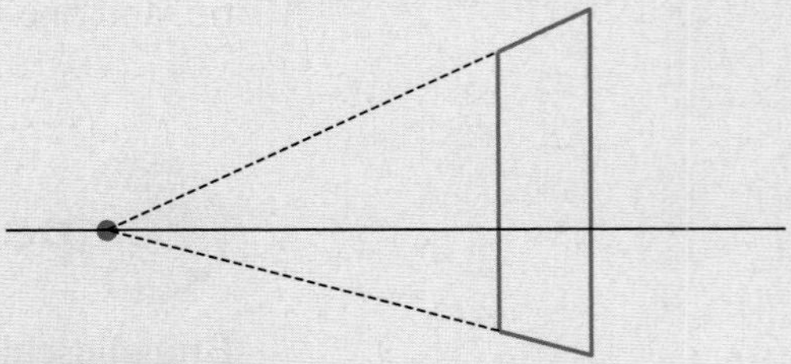

FIGURE 8.110 Steps 1, 2 and 3.

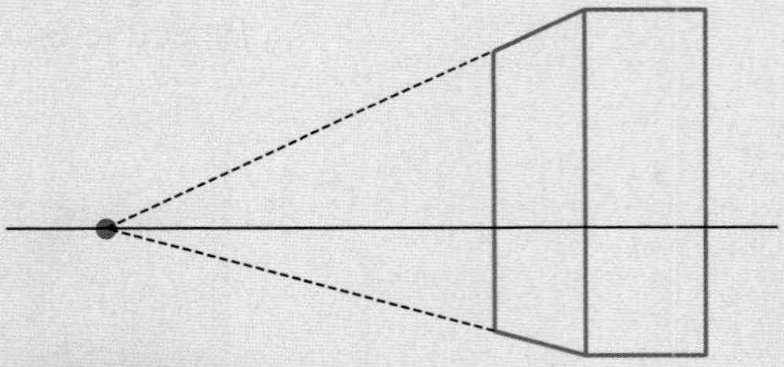

4. Draw a horizontal line that starts at the top of the vertical line farthest from the vantage point and goes away from the vanishing point.
5. Draw another horizontal line like the one in step 4 that starts at the bottom rather than the top.
6. Draw a vertical line that connects the lines from steps 4 and 5.
7. Erase unneeded parts of lines.

FIGURE 8.111 Steps 4, 5, and 6.

Two-Point Perspective

Two-point perspective is used if the building is rotated so that its face is not parallel to the painting's surface. This allows the viewer to look directly at the corner of a building. With two point perspective, there are two vanishing points.

DRAWING A BOX IN TWO-POINT PERSPECTIVE

1. Draw a horizon line with *two* vanishing points.
2. Draw two upper perspective lines, one coming out of each vanishing point.
3. Draw a vertical line that starts where the perspective lines intersect and goes down.

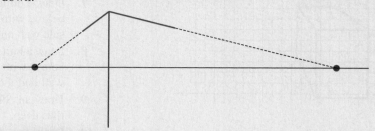

FIGURE 8.112 Steps 1, 2, and 3.

4. Draw two lower perspective lines, one coming out of each vanishing point. Have them intersect at the vertical line from step 3.

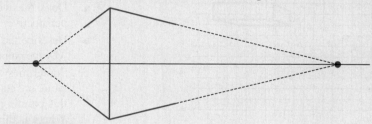

FIGURE 8.113 Step 4.

5. Draw two vertical lines, one on each side of the vertical line from step 3.
6. Erase unneeded parts of lines.

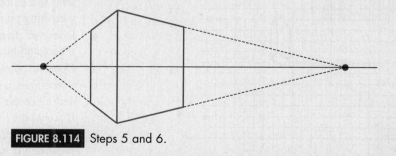

FIGURE 8.114 Steps 5 and 6.

8.6 EXERCISES

In Exercises 1–4, answer the following questions.

a. Is the box drawn in one-point perspective or two-point perspective?

b. Is the box viewed from above or below?

c. At what approximate ordered pair(s) is/are the vanishing point(s)?

d. What is the equation of the horizon line?

HINT: In Exercise 1, the box has a corner (approximately) at the ordered pair (3, 6) and an edge on the line with equation $y = 8$.

1.

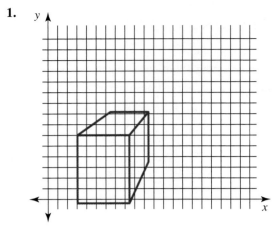

2.

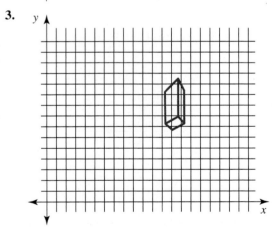

3.

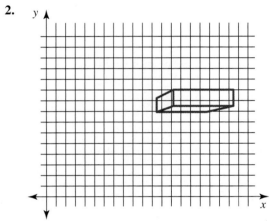

4.

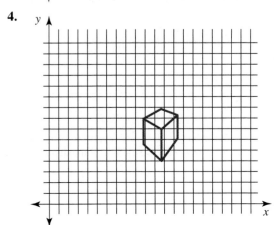

5. Draw a building in one-point perspective, viewed from above, with two rows of three windows each on the side wall.

6. Draw a building in one-point perspective, viewed from below, with three rows of two windows each on the side wall.

7. Draw a building in two-point perspective, viewed from below, with three rows of two windows each on one side wall and a door on the other side wall.

8. Draw a building in two-point perspective, viewed from below, with two rows of two windows each on one side wall and a door and a window on the other side wall.

9. Draw an Albertian grid that is five tiles wide and four tiles deep.

10. Draw an Albertian grid that is six tiles wide and three tiles deep.

In Exercises 11–20, answer the following questions. If you think it would be easier to use a larger image, you can find one by googling the painter's name and the name of the painting.

a. Does the work use one-point perspective, two-point perspective, or both? Justify your answer.

b. Describe the location of the vanishing point (s). Discuss why the painter might have chosen that/those location(s).

c. Sometimes a small feature of the painting, such as a table, has its own vanishing point. Does this happen in this painting? If so, what has its own vanishing point? Where is this vanishing point?

d. Sometimes two different main parts of a painting have separate vanishing points. Does this happen in this painting? If so, describe what has separate vanishing points. Where are the vanishing points? Also, discuss why the painter might have chosen to have multiple vanishing points.

e. Does the painting use foreshortening? If so, describe where and how.

f. Does the painting have pavement? If so, describe where.

g. Sometimes a painting uses an Albertian grid on the ceiling or the walls. Does this happen in this painting? If so, where?

h. Is the painting's horizon line an actual part of the painting? If so, where?

11.

Raphael, *School of Athens.*

12.

Masaccio, *Trinity.*

13.

Mantegna, *St. James on his Way to the Execution.*

14.

Masaccio, *Tribute Money.*

15.

Pissaro, *Road to Louveciennes,* 1872.

16.

Leonardo da Vinci, *The Annunciation.*

17.

The Flagellation of Christ, c.1463–4 (tempera on panel), Piero della, Francesca, Piero della (c.1415–92)/Galleria Nazionale delle Marche, Urbino, Italy/The Bridgeman Art Library International

Piero della Francesca, *The Flagellation of Christ.*

18.

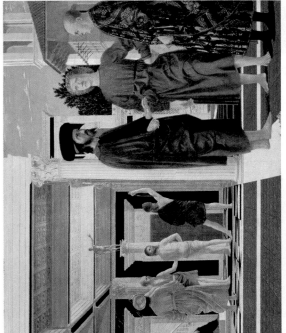

Train in the Snow or The Locomotive, 1875 (oil on canvas), Monet, Claude (1840–1926)/Musee Marmottan, Paris, France/Giraudon/The Bridgeman Art Library International

Claude Monet, *The Train in the Snow,* 1875.

19.

Digital Image © The Museum of Modern Art/Licensed by SCALA/Art

Andrew Wyeth, *Christina's World,* 1948.

20.

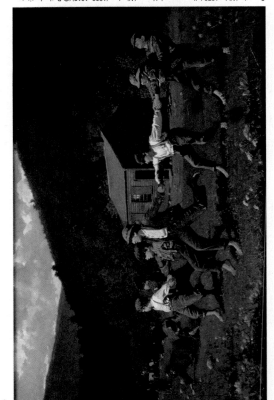

Snap the Whip, 1872 (oil on canvas), Homer, Winslow (1836–1910)/© Butler Institute of American Art, Youngstown, OH, USA/Museum Purchase 1918/The Bridgeman Art Library International

Winslow Homer, *Snap the Whip,* 1872.

21. After comparing the following perspective study with the work itself, thoroughly discuss how Eakins used perspective in *The Pair-Oared Shell.*

Thomas Eakins, *The Pair-Oared Shell.*

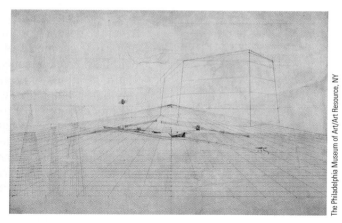

Thomas Eakins, perspective study.

In Exercises 22–25, find the work in an art book or by googling the painter's name and the name of the painting, and answer the questions from Exercises 11–20.

22. Leonardo da Vinci, *The Last Supper.*

23. John Singer Sargent, *A Street in Venice.*
24. A screen shot from Pixar Studies.
25. M.C. Escher, *Ascending and Descending.*

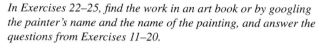

8.7 Conic Sections and Analytic Geometry

OBJECTIVES

- Find the center and radius of a circle, and sketch its graph
- Find the focus of a parabola, and sketch its graph
- Find the foci of an ellipse, and sketch its graph
- Find the foci of a hyperbola, and sketch its graph

In order to seek truth it is necessary once in the course of our life to doubt as far as possible all things.
—RENÉ DESCARTES

It is easy to take our system of mathematical notation and algebraic manipulation for granted or even to regard the study of mathematics as an inconvenience and to doubt its relevance to our individual lives and goals. René Descartes, one of the great philosophers and mathematicians of the seventeenth century, felt the same way. In doubting the "system" of mathematics of his predecessors, Descartes contributed to the creation of one of the major foundations of modern mathematics: analytic geometry. Simply put, **analytic geometry** is the marriage of algebra and geometry.

Most mathematics from ancient times through the Middle Ages consisted of a verbal description of a geometric method for solving a single problem. (The empirical geometry of the Egyptians is a case in point.) In the Europe of the Middle Ages, "advanced" mathematics was available only to people who knew Latin, as all major works were written in this "language of scholars." Rather than relying on language (which differs from culture to culture), Descartes pioneered the modern notion of using single letters to represent known and unknown quantities. He used the last letters of the alphabet—x, y, and z—to represent variables and the first letters—a, b, and c—to represent constants.

The Greeks verbally defined a parabola to be "the set of all points in a plane equidistant from a given line and a point not on the line"; modern mathematicians, by contrast, favor the equation $y = ax^2 + bx + c$. When we say that the graph of the equation $y = ax^2 + bx + c$ is a parabola, we are operating in the realm of analytic geometry. No doubt your initial exposure to this field was in your beginning or intermediate algebra course, when you were asked to graph an equation such as $y = 2x + 1$ or $y = x^2$.

Analytic geometry was not created overnight. Many individuals from diverse cultures made significant contributions. Some historians take the easy way out and say that Descartes invented it, but the story of the creation of analytic geometry is much more complicated than that. Because of its intimate association with the creation of calculus, the development of analytic geometry is discussed in Chapter 13.

Conic Sections

A **conic section** is the figure formed when a plane intersects a cone. What do the cross sections of a cone look like? The ancient Greeks studied this problem extensively. Recall that their approach was verbal, involving statements such as "a parabola is the set of all points in a plane equidistant from a given line and a point not on the line" or "a circle is the set of all points in a plane equidistant from a given point." When we study the various conic sections, we use a "double cone" like the one shown in Figure 8.115. Depending on the inclination of the plane of intersection, a circle, parabola, ellipse, or hyperbola is formed. (See Figure 8.116.)

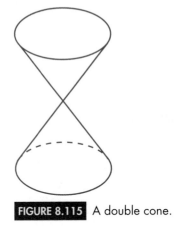

FIGURE 8.115 A double cone.

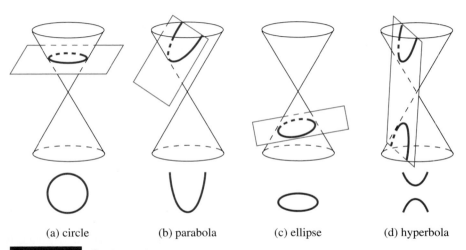

(a) circle　　　(b) parabola　　　(c) ellipse　　　(d) hyperbola

FIGURE 8.116 Conic sections.

©Bettmann/CORBIS

HISTORICAL NOTE

HYPATIA, A.D. 370–415

Hypatia, one of the first women to be recognized for her mathematical accomplishments, lived in Alexandria, one of the largest and most academically prominent cities on the Mediterranean Sea. This Hellenistic city was famed for its university and its library, said to be the largest of its day.

Hypatia's early environment was filled with intellectual challenge and stimulation. Her father, Theon, was a professor of mathematics and director of the museum and library at the University of Alexandria. He gave his daughter a classic education in the arts, literature, mathematics, science, and philosophy. In addition to her studies at the University of Alexandria, Hypatia traveled the Mediterranean. While in Athens, she attended a school conducted by the famed writer Plutarch.

On her return to Alexandria, Hypatia continued in her father's footsteps: She lectured on mathematics and philosophy at the university and directed the museum and library. Her lectures were well received by enthusiastic students and scholars alike; many considered Hypatia to be an oracle. Although none of her writings remains intact, historians attribute several mathematical treatises to Hypatia, including commentaries on the astronomical works of Diophantus and Ptolemy, on the conics of Apollonius, and on the geometry of Euclid. In addition to her insightful lectures and mathematical works, letters written by Hypatia's contemporaries credit her with the invention of devices used in the study of astronomy.

At this time, Alexandria was part of the Roman Empire and was undergoing a power struggle between Christians and pagans (worshippers of Greek and Roman gods). Bishop Cyril was using his position in the Christian church to usurp the power of the Alexandrian government, which was under the rule of the Roman prefect Orestes. Orestes was known to have attended many of Hypatia's lectures and was believed to have been Hypatia's lover. Since Hypatia was a symbol of classic Greek culture, Bishop Cyril associated her with paganism and viewed her as a threat to his quest for Christian power.

In a frenzied attempt to eradicate the pagan influences of Alexandria, a Christian mob incited by Bishop Cyril attacked Hypatia while she was riding in her chariot. She was stripped naked and dragged through the streets, then tortured and murdered.

Hypatia was the last symbol of the ancient culture of Alexandria. She preserved and carried forward the knowledge and wisdom of the Golden Age of Greek civilization. It is tragic that her intelligence and devotion led to her violent death.

The Circle

When a cross section is taken parallel to the base of the cone, as in Figure 8.116(a), a **circle** is obtained. The Greeks defined a circle as "the set of all points in a plane equidistant from a given fixed point." That is, no matter which point P on the circle you select, its distance from the fixed point (the center) is always the same, as shown in Figure 8.117.

By using modern algebra, a circle can also be described as the graph of an algebraic equation, as shown in Figure 8.118.

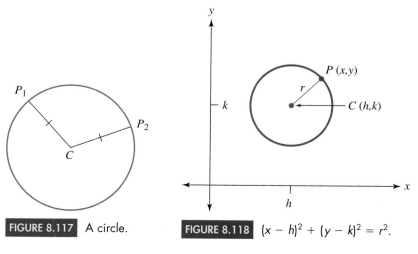

FIGURE 8.117   A circle.

FIGURE 8.118 $(x - h)^2 + (y - k)^2 = r^2$.

EQUATION OF A CIRCLE

A **circle** of radius r, centered at the point (h, k), is the set of all points satisfying the equation

$$(x - h)^2 + (y - k)^2 = r^2$$

EXAMPLE 1

COMPLETING THE SQUARE TO FIND THE CENTER AND RADIUS OF A CIRCLE Find the center and radius of the circle $x^2 + y^2 + 6x - 8y + 21 = 0$.

SOLUTION

To find the coordinates of the center (that is, to find h and k), we must put the given equation in the form $(x - h)^2 + (y - k)^2 = r^2$. The expressions $(x - h)^2$ and $(y - k)^2$ are called *perfect squares;* each is the square of a binomial.

First, group the terms containing x and those containing y and put the constant on the right side of the equation:

$$x^2 + y^2 + 6x - 8y + 21 = 0$$
$$(x^2 + 6x) + (y^2 - 8y) = -21$$

The given expression $(x^2 + 6x)$ is *not* a perfect square; we must add the appropriate term to complete the square. To find this missing term, take half the coefficient of x and square it; we must add $\left(\frac{6}{2}\right)^2 = 9$. Likewise, we must add $\left(\frac{-8}{2}\right)^2 = 16$ to complete the square of y.

$$(x^2 + 6x + 9) + (y^2 - 8y + 16) = -21 + 9 + 16$$
$$(x + 3)^2 + (y - 4)^2 = 4$$
$$[x - (-3)]^2 + (y - 4)^2 = 2^2$$

Therefore, the center is the point $C(-3, 4)$, and the radius is $r = 2$, as shown in Figure 8.119.

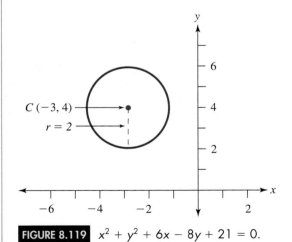

 **FIGURE 8.119** $x^2 + y^2 + 6x - 8y + 21 = 0$.

The Parabola

When a cross section passes through the base of the cone and intersects only one portion, as in Figure 8.116(b), the resulting figure is a **parabola.** The Greeks defined a parabola as "the set of all points in a plane equidistant from a given line and a given point not on the line." No matter which point P on the parabola you select, its distance from the given line (called the *directrix*) is the same as its distance from the given point (called the *focus*), as shown in Figure 8.120. Every parabola can be divided into two equal pieces. The line that cuts the parabola into symmetric halves is called the **line of symmetry.** The point at which the line of symmetry intersects the parabola is called the **vertex** of the parabola.

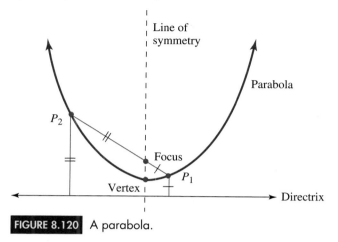

FIGURE 8.120 A parabola.

Parabolas have many applications. In particular, satellite dish antennas and the dishes used with sports microphones are designed so that their cross sections are parabolas; that is, they are parabolic reflectors. When incoming radio waves, microwaves, or sound waves hit the dish, they are reflected to the focus, where they are collected and converted to electronic pulses. The focusing ability of a parabola is also applied to the construction of mirrors used in telescopes and in solar heat collection. (See Figure 8.121.)

Our study of parabolas concentrates on locating the focus. In this respect, we will consider only a parabola whose vertex is at the origin (0, 0) of a rectangular coordinate system.

A PARABOLA AND ITS FOCUS

A **parabola,** with vertex at the point (0, 0) and **focus** at the point (0, p), is the set of all points (x, y) satisfying the equation

$$4py = x^2$$

A parabola and its focus, located on a rectangular coordinate system, are shown in Figure 8.122.

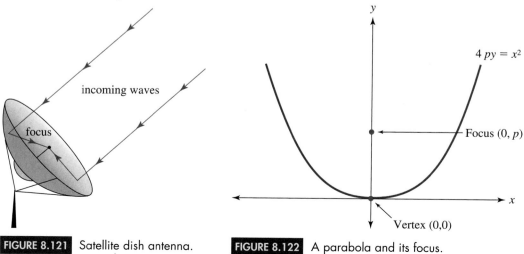

FIGURE 8.121 Satellite dish antenna.

FIGURE 8.122 A parabola and its focus.

EXAMPLE **2**

SOLUTION

FINDING THE FOCUS OF A PARABOLA A Peace Corps worker has found a way for villagers to obtain Mylar-coated parabolic dishes that will concentrate the sun's rays for the solar heating of water. If a dish is 7 feet wide and 1.5 feet deep, where should the water container be placed?

The water will be heated most rapidly if the container is placed at the focus. (The incoming solar radiation will be reflected by the parabolic dish and concentrated at the focus of the parabola.)

Draw a parabola with its vertex at the origin. As Figure 8.123 demonstrates, the given dimensions tell us that the point $Q(3.5, 1.5)$ is on the parabola.

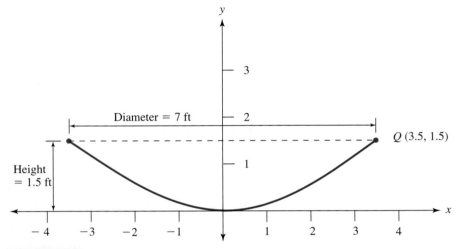

FIGURE 8.123 Solar parabolic dish.

Substituting $x = 3.5$ and $y = 1.5$ into the general equation $4py = x^2$, we can find p, the location of the focus:

$$4py = x^2$$
$$4p(1.5) = (3.5)^2$$
$$6p = 12.25$$
$$p = 2.04166666\ldots$$
$$p \approx 2$$

The focus is located at the point $(0, 2)$. Therefore, the water container should be placed 2 feet above the bottom of the parabolic dish.

Besides collecting incoming waves, parabolic reflectors also concentrate outgoing waves. The reflectors in flashlights and classic automotive headlights are parabolic. If the bulb is positioned at the focus, then the outgoing light will be more intensely concentrated in a forward beam.

The Ellipse

When a cross section of a cone is not parallel to the base and does not pass through the base, as in Figure 8.116(c), the resulting figure is an **ellipse.** An ellipse is a symmetric oval. The Greeks defined an ellipse as "the set of all points in a plane the *sum* of whose distances from two given points is constant." That is, if C_1 and C_2

represent the given points (called the **foci**), then no matter which point P on the ellipse you select, the distance from P to C_1 plus the distance from P to C_2 is always the same, as shown in Figure 8.124.

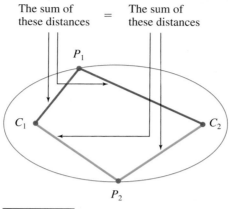

The sum of these distances = The sum of these distances

FIGURE 8.124   Ellipse.

Ellipses are commonly used in astronomy. The planets revolve around the sun, each following its own elliptical path with the sun at one of the foci. The elliptical motion of the planets was first hypothesized by Johann Kepler in 1609.

An ellipse has an interesting reflective property. Anything that is emitted from one focus and strikes the ellipse will be directed to the other focus. For example, suppose you had an elliptical billiard table with a hole located at one of the foci. A cue ball located at the other focus would always go in the hole, regardless of the direction in which it was shot. This principle is used in the design of whispering galleries. If the walls and ceiling of a room are elliptical, a person standing at one focus will be able to hear the whispering of someone standing at the other focus; the sound waves are reflected and directed from one focus to the other. The world's most famous whispering gallery is located in St. Paul's Cathedral in London, England, which was completed in 1710. Other galleries are located in the U.S. Capitol in Washington, D.C.; in the Mormon Tabernacle in Salt Lake City, Utah; and in Gloucester Cathedral (built in the eleventh century) in Gloucester, England.

Although they can be located anywhere in a plane, we will consider only ellipses that are centered at the origin of a rectangular coordinate system. (See Figure 8.125.) (If the center is not at the origin, the method of completing the square can be used to locate the center, as in Example 1.)

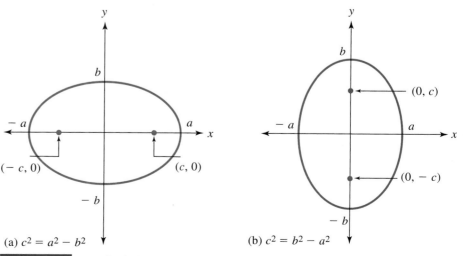

(a) $c^2 = a^2 - b^2$ (b) $c^2 = b^2 - a^2$

FIGURE 8.125   Standard ellipses.

EQUATION OF AN ELLIPSE

An **ellipse,** centered at the origin, is the set of all points (x, y) satisfying the equation

$$\frac{x^2}{a^2} + \frac{y^2}{b^2} = 1 \quad a \text{ and } b \text{ are positive constants.}$$

The **foci** are located on the longer axis of the ellipse and are found by solving the equation $c^2 = $ difference between a^2 and b^2.

EXAMPLE **3**

SOLUTION

GRAPHING AN ELLIPSE   Sketch the graph and find the foci of the ellipse $9x^2 + 25y^2 = 225$.

First, we put the equation in the general formula $x^2/a^2 + y^2/b^2 = 1$. To obtain this form, we divide each side of the equation by 225 so that the right side will equal 1:

$$9x^2 + 25y^2 = 225$$

$$\frac{9x^2}{225} + \frac{25y^2}{225} = \frac{225}{225} \quad \textbf{dividing by 225}$$

$$\frac{x^2}{25} + \frac{y^2}{9} = 1 \quad \textbf{simplifying the fractions}$$

(Note that $a^2 = 25$ and $b^2 = 9$.) We can now sketch the ellipse by finding the x- and y-intercepts.

Finding the x-Intercepts:	*Finding the y-Intercepts:*	*Finding the Foci:*
(Substitute $y = 0$ and solve for x.)	(Substitute $x = 0$ and solve for y.)	$c^2 = a^2 - b^2$
$\dfrac{x^2}{25} + \dfrac{0^2}{9} = 1$	$\dfrac{0^2}{25} + \dfrac{y^2}{9} = 1$	$= 25 - 9$
$\dfrac{x^2}{25} = 1$	$\dfrac{y^2}{9} = 1$	$= 16$
$x^2 = 25$	$y^2 = 9$	Therefore, $c = \pm4$.
$x = \pm5$	$y = \pm3$	

Plotting the intercepts and connecting them with a smooth oval, we sketch the ellipse and its foci as shown in Figure 8.126.

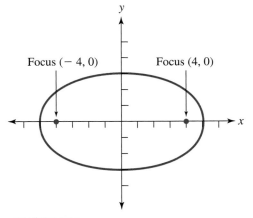

Focus $(-4, 0)$ Focus $(4, 0)$

FIGURE 8.126 $9x^2 + 25y^2 = 225$.

The Hyperbola

When a cross section intersects both portions of the double cone, as in Figure 8.116(d), the resulting figure is a **hyperbola.** A hyperbola consists of two separate, symmetric branches. The Greeks defined a hyperbola as "the set of all points in a plane the *difference* of whose distances from two given points is constant." That is, if C_1 and C_2 represent the given points (called the **foci**), then no matter which point P on the hyperbola you select, subtracting the smaller distance (between the point and each focus) from the larger distance always results in the same value, as shown in Figure 8.127.

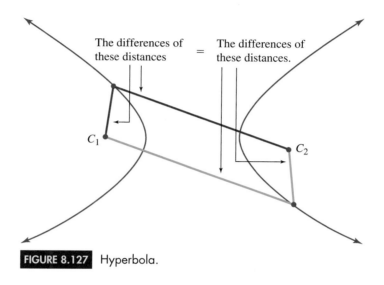

FIGURE 8.127   Hyperbola.

As with the ellipse, we will consider only hyperbolas centered at the origin of a rectangular coordinate system, as shown in Figure 8.128.

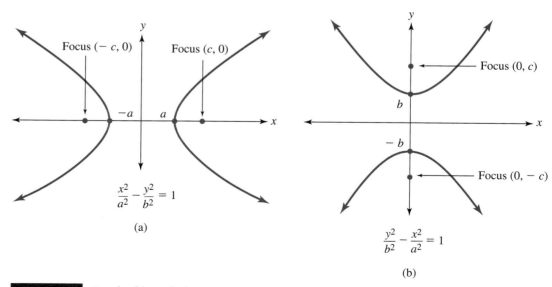

FIGURE 8.128   Standard hyperbolas.

EQUATION OF A HYPERBOLA

A **hyperbola**, centered at the origin, is the set of all points (x, y) satisfying either the equation

$$\frac{x^2}{a^2} - \frac{y^2}{b^2} = 1 \quad \text{or} \quad \frac{y^2}{b^2} - \frac{x^2}{a^2} = 1 \quad a \text{ and } b \text{ are positive constants.}$$

The **foci** are found by solving the equation $c^2 = a^2 + b^2$.

EXAMPLE **4**

GRAPHING A HYPERBOLA   Sketch the graph and find the foci of the hyperbola $y^2 - 4x^2 = 4$.

SOLUTION

First, put the equation in the general form, with the right side of the equation equal to 1:

$$y^2 - 4x^2 = 4$$

$$\frac{y^2}{4} - \frac{4x^2}{4} = \frac{4}{4} \quad \text{dividing by 4}$$

$$\frac{y^2}{4} - \frac{x^2}{1} = 1 \quad \text{simplifying the fractions}$$

(Note that $a^2 = 1$ and $b^2 = 4$.)

Finding the x-Intercepts:
(Substitute $y = 0$ and solve for x.)

$$\frac{0^2}{4} - \frac{x^2}{1} = 1$$

$$-\frac{x^2}{1} = 1$$

$$x^2 = -1$$

no solution; no x-intercepts

Finding the y-Intercepts:
(Substitute $x = 0$ and solve for y.)

$$\frac{y^2}{4} - \frac{0^2}{1} = 1$$

$$\frac{y^2}{4} = 1$$

$$y^2 = 4$$

$$y = \pm 2$$

Because there are no x-intercepts, the branches of the hyperbola open up and down. (Whenever the y^2 term comes first in the general equation, the branches of the hyperbola will open up and down.)

The values of a and b determine the shape of the hyperbola. Locate the points $a = \pm 1$ on the x-axis. Locate the y-intercepts $b = \pm 2$.

Using a dashed line, draw the rectangle (with sides parallel to the axes) formed by these points. Lightly draw in the diagonals of this rectangle. Now draw in the two branches of the hyperbola, as shown in Figure 8.129.

Finding the Foci:
$$c^2 = a^2 + b^2$$
$$= 1 + 4 = 5$$

Therefore, $c = \pm\sqrt{5}$ (≈ 2.24). The foci are the points $(0, \sqrt{5})$ and $(0, -\sqrt{5})$.

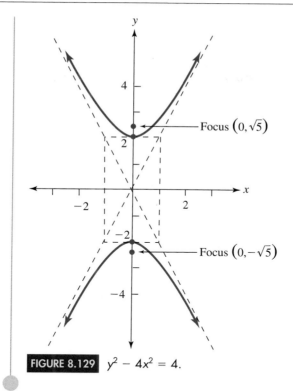

FIGURE 8.129 $y^2 - 4x^2 = 4$.

The most common application of the hyperbola is in radio-assisted-navigation, or LORAN (LOng-RAnge Navigation). Various national and international transmitters continually emit radio signals at regular time intervals. By measuring the *difference* in reception times of a pair of these signals, the navigator of a ship at sea knows that the vessel is somewhere on the hyperbola that has the transmitters as foci. Using a second pair of transmitters, the navigator knows that the ship is somewhere on a second hyperbola. By plotting both hyperbolas on a map, the navigator knows that the ship is located at the intersection of the two hyperbolas, as shown in Figure 8.130.

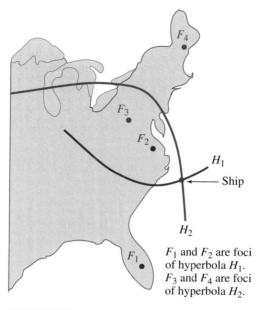

F_1 and F_2 are foci of hyperbola H_1.
F_3 and F_4 are foci of hyperbola H_2.

FIGURE 8.130 Hyperbolas are used in radio-assisted navigation.

In Exercises 1–8, find the center and radius of the given circle and sketch its graph.

1. $x^2 + y^2 = 1$
2. $x^2 + y^2 = 4$
3. $x^2 + y^2 - 4x - 5 = 0$
4. $x^2 + y^2 + 8y + 12 = 0$
5. $x^2 + y^2 - 10x + 4y + 13 = 0$
6. $x^2 + y^2 + 2x - 6y - 15 = 0$
7. $x^2 + y^2 - 10x - 10y + 25 = 0$
8. $x^2 + y^2 + 8x + 8y + 16 = 0$

In Exercises 9–12, sketch the graph of the given parabola and find its focus.

9. $y = x^2$
10. $y = 2x^2$
11. $y = \frac{1}{2}x^2$
12. $y = \frac{1}{4}x^2$

13. A Mylar-coated parabolic reflector dish (used for the solar heating of water) is 9 feet wide and $1\frac{3}{4}$ feet deep. Where should the water container be placed to heat the water most rapidly?

14. A satellite dish antenna has a parabolic reflector dish that is 18 feet wide and 4 feet deep. Where is the focus located?

15. A flashlight has a diameter of 3 inches. If the reflector is 1 inch deep, where should the light bulb be located to concentrate the light in a forward beam?

16. An automotive headlight has a diameter of 7 inches. If the reflector is 3 inches deep, where should the light bulb be located to concentrate the light in a forward beam?

In Exercises 17–24, sketch the graph and find the foci of the given ellipse.

17. $4x^2 + 9y^2 = 36$
18. $x^2 + 4y^2 = 16$
19. $25x^2 + 4y^2 = 100$
20. $x^2 + 9y^2 = 9$
21. $16x^2 + 9y^2 = 144$
22. $4x^2 + y^2 = 36$
23. $4x^2 + y^2 - 8x - 4y + 4 = 0$

   HINT: Complete the square for x and for y.

24. $25x^2 + 9y^2 + 150x - 90y + 225 = 0$

   HINT: Complete the square for x and for y.

25. An elliptical room is designed to function as a whispering gallery. If the room is 30 feet long and 24 feet wide, where should two people stand to optimize the whispering effect?

26. An elliptical billiard table is 8 feet long and 5 feet wide. Where are the foci located?

27. During the earth's elliptical orbit around the sun (the sun is located at one of the foci), the earth's greatest distance from the sun is 94.51 million miles, and the shortest distance is 91.40 million miles. Find the equation (in the form $x^2/a^2 + y^2/b^2 = 1$) of the earth's elliptical orbit.

   HINT: Use $a + c$ = greatest distance and $a - c$ = shortest distance, as shown in Figure 8.131, to solve for a and then c; then find b.

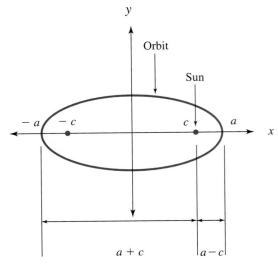

FIGURE 8.131 Elliptical orbit of the earth around the sun.

28. During Mercury's elliptical orbit around the sun (the sun is located at one of the foci), the planet's greatest distance from the sun is 43.38 million miles, and the shortest distance is 28.58 million miles. Find the equation (in the form $x^2/a^2 + y^2/b^2 = 1$) of Mercury's elliptical orbit.

   HINT: Use $a + c$ = greatest distance and $a - c$ = shortest distance, as shown in Figure 8.131, to solve for a and c; then find b.

In Exercises 29–36, sketch the graph and find the foci of the given hyperbola.

29. a. $x^2 - y^2 = 1$
    b. $y^2 - x^2 = 1$
30. a. $x^2 - y^2 = 4$
    b. $y^2 - x^2 = 4$
31. $4x^2 - 9y^2 = 36$
32. $x^2 - 4y^2 = 16$
33. $25y^2 - 4x^2 = 100$
34. $y^2 - 9x^2 = 9$

35. $x^2 - 4y^2 - 6x + 5 = 0$

HINT: Complete the square for x and for y.

36. $9y^2 - x^2 - 36y - 2x + 26 = 0$

HINT: Complete the square for x and for y.

 Answer the following questions using complete sentences and your own words.

• HISTORY QUESTIONS

37. Who is one of the first women to be mentioned in the history of mathematics? Why was she murdered?

38. Write a research paper on a historical topic referred to in this section or on a related topic. Following is a list of possible topics:
  • Apollonius of Perga (What contributions did he make to the study of conic sections?)
  • Hypatia (Discuss her accomplishments.)
  • René Descartes (What contributions did he make to the study of conic sections?)
  • Edmond Halley (What is Halley's comet? How did Halley predict its arrival?)
  • Johann Kepler (How did he develop his theory of elliptical orbits?)
  • LORAN (How is LORAN used in navigation? When and where did it originate?)
  • Parabolic reflectors and antennas (How have parabolic reflectors changed the dissemination of information around the world?)

8.8 Non-Euclidean Geometry

OBJECTIVES

  • Understand the Parallel Postulate
  • Explore alternatives to the Parallel Postulate
  • Explore the history and contributions of prominent mathematicians in the development of non-Euclidean geometry
  • Explore models and applications of non-Euclidean geometries

Mathematics is that subject in which we do not know what we are talking about, or whether what we are saying is true.
—BERTRAND RUSSELL

From the time of Thales of Miletus, mathematics has been based on the axiomatic system of deduction and proof. Euclid's monumental work, *Elements,* was the epitome of this effort. By the careful choice of five geometric postulates (things that are assumed to be true), Euclid proceeded to build the entire structure of all geometric knowledge of the day. The ancient Greeks' attitude was "the truth of these five axioms is obvious; therefore, everything that follows from them is true also." Over 2,000 years and many headaches, heartbreaks, and humiliations later, the attitude of a current mathematician is more "*if* we *assume* these axioms to be valid, then everything that follows from them is valid also." While the Greeks considered geometry (and all of mathematics) to be a system of absolute truth, Bertrand Russell's twentieth-century comment implies that results that are based on *assumptions* cannot be considered true or false; the most one can hope for is *consistency.*

The Parallel Postulate

Euclid's empire was based on his perception of reality. This reality in turn consisted of five axiomatic concepts so simple and so obvious that no one could deny that they were in fact true. After all, reality is what is confirmed by experience, and experience had never shown the following statements to be false:

1. A straight line can be drawn from any point to any other point.
2. A (finite) straight line segment can be extended continuously in an (infinite) straight line.
3. A circle may be described with any center and any radius.
4. All right angles are equal to one another.
5. Given a line and a point not on that line, there is one and only one line through the point parallel to the original line.

Taking the axiomatic method to heart, scholars scrutinized even the five "obvious" postulates for any possible overlap or dependence. Could one of the five be deduced logically from the others? The search was on for the smallest possible set of postulates that could serve as the basis of all geometry.

The most likely candidate for a dependent postulate was number 5, the so-called **Parallel Postulate** (see Figure 8.132). From the beginning, it aroused the curiosity and inquiry of mathematicians the world over. Many "proofs" of this postulate were put forth, but each was dismissed upon further scrutiny by later investigations. Each "proof" had its flaw, its logical demise. Even Euclid himself had tried to prove it but couldn't, and he consequently called it a postulate (it was obviously true).

The more mathematicians studied the postulate, the more elusive it became. Indeed, the competition to prove the Parallel Postulate became so intense that the French mathematician d'Alembert called it "the scandal of elementary geometry." On one occasion, the highly admired Joseph-Louis Lagrange presented a paper on the postulate to the French Academy. However, halfway through his presentation, he stopped, with the comment "I must meditate further on this." He put the paper away and never spoke of it again.

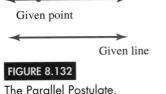

One parallel ?

Given point

Given line

FIGURE 8.132

The Parallel Postulate.

Girolamo Saccheri

In all the furor to prove the Parallel Postulate, one particular attempt is worth mentioning. In 1733, a Jesuit priest named Girolamo Saccheri published *Euclides ab Omni Naevo Vindicatus* (*Euclid Vindicated of Every Blemish*). This work contained Saccheri's "proof" of the Parallel Postulate. Saccheri was a professor of mathematics, logic, and philosophy at the University of Pavia, Italy. He is said to have had an incredible memory; commonly mentioned was his ability to play three games of chess simultaneously, without looking at the chessboards!

Saccheri was the first to examine the consequences of assuming the Parallel Postulate to be false; that is, he attempted a proof by contradiction. Proof by contradiction is an indirect means of proving a statement. Simply put, to prove something, you assume it to be false. If you can show that this assumption leads to a contradiction, then the assumed falsity of the statement is itself false, implying that the statement is true. This is an acceptable line of reasoning that has been used since the days of classic Greek logic. Euclid himself used proof by contradiction in *Elements,* though not in reference to the Parallel Postulate.

In Saccheri's quest to prove the Parallel Postulate, he assumed it to be false, thereby hoping to obtain some sort of logical contradiction. The denial of the Parallel Postulate consists of two alternatives:

1. Given a line and a point not on the line, there are no lines through the point parallel to the original line. (Parallels don't exist.)

2. Given a line and a point not on the line, there are at least two lines through the point parallel to the original line.

Surely, each of these assumptions would lead to a contradiction. To his surprise, Saccheri couldn't find one! How could this be? Were the foundations of Euclid's world built on rock or on sand? Rather than accepting the implications of his work, Saccheri manufactured a spurious contradiction in order to be able to proclaim support of the Parallel Postulate.

Saccheri's work had little impact at the time. The implication that the Parallel Postulate was independent of the other four postulates (that is, that it could not be proved) was too far out of line for the thinking of the day. Mathematicians were so convinced of the dependence of the Parallel Postulate that little interest was given to any other alternative. However, less than 100 years later, three men (Carl Gauss, Janos Bolyai, and Nikolai Lobachevsky) independently resurrected the denial of the Parallel Postulate and took roads similar to the one that had led Saccheri to question the "truth" of Euclidean geometry.

 Carl Friedrich Gauss

Carl Friedrich Gauss.

Carl Friedrich Gauss, the "Prince of Mathematicians," first became interested in the study of parallels in 1792, at the age of fifteen (see Historical Note on page 282). Although he never published any theory that went counter to Euclid's Parallel Postulate, his diaries and his letters to close friends over the years affirmed his realization that geometries based on axioms different from Euclid's could exist.

Having failed to prove the Parallel Postulate, Gauss began to ponder its sanctity. According to his diaries, he put forth an axiom that contradicted Euclid's; specifically, he assumed that more than one parallel could be drawn through a point not on a line. Rather than trying to obtain a contradiction as Saccheri had attempted (and thus prove the Parallel Postulate via reductio ad absurdum), Gauss began to see that a geometry that was at odds with Euclid's but internally consistent could be developed. However, Gauss never went public; he didn't want to face the inevitable public controversy.

 Janos Bolyai

Janos Bolyai was a Hungarian army officer and the son of a respected mathematician. From an early age, Janos was exposed to the world of mathematics. Janos's father, Wolfgang, had studied with Gauss at Göttingen, and their sporadic correspondence lasted a lifetime. In fact, in their correspondence, Gauss had pointed out the fallacy in the elder Bolyai's "proof" of the Parallel Postulate.

In 1817, Janos entered the Imperial Engineering Academy in Vienna at the age of fifteen. After completing his studies in 1823, he embarked upon a military career. An expert fencer with numerous successful duels to his credit, Bolyai was also an accomplished violinist.

The early teachings of his father prompted Bolyai to continue his study of mathematics. However, when Janos informed his father of his interest in the study of parallels, the elder Bolyai wrote, "Do not waste an hour's time on that problem. It does not lead to any result; instead it will come to poison all your life." Ignoring his father's decree, the younger Bolyai pursued the elusive Parallel Postulate.

After several failed attempts to prove the Parallel Postulate, Bolyai, too, proceeded down the path of denying Euclid's assumption of the existence of a unique parallel. In a manner not unlike that of Gauss, even though he was unaware of Gauss's work, Bolyai assumed that more than one parallel existed through a point not on a line. He developed several theorems in this new, consistent geometry. He was so excited over his discovery that he wrote, "Out of nothing I have created a strange new world."

Hastening to publish his work, the younger Bolyai published his theory as a twenty-four-page appendix to the elder Bolyai's two-volume *Tentamen Juventutem Studiosam in Elementa Matheseos Purae* (*An Attempt to Introduce Studious Youth to the Elements of Pure Mathematics*). Although the book is dated 1829, it was actually printed in 1832.

When Wolfgang Bolyai sent a copy to his old friend Carl Gauss, the reply was less than heartening to Janos. Gauss responded, "If I begin by saying that I dare not praise this work, you will of course be surprised for a moment; but I cannot do otherwise. To praise it would amount to praising myself. For the entire content of the work, the approach which your son has taken, and the results to which he is led, coincide almost exactly with my own meditations which have occupied my mind for the past thirty or thirty-five years. It was my plan to put it all down on paper eventually, so that at least it would not perish with me. So I am greatly surprised to be spared this effort, and am overjoyed that it happens to be the son of my old friend who outstrips me in such a remarkable way."

Not seeing the true compliment that the great Gauss had bestowed on him, Bolyai feared that his work was being stolen. His feelings were compounded by the total lack of interest on the part of other mathematicians, and deep periods of depression set in. Janos Bolyai never published again.

Nikolai Lobachevsky.

© The Granger Collection, New York

Lobachevsky's *On the Foundations of Geometry* was published in this 1829 issue of the academic journal of the University of Kazan.

© The Granger Collection, New York

Nikolai Lobachevsky

Labeling it "imaginary geometry," Nikolai Ivanovitch Lobachevsky published the first complete text on non-Euclidean geometry in 1829. Many people have since heralded him as the Copernicus of geometry. Just as Copernicus had challenged the long-established theory that the earth was the center of the universe, Lobachevsky's alternative view of geometry was in direct contradiction to the long-revered work of Euclid. Referring to Lobachevsky, Einstein said, "He dared to challenge an axiom."

Lobachevsky spent most of his life at the University of Kazan, near Siberia. Founded by Czar Alexander I in 1804, Kazan was Europe's easternmost center of higher education. Among its first students, Lobachevsky received his master's degree in mathematics and physics in 1811 at the age of eighteen. He remained at Kazan teaching courses for civil servants until 1816, when he was promoted to a full professorship.

Geometry received Lobachevsky's special attention. Being inquisitive, he made several attempts to prove the Parallel Postulate; he failed at each. He then proceeded to examine the consequences of substituting an alternative to Euclid's postulate of a unique parallel; Lobachevsky assumed that more than one parallel could be drawn through a point. Living in distant isolation from the learning capitals of

Europe, Lobachevsky was unaware of the similar approaches taken by Gauss and Bolyai.

Lobachevsky's idea of a geometry based on an axiom in opposition to Euclid began to take form in 1823, when he drew up an outline for a geometry course he was teaching. In 1826, he gave a lecture and presented a paper incorporating his belief in the feasibility of a geometry based on axioms different from Euclid's. Although this paper has been lost (as have so many in the history of mathematics), it was the first recorded attempt to breach Euclid's bastion. In 1829, the monthly academic journal of the University of Kazan printed a series of his works titled *On the Foundations of Geometry;* this publication is considered by many to be the official birth of the radically new, non-Euclidean geometry.

As with anything new and at odds with the status quo, Lobachevsky's work was not greeted with open arms. The St. Petersburg Academy rejected it for publication in its scholarly journal and printed an uncomplimentary review. In contrast to Gauss, who did not have the courage to print, and Bolyai, who did not have the fortitude to face his opponents, Lobachevsky remained undaunted. He proceeded in his work, expanding *Foundations* into *New Elements of Geometry, with a Complete Theory of Parallels,* which was also published in Kazan's academic journal of 1835. Shortly thereafter, his work began to be recognized outside of Kazan; he was published in Moscow, Paris, and Berlin.

In 1846, when Gauss received a copy of Lobachevsky's latest book, *Geometrical Investigations on the Theory of Parallels* (which contained only sixty-one pages), he wrote to a colleague, "I have had occasion to look through again that little volume by Lobachevsky. You know that for fifty-four years now I have held the same conviction. I have found in Lobachevsky's work nothing that is new to me, but the development is made in a way different from that which I have followed, and certainly by Lobachevsky in a skillful way and a truly geometrical spirit." However, Gauss did not give public approval to Lobachevsky's work.

As is the case with those who pioneer ideas and art forms that are incomprehensible to the world at large, the radicals of geometry, Lobachevsky and Bolyai, never received full recognition of the value of their work during their lifetimes. However, they opened the door for a host of new ideas in geometry and in the axiomatic system in general.

The Granger Collection, New York

Bernhard Riemann.

Bernhard Riemann

The groundbreaking work of Lobachevsky and Bolyai was ignored for many years. The mathematician who finally convinced the academic world of the merits of non-Euclidean geometry (that is, geometries based on the denial of Euclid's Parallel Postulate) was born in 1826, at the time when Lobachevsky and Bolyai were initially presenting their ideas. Although his life was cut short (he died of tuberculosis at the age of thirty-nine), Georg Friedrich Bernhard Riemann's contributions to the world of modern mathematics were monumental.

At the age of nineteen, Bernhard Riemann enrolled at the University of Göttingen with the intention of pursuing the study of theology and philosophy. However, he became so interested in the study of mathematics that he devoted his life entirely to that field. As a graduate student at Göttingen, Riemann studied under Gauss, who was unusually impressed with his protégé's abilities. In his report to the faculty, Gauss said that Riemann's dissertation came from "a creative, active, truly mathematical mind, and of a gloriously fertile originality." Coming from the Prince of Mathematicians, that was a compliment indeed!

After receiving his degree, Riemann wanted to stay at Göttingen as a member of the faculty. To prove himself, he first had to deliver a "probationary" lecture to his former teachers. When he submitted three possible topics for his oration, Gauss chose the third, "On the Hypotheses That Underlie the Foundation of Geometry." Delivered in 1854 and published in 1868 (two years after his death), Riemann's lecture is considered by many to be one of the highlights of modern mathematical history.

Unlike Gauss, Lobachevsky, and Bolyai (each of whom assumed that through a point not on a line more than one parallel existed), Riemann denied the existence of all parallel lines! He philosophized that we could just as well assume that all lines eventually intersect as assume that some (parallel lines) do not intersect.

Geometric Models

One reason for the initial resistance to non-Euclidean geometry was practical experience. In our perception of the "flatness" of the immediate world in which we live, we naturally accept the Parallel Postulate as true. It is easy to envision a rectangular grid of city blocks with straight streets that do not intersect. However, our world is not flat; the earth is spherical. An age-old problem associated with map making has been the projection of the curved earth (a globe) into a plane (a flat map) and vice versa. (What happens if we try to peel the "skin" off of a globe or stretch a piece of rectangular graph paper around a sphere?)

Granted, non-Euclidean geometries run counter to common sense. To understand them, we need to draw pictures; we must create models. Each model will have its own unique definition of a line and a plane. In all cases, **parallel lines** are lines in a plane that do not intersect one another. Geometries can be categorized by three types, depending on what is assumed concerning parallel lines:

1. *Euclidean geometry:* Through a point not on a given line, there exists exactly one line parallel to the given line (Euclid's Parallel Postulate).
2. *Lobachevskian geometry:* Through a point not on a given line, there exist more than one line parallel to the given line.
3. *Riemannian geometry:* Through a point not on a given line, there exists no line parallel to the given line.

Because they are based on postulates that contradict Euclid's Parallel Postulate, Lobachevskian and Riemannian geometries are called non-Euclidean geometries.

The model that is used to express Euclidean geometry is the plane. Lying in a flat plane, lines extend indefinitely into space and have infinite length; they do not "wrap" around the earth. See Figure 8.133.

In contrast, the model used to express Riemannian geometry is the sphere. Using this model, we define lines to be the great circles that encompass the sphere. A **great circle** is a circle whose center lies at the center of the sphere, as shown in Figure 8.134. No matter how they are drawn, each pair of great circles will always intersect in two points.

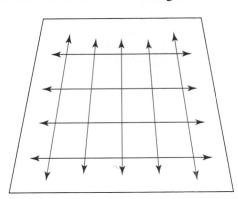

FIGURE 8.133

The Euclidean planar model.

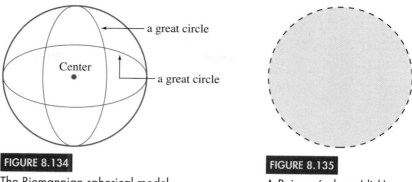

FIGURE 8.134

The Riemannian spherical model.

FIGURE 8.135

A Poincaré plane (disk).

Consequently, parallel lines do not exist! Riemannian geometry is important in navigation, because the shortest distance between two points on a sphere is the path along a great circle.

A model that is used to express Lobachevskian geometry is the interior of a circle. This model was described by the Frenchman Henri Poincaré (1854–1912). The Poincaré model defines a plane to be all points "inside" a circle. (The points *on* the circle are excluded.) This region is also called a **disk**; it is shown in Figure 8.135.

Poincaré's model of a Lobachevskian geometry defines a line to be either of the following:

1. a diameter of the disk

2. a circular arc connecting two points on the boundary of the disk[*]

The two types of Poincaré lines are shown in Figure 8.136. Because the boundary of the disk is excluded, a (Poincaré) line has no endpoints.

Using Poincaré's disk model, we can construct more than one parallel line through a given point, as shown in Figure 8.137. Notice that lines *CD* and *EF* pass through the point *Q* and that each line is parallel to line *AB*. (*CD* and *AB* are parallel because they do not intersect. Likewise, *EF* and *AB* are parallel because they do not intersect.)

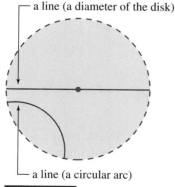

a line (a diameter of the disk)

a line (a circular arc)

FIGURE 8.136

The Poincaré disk model for Lobachevskian geometry.

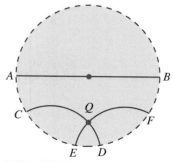

FIGURE 8.137

More than one parallel line through Q.

---

[*] Technically, the arc is the intersection of the disk with an orthogonal circle. Orthogonal circles are circles whose radii are perpendicular at the points where the circles intersect.

A Comparison of Triangles

Owing to the difference in the number of parallel lines, non-Euclidean geometries have properties that differ from those of Euclidean geometry. As you know, the sum of the angles of a triangle always *equals* 180° in Euclidean geometry, as shown in Figure 8.138(a). In contrast, when a triangle is drawn on a sphere, the sum of the angles is *greater* than 180°, as shown in Figure 8.138(b). Finally, the sum of the angles of a triangle drawn on a Poincaré disk is *less* than 180°, as shown in Figure 8.138(c).

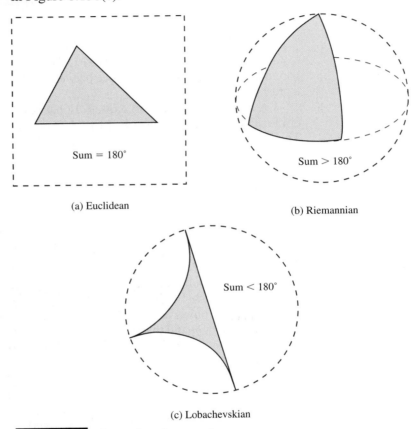

(a) Euclidean   (b) Riemannian

(c) Lobachevskian

FIGURE 8.138 The angles of a triangle.

Geometry in Art

The study of geometry has always been an integral part of an education in art. Whether an artist leans toward realism or surrealism, a fundamental knowledge of geometric relationships is essential. No artist has been more successful in combining art and geometry than the Dutch artist Maurits Cornelis Escher (1898–1972). Escher is known for his repetitious plane-filling patterns, which often depict the metamorphosis of one figure into another.

The precise geometric appearance of Escher's work is not accidental; he possessed the knowledge, the talent, and the creativity to manipulate perspective and dimension in both Euclidean and non-Euclidean geometry. In fact, several of Escher's images were based on the non-Euclidean geometry of a Poincaré disk. A sketch from one of his many notebooks (Figure 8.139) shows Escher's study of a disk in his preparation to work in non-Euclidean space. Notice the triangles formed by the circular arcs.

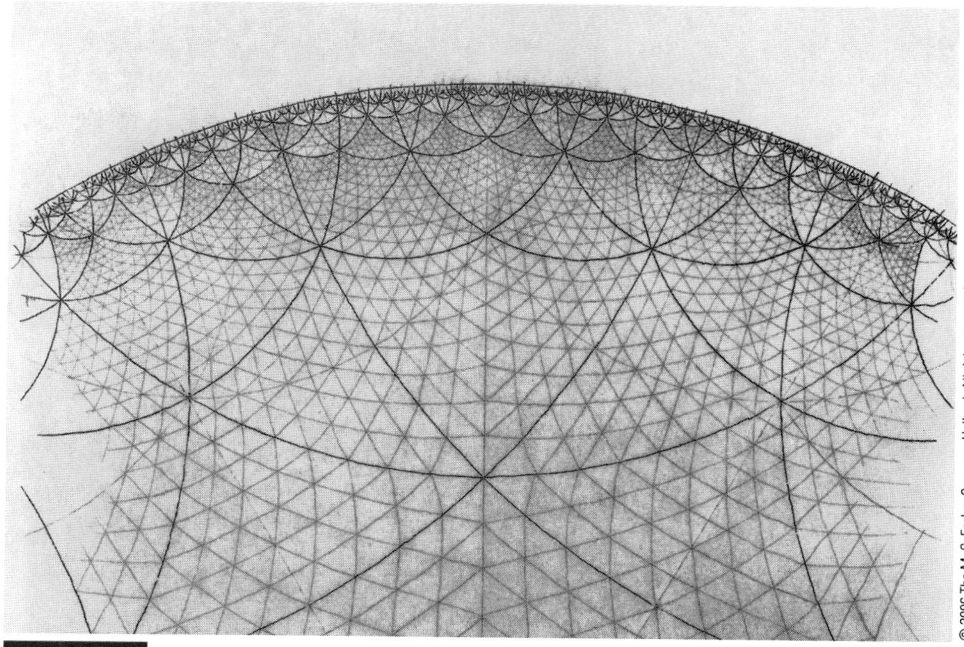

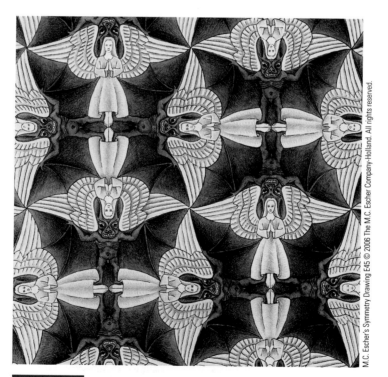

FIGURE 8.139

Escher drew this sketch of a Poincaré disk before creating the work of art shown in Figure 8.141.

FIGURE 8.140

A repetitive pattern that completely fills a plane is known as a tiling; these complementary angels and devils form a tiling of the Euclidian plane.

Escher often created the same image on different geometric models. For instance, after he developed a pattern of complementary angels and devils on a "flat" Euclidean plane (Figure 8.140), Escher then employed Lobachevskian geometry to map the pattern onto a disk (Figure 8.141) and Riemannian geometry to transpose the pattern onto a sphere (Figure 8.142).

FIGURE 8.141

Utilizing Lobachevskian geometry, Escher tiled a Poincaré disk with the angels and devils of Figure 8.140.

FIGURE 8.142

Utilizing Riemannian geometry, Escher transposed the angels and devils of Figure 8.140 onto a sphere; the design is carved in solid maple.

In Exercises 1–6, answer the questions for the planar model of Euclidean geometry.

1. Through a point not on a given line, how many parallels to the given line are there?

2. What can be said about the sum of the angles of a triangle?

3. How many right angles can a triangle have?

4. If two different lines are each parallel to a third line, are they necessarily parallel to each other?

5. In how many points can a pair of distinct lines intersect?

6. Do lines have finite or infinite length?

In Exercises 7–12, answer the questions for the spherical model of Riemannian geometry.

7. Through a point not on a given line, how many parallels to the given line are there?

8. What can be said about the sum of the angles of a triangle?

9. How many right angles can a triangle have?

10. Do lines have finite or infinite length?

11. In how many points can a pair of distinct lines intersect?

12. What is the minimum number of sides required to form a polygon?

In Exercises 13–18, answer the questions for Poincaré's model of Lobachevskian geometry.

13. Through a point not on a given line, how many parallels to the given line are there?

14. What can be said about the sum of the angles of a triangle?

15. How many right angles can a triangle have?

16. If two different lines are each parallel to a third line, are they necessarily parallel to each other?

17. In how many points can a pair of distinct lines intersect?

18. Do lines have finite or infinite length?

 Answer the following questions using complete sentences and your own words.

• HISTORY QUESTIONS

19. What is Euclid's Parallel Postulate?

20. What was "the scandal of elementary geometry"?

21. Who was the first mathematician to examine the consequences of assuming the Parallel Postulate to be false? What were his findings? What was his reaction?

22. Who was the first mathematician to affirm the existence of a non-Euclidean geometry? Did he publish his work? Why or why not?

23. Who stopped publishing his mathematical ideas after he had published his work on non-Euclidean geometry? Why did he stop?

24. Who is called the Copernicus of geometry? Why?

25. Who introduced a geometry in which parallel lines do not exist?

26. Write a research paper on a historical topic referred to in this section or on a related topic. Following is a list of possible topics:

   - Janos Bolyai (What influence did he have on the development of non-Euclidean geometry?)
   - M. C. Escher (How did he combine art and geometry?)
   - Carl Friedrich Gauss (What influence did he have on the development of non-Euclidean geometry?)
   - Immanuel Kant (How did his philosophy affect the development of non-Euclidean geometry?)
   - Felix Klein (What model did he create to represent Lobachevskian geometry?)
   - Adrien-Marie Legendre (How did he attempt to prove the Parallel Postulate?)
   - Nikolai Lobachevsky (Why is he given most of the credit for the creation of non-Euclidean geometry?)
   - Henri Poincaré (What did he contribute to the study of geometry?)
   - Georg Friedrich Bernhard Riemann (What did he contribute to the study of geometry?)
   - Girolamo Saccheri (What is the connection between Saccheri quadrilaterals and non-Euclidean geometry?)

WEB PROJECT

27. There are numerous web sites devoted to the works of M. C. Escher; many are animated and interactive. Investigate some of these web sites, and write an essay and/or make a presentation about your findings. Some useful links for this web project are listed on the text web site: **www.cengage.com/math/johnson**

8.9 Fractal Geometry

OBJECTIVES

- Understand how a recursive process can be used to generate a fractal
- See how self-similarity occurs in both the real world and fractal geometry
- Comprehend and computing a fractal's fractional dimension
- Grasp that some real-world objects have fractional dimensions

The shapes of the classic geometry of the Greeks—the triangle, square, rectangle, circle, ellipse, parabola and hyperbola—do not seem natural to us. Although these idealized shapes occasionally exist in nature, we more commonly see complex, less perfect shapes such as those of mountains and plants, rather than the simple, Euclidean shapes of circles and rectangles. As mathematician Benoit Mandelbrot said, "Clouds are not spheres, mountains are not cones, coastlines are not circles, and bark is not smooth, nor does lightning travel in a straight line" (Benoit Mandelbrot, *The Fractal Geometry of Nature*).

Amazingly, the fern and the mountain in Figure 8.143 are neither photographs nor drawings; they are shapes formed with fractal geometry, a new alternative

A fractal forgery of a landscape.

FIGURE 8.143 A fractal forgery of a fern.

geometry. In many respects, fractal geometry has more in common with nature than does the classic geometry of the Greeks.

The Sierpinski Gasket

The **Sierpinski gasket** is one of the older examples of fractal geometry. It was first studied by Polish mathematician Waclaw Sierpinski in 1915. To understand this shape, we will create one. Take the following steps, which are illustrated in Figure 8.144.

1. Draw a triangle and fill in its interior.

2. Put a hole in the triangle:
    *   Find the midpoints of the sides of the triangle.
    *   Connect the midpoints and form a new triangle.
    *   Form a hole by removing the new triangle.

    This leaves three filled in triangles, each a smaller version of the original triangle from step 1.

3. Put a hole in each of the three triangles from step 2 by applying the procedure from step 2 to each triangle.

4. Put a hole in each of the triangles from step 3 by applying the procedure from step 2 to each triangle.

5 (and beyond). Put a hole in each of the triangles from the previous step by applying the procedure from step 2 to each triangle. Then do it again and again.

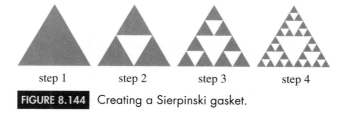

step 1   step 2   step 3   step 4

FIGURE 8.144   Creating a Sierpinski gasket.

The Sierpinski gasket (Figure 8.145) is the result of continuing the process described above through steps 6 and 7 and 8 and beyond without stopping.

After several steps, it becomes impossible to visually tell the difference between the result of one step and the result of the next step. That is, you probably couldn't tell the difference between the result of stopping at step 10 and the result of stopping at step 11 unless you used a magnifying glass. However, neither of these shapes would be a Sierpinski gasket; the only way to get a Sierpinski gasket is to continue the process without stopping, forming more and more ever-tinier triangles, each with holes and even tinier triangles inside it.

FIGURE 8.145   Step 5 of a Sierpinski gasket.

Self-Similarity

If you did use a magnifying glass to enlarge a portion of a Sierpinski gasket, what would you see? The enlarged portion would look exactly the same as the non-enlarged portion, as shown in Figure 8.146. This is called *self-similarity*. A shape has **self-similarity** if parts of that shape appear within itself at different scales. Fractal geometry utilizes shapes that have self-similarity. Some shapes have **exact self-similarity** (in which the enlarged portion is exactly the same as the nonenlarged portion), and some shapes have **approximate self-similarity.**

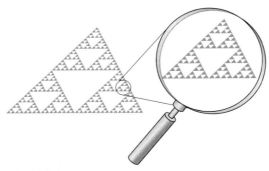

FIGURE 8.146 The self-similarity of a Sierpinski gasket.

Notice that the fern in Figure 8.143 has self-similarity. One branch of the fern looks just like the fern itself. And that one branch consists of many smaller branches, each of which looks just like the fern itself. Approximate self-similarity is a common feature of shapes that exist in nature. For example, trees and lightning can have a self-similar structure (see Figures 8.147 and 8.148). The fact that fractal geometry utilizes self-similarity is one reason why fractal geometry generates shapes that could exist in the real world.

FIGURE 8.147 A small part of this oak tree looks like the whole tree.

FIGURE 8.148 Is one fork of the lightning bolt very different in appearance from the entire bolt?

Self-similarity can be found in art, too. For example, consider M. C. Escher's *Smaller and Smaller* (Figure 8.149), and Jonathan Swift's *On Poetry. A Rhapsody*. These two works certainly illustrate self-similarity.

FIGURE 8.149 M. C. Escher's *Smaller and Smaller.*

So Nat'ralists observe, a Flea
Hath smaller fleas that on him prey;
And these have smaller Fleas to bite 'em
And so proceed ad infinitum.

—JONATHAN SWIFT
ON POETRY. A RHAPSODY

Recursive Processes

Step 2 of building a Sierpinski gasket gives a procedure for putting a hole in a triangle. Steps 3 and beyond instruct us to apply that procedure to the result, over and over again. Such a process, in which a procedure is applied and the same procedure is applied to the result, over and over again, is called a **recursive process.**

Two of the key elements of fractal geometry are self-similarity and recursive processes. We will not attempt a more formal definition of fractal geometry, since it goes beyond the scope of this text. Benoit Mandelbrot, who coined the term **fractal,** described a fractal as a mathematical object whose form is extremely irregular and/or fragmented at all scales. Mandelbrot derived the term *fractal* from the Latin *fractus* (to break, to create irregular fragments).

The Skewed Sierpinski Gasket

The **skewed Sierpinski gasket** is a simple variation of the Sierpinski gasket. The steps are the same as before, except for a twist.

1. Draw a triangle and fill in its interior (as before).

2. Put a twisted hole in the triangle: (See Figure 8.150.)

   • Find the midpoints of the sides of the triangle (as before).
   • For each midpoint, find a nearby point (this is the new twist).

- Connect the nearby points and the vertices of the original triangle.
- Form a twisted hole by removing the center triangle.

3. Put a twisted hole in each of the three triangles from step 2 by applying the procedure from step 2 to each triangle.

4. Put a twisted hole in each of the triangles from step 3 by applying the procedure from step 2 to each triangle.

5 (and beyond). Put a twisted hole in each of the triangles from the previous step by applying the procedure from step 2 to each triangle. Continue this process indefinitely.

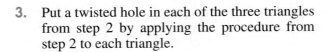

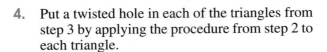

The midpoints and nearby points in step 2

Finishing step 2

FIGURE 8.150 Creating a skewed Sierpinski gasket.

Figure 8.151 shows a skewed Sierpinski gasket. Surprisingly, this fractal looks somewhat like a mountain! It has both the shape and the texture of a real mountain. Fractal geometry generates shapes that could exist in the real world.

FIGURE 8.151 A skewed Sierpinski gasket and a real mountain.

The Sierpinski Gasket and the Chaos Game

There is an interesting connection between probability (as studied in Chapter 3) and the Sierpinski gasket. This connection involves the **chaos game,** which is similar to a game of chance.

CHAOS GAME EQUIPMENT:

- A sketch of a triangle with the vertices labeled A, B, and C. (See Figure 8.152.)
- A die
  - If you roll a 1 or a 2, vertex A "wins."
  - If you roll a 3 or a 4, vertex B "wins."
  - If you roll a 5 or a 6, vertex C "wins."

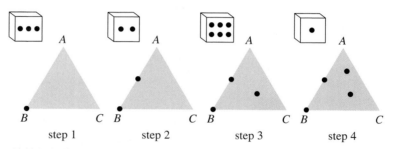

FIGURE 8.152   The beginnings of the chaos game.

CHAOS GAME PROCEDURE:

To play this game, follow these steps, which are illustrated in Figure 8.152.

1. Roll the die. Put a dot at the winning vertex. For example, if you roll a 3, put a dot at vertex B.

2. Roll the die. Put a dot halfway between the last dot and the winning vertex. For example, if you roll a 2, put a dot halfway between vertex B (the last dot) and vertex A (the winning vertex).

3. Roll the die. Put a dot halfway between the last dot and the winning vertex, as determined by the die. For example, if you roll a 6, put a dot halfway between the last dot and vertex C (the winning vertex).

4 (and beyond). Continue rolling the die. Each time, put a dot halfway between the last dot and the winning vertex, as determined by the die. For example, if you roll a 1, put a dot halfway between the last dot and vertex A (the winning vertex). (See Figure 8.152.) Then roll again.

Figure 8.153(a) shows the chaos game after 100 rolls. Not surprisingly, it's just a scattering of dots. Figure 8.153(d) shows the chaos game after 10,000 rolls. Quite surprisingly, it looks like the Sierpinski gasket! After all, the chaos game is a sequence of random events, so you would think that you would get a random result, and if you play the game twice, you would think that you'd get two different results. But this is not the case; no matter how many times you play the chaos game, if you roll the die enough, you get the same result: the Sierpinski gasket.

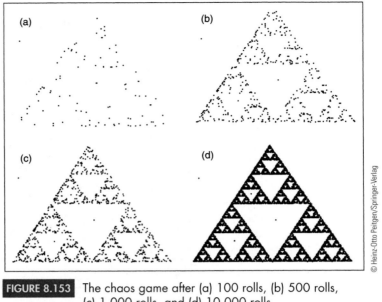

FIGURE 8.153 The chaos game after (a) 100 rolls, (b) 500 rolls, (c) 1,000 rolls, and (d) 10,000 rolls

Pascal's Triangle and the Sierpinski Gasket (*for those who have read Section 2.4: Permutations and Combinations.*)

There is an interesting connection between Pascal's triangle (as studied in Section 2.4) and the Sierpinski gasket. If you color the odd numbers of the first eight rows of Pascal's triangle black and the even numbers white, the result looks like the result of step 3 in creating the Sierpinski gasket. This is shown in Figure 8.154.

The first eight rows of Pascal's Triangle

The first eight rows of Pascal's Triangle with the odd numbers colored black and the even numbers colored white

FIGURE 8.154

Figure 8.155 shows the result of coloring the first 64 rows of Pascal's triangle. Is this a Sierpinski gasket? (See Exercise 20.)

© Heinz-Otto Peitgen/Springer-Verlag

FIGURE 8.155   The first 64 rows of Pascal's Triangle with the odd numbers colored black and the even numbers colored white.

The Koch Snowflake

The **Koch snowflake** is another example of fractal geometry. It was first studied by Swedish mathematician Helge von Koch in 1904. To create this shape, follow these steps, which are illustrated in Figure 8.156.

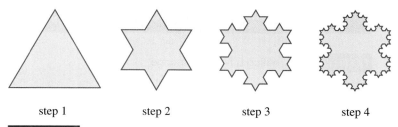

step 1   step 2   step 3   step 4

FIGURE 8.156   Creating a Koch snowflake.

1. Draw an equilateral triangle (that is, a triangle whose sides are equal and whose angles are equal).

2. In the middle of each side of the equilateral triangle, attach a smaller equilateral triangle (pointing outwards). The smaller equilateral triangle's sides should be one third as long as the original larger equilateral triangle's sides. This forms a Star of David, with 12 sides.

3. In the middle of each side of the Star of David, attach an equilateral triangle (pointing outward). This equilateral triangle's sides should be one third as long as the Star of David's sides.

4 (and beyond). Alter the result of the previous step by attaching an equilateral triangle to the middle of each side. Continue this process indefinitely. The resulting curve, shown in Figure 8.157, is called a Koch snowflake. Notice its similarity to a real snowflake, such as that in Figure 8.158.

FIGURE 8.157 A Koch snowflake.

FIGURE 8.158 Notice the similarity between the Koch snowflake and the real snowflake shown here.

The Menger Sponge

The **Menger sponge** is closely related to the Sierpinski gasket. To create this shape, follow these steps, which are illustrated in Figure 8.159.

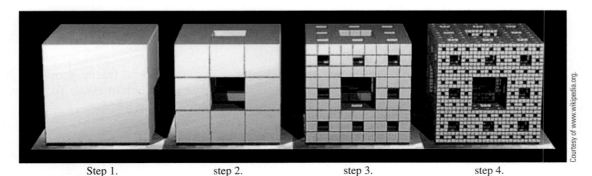

Step 1.   step 2.   step 3.   step 4.

FIGURE 8.159 Creating a Menger sponge.

1. Draw a cube.

2. Remove some of the cube:
   - Subdivide each of the cube's faces into nine equal squares. This subdivides the cube itself into 27 small cubes (nine in the front, nine in the middle, and nine in the back).
   - Remove the small cube at the center of each face. There are six such small cubes.
   - Remove the small cube at the very center of the original cube.

3. A number of small cubes remain at the end of step 2. Remove some of each of these cubes by applying the procedure from step 2 to them.

4 (and beyond). A number of cubes remain at the end of the previous step. Remove some of each of these cubes by applying the procedure from step 2 to them. Continue this process indefinitely. The resulting shape is called a Menger sponge.

Some Applications of Fractals

Biologists use *diffusion fractals* to analyze how bacteria cultures grow. Biologists also use *l-systems,* a method of generating fractals, to study the structures of plants. Chemists use *strange attractors,* a type of fractal, to study chaotic behavior in chemical reactions. Anatomists use *fractal canopies* to study the structure of lungs. Computer scientists use *fractal image compression* to squeeze 7,000 photographs onto the Microsoft Encarta encyclopedia CD. And moviemakers use fractals to create special effects in movies.

Dimension

You're probably familiar with the term *dimension.* A box has three dimensions: up/down, back/forth, and left/right. A rectangle has two dimensions: up/down and left/right. A line has one dimension: left/right. A circle (one that's not filled in) also has one dimension: clockwise/counterclockwise. See Figure 8.160.

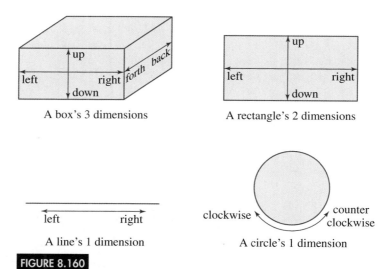

A box's 3 dimensions

A rectangle's 2 dimensions

A line's 1 dimension

A circle's 1 dimension

FIGURE 8.160

What is the dimension of a piece of paper? Many would say that it has two dimensions: up/down and left/right. Actually, a piece of paper has a third dimension, which is thickness. Tissue paper is very thin, typing paper is somewhat thicker, and card stock is thicker still. It's just that we tend to forget about this third dimension, the thickness, because it's so small relative to the paper's length and width. The only way a piece of paper would have two dimensions would be if it were infinitely thin. And paper isn't infinitely thin.

The Dimension of a Fractal

What is the dimension of a Koch snowflake? Many would say that, like a circle, it has one dimension: clockwise/counterclockwise. But a Koch snowflake wiggles around a lot; it's filled with little tiny crinkles. And those crinkles seem to make it somewhat thicker than a circle. The fact that a piece of paper has some thickness makes the paper's dimension 3 rather than 2. Does the fact that the Koch snowflake has some thickness affect its dimension?

In the late 1970s, Loren Carpenter was using computer-aided design to create images of airplanes at Boeing. His personal goal was to use computers to help make science fiction movies. He worked at Boeing to gain access to the computers that might help him accomplish that goal.

Carpenter read Benoit Mandelbrot's *The Fractal Geometry of Nature* (see the historical note on Mandelbrot on page 642 of this section). The book's fractal landscapes appealed to Carpenter. In a recent interview, he said, "I wanted to be able to create landscapes for my airplanes. But the method described (in Mandelbrot's book) did not allow you to stand *in* the landscape and have a level of detail in the background and foreground that was consistent with reality, to have a situation where the magnification factor can vary by 1000 in the same object. And that is what I needed." After some time, he figured out how to write computer algorithms "to make lightning bolts, landscapes, clouds, and a host of other things that had an infinite variety of detail and scale and would also animate, because they had a consistent geometry."*

Carpenter wanted to work at Lucasfilm's new computer graphics department, but he had to find a way to be noticed. So he made a movie to demonstrate his fractal discoveries. He used Boeing's computers at night and on weekends for four months to make the two-minute film *Vol Libre* ("Free Flight"), in which the viewer flies through a range of mountains. You can watch it on YouTube. Search for "Vol Libre."

In *Vol Libre*, Carpenter used fractals to create the landscapes. These landscapes were different from those in Mandelbrot's book in that the viewer could move through them and see them from different perspectives and different distances.

SIGGRAPH is a professional organization of computer scientists, artists and filmmakers who are interested in computer graphics. Carpenter showed his 16 mm film at their annual conference in 1980. He reported that "the audience erupted. And Ed Catmull, then head of the computer division at Lucasfilm, and Alvy Ray Smith, then director of the computer graphics group, were in the front row. They made me a job offer on the spot."

At Lucasfilm, Carpenter used fractals in the Genesis Sequence of *Star Trek II: The Wrath of Khan*. This was the first completely computer-generated sequence. You can watch it on YouTube. Search for "Genesis effect for Star Trek II The Wrath of Khan." The result was overwhelmingly successful. The Genesis Sequence was used in the next three Star Trek movies.

In 1986, Steve Jobs of Apple Computer fame bought Lucasfilm's computer division and renamed it Pixar. Carpenter became the company's first Senior Scientist, and he still holds that position. In 1993, he received a technical Academy Award for his invention of RenderMan image synthesis software system. RenderMan was used in rendering the dinosaurs in *Jurassic Park* and in making many other movies. See Exercise 35.

In 2001, Carpenter was awarded an Oscar for his fundamental contributions to the technology of motion pictures.

Thanks to Loren Carpenter for the information in this article and for the *Vol Libre* and *Star Trek* images.

*Source for all quotes: Perry, Tekla S. "And The Oscar Goes To . . . ," IEEE Spectrum, April 2001.

Carpenter's fractal mountains in *Vol Libre*

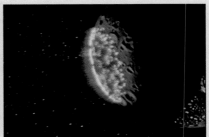
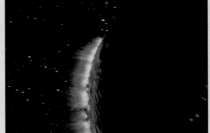

From the genesis sequence in *Star Trek II: The Wrath of Khan*.

The dimension of a fractal is certainly less obvious than the dimension of a box, a rectangle, or a line. Because of this, it must be determined with a new approach. We'll start by applying this new approach to a square.

If we take a square and double both its length and its width, we get four new squares, each equal to the original square. (See Figure 8.161.) We say that the scale factor is 2, because we multiplied the length and width by 2.

What happens if we apply a scale factor of 3 to the original square? If we multiply both the length and the width by 3, we get nine new squares, each equal to the original square. (See Figure 8.162.)

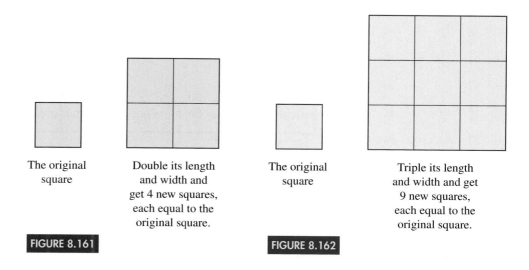

The original square

Double its length and width and get 4 new squares, each equal to the original square.

FIGURE 8.161

The original square

Triple its length and width and get 9 new squares, each equal to the original square.

FIGURE 8.162

In each case, the scale factor, raised to the second power, gives the number of new squares:

- When the scale factor is 2, we get $2^2 = 4$ new squares.
- When the scale factor is 3, we get $3^2 = 9$ new squares.

What happens with cubes? If we apply a scale factor of 2 to a cube, then we get eight new cubes, each equal to the original cube. And if we apply a scale factor of 3, then we get twenty-seven new cubes, each equal in size to the original cube. See Figure 8.163.

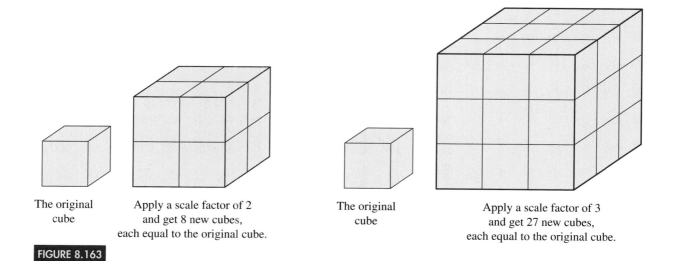

The original cube

Apply a scale factor of 2 and get 8 new cubes, each equal to the original cube.

FIGURE 8.163

The original cube

Apply a scale factor of 3 and get 27 new cubes, each equal to the original cube.

With the square, the scale factor raised to the *second* power gives the number of new squares. But with the cube, the scale factor raised to the *third* power gives the number of new cubes:

- When the scale factor is 2, we get $2^3 = 8$ new cubes.
- When the scale factor is 3, we get $3^3 = 27$ new cubes.

Clearly, there is a relationship between the scale factor, the number of new objects, and the dimension. The scale factor, raised to the power of the dimension, equals the number of new objects. It turns out that this relationship can be applied to fractals just as well as it can be applied to squares and cubes.

DIMENSION FORMULA

The dimension of an object is the number d that satisfies the equation

$$s^d = n$$

where s is the scale factor and n is the number of new objects that result, where each new object is equal in size to the original object.

EXAMPLE **1**

SOLUTION

A FRACTAL'S DIMENSION   Find the dimension of a Koch snowflake.

We will start by looking at one piece of a Koch snowflake. If we apply a scale factor of $s = 3$ to this piece, we get a larger piece of a Koch snowflake. This larger piece has $n = 4$ parts, each of which is equal in size to the original piece. See Figure 8.164.

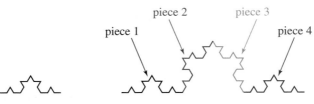

The original piece

Apply a scale factor of 3 and get 4 new pieces, each equal to the original piece.

FIGURE 8.164

This means that the dimension of the piece of a Koch snowflake is the number d that solves the equation $3^d = 4$:

$s^d = n$ **the dimension formula**

$3^d = 4$ **substituting**

If we apply a scale factor of $s = 3$ to an entire Koch snowflake, rather than to just one piece of the snowflake, each piece turns into $n = 4$ pieces, each of which is equal to the original piece. So the dimension of the entire snowflake is the same as that of the piece of a snowflake. It is the number d that solves the equation $3^d = 4$.

What number would d be? Solving this equation involves mathematics that will be covered in Chapter 10. Instead, we'll approximate d to the nearest tenth, using a trial and error process.

Clearly, $d \neq 1$, because $3^1 = 3 \neq 4$. Also, $d \neq 2$, because $3^2 = 9 \neq 4$. Substituting 1 for d gives a result that's too small, and substituting 2 for d gives a result that's too big. Instead, d is a number between 1 and 2. We'll approximate d to the nearest tenth, using a trial and error process. Using a calculator, we get the following results:

$$3^{1.1} = 3.348 \ldots$$
$$3^{1.2} = 3.737 \ldots$$
$$3^{1.3} = 4.171 \ldots$$

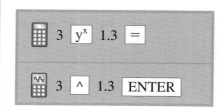

Substituting 1.2 for d gives a result that's a little too small, and substituting 1.3 gives a result that's a little too big. This means that d is between 1.2 and 1.3. And since 4.171 . . . is closer to 4 than is 3.737 . . . , we can tell that d is closer to 1.3 than to 1.2. Thus, $d \approx 1.3$.

So the dimension of a Koch snowflake is a little bigger than the dimension of a circle. The Koch snowflake has dimension $d \approx 1.3$, while the circle has dimension $d = 1$. The little tiny crinkles do make the snowflake somewhat thicker than a circle. And this extra thickness makes the snowflake's dimension somewhat bigger than that of a circle. But it's quite strange that a fractal can have a dimension that is not a whole number.

In discussing squares and cubes, it didn't matter what scale factor we used. We got the same result for the dimension regardless of whether we used $s = 2$ or $s = 3$. In fact, we would get the same result for any scale factor. But in Example 1, we deliberately used a scale factor of 3 because that scale factor gave us a larger piece of a Koch snowflake with four parts, *each of which is equal to the original piece*. If we used a scale factor of 2, we would still get a larger piece of a Koch snowflake with four parts, *but those parts would not be equal to the original piece*. They would be too small to equal the original piece. See Figure 8.165. With fractals, it's important to choose a scale factor carefully so that each of the new pieces is equal in size to the original piece.

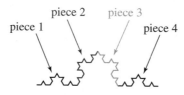

piece 2   piece 3

piece 1   piece 4

The original piece

Apply a scale factor of 2 and get 4 new pieces, but they are <u>not</u> equal in size to the original piece.

FIGURE 8.165

Some Applications of Fractal Dimension

A country's coastline has self-similarity. If you took two photographs of an uninhabited coastline, one from an elevation of 1,000 feet and the other from an elevation of 5,000 feet, you would have difficulty determining which photo was which. This self-similarity means that fractal geometry can be used to analyze the complexity of coastlines. In fact, geographers classify coastlines according to their

HISTORICAL NOTE

BENOIT MANDELBROT 1924–

© Roger Ressmeyer/CORBIS

Benoit Mandelbrot was born in Warsaw, Poland. His father made his living buying and selling clothes. His mother was a doctor. When Mandelbrot was 11, he emigrated with his family to France. There, his uncle, who was a professor of mathematics at the Collège de France, took responsibility for his education. After World War II started, poverty and the need to survive kept Mandelbrot away from school. As a result, he educated himself. He now attributes much of his success to this self-education. It allowed him to think in different ways and to avoid the standard modes of thought encouraged by a conventional education.

After obtaining a master's degree in aeronautics from the California Institute of Technology and a Ph.D. at the Université de Paris, Mandelbrot went to the Institute for Advanced Study at Princeton. Later, he became an IBM Fellow at their Watson Research Center in New York. IBM allowed him great latitude in choosing the direction of his research. At IBM, he started his famous work on fractals, which led to his being called "the father of fractal geometry." Also at IBM, he developed one particularly important fractal called the Mandelbrot set, and he developed some of the first computer programs that print graphics.

In part because of his unique education, he decided early on to make contributions to many different branches of science. His academic posts indicate this; they include professor of mathematics at Yale, Harvard, and the École Polytechnique and visiting professor at Harvard (first in economics and later in applied mathematics), Yale (engineering), and the Einstein College of Medicine (physiology). He has received many honors and prizes, including the Barnard Medal for Meritorious Service to Science, the Franklin Medal, the Alexander von Humbolt Prize, the Steinmetz Medal, the Science for Art Prize, the Harvey Prize for Science and Technology, the Nevada Medal, the Wolf Prize in physics, and the Honda Prize.

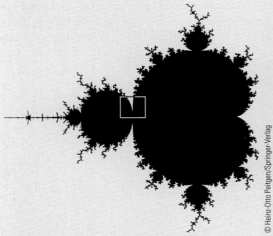

© Heinz-Otto Peitgen/Springer-Verlag

This is the Mandelbrot set, shown in black and white. It can be intriguingly beautiful along it's edge. The result of zooming in on the boxed portion is shown to the right.

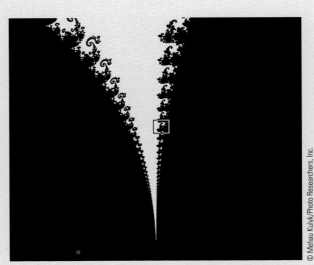

© Mehau Kulyk/Photo Researchers, Inc.

We'll zoom in again. The result is shown on page 643.

The edge has many fascinating details. We'll zoom in on the spiral.

Now we'll zoom in on a detail of the spiral.

BENOIT MANDELBROT 1924– (*CONTINUED*)

Another zoom.

One more zoom, and what do we see?

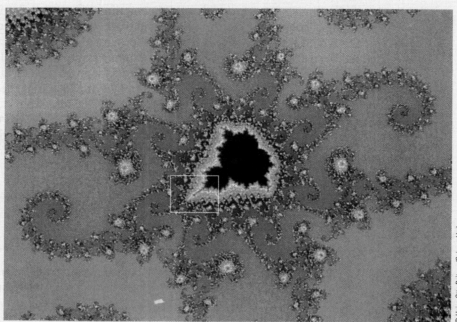

We see the Mandelbrot set again, surrounded by beautiful lace filigree. What an extreme example of self-similarity! And, if we zoom in one final time . . .

. . . we find another Mandelbrot set, slightly to the left and below the center, surrounded by ornate details. Even more self-similarity.

fractal dimension. South Africa's coastline is quite smooth, with a fractal dimension close to 1. On the other hand, Norway's coastline is heavily indented by glacier-carved fjords; its fractal dimension is about 1.5.

It's interesting to note that a Koch snowflake is sometimes called a *Koch island*, because the snowflake's outline looks somewhat like the coastline of an island. (See Exercise 34.)

Geologists have found that the fractal dimension of geologic features such as mountains, coastlines, faults, valleys, and river courses are related to the conditions under which they were formed and the processes to which they have been subjected. The fractal dimension of a mountain's contour line will be higher if the mountain is made of easily eroded rock or if it is in an area of high rainfall.

Scientists have also found that the fractal dimension of the surfaces of small particles such as soot and pharmaceuticals is related to their physical characteristics. For example, the tendency of soot to adhere to lung surfaces and the rate of absorption of ingested pharmaceuticals are both related to their fractal dimensions.

8.9 EXERCISES

1. *The Sierpinski gasket*
   a. Using a straightedge, carefully and accurately draw five equilateral triangles, each on a separate piece of paper. Make each side 1 foot in length (or, optionally, make your triangle as large as the paper will allow and label each side as being 1 foot in length).
   b. With the first equilateral triangle, complete the first step of a Sierpinski gasket. With the second equilateral triangle, complete the first two steps of a Sierpinski gasket. Continue in this fashion until you have completed the first five steps with the fifth equilateral triangle. With each triangle, shade in the appropriate regions.

2. *The Sierpinski carpet.* The Sierpinski carpet is closely related to the Sierpinski gasket and the Menger sponge. It is the square version of the Sierpinski gasket and the flat version of the Menger sponge. To create this shape, use the following steps. (See Figure 8.166.)

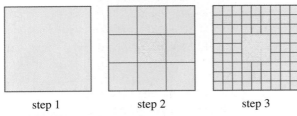

step 1　　　　step 2　　　　step 3

FIGURE 8.166 Creating a Sierpinski carpet.

1. Draw a square.
2. Subdivide the square into nine equal squares. Remove the central square.
3. The original square now has a square hole in its center, surrounded by eight other squares. Subdivide

each of these eight squares into nine smaller squares. Remove each of the central squares.

4 (and beyond). Continue this process indefinitely.
   a. Using a straightedge, carefully and accurately draw four squares, each on a separate piece of paper. Make each side 1 foot in length (or, optionally, make your triangle as large as the paper will allow and label each side as being 1 foot in length).
   b. With the first square, complete the first step of a Sierpinski carpet. With the second square, complete the first two steps of a Sierpinski carpet. Continue in this fashion until you have completed the first four steps with the fourth square. With each square, shade in the appropriate regions.

3. *The Mitsubishi gasket.* Carefully and accurately complete the first four steps of a Mitsubishi gasket. To create this shape, use the following steps. (See Figure 8.167.)

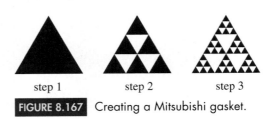

step 1　　　　step 2　　　　step 3

FIGURE 8.167 Creating a Mitsubishi gasket.

1. Draw a triangle and fill in its interior.
2. Put three holes in the triangle:
   • Find the points that divide the sides of the triangle into thirds.

- Form nine new triangles by connecting those points with lines parallel to the sides.
- Form three holes by removing three of these new triangles, as shown in Figure 8.167.

This leaves six filled in triangles, each a smaller version of the original triangle from step 1.

3. Put three holes in each of the six triangles from step 2 by applying the procedure from step 2 to each triangle.

4. Put three holes in each of the triangles from step 3 by applying the procedure from step 2 to each triangle.

5 (and beyond). Put three holes in each of the triangles from the previous step by applying the procedure from step 2 to each triangle.

4. *The square snowflake.* Carefully and accurately complete the first four steps of a square snowflake. To create this shape, use the following steps. (See Figure 8.168.)

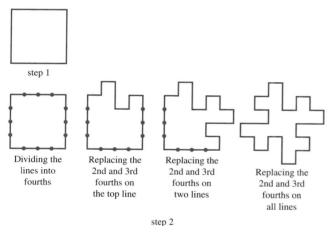

Dividing the lines into fourths | Replacing the 2nd and 3rd fourths on the top line | Replacing the 2nd and 3rd fourths on two lines | Replacing the 2nd and 3rd fourths on all lines

step 2

FIGURE 8.168 Creating a square snowflake.

1. Draw a square.

2. Divide each of the square's straight lines into fourths. In a clockwise order:
   - Leave the first fourth alone.
   - Replace the second fourth with a square that's on the outer side of the line.
   - Replace the third fourth with a square that's on the inner side of the line.
   - Leave the last fourth alone.

   Apply the above procedure to each of the square's straight lines. *Be sure to do the above in a clockwise order!*

3. Apply the procedure from step 2 to each of the straight lines in step 2.

4 (and beyond). Apply the procedure from step 2 to each of the previous step's straight lines.

5. Find the exact dimension of a fractal that has $n = 25$ and $s = 5$. Do not approximate.

6. Find the exact dimension of a fractal that has $n = 5$ and $s = 25$. Do not approximate.

7. Find the exact dimension of a fractal that has $n = 3$ and $s = 27$. Do not approximate.

8. Find the exact dimension of a fractal that has $n = 2$ and $s = 32$. Do not approximate.

9. Find the approximate dimension of the Sierpinski gasket. Round off to the nearest tenth. What does this number say about the fractal?

10. Find the approximate dimension of the Sierpinski carpet from Exercise 2. Round off to the nearest tenth. What does this number say about the fractal?

11. Find the approximate dimension of the Mitsubishi gasket from Exercise 3. Round off to the nearest tenth. What does this number say about the fractal?

12. Find the approximate dimension of the Menger sponge. Round off to the nearest tenth. What does this number say about the fractal?

13. Find the approximate dimension of the square snowflake from Exercise 4. Round off to the nearest tenth. What does this number say about the fractal?

14. Of the assigned problems from 9 through 13, which fractal is the least complex? Which is the most complex? Why?

15. Use a scale factor of 4 to find the dimension of a cube. What do you observe?

16. Use a scale factor of 5 to find the dimension of a square. What do you observe?

17. Use scale factors to find the dimension of a rectangle.

18. Play the chaos game through 40 rolls of a die. Label the points "P_1," "P_2," "P_3," etc. Discuss your results.

Answer the following using complete sentences and your own words.

- **CONCEPTS QUESTIONS**

19. When playing the chaos game, why is it impossible to ever land in the hole formed in step 2 of the Sierpinski gasket process? Why is this an important consideration in seeing that the chaos game yields the Sierpinski gasket?

20. Figure 8.153(d) shows the result of the chaos game after 10,000 rolls. Is this a Sierpinski gasket, or is it the result of one of the steps in forming a Sierpinski gasket or is it merely similar to the Sierpinski gasket or one of the steps in forming the gasket? Explain your answer.

21. *For those who have read Section 2.4: Permutations and Combinations.* Figure 8.155 shows the result of coloring the first 64 rows of Pascal's triangle. Is this a Sierpinski gasket, or is it the result of one of the steps in forming a Sierpinski gasket? Explain your answer. If it is the result of one of the steps in forming a Sierpinski gasket, determine which step.

22. Describe the self-similarity of the given fractal. Also determine whether the self-similarity is exact or approximate.
    a. the Sierpinski gasket
    b. the skewed Sierpinski gasket
    c. the Koch snowflake
    d. the Menger sponge

23. Describe the self-similarity for the fractal described in each of the assigned exercises (see Exercises 2–4). In each case, describe the self-similarity. Also determine whether the self-similarity is exact or approximate.

24. Discuss the recursive nature of:
    a. the Sierpinski gasket
    b. the skewed Sierpinski gasket
    c. the Koch snowflake
    d. the Menger sponge

25. Discuss the recursive nature of the fractal described in each of the assigned exercises (see Exercises 2–4).

26. Name three naturally occurring objects (other than those specifically mentioned in the text) that have self-similarity. Do these objects have exact or approximate self-similarity?

27. Which type of self-similarity is most typically found in nature: exact or approximate? Which type is most typically found in fractal geometry?

28. Name a naturally occurring object that has a recursive nature. Describe its recursive nature.

29. In fractal geometry, self-similarity occurs at ever-smaller scales. That is, parts of a shape appear within itself at infinitely many different scales. Is this the case with the self-similarity found in nature? Why?

30. Describe two applications of fractal dimension.

• HISTORY QUESTIONS

31. Who is known as the father of fractal geometry?

32. What was different about the education of the father of fractal geometry? What impact did this difference have on him?

• PROJECTS

33. *Design your own fractal.* Discuss its self-similarity and its recursive nature. Find its dimension

34. *Create your own coastline.* Draw a line segment. Roll a die. If either a 1, 2, or 3 comes up, apply step 2 of the Koch snowflake procedure to that line segment, with the triangle(s) pointing upwards. If a 4, 5, or 6 comes up, apply step 2 of the Koch snowflake procedure to that line segment, with the triangle(s) pointing downwards. Continue this process until your fractal resembles a coastline. Include in your work the final sketch and a description of exactly how you created that shape. In your description, list each roll of the die and describe how that roll changed your image.

WEB PROJECTS

35. Write an essay on one of the uses fractals by the film industry in creating computer graphics.

36. Investigate some fractal-related web sites and write an essay about your findings.

    Some useful links for these web projects are listed on the text companion web site: Go to **www.cengage.com/math/johnson** to access the web site.

8.10 The Perimeter and Area of a Fractal

OBJECTIVES

- Determining the perimeter and area of a fractal
- Appreciating how some real-world objects' perimeters and areas are like those of fractals

There is something both puzzling and useful about the perimeter and area of a fractal. We're going to find the perimeter of a Koch snowflake. To do so, we'll first find the perimeter of the first few steps of the process of creating a Koch snowflake. Then we'll use that work to find the perimeter of a Koch snowflake.

The Perimeter of the Koch Snowflake

EXAMPLE **1**

A KOCH SNOWFLAKE'S PERIMETER, PART I   Find the perimeter P_1 of the equilateral triangle formed in step 1 of the process of creating a Koch snowflake, if each side is of length 1 foot. (See Figure 8.169.)

SOLUTION

The perimeter is just the sum of the lengths of the sides, so it is

$$P_1 = 1 \text{ ft} + 1 \text{ ft} + 1 \text{ ft} = 3 \text{ ft}$$

Alternatively, there are three sides, each of length 1 foot, so the perimeter is

$$P_1 = (3 \text{ sides}) \cdot (1 \text{ foot/side}) = 3 \text{ ft}$$

(Notice that we are using dimensional analysis, as discussed in Appendix E.) We call this P_1 as a memory device; P stands for perimeter, and the subscript 1 indicates that it is the perimeter of the triangle formed in step one.

FIGURE 8.169

Finding the perimeter P_1.

EXAMPLE **2**

A KOCH SNOWFLAKE'S PERIMETER, PART II   Find the perimeter P_2 of the Star of David that is formed in step 2 of the process of creating a Koch snowflake. (See Figure 8.170.)

SOLUTION

The sides of the new smaller triangles are one third as long as the sides of the larger equilateral triangle from Example 1, so each side is

$$1/3 \cdot 1 \text{ ft} = 1/3 \text{ ft}$$

There are 12 sides, so the perimeter is

$$P_2 = (12 \text{ sides}) \cdot (1/3 \text{ ft/side}) = 4 \text{ ft} \quad \textbf{using dimensional analysis}$$

$\frac{1}{3}$ ft

FIGURE 8.170   Finding the perimeter P_2.

EXAMPLE **3**

A KOCH SNOWFLAKE'S PERIMETER, PART III   Find the perimeter P_3 of the shape that is formed in step 3 of the process of creating a Koch snowflake. (See Figure 8.171.)

SOLUTION

The sides of the new smaller triangles are one-third as long as the sides of the triangle from Example 2, so each side is

$$1/3 \cdot 1/3 \text{ ft} = 1/9 \text{ ft}$$

There are 48 sides (count them!), so the perimeter is

$$P_3 = (48 \text{ sides}) \cdot (1/9 \text{ ft/side})$$
$$= 48/9 \text{ ft} = 16/3 \text{ ft} \quad \textbf{using dimensional analysis}$$

The results of Examples 1, 2, and 3 are summarized in Figure 8.172.

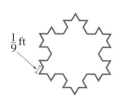

$\frac{1}{9}$ ft

FIGURE 8.171

Finding the perimeter P_3.

Step	Picture	Number of Sides	Length of Each Side	Perimeter
1		3 sides	1 ft/side	$P_1 = 3 \text{ sides} \cdot 1 \text{ ft/side}$ $= 3 \text{ ft}$
2		12 sides	1/3 ft/side	$P_2 = 12 \text{ sides} \cdot 1/3 \text{ ft/side}$ $= 4 \text{ ft}$
3		48 sides	1/9 ft/side	$P_3 = 48 \text{ sides} \cdot 1/9 \text{ ft/side}$ $= 16/3 \text{ ft}$

FIGURE 8.172   Finding the perimeter P.

EXAMPLE **4**

A KOCH SNOWFLAKE'S PERIMETER   Find the perimeter P of the Koch snowflake.

SOLUTION

To find the perimeter P, we have to generalize from the solutions of Examples 1, 2, and 3. There are several patterns to observe.

First notice that the number of sides is increasing by a factor of 4. In Figure 8.172, under "Number of Sides," each entry is 4 times the previous entry. In particular, $4 \cdot 3$ sides = 12 sides and $4 \cdot 12$ sides = 48 sides. This should make sense, because in each step in the Koch snowflake procedure, you replace every _____ with a four-sided __/__.

This means that every single side is replaced with four sides. So naturally, the number of sides increases by a factor of 4. (The word *factor* means "multiplier," so saying that the number of sides increases by a factor of 4 means that the number of sides is multiplied by 4.)

The second pattern to observe is that the length of each side is decreasing by a factor of 1/3. In Figure 8.172, under "Length of Each Side," each entry is 1/3 times the previous entry. In particular, $1/3 \cdot 1$ ft = 1/3 ft and $1/3 \cdot 1/3$ ft = 1/9 ft. This too should make sense, because each step in the Koch snowflake procedure makes each new side 1/3 as long as each side from the previous step.

The last pattern to observe is that the perimeter is increasing by a factor of 4/3. In Figure 8.172, under "Perimeter," each entry is 4/3 times the previous entry. In particular, $4/3 \cdot 3$ ft = 4 ft and $4/3 \cdot 4$ ft = 16/3 ft. This should also make sense, because there are four times as many sides, and each side is 1/3 as long, so the perimeter increases by a factor of $4 \cdot 1/3 = 4/3$.

These observations are summarized and extended in Figure 8.173.

Notice that for each step number, the perimeter is 3 times 4/3 to a power, where the power is 1 less than the step number. In other words, if we let n be the step number, then the perimeter at step n is

$$P_n = 3 \cdot (4/3)^{n-1}$$

Let's check this formula by using it to find P_5:

$$P_5 = 3 \cdot (4/3)^{5-1} = 3 \cdot (4/3)^4$$

This checks with the last entry in Figure 8.173.

The Koch snowflake is the result of continuing the process illustrated in Figure 8.173 indefinitely. That means that the perimeter of the Koch snowflake is the result of multiplying previous perimeters by 4/3, indefinitely. At step 100, the perimeter is huge:

$$P_{100} = 3 \cdot (4/3)^{100-1} = 3 \cdot (4/3)^{99} \text{ ft}$$

With a calculator, we find that this is approximately 7×10^{12} ft = 7,000,000,000,000 ft. At step 1,000, the perimeter is incredibly huge:

$$P_{1000} = 3 \cdot (4/3)^{1000-1} = 3 \cdot (4/3)^{999} \text{ ft}$$

On the calculator, this is approximately 2×10^{125} ft. (Remember, 2×10^{125} is a 2 followed by 125 zeros!)

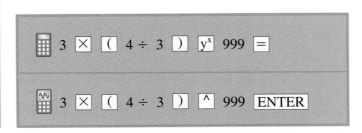

Step	Picture	Number of Sides	Length of Each Side	Perimeter
1	*(equilateral triangle, sides 1 ft, base 1 ft)*	3 sides	1 ft/side	$P_1 = 3$ ft
2	*(six-pointed star, $\frac{1}{3}$ ft)*	$4 \cdot 3$ sides $= 12$ sides	$1/3 \cdot 1$ ft/side $= 1/3$ ft/side	$P_2 = 4/3 \cdot P_1$ $= 4/3 \cdot 3$ ft $= 4$ ft
3	*(snowflake shape, $\frac{1}{9}$ ft)*	$4 \cdot 12$ sides $= 48$ sides	$1/3 \cdot 1/3$ ft/side $= 1/9$ ft/side	$P_3 = 4/3 \cdot P_2$ $= 4/3 \cdot 4/3 \cdot 3$ ft $= (4/3)^2 \cdot 3$ ft ≈ 5.3 ft
4		$4 \cdot 48$ sides $= 192$ sides	$1/3 \cdot 1/9$ ft/side $= 1/27$ ft/side	$P_4 = 4/3 \cdot P_3$ $= 4/3 \cdot (4/3)^2 \cdot 3$ ft $= (4/3)^3 \cdot 3$ ft ≈ 7.1 ft
5		$4 \cdot 192$ sides $= 768$ sides	$1/3 \cdot 1/27$ ft/side $= 1/81$ ft/side	$P_5 = 4/3 \cdot P_4$ $= 4/3 \cdot (4/3)^3 \cdot 3$ ft $= (4/3)^4 \cdot 3$ ft ≈ 9.5 ft

FIGURE 8.173   Continuing to find the perimeter P.

We must conclude that the perimeter of the Koch snowflake is infinite! This is hard to make sense of, because the perimeter just doesn't look that long. But remember, after several steps, it becomes impossible to visually tell the difference between the result of one step and the result of the next step. So you can't visually tell the difference between the result of step 5 (which has perimeter $P_5 \approx 9.48$ ft) and a Koch snowflake (which has infinite perimeter). Most of the perimeter is in all the little tiny crinkles.

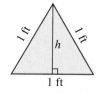

FIGURE 8.174

Finding the area A_1.

The Area of the Koch Snowflake

EXAMPLE **5**   **A KOCH SNOWFLAKE'S AREA, PART I**   Find the area A_1 of the equilateral triangle formed in step 1 of the process of creating a Koch snowflake, if each side is of length 1 foot. (See Figure 8.174.)

SOLUTION   To find the area, we need the base and the height. It's easy to see that the base is 1 foot, but we have to calculate the height h. The height h splits the equilateral triangle into two identical triangles, one on the each side of the height. Because the two triangles are identical, their bases are equal, so each must be 1/2 ft. We can

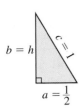

FIGURE 8.175
Finding the height h.

apply the Pythagorean Theorem to either of these triangles, because they are right triangles. (See Figure 8.175.)

$$a^2 + b^2 = c^2 \qquad \text{\textbf{Pythagorean Theorem}}$$
$$(1/2)^2 + h^2 = 1^2 \qquad \text{\textbf{substituting}}$$
$$1/4 + h^2 = 1$$
$$h^2 = 1 - 1/4 = 3/4$$
$$h = \sqrt{\frac{3}{4}} = \frac{\sqrt{3}}{\sqrt{4}} = \frac{\sqrt{3}}{2} \text{ ft}$$

This is the height of the right triangle, but it is also the height of the triangle formed in step 1 of the Koch snowflake process. This means that the area of that triangle is

$$A_1 = \frac{1}{2}bh = \frac{1}{2} \cdot 1 \text{ ft} \cdot \frac{\sqrt{3}}{2} \text{ ft} = \frac{\sqrt{3}}{4} \text{ sq ft}$$

EXAMPLE **6**

A KOCH SNOWFLAKE'S AREA, PART II   Find the area A_2 of the Star of David formed in step 2 of the process of creating a Koch snowflake. (See Figure 8.176.)

SOLUTION

The area is the sum of the area of the original equilateral triangle plus the areas of the three new smaller triangles. The original equilateral triangle's area was found in Example 5. It is

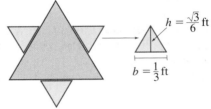

$$A_1 = \frac{\sqrt{3}}{4} \text{ sq ft} \approx 0.43 \text{ sq ft}$$

FIGURE 8.176   Finding the area A_2.

The new smaller triangles are identical, so they have the same area. In step 2, all sides are $1/3$ of what they were in step 1, so the base is $b = 1/3 \cdot 1 \text{ ft} = 1/3 \text{ ft}$, and the height is $h = 1/3 \cdot \sqrt{3}/2 = \sqrt{3}/6 \text{ ft}$. This means that the area of each smaller triangle is

$$A = \frac{1}{2}bh = \frac{1}{2} \cdot \frac{1}{3} \text{ ft} \cdot \frac{\sqrt{3}}{6} \text{ ft} = \frac{\sqrt{3}}{36} \text{ sq ft}$$

There are three of these new smaller triangles. Thus, the area A_2 of the Star of David is

$$A_2 = \text{(area of original equilateral triangle)} + 3 \cdot \text{(area of new smaller triangle)}$$
$$= \frac{\sqrt{3}}{4} + 3 \cdot \frac{\sqrt{3}}{36} \text{ sq ft}$$
$$= \frac{\sqrt{3}}{4} + \frac{\sqrt{3}}{12} \text{ sq ft} \qquad \text{\textbf{canceling}}$$
$$\approx 0.58 \text{ sq ft}$$

EXAMPLE **7**

A KOCH SNOWFLAKE'S AREA, PART III   Find the area A_3 of the shape formed in step 3 of the process of creating a Koch snowflake. (See Figure 8.177.)

SOLUTION

The area is the sum of the area of the Star of David from Example 6 plus the areas of the new smaller triangles. The Star of David's area is

$$A_2 = \frac{\sqrt{3}}{4} + \frac{\sqrt{3}}{12} \text{ sq ft}$$

The new smaller triangles are identical, so they have the same area. The base is $b = 1/3 \cdot 1/3$ ft $= 1/9$ ft, and the height is $h = 1/3 \cdot \sqrt{3}/6 = \sqrt{3}/18$ ft. This means that the area of each smaller triangle is

$$A = \frac{1}{2}bh = \frac{1}{2} \cdot \frac{1}{9} \text{ ft} \cdot \frac{\sqrt{3}}{18} \text{ ft} = \frac{\sqrt{3}}{324} \text{ sq ft}$$

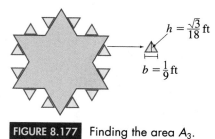

FIGURE 8.177 Finding the area A_3.

Also, there are 12 new smaller triangles (count them!). Thus, the area A_3 is

A_3 = (area of Star of David) + 12 · (area of new smaller triangle)

$$= \frac{\sqrt{3}}{4} + \frac{\sqrt{3}}{12} + 12 \cdot \frac{\sqrt{3}}{324} \text{ sq ft}$$

$$= \frac{\sqrt{3}}{4} + \frac{\sqrt{3}}{12} + \frac{\sqrt{3}}{27} \text{ sq ft} \qquad \textbf{canceling}$$

$$\approx 0.64 \text{ sq ft}$$

The areas are as follows:

- $A_1 \approx 0.43$ sq ft
- $A_2 \approx 0.58$ sq ft
- $A_3 \approx 0.64$ sq ft

While these areas are increasing, they don't seem to be getting that big, as the perimeters did in Example 4.

Notice that the result of each step of the process of creating a Koch snowflake can be completely enclosed in a square that is 2 feet on each side, with plenty of room left over, as shown in Figure 8.178. This means that the area of the Koch snowflake is significantly less than the area of the square, which is 4 square feet. It is not infinite.

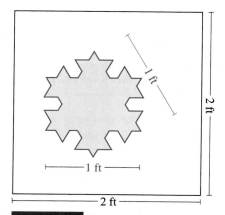

FIGURE 8.178 Approximating the area A of a Koch snowflake.

While there are a number of patterns that we could observe, these patterns lead to a more complicated result than do the perimeter patterns discussed in Example 4. Utilizing these patterns to compute the exact area of the Koch snowflake is beyond the scope of this text. In Exercise 11, we will use Figure 8.177 to approximate the area.

The Perimeter and Area of a Fractal

At the beginning of this section, we said, "There is something both *puzzling* and *useful* about the perimeter and area of a fractal." Now that we've discussed the perimeter and area of the Koch snowflake, we can discuss the puzzling and useful aspect of this topic.

The Koch snowflake has an infinite perimeter but an area that is significantly less than 4 square feet. This means that the Koch snowflake fits an infinitely long curve (the perimeter) inside a small space. The Sierpinski gasket also has an infinite perimeter but an area of zero! (The area is zero because of all of the holes—they just take over the Sierpinski gasket.) So the Sierpinski gasket also fits an infinitely long

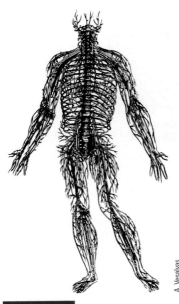

A. Vesalvas

FIGURE 8.179

The human body's system of veins, arteries and capillaries.

curve inside a small space. The Menger sponge has an infinite surface area but a volume of zero! This means that the Menger sponge fits an infinitely large surface inside a small space.

An Application of Fractals

It is certainly puzzling that a Koch snowflake has an infinite perimeter but a small area, and it's certainly puzzling that a Menger sponge has an infinite surface area but a volume of zero. But why are these facts useful?

Consider the human body's system of veins, arteries, and capillaries (see Figure 8.179). This system takes up a small part of the body's space, but its length is incredibly large; laid end to end, the system of veins and arteries would be more than 40,000 miles long. Like a Koch snowflake, the length is huge, but it fits inside a relatively small space.

Also, consider the human kidney. Its surface area is practically infinite (and for good reason, since the kidney's surface area determines how much blood it can cleanse), but its volume is small—the kidney is only 4 inches long.

Scientists use fractal geometry to analyze these various parts of the body. It may well be that fractal geometry will help us better understand the functioning of the human body.

8.10 EXERCISES

1. *Finding the perimeter of a Sierpinski gasket.* (By "perimeter," we mean the total distance around all of the filled in regions, not the distance around the outside only.)

   a. Use your drawings from Exercise 1 of Section 8.9 to explain each of the calculations in the chart in Figure 8.180, where "triangle" refers to a filled-in triangle, not a triangular hole.

Step	Number of Triangles	Length of Each Side	Perimeter of One Triangle	Total Perimeter of All Triangles
1	1	1 ft	$3 \cdot 1$ ft = 3 ft	$1 \cdot 3$ ft = 3 ft
2	3*	$1/2 \cdot 1$ ft = 1/2 ft	$3 \cdot 1/2$ ft = 3/2 ft	$3 \cdot 3/2$ ft = 9/2 ft
3				
4				
5				
6				

   *3 not 4, because only three triangles are filled in.

   FIGURE 8.180 For Exercise 1.

   b. Use your drawings from Exercise 1 of Section 8.9 to complete the chart in Figure 8.180. Include units.

   c. By what factor does the number of triangles increase? Explain why your answer makes sense.

   d. By what factor does the length of each side decrease? Explain why your answer makes sense.

   e. By what factor does the perimeter of one triangle change? Explain why your answer makes sense. Is the perimeter of one triangle increasing or decreasing? Why?

   f. By what factor does the total perimeter of all triangles change? Explain why your answer makes sense. Is the total perimeter of one triangle increasing or decreasing? Why?

   g. Use your answers to parts (a)–(f) to find a formula for the total perimeter of all triangles at step n.

   h. Use your answer to part (g) and your calculator to find the perimeter of the Sierpinski gasket. Explain your work. (By "perimeter," we mean the total distance around all of the filled in regions, not the distance around the outside only.)

2. *Finding the area of a Sierpinski gasket.* (By "area," we mean the total area of all the filled in regions.)

   a. Use your drawings in Exercise 1 of Section 8.9 to complete the chart in Figure 8.181, where "triangle" refers to a filled in triangle, not to a triangular hole. Include units.

   b. By what factor does the number of triangles increase? Explain why your answer makes sense.

   c. By what factor does the length of the base decrease? Explain why your answer makes sense.

Step	Number of Triangles	Length of the Base	Length of the Height	Area of one Triangle	Total area of all Triangles
1	1				
2	3*				
3					
4					
5					
6					

*3 not 4, because only three triangles are filled in.

FIGURE 8.181 For Exercise 2.

Step	Number of New Squares*	Length of Each Side	Perimeter of One New Square*	Total Perimeter of All New Squares*	Total Perimeter of All Squares*
1	1	1 ft	4 · 1 ft = 4 ft	1 · 4 ft = 4 ft	4 ft
2	1	1/3 · 1 ft = 1/3 ft	4 · 1/3 ft = 4/3 ft	1 · 4/3 ft = 4/3 ft	4 ft + 4/3 ft
3	8				
4					
5					
6					

*"New square" refers to the squares that are both new to this step and that contribute to the total perimeter. "All squares" refers to the squares that are new to this step as well as those from previous steps.

FIGURE 8.182 For Exercise 3.

d. By what factor does the length of the height decrease? Explain why your answer makes sense.

e. By what factor does the area of each triangle decrease? Explain why your answer makes sense.

f. By what factor does the total area of all triangles change? Explain why your answer makes sense. Is the total area increasing or decreasing? Why?

g. Use your answers to parts (a)–(f) to find a formula for the total area of all triangles at step n.

h. Use your answer to part (g) and your calculator to find the area of the Sierpinski gasket. (By "area," we mean the total area of all of the filled in regions.) Explain your work.

3. *Finding the perimeter of a Sierpinski carpet.* (See Exercise 2 in Section 8.9 for a description of this fractal.) (By "perimeter," we mean the total distance around all of the filled in regions.)

a. Use your drawings in Exercise 2 of Section 8.9 to explain each of the calculations in the chart in Figure 8.182. These calculations are illustrated in Figure 8.183.

b. Use your drawings from Exercise 2 of Section 8.9 to complete the chart in Figure 8.182. Include units.

step 1: Each side is of length 1 ft.

step 2: There is one new square. Each of its sides is of length $\frac{1}{3}$ ft.

step 3: There are eight new squares. What is the length of each of their sides?

Finding the perimeter and area of a Sierpinski carpet.

FIGURE 8.183 For Exercise 3.

Step	Number of New Squares*	Length of Each Side	Area of One New Square	Total Area of All New Squares*	Total area after Removing All New Squares*
1	1	1 ft	$(1 \text{ ft})^2$ $= 1$ sq ft	$1 \cdot 1$ sq ft $= 1$ sq ft	1 sq ft $- 0$ $= 1$ sq ft (Nothing is removed in step 1.)
2	1	$1/3 \cdot 1$ ft $= 1/3$ ft	$(1/3 \text{ ft})^2$ $= 1/9$ sq ft	$1 \cdot 1/9$ sq ft $= 1/9$ sq ft	1 sq ft $- 1/9$ sq ft $= 8/9$ sq ft
3	8				
4					
5					
6					

*"New square" refers to the squares that are both new to this step and that contribute to the area.

FIGURE 8.184 For Exercise 4.

c. By what factor does the number of squares increase (starting at step 2)? Explain why your answer makes sense.

d. By what factor does the length of each side decrease? Explain why your answer makes sense.

e. By what factor does the perimeter of one new square change? Explain why your answer makes sense. Is the perimeter of one new square increasing or decreasing? Why?

f. By what factor does the total perimeter of all new squares change? Explain why your answer makes sense. Is the total perimeter of all new squares increasing or decreasing? Why?

g. Use your answers to parts (a)–(f) to find a formula for the total perimeter of all new squares at step n.

h. Use your answer to part (g) and your calculator to determine what happens to the total perimeter of all new squares as we move from step to step.

i. Use your answer to part (h) to find the perimeter of the Sierpinski carpet. Explain your work. (By "perimeter," we mean the total distance around all of the filled in regions, not the distance around the outside only.)

HINT: How must the total perimeter of all squares compare with the total perimeter of all new squares?

4. *Finding the area of a Sierpinski carpet.* (See Exercise 2 in Section 8.9 for a description of this fractal.) (By "area," we mean the total area of all the filled in regions.)

a. Use your drawings in Exercise 2 of Section 8.9 to explain each of the calculations in the chart in Figure 8.184. These calculations are illustrated in Figure 8.183.

b. Use your drawings from Exercise 2 of Section 8.9 to complete the chart in Figure 8.184.

c. By what factor does the number of squares increase (starting at step 2)? Explain why your answer makes sense.

d. By what factor does the length of each side decrease? Explain why your answer makes sense.

e. By what factor does the area of one square change? Explain why your answer makes sense. Is the area of one square increasing or decreasing? Why?

f. By what factor does the total area of all new squares change? Explain why your answer makes sense. Is the total area of all new squares increasing or decreasing? Why?

g. By what factor is the total area after removing all new squares changing? Is the total area after removing all new squares increasing or decreasing? Why?

h. Use your answers to parts (a)–(g) to find a formula for the total area after removing all new squares at step n.

i. Use your answer to part (h) and your calculator to find the area of the Sierpinski carpet. Explain your answer.

5. *The Mitsubishi Gasket.* (See Exercise 3 in Section 8.9 for a description of this fractal.)

a. Find a formula for the total perimeter of step n of the process described in Exercise 3, Section 8.9.

b. Use the formula from part (a) to find the perimeter of the Mitsubishi gasket.

c. Find a formula for the total area of step n of the above process.

Step	Number of New Triangles	Base of New Triangle	Height of New Triangle	Area of each New Triangle	Area of all New Triangles	Total Area
1	1	1 ft	$\frac{\sqrt{3}}{2}$ ft	$A_1 = \frac{1}{2}\cdot 1 \cdot \frac{\sqrt{3}}{2}$ $= \frac{\sqrt{3}}{4}$ sq ft	$1\cdot\frac{\sqrt{3}}{4}$ sq ft $= \frac{\sqrt{3}}{4}$ sq ft	$A_1 = \frac{\sqrt{3}}{4}$ sq ft
2	3	$\frac{1}{3}\cdot 1$ ft $= \frac{1}{3}$ ft	$\frac{1}{3}\cdot\frac{\sqrt{3}}{2}$ ft $= \frac{\sqrt{3}}{6}$ ft	$\frac{1}{2}\cdot\frac{1}{3}\cdot\frac{\sqrt{3}}{6}$ $= \frac{\sqrt{3}}{36}$ sq ft	$3\cdot\frac{\sqrt{3}}{36}$ sq ft $= \frac{\sqrt{3}}{12}$ sq ft	$A_2 = \frac{\sqrt{3}}{4}$ $+ \frac{\sqrt{3}}{12}$ sq ft
3						

FIGURE 8.185 For Exercise 11.

d. Use the formula from part (c) to find the area of the Mitsubishi gasket.

6. *A square snowflake.* (See Exercise 4 in Section 8.9 for a description of this fractal.)

   a. Find a formula for the total perimeter of step n of the process described in Exercise 4, Section 8.9.

   b. Use the formula from part (a) to find the perimeter of a square snowflake.

   c. Find a formula for the total area of step n of the above process.

   HINT: It's an incredibly easy formula. Don't make it hard.

   d. Use the formula from part (c) to find the area of a square snowflake.

7. *For those who have read Section 1.1: Deductive versus Inductive Reasoning.* In Example 4, we discussed the perimeter of the results of the first few steps of the process of creating a Koch snowflake, and we concluded that $P_4 = 64/9$ and $P_5 = 256/27$. In arriving at these conclusions, some of the reasoning was inductive and some was deductive. Which reasoning was inductive? Which was deductive? Why?

8. *For those who have read Section 8.5: Right Triangle Trigonometry.* In part (b) of Example 1, we found the base and height of a triangle. Discuss how to use the information developed in Section 8.5, under *Special Triangles,* to find this base and height.

Answer the questions 9 and 10 using complete sentences and your own words.

• CONCEPT QUESTIONS

9. Describe an application of fractal area, perimeter, and volume.

• PROJECTS

10. *Design your own fractal.* Discuss its self-similarity and its recursive nature. Find its perimeter and area (or its surface area and volume, if appropriate).

11. *Approximating the area of a Koch snowflake.*

   a. Use Example 7 to complete the chart in Figure 8.185. Include units.

   b. Expand the chart to include steps 4–8.

   c. Use your calculator and the "Total Area" entries in the chart you completed in part (b) to compute decimal approximations of A_1 through A_8. Round off to four decimal places.

   d. So far, we haven't found the area of the Koch snowflake. Instead, we've found the areas of the first few steps of creating a Koch snowflake. Would the area of the Koch snowflake be larger or smaller than A_8? Why? Use the result of part (c) to estimate the area of the Koch snowflake.

8 CHAPTER REVIEW

TERMS

Albertian grid
analytic geometry
angle of depression

angle of elevation
approximate self-
   similarity
area
astronomical unit (AU)

center
chaos game
circle
circumference
circumscribed polygon

cone
congruent triangles
conic section
cosine
cylinder

degree	hypotenuse	parsec (pc)	similar triangles
diameter	inscribed polygon	perimeter	sine
dimension	Koch snowflake	π (pi)	skewed Sierpinski gasket
disk	leg	polygon	sphere
ellipse	light-year (ly)	pyramid	surface area
exact self-similarity	linear perspective	Pythagorean Theorem	tangent
foci	line of symmetry	radius	trapezoid
focus	Menger sponge	rectangle	triangle
foreshortening	minutes	recursive process	trigonometry
fractal	parabola	right triangle	vanishing point
great circle	parallel lines	seconds	vertex
height	Parallel Postulate	self-similarity	volume
hyperbola	parallelogram	Sierpinski gasket	

REVIEW EXERCISES

1. Find the area and perimeter of Figure 8.186.

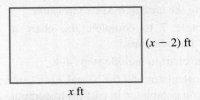

FIGURE 8.186 Diagram for Exercise 1.

2. Find the area and perimeter of Figure 8.187.

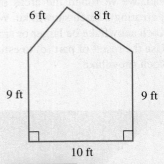

FIGURE 8.187 Diagram for Exercise 2.

3. Find the area and perimeter of Figure 8.188.

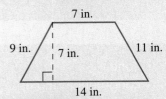

FIGURE 8.188 Diagram for Exercise 3.

4. Find the area and perimeter of Figure 8.189.

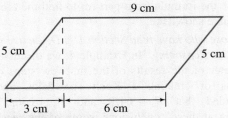

FIGURE 8.189 Diagram for Exercise 4.

5. Find the area and perimeter of Figure 8.190.

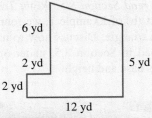

FIGURE 8.190 Diagram for Exercise 5.

6. Saturn (the ringed planet) revolves around the sun in an elliptical orbit. Consequently, the distance between Saturn and the sun varies from approximately 840.4 million to 941.1 million miles.

   a. What is the (approximate) average distance from Saturn to the sun?

   b. On average, how far from the sun is Saturn in AU?

   c. Interpret the answer to part (b) in terms of the earth's distance from the sun.

7. Scientists believe that the star Vega (the brightest star in the constellation Lyra) is approximately 239.7 trillion kilometers from the earth.

a. How far, in astronomical units, is Vega from the earth?

b. How far, in light-years, is Vega from the earth?

c. How far, in parsecs, is Vega from the earth?

8. Find the volume of Figure 8.191.

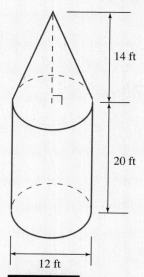

14 ft

20 ft

12 ft

FIGURE 8.191 Diagram for Exercise 8.

9. Find the volume and surface area of Figure 8.192.

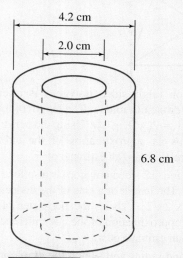

4.2 cm

2.0 cm

6.8 cm

FIGURE 8.192 Diagram for Exercise 9.

10. Find the volume and surface area of Figure 8.193.

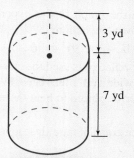

3 yd

7 yd

FIGURE 8.193 Diagram for Exercise 10.

11. Find the volume and surface area of Figure 8.194.

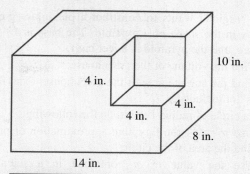

10 in.

4 in.

4 in.

4 in.

8 in.

14 in.

FIGURE 8.194 Diagram for Exercise 11.

12. You jog $1\frac{3}{4}$ miles due south, then jog $\frac{1}{2}$ mile due west, and then return to your starting point via a straight-line path. How many miles have you run?

13. Tony Endres wants to fertilize his backyard. The yard is a triangle whose sides are 100 feet, 130 feet, and 160 feet. If one bag of fertilizer will cover 800 square feet, how many bags does he need?

14. An oval athletic field is the combination of a square and semicircles at opposite ends, as shown in Figure 8.195. If the total area of the field is 1,200 square yards, find the area of the square.

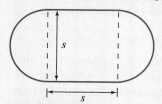

s

s

FIGURE 8.195 Diagram for Exercise 14.

15. From a 12-inch-by-18-inch piece of cardboard, 2-inch-square corners are cut out as shown in Figure 8.196, and the resulting flaps are folded up to form an open box. Find (a) the volume and (b) the external surface area of the box.

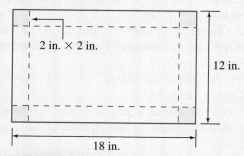

2 in. × 2 in.

12 in.

18 in.

FIGURE 8.196 Diagram for Exercise 15.

16. A regulation baseball (hardball) has a circumference of 9 inches. A regulation tennis ball has a diameter of $2\frac{1}{2}$ inches.

a. Find and compare the volumes of the two types of balls.

b. Find and compare the surface areas of the two types of balls.

17. Ted Nirgiotis wants to construct a plexiglass greenhouse in the shape of a pyramid. The base is 13 feet square, and the pyramid is 8 feet high.
    a. Find the volume of the greenhouse.
    b. Find the surface area of the greenhouse. (The floor is not included.)

18. For a circle of radius 1 cubit, do the following:
    a. Use $\pi = \frac{256}{81}$ (the Egyptian approximation of pi) to find the area of the circle.
    b. Use the value of π contained in a scientific calculator to find the area of the circle.
    c. Find the error of the Egyptian calculation relative to the calculator value.

19. For a sphere of radius 1 cubit, do the following.
    a. Use $\pi = \frac{256}{81}$ (the Egyptian approximation of pi) to find the volume of the sphere.
    b. Use the value of π contained in a scientific calculator to find the volume of the sphere.
    c. Find the error of the Egyptian calculation relative to the calculator value.

20. Which of the following structures has the larger storage capacity?
    a. a truncated pyramid 30 cubits high with a 6-cubit-by-6-cubit top and a 40-cubit-by-40-cubit bottom
    b. a regular pyramid 50 cubits tall with a 40-cubit-by-40-cubit base

21. Following the method of Problem 48 of the Rhind Papyrus, find a square, with a whole number as the length of its side, that has the same (approximate) area as a circle of diameter 18 units.
    HINT: Divide the sides of an 18-by-18 square into three equal segments.

22. Use the result of Exercise 21 to obtain an approximation of π.

23. Find the length of the longest object that will fit inside a rectangular trunk 5 feet long, 3.5 feet wide, and 2.5 feet high.

24. Find the length of the longest object that will fit inside a garbage can (a cylinder) that has a diameter of 2.5 feet and a height of 4 feet.

25. Use the diagrams in Figure 8.197 to prove the Pythagorean Theorem.
    HINT: Use the method of rearrangement and label the sides of each square in the right-hand diagram in Figure 8.197(b).

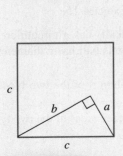
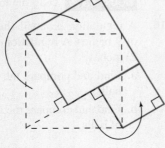

FIGURE 8.197 Diagram for Exercise 25.

In Exercises 26 and 27, use the two-column method to prove the indicated result.

26. Given: $AD = CD$
    $\qquad\quad AB = CB$
    Prove: $AE = CE$

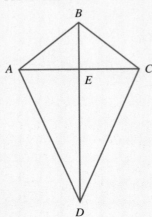

27. Given: $\angle CBA = \angle DAB$
    $\qquad\quad BC = AD$
    Prove: $\angle CAD = \angle DBC$

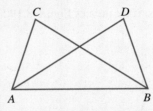

28. A regular octagon (eight-sided polygon) is circumscribed around a circle of radius r. The length of a side of the octagon is $s = 2(\sqrt{2} - 1)r$. Using the perimeter of the octagon as an approximation of the circumference of the circle, obtain an estimate of π.

29. A regular dodecagon (12-sided polygon) is inscribed in a circle of radius r. The length of a side of the dodecagon is $s = [(\sqrt{6} - \sqrt{2})/2]r$. Using the perimeter of the dodecagon as an approximation of the circumference of the circle, obtain an estimate of π.

30. Find the center and radius and sketch the graph of the circle given by
    $$x^2 + y^2 - 2x - 2y + 1 = 0$$

31. Find the foci and sketch the graph of the hyperbola given by each of the following equations.
    a. $9x^2 - 4y^2 = 36$ b. $4y^2 - 9x^2 = 36$

32. A Mylar-coated parabolic reflector is used to solar-heat water. The disk is 8 feet wide and 2 feet deep. Where should the water container be placed to heat the water most rapidly?

33. Find the foci and sketch the graph of the ellipse given by
    $$9x^2 + 4y^2 = 36$$

34. An elliptical room is designed to function as a whispering gallery. If the room is 40 feet long and 20 feet wide, where should two people stand to optimize the whispering effect?

In Exercises 35 and 36, draw and label a right triangle like the one shown in Figure 8.198, and use trigonometric ratios and a calculator to find the unknown sides and angles. Round off your answers to one decimal place.

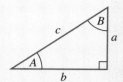

FIGURE 8.198 Diagram for Exercises 35 and 36.

35. $A = 25.4°$ and $c = 56.1$ ft.

36. $a = 45.2$ cm and $b = 67.8$ cm

37. An electrician is at the top of a 50.0-foot utility pole. She measures the angle of depression of the base of another utility pole on the opposite side of a river. If the angle is 13.2°, how far apart are the poles?

38. While sightseeing in Paris, you visit the Eiffel Tower. From an unknown distance, you measure the angle of elevation of the top of the tower. You then move 120 feet backwards (directly away from the tower) and measure the angle of elevation of the top of the tower. If the angles are 79.7° and 73.1°, respectively, how tall (to the nearest foot) is the Eiffel tower?

39. Calculate the distance (in parsecs) from the sun to an object in space if the angle between the sun and the earth (as seen from the object) is 3 seconds and the distance between the earth and the sun is 1 AU.

40. Calculate the distance (in parsecs) from the sun to an object in space if the angle between the sun and the earth (as seen from the object) is $\frac{2}{3}$ second and the distance between the earth and the sun is 1 AU.

41. Suppose the distance from the sun to an object in space is 0.85 parsec and the distance between the earth and the sun is 1 AU.

   a. How far from the sun is the object in terms of AU?

   b. Calculate the angle between the sun and the earth, as seen from the object. Express your answer in terms of seconds, rounded off to the nearest hundredth of a second.

42. **a.** What is the Parallel Postulate?

   b. What are the two alternatives if we assume the Parallel Postulate to be false?

   c. What geometric models can be used to describe a geometry based on each of the alternatives listed in part (b)?

   d. Draw a triangle using each of the geometric models listed in part (c).

In Exercises 43 and 44, answer each question for (a) Euclidean, (b) Lobachevskian, and (c) Riemannian geometry.

43. Through a point not on a given line, how many parallels to the given line are there?

44. What can be said about the sum of the angles of a triangle?

45. *The box fractal.* To create this shape, use the following steps:

   1. Draw a square and fill in its interior.

   2. Remove some of the square:

     • Subdivide the square into nine equal squares.

     • Remove the four small squares at the middle of each edge.

This leaves five filled in squares, one at each corner and one at the very center. See Figure 8.199.

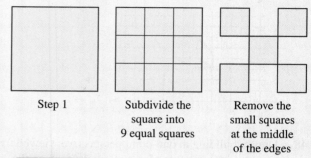

Step 1   Subdivide the square into 9 equal squares   Remove the small squares at the middle of the edges

FIGURE 8.199 Creating a box fractal.

   3. Remove some of the five squares from step 2 by applying the procedure from step 2 to each filled in square.

   4. Remove some of the squares from step 3 by applying the procedure from step 2 to each filled in square.

   5 (and beyond). Remove some of the five squares from the previous step by applying the procedure from step 2 to each filled in square.

   a. Carefully and accurately complete the first five steps of the box fractal.

   b. Find the total perimeter of steps 1–5 of the above process.

   c. Find a formula for the total perimeter of step n of the above process.

   d. Use the formula from part (c) to find the perimeter of the box fractal.

   e. Find a formula for the total area of step n of the above process.

   f. Use the formula from part (e) to find the area of the box fractal.

   g. Find the dimension of the box fractal.

   h. Discuss the self-similarity of the box fractal.

   i. Discuss the recursive nature of the box fractal.

In Exercises 46 and 47, answer the following questions:

a. Is the box drawn in one-point perspective or two-point perspective?

b. Is the box viewed from above or below?

c. At what approximate ordered pair is/are the vanishing point(s)?

d. What is the equation of the horizon line?

46.

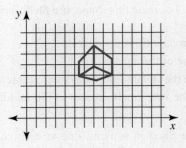

47.

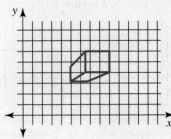

48. Draw a building in one-point perspective, viewed from above, with two rows of two windows each on the side wall and a rectangular skylight on the roof.

49. Draw a building in two-point perspective, viewed from above, with two rows of two windows each on each side wall.

50. Draw an Albertian grid that is four tiles wide and six tiles deep.

In Exercises 51 and 52, answer the following questions:

a. Does the work use one-point perspective, two-point perspective, or three-point perspective? Justify your answer.

b. Describe the location of the vanishing point(s). Discuss why the painter might have chosen that/those location(s).

c. Sometimes a small feature of the painting has its own vanishing point. Does this happen in this painting? If so, what has its own vanishing point? Where is the vanishing point?

d. Sometimes two different main parts of a painting have separate vanishing points. Does this happen in this painting? If so, describe what has separate vanishing points. Where are the vanishing points? Discuss why the painter might have chosen to have multiple vanishing points.

e. Does the painting use foreshortening? If so, describe where and how.

f. Does the painting have pavement? If so, describe where.

g. Sometimes a painting uses an Albertian grid on the ceiling or the walls. Does this happen in this painting? If so, where?

h. Is the painting's horizon line an actual part of the painting? If so, where?

51.

Jan Vermeer, *The Music Lesson*, 1665.

52.

Edward Hopper, *The Lighthouse at Two Lights*, 1929.

 Answer the following questions using complete sentences and your own words.

• HISTORY QUESTIONS

53. What roles did the following people play in the development and application of geometry?
- Archimedes of Syracuse
- Euclid of Alexandria
- Hypatia
- Johann Kepler
- Pythagoras of Samos
- Girolamo Saccheri
- Janos Bolyai
- Carl Friedrich Gauss
- Nikolai Lobachevsky
- Bernhard Riemann
- Thales of Miletus
- Benoit Mandelbrot

GRAPH THEORY

9

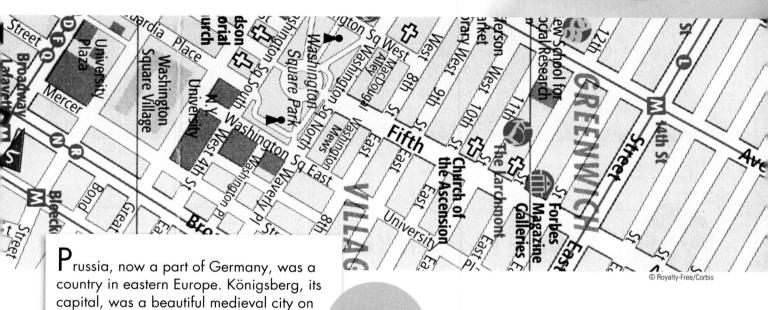

© Royalty-Free/Corbis

Prussia, now a part of Germany, was a country in eastern Europe. Königsberg, its capital, was a beautiful medieval city on the Pregel River. The Prussian king lived there in a castle.

Königsberg's residents often strolled through their city, enjoying its beauty. These strolls usually involved crossing several of the seven bridges over the Pregel River. The walkers tried to find a route that allowed them to cross each of seven bridges once. They never succeeded.

Leonhard Euler, a prominent eighteenth-century mathematician, got involved in this quest. He discovered that the residents' lack of success wasn't due to bad luck or a lack of ingenuity. Instead, he logically proved that such a walk was impossible. In doing this, Euler founded graph theory.

In this chapter, we will use graph theory in several interesting contexts.

WHAT WE WILL DO IN THIS CHAPTER

WE WILL FIND SCENIC ROUTES:

- We'll find a way to island-hop through the Puget Sound in the Pacific Northwest.
- We'll find a way to tour Manhattan and its neighboring boroughs in New York City.
- We'll find a way to tour the San Francisco Bay Area on BART, its rapid transit system.

WE WILL SCHEDULE MULTIPLE-LEG AIRPLANE TRIPS:

- Have you ever taken an airplane trip that involved flying between several different cities? If so, you know that it can be surprisingly difficult to schedule such a trip without crisscrossing the country and paying exorbitant airplane charges. We'll use graph theory to create an efficient schedule for such a trip.

WE WILL DISCUSS HOW TO SET UP A SECURE COMPUTER NETWORK:

- Do you use wi-fi networks at coffee shops, on campus, or in airports? When you log onto a public wi-fi network, others can easily access personal information such as a user name, password, or a

continued

credit card number. That's over the *official* network, and at some airports, there are rogue networks waiting to ensnare the unsuspecting.

- Do you have a wi-fi network at your home? When you installed it, you probably turned it on and started to use it *without any security*. To turn on the router's security features, you have to *configure the device*. Doing this is notoriously difficult and frequently involves several calls to technical support. People often give up after a while.

- The alternative is a hardwired network. Such a network is very secure. Many corporations use hardwired networks because of security concerns. Designing a reasonable network can be a daunting task, but graph theory makes it easier.

9.1 A Walk through Königsberg

OBJECTIVES

- Comprehend Euler's solution to the Königsberg bridge problem
- Understand the basic terms and concepts of graph theory
- See that graph theory has many, varied applications

Königsberg, the Pregel River, and the king's castle.

The residents of the Prussian city of Königsberg used to stroll through their city, crossing bridges across the Pregel River and enjoying the city's beauty. Many of them tried to find a route that allowed them to cross each of seven bridges just once.

Figure 9.1 shows a map of the city and a diagram of an unsuccessful attempt. Someone who followed this route, starting at the upper left bridge, would end up on the island. The walker had already crossed each of the island's bridges, so there was no way to get off the island and cross the remaining bridge at the lower right.

The Solution

The problem was solved by Swiss mathematician Leonhard Euler (pronounced "Oiler"). Euler reasoned like this:

- A walker uses up a bridge when he crosses the first bridge on his walk.
- A walker uses up two bridges when he walks across a piece of land: one in approaching the land and one in leaving the land.
- A walker uses up a bridge when he crosses the last bridge on his walk.

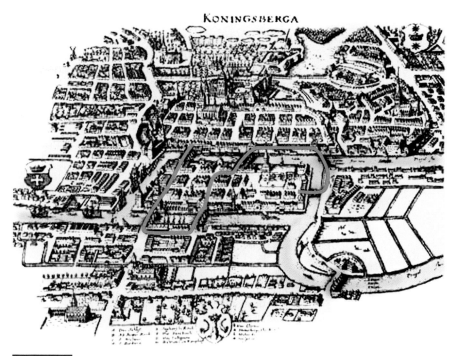

FIGURE 9.1 Königsberg's seven bridges and a diagram of an unsuccessful walking tour.

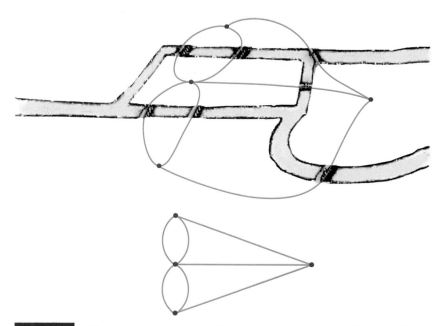

FIGURE 9.2 Euler's points and connecting lines, both with (above) and without (below) the river and bridges.

Euler drew a simple picture to replace the map, as shown in Figure 9.2. In that picture, a line represents a bridge and a point represents land. Each line connects two points because each bridge connects two pieces of land. Euler's thoughts about using up bridges turned into thoughts about using up lines.

- A single line gets used up when the walker crosses his first bridge.
- Two lines get used up when the walker walks across land: the line that represents the bridge that approaches the land and the line that represents the bridge that leaves the land.
- A line gets used up when the walker crosses his last bridge.

This means that if the walker were to start and stop the walk at different places, the starting point must have an odd number of lines coming out of it: one line to start the walk and two lines (or four or six . . .) to revisit the point when crossing other bridges. And the ending point must have an odd number of lines coming out of it for the same reason. But all other points must have an even number of lines so that the walker can both approach the point and depart the point. See Figure 9.3.

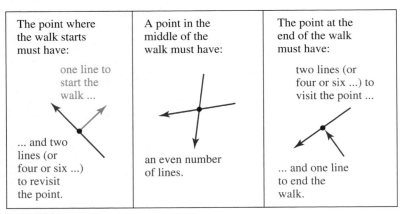

The point where the walk starts must have:	A point in the middle of the walk must have:	The point at the end of the walk must have:
one line to start the walk ... ... and two lines (or four or six ...) to revisit the point.	an even number of lines.	two lines (or four or six ...) to visit the point ... ... and one line to end the walk.

FIGURE 9.3 If the walker were to start and stop the walk at different places.

If the walker were to start and stop the walk at the same point, then that point must have an even number of lines: one line to start the walk, an even number to revisit the point when crossing other bridges, and one line to return to the starting point. This means that all points must have an even number of lines. See Figure 9.4.

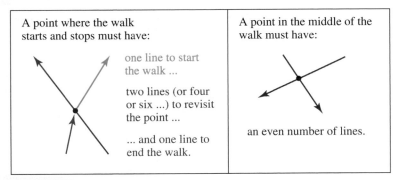

A point where the walk starts and stops must have:	A point in the middle of the walk must have:
one line to start the walk ... two lines (or four or six ...) to revisit the point ... ... and one line to end the walk.	an even number of lines.

FIGURE 9.4 If the walker were to start and stop the walk at different places.

In the graph of Königsberg, every point has an odd number of lines. See Figure 9.5. This means that each point could be used as a starting point or as a stopping point. But none of the points could be visited without starting or stopping there. So there is no way to walk across each of Königsberg's seven bridges once, much to the chagrin of its residents.

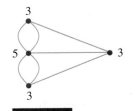

FIGURE 9.5

Each of Königsberg's points has an odd number of lines.

Graph Theory

In solving the Königsberg puzzle, Euler started a new type of geometry called *graph theory*. Despite its lighthearted beginnings, graph theory has important applications in business, psychology, sociology, computer science, chemistry, and biology.

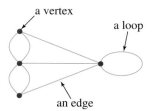

FIGURE 9.6

The Königsberg graph with an added loop.

A **graph** is a diagram that consists of points, called **vertices** (the singular is **vertex**) and connecting lines, called **edges.** An edge that connects a vertex with itself is called a **loop.** Euler's simplified picture of Königsberg is a graph that has edges and vertices but no loops. Figure 9.6 shows that graph, with an added loop.

Euler's graph representing Königsberg and its bridges was rather novel at the time. Now, it's quite common to use a graph to represent a subway system or a rapid transit system. In such a graph, the vertices represent stations and the edges represent subway or rapid transit lines. See Figure 9.7.

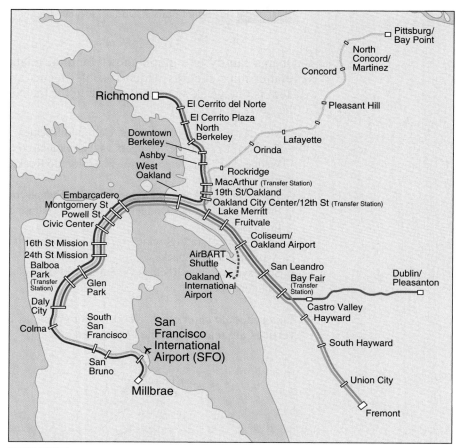

FIGURE 9.7 A graph of San Francisco's Bay Area Rapid Transit System (BART).

A graph lacks details that are common to most maps. For example, you can't look at the Königsberg graph and tell how far it is from one bridge to another or what streets to follow to get to a bridge. And the shape of an edge does not necessarily resemble the shape of the connection that it represents. But these details don't matter to subway riders.

Graphs Describe Relationships

A family tree is a graph. Figure 9.8 shows a part of President Barack Obama's family tree. Its vertices are family members, and its edges are parent/child relationships.

Graphs can be used to describe relationships *of any type*. The Königsberg graph and the Bay Area Rapid Transit graph show geographic relationships. The

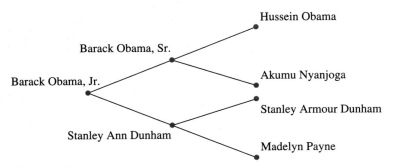

FIGURE 9.8 President Barack Obama's family tree.

Obama family tree graph shows familial relationships. Graphs can even show relationships such as which baseball teams are scheduled to play each other. Here is one week's baseball schedule for the National League West, right out of the newspaper:

- Monday: San Diego Padres vs. Colorado Rockies
- Tuesday: Los Angeles Dodgers vs. San Francisco Giants
- Wednesday: San Diego Padres vs. Colorado Rockies, Los Angeles Dodgers vs. San Francisco Giants
- Thursday: Los Angeles Dodgers vs. San Francisco Giants
- Friday: Los Angeles Dodgers vs. Arizona Diamondbacks, Colorado Rockies vs. San Francisco Giants
- Saturday: Colorado Rockies vs. San Francisco Giants, Los Angeles Dodgers vs. Arizona Diamondbacks
- Sunday: Colorado Rockies vs. San Francisco Giants, Los Angeles Dodgers vs. Arizona Diamondbacks

This schedule can be described with a graph, as in Figure 9.9. With this graph, the vertices are teams, and the edges are games.

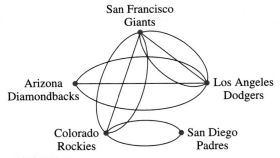

FIGURE 9.9 One week's baseball schedule.

Notice that depending on what you want to know, Figure 9.9 can be easier to read than is the written schedule. For example, a glance at Figure 9.9 is sufficient to determine that the Giants play the Dodgers three times and the Dodgers don't play the Padres at all. You would have to inspect the written schedule rather closely to determine these things.

Also notice that in Figure 9.9 the edges cross each other at many places, only some of which are vertices. Some graphs have this quality. Others, like the Königsberg graph, don't. It's important to understand that "vertex" isn't the same as "where two edges cross."

In Exercises 1–3, use the following information: After much discussion, the Königsberg city council decided that it would be a good thing if the city's residents succeeded in finding a bridge walk, that is, a walk that would take the walker across each of the city's bridges once. Council members reasoned that a bridge walk would foster civic pride and attract tourists. They voted to close one bridge if doing so would allow the creation of a bridge walk. (Note: The Königsberg story presented earlier in this section is true. However, the details presented in the exercises are not.)

1. Is there a bridge whose closure would create a bridge walk if the walk's starting point had to be different from its stopping point? If your answer is "yes," specify the bridge, draw a graph of the new six-bridge system, and describe the walk. If your answer is "no," explain why.

2. Is there a bridge whose closure would create a bridge walk if the walk's starting point had to be the same as its stopping point? If your answer is "yes," specify the bridge, draw a graph of the new six-bridge system, and describe the walk. If your answer is "no," explain why.

3. Is there a bridge whose closure would not create a bridge walk? If so, specify the bridge and explain why a walk would not be created.

In Exercises 4–6, use the following information: After closing a bridge, as discussed in Exercises 1–3, a number of Königsberg merchants complained to the city council that there was insufficient access to their stores. The council voted to reopen the closed bridge and to build a new bridge if doing so would allow the creation of a bridge walk.

4. Is there a site for a new bridge whose opening would create a bridge walk if the walk's starting point had to be the same as its stopping point? If your answer is "yes," specify the site, draw a graph of the new eight-bridge system, and describe the walk. If your answer is "no," explain why.

5. Is there a site for a new bridge whose opening would create a bridge walk if the walk's starting point had to be different from its stopping point? If your answer is "yes," specify the site, draw a graph of the new eight-bridge system, and describe the walk. If your answer is "no," explain why.

6. Is there a site for a new bridge whose opening would not create a bridge walk? If so, specify the site and explain why a walk would not be created.

In Exercises 7–10, determine the number of vertices, edges, and loops in the given graph.

7.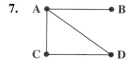

8.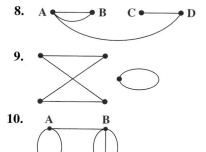

9.

10.

11. **a.** Determine the number of vertices, edges and loops in the Obama family tree graph in Figure 9.8.

   b. Could a family tree graph ever have a loop? Why or why not?

12. **a.** Determine the number of vertices, edges and loops in the baseball schedule graph in Figure 9.9.

   b. Could a baseball schedule graph ever have a loop? Why or why not?

13. Draw a graph with three vertices and six edges, one of which is a loop.

14. Draw a graph with four vertices and eight edges, two of which are loops.

15. **a.** Why do the two diagrams in Figure 9.10 represent the same graph?

   b. Draw another representation of the same graph that looks significantly different from either of these.

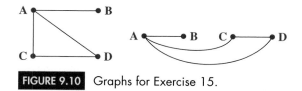

FIGURE 9.10 Graphs for Exercise 15.

16. **a.** Why do the two diagrams in Figure 9.11 represent the same graph?

   b. Draw another representation of the same graph that looks significantly different from either of these.

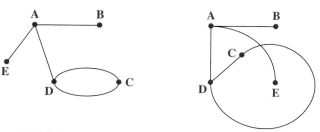

FIGURE 9.11 Graphs for Exercise 16.

17. Draw another representation of the graph in Figure 9.12 that looks significantly different.

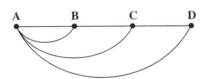

FIGURE 9.12 A graph for Exercise 17.

18. Draw another representation of the graph in Figure 9.13 that looks significantly different.

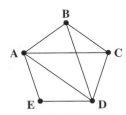

FIGURE 9.13 A graph for Exercise 18.

• CONCEPT QUESTION

19. The Königsberg graph shows the relationship between bridges and landmasses. The Bay Area Rapid Transit graph shows the relationship between transit stations and transit lines. The Obama family tree graph shows familial relationships. The baseball schedule graph shows which baseball teams are scheduled to play each other. Give an example of another type of relationship that could be displayed with a graph, and describe how that relationship could be displayed.

• PROJECT

20. Find some information in a newspaper, magazine, or other similar source, and present that information as a graph.

9.2 Graphs and Euler Trails

OBJECTIVES

- Be able to utilize Euler's Theorems
- Apply Fleury's algorithm
- Understand Eulerization

Terminology

Before going any further, we need to discuss some of the terms that are used in studying graphs.

Two vertices are **adjacent** if they are joined by an edge. In Figure 9.14, the Arizona Diamondbacks and Los Angeles Dodgers vertices are adjacent vertices, because they are joined by an edge. Notice that the Arizona Diamondbacks and San Francisco Giants vertices are not adjacent, even though they are next to each other on the graph.

Two edges are **adjacent** if they have a vertex in common. In Figure 9.14, the two edges coming out of the San Diego Padres vertex are adjacent edges, because they have the San Diego Padres vertex in common.

The graph in Figure 9.15 has a different appearance than does the graph in Figure 9.14. However, it describes the same schedule of games. Edges that are

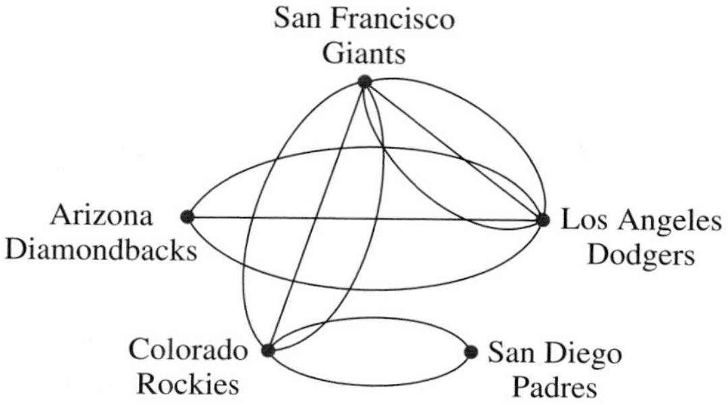

San Francisco
Giants

Arizona
Diamondbacks

Los Angeles
Dodgers

Colorado
Rockies

San Diego
Padres

FIGURE 9.14 One week's baseball schedule.

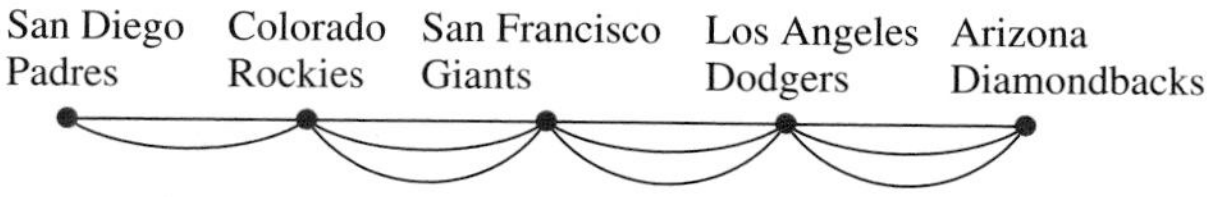

San Diego   Colorado   San Francisco   Los Angeles   Arizona
Padres   Rockies   Giants   Dodgers   Diamondbacks

FIGURE 9.15 A different description of the same schedule, with an identical graph.

adjacent in Figure 9.14 are adjacent in Figure 9.15, and vertices that are adjacent in Figure 9.14 are adjacent in Figure 9.15. For this reason, the graphs in Figures 9.14 and 9.15 are said to be **identical.**

A **trail** is a sequence of adjacent vertices and the distinct edges that connect them. Trails are most easily described by listing the distinct edges in order. Figure 9.16 shows the Königsberg graph, with the edges (that is, the bridges) labeled. One trail is a, d. That trail describes a walk that starts at the upper bank, crosses a bridge (edge a) to the island, and then crosses another bridge (edge d) to the lower bank.

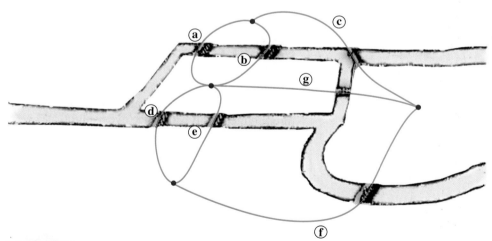

FIGURE 9.16 The Königsberg graph with the edges labeled.

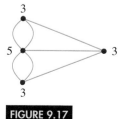

FIGURE 9.17

The vertices' degrees.

A **circuit** is a trail that begins and ends at the same vertex. One circuit is a, d, f, c.

The **degree of a vertex** is the number of edges that connect to that vertex. A loop connects to a vertex twice, so a loop contributes to the degree twice. An **odd vertex** is a vertex with an odd degree, and an **even vertex** is a vertex with an even degree. Figure 9.17 shows each vertex's degree in the Königsberg graph.

An **Euler trail** is a trail that travels through *every* edge of a graph *exactly once*. An **Euler circuit** is an Euler trail that is a circuit.

Euler's Theorems

In Section 9.1, we stated, "In the graph of Königsberg, every point has an odd number of lines . . . [so] each point could be used as a starting point or as a stopping point. But none of the points could be visited without starting or stopping there. So there is no way to walk across each of Königsberg's seven bridges once." The same statement, phrased with the terminology of graph theory, is "The Königsberg graph has more than two odd vertices, so that graph has no Euler trails and no Euler circuits."

If we generalize from the above statement, we obtain Euler's Theorems.

EULER'S THEOREMS

- A connected graph with *only even vertices* has at least one Euler trail, which is also an Euler circuit.
- A connected graph with *exactly two odd vertices* and any number of even vertices has at least one Euler trail. Each of these trails will start at one odd vertex and end at the other odd vertex.
- A graph with *more than two odd vertices* has no Euler trails and no Euler circuits.
- It's impossible for a connected graph to have only one odd vertex.

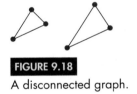

FIGURE 9.18

A disconnected graph.

Euler's Theorems refer to *connected graphs*. A graph is **connected** if every pair of vertices is connected by a trail. Figure 9.18 shows a **disconnected graph,** or a graph that is not connected. This graph is disconnected because there is no trail connecting any of the three vertices on the left triangle with any of the three vertices on the right triangle. It's easy to see that disconnected graphs never have Euler trails and never have Euler circuits.

Using Euler's Theorems

EXAMPLE **1**

USING EULER'S THEOREMS   Figure 9.19 shows a map of the Serene Oaks gated community. After several burglaries, the residents hired a security guard to walk through the subdivision once every night.

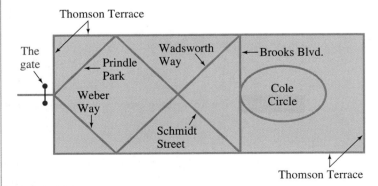

FIGURE 9.19   The Serene Oaks gated community.

a. Draw a graph of the neighborhood, where vertices are corners or intersections and edges are streets. In that graph, show the degree of each vertex.

b. The guard, who is not paid by the hour, would prefer to park his car, walk a route that allows him to walk through every block just once, and then return to his car. Is this possible?

SOLUTION

a. The map itself is almost a graph. It lacks only vertices. The graph is shown in Figure 9.20. Notice that the vertex that is the intersection of Brooks Blvd. and Cole Circle is of degree 4. A loop connects to a vertex twice, so a loop contributes to the degree twice.

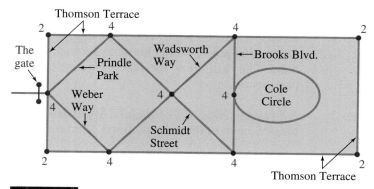

FIGURE 9.20 A graph of the Serene Oaks neighborhood, showing vertices and their degrees.

b. This is a connected graph with only even vertices, so it has at least one Euler trail, which is also an Euler circuit. This means that it is possible for the guard to walk a route that allows him to walk through every block just once.

Using Fleury's Algorithm to Find Euler Trails and Euler Circuits

With some graphs, it's easy to find an Euler trail or an Euler circuit, once you have used Euler's Theorem to verify that the graph has such a trail or circuit. However, there are some traps that can be easily avoided, if you know what to look for. *Fleury's algorithm* describes those traps and specifies how to avoid them. An **algorithm** is a logical step-by-step procedure, or recipe, for solving a problem. Fleury's algorithm is a logical step-by-step procedure for finding Euler circuits and Euler trails.

FLEURY'S ALGORITHM FOR FINDING EULER TRAILS AND EULER CIRCUITS

1. *Verify that the graph has an Euler trail or Euler circuit,* using Euler's Theorem.
2. *Choose a starting point.*
   - If the graph has two odd vertices, we can find an Euler trail. Pick either of the odd vertices as the starting point.
   - If the graph has no odd vertices, we can find an Euler circuit. Pick any point as the starting point.
3. *Label each edge alphabetically* as you travel that edge.
4. When choosing edges:
   - *Never choose an edge that would make the yet-to-be-traveled part of the graph disconnected,* because you won't be able to get from one portion of the graph to the other.
   - *Never choose an edge that has already been followed,* since you can't trace any edges twice in Euler trails and Euler circuits.
   - *Never choose an edge that leads to a vertex that has no other yet-to-be-traveled edges,* because you won't be able to leave that vertex.

EXAMPLE **2**

APPLYING FLEURY'S ALGORITHM Use Fleury's algorithm to find a route through the Serene Oaks neighborhood for the security guard.

SOLUTION

Step 1

Verify that the graph has an Euler trail or Euler circuit. We did this in Example 1.

Step 2

Choose a starting point. The graph has no odd vertices, so we can pick any point as the starting point. We'll choose the corner at the lower left corner of the neighborhood. This will allow the security guard to drive through the gate, turn right and park his car.

Step 3

Label each edge alphabetically. We'll start by following Thomson Terrace along the perimeter of the neighborhood, from edge a to b to c and so on. See Figure 9.21.

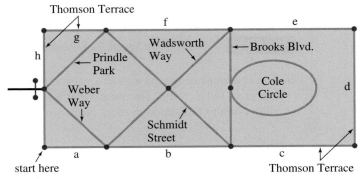

FIGURE 9.21 Edges a through h in orange.

It's only when we get through edge h and arrive at the vertex that is the intersection of Thomson Terrace, Prindle Park, and Weber Way that we have to choose an edge with step 4 in mind. If we continue along Thomson Terrace, we will arrive at a vertex (our starting part, actually) that *has no yet-to-be traveled edges.* That route is prohibited by Fleury's algorithm, so we will take Weber Way instead.

At the next vertex, we must take Wadsworth Way rather than Thomson Terrace, because *we cannot choose an edge that has already been traveled.* For the same reason, after turning onto Schmidt Street, we must choose Brooks Blvd. rather than Thomson Terrace. See Figure 9.22.

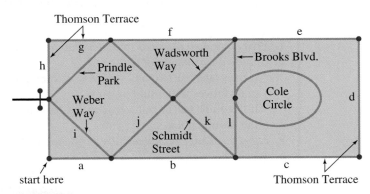

FIGURE 9.22 Edges a through l.

Now we're at the intersection of Brooks Blvd. and Cole Circle. If we continue on Brooks Blvd., we would be choosing an edge that would *make the yet-to-be-traveled part of the graph disconnected.* Specifically, Cole Circle would be disconnected from the rest of the yet-to-be-traveled part of the graph. See Figure 9.23.

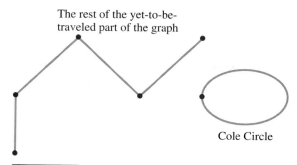

The rest of the yet-to-be-traveled part of the graph

Cole Circle

FIGURE 9.23 Choosing Brooks Blvd. would make the yet-to-be-traveled part of the graph disconnected.

Continuing on Brooks Blvd. is prohibited by Fleury's algorithm, so we will take Cole Circle instead. See Figure 9.24.

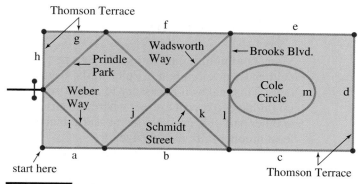

FIGURE 9.24 Edges a through m.

The remaining edge choices are straightforward. The finished Euler circuit is shown in Figure 9.25.

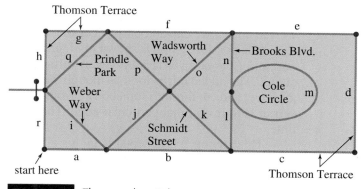

FIGURE 9.25 The complete Euler circuit.

Applications of Euler Trails and Circuits

Euler trails and circuits have many practical applications. Any time services must be delivered along streets, Euler's theorem and Fleury's algorithm can make the job more efficient. The post office must deliver mail six days a week. Utility companies must read meters every month. Garbage companies must pick up garbage every week. Inefficient routes create unnecessary expenses for these firms—expenses that must be passed on to their customers: you and me.

Notice that a utility company must visit each side of the street, because each house has its own meter. This creates a different type of graph. Figure 9.26 shows the effect of this on the lower left part of the Serene Oaks gated community. The right-hand graph has **multiple edges,** which are edges that connect the same two vertices.

If each street must be visited once.

If each *side* of the street must be visited once.

FIGURE 9.26

Revisiting Edges and Eulerization

Many communities are unlike the Serene Oaks gated community in that they have neither Euler circuits nor Euler trails, because they have more than two odd vertices. This makes it necessary to revisit some edges. For example, a postal carrier might have to revisit a street where she has already delivered the mail, just to get to another part of her route. The most efficient trails and circuits are those that have the least number of revisited edges.

Consider the Secluded Glen community, whose graph is shown in Figure 9.27. This community is typical of many in that it is laid out like a grid, with parallel streets. The graph has six odd vertices, shown in blue, so it has no Euler circuit.

The goal here is to create an efficient route, so we want to add as few new edges as possible. The most efficient thing to do would be to add three new edges, each connecting a pair of odd vertices. We have done this in Figure 9.28.

While this seems efficient, the new diagonal edges require the postal carrier to leave the street and cut through people's backyards, which is not feasible. When we add edges, we can add only edges that are duplicates of existing edges.

One possible graph is shown in Figure 9.29. The added edges (shown in blue) turn all of the odd vertices into even vertices, so the resulting graph has an Euler circuit. The added edges look as though we're requiring the postal carrier to leave the street and cut through somebody's front yard. Instead, they mean that the postal carrier is to revisit that particular edge. That is, she is to walk that block a second time.

This approach is acceptable in that all added edges are duplicates of existing edges. However, it has seven added edges. Wherever the postal carrier starts and

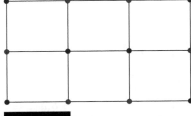

FIGURE 9.27

A graph of the Secluded Glen community. Odd vertices are shown in blue.

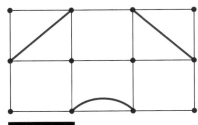

FIGURE 9.28

A graph of the Secluded Glen community, with three illegal new edges.

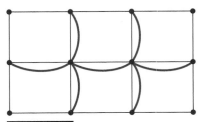

FIGURE 9.29

A graph of the Secluded Glen community with some added edges.

stops, this plan would require her to walk seven blocks a second time. Let's try to find a more efficient plan.

The process of adding duplicate edges so that all odd vertices are turned into even vertices is called **Eulerization** (or **Eulerizing** a graph) because it results in a graph that has an Euler circuit. Figure 9.29 shows one Eulerization of the graph shown in Figure 9.27.

Whenever a graph is laid out like a grid, with parallel edges, the odd vertices are always along the outer perimeter of the graph. The following Eulerization algorithm results in efficient Eulerizations because it recognizes that fact and adds duplicate edges along the outer perimeter only.

EULERIZATION ALGORITHM

To efficiently Eulerize a graph that is laid out like a grid:

1. Choose a vertex along the outer perimeter of the graph.
2. If a vertex is an even vertex, add no edges and move on to the next vertex along the outer perimeter.
3. If a vertex is an odd vertex, add a duplicate edge that connects it to the next adjacent vertex along the outer perimeter, and move on to the next vertex along the outer perimeter.
4. Repeat these steps until you return to the vertex in step 1.

EXAMPLE 3

EULERIZING TO OBTAIN AN EULER CIRCUIT

a. Apply the Eulerization algorithm to the Secluded Glen community.
b. Use the result of part (a) to find an efficient route through the community for the postal carrier. All mailboxes are on one side of the street.

SOLUTION

a. *Applying the Eulerization algorithm.*

Step 1   *Choose a vertex along the outer perimeter of the graph.* We'll choose the vertex in the upper left corner.

Step 2   *If a vertex is an even vertex, add no edges and move on to the next vertex along the outer perimeter.* The vertex in the upper left corner is indeed an even vertex, so we add no edges and move to the right to the next vertex.

Step 3   *If a vertex is an odd vertex, add a duplicate edge that connects it to the next adjacent vertex along the outer perimeter.* The second vertex is an odd vertex, so we add a duplicate edge that connects it to the next vertex. See Figure 9.30.

Step 4   *Repeat these steps until you return to the vertex in step 1.* This results in the graph shown in Figure 9.31.

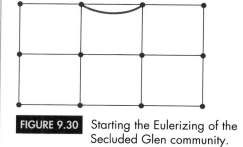

FIGURE 9.30 Starting the Eulerizing of the Secluded Glen community.

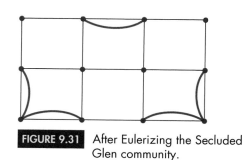

FIGURE 9.31 After Eulerizing the Secluded Glen community.

This scheme has five added edges. Wherever the postal carrier starts and stops, this scheme would require her to walk five blocks a second time. This is the most efficient scheme.

b. *Using the result to find an efficient route through the community for the postal carrier.* We'll do this with Fleury's algorithm. The result is shown in Figure 9.32.

Remember that the added edges are supposed to be duplicates of existing edges. For example, edge q is supposed to be a duplicate of edge b. We draw them this way so that we can tell when we're traveling an edge for the second time, and so we can apply Euler's theorem and Fleury's algorithm.

In Figure 9.33, we show a map of the Secluded Glen community without the added edges but with street names.

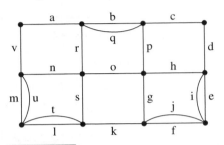

FIGURE 9.32

Applying Fleury's algorithm to Figure 9.31.

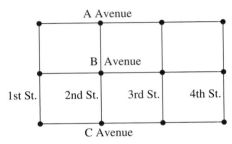

FIGURE 9.33

Applying Fleury's algorithm to Figure 9.32.

The beginning of the route described in Figure 9.32 with the letters a, b, c, and so on translates like this:

- Start at the corner of A Avenue and 1st Street.
- Go along A until you come to 4th, and turn onto 4th.
- Go along 4th until you come to C, and turn onto C.
- From C, turn onto 3rd.
- From 3rd, turn right onto B.
- From B, turn right onto 4th and then onto C. Revisit two blocks.
- From C, turn onto 1st and then onto B.
- Turn left onto 3rd.
- Turn left onto A, and revisit one block.
- Turn onto 2nd, and then right onto C, revisiting one block.
- Turn onto 1st, revisit one block, and continue to the starting point.

9.2 EXERCISES

In Exercises 1–6, do the following:

a. Find two adjacent edges.

b. Find two adjacent vertices.

c. Find the degree of each vertex.

d. Determine whether the graph is connected.

1.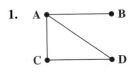

2.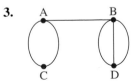

3.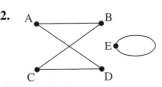

4.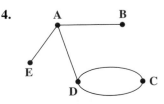

5.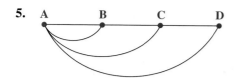

6.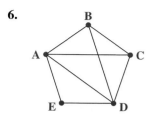

7. Draw a graph with four vertices in which:

  a. each vertex is of degree 2.

  b. each vertex is of degree 3.

  c. two vertices are of degree 2 and two are of degree 3.

8. Draw a graph with five vertices in which:

  a. each vertex is of degree 2.

  b. each vertex is of degree 4.

  c. one vertex is of degree 2 and four are of degree 3.

9. Draw a graph with six vertices in which:

  a. each vertex is of degree 1.

  b. each vertex is of degree 0.

10. Draw a graph with eight vertices in which each vertex is of degree 3.

11. Draw a graph with six vertices in which each vertex is of degree 3 and that has:

  a. loops but no multiple edges.

  b. multiple edges but no loops.

  c. neither loops nor multiple edges.

  d. both loops and multiple edges.

12. Draw a graph with five vertices in which each vertex is of degree 4 and that has:

  a. loops but no multiple edges.

  b. multiple edges but no loops.

  c. neither loops nor multiple edges.

  d. both loops and multiple edges.

In Exercises 13–18, do the following.

a. *Use Euler's Theorems to determine whether the graph has an Euler trail or an Euler circuit.*

b. *If it does, find it. If it does not, say why. Then Eulerize the graph if possible.*

13. The graph in Exercise 1

14. The graph in Exercise 2

15. The graph in Exercise 3

16. The graph in Exercise 4

17. The graph in Exercise 5

18. The graph in Exercise 6

19. Find an Euler circuit for the Serene Oaks gated community other than that created in Example 2.

20. Find an Eulerization for the Secluded Glen community other than that created in Example 3.

21. Once a month, the utility company sends an agent through the Serene Oaks gated community (see Figure 9.19) to read the meters. Each house has its own meter. Find a route for the meter reader:

  a. if there are houses on both sides of each street.

  b. if there are houses on both sides of each street except that there is a greenbelt consisting of lawns and playgrounds along the outer edge of the community, outside of Thomson Terrace.

  c. there are houses on both sides of each street except that the inside of Cole Circle is a park.

22. Once a month, the utility company sends an agent through the Secluded Glen community (see Figure 9.33) to read the meters. Each house has its own meter. Find a route for the meter reader:

  a. if there are houses on both sides of each street.

  b. if there are houses on both sides of each street except that there is a greenbelt consisting of lawns and playgrounds along the outer edge of the community, outside of A Avenue, C Avenue, First Street, and Fourth Street.

  c. if there are houses on both sides of each street except that the area bounded by A Avenue, B Avenue, First Street, and Second Street is a park.

23. Find an efficient route through the community shown in Figure 9.34 for the postal carrier if mailboxes are all on one side of each street.

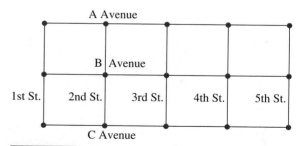

FIGURE 9.34 A map for Exercises 23 and 24.

24. Once a month, the utility company sends an agent through the community shown in Figure 9.34 to read the meters. Each house has its own meter. Find a route for the meter reader:

  a. if there are houses on both sides of each street.

  b. if there are houses on both sides of each street except that there is a greenbelt consisting of lawns and playgrounds along the outer edge of the community, outside of A Avenue, C Avenue, First Street, and Fifth Street.

c. if there are houses on both sides of each street except the area bounded by A Avenue, B Avenue, First Street, and Second Street is a park.

25. Figure 9.35 shows a map of New York City's bridges and tunnels.

   a. Draw a graph of this. Use "Manhattan," "the Bronx," "Queens," "Brooklyn," and "New Jersey" as vertices. Consider the Triborough bridge to be three separate bridges: one from Manhattan to the Bronx, one from Manhattan to Queens, and one from Queens to the Bronx.

   b. Does New York have a bridge and tunnel drive, that is, a drive that would traverse each of the city's bridges and tunnels once if the drive's starting point must be the same as its stopping point? If your answer is "yes," describe the drive. If your answer is "no," explain why. Then Eulerize the graph if possible.

   c. Does New York have a bridge and tunnel drive, that is, a drive that would traverse each of the city's bridges and tunnels once if the drive's starting point must be different from its stopping point? If your answer is "yes," describe the drive. If your answer is "no," explain why. Then Eulerize the graph if possible.

26. Figure 9.36 shows a map of the San Francisco Bay Area's bridges.

   a. Draw a graph of this. Use "the north bay," "the east bay," and "the peninsula" as vertices.

   b. Does the Bay Area have a bridge drive, that is, a drive that would traverse each of the area's bridges

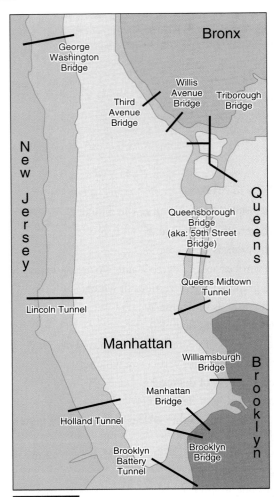

FIGURE 9.35 New York City's Bridges and Tunnels.

New York's Brooklyn Bridge.

once if the drive's starting point must be the same as its stopping point? If your answer is "yes," describe the drive. If your answer is "no," explain why. Then Eulerize the graph if possible.

c. Does the Bay Area have a bridge drive, that is, a drive that would traverse each of the area's bridges once if the drive's starting point must be different from its stopping point? If your answer is "yes," describe the drive. If your answer is "no," explain why. Then Eulerize the graph if possible.

27. Figure 9.37 shows a map of the bridges and ferry routes in the Seattle/Puget Sound area. These ferries take cars as well as passengers.

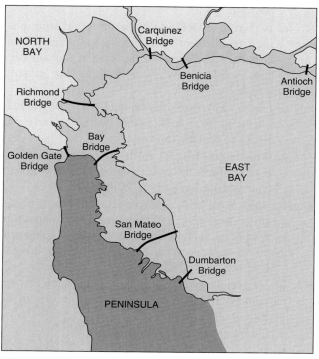

FIGURE 9.36 The San Francisco Bay Area's bridges.

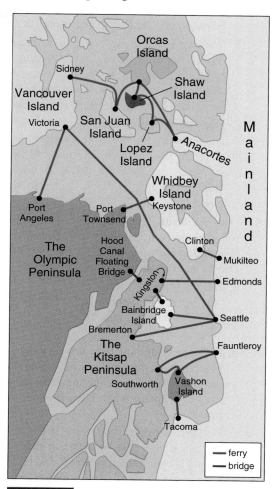

FIGURE 9.37 The Seattle/Puget Sound area.

San Francisco's Golden Gate Bridge.

a. Draw a graph of this. Use "mainland," "Whidbey Island," "Vancouver Island," Orcas Island," "Shaw Island," "San Juan Island," "Lopez Island," "the Olympic Peninsula," "Bainbridge Island," "Vashon Island," and "the Kitsap Peninsula" as vertices.

b. Does the Seattle/Puget Sound area have a bridge and ferry drive if the drive's starting point must be the same as its stopping point? If your answer is "yes," describe the drive. If your answer is "no," explain why. Then Eulerize the graph if possible.

c. Does the Seattle/Puget Sound area have a bridge and ferry drive if the drive's starting point must be different from its stopping point? If your answer is "yes," describe the drive. If your answer is "no," explain why. Then Eulerize the graph if possible.

28. Does the Seattle area have a bridge and ferry drive if we exclude Vancouver Island, Orcas Island, Shaw Island, San Juan Island, and Lopez Island? See Exercise 27 and Figure 9.37.

a. Draw a graph. Use "mainland," "Whidbey Island," "the Olympic Peninsula," "Bainbridge Island," "Vashon Island," and "the Kitsap Peninsula" as vertices.

b. Is there a bridge and ferry drive if the drive's starting point must be the same as its stopping point? If your answer is "yes," describe the drive. If your answer is "no," explain why. Then Eulerize the graph if possible.

c. Is there a bridge and ferry drive if the drive's starting point must be different from its stopping point? If your answer is "yes," describe the drive. If your answer is "no," explain why. Then Eulerize the graph if possible.

29. Figure 9.7 on page 667 shows a map of the BART (Bay Area Rapid Transit) system. On that map,

different rapid transit lines are colored differently. For example, the rapid transit line between Millbrae and Pittsburg/Bay Point is colored yellow.

a. Draw a graph of this. If two adjacent stations have three different lines connecting them, then those stations' vertices should have three different edges connecting them.

b. Is there a BART ride if the drive's starting point must be the same as its stopping point? If your answer is "yes," describe the drive. If your answer is "no," explain why. Then Eulerize the graph if possible.

c. Is there a BART ride if the drive's starting point must be different from its stopping point? If your answer is "yes," describe the drive. If your answer is "no," explain why. Then Eulerize the graph if possible.

d. Explain why the existence of a BART ride would be important to BART police, who regularly ride BART to provide security.

e. Which type of BART ride would be more useful to BART police: the type described in part (b) or the type described in part (c)? Why?

30. Is there a BART ride (see Exercise 29) if we disregard the difference between different rapid transit lines (that is, if adjacent stations have only one edge connecting them)? If your answer is "yes," describe the drive. If your answer is "no," explain why.

31. In this exercise, you will find the relationship between the number of edges in a graph and the sum of the degrees of all of the vertices.

a. Complete the chart in Figure 9.38.

b. Using the result of part (a), make a guess as to the relationship between the number of edges in a graph and the sum of the degrees of all of the vertices.

c. How many vertices must an edge connect?

A ferry in the Puget Sound.

Graph	Number of edges in that graph	Sum of the degrees of all of the vertices
Figure 9.15	11	2 + 5 + 6 + 6 + 3 = 22
Figure 9.17		
Figure 9.20		

FIGURE 9.38 Chart for Exercise 31.

d. How much does an edge contribute to the degree of each of the vertices it connects?

e. Use the results of parts (c) and (d) to explain why your guess in part (a) must be correct.

32. In this exercise, you will find why it is impossible for a graph to have only one odd vertex.

a. Use the results of Exercise 31 to explain why the sum of the degrees of all of the vertices must be an even number for any graph.

b. Use the results of part (a) to explain why a graph must have an even number of odd vertices.

WEB PROJECTS

33. There are many interesting problems in graph theory. Some of these problems have been solved, and some remain unsolved. Many of these problems are discussed on the web. Visit several web sites and choose a specific problem. Describe the problem, its history, and its applications. If it has been solved, describe the method of solution if possible.

34. The Four Color Map Problem started in 1852 when Francis Guthrie was coloring a map of the countries that England ruled. He noticed that if he used four colors, he was able to color the map so that no two adjacent countries were the same color. He and his brother Frederick went to mathematician Augustus DeMorgan and asked whether this was always true. Investigate this problem on the Web. Describe its connection with graph theory, its history, its proof in 1976 by Wolfgang Haken and Kenneth Appel, and the nature of the proof.

Some useful links for these web projects are listed on the text companion web site. Go to **www.cengage.com/math/johnson** to access the web site.

9.3 Hamilton Circuits

OBJECTIVES

● Understand the concept of a minimum Hamilton circuit

● Be able to apply the nearest neighbor algorithm, the repetitive nearest neighbor algorithm, and the cheapest edge algorithm

Finding the Best Route

Kim Blum owns her own business in New York City. One month, she had to fly to Los Angeles and Chicago to meet with clients. Her firm was a recent startup, and money was tight, so she made a point of finding the least expensive flights. She logged onto a travel web site and found the following costs of one-way flights between the cities:

New York to Chicago	$204
Chicago to New York	$219
New York to Los Angeles	$177
Los Angeles to New York	$199
Los Angeles to Chicago	$314
Chicago to Los Angeles	$314

Source: www.travelocity.com

There are two ways to make this trip: She could go to Chicago first and then Los Angeles, or she could go to Los Angeles first and then Chicago. See Figure 9.39.

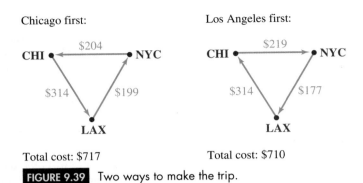

Total cost: $717   Total cost: $710

FIGURE 9.39 Two ways to make the trip.

Kim flew to Los Angeles first and saved $7.

Weighted Graphs and Digraphs

We described each of Kim's possible routes with a graph. However, these graphs are different from those in Sections 9.1 and 9.2 in that each edge has a number associated with it. For example, the edge from CHI to NYC in the "Los Angeles route first" graph above has the cost $219 associated with it. These numbers are called **weights.** A graph that includes weights is called a **weighted graph.**

Kim's circuits are different from those in Sections 9.1 and 9.2 in another way: Each edge has a direction associated with it. Here, it is necessary to distinguish between directions, because the edge from CHI to NYC has a cost of $219 associated with it, while the edge from NYC and CHI has a cost of $204 associated with it. A graph whose edges have directions is called a **directed graph,** or a **digraph.**

So each of Kim's possible routes is illustrated above with a particular type of graph—a *weighted digraph.*

A Weighted Digraph with Four Vertices

A month later, Kim had to visit the same clients and a new client in Miami. She wasn't impressed with her $7 savings a month earlier, so she priced out only two routes. To her surprise, she found that the total cost of her two routes varied quite a bit, so she checked out all the possibilities. The old airfares hadn't changed. She found the following new fares:

New York to Miami	$98
Miami to New York	$286
Los Angeles to Miami	$179
Miami to Los Angeles	$514
Chicago to Miami	$124
Miami to Chicago	$244

Source: www.travelocity.com

Route 1:

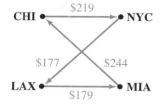

Total trip cost: $819

Route 2:

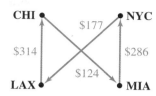

Total trip cost: $901

Route 3:

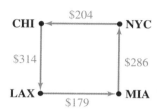

Total trip cost: $983

Route 4:

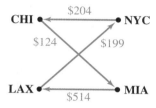

Total trip cost: $1041

Route 5:

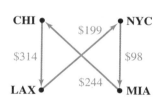

Total trip cost: $855

Route 6:

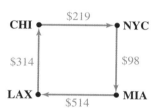

Total trip cost: $1145

FIGURE 9.40 Six different routes.

Then Kim started listing routes. She quickly realized that there were several. She figured that she was going to three different cities, so there were three different choices for her trip's first leg. That left two choices for the trip's second leg and only one possibility for the third leg. All together, there are $3 \cdot 2 \cdot 1 = 3! = 6$ different routes. See Figure 9.40.

Kim flew route 1 and paid $819. This saved $326.

Hamilton Circuits

Kim's route through four cities forms a circuit, because it is a trail that begins and ends at the same vertex. Kim's circuit is different from the Euler circuits that we studied in Sections 9.1 and 9.2 in that her goal is not to pass through every edge exactly once. In fact, none of her possible circuits passed through every edge. For example, route 6 in Figure 9.40 did not pass through the edge joining NYC and LAX, and it did not pass through the edge joining CHI and MIA. (The illustrations show only the edges that were used.)

Instead, Kim's goal is to pass through every *vertex* exactly one time. This type of circuit is called a **Hamilton circuit.**

Many Possibilities Create a Need for Strategies

A few months later, Kim had to visit the same clients and two new clients, one in Seattle and one in Houston. Kim instructed her newly hired assistant to start pricing all the possibilities. After some time, the assistant told her that there were just too many possibilities to try.

Kim calculated that she was going to five different cities, so there were five different choices for her trip's first leg. That left four choices for the trip's second leg, three choices for the third leg, two for the fourth leg, and only one possibility for the fifth leg. All together, there are $5 \cdot 4 \cdot 3 \cdot 2 \cdot 1 = 5! = 120$ different routes. Pricing out every route could end up costing more money (in the form of the assistant's wages) than it would save.

Kim's original approach was to list every possible circuit, add up each circuit's total weight, and choose the circuit with the least total weight. This *brute force algorithm* will find the minimum Hamilton circuit, but most real-world problems have too many circuits for this approach to be realistic.

Nonetheless, past experience had taught Kim that different routes could have wildly different prices. And her booming business was going to require a significant amount of traveling. She needed some strategies.

Figure 9.41 shows a complete graph in that it shows every possible leg. In fact, a graph in which every pair of vertices is joined with one edge is called a **complete graph.**

Figure 9.42 shows the various costs.

FIGURE 9.41

The complete graph.

		From:				
		CHI	HOU	LAX	MIA	NYC	SEA
	CHI		$305	$314	$244	$204	$179
	HOU	$288		$324	$214	$336	$319
To:	**LAX**	$314	$284		$514	$177	$242
	MIA	$124	$190	$179		$98	$307
	NYC	$219	$339	$199	$286		$169
	SEA	$179	$315	$242	$307	$132	

FIGURE 9.42 Flight costs. *Source:* www.travelocity.com.

The Nearest Neighbor Algorithm

Kim decided that it would be a good beginning to choose the cheapest flight out of New York. That was a $98 flight to Miami. From Miami, the cheapest flight is a $214 flight to Houston. Figure 9.43 continues to choose cheap flights, avoiding flights that take her to a city she has already visited and avoiding flights that take her home before she's visited all the necessary cities. This method of finding a cheap circuit is called the *nearest neighbor algorithm.*

From:	*The cheapest and most usable flight is:*
New York City	NYC to MIA at $98
Miami	MIA to HOU at $214
Houston	HOU to MIA at $190—but she'd already been to Miami so use the second cheapest which is HOU to LAX at $284
Los Angeles	LAX to MIA—no, been to Miami LAX to NYC—no, can't go home yet LAX to SEA at $242
Seattle	SEA to NYC—no, can't go home yet SEA to CHI at $179
Chicago	Time to go home so CHI to NYC at $219

FIGURE 9.43 Using the nearest neighbor algorithm.

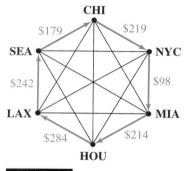

FIGURE 9.44

The result of the nearest neighbor algorithm.

The total cost is $98 + $214 + $284 + $242 + $179 + $219 = $1,236. See Figure 9.44.

THE NEAREST NEIGHBOR ALGORITHM FOR FINDING A MINIMUM HAMILTON CIRCUIT

1. From the starting vertex, go to the "nearest neighbor"—that is, use the edge with the smallest weight. If there are any ties, explore each option if the number of options is acceptably small.
2. From all other vertices, use the edge that:
   • has the smallest weight and
   • leads to a vertex that hasn't been visited yet and
   • does not lead to the starting vertex.
   If there are any ties, explore each option if the number of options is acceptably small.
3. The last edge must lead to the starting vertex.

Applications of Hamilton Circuits

Finding the cheapest route for Kim's trip is called a **traveling salesman problem.** The goal of any traveling salesman problem is to find a **minimum Hamilton circuit.** That is, the goal is to find a circuit that visits every vertex exactly once and that does so *with the least total weight.*

Finding cheap airplane routes is not the only application of Hamilton circuits. Here are two more:

• *Delivery services.* Companies such as Fed Ex (Federal Express) and UPS (United Parcel Service) deliver millions of packages every day. Using less efficient routes can be prohibitively expensive. Using Hamilton circuits can make a huge impact on a firm's profitability. Here, a vertex is a place where a package must be delivered. An edge's weight is the time it takes to drive between that edge's vertices. Typically, each truck makes between 100 and 200 stops, so one truck's graph would have between 100 and 200 vertices. And a graph with 200 vertices has $200! \approx 8 \times 10^{374}$ different Hamilton circuits. (Remember, 8×10^{374} means an 8 followed by 374 zeros.) So finding an efficient route by listing them all would be impossibly time consuming, even for the fastest of computers.

• *Circuit boards.* Computers and other electronic gadgets have circuit boards inside of them. A circuit board has thousands of small holes drilled at precise locations so that connecting electrical paths can be made between the board's two surfaces. The holes are drilled robotically. The cost of producing a circuit board is least if the order in which the holes are drilled is such that the entire drilling sequence is completed in the least amount of time. Here, a vertex is a hole, and an edge's weight is the distance between the edge's vertices.

Delivery services such as FedEx and circuit board manufacturers benefit if they use minimum Hamilton circuits.

Approximation Algorithms

The nearest neighbor algorithm is not guaranteed to find the minimum Hamilton circuit. Instead, it only approximates the minimum Hamilton circuit. For this reason, it is called an **approximation algorithm.**

There is no known algorithm that is guaranteed to find the minimum Hamilton circuit in an acceptable amount of time. Instead, there are only approximation algorithms. The brute force algorithm is usually unacceptable, because most real-world problems have too many circuits to do this. Recall that the brute force algorithm involves listing every possible circuit, adding up each circuit's total weight, and choosing the circuit with the least total weight. We used this algorithm earlier in this section when Kim visited three cities and when she visited four cities.

The Repetitive Nearest Neighbor Algorithm

Kim regretted not being able to take advantage of the $190 flight out of Houston to Miami. The nearest neighbor algorithm required that she instead take the $284 flight out of Houston to Los Angeles, because she had already been to Miami. For a moment, she entertained the absurd idea that she should move her business to Houston when she had a flash of insight: Once you have a circuit, it doesn't matter where you start because the circuit goes through all of the vertices once. So she could use the nearest neighbor algorithm with Houston as a starting vertex. Then, once the complete circuit is built, she could use that circuit, with New York City as the starting vertex. This process is illustrated in Figure 9.45.

From:	The cheapest and most usable flight is:
Houston	HOU to MIA at $190
Miami	MIA to HOU—no, been to MIA MIA to CHI at $244
Chicago	CHI to MIA—no, been to MIA CHI to SEA at $179
Seattle	SEA to NYC—no, can't go home yet SEA to CHI—no, been to Chicago SEA to LAX at $242
Los Angeles	LAX to MIA—no, been to Miami LAX to NYC at $199
New York City	NYC to HOU at $336

FIGURE 9.45 Starting to use the repetitive nearest neighbor algorithm.

The total cost is $1,390. See Figure 9.46.

Kim was disappointed because this is worse than the $1,236 that the nearest neighbor algorithm gave, starting from New York City. But this approach could still yield a better price. We have yet to try the other cities as starting vertices, and at least one of those other starting vertices yields a very good total price. We'll save the details for the exercises, but one starting city yields a total price of $1,132. See Exercise 31.

The *repetitive nearest neighbor algorithm* involves applying the nearest neighbor algorithm repeatedly, with each vertex as a starting point. Kim's trip from

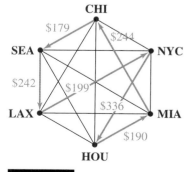

FIGURE 9.46

You can start this circuit at Houston or at New York City for a total cost of $1,390.

New York City to five other cities involves six separate applications of the nearest neighbor algorithm—one with New York City as the starting vertex, one with Houston as the starting vertex, one with Chicago as the starting vertex, and so on. This seems inefficient, but remember, there are $5! = 120$ different possible circuits. We would be checking only six circuits, not all 120 circuits. This is efficient.

THE REPETITIVE NEAREST NEIGHBOR ALGORITHM FOR FINDING A MINIMUM HAMILTON CIRCUIT

1. Apply the nearest neighbor algorithm, using any particular vertex as the starting point, and compute the total cost of the circuit obtained.
2. Apply the nearest neighbor algorithm, using another vertex as the starting point, and compute the total cost of the circuit obtained.
3. Repeat the above process until you have applied the nearest neighbor algorithm, using each vertex as the starting point. If there are any ties, explore each option if the number of options is acceptably small.
4. Use the best of the Hamilton circuits obtained in steps 1–3. If this circuit starts at an inappropriate vertex, resequence the circuit so that it starts at the appropriate vertex.

The Cheapest Edge Algorithm

The key idea behind the repetitive nearest neighbor algorithm is that you can build a circuit with any city as the starting vertex, regardless of the fact that you really need to start at one particular city. You just keep adding cheap flights. And you add them in order so that they connect up and make a circuit.

The key idea behind the cheapest edge algorithm is that you can just keep adding cheap flights so that they connect up and make a circuit, but you don't have to add them in the order that you travel.

We'll start by listing flights in order of cost. For the time being, we'll list only the flights that cost under \$300. See Figure 9.47. With any luck, we won't have to use flights that cost more than \$300.

Next, we'll start building a circuit, using the cheapest flights first. The first flight we can use is the \$98 special from NYC to MIA. See Figure 9.48.

Cost	Flight
\$98	NYC to MIA
\$124	CHI to MIA
\$132	NYC to SEA
\$169	SEA to NYC
\$177	NYC to LAX
\$179	LAX to MIA
\$179	SEA to CHI
\$179	CHI to SEA
\$190	HOU to MIA
\$199	LAX to NYC
\$204	NYC to CHI
\$214	MIA to HOU
\$219	CHI to NYC
\$242	SEA to LAX
\$242	LAX to SEA
\$244	MIA to CHI
\$284	HOU to LAX
\$286	MIA to NYC
\$288	CHI to HOU

FIGURE 9.47

Flights in order of cost.

Cost	Flight	Comment
\$98	NYC to MIA	Use it!
\$124	CHI to MIA	Can't use—already flew in to MIA.
\$132	NYC to SEA	Can't use—already flew out of NYC.

FIGURE 9.48 Starting the cheapest edge algorithm.

The second flight we can use is the $169 SEA to NYC flight. See Figure 9.49.

Cost	Flight	Comment
$169	SEA to NYC	Use it!
$177	NYC to LAX	Can't use—already flew out of NYC.
$179	LAX to MIA	Can't use—already flew in to MIA.
$179	SEA to CHI	Can't use—already flew out of SEA.

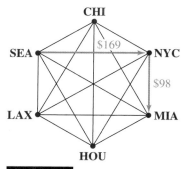

FIGURE 9.49 Continuing the cheapest edge algorithm.

The third flight we can use is the $179 CHI to SEA flight. See Figure 9.50.

Cost	Flight	Comment
$179	CHI to SEA	Use it!
$190	HOU to MIA	Can't use—already flew in to MIA.
$199	LAX to NYC	Can't use—already flew in to NYC.
$204	NYC to CHI	Can't use—already flew out of NYC.

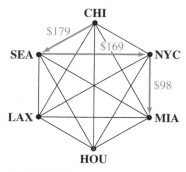

FIGURE 9.50 Continuing the cheapest edge algorithm.

The fourth flight we can use is the $214 MIA to HOU flight. See Figure 9.51.

Cost	Flight	Comment
$214	MIA to HOU	Use it!
$219	CHI to NYC	Can't use—already flew out of CHI.
$242	SEA to LAX	Can't use—already flew out of SEA.
$242	LAX to SEA	Can't use—already flew in to SEA.
$244	MIA to CHI	Can't use—already flew out of MIA.

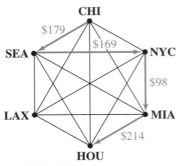

FIGURE 9.51 Continuing the cheapest edge algorithm.

The fifth flight we can use is the $284 HOU to LAX flight. See Figure 9.52. Now we have to complete the circuit with the $314 LAX to CHI flight. See Figure 9.53.

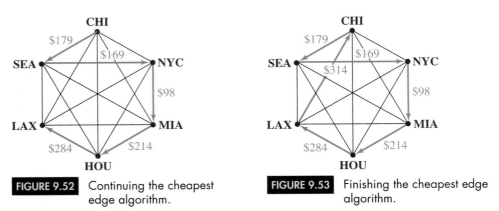

FIGURE 9.52 Continuing the cheapest edge algorithm.

FIGURE 9.53 Finishing the cheapest edge algorithm.

The total cost of this route is $1,258. For this particular problem, the repetitive nearest neighbor algorithm gives the cheaper route at $1,132. You never know which algorithm will work best until you try each one. And you never know if there is a better route unless you use the brute force algorithm and actually find the total weight for every possible route.

THE CHEAPEST EDGE ALGORITHM FOR FINDING A MINIMUM HAMILTON CIRCUIT

1. List the weights, in order, from smallest to largest.
2. Draw a complete graph for the problem.
3. Pick the edge with the smallest weight. Mark that edge on the graph from step 2. If there are any ties, explore each options if the number of options is acceptably small.
4. Continue picking edges, in order, from the list from step 1. Do not pick an edge if:
   - it leads to a vertex that you've already "flown in to."
   - it departs from a vertex that you've already "flown out of."
   - it closes a circuit that shouldn't be closed yet.

   If there are any ties, explore each options if the number of options is acceptably small.
5. When all of the vertices are joined, close the circuit.

If more than one edge will close the circuit, choose the one with the smallest weight.

Choosing an Algorithm

We have discussed four different methods:

- the brute force method
- the nearest neighbor algorithm
- the repetitive nearest neighbor algorithm
- the cheapest edge algorithm

The first method is realistic only if there aren't too many circuits to check. The last three work well if there are a lot of circuits, but they're not guaranteed to find the minimum Hamilton circuit. They are approximation algorithms, and they approximate the minimum Hamilton circuit.

It's easy to tell how many different circuits there are. Think of our experience with Kim Blum's various flights. See Figure 9.54. If Kim had to fly between n cities, there would be $(n - 1)!$ different circuits. And a complete digraph with n vertices has $(n - 1)!$ different Hamilton circuits.

When Kim had to fly between:	There were:
3 cities	$2! = 2 \cdot 1 = 2$ different circuits
4 cities	$3! = 3 \cdot 2 \cdot 1 = 6$ different circuits
6 cities	$5! = 5 \cdot 4 \cdot 3 \cdot 2 \cdot 1 = 120$ different circuits

FIGURE 9.54 Kim's travels.

When we graphed Kim's flight options, we gave each edge a direction. This was necessary because a flight's direction can affect its cost. For example, a flight from Miami to Chicago would cost Kim $244, but a flight from Chicago to Miami would cost $124. Recall that a graph with directed edges is called a *directed graph* or a *digraph*.

If the edges' weights were flight times rather than flight costs, we wouldn't give the edges directions, because a flight from Miami to Chicago takes the same amount of time as does a flight from Chicago to Miami. This means that there would be half as many different Hamilton circuits. And a complete graph with n vertices has $\frac{(n-1)!}{2}$ different Hamilton circuits.

CHOOSING AN ALGORITHM

1. Find how many different circuits your graph has.
   - A complete directed graph with n vertices has $(n - 1)!$ different Hamilton circuits.
   - A complete nondirected graph with n vertices has $\frac{(n-1)!}{2}$ different Hamilton circuits.
2. Determine the appropriate method(s).
   - If the number of different circuits is acceptably small, then use the brute force algorithm, and be certain that you've found the Hamilton circuit with the least total weight.
   - If the number of different circuits is not acceptably small, then use any (or, better yet, all) of the approximation algorithms:
     - the nearest neighbor algorithm
     - the repetitive nearest neighbor algorithm
     - the cheapest edge algorithm

Note: In the following exercises, answers will vary if a tie is encountered.

Exercises 1–4 use the flight costs shown in Figure 9.55.

1. You live in Atlanta, and you need to visit Boston, Denver, and Phoenix.
   a. How many different circuits does your graph have?
   b. Approximate the cheapest route, using the nearest neighbor algorithm. Draw this route's graph.
   c. Approximate the cheapest route, using the repetitive nearest neighbor algorithm. Draw this route's graph.
   d. Approximate the cheapest route, using the cheapest edge algorithm. Draw this route's graph.
   e. Find the cheapest route, using the brute force method. Draw this route's graph.

2. You live in Phoenix, and you need to visit Portland, San Francisco, and Washington, D.C.
   a. How many different circuits does your graph have?
   b. Approximate the cheapest route, using the nearest neighbor algorithm. Draw this route's graph.
   c. Approximate the cheapest route, using the repetitive nearest neighbor algorithm. Draw this route's graph.
   d. Approximate the cheapest route, using the cheapest edge algorithm. Draw this route's graph.
   e. Find the cheapest route, using the brute force method. Draw this route's graph.

3. You live in Atlanta, and you need to visit Boston, Denver, Phoenix, and Portland.
   a. How many different circuits does your graph have?
   b. Approximate the cheapest route, using the nearest neighbor algorithm. Draw this route's graph.
   c. Approximate the cheapest route, using the repetitive nearest neighbor algorithm. Draw this route's graph.
   d. Approximate the cheapest route, using the cheapest edge algorithm. Draw this route's graph.

   e. Why would it not be appropriate to find the cheapest route, using the brute force method?

4. You live in Denver, and you need to visit Phoenix, Portland, San Francisco, and Washington D.C.
   a. How many different circuits does your graph have?
   b. Approximate the cheapest route, using the nearest neighbor algorithm. Draw this route's graph.
   c. Approximate the cheapest route, using the repetitive nearest neighbor algorithm. Draw this route's graph.
   d. Approximate the cheapest route, using the cheapest edge algorithm. Draw this route's graph.
   e. Why would it not be appropriate to find the cheapest route, using the brute force method?

Exercises 5–12 use the FedEx travel times shown in Figure 9.56.

Exercises 5–8 involve deliveries to Kinko's, City Hall, the insurance office, the lawyer's office, and the bank.

5. You must make deliveries from the FedEx warehouse to Kinko's, City Hall, the insurance office, the lawyer's office, and the bank. Use the nearest neighbor algorithm to determine the best sequence for those deliveries.

6. You must make deliveries from the FedEx warehouse to Kinko's, City Hall, the insurance office, the lawyer's office, and the bank. Use the repetitive nearest neighbor algorithm to determine the best sequence for those deliveries.

7. You must make deliveries from the FedEx warehouse to Kinko's, City Hall, the insurance office, the lawyer's office, and the bank. Use the cheapest edge algorithm to determine the best sequence for those deliveries.

8. If you must make deliveries from the FedEx warehouse to Kinko's, City Hall, the insurance office, the lawyer's office, and the bank, how many different routes are possible? Of Exercises 5, 6, and 7, which assigned exercise generates the best delivery sequence?

			From:				
		ATL	BOS	DEN	PHX	PORT	SFO	WASH
	ATL		$104	$144	$357	$467	$154	$74
	BOS	$104		$312	$446	$179	$134	$54
To:	DEN	$144	$310		$156	$122	$192	$175
	PHX	$447	$444	$156		$104	$136	$182
	PORT	$216	$177	$122	$104		$84	$492
	SFO	$154	$132	$192	$136	$84		$124
	WASH	$74	$52	$175	$182	$492	$144	

FIGURE 9.55 Flight costs between Atlanta, Boston, Denver, Phoenix, Portland, San Francisco, and Washington, D.C. *Source:* www.travelocity.com.

	Kinko's	City Hall	Insurance office	Lawyer's office	Bank	Real estate office
FedEx warehouse	14	22	10	35	27	11
Kinko's		17	8	32	15	8
City Hall			14	37	18	11
Insurance office				15	25	17
Lawyer's office					19	30
Bank						29

FIGURE 9.56 FedEx travel times in minutes.

Exercises 9–12 involve deliveries to City Hall, the insurance office, the lawyer's office, the real estate office, and the bank.

9. You must make deliveries from the FedEx warehouse to City Hall, the insurance office, the lawyer's office, the real estate office, and the bank. Use the nearest neighbor algorithm to determine the best sequence for those deliveries.

10. You must make deliveries from the FedEx warehouse to City Hall, the insurance office, the lawyer's office, the real estate office, and the bank. Use the repetitive nearest neighbor algorithm to determine the best sequence for those deliveries.

11. You must make deliveries from the FedEx warehouse to City Hall, the insurance office, the lawyer's office, the real estate office, and the bank. Use the cheapest edge algorithm to determine the best sequence for those deliveries.

12. If you must make deliveries from the FedEx warehouse to City Hall, the insurance office, the lawyer's office, the real estate office, and the bank, how many different routes are possible? Of Exercises 9, 10, and 11, which assigned exercise generates the best delivery sequence?

Use the following information in Exercises 13–22.

You are programming a robotic drill to drill holes in a circuit board. After the holes have been drilled, electronic components will be inserted into them. You use an ordered pair to specify a hole's location. For example, the ordered pair (1, 2) specifies a location that is 1 mm to the right and 2 mm above the board's lower left corner, as shown in Figure 9.57. The drill is programmed to return to (0, 0) after completing a circuit board. The cost of drilling a board is determined by the distance the drill must travel, and the drill moves only horizontally and vertically. For example, if the drill starts at (0, 0) and then travels to (1, 2) to drill a hole, it travels horizontally 1 mm and vertically 2 mm, for a total of 3 mm.

13. The machine drills holes at (4, 1) and then (3, 7).

    a. How far does the drill travel when it moves from (0, 0) to (4, 1)?

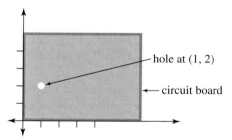

FIGURE 9.57 A hole in a circuit board at (1, 2).

    b. How far does the drill travel when it moves from (4, 1) to (3, 7)?

    c. How far does the drill travel when it moves from (3, 7) to (0, 0)?

    d. How far does the drill travel in total?

14. The machine drills holes at (5, 2) and then (1, 1).

    a. How far does the drill travel when it moves from (0, 0) to (5, 2)?

    b. How far does the drill travel when it moves from (5, 2) to (1, 1)?

    c. How far does the drill travel when it moves from (1, 1) to (0, 0)?

    d. How far does the drill travel in total?

15. A board needs holes at (4, 5), (3, 1), (6, 4), (7, 5), and (2, 2). Use the nearest neighbor algorithm to determine a sequence for the drilling of those holes.

16. A board needs holes at (4, 5), (3, 1), (6, 4), (7, 5), and (2, 2). Use the repetitive nearest neighbor algorithm to determine a sequence for the drilling of those holes.

17. A board needs holes at (4, 5), (3, 1), (6, 4), (7, 5), and (2, 2). Use the cheapest edge algorithm to determine a sequence for the drilling of those holes.

18. If a board needs holes at (4, 5), (3, 1), (6, 4), (7, 5), and (2, 2), how many different routes are possible? Of Exercises 15, 16, and 17, which assigned exercise generates the best drilling sequence?

19. A board needs holes at (1, 8), (2, 7), (5, 1), (3, 4), and (4, 3). Use the nearest neighbor algorithm to determine a sequence for the drilling of those holes.

20. A board needs holes at (1, 8), (2, 7), (5, 1), (3, 4), and (4, 3). Use the repetitive nearest neighbor algorithm to determine a sequence for the drilling of those holes.

21. A board needs holes at (1, 8), (2, 7), (5, 1), (3, 4), and (4, 3). Use the cheapest edge algorithm to determine a sequence for the drilling of those holes.

22. If a board needs holes at (1, 8), (2, 7), (5, 1), (3, 4), and (4, 3), how many different routes are possible? Of Exercises 19, 20, and 21, which assigned exercise generates the best drilling sequence?

23. Use the information in Figure 9.55 and the nearest neighbor algorithm to find the cheapest route between all seven cities.

24. Use the information in Figure 9.56 and the nearest neighbor algorithm to find the best delivery sequence from the FedEx warehouse to Kinko's, City Hall, the insurance office, the lawyer's office, the bank, and the real estate office.

25. Use the information in Figure 9.55 and the repetitive nearest neighbor algorithm to find the cheapest route between all seven cities.

26. Use the information in Figure 9.56 and the repetitive nearest neighbor algorithm to find the best delivery sequence from the FedEx warehouse to Kinko's, City Hall, the insurance office, the lawyer's office, the bank, and the real estate office.

27. Use the information in Figure 9.55 and the cheapest edge algorithm to find the cheapest route between all seven cities.

28. Use the information in Figure 9.56 and the cheapest edge algorithm to find the best delivery sequence from the FedEx warehouse to Kinko's, City Hall, the insurance office, the lawyer's office, the bank, and the real estate office.

29. Of Exercises 23, 25, and 27, which assigned exercise generates the cheapest route?

30. Of Exercises 24, 26, and 28, which assigned exercise generates the best delivery sequence?

31. Find the result of applying the repetitive nearest neighbor algorithm to Kim Blum's trip from New York to all five cities: Chicago, Miami, Houston, Los Angeles, and Seattle.

WEB PROJECT

32. Investigate traveling salesman problems on the web. What do they have to do with Hamilton circuits? Summarize their history. Describe specific problems that led to progress.

Some useful links for this web project are listed on the text companion web site. Go to **www.cengage.com/math/johnson** to access the web site.

9.4 Networks

OBJECTIVES

- Understand what a *tree* and a *spanning tree* are
- Be able to apply Kruskal's algorithm
- Be able to locate and use Steiner points to find a minimum spanning tree

Installing a Network

Sarah and Miles are joining the growing ranks of telecommuters. Each has an office in their home. They are going to install a network so that they can share one internet connection and one color laser printer. For security reasons, their employers have insisted that they have a hardwired network rather than a wireless network.

Their daughter Liz and their son Pete are working on degrees at State University. Each has a laptop computer, which they use in their bedrooms. So altogether, the network should connect four computers.

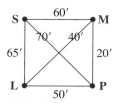

FIGURE 9.58

Amounts of wire needed.

Sarah and Miles looked into having the network installed by a professional. However, the cost was rather excessive—labor ran over $100 an hour. After buying a book on networks, they realized that they could do the work themselves.

They measured the various distances between the rooms. They found that they would need 60 feet of wire to form a link between Sarah's and Miles's offices. These and other measurements are shown in the complete weighted graph in Figure 9.58.

Avoiding Unnecessary Connections

Sarah and Miles quickly realized that it is not necessary to wire up every edge of the graph in Figure 9.58. In particular, there is no reason to form a circuit. For example, if they wired up the 60' edge joining Sarah's and Miles's office, and the 40' edge joining Miles's office and Liz's room, then Liz's room and Sarah's office would be connected through Miles's office. In this case, there would be no need to wire up the 65' edge joining Liz's room and Sarah's office. See Figure 9.59. And avoiding extra connections means saving money.

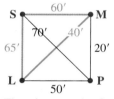

There is no need to form this blue circuit . . .

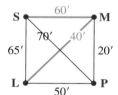

because these two blue edges connect S, M and L.

FIGURE 9.59

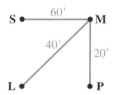

FIGURE 9.60

A partially completed network.

Choose the Shortest Edges First

Sarah and Miles decided that their network should definitely involve wiring up the 20' edge joining Miles's office and Pete's room, because that's the shortest (and therefore cheapest) connection possible.

They also decided that their network should include the second-shortest connection, which is the 40' edge joining Miles's office and Liz's room. These two decisions gave them the partially completed network shown in Figure 9.60. The network isn't complete yet, so we'll continue to show all the edges.

The third-shortest connection is the 50' edge joining Pete's room and Liz's room. But there's no need to wire up this edge because Pete's room and Liz's room are already connected through Miles's office.

The fourth-shortest connection is the 60' edge joining Sarah's office and Miles's office. This connection completes the network, because all four rooms are now connected. See Figure 9.61.

FIGURE 9.61

Sarah's and Miles's network.

Trees

The network in Figure 9.61 is an example of a *tree*. A **tree** is a connected graph that has no circuits. The graph of Sarah's and Miles's network must be connected because if it weren't, someone's computer wouldn't have access to the internet connection. And as was discussed previously, it's unnecessary to form a circuit.

There are many other trees that would describe a functional network for Sarah and Miles. For example, the tree in Figure 9.62 would work. However, it involves

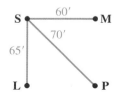

FIGURE 9.62

A tree describing a more expensive network.

60' + 70' + 65' = 195' of wire, so it would be more expensive and probably more time consuming to install.

The trees in Figure 9.61 and 9.62 are both **spanning trees** because they both include all of the vertices in the original graph. (Here, *span* means "to reach over or across something.") However, the spanning tree in Figure 9.61 has less total weight than does the spanning tree in Figure 9.62. In fact, it has the least possible weight of all spanning trees. For this reason, it is called the **minimum spanning tree.**

Kruskal's Algorithm

While it seems to be just common sense, the method that Sarah and Miles used to design their network guarantees a minimum spanning tree. It is called *Kruskal's algorithm,* named after the Bell Labs mathematician who described it in 1956. Kruskal's algorithm is very similar to the cheapest edge algorithm discussed in Section 9.3. However the cheapest edge algorithm is an approximation algorithm, so it does not guarantee a minimum Hamilton circuit.

KRUSKAL'S ALGORITHM FOR FINDING A MINIMUM SPANNING TREE

1. List the weights, in order, from smallest to largest.
2. Draw a complete graph for the problem.
3. Pick the edge with the smallest weight. (In case of a tie, pick one at random.) Mark that edge on the graph from step 2.
4. Continue picking edges, in order, from the list from step 1. Do not pick an edge if it closes a circuit.
5. Stop adding edges when all the vertices are joined.

EXAMPLE **1**

FINDING A MINIMUM SPANNING TREE   You have been hired to design a network linking Atlanta, Boston, Chicago, and Kansas City. Find the minimum spanning tree connecting the cities and its total length. See Figure 9.63.

	Boston	**Chicago**	**Kansas City**
Atlanta	932	585	678
Boston		851	1247
Chicago			407

FIGURE 9.63   Distances between cities, in miles.

SOLUTION

Step 1

List the weights, in order, from smallest to largest.
- 407: Chicago/Kansas City
- 585: Atlanta/Chicago
- 678: Atlanta/Kansas City
- 851: Boston/Chicago
- 932: Atlanta/Boston
- 1,247: Boston/Kansas City

Step 2

Draw a complete graph for the problem. See Figure 9.64.

Step 3

Pick the edge with the smallest weight.

- 407: Chicago/Kansas City

See Figure 9.65.

Step 4

Continue picking edges, in order.

- 585: Atlanta/Chicago
- 678: Atlanta/Kansas City—do not use because it closes a circuit.
- 851: Boston/Chicago

See Figure 9.65.

Step 5

Stop adding edges when all the vertices are joined. We're done. The minimum spanning tree links Chicago/Boston, Chicago/Atlanta, and Chicago/Kansas City. Its total length is 851 + 585 + 407 = 1,742 miles.

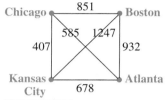

FIGURE 9.64

Step 2: Draw a complete graph.

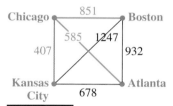

FIGURE 9.65

Steps 3 and 4.

Adding a Vertex

With Sarah's and Miles's network, all lines come together at Miles's office, as shown in Figure 9.61. This means that they would install a *router* in Miles's office. A router is a small, specialized computer that controls the various computer and internet connections. The four computers, the color laser printer, and the internet all plug into the router.

Sarah suggested to Miles that if they put the router in the attic, they might end up with shorter connections. They measured the appropriate distances and came up with a new design. That new design is shown in Figure 9.66.

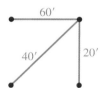

By adding a vertex they could replace this . . .

with an even shorter minimum spanning tree.

FIGURE 9.66

Using the language of graph theory, putting the router in the attic means adding a vertex to the graph. And adding a vertex might allow Sarah and Miles to replace their minimum spanning tree with a new minimum spanning tree that has less total weight.

Where to Add a Vertex

Suppose you were designing a telephone network that would connect three desert towns: Badwater, Dry Springs, and Hotspot. Each of these towns is 300 miles from each of the other two towns.

A minimum spanning tree consists of any two of the three edges, with a total length of 600 miles. See Figure 9.67 on the next page.

There is nothing but desert between the towns, so we are free to run the telephone cables anywhere we want. In particular, we could shorten the edge that connects Hotspot to the rest of the network by adding a vertex somewhere along the Dry Springs/Badwater edge and connecting there instead of at Dry Springs. See Figure 9.68.

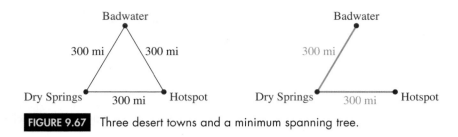

FIGURE 9.67   Three desert towns and a minimum spanning tree.

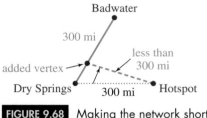

FIGURE 9.68   Making the network shorter by adding a vertex and shortening an edge.

A 90° angle gives the shortest Hotspot edge, as shown in Figure 9.69. This edge works out to be 259.8 miles long, so the network is $300 + 259.8 = 559.8$ miles. By not connecting Hotspot directly to Dry springs, we shortened the network by 40 miles.

Perhaps we can do even better by not connecting Dry Springs directly to Badwater. Lets move the newly added vertex closer to Hotspot as shown in Figure 9.69.

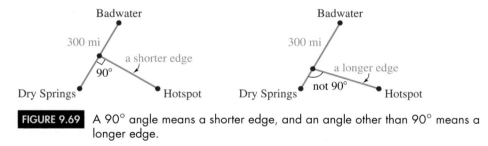

FIGURE 9.69   A 90° angle means a shorter edge, and an angle other than 90° means a longer edge.

One possibility is shown in Figures 9.70 and 9.71. Here, we moved the added vertex so that the edges form three 120° angles.

FIGURE 9.70   Moving the vertex might result in an even shorter minimum spanning tree.

FIGURE 9.71   Adding a vertex so we get three 120° angles.

Steiner Points

The spanning tree in Figure 9.71 is the shortest network connecting the three cities. It is 519.6 miles long. It is impossible to find a shorter network. The added vertex at the center of this network is called a *Steiner point*. A **Steiner point** is a vertex where three edges come together, forming three 120° angles. Usually, but not always, shortest networks involve one or more Steiner points.

EXAMPLE 2

SOLUTION

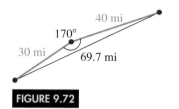

FIGURE 9.72

For Example 2.

USING A STEINER POINT Determine whether adding a Steiner point forms the shortest network connecting the points in Figure 9.72.

If we don't add a Steiner point, we get a spanning tree of length 30 mi + 40 mi = 70 mi.

As we see in Figure 9.73, it is impossible to add a Steiner point vertex. The 170° angle is just too big for a Steiner point to work.

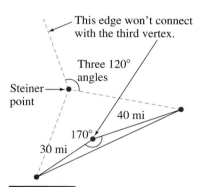

FIGURE 9.73 Unsuccessfully trying to add a Steiner point vertex.

In Example 2, we were unsuccessful in adding a Steiner point vertex because of the 170° angle. It's easy to see that we would have the same difficulty if the angle were 160° or 150°. In fact, this difficulty arises whenever a triangle has an angle of 120° or more.

THE SHORTEST NETWORK CONNECTING THREE POINTS

If the three points form a triangle with angles that are less than 120°, then the shortest network linking the three points is obtained by adding a Steiner point inside the triangle, as shown in blue in Figure 9.74.

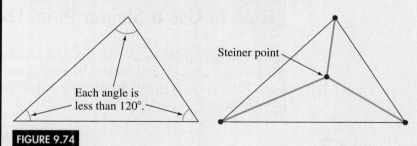

FIGURE 9.74

If the three points form a triangle with an angle that is 120° or more, then the shortest network consists of the two shortest sides of the triangle, as shown in blue in Figure 9.75.

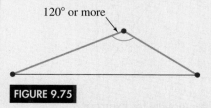

FIGURE 9.75

Locating Steiner Points

We'll use a *Steiner point locator* to find Steiner points. To make a Steiner point locator, you will need a protractor and either a piece of thin but durable paper or a transparency.

HOW TO MAKE A STEINER POINT LOCATOR

1. Use the ruler part of a protractor to carefully draw a horizontal line from the center of the paper to one edge of the paper, as shown in Figure 9.76.
2. Use the protractor to draw a second line that forms a 120° angle with the first line, as shown in Figure 9.76.
3. Use the protractor to draw a third line that forms 120° angles with the lines from steps 1 and 2, as shown in Figure 9.76.
4. Make a small hole where the three lines come together.

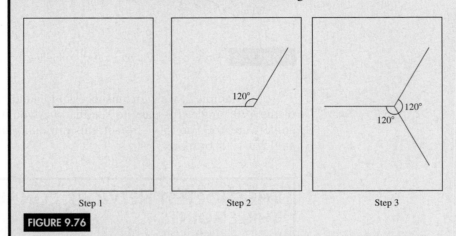

Step 1   Step 2   Step 3

FIGURE 9.76

How to Use a Steiner Point Locator

Place a Steiner point locator on top of a drawing of a triangle. Move the Steiner point locator around until each of its edges goes through one of the triangle's vertices. Then use your pencil to put make a mark through the small hole in the Steiner point locator.

EXAMPLE **3**

USING A STEINER POINT LOCATOR Use a Steiner point locator to locate the Steiner point for the triangle shown in Figure 9.77.

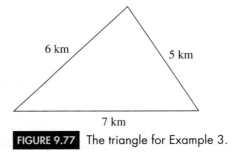

FIGURE 9.77 The triangle for Example 3.

SOLUTION

First, make an accurate drawing of the triangle. Let 1 km correspond to 1 in. A compass can be helpful in doing this. Next, place your Steiner point locator on top of your drawing, as shown in Figure 9.78.

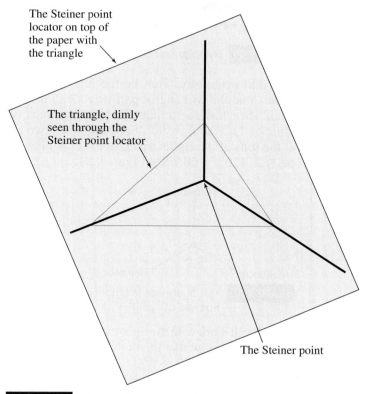

The Steiner point locator on top of the paper with the triangle

The triangle, dimly seen through the Steiner point locator

The Steiner point

FIGURE 9.78   Using a Steiner point locator.

Once the Steiner point has been found, it's easy to use a ruler to measure the distances.

For Those Who Have Read Section 8.5, "Right Triangle Trigonometry"

EXAMPLE **4**

USING TRIGONOMETRY   Find the total length of the Badwater/Dry Springs/ Hotspot network with three 120° angles, as shown in Figure 9.79.

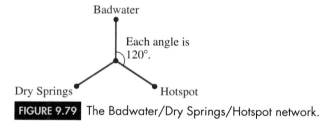

Badwater

Each angle is 120°.

Dry Springs   Hotspot

FIGURE 9.79   The Badwater/Dry Springs/Hotspot network.

SOLUTION

To use trigonometry, we need to find a right triangle. Reinserting the Dry Springs/ Hotspot edge gives us two triangles. Continuing the Badwater edge straight down gives us what appears to be two right triangles. See Figure 9.80.

The fact that they *appear* to be right triangles doesn't mean that they *are* right triangles. Let's see if they are. The minimum spanning tree in Figure 9.80 has

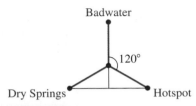

FIGURE 9.80 Trying to find a right triangle.

left/right symmetry. That is, the left side and the right side are identical. This means that the two angles part way along the Dry Springs/Hotspot edge are identical. They form a straight line, so they must add to 180°. This means that each must be $1/2 \cdot 180° = 90°$. So we do have two right triangles. Also, the two angles at the tops of the rights triangles are equal in size. They add to 120°, so each must be $1/2 \cdot 120° = 60°$. See Figure 9.81.

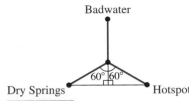

FIGURE 9.81 This figure has left/right symmetry, so the triangle has two 60° angles and two 90° angles.

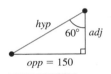

FIGURE 9.82

The right triangle on the Dry Springs side.

We'll work with the right triangle on the Dry Springs side, shown in Figure 9.82. We already know that the bottom edge's length is 300, so the left and right halves must each be $\frac{1}{2} \cdot 300 = 150$.

We need to know the length of the Dry Springs edge, which is the hypotenuse. And we know that the opposite is 150. So we'll use the sine ratio.

$$\sin \theta = \frac{opp}{hyp} \qquad \textbf{definition of the sine ratio}$$

$$\sin 60° = \frac{150}{hyp} \qquad \textbf{substituting for } \textit{opp}$$

$$hyp \cdot \sin 60° = 150 \qquad \textbf{multiplying by } \textit{hyp}$$

$$hyp \cdot \frac{\sqrt{3}}{2} = 150 \qquad \textbf{since } \sin 60° = \frac{\sqrt{3}}{2}$$

$$hyp \cdot \frac{\sqrt{3}}{2} \cdot \frac{2}{\sqrt{3}} = 150 \cdot \frac{2}{\sqrt{3}} \qquad \textbf{solving for } \textit{hyp}$$

$$hyp = \frac{300}{\sqrt{3}} \approx 173.2$$

This makes each edge 173.2 miles, and the network has a total length of $3 \cdot 173.2 = 519.6$ miles, which is significantly shorter than the earlier networks of 559.8 miles and 600 miles. See Figure 9.83.

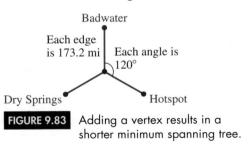

FIGURE 9.83 Adding a vertex results in a shorter minimum spanning tree.

FEATURED IN THE NEWS

SOLUTION TO AN OLD PUZZLE: HOW SHORT A SHORTCUT?

Two mathematicians have solved an old problem in the design of networks that has enormous practical importance but has baffled some of the sharpest minds in the business.

Dr. Frank Hwang of A.T. & T. Bell Laboratories in Murray Hill, N. J., and Dr. Ding Xhu Du, a postdoctoral student at Princeton University, announced at a meeting of theoretical computer scientists last week that they had found a precise limit to the design of paths connecting three or more points.

Designers of things like computer circuits, long-distance telephone lines, or mail routings place great stake in finding the shortest path to connect a set of points, like the cities an airline will travel to or the switching stations for long-distance telephone lines. Dr. Hwang and Dr. Du proved, without using any calculations, that an old trick of adding extra points to a network to make the path shorter can reduce the length of the path by no more than 13.4 percent.

"This problem has been open for 22 years," said Dr. Ronald L. Graham, a research director at A. T. & T. Bell Laboratories who spent years trying in vain to solve it. "The problem is of tremendous interest at Bell Laboratories, for obvious reasons."

In 1968, two other mathematicians guessed the proper answer, but until now no one had proved or disapproved their conjecture.

In the tradition of Dr. Paul Erdos, an eccentric Hungarian mathematician who offers prizes for solutions to hard problems that interest him, Dr. Graham offered $500 for the solution to this one. He said he is about to mail his check.

ADD POINT, REDUCE PATH

The problem has its roots in an observation made by mathematicians in 1640.

They noticed that if you want to find a path connecting three points that form the vertices of an equilateral triangle, the best thing to do is to add an extra point in the middle of the set. By connecting these four points, you can get a shorter route around the original three than if you had simply drawn lines between those three points in the first place.

Later, mathematicians discovered many more examples of this paradoxical phenomenon. It turned out to be generally true that if you want to find the shortest path connecting a set of points, you can shorten the path by adding an extra point or two.

But it was not always clear where to add the extra points, and even more important, it was not always clear how much you would gain. In 1977, Dr. Graham and his colleagues, Dr. Michael Garey and Dr. David Johnson of Bell Laboratories, proved that there was no feasible way to find where to place the extra points, and since there are so many possibilities, trial and error is out of the question.

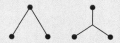

The Shorter Route

The length of string needed to join the three corners of a triangle is shorter if a point is added in the middle, as in the figure at right.

Source: F. K. Hwang

These researchers and others found ways to guess where an extra point or so might be beneficial, but they were always left with the nagging question of whether this was the best they could do.

The conjecture, made in 1968 by Dr. Henry Pollack, who was formerly a research director at Bell Laboratories, and Dr. Edger Gilbert, a Bell mathematician, was that the best you could

ever do by adding points was to reduce the length of the path by the square root of three divided by two, or about 13.4 percent. This was exactly the amount the path was reduced in the original problem with three vertices of an equilateral triangle.

KNOWING YOUR LIMITS

In the years since, researchers proved that the conjecture was true if you wanted to connect four points, then they proved it for five.

"Everyone who contributed devised some special trick," Dr. Hwang said. "But still, if you wanted to generalize to the next number of points, you couldn't do it."

Dr. Du and Dr. Hwang's proof relied on converting the problem into one involving mathematics of continuously moving variables. The idea was to suppose that there is an arrangement for which added points make you do worse than the square root of three over two. Then, Dr. Du and Dr. Hwang showed, you can slide the points in your network around and see what happens to your attempt to make the connecting path shorter. They proved that there was no way you could slide points around to do any better than a reduction in the path length of the square root of three over two.

Dr. Graham said the proof should allow network designers to relax as they search for the shortest routes. "If you can never save more than the square route of three over two, it may not be worth all the effort it takes to look for the best solution" he said.

By GINA KOLATA
Source: New York Times, Oct. 30, 1990. Copyright © 1990 by the New York Times Co. Reprinted with permission.

In Exercises 1–12, determine whether the given graph is a tree. If not, say why.

1.

2.

3.

4.

5.

6.

7.

8.

9.

10.

11.

12.

In Exercises 13–18, find a spanning tree for the given graph. Describe it by listing its edges. Note: Answers will vary, so answers are not given in the back of the book.

13.

14.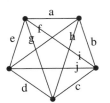

15.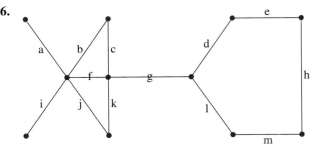

16.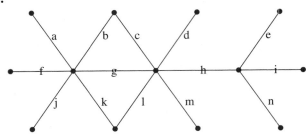

17.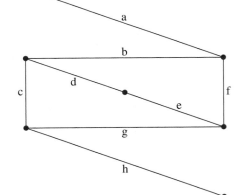

18.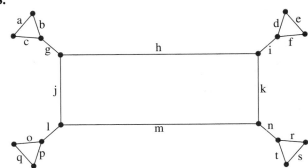

In Exercises 19–24, use Kruskal's algorithm to find the minimum spanning tree for the given weighted graph. Give the total weight of the minimum spanning tree.

19.
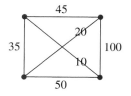

20.
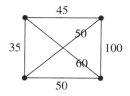

21.
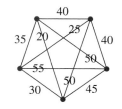

22.
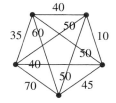

23.
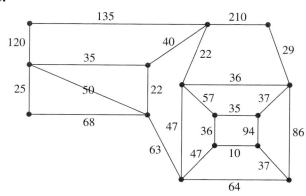

24.
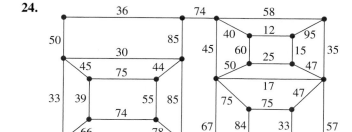

In Exercises 25–30, use the information in Figure 9.84. The distances given are not along roads. Rather, they are "as the crow flies" distances.

25. You have been hired to design a network linking New York City, Chicago, Boston, Kansas City, and San Francisco. Using only those cities as vertices, find the minimum spanning tree connecting the cities and its total length.

26. You have been hired to design a network linking Atlanta, Denver, Philadelphia, Portland, and San Francisco. Using only those cities as vertices, find the minimum spanning tree connecting the cities and its total length.

27. You have been hired to design a network linking Atlanta, Denver, Philadelphia, Portland, New York City, and San Francisco. Using only those cities as vertices, find the minimum spanning tree connecting the cities and its total length.

28. You have been hired to design a network linking New York City, Chicago, Boston, Kansas City, Philadelphia,

	Boston	Chicago	Denver	Kansas City, MO	New York City	Philadelphia	Portland, OR	San Franciso
Atlanta	932	585	1213	678	744	658	2170	2138
Boston		851	1767	1247	187	273	2536	2696
Chicago			917	407	714	664	1755	1855
Denver				558	1630	1574	979	947
Kansas City, MO					1094	1032	1493	1500
New York						86	2442	2570
Philadel-phia							2407	2518
Portland, OR								538

FIGURE 9.84   Distances between cities, in miles.

and San Francisco. Using only those cities as vertices, find the minimum spanning tree connecting the cities and its total length.

29. You have been hired to design a network linking Atlanta, New York City, Chicago, Boston, Kansas City, Philadelphia, Portland, and San Francisco. Using only those cities as vertices, find the minimum spanning tree connecting those cities and its total length.

30. You have been hired to design a network linking Atlanta, New York City, Boston, Denver, Kansas City, Philadelphia, Portland, and San Francisco. Using only those cities as vertices, find the minimum spanning tree connecting those cities and its total length.

31. Figure 9.85 shows the floor plan of Pinstripe National Bank's new information technologies office. Each vertex represents a cubicle, and the edges' weights represent the distances between the cubicles. The cubicles are laid out in a grid pattern, so edges are parallel. You have been asked to design a computer network linking the cubicles. Without adding a vertex, find the minimum spanning tree connecting the cubicles and its total length.

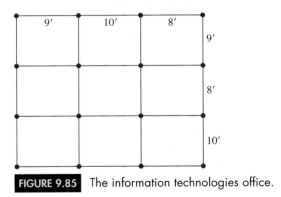

FIGURE 9.85 The information technologies office.

32. Figure 9.86 shows the floor plan of Second National Bank's new web development office. Each vertex represents a cubicle, and the edges' weights represent the distances between the cubicles. The cubicles are laid out in a grid pattern, so edges are parallel. You have been asked to design a computer network linking the cubicles. Without adding a vertex, find the minimum spanning tree connecting the cubicles and its total length.

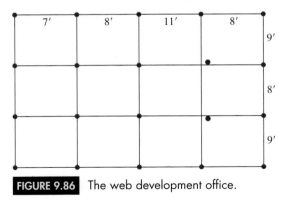

FIGURE 9.86 The web development office.

33. A triangle has three vertices: X, Y, and Z. Inside the triangle are three points: A, B, and C, one of which is a Steiner point. The distances between various points are given in Figure 9.87. Which point, A, B, or C, is a Steiner point? Why?

	A	B	C
X	30	40	50
Y	40	30	50
Z	80	70	60

FIGURE 9.87 Data for Exercise 33.

34. A triangle has three vertices: X, Y, and Z. Inside the triangle are three points: A, B, and C, one of which is a Steiner point. The distances between various points are given in Figure 9.88. Which point, A, B, or C, is a Steiner point? Why?

	A	B	C
X	50	60	70
Y	50	70	40
Z	60	40	60

FIGURE 9.88 Data for Exercise 34.

35. Find the length of the network in Example 3.

36. Use a Steiner point locator to find the shortest network for a triangle with sides 7 mi, 8 mi, and 10 mi and its length.

37. Use a Steiner point locator to find the shortest network for a triangle with sides 8 mi, 9 mi, and 10 mi and its length.

38. Use a Steiner point locator to find the shortest network for a triangle with sides 9 mi, 10 mi, and 12 mi and its length.

39. Use a Steiner point locator and the map of Nevada in Figure 9.89 on page 709 to find the shortest network joining Reno, Las Vegas, and Wells and its length.

40. Use a Steiner point locator and the map of Arizona in Figure 9.90 on page 709 to find the shortest network joining Phoenix, Winslow, and Prescott and its length.

41. Use a Steiner point locator and the map of Florida in Figure 9.91 on page 710 to find the shortest network joining Tampa, Orlando, and Miami and its length.

42. Use a Steiner point locator and the map of New York in Figure 9.92 on page 710 to find the shortest network joining Poughkeepsie, Rochester, and Troy and its length.

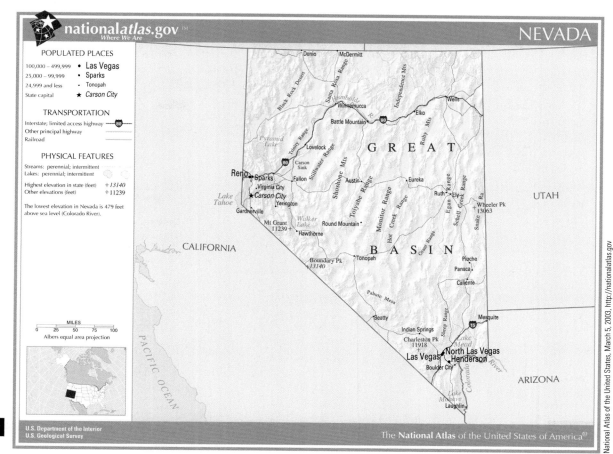

FIGURE 9.89

Nevada.

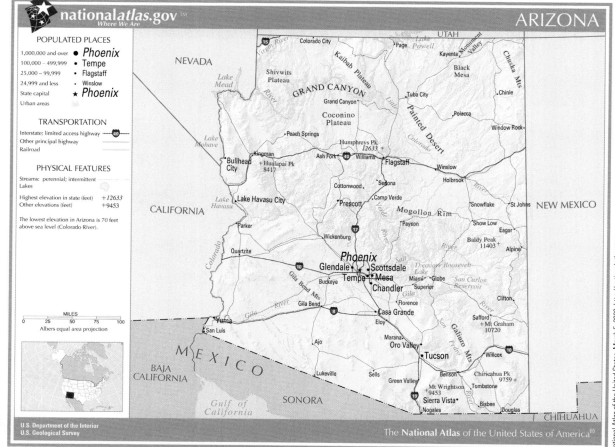

FIGURE 9.90

Arizona.

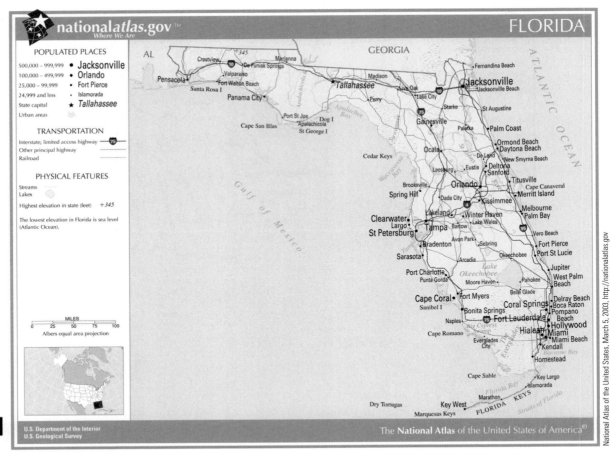

FIGURE 9.91

Florida.

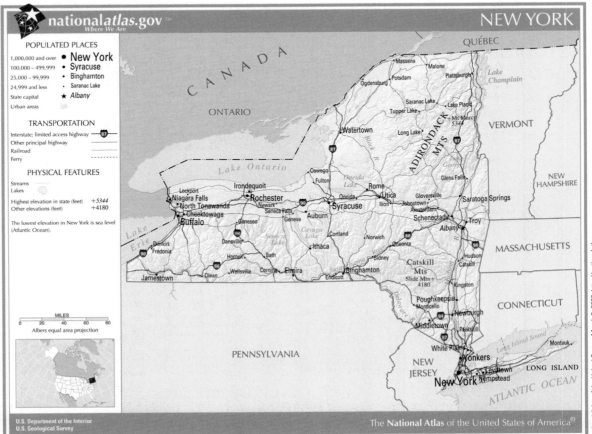

FIGURE 9.92

New York.

Exercises 43–60 are for those who have read Section 8.5: Right Triangle Trigonometry.

43. a. Use trigonometry and the combined $20° + 20° = 40°$ angle at the left of the big triangle in Figure 9.93 to find the size of $x + y$.

b. Use trigonometry and the $20°$ angle at the left of the lower triangle to find the size of x.

c. Use the answers to parts (a) and (b) to find the size of y.

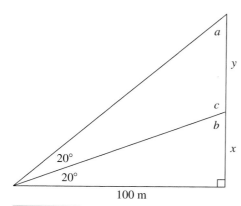

FIGURE 9.93 A triangle for Exercises 43 and 45.

44. a. Use trigonometry and the combined $10° + 15° = 25°$ angle at the left of the big triangle in Figure 9.94 to find the size of $x + y$.

b. Use trigonometry and the $15°$ angle at the left of the lower triangle to find the size of x.

c. Use the answers to parts (a) and (b) to find the size of y.

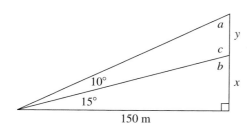

FIGURE 9.94 A triangle for Exercises 44 and 46.

45. a. Use the combined $20° + 20° = 40°$ angle at the left of the big triangle in Figure 9.93 and the fact that the three angles of a triangle always add to $180°$ to find the size of angle a.

b. Use the $20°$ angle at the left of the lower triangle and the fact that the three angles of a triangle always add to $180°$ to find the size of angle b.

c. Use the answer to part (b) to find the size of angle c.

46. a. Use the combined $10° + 15° = 25°$ angle at the left of the big triangle in Figure 9.94 and the fact that the three angles of a triangle always add to $180°$ to find the size of angle a.

b. Use the $15°$ angle at the left of the lower triangle and the fact that the three angles of a triangle always add to $180°$ to find the size of angle b.

c. Use the answer to part (b) to find the size of angle c.

47. a. Use trigonometry and the combined $20° + 25° = 45°$ angle at the left of the big triangle in Figure 9.95 to find the size of $x + y$.

b. Use trigonometry and the $25°$ angle at the left of the lower triangle to find the size of x.

c. Use the answers to parts (a) and (b) to find the size of y.

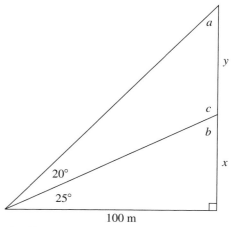

FIGURE 9.95 A triangle for Exercises 47 and 49.

48. a. Use trigonometry and the combined $20° + 15° = 35°$ angle at the left of the big triangle in Figure 9.96 to find the size of $x + y$.

b. Use trigonometry and the $15°$ angle at the left of the lower triangle to find the size of x.

c. Use the answers to parts (a) and (b) to find the size of y.

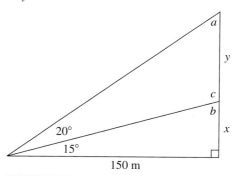

FIGURE 9.96 A triangle for Exercises 48 and 50.

49. a. Use the combined $20° + 25° = 45°$ angle at the left of the big triangle in Figure 9.95 and the fact that the three angles of a triangle always add to $180°$ to find the size of angle a.

b. Use the $25°$ angle at the left of the lower triangle and the fact that the three angles of a triangle always add to $180°$ to find the size of angle b.

c. Use the answer to part (b) to find the size of angle c.

50. a. Use the combined $20° + 15° = 35°$ angle at the left of the big triangle in Figure 9.96 and the fact that the three angles of a triangle always add to $180°$ to find the size of angle a.

b. Use the $15°$ angle at the left of the lower triangle and the fact that the three angles of a triangle always add to $180°$ to find the size of angle b.

c. Use the answer to part (b) to find the size of angle c.

In Exercises 51–54, use symmetry, trigonometry, and Steiner points to find the shortest network connecting the three cities, and the length of that network.

51.

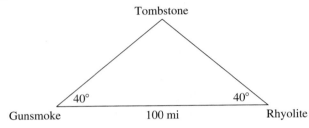

52.

53.

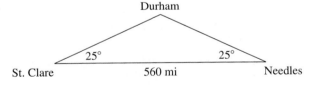

54.

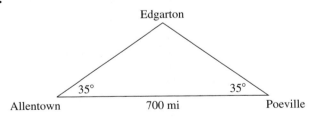

55. Which is the shortest network? What is its length? All interior points are Steiner points, and the object is a rectangle.

a. 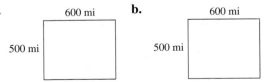 **b.**

c. 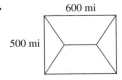 **d.**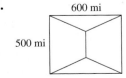

56. Which is the shortest network? What is its length? All interior points are Steiner points, and the object is a rectangle.

a.

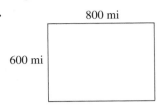

b.

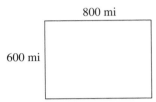

c.

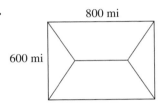

d.

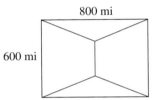

In Exercises 57–60, use symmetry, trigonometry, and Steiner points to find the shortest network and the length of that network.

57. Find the shortest network and its length for a rectangle that is 700 miles wide and 600 miles long.

58. Find the shortest network and its length for a rectangle that is 700 miles wide and 800 miles long.

59. Which is the shortest network? What is its length? The interior points in (c) are Steiner points, and the object is a square.

a.

b.

c.

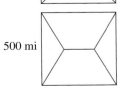

60. Find the shortest network and its length for a square that is 800 miles wide.

Answer the following questions using complete sentences and your own words.

• CONCEPT QUESTIONS

61. In the *New York Times* article on page 705, Dr. Ronald Graham is quoted as saying "The problem is of tremendous interest . . . for obvious reasons." What are the obvious reasons?

62. Figure 9.97 shows a Japanese stamp issued in honor of TPC-3 ("Trans-Pacific Cable number 3"). This was the first fiber-optic telephone cable across the Pacific. It links Japan, Guam, and Hawaii. It was constructed by a consortium of what were then the worlds largest telephone companies: AT&T, Sprint, MCI, and British Telephone.

   a. TPC-3, the network colored purple on the stamp, links Hawaii, Japan, and Guam. What type of network is this? Why was this type of network used?

FIGURE 9.97 A Japanese stamp in honor of TPC-3.

   b. HAW-4, the network colored blue on the stamp, links Hawaii and the west coast of the United States. What type of network is this? Why was this type of network used?

   c. Japan is the purple island on the stamp. It is 3,910 miles from Hawaii and 1,620 miles from Guam. Guam is the blue dot directly below Japan. It is 3,820 miles from Hawaii. Use this information and a Steiner point locator to approximate the length of the network linking Hawaii, Japan, and Guam.

9.5 Scheduling

OBJECTIVES

- Understand the terms *sequential tasks*, *parallel tasks*, and *limiting tasks*
- Be able to use a PERT chart to efficiently schedule a project
- Be able to use a Gantt chart to display an efficient schedule

Project managers and other people involved in scheduling projects typically use PERT charts and Gantt charts in producing and maintaining schedules. These charts make it easier to schedule a project's tasks efficiently. They can have a huge impact on the project's cost and completion time.

A Group Project

You have been assigned to work with three other people to write a large research paper. You have been given five weeks to complete the task.

    At the group's first planning meeting, one group member pointed out that the assigned topic naturally breaks down to two separate parts. That person went on to suggest that two group members could work on one topic, and two could work on the other. Furthermore, of the two people working on one topic, one could do the research, and one could do the writing. So this plan involved four distinct jobs, one for each of the group's members:

Topic 1:

> Al would research.
> Betty would write.

Topic 2:

> Carla would research.
> Dave would write.

Dependencies, Milestones, and PERT Charts

Another group member suggested that since the workload would be split among four people, the group could wait four weeks before starting. He reasoned that in the last week, Betty and Dave could do their writing while Al and Carla did their research. The others politely pointed out that the writing couldn't happen until after the research was done. In scheduling, this is called a **dependency.**

The group decided that Al's research would take two weeks but Carla's research would take only one week. They allotted two weeks for Betty and Dave to write, but they decided that the writing couldn't start until all the research was finished. They also realized that someone should make sure that the two topics were put together well when the writing was over. Carla agreed to do this, since her research task was less time consuming.

The weighted digraph in Figure 9.98 is an easy way of describing the proposed schedule. This type of graph is called a **PERT chart.** *PERT* stands for "program evaluation and review technique." PERT was developed in the 1950s by the U.S. Navy to aid in developing and managing the Polaris submarine project. It is used frequently today by project managers.

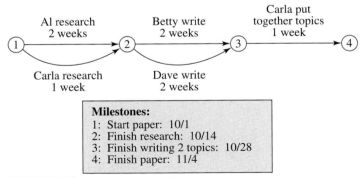

FIGURE 9.98 The group's proposed schedule, displayed on a PERT chart.

A PERT chart's vertices are called **milestones.** They are numbered, and the numbers are used to list each milestone and its date.

A PERT chart's edges are the project's tasks. The tasks "Al research" and "Betty write" are called **sequential tasks** because they must happen in sequence, one after the other. The tasks "Al research" and "Carla research" are called **parallel tasks** because there is no dependency between them and they can happen simultaneously.

Limiting Tasks and Critical Paths

Of the two parallel research tasks, Al's research takes two weeks, and Carla's takes only one week. This means that the task "Al research" limits the total time to complete the project, and "Carla research" does not limit the total time. For this reason, "Al research" is called a *limiting task*. When parallel paths connect consecutive milestones, the task that takes the longest is called a **limiting task.**

The project's **critical path** is the path that includes all of the limiting tasks. The length of the critical path is the project's **completion time.** Figure 9.99 uses heavier arrows to show the research paper's critical path. The paper's completion time is

$$2 \text{ weeks } + 2 \text{ weeks } + 1 \text{ week } = 5 \text{ weeks}$$

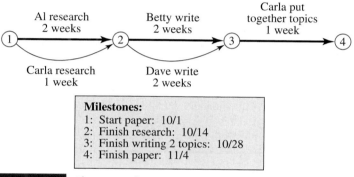

Milestones:
1: Start paper: 10/1
2: Finish research: 10/14
3: Finish writing 2 topics: 10/28
4: Finish paper: 11/4

FIGURE 9.99   The research paper's critical path.

Gantt Charts

Gantt charts are also used in scheduling. A Gantt chart describing the proposed schedule is shown in Figure 9.100. Tasks are described along the side, and dates are given along the top. There is a horizontal bar for each task. Each bar shows the task's start, duration, and completion. The dependencies are shown with arrows.

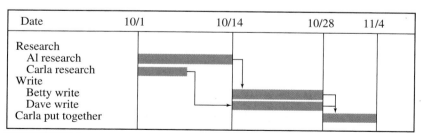

FIGURE 9.100   The group's proposed schedule, displayed on a Gantt chart.

Gantt charts were invented by Henry Gantt (1861–1919), a mechanical engineer and management consultant. The Gantt chart was an important innovation in the 1920s. Gantt charts were used on large construction projects such as the Hoover Dam in 1931 and the interstate highway network in 1956. Today, they are frequently used by project managers, often in conjunction with PERT charts.

A More Extensive Project

Projects that are more extensive than the research paper require a more organized approach. We will use the following steps.

PERT STEPS

1. *List all tasks.*
   - Give each task's estimated time length and dependencies.
   - In this book, this step will be done for you.
2. *Draw an edge for each task that has no dependencies.*
   - Start these edges at the beginning vertex, because they lack dependencies.
   - Weight them with the task's time length.
   - Briefly describe each task, either on the task's edge or in a key.
3. *Draw an edge for each task that depends on the tasks in step 2.*
   - Position these edges so that dependencies are clear.
4. *Draw an edge for each task that depends on the tasks that are already drawn.*
   - Position these edges so that dependencies are clear.
5. *Create a milestone key and possibly a task key.*
   - Number each vertex (or milestone).
   - Include a key that describes the vertices' dates and the events to occur on those dates.
6. *Determine the project's critical path and completion time.*

EXAMPLE 1

MAKING A PERT CHART   Producing a film is a very complex project. A general discussion of such a project, starting from when a script is obtained, is given in Figure 9.101. Draw a PERT chart for the project, and determine its critical path and completion time.

SOLUTION

Step 1

List all tasks. This step is done for us in Figure 9.101.

Task	Length	Dependency
a. Create budget, obtain funds	8 weeks	
b. Hire director	4 weeks	
c. Hire actors	4 weeks	b
d. Hire production staff	2 weeks	a
e. Advertise and contact agents	3 weeks	
f. Scout locations	2 weeks	c, d, e
g. Build sets	4 weeks	f
h. Film non-set-dependent scenes	12 weeks	f
i. Film set-dependent scenes	9 weeks	g
j. Pick conductor and music	3 weeks	h, i
k. Edit film	4 weeks	h, i
l. Write press releases, create preview	2 weeks	h, i
m. Prescreen with audiences	2 weeks	k
n. Create soundtrack CD	1 weeks	j

FIGURE 9.101   A film project.

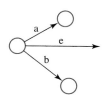

FIGURE 9.102

Tasks a, b, and e.

Step 2 *Draw an edge for each task that has no dependencies.* These are tasks a, b, and e. See Figure 9.102.
- We do not describe the tasks on the edges, because of space constraints.
- Instead, we will create a task key at Step 5.

Step 3 *Draw an edge for each task that depends on the tasks in step 2.* These are tasks c and d. See Figure 9.103.

Step 4 *Draw an edge for each task that depends on the tasks that are already drawn.*
- Task f depends on tasks c, d, and e.
- Tasks g and h depend on task f. See Figure 9.104.
- Task i depends on task g, and tasks j, k, and l depend on tasks h and i. See Figure 9.105.

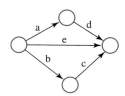

FIGURE 9.103

Tasks a through e.

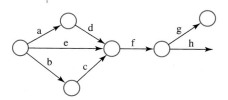

FIGURE 9.104

Tasks a through h.

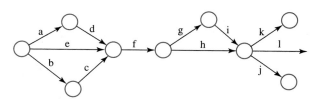

FIGURE 9.105

Tasks a through l.

- We've inserted the remaining tasks in Figure 9.106.

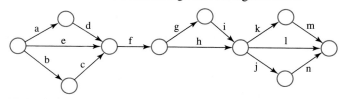

FIGURE 9.106 All tasks are inserted.

Step 5 *Create a milestone key* and possibly a task key. We created both a task key and a milestone key, as discussed in step 2. See Figure 9.107.

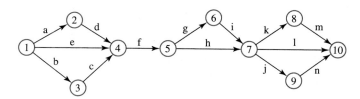

Tasks:	Milestones:
a. Create budget, obtain funds: 8 wks	1. Script obtained
b. Hire director: 4 wks	2. Funding acquired
c. Hire actors: 4 wks	3. Director signed
d. Hire production staff: 2 wks	4. Actors and production staff signed
e. Advertise and contact agents: 3 wks	5. Locations picked
f. Scout locations: 2 wks	6. Sets finished
g. Build sets: 4 wks	7. Filming completed
h. Film non-set-dependent scenes: 12 wks	8. Film edited
i. Film set-dependent scenes: 9 wks	9. Soundtrack completed
j. Pick conductor, music: 3 wks	10. Film released
k. Edit film: 4 wks	
l. Write press releases, create preview: 2 wks	
m. Prescreen with audiences: 2 wks	
n. Create soundtrack CD: 1 wk	

FIGURE 9.107 The complete PERT chart.

Step 6

Determine the project's critical path and completion time.

Tasks a and d, e, and b and c are parallel tasks.

Tasks a and d take 8 wks + 2 wks = 10 wks
Task e takes 3 wks
Tasks b and c take 4 wks + 4 wks = 8 wks

So tasks a and d are limiting tasks.
Task f has no parallel tasks, so it is a limiting task.
Tasks g and i, and h are parallel.

Tasks g and i take 4 wks + 9 wks = 13 wks
Task h takes 12 wks

So tasks g and i are limiting tasks.
Tasks k and m, l, and j and n are parallel tasks.

Tasks k and m take 4 wks + 2 wks = 6 wks
Task l takes 2 wks
Tasks j and n take 3 wk + 1 wks = 4 wks

So tasks k and m are limiting tasks.
The critical path is shown in Figure 9.108.

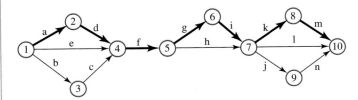

FIGURE 9.108   The critical path.

The completion time is

8 wks + 2 wks + 2 wks + 4 wks + 9 wks + 4 wks + 2 wks = 31 wks

Once a PERT chart has been made, it's easy to make a Gantt chart.

GANTT STEPS

1. *Draw a PERT chart.*
2. *Draw a bar for each task that has no dependencies.*
   - Start these bars at the beginning edge of the chart.
   - Label the bars appropriately.
   - Put a timeline at the chart's top, and adjust the bars' lengths to the timeline.
3. *Draw a bar for each task that depends on the tasks in step 2.*
   - As much as possible, position each of these bars so that each is immediately below the task it depends on.
   - Position each of these bars to start at the time when the task it depends on stops.
   - Use arrows to make the dependencies clear.
   - Label the bars appropriately
4. *Draw an edge for each task that depends on the tasks that are already drawn.*
   - Proceed as in step 3.

EXAMPLE **2**

MAKING A GANTT CHART

a. Draw a Gantt chart for the project described in Example 1.
b. If the film proceeds on schedule, how much time should have elapsed before the actors are hired?

SOLUTION

a.

Step 1
Draw a PERT chart. This was done in Example 1.

Step 2
Draw a bar for each task that has no dependencies.
- We have positioned tasks a, b, and e so that they start at the beginning edge.
- We have labeled the tasks with brief descriptions.
- We have also put a timeline at the top of the chart and adjusted the bars' lengths to this timeline.

See Figure 9.109.

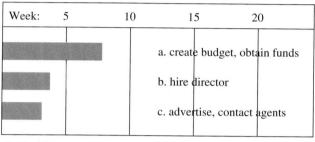

FIGURE 9.109 Step 2.

Step 3
Draw a bar for each task that depends on the tasks in step 2. These are tasks c and d.
- We have positioned each immediately below the task it depends on.
- We have positioned each to start at the time when the task it depends on stops.
- We have continued to label the tasks with brief descriptions.
- Notice the dependency arrows connecting tasks a and d; b and c; and d, c, and f.

See Figure 9.110.

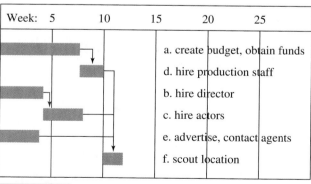

FIGURE 9.110 Step 3.

Step 4
Draw an edge for each task that depends on the tasks that are already drawn. We have inserted the remaining tasks in Figure 9.111.

b. The Gantt chart indicates that the actors could be hired as early as week 4. However, the only task that is immediately dependent on this is task f, "scout location." This task cannot start until week 10, because of other dependencies. So the actors should be hired between weeks 4 and 10.

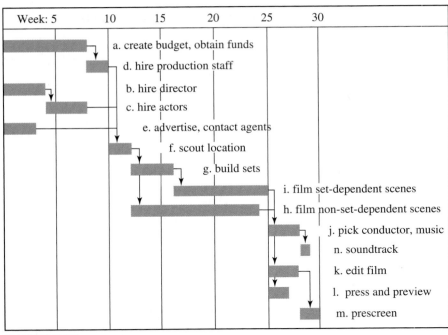

Week: 5	10	15	20	25	30

a. create budget, obtain funds
d. hire production staff
b. hire director
c. hire actors
e. advertise, contact agents
f. scout location
g. build sets
i. film set-dependent scenes
h. film non-set-dependent scenes
j. pick conductor, music
n. soundtrack
k. edit film
l. press and preview
m. prescreen

FIGURE 9.111 A Gantt chart for Example 2.

Advantages and Disadvantages

The Helios aircraft.

A PERT chart shows a project's critical path. This in turn makes it easier to use a PERT chart to determine which tasks must be finished on time and which can be delayed for a while. It also makes it easier to use a PERT chart to determine which tasks would be good candidates for acceleration if the project's completion time must be shortened.

It's easier to use a Gantt chart to determine whether the project is on schedule. If the project is not on schedule, it's easier to use a Gantt chart to find remedial actions that would allow it to return to schedule.

TOPIC X   NASA: PERT CHARTS IN THE REAL WORLD

The following is excerpted from a NASA article on the Helios Solar aircraft project: "Most program/project and task managers use the Gantt chart format for their graphic display of project plans and actual accomplishments. It is a simple tool to use, and displays a lot of information on a computer screen. . . . From the standpoint of communicating the overall picture of what needs to be done, when and why, to both the project team and our customers, however, I've found the

PERT chart to be better. . . . In our solar aircraft development program, we used two types of precedence charts extensively for communication of program/project plans. . . . A top-level program chart, spanning eight years, is shown in Figure 1 [Figure 9.112 in this text].

"The chart was much more than window dressing, as we often referred back to it in team meetings to help redefine the importance of a current task and to see how it fit into 'the big picture.'" This

became a very valuable tool for the team.

"With pride, we saw blocks filled in with actual pictures of our accomplishments (as well as programmatic re-adjustments when necessitated by problems). Enthusiasm for accomplishing the next goal was reborn each time we looked at the graphics on our wall."

See Exercises 28–34.

(Box continued on next page)

TOPIC X NASA: PERT CHARTS IN THE REAL WORLD (CONTINUED)

NASA ERAST* Solar Aircraft Development Plan

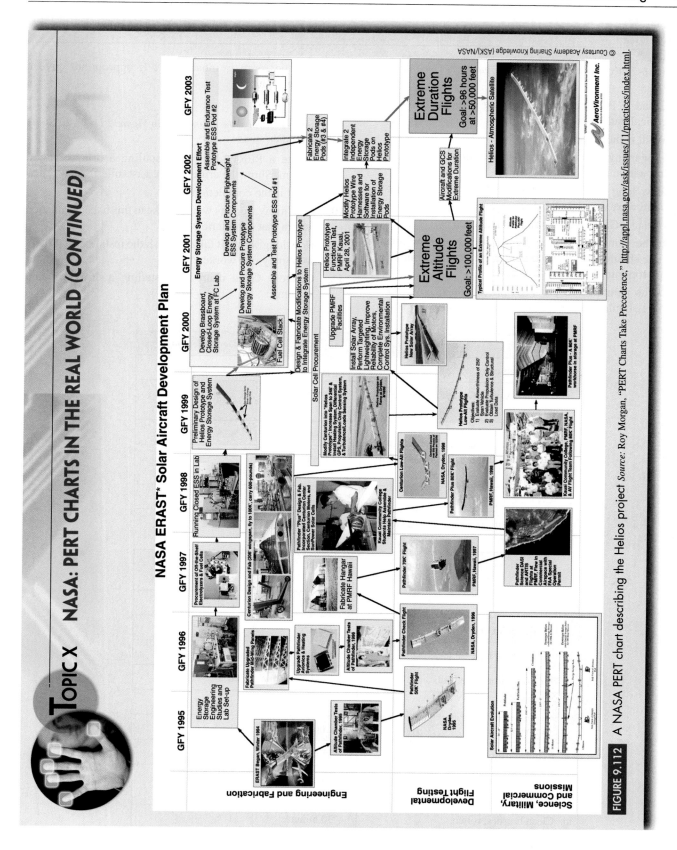

FIGURE 9.112 A NASA PERT chart describing the Helios project *Source:* Roy Morgan, "PERT Charts Take Precedence," http://appl.nasa.gov/ask/issues/11/practices/index.html.

Exercises 1–7 involve the information in Figure 9.113 on painting a room.

1. Create a PERT chart for painting a room. Show the critical path. List milestones.
2. What is the completion time?
3. How many workers would it take to finish the painting job within the completion time?
4. How long would it take to finish the painting job if there were only one worker?
5. Create a Gantt chart for painting a room.
6. If the painting job proceeds on schedule, how much time should have elapsed before the trim is painted? If appropriate, give a range of time.
7. If the painting job proceeds on schedule, how much time should have elapsed before the wall paint is dry and the room is inhabitable? If appropriate, give a range of time.

Exercises 8–14 involve the information in Figure 9.114 on installing a drip irrigation system.

8. Create a PERT chart for installing a drip irrigation system. Show the critical path. List milestones.
9. What is the completion time?
10. How many workers would it take to finish the installation within that completion time?
11. How long would it take to finish the installation if there were only one worker?
12. Create a Gantt chart for installing a drip irrigation system.

Task	Length	Dependencies
a. Select paint color	2 days	
b. Calculate square footage of area to be painted.	30 minutes	
c. Buy paint	1 hour	a, b
d. Buy drop cloths, brushes, rollers, TSP, tape	1 hour	
e. Wash walls with TSP	2 hours	d
f. Remove lights, plug covers and switch covers	30 minutes	
g. Tape walls and windows alongside trim	1 hour	d
h. Paint trim	3 hours	c, d, e, g
i. Let trim paint dry	2 hours	h
j. Remove wall tape	15 minutes	i
k. Tape trim alongside walls	1 hour	d, i
l. Paint walls	3 hours	c, d, e, f, g, h, k
m. Let wall paint dry	2 hours	l
n. Remove trim tape and window tape	15 minutes	m
o. Replace lights, plug covers and switch covers	30 minutes	m
p. Clean brushes, rollers	30 minutes	l
q. Dispose of drop cloths, tape	15 minutes	n

FIGURE 9.113 Painting a room.

Task	Length	Dependencies
a. Determine water needs of plants	3 hours	
b. Lay out system	1 hour	a
c. Determine length of tubing needed	15 minutes	a, b
d. Determine number and type of emitters needed	15 minutes	a, b
e. Determine location of control panel	15 minutes	
f. Determine number of separately controlled lines	15 minutes	a, b, d
g. Buy emitters, tubing, valves, control panel, wire	1 hour	b, c, d, e, f
h. Get wire from control panel to valves	3 hours	e, f, g
i. Put tubing in place	1 hours	b, c, f, g
j. Cut tubing	30 minutes	g, i
k. Connect cut sections of tubing	30 minutes	g, j
l. Install emitters	1 hour	d, g, i, j, k
m. Test system	15 minutes	k, l

FIGURE 9.114 Installing a drip irrigation system.

13. If the installation proceeds on schedule, how much time should have elapsed before all necessary purchases are made? If appropriate, give a range of time.

14. If the installation proceeds on schedule, how much time should have elapsed before the valves are wired to the control panel? If appropriate, give a range of time.

Exercises 15–21 involve the information in Figure 9.116 on page 725 on having a wedding reception or anniversary celebration.

15. Create a PERT chart for having a wedding reception or anniversary celebration. Show the critical path. List milestones.

16. What is the completion time?

17. How many workers would it take to finish the preparations within that completion time?

18. How long would it take to finish the preparations if there were only one worker?

19. Create a Gantt chart for having a wedding reception or anniversary party.

20. If the project proceeds on schedule, how much time should have elapsed before the caterer is hired? If appropriate, give a range of time.

21. If the project proceeds on schedule, how much time should have elapsed before needed clothing is purchased? If appropriate, give a range of time.

Exercises 22–27 involve the information in Figure 9.116 on constructing a home.

22. Create a PERT chart for constructing a home. Show the critical path. List milestones.

23. What is the completion time?

24. Create a Gantt chart for constructing a home.

25. If the project proceeds on schedule, how much time should have elapsed before the roof is built? If appropriate, give a range of time.

26. If the roof is late, what other tasks could be accelerated to keep the job on schedule?

27. If the project proceeds on schedule, how much time should have elapsed before the inspection? If appropriate, give a range of time.

Use the PERT chart in Figure 9.112 to answer Exercises 28–34.

28. When did the Helios program begin? What were the first tasks?

29. The task "Preliminary design of Helios prototype" was immediately dependent on what other tasks? What is the task's approximate length? What is the task's approximate completion date?

30. The task "Centurion low altitude flights" was immediately dependent on what other tasks? What is

Task	Length	Dependencies
a. Create guest list	1 week	
b. Check out different room options	6 hours	
c. Select invitations and place cards	2 hours	
d. Wait for invitations to return from printer	1 week	c
e. Mail invitations	1 hour	d
f. Wait for RSVPs to return from invited guests	2 weeks	d
g. Rent a room	1 hour	a
h. Check out different hotel options for out of town guests	4 hours	
i. Make hotel arrangements	1 hour	h
j. Check out catering options	6 hours	
k. Hire caterer	1 hour	j
l. Determine menu	2 hours	k
m. Give caterer info on number of guests and their food choices	1 hour	f, k
n. Give printer list of attendees for place cards	30 mins	f
o. Wait for place cards to return from printer	1 week	n
p. Check out band options	3 hours	
q. Hire band	1 hour	g, p
r. Check out table decoration options	4 hours	
s. Hire florist	1 hour	r
t. Determine bar needs	2 hours	f
u. Arrange for bar	1 hour	t
v. Determine clothing needs of immediate family	1 hour	
w. Purchase needed clothing	8 hours	v

FIGURE 9.115 Having a wedding reception or anniversary celebration.

the task's approximate length? What is the task's approximate completion date?

31. The task "Extreme altitude flights" was immediately dependent on what other tasks? What is the task's approximate length? What is the task's approximate completion date?

32. Of all the tasks, which has the largest number of immediate dependencies?

33. What is the project's approximate completion time? What is its approximate completion date?

34. How does the NASA PERT chart differ from those in the text? Why do you think it has this difference?

PROJECT

35. Identify a project (such as painting a room, installing a drip irrigation system, or planning a wedding reception) that you are familiar with. Do not use the projects discussed in the text. Create a detailed PERT chart and Gantt chart for the project.

Task	Length	Dependency
a. Write and agree on contracts	4 weeks	
b. Secure financing	4 weeks	a
c. Obtain permits	6 weeks	a, b
d. Site work	4 weeks	a, b, c
e. Build foundation	3 weeks	b, c, d
f. Erect walls	3 weeks	d, e
g. Build roof	4 weeks	f
h. Rough in plumbing	2 weeks	f
i. Rough in electrical	2 weeks	f
j. Rough in furnace and air conditioning	1 week	f
k. Finish exterior	6 weeks	f, g, i
l. Finish interior	6 weeks	f, g, h, i, j
m. Finish plumbing	2 weeks	h, i
n. Finish electrical	1 week	i, l
o. Finish furnace and air conditioning	1 day	j, l
p. Wait for inspection	2 weeks	k, l, m, n, o

FIGURE 9.116 Constructing a home.

9 CHAPTER REVIEW

TERMS

adjacent edges
adjacent vertices
algorithm
approximation
    algorithm
circuit
complete graph
completion time
connected graph
critical path

degree of a vertex
dependency
digraph
directed graph
disconnected graph
edge
Euler circuit
Euler trail
Eulerization
even vertex
Gantt chart
graph

Hamilton circuit
identical graphs
limiting task
loop
milestones
minimum Hamilton
    circuit
minimum spanning tree
multiple edges
odd vertex
parallel tasks
PERT chart

sequential tasks
spanning tree
Steiner point
trail
traveling salesman
    problem
tree
vertex
weighted graph
weights

THEOREMS AND ALGORITHMS

Euler's Theorems

- A connected graph with only even vertices has at least one Euler trail, which is also an Euler circuit.
- A connected graph with exactly two odd vertices and any number of even vertices has at least one Euler trail.
- A graph with more than two odd vertices has no Euler trails and no Euler circuits.
- It's impossible for a connected graph to have only one odd vertex.

Fleury's algorithm for finding Euler circuits

1. Verify that the graph has an Euler trail or Euler circuit, using Euler's theorem.
2. Choose a starting point.
   - If the graph has two odd vertices, pick either of the odd vertices as the starting point.
   - If the graph has no odd vertices, pick any point as the starting point.
3. Label each edge alphabetically as you travel that edge.
4. When choosing edges:
   - Never choose an edge that would make the yet-to-be-traveled part of the graph disconnected.
   - Never choose an edge that has already been followed.
   - Never choose an edge that leads to a vertex which has no other yet-to-be-traveled edges.

Eulerization algorithm

To efficiently Eulerize a graph that is laid out like a grid:

1. Choose a vertex along the outer perimeter of the graph.
2. If a vertex is an even vertex, add no edges and move on to the next vertex along the outer perimeter.
3. If a vertex is an odd vertex, add a duplicate edge that connects it to the next adjacent vertex along the outer perimeter, and move on to the next vertex along the outer perimeter.
4. Repeat these steps until you return to the vertex in Step 1.

The nearest neighbor algorithm for finding a minimum Hamilton circuit

1. From the starting vertex, go to the "nearest neighbor"
2. From all other vertices, use the edge that:
   - has the smallest weight and
   - leads to vertex that hasn't been visited yet and
   - does not lead to the starting vertex.

   (If there are any ties in steps 1 or 2, choose at random. Different choices will result in different circuits, which may have different totals.)
3. The last edge must lead to the starting vertex.

The repetitive nearest neighbor algorithm for finding a minimum Hamilton circuit

1. Apply the nearest neighbor algorithm repeatedly, using each vertex as the starting point, and compute the total cost of the circuit obtained.
2. Choose the best of the Hamilton circuits obtained in step 1. If this circuit starts at an inappropriate vertex, resequence the circuit so that it starts at the appropriate vertex.

   (If there are any ties in steps 1 or 2, explore each option if the number of options is acceptably small.)

The cheapest edge algorithm for finding a minimum Hamilton circuit

1. List the weights, in order, from smallest to largest.
2. Draw a complete graph for the problem.
3. Pick edges, in order, from the list from step 1. Do not pick an edge if:
   - it leads to a vertex that you've already "flown in to."
   - it departs from a vertex that you've already "flown out of."
   - it closes a circuit that shouldn't be closed yet.

   (If there is a tie, choose at random. Different choices will result in different circuits, which may result in different totals.)
4. When all of the vertices are joined, close the circuit.

Choosing a Hamilton circuit algorithm

1. Find how many different circuits your graph has.
   - A complete directed graph with n vertices has $(n-1)!$ different Hamilton circuits.
   - A complete nondirected graph with n vertices $\frac{(n-1)!}{2}$ different Hamilton circuits.
2. Determine the appropriate method(s).
   - If the number of different circuits is acceptably small, then use the brute force method and be certain that you've found the Hamilton circuit with the least total weight.
   - If the number of different circuits is not acceptably small, then use any (or, better yet, all) of the approximation algorithms: the nearest neighbor algorithm, the repetitive nearest neighbor algorithm, or the cheapest edge algorithm.

Kruskal's algorithm for finding a minimum spanning tree

1. List the weights, in order, from smallest to largest.
2. Draw a complete graph for the problem.
3. Pick edges, in order, from the list from step 1. Do not pick an edge if it closes a circuit.
4. Stop adding edges when all the vertices are joined.

REVIEW EXERCISES

In Exercises 1–6, do the following:

a. Determine the number of vertices, edges, and loops in the given graph.

b. Draw another representation of the graph that looks significantly different.

c. Find two adjacent edges.

d. Find two adjacent vertices.

e. Find the degree of each vertex.

f. Determine whether the graph is connected.

1.
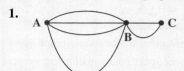

2.
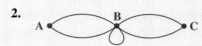

3.
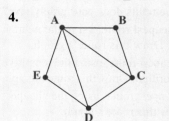

4.
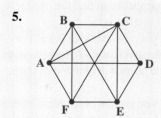

5.
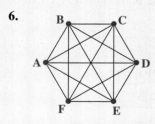

6.
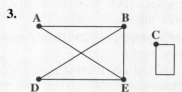

7. Draw a graph with five vertices: three of degree 2 and two of degree 3.

8. Draw a graph with six vertices: five of degree 1 and one of degree 3.

9. Draw a graph with four vertices: two of degree 3 and two of degree 4.

10. Draw a graph with four vertices: all of degree 3.

11. Draw a graph with four vertices in which each is of degree 3 and that has:

   a. loops but no multiple edges.

   b. multiple edges but no loops.

   c. neither loops nor multiple edges.

   d. both loops and multiple edges.

12. Draw a graph with six vertices in which each is of degree 4 and that has:

   a. loops but no multiple edges.

   b. multiple edges but no loops.

   c. both loops and multiple edges.

In Exercises 13–20, do the following.

a. Use Euler's theorem to determine whether the graph has an Euler trail or an Euler circuit.

b. If it does, find it. If it does not, say why. Then Eulerize the graph if possible.

13. The graph in Exercise 1

14. The graph in Exercise 2

15. The graph in Exercise 3

16. The graph in Exercise 4

17. The graph in Exercise 5

18. The graph in Exercise 6

19.

    20.
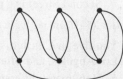

21. Figure 9.117 shows a map of the Hidden Circles gated community. After several burglaries, the residents hired a security guard to walk through the subdivision once every night.

   a. The guard would prefer to park his car, walk a route that allows him to walk through every block just once, and then return to his car. Is this possible? Explain why or why not.

   b. If such a route is possible, find it. If such a route is not possible, find an efficient route.

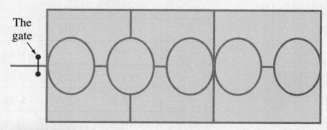

FIGURE 9.117 The Hidden Circles gated community.

22. Once a month, the utility company sends an agent through the Hidden Circles gated community to read the meters. (See Exercise 21 and Figure 9.117.) Each house has its own meter. Find a route for the meter reader if there are houses on both sides of each street.

23. Find an efficient route through the community shown in Figure 9.118 for the postal carrier it mailboxes are all on one side of each street.

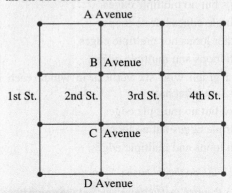

FIGURE 9.118 A graph for Exercise 23.

24. Once a month, the utility company sends an agent through the community shown in Figure 9.118 to read the meters. Each house has its own meter.
   a. Find a route for the meter reader if there are houses on both sides of each street.
   b. Find a route for the meter reader if there are houses on both sides of each street except that the area bounded by A Avenue, B Avenue, First Street, and Second Street is a park.

Exercises 25–31 use the flight costs shown in Figure 9.119.

25. You live in Chicago, and you need to visit New York, Los Angeles, and Miami.
   a. How many different circuits does your graph have?
   b. Approximate the cheapest route, using the nearest neighbor algorithm. Draw this route's graph.
   c. Approximate the cheapest route, using the repetitive nearest neighbor algorithm. Draw this route's graph.
   d. Approximate the cheapest route, using the cheapest edge algorithm. Draw this route's graph.
   e. Find the cheapest route, using the brute force method. Draw this route's graph.

26. You live in Los Angeles, and you need to visit Miami, Seattle, and Houston.
   a. How many different circuits does your graph have?
   b. Approximate the cheapest route, using the nearest neighbor algorithm. Draw this route's graph.
   c. Approximate the cheapest route, using the repetitive nearest neighbor algorithm. Draw this route's graph.
   d. Approximate the cheapest route, using the cheapest edge algorithm. Draw this route's graph.
   e. Find the cheapest route, using the brute force method. Draw this route's graph.

27. You live in Chicago, and you need to visit New York, Los Angeles, Miami, and Seattle.
   a. How many different circuits does your graph have?
   b. Approximate the cheapest route, using the nearest neighbor algorithm. Draw this route's graph.
   c. Approximate the cheapest route, using the repetitive nearest neighbor algorithm. Draw this route's graph.
   d. Approximate the cheapest route, using the cheapest edge algorithm. Draw this route's graph.
   e. Why would it not be appropriate to find the cheapest route, using the brute force method?

28. You live in New York, and you need to visit Los Angeles, Miami, Seattle, and Houston.
   a. How many different circuits does your graph have?
   b. Approximate the cheapest route, using the nearest neighbor algorithm. Draw this route's graph.
   c. Approximate the cheapest route, using the repetitive nearest neighbor algorithm. Draw this route's graph.
   d. Approximate the cheapest route, using the cheapest edge algorithm. Draw this route's graph.
   e. Why would it not be appropriate to find the cheapest route, using the brute force method?

29. You live in Chicago, and you need to visit New York, Los Angeles, Miami, Seattle, and Houston. Approximate the cheapest route, using the nearest neighbor algorithm. Draw this route's graph.

30. You live in Chicago, and you need to visit New York, Los Angeles, Miami, Seattle, and Houston. Approximate

		From:					
		CHI	NYC	LAX	MIA	SEA	HOU
	CHI		$194	$299	$74	$129	$133
	NYC	$243		$119	$79	$109	$148
To:	LAX	$299	$119		$124	$89	$154
	MIA	$74	$79	$124		$169	$94
	SEA	$129	$109	$89	$169		$294
	HOU	$133	$148	$154	$94	$294	

FIGURE 9.119 Flight costs between Chicago, New York, Los Angeles, Miami, Seattle, and Houston *Source:* www.travelocity.com.

the cheapest route, using the repetitive nearest neighbor algorithm. Draw this route's graph.

31. You live in Chicago, and you need to visit New York, Los Angeles, Miami, Seattle, and Houston. Approximate the cheapest route, using the cheapest edge algorithm. Draw this route's graph.

Use the following information in Exercises 32–34.

You are programming a robotic drill to drill holes in a circuit board. The cost of drilling a board is determined by the distance the drill must travel, and the drill moves only horizontally and vertically.

32. A board needs holes at (5, 7), (1, 1), (6, 3), (7, 2), and (2, 4). Use the nearest neighbor algorithm to determine a sequence for the drilling of those holes.

33. A board needs holes at (5, 7), (1, 1), (6, 3), (7, 2), and (2, 4). Use the repetitive nearest neighbor algorithm to determine a sequence for the drilling of those holes.

34. A board needs holes at (5, 7), (1, 1), (6, 3), (7, 2), and (2, 4). Use the cheapest edge algorithm to determine a sequence for the drilling of those holes.

In Exercises 35–42, find a spanning tree for the given graph. Note: Answers will vary.

35. The graph in Exercise 1
36. The graph in Exercise 2
37. The graph in Exercise 3
38. The graph in Exercise 4
39. The graph in Exercise 5
40. The graph in Exercise 6
41. The graph in Exercise 19
42. The graph in Exercise 20

In Exercises 43–48, use Kruskal's algorithm to find the minimum spanning tree for the given weighted graph. Give the total weight of the minimum spanning tree.

43.
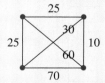

44.
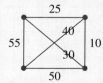

45.
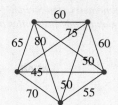

46.
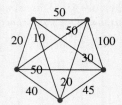

47.
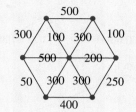

48.
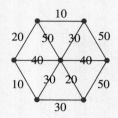

In Exercises 49–52, use the information in Figure 9.120.

49. You have been hired to design a network linking Boston, Chicago, Denver, Kansas City, and New York. Using only those cities as vertices, find the minimum spanning tree connecting the cities, and its total length.

50. You have been hired to design a network linking Kansas City, New York, Philadelphia, Portland, and San Francisco. Using only those cities as vertices, find the minimum spanning tree connecting the cities and its total length.

51. You have been hired to design a network linking Boston, Chicago, Denver, Kansas City, New York, Philadelphia, and San Francisco. Using only those cities as vertices, find the minimum spanning tree connecting the cities and its total length.

52. You have been hired to design a network linking Boston, Chicago, Denver, Kansas City, New York, Philadelphia, Portland, and San Francisco. Using only those cities as vertices, find the minimum spanning tree connecting the cities and its total length.

	Boston	Chicago	Denver	Kansas City, MO	New York City	Philadelphia	Portland, OR	San Francisco
Atlanta	932	585	1213	678	744	658	2170	2138
Boston		851	1767	1247	187	273	2536	2696
Chicago			917	407	714	664	1755	1855
Denver				558	1630	1574	979	947
Kansas City, MO					1094	1032	1493	1500
New York						86	2442	2570
Philadelphia							2407	2518
Portland, OR								538

FIGURE 9.120 Distances between cities, in miles.

In Exercises 53–55, use symmetry, trigonometry, and Steiner points to find the shortest network connecting the three cities and the length of that network.

53.

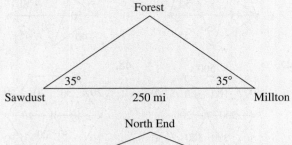

Forest

35°   35°

Sawdust   250 mi   Millton

54.

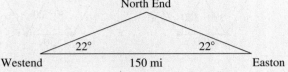

North End

22°   22°

Westend   150 mi   Easton

55.

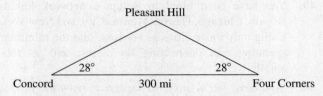

Pleasant Hill

28°   28°

Concord   300 mi   Four Corners

56. In Figure 9.121, four different networks are drawn for the same rectangle. Which is the shortest network: a, b, c, or d? What is its length? All interior points are Steiner points.

57. Use symmetry, trigonometry, and Steiner points to find the shortest network and its length for a rectangle that is 700 miles wide and 900 miles long.

58. Use symmetry, trigonometry, and Steiner points to find the shortest network and its length for a rectangle that is 700 miles wide and 1,000 miles long.

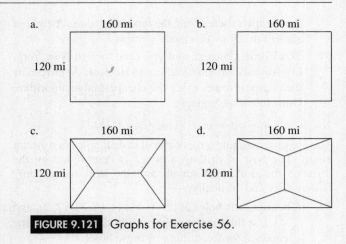

FIGURE 9.121 Graphs for Exercise 56.

Exercises 59–64 involve the information in Figure 9.122 on staining a deck.

59. Create a PERT chart for staining a deck. Show the critical path. List milestones.

60. What is the completion time?

61. How many workers would it take to finish the painting job within the completion time?

62. How long would it take to finish the painting job if there were only one worker?

63. Create a Gantt chart for staining a deck.

64. If the staining job proceeds on schedule, how much time should have elapsed before the stain is purchased? It appropriate, give a range of time.

Task	Length	Dependency
a. Rent power washer	1 hour	
b. Remove patio furniture from deck	15 minutes	
c. Power wash deck	16 hours	a, b
d. Select type of stain (opaque or semitransparent) and stain color	2 days	
e. Calculate square footage of area to be stained	15 minutes	
f. Buy stain	1 hour	d, e
g. Buy drop cloths, brushes, and rollers	1 hour	
h. Allow deck to dry	2 days	c
i. Stain deck	24 hours	f, h
j. Allow deck to dry	2 days	i

FIGURE 9.122 Staining a deck.

EXPONENTIAL AND LOGARITHMIC FUNCTIONS

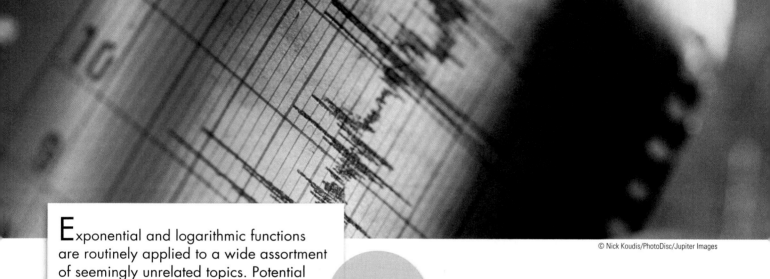

© Nick Koudis/PhotoDisc/Jupiter Images

Exponential and logarithmic functions are routinely applied to a wide assortment of seemingly unrelated topics. Potential home purchasers use them to analyze the effect of inflation on home prices. Residents of earthquake-prone areas use them and the Richter scale to compare earthquakes of different strengths. Musicians, as well as people who are interested in the effect of sounds on their ears, use them and the decibel scale to compare sounds of different volumes. Biologists and social scientists use them to predict the future sizes of human and animal populations. Archaeologists use them to determine the ages of artifacts. Medical technicians use them to monitor the decay of radioactive material used in various diagnostic tests and internal imaging procedures.

WHAT WE WILL DO IN THIS CHAPTER

WE'LL ANSWER QUESTIONS SUCH AS THE FOLLOWING:

- When will the world's population exceed the earth's ability to support it?

- How long will it take for my home to double in value?

- How long must the radioactive waste from a nuclear plant be stored?

- What does radiocarbon dating say about the authenticity of the Dead Sea Scrolls and the Shroud of Turin?

- How long will a typical dose of radioactive tracer remain in your body?

- How does a 7.0 earthquake compare with a 8.4 earthquake?

OBJECTIVES

- Define and graph an exponential function; define the natural exponential function

- Use a calculator to find values of the exponential function 10^x and the natural exponential function e^x

- Define and understand the meaning of a logarithm

- Rewrite a logarithm as an exponential equation and vice versa

- Use a calculator to find values of the common and natural logarithmic functions $\log x$ and $\ln x$

Functions

An equation is said to be a **function** if to each value of x there corresponds one and only one value of y. For example, consider the equation $y = 3x + 1$; $x = 1$ corresponds to one and only one value of y (in particular, to $y = 4$). Other values of x also correspond to one and only one value of y, so the equation $y = 3x + 1$ is a function. When the value of y depends in this manner on the value of x, we say that y is a function of x; x is called the **independent variable,** and y is called the **dependent variable.**

Our equation $y = 3x + 1$ is called a **linear function,** because it is a function whose graph is a line. The graph of the equation $y = 3x + 1$ is shown in Figure 10.1. The slope of this line is 3; for every one-unit increase in x, the value of y increases by three units [slope = rise/run = (change in y)/(change in x)].

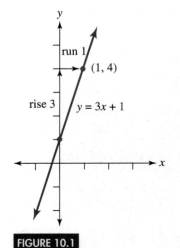

FIGURE 10.1

A linear function.

Exponential Functions

A function in which x appears only in the exponent is called an **exponential function.** For example, $y = 2^x$ is an exponential function. Again, y is the dependent variable and x is the independent variable.

EXPONENTIAL FUNCTION

An equation of the form $y = b^x$, where b is a positive constant, ($b \neq 1$) is called an **exponential function.** The positive constant b is called the **base.**

As Example 1 shows, the graph of an exponential function is quite different from the graph of a linear function.

EXAMPLE **1**

SOLUTION

GRAPHING AN EXPONENTIAL FUNCTION Sketch the graph of $y = 2^x$.

We can graph this exponential function by finding several ordered pairs, as in Figure 10.2.

The graph of all ordered pairs (x, y) satisfying the equation $y = 2^x$ is the curve shown in Figure 10.3.

x	$y = 2^x$	Ordered Pair (x, y)
3	$2^3 = 8$	$(3, 8)$
2	$2^2 = 4$	$(2, 4)$
1	$2^1 = 2$	$(1, 2)$
0	$2^0 = 1$	$(0, 1)$
-1	$2^{-1} = \dfrac{1}{2}$	$\left(-1, \dfrac{1}{2}\right)$
-2	$2^{-2} = \dfrac{1}{4}$	$\left(-2, \dfrac{1}{4}\right)$
-3	$2^{-3} = \dfrac{1}{8}$	$\left(-3, \dfrac{1}{8}\right)$

FIGURE 10.2 Finding ordered pairs.

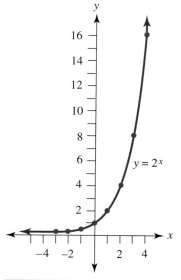

FIGURE 10.3 An exponential function.

The graph of $y = 2^x$ is not a straight line. Typically, the graph of an exponential function is nearly horizontal over a large portion of the x-axis and then turns upward rather abruptly, increasing without bound. Because of this curvature, the slope is not constant, as it is with a linear function; the vertical change is minimal where the graph is nearly horizontal, whereas the vertical change is quite extreme where the graph is steep.

In this chapter, we will study various applications of exponential functions. These applications include the prediction of the size of human and animal populations, inflation, the decay of radioactive materials, the dating of archeological artifacts, the Richter scale (which rates earthquakes), and the decibel scale (which rates volumes of sounds).

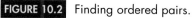

Rational and Irrational Numbers

Real numbers are either rational or irrational. A real number that is a terminating decimal, such as $\frac{3}{4} = 0.75$, or a repeating decimal, such as $\frac{2}{3} = 0.6666\ldots$, is called a ***ratio*nal number** because it can be written as the *ratio* of two integers—that is, as a fraction. A real number that is neither a terminating decimal nor a repeating decimal is called an **irrational number.** The irrational numbers with which you are probably most familiar are square roots. For example, $\sqrt{2} = 1.414213562\ldots$ and $\sqrt{23} = 4.795831523\ldots$ are irrational.

One important irrational number is π. Whenever the circumference of any circle is divided by its diameter, the resulting number is *always* $\pi = 3.141592654\ldots.$

Another "famous" irrational number is $e = 2.71828182845\ldots$ The number e is used in most of the applications of mathematics in this chapter, including population prediction, inflation, radioactive decay, and the dating of archeological artifacts.

The Natural Exponential Function

The **natural exponential function** is $y = e^x$. Calculations involving e^x will be performed on a calculator. Some scientific calculators have a button that is labeled "e^x" on the button itself and "ln" or "ln x" above the button; to calculate e^1 with such a calculator, press

$$1 \quad \boxed{e^x}$$

Other scientific calculators have a button that is labeled "ln" or "ln x" on the button itself and "e^x" above the button; to calculate e^1 with such a calculator, press either

$$1 \quad \boxed{2\text{nd}} \quad \boxed{e^x} \qquad \text{or} \qquad 1 \quad \boxed{\text{shift}} \quad \boxed{e^x} \qquad \text{or} \qquad 1 \quad \boxed{\text{INV}} \quad \boxed{e^x}$$

In the future, we will simply write $\boxed{e^x}$ to refer to each of these sequences of keystrokes.

Most graphing calculators have a button that is labeled "LN" on the button itself and "e^x" above the button; to calculate e^1 with such a calculator, press

$$\boxed{2\text{nd}} \quad \boxed{e^x} \quad 1 \quad \boxed{\text{ENTER}}$$

EXAMPLE **2**

CALCULATING VALUES CONTAINING e Use a calculator to find the following values:

a. e **b.** $1/e$ **c.** e^2 **d.** $1/e^2$ **e.** e^3

f. Using the above values, sketch the graph of $y = e^x$.

SOLUTION

Finding the Values.

a. $e = e^1 = 2.718281828\ldots$

b. $1/e = 0.367879441\ldots$

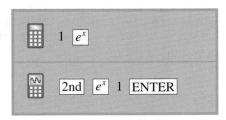

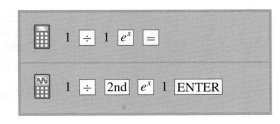

An alternative way to find $1/e$ is to realize that $1/e = e^{-1}$.

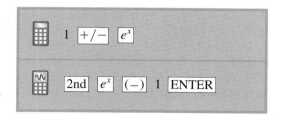

c. $e^2 = 7.389056099\ldots$

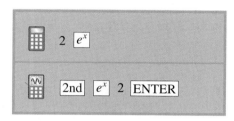

d. $1/e^2 = 0.135335283\ldots$

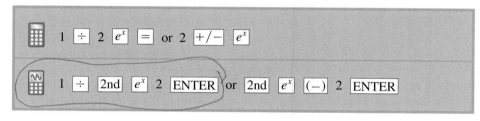

e. $e^3 = 20.08553692\ldots$

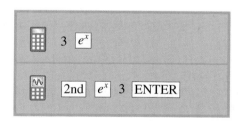

f. *Sketching the graph of* $y = e^x$. First, express the above values as ordered pairs, rounding off y-coordinates to the nearest tenth (which is sufficient for plot plotting), as shown in Figure 10.4. Then plot the ordered pairs and connect them with a smooth curve, as shown in Figure 10.5.

x	$y = e^x$	Ordered Pair (x, y)
3	$e^3 \approx 20.1$	$(3, 20.1)$
2	$e^2 \approx 7.4$	$(2, 7.4)$
1	$e^1 \approx 2.7$	$(1, 2.7)$
0	$e^0 = 1$	$(0, 1)$
-1	$e^{-1} \approx 0.4$	$(-1, 0.4)$
-2	$e^{-2} \approx 0.1$	$(-2, 0.1)$

FIGURE 10.4   Finding ordered pairs.

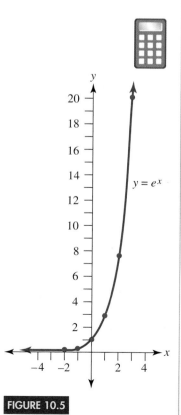

FIGURE 10.5

The natural exponential function.

The graph of $y = e^x$ has the same shape as the graph of $y = 2^x$. Both are nearly horizontal over a portion of the x-axis and then turn upward, increasing without bound; both graphs are entirely above the x-axis. Exponential functions (functions of the form $y = b^x$) always have these qualities. The independent variable x can have any value, but the dependent variable y can have only a positive value.

Because $y = e^x$ is an exponential function, its graph is not a straight line, and the slope is not constant; going from $x = -2$ to $x = -1$, the slope is

$$m \approx \frac{0.4 - 0.1}{-1 - (-2)} = 0.3 \quad \text{using the ordered pairs generated in Figure 10.4}$$

However, going from $x = 2$ to $x = 3$, the slope is

$$m \approx \frac{20.1 - 7.4}{3 - 2} = 12.7$$

This means that e^x grows much more quickly when x is a larger number. When we are dealing with population growth, this trait can become quite disturbing; the larger the population, the faster it grows. The use of exponential functions to analyze population growth is explored in Section 10.1.

Logarithms

Ten raised to what power gives 100? 1,000? 346? The answers to the first and second questions are easily found; the solution to $10^x = 100$ is $x = 2$, and the solution to $10^x = 1,000$ is $x = 3$. However, the solution to $10^x = 346$ is not so obvious! We could safely say that because $10^2 = 100$ and $10^3 = 1,000$, and because 346 is between 100 and 1,000, the solution to $10^x = 346$ must be between 2 and 3. To be more accurate would be difficult.

The idea of finding the exponent to which a number must be raised in order to get some particular number is the central concept of a **logarithm.** The question "3 raised to what power gives 9?" is the same question as "What is the logarithm (base 3) of 9?" (The answer to either question is 2.) Algebraically, the question "3 raised to what power gives 9?" is written as "$3^x = 9$," and the question "What is the logarithm (base 3) of 9?" is written "$\log_3 9 = x$." When we say $x = \log_3 9$, we are saying that x is the exponent to which we must raise 3 in order to get 9.

LOGARITHM DEFINITION

$\log_b u = v$ (or the **logarithm** of u)

means the same as

$b^v = u$

In either equation, b is called the **base** and must be a positive number ($b \neq 1$).

This definition of a logarithm allows us to rewrite a logarithmic equation as an exponential equation. The next example uses this fact.

HISTORICAL NOTE

LEONHARD EULER, 1707–1783

Leonhard Euler (pronounced "Oiler") was probably the most versatile and prolific writer in the history of mathematics. Euler produced over 700 books and papers, many of which were created during the last seventeen years of his life while he was totally blind. He wrote an average of 800 pages of mathematics each year. Euler's writings spanned pure mathematics, astronomy, annuities, life expectancy, lotteries, and music.

Euler was the son of a Swiss minister and mathematician who studied under Jacob Bernoulli. Euler was schooled in theology and mathematics and was to become a minister. However, his mathematical curiosity and ability prevailed, and his father eventually allowed him to concentrate on mathematics. He was awarded his master's degree at the age of sixteen! At the age of twenty-six, at the invitation of Catherine I, he was appointed to the Academy of St. Petersburg, a major center of scientific research. Later he was invited to Berlin by Frederick the Great.

Euler's phenomenal memory has been the subject of many legends. Having memorized Virgil's *Aeneid*, Euler could recite the first and last lines on any page of his copy. When confronted by two students who asked him to settle a disputed mathematical calculation in the fiftieth place, Euler successfully calculated the true result in his head!

One of Euler's skills was the creation of notations that were useful and compact. His writings were very popular, and the notations he introduced have endured. He introduced the notation $f(x)$ for a function, and he was the first to use π for $3.14159\ldots$, i for $\sqrt{-1}$, and e for $2.718281828.\ldots$

EXAMPLE 3

FINDING MISSING VALUES IN A LOGARITHMIC EQUATION

a. Find v if $v = \log_2 8$.   **b.** Find u if $\log_5 u = -2$.   **c.** Find b if $\log_b 9 = 2$.

SOLUTION

a. $v = \log_2 8$ can be rewritten as $2^v = 8$. By inspection, $v = 3$.

b. $\log_5 u = -2$ can be rewritten as $5^{-2} = u$. Therefore, $u = 1/5^2 = 1/25$.

c. $\log_b 9 = 2$ can be rewritten as $b^2 = 9$. By inspection, $b = \pm 3$. However, the base b of a logarithm is required to be positive. Thus, $b = 3$.

Common Logarithms

$\log_{10} x$, in which 10 is the base, is called the **common logarithm.** Base 10 logarithms are used so commonly that the base is understood to be 10 when no base is written. Thus, $\log 346$ is an abbreviation for $\log_{10} 346$; $\log 346$ is the exponent to which 10 must be raised in order to get 346.

COMMON LOGARITHM DEFINITION

$$y = \log x \qquad (\text{or} \quad y = \log_{10} x)$$

means the same as

$$10^y = x$$

EXAMPLE **4**

CALCULATING COMMON LOGARITHMS   Find the following values by using a calculator:

a. log 346
b. log(0.82)
c. log(−10)
d. log(10^5)

SOLUTION

a. The ⬛log⬛ button on your calculator is used to find the common logarithm of a number. Simply enter the number you want to find the logarithm of and press ⬛log⬛. Some calculators require that you press ⬛log⬛ first and then enter the desired number.

$$\log 346 = 2.539076099\ldots$$

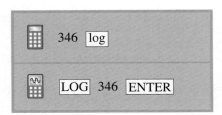

Therefore, $10^{2.539076099} \approx 346$. (This is the answer to the earlier question "Ten raised to what power gives 346?" We had estimated that the power must be between 2 and 3; now we know that the power 2.539076099 gives 346.)

b. $\log(0.82) = -0.086186147\ldots$

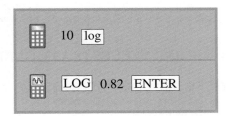

Therefore, $10^{-0.086186147} \approx 0.82$

c. $\log(-10) = \text{ERROR}$

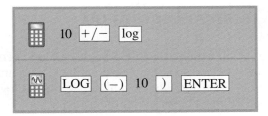

Your calculator cannot find the common logarithm of −10 because negative numbers do not have logarithms! To see why, let $x = \log(-10)$. The corresponding exponential equation would be $10^x = -10$. This equation has no solution, because 10 raised to *any* power *always* gives a positive result; it is impossible for 10^x to equal −10. Consequently, $\log(-10)$ is undefined.

d. $\log(10^5) = \log 100{,}000 = 5$

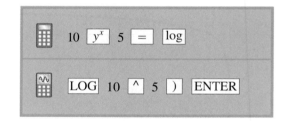

The values of the independent variable x in the common logarithm function $y = \log x$ must always be positive, as was discussed in part (c) of Example 4, and the values of the dependent variable y can be either positive or negative, as was shown in parts (a) and (b). This situation is exactly the reverse of that for an exponential function, in which the independent variable can have any value and the dependent variable must be positive.

It should be noted that the explanation given in part (c) of Example 4 is an oversimplification; technically, it should be stated that negative numbers do not have logarithms that are *real numbers*. Negative numbers *do* have logarithms; however, the logarithm of a negative number is a complex (or imaginary) number. Some graphing calculators will compute the (complex) logarithm of a negative number; however, we will not study complex numbers in this textbook.

EXAMPLE 5

SOLUTION

GRAPHING A LOGARITHMIC FUNCTION   Sketch the graph of $y = \log x$.

Rather than computing the value of y for specific values of x, recall that $y = \log x$ means the same as $x = 10^y$. This exponential form allows for easier computations than does the original logarithmic form. Therefore, we compute the value of x for specific integer values of y. See Figure 10.6. Now plot the ordered pairs and connect them with a smooth curve, as shown in Figure 10.7.

y	$x = 10^y$	(x, y)
1	$10^1 = 10$	$(10, 1)$
0	$10^0 = 1$	$(1, 0)$
-1	$10^{-1} = \dfrac{1}{10}$	$\left(\dfrac{1}{10}, -1\right)$
-2	$10^{-2} = \dfrac{1}{100}$	$\left(\dfrac{1}{100}, -2\right)$

FIGURE 10.6   Finding ordered pairs.

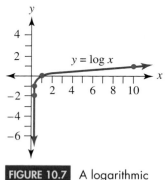

FIGURE 10.7   A logarithmic function.

Notice that when x is between 0 and 1, the graph rises very steeply and the slope is large. In this interval, small changes in x produce large changes in y; for example, as x changes from $x = \frac{1}{100}$ to $x = \frac{1}{10}$, y changes from $y = -2$ to $y = -1$. On the other hand, when x is greater than 1, the graph rises very slowly and the slope is small. Here, large changes in x produce very small changes in y; for example, as x changes from $x = 1$ to $x = 10$, y changes only from $y = 0$ to $y = 1$. Consequently, logarithms can be used to expand small variations and compress large ones (see Figure 10.8). This characteristic will be important in our study of the Richter scale and the decibel scale in Section 10.3.

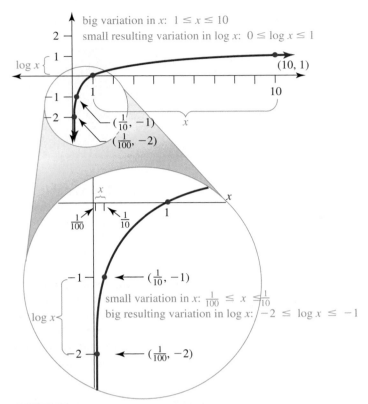

FIGURE 10.8 Logarithms expand small variations and compress large ones.

The Natural Logarithm Function

The number e is often used as the base of a logarithm. $\text{Log}_e\, x$ is called the **natural logarithm function** and is abbreviated $\ln x$. Thus, $\ln 2$ is an abbreviation for $\log_e 2$; $\ln 2$ is the exponent to which e must be raised in order to get 2.

NATURAL LOGARITHM DEFINITION

$$y = \ln x \qquad (\text{or} \quad y = \log_e x)$$

means the same as

$$e^y = x$$

The common logarithm (base 10) and the natural logarithm (base e) are the most frequently used types of logarithms.

EXAMPLE 6

CALCULATING NATURAL LOGARITHMS Find the following values by using a calculator:

a. $\ln 5.2$ **b.** $\ln 0.4$ **c.** $\ln(-1.2)$ **d.** $\ln(e^2)$

SOLUTION

a. The $\boxed{\ln x}$ button on your calculator is used to find the natural logarithm of a number. Simply enter the number you want to find the logarithm of and press

$\boxed{\ln x}$. On a graphing calculator, you must press $\boxed{\text{LN}}$ first and then enter the desired number.

$$\ln 5.2 = 1.648658626 \ldots$$

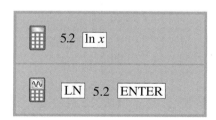

Therefore, $e^{1.648658626} \approx 5.2$.

b. $\ln 0.4 = -0.916290731 \ldots$

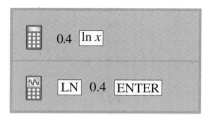

Therefore, $e^{-0.916290731} \approx 0.4$.

c. $\ln(-1.2) = \text{ERROR}$

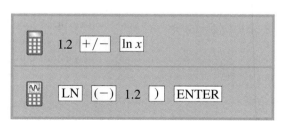

Your scientific calculator cannot find the natural logarithm of -1.2 because negative numbers do not have logarithms (in the real number system). If we let $y = \ln(-1.2)$, then the corresponding exponential equation would be $e^y = -1.2$. However, this equation has no solution (in the real number system) because e raised to any power always gives a positive result. Because we are working with real numbers only, $\ln(-1.2)$ is undefined.

d. To find $\ln(e^2)$, you must first calculate e^2: $e^2 = 7.389056099\ldots$. Now find the natural logarithm of this number.

$$\ln(e^2) = \ln(7.389056099\ldots) = 2$$

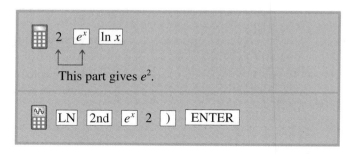

In Exercises 1–12, find the value of u, v, or b.

1. $v = \log_2 4$ **2.** $v = \log_2 32$

3. $v = \log_2\left(\dfrac{1}{16}\right)$ **4.** $v = \log_2\left(\dfrac{1}{2}\right)$

5. $\log_5 u = 2$ **6.** $\log_5 u = 3$

7. $\log_3 u = 0$ **8.** $\log_3 u = 1$

9. $\log_b 16 = 2$ **10.** $\log_b 36 = 2$

11. $\log_b 8 = -3$ **12.** $\log_b 27 = -3$

In Exercises 13–18, rewrite the logarithm as an exponential equation.

13. $P = \log_b Q$ **14.** $S = \log_b T$

15. $M = \log_b(N + T)$ **16.** $J = \log_b(K - L)$

17. $M + R = \log_b(N + T)$ **18.** $J + E = \log_b(K - L)$

In Exercises 19–26, rewrite the exponential equation as a logarithm.

19. $b^F = G$ **20.** $b^w = Y$

21. $b^{F+2} = G$ **22.** $b^{w-1} = Y$

23. $b^{CD} = E - F$ **24.** $b^{EF} = C - D$

25. $b^{2-H} = Z + 3$ **26.** $b^{2+K} = Q - 9$

In Exercises 27–56, use a calculator to find each value. Give the entire display of your calculator.

27. a. $e^{1.4}$ **b.** $10^{1.4}$

28. a. $e^{3.2}$ **b.** $10^{3.2}$

29. a. $2e^{0.07}$ **b.** $2(10^{0.07})$

30. a. $3e^{0.09}$ **b.** $3(10^{0.09})$

31. a. $\dfrac{1}{e^{1.2}}$ **b.** $\dfrac{1}{10^{1.2}}$

32. a. $\dfrac{1}{e^{2.1}}$ **b.** $\dfrac{1}{10^{2.1}}$

33. a. $\dfrac{5}{e^{0.24}}$ **b.** $\dfrac{5}{10^{0.24}}$

34. a. $\dfrac{13}{e^{0.68}}$ **b.** $\dfrac{13}{10^{0.68}}$

35. a. $\dfrac{e^{5.6}}{2e^{3.4}}$ **b.** $\dfrac{10^{5.6}}{2(10^{3.4})}$

36. a. $\dfrac{e^{0.078}}{3e^{0.065}}$ **b.** $\dfrac{10^{0.078}}{3(10^{0.065})}$

37. a. $\dfrac{2e^{5.6}}{e^{3.4}}$ **b.** $\dfrac{2(10^{5.6})}{10^{3.4}}$

38. a. $\dfrac{3e^{0.078}}{e^{0.065}}$ **b.** $\dfrac{3(10^{0.078})}{10^{0.065}}$

39. a. $\dfrac{e^{4.5}}{10^{1.2}}$ **b.** $\dfrac{10^{1.2}}{e^{4.5}}$

40. a. $\dfrac{e^{3.9}}{10^{1.5}}$ **b.** $\dfrac{10^{1.5}}{e^{3.9}}$

41. a. $\ln 2.67$ **b.** $\log 2.67$

42. a. $\ln 8.76$ **b.** $\log 8.76$

43. a. $\ln 0.85$ **b.** $\log 0.85$

44. a. $\ln 0.33$ **b.** $\log 0.33$

45. a. $\ln(e^{4.1})$ **b.** $\log(10^{4.1})$

46. a. $\ln(e^{3.8})$ **b.** $\log(10^{3.8})$

47. a. $\ln(2e^{4.1})$ **b.** $\log[2(10^{4.1})]$

48. a. $\ln(2e^{3.8})$ **b.** $\log[2(10^{3.8})]$

49. a. $e^{\ln 2.3}$ **b.** $10^{\log 2.3}$

50. a. $e^{\ln 3.2}$ **b.** $10^{\log 3.2}$

51. a. $e^{3\ln 2}$ **b.** $10^{3\log 2}$

52. a. $e^{2\ln 2}$ **b.** $10^{2\log 2}$

53. a. $\ln 10$ **b.** $\ln(10^2)$ **c.** $\ln(10^3)$

54. a. $\ln 5$ **b.** $\ln(5^2)$ **c.** $\ln(5^3)$

55. a. $\log e$ **b.** $\log(e^2)$ **c.** $\log(e^3)$

56. a. $\log 5$ **b.** $\log(5^2)$ **c.** $\log(5^3)$

57. Put the following in numerical order, from smallest to largest: $e^{2.9}$, $\ln 2.9$, $10^{2.9}$, $\log 2.9$.

58. Put the following in numerical order, from smallest to largest: $e^{1.6}$, $\ln 1.6$, $10^{1.6}$, $\log 1.6$.

Answer the following questions using complete sentences and your own words.

• CONCEPT QUESTIONS

59. What is an exponential function?

60. What is a logarithmic function?

61. If $x > 0$, which value is larger, 10^x or $\log x$? Why?

62. If $x > 0$, which value is larger, e^x or $\ln x$? Why?

63. If $x > 1$, which value is larger, $\log x$ or $\ln x$? Why?

64. If $0 < x < 1$, which value is larger, $\log x$ or $\ln x$? Why?

65. If $x > 0$, which value is larger, e^x or 10^x? Why?

66. If $x < 0$, which value is larger, e^x or 10^x? Why?

• HISTORY QUESTION

67. Who invented the symbol e? What other symbols did this person introduce?

10.0B Review of Properties of Logarithms

BJECTIVES

- Understand and apply the inverse properties of logarithms and exponentials
- Understand and apply the arithmetic properties (addition, subtraction, and multiplication) of logarithms
- Solve exponential equations
- Solve logarithmic equations

Logarithms have several properties that can be used in simplifying complicated expressions and solving equations. We will use these properties extensively in the applications of logarithms and exponentials in this chapter. Perhaps the most important of these properties are the *Inverse Properties*.

The Inverse Properties

In Section 10.0A, we used the calculator to compute $\log(10^5) = 5$ and $\ln(e^2) = 2$ (Examples 4d and 6d). Note that in either case, the base of the logarithm is the same as the base of the exponential and that the logarithm appears to have "canceled" the exponential. Does this "canceling" always work? Does $\log_b(b^v) = v$ for all bases b and all numbers v?

Recall that by the definition of a logarithm, $\log_b u = v$ means the same as $b^v = u$. Since b^v is equal to u, we can substitute b^v for u in the equation $\log_b u = v$. This substitution results in $\log_b(b^v) = v$, which indicates that a base b logarithm will always "cancel" a base b exponential. *Cancel* is not really the right word here; *canceling* refers to what happens when you reduce a fraction. Instead, we say that a base b logarithm is the **inverse** of a base b exponential; whatever an exponential function does to a number, a logarithm undoes it and gives back the number.

Because $\log_b(b^x) = x$ for *any* base b, it is true for bases $b = 10$ and $b = e$. In other words, $\log_{10}(10^x) = x$ and $\log_e(e^x) = x$. Using the alternative notations, $\log(10^x) = x$ and $\ln(e^x) = x$.

INVERSE PROPERTIES

$$\log(10^x) = x \qquad (\text{or} \qquad \log_{10}(10^x) = x)$$
$$\ln(e^x) = x \qquad (\text{or} \qquad \log_e(e^x) = x)$$

EXAMPLE **1** | **APPLYING THE INVERSE PROPERTIES OF LOGARITHMS** Simplify the following by using the Inverse Properties:

a. $\log(10^{3x})$ **b.** $\ln(e^{-0.012x})$ **c.** $\ln(10^{5x})$

743

SOLUTION

a. We are applying a base 10 logarithm to a base 10 exponential, so we can apply the Inverse Property $\log(10^x) = x$ and obtain

$$\log(10^{3x}) = \log_{10}(10^{3x}) = 3x$$

b. $\ln(e^{-0.012x}) = \log_e(e^{-0.012x}) = -0.012x$ **using the Inverse Property $\ln(e^x) = x$**

c. The Inverse Properties do *not* apply to $\ln(10^{5x}) = \log_e(10^{5x})$, because the logarithm and the exponential are different bases (base e and base 10, respectively). Thus, $\ln(10^{5x})$ cannot be simplified by using the Inverse Properties.

We have just seen that a logarithm "undoes" an exponential when the logarithm is applied to the exponential and the bases are the same. Will the same thing happen when an exponential is applied to a logarithm? Does $b^{(\log_b x)}$ simplify to just x?

Recall that by the definition of a logarithm, $b^v = u$ means the same as $\log_b u = v$. Because $\log_b u$ is equal to v, we can substitute it for v in the equation $b^v = u$. This substitution results in $b^{(\log_b u)} = u$. This means that a base b exponential will always "undo" a base b logarithm; an exponential (of the same base) is the **inverse** of a logarithm.

Because $b^{(\log_b x)} = x$ for *any* base b, it is true for bases $b = 10$ and $b = e$. In other words, $10^{(\log_{10} x)} = x$ and $e^{(\log_e x)} = x$. Using the alternative notations, $10^{(\log x)} = x$ and $e^{(\ln x)} = x$.

INVERSE PROPERTIES

$$10^{\log x} = x \qquad (\text{or} \qquad 10^{\log_{10} x} = x)$$
$$e^{\ln x} = x \qquad (\text{or} \qquad e^{\log_e x} = x)$$

EXAMPLE **2**

APPLYING THE INVERSE PROPERTIES OF EXPONENTIALS   Simplify the following.

a. $10^{\log(2x-3)}$ **b.** $e^{\ln(0.023x)}$

SOLUTION

a. We are raising 10 to a common log (base 10) power, so we can apply an Inverse Property.

$$10^{\log(2x-3)} = 2x - 3 \quad \text{using the Inverse Property } 10^{\log x} = x$$

b. We are raising e to a natural log (base e) power, so we can apply an Inverse Property.

$$e^{\ln(0.023x)} = 0.023x \quad \text{using the Inverse Property } e^{\ln x} = x$$

Solving Exponential Equations

Recall that an exponential equation is one in which x appears only in the exponent. If the variable can be removed from the exponent, then the resulting equation can be solved in a manner that is familiar to you from algebra. To solve an exponential equation, apply a logarithm and use the properties of logarithms to remove the variable from the exponent.

EXAMPLE **3**

SOLVING AN EXPONENTIAL EQUATION, BASE 10   Solve $10^x = 0.47$.

SOLUTION

We have a base 10 exponential, so we apply the common logarithm to both sides and simplify.

$$10^x = 0.47$$
$$\log(10^x) = \log(0.47)$$
$$x = \log 0.47 \qquad \textbf{using the Inverse Property } \log(10^x) = x$$
$$x = -0.327902142 \ldots$$

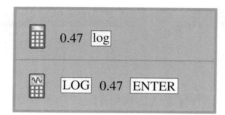

As a rough check, we know that $x = -0.33$ is between 0 and -1, so we should expect $10^x = 10^{-0.33}$ to be between $10^0 = 1$ and $10^{-1} = \frac{1}{10} = 0.1$. As a more accurate check, use your calculator to verify that $10^{-0.327902142} \approx 0.47$.

10 $\boxed{y^x}$ 0.327902142 $\boxed{+/-}$ $\boxed{=}$

10 $\boxed{\wedge}$ $\boxed{(}$ $\boxed{(-)}$ 0.327902142 $\boxed{)}$ $\boxed{\text{ENTER}}$

EXAMPLE 4

SOLUTION

SOLVING AN EXPONENTIAL EQUATION, BASE e   Solve $5e^{0.01x} = 8$.

Before we apply a logarithm, we must first "isolate" the exponential by dividing by 5.

$$5e^{0.01x} = 8$$

$$e^{0.01x} = \frac{8}{5}$$

$$\ln(e^{0.01x}) = \ln\frac{8}{5} \qquad\qquad \textbf{taking ln of each side}$$

$$0.01x = \ln\frac{8}{5} \qquad\qquad \textbf{using the Inverse Property } \ln(e^x) = x$$

$$x = \frac{\ln(8/5)}{0.01}$$

$$x = 47.00036292\ldots$$

8 $\boxed{\div}$ 5 $\boxed{=}$ $\boxed{\ln x}$ $\boxed{\div}$ 0.01 $\boxed{=}$

$\boxed{\text{LN}}$ 8 $\boxed{\div}$ 5 $\boxed{)}$ $\boxed{\div}$ 0.01 $\boxed{\text{ENTER}}$

Note: A common mistake is to improperly apply the natural logarithm at the first step; that is, if $5e^{0.01x} = 8$, then $5\ln(e^{0.01x}) \neq \ln 8$. You should always isolate the exponential as your first step.

Use your calculator to verify that $5e^{(0.01)(47.00036292)}$ is 8.

0.01 $\boxed{\times}$ 47.00036292 $\boxed{=}$ $\boxed{e^x}$ $\boxed{\times}$ 5 $\boxed{=}$

5 $\boxed{\times}$ $\boxed{\text{2nd}}$ $\boxed{e^x}$ 0.01 $\boxed{\times}$ 47.00036292 $\boxed{)}$ $\boxed{\text{ENTER}}$

STEPS FOR SOLVING EXPONENTIAL EQUATIONS

1. Isolate the exponential; that is, rewrite the problem in the form $e^A = B$.
2. Take the natural logarithm of each side.
3. Use the Inverse Property $\ln(e^x) = x$ to simplify.
4. Solve.
5. Use your calculator to check the answer.

(These steps also apply to solving a base 10 exponential equation. However, instead of applying the natural logarithm to each side, apply the common log.)

The Exponent-Becomes-Multiplier Property

A property of logarithms that is useful in solving exponential equations is the *Exponent-Becomes-Multiplier* Property.

EXPONENT-BECOMES-MULTIPLIER PROPERTY

$$\log(A^n) = n \cdot \log A$$
$$\ln(A^n) = n \cdot \ln A$$

When we take the logarithm of a base raised to a power, the exponent can be brought down in front of the logarithm, thus becoming a multiplier of the logarithm. Of course, these equations can be "reversed" to obtain $n \cdot \log A = \log(A^n)$ and $n \ln A = \ln(A^n)$; a multiplier of a logarithm can become an exponent.

This property seems rather strange—after all, exponents don't normally turn into multipliers. Or do they? Consider the Exponent Property:

$$(x^m)^n = x^{m \cdot n}$$

In the left-hand side of the equation, n is an exponent; in the right-hand side, n is a multiplier. This property is actually the basis of the Exponent-Becomes-Multiplier Property.

To see why $\log(A^n) = n \cdot \log A$, let $\log A = a$ and rewrite this logarithm as an exponential.

$$\log A = a$$
$$A = 10^a \quad \text{using the common logarithm definition}$$

Raising each side to the nth power, we get

$$A^n = (10^a)^n$$
$$= 10^{an} \quad \text{using the exponent law } (x^m)^n = x^{mn}$$

HISTORICAL NOTE

JOHN NAPIER, 1550–1617

The Granger Collection, New York

John Napier was a Scottish landowner and member of the upper class. As such, he had a great deal of leisure time, much of which he devoted to mathematics, politics, and religion.

As is the case today, scientists of Napier's time frequently had to multiply and divide large numbers. Of course, at that time no calculators (or even slide rules) existed; a scientist had to make all calculations by hand. Such work was tedious and prone to errors. Napier invented logarithms as a system that would allow the relatively easy calculation of products and quotients, as well as powers and roots.

With this system, the product of two numbers was calculated by finding the logarithms of those numbers in a book of tables and adding the results. Napier spent *twenty years* creating these tables. This procedure represented a real shortcut: Addition replaced multiplication, and addition is much simpler than multiplication when done by hand. This method of multiplying by adding was an especially useful application of what we know as the Multiplication-Becomes-Addition Property, $\log(A \cdot B) = \log A + \log B$. Napier did not invent the notation $\log_b x$; this notation was invented much later by Leonhard Euler. Napier did not use any notation in his writings on logarithms; he wrote everything out verbally.

Although Napier did not invent the decimal point, he is responsible for its widespread use. Napier's calculating system was very popular, and the logarithms in his tables were decimal numbers written with a decimal point.

A similar system was developed independently by Joost Bürgi, a Swiss mathematician and watchmaker. Napier is generally credited with the invention

of logarithms, because he published his work before Bürgi.

As a member of the Scottish aristocracy, Napier was an active participant in local and national affairs. He was also a very religious man and belonged to the Church of Scotland. These two interests led to an interesting story about Napier.

In sixteenth-century Scotland, politics and religion were inexorably entwined. Mary, Queen of Scots, was a Roman Catholic; Elizabeth, Queen of England, was a Protestant; and both the Church of Scotland and the Church of England were Protestant churches. Furthermore, Mary had a strong claim to the throne of England. Catholic factions wanted Scotland to form an alliance with France; Protestant factions wanted an alliance with England. After a series of power struggles, Mary was forced to abdicate her throne to her son, who then became James VI, King of Scotland. Mary was eventually beheaded for plotting against the English throne.

It was well known that James VI wished to succeed Elizabeth to the English throne. It was suspected that he had enlisted the help of Phillip II, King of Spain and a Catholic, to attain this goal. It was also suspected that James VI was arranging an invasion of Scotland by Spain. John Napier was a member of the committee appointed by the Scottish church to express its concern to James.

Napier was not content with expressing his concerns through the church. He wrote one of the earliest Scottish interpretations of the scriptures, *A Plaine Discovery of the Whole Revelation of Saint John*, which was clearly calculated to influence contemporary events. In this work, Napier urged James VI to see that "justice be done

against the enemies of God's Church." It also declared, "Let it be your Majesty's continual study to reform . . . your country, and first to begin at your Majesty's owne house, familie and court, and purge the same of all suspicion of Papists and Atheists and Newtrals."

This tract was widely read in Europe as well as in Scotland, and Napier earned a considerable reputation as a scholar and theologian. It has been suggested that the tract saved him from persecution as a warlock; Napier had previously been suspected of being in league with the devil.

Napier also invented several secret weapons for the defense of his faith and country against a feared Spanish invasion. These inventions included two kinds of burning mirrors designed to set fire to enemy ships at a distance and an armored chariot that would allow its occupants to fire in all directions. Whether any of these devices were ever constructed is not known.

MIRIFICI

LOGARITHMORVM

CANONIS CON-
STRVCTIO;

Et eorum ad naturales ipforum numeros habitudines;

VNÀ CVM

Appendice, de aliâ eâque præftantiore Logarithmorum fpecie condenda.

QVIBVS ACCESSERE

Propofitiones ad triangula fphærica faciliore calculo refolvenda:
Vnà cum Annotationibus aliquot doctiffimi D. HENRICI
BRIGGII, in eas & memoratam appendicem.

Authore & Inventore *Ioanne Nepero*, Barone
Merchiftonii, &c. Scoto.

EDINBVRGI,
Excudebat ANDREAS HART.
ANNO DOMINI 1619.

The Granger Collection, NY

Taking the log of each side and simplifying, we have

$$\log(A^n) = \log(10^{an})$$
$$= an \qquad \text{using the Inverse Property } \log(10^x) = x.$$

Finally, we substitute the original expression $\log A = a$ to get the desired result.

$$\log(A^n) = n \cdot \log A$$

A similar method can be used to obtain the natural logarithm version of the Exponent-Becomes-Multiplier Property. (See Exercise 56.)

EXAMPLE 5

APPLYING THE EXPONENT-BECOMES-MULTIPLIER PROPERTY

a. Rewrite $\log(1.0125^x)$ so that the exponent is eliminated.
b. Rewrite $\ln(1 + 0.09/12)^x$ so that the exponent is eliminated.

SOLUTION

a. We are taking the log of an exponential, so we can apply the Exponent-Becomes-Multiplier Property.

$$\log(A^n) = n \cdot \log A \qquad \text{Exponent-Becomes-Multiplier Property}$$
$$\log(1.0125^x) = x \cdot \log 1.0125 \qquad \text{substituting 1.0125 for } A \text{ and } x \text{ for } n$$

b. The Exponent-Becomes-Multiplier Property applies to natural logs too.

$$\ln(A^n) = n \cdot \ln A \qquad \text{Exponent-Becomes-Multiplier Property}$$

$$\ln\left(1 + \frac{0.09}{12}\right)^x = x \cdot \ln\left(1 + \frac{0.09}{12}\right) \qquad \text{substituting } \frac{0.09}{12} \text{ for } A \text{ and } x \text{ for } n$$
$$= x \cdot \ln(1 + 0.0075)$$
$$= x \cdot \ln(1.0075)$$

EXAMPLE 6

SOLVING AN EXPONENTIAL EQUATION, NONSTANDARD BASE
Solve $1.03^x = 2$.

SOLUTION

This equation is different from those in the previous examples in that the base is neither 10 nor e, so the Inverse Properties cannot be used to simplify the equation. However, by taking the log (or ln) of each side, we can use the Exponent-Becomes-Multiplier Property of logarithms to change the exponent x into a multiplier, which gives us an easier equation to work with.

$$1.03^x = 2$$
$$\log(1.03^x) = \log 2 \qquad \text{taking the log of each side}$$
$$x \cdot \log 1.03 = \log 2 \qquad \text{using the Exponent-Becomes-Multiplier Property } \log(A^n) = n \cdot \log A$$
$$x = \frac{\log 2}{\log 1.03} \qquad \text{dividing by log 1.03}$$
$$x = 23.44977225 \ldots$$

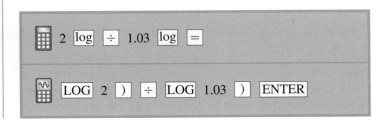

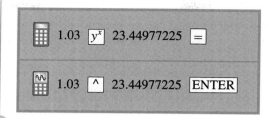

Use your calculator to verify that $1.03^{23.44977225}$ is 2.

| $\boxplus$ 1.03 $\boxed{y^x}$ 23.44977225 $\boxed{=}$ |
| $\boxplus$ 1.03 $\boxed{\wedge}$ 23.44977225 $\boxed{\text{ENTER}}$ |

The Division-Becomes-Subtraction Property

There are times when rewriting one logarithm as two logarithms, or vice versa, is advantageous. To this end, we have the following property.

DIVISION-BECOMES-SUBTRACTION PROPERTY

$$\log \frac{A}{B} = \log A - \log B$$

$$\ln \frac{A}{B} = \ln A - \ln B$$

The logarithm of a quotient can be rewritten as the logarithm of the numerator *minus* the logarithm of the denominator, hence the title *Division-Becomes-Subtraction*. Of course, these equations can be "reversed" to obtain

$$\log A - \log B = \log \frac{A}{B} \quad \text{and} \quad \ln A - \ln B = \ln \frac{A}{B}$$

These properties seem rather strange; after all, division doesn't normally turn into subtraction. Or does it? Consider the exponent property

$$\frac{x^m}{x^n} = x^{m-n}$$

When we use exponents, division *does* become subtraction. Furthermore, logarithms are closely related to exponents, so it shouldn't be surprising that both logarithms and exponents have a Division-Becomes-Subtraction Property.

To see why $\log A - \log B = \log \frac{A}{B}$, let $\log A = a$ and $\log B = b$. Now rewrite each expression as an exponential.

$$A = 10^a \quad \text{and} \quad B = 10^b \quad \text{using the Common Logarithm Definition}$$

Dividing the first equation by the second yields the following:

$$\frac{A}{B} = \frac{10^a}{10^b}$$

$$\frac{A}{B} = 10^{a-b} \qquad \text{using the exponent law } \frac{x^m}{x^n} = x^{m-n}$$

$$\log \frac{A}{B} = \log(10^{a-b}) \quad \text{taking the log of each side}$$

$$\log \frac{A}{B} = a - b \qquad \text{using the Inverse Property } \log(10^x) = x$$

Finally, substitute the original expressions for a and b to get the desired result:

$$\log \frac{A}{B} = \log A - \log B$$

A similar method can be used to obtain the natural logarithm version of the Division-Becomes-Subtraction Property. (See Exercise 55.)

EXAMPLE 7

APPLYING THE DIVISION-BECOMES-SUBTRACTION PROPERTY

a. Rewrite $\log \frac{x}{3}$ so that the fraction is eliminated.
b. Rewrite $\log(2x) - \log 8$ as one logarithm.

SOLUTION

a. We have the log of a fraction, so we can apply the Division-Becomes-Subtraction Property.

$$\log \frac{A}{B} = \log A - \log B \qquad \textbf{Division-Becomes-Subtraction Property}$$

$$\log \frac{x}{3} = \log x - \log 3 \qquad \textbf{substituting } x \textbf{ for } A \textbf{ and } 3 \textbf{ for } B$$

b. We are subtracting two logs, so we can apply the reverse of the Division-Becomes-Subtraction Property.

$$\log A - \log B = \log \frac{A}{B} \qquad \textbf{Division-Becomes-Subtraction Property}$$

$$\log 2x - \log 8 = \log \frac{2x}{8} \qquad \textbf{substituting } 2x \textbf{ for } A \textbf{ and } 8 \textbf{ for } B$$

$$= \log \frac{x}{4} \qquad \textbf{canceling}$$

In the last step of Example 6, we had

$$x = \frac{\log 2}{\log 1.03}$$

Frequently, students will attempt to rewrite this as

$$x = \log(2 - 1.03)$$

This is incorrect; it is the result of misremembering the Division-Becomes-Subtraction Property. That property says that $\log \frac{A}{B} = \log A - \log B$; it does *not* refer to $\frac{\log A}{\log B}$ or to $\log(A - B)$.

COMMON ERRORS

You might be tempted to "simplify" the expression $\dfrac{\log A}{\log B}$. However,

1. $\dfrac{\log A}{\log B} \neq \log A - \log B$

    (Subtracting two logs is not the same as dividing two logs.)

2. $\dfrac{\log A}{\log B} \neq \dfrac{A}{B}$

    (You cannot cancel the logs in the numerator and denominator of a fraction.)

3. $\dfrac{\log A}{\log B} \neq \log(A - B)$

    (The two logs on the left cannot be reduced into one log.)

The Multiplication-Becomes-Addition Property

We have seen that when logarithms are subtracted, they can be combined into one logarithm via the Division-Becomes-Subtraction Property. There is a similar property when logarithms are added.

MULTIPLICATION-BECOMES-ADDITION PROPERTY

$$\log(A \cdot B) = \log A + \log B$$
$$\ln(A \cdot B) = \ln A + \ln B$$

The logarithm of a product can be rewritten as the sum of two logarithms, hence the title *Multiplication-Becomes-Addition*. Of course, these equations can be "reversed" to obtain $\log A + \log B = \log(A \cdot B)$ and $\ln A + \ln B = \ln(A \cdot B)$.

EXAMPLE 8

APPLYING THE MULTIPLICATION-BECOMES-ADDITION PROPERTY

a. Rewrite $\log(3x)$ as two logarithms.
b. Rewrite $\ln(2x) + \ln 8$ as one simplified logarithm.

SOLUTION

a. We have the log of a product, so we can apply the Multiplication-Becomes-Addition Property.

$\log(A \cdot B) = \log A + \log B$ **multiplication-Becomes-Addition Property**

$\log(3x) = \log 3 + \log x$ **substituting 3 for A and x for B**

b. We are adding two logs, so we can apply the reverse of the Multiplication-Becomes-Addition Property.

$\ln A + \ln B = \ln(A \cdot B)$ **multiplication-Becomes-Addition Property**

$\ln(2x) + \ln 8 = \ln(2x \cdot 8)$ **substituting $2x$ for A and 8 for B**

$= \ln(16x)$ **multiplying**

Solving Logarithmic Equations

An equation in which x appears "inside" a logarithm is called a **logarithmic equation**. To solve an equation of this type, we apply an exponential function to each side and use the various properties that we have developed.

EXAMPLE 9

SOLUTION

SOLVING A LOGARITHMIC EQUATION   Solve $\log x = 5 + \log 3$.

To solve for x, we have to eliminate the common logarithm. We can accomplish this by using the Inverse Property $10^{\log x} = x$. However, for this property to apply, the equation must contain only one logarithm. Therefore, our first step is to get all the log terms on one side and then combine them into one logarithm.

$\log x = 5 + \log 3$

$\log x - \log 3 = 5$

$\log\left(\dfrac{x}{3}\right) = 5$ **using the Division-Becomes-Subtraction Property**

We can now apply the base 10 exponential function to each side—that is, exponentiate each side (base 10), and simplify.

$$10^{\log(x/3)} = 10^5$$

$$\frac{x}{3} = 100{,}000 \quad \text{using the Inverse Property } 10^{\log x} = x$$

$$x = 300{,}000$$

Use your calculator to verify that log 300,000 = 5 + log 3 by computing the right and left sides separately and verifying that they are equal.

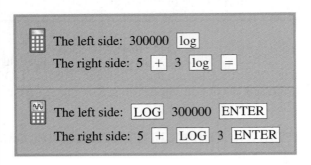

The left side: 300000 [log]
The right side: 5 [+] 3 [log] [=]

The left side: [LOG] 300000 [ENTER]
The right side: 5 [+] [LOG] 3 [ENTER]

pH: An Application of Logarithms

Chemists define pH by the formula pH $= -\log[\text{H}^+]$, where $[\text{H}^+]$ is the hydrogen ion concentration measured in moles per liter. Solutions with a pH of 7 are said to be neutral, whereas a pH less than 7 is classified as an acid and a pH greater than 7 is classified as a base.

EXAMPLE **10**

WORKING WITH PH IN CHEMISTRY

a. An unknown substance has a hydrogen ion concentration of $[\text{H}^+] = 1.7 \times 10^{-5}$ moles per liter. Determine the pH and classify the substance as an acid or a base.
b. If a solution has a pH of 6.2, find the hydrogen ion concentration of the solution.

SOLUTION

a. Because we know the hydrogen ion concentration, we substitute it into the pH formula and use a calculator.

$$\begin{aligned}
\text{pH} &= -\log[\text{H}^+] && \textbf{pH formula} \\
&= -\log(1.7 \times 10^{-5}) && \textbf{substituting } 1.7 \times 10^{-5} \textbf{ for } [\text{H}^+] \\
&= 4.769551079 && \textbf{using a calculator} \\
&= 4.8 && \textbf{rounding to one decimal place}
\end{aligned}$$

The pH of the substance is 4.8 (less than 7), so the substance is classified as an acid.
b. Because we know the pH, we substitute it into the pH formula and solve for $[\text{H}^+]$, the hydrogen ion concentration.

$$\begin{aligned}
\text{pH} &= -\log[\text{H}^+] && \textbf{pH formula} \\
6.2 &= -\log[\text{H}^+] && \textbf{substituting 6.2 for pH} \\
-6.2 &= \log[\text{H}^+] && \textbf{multiplying by } -1 \\
10^{-6.2} &= 10^{\log[\text{H}^+]} && \textbf{exponentiating with base 10} \\
10^{-6.2} &= [\text{H}^+] && \textbf{using the Inverse Property } 10^{\log x} = x \\
[\text{H}^+] &= 6.309573445 \times 10^{-7} && \textbf{using a calculator}
\end{aligned}$$

The hydrogen ion concentration is 6.3×10^{-7} moles per liter.

STEPS FOR SOLVING LOGARITHMIC EQUATIONS

1. Get all the log terms on one side and all the nonlog terms on the other.
2. Combine the log terms into one term, using the **Division-Becomes-Subtraction Property**

$$\log\left(\frac{A}{B}\right) = \log A - \log B$$

and the **Multiplication-Becomes-Addition Property**

$$\log(A \cdot B) = \log A + \log B$$

3. Exponentiate each side (base 10).
4. Use the **Inverse Property** $10^{\log x} = x$ and simplify.
5. Solve.

(These steps also apply to solving an equation that contains natural logarithms. However, instead of exponentiating base 10, use the base e exponential and the inverse property $e^{\ln x} = x$ to simplify.)

For Those Who Have Read Section 8.8, "Fractal Geometry"

EXAMPLE 11

CALCULATING A LOGARITHM   In Example 1 of Section 8.8 we found that the dimension of a Koch snowflake is the number d that satisfies the equation $3^d = 4$. We used a trial-and-error process to approximate the value of d because we were unable to solve this equation.

a. Use logarithms to solve the equation and find the exact value of d.
b. Use the result of part (a) to approximate d to the nearest hundredth.
c. Use your calculator to check your answer.

SOLUTION

a. *Finding the exact value of d.*

$$3^d = 4$$
$$\log(3^d) = \log 4 \quad \text{taking the log of each side}$$
$$d \cdot \log 3 = \log 4 \quad \text{using the Exponent-Becomes-Multiplier Property}$$
$$d = \frac{\log 4}{\log 3} \quad \text{dividing by log 3}$$

The exact value of d is

$$d = \frac{\log 4}{\log 3}$$

b. *Approximating d.*   Using a calculator, we get

$$d = 1.26185950 \ldots \approx 1.26$$

Notice that this result agrees with that found in Example 1 of Section 8.8.

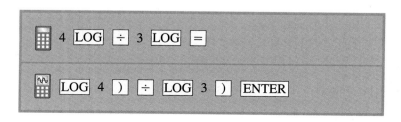

c. *Checking the answer.* You can check your answer by verifying that $3^{1.26185950\cdots}$ is 4.

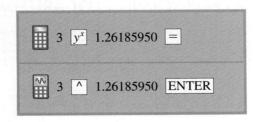

10.0B   EXERCISES

In Exercises 1–12, simplify by using the Inverse Properties.

1. $\log(10^{6x})$

2. $\log(10^{-2x})$

3. $\ln(e^{-0.036x})$

4. $\ln(e^{0.114x})$

5. $10^{\log(2x+5)}$

6. $10^{\log(4-3x)}$

7. $e^{\ln(1-x)}$

8. $e^{\ln(1.75x)}$

9. a. $e^{2\ln x}$ **b.** $10^{2\log x}$

10. a. $e^{3\ln x}$ **b.** $10^{3\log x}$

11. a. $e^{2\ln(3x)}$ **b.** $10^{2\log(3x)}$

12. a. $e^{3\ln(2x)}$ **b.** $10^{3\log(2x)}$

In Exercises 13–22, rewrite the given logarithm so that all products, quotients, and exponents are eliminated.

13. $\log\left(\dfrac{x}{4}\right)$

14. $\ln\left(\dfrac{x}{7}\right)$

15. $\ln(1.8x)$

16. $\log(12x)$

17. $\log(1.225^{x})$

18. $\ln(1.005^{x})$

19. $\ln(3x^{4})$

20. $\log(6x^{5})$

21. $\log\left(\dfrac{5x^{2}}{7}\right)$

22. $\ln\left(\dfrac{7x^{3}}{11}\right)$

In Exercises 23–32, rewrite as one simplified logarithm.

23. $\ln(4x) + \ln 5$

24. $\log(6x) + \log 2$

25. $\log(6x) - \log 2$

26. $\ln(5x) - \ln 5$

27. $\ln x - \ln 3 + \ln 6$

28. $\log x - \log 4 + \log 22$

29. $3\log(2x) - \log 8$

30. $2\ln(5x) - \ln 5$

31. $\ln(9x) + \ln(4x) - 2\ln(6x)$

32. $2\log(3x) + 4\log(2x) - 2\log(12x)$

In Exercises 33–54, solve the given equation. (The solution of every other odd-numbered exercise is given in the back of the book; check the other exercises as shown in the text.)

33. a. $e^{x} = 0.35$ **b.** $10^{x} = 0.35$

34. a. $e^{x} = 0.75$ **b.** $10^{x} = 0.75$

35. a. $145e^{0.024x} = 290$ **b.** $145(10)^{0.024x} = 290$

36. a. $72e^{0.068x} = 144$ **b.** $72(10)^{0.068x} = 144$

37. a. $2,000e^{0.004x} = 8,500$ **b.** $2,000(10)^{0.004x} = 8,500$

38. a. $5,000e^{0.15x} = 18,000$ **b.** $5,000(10)^{0.15x} = 18,000$

39. a. $50e^{-0.035x} = 25$ **b.** $50(10)^{-0.035x} = 25$

40. a. $13e^{-0.001x} = 6.5$ **b.** $13(10)^{-0.001x} = 6.5$

41. a. $80e^{-0.0073x} = 65$ **b.** $80(10)^{-0.0073x} = 65$

42. a. $42e^{-0.0037x} = 35$ **b.** $42(10)^{-0.0037x} = 35$

43. a. $\ln x = 0.66$ **b.** $\log x = 0.66$

44. a. $\ln x = 0.86$ **b.** $\log x = 0.86$

45. a. $\ln x = 3.66$ **b.** $\log x = 3.66$

46. a. $\ln x = 2.86$ **b.** $\log x = 2.86$

47. a. $\ln x + \ln 6 = 2$ **b.** $\log x + \log 6 = 2$

48. a. $\ln x + \ln 4 = 3$
  b. $\log x + \log 4 = 3$

49. a. $\ln x - \ln 6 = 2$
  b. $\log x - \log 6 = 2$

50. a. $\ln x - \ln 4 = 3$
  b. $\log x - \log 4 = 3$

51. a. $\ln x = 4.8 + \ln 6.9$
  b. $\log x = 4.8 - \log 6.9$

52. a. $\ln x = 2.7 + \ln 3.5$
  b. $\log x = 2.7 + \log 3.5$

53. a. $\ln 0.9 = 3.1 - \ln(4x)$
  b. $\log 0.9 = 3.1 - \log(4x)$

54. a. $\ln 0.8 = 2.4 - \ln(3x)$
  b. $\log 0.8 = 2.4 - \log(3x)$

55. Show that $\ln \frac{A}{B} = \ln A - \ln B$.

HINT: Use the same method as in the proof of $\log \frac{A}{B} = \log A - \log B$.

56. Show that $\ln(A^n) = n \cdot \ln A$.

HINT: Use the same method as in the proof of $\log (A^n) = n \cdot \log A$.

57. Show that $\log(A \cdot B) = \log A + \log B$.

HINT: Use a method similar to that used in Exercise 49.

58. Show that $\ln(A \cdot B) = \ln A + \ln B$.

Exercises 59–70 refer to Example 10.

59. For a certain fruit juice, $[H^+] = 3.0 \times 10^{-4}$. Determine the pH and classify the juice as acid or base.

60. For milk, $[H^+] = 1.6 \times 10^{-7}$. Determine the pH and classify milk as acid or base.

61. An unknown substance has a hydrogen ion concentration of 3.7×10^{-8} moles per liter. Determine the pH and classify the substance as acid or base.

62. An unknown substance has a hydrogen ion concentration of 2.4×10^{-5} moles per liter. Determine the pH and classify the substance as acid or base.

63. Fresh-brewed coffee has a hydrogen ion concentration of about 1.3×10^{-5} moles per liter. Determine the pH of fresh-brewed coffee.

64. Normally, human blood has a hydrogen ion concentration of about 3.98×10^{-8} moles per liter. Determine the normal pH of human blood.

65. When the pH of a person's blood drops below 7.4, a condition called *acidosis* sets in. Acidosis can result in death if pH reaches 7.0. What would the hydrogen ion concentration of a person's blood be at that point?

66. When the pH of a person's blood rises above 7.4, a condition called *alkalosis* sets in. Alkalosis can result in death if pH reaches 7.8. What would the hydrogen ion concentration of a person's blood be at that point?

67. You want to plant tomatoes in your backyard. Tomatoes prefer soil that has a pH range of 5.5 to 7.5. Using a testing kit, you determine the hydrogen ion concentration of your soil to be
a. 3.5×10^{-7} moles per liter. Should you plant tomatoes? Why or why not?
b. 3.5×10^{-4} moles per liter. Should you plant tomatoes? Why or why not?

68. You want to plant potatoes in your backyard. Potatoes prefer soil that has a pH range of 4.5 to 6.0. Using a testing kit, you determine the hydrogen ion concentration of your soil to be
a. 2.1×10^{-7} moles per liter. Should you plant potatoes? Why or why not?
b. 2.1×10^{-5} moles per liter. Should you plant potatoes? Why or why not?

69. Paprika prefers soil that has a pH range of 7.0 to 8.5. What range of hydrogen ion concentration does paprika prefer?

70. Spruce prefers soil that has a pH range of 4.0 to 5.0. What range of hydrogen ion concentration does spruce prefer?

Exercises 71–76 are for those who have covered Section 5.2: Compound Interest.

71. Use logarithms to solve Exercise 76 on page 355 in Section 5.2.

72. Use logarithms to solve Exercise 77 on page 355 in Section 5.2.

73. Use logarithms to solve Exercise 78 on page 355 in Section 5.2.

74. Use logarithms to solve Exercise 79 on page 355 in Section 5.2.

75. Use logarithms to solve Exercise 80 on page 355 in Section 5.2.

76. Use logarithms to solve Example 9 on page 355 in Section 5.2.

For those who have read Section 8.8, "Fractal Geometry," in Exercises 77–81, (a) Use logarithms to find the exact value of the dimension of the given fractal. (b) Use the result of part (a) to approximate d to the nearest hundredth. (The solution to part (b) is not given in the back of the book; instead, check as shown in the text.)

77. The Sierpinski gasket (See Exercise 9 in Section 8.9.)

78. The Sierpinski carpet (See Exercises 2 and 10 in Section 8.9.)

79. The Mitsubishi gasket (See Exercises 3 and 11 in Section 8.9.)

80. The Menger sponge (See Exercise 12 in Section 8.9.)

81. The square snowflake (See Exercises 4 and 13 in Section 8.9.)

 Answer the following questions using complete sentences and your own words.

● HISTORY QUESTIONS

82. Who invented logarithms?

83. What motivated the invention of logarithms?

84. Historically, how were logarithms used to calculate quotients?

HINT: Alter the discussion of how logarithms were used to calculate products.

85. Was Napier in favor of or opposed to the attempt of James VI to succeed Elizabeth to the English throne? Why?

86. From what possible catastrophe did Napier's political pursuits save him?

OBJECTIVES

- Understand average growth rates and what they measure
- Understand relative growth rates and what they measure
- Distinguish between average and relative growth rates
- Use the exponential model to analyze growth

In Sections 10.0A and 10.0B, we reviewed the algebra that is used in connection with the exponential function $y = ae^{bx}$. In this and the following section, we will see how this function is used in topics as diverse as population growth, inflation, the decay of nuclear wastes, and radiocarbon dating. Each of these quantities grows (or decays) at a rate that reflects its current size: A larger population grows faster than a smaller one, and a larger amount of uranium decays faster than a smaller amount. This common relationship yields a powerful model that allows us to predict what the world population will be in the year 2100, how long it takes a home to double in value, and how long nuclear wastes must be stored. Or, looking back in time, we can use the model to determine the age of the Dead Sea Scrolls.

The first thing we need to do is determine why some things grow at a rate that reflects their current size.

EXAMPLE 1

HOW DO POPULATIONS GROW?   A farmer has been studying the aphids in her alfalfa so that she can determine the best time to spray or release natural predators. In each of her two fields, she marked off one square foot of alfalfa and counted the number of aphids in that square foot. In her first field, she counted 100 aphids. One week later, the population had increased to 140; that is, its average growth rate was 40 aphids per week. In her second field, she counted 200 aphids. If the two fields provide the aphids with similar living conditions (favorable temperature, an abundance of food and space, and so on), what would be the most likely average growth rate for the second field?

SOLUTION

Forty aphids per week would not be realistic. Certainly, the larger population would have more births per week than the smaller one. The most likely average growth rate for the second field would be 80 aphids per week; this reflects the difference in population size. Either population, then, would grow by the same percent; that is, the growth rate would be the same percentage of the population (see Figure 10.9).

	Average Growth Rate	Population	Growth as a Percent of (or Relative to) the Population
First Field	40 aphids/week	100 aphids	$\frac{40}{100} = 0.40 = 40\%$
Second Field	80 aphids/week	200 aphids	$\frac{80}{200} = 0.40 = 40\%$

FIGURE 10.9 Aphid growth.

Delta Notation

The symbol "Δ" is the Greek letter *delta*. **Delta notation** is frequently used to describe changes in quantities, with the symbol Δ meaning "change in." In Example 1, the change in population p could be written Δp, and the change in time t could be written Δt. Expressed in delta notation, the **average growth rate** of 40 aphids per week would be written as

$$\frac{\Delta p}{\Delta t} = \frac{40 \text{ aphids}}{1 \text{ week}}$$

while the **relative growth rate** (that is, the growth rate *relative to*, or as a percentage of, the population) would be written as

$$\frac{\Delta p / \Delta t}{p} = \frac{40 \text{ aphids}/1 \text{ week}}{100 \text{ aphids}} = 40\% \text{ per week}$$

using dimensional analysis, as discussed in Appendix E, to cancel "aphids" with "aphids"

A **rate of change** always means one change divided by another change; the average growth rate

$$\frac{\Delta p}{\Delta t} = \frac{40 \text{ aphids}}{1 \text{ week}} = 40 \text{ aphids/week}$$

is an example of a rate of change.

EXAMPLE **2**

UNDERSTANDING AVERAGE GROWTH RATES Two years ago, Anytown, U.S.A., had a population of 30,000. Last year, there were 900 births and 300 deaths, for a net growth of 600. There was no immigration to or emigration from the town, so births and deaths were the only sources of population change. The average growth rate was therefore

$$\frac{\Delta p}{\Delta t} = \frac{600 \text{ people}}{1 \text{ year}}$$

What is the most likely average growth rate for Anytown this year, assuming no change in living conditions?

SOLUTION

Anytown should grow by the same percent each year, but by a percent of a changing amount. Because Anytown's population has increased slightly, the average growth rate would be a percent of a larger amount and would therefore be slightly higher. During the first year, the relative growth rate was

$$\frac{\Delta p / \Delta t}{p} = \frac{600 \text{ people}/1 \text{ year}}{30,000 \text{ people}} = 2\% \text{ per year}$$

using dimensional analysis to cancel "people" with "people"

In the following year, $\dfrac{\Delta p / \Delta t}{p}$ should also be 2%, but p is now 30,600.

$$\frac{\Delta p / \Delta t}{p} = 2\%$$

$$\frac{\Delta p / \Delta t}{30,600} = 2\%$$

$$\Delta p / \Delta t = 2\% \text{ of } 30,600 = 612$$

This means that the most likely average growth rate for Anytown this year would be 612 people/year (see Figure 10.10). It is important to realize that this is only a prediction; it might be a good prediction, but it is not a guarantee. For example, there might be fewer residents in their childbearing years this year than previously. This would have a significant effect on the average growth rate, as would changes in living conditions.

	Average Growth Rate $\Delta p/\Delta t$	**Population** p	**Relative Growth Rate** $\dfrac{\Delta p/\Delta t}{p}$
First Year	600 people/year	30,000 people	$\dfrac{600}{30,000} = 0.02 = 2\%$
Second Year	612 people/year	30,600 people	$\dfrac{612}{30,600} = 0.02 = 2\%$

FIGURE 10.10 Anytown's growth.

In Examples 1 and 2, we saw that populations tend to grow at rates that reflect their size; it is most likely that a population will have an average growth rate that is a fixed percent of the current population size. In the next example, we will see that populations are not the only things that grow in this manner.

EXAMPLE **3**

UNDERSTANDING RELATIVE GROWTH RATES   A house is purchased for $140,000 in January 2004. A year later, the house next door is sold for $149,800. The two houses are of the same style and size and are in similar condition, so they should have equal value. What would either house be worth in January 2006?

SOLUTION

Another increase of $9,800, for a total value of $159,600, would not be realistic. Experience has shown that more expensive homes tend to increase in value by a larger amount than do less expensive homes. The value would be more likely to increase by the same percent each year than by the same dollar amount. During 2004, the relative growth rate was

$$\frac{\Delta v/\Delta t}{v} = \frac{\$9,800/1\ \text{year}}{\$140,000} = 7\% \text{ per year}$$

In the following year, $\dfrac{\Delta v/\Delta t}{v}$ should also be 7% per year, but v is now $149,800.

$$\frac{\Delta v/\Delta t}{v} = 7\% \text{ per year}$$

$$\frac{\Delta v/\Delta t}{149,800} = 7\% \text{ per year}$$

$$\Delta v/\Delta t = 7\% \text{ of } \$149,800 \text{ per year} = \$10,486 \text{ per year}$$

Therefore, the average growth rate in 2005 should be

$$\frac{\Delta v}{\Delta t} = \frac{\$10,486}{1\ \text{year}} = \$10,486 \text{ per year}$$

and the best guess of the value in January 2006 would be

$$\$149,800 + \$10,486 = \$160,286$$

However in 2006, the real estate market crashed and houses did not appreciate as they had in 2004 and 2005. Calculations like those above would not accurately predict 2006 and 2007 housing values. These predictions are based on the assumption that the pattern set previously continues.

The Exponential Model

In the examples on populations, the average growth rate $\Delta p/\Delta t$ was a constant percent of the current population. In the example on inflation, the average growth rate $\Delta v/\Delta t$ was a constant percent of the current value. More generally, we are looking at situations in which *the average growth rate $\Delta y/\Delta t$ is a constant percent of the current value of y*, that is, in which

$$\frac{\Delta y}{\Delta t} = k \cdot y$$

In the examples above, we used p for population and v for value in place of y, as a memory device, and k was a specific number, such as $7\% = 0.07$. In calculus, it is determined that if a quantity y behaves in such a way that the average growth rate $\Delta y/\Delta t$ is a constant percent of the current value of y, then the size of y at some later time can be predicted by using the equation

$$y = ae^{bt}$$

where t is time and a and b are constants. We used equations like this in Sections 10.0A and 10.0B, with specific constants such as 3 and 5 in place of a and b.

A **mathematical model** is an equation that is used to describe a real-world subject. The exponential equation

$$y = ae^{bt}$$

is called the **exponential model** because it is an exponential equation and is used to describe subjects such as populations and real estate appreciation. This model is extremely powerful—it allows us to predict the value of y and its growth rate with just a few data. Because many different quantities have a growth rate proportional to their size, the applications of this model are wide and varied. The model's use is illustrated in Examples 4–6, which are continuations of Examples 1–3.

EXAMPLE 4

USING THE EXPONENTIAL MODEL WITH AN INSECT POPULATION
One square foot of alfalfa has a population of 100 aphids. In one week, the population increases to 140; that is, its average growth rate is 40 aphids per week.

a. Develop the model that represents the population of aphids.
b. Predict the aphid population after three weeks.

SOLUTION

a. *Developing the model.*   As we discovered in Examples 1 and 2, the rate at which a population changes is a constant percent of the size of the population (as long as the population's growth is not limited by a lack of food, space, or other constraints). Thus, we can use the model $y = ae^{bt}$ or, with a more appropriate letter, $p = ae^{bt}$. To develop this model for our situation, we need to find the constants a and b.

Step 1   *Write the given information as two ordered pairs (t, p).*   At the beginning of the experiment (at "time 0"), there were 100 aphids; that is, when $t = 0$, $p = 100$.

$(t, p) = (0, 100)$

After one week, there were 140 aphids, that is, when $t = 1$, $p = 140$:

$(t, p) = (1, 140)$

Step 2   *Substitute the first ordered pair into the model $p = ae^{bt}$ and simplify.*

$(t, p) = (0, 100)$	**the first ordered pair**
$100 = ae^{b \cdot 0}$	**substituting**
$100 = ae^{0}$	**simplifying**
$100 = a \cdot 1$	
$a = 100$	

Because a is a constant, we can rewrite our model $p = ae^{bt}$ as $p = 100\,e^{bt}$. To totally develop this model to fit our situation, we need to determine the value of the remaining constant b.

Step 3

Substitute the second ordered pair into the model $p = 100\,e^{bt}$, simplify, and solve for b.

$$(t, p) = (1, 140) \quad \text{the second ordered pair}$$

$$140 = 100\,e^{b \cdot 1} \quad \text{substituting}$$

$$140 = 100\,e^{b} \quad \text{simplifying}$$

$$\frac{140}{100} = e^{b} \quad \text{solving for } b$$

$$1.40 = e^{b}$$

$$\ln 1.40 = \ln e^{b}$$

$$\ln 1.40 = b \quad \text{using the Inverse Property } \ln(e^x) = x$$

$$b \approx 0.3364722366$$

1.40 [ln x]

[LN] 1.40 [ENTER]

We have now developed the model $p = ae^{bt}$ to fit our situation. After substituting $a = 100$ and $b = 0.3364722366$, we get

$$p = 100\,e^{0.3364722366t}$$

where t is the number of weeks after the beginning of the experiment. Notice that a is the initial size of the population and that b is somewhat close to the relative growth rate of $40\% = 0.40$. The values of a and b always have these characteristics. The graph of the model is shown in Figure 10.11.

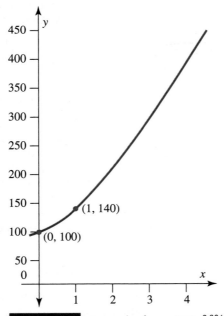

FIGURE 10.11 A graph of $p = 100e^{0.3364722366t}$, showing the population of aphids.

Our model will be most accurate if we do not round off $b = 0.3364722366$ at this point—that is, if we use all the decimal places that the calculator displays.

(Your calculator might display more or fewer decimal places than shown here.) One way to do this is to store the number in your calculator's memory. To do so, press the following when the number is on your calculator's screen.

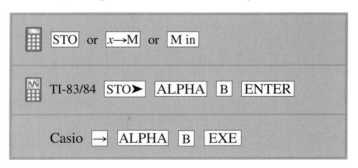

b. *Predicting the population after three weeks.* Now that the model has been developed, predictions can be made. At the end of three weeks, the aphid population would probably be as follows:

$$p = 100\, e^{(0.3364722366)(3)} \quad \textbf{substituting 3 for } t$$
$$= 274.3999\ldots$$
$$\approx 274 \qquad\qquad \textbf{rounding off to the nearest whole number of aphids}$$

Therefore, at the end of three weeks, we would expect 274 aphids.

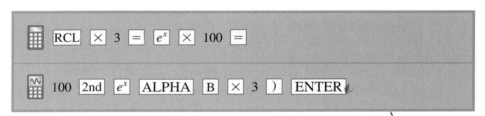

EXAMPLE 5

USING THE EXPONENTIAL MODEL WITH A HUMAN POPULATION
In the year 2009, Anytown, U.S.A., had a population of 30,000. In the following year, there were 900 births and 300 deaths, for a net growth of 600. There was no immigration to or emigration from the town, so births and deaths were the only sources of population change.

a. Develop the model that represents Anytown's population.
b. Predict Anytown's population in the year 2015.

SOLUTION

a. *Developing the model.* As we discovered, the rate at which a population changes is a constant percent of the size of the population. Thus, we can use the model $p = ae^{bt}$. To develop this model for our situation, we need to find the constants a and b.

Step 1 *Write the given information as two ordered pairs* (t, p). For ease in calculating, let 2009 be $t = 0$; then 2010 is $t = 1$ and 2015 is $t = 6$. Also, we will count the population in thousands. The ordered pairs are $(0, 30)$ and $(1, 30.6)$.

Step 2 *Substitute the first ordered pair into the model* $p = ae^{bt}$ *and simplify.*

$$(t, p) = (0, 30) \quad \textbf{the first ordered pair}$$
$$30 = ae^{b \cdot 0} \quad \textbf{substituting}$$
$$30 = ae^0 \quad \textbf{simplifying}$$
$$30 = a \cdot 1$$
$$a = 30$$

As a shortcut, recall that a is always the initial population.

Step 3 | *Substitute the second ordered pair into the model $p = 30e^{bt}$, simplify, and solve for b.*

$(t, p) = (1, 30.6)$ **the second ordered pair**

$30.6 = 30\, e^{b \cdot 1}$ **substituting**

$\dfrac{30.6}{30} = e^{b}$ **simplifying**

$1.02 = e^{b}$ **solving for b**

$\ln 1.02 = \ln e^{b}$

$\ln 1.02 = b$ **using the Inverse Property $\ln(e^x) = x$**

$b \approx 0.0198026273$

Store this value of b in your calculator's memory.

> Notice that the value of b is close to the relative growth rate of $2\% = 0.02$, as it should be.

Thus, our model for Anytown's growth is

$$p = 30\, e^{0.0198026273t}$$

where t is years beyond 2009 and p is population in thousands.

b. *Predicting Anytown's population in the year 2015.* Now that the model has been determined, predictions can be made. In the year 2015, $t = 6$ and

$p = 30\, e^{(0.0198026273)(6)}$ **substituting 6 for t**

≈ 33.78487258

The population would probably be 33,785.

At this point, you might ask, "Why can't I just substitute the relative growth rate for b rather than use the second ordered pair to actually calculate b? After all, b is close to the relative growth rate." You can, but the resulting calculation will *not* necessarily be close to the true figure. In Example 5, if we use the relative growth rate for b, the resulting calculation is off by 40 people. Using the relative growth rate for b is equivalent to rounding off b; whenever you round off early, you lose accuracy.

Furthermore, while b is always somewhat close to the relative growth rate, it is not necessarily extremely close to that rate. In Example 4 on aphids, b was 0.3364722, and the relative growth rate was somewhat close at $40\% = 0.40$. In Example 5 on Anytown, b was 0.0198026273, and the relative growth rate was extremely close at $2\% = 0.02$. In general, b is always close to the relative growth rate, and the smaller these two numbers are, the closer they are.

In the previous two examples, we saw how to use the exponential model to predict the size of a population at some future date. The model can also be used to determine the time at which a quantity will have grown to any specified value.

EXAMPLE **6**

USING THE EXPONENTIAL MODEL WITH REAL ESTATE VALUES   A house is purchased for $140,000 in January 2004. A year later, the house next door is sold for $149,800. The two houses are of the same style and size and are in similar condition, so they should have equal value.

a. Develop the mathematical model that represents the home's value.
b. Find when the house would be worth $200,000 (assuming that the rate of appreciation for houses actually continued unchanged).

SOLUTION

a. *Developing the model.*

Step 1

Write the given information as two ordered pairs (t, v). If we let t measure years after January 2004 and v measure value in thousands of dollars, then the ordered pairs are $(t, v) = (0, 140)$ and $(t, v) = (1, 149.8)$.

Step 2

Substitute the first ordered pair into the model $y = ae^{bt}$ and simplify to find a. (Or remember that a is always the initial value of y.) Since a is the initial value of y (or here, v), $a = 140$. Alternatively, you could substitute $(0, 140)$ into $v = ae^{bt}$.

$$140 = ae^{b \cdot 0}$$
$$140 = ae^{0}$$
$$140 = a \cdot 1$$
$$a = 140$$

The model is now

$$v = 140e^{bt}$$

Step 3

Substitute the second ordered pair into the model, simplify, and solve for b.

$$(t, v) = (1, 149.8)$$
$$149.8 = 140e^{b \cdot 1}$$
$$149.8 = 140e^{b}$$
$$\frac{149.8}{140} = e^{b}$$
$$1.07 = e^{b}$$
$$\ln 1.07 = \ln e^{b}$$
$$\ln 1.07 = b$$
$$b \approx 0.0676586485$$

Store this value of b in your calculator's memory.

Notice that b is close to the relative growth rate of $7\% = 0.07$.

The model is

$$v = 140\, e^{0.0676586485t}$$

where t is the number of years after January 2004 and v is the value in thousands of dollars.

b. *Finding when the house will be worth $200,000. The problem is to find t when $v = 200$, so we substitute 200 for v in our model:*

$$200 = 140e^{0.0676586485t} \qquad \textbf{substituting}$$

$$\frac{200}{140} = e^{0.0676586485t} \qquad \textbf{solving for } t$$

$$\ln \frac{200}{140} = \ln(e^{0.0676586485t}) \qquad \textbf{taking the natural log of each side}$$

$$\ln \frac{200}{140} = 0.0676586485t \qquad \text{using the Inverse Property } \ln(e^x) = x$$

$$t = \frac{\ln \dfrac{200}{140}}{0.0676586485} \qquad \text{dividing}$$

$$t = 5.27168 \ldots$$

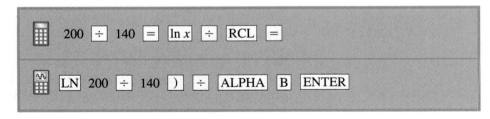

Therefore, our prediction is that if the real estate market hadn't crashed, the house would be worth \$200,000 about $5\frac{1}{4}$ years after January 2004—that is, in April 2009.

Real estate values are affected by many things, and the rate changes frequently. The model $y = ae^{bt}$ allows us to make predictions, but it is important to remember that they are *only* predictions.

STEPS IN DEVELOPING AN EXPONENTIAL MODEL

1. *Write the information given as to the value of the quantity at two points in time as two ordered pairs* (t, y). (You might want to use another letter in place of y as a memory device.)
2. *Substitute the first ordered pair into the model* $y = ae^{bt}$ *and simplify to find* a. (Or remember that a is always the initial value of y.)
3. *Substitute the second ordered pair into the model, simplify and solve for* b. (As a check, recall that b is close to the relative growth rate.)

Exponential Growth and Compound Interest (For Those Who Have Read Chapter 5: Finance)

You might be wondering whether there is a relationship between the inflation of real estate, discussed in Examples 3 and 6, and compound interest, as discussed in Section 5.2. In Example 3, we found that a \$140,000 home increased in value by 7% from January 2004 to January 2005 (that is, its relative growth rate is 7%). Is this equivalent to depositing \$140,000 in an account that earns 7% interest compounded annually?

EXAMPLE 7

COMPOUND INTEREST IS EXPONENTIAL GROWTH   In January 2004, \$140,000 is deposited in an account that earns 7% interest compounded annually. That interest rate is guaranteed for 10 years.

a. Find the future value in January 2005.
b. Find when the account would hold \$200,000.
c. Develop a compound interest model that can be used to answer questions involving the future value of the account.

SOLUTION

a. $FV = P(1 + i)^n$ the Compound Interest Formula
      $= 140{,}000(1 + 0.07)^1$ substituting
      $= \$149{,}800$

b. $FV = P(1 + i)^n$ the Compound Interest Formula
   $200{,}000 = 140{,}000(1 + 0.07)^n$ substituting

We will solve this exponential equation by following the steps developed in Section 10.0B. First, isolate the exponential by dividing by 140,000.

$$\frac{200000}{140000} = (1.07)^n$$

$$\frac{20}{14} = (1.07)^n \qquad \text{reducing}$$

$\ln(20/14) = \ln 1.07^n$ **taking ln of each side**

$\ln(20/14) = n \ln 1.07$ **using the Exponent-Becomes-Multiplier Property $\ln(A^n) = n \cdot \ln A$**

$\dfrac{\ln(20/14)}{\ln 1.07} = n$ **dividing by ln 1.07**

$$n = 5.27168\ldots$$

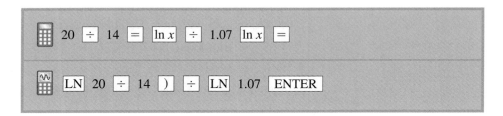

Since n is the number of compounding periods and interest is compounded annually, $n = 5.27168\ldots$ means that it will take $5.27168\ldots$ years for the account to hold $200,000.

This solution is mathematically correct. Practically speaking, however, the ".27168 . . ." part of the solution doesn't make sense. If interest is compounded annually, then at the end of each year, your account is credited with that year's interest. After five years, your account balance would be $140{,}000(1 + 0.07)^5 \approx \$196{,}357.24$. After $5.27168\ldots$ years, your account balance would not have changed, since interest won't be credited until the end of the year. After six years, your account balance would be $140{,}000(1 + 0.07)^6 \approx \$210{,}102.25$. Thus, the best answer to the question is that the account will never hold exactly $200,000, but after six years (i.e., in January of 2010), the account will hold more than $200,000.

c. To develop a compound interest model that can be used to answer questions involving the future value of the account, substitute 140,000 for P and $7\% = 0.07$ for i into the Compound Interest Formula.

     $FV = P(1 + i)^n$ the Compound Interest Formula
     $FV = 140{,}000(1.07)^n$ substituting

In this model, n must be a whole number of years.

Notice that the future value in part (a) of Example 7 matches the future value in Example 3. That is, a $140,000 home that appreciates 7% in 1 year has the same future value as does a $140,000 deposit that earns 7% interest in 1 year.

Notice also that the mathematically correct (but practically incorrect) answer to part (b) of Example 7 matches that of part (b) of Example 6. That is, the home and the bank account will each be worth $200,000 in 5.27168 years.

These two observations imply that $v = 140 \, e^{0.0676586485t}$, the exponential model developed in Example 6, and $FV = 140{,}000(1.07)^n$, the compound interest model developed in Example 7, are mathematically interchangeable. Regardless of which model is used, only a whole number of years makes sense in a compound interest problem if the interest is compound annually.

EXAMPLE **8**

MORE ON COMPOUND INTEREST AND EXPONENTIAL GROWTH
Show that the exponential model $v = 140 \, e^{0.0676586485t}$ from Example 6 and the compound interest model $v = 140{,}000(1.07)^n$ from Example 7 are mathematically interchangeable.

SOLUTION

We will use the properties of logarithms to convert the exponential model to the compound interest model. In Example 6, we modeled the value of a house with

$$v = 140 \, e^{0.0676586485t}$$

where v is the value in thousands of dollars. To convert this to dollars, as used in Example 7, multiply by 1,000:

$$v = 1{,}000 \cdot 140 \, e^{0.067586485t}$$
$$= 140{,}000 \, e^{0.0676586485t}$$

In Example 6, we found that $b = \ln 1.07 \approx 0.0676586485$, so we can replace 0.0676586485 with $\ln 1.07$. This gives us

$$v = 140{,}000 \, e^{0.0676586485t}$$
$$= 140{,}000 \, e^{(\ln 1.07)(t)}$$

For the moment, focus on e's exponent, $(\ln 1.07)(t)$:

$$(\ln 1.07)(t) = t \cdot \ln 1.07$$
$$= \ln(1.07^t) \quad \textbf{using the Exponent-Becomes-Multiplier}$$
$$\textbf{Property } n \cdot \ln A = \ln(A^n)$$

Thus, we can replace e's exponent, $(\ln 1.07)(t)$, with $\ln(1.07^t)$:

$$v = 140{,}000 \, e^{(\ln 1.07)(t)}$$
$$= 140{,}000 \, e^{\ln(1.07^t)}$$
$$= 140{,}000(1.07^t) \quad \textbf{using the Inverse Property } e^{\ln x} = x$$
$$= 140{,}000(1.07^n) \quad t \textbf{ and } n \textbf{ both measure number of years}$$

Since we were able to convert the exponential model $v = 140 \, e^{0.0676586485t}$ to the compound interest model $v = 140{,}000(1.07)^n$, we know that they are algebraically equivalent and therefore are mathematically interchangeable. Regardless of which model is used, only a whole number of years makes sense in a compound interest problem (if the interest is compounded annually).

10.1 EXERCISES

1. Use the model $p = 30 \, e^{0.0198026273t}$ developed in Example 5 to predict the population of Anytown in the year 2013.

2. Use the model $p = 30 \, e^{0.0198026273t}$ developed in Example 5 to predict the population of Anytown in the year 2032.

3. Use the model $v = 140 \, e^{0.0676586485t}$ developed in Example 6 to predict when the house would be worth $250,000.

4. Use the model $v = 140 \, e^{0.0676586485t}$ developed in Example 6 to predict when the house would be worth $300,000.

Exercises 5–12 deal with data from the U.S. Bureau of the Census on populations of metropolitan areas.* These data allow us to find how fast the population is growing and when it will reach certain levels. Such calculations are very important, because they indicate the future needs of the population for goods and services and how well the area can support the population.

5. The third largest metropolitan area in the United States is the Chicago/Naperville/Joliet metropolitan area. Its population in 2004 was 9,392 (in thousands); in 2008, it was 9,786.

    a. Convert this information to two ordered pairs (t, p), where t measures years since 2004 and p measures population in thousands.

    b. Find Δt, the change in time.

    c. Find Δp, the change in population.

    d. Find $\Delta p/\Delta t$, the average growth rate.

    e. Find $(\Delta p/\Delta t)/p$, the relative growth rate.

Chicago is the third largest metropolitan area in the United States.

6. The second largest metropolitan area in the United States is the Los Angeles/Long Beach/Santa Ana metropolitan area. Its population in 2004 was 12,858 (in thousands); in 2007, it was 12,876.

    a. Convert this information to two ordered pairs (t, p), where t measures years since 2004 and p measures population in thousands.

    b. Find Δt, the change in time.

    c. Find Δp, the change in population.

    d. Find $\Delta p/\Delta t$, the average growth rate.

    e. Find $(\Delta p/\Delta t)/p$, the relative growth rate.

*U.S. Department of Commerce, Bureau of the Census, Annual Estimates of the Population.

7. The largest metropolitan area in the United States is the New York/northern New Jersey/Long Island metropolitan area. Its population in 2004 was 18,731 (in thousands); in 2008, it was 19,007.

New York is the largest metropolitan area in the United States.

    a. Convert this information to two ordered pairs (t, p), where t measures years since 2004 and p measures population in thousands.

    b. Find Δt, the change in time.

    c. Find Δp, the change in population.

    d. Find $\Delta p/\Delta t$, the average growth rate.

    e. Find $(\Delta p/\Delta t)/p$, the relative growth rate.

8. The twelfth largest metropolitan area in the United States is the San Francisco/Oakland/Fremont metropolitan area. Its population in 2004 was 4,154 (in thousands); in 2007, it was 4,204.

    a. Convert this information to two ordered pairs (t, p), where t measures years since 2004 and p measures population in thousands.

    b. Find Δt, the change in time.

    c. Find Δp, the change in population.

    d. Find $\Delta p/\Delta t$, the average growth rate.

    e. Find $(\Delta p/\Delta t)/p$, the relative growth rate.

9. **a.** Develop the model that represents the population of the Chicago/Naperville/Joliet metropolitan area (see Exercise 5).

    b. Predict the population in 2012.

    c. Predict the population in 2017.

    d. Predict when the population will be double its 2004 population.

10. **a.** Develop the model that represents the population of the Los Angeles/Long Beach/Santa Ana metropolitan area (see Exercise 6).

    b. Predict the population in 2013.

    c. Predict the population in 2017.

    d. Predict when the population will be 50% more than its 2004 population.

11. **a.** Develop the model that represents the population of the New York/northern New Jersey/Long Island metropolitan area (see Exercise 7).

    b. Predict the population in 2010.

c. Predict the population in 2017.

d. Predict when the population will be 50% more than it was in 2004.

12. a. Develop the model that represents the population of the San Francisco/Oakland/Fremont metropolitan area (see Exercise 8).

b. Predict the population in 2015.

c. Predict the population in 2024.

d. Predict when the population will be double what it was in 2004.

13. A biologist is conducting an experiment that involves a colony of fruit flies. (Biologists frequently study fruit flies because their short life span allows the experimenters to easily study several generations.) One day, there were 2,510 flies in the colony. Three days later, there were 5,380.

a. Develop the mathematical model that represents the population of flies.

b. Use the model to predict the population after one week.

c. Use the model to predict when the population will be double its initial size.

14. A university keeps a number of mice for psychology experiments. One day, there were 89 mice. Three weeks later, there were 127.

a. Develop the mathematical model that represents the population of mice.

b. Use the model to predict the population after two months.

c. Use the model to predict when the population will be double its initial size.

15. In August 2009, Buck Meadows bought a house for $230,000. In February 2010, a nearby house with the same floor plan was sold for $310,000.

a. Develop the mathematical model that represents the value of the house.

b. Use the model to predict when the house will double its 2009 value.

c. Use the model to determine the value of the house one year after Mr. Meadows purchased it.

d. Use part (c) to determine $\Delta v / \Delta t$, that is, the rate at which the value of the house grew.

16. In July 2008, Alvarado Niles bought a house for $189,000. In September 2009, a nearby house with the same floor plan was sold for $207,000.

a. Develop the mathematical model that represents the value of the house.

b. Use the model to predict when the house will double its 2008 value.

c. Use the model to determine the value of the house one year after Mr. Niles purchased it.

d. Use part (c) to determine $\Delta v / \Delta t$, that is, the rate at which the value of the house grew.

17. An October 2009 article in *The Industry Standard* states that "independent Twitter data shows exponential tweet growth." GigaTweet, an independent tweet-counting service, reports that the number of tweets was 5.0 billion in October 2009. In April 2009, the number of tweets was 1.6 billion.

a. Develop the exponential growth model that fits with these data, where t is the number of months after April, 2009.

b. Use the model to predict when there will be 10.0 billion tweets.

18. According to a September 9, 2007, article in *ZDNet*, the number of Facebook users "is growing at 3 to 4 percent compounded weekly, with about 38 million members currently."

a. Why does this statement imply exponential growth?

b. Use these data (and a 3.5% weekly growth rate) to develop the exponential growth model for the number of users, where t is the number of weeks after September 9, 2007.

c. Use the model to predict the number of Facebook users in the first week of January 2011.

d. The article goes on to say that "In three years, we could have everybody on the planet, but that's not going to happen." Discuss why it's not going to happen and what this means about using an exponential model to make predictions.

19. The number of cell phone subscribers has been growing exponentially for some time. According to *Information Please Almanac*, there were 109,478 thousand cellular phone subscriptions in the United States in 2000. In 2002, there were 140,766.

a. Develop the exponential model that represents the nations cellular phone subscriptions.

b. Use the model to predict the number of subscriptions in 2004. (The actual number in 2004 was 182,140 thousand.)

c. Use the model to predict the number of subscriptions in 2006. (The actual number in 2006 was 233,000 thousand.)

20. Which has been growing faster: Twitter, Facebook, cell phones, or YouTube?

a. Use the exponential model in Exercise 17 to find how long it takes for the number of tweets to double.

b. Use the exponential model in Exercise 18 to find how long it takes for the number of Facebook users to double.

c. Use the exponential model in Exercise 19 to find how long it takes for the number of cell phone subscribers to double.

d. YouTube does not release usage information. However, "Mobile Uploads to YouTube Increase Exponentially," a June 2009 article on *The Official*

YouTube Blog, claims exponential growth in the number of YouTube uploads from mobile phones to YouTube, with an increase of 1700% (yes, that's really 1700%) from December 2008 to June 2009. Develop the exponential growth model that fits with these data, where t is the number of months after October 2009.

  e. Use the model from part (d) to find how long it takes for the number of mobile phone uploads to YouTube to double.

  f. Which has been growing faster: Twitter, Facebook, cell phone, or YouTube?

21. The first census of the United States was taken on August 2, 1790; the population then was 3,929,214. Since then, a census has been taken every ten years. The next census gave a U.S. population of 5,308,483.*

  a. Use these data to develop the mathematical model that represents the population of the United States.

  b. Which prediction would be more accurate: U.S. population in 1810 or in 1990? Why?

  c. Use the model to predict the U.S. population in 1810 and in 2000. (The actual populations were 7,239,881, and 281,421,906.)

22. In 1969, the National Academy of Sciences published a study titled "Resources and Man." That study places "the earth's ultimate carrying capacity at about 30 billion people, at a level of chronic near-starvation for the great majority (and with massive immigration to the now less-densely populated lands)!" The study goes on to state that 10 billion people is "close to (if not above) the maximum that an intensively managed world might hope to support with some degree of comfort and individual choice." The world population in 1990 was 5.283 billion; in 2000, it was 6.082 billion.

  a. Develop the mathematical model that represents world population.

  b. Use the model to predict when world population would reach the "somewhat comfortable" level of 10 billion.

  c. Use the model to predict when world population would reach "the earth's ultimate carrying capacity" of 30 billion.

23. Perhaps the most intuitive way to compare how fast different populations are growing is to calculate their "doubling times," that is, how long it takes for the populations to double. Find the doubling times for Africa, North America, and Europe, using the data in Figure 10.12.

	Africa	North America	Europe
1990 population	643 million	277 million	509 million
2000 population	794 million	314 million	727 million

FIGURE 10.12

24. China's population in 1980 was 983 million; in 1990, it was 1,154 million.

  a. Develop the mathematical model that represents China's population.

  b. Use the model developed in Exercise 23 that represents Africa's population and the model in part (a) to predict when Africa's population will exceed China's.

25. Use the data in Exercise 22 to compute the doubling time for the world's population.

26. Use the data in Exercise 24 to compute the doubling time for China's population.

27. A real estate investor bought a house for $125,000. He estimated that houses in that area increase their value by 10% per year.

  a. If that estimate is accurate, why would it not take five years for the house to increase its value by 50%?

  b. How long would it take for the house to increase its value by 50%?

28. The average home in Metropolis in 2008 was $235,600. In 2009, it was $257,400. The average home price in Smallville in 2008 was $112,100. In 2009, it was $137,600. If these trends continue, when will Smallville's average home price exceed that of Metropolis?

Exercises 29–37 are only for those who have read Chapter 5.

29. Develop a compound interest model that represents the population of the New York/northern New Jersey/Long Island metropolitan area. Use the information in Exercise 7 and the method of Example 7.

30. Develop a compound interest model that represents the population of the San Francisco/Oakland/Fremont metropolitan area. Use the information in Exercise 8 and the method of Example 7.

31. Develop a compound interest model that represents the population of the Chicago/Naperville/Joliet metropolitan area. Use the information in Exercise 5 and the method of Example 7.

32. Develop a compound interest model that represents the population of the Los Angeles/Long Beach/Santa Ana metropolitan area. Use the information in Exercise 6 and the method of Example 7.

33. Show that the exponential model developed in Exercise 9 and the compound interest model developed in Exercise 31 are mathematically interchangeable.

34. Show that the exponential model developed in Exercise 10 and the compound interest model developed in Exercise 32 are mathematically interchangeable.

35. In Section 5.2, we found the doubling time of an account that pays 5% interest compounded daily by using the TVM feature of a TI graphing calculator. Instead, we could use logarithms to solve the equation:

$$2P = P\left(1 + \frac{0.05}{365}\right)^n$$

Solve this equation.

36. **a.** What equation would we solve to answer Exercise 78 in Section 5.2?
    b. Use logarithms to solve that equation.

37. **a.** What equation would we solve to answer Exercise 79 in Section 5.2?
    b. Use logarithms to solve that equation.

 Answer the following questions using complete sentences and your own words.

• CONCEPT QUESTIONS

38. The exponential model is a powerful predictor of population growth, but it is based on the assumption that the two ordered pairs used to find a and b describe a steady tendency in the population's growth. In fact, that tendency may not be so steady, and the model's prediction may not be a good one.
    a. What information were you given in Example 1 that implied a steady tendency in the aphid population's growth?
    b. What information were you given in Example 2 that implied a steady tendency in Anytown's growth?
    c. Discuss the factors of an animal population's growth that affect the accuracy of a prediction based on the exponential model. What other factors would be involved if the population were human?

39. In which of the following four situations would the exponential model give the most accurate prediction? The least accurate? Why?
    • p is the population of a specific country; the two ordered pairs that are used to find a and b cover a time span of five years; you are asked to predict the population five years later.
    • p is the population of a specific country; the two ordered pairs that are used to find a and b cover a

time span of 100 years; you are asked to predict the population five years later.
    • p is the population of the world; the two ordered pairs that are used to find a and b cover a time span of five years; you are asked to predict the population five years later.
    • p is the population of the world; the two ordered pairs that are used to find a and b cover a time span of 100 years; you are asked to predict the population five years later.

• PROJECTS

40. Do some research and find out the population of your community at two fairly recent points in time. Use the data to:
    a. Find $\Delta p / \Delta t$, the average growth rate.
    b. Find $(\Delta p / \Delta t)/p$, the relative growth rate.
    c. Develop the model that represents the population of your community.
    d. Predict the population of your community in 10 years.
    e. Determine your community's doubling time.
    f. Also, discuss the accuracy of your model. Be certain to include in your discussion the prominence of emigration and immigration in your community.

WEB PROJECT

41. This web project will take approximately two weeks to complete. The U.S. Census Bureau has two population clocks on its web page (www.census .gov). One gives the estimated current U.S. population, and the other gives the estimated current world population.
    a. Visit this site at two different times approximately one week apart, and record the time and the population (either U.S. or world population). Use these two different ordered pairs to create an exponential model, where t is time in hours.
    b. Use the exponential model to predict the population approximately one week into the future.
    c. Revisit the Census Bureau's web site at the appropriate time to compare your prediction from part (b) with the actual population. What is the error in your prediction? That is, what is the difference between your prediction and the actual population?
    d. What is the percent error in your prediction? To find this percent error, take the error from part (c), divide it by the actual population, and convert the result to a percent.

OBJECTIVES

- Create an exponential model to represent the decay of a radioactive substance
- Calculate the amount of radioactive substance remaining after a specified period of time
- Calculate the amount of time required for a substance to lose a specified portion of its radioactivity
- Find average and relative decay rates
- Use the carbon-14 model in determining the age of artifacts
- Explore the use of radioactive substances in the field of medicine

What does the first article below mean when it says that tritium gas "decays at a rate of 5.5% a year"? Why do scientists claim that the frozen corpse found in 1991 in the Austrian Alps "is 5,000 to 5,500 years old"? (See the second article on the next page.) The answers have to do with radioactivity. Radioactive materials have become more and more prevalent and useful since Marie Curie introduced them to the world in 1898.

A radioactive substance is not stable; over time, it transforms itself into another substance. This is called **radioactive decay,** and it is due to the interaction between nuclear particles (protons, neutrons, and electrons) in the radioactive substance. Because a larger quantity of a radioactive substance has more nuclear particles (and hence more interactions), we might guess that it decays faster, just as a larger population produces more offspring. This, in fact, is the case; a larger amount of radioactive material does experience more decay. That is, *the rate of decay is proportional to the amount of radioactive substance present,* just as the

FEATURED IN THE NEWS

RADIOACTIVE GAS USED IN A-BOMB MAY BE MISSING

WASHINGTON—The Energy Department and the Nuclear Regulatory Commission are investigating the possible loss of enough tritium to help make a nuclear bomb, and authorities have suspended sales of the radioactive gas during the probe.

Tritium is used in nuclear weapons to increase their power. Since it decays at the rate of 5.5 percent a year, the gas must be replenished if weapons are to maintain their full explosive potential.

The federal government also sells 200 to 300 grams of tritium a year to American and foreign companies for use in biological and energy research and to make self-luminous lights, signs and dials.

The incident has led to fears in Congress that some of the gas is missing and has fallen into unfriendly hands. But federal officials say that there are no signs of a diversion and that the discrepancy may have resulted from errors in measuring the tritium.

From William Broad, "U.S. Halts Sale of Tritium After Loss Of Enough to Make a Nuclear Bomb," *New York Times, 7/26/89.* Copyright © 1989 by the New York Times Co. Reprinted by permission.

TESTS ON ICEMAN MAKE HIM PART OF STONE AGE

NEW YORK—Carbon dating tests show that the well-preserved body of a prehistoric human hunter found in an Alpine glacier last year is 5,000 to 5,500 years old, scientists reported yesterday.

The first scientifically established age for the frozen corpse is more than 1,000 years older than original estimates. It means the man lived and presumably froze to death well before the Bronze Age replaced the late Stone Age in Europe.

"Now we know he's not from the Bronze Age, but much older," said Dr. Werner Platzer, head of the anatomy department at Innsbruck University in Austria, who is directing research on the mummified corpse. "It means, I believe, that this is the only corpse we have from the Stone Age."

The tests on bones and skin tissue were conducted by scientists at Oxford University in England and a Swiss physics institute in Zurich. The body is being kept in cold storage at Innsbruck University.

The corpse was found in September in a glacier 10,500 feet up in the Austrian Alps, close to the Italian border. The body was mummified, and its leather and fur clothing was badly deteriorated. But scholars of prehistoric people were fascinated by a leather quiver with 14 arrows and an ax found by the iceman's side.

On first examination, the ax was thought to be made of bronze, which led to the assumption that this was an early Bronze Age man who lived about 4,000 years ago. More careful study showed the ax to be made of copper, thus a product of a simpler technology that seemed to place the iceman in the late Stone Age. This seemed to be confirmed by previous tests on samples of grasses taken from the hunter's clothing, which indicated that the body might be 5,000 years old.

rate of growth of a population is proportional to the size of the population. Consequently, we can use the exponential model $y = ae^{bt}$, developed in Section 10.1. As a memory device, we will use the variable Q instead of y to represent "quantity"; therefore, our model will be $Q = ae^{bt}$.

EXAMPLE 1

APPLYING AN EXPONENTIAL MODEL TO RADIOACTIVE DECAY
Hospitals use the radioactive substance iodine-131 in research. It is effective in locating brain tumors and in measuring heart, liver, and thyroid activity. A hospital purchased 20 grams of the substance. Eight days later, when a doctor wanted to use some of the iodine-131, he observed that only 10 grams remained (the rest had decayed).

a. Develop a mathematical model that represents the amount of iodine-131 present.
b. Predict the amount remaining two weeks after purchase.

SOLUTION

a. *Developing the model.* Because we're using the exponential model $Q = ae^{bt}$, we follow the steps from Section 10.1.

Step 1
Write the given information as two ordered pairs (t, Q). Originally there were 20 grams, so we have $(t, Q) = (0, 20)$. Eight days later, there were 10 grams, so we have $(t, Q) = (8, 10)$.

Step 2
Recall that *a is always the initial value of Q, so* a $= 20$. Alternatively, we could *substitute* $(0, 20)$ *into the model* $Q = ae^{bt}$ *and simplify.*

$$Q = ae^{bt}$$
$$20 = ae^{b(0)}$$
$$20 = ae^0$$

$$20 = a \cdot 1$$
$$a = 20$$

The model is now $Q = 20e^{bt}$.

Step 3   *Substitute the second ordered pair* (8, 10) *into this new model and simplify.*

$$Q = 20\, e^{bt}$$
$$10 = 20\, e^{b(8)}$$
$$\frac{10}{20} = e^{8b}$$
$$0.5 = e^{8b}$$
$$\ln 0.5 = \ln e^{8b} \qquad \textbf{taking ln of each side}$$
$$\ln 0.5 = 8b \qquad \textbf{using the Inverse Property } \ln(e^x) = x$$
$$b = \frac{\ln 0.5}{8}$$
$$b \approx -0.086643397$$

Store this value of b in your calculator's memory.

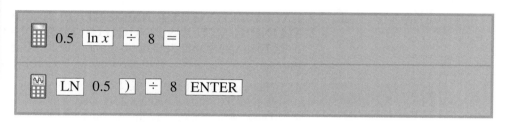

Thus, our model for the amount Q of iodine-131 remaining t days after purchase is $Q = 20\, e^{-0.086643397t}$.

b. *Predictions.*   Now that the model has been determined, predictions can be made. Two weeks after purchasing the iodine-131, $t = 14$ days, and

$$Q = 20\, e^{-0.086643397t}$$
$$= 20\, e^{-0.086643397(14)}$$
$$= 5.946035575 \ldots$$
$$\approx 5.9$$

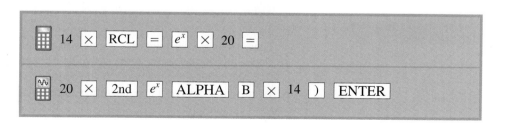

After two weeks (14 days), we would expect approximately 5.9 grams to remain.

Notice that b is negative in Example 1. This indicates that the amount of iodine-131 is *decreasing*. In contrast, b was positive in Section 10.1, because the quantities under study were *increasing*.

EXPONENTIAL GROWTH OR EXPONENTIAL DECAY

In general, any quantity for which *the rate of change is proportional to the amount present* can be modeled by the formula $y = ae^{bt}$.

1. If $b > 0$, then y is growing exponentially.
2. If $b < 0$, then y is decaying exponentially.

Half-Life

A radioactive substance is not stable. One way to measure its instability is to determine the **half-life** of the substance, that is, the amount of time required for a quantity to reduce to one-half its initial size. In the previous example, the half-life of iodine-131 was 8 days; it took 8 days for 20 grams of the substance to decay into 10 grams. As we will see, it will take an additional 8 days for those 10 grams to decay into 5 grams.

EXAMPLE 2

CALCULATING THE AMOUNT OF RADIOACTIVE SUBSTANCE REMAINING AFTER A PERIOD OF TIME Use the model $Q = 20e^{-0.086643397t}$ from Example 1 to calculate the amount of iodine-131 remaining at the following times.

a. 16 days after purchase   **b.** 24 days after purchase

SOLUTION

a. Sixteen days after purchasing the 20 grams of iodine-131 (or eight days after observing that half of it had decayed), $t = 16$; therefore,

$$Q = 20\, e^{-0.086643397t}$$
$$= 20\, e^{-0.086643397(16)}$$
$$= 5$$

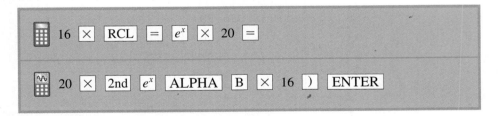

Thus, 16 days after purchase, 5 grams of the initial 20 grams of iodine-131 will remain. Notice that 16 days is the same as two half-life periods. After one half-life period (8 days), 10 grams remained; after two half-life periods (16 days), 5 grams remained.

b. Twenty-four days (or three half-life periods) after purchasing the 20 grams of iodine-131, $t = 24$; therefore,

$$Q = 20\, e^{-0.086643397t}$$
$$= 20\, e^{-0.086643397(24)}$$
$$= 2.5$$

Thus, 24 days after purchase, 2.5 grams of the initial 20 grams of iodine-131 will remain. After another 8 days (or a total of 32 days), only 1.25 grams will remain. Every 8 days, half of the material decays.

We can obtain a graph showing the exponential decay of the initial 20 grams of iodine-131 from the example above by plotting the ordered pairs (0, 20), (8, 10), (16, 5), and (24, 2.5) and continuing in this manner (after each 8-day period, only half the previous quantity remains). This graph is shown in Figure 10.13.

The half-life of a radioactive substance does not depend on the amount of substance present. For any given radioactive substance, the half-life is intrinsic to the substance itself; each radioactive substance has its own half-life. Figure 10.14 lists the half-life for various substances. Notice that half-lives have an incredibly wide range of values—from a few seconds to well over 20,000 years.

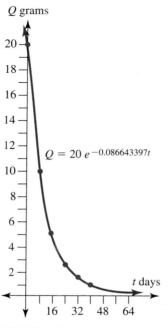

Q grams

$Q = 20 \, e^{-0.086643397t}$

t days

FIGURE 10.13 A graph showing the amount of iodine-131 remaining t days after purchasing 20 grams.

Radioactive Substance	Half-life
krypton-91	10 seconds
silicon-31	2.6 hours
cobalt-55	18.2 hours
magnesium-28	21.0 hours
iodine-124	4.5 days
iodine-131	8.0 days
cobalt-60	5.3 years
plutonium-241	13 years
plutonium-238	86 years
carbon-14	5,730 years
plutonium-239	24,400 years

FIGURE 10.14 Selected half-lives.

EXAMPLE **3**

USING HALF-LIFE TO CREATE A MODEL AND MAKE A PREDICTION
Professor Frank Stein received 8.2 grams of plutonium-241 and stored it in his laboratory for future use in experiments. If he does not use any of the substance for experimental purposes, how much will remain one year later?

SOLUTION

Rather than focusing on the *specific* initial amount of 8.2 grams, we will determine the *general* model $Q = ae^{bt}$ for *any* initial amount and substitute our given values later. We do this here to emphasize the general approach of developing a model, rather than dwelling on the specific numerical calculations.

Step 1

Express the information as ordered pairs (t, Q). Because a is the initial amount of plutonium-241, $(t, Q) = (0, a)$. To obtain a second ordered pair, we find the half-life of plutonium-241 in Figure 10.14 (the half-life is $t = 13$ years). In 13 years, half of a, or $\frac{a}{2}$, will remain, and we obtain $(t, Q) = (13, \frac{a}{2})$.

Step 2

Substitute the first ordered pair into the model, and simplify to find a. Because we are not using a specific value for the initial amount a, we can omit this step. (Our model is still $Q = ae^{bt}$.)

Step 3   *Substitute* $(t, Q) = (13, \frac{a}{2})$ *into the model and simplify.*

$$Q = ae^{bt}$$

$$\frac{a}{2} = ae^{b(13)}$$

$$\frac{1}{2} = e^{13b} \qquad \textbf{dividing by } a$$

$$0.5 = e^{13b}$$

$$\ln 0.5 = \ln e^{13b} \qquad \textbf{taking the natural log}$$

$$\ln 0.5 = 13b \qquad \textbf{using the Inverse Property } \ln(e^x) = x$$

$$b = \frac{\ln 0.5}{13}$$

$$b \approx -0.053319013$$

Store this value of b in your calculator's memory.

Thus, our model for the amount Q of plutonium-241 remaining t years after receiving an initial quantity of a grams is $Q = ae^{-0.053319013t}$. To predict the amount of plutonium-241 that will remain one year after Professor Stein receives his 8.2 grams, we substitute 8.2 for a and 1 for t.

$$Q = ae^{-0.053319013t}$$

$$= 8.2 \cdot e^{-0.053319013(1)}$$

$$= 7.774235618 \ldots$$

$$\approx 7.8$$

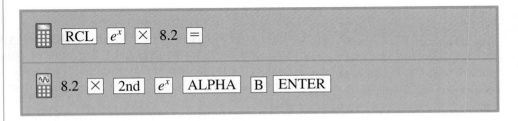

On the basis of the model $Q = 8.2\, e^{-0.053319013t}$, Professor Stein should expect approximately 7.8 grams of plutonium-241 to remain one year after receiving the 8.2 grams.

Relative Decay Rate

In Section 10.1, we discussed the average growth rate $\Delta p / \Delta t$, which indicates how fast a quantity is growing, and the relative growth rate $(\Delta p / \Delta t)/p$, which indicates how fast the quantity is growing *as a percent*. When calculating the constant b in the model $p = ae^{bt}$, we were able to check our work by seeing whether b was close to the relative growth rate.

This same concept can be applied to exponential decay. When using the exponential decay model $Q = ae^{bt}$, we can compute the **average decay rate** $\Delta Q / \Delta t$ by dividing the change in the amount of radioactive material by the corresponding change in time. We can also compute the **relative decay rate** by calculating $(\Delta Q / \Delta t)/Q$. Because Q is decreasing, ΔQ will be negative, as will the relative decay rate.

EXAMPLE 4

FINDING AVERAGE AND RELATIVE DECAY RATES   Find (a) the average decay rate and (b) the relative decay rate for the data given in Example 3.

SOLUTION

a. *Finding the average decay rate.* The initial quantity was 8.2 grams, and 7.774235625 grams remained after $t = 1$ year. (Using the rounded-off figure of 7.8 in place of 7.774235625 would result in a less accurate answer.)

$$\Delta Q = \text{final quantity} - \text{initial quantity}$$
$$= 7.774235625 - 8.2 = -0.425764375 \text{ gram}$$
$$\Delta t = 1 \text{ year}$$

Therefore, the average decay rate is

$$\frac{\Delta Q}{\Delta t} = \frac{-0.425764375 \text{ gram}}{1 \text{ year}} \approx -0.4 \text{ gram per year}$$

The following year, there will not be as much material left, and what remains will decay at a slower rate. The average decay rate will not remain constant.

b. *Finding the relative decay rate.*

$$\frac{\Delta Q / \Delta t}{Q} = \frac{-0.425764375 \text{ gram per year}}{8.2 \text{ grams}}$$
$$= -0.051922484 \text{ per year} = -5.1922484\% \text{ per year}$$

This implies that the amount of radioactive material is decreasing by about 5.2% per year. While the average decay rate will change from year to year (because the material decays more slowly as the amount decreases), the relative decay rate will remain constant. Notice that the relative decay rate is close to b, which was -0.053319013.

EXAMPLE 5

CALCULATING THE AMOUNT OF TIME REQUIRED FOR A SUBSTANCE TO LOSE A SPECIFIED PORTION OF ITS RADIOACTIVITY Plutonium-239 is a waste product of nuclear reactors. How long will it take for this waste to lose 99.9% of its radioactivity and therefore be considered relatively harmless to the biosphere?*

SOLUTION

Regardless of the initial amount a of plutonium-239, we need to determine the time t required for 0.1% of the radioactivity to *remain*. (Remember, the model $Q = ae^{bt}$ determines the amount Q *remaining* after a time period t.) We first need to determine the model for plutonium-239.

Step 1

Express the initial data. Let $a = $ the initial amount of plutonium-239, and note that the half-life is 24,400 years (from Figure 10.14). *Express this information as ordered pairs* (t, Q). We have $(0, a)$ and $(24{,}400, \frac{a}{2})$.

---

*H. A. Bethe, "The Necessity of Fission Power," *Scientific American,* January 1976.

Step 2   *Substitute the first ordered pair into the model, and simplify to find a.* Because we are not using a specific value for the initial amount *a*, we can omit this step. (Our model is still $Q = ae^{bt}$.)

Step 3   *Substitute $(t, Q) = (24{,}400, \frac{a}{2})$ into the model and simplify.*

$$Q = ae^{bt}$$

$$\frac{a}{2} = ae^{b(24{,}400)}$$

$$\frac{1}{2} = e^{24{,}400b}$$

$$0.5 = e^{24{,}400b}$$

$$\ln 0.5 = \ln(e^{24{,}400b}) \quad \textbf{taking ln of each side}$$

$$\ln 0.5 = 24{,}400b \qquad \textbf{using an Inverse Property}$$

$$b = \frac{\ln 0.5}{24{,}400}$$

$$b \approx -0.000028407$$

Store this value of *b* in your calculator's memory.

For plutonium-239, therefore, our model is $Q = ae^{-0.000028407t}$. Because the initial amount of plutonium-239 is *a*, we need to determine the time *t* when 0.1% of *a* remains—that is, when the amount *Q* of plutonium-239 will equal 0.001*a*. We substitute 0.001*a* for *Q* in the model $Q = ae^{-0.000028407t}$ and solve.

$$Q = ae^{-0.000028407t}$$

$$0.001a = ae^{-0.000028407t}$$

$$0.001 = e^{-0.000028407t} \qquad \textbf{dividing by } a$$

$$\ln 0.001 = \ln(e^{-0.000028407t}) \qquad \textbf{taking ln of each side}$$

$$\ln 0.001 = -0.000028407t \qquad \textbf{using an Inverse Property}$$

$$t = \frac{\ln 0.001}{-0.000028407}$$

$$= 243{,}165.1365\ldots$$

$$\approx 240{,}000$$

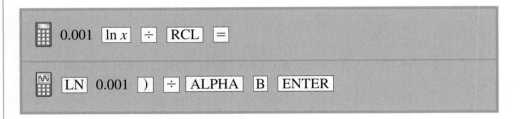

It will take approximately 240,000 years for any given quantity of plutonium-239 to lose 99.9% of its radioactivity and thus be considered relatively harmless to the biosphere. How long a time span is this, relative to all of human history?

The United States has assembled a vast stockpile of about 50 tons of plutonium. Because plutonium is so deadly, its storage and disposal are of major concern, as indicated by the 1993 news article shown on the next page. However, not every country in the world shares the United States' concern over the perils of plutonium. The Japanese Nuclear Agency has been promoting the use of plutonium in its power plants, as indicated by the 1994 news article.

FEATURED IN THE NEWS

PLUTONIUM STOCKPILE COULD LAST FOREVER
U.S. HAS 50 TONS TO THINK ABOUT DISPOSING

AMARILLO, TEXAS—In 16 unremarkable concrete bunkers built by the Army for a war with Hirohito and Hitler, the United States has begun assembling about 50 tons of plutonium, a vast stockpile of one of the most expensive materials ever produced and perhaps the most important to safeguard.

The Energy Department says the bunkers, each about the size of a two-car garage, are going to be used for interim storage, meaning six or seven years.

But plutonium, which was invented by the Energy Department's predecessor, the Manhattan Project, may turn out to be the hardest thing on Earth to dispose of. And at the Energy Department, "interim" can have an elastic meaning. "Immediate" tends to mean several years. "Several years" can mean never.

President Clinton announced the formation of an inter-agency task force in September to consider how much plutonium the nation needs in the post-Soviet era and how to dispose of what is surplus. The Energy Department is also drawing up a plan for its weapons production complex for the next century.

But officials are choosing among a short list of unattractive options.

At the Pantex plant near Amarillo, where the plutonium is piling up, general manager Rich Loghry said when asked what would happen next, "I don't think people have a really good answer for what is going to happen to the plutonium."

Storage may be the leading option, but even this is tricky. Plutonium loses half its radioactivity every 24,000 years, so it will reach background levels of radioactivity in 10 "half-lives," or 240,000 years; the U.S. political system is focused largely on problems that can be solved in less than four.

FEATURED IN THE NEWS

JAPANESE MAKE PLUTONIUM CUTE
VIDEO CLAIMS IT'S SAFE TO DRINK

TOKYO—Meet Mr. Pluto, the Japanese nuclear agency's round-faced, rosy-cheeked, animated answer to the public's concern about its plan to import 30 tons of plutonium as fuel for power plants.

In the country that best knows the dark side of atomic energy, not everyone is charmed by Mr. Pluto, who is featured in a promotional videotape prepared by the Power Reactor and Nuclear Fuel Development Corp.

Anti-nuclear groups said yesterday that they will campaign against distribution of the video, entitled "The Story of Plutonium: That Dependable Fellow, Mr. Pluto." They contend that it irresponsibly plays down the dangers plutonium poses.

Perky and pint-sized, Mr. Pluto is childlike, with cute red boots and a

BY ASSOCIATED PRESS

The agency's video shows 'Mr. Pluto' greeting a boy drinking soda dosed with plutonium.

green helmet with antennae. On the front of the helmet is the chemical symbol for plutonium, Pu.

In one scene in the video, he shakes the hand of a cheerful youngster who is drinking a mug of plutonium-laced soda. The narration says that if plutonium were ingested, most of it would pass through the body without harm.

"The most fundamental lie in this video is the idea that plutonium is not dangerous," said Jinzaburo Takagi, a former nuclear chemist who heads the Citizens' Nuclear Information Center.

"Of course, it's very dangerous to drink plutonium," Takagi said. "To say otherwise, as they do in this video, is completely outrageous."

The highly radioactive, silvery metal is toxic to humans because it is absorbed by bone marrow. The inhalation of .0001 of a gram can induce lung cancer.

Radiocarbon Dating

Radioactive substances are used to determine the age of fossils and artifacts. The procedure is based on the fact that two types of carbon occur naturally: carbon-12, which is stable, and carbon-14, which is radioactive. The total amount of carbon-14 in the environment is rather small; there is only one atom of carbon-14 for every 1 trillion atoms of carbon-12! Living organisms maintain this ratio due to their intake of water, air, and nutrients. However, when an organism dies, the amount of carbon-14 decreases exponentially due to radioactive decay. By making extremely delicate measurements of the amounts of carbon-14 and carbon-12 in a fossil or artifact, scientists are able to estimate the age of the item under investigation.

EXAMPLE **6**

SOLUTION

Step 1

CREATING THE CARBON-14 DECAY MODEL Determine the model representing the amount Q of carbon-14 remaining t years after the death of an organism.

Let a represent the original quantity of carbon-14 present in a living organism. The half-life of carbon-14 is 5,730 years (from Figure 10.14), so at time $t = 5{,}730$, the quantity of carbon-14 present would be $\frac{a}{2}$. Expressing this as ordered pairs (t, Q), we have $(0, a)$ and $(5730, \frac{a}{2})$.

HISTORICAL NOTE

MARIE CURIE, 1867–1934

Born Marie Sklodowska in Warsaw, Poland, Marie Curie moved to Paris and enrolled as a student of physics at the Sorbonne in 1891. While researching the magnetic properties of various steel alloys, she met Pierre Curie, and they married in 1895. In the following year, Antoine Henri Becquerel discovered radioactivity in uranium. As a team, the Curies further investigated uranium and discovered the elements radium and polonium (named after Marie's native country). They also discovered that diseased, tumor-forming cells were destroyed faster than healthy cells when exposed to radium, laying the groundwork for modern radiation therapy. The word *radioactivity* was coined by Madame Curie.

In addition to publishing many important scientific papers and books, Madame Curie received many honors. She was the first person to win two Nobel Prizes, one in 1903 for the discovery of radioactivity (which she shared with her husband and Becquerel) and one in 1911 in chemistry. In 1908, at the University of Paris, she taught the first course on radioactivity ever offered. The Sorbonne created a special chair in physics for Pierre Curie; Marie was appointed his successor after he died in a street accident.

During World War I, Madame Curie devoted much of her time to providing radiological services to hospitals. On her death, Albert Einstein said, "Marie Curie is, of all celebrated beings, the only one whom fame has not corrupted."

THÈSES

PRÉSENTÉES

A LA FACULTÉ DES SCIENCES DE PARIS

POUR OBTENIR

LE GRADE DE DOCTEUR ÈS SCIENCES PHYSIQUES,

PAR

Mᵐᵉ SKLODOWSKA CURIE.

1ʳᵉ THÈSE. — Recherches sur les substances radio-actives.

2ᵉ THÈSE. — Propositions données par la Faculté.

Soutenues le juin 1903, devant la Commission d'Examen.

MM. LIPPMANN, *Président*.
BOUTY,
MOISSAN, } *Examinateurs*.

PARIS,

GAUTHIER-VILLARS, IMPRIMEUR-LIBRAIRE

DU BUREAU DES LONGITUDES, DE L'ÉCOLE POLYTECHNIQUE,

Quai des Grands-Augustins, 55.

1903

HISTORICAL NOTE

WILLARD FRANK LIBBY, 1908–1980

© Bettmann/Corbis

Willard Frank developed the radiocarbon dating technique in the mid-1940s. Carbon-14 was known to exist in nature, but little was known of its origins and properties. In 1939, Libby discovered that cosmic rays interacting with nitrogen at high altitudes produced a rapid formation of carbon-14. This high-altitude formation is the basis of the claim that the current ratio of carbon-14 to carbon-12 has been constant throughout history.

While working at the Enrico Fermi Institute of Nuclear Studies in Chicago,

Libby was able to artificially produce carbon-14 and accurately determine its half-life. In addition, he devised a relatively simple device that measures the amount of carbon-14 in an organic sample. Before the creation of this device, measuring carbon-14 was a very expensive and difficult process. Libby's method made radiocarbon dating a practical possibility and revolutionized the fields of archeology and geology.

Libby received his doctorate degree in chemistry from the University of California at Berkeley in 1933 and

taught there until 1945. During World War II, Libby also worked on the Manhattan Project, which developed the atomic bomb. During the years 1955–1959, he served on the U.S. Atomic Energy Commission, where he was instrumental in the formulation of many aspects of the commission.

After many years of dedicated research and numerous discoveries, Libby was awarded the Nobel Prize in chemistry in 1960 "for his method of using carbon-14 as a measurer of time in archeology, geology, geophysics, and other sciences."

Step 2 | Because we are not using a specific value for the initial amount a, we can omit this step. (Our model is still $Q = ae^{bt}$.)

Step 3 | *We substitute $(t, Q) = (5730, \frac{a}{2})$ into the model and simplify.*

$$Q = ae^{bt}$$

$$\frac{a}{2} = ae^{b(5,730)}$$

$$\frac{1}{2} = e^{5,730b} \qquad \textbf{dividing by } a$$

$$0.5 = e^{5,730b}$$

$$\ln 0.5 = \ln(e^{5,730b}) \qquad \textbf{taking ln of each side}$$

$$\ln 0.5 = 5,730b \qquad \textbf{using an Inverse Property}$$

$$b = \frac{\ln 0.5}{5,730}$$

$$b \approx -0.000120968$$

© Mary Evans Picture Library/ The Image Works

Store this value of b in your calculator's memory.

We now have our model, $Q = ae^{-0.000120968t}$

The calculations in Example 6 produce the following general model.

RADIOCARBON DATING MODEL

The quantity Q of carbon-14 remaining t years after the death of an organism (that had an initial amount a) is

$$Q = ae^{-0.000120968t} \qquad \text{or} \qquad Q = ae^{bt} \quad \text{where } b = \frac{\ln 0.5}{5,730}$$

EXAMPLE **7**

USING THE CARBON-14 MODEL TO DETERMINE THE AGE OF AN ARTIFACT The first of the Dead Sea Scrolls were discovered in 1947 in caves near the northwestern shore of the Dead Sea in the Middle East. The parchment scrolls, which were wrapped in linen and leather, contained all the books of the Old Testament (except Esther). An analysis showed that the scrolls contained approximately 79% of the expected amount of carbon-14 found in a living organism. Determine the age of the scrolls.

SOLUTION

Using the model $Q = ae^{-0.000120968t}$, we need to determine the amount of time t required so that 79% of a remains. To do this, we substitute $0.79a$ for Q in the model and solve.

$$Q = ae^{-0.000120968t}$$
$$0.79a = ae^{-0.000120968t} \qquad \textbf{substituting } Q = 0.79a$$
$$0.79 = e^{-0.000120968t} \qquad \textbf{dividing by } a$$
$$\ln 0.79 = \ln(e^{-0.000120968t}) \qquad \textbf{taking ln of both sides}$$
$$\ln 0.79 = -0.000120968t \qquad \textbf{using an Inverse Property}$$
$$t = \frac{\ln 0.79}{-0.000120968} \qquad \textbf{dividing by } -0.000120968$$
$$t = 1{,}948.22 \ldots$$
$$t \approx 1{,}950$$

A fragment of the Habbakuk Commentary of the Old Testament.

THE FAR SIDE BY GARY LARSON

Early archaeologists

People have always been curious about the ages of artifacts.

Thus, the Dead Sea Scrolls were approximately 1,950 years old when they were discovered and were therefore apparently created around the time when Jesus was alive. This result has been used to support the authenticity of the scrolls.

It should be noted that an underlying assumption in radiocarbon dating is that the current ratio of carbon-14 to carbon-12 in the biosphere remains constant over time and location. There have been disagreements within the scientific community over the validity of this assumption. Therefore, the dates obtained from radiocarbon dating are not guaranteed to be 100% reliable. Some of the results of radiocarbon dating have been controversial, as the article above indicates.

EXAMPLE **8**

EXPLORING THE USE OF TECHNETIUM (TC-99 M)   Suppose that a dose of a mCu of Tc-99m is administered to a patient so that a blood loss imaging scan can be conducted.

a. Develop a mathematical model that represents the amount of Tc-99m remaining in the patient t hours after the injection.
b. What portion of the initial dose will remain in the patient 1 hour after injection?
c. What portion of the initial dose will remain in the patient 10 hours after injection?
d. How long will it take so that 99% of the initial amount of Tc-99m has decayed?

SOLUTION

a. To develop the mathematical model $Q = ae^{bt}$, we use the same procedure as in Example 3. That is, we first express the information as ordered pairs.

$$(t, Q) = (0, a) \text{ and } (t, Q) = \left(6.02, \frac{a}{2} \right) \quad \begin{array}{l} \textbf{the initial amount is } a \textbf{, and the} \\ \textbf{half-life is 6.02 hours.} \end{array}$$

Substituting $(t, Q) = (6.02, \frac{a}{2})$ into the model, we have

$$Q = ae^{bt}$$
$$\frac{a}{2} = ae^{b(6.02)}$$

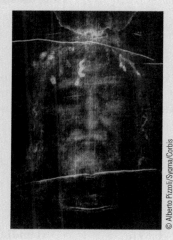

FEATURED IN THE NEWS

SHROUD OF TURIN WAS CREATED IN 14TH CENTURY, OFFICIAL SAYS

ROME—Laboratory tests show that the Shroud of Turin was made in the 14th century and could not be the burial cloth of Christ, the scientific adviser to the archbishop of Turin said he learned yesterday.

Professor Luigi Gonella said he has not yet seen the official report from the three laboratories that conducted the carbon-14 dating tests, but that all the leaks to the press dated it to the 14th century, and "somebody let me understand that the rumors were right."

He refused to identify who had told him about the results of the tests at Oxford University, the University of Arizona and the Swiss Federal Institute of Technology at the University of Zurich.

Topic X NUCLEAR MEDICINE
RADIOACTIVITY IN THE REAL WORLD

If a person has the misfortune to experience internal bleeding, how can a doctor determine the source and severity of the problem? A common investigative procedure (known as blood loss imaging) that involves the introduction of a radioactive material into the patient's bloodstream can assist in the diagnosis. Typically, a small amount of blood is drawn from the patient, a radioactive material (called a *tracer*) is combined with the blood sample, and the mixture is injected back into the bloodstream. Then the person's body is scanned with highly sensitive electronic sensors. These sensors detect the presence of gamma rays. Gamma rays result from the decay of radioactive elements. By monitoring the location and concentration of the gamma rays, doctors can pinpoint internal pools of blood and their sources. Once the bleeding sources have been located, an appropriate course of medical action or surgery can be initiated. The use of radioactive materials in the medical field is known as *nuclear medicine*.

One of the radioactive elements that is frequently used in blood loss imaging is technetium, commonly referred to as Tc-99m. Technetium was discovered in the early twentieth century. However, there is a debate as to when and where it was discovered; some say that German scientists discovered technetium in 1925, while others dispute this "finding" (their experimental data could not be reproduced) and say that technetium was "officially" discovered in 1937 in Italy.

Tc-99m is used in nuclear medicine because it has a relatively short half-life of 6.02 hours and is easily detectable in small amounts. When measuring the amount of radioactivity in a substance, technicians use units called *curies* (named after Pierre Curie) and *millicuries;* one curie equals 1000 millicuries. Curies and millicuries are abbreviated as "Cu" and "mCu," respectively; hence, 1 Cu = 1000 mCu. Unlike a gram or milligram, a curie does not measure weight (or mass). Instead, a curie measures the actual amount of radioactivity in a substance.

The definition of a curie is based on the element radium (discovered by Pierre and Marie Curie); 1 curie was originally defined as the amount of radioactivity in 1 gram of pure radium. However, radioactivity is now measured in terms of the number of atomic disintegrations per second; in 1953, scientists agreed that 1 curie would be defined as 37 billion (3.7×10^{10}) atomic disintegrations per second.

When Tc-99m is administered for blood loss imaging, the suggested dose ranges from 10 to 20 mCu. Therefore, if a patient receives 10 mCu of Tc-99m, then 5 mCu will remain in the patient approximately 6 hours after being injected, 2.5 mCu will remain approximately 12 hours after being injected, approximately 1.25 mCu will remain 18 hours after being injected, and so on (owing to Tc-99m's half-life of 6.02 hours). See Example 8.

$$\frac{1}{2} = e^{b(6.02)} \qquad \textbf{dividing by } a$$

$$0.5 = e^{6.02b}$$

$$\ln 0.5 = \ln e^{6.02b} \qquad \textbf{taking the natural log}$$

$$\ln 0.5 = 6.02b \qquad \textbf{using the Inverse Property } \ln(e^x) = x$$

$$b = \frac{\ln 0.5}{6.02}$$

$$b \approx -0.1151407277$$

Store this value in your calculator's memory. Thus, our model for the amount of Tc-99m remaining t hours after an injection of a mCu is $Q = ae^{-0.1151407277t}$.

b. To calculate what portion of the initial dose remains 1 hour after injection, we substitute $t = 1$ and obtain

$$Q = ae^{-0.1151407277t}$$

$$Q = ae^{-0.1151407277(1)}$$

$$Q = ae^{-0.1151407277}$$

$$Q = a(0.8912407129)$$

$$Q = 0.891a \qquad \textbf{rounding off to three decimal places}$$

Therefore, 1 hour after injection, 0.891 (or 89.1%) of initial amount of Tc-99m will remain.

c. To calculate what portion of the initial dose remains 10 hours after injection, we substitute $t = 10$ and obtain

$$Q = ae^{-0.1151407277t}$$

$$Q = ae^{-0.1151407277(10)}$$

$$Q = ae^{-1.151407277}$$

$$Q = a(0.3161914872)$$

$$Q = 0.316a \qquad \text{rounding off to three decimal places}$$

Therefore, 10 hours after injection, 0.316 (or 31.6%) of the initial amount of Tc-99m will remain.

d. If 99% of the Tc-99m has decayed, 1% will still remain in the patient. To determine how long it will take so that only 1% remains, we substitute $Q = 0.01a$ into the model and solve for t.

$$Q = ae^{-0.1151407277t}$$

$$0.01a = ae^{-0.1151407277t}$$

$$0.01 = e^{-0.1151407277t} \qquad \text{dividing by } a$$

$$\ln 0.01 = \ln e^{-0.1151407277t} \qquad \text{taking the natural log}$$

$$\ln 0.01 = -0.1151407277t \qquad \text{using the Inverse Property } \ln(e^x) = x$$

$$t = \frac{\ln 0.01}{-0.1151407277}$$

$$t = 39.99601426$$

Therefore, 40.0 hours after injection, 1% of the initial amount of Tc-99m remains in the patient, that is, 99% has decayed.

10.2 EXERCISES

1. Using the model $Q = 20\, e^{-0.086643397t}$ developed in Example 1, predict how much iodine-131 will remain after three weeks.

2. Using the model $Q = 20\, e^{-0.086643397t}$ developed in Example 1, predict how much iodine-131 will remain after thirty days.

3. Using the model $Q = 8.2\, e^{-0.053319013t}$ developed in Example 3, predict how much plutonium-241 will remain after two years.

4. Using the model $Q = 8.2\, e^{-0.053319013t}$ developed in Example 3, predict how much plutonium-241 will remain after ten years.

In Exercises 5–16, use Figure 10.14.

5. Silicon-31 is used to diagnose certain medical ailments. Suppose a patient is given 50 milligrams.

   a. Develop the mathematical model that represents the amount of silicon-31 present at time t.

   b. Predict the amount of silicon-31 remaining after one hour.

   c. Predict the amount of silicon-31 remaining after one day.

   d. Find $\Delta Q / \Delta t$, the average decay rate, for the first hour.

e. Find $(\Delta Q/\Delta t)/Q$, the relative decay rate, for the first hour.

f. Find $\Delta Q/\Delta t$, the average decay rate, for the first day.

g. Find $(\Delta Q/\Delta t)/Q$, the relative decay rate, for the first day.

h. Why is the absolute value of the answer to part (e) greater than the answer to part (g)?

6. Plutonium-238 is used as a compact source of electrical power in many applications, ranging from pacemakers to spacecraft. Suppose a power cell initially contains 1.6 grams of plutonium-238.

   a. Develop the mathematical model that represents the amount of plutonium-238 present at time t.

   b. Predict the amount of plutonium-238 remaining after one year.

   c. Predict the amount of plutonium-238 remaining after 20 years.

   d. Find $\Delta Q/\Delta t$, the average decay rate, for the first year.

   e. Find $(\Delta Q/\Delta t)/Q$, the relative decay rate, for the first year.

   f. Find $\Delta Q/\Delta t$, the average decay rate, for the first 20 years.

   g. Find $(\Delta Q/\Delta t)/Q$, the relative decay rate, for the first 20 years.

   h. Why is the absolute value of the answer to part (e) greater than the answer to part (g)?

7. How long will it take 64 grams of magnesium-28 to decay into the following amounts?

   a. 32 grams

   b. 16 grams

   c. 8 grams

8. How long will it take 56 milligrams of cobalt-55 to decay into the following amounts?

   a. 28 milligrams

   b. 14 milligrams

   c. 7 milligrams

9. How long will it take 500 grams of plutonium-241 to decay into 100 grams?

10. How long will it take 300 grams of cobalt-60 to decay into 10 grams?

11. How long will it take 30 milligrams of plutonium-238 to decay into 20 milligrams?

12. How long will it take 900 milligrams of silicon-31 to decay into 700 milligrams?

13. How long will it take a given quantity of plutonium-239 to lose 90% of its radioactivity?

14. How long will it take a given quantity of plutonium-239 to lose 95% of its radioactivity?

15. How long will it take a given quantity of krypton-91 to lose 99.9% of its radioactivity?

16. How long will it take a given quantity of carbon-14 to lose 99.9% of its radioactivity?

17. The article on the "Iceman" in this section stated that the frozen corpse was 5,000 to 5,500 years old. Assuming the corpse was 5,250 years old, how much carbon-14 should it contain?

18. The article on radioactive gas in this section at the beginning of this section stated that tritium decays at the rate of 5.5% a year; that is, 94.5% of any given quantity will remain at the end of 1 year.

   a. Find the value of b in the mathematical model for the exponential decay of tritium.

   b. Determine the half-life of tritium gas.

19. A lab technician had 58 grams of a radioactive substance. Ten days later, only 52 grams remained. Find the half-life of the radioactive substance.

20. A lab technician had 32 grams of a radioactive substance. Eight hours later, only 30 grams remained. Find the half-life of the radioactive substance.

21. In 1989, a Mayan codex (a remnant of ancient written records) was found in the remains of a thatched hut that had been buried under 15 feet of volcanic ash after a prehistoric eruption of Laguna Caldera Volcano, just north of San Salvador. An analysis of the roof material concluded that the material contained 84% of the expected amount of carbon-14 found in a living organism. Determine the age of the roof material and hence the age of the codex.

22. Two Ohlone Indian skeletons, along with burial goods such as quartz crystals, red ocher, mica ornaments, and olivella shell beads, were dug up at a construction site in San Francisco in 1989. An analysis of the skeletons revealed that they contained 88% of the expected amount of carbon-14 found in a living person. Determine the age of the skeletons.

23. In 1940, beautiful prehistoric cave paintings of animals, hunters, and abstract designs were found in the Lascaux cave near Montignac, France. Analyses revealed that charcoal found in the cave had lost 83% of the expected amount of carbon-14 found in living plant material. Determine the age of the charcoal and the paintings.

24. To determine the onset of the last Ice Age, scientists analyzed fossilized wood found in Two Creeks Forest, Wisconsin. They theorized that the trees from which the wood was taken had been killed by the glacial advance. If the wood had lost 75% of its carbon-14, determine its age.

25. An ancient parchment contained 70% of the expected amount of carbon-14 found in living matter. Estimate the age of the parchment.

26. When analyzing wood found in an Egyptian tomb, scientists determined that the wood contained 55% of the expected amount of carbon-14 found in living matter. Estimate the age of the wood.

27. As noted in the article on the Shroud of Turin in this section, radiocarbon dating analysis concluded that the shroud was made in the fourteenth century. Assuming that the shroud was made in 1350 A.D., how much carbon-14 did it contain when it was tested in 1988? If the shroud was made at the time of Christ's death in 33 A.D., how much carbon-14 should it have contained when it was tested in 1988?

28. In 1993, archeologists digging in southern Turkey found an ancient piece of cloth. Radiocarbon dating determined that the cloth was 9,000 years old. How much carbon-14 did the cloth have when it was tested?

29. How much carbon-14 would you expect to find in a 5,730-year-old relic?

30. How much carbon-14 would you expect to find in an 11,460-year-old relic?

31. A museum claims that one of its mummies is 5,000 years old. An analysis reveals that the mummy contains 62% of the expected amount of carbon-14 found in living organisms. Is the museum's claim justified?

32. A museum claims that one of its skeletons is 9,000 years old. An analysis reveals that the skeleton contains 42% of the expected amount of carbon-14 found in living organisms. Is the museum's claim justified?

33. A flute carved from the wing bone of a crane was discovered in Jiahu, an excavation site of Stone Age artifacts in China's Yellow River Valley in the 1980s. Scientists estimate that the artifact has lost 63.5% of its carbon-14. Estimate the age of the flute.

34. Suppose a patient receives 20.0 mCu of Tc-99m.
   a. Complete the chart in Figure 10.15 by determining the amount of Tc-99m remaining in the patient 5, 10, 15, 20, 25, and 30 hours after the injection. (The half-life of Tc-99m is 6.02 hours.)
   b. Sketch a graph of the results obtained in part (a).

Hours after Injection	0	5	10	15	20	25	30
Amount Remaining (mCu)	20.0						

FIGURE 10.15 Chart for Exercise 34.

35. In Example 8, we determined that 0.891 (or 89.1%) of an initial dose of Tc-99m would remain in a patient 1 hour after injection. In a similar fashion, complete the chart in Figure 10.16.

Hours after Injection	1	2	3	4	5
Portion Remaining	0.891				

FIGURE 10.16 Chart for Exercise 35.

36. In Example 8, we determined that 0.316 (or 31.6%) of an initial dose of Tc-99m would remain in a patient 10 hours after injection. In a similar fashion, complete the chart in Figure 10.17.

Hours after Injection	6	7	8	9	10
Portion Remaining					0.316

FIGURE 10.17 Chart for Exercise 36.

37. How long will it take so that 99.5% of an initial dose of Tc-99m has decayed?

38. How long will it take so that 99.9% of an initial dose of Tc-99m has decayed?

39. **a.** How many atomic disintegrations per second occur in 1 mCu of radioactivity?
   b. How many atomic disintegrations per second occur in 10 mCu of radioactivity?
   c. How many atomic disintegrations per second occur in 20 mCu of radioactivity?

 Answer the following questions using complete sentences and your own words.

• **CONCEPT QUESTIONS**

40. What does the phrase *exponential decay* mean?

41. What does the phrase *half-life* mean?

42. In terms of mathematical models, what is the basic difference between exponential growth and exponential decay?

43. In terms of units, what is the difference between the average decay rate and the relative decay rate?

44. How does radiocarbon dating work?

45. What aspect of radiocarbon dating is controversial?

46. What is nuclear medicine?

47. What basic unit is used in measuring radioactivity? How is it defined?

• HISTORY QUESTIONS

48. Who invented radiocarbon dating? When? What other famous project did this person work on?

49. Who received a Nobel Prize for the discovery of radioactivity? When?

50. Who was the first person to receive two Nobel Prizes?

WEB PROJECT

51. Although it is widely used, radiocarbon dating is not the only method employed when determining the age of artifacts. Write a report, or make a presentation, on an alternative method used in archaeometry (the science of determining the age of artifacts).

Some useful links for this web project are listed on the text web site: **www.cengage.com/math/johnson**

10.3 Logarithmic Scales

OBJECTIVES

- Understand earthquake amplitude and what it measures
- Understand earthquake magnitude and what it measures
- Compare two earthquakes
- Convert between amplitude and magnitude
- Understand a sound's intensity and what it measures
- Understand a sound's decibel rating and what it measures
- Convert between intensity and decibel rating
- Compare two sounds

Until now, we have been studying exponential models that are based on the function $y = ae^{bt}$. Now let's look at another topic closely related to exponential models: logarithmic scales. A **logarithmic scale** is a scale in which logarithms serve to make data more manageable by expanding small variations and compressing large ones, as shown in Figure 10.18. We will look at the Richter scale, which is used to rate earthquakes, and the decibel scale, which is used to rate the loudness of sounds.

Earthquakes

Most earthquakes are mild and cause little or no damage, but some are incredibly devastating. San Francisco was almost totally destroyed in 1906 by a large earthquake (and the fires it caused). Over 200,000 people were killed by a large quake in the Indian Ocean in 2004. However, engineers have made great progress in designing structures that can withstand major

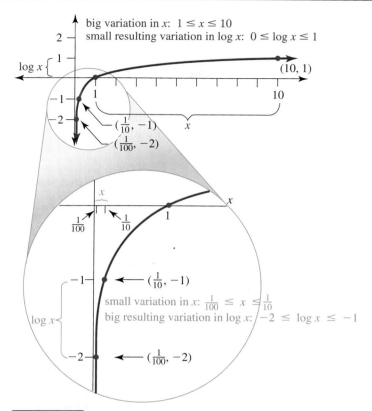

big variation in x: $1 \le x \le 10$
small resulting variation in $\log x$: $0 \le \log x \le 1$

$\log x \Big\{$

$(10, 1)$

1

10

$(\frac{1}{10}, -1)$

$(\frac{1}{100}, -2)$

x

$\frac{1}{100}$ $\frac{1}{10}$ 1

$\longleftarrow (\frac{1}{10}, -1)$

small variation in x: $\frac{1}{100} \le x \le \frac{1}{10}$
big resulting variation in $\log x$: $-2 \le \log x \le -1$

$\log x \Big\{$

$\longleftarrow (\frac{1}{100}, -2)$

FIGURE 10.18 With logarithmic scales, logarithms expand small variations and compress large ones.

San Francisco was almost totally destroyed by the great quake of 1906 (above left). Only limited damage was sustained in the quake of 1989 (above right). The few buildings that collapsed were old ones, and most were made of unreinforced masonry or stucco or were built over first-floor garages, as was the building shown above.

earthquakes. San Francisco survived its 1989 quake with very limited damage; although there was some major damage, most of the city survived without a scratch.

Most earthquakes in recent times have been in the Middle East and along the Pacific Rim, but they are not limited to those regions. Among the most powerful earthquakes in North America's history were a series of three temblors in 1811–1812 on the New Madrid fault in Missouri. These quakes rerouted the Mississippi River, cracked plaster in Boston, and made church bells ring in

Montreal. The U.S. Geological Survey now lists 39 states as having earthquake potential. (See Figure 10.19.) Geologists say that the eastern United States is certain to receive a violent jolt in the near future.* Perhaps the greatest current threat is posed by the New Madrid fault, where experts say there is a 90% chance of a serious quake in the next 50 years. Because of the region's geology, such a quake would cause damage over a huge area, including parts of more than a dozen states. (See Figure 10.20.) And it has only been in the last ten years that these states have started to adopt seismic building codes.

Seismicity of the United States: 1990 - 2000

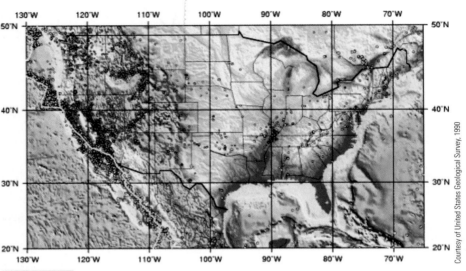

Courtesy of United States Geological Survey, 1990

FIGURE 10.19 California is not the only state with tectonic stresses and thus prone to earthquakes. Each dot represents an earthquake between 1990 and 2000.

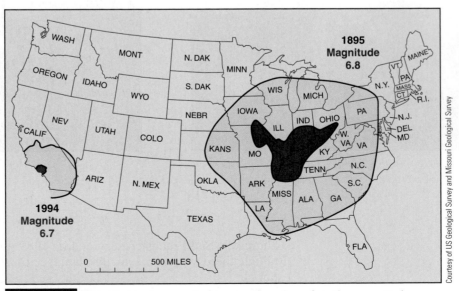

Courtesy of US Geological Survey and Missouri Geological Survey

FIGURE 10.20 Comparing the extent of two earthquakes of similar magnitudes, one in Southern California and one in Missouri.

Newsweek, 30 October 1989, "East of the Rockies: A Lot of Shaking Going On."

Most earthquakes are related to **tectonic stress** (that is, stress in the earth's crust). A group of more than 30 scientists from 19 different countries is compiling a global data base of stresses. A result of their efforts is the *World Stress Map,* shown in Figure 10.21, which indicates the presence of tectonic stresses throughout most of North America as well as much of Asia and Europe.

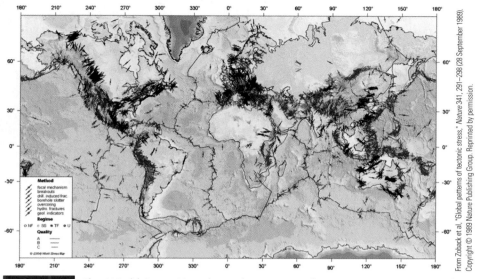

From Zoback et al, "Global patterns of tectonic stress," *Nature* 341, 291–298 (28 September 1989). Copyright © 1989 Nature Publishing Group. Reprinted by permission.

FIGURE 10.21   The *World Stress Map* shows the pressure of tectonic stresses in most of North America, Europe, and Asia.

A **seismograph** (from *seismos,* a Greek word meaning "earthquake") is an instrument that records the amount of earth movement generated by an earthquake's seismic wave; the recording is called a **seismogram.** The **amplitude of a seismogram** is the vertical distance between the peak or valley of the recording of the seismic wave and a horizontal line formed if there is no earth movement; the amplitude is usually measured in micrometers (μm).

The Richter Scale

The most common method of comparing earthquakes was developed by Charles F. Richter, a seismologist at the California Institute of Technology, in 1935. Prior to Richter, it was known that the amplitude of a recording of a seismic wave is affected by the strength of the earthquake and by the distance between the earthquake and the seismograph. Richter wanted to develop a scale that would reflect only the actual strength of the earthquake and not the distance between the source (or **epicenter**) and the seismograph. Because larger earthquakes have amplitudes millions of times greater than those of smaller quakes, Richter used the common logarithm of the amplitude of the quake in developing his scale in order to compress the enormous variation inherent in earth movement down to a more manageable range of numbers.

Examining data from many earthquakes, Richter discovered an interesting pattern: If A_{10} and A_{20} are the amplitudes of one earthquake measured 10 and 20 kilometers, respectively, from the epicenter and if B_{10} and B_{20} are similar 10- and 20-kilometer measurements for a second earthquake, as shown in Figure 10.22, then

$$\log A_{10} - \log B_{10} = \log A_{20} - \log B_{20}$$

Because the epicenter of the 1994 Northridge quake was within Los Angeles itself, the city suffered more extensive damage than did San Francisco in 1989.

or, equivalently, by the Division-Becomes-Subtraction Property,

$$\log \frac{A_{10}}{B_{10}} = \log \frac{A_{20}}{B_{20}}$$

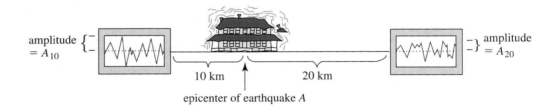

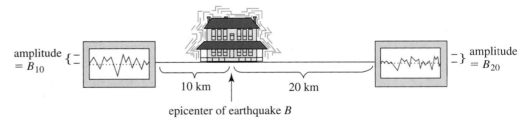

FIGURE 10.22 The concept behind the Richter scale.

This pattern continues *regardless of the distance*. That is, if A_{95} and B_{95} are the amplitudes of two quakes measured 95 kilometers from their epicenters, then

$$\log A_{10} - \log B_{10} = \log A_{20} - \log B_{20} = \log A_{95} - \log B_{95}$$

Richter's goal was to develop a scale that would measure the actual strength of an earthquake *without being biased by the distance between the quake's epicenter and the seismograph*. This pattern went a long way toward fulfilling Richter's goal; the only difficulty was that the pattern is a comparison of two quakes, not a measurement of a single quake. Richter remedied that problem by creating a "standard earthquake" of a certain fixed strength; this standard earthquake would be a yardstick against which all future earthquakes would be measured. Richter's standard earthquake is an average of a large number of extremely small southern California earthquakes, and the magnitude of an earthquake is a measure of how much stronger than the standard a given earthquake is. Figure 10.23 gives logs of amplitudes at different distances for the standard quake of "magnitude 0."

Distance (km)	20	60	100	140	180	220	260	300	340	380	420	460	500	540	580
Log of amplitude (πm)	−1.7	−2.8	−3.0	−3.2	−3.4	−3.6	−3.8	−4.0	−4.2	−4.4	−4.5	−4.6	−4.7	−4.8	−4.9

FIGURE 10.23 Richter's standard earthquake.

RICHTER'S DEFINITION OF EARTHQUAKE MAGNITUDE

The magnitude M of an earthquake of amplitude A is

$$M = \log A - \log A_0$$

where A_0 is the amplitude of the "standard earthquake" measured at the same distance.

EXAMPLE 1

THE RELATIONSHIP BETWEEN AN EARTHQUAKE'S AMPLITUDE AND ITS MAGNITUDE A seismograph 20 kilometers from an earthquake's epicenter recorded a maximum amplitude of 5.0×10^6 micrometers. Find the earthquake's magnitude.

SOLUTION

$$M = \log A - \log A_0$$
$$= \log(5.0 \times 10^6) - \log A_0 \quad \text{substituting for } A$$
$$= \log(5.0 \times 10^6) - (-1.7) \quad \text{using Figure 10.23 to find } \log A_0$$
$$\approx 6.699 + 1.7$$
$$= 8.399$$
$$\approx 8.4$$

The earthquake was 8.4 on the Richter scale. (By tradition, earthquake magnitudes are rounded to the nearest tenth.)

An earthquake is usually measured by seismographs in at least three different locations in order to accurately determine its Richter scale rating and the location of its epicenter. The Richter scale ratings of some major earthquakes are given in Figure 10.24 on the next page.

In addition to comparing the amplitude of an earthquake with that of Richter's artificial "standard earthquake," we can also use Richter's definition to compare the amplitudes of two actual earthquakes.

1811–1812	New Madrid, MO	8.7	1994	Kuril Islands (between Russia and Japan)	8.2
1906	San Francisco	8.3	1997	Western Pakistan	7.3
1933	Japan	8.9	1999	Turkey	7.8
1950	India	8.7	1999	Taiwan	7.6
1960	Chile (the largest earthquake ever recorded)	9.5	1999	Oaxaca, Mexico	7.6
1964	Alaska	9.2	2001	El Salvador	7.7
1968	22 midwestern states	5.5	2001	India	7.7
1976	China	8.0	2001	Peru	8.4
1983	Coalinga, CA	6.5	2004	Indian Ocean (the second largest ever recorded)	9.2
1985	Mexico City	8.1	2005	Indonesia	8.6
1989	San Francisco/Loma Prieta (the "world series quake")	7.1	2006	Russia	8.3
1990	Iran	7.7	2007	Sumatra	8.5
1990	Philippines	7.7	2008	China	7.9
1991	India and Nepal	7.7	2009	Samoa	8.0
1992	Yucca Valley, CA	7.4	2010	Chile	8.8
1994	Los Angeles (Northridge)	6.8	2010	Haiti	7.0
1995	Kobe, Japan	7.2			

FIGURE 10.24 Magnitudes of some major earthquakes.

MAGNITUDE COMPARISON FORMULA

If M_1 and M_2 are the magnitudes of two earthquakes, and if A_1 and A_2 are their amplitudes measured at equal distances, then

$$M_1 - M_2 = \log\left(\frac{A_1}{A_2}\right)$$

The Magnitude Comparison Formula is true because

$$M_1 - M_2 = (\log A_1 - \log A_0) - (\log A_2 - \log A_0) \quad \textbf{using Richter's definition to rewrite } M_1 \textbf{ and } M_2$$

$$= \log A_1 - \log A_2 \quad \textbf{simplifying}$$

$$= \log\left(\frac{A_1}{A_2}\right) \quad \textbf{using the Division-Becomes-Subtraction Property}$$

Notice that if the second earthquake in the formula above is Richter's standard earthquake, $M_2 = 0$ (Richter's standard quake is one of magnitude zero), and if A_2 is A_0, the magnitude comparison formula turns into Richter's definition of earthquake magnitude.

EXAMPLE **2**

SOLUTION

COMPARING DIFFERENT MAGNITUDES If one earthquake has magnitude 6 and another has magnitude 3, it does not mean that the first caused twice as much earth movement, as many people believe. Find how the stronger earthquake actually compares to the weaker one.

We use the Magnitude Comparison Formula:

$$M_1 - M_2 = \log\left(\frac{A_1}{A_2}\right)$$

$$6 - 3 = \log\left(\frac{A_1}{A_2}\right)$$

$$3 = \log\left(\frac{A_1}{A_2}\right)$$

$$10^3 = 10^{\log(A_1/A_2)}$$

$$10^3 = \frac{A_1}{A_2} \qquad \textbf{using the Inverse Property } \mathbf{10^{\log x} = x}$$

$$A_1 = 10^3 \, A_2$$

The stronger earthquake's amplitude is actually $10^3 = 1{,}000$ times that of the weaker earthquake; thus, the stronger earthquake would cause about 1,000 times as much earth movement (not twice as much)!

Earthquake Magnitude and Energy

The magnitude of an earthquake is determined by the amplitude of the earthquake's seismic wave and thus by the amount of ground movement caused by the quake. The amount of ground movement does not depend exclusively on the amount of energy radiated by the earthquake, because different geological compositions will transmit the energy in different ways. However, ground movement and energy are closely related.

ENERGY FORMULA

The energy E (in ergs) released by an earthquake of magnitude M is approximated by

$$\log E \approx 11.8 + 1.45M$$

EXAMPLE **3**

SOLUTION

ENERGY AND MAGNITUDE Find approximately how much more energy is released by an earthquake of magnitude 6 than by an earthquake of magnitude 3.

$M_1 = 6$	$M_2 = 3$
$\log E_1 \approx 11.8 + 1.45M_1$	$\log E_2 \approx 11.8 + 1.45M_2$
$\quad = 11.8 + 1.45 \cdot 6$	$\quad = 11.8 + 1.45 \cdot 3$
$\quad = 20.5$	$\quad = 16.15$
$10^{\log E_1} \approx 10^{20.5}$	$10^{\log E_2} \approx 10^{16.15}$
$\quad E_1 \approx 10^{20.5}$	$\quad E_2 \approx 10^{16.15}$

$$\frac{E_1}{E_2} \approx \frac{10^{20.5}}{10^{16.15}}$$
$$= 10^{20.5-16.15}$$
$$= 10^{4.35}$$
$$= 22{,}387.21139\ldots$$
$$\approx 22{,}000$$
$$E_1 \approx 22{,}000 \cdot E_2$$

The earthquake of magnitude 6 releases approximately 22,000 times as much energy as the earthquake of magnitude 3. (Recall from Example 2 that the magnitude 6 quake causes 1,000 times as much earth movement as the magnitude 3 quake.)

The Decibel Scale

A sound is a vibration received by the ear and processed by the brain. The **intensity of a sound** is a measure of the "strength" of the vibration; it is determined by placing a surface in the path of the sound and measuring the amount of energy in that surface per unit of area per second. This surface acts like an eardrum.

Intensity is not a good measure of loudness as perceived by the human brain, for two reasons. First, experiments have shown that humans perceive loudness on the basis of the ratio of intensities of two different sounds. For example, in order for the human ear and brain to distinguish an increase in loudness, one sound must typically have 25% more energy than another. If two sounds have intensities I_1 and I_2, and if I_1 is 25% greater than I_2, then

$$I_1 = 125\% \text{ of } I_2$$
$$= 1.25 I_2$$

and the ratio of I_1 to I_2 is

$$\frac{I_1}{I_2} = 1.25$$

and the first sound would be perceived as slightly louder than the second. To determine loudness (as humans perceive it), we must look at the ratio of sound intensities.

The second reason why intensity is not a good measure of loudness has to do with how the brain processes sound. The human ear registers an amazing range of sound intensities. The intensity of a painfully loud sound is 100 trillion ($10^{14} = 100{,}000{,}000{,}000{,}000$) times that of a barely audible sound. More precisely, if I_1 is the intensity of a painfully loud sound and I_2 is the intensity of a barely audible sound, then $I_1 = 10^{14} \cdot I_2$, or $I_1/I_2 = 10^{14}$. The ear registers this range of

DECIBEL RATING DEFINITION

If a sound has intensity I (in watts per square centimeter, measured at a standard distance), then its decibel rating is

$$D = 10 \log\left(\frac{I}{I_0}\right)$$

where I_0 is a "standard intensity" ($I_0 \approx 10^{-16}$ watts/cm^2, the intensity of a barely audible sound).

HISTORICAL NOTE

ALEXANDER GRAHAM BELL, 1847–1922

The Granger Collection, New York

Alexander Melville Bell lectured at the University of Edinburgh (Scotland), at the University of London, and in Boston. He developed a system of visible speech for the deaf, with symbols for every sound of the human voice.

Alexander Graham Bell, the son of Alexander Melville Bell and the husband of a deaf woman, had his own school of vocal physiology in Boston and was very active in issues related to education of the deaf. In 1875, he conceived the idea of the telephone. On March 10, 1876, he used his experimental apparatus to transmit the now famous sentence "Watson, come here; I want you" to his assistant. Later that same year, the telephone was introduced to the world at the Philadelphia Centennial Exposition. The Bell Telephone Company was organized a year later.

Bell also established the Volta Laboratory in Washington, D.C., where the first successful phonograph record was produced. Bell invented both the flat and cylindrical wax recorders for phonographs, as well as the photophone, which transmits speech by light rays, and the audiometer, which measures a person's hearing ability. He investigated the nature and causes of deafness and studied its heredity. He helped found the magazine *Science*, was president of the National Geographic Society, and was a regent of the Smithsonian Institution.

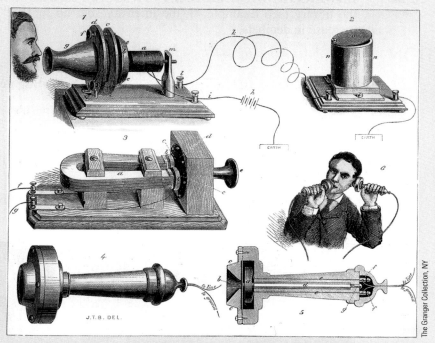

The Granger Collection, NY

J.T.B. DEL.

Bell's telephone, as described in an 1877 newspaper. The first two figures are the transmitter and receiver that were shown at the Philadelphia Exposition.

sound intensities, but the brain condenses it to a smaller, more manageable range. The brain processes sound in a roughly logarithmic fashion.

The **decibel scale** approximates loudness as perceived by the human brain. It is based on the ratio of intensities of sounds, and its range is roughly on a par with the range in loudness that a human perceives rather than with the range in energy intensities that the ear or a mechanical device would actually measure.

The word *decibel* is abbreviated *dB*. The prefix *deci* is the metric system's prefix meaning "tenth." One decibel, therefore, is one-tenth of a bel. The *bel* is named for Alexander Graham Bell, the inventor of the telephone.

EXAMPLE **4**

THE RELATIONSHIP BETWEEN A SOUND'S INTENSITY AND ITS DECIBEL RATING The background noise of a quiet library has a measured intensity of 10^{-12} watts/cm^2. Find the decibel rating of the sound.

SOLUTION

$$I_1 = 10^{-12} \qquad \text{\textbf{the given intensity}}$$

$$D_1 = 10 \log\left(\frac{I_1}{I_0}\right) \qquad \text{\textbf{the decibel rating definition}}$$

$$= 10 \log\left(\frac{10^{-12}}{10^{-16}}\right) \qquad \text{\textbf{substituting}}$$

$$= 10 \log (10^{-12 - (-16)}) \qquad \text{\textbf{using an Exponent Property}}$$

$$= 10 \log 10^4$$

$$= 10 \cdot 4 \qquad \text{\textbf{using the Inverse Property } } \log 10^x = x$$

$$= 40$$

The decibel rating of the background noise is 40 dB.

EXAMPLE **5**

UNDERSTANDING dB GAIN When two students started a conversation in the library (see Example 4), the intensity shot up to 10^{-10} watts/cm^2. Find the increase in decibels.

SOLUTION

$$I_2 = 10^{-10} \qquad \text{\textbf{the given intensity}}$$

$$D_2 = 10 \log\left(\frac{I_2}{I_0}\right) \qquad \text{\textbf{the decibel rating definition}}$$

$$= 10 \log\left(\frac{10^{-10}}{10^{-16}}\right) \qquad \text{\textbf{substituting}}$$

$$= 10 \log(10^{-10 - (-16)}) \qquad \text{\textbf{using an Exponent Property}}$$

$$= 10 \log 10^6$$

$$= 10 \cdot 6 \qquad \text{\textbf{using the Inverse Property } } \log 10^x = x$$

$$= 60$$

Recording engineers monitor the music's decibel level so as to get a high-quality recording.

AP Photo/Richard Drew

$$D_2 - D_1 = 60 - 40 \qquad\qquad D_1 = 40, \text{from Example 4}$$
$$= 20$$

The increase in decibels, or **dB gain,** is 20 dB.

An alternative approach to finding the dB gain is to use a slightly altered version of the Decibel Formula.

---

dB GAIN FORMULA

If I_1 and I_2 are the intensities of two sounds, then the dB gain is

$$D_1 - D_2 = 10 \log \left(\frac{I_1}{I_2} \right)$$

The dB Gain Formula is true because

$$D_1 - D_2 = 10 \log \left(\frac{I_1}{I_0} \right) - 10 \log \left(\frac{I_2}{I_0} \right)$$ **using the decibel rating definition to rewrite D_1 and D_2**

$$= 10 \left[\log\left(\frac{I_1}{I_0}\right) - \log\left(\frac{I_2}{I_0}\right) \right]$$ **factoring**

$$= 10[(\log I_1 - \log I_0) - (\log I_2 - \log I_0)]$$ **using the Division-Becomes-Subtraction Property**

$$= 10[\log I_1 - \log I_2]$$

$$= 10 \log \left(\frac{I_1}{I_2} \right)$$ **using the Division-Becomes-Subtraction Property**

EXAMPLE 6

USING THE dB GAIN FORMULA   Use the dB Gain Formula to find the increase in decibels when the quiet library in Example 4 became somewhat less quiet owing to the students' conversation in Example 5.

SOLUTION

$$I_1 = 10^{-10} \text{ watts/cm}^2 \qquad \text{\textbf{the conversation's intensity}}$$
$$I_2 = 10^{-12} \text{ watts/cm}^2 \qquad \text{\textbf{the background noise's intensity}}$$
$$D_1 - D_2 = 10 \log \left(\frac{I_1}{I_2} \right) \qquad \text{\textbf{the dB Gain Formula}}$$

$$= 10 \log \left(\frac{10^{-10}}{10^{-12}} \right) \qquad \text{\textbf{substituting}}$$

$$= 10 \log(10^{-10 -(-12)}) \qquad \text{\textbf{using an Exponent Property}}$$

$$= 10 \log 10^2$$

$$= 10 \cdot 2 = 20 \text{ dB gain}$$

Notice that $I_1/I_2 = 10^{-10}/10^{-12} = 10^2$, so the conversation had a sound intensity $10^2 = 100$ times that of the background noise. This means that there was 100 times as much vibration.

EXAMPLE 7

USING THE dB GAIN FORMULA   The noise on a busy freeway varies from 81 dB to 92 dB. Find the corresponding variation in intensities.

SOLUTION

We are given $D_1 = 92$ dB and $D_2 = 81$ dB, and we are asked to compare I_1 and I_2.

$$D_1 - D_2 = 10 \log\left(\frac{I_1}{I_2}\right) \qquad \text{the dB Gain Formula}$$

$$92 - 81 = 10 \log\left(\frac{I_1}{I_2}\right) \qquad \text{substituting}$$

$$11 = 10 \log\left(\frac{I_1}{I_2}\right)$$

$$1.1 = \log\left(\frac{I_1}{I_2}\right) \qquad \text{solving for } I_1$$

$$10^{1.1} = 10^{\log(I_1/I_2)}$$

$$10^{1.1} = \frac{I_1}{I_2} \qquad \text{using the Inverse Property } 10^{\log x} = x$$

$$I_1 = 10^{1.1} I_2 \approx 12.5893 I_2$$

The louder freeway noise has almost 13 times the sound intensity of the quieter freeway noise. This means that the louder freeway noise has almost 13 times as much vibration as does the quieter freeway noise.

FEATURED IN THE NEWS

MUSIC MAKING FANS DEAF?
HOW THE IPOD GENERATION MAY BE LOSING ITS HEARING WITHOUT EVEN KNOWING IT

Hearing loss is one of the dirty secrets of the music business, and everyone involved—from musicians onstage to fans who crank MP3s through headphones—is at risk. "We turn it up without realizing that we're doing damage," says Brian Fligor, an audiologist at Boston Children's Hospital. "Noise-induced hearing loss develops so slowly and insidiously that we don't know it's happened until it's too late."

In 1989, Pete Townshend admitted that he had sustained "very severe hearing damage." Since then, Neil Young, Beatles producer George Martin, Sting, Ted Nugent and Jeff Beck have all discussed their hearing problems.

Twenty-two million American adults own an iPod or other digital-music player, and studies show that sustained listening, even at moderate volume, can cause serious harm.

In 2001, Fligor began a study to determine how loud—and for how long—you can safely listen to a portable music player through headphones. He found that the kind of headphones you use greatly affects the risk. "The closer to the eardrum, the higher the sound levels the system is capable of producing," Fligor

says. On average, Fligor found that you can safely listen to over-the-ear headphones with a player set at level six (out of ten) for an hour a day. For most in-the-ear headphones, like the earbuds that come with most MP3 players, the acceptable time at that level is less—around thirty minutes for some models before you've exceeded your safe daily dose. Fligor's preliminary findings indicate that the iPod is comparable to a Sony CD Walkman with earbuds, which can go as high as 130 decibels—equivalent to a jackhammer. European iPods, in contrast, are capped at 100 decibels by law.

"Every other musician I know has some form of problem with their hearing," says Garbage singer Shirley Manson. "All the guys in my band have weird schisms that come from being around really loud music." Pearl Jam guitarist Stone Gossard says he isn't affected but that bassist Jeff Ament, guitarist Mike McCready and drummer Matt Cameron all have hearing loss or tinnitus. "It's hard for them to deal with it, so we've made adjustments," Gossard says. "We're trying to keep the volume at a level that allows everyone to keep playing music."

FIVE WAYS TO SAVE YOUR EARS

1. Wear earplugs: Coldplay and Dave Matthews Band wear ear protection. You should too.
2. Turn it down: Don't crank up your portable music player too loud, especially to compensate for other noise around you.
3. Get better headphones: Those that shut out external noise allow you to turn down the tunes.
4. Give your ears a rest: "There's nothing wrong with going to a rock concert on Friday night," says Marshall Chasin. "Just don't mow your lawn on Saturday." Your ears need about eighteen hours after exposure to sustained high volumes before they return to normal.
5. Quit smoking: It doubles the risk of noise-induced hearing loss.

Topic X   THE NEGATIVE IMPACT OF SOUNDS: LOGARITHMS AND THE REAL WORLD

Sound has an incredibly positive impact on our lives. It provides us with music, and it provides us with oral communication. However, sound can have a negative impact as well. Loud concerts and iPods are damaging our hearing. iPods and other personal listening devices can be especially damaging, for several reasons:

- The earbud is placed in the ear, close to the eardrum. This proximity makes loud music more damaging than it would be if it were produced by a band at a concert or even over-the-ear headphones.
- Some people listen to their iPods for hours without interruption, because of the iPod's portability and its high-quality sound. According to federal government safety standards, workers should not be exposed to noise above 90 decibels for more than

eight hours. For every 5-decibel increase, the exposure time is cut in half. But iPod users listen at 130 decibels for hours.

- Hearing loss comes on very slowly. If it is due to loud music, the loss starts in the high frequencies, above the pitch of most conversations, and people don't notice. Instead, they continue to

abuse their ears with overly loud music until it's too late.

See the *Rolling Stone* article on page 800, Figure 10.25, and Exercise 33.

Loud music isn't the only cause of harmful sounds. The U.S. Navy's sonar system is harming marine mammals. See the articles on these topics on page 804 and Exercises 35–36.

William Thomas Cain/Getty Images

Decibels	Example of Sound at This Level	OSHA Maximum Exposure Time
0 dB	Threshold of hearing	
10 dB	Soundproof room	
20 dB	Radio, television or recording studio	
30–40 dB	Quiet room	Safe listening zone
50 dB	Private office	
60 dB	Normal conversation	
70 dB	Street noise	
80 dB	Telephone dial tone	
90 dB	Train whistle at 500 ft	
95 dB	Subway train at 200 ft	4 hours
100 dB	Chain saw	2 hours
105 dB	Power mower	1 hour
110 dB	Dance clubs	30 minutes
120 dB	Band practice	7.5 minutes
130 dB	Threshold of pain	Less than 2 minutes
140 dB	Jet engine at 100 ft	

FIGURE 10.25   When it's safe to listen.

10.3 EXERCISES

In Exercises 1–4, find the magnitude of the given earthquake.

1. An earthquake was measured by a seismograph 100 kilometers from the epicenter that recorded a maximum amplitude of 3.9×10^4 μm.

2. An earthquake was measured by a seismograph 60 kilometers from the epicenter that recorded a maximum amplitude of 6.3×10^5 μm.

3. An earthquake was measured by three seismographs. The closest, 60 kilometers from the epicenter, recorded a maximum amplitude of 250 μm. A second seismograph 220 kilometers from the epicenter recorded a maximum amplitude of 40 μm. The third seismograph, 460 kilometers from the epicenter, recorded a maximum amplitude of 4 μm.

4. An earthquake was measured by three seismographs. The closest, 20 kilometers from the epicenter, recorded a maximum amplitude of 25 μm. A second seismograph 60 kilometers from the epicenter recorded a maximum amplitude of 2 μm. The third seismograph, 180 kilometers from the epicenter, recorded a maximum amplitude of 0.50 μm.

In Exercises 5–12, use the information in Figure 10.24 (page 794) to compare the two earthquakes given by (a) finding how much more earth movement was caused by the larger and (b) finding how much more energy was released by the larger.

5. San Francisco (1989) and San Francisco (1906)
6. New Madrid and Los Angeles
7. San Francisco (1906) and Los Angeles
8. San Francisco (1989) and Los Angeles
9. Indian Ocean and San Francisco (1989)
10. Turkey and Taiwan
11. New Madrid and Coalinga
12. Los Angeles and Indian Ocean

13. Shortly after the 1989 San Francisco quake, it was announced that the quake was of magnitude 7.0. Later, after the data from more seismographs had been analyzed, the rating was increased to 7.1.

    a. How large an increase of earth movement corresponds to this increase in magnitude?

    b. How large an increase of energy released corresponds to this increase in magnitude?

14. Find a rule of thumb for the increase in earth movement that corresponds to an increase in Richter magnitude of one unit.

    HINT: Follow Example 2, using magnitudes $x + 1$ and x.

15. Find a rule of thumb for the increase in energy released that corresponds to an increase in Richter magnitude of one unit.

HINT: Follow Example 3, using magnitudes $x + 1$ and x.

In Exercises 16–20, find the decibel rating of the given sound.

16. a very faint whisper at 10^{-14} watts/cm^2
17. a television at 10^{-9} watts/cm^2
18. a quiet residence at 3.2×10^{-12} watts/cm^2
19. a vacuum cleaner at 2.5×10^{-9} watts/cm^2
20. an outboard motor at 1.6×10^{-6} watts/cm^2

In Exercises 21–24, find the dB gain for the given sound.

21. a whisper increasing from 10^{-14} watts/cm^2 to 10^{-13} watts/cm^2

22. noise in a dormitory increasing from 3.2×10^{-12} watts/cm^2 to 2.1×10^{-11} watts/cm^2

23. a stereo increasing from 3.9×10^{-9} watts/cm^2 to 3.2×10^{-8} watts/cm^2

24. a motorcycle increasing from 6.3×10^{-8} watts/cm^2 to 3.1×10^{-6} watts/cm^2

25. If a single singer is singing at 74 dB, how many singers have joined him if the level increases to 81 dB and each singer is equally loud?

26. If a single singer is singing at 74 dB, how many singers have joined him if the level increases to 83 dB and each singer is equally loud?

27. If a single singer is singing at 74 dB, how many singers have joined her if the level increases to 77 dB and each singer is equally loud?

28. If a single singer is singing at 74 dB, how many singers have joined her if the level increases to 82.5 dB and each singer is equally loud?

29. If a single trumpet is playing at 78 dB, how many trumpets have joined in if the level increases to 85.8 dB and each trumpet is equally loud?

30. If a single trumpet is playing at 78 dB, how many trumpets have joined in if the level increases to 88 dB and each trumpet is equally loud?

31. Find a rule of thumb for the dB gain if the number of sound sources doubles (where each source produces sounds at the same level).

32. Find a rule of thumb for the dB gain if the number of sound sources increases tenfold (where each source produces sounds at the same level).

33. According to the article from *Rolling Stone* on page 801, an American iPod can be turned up louder than a European iPod.

    a. How much more impact does an American iPod have on the eardrum than does a European iPod if they are both turned up all the way?

FEATURED IN THE NEWS

VENICE BALKS AT CONCERT BY FLOATING PINK FLOYD

VENICE—The British rock band Pink Floyd arrived yesterday to prepare for a weekend world television concert as promoters and city leaders argued over whether the group's decibels would damage ancient monuments.

Pink Floyd originally was scheduled to perform tomorrow on a giant floating platform moored alongside the lagoon city's world famous St. Mark's Square.

But the city's superintendent for architectural and environmental property vetoed the idea on grounds that vibrations from the group's amplified music could damage the structures of such historic buildings as St. Mark's Basilica and the Ducal Palace.

Yesterday morning, the organizers and city officials reached agreement that the concert could go ahead provided the floating platform is moored far enough out in the Venice lagoon to restrict the sound hitting the famed buildings to 60 decibels.

Scientists calculated that could be achieved if the huge platform—which measures 295 feet by 88 feet—is anchored 150 yards away from the St. Mark's Square shore.

UNITED PRESS INTERNATIONAL
Reprinted with permission.

FEATURED IN THE NEWS

PINK FLOYD'S VENICE CONCERT DRAWS 200,000

VENICE—Fans of the rock group Pink Floyd, including Woody Allen and Tom Cruise, more than doubled the population of this canal city to see the band perform on a floating stage near St. Mark's Square.

100 MILLION LISTENERS

It was broadcast live in Italy and 23 other countries, reaching a total audience of 100 million people, according to the state-run RAI network. It was not broadcast in the United States.

The concert set an attendance record as Italy's largest rock event and attracted more visitors to the historic city than did the annual Carnival festival.

City preservation officials had threatened to deny permission for the concert. But on Thursday they dropped their objections when concert organizers agreed to measures to reduce potential damage to the surrounding buildings.

A QUIETER SOUND

The stage was pulled farther from the shore than previously planned and the rock group reduced its volume to a maximum of 60 decibels, instead of 105 as planned.

Italy's vice premier, Gianni De Michelis, spoke out in favor of the concert.

"Venice must give space to all cultural languages, including that of rock music," he said.

ASSOCIATED PRESS
Reprinted with permission.

---

b. Use the information in the box on page 801 and Figure 10.25 to determine the maximum exposure time for an American iPod and for a European iPod if they are both turned up all the way.

34. According to the above articles, the rock group Pink Floyd reduced its volume to protect Venice's buildings.

   a. How much effect would the stated decrease in volume have?

   b. Use the information in Figure 10.25 to determine the maximum exposure time for the concert at both the proposed and actual decibel levels.

35. According to the article on the next page "Suit Filed to Halt Navy Sonar System" from the *San Francisco Chronicle*, the Navy will restrict its sonar operators to a lower sound level to protect marine mammals. How much effect would the stated decrease in volume have?

36. According to the article on the next page "Whales at Risk" from *National Geographic News*, mid-frequency sonar can emit continuous sound at 235 dB or higher. How many times as intense is this sound than an American iPod turned up all the way to 130 dB?

FEATURED IN
THE NEWS

SUIT FILED TO HALT NAVY SONAR SYSTEM

Environmental groups sued . . . to stop the U.S. Navy from deploying a new anti-submarine sonar system that they say could harm sea creatures wherever it's used in the world's oceans. The federal lawsuit seeks an injunction against using the low-frequency active sonar system, which critics fear will threaten entire populations of whales,

dolphins, seals and other marine mammals. . . . The Navy says active sonar arrays are capable of generating 215 decibels of sound—dangerous for marine mammals—but that operators will be restricted to 180 decibels, a sound level emitted by some whales. . . . Naomi Rose, marine mammal scientist for the New York-based Humane Society of the United States, . . . called a safety exposure level of 180 decibels

"completely unacceptable and not substantiated by data." . . . "It's not precautionary, as the Marine Mammal Protection Act demands," Rose said, "and is out of step with the standards that other countries are setting."

From Jane Kay, "Environmental groups say anti-sub device harms whales and other ocean mammals," *San Francisco Chronicle*, August 8, 2002. Reprinted by permission.

FEATURED IN
THE NEWS

WHALES AT RISK FROM NEW U.S. NAVY SONAR RANGE, ACTIVISTS SAY

The U.S. Navy is moving ahead with plans to build an undersea warfare training range on the U.S. East Coast despite fierce opposition from conservation and animal welfare organizations. Groups opposed to the military project say endangered North Atlantic right whales, dolphins, and sea turtles could potentially be injured or killed from powerful sonar blasts emitted during training exercises. "Protecting whales and preserving national security are not mutually exclusive," said Fred O'Regan, president of the International Fund for Animal Welfare.

Mid-frequency sonar used during training exercises can emit continuous sound well above 235 decibels—an intensity roughly comparable to a rocket

blastoff, according to the Natural Resources Defense Council (NRDC), a conservation nonprofit group. Experts say that's a problem for marine mammals and other aquatic animals. Sound is their primary means of learning about their environment, communicating, and navigating. "Military sonar needlessly threatens whole populations of whales and other marine animals," said Joel Reynolds, an attorney for NRDC. "The Navy refuses to take basic precautions that could spare these majestic creatures." The Washington, D.C.-based nonprofit, along with several animal welfare and environmental groups, sued the Navy last week, citing harm to whales caused by mid-frequency sonar.

The technology is associated with strandings—when marine animals swim or float onto shore and become

beached or stuck in shallow water. In 2000 17 whales from three species beached themselves in the Bahamas after Navy ships conducted mid-range sonar exercises. Of those stranded, seven animals died. A federal investigation concluded sonar was the most likely cause of the whales' stranding. Adding to the controversy, the military's preferred training site in North Carolina is near an area where a mass whale stranding took place in January, not long after the Navy conducted sonar exercises in the region.

Maryann Mott, "Whales at Risk from New U.S. Navy Sonar Range, Activists Say," National Geographic News, Nov. 3, 2005. Reprinted by permission.

 Answer the following questions using complete sentences and your own words.

• CONCEPT QUESTION

37. Compare and contrast the Richter scale and the decibel scale. Why are the two formulas so similar? Why does each formula involve logarithms?

• HISTORY QUESTIONS

38. Name three of Bell's other inventions besides the telephone.

39. The inventor of the Richter scale performed his research at what school?

40. In addition to electronics, what field of study interested Alexander Graham Bell?

TERMS

amplitude of a
  seismogram
average decay rate
average growth
  rate
base of a logarithmic
  function
base of an exponential
  function

common logarithm
dB gain
decibel scale
delta notation
dependent variable
epicenter
exponential function
exponential model
  function
half-life

independent
  variable
intensity of a sound
  (versus its perceived
  loudness)
irrational number
logarithm
logarithmic equation
logarithmic scale
mathematical
  model

natural exponential
  function
natural logarithm
radioactive decay
rate of change
rational number
relative decay rate
relative growth rate
seismogram
seismograph
tectonic stress

PROCEDURES

Solving exponential equations

1. Isolate the exponential.
2. Take the natural logarithm of each side.
3. Use an Inverse Property to simplify.
4. Solve.

Solving logarithmic equations

1. Get all the log terms on one side and all the nonlog terms on the other.
2. Combine the log terms into one terms using the Division-Becomes-Subtraction Property and the Multiplication-Becomes-Addition Property.
3. Exponentiate each side.
4. Use an Inverse Property to simplify.
5. Solve.

Developing an exponential model

1. Write the information given as to the value of the quantity at two points in time as two ordered pairs (t, y).
2. Substitute the first ordered pair into the model $y = ae^{bt}$ and simplify to find a.
3. Substitute the second ordered pair into the model and simplify to find b.

APPLICATION DEFINITIONS AND FORMULAS

Richter's definition of earthquake magnitude:
$M = \log A - \log A_0$ (get $\log A_0$ from the table on page 793)

Magnitude comparison formula: $M_1 - M_2 = \log\left(\dfrac{A_1}{A_2}\right)$

Energy formula: $\log E \approx 11.8 + 1.45M$

Decibel rating definition: $D = 10 \log\left(\dfrac{I}{I_0}\right)$

dB gain formula: $D_1 - D_2 = 10 \log\left(\dfrac{I_1}{I_2}\right)$

LOGARITHM DEFINITIONS AND FORMULAS

For any base b, $\log_b u = v$ means the same as $b^v = u$. For base 10 and base e:

$y = \log x$ means the same as $10^y = x$

$y = \ln x$ means the same as $e^y = x$

Division-Becomes-Subtraction Properties

$$\log \frac{A}{B} = \log A - \log B \qquad \ln \frac{A}{B} = \ln A - \ln B$$

Multiplication-Becomes-Addition Property

$$\log(A \cdot B) = \log A + \log B \qquad \ln(A \cdot B) = \ln A + \ln B$$

Exponent-Becomes-Multiplier Properties

$$\log(A^n) = n \cdot \log A \qquad \ln(A^n) = n \cdot \ln A$$

Inverse Properties

$$\log(10^x) = x \qquad \ln(e^x) = x$$
$$10^{\log x} = x \qquad e^{\ln x} = x$$

805

REVIEW EXERCISES

In Exercises 1–3, find the value of x.

1. $x = \log_3 81$
2. $2 = \log_x 16$
3. $5 = \log_5 x$
4. State the Inverse Properties of the common logarithm.
5. State the Inverse Properties of the natural logarithm.
6. State the Division-Becomes-Subtraction Property of the common logarithm.
7. State the Division-Becomes-Subtraction Property of the natural logarithm.
8. State the Exponent-Becomes-Multiplier Property of the common logarithm.
9. State the Exponent-Becomes-Multiplier Property of the natural logarithm.
10. State the Multiplication-Becomes-Addition Property of the common logarithm.
11. State the Multiplication-Becomes-Addition Property of the natural logarithm.

In Exercises 12–14, rewrite each of the expressions (if possible) using properties of logarithms.

12. $\log 3x - \log 4 + \log 7x^2$
13. $\log(x + 2)$
14. $\log(\frac{x}{2})$
15. Solve $520\,e^{0.03x} = 730$.
16. Solve $\ln x - \ln 4 = \ln 2$.
17. Solve $\log 5x + \log x^2 - \log x = 12$.
18. A bacteria culture had a population of about 10,000 at 12 noon. At 2 P.M., the population had grown to 25,000.
    a. Develop the mathematical model that represents the population of the bacteria culture.
    b. Use the model to predict the population at 6 A.M. the next day.
    c. Use the model to predict when the population will double its original size.
    d. On what assumptions is the model based?
19. Cobalt-60 is used by hospitals in radiation therapy. Suppose a hospital purchases 300 grams.
    a. Develop the mathematical model that represents the amount of cobalt-60 at time t.
    b. Predict the amount *lost* if delivery takes two weeks.
    c. Predict the amount *remaining* after one year (if none is used).
    d. Determine how long it will take for 90% of the cobalt-60 to decay.
20. Archeologists found a wooden bowl in an Indian burial mound. The bowl contained 73% of the amount of

carbon-14 found in living matter. Estimate the age of the burial mound.

21. An earthquake was measured by two seismographs. The closest, 20 kilometers from the epicenter, recorded a maximum amplitude of 25 μm. The second seismograph, 60 km from the epicenter, recorded a maximum amplitude of 2 μm. Find the magnitude of the earthquake.

22. A 1989 earthquake in Newcastle, Australia measured 5.5 on the Richter scale. The 1989 San Francisco earthquake measured 7.1.
    a. How much more earth movement was caused by the San Francisco quake?
    b. How much more energy was released by the San Francisco quake?

23. Find the decibel rating of a scooter at 1.6×10^{-6} watts/cm^2.

24. The scooter in Exercise 23 increases to 1.3×10^{-5} watts/cm^2. Find the dB gain.

25. If a single trumpet is playing at 78 dB, how many trumpets have joined in if the level increases to 84 dB?

 Answer the following questions using complete sentences and your own words.

● CONCEPT QUESTIONS

26. What is an exponential function?
27. What is a logarithmic function?
28. In terms of mathematical models, what is the basic difference between exponential growth and exponential decay?
29. How does radiocarbon dating work? What aspect of radiocarbon dating is controversial?

● HISTORY QUESTIONS

30. What underlying theme unites all of Alexander Graham Bell's professional interests?
31. Name three of Bell's other inventions besides the telephone.
32. Who introduced the symbol e? What other symbols did this person invent?
33. Who invented logarithms? What motivated their invention?
34. Who invented radiocarbon dating?
35. Who was the first person to receive two Nobel Prizes? For what was the first prize awarded?

MATRICES AND MARKOV CHAINS

11

Andy Warhol's *100 Cans* illustrates America's fascination with mass-produced products.

© The Andy Warhol Foundation for the Visual Arts/ARS, NY.

Markov chains use probabilities to analyze trends and to predict the outcomes of those trends. Markov chains have many applications in business, sociology, the physical sciences, and biology.

IN BUSINESS:

* "Retro Coke" uses sugar rather than high-fructose corn syrup. It is designed to taste more like the original Coca-Cola and to appeal to people who are concerned about what they eat. It was test-marketed in parts of the United States during the summer of 2009.

* Test-marketing helps manufacturers determine whether they should mass-market a new product and, if so, how to tweak the product or its advertising.

* Businesses such as Coca-Cola use Markov chains to analyze test-marketing results.

continued

WHAT WE WILL DO IN THIS CHAPTER

* We will discuss how businesses use Markov checking to analyze test-marketing results.

* We will learn how businesses use Markov chains to predict short-term and long-term results of competitive changes.

* We will investigate how sociologists use Markov chains to predict the future results of ongoing changes, such as urban flight and the shrinking of the middle class.

- Sales of the Honda Fit were greatly reduced when the retooled Ford Focus was released in 2010. An already established product's success can be affected by a new or redesigned competing product. Businesses such as Ford and Honda use Markov chains to predict short-term and long-term results of competitive changes.

IN SOCIOLOGY:

- The American middle class has been shrinking. Our society has larger proportions of poor people and of wealthy people than it used to.
- "Urban flight" is a reality. People have been moving out of the cities into the suburbs, and the suburbs are encroaching on rural and agricultural areas.
- Markov chains are used to analyze trends such as the shrinking of the American middle class and the growth of the suburbs and to predict the eventual outcome of those trends.

11.0 Review of Matrices

OBJECTIVES

- Become familiar with matrix terminology
- Learn how to multiply matrices
- Be able to use the associative property and the identity property when multiplying matrices

Markov chain calculations use matrices, so we will start by reviewing matrices.

Terminology and Notation

A **matrix** (plural: *matrices*) is a rectangular arrangement of numbers enclosed by brackets. For example,

$$\begin{bmatrix} 3 & -2 & 0 \\ 1 & -27 & 5 \end{bmatrix}$$

is a matrix. Each number in the matrix is called an **element** or an **entry** of the matrix; 3 is an element of the above matrix, as is -27. The horizontal listings of elements are called **rows;** the vertical listings are called **columns.** The first row of the matrix

$$\begin{bmatrix} 3 & -2 & 0 \\ 1 & -27 & 5 \end{bmatrix} \leftarrow \textbf{first row}$$

$$\uparrow$$
$$\textbf{last column}$$

is "3 -2 0," and the last column is "0 5." The **dimensions of a matrix** with m rows and n columns are $m \times n$ (read "m by n"). The matrix above has two rows and three columns, so its dimensions are 2×3; that is, it is a 2×3 matrix.

The dimensions of a matrix are like the dimensions of a room. If a room is 7 feet on one side and 10 feet on the other, then the dimensions of that room are 7×10. We do not actually multiply 7 times 10 to determine the dimensions (although we would to find the area). Likewise, the dimensions of the above matrix are 2×3; we do not actually multiply 2 times 3 and get 6. If we did, we would have the *number of elements* in the matrix rather than the *dimensions* of the matrix.

A matrix with only one row is called a **row matrix.** A matrix with only one column is called a **column matrix.** A matrix is **square** if it has as many rows as columns. For example,

$[3 \quad -2 \quad 0]$ is a row matrix.

$\begin{bmatrix} 26 \\ 89 \end{bmatrix}$ is a column matrix.

$\begin{bmatrix} 0 & -1 \\ 5 & 3 \end{bmatrix}$ is a square matrix.

$\begin{bmatrix} 3 & -4 & 5 \\ 0 & 18 & 2 \end{bmatrix}$ is neither a row matrix, a column matrix, nor a square matrix.

A matrix is usually labeled with a capital letter, and the entries of that matrix are labeled with the same letter in lowercase with a double subscript. The first subscript refers to the entry's row; the second subscript refers to its column. For example, b_{23} refers to the element of matrix B that is in the second row and the third column. Similarly, if

$$A = \begin{bmatrix} 5 & 7 \\ 9 & 11 \end{bmatrix}$$

then $a_{11} = 5$, $a_{12} = 7$, $a_{21} = 9$, and $a_{22} = 11$.

Matrix Multiplication

In your intermediate algebra class, you might have learned to add, subtract, and multiply matrices and to take the determinant of a matrix. In this chapter, we will be studying applications that involve the multiplication of matrices but not their addition or subtraction or the taking of determinants. Thus, we limit our discussion to matrix multiplication.

Matrix multiplication is a process that at first seems rather strange and obscure. Its usefulness will become apparent when we apply it to probability trees in the method of Markov chains later in this chapter.

EXAMPLE 1

MULTIPLYING TWO MATRICES   If $A = [3 \quad -2]$ and $B = \begin{bmatrix} 1 \\ 0 \end{bmatrix}$, find the product AB.

SOLUTION

$$AB = [3 \quad -2]\begin{bmatrix} 1 \\ 0 \end{bmatrix} = ?$$

To find AB,

- multiply the first element of A by the first element of B,
- multiply the second element of A by the second element of B,
- add the results:

$$AB = [3 \quad -2]\begin{bmatrix} 1 \\ 0 \end{bmatrix} = [3 \cdot 1 \ + \ -2 \cdot 0] = [3]$$

EXAMPLE **2**

MULTIPLYING TWO BIGGER MATRICES If $A = \begin{bmatrix} 3 & -2 \\ 4 & 1 \end{bmatrix}$ and $B = \begin{bmatrix} 1 & 5 \\ 0 & -7 \end{bmatrix}$, find the product AB.

SOLUTION

$$AB = \begin{bmatrix} 3 & -2 \\ 4 & 1 \end{bmatrix}\begin{bmatrix} 1 & 5 \\ 0 & -7 \end{bmatrix} = ?$$

Start with row 1 of A and column 1 of B. Multiply corresponding elements together, add the products, and put the sum in row 1, column 1 of the product matrix, as we did in Example 1.

$$AB = \begin{bmatrix} 3 & -2 \\ 4 & 1 \end{bmatrix}\begin{bmatrix} 1 & 5 \\ 0 & -7 \end{bmatrix} = \begin{bmatrix} 3 \cdot 1 + -2 \cdot 0 & ? \\ ? & ? \end{bmatrix} = \begin{bmatrix} 3 & ? \\ ? & ? \end{bmatrix}$$

Now multiply row 1 of A and column 2 of B, and put the result in row 1, column 2 of the product matrix:

$$AB = \begin{bmatrix} 3 & -2 \\ 4 & 1 \end{bmatrix}\begin{bmatrix} 1 & 5 \\ 0 & -7 \end{bmatrix} = \begin{bmatrix} 3 & 3 \cdot 5 + -2 \cdot -7 \\ ? & ? \end{bmatrix} = \begin{bmatrix} 3 & 29 \\ ? & ? \end{bmatrix}$$

We're finished with row 1; what remains is to multiply row 2 of A first by column 1 of B and then by column 2. Multiplying row 2 of A by column 1 of B gives

$$AB = \begin{bmatrix} 3 & -2 \\ 4 & 1 \end{bmatrix}\begin{bmatrix} 1 & 5 \\ 0 & -7 \end{bmatrix} = \begin{bmatrix} 3 & 29 \\ 4 \cdot 1 + 1 \cdot 0 & ? \end{bmatrix} = \begin{bmatrix} 3 & 29 \\ 4 & ? \end{bmatrix}$$

Finally, multiply row 2 of A by column 2 of B and put the result in row 2, column 2 of the product matrix:

$$AB = \begin{bmatrix} 3 & -2 \\ 4 & 1 \end{bmatrix}\begin{bmatrix} 1 & 5 \\ 0 & -7 \end{bmatrix} = \begin{bmatrix} 3 & 29 \\ 4 & 4 \cdot 5 + 1 \cdot -7 \end{bmatrix} = \begin{bmatrix} 3 & 29 \\ 4 & 13 \end{bmatrix}$$

The product of A and B is

$$AB = \begin{bmatrix} 3 & 29 \\ 4 & 13 \end{bmatrix}$$

EXAMPLE **3**

USING MATRIX MULTIPLICATION TO COMPUTE TOTAL PRICE A matrix is really just a table. Consider the table in Figure 11.1, which gives prices of CDs and DVDs.

    If we were to purchase three CDs and two DVDs, we would calculate the sale price as

$$3 \cdot \$14 + 2 \cdot \$20 = \$82$$

and the regular price as

$$3 \cdot \$19 + 2 \cdot \$25 = \$107$$

	Sale Price	Regular Price
CDs	$14	$19
DVDs	$20	$25

FIGURE 11.1 Disc prices.

This calculation is matrix multiplication. It's the same as

$$[3 \quad 2]\begin{bmatrix} 14 & 19 \\ 20 & 25 \end{bmatrix} = [3 \cdot 14 + 2 \cdot 20 \quad 3 \cdot 19 + 2 \cdot 25]$$

$$= [82 \quad 107]$$

It's not always possible to multiply two matrices together. If the first matrix in Example 3 had one extra element, we wouldn't be able to multiply.

$$[3 \quad 2 \quad 77]\begin{bmatrix} 14 & 19 \\ 20 & 25 \end{bmatrix} = [3 \cdot 14 + 2 \cdot 20 + 77 \cdot ? \quad 3 \cdot 19 + 2 \cdot 25 + 77 \cdot ?]$$

It's easy to tell when two matrices cannot be multiplied: If you run out of elements to pair together when you're multiplying the first row by the first column, then the two matrices cannot be multiplied. In the above problem, we paired 3 and 14, and 2 and 20, but we had nothing to pair with 77. This tells us that the two matrices cannot be multiplied together.

Some people prefer to determine whether it's possible to multiply two matrices before they actually start multiplying. In the above problem, we could not multiply a 1×3 matrix by a 2×2 matrix.

$$[3 \quad 2 \quad 77]\begin{bmatrix} 14 & 19 \\ 20 & 25 \end{bmatrix}$$

$$1 \times 3 \qquad 2 \times 2$$

$$\uparrow \qquad \uparrow$$

different, so we can't multiply.

Previously, when 77 wasn't there, we *were* able to multiply.

$$[3 \quad 2]\begin{bmatrix} 14 & 19 \\ 20 & 25 \end{bmatrix}$$

$$1 \times 2 \qquad 2 \times 2$$

$$\uparrow \qquad \uparrow$$

the same, so we can multiply.

EXAMPLE **4**

NOT ALL MATRICES CAN BE MULTIPLIED $A = \begin{bmatrix} 3 & -2 \\ 4 & 1 \end{bmatrix}$, $B = [1 \quad 0]$, and $C = \begin{bmatrix} 1 \\ 0 \end{bmatrix}$.

a. Find AB, if it exists.
b. Find AC, if it exists.

SOLUTION

a. $AB = \begin{bmatrix} 3 & -2 \\ 4 & 1 \end{bmatrix}[1 \quad 0] = \begin{bmatrix} 3 \cdot 1 + -2 \cdot ? & 3 \cdot 0 + -2 \cdot ? \\ 4 \cdot 1 + 1 \cdot ? & 4 \cdot 0 + 1 \cdot ? \end{bmatrix}$

The product AB does not exist, because we ran out of elements to pair together. Alternatively, A is a 2×2 matrix, and B is a 1×2 matrix.

$$\begin{bmatrix} 3 & -2 \\ 4 & 1 \end{bmatrix}[1 \quad 0]$$

$$2 \times 2 \qquad 1 \times 2$$

$$\uparrow \qquad \uparrow$$

different, so we can't multiply.

b. $AC = \begin{bmatrix} 3 & -2 \\ 4 & 1 \end{bmatrix}\begin{bmatrix} 1 \\ 0 \end{bmatrix} = \begin{bmatrix} 3 \cdot 1 + -2 \cdot 0 \\ 4 \cdot 1 + 1 \cdot 0 \end{bmatrix} = \begin{bmatrix} 3 \\ 4 \end{bmatrix}$

The product of A and C is

$$\begin{bmatrix} 3 \\ 4 \end{bmatrix}$$

If we had checked the dimensions in advance, we would have found that A and C can be multiplied:

$$\begin{bmatrix} 3 & -2 \\ 4 & 1 \end{bmatrix} \begin{bmatrix} 1 \\ 0 \end{bmatrix}$$
$$\qquad 2 \times 2 \qquad 2 \times 1$$
$$\qquad\quad \uparrow \qquad\quad \uparrow$$

the same, so we can multiply.

This dimension check also gives the dimensions of the product matrix. The matching twos "cancel" and leave a 2×1 matrix.

MATRIX MULTIPLICATION

If matrix A is an $m \times n$ matrix and matrix B is an $n \times p$ matrix, then the product AB exists and is an $m \times p$ matrix.

(*Think: $m \times n \cdot n \times p$ yields $m \times p$.*)

To find the entry in the ith row and the jth column of the product matrix AB, multiply each element of A's ith row by the corresponding element of B's jth column and add the results.

Properties of Matrix Multiplication

When you multiply numbers together, you use certain properties so automatically that you don't even realize you are using them. In particular, multiplication of real numbers is *commutative*; that is, $ab = ba$ (the order in which you multiply doesn't matter). Multiplication of real numbers is also *associative*; that is, $a(bc) = (ab)c$ (if you multiply three numbers together, it doesn't matter which two you multiply first).

Matrix multiplication is certainly different from real number multiplication. It's important to know whether matrix multiplication is commutative and associative, because people tend to use those properties without being aware of doing so.

EXAMPLE 5

IS MATRIX MULTIPLICATION COMMUTATIVE? Determine whether matrix multiplication is commutative by finding the product BA for the matrices B and A given in Example 2 and comparing BA with AB.

SOLUTION

$$BA = \begin{bmatrix} 1 & 5 \\ 0 & -7 \end{bmatrix} \begin{bmatrix} 3 & -2 \\ 4 & 1 \end{bmatrix} = \begin{bmatrix} 1 \cdot 3 + 5 \cdot 4 & 1 \cdot -2 + 5 \cdot 1 \\ 0 \cdot 3 + -7 \cdot 4 & 0 \cdot -2 + -7 \cdot 1 \end{bmatrix}$$

$$= \begin{bmatrix} 23 & 3 \\ -28 & -7 \end{bmatrix}$$

The product of B and A is $BA = \begin{bmatrix} 23 & 3 \\ -28 & -7 \end{bmatrix}$.

In Example 2, we found that $AB = \begin{bmatrix} 3 & 29 \\ 4 & 13 \end{bmatrix}$.

Clearly, $AB \neq BA$.

Matrix multiplication is not commutative; in general, $AB \neq BA$. This means that you have to be careful to multiply in the right order—it's easy to be careless and find BA when you're asked to find AB.

EXAMPLE 6

IS MATRIX MULTIPLICATION ASSOCIATIVE?   Check the associative property for matrix multiplication by finding $A(BC)$ and $(AB)C$ for matrices A, B, and C given below.

$$A = \begin{bmatrix} 2 & -3 \end{bmatrix} \quad B = \begin{bmatrix} 4 & 0 & -3 \\ 2 & -1 & 5 \end{bmatrix} \quad C = \begin{bmatrix} -3 & 5 \\ 2 & 0 \\ 1 & -1 \end{bmatrix}$$

SOLUTION

To find $A(BC)$, we first find BC and then multiply it by A. Notice that B's dimensions are 2×3 and C's dimensions are 3×2. This tells us that B and C can be multiplied and that the product will be a 2×2 matrix.

$$BC = \begin{bmatrix} 4 & 0 & -3 \\ 2 & -1 & 5 \end{bmatrix} \begin{bmatrix} -3 & 5 \\ 2 & 0 \\ 1 & -1 \end{bmatrix}$$

$$= \begin{bmatrix} 4 \cdot -3 + 0 \cdot 2 + -3 \cdot 1 & 4 \cdot 5 + 0 \cdot 0 + -3 \cdot -1 \\ 2 \cdot -3 + -1 \cdot 2 + 5 \cdot 1 & 2 \cdot 5 + -1 \cdot 0 + 5 \cdot -1 \end{bmatrix}$$

$$= \begin{bmatrix} -15 & 23 \\ -3 & 5 \end{bmatrix}$$

To find $A(BC)$, we place A on the *left* of BC and multiply. Notice that A's dimensions are 1×2 and BC's dimensions are 2×2, so A and BC can be multiplied, and the product will be a 1×2 matrix.

$$A(BC) = \begin{bmatrix} 2 & -3 \end{bmatrix} \begin{bmatrix} -15 & 23 \\ -3 & 5 \end{bmatrix}$$

$$= \begin{bmatrix} 2 \cdot -15 + -3 \cdot -3 & 2 \cdot 23 + -3 \cdot 5 \end{bmatrix}$$

$$= \begin{bmatrix} -21 & 31 \end{bmatrix}$$

Now that we've found $A(BC)$, we need to find $(AB)C$ and then see whether the products are the same. To find $(AB)C$, we first find AB and then multiply it by C. Notice that A's dimensions are 1×2 and B's dimensions are 2×3, so A and B can be multiplied.

$$AB = \begin{bmatrix} 2 & -3 \end{bmatrix} \begin{bmatrix} 4 & 0 & -3 \\ 2 & -1 & 5 \end{bmatrix}$$

$$= \begin{bmatrix} 2 \cdot 4 + -3 \cdot 2 & 2 \cdot 0 + -3 \cdot -1 & 2 \cdot -3 + -3 \cdot 5 \end{bmatrix}$$

$$= \begin{bmatrix} 2 & 3 & -21 \end{bmatrix}$$

To find $(AB)C$, we place AB on the *left* of C and multiply.

$$(AB)C = \begin{bmatrix} 2 & 3 & -21 \end{bmatrix} \begin{bmatrix} -3 & 5 \\ 2 & 0 \\ 1 & -1 \end{bmatrix}$$

$$= \begin{bmatrix} 2 \cdot -3 + 3 \cdot 2 + -21 \cdot 1 & 2 \cdot 5 + 3 \cdot 0 + -21 \cdot -1 \end{bmatrix}$$

$$= \begin{bmatrix} -21 & 31 \end{bmatrix}$$

Therefore, $A(BC) = (AB)C$.

In Example 5, $A(BC) = (AB)C$, even though the work involved in finding $A(BC)$ was different from the work involved in finding $(AB)C$. This always happens;

HISTORICAL NOTE

ARTHUR CAYLEY, 1821–1895, & JAMES JOSEPH SYLVESTER, 1814–1897

Arthur Cayley

The Granger Collection, New York

James Sylvester

The Granger Collection, New York

The theory of matrices was a product of the unique partnership of Arthur Cayley and James Sylvester. They met in their twenties and were friends, colleagues, and coauthors for the rest of their lives.

Cayley's mathematical ability was recognized at an early age, and he was encouraged to study the subject. He graduated from Cambridge University at the top of his class. After graduation, Cayley was awarded a three-year fellowship that allowed him to do as he pleased. During this time, he made several trips to Europe, where he spent his time taking walking tours, mountaineering, painting, reading novels, and studying architecture, as well as reading and writing mathematics. He wrote twenty-five papers in mathematics, papers that were well received by the mathematical community.

When Cayley's fellowship expired, he found that no position as a mathematician was open to him unless he entered the clergy, so he left mathematics and prepared for a legal career. When he was admitted to the bar, he met James Joseph Sylvester.

Sylvester's mathematical ability had also been recognized at an early age. He studied mathematics at the University of London at the age of fourteen, under Augustus De Morgan (see Chapter 2 for more information on this famous mathematician). He entered Cambridge University at the age of seventeen and won several prizes. However, Cambridge would not award Sylvester his degrees because he was Jewish. He completed his bachelor's and master's degrees at Trinity College in Dublin. Many years

later, Cambridge changed its discriminatory policy and gave Sylvester his degrees.

Sylvester taught science at University College in London for two years, found that he didn't like teaching science, and quit. He went to the United States, where he got a job teaching mathematics at the University of Virginia. After three months, he quit when the administration refused to discipline a student who had insulted him. After several unsuccessful attempts to obtain a teaching position, he returned to England and worked for an insurance firm as an actuary, retaining his interest in mathematics only through tutoring. Florence Nightingale was one of his private pupils. When Sylvester became thoroughly bored with insurance work, he studied for a legal career and met Cayley.

Cayley and Sylvester revived and intensified each other's interest in mathematics, and each started to write mathematics again. During his fourteen years spent practicing law, Cayley wrote almost 300 papers. Cayley and Sylvester frequently expressed gratitude to each other for assistance and inspiration. In one of his papers, Sylvester wrote that "the theorem above enunciated was in part suggested in the course of a conversation with Mr. Cayley (to whom I am indebted for my restoration to the enjoyment of mathematical life)." In another, he said, "Mr. Cayley habitually discourses pearls and rubies."

Cayley joyfully departed from the legal profession when Cambridge offered him a professorship in mathematics, even though his income suffered as a result. He was finally able to spend his life studying, teaching, and writing mathematics. He became quite famous as a mathematician, writing almost 1,000 papers in algebra and geometry, often in collaboration with Sylvester. Many of these papers are pioneering works of scholarship. Cayley also played an important role in changing Cambridge's policy that had prohibited the admission of women as students.

Sylvester was repeatedly honored for his pioneering work in algebra. He left the law but was unable to obtain a professorship in mathematics at a prominent institution until late in his life. At the age of sixty-two, he accepted a position at the newly founded Johns Hopkins University in Baltimore as its first professor of mathematics. While there, he founded the *American Journal of Mathematics*, introduced graduate work in mathematics into American universities, and generally stimulated the development of mathematics in America. He also arranged for Cayley to spend a semester at Johns Hopkins as guest lecturer. At the age of seventy, Sylvester returned to England to become Savilian Professor of Geometry at Oxford University.

Cayley and Sylvester were responsible for the theory of matrices, including the operation of matrix multiplication. Sixty-seven years after the invention of matrix theory, Heisenberg recognized it as the perfect tool for his revolutionary work in quantum mechanics. The work of Cayley and Sylvester in algebra became quite important for modern physics, particularly in the theory of relativity. Cayley also wrote on non-Euclidean geometry (see Chapter 8 for information on non-Euclidean geometry).

$A(BC) = (AB)C$, provided that the dimensions of A, B, and C are such that they can be multiplied together. *Matrix multiplication is associative.*

Identity Matrices

EXAMPLE **7**

A MATRIX THAT IS LIKE THE NUMBER 1 If $I = \begin{bmatrix} 1 & 0 \\ 0 & 1 \end{bmatrix}$ and $A = \begin{bmatrix} 2 & 3 \\ 4 & 5 \end{bmatrix}$, find IA.

SOLUTION

$$IA = \begin{bmatrix} 1 & 0 \\ 0 & 1 \end{bmatrix} \begin{bmatrix} 2 & 3 \\ 4 & 5 \end{bmatrix}$$
$$= \begin{bmatrix} 1 \cdot 2 + 0 \cdot 4 & 1 \cdot 3 + 0 \cdot 5 \\ 0 \cdot 2 + 1 \cdot 4 & 0 \cdot 3 + 1 \cdot 5 \end{bmatrix}$$
$$= \begin{bmatrix} 2 & 3 \\ 4 & 5 \end{bmatrix}$$

Notice that $IA = A$. If you were to multiply in the opposite order, you would find that $AI = A$. The matrix I is similar to the number 1, because $I \cdot A = A \cdot I = A$, just as $1 \cdot a = a \cdot 1 = a$.

The matrix $\begin{bmatrix} 1 & 0 \\ 0 & 1 \end{bmatrix}$ is called an *identity matrix*, because multiplying this matrix by any other 2×2 matrix A yields a product *identical* to A. The matrix $\begin{bmatrix} 1 & 0 & 0 \\ 0 & 1 & 0 \\ 0 & 0 & 1 \end{bmatrix}$ is also called an identity matrix, because multiplying this 3×3 matrix by any other 3×3 matrix A yields a product identical to A. An **identity matrix** is a square matrix I that has ones for each entry in the **diagonal** (the diagonal that starts in the upper left corner) and zeros for all other entries. The product of any matrix A and an identity matrix is always A, as long as the dimensions of the two matrices are such that they can be multiplied.

PROPERTIES OF MATRIX MULTIPLICATION

1. *There is no commutative property:* In general, $AB \neq BA$. You must be careful about the order in which you multiply.
2. **Associative property:** $A(BC) = (AB)C$, provided the dimensions of A, B, and C are such that they can be multiplied together.
3. **Identity property:** An **identity matrix** is a square matrix I that has ones for each entry in the diagonal (the diagonal that starts in the upper left corner) and zeros for all other entries. If I and A have the same dimensions, then $IA = AI = A$.

11.0 EXERCISES

In Exercises 1–10, (a) find the dimensions of the given matrix and (b) determine whether the matrix is a row matrix, a column matrix, a square matrix, or none of these.

1. $A = \begin{bmatrix} 5 & 0 \\ 22 & -3 \\ 18 & 9 \end{bmatrix}$

2. $B = \begin{bmatrix} 1 & 13 & 207 \\ -4 & 8 & 100 \\ 0 & 1 & 5 \end{bmatrix}$

3. $C = \begin{bmatrix} 23 \\ 41 \end{bmatrix}$

4. $D = \begin{bmatrix} 2 & 0 & 19 & -3 \\ 62 & 13 & 44 & 1 \\ 5 & 5 & 30 & 12 \\ 0 & 0 & 0 & 0 \end{bmatrix}$

5. $E = \begin{bmatrix} 3 & 0 \end{bmatrix}$

6. $F = \begin{bmatrix} -2 & 10 \\ 4 & -3 \end{bmatrix}$

7. $G = \begin{bmatrix} 12 & -11 & 5 \\ -9 & 4 & 0 \\ 1 & 9 & 5 \end{bmatrix}$

8. $H = \begin{bmatrix} 1 & 0 \\ 0 & -1 \end{bmatrix}$

9. $J = \begin{bmatrix} 5 \\ -3 \\ 11 \end{bmatrix}$

10. $K = \begin{bmatrix} 2 & -5 & 13 & 0 \\ -1 & 4 & 3 & 6 \\ 8 & -10 & 4 & 0 \end{bmatrix}$

In Exercises 11–20, find the indicated elements of the matrices given in Exercises 1–10.

11. a_{21}

(*HINT:* You're asked to find the entry in row 2, column 1 of the matrix A, which is given in Exercise 1.)

12. b_{23} **13.** c_{21} **14.** d_{34}
15. e_{11} **16.** f_{22} **17.** g_{12}
18. h_{21} **19.** j_{21} **20.** k_{23}

In Exercises 21–30, find the indicated products (if they exist) of the matrices given in Exercises 1–10.

21. a. AC **b.** CA **22. a.** AE **b.** EA
23. a. AD **b.** DA **24. a.** GH **b.** HG
25. a. CG **b.** GC **26. a.** DK **b.** KD
27. a. JB **b.** BJ **28. a.** HK **b.** KH
29. a. AF **b.** FA **30. a.** AB **b.** BA

31. Use matrix multiplication and Figure 11.2 to find the sale price and the regular price of purchasing five CDs and three DVDs.

	Sale Price	Regular Price
CDs	$14	$19
DVDs	$20	$25

FIGURE 11.2 Disc data for Exercises 31 and 32.

32. Use matrix multiplication and Figure 11.2 to find the sale price and the regular price of purchasing four CDs and one DVD.

33. Use matrix multiplication and Figure 11.3 to find the price of purchasing two slices of pizza and one cola at Blondie's and at SliceMan's.

	Blondie's Price	SliceMan's Price
Pizza	$1.25	$1.30
Cola	$ .95	$1.10

FIGURE 11.3 Pizza data for Exercises 33 and 34.

34. Use matrix multiplication and Figure 11.3 to find the price of purchasing six slices of pizza and two colas at Blondie's and at SliceMan's.

35. Jim, Eloise, and Sylvie are enrolled in the same four courses: English, Math, History, and Business. English and Math are four-unit courses, while History and Business are three-unit courses. Their grades are given in Figure 11.4.

	English	Math	History	Business
Jim	A	B	B	C
Eloise	C	A	A	C
Sylvie	B	A	B	A

FIGURE 11.4 Grade data for Exercise 35.

Counting A's as four grade points, B's as three grade points, and C's as two grade points, use matrix multiplication to compute Jim's, Eloise's, and Sylvie's grade point averages (GPAs).

HINT: Start by computing Jim's total grade points without using matrices. Then set up two matrices: one that is a matrix version of Figure 11.4 with grade points rather than letter grades and another that gives the four courses' units. Set up these two matrices so that their product mirrors your by-hand calculation of Jim's total grade points. Finally, multiply by 1 over the number of units taken so that the result is GPA, not total grade points.

36. Dave, Jay, Conan, and Jimmy are each enrolled in the same four courses: Communications, Broadcasting, Marketing, and Contracts. Communications and Contracts are four-unit courses, while Broadcasting and Marketing are three-unit courses. Their grades are given in Figure 11.5.

Counting A's as 4 grade points, B's as 3 grade points, and C's as 2 grade points, use matrix multiplication to

	Communi-cations	Broad-casting	Marketing	Contracts
Dave	A	B	C	A
Jay	A	B	C	B
Conan	A	A	A	A
Jimmy	B	A	B	C

FIGURE 11.5 Grade data for Exercise 36.

compute Dave's, Jay's, Conan's, and Jimmy's grade point averages. (See *HINT,* in Exercise 35.)

37. **Tourism Sales.** Figure 11.6 gives the total sales in the United States for two tourism-related industries: hotels and restaurants.

	2002	2003
Hotels	$105.8 billion	$105.5 billion
Restaurants	$ 61.9 billion	$ 66.3 billion

FIGURE 11.6 Tourism data for Exercise 37. *Source:* Bureau of Economic Analysis, *News Release: Sales of Tourism Industries,* www.bea.gov.

a. Convert this chart into a 2×2 matrix S, and find the product $S\begin{bmatrix} -1 \\ 1 \end{bmatrix}$.

b. Interpret the solution to part (a) in the context of the data.

38. **Tourism Sales.** Figure 11.7 gives the total sales in the United States for two tourism-related industries: air transportation and auto rental.

a. Convert this chart into a 2×2 matrix S, and find the product $S\begin{bmatrix} -1 \\ 1 \end{bmatrix}$.

b. Interpret the solution to part (a) in the context of the data.

	2001	2002
Air Transportation	$103.1 billion	$93.3 billion
Auto Rental	$ 21.6 billion	$23.5 billion

FIGURE 11.7 Tourism data for Exercise 38. *Source:* Bureau of Economic Analysis, *News Release: Sales of Tourism Industries,* www.bea.gov.

39. **Fence Construction.** Borg Fences builds wood fences. They purchase their materials from two suppliers: Piedmont Lumber and Truitt & White. Figure 11.8 shows current prices for their materials and the materials needed to build one 6-foot section of fence.

a. Make two matrices out of the given data, one showing prices and the other showing materials needed.

b. Use matrix multiplication and the two matrices from part (a) to determine the materials cost of one 6-foot section of fence at each of the two suppliers.

40. **Fence Construction.** Piedmont Lumber increased all their prices by 10%, and Truitt & White increased theirs by 9%. Use matrix multiplication and the solutions to Exercise 39 to find the materials cost of one 6-foot section of fence at each of the two suppliers after the price increase.

41. **U.S. Population Movement.**

a. In 2000, the regional population of the United States was 53,594 in the Northeast, 64,393 in the Midwest, 100,237 in the South, and 63,198 in the West. Make a 1×4 matrix A out of these data.

b. Figure 11.9 shows the population movement in the United States during the period from 2000 to 2001. For example, 98.85% = 0.9885 of the residents of the Northeast stayed in the Northeast, and 0.15% = 0.0015 of the residents of the Northeast moved to

	Piedmont Lumber	Truitt & White	Materials Needed for a 6-Foot Section
50 lb. Bag of Concrete	$6.25	$6.10	2
Pressure-Treated 4″ × 4″ × 8′	$8.97	$8.75	1
Redwood 2″ × 4″ × 6′	$4.97	$5.25	2
Framed Redwood lattice 2′ × 6′	$24.85	$22.12	1
Redwood 5/8″ × 5 1/2″ × 8′	$6.98	$6.98	9
Redwood 1″ × 1″ × 6′	$3.88	$3.75	4

FIGURE 11.8 Fence data for Exercise 39.

Region Moved From	Region Moved To			
	Northeast	Midwest	South	West
Northeast	0.9885	0.0015	0.0076	0.0024
Midwest	0.0011	0.9901	0.0057	0.0032
South	0.0018	0.0042	0.9897	0.0043
West	0.0017	0.0035	0.0077	0.9870

FIGURE 11.9 Population data for Exercise 41. *Source:* U.S. Census Bureau, *Statistical Abstract of the United States, 2000 and Current Population Reports, P20–538: Geographical Mobility.*

the Midwest. Make a 4 × 4 matrix *B* out of these data.

c. Compute *AB*. What do its entries measure?

42. U.S. Population Movement. Assume that the data in Figure 11.9 also describe the population movement from 2001 to 2002. Compute (*AB*)*B*, using the matrices *A* and *B* from Exercise 41. What do its entries measure?

43. Verify the associative property by finding *A*(*BC*) and (*AB*)*C*.

$$A = \begin{bmatrix} -4 & 5 \\ 2 & 3 \end{bmatrix} \qquad B = \begin{bmatrix} -3 & 0 & -1 \\ 1 & 4 & -2 \end{bmatrix}$$

$$C = \begin{bmatrix} 0 \\ 5 \\ -1 \end{bmatrix}$$

44. Verify the associative property by finding *A*(*BC*) and (*AB*)*C*.

$$A = \begin{bmatrix} 2 & 0 \\ -1 & 1 \\ 3 & 2 \end{bmatrix} \qquad B = \begin{bmatrix} 5 & 3 & -2 \\ 2 & 0 & 1 \end{bmatrix}$$

$$C = \begin{bmatrix} 1 \\ 2 \\ 0 \end{bmatrix}$$

In Exercises 45–50, find the product (if it exists).

45. $\begin{bmatrix} 1 & 0 \\ 0 & 1 \end{bmatrix} \cdot \begin{bmatrix} 3 & -2 \\ 4 & 0 \end{bmatrix}$

46. $\begin{bmatrix} 1 & 0 & 0 \\ 0 & 1 & 0 \\ 0 & 0 & 1 \end{bmatrix} \cdot \begin{bmatrix} -4 & 5 & 2 \\ 8 & -1 & 9 \\ -2 & 27 & 4 \end{bmatrix}$

47. $\begin{bmatrix} 1 & 0 & 0 \\ 0 & 1 & 0 \\ 0 & 0 & 1 \end{bmatrix} \cdot \begin{bmatrix} 4 & 1 & -1 \\ 5 & 12 & 3 \end{bmatrix}$

48. $\begin{bmatrix} 27 & 19 \\ 42 & 25 \end{bmatrix} \cdot \begin{bmatrix} 1 & 0 \\ 0 & 1 \end{bmatrix}$

49. $\begin{bmatrix} 19 & 7 & 34 \\ 74 & 0 & -11 \\ 13 & -2 & 44 \end{bmatrix} \cdot \begin{bmatrix} 1 & 0 & 0 \\ 0 & 1 & 0 \\ 0 & 0 & 1 \end{bmatrix}$

50. $\begin{bmatrix} 1 & 0 \\ 0 & 1 \end{bmatrix} \cdot \begin{bmatrix} 6 & 18 & -3 \\ 0 & 1 & 5 \\ -14 & 5 & -2 \end{bmatrix}$

 Answer the following questions using complete sentences and your own words.

• CONCEPT QUESTIONS

51. Give an example of two matrices that cannot be multiplied.

52. Describe two ways of determining whether two matrices can be multiplied.

53. Why is an identity matrix so named?

54. Give an example of two matrices *A* and *B* such that *AB* = *BA*.

55. Give an example of two matrices *AB* such that *AB* exists but *BA* does not exist.

56. What must be true about matrices *A* and *B* if the products *AB* and *BA* both exist?

57. Make up a realistic word problem that involves matrix multiplication, as Exercises 31 through 42 do.

• HISTORY QUESTIONS

58. Compare and contrast Arthur Cayley and James Joseph Sylvester. Be sure to discuss both their similarities and their differences. In particular, compare and contrast why each turned his back on mathematics, why each returned to mathematics, their relative success in schooling, and their relative success in employment.

59. Why did Cambridge University not award Sylvester his degrees?

60. Why was no position as a mathematician open to Cayley early in his career?

61. What did Sylvester do in the United States?

MATRIX MULTIPLICATION ON A GRAPHING CALCULATOR

The TI-83, TI-84, and Casio graphing calculators can multiply matrices. Consider the matrices

$$A = \begin{bmatrix} 1 & 2 \\ 3 & 4 \end{bmatrix} \quad \text{and} \quad B = \begin{bmatrix} 1 & 5 \\ 0 & -7 \end{bmatrix}$$

The TI-83 and TI-84 require that you name these matrices "[A]" and "[B]" with brackets. The Casio requires that you name them "Mat A" and "Mat B."

Entering a Matrix

To enter matrix A on a TI, select "EDIT" from the "MATRIX" menu and then matrix A from the "MATRX EDIT" menu:

	TI-83/84	• Press MATRX
		• Use the ► button to highlight "EDIT"
		• Highlight option 1 "[A]"*
		• Press ENTER
	Casio	• Press MENU
		• Use the arrow buttons to highlight "MAT"
		• Press EXE
		• Use the arrow buttons to highlight "Mat A"

Then enter the dimensions of our 2×2 matrix A by typing

2 ENTER 2 ENTER

To enter the elements, type

1 ENTER 2 ENTER 3 ENTER 4 ENTER

In a similar manner, enter matrix B. When you're done entering, type 2nd QUIT.

Note that the calculator uses double subscript notation. When we entered the "4", the calculator showed that this was entered as a_{22}.

Viewing a Matrix

To view matrix A, type the following:

	TI-83/84	• Type MATRX
		• Highlight option 1: "[A]"
		• Press ENTER, and "[A]" will appear on the screen
		• Press ENTER, and the matrix itself will appear on the screen, as shown in Figure 11.10.

```
[A]
        [[1 2]
         [3 4]]
■
```

FIGURE 11.10

Matrix A.

---

*Option 1 is automatically highlighted. If we were selecting some other option, we would use the ▲ and ▼ buttons to highlight it.

Casio
- Press MENU
- Use the arrow buttons to highlight "MAT" and press EXE
- Use the arrow buttons to highlight "Mat A"
- Press EXE , and the matrix will appear on the screen, similar to Figure 11.10

Multiplying Two Matrices

To calculate AB, make the screen read "$[A]*[B]$" or "MatA $\times$ MatB" by typing the following:

TI-83/84   MATRX 1 $\times$ MATRX 2 ENTER

Casio
- Press MENU
- Use the arrow buttons to highlight "RUN" and press EXE
- Press OPTN
- Press MAT (i.e., F2)
- Make the screen read "Mat A $\times$ Mat B" by typing
  MAT ALPHA A $\times$ MAT ALPHA B
- Press EXE

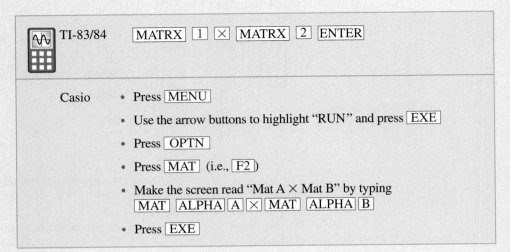

[A]*[B]

$$\begin{bmatrix} 1 & -9 \\ 3 & -13 \end{bmatrix}$$

FIGURE 11.11

Multiplying two matrices.

The result on a TI-83/84 is shown in Figure 11.11.

EXERCISES

In Exercises 62–70, use the following matrices.

$$A = \begin{bmatrix} 5 & -7 \\ 3 & 9 \end{bmatrix} \qquad B = \begin{bmatrix} 7 & 9 \\ -8 & 0 \end{bmatrix}$$

$$C = \begin{bmatrix} 8 & -12 & 13 \\ 52 & 17 & -31 \\ 72 & 28 & -15 \end{bmatrix}$$

$$D = \begin{bmatrix} 14 & -22 & 53 \\ 94 & -15 & -35 \\ 83 & 0 & 7 \end{bmatrix} \qquad E = \begin{bmatrix} 5 & -31 \\ 83 & -33 \\ 60 & 0 \end{bmatrix}$$

$$F = \begin{bmatrix} 93 & -11 & 39 \\ 53 & 66 & 83 \end{bmatrix}$$

In Exercises 62–63, find the indicated product (a) by hand and (b) with a graphing calculator. Check your work by comparing the solutions to parts (a) and (b); answers are not given in the back of the book.

62. AB **63.** BA

In Exercises 64–70, find the indicated product with a graphing calculator. (When a matrix is too big for the screen, use the arrow keys to view it.)

64. a. CD b. DC
65. a. EF b. FE
66. a. DE b. ED
67. a. $(BF)C$ b. $B(FC)$
68. a. $(DE)A$ b. $D(EA)$
69. a. C^2 b. C^5
70. a. D^2 b. D^6

OBJECTIVES

- Be able to use probabilities to make a probability matrix and a transition matrix
- Use a probability matrix and a transition matrix to make predictions
- Understand how a prediction can be made with either a tree or matrix multiplication

Suppose a college student observes her own class standing at the beginning of each school year. These observations vary among elements of the finite set {freshman, sophomore, junior, senior, quit, graduated}. If the student is a freshman at the beginning of the 2010–2011 school year, then she could be a freshman or a sophomore or quit at the beginning of the next school year, each with a certain probability. Her class standing at the beginning of any one year is dependent on her class standing in the preceding school year and not on her class standing in any earlier year.

We're talking about repeatedly observing a certain changeable quality (such as class standing, observed at the beginning of the school year). If the probability of making a certain observation at a certain time depends only on the immediately preceding observation and if the observations' outcomes are elements of a finite set, then these observations and their probabilities form a **Markov chain.**

Our college student's observations form a Markov chain. The probability of having a certain class standing at any one year depends only on her class standing in the preceding year. And her possible class standings are elements of the finite set {freshman, sophomore, junior, senior, quit, graduated}.

Markov chains were developed in the early 1900s by Andrei Markov, a Russian mathematician. They have many applications in the physical sciences, business, sociology, and biology. In business, Markov chains are used to analyze data on customer satisfaction with a product and the effect of the product's advertising and to predict what portion of the market the product will eventually command. In sociology, Markov chains are used to analyze sociological trends such as the shrinking of the U.S. middle class and the growth of the suburbs at the expense of the central cities and the rural areas and to predict the eventual outcome of such trends. Markov chains are also used to predict the weather, to analyze genetic inheritance, and to predict the daily fluctuation in a stock's price.

EXAMPLE **1**   **NUTRITION BARS**   Brute is a nutrition bar for men. Manfood, the company that makes Brute, has just launched a new advertising campaign. A market analysis indicates that as a result of this campaign, 40% of the men who currently *do not* use Brute will buy it the next time they buy a nutrition bar. Another market analysis has studied customer loyalty toward Brute. It indicates that 80% of the consumers who currently *do* use Brute will buy it again the next time they buy a nutrition bar. Rewrite these data in probability form and find the complements of these events.

SOLUTION   We are given a number of probabilities. Since "40% of the men who currently do not use Brute will buy it the next time they buy a nutrition bar," we know that

$$p(\text{next purchase is Brute} \mid \text{current purchase is not Brute}) = 40\% = 0.4$$

Also, "80% of the consumers who currently do use Brute will buy it again the next time they buy a nutrition bar," so

$$p(\text{next purchase is Brute} \mid \text{current purchase is Brute}) = 80\% = 0.8$$

Each of these two events has a complement. If 40% of the men who currently do not use Brute *will* buy it the next time they buy a nutrition bar, then the other 60% of the men who currently do not use Brute *will not* buy it the next time they buy a nutrition bar. That is,

$$p(\text{next purchase is not Brute} \mid \text{current purchase is not Brute})$$
$$= 100\% - 40\% = 60\% = 0.6$$

Similar reasoning gives us

$$p(\text{next purchase is not Brute} \mid \text{current purchase is Brute}) = 1 - 0.8 = 0.2$$

These relationships are summarized in Figure 11.12.

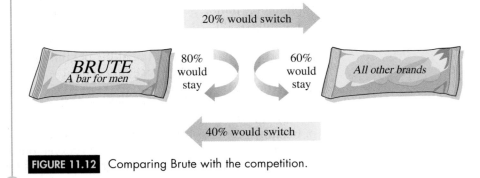

FIGURE 11.12 Comparing Brute with the competition.

Notice that Example 1 describes a Markov chain because:

• the observations are made at successive points in time (shopping trips during which a nutrition bar is purchased);
• the observations' outcomes are elements of the set {Brute is purchased, something other than Brute is purchased}; and
• the probability that the next purchase is Brute depends only on the current purchase and not on any previous purchases—or, more generally, the probability of making a certain observation at a certain time depends only on the outcome of the immediately preceding observation.

The observations' outcomes are called **states.** The states in Example 1 are "a consumer purchases Brute ("Brute," or B for short) and "a consumer purchases something other than Brute ("not Brute," or B′ for short).

All states come as both **current states** and **following states.** Our *current* states are "a consumer currently purchases Brute" and "a consumer currently purchases something other than Brute," and the *next following* states are "a consumer will purchase Brute the next time he purchases a nutrition bar" and "a consumer will purchase something other than Brute the next time he purchases a nutrition bar."

Transition Matrices

A **transition matrix** is a matrix whose entries are the probabilities of passing from current states to following states. A transition matrix has a row and a column for each state. The rows refer to current states, and the columns refer to later (or following) states. A transition matrix is read like a chart. An entry in a certain row and column represents the probability of making a transition from the current state represented by that row to the following state represented by that column.

EXAMPLE 2

NUTRITION BARS AND A TRANSITION MATRIX Write the transition matrix T for the data in Example 1.

SOLUTION

The states are "Brute" and "not Brute" (or just B and B′). Therefore our transition matrix T is the following 2×2 matrix:

a column for each following state

$$
T = \begin{array}{cc} & \begin{array}{cc} \text{B} & \text{B}' \end{array} \\ \begin{array}{c} \\ \end{array} & \left[\begin{array}{cc} 0.8 & 0.2 \\ 0.4 & 0.6 \end{array} \right] \begin{array}{l} \text{B} \\ \text{B}' \end{array} \end{array} \quad \longleftarrow \quad \textbf{a row for each current state}
$$

The entry 0.4 is in the B′ row and the B column, so it is the probability that a consumer makes a transition from buying something other than Brute (current state is B′) to buying Brute (first following state is B).

We can make several observations about transition matrices.

1. *A transition matrix must be square,* because there is one row and one column for each state. If you create a transition matrix that is not square, go back and see which state you did not list as both a current state and a following state.
2. *Each entry in a transition matrix must be between 0 and 1 (inclusive),* because the entries are probabilities. If you create a transition matrix that has an entry that is less than 0 or greater than 1, go back and find your error.
3. *The sum of the entries of any row must be 1,* because the entries of a row are the probabilities of changing from the state represented by that row to *all* of the possible following states. If you create a transition matrix that has a row that does not add to 1, go back and find your error.

TRANSITION MATRIX OBSERVATIONS

1. A transition matrix must be square.
2. Each entry in a transition matrix must be between 0 and 1 (inclusive).
3. The sum of the entries of any row must be 1.

The transition matrix in Example 2 shows the transitional trends in the consumers' selection of their next nutrition bar purchases. What effect will these trends have on Brute's success as a product? Will Brute's market share increase or decrease?

Probability Matrices

EXAMPLE 3

NUTRITION BARS REVISITED A marketing analysis for Brute shows that Brute currently commands 25% of the market. Write this information and its complement in probability form.

SOLUTION

We are given

$p(\text{current purchase is Brute}) = 0.25$

Its complement is

$p(\text{current purchase is not Brute}) = 1 - 0.25 = 0.75$

A **probability matrix** is a row matrix, with a column for each state. The entries are the probabilities of the different states. The columns of a probability matrix must be labeled in the same way, as are the rows and columns of the transition matrices.

EXAMPLE **4**

NUTRITION BARS AND A PROBABILITY MATRIX   Write the probability matrix P for the data in Example 3.

SOLUTION

The probability matrix is a row matrix, with a column for B and a column for B′.

$$P = \begin{bmatrix} \overset{\text{B}}{0.25} & \overset{\text{B}'}{0.75} \end{bmatrix}$$

We can make several observations about probability matrices that parallel the observations made earlier about transition matrices.

> ## PROBABILITY MATRIX OBSERVATIONS
>
> **1.** A probability matrix is a row matrix.
> **2.** Each entry in a probability matrix must be between 0 and 1 (inclusive).
> **3.** The sum of the entries of the row must be 1.

Using Markov Chains to Predict the Future

EXAMPLE **5**

NUTRITION BARS' FUTURE MARKET SHARE, USING A TREE   Use a probability tree to predict Brute's market share after the first following purchase. In other words, find p(1st following purchase is Brute) and p(1st following purchase is not Brute).

SOLUTION

We will use the tree in Figure 11.13.

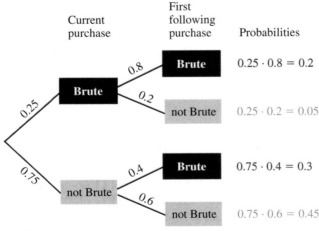

FIGURE 11.13 Predicting Brute's future success.

We want p(1st following purchase is Brute), which the sum of the probabilities of the limbs that stop at Brute. Similarly, p(1st following purchase is not Brute) is the sum of the probabilities of the limbs that stop at not Brute.

$$p(\text{1st following purchase is Brute}) = (0.25 \cdot 0.8) + (0.75 \cdot 0.4)$$
$$= 0.2 + 0.3 = 0.5$$
$$p(\text{1st following purchase is not Brute}) = (0.25 \cdot 0.2) + (0.75 \cdot 0.6)$$
$$= 0.05 + 0.45 = 0.5$$

In other words, after the first following purchase, Brute will command 50% of the market (up from a previous 25%). Remember, though, that this is only a prediction, not a guarantee. And it's based on the assumption that current trends continue unchanged.

EXAMPLE 6

FUTURE MARKET SHARE, USING MATRIX MULTIPLICATION   The calculations done in Example 5 can be done by computing the product of the probability matrix P and the transition matrix T:

$$PT = \begin{bmatrix} 0.25 & 0.75 \end{bmatrix} \begin{bmatrix} 0.8 & 0.2 \\ 0.4 & 0.6 \end{bmatrix}$$
$$= \begin{bmatrix} 0.25 \cdot 0.8 + 0.75 \cdot 0.4 & 0.25 \cdot 0.2 + 0.75 \cdot 0.6 \end{bmatrix}$$
$$= \begin{bmatrix} 0.2 + 0.3 & 0.05 + 0.45 \end{bmatrix}$$
$$= \begin{bmatrix} 0.5 & 0.5 \end{bmatrix}$$

The parallelism between trees (in Example 5) and matrices (in Example 6) is illustrated in Figure 11.14.

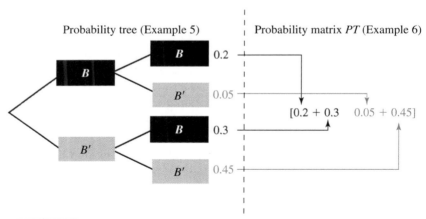

Probability tree (Example 5)   Probability matrix PT (Example 6)

FIGURE 11.14 Comparing tree results and matrix results.

EXAMPLE 7

NUTRITION BARS AND A LATER MARKET SHARE   Use a probability tree to predict Brute's market share after the second following purchase.

SOLUTION

We'll use the tree in Figure 11.15. We want $p(\text{2nd following purchase is Brute})$, which is the sum of the probabilities of the limbs that stop at Brute under "2nd following purchase."
This tree gives us

$$p(\text{2nd following purchase is B}) = (0.25 \cdot 0.8 \cdot 0.8) + (0.25 \cdot 0.2 \cdot 0.4)$$
$$+ (0.75 \cdot 0.4 \cdot 0.8) + (0.75 + 0.6 + 0.4)$$
$$= 0.6$$

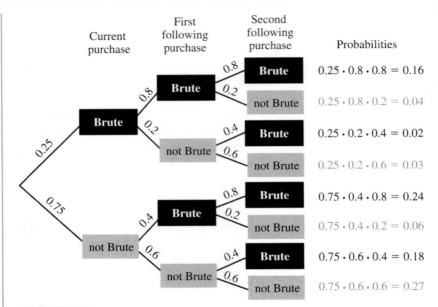

FIGURE 11.15 Predicting Brute's future success after the second following purchase.

In other words, after the second following purchase, Brute will command 60% of the market. Remember, though, that this is only a prediction, not a guarantee. It is based on the assumption that current trends continue unchanged.

EXAMPLE 8

A LATER MARKET SHARE BY MATRIX MULTIPLICATION   The calculations done in Example 7 can be done by computing the product of the probability matrix PT and the transition matrix T.

$$PT^2 = PT \cdot T$$
$$= \begin{bmatrix} 0.5 & 0.5 \end{bmatrix} \cdot \begin{bmatrix} 0.8 & 0.2 \\ 0.4 & 0.6 \end{bmatrix}$$
$$= \begin{bmatrix} 0.5 \cdot 0.8 + 0.5 \cdot 0.4 & 0.5 \cdot 0.2 + 0.5 \cdot 0.6 \end{bmatrix}$$
$$= \begin{bmatrix} 0.6 & 0.4 \end{bmatrix}$$

This tells us that $P(\text{2nd following purchase is } B) = 0.6$ and $P(\text{2nd following purchase is } B') = 0.4$.

Predictions of Brute's future market shares can be made either with trees or with matrices. The matrix work is certainly simpler, both to set up and to calculate.

EXAMPLE 9

SOLUTION

MORE PREDICTIONS USING MATRICES   Use matrices to predict Brute's market share after the third following purchase.

Brute's market share will be an entry in PT^3. To do this, observe that

$$PT^3 = P(T^2 T) = (PT^2)T \quad \textbf{associative property}$$

and use the previously calculated PT^2.

$$PT^3 = (PT^2)T = \begin{bmatrix} 0.6 & 0.4 \end{bmatrix}\begin{bmatrix} 0.8 & 0.2 \\ 0.4 & 0.6 \end{bmatrix} = \begin{bmatrix} 0.64 & 0.36 \end{bmatrix}$$

This tells us that $p(\text{third following purchase is Brute}) = .64$, and $p(\text{third following purchase is not Brute}) = .36$. This means that three purchases after the time when the market analysis was done, 64% of the men's nutrition bars purchased will be

Brute (*if current trends continue*). In other words, Brute's market share will be 64%. Brute will be doing quite well. Recall that it had a market share of only 25% at the beginning of the new advertising campaign.

PREDICTIONS WITH MARKOV CHAINS

1. *Create the probability matrix P. P* is a row matrix whose entries are the initial probabilities of the states.
2. *Create the transition matrix T. T* is the square matrix whose entries are the probabilities of passing from current states to first following states. The rows refer to the current states, and the columns refer to the next following states.
3. *Calculate PT^n. PT^n* is a row matrix whose entries are the probabilities of the nth following states. Be careful that you multiply in the correct order: $PT^n \neq T^nP$.

EXAMPLE **10**

IS THE MIDDLE CLASS SHRINKING? Sociologists have found that a strong determinant of an individual's income is the income of his or her parents.

a. Convert the data given in Figure 11.16 on family incomes in 1999 into a probability matrix.

b. Census data suggest the following (illustrated in Figure 11.17):

- Of those individuals whose parents belong to the lower income group, 21% will become members of the middle income group, and 1% will become members of the upper income group.
- Of those individuals whose parents belong to the middle income group, 6% will become members of the lower income group, and 4% will become members of the upper income group.
- Of those individuals whose parents belong to the upper income group, 1% will become members of the lower income group, and 10% will become members of the middle income group.

Convert this information into a transition matrix.

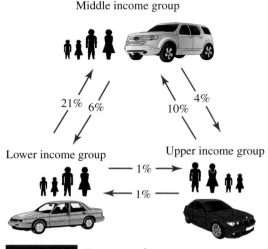

Family Income	Percent of Population
Under $15,000	12%
$15,000–$74,999	62%
$75,000 or more	26%

FIGURE 11.16 Income levels in the United States.
Source: U.S. Bureau of the Census, *Statistical Abstract of the United States, 2001.*

FIGURE 11.17 Transitions between income groups.

c. Predict the percent of U.S. families in the lower, middle, and upper income groups after one generation.

d. Predict the percent of U.S. families in the lower, middle, and upper income groups after two generations.

SOLUTION

a. *Creating the probability matrix P.* Recall that a probability matrix has one row, with a column for each state. Our states are "lower," "middle," and "upper." Our probability matrix is

$$
\begin{array}{ccc}
\text{low} & \text{mid} & \text{up}
\end{array}
$$
$$
P = \begin{bmatrix} 0.12 & 0.62 & 0.26 \end{bmatrix}
$$

b. *Creating the transition matrix T.* Because we have three income groups, the transition matrix T is a 3×3 matrix, with rows referring to the current income group of the parents, and columns referring to the income group of the child. We are given the following portion of T:

income group of child

$$
T = \begin{bmatrix}
? & 0.21 & 0.01 \\
0.06 & ? & 0.04 \\
0.01 & 0.10 & ?
\end{bmatrix}
\begin{array}{l}
\text{low} \\
\text{mid} \\
\text{up}
\end{array}
\quad \textbf{income group of parents}
$$

with column headers: low, mid, up.

Filling in the blanks is easy. If 21% of the children of lower income parents become members of the middle income group and 1% become members of the upper income group, that leaves 78% to remain members of the lower income group. Alternatively, the entries in any row must add to 1, so the missing entry is $1 - (0.21 + 0.01) = 0.78$. Similar calculations fill in the other blanks. Thus, we have

$$
\begin{array}{ccc}
\text{low} & \text{mid} & \text{up}
\end{array}
$$
$$
T = \begin{bmatrix}
0.78 & 0.21 & 0.01 \\
0.06 & 0.90 & 0.04 \\
0.01 & 0.10 & 0.89
\end{bmatrix}
\begin{array}{l}
\text{low} \\
\text{mid} \\
\text{up}
\end{array}
$$

c. *Predicting the distribution after one generation.* We are asked to calculate PT^1, the probabilities of the first following generation.

$$
PT^1 = PT = \begin{bmatrix} 0.12 & 0.62 & 0.26 \end{bmatrix}
\begin{bmatrix}
0.78 & 0.21 & 0.01 \\
0.06 & 0.90 & 0.04 \\
0.01 & 0.10 & 0.89
\end{bmatrix}
$$
$$
= \begin{bmatrix} 0.1334 & 0.6092 & 0.2574 \end{bmatrix}
$$
$$
\begin{array}{ccc}
\text{low} & \text{mid} & \text{up}
\end{array}
$$
$$
\approx \begin{bmatrix} 0.13 & 0.61 & 0.26 \end{bmatrix}
$$

This means that if current trends continue, in one generation's time, the lower income group will grow from 12% of the population to 13%, the middle income group will shrink from 62% to 61%, and the upper income group will stay at 26%.

Notice that the row adds to $0.13 + 0.61 + 0.26 = 1$, as it should.

d. *Predicting the distribution after two generations.* We are asked to calculate PT^2, the probabilities of the second following generation. The easiest way to do this calculation is to observe that $PT^2 = (PT)T$ and to use the previously calculated PT. Our answers will be more accurate if we do not round off PT before multiplying it by T.

$$
PT^2 = (PT)T = \begin{bmatrix} 0.1334 & 0.6092 & 0.2574 \end{bmatrix} \cdot
\begin{bmatrix}
0.78 & 0.21 & 0.01 \\
0.06 & 0.90 & 0.04 \\
0.01 & 0.10 & 0.89
\end{bmatrix}
$$
$$
= \begin{bmatrix} 0.143178 & 0.602034 & 0.254788 \end{bmatrix}
$$
$$
\begin{array}{ccc}
\text{low} & \text{mid} & \text{up}
\end{array}
$$
$$
\approx \begin{bmatrix} 0.14 & 0.60 & 0.25 \end{bmatrix}
$$

This means that if current trends continue, in two generations' time, the lower income group will grow from 12% of the population to 14%, the middle income group will shrink from 62% to 60%, and the upper income group will shrink from 26% to 25%.

> Notice that the row adds to 1 before rounding.
>
> $$0.143178 + 0.602034 + 0.254788 = 1$$
>
> If it didn't, we would have to find our error.

11.1 EXERCISES

1. $P = \begin{bmatrix} 0.8 & 0.2 \end{bmatrix}$, and $T = \begin{bmatrix} 0.1 & 0.9 \\ 0.7 & 0.3 \end{bmatrix}$.

    a. What is the probability of moving from state 1 (as a current state) to state 2 (as a first following state)?

    b. What is state 2's current market share?

    c. Find PT.

    d. Find PT^2.

    e. Draw a tree diagram illustrating P and T.

2. $P = \begin{bmatrix} 0.75 & 0.25 \end{bmatrix}$, and $T = \begin{bmatrix} 0.6 & 0.4 \\ 0.5 & 0.5 \end{bmatrix}$.

    a. What is the probability of moving from state 1 (as a current state) to state 2 (as a first following state)?

b. What is state 2's current market share?

c. Find PT.

d. Find PT^2.

e. Draw a tree diagram illustrating P and T.

3. $P = \begin{bmatrix} 0.6 & 0.4 \end{bmatrix}$, and $T = \begin{bmatrix} 0.2 & 0.8 \\ 0 & 1 \end{bmatrix}$.

a. What is the probability of moving from state 1 (as a current state) to state 2 (as a first following state)?

b. What is state 2's current market share?

c. Find PT.

d. Find PT^2.

e. Draw a tree diagram illustrating P and T.

4. $P = \begin{bmatrix} 0.6 & 0.4 \end{bmatrix}$, and $T = \begin{bmatrix} 0.3 & 0.7 \\ 0.4 & 0.6 \end{bmatrix}$.

a. What is the probability of moving from state 1 (as a current state) to state 2 (as a first following state)?

b. What is state 2's current market share?

c. Find PT.

d. Find PT^2.

e. Draw a tree diagram illustrating P and T.

In Exercises 5–8, (a) rewrite the given data (and, if appropriate, their complements) in probability form, and (b) convert these probabilities into a probability matrix.

5. A marketing analysis shows that KickKola currently commands 14% of the Cola market.

6. A marketing analysis shows that SoftNWash currently commands 26% of the fabric softener market.

7. Silver's Gym currently commands 48% of the health club market in Metropolis, Fitness Lab commands 37%, and ThinNFit commands the balance of the market.

8. Smallville has three Chinese restaurants: Asia Gardens, Chef Chao's, and Chung King Village. Currently, Asia Gardens gets 41% of the business, Chef Chao's gets 33%, and Chung King Village gets the balance.

In Exercises 9–12, (a) rewrite the given data in probability form, and (b) convert these probabilities into a transition matrix.

9. A marketing analysis for KickKola indicates that 12% of the consumers who do not currently drink KickKola will purchase KickKola the next time they buy a cola (in response to a new advertising campaign) and that 63% of the consumers who currently drink KickKola will purchase it the next time they buy a cola.

10. A marketing analysis for SoftNWash indicates that 9% of the consumers who do not currently use SoftNWash will purchase SoftNWash the next time they buy a fabric softener (in response to a free sample sent to selected consumers) and that 29% of the consumers who currently use SoftNWash will purchase it the next time they buy a fabric softener.

11. An extensive survey of gym users in Metropolis indicates that 71% of the current members of Silver's will continue their annual membership when it expires, 12% will quit and join Fitness Lab, and the rest will quit and join ThinNFit. Fitness Lab has been unable to keep its equipment in good shape, and as a result 32% of its members will defect to Silver's and 34% will leave for ThinNFit. ThinNFit's members are quite happy, and as a result, 96% plan on renewing their annual membership, with half of the rest planning on moving to Silver's and half to Fitness Lab.

12. Chung King Village recently mailed a coupon to all residents of Smallville offering two dinners for the price of one. As a result, 67% of those who normally eat at Asia Gardens plan on trying Chung King Village within the next month they eat Chinese food, and 59% of those who normally eat at Chef Chao's plan on trying Chung King Village. Also, all of Chung King Village's regular customers will return to take advantage of the special. And 30% of the people who normally eat at Asia Gardens will eat there within the next month, because of its convenient location. Chef Chao's has a new chef who isn't doing very well, and as a result, only 15% of the people who normally eat at there are planning on returning next time.

13. Use the information in Exercises 5 and 9 to predict KickKola's market share at:

a. the first following purchase, using probability trees

b. the second following purchase, using probability trees

c. the first following purchase, using matrices

d. the second following purchase, using matrices

e. the sixth following purchase

f. Also, discuss the advantages and disadvantages of using trees as well of those of using matrices.

14. Use the information in Exercises 6 and 10 to predict SoftNWash's market share at:

a. the first following purchase, using probability trees

b. the second following purchase, using probability trees

c. the first following purchase, using matrices

d. the second following purchase, using matrices

e. the seventh following purchase

f. Also, discuss the advantages and disadvantages of using trees as well of those of using matrices.

15. Use the information in Exercises 7 and 11 to predict the market shares of Silver's Gym, Fitness Lab, and ThinNFit:

a. in one year, using probability trees

b. in one year, using matrices

c. in two years, using matrices

d. in three years

e. in five years

Note: The answer to part (b) is not given in the back of the book, but it is the same as that of part (a).

16. Use the information in Exercises 8 and 12 to predict the market shares of Asia Gardens, Chef Chao's, and Chung King Village:

  a. in one month, using probability trees

  b. in one month, using matrices

  c. in two months, using matrices

  d. in three months

  e. in one year

17. A census report shows that currently 32% of the residents of Metropolis own their own home and that 68% rent. The census report also shows that 12% of the renters plan to buy a home in the next 12 months and that 3% of the homeowners plan to sell their home and rent instead. If residents follow their plans, find what percentage of Metropolis residents will be home owners and what percent will be renters:

  a. in 1 year   **b.** in 2 years

  c. in 3 years

18. A survey shows that 23% of the shoppers in seven Midwestern states regularly buy their groceries at Safe Shop, 29% regularly shop at PayNEat, and the balance shops at any one of several smaller markets. The survey also indicates that 8% of the consumers who currently shop at Safe Shop will purchase their groceries at PayNEat the next time they shop and that 5% will switch to some other store. Also, 12% of the consumers who currently shop at PayNEat will purchase their groceries at Safe Shop the next time they shop, and 2% will switch to some other store. In addition, 13% of the consumers who currently shop at neither store will purchase their groceries at Safe Shop the next time they shop, and 10% will shop at PayNEat. A typical shopper goes to a grocery store once every week. Find the percentage of the shoppers that will shop at Safe Shop and at PayNEat:

  a. in 1 week   **b.** in 2 weeks

  c. in 4 weeks

19. Sierra Cruiser currently commands 41% of the mountain bike market. A nationwide survey performed by *Get Out of My Way* magazine indicates that 31% of the bike owners who do not currently own a Sierra Cruiser will purchase a Sierra Cruiser the next time they buy a mountain bike and that 12% of the bikers who currently own a Sierra Cruiser will purchase one the next time they buy a mountain bike. If the average customer buys a new mountain bike every two years, predict Sierra Cruiser's market share in four years.

20. A marketing analysis shows that Clicker Pens currently commands 46% of the pen market. The analysis also indicates that 46% of the consumers who do not currently own a Clicker pen will purchase a Clicker the next time they buy a pen and that 37% of the consumers who currently own a Clicker will not purchase one the next time they buy a pen. If the average consumer buys a new pen every three weeks, predict Clicker's market share after six weeks.

● PROJECTS

21. **a.** The Census Bureau classifies all residents of the United States as residents of a central city, a suburb, or a nonmetropolitan area. In the 1990 census, the bureau reported that the central cities held 77.8 million people, the suburbs 114.9 million people, and the nonmetropolitan areas 56.0 million people. Compute the proportion of U.S. residents for each category and present those proportions as a probability matrix. (Round off to three decimal places.)

  b. The Census Bureau (in Current Population Reports, P20–463, *Geographic Mobility, March 1990 to March 1991*) also reported the information in Figures 11.18 and 11.19 regarding migration between these three areas from 1990 to 1991. (Numbers are in thousands.)

Moved from	Moved to		
	Central City	Suburb	Nonmetropolitan Area
Central City	(x)	4946	736
Suburb	2482	(x)	964
Nonmetropolitan Area	741	1075	(x)

FIGURE 11.18   Migration data for Exercise 21. *Note:* Persons moving from one central city to another were not considered. This is represented by "(x)."

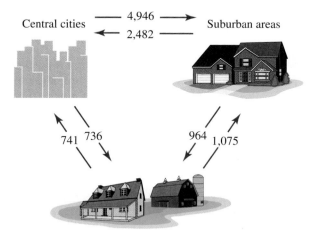

Nonmetropolitan areas

FIGURE 11.19   Migration data in graphic form. (Numbers are in thousands.)

Use the data in Figure 11.18 and the data in part (a) to compute the probabilities that a resident of any one type of environment will move to any of the other types. Present these probabilities in the form of a transition matrix. (Round off to three decimal places.)

c. Predict the percent of U.S. residents who will reside in a central city, a suburb, and a nonmetropolitan area in 1991.

d. Predict the percent of U.S. residents who will reside in a central city, a suburb, and a nonmetropolitan area in 1992.

e. Which prediction is the stronger one; that in part (c) or that in part (d)? Why?

22. a. The Census Bureau (in *Statistical Abstract of the United States, 2000*) reported the 2000 national and regional populations given in Figure 11.20.

Region	Population (in thousands)
U.S.	281,422
Northeast	53,594
Midwest	64,393
South	100,237
West	63,198

FIGURE 11.20 Regional population data for Exercise 22.

Compute the proportion of U.S. residents of each region, and present those proportions as a probability matrix. (Round off to three decimal places.)

b. The Census Bureau (in Current Population Reports, P20–465, *Geographic Mobility: March 1999 to March 2000*) also reported the information in Figures 11.21 and 11.22 regarding movement between regions from 1999 to 2000. (All numbers are in thousands.)

Compute the probabilities that a resident of any one region will move to any of the other regions. Present these probabilities in the form of a transition matrix. Assume that the transitional probabilities

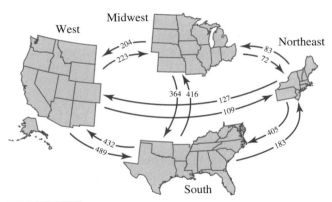

FIGURE 11.22 Interregional migration in graphic form.

for 2000–2001 are equal to those for 1999–2000. (Round off to three decimal places.)

c. Predict the percent of U.S. residents who resided in the Northeast, the Midwest, the South, and the West in 2001.

d. Predict the percent of U.S. residents who resided in the Northeast, the Midwest, the South, and the West in 2002.

23. Land Use. Neil Sampson, executive vice president of the National Association of Conservation Districts, compiled data on shifting land use in the United States (*Source:* R. Neil Sampson *Farmland or Wasteland: A Time To Choose.* Emmaus, PA: Rodale Press, 1981). He stated that in 1977, there were 413 million acres of cropland, 127 million acres of land that had a high or medium potential for conversion to cropland (for example, grasslands and forests), and 856 million acres of land that had little or no potential for conversion to cropland (such as urban land or land with heavily deteriorated soil).

a. Compute the proportion of land in each category, and present those proportions as a probability matrix. (Round off to three decimal places.)

b. He also estimated that, between 1967 and 1977, the shifts in land use shown in Figures 11.23 and Figure 11.24 occurred (in millions of acres).

Region Moved from	Region Moved to			
	Northeast	**Midwest**	**South**	**West**
Northeast	(x)	83	405	127
Midwest	72	(x)	364	204
South	183	416	(x)	432
West	109	223	489	(x)

FIGURE 11.21 Interregional migration. *Note:* The Census data involved interregional movement. Thus, persons moving from one part of a region to another part of the same region were not considered. This is represented in the above chart by "(x)."

Previous Use of Land	New Use of Land		
	Cropland	**Potential Cropland**	**Non-cropland**
Cropland	(x)	17	35
Potential Cropland	34	(x)	0[1]
Non-cropland	0	0[1]	(x)

[1]Unknown but assumed to cancel each other out.

FIGURE 11.23 Shifting land use. *Source:* R. Neil Sampson, *Farmland or Wasteland: A Time to Choose.* Emmaus, PA: Rodale Press, 1981.

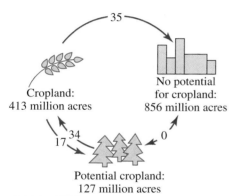

35

Cropland:
413 million acres

No potential
for cropland:
856 million acres

17 $\overset{34}{}$ 0

Potential cropland:
127 million acres

FIGURE 11.24 Shifting land use in graphic form.

Use these data and the data in part (a) to compute the probabilities that an acre of land of any of the three categories will shift to another use (in ten year's time). Present these probabilities in the form of a transition matrix. (Round off to three decimal places.) Assume that the transitional probabilities for 1967–1977 are equal to those for 1977–1987.

c. Predict the amount of land that will be cropland, potential cropland, and noncropland in 1987, assuming that the trend described above continues.

d. Predict the amount of land that will be cropland, potential cropland, and noncropland in 1997, assuming that the trend described above continues.

Answer the following questions using complete sentences and your own words.

• CONCEPT QUESTIONS

24. Why does the information in Exercises 5 and 9 describe a Markov chain?

25. A coin is tossed ten times. Does this experiment and its probabilities form a Markov chain? Why or why not?

26. Are PT, PT^2, and PT^3 transition matrices or probability matrices? Why?

27. Are T^2 and T^3 transition matrices or probability matrices? Why?

28. What are the differences between a transition matrix and a probability matrix?

29. What information would a marketing analyst need to predict the future market share of a product? How would he or she obtain such information?

30. When a prediction is made using Markov chains, that prediction is based on a number of assumptions, one involving the trend, one involving the data summarized in the probability matrix P, and one involving the data summarized in the transition matrix T. What are these assumptions? What would make the assumption invalid?

31. Compare and contrast the tree method of Markov chains with the matrix method of Markov chains. What are the advantages of the two different methods?

32. Choose one of Exercises 13–20 and discuss why the given data describes a Markov chain.

33. What is the difference between a current state and a following state?

34. Suppose a transition matrix contains a row where one entry is a 1 and all of the other entries are zeros. What does that tell you?

• HISTORY QUESTIONS

35. In creating Markov chains, was Markov motivated by theoretical concerns or by specific applications?

36. Was Markov a supporter of the czar or of the revolution?

37. How did Markov celebrate the 300th anniversary of the House of Romanov?

38. What did Markov do when the czar abdicated?

11.2 Systems of Linear Equations

OBJECTIVES

- Learn how to use the elimination method to solve a system of linear equations
- Understand the relationship between the number of equations in a system, the number of variables, and the number of solutions

To go further with Markov chains, we have to solve systems of linear equations. We'll return to Markov chains after reviewing systems.

Linear Equations

To graph the equation $2x + 3y = 8$, first solve the equation for y.

$$2x + 3y = 8$$
$$3y = -2x + 8$$
$$y = -\frac{2}{3}x + \frac{8}{3}$$

Comparing this equation with the slope-intercept formula $y = mx + b$, we find that this is the equation of a line with slope $m = -\frac{2}{3}$ and y-intercept $b = \frac{8}{3}$, as shown in Figure 11.25.

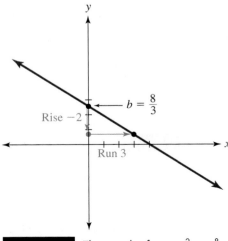

FIGURE 11.25 The graph of $y = -\frac{2}{3}x + \frac{8}{3}$.

The graph of the equation $5x + 7y = 12$ is also a line, because we could follow the above procedure and rewrite the equation in the form $y = mx + b$. In fact, the graph of any equation of the form $ax + by = c$ is a line. (If b were 0, the above procedure wouldn't work, but the graph would still be a line.) For this reason, an equation of the form $ax + by = c$ is called a *linear equation*.

To graph an equation of the form $ax + by + cz = d$, we would need a z-axis in addition to the usual x- and y-axes; the graph of such an equation is always a plane. Despite its geometrical appearance, however, such an equation is called a linear equation owing to its algebraic similarity to $ax + by = c$.

A **linear equation** is an equation that can be written in the form $ax + by = c$, the form $ax + by + cz = d$, or a similar form with more unknowns. In Section 11.3, we will use linear equations to solve Markov chain problems.

Systems of Equations

A **system of equations** is a set of more than one equation. **Solving a system of equations** means finding all ordered pairs (x, y) [or ordered triples (x, y, z) and so on] that will satisfy each equation in the system. Geometrically, this means finding all points that are on the graph of each equation in the system.

The system

$$x + y = 3$$
$$2x + 3y = 8$$

is a system of linear equations because it is a set of two equations and each equation is of the form $ax + by = c$. The ordered pair $(1, 2)$ is a solution to the system because $(1, 2)$ satisfies each equation:

$$x + y = 1 + 2 = 3$$
$$2x + 3y = 2 \cdot 1 + 3 \cdot 2 = 8$$

Does this system have any other solutions? If we solve each equation for y, we get

$$y = -1x + 3$$
$$y = \frac{-2}{3}x + \frac{8}{3}$$

The first equation is the equation of a line with slope -1 and y-intercept 3; the second is the equation of a line with slope $\frac{-2}{3}$ and y-intercept $\frac{8}{3}$. Because the slopes are different, the lines are not parallel and intersect in only one point. Thus, $(1, 2)$ is the only solution to the system; it is the point at which the two lines intersect. This system is illustrated in Figure 11.26.

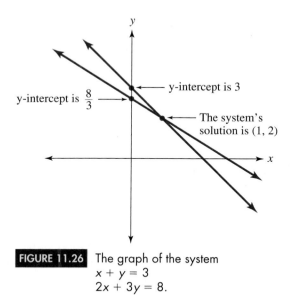

FIGURE 11.26 The graph of the system
$x + y = 3$
$2x + 3y = 8$.

Not all systems have only one solution. A system with two linear equations in two unknowns describes a pair of lines. The two lines can be parallel and not intersect (in which case the system has no solution), or the two lines can be the same (in which case the system has an infinite number of solutions), as shown in Figure 11.27.

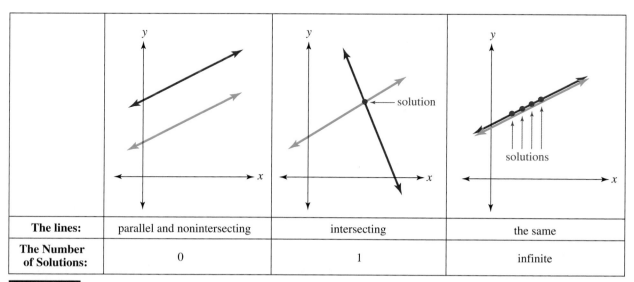

The lines:	parallel and nonintersecting	intersecting	the same
The Number of Solutions:	0	1	infinite

FIGURE 11.27 A system of two linear equations in two unknowns can have no solutions, one solution, or an infinite number of solutions.

A system of two linear equations in *three* unknowns (x, y, and z) describes a set of two planes. Such a system can have no solution or an infinite number of solutions, but it cannot have only one solution, as illustrated in Figure 11.28. In fact, *any system of linear equations that has fewer equations than unknowns cannot have a unique solution.*

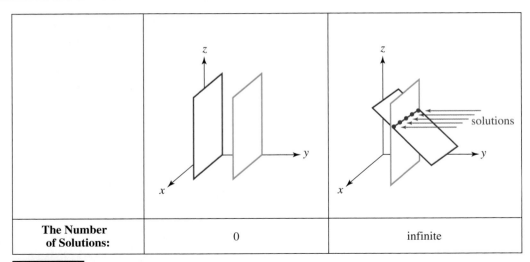

The Number of Solutions:	0	infinite

FIGURE 11.28 A system of two linear equations in three unknowns can have no solutions or an infinite number of solutions.

A system of *three* linear equations in three unknowns describes a set of three planes. Such a system can have no solution, only one solution, or an infinite number of solutions, as illustrated in Figure 11.29. In fact, *any system of linear equations that has as many equations as unknowns can have either no solution, one unique solution, or an infinite number of solutions.*

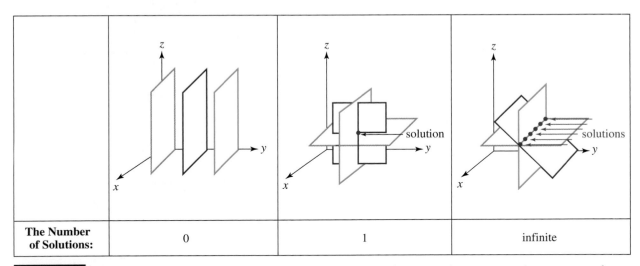

The Number of Solutions:	0	1	infinite

FIGURE 11.29 A system of three linear equations in three unknowns can have no solutions, one solution, or an infinite number of solutions.

NUMBER OF SOLUTIONS OF A SYSTEM OF LINEAR EQUATIONS

A SYSTEM THAT HAS:	**WILL HAVE:**
fewer equations than unknowns	either no solution or an infinite number of solutions
as many unique equations as unknowns	either no solution, one solution, or an infinite number of solutions

EXAMPLE **1**

FINDING THE NUMBER OF SOLUTIONS How many solutions could the following system have?

$$3x - \quad y + \quad z = 5$$
$$2x + \quad y - \quad 3z = 1$$
$$20x + 10y - 30z = 10$$

SOLUTION

This system seems to have as many equations as unknowns; however, this is not the case. The third equation is 10 times the second equation, so its presence is unnecessary. The above system is equivalent to the following system:

$$3x - y + \quad z = 5$$
$$2x + y - 3z = 1$$

This system has three unknowns but only two *unique* equations, so it can have either no solution or an infinite number of solutions. It cannot have one solution.

Solving Systems: The Elimination Method

Our work with Markov chains will involve solving systems of equations. Many methods are used to solve systems of equations. Perhaps the easiest method for solving a small system is the elimination method (also called the addition method). With this method, you add together equations or multiples of equations in such a way that a variable is eliminated. You probably saw this method in your intermediate algebra class.

EXAMPLE **2**

USING THE ELIMINATION METHOD Solve the following system:

$$3x - 2y = 5$$
$$2x + 3y = 4$$

SOLUTION

We'll eliminate x if we multiply the first equation by 2 and the second equation by -3 and add the results.

$$\begin{array}{rl} 6x - \quad 4y = 10 & \textbf{2 times equation 1} \\ + \ -6x - \quad 9y = -12 & \textbf{-3 times equation 2} \\ \hline 0x - 13y = -2 & \textbf{adding} \\ y = \dfrac{2}{13} & \textbf{solving for } y \end{array}$$

We can find x by substituting $y = \frac{2}{13}$ into either of the original equations.

$$3x - 2y = 5 \qquad \textbf{one of the original equations}$$

$$3x - 2 \cdot \frac{2}{13} = 5 \qquad \textbf{substituting for } y$$

$$3x - \frac{4}{13} = \frac{65}{13} \qquad \textbf{rewriting 5 with a common denominator}$$

$$3x = \frac{69}{13} \qquad \textbf{solving for } x$$

$$x = \frac{23}{13}$$

Thus, the solution to the system of equations is the ordered pair $\left(\frac{23}{13}, \frac{2}{13}\right)$.

We can easily check our work by substituting the solution back into the original equations.

$$3x - 2y = 3 \cdot \frac{23}{13} - 2 \cdot \frac{2}{13} = \frac{69}{13} - \frac{4}{13} = \frac{65}{13} = 5 \checkmark$$

$$2x + 3y = 2 \cdot \frac{23}{13} + 3 \cdot \frac{2}{13} = \frac{46}{13} + \frac{6}{13} = \frac{52}{13} = 4 \checkmark$$

Naturally, these calculations can be performed on a calculator.

11.2 EXERCISES

In Exercises 1–6, determine whether the given ordered pair or ordered triple solves the given system of equations.

1. $(4, 1)$
 $3x - 5y = 7$
 $2x + 2y = 10$

2. $(7, -2)$
 $2x - 4y = 30$
 $x + y = 5$

3. $(-5, 3)$
 $3x + y = 4$
 $10x - 4y = -62$

4. $(-1, -2)$
 $2x + 2y = -5$
 $3x - 4y = 2$

5. $(4, -1, 2)$
 $2x + 3y - z = 3$
 $x + y - z = 5$
 $10x - 2y = 3$

6. $(0, 5, -1)$
 $3x - 2y + z = -11$
 $2x + 4y + z = 19$
 $x - z = 1$

In Exercises 7–12, do the following: (a) Find each line's slope and y-intercept. (b) Use the slopes and y-intercepts to determine whether the given system has no solution, one solution, or an infinite number of solutions. Do not actually solve the system.

7. $5x + 2y = 4$
 $6x - 19y = 72$

8. $3x + 132y = 19$
 $45x + 17y = 4$

9. $4x + 3y = 12$
 $8x + 6y = 24$

10. $19x - 22y = 1$
 $190x - 220y = 10$

11. $x + y = 7$
 $3x + 2y = 8$
 $2x + 2y = 14$

12. $3x - y = 12$
 $2x + 3y = 5$
 $5x + 2y = 17$

In Exercises 13–18, determine whether the given system could have a single solution. Do not actually solve the system.

13. $3x - 2y + 5z = 1$
 $2x + y = 2$
 $5x + 7y - z = 0$

14. $8x - 4y + 2z = 10$
 $3x + y + z = 1$
 $4x - 2y + z = 5$

15. $x + y + z = 1$
 $2x + 2y + 2z = 2$
 $3x - y + 10z = 45$

16. $x + y + z = 34$
 $5x - y + 2z = 9$
 $3x + 3y + 3z = 102$

17. $x + 2y + 3z = 4$
 $5x - y = 2$

18. $9x - 21y = 476$
 $x + 3y + z = 12$

Solve the systems in Exercises 19–24 with the elimination method. Check your answers by substituting them back in. (Answers are not given at the back of the book.)

19. $2x + 3y = 5$
 $4x - 2y = 2$

20. $5x + 3y = 11$
 $2x + 7y = 16$

21. $3x - 7y = 27$
 $4x - 5y = 23$

22. $5x - 2y = -23$
 $x + 2y = 5$

23. $5x - 9y = -12$
 $3x + 7y = -1$

24. $5x - 12y = 9$
 $3x + 3y = 2$

Answer the following questions using complete sentences and your own words.

• CONCEPT QUESTIONS

25. By sketching lines, determine whether a system of three equations in two unknowns could have no solution, one solution, or an infinite number of solutions. What if the three equations were unique?

26. By sketching planes, determine whether a system of four equations in three unknowns could have no solution, one solution, or an infinite number of solutions. What if the four equations were unique?

27. Solving a system of two equations in two unknowns is the same as finding all points that are on the graph of both of the two equations. Why do we not solve systems by graphing them (by hand) and locating the points that are on both graphs?

28. What does the graph of a linear equation in two unknowns look like? What does the graph of a linear equation in three unknowns look like? Why is a linear equation in three unknowns called a linear equation?

29. Suppose you have a linear equation in two unknowns, and you multiply both sides of that equation by a constant (not zero). How would the graph of the new equation compare to that of the original equation? Would they be parallel lines, perpendicular lines, the same lines, or what? Why?

SOLVING SYSTEMS OF TWO LINEAR EQUATIONS IN TWO UNKNOWNS ON A GRAPHING CALCULATOR

A system of two linear equations in two unknowns can be solved nicely on a graphing calculator.

> ## SOLVING SYSTEMS OF TWO LINEAR EQUATIONS IN TWO UNKNOWNS ON A GRAPHING CALCULATOR
>
> **Step 1** Solve each equation for y.
>
> **Step 2** Enter the results into the graphing calculator as Y_1 and Y_2.
>
> **Step 3** Graph the two equations on the graphing calculator.
>
> **Step 4** Use the calculator to find the point of intersection. (The details are given in Appendices C and D.)
>
> **Step 5** Check your solution by substituting it into each of the original equations.

EXAMPLE 3

USING A GRAPHING CALCULATOR Solve the following system:

$$3x - 2y = 5$$
$$2x + 3y = 4$$

SOLUTION

Step 1   *Solve each equation for y.*

$$3x - 2y = 5 \Rightarrow -2y = -3x + 5 \Rightarrow \frac{-2y}{-2} = \frac{-3x + 5}{-2} \Rightarrow y = \frac{3}{2}x - \frac{5}{2}$$

$$2x + 3y = 4 \Rightarrow 3y = -2x + 4 \Rightarrow \frac{3y}{3} = \frac{-2x + 4}{3} \Rightarrow y = -\frac{2}{3}x + \frac{4}{3}$$

Step 2   *Enter the results into the graphing calculator as Y_1 and Y_2. See Figure 11.30.*

FIGURE 11.30 A TI's "Y=" screen.

Step 3   *Graph the two equations on the graphing calculator. See Figure 11.31.*

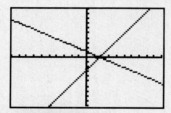

FIGURE 11.31 The graph of the system
$$3x - 2y = 5$$
$$2x + 3y = 4.$$

839

Step 4 | *Use the calculator to find the point of intersection.* See Figure 11.32.

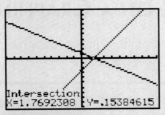

FIGURE 11.32   The solution of the system
$$3x - 2y = 5$$
$$2x + 3y = 4.$$

The system's solution is approximately (1.7692308, 0.15384615).

Step 5 | *Check your solution by substituting it into each of the original equations.*

$$3x - 2y = 3 \cdot 1.769230 - 2 \cdot 0.15384615 = 4.9999167 \approx 5$$
$$2x + 3y = 2 \cdot 1.769230 + 3 \cdot 0.15384615 = 3.99994445 \approx 4$$

The solution checks, but not perfectly. This is because the calculator rounded off the point of intersection in step 4.

EXERCISES

Solve the systems in Exercises 30–35 with a graphing calculator. Check your answers by substituting them back in. (Answers are not given at the back of the book.)

30. Exercise 20   **31.** Exercise 21

32. Exercise 22   **33.** Exercise 23

34. Exercise 24   **35.** Exercise 19

• **CONCEPT QUESTIONS**

36. Describe how you could use a graphing calculator to solve a system of three equations in two unknowns.

37. Why could you not use a graphing calculator to solve a system of two equations in three unknowns?

11.3   Long-Range Predictions with Markov Chains

OBJECTIVES

● Understand how to find the equilibrium matrix L

● Use L to find the long-term result of a trend

In Section 11.1, we were given Brute's current market share (25%), and we predicted Brute's market share after one, two, and three purchases (that is, we found the probabilities of the first, second, and third following states) by computing

PT, PT^2, and PT^3. In a similar way, we could predict Brute's market share after four purchases by computing PT^4. These predictions are shown in Figure 11.33.

Brute's Projected Market Share	Brute	Not Brute	Probability Matrix
Current purchase	25%	75%	P
First following purchase	50%	50%	$PT = PT^1$
Second following purchase	60%	40%	PT^2
Third following purchase	64%	36%	PT^3
Fourth following purchase	65.6%	34.4%	PT^4

FIGURE 11.33   Brute is gaining market share.

It appears that the trend in increasing market share created by Brute's new advertising campaign and customer satisfaction with the product will stabilize and that Brute will ultimately command close to 70% of the market.

If Brute started out with a much smaller initial market share, say, only 10% rather than 25%, a similar pattern emerges, as shown in Figure 11.34.

$$P = \begin{bmatrix} 0.1 & 0.9 \end{bmatrix} \quad \textbf{a new probability matrix}$$

$$T = \begin{bmatrix} 0.8 & 0.2 \\ 0.4 & 0.6 \end{bmatrix} \quad \textbf{the same transition matrix as before}$$

Brute's Projected Market Share	Brute	Not Brute	Probability Matrix
Current purchase	10%	90%	P
First following purchase	44%	56%	$PT = PT^1$
Second following purchase	57.6%	42.4%	PT^2
Third following purchase	63.04%	36.96%	PT^3
Fourth following purchase	65.22%	34.78%	PT^4

FIGURE 11.34   The same result, even with a different starting point.

Surprisingly, it appears that Brute's new advertising campaign and customer satisfaction with the product would have the same ultimate effect, regardless of Brute's initial market share. Under certain circumstances, PT^n will stabilize, and it will do so in a way that is unaffected by the value of P. Our goal in this section is to find the level at which the trend will stabilize.

The **equilibrium matrix L,** the matrix at which the trend stabilizes, is the probability matrix $L = PT^n$ such that all following probability matrices are equal. In other words, multiplying the equilibrium matrix L by T would have no effect, and $LT = L$. In fact, we find the equilibrium matrix by solving the matrix equation $LT = L$.

EXAMPLE **1** | **FINDING THE EQUILIBRIUM MATRIX L**   Make a long-range forecast for Brute's market share given the transition data from Section 11.1.

SOLUTION

We need to find the equilibrium matrix L. We find L by solving the matrix equation $LT = L$.

$$L = [x \quad y] \qquad \text{\textbf{L is a row matrix, since L is a probability matrix}}$$

$$T = \begin{bmatrix} 0.8 & 0.2 \\ 0.4 & 0.6 \end{bmatrix} \qquad \text{\textbf{from Section 11.1}}$$

$$\begin{array}{ccc} L & T & = & L \end{array} \qquad \text{\textbf{the equilibrium equation}}$$

$$[x \quad y]\begin{bmatrix} 0.8 & 0.2 \\ 0.4 & 0.6 \end{bmatrix} = [x \quad y] \qquad \text{\textbf{substituting for L and T}}$$

$$[0.8x + 0.4y \quad 0.2x + 0.6y] = [x \quad y] \qquad \text{\textbf{multiplying}}$$

This yields the system

$$0.8x + 0.4y = x$$
$$0.2x + 0.6y = y$$

Combining like terms gives

$$-0.2x + 0.4y = 0 \qquad \text{\textbf{subtracting x from each side of the first equation}}$$
$$0.2x - 0.4y = 0 \qquad \text{\textbf{subtracting y from each side of the second equation}}$$

These two equations are equivalent (we can multiply one equation by -1 to get the other), so we may discard one. This leaves us in a difficult spot: our single equation has an infinite number of solutions, and we are looking for a single solution. However, recall that $L = [x \quad y]$ is a probability matrix, so its entries must add to 1. Thus, $x + y = 1$ is our needed second equation. Now we can solve the system

$$0.2x - 0.4y = 0 \qquad \text{\textbf{either of the above two equations}}$$
$$x + y = 1 \qquad \text{\textbf{since [x \quad y] is a probability matrix}}$$

This system can easily be solved with the elimination method or a graphing calculator.
    Proceeding with the elimination method, we multiply the second equation by 0.4. The resulting system is

$$\begin{array}{ll} 0.2x - 0.4y = 0 & \text{\textbf{the first equation from above}} \\ \underline{+\ 0.4x + 0.4y = 0.4} & \text{\textbf{multiplying the second equation by 0.4}} \\ 0.6x + \quad 0y = 0.4 & \text{\textbf{adding}} \end{array}$$

$$\frac{0.6x}{0.6} = \frac{0.4}{0.6} \qquad \text{\textbf{solving for x}}$$

$$x = \frac{2}{3}$$

Substituting into $x + y = 1$ gives

$$\frac{2}{3} + y = 1 \qquad \text{\textbf{substituting } \tfrac{2}{3} \text{ for x}}$$

$$y = \frac{1}{3} \qquad \text{\textbf{solving for y}}$$

Thus, $L = [x \quad y] = [\tfrac{2}{3} \quad \tfrac{1}{3}]$, and Brute's market share should stabilize at $\tfrac{2}{3}$, or approximately 67%, as shown in Figure 11.35.

As a check, multiply L by T and see whether you get L.

$$LT = \begin{bmatrix} \dfrac{2}{3} & \dfrac{1}{3} \end{bmatrix}\begin{bmatrix} 0.8 & 0.2 \\ 0.4 & 0.6 \end{bmatrix}$$

$$= \begin{bmatrix} \dfrac{2}{3} & \dfrac{1}{3} \end{bmatrix}$$

FIGURE 11.35 As a result of a new advertising campaign and customers' satisfaction with the product, Brute is predicted to increase its market share from 25% . . .

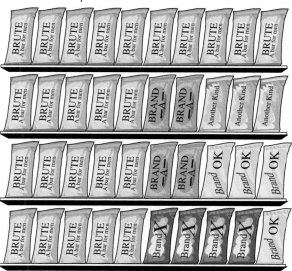

ultimately to 67%.

LONG-RANGE PREDICTIONS WITH REGULAR MARKOV CHAINS

Step 1 *Create the transition matrix T. T* is a square matrix, discussed in Section 11.1, whose entries are the probabilities of passing from current states to next following states. The rows refer to the current states and the columns refer to the next following states.

Step 2 *Create the equilibrium matrix L.* The equilibrium matrix L is the long-range prediction. L is the matrix that solves the matrix equation by $LT = L$. If there are two states, then $L = \begin{bmatrix} x & y \end{bmatrix}$.

Step 3 *Find and simplify the system of equations described by $LT = L$.*

Step 4 *Discard any redundant equations, and include the equation $x + y = 1$.*

Step 5 *Solve the resulting system.* Use the elimination method or a graphing calculator.

Step 6 ✔ *Check your work by verifying that $LT = L$.*

In Exercises 1–4, do the following.

a. *Find the equilibrium matrix L by solving LT = L for L.*

b. *Check your solution to part (b) by verifying that LT = L. (Answers are not given in the back of the book.)*

1. $T = \begin{bmatrix} 0.1 & 0.9 \\ 0.2 & 0.8 \end{bmatrix}$

2. $T = \begin{bmatrix} 0.5 & 0.5 \\ 0.6 & 0.4 \end{bmatrix}$

3. $T = \begin{bmatrix} 0.3 & 0.7 \\ 0.8 & 0.2 \end{bmatrix}$

4. $T = \begin{bmatrix} 0.6 & 0.4 \\ 0.25 & 0.75 \end{bmatrix}$

5. A marketing analysis shows that 12% of the consumers who do not currently drink KickKola will purchase KickKola the next time they buy a cola and that 63% of the consumers who currently drink KickKola will purchase it the next time they buy a cola. Make a long-range prediction of KickKola's ultimate market share, assuming that current trends continue. (See Exercise 9 in Section 11.1.)

6. A marketing analysis shows that 9% of the consumers who do not currently use SoftNWash will purchase SoftNWash the next time they buy a fabric softener and that 29% of the consumers who currently use SoftNWash will purchase it the next time they buy a fabric softener. Make a long-range prediction of SoftNWash's ultimate market share, assuming

The Monopoly board.

that current trends continue. (See Exercise 10 in Section 11.1.)

7. A census report shows that 32% of the residents of Metropolis own their own home and that 68% rent. The report shows that 12% of the renters plan on buying a home in the next 12 months and that 3% of the homeowners plan on selling their home and renting instead. Make a long-range prediction of the percentage of Metropolis residents that will own their own home and the percentage that will rent. Give two assumptions on which this prediction is based. (See Exercise 17 in Section 11.1.)

8. A marketing analysis indicates that 46% of the consumers who do not currently own a Clicker pen will purchase a Clicker the next time they buy a pen and that 37% of the consumers who currently own a Clicker will not purchase one the next time they buy a pen. If the average consumer buys a new pen every three weeks, make a long-range prediction of Clicker's ultimate market share, assuming that current trends continue. (See Exercise 20 in Section 11.1.)

9. A nationwide survey preformed by *Get Out of My Way* magazine indicates that 31% of the bike owners who do not currently own a Sierra Cruiser will purchase a Sierra Cruiser the next time they buy a mountain bike and that 12% of the bikers who currently own a Sierra Cruiser will purchase one the next time they buy a mountain bike. If the average customer buys a new mountain bike every two years, make a long-range prediction of Sierra Cruiser's ultimate market share, assuming that current trends continue. (See Exercise 19 in Section 11.1.)

WEB PROJECT

10. Monopoly is "the most played board game in the world," according to Hasbro, its manufacturer. Players take turns moving around the board, according to a roll of a pair of dice. The board consists of 40 squares, 28 of which are properties such as Marvin Gardens, B&O Railroad, and Water Works. There are also three Chance squares, three Community Chest squares, Luxury Tax, Income Tax, Jail, Free Parking, Go, and Go to Jail.

   This game can be analyzed by using Markov chains. There are 40 states, one for each square on the board. Read about the game's rules, and create a transition matrix for the game. Do not try to make a long-range prediction. Instead, discuss what information would be contained in such a prediction.

   HINT: The probabilities of moving from "Go" to Mediterranean Avenue, Community Chest, Baltic Avenue, Income Tax, Reading Railroad, or Oriental Avenue are each $\frac{1}{6}$.

   Some useful links for this web project are listed on the text companion web site. Go to **www.cengage .com/math/johnson** to access the web site.

11.4 Solving Larger Systems of Equations

OBJECTIVE

• Understand how to apply the elimination method to a larger system of equations

In this section, we will discuss how to use the elimination method to solve larger systems of equations (i.e., systems of more than two equations with more than two unknowns).

In the technology portion of this section, we will discuss how to use a graphing calculator and the Gauss-Jordan method to solve larger systems of equations. The Gauss-Jordan method is a technology-friendly version of the elimination method that uses matrices rather than equations.

Solving Larger Systems with the Elimination Method

Solving a system of three equations in three unknowns with the elimination method is quite similar to solving a system of two equations in two unknowns. There are a few more steps because there are more variables to eliminate.

> # SOLVING A SYSTEM OF THREE EQUATIONS IN THREE UNKNOWNS WITH THE ELIMINATION METHOD
>
> **Step 1** Combine two equations so that one variable is eliminated.
>
> **Step 2** Combine two different equations so that the same variable is eliminated.
>
> **Step 3** Steps 1 and 2 result in a system of two equations in two unknowns. Solve that system with the elimination method, as discussed in Section 11.2.
>
> **Step 4** Substitute back into various equations to find the values of the remaining variables.
>
> **Step 5** Check your answer by substituting it into the original equations.

EXAMPLE 1

SOLVING A LARGER SYSTEM Use the elimination method to solve the following system:

$$5x - 7y + 2z = 25$$
$$3x + 2y - 5z = -16$$
$$7x + 3y - 9z = -25$$

SOLUTION

Step 1

Combine two equations so that one variable is eliminated.
We'll combine the first two equations, and eliminate z. To do this, multiply the first equation by 5 and the second by 2, and add the results.

$$
\begin{array}{rl}
25x - 35y + 10z = 125 & \textbf{5 times equation 1} \\
+ 6x + 4y - 10z = -32 & \textbf{2 times equation 2} \\
\hline
31x - 31y = 93 & \textbf{adding}
\end{array}
$$

Step 2

Combine two different equations so that the same variable is eliminated.
We'll combine the last two equations and eliminate z. To do this, multiply the second equation by 9 and the third by -5, and add the results.

$$
\begin{array}{rl}
27x + 18y - 45z = -144 & \textbf{9 times equation 2} \\
+ {-35x} - 15y + 45z = 125 & \textbf{-5 times equation 3} \\
\hline
-8x + 3y = -19 & \textbf{adding}
\end{array}
$$

Step 3

Steps 1 and 2 result in a system of two equations in two unknowns. Solve that system with the elimination method, as discussed in Section 11.2. We are left with a smaller system to solve:

$$
\begin{array}{rl}
31x - 31y = 93 & \textbf{the result of step 1} \\
-8x + 3y = -19 & \textbf{the result of step 2}
\end{array}
$$

First, simplify the first equation of this new system by dividing through by 31:

$$
\begin{array}{rl}
x - y = 3 & \textbf{dividing the first equation by 31} \\
-8x + 3y = -19 & \textbf{the second equation}
\end{array}
$$

We'll eliminate y if we multiply the first equation by 3 and add the result to the second equation.

$$
\begin{array}{rl}
3x - 3y = 9 & \textbf{3 times equation 1} \\
+ {-8x} + 3y = -19 & \textbf{equation 2} \\
\hline
-5x = -10 & \textbf{adding} \\
-5x/{-5} = -10/{-5} & \textbf{solving for } x \\
x = 2 &
\end{array}
$$

Step 4 | *Substitute back into various equations to find the values of the remaining variables.*
We can find y by substituting $x = 2$ into $x - y = 3$

$$2 - y = 3 \quad \text{substituting 2 for } x$$
$$-y = 1 \quad \text{solving for } y$$
$$y = -1$$

We can find z by substituting $x = 2$ and $y = -1$ into $5x - 7y + 2z = 25$:

$$5x - 7y + 2z = 25$$
$$5 \cdot 2 - 7 \cdot (-1) + 2z = 25 \quad \text{substituting 2 for } x \text{ and } -1 \text{ for } y$$
$$17 + 2z = 25 \quad \text{simplifying}$$
$$2z = 8 \quad \text{solving for } z$$
$$z = 8/2 = 4$$

The system's solution is $(x, y, z) = (2, -1, 4)$.

Step 5 | *Check your answer by substituting it into the original equations.* It's not necessary to substitute this solution into the original system's first equation, because in step 4 we just finished forcing that equation to work.

$$3x + 2y - 5z = -16 \quad \Rightarrow \quad 3 \cdot 2 + 2 \cdot (-1) - 5 \cdot 4 = -16 \quad \checkmark$$
$$7x + 3y - 9z = -25 \quad \Rightarrow \quad 7 \cdot 2 + 3 \cdot (-1) - 9 \cdot 4 = -25 \quad \checkmark$$

11.4 EXERCISES

Solve the given system with the elimination method.

1. $5x + y - z = 17$
$2x + 5y + 2z = 0$
$3x + y + z = 11$

2. $9x - 2y + 4z = 29$
$2x + 3y - 4z = 3$
$x + y + z = 1$

3. $x + y + z = 14$
$3x - 2y + z = 3$
$5x + y + 2z = 29$

4. $2x + y - z = 4$
$3x + 2y - 7z = 0$
$5x - 3y + 2z = 20$

5. $x - y + 4z = -13$
$2x - z = 12$
$3x + y = 25$

6. $2x + 2y + z = 5$
$3x - y = 5$
$2y + 7z = -9$

7. $8x + 7y - 3z = 38$
$4x - 3y + 2z = 11$
$9x - 11y + 5z = 7$

8. $4x - 3y - 2z = 42$
$5x + 5y + 4z = 78$
$2x - 9y - 6z = 6$

9. $4x + 4y + 2z = 10$
$3x - 2y - 2z = 18$
$7x - y + 8z = 4$

10. $11x + 17y - 13z = -186$
$12x + 32y + 64z = -92$
$15x - 41y + 88z = 1132$

11. $47x + 58y + 37z = -113$
$22x - 37y + 27z = 332$
$47x + 15y + 52z = 145$

12. $74x - 22y - 48z = -1052$
$49x - 75y - 47z = 506$
$53x + 10y + 77z = 4608$

SOLVING LARGER SYSTEMS OF LINEAR EQUATIONS ON A GRAPHING CALCULATOR

We can't solve a system of equations in three unknowns by graphing the system, as we did in Section 11.2, because a graphing calculator cannot graph an equation in three unknowns.

We can, however, solve such a system using a graphing calculator and the Gauss-Jordan method. The *Gauss-Jordan method* is a technology-friendly version of the elimination method that uses matrices rather than equations.

Solving Systems on a Graphing Calculator

EXAMPLE 2

USING TECHNOLOGY Use the Gauss-Jordan method and a graphing calculator to solve the following system:

$$2x + z = y + 9$$
$$x + 2y + z = 4$$
$$4x - 3z = -4$$

SOLUTION

Step 1

Write the system with all variables on the left side of the equations and with all coefficients showing. Specifically, what needs to be done?

- The first equation has the variable y on the right side, and this method requires that all variables be on the left side. We'll move y to the left side by subtracting y from each side.
- In the first equation, both z and y lack coefficients. We'll insert coefficients of 1. We'll do the same thing with the second equation.
- Finally, the third equation lacks a y term. We'll insert $0y$.

This gives us

$$2x - 1y + 1z = 9$$
$$1x + 2y + 1z = 4$$
$$4x + 0y - 3z = -4$$

Step 2

Rewrite the system in matrix form. Eliminate the letters, addition symbols, and equal symbols.

$$\begin{matrix} x & y & z & \\ \begin{bmatrix} 2 & -1 & 1 & 9 \\ 1 & 2 & 1 & 4 \\ 4 & 0 & -3 & -4 \end{bmatrix} \end{matrix}$$

Notice that we've labeled the first three columns x, y, and z. This is an optional way of keeping track of the origins of the numbers.

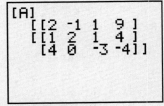

FIGURE 11.36 Matrix [A].

Step 3

Enter the matrix into your graphing calculator, as matrix [A]. This was discussed in Section 11.0.

Step 4

Have your calculator put the matrix in reduced row echelon form (or "rref"). To do this, make the screen read "rref([A]" on a TI-83/84, or "rref A" on a TI-86 by typing:

TI-83:	MATRX, scroll to "MATH" and select "rref(" (You will have to use the down arrow button to find it.)	MATRX 1 ENTER
TI-84:	2ND MATRX, scroll to "MATH" and select "rref(" (You will have to use the down arrow button to find it.)	MATRX 1 ENTER
	This generates "rref(" or "rref".	*This generates "[A]" or "A".*

Once "rref([A]" is on the screen, press ENTER, and the screen will appear as in Figure 11.37.

The "rref" command has the calculator add multiples of equations together to eliminate variables and solve the system.

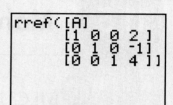

Step 5   *Rewrite the matrix in system form.* This is the reverse of step 2, in which we put the system in matrix form.

The matrix is

$$\begin{array}{ccc} x & y & z \\ \begin{bmatrix} 1 & 0 & 0 & 2 \\ 0 & 1 & 0 & -1 \\ 0 & 0 & 1 & 4 \end{bmatrix} \end{array}$$

FIGURE 11.37   Matrix [A] in rref (reduced row echelon form).

This translates into the system

$$1x + 0y + 0z = 2$$
$$0x + 1y + 0z = -1$$
$$0x + 0y + 1z = 4$$

The system simplifies to

$x = 2$ **simplifying the above first equation**
$y = -1$ **simplifying the second equation**
$z = 4$ **simplifying the third equation**

So the solution is $(x, y, z) = (2, -1, 4)$.

Step 6   *Check your answer by substituting it into the system.*

$$2x - 1y + 1z = 2 \cdot (2) - 1 \cdot (-1) + 1 \cdot (4) = 9 \checkmark$$
$$1x + 2y + 1z = 1 \cdot (2) + 2 \cdot (-1) + 1 \cdot (4) = 4 \checkmark$$
$$4x + 0y - 3z = 4 \cdot (2) + 0 \cdot (-1) - 3 \cdot (4) = -4 \checkmark$$

EXERCISES

In Exercises 13–24, solve the given system with the Gauss-Jordan method and a graphing calculator.

13.   The system in Exercise 1
14.   The system in Exercise 2
15.   The system in Exercise 3
16.   The system in Exercise 4
17.   The system in Exercise 5
18.   The system in Exercise 6
19.   The system in Exercise 7
20.   The system in Exercise 8
21.   The system in Exercise 9
22.   The system in Exercise 10
23.   The system in Exercise 11
24.   The system in Exercise 12

In Exercises 25–28, solve the given system with the Gauss-Jordan method but without using a graphing calculator.

25. $2x - 3y = 22$
$6x + 7y = 2$

26. $3x + 5y = -17$
$x + 8y = -31$

27. $4x - 6y = -10$
$2x + 9y = 31$

28. $3x + 5y = 7$
$x - 7y = -15$

11.5 More on Markov Chains

OBJECTIVE

- Understand how to find L in a problem with more than two states

In this section, we will continue to make long-range predictions with Markov chains, using the same steps that we used in Section 11.3. However, the problems in this section will involve more than two unknowns. The resulting systems can be solved by using either the elimination method or the graphing calculator and the Gauss-Jordan method, as discussed in Section 11.4.

EXAMPLE 1

IS THE MIDDLE CLASS SHRINKING? Sociologists have found that a strong determinant of an individual's income is the income of his or her parents. In Example 10 of Section 11.1, we used transitional data on the lower income group, the middle income group, and the upper income group to create a transition matrix T. Use T to make a long-range prediction of the levels at which the lower, middle, and upper income groups will stabilize.

SOLUTION

We are to find the equilibrium matrix L.

Step 1 *Create the transition matrix T. We did this in Section 11.1.*

$$T = \begin{bmatrix} 0.78 & 0.21 & 0.01 \\ 0.06 & 0.90 & 0.04 \\ 0.01 & 0.10 & 0.89 \end{bmatrix}$$

Step 2 *Create the equilibrium matrix L. L is a row matrix, and in this problem, L has three entries: one for lower income group, one for middle income group, and one for upper income group.*

$$L = \begin{bmatrix} x & y & z \end{bmatrix}$$

Step 3 *Find and simplify the system of equations described by $LT = L$.*

$$\begin{array}{ccc} L & T & = & L \end{array}$$

$$\begin{bmatrix} x & y & z \end{bmatrix} \begin{bmatrix} 0.78 & 0.21 & 0.01 \\ 0.06 & 0.90 & 0.04 \\ 0.01 & 0.10 & 0.89 \end{bmatrix} = \begin{bmatrix} x & y & z \end{bmatrix} \quad \textbf{substituting for } L \textbf{ and } T$$

$$\begin{bmatrix} 0.78x + 0.06y + 0.01z & 0.21x + 0.90y + 0.10z & 0.01x + 0.04y + 0.89z \end{bmatrix}$$
$$= \begin{bmatrix} x & y & z \end{bmatrix} \quad \textbf{multiplying matrices}$$

This matrix equation describes the following system:

$$0.78x + 0.06y + 0.01z = x$$
$$0.21x + 0.90y + 0.10z = y$$
$$0.01x + 0.04y + 0.89z = z$$

This system can be simplified by combining like terms.

$$-0.22x + 0.06y + 0.01z = 0 \quad \textbf{subtracting } x \textbf{ in the first equation above}$$
$$0.21x - 0.10y + 0.10z = 0 \quad \textbf{subtracting } y \textbf{ in the second equation}$$
$$0.01x + 0.04y - 0.11z = 0 \quad \textbf{subtracting } z \textbf{ in the third equation}$$

The system can be further simplified by multiplying by 100.

$$-22x + 6y + 1z = 0$$
$$21x - 10y + 10z = 0$$
$$1x + 4y - 11z = 0$$

Step 4   *Discard any redundant equations, and include the equation $x + y + z = 1$.* The third equation is the negative of the sum of the first two equations, so its presence is redundant. Thus, any one of the three equations can be dropped. Also, $L = [x \quad y \quad z]$ is a probability matrix, so the sum of its entries must be 1. That is, $x + y + z = 1$.

$$-22x + 6y + 1z = 0 \quad \text{the above first equation}$$
$$21x - 10y + 10z = 0 \quad \text{the above second equation}$$
$$x + y + z = 1 \quad \text{since } [x \quad y \quad z] \text{ is a probability matrix}$$

Step 5   *Solve the resulting system.*
We'll combine the first and third equations and eliminate x. To do this, multiply the third equation by 22, and add the result to the first equation.

$$
\begin{array}{rl}
-22x + 6y + 1z = 0 & \text{the first equation} \\
+ \quad 22x + 22y + 22z = 22 & \text{22 times the third equation} \\
\hline
28y + 23z = 22 & \text{adding}
\end{array}
$$

Next, we'll combine two different equations so that x is eliminated. We'll multiply the third equation by -21 and add the result to the second equation.

$$
\begin{array}{rl}
21x - 10y + 10z = 0 & \text{the second equation} \\
+ \quad -21x - 21y - 21z = -21 & -21 \text{ times the third equation} \\
\hline
-31y - 11z = -21 & \text{adding}
\end{array}
$$

Finally, we'll solve the system created by the above two calculations.

$$28y + 23z = 22 \quad \text{the result of the first calculation}$$
$$-31y - 11z = -21 \quad \text{the result of the second calculation}$$

We'll eliminate z if we multiply the first equation by 11 and the second by 23, and we'll add the results.

$$
\begin{array}{rl}
308y + 253z = 242 & \textbf{11 times the first equation} \\
+ \quad -713y - 253z = -483 & \textbf{23 times the second calculation} \\
\hline
-405y = -241 & \textbf{adding} \\
-405y/-405 = -241/-405 & \textbf{solving for } y \\
y = 241/405 = 0.59506\ldots \approx 0.5951
\end{array}
$$

Substitute this into one of the above equations with only y and z in it to get z.

$$
\begin{array}{rl}
28y + 23z = 22 & \textbf{an earlier result} \\
28 \cdot (241/405) + 23z = 22 & \textbf{substituting for } y \\
6748/405 + 23z = 8910/405 & \textbf{writing with a common denominator} \\
23z = 8910/405 - 6748/405 & \textbf{solving for } z \\
23z = 2162/405 & \textbf{dividing} \\
z = \dfrac{2162}{405} \cdot \dfrac{1}{23} = \dfrac{94}{405} = 0.23209\ldots \approx 0.2321
\end{array}
$$

Substitute this into one of the original equations to get x.

$$
\begin{array}{rl}
x + y + z = 1 & \textbf{an original equation} \\
x + 241/405 + 94/405 = 1 & \textbf{substituting for } y \text{ and } z \\
x + 335/405 = 405/405 & \textbf{writing with a common denominator} \\
x = 70/405 = 0.17283\ldots \approx 0.1728 & \textbf{solving for } x
\end{array}
$$

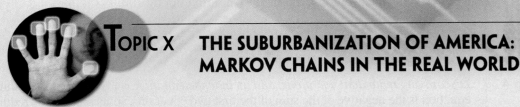

TOPIC X THE SUBURBANIZATION OF AMERICA: MARKOV CHAINS IN THE REAL WORLD

The suburbs are a big part of American life. This wasn't always the case, though. The suburbanization of America began immediately after World War II, when returning war veterans needed housing. Cheaper building materials and government-backed mortgages allowed homebuilders to meet this demand by building new housing developments where land was cheap and plentiful—in the suburbs. In the 1950s, 20 million people moved to developments such as the three Levittowns: Levittown, New York; Levittown, Pennsylvania; and Willingboro, New Jersey.

People are still moving from the cities to the suburbs. How big is this trend? How important is it? We'll find out in Exercise 9.

In the 1970s, the Midwest hit the economic skids. Factories were abandoned, and unemployment was rampant. Manufacturers were relocating in the South. As a result, there was a great migration from the Rust Belt to the Sun Belt. We'll find out in Exercise 12 whether this is still an important trend.

From Electrical Merchandising, July 1957

Courtesy General Electric

The resulting solution is

$$L = [70/405 \quad 241/405 \quad 94/405] \approx [0.1728 \quad 0.5951 \quad 0.2321]$$

Step 6 *Check your work by verifying that LT = L.*

$$
\begin{array}{ccc}
L & T & = \quad L
\end{array}
$$

$$
[70/405 \quad 241/405 \quad 94/405] \cdot
\begin{bmatrix}
0.78 & 0.21 & 0.01 \\
0.06 & 0.90 & 0.04 \\
0.01 & 0.10 & 0.89
\end{bmatrix}
\approx [0.1728 \quad 0.5951 \quad 0.2321]
$$

Our solution checks. ✓

The solution, rounded to the nearest 1%, is [17% 60% 23%]. This means that if current trends continue, the lower income group will eventually stabilize at

17% of the population, the middle income group will stabilize at 60%, and the upper income group will stabilize at 23%. (Recall that in 1999, the lower income group was 12%, the middle income group was 62%, and the upper income group was 26%, according to the U.S. Census Bureau.) It is important to remember that this prediction is based on the assumption that the current trends in the shift of people between the income groups will continue.

11.5 EXERCISES

In Exercises 1– 4, do the following.

a. Find the equilibrium matrix L by solving LT = L for L.
b. Check your solution to part (b) by verifying that LT = L.

1. $T = \begin{bmatrix} 0.3 & 0.2 & 0.5 \\ 0.1 & 0.8 & 0.1 \\ 0.4 & 0.3 & 0.3 \end{bmatrix}$

2. $T = \begin{bmatrix} 0.4 & 0.3 & 0.3 \\ 0.2 & 0.7 & 0.1 \\ 0.3 & 0.3 & 0.4 \end{bmatrix}$

3. $T = \begin{bmatrix} 1/2 & 1/4 & 1/4 \\ 2/3 & 1/6 & 1/6 \\ 0 & 0 & 1 \end{bmatrix}$

4. $T = \begin{bmatrix} 1/8 & 2/8 & 5/8 \\ 3/8 & 2/8 & 3/8 \\ 1/2 & 1/8 & 3/8 \end{bmatrix}$

5. An extensive survey of gym users in Metropolis indicates that 71% of the current members of Silver's will continue their annual membership when it expires, 12% will quit and join Fitness Lab, and the rest will quit and join ThinNFit. Fitness Lab has been unable to keep its equipment in good shape, and as a result 32% of its members will defect to Silver's, and 34% will leave for ThinNFit. ThinNFit's members are quite happy, and as a result, 96% plan on renewing their annual membership, with half of the rest planning on moving to Silver's and half to Fitness Lab. Make a long-range prediction of the ultimate market shares of the three health clubs, assuming that current trends continue. (See Exercise 11 in Section 11.1.)

6. Smallville has three Chinese restaurants: Asia Gardens, Chef Chao's, and Chung King Village. Chung King Village recently mailed a coupon to all residents of Smallville offering two dinners for the price of one. As a result, 67% of those who normally eat at Asia Gardens plan on trying Chung King Village within the next month, and 59% of those who normally eat at Chef Chao's plan on trying Chung King Village. Also, all of Chung King Village's regular customers will return to

take advantage of the special. And 30% of the people who normally eat at Asia Gardens will eat there within the next month because of its convenient location. Chef Chao's has a new chef who isn't doing very well, and as a result, only 15% of the people who normally eat at there are planning on returning next time. Make a long-range prediction of the ultimate market shares of the three restaurants, assuming that current trends continue. (See Exercise 12 in Section 11.1.)

7. A survey shows that 23% of the shoppers in seven Midwestern states regularly buy their groceries at Safe Shop, 29% regularly shop at PayNEat, and the balance shop at any one of several smaller markets. The survey indicates that 8% of the consumers who currently shop at Safe Shop will purchase their groceries at PayNEat the next time they shop and that 5% will switch to some other store. Also, 12% of the consumers who currently shop at PayNEat will purchase their groceries at Safe Shop the next time they shop, and 2% will switch to some other store. In addition, 13% of the consumers who currently shop at neither store will purchase their groceries at Safe Shop they next time they shop, and 10% will shop at PayNEat. Predict the percentage of Midwestern shoppers that will be regular Safe Shop customers and the percent that will be regular PayNEat customers as a result of this trend. (See Exercise 18 in Section 11.1.)

8. In October 1990, a New York Times/CBS News poll contained the information shown in Figure 11.38 on page 854. For example, 78% of those registered as a Democrat reported that they would vote for a Democrat in the next election.

   a. Complete the chart.

   b. What assumption must be made in completing the chart?

   c. If the poll were interpreted as predicting a change in registered party affiliation, make a long-range prediction of the percentage of voters who will be registered as Democrats, as Republicans, and as Independents.

   d. Do you think that the prediction in part (b) would be a good prediction? Why?

	Next Vote Will Be for a Democrat	Next Vote Will Be for a Republican	Next Vote Will Be for an Independent
Registered as a Democrat	78%	6%	?
Registered as a Republican	6%	75%	?
Registered as an Independent	30%	29%	?

FIGURE 11.38 Party membership changes. *Source: New York Times,* October 12 1990, page A21.

• PROJECTS

9. The U.S. Census Bureau classifies all residents of the United States as residents of a central city, a suburb, or a nonmetropolitan area. In the 1990 census, the bureau reported that the central cities held 77.8 million people, the suburbs 114.9 million people, and the non-metropolitan areas 56.0 million people. The Census Bureau (in Current Population Reports, P20–463, *Geographic Mobility, March 1990 to March 1991*) also reported the information in Figure 11.39 regarding migration between these three areas from 1990 to 1991. (Numbers are in thousands.)

Moved From	Moved to		
	Central City	Suburb	Nonmetropolitan Area
Central City	(x)	4946	736
Suburb	2482	(x)	964
Nonmetropolitan Area	741	1075	(x)

Note: Persons moving from one central city to another were not considered. This is represented by the "(x)" in the upper-left corner. The other "(x)'s" have similar meanings.

FIGURE 11.39 Migration data for Exercise 9.

a. Compute the probabilities that a resident of any one type of environment will move to any of the other types. Present these probabilities in the form of a transition matrix. (Round off to three decimal places.) (See Exercise 21 in Section 11.1.)

b. Make a long-range prediction of the ultimate percentage of U.S. residents who will reside in a central city, a suburb, and a nonmetropolitan area, assuming that the trend indicated by the above data continues.

10. Snapdragons have no color dominance. A snapdragon with two red genes will have red flowers, a snapdragon with one red gene and one white gene will have pink flowers, and a snapdragon with two white genes will have white flowers. A commercial nursery crosses many snapdragons (some with red flowers, some with pink flowers, and some with white flowers) with pink flowered plants.

a. Use Punnett squares to create a transition matrix for this situation. (Punnett squares are covered in Section 3.2.)

b. The nursery crosses snapdragons from each generation with pink-flowered plants to produce the next generation, over and over again. Find the percentages at which the snapdragon will stabilize.

11. In May 1992, the *New York Times* reported on a University of Michigan study on intergenerational transition between income groups. That study contained the information shown in Figure 11.40. For example, 30% of those whose fathers were in the lowest income group stayed in the lowest income group.

a. Make a long-range prediction of the percent of males in the various income groups.

b. The predictions in part (a) are not 10%, 40%, 40%, and 10%. Should they be? Explain.

HINT: Does "the bottom 10%" mean the same income bracket for the son as it does for the father?

	Son's Income in Bottom 10%	Son's Income from 10% to 50%	Son's Income from 50% to 90%	Son's Income in Top 10%
Father's Income in Bottom 10%	30%	52%	17%	1%
Father's Income from 10% to 50%	10%	48%	38%	4%
Father's Income from 50% to 90%	4%	38%	48%	10%
Father's Income in Top 10%	1%	17%	52%	30%

FIGURE 11.40 Intergenerational income group transition data for Exercise 11. *Source: New York Times,* May 18 1992, page D5. Data have been adjusted so that percentages add to 100%.

c. What assumptions are made in the prediction in part (a)? Consider the fact that some men have no sons, some have one son, and some have many sons. Also consider the fact that only males are involved in the study.

12. The Census Bureau (in *Statistical Abstract of the United States, 2000*) reported the 2000 national and regional populations in Figure 11.41.

Region	Population (in thousands)
United States	281,422
Northeast	53,594
Midwest	64,393
South	100,237
West	63,198

FIGURE 11.41 U.S. regional population. *Source:* U.S. Census Bureau, *Statistical Abstract of the United States, 2000.*

The Census Bureau also reported (*Geographical Mobility: March 1999 to March 2000*) the information in Figure 11.42 regarding the movement between the regions from 1999 to 2000 (numbers are in thousands).

Region Moved from	Region Moved to			
	Northeast	Midwest	South	West
Northeast	(x)	83	405	127
Midwest	72	(x)	364	204
South	183	416	(x)	432
West	108	223	489	(x)

FIGURE 11.42 Geographical mobility. *Source:* U.S. Census Bureau, *Current Population Reports P20–538, Geographical Mobility: March 1999 to 2000.*

a. Compute the probabilities that a resident of any one region will move to any of the other regions. Present these probabilities in the form of a transition matrix. Assume that the transitional probabilities for 2000–2001 are equal to those for 1999–2000. (Round off to three decimal places.) (See Exercise 22 in Section 11.1.)

b. Make a long-range prediction of the ultimate percent of U.S. residents who will reside in the Northeast, the Midwest, the South, and the West, assuming that the trend described by the above data continues.

11 CHAPTER REVIEW

TERMS

associative property of matrix multiplication
column
column matrix
current states
diagonal

dimensions of a matrix
element
entry
equilibrium matrix
following states
identity matrix
identity property of matrix multiplication

linear equation
Markov chain
matrix
matrix multiplication
probability matrix
row
row matrix

solving a system of equations
square matrix
states
system of equations
transition matrix

PROCEDURES

To make **short-range predictions** with Markov chains:

1. *Create the probability matrix P*, which is a row matrix whose entries are the initial probabilities of the various states.

2. *Create the transition matrix T*, which is a square matrix whose entries are the probabilities of passing from current states to first following states.

3. *Calculate PT^n*, which is a row matrix whose entries are the probabilities of the nth following states.

To make **long-range-predictions** with Markov chains:

1. *Create the transition matrix T.*
2. *Create the equilibrium matrix L, using variables.*
3. *Find and simplify the system of equations described by LT = L.*

4. *Discard any redundant equations and include the equation x + y = 1 (or x + y + z = 1 and so on).*
5. *Solve the resulting system.*
6. *Check your work by verifying that LT = L.*

REVIEW EXERCISES

In Exercises 1–4, determine whether the given matrix is a row matrix, a column matrix, a square matrix, or neither.

1. $\begin{bmatrix} 5 & 7 \\ 3 & 9 \\ 1 & 0 \end{bmatrix}$

2. $\begin{bmatrix} 6 & 3 & 8 \\ 1 & 6 & 3 \\ 4 & 9 & 6 \end{bmatrix}$

3. $\begin{bmatrix} 8 \\ 2 \\ 4 \\ 7 \end{bmatrix}$

4. $\begin{bmatrix} 4 & 67 & 43 & 27 \end{bmatrix}$

In Exercises 5–10, find the indicated product if it exists.

5. $\begin{bmatrix} 1 & 6 \\ 8 & 4 \end{bmatrix} \cdot \begin{bmatrix} -3 & 6 \\ 0 & -5 \end{bmatrix}$

6. $\begin{bmatrix} -3 & 6 \\ 0 & -5 \end{bmatrix} \cdot \begin{bmatrix} 1 & 6 \\ 8 & 4 \end{bmatrix}$

7. $\begin{bmatrix} 6 & 3 & 5 \\ 0 & 2 & 5 \end{bmatrix} \cdot \begin{bmatrix} -3 & 6 \\ 0 & -5 \end{bmatrix}$

8. $\begin{bmatrix} -3 & 6 \\ 0 & -5 \end{bmatrix} \cdot \begin{bmatrix} 6 & 3 & 5 \\ 0 & 2 & 5 \end{bmatrix}$

9. $\begin{bmatrix} 5 & 3 & -9 \\ 1 & 0 & 5 \end{bmatrix} \cdot \begin{bmatrix} 7 & 4 & 6 & 9 \\ 3 & 5 & -7 & 2 \\ 1 & 6 & 9 & 3 \end{bmatrix}$

10. $\begin{bmatrix} 7 & 4 & 6 & 9 \\ 3 & 5 & -7 & 2 \\ 1 & 6 & 9 & 3 \end{bmatrix} \cdot \begin{bmatrix} 5 & 3 & -9 \\ 1 & 0 & 5 \end{bmatrix}$

11. Johnco Industries is planning a new promotional campaign for its Veg-O-Slicer. Use matrix multiplication and Figure 11.43 to find the cost of three newspaper ads and four television ads in New York City and the cost of a similar campaign in Washington, D.C.

	Newspaper Ad	Television Ad
New York City	$5000	$7000
Washington, D.C.	$5500	$6200

FIGURE 11.43 Ad data for Exercise 11.

12. Is matrix multiplication commutative? Is it associative?

In Exercises 13–18, do the following.

a. *Solve the system of equations.*
b. *Check your answer by substituting it back in.*

13. $3x + 5y = -14$
    $4x - 7y = 36$

14. $10x - 11y = 28$
    $5x + 7y = 39$

15. $5x - 7y = -29$
    $2x + 4y = 2$

16. $8x - 7y = 54$
    $3x + 15y = -15$

17. $7x - 8y + 3z = -43$
    $5x + 3y - z = 24$
    $11x - 5y + 12z = 0$

18. $5x - 4y + 3z = 10$
    $6x - 3y - 3z = 9$
    $11x - 15y + 10z = 13$

19. $P = \begin{bmatrix} 0.63 & 0.37 \end{bmatrix}$, and $T = \begin{bmatrix} 0.3 & 0.7 \\ 0.1 & 0.9 \end{bmatrix}$.

    a. What is the probability of moving from state 1 (as a current state) to state 2 (as a first following state)?
    b. What is state 2's current market share?
    c. Find *PT*.
    d. Find PT^2.
    e. Draw a tree diagram illustrating *P* and *T*.

20. $P = \begin{bmatrix} 0.55 & 0.45 \end{bmatrix}$, and $T = \begin{bmatrix} 0.4 & 0.6 \\ 0.5 & 0.5 \end{bmatrix}$.

    a. What is the probability of moving from state 1 (as a current state) to state 2 (as a first following state)?
    b. What is state 2's current market share?
    c. Find *PT*.
    d. Find PT^2.
    e. Draw a tree diagram illustrating *P* and *T*.

In Exercises 21–26, do the following.

a. *Find the equilibrium matrix L by solving LT = L for L.*
b. *Check your answer by verifying that LT = L.*

21. $T = \begin{bmatrix} 0.7 & 0.3 \\ 0.2 & 0.8 \end{bmatrix}$

22. $T = \begin{bmatrix} 0.8 & 0.2 \\ 0.2 & 0.8 \end{bmatrix}$

23. $T = \begin{bmatrix} 0.4 & 0.6 \\ 0.1 & 0.9 \end{bmatrix}$

24. $T = \begin{bmatrix} 0.8 & 0.1 & 0.1 \\ 0.2 & 0.3 & 0.5 \\ 0.4 & 0.3 & 0.3 \end{bmatrix}$

25. $T = \begin{bmatrix} 0.5 & 0.1 & 0.4 \\ 0.2 & 0.2 & 0.6 \\ 0.4 & 0.3 & 0.3 \end{bmatrix}$ **26.** $T = \begin{bmatrix} 0.4 & 0.4 & 0.2 \\ 0.2 & 0.3 & 0.5 \\ 0.1 & 0.3 & 0.6 \end{bmatrix}$

In Exercises 27–30, use the following information. Department of Motor Vehicles records indicate that 12% of the automobile owners in the state of Jefferson own a Toyonda. A recent survey of Jefferson automobile owners, commissioned by Toyonda Motors, shows that 62% of the Toyonda owners would buy a Toyonda for their next car and that 16% of automobile owners who do not own a Toyonda would buy a Toyonda for their next car. The same survey indicates that automobile owners buy a new car an average of once every three years.

27. Predict Toyonda Motors' market share after three years.

28. Predict Toyonda Motors' market share after six years.

29. Make a long-range prediction of Toyonda Motors' market share.

30. On what assumptions are these predictions based?

In Exercises 31–33, use the following information. Currently, 23% of the residences of Foxtail County are apartments, 18% are condominiums or townhouses, and the balance are single-family houses. The Foxtail County Contractors' Association has commissioned a survey that shows that 3% of the apartment residents plan on moving to a condominium or townhouse within the next two years, and 5% plan on moving to a single family house. The same survey indicates that 1% of those who currently reside in a condominium or townhouse plan on moving to an apartment, and 11% plan on moving to a single family house; 2% of the single family house dwellers plan on moving to an apartment, and 4% plan on moving to a condominium or townhouse.

31. What recommendation should the FCCA make regarding the construction of apartments, condominiums, and single-family houses in the next four years?

32. What long-term recommendation should the FCCA make?

33. What important factors does the survey ignore?

LINEAR PROGRAMMING

© British Retail Photography / Alamy

*L*inear programming was invented in the 1940s by George Dantzig as a result of a military research project on planning how to distribute men, weapons, and supplies efficiently to the various fronts during World War II. (Here, the word *programming* means creating a plan that solves a problem; it is not a reference to computer programming.)

Many businesses, industries, and government agencies use linear programming successfully.

- Airlines use it to schedule flights in a way that minimizes costs without overloading a pilot or crew with too many hours.
- Ecological organizations and government agencies use it to determine the best way to keep our water and air clean.

continued

WHAT WE WILL DO IN THIS CHAPTER

- You may well find yourself working in a small business. Over half of the American workforce does, according to the U.S. Small Business Administration. Small businesses have limited resources, and how they allocate their resources can make or break the business. We will look at how businesses can use linear programming to find the best way to do this.

- Starbucks stores and smaller local coffee shops seem to be everywhere. They blend coffee beans from different parts of the world. Like them, many businesses sell a product that involves blending resources. These businesses have many factors to consider: the desired quality of the blend, the availability of the raw materials, and the cost. We will investigate how linear programming is used do this.

- Many businesses use the same resources to make different products. A craftsman uses the same wood to make different tables and chairs. A fashion designer uses the same fabrics to make different garments. We will learn how linear programming is used to find the best way to allocate resources to different products.

859

- Businesses use it to find the best way to assign personnel to jobs.
- Supermarket chains use it to determine which warehouses should ship which products to which stores.
- Investment companies use it to create portfolios with the best mix of stocks and bonds.
- Refineries use it to decide what crude oil to buy and to determine what products to produce with the oil.

12.0 Review of Linear Inequalities

OBJECTIVES

- Review how to graph a linear inequality
- Learn how to graph a system of linear inequalities
- Understand how to find corner points

A **linear equation** in two variables x and y is an equation that can be written in the form $ax + by = c$, where a, b, and c are constants. Such an equation is referred to as a *linear* equation because its graph is a line. We can graph most linear equations by solving for y and using the slope-intercept formula $y = mx + b$ to find the slope m and the y-intercept b of the line.

A **linear inequality** in two variables x and y is an inequality that can be written in the form $ax + by < c$ (or with $>$, $\leq$, or $\geq$ instead of $<$). In other words, a linear inequality is the result of replacing a linear equation's equals symbol with an inequality symbol.

Our goal in this chapter is to **optimize** a quantity—that is, to maximize or minimize a quantity. A craftsman who manufactures coffee tables and end tables by hand would want to know how to invest his resources of time and money to maximize his profit. His resources are restricted; he wants to work *no more than* 40 hours each week, and he has *at most* $1,000 to spend on materials each week. These restrictions are described mathematically with inequalities, because they involve the phrases "no more than" and "at most." To analyze the craftsman's restricted resources and the effect of these restrictions on his profit, we must be able to graph the inequalities that describe the restrictions.

Each restriction in a linear programming problem must be expressible as a linear inequality; this is the meaning of the word *linear* in the term *linear programming*. In this section, we discuss the graphing of linear inequalities. We can graph a linear inequality by solving the inequality for y, graphing the line described by the associated equation, and then shading the region to one side of that line.

EXAMPLE **1**

SOLUTION

Step 1

GRAPHING A LINEAR INEQUALITY   Graph $2x + y \leq 6$.

Solve the linear inequality for y.

$$2x + y \leq 6 \rightarrow y \leq -2x + 6$$

Step 2

Graph the line. The equation associated with the inequality is $y = -2x + 6$. Comparing this equation with the slope-intercept formula $y = mx + b$, we find that this is the equation of a line with slope $m = -2$ and y-intercept $b = 6$. Because the slope is -2 and because slope means "rise over run," we have

$$\frac{\text{rise}}{\text{run}} = -2 = \frac{-2}{1}$$

To graph the line, we place a point at 6 on the y-axis (since the y-intercept is 6) and then from that point rise -2 (that is, move two units down) and run 1 (that is, move one unit to the right). This takes us to a new point. We connect the points with a line.

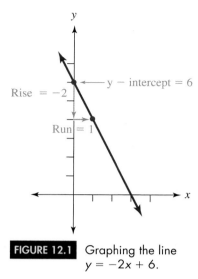

FIGURE 12.1   Graphing the line $y = -2x + 6$.

Because "=" is a part of "≤," any point on the line $y = -2x + 6$ must also be a point on the graph of the inequality $y \leq -2x + 6$. We show this by using a solid line, as in Figure 12.1. (If our inequality were $y < -2x + 6$, then a point on the line would *not* be a point on the graph of the inequality. We would show this by using a dashed line.)

Step 3

Shade in one side of the line. Two types of points satisfy the inequality $y \leq -2x + 6$: points that satisfy $y = -2x + 6$ and points that satisfy $y < -2x + 6$.

Points that satisfy $y = -2x + 6$ were graphed in step 2, when we graphed the line. Points that satisfy y *is less than* $-2x + 6$ are the points *below* the line, because values of y decrease if we move down and increase if we move up, as shown in Figure 12.2.

Thus, to graph the inequality $y \leq -2x + 6$, we make the line solid and shade in the region *below* the line, as shown in Figure 12.3. The solution of $y \leq -2x + 6$ is the set of all points on or below the line; any point on or below the line will successfully substitute into the inequality, and any point above the line will not. This region is called the **region of solutions** of the inequality.

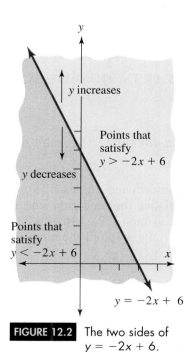

FIGURE 12.2   The two sides of $y = -2x + 6$.

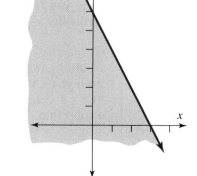

FIGURE 12.3   The region of solution of $y \leq -2x + 6$ (or equivalently of $2x + y \leq 6$).

EXAMPLE **2**

GRAPHING ANOTHER LINEAR INEQUALITY Graph $3x - 2y < 12$.

SOLUTION

Step 1

Solve the linear inequality for y.

$$3x - 2y < 12$$
$$-2y < -3x + 12 \qquad \text{solving for } y$$
$$\frac{-2y}{-2} > \frac{-3x + 12}{-2} \qquad \text{multiplying or dividing by a negative reverses the direction of an inequality}$$
$$y > \frac{-3x}{-2} + \frac{12}{-2} \qquad \text{distributing } -2$$
$$y > \frac{3}{2}x - 6 \qquad \text{simplifying}$$

Step 2

Graph the line. The associated equation is $y = \frac{3}{2}x - 6$, which is a line with slope $m = \frac{3}{2}$ and y-intercept $b = -6$.

$$m = \frac{3}{2} \rightarrow \frac{\text{rise}}{\text{run}} = \frac{3}{2}$$

To graph the line, we place a point at -6 on the y-axis and from that point rise 3 and run 2.

Because "$=$" is *not* part of "$>$," a point on the line $y = \frac{3}{2}x - 6$ is *not* a point on the graph of the inequality $y > \frac{3}{2}x - 6$. We show this by using a dashed line.

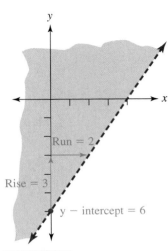

FIGURE 12.4

The region of solutions of $y > \frac{3}{2}x - 6$ (or equivalently of $3x - 2x < 12$).

Step 3

Shade in one side of the line. Because values of y *increase* if we move upward and because we want to graph where *y is greater than* $\frac{3}{2}x - 6$, we shade in the region above the dashed line. The region of solutions of the inequality is the set of all points above (but not on) the line, as shown in Figure 12.4.

EXAMPLE **3**

GRAPHING A LINEAR INEQUALITY WITH NO *y*-TERM Graph $x \geq 3$.

SOLUTION

Step 1

Solve the linear inequality for y. This can't be done, since there is no y in the inequality $x \geq 3$.

Step 2

Graph the line. The associated equation is $x = 3$. Because any point with an x-coordinate of 3 satisfies this equation, the graph of $x = 3$ is a vertical line through $(3, 0)$ and $(3, 1)$ and $(3, 2)$. Also, "$=$" is part of "$\geq$," so any point on the line $x = 3$ is a point on the graph of the inequality $x \geq 3$. We show this by using a solid line.

Step 3

Shade in one side of the line. Values of x increase if we move to the right and decrease if we move to the left. The points on the line are the points where x equals 3, and the points to the right of the line are the points where x is greater than 3. Thus, the region of solutions of the inequality is the set of all points on or to the right of the line, as shown in Figure 12.5.

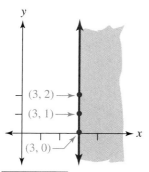

FIGURE 12.5

The region of solutions of $x \geq 3$.

GRAPHING THE REGION OF SOLUTIONS OF A LINEAR INEQUALITY

1. *Solve the linear inequality for y.* This puts the inequality in slope-intercept form. If the inequality has no y, then solve the inequality for x. If you multiply or divide each side by a negative number, reverse the direction of the inequality.

2. *Graph the line.* The line is described by the equation associated with the inequality.

If the equation is:	then the line:
in slope-intercept form ($y = mx + b$) ...	has slope m and y-intercept b.
in the form $x = a$	is a vertical line through $(a, 0)$.

If the inequality is:	then the line:
$\leq$ or $\geq$	is part of the region of solutions—use a solid line.
$<$ or $>$	is not part of the region of solutions—use a dashed line.

3. *Shade in one side of the line.*
   y increases as you move up, and decreases as you move down.
   x increases as you move to the right, and decreases as you move to the left.

Systems of Linear Inequalities

The inequalities we deal with in this chapter come from restrictions. Usually, a linear programming problem involves more than one restriction and therefore more than one inequality. A **system of linear inequalities** is a set of more than one linear inequality. The **region of solutions** of a system of linear inequalities is the set of all points that simultaneously satisfy each inequality in the system.

To graph the region of solutions of a system of linear inequalities, we graph each inequality on the same axes and shade in the intersection of their solutions.

EXAMPLE 4

GRAPHING A SYSTEM OF INEQUALITIES   Graph the following system:

$$x + \quad y \geq 3$$
$$-x + 2y \leq 0$$

SOLUTION

With the aid of the chart in Figure 12.6, we can graph the two inequalities on the same axes, as shown in Figure 12.7.

Original Inequality	Slope-Intercept Form	Associated Equation	Graph of the Inequality
$x + y \geq 3$	$y \geq -x + 3$	$y = -1x + 3$	all points on or above the line with slope -1 and y-intercept 3
$-x + 2y \leq 0$	$2y \leq x \rightarrow y \leq \frac{1}{2}x$	$y = \frac{1}{2}x + 0$	all points on or below the line with slope $\frac{1}{2}$ and y-intercept 0

FIGURE 12.6

The region of solutions of the system is the set of all points that satisfy both the first *and* the second inequality. The region of solutions is the intersection of the graphs of the two inequalities. This is the region shaded in Figure 12.8; it includes the solid lines bounding that region.

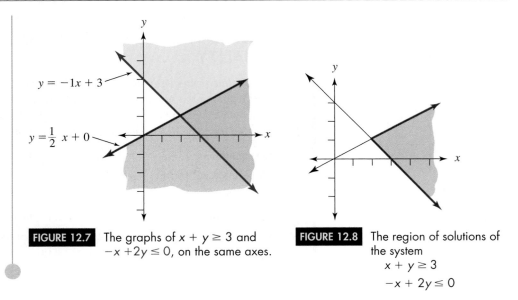

FIGURE 12.7   The graphs of $x + y \geq 3$ and $-x + 2y \leq 0$, on the same axes.

FIGURE 12.8   The region of solutions of the system
$$x + y \geq 3$$
$$-x + 2y \leq 0$$

EXAMPLE **5**

GRAPHING A BIGGER SYSTEM   Graph the following system of inequalities:

$$2x + y \leq 8$$
$$x + 2y \leq 10$$
$$x \geq 0$$
$$y \geq 0$$

SOLUTION

With the aid of the chart in Figure 12.9, we can graph the two inequalities on the same axes.

The inequalities $x \geq 0$ and $y \geq 0$ appear frequently in linear programming problems. They tell us that our graph is in the first quadrant.

The region of solutions of the system is the intersection of the graph of each individual inequality. It is shown in Figure 12.10.

Original Inequality	Slope-Intercept Form	Associated Equation	Graph of the Inequality
$2x + y \leq 8$	$y \leq -2x + 8$	$y = -2x + 8$	all points on or below the line with slope -2 and y-intercept 8
$x + 2y \leq 10$	$2y \leq -x + 10 \rightarrow$ $y \leq -\frac{1}{2}x + 5$	$y = -\frac{1}{2}x + 5$	all points on or below the line with slope $-\frac{1}{2}$ and y-intercept 5
$x \geq 0$	(not applicable)	$x = 0$	all points on or to the right of the y-axis
$y \geq 0$	$y \geq 0$	$y = 0$	all points on or above the x-axis

FIGURE 12.9   Getting ready to graph the region of solutions.

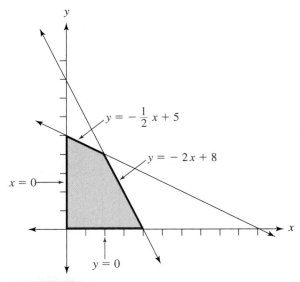

FIGURE 12.10   The region of solutions of the system
$$2x + y \leq 8$$
$$x + 2y \leq 10$$
$$x \geq 0$$
$$y \geq 0$$

Bounded and Unbounded Regions

The region of solutions in Example 4 is different from that of Example 5 in that the former is not totally enclosed. For that reason, it is called an **unbounded region.** Regions that are totally enclosed, like that in Example 5, are called **bounded regions.** See Figure 12.11. When we graph a system of inequalities as part of a linear programming problem, we must analyze unbounded regions differently than we do bounded regions.

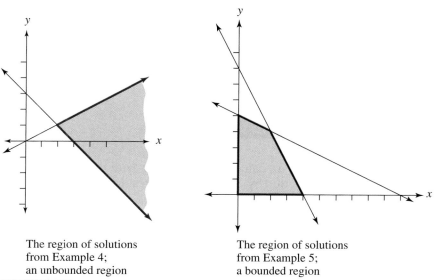

The region of solutions from Example 4; an unbounded region

The region of solutions from Example 5; a bounded region

FIGURE 12.11   Comparing bounded and unbounded regions.

Finding Corner Points

Our work in linear programming will involve graphing the region of solutions of a system of linear inequalities, as we did in Example 5. It will also involve finding the region's corner points. A **corner point** is a point that is at a corner of the region of solutions.

EXAMPLE 6

FINDING A REGION'S CORNER POINTS   Find the corner points of the region of solutions in Example 5 on page 866.

SOLUTION

The region has four corner points, as shown in Figure 12.12 on the next page.

- P_1 at the origin
- P_2 where $y = -\frac{1}{2}x + 5$ intersects $x = 0$ (the y-axis)
- P_3 where $y = -\frac{1}{2}x + 5$ intersects $y = -2x + 8$
- P_4 where $y = -2x + 8$ intersects $y = 0$ (the x-axis)

There is no work to do to find points P_1 and P_2. Point P_1 is clearly $(0, 0)$. Point P_2 is the y-intercept of $y = -\frac{1}{2}x + 5$; that y-intercept is $b = 5$, so P_2 is $(0, 5)$.

We can find P_3 by solving the system of equations

$$y = -\frac{1}{2}x + 5$$
$$y = -2x + 8$$

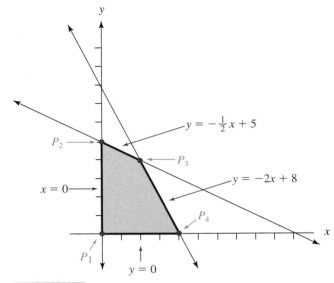

FIGURE 12.12   Corner points.

We can solve this system with the elimination method. Multiply the second equation by -1 and add the results.

$$y = -\frac{1}{2}x + 5$$

$$+ \; -y = \quad 2x - 8 \quad \textbf{-1 times the second equation}$$

$$0 = \quad \frac{3}{2}x - 3 \quad \textbf{adding}$$

$$\frac{3}{2}x = 3 \qquad\qquad \textbf{solving for } x$$

$$\frac{2}{3} \cdot \frac{3}{2}x = \frac{2}{3} \cdot 3$$

$$x = 2$$

We can find y by substituting $x = 2$ into either of the original equations.

$$y = -2x + 8 \qquad\qquad \textbf{the second original equation}$$
$$y = -2 \cdot 2 + 8 = 4 \quad \textbf{substituting 2 for } x$$

Thus, P_3 is $(2, 4)$.

We can find P_4 by solving the system of equations

$$y = -2x + 8$$
$$y = 0 \qquad\qquad\quad \textbf{the equation of the } x\textbf{-axis}$$

To proceed with the elimination method, multiply the second equation by -1 and add the results.

$$y = -2x + 8$$
$$+ \; -y = \qquad\quad 0 \quad \textbf{-1 times the second equation}$$
$$0 = -2x + 8 \quad \textbf{adding}$$
$$2x = 8 \qquad\qquad \textbf{solving for } x$$
$$x = 4$$

Since P_4 is on the x-axis, its y-coordinate must be 0. Thus, P_4 is $(4, 0)$. The region of solutions' four corner points are

- P_1 at $(0, 0)$ • P_3 at $(2, 4)$
- P_2 at $(0, 5)$ • P_4 at $(4, 0)$

In Exercises 1–8, graph the region of solutions of the given linear inequality.

1. $3x + y < 4$
2. $8x + y > 2$
3. $4x - 3y \leq 9$
4. $5x - 2y \geq 6$
5. $x \geq 4$
6. $x \leq -3$
7. $y \leq -4$
8. $y \geq -2$

In Exercises 9–21, (a) graph the region of solutions of the given system of linear inequalities, (b) determine whether the region of solutions is bounded or unbounded, and (c) find all of the region's corner points.

9. $y > 2x + 1$
   $y \leq -x + 4$
10. $y < -2x + 6$
    $y \geq -x + 7$
11. $2x + 3y < 17$
    $3x - y \geq -2$
12. $5x - y \geq 7$
    $2x - 3y < -5$
13. $x + 2y \leq 4$
    $3x - 2y \leq -12$
    $x - y < -7$
14. $x - y + 1 \geq 0$
    $3x + 2y + 8 \geq 0$
    $3x - y < 6$
15. $2x + 5y \leq 70$
    $5x + y \leq 60$
    $x \geq 0$
    $y \geq 0$
16. $x + 20y \leq 460$
    $21x + y \leq 861$
    $x \geq 0$
    $y \geq 0$
17. $15x + 22y \leq 510$
    $35x + 12y \leq 600$
    $x + y > 10$
    $x \geq 0$
    $y \geq 0$
18. $3x + 20y \leq 2,200$
    $19x + 9y \leq 2,755$
    $2x + y \leq 120$
    $x \geq 0$
    $y \geq 0$
19. $0.50x + 1.30y \leq 2.21$
    $6x + y \leq 9$
    $0.7x + 0.6y < 3.00$
    $x \geq 0$
    $y \geq 0$
20. $3.70x + 0.30y \leq 1.17$
    $0.10x + 2.20y \leq 0.47$
    $x \geq 0$
    $y \geq 0$
21. $x - 2y + 16 \geq 0$
    $3x + y \leq 30$
    $x + y \leq 14$
    $x \geq 0$
    $y \geq 0$

• CONCEPT QUESTIONS

22. Make up a system of inequalities with a region of solutions that is bounded. Do not use any of the systems discussed in the text.

23. Make up a system of inequalities with a region of solutions that is unbounded. Do not use any of the systems discussed in the text.

24. Make up a system of inequalities with a region of solutions that is the triangle with corner points $(0, 0)$, $(10, 0)$, and $(0, 20)$.

25. Make up a system of inequalities with a region of solutions that has corner points $(0, 0)$, $(20, 0)$, $(10, 40)$, and $(0, 50)$.

Answer the following questions with complete sentences and your own words.

• CONCEPT QUESTIONS

26. Why do we describe the solution of a system of linear inequalities with a graph, rather than a list of points?

27. List three industries that routinely use linear programming, and give examples of how they use it.

28. Why do large corporations have to be concerned about "limited" resources?

GRAPHING LINEAR INEQUALITIES ON A GRAPHING CALCULATOR

A Texas Instruments graphing calculator can perform all of the specific tasks that are part of graphing the region of solutions of a system of linear inequalities: It can graph the lines associated with the system, shade in the appropriate sides of those lines, and locate corner points. However, the shading makes the graph hard to see if there are more than two inequalities. It is much easier to do the shading by hand and to use the calculator to graph the lines associated with the inequalities and to find the corner points. If you wish to explore the use of the shading feature command, consult your calculator's operating manual.

Graphing a Linear Inequality

EXAMPLE 7

USING TECHNOLOGY   Graph $3x - 2y < 12$.

SOLUTION

Step 1   *Solve the linear inequality for y.*

$$3x - 2y < 12 \rightarrow -2y < -3x + 12 \rightarrow y > \frac{3}{2}x - 6$$ **Multiplying or dividing by a negative reverses the direction of the inequality.**

Step 2   *Graph the line,* using the procedure discussed in Appendix C. The line associated with the inequality is $y = \frac{3}{2}x - 6$. Enter $3/2*x - 6$ for Y_1. Be sure to include the "*." Some calculators would interpret $3/2x - 6$ as $\frac{3}{2x} - 6$. Be sure that no other equations are selected. Use the standard viewing window. The line's graph is shown in Figure 12.13.

Step 3   *Copy the line's graph onto paper.* The inequality is a ">" inequality, so use a dashed line.

Step 4   *By hand, shade in one side of the line.* To graph $y > \frac{3}{2}x - 6$, shade in the region above the line $y = \frac{3}{2}x - 6$. The region's graph is shown in Figure 12.14.

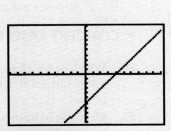

FIGURE 12.13   The graph of $y = \frac{3}{2}x - 6$.

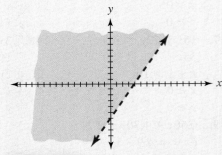

FIGURE 12.14   The graph of $y > \frac{3}{2}x - 6$.

Graphing a System of Linear Inequalities

EXAMPLE 8

GRAPHING A SYSTEM WITH TECHNOLOGY   Graph the following system of linear inequalities:

$$x + y \geq 3$$
$$-x + 2y \leq 0$$

SOLUTION

Step 1   *Solve the linear inequalities for y.*

$$x + y \geq 3 \quad \rightarrow \quad y \geq -x + 3$$
$$-x + 2y \leq 0 \quad \rightarrow \quad y \leq \frac{1}{2}x$$

Step 2   *Graph the lines,* using the procedure discussed in Appendix C. Enter $-x + 3$ for Y_1 and $1/2*x$ for Y_2. Be sure that no other equations are selected. Use the standard viewing window. See Figure 12.15.

Step 3   *Copy the lines' graphs onto paper.* The inequalities are both "≤ or ≥" inequalities, so use solid lines.

Step 4   *By hand, shade in the region of solutions.* The region of solutions is above the line $Y_1 = -x + 3$, because the associated inequality is $y \geq -x + 3$, and y increases as you move up. The region of solutions is also below the line $Y_2 = \frac{1}{2}x$, because the associated inequality is $y \leq \frac{1}{2}x$, and y decreases as you move down. The region of solutions is the region that is both above the line $Y_1 = -x + 3$ and below the line $Y_2 = \frac{1}{2}x$. It is shown in Figure 12.16.

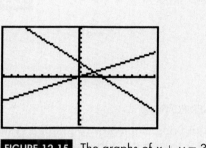

FIGURE 12.15 The graphs of $x + y = 3$ and $-x + 2y = 0$.

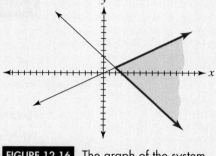

FIGURE 12.16 The graph of the system
$$x + \ y \geq 3$$
$$-x + 2y \leq 0$$

NONSTANDARD VIEWING WINDOWS

The previous examples were chosen so that the standard viewing window would be appropriate. In the following example, we must determine an appropriate viewing window.

EXAMPLE 9   **NONSTANDARD VIEWING WINDOWS**   Graph the system of inequalities:

$$2x + y \leq 8$$
$$x + 2y \leq 10$$
$$x \geq 0$$
$$y \geq 0$$

SOLUTION
$$2x + y \leq 8 \quad \rightarrow \quad y \leq -2x + 8$$
$$x + 2y \leq 10 \quad \rightarrow \quad 2y \leq -x + 10 \quad \rightarrow \quad y \leq -\frac{1}{2}x + 5$$

$x \geq 0$ cannot be solved for y, and $y \geq 0$ is already solved for y

Enter $-2x + 8$ for Y_1 and $-1/2*x + 5$ for Y_2. The last pair of inequalities tells us that the region of solutions is in the first quadrant; if we set Xmin and Ymin equal to -1, we'll leave ourselves a little extra room. By inspecting the two lines' equations, we can tell that the largest y-intercept is 8, so we'll set Ymax equal to 8. Substituting 0 for y in each equation gives x-intercepts of $(4, 0)$ and $(10, 0)$, so we'll set Xmax equal to 10. The resulting graph is shown in Figure 12.17.

The region of solutions is below the line $Y_1 = -2x + 8$, because the associated inequality is $y \leq -2x + 8$, and y decreases as you move down. The region of solutions is also below the line $Y_2 = -\frac{1}{2}x + 5$, because the associated inequality is $y \leq -\frac{1}{2}x + 5$. The region of solutions is the part of the first quadrant that is both on or below the line $Y_1 = -2x + 8$ and on or below the line $Y_2 = -\frac{1}{2}x + 5$. It is shown in Figure 12.18.

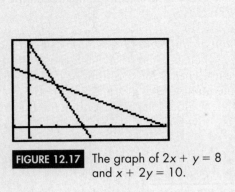

FIGURE 12.17 The graph of $2x + y = 8$ and $x + 2y = 10$.

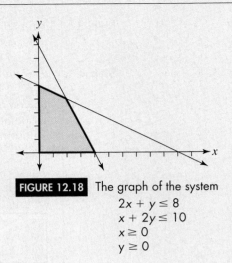

FIGURE 12.18 The graph of the system
$$2x + y \leq 8$$
$$x + 2y \leq 10$$
$$x \geq 0$$
$$y \geq 0$$

FINDING CORNER POINTS

After using a graphing calculator to graph a system's region of solutions, you can use the calculator to find the corner points as well.

EXAMPLE 10

CORNER POINTS WITH TECHNOLOGY Find the corner points for the region graphed in Example 9.

SOLUTION

This region has four corner points. One is clearly located at $(0, 0)$. A second is the y-intercept of the line $y = -\frac{1}{2}x + 5$; by inspecting the equation, we can tell that it is $(0, 5)$.

A third corner point is the intersection of the two lines; it can be found using the procedure discussed in Appendix D (summarized below).

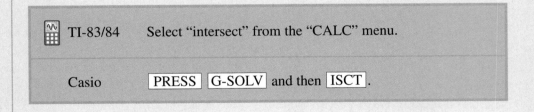

TI-83/84   Select "intersect" from the "CALC" menu.

Casio   PRESS G-SOLV and then ISCT .

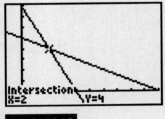

FIGURE 12.19

Finding the intersection.

This third corner point is $(2, 4)$, as shown in Figure 12.19.

The fourth corner point is the x-intercept or root of the line $y = -2x + 8$; it can be found using the procedure discussed in Appendix C, Exercises 8 and 9 (summarized below).

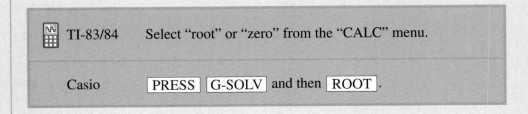

TI-83/84   Select "root" or "zero" from the "CALC" menu.

Casio   PRESS G-SOLV and then ROOT .

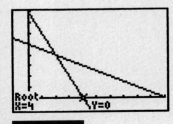

FIGURE 12.20

Finding the root.

This fourth corner point is $(4, 0)$, as shown in Figure 12.20.

EXERCISES

In Exercises 29–36, use a graphing calculator to graph the region of solutions of the inequality given earlier in this section.

29. Exercise 1   **30.** Exercise 2
31. Exercise 3   **32.** Exercise 4
33. Exercise 5   **34.** Exercise 6
35. Exercise 7   **36.** Exercise 8

In Exercises 37–49, use a graphing calculator to graph the region of solutions of the system of inequalities given earlier in this section, and to find the corner points.

37. Exercise 9   **38.** Exercise 10
39. Exercise 11   **40.** Exercise 12
41. Exercise 13   **42.** Exercise 14
43. Exercise 15   **44.** Exercise 16
45. Exercise 17   **46.** Exercise 18
47. Exercise 19   **48.** Exercise 20
49. Exercise 21

12.1   The Geometry of Linear Programming

OBJECTIVES

- Understand how to create a model
- Be able to use a model to generate a region
- Learn how to use a region and its corner points to solve a problem

As we stated in the introduction to this chapter, linear programming is a method for solving problems in which a quantity is to be maximized or minimized, when that quantity is subject to various restrictions. The following is a typical linear programming problem.

A craftsman produces two products: coffee tables and end tables. Production of one coffee table requires six hours of his labor, and the materials cost him $200. Production of one end table requires five hours of labor, and the materials cost him $100. The craftsman wants to work no more than 40 hours each week, and his financial resources allow him to pay no more than $1,000 for materials each week. If he can sell as many tables as he can make and if his profit is $240 per coffee table and $160 per end table, how many coffee tables and how many end tables should he make each week to maximize weekly profit?

Any linear programming problem has three features: *variables,* an *objective,* and *constraints.* In the problem above, the **variables** (or quantities that can vary) are the following:

- The number of coffee tables made each week
- The number of end tables made each week
- The number of hours the craftsman works each week
- The amount of money he spends on materials each week
- The weekly profit

How should a craftsman allocate his time and money to maximize profit?

© Owaki-Kulla/CORBIS

The last three variables depend on the first two, so they are called the **dependent variables,** whereas the first two are called the **independent variables.**

The craftsman's objective is to maximize profit. The **objective function** is a function that mathematically describes the profit.

The **constraints** (or restrictions) are as follows:

- The craftsman's weekly hours ≤ 40
- The craftsman's weekly expenses $\leq \$1,000$

The constraints form a system of inequalities. To analyze the effect of these constraints on the craftsman's profit, we must graph the system of inequalities. The resulting graph is called the **region of possible solutions,** because it contains all the points that could *possibly* solve the craftsman's problem.

Creating a Model

A **model** is a mathematical description of a real-world situation. In this section, we discuss how to model a linear programming problem (that is, how to translate it into mathematical terms), how to find and graph the region of possible solutions, and how to analyze the effect of the constraints on the objective and solve the problem.

EXAMPLE **1**

GRAPHING A LINEAR PROGRAMMING PROBLEM'S REGION OF POSSIBLE SOLUTIONS Model the linear programming problem from the beginning of this section and graph the region of possible solutions. The problem is summarized as follows:

A craftsman produces two products: coffee tables and end tables. Production data are given in Figure 12.21. If the craftsman wants to work no more than 40 hours each week and if his financial resources allow him to pay no more than $1,000 for materials each week, how many coffee tables and how many end tables should he make each week to maximize weekly profit?

	Labor (per table)	Cost of Materials (per table)	Profit (per table)
Coffee Tables	6 hours	$200	$240
End Tables	5 hours	$100	$160

FIGURE 12.21 Data for Example 1.

SOLUTION

Step 1

List the independent variables. We have already done this. If we call them x and y, the independent variables are

x = number of coffee tables made each week

y = number of end tables made each week

Step 2

List the constraints and translate them into linear inequalities. We have already determined that the constraints (or restrictions) are as follows:

the craftsman's weekly hours ≤ 40

the craftsman's weekly expenses $\leq \$1,000$

We need to translate these constraints into linear inequalities. First, let's translate the time constraint.

$$\text{hours} \leq 40$$

$$(\text{coffee table hours}) \quad + \quad (\text{end table hours}) \qquad \leq 40$$

$$\left(\begin{array}{c}6 \text{ hours per} \\ \text{coffee table}\end{array}\right) \cdot \left(\begin{array}{c}\text{number of} \\ \text{coffee tables}\end{array}\right) + \left(\begin{array}{c}5 \text{ hours per} \\ \text{end table}\end{array}\right) \cdot \left(\begin{array}{c}\text{number of} \\ \text{end tables}\end{array}\right) \leq 40$$

$$6 \qquad\quad x \quad + \quad 5 \qquad\quad y \qquad \leq 40$$

Next, we'll translate the money constraint.

$$\text{money spent} \leq 1{,}000$$

$$(\text{coffee table money}) \quad + \quad (\text{end table money}) \qquad \leq 1{,}000$$

$$\left(\begin{array}{c}\$200 \text{ per} \\ \text{coffee table}\end{array}\right) \cdot \left(\begin{array}{c}\text{number of} \\ \text{coffee tables}\end{array}\right) + \left(\begin{array}{c}\$100 \text{ per} \\ \text{end table}\end{array}\right) \cdot \left(\begin{array}{c}\text{number of} \\ \text{end tables}\end{array}\right) \leq 1{,}000$$

$$200 \qquad\quad x \quad + \quad 100 \qquad\quad y \qquad \leq 1{,}000$$

There are two more constraints. Both x and y count things (the number of tables), so neither can be negative; therefore,

$$x \geq 0 \qquad \text{and} \qquad y \geq 0$$

Our constraints are

$$6x + 5y \leq 40 \qquad\qquad \textbf{the time constraint}$$
$$200x + 100y \leq 1{,}000 \qquad \textbf{the money constraint}$$
$$x \geq 0 \qquad \text{and} \qquad y \geq 0$$

Step 3   *Find the objective and translate it into a linear equation.* The objective is to maximize profit. If we let $z = $ profit, we get

$$z = (\text{coffee table profit}) \qquad\qquad + \quad (\text{end table profit})$$

$$= \left(\begin{array}{c}\$240 \text{ per} \\ \text{coffee table}\end{array}\right) \cdot \left(\begin{array}{c}\text{number of} \\ \text{coffee tables}\end{array}\right) + \left(\begin{array}{c}\$160 \text{ per} \\ \text{end table}\end{array}\right) \cdot \left(\begin{array}{c}\text{number of} \\ \text{end tables}\end{array}\right)$$

$$= \qquad 240 \quad\cdot\quad x \quad + \quad 160 \quad\cdot\quad y$$

This equation is our objective function.

Steps 1 through 3 yield the model, or mathematical description, of the problem.

Independent Variables

$x = $ number of coffee tables

$y = $ number of end tables

Constraints

$$6x + 5y \leq 40 \qquad\qquad \textbf{the time constraint}$$
$$200x + 100y \leq 1{,}000 \qquad \textbf{the money constraint}$$
$$x \geq 0 \qquad \text{and} \qquad y \geq 0$$

Objective Function

$$z = 240x + 160y \qquad\qquad z \text{ measures profit}$$

Step 4   *Graph the region of possible solutions.* With the aid of the chart in Figure 12.22, we can graph the two inequalities on the same axes, as shown in Figure 12.23.

The last two constraints tell us that the region of possible solutions is in the first quadrant. The region of possible solutions is shown in Figure 12.23.

Original Inequality	Slope-Intercept Form	Associated Equation	Graph of the Inequality
$6x + 5y \leq 40$	$5y \leq -6x + 40$ $\rightarrow y \leq -\frac{6}{5}x + 8$	$y = -\frac{6}{5}x + 8$	all points on or below the line with slope $-6/5$ and y-intercept 8
$200x + 100y \leq 1,000$	$100y \leq -200x + 1,000$ $\rightarrow y \leq -2x + 10$	$y = -2x + 10$	all points on or below the line with slope -2 and y-intercept 10
$x \geq 0$	(not applicable)	$x = 0$	all points on or to the right of the y-axis
$y \geq 0$	$y \geq 0$	$y = 0$	all points on or above the x-axis

FIGURE 12.22 Getting ready to graph the region of possible solutions.

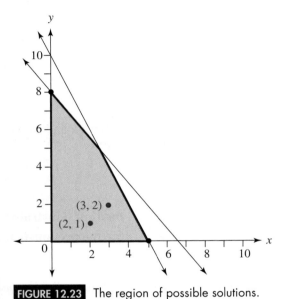

FIGURE 12.23 The region of possible solutions.

Analyzing the Graph

The region of solutions in Figure 12.23 is a graph of the possible numbers of coffee tables and end tables the craftsman could make in one week without violating his constraints. That means that it consists of all the points at which the craftsman's weekly hours are no more than 40 and his weekly expenses are no more than $1000.

By inspecting the graph in Figure 12.23, we can see that $(2, 1)$ and $(3, 2)$ are clearly in the system's region of solutions. That means that each point should satisfy all of the constraints. Let's verify that that happens.

The first constraint is that the time used, $6x + 5y$, be no more than 40 hours per week. The second constraint is that the money used, $200x + 100y$, be no more than $1000 each week. The third constraint is that x is not negative (because x counts tables), and the fourth constraint is that y is not negative. Figure 12.24 shows the calculations for our two points.

The table in Figure 12.24 shows that if the craftsman makes two coffee tables and one end table each week, he will use 17 hours (of 40 available hours), spend $500 (of the $1,000 available), and profit $640. If he makes three coffee tables and two end tables each week, he will use 28 hours, spend $800, and profit $1,040. Each of these points represents a *possible* solution to the craftsman's problem, because each satisfies the time constraint and the money constraint. Neither represents the *actual* solution, because neither maximizes his profit; he has both money and time left over, so he should be able to increase his profit by building more tables. This is why the region in Figure 12.23 is called the region of *possible* solutions.

Point	(2, 1)	(3, 2)
Time Used	$6x + 5y$ $= 6 \cdot 2 + 5 \cdot 1$ $= 17 \le 40$	$6x + 5y$ $= 6 \cdot 3 + 5 \cdot 2$ $= 28 \le 40$
Money Spent	$200x + 100y$ $= 200 \cdot 2 + 100 \cdot 1$ $= 500 \le 1000$	$200x + 100y$ $= 200 \cdot 3 + 100 \cdot 2$ $= 800 \le 1000$
x Not Negative	$x = 2 \ge 0$	$x = 3 \ge 0$
y Not Negative	$y = 1 \ge 0$	$y = 2 \ge 0$

FIGURE 12.24 Checking the craftsman's constraints.

Common sense tells us that to maximize profit, our craftsman must use all his time and/or money and make more tables. To make more tables means to increase the value of x and/or y, which implies that we should choose points on the boundary of the region of feasible solutions. There are quite a few points on the boundary—too many to find and substitute into the equation for profit. Fortunately, the **Corner Principle** comes to our rescue. (We discuss why the Corner Principle is true later in this section.)

CORNER PRINCIPLE

The maximum and minimum values of an objective function occur at corner points of the region of possible solutions if that region is bounded.

Our region is a bounded region, so the Corner Principle applies. The region has four corner points (see Figure 12.25):

- P_1 at the origin
- P_2 where $y = -\frac{6}{5}x + 8$ intersects the y-axis
- P_3 where $y = -\frac{6}{5}x + 8$ intersects $y = -2x + 10$
- P_4 where $y = -2x + 10$ intersects the x-axis

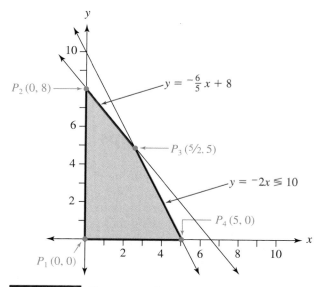

FIGURE 12.25 The region of possible solutions and its corner points.

We can find each of these points by solving a system of equations. For example, P_3 can be found by solving the following system:

$$y = -\tfrac{6}{5}x + 8$$
$$y = -2x + 10$$

P_4 can be found by solving the following system:

$$y = -2x + 10$$
$$y = 0 \qquad \textbf{the } x\textbf{-axis}$$

The corner points, shown in Figure 12.25 are as follows:

$$P_1 \, (0, 0)$$
$$P_2 \, (0, 8)$$
$$P_3 \, (\tfrac{5}{2}, 5)$$
$$P_4 \, (5, 0)$$

Let's verify that the constraints are satisfied at each of these points and, more important, find the profit at each point, as shown in Figure 12.26.

Point	$P_1 \, (0, 0)$	$P_2 \, (0, 8)$	$P_3 \left(\dfrac{5}{2}, 5\right)$	$P_4 \, (5, 0)$
Time Used	$6 \cdot 0 + 5 \cdot 0 = 0$	$6 \cdot 0 + 5 \cdot 8 = 40$	$6 \cdot \dfrac{5}{2} + 5 \cdot 5 = 40$	$6 \cdot 5 + 5 \cdot 0 = 30$
Money Spent	$200 \cdot 0 + 100 \cdot 0 = 0$	$200 \cdot 0 + 100 \cdot 8 = 800$	$200 \cdot \dfrac{5}{2} + 100 \cdot 5 = 1{,}000$	$200 \cdot 5 + 100 \cdot 0 = 1{,}000$
$z = $ **Profit**	$240 \cdot 0 + 160 \cdot 0 = 0$	$240 \cdot 0 + 160 \cdot 8 = 1{,}280$	$240 \cdot \dfrac{5}{2} + 160 \cdot 5 = 1{,}400$	$240 \cdot 5 + 160 \cdot 0 = 1{,}200$

FIGURE 12.26   Verifying constraints and finding profit.

At each corner point, each constraint is satisfied: the time used is at most 40 hours, and the money spent is at most \$1,000. The highest profit, \$1,400, occurs at $P_3 \, (\tfrac{5}{2}, 5)$. The Corner Principle tells us that this is the point in the region of possible solutions at which the highest profit occurs. The craftsman will maximize his profit if he makes $2\tfrac{1}{2}$ coffee tables and 5 end tables each week (finishing that third coffee table during the next week), working his entire 40 hours per week and spending his entire \$1,000 per week on materials.

Some Graphing Tips

In solving a linear programming problem, an accurate graph of the region of possible solutions is important. A less-than-accurate graph can be the source of errors and frustration. If you are graphing without the aid of technology, you can avoid some difficulty by following these recommendations:

- Find the x-intercept of each line and use it (as well as the slope and y-intercept) to graph the line.
- Use graph paper and a straightedge.
- Use your graph to check your corner point computations—the computed location of a corner point should fit with the graph.

(Why the Corner Principle Works

Each point in the region of possible solutions has a value of z associated with it. For example, we found that $P_3(\frac{5}{2}, 5)$ has a z-value of 1,400. Think of each point in the region as being a light bulb, with the brightness of the light given by the z-value. The point at which the craftsman's profit is maximized is the point with the brightest light.

Some points are just as bright as others. For example, all points that satisfy the equation $320 = 240x + 160y$ have a brightness of $z = 320$. If we solve this equation for y, we get

$$z = 240x + 160y \quad \text{the profit function}$$
$$320 = 240x + 160y \quad \text{substituting 320 for } z$$
$$-160y = 240x - 320 \quad \text{solving for } y$$
$$y = \frac{-3}{2}x + 2$$

This is a line with slope $\frac{-3}{2}$ and y-intercept 2.

Points that satisfy the equation $640 = 240x + 160y$ have a brightness of $z = 640$. If we solve this equation for y, we get

$$z = 240x + 160y \quad \text{the profit function}$$
$$640 = 240x + 160y \quad \text{substituting 640 for } z$$
$$-160y = 240x - 640 \quad \text{solving for } y$$
$$y = \frac{-3}{2}x + 4$$

Points that satisfy the equation $960 = 240x + 160y$ have a brightness of 960; solving for y gives $y = (\frac{-3}{2})x + 6$.

These three lines have the same slope, so they are parallel, as shown in Figure 12.27.

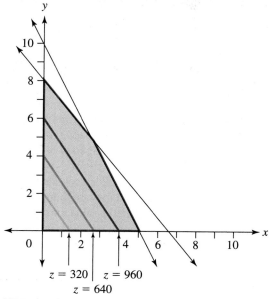

FIGURE 12.27 Why the corner principle works.

The light bulbs that fill the region of possible solutions form parallel rows, and all bulbs in any one row are equally bright. Rows closer to the upper right corner of the region are brighter, and rows closer to the lower left corner are dimmer. The brightest bulb is at corner point $P_3(\frac{5}{2}, 5)$, and the dimmest is at corner point $P_1(0, 0)$.

Any linear programming problem (including our problem about the craftsman) must have constraints that are expressible as linear inequalities and an objective function that is expressible as a linear equation. This means that when a problem has two independent variables, the region of possible solutions is bounded by lines, and each value of z will correspond to a line. Thus, the graph must always look something like one of the three possibilities shown in Figure 12.28.

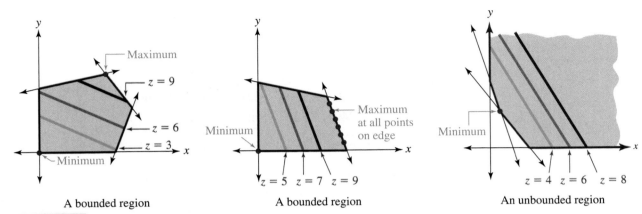

FIGURE 12.28 Where maximums and minimums occur.

If the region is bounded, the maximum and minimum must be at corner points or at all points on an edge. Therefore, we can find all corner points, substitute them into the objective function, and choose the biggest or smallest. If two corners yield the same maximum z-value, we know that the maximum occurs at all points on the boundary line between those corners.

An unbounded region does not necessarily have a maximum or a minimum. Unbounded regions are explored further in the exercises.

LINEAR PROGRAMMING STEPS

1. *List the independent variables.*
2. *List the constraints and translate them into linear inequalities.*
3. *Find the objective and translate it into a linear equation.* This equation is called the objective function.
4. *Graph the region of possible solutions.* This is the region described by the constraints. Graph each line carefully, using the x-intercept as well as the slope and y-intercept, if you are graphing without the aid of technology.
5. *Find all corner points and the z-values associated with these points.*
   ✔ Check your corner point computations by verifying that a point's computed location fits with the graph, if you are graphing without the aid of technology.
6. *Find the maximum/minimum.* For a bounded region, the maximum occurs at the corner with the largest z-value, and the minimum occurs at the corner with the smallest z-value. If two corners give the same maximum (or minimum) value, then the maximum (or minimum) occurs at all points on the boundary line between those corners.

In Example 1, the craftsman maximized his profit by exhausting all of his resources. In Example 2, we will find that maximizing profit does not necessarily entail exhausting all resources.

EXAMPLE 2 | **SOLVING A LINEAR PROGRAMMING PROBLEM** Pete's Coffees sells two blends of coffee beans, Rich Blend and Hawaiian Blend. Rich Blend is one-half Colombian beans and one-half Kona beans, and Hawaiian Blend is one-quarter Colombian beans and three-quarters Kona beans. Profit on the Rich Blend is $2 per pound, while profit on the Hawaiian Blend is $3 per pound. Each day, the shop can obtain 200 pounds of Colombian beans and 60 pounds of Kona beans, and it uses that coffee only in the two blends. If the shop can sell all that it makes, how many pounds of Rich Blend and of Hawaiian Blend should Pete's Coffees prepare each day to maximize profit?

How should Pete's Coffees blend its beans to maximize its profit?

SOLUTION

Step 1

List the independent variables. The variables are as follows:

- The amount of Rich Blend to be prepared each day
- The amount of Hawaiian Blend to be prepared each day
- The daily profit

Profit depends on the amount of the two blends prepared, so profit is the dependent variable, and the amounts of Rich Blend and Hawaiian Blend are the independent variables. If we call them x and y, then

x = pounds of Rich Blend to be prepared each day

y = pounds of Hawaiian Blend to be prepared each day

Step 2

List the constraints and translate them into linear inequalities. The constraints, or restrictions, are that there are 200 pounds of Colombian beans available each day and only 60 pounds of Kona beans. In the blends, no more than the amount available can be used. First, let's translate the Colombian bean constraint:

Colombian beans used ≤ 200

(Colombian in Rich Blend) + (Colombian in Hawaiian Blend) ≤ 200

(one-half of Rich Blend) + (one-fourth of Hawaiian Blend) ≤ 200

$$\frac{1}{2}x \qquad + \qquad \frac{1}{4}y \qquad \leq 200$$

Next, we'll translate the Kona bean constraint:

Kona beans used ≤ 60

(Kona in Rich Blend)	+	(Kona in Hawaiian Blend)	≤ 60
(one-half of Rich Blend)	+	(three-quarters of Hawaiian Blend)	≤ 60
$\frac{1}{2}x$	+	$\frac{3}{4}y$	≤ 60

Also, x and y count things (pounds of coffee), so neither can be negative.

$$x \geq 0 \quad \text{and} \quad y \geq 0$$

There is another implied constraint in this problem. Pete's sells its coffee in 1-pound bags, so x and y must be whole numbers. Special methods are available for handling such constraints, but these methods are beyond the scope of this class. We will ignore such constraints and accept fractional answers should they occur.

Step 3 *Find the objective and translate it into a linear equation.* The objective is to maximize profit. If we let $z = $ profit, we get

$z = $ (Rich Blend profit) + (Hawaiian Blend profit)

$\quad = $ ($2 per pound) (pounds of Rich Blend) + ($3 per pound) (pounds of Hawaiian Blend)

$\quad = 2x + 3y$

Steps 1 through 3 yield the mathematical model:

Independent Variables

$x = $ pounds of Rich Blend to be prepared each day

$y = $ pounds of Hawaiian Blend to be prepared each day

Constraints

$$\frac{1}{2}x + \frac{1}{4}y \leq 200 \quad \textbf{the Colombian bean constraint}$$

$$\frac{1}{2}x + \frac{3}{4}y \leq 60 \quad \textbf{the Kona bean constraint}$$

$$x \geq 0 \quad \text{and} \quad y \geq 0$$

Objective Function

$$z = 2x + 3y \quad \textbf{z measures profit}$$

Step 4 *Graph the region of possible solutions.* With the aid of the chart in Figure 12.29, we can graph the two inequalities on the same axes, as shown in Figure 12.30.

Original Inequality	Slope-Intercept Form	Associated Equation	Graph of the Inequality
$\frac{1}{2}x + \frac{1}{4}y \leq 200$	$\frac{1}{4}y \leq -\frac{1}{2}x + 200 \rightarrow$ $y \leq -2x + 800$	$y = -2x + 800$	all points on or below the line with slope -2 and y-intercept 800
$\frac{1}{2}x + \frac{3}{4}y \leq 60$	$\frac{3}{4}y \leq -\frac{1}{2}x + 60 \rightarrow$ $y \leq -\frac{2}{3}x + 80$	$y = -\frac{2}{3}x + 80$	all points on or below the line with slope $-2/3$ and y-intercept 80
$x \geq 0$	(not applicable)	$x = 0$	all points on or to the right of the y-axis
$y \geq 0$	$y \geq 0$	$y = 0$	all points on or above the x-axis

FIGURE 12.29 Getting ready to graph the region of possible solutions.

The region of possible solutions is shown in Figure 12.30. It is a bounded region. Notice that $y = -2x + 800$ is not a boundary of the region of possible solutions; if we were not given the first constraint, we would have the same region of possible solutions. That constraint, which describes the limited amount of Colombian beans available to Pete's Coffees, is not really a limitation.

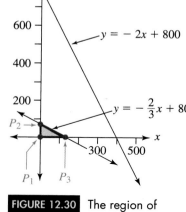

FIGURE 12.30 The region of possible solutions.

Step 5 *Find all corner points and the z-values associated with these points.* Clearly, $P_1 = (0, 0)$. P_2 has already been found; it is the y-intercept of $y = -\frac{2}{3}x + 80$. P_2 is $(0, 80)$.

P_3 is the x-intercept of $y = -\frac{2}{3}x + 80$, solve the system

$$y = -\frac{2}{3}x + 80$$
$$y = 0 \qquad \textbf{the equation of the } x\textbf{-axis}$$

P_3 is the point $(120, 0)$.

Step 6 *Find the maximum.* The result of substituting the corner points into the objective function $z = 2x + 3y$ is given in Figure 12.31.

Point	Value of $z = 2x + 3y$
P_1 (0, 0)	$z = 2 \cdot 0 + 3 \cdot 0 = 0$
P_2 (0, 80)	$z = 2 \cdot 0 + 3 \cdot 80 = 240$
P_3 (120, 0)	$z = 2 \cdot 120 + 3 \cdot 0 = 240$

FIGURE 12.31 Finding the maximum.

The two corner points P_2 and P_3 give the same maximum value, $z = 240$. Thus, the maximum occurs at P_2 and P_3 and at all points between them on the boundary line $y = -\frac{2}{3}x + 80$, as shown in Figure 12.32. This line includes points such as $(30, 60)$, $(60, 40)$, and $(90, 20)$. (Find points like these by substituting appropriate

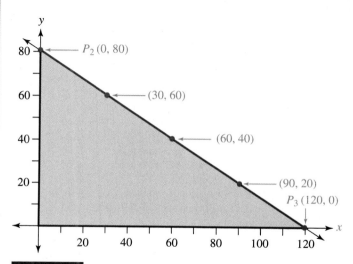

FIGURE 12.32 The maximum occurs at many points.

values of x in the equation.) The meaning of these points is given in Figure 12.33. Pete's Coffees can choose to produce its Rich Blend and Hawaiian Blend in any of the amounts given in Figure 12.33 (or any other amount given by a point on the line). Because each of these choices will maximize Pete's profit at $240 per day, the choice must be made using criteria other than profit. Perhaps the Hawaiian Blend tends to sell out earlier in the day than the Rich Blend. In this case, Pete might choose to produce 30 pounds of Rich Blend and 60 pounds of Hawaiian Blend.

Point	Interpretation
P_2 (0, 80)	Each day, prepare no Rich Blend and 80 pounds of Hawaiian Blend for a profit of $240.
P_3 (120, 0)	Each day, prepare 120 pounds of Rich Blend and no Hawaiian Blend for a profit of $240.
(30, 60)	Each day, prepare 30 pounds of Rich Blend and 60 pounds of Hawaiian Blend for a profit of $240.
(60, 40)	Each day, prepare 60 pounds of Rich Blend and 40 pounds of Hawaiian Blend for a profit of $240.
(90, 20)	Each day, prepare 90 pounds of Rich Blend and 20 pounds of Hawaiian Blend for a profit of $240.

FIGURE 12.33 Interpreting some typical boundary line maximums.

TOPIC X   PILKINGTON LIBBEY-OWENS-FORD LINEAR PROGRAMMING IN THE REAL WORLD

We have been looking at how linear programming can be used in making product mix decisions. Our friend the craftsman had to make a product mix decision: He had to decide what mix of his products was best. And Pete's Coffees had to decide what mix of their products were best.

Pilkington Libbey-Owens-Ford, a large plate glass manufacturer, has used linear programming quite successfully in making product mix decisions. Their linear programming model deals with more than 200 products made at four different plants. The model's constraints include market demand, production capabilities, inventory storage capabilities, and the availability of railroad shipping, as well as production, inventory, and shipping costs. The model is run ten to twenty times per month and makes recommendations regarding the amount of each product to produce at each plant, the amount of inventory to hold at each plant, and the amount of each product to ship between the plants. It has also been used to determine what new products to introduce and what old products to eliminate and to determine how, when, and where to increase manufacturing capabilities. The model, which took two years to complete, has resulted in savings of more than $2 million each year.

Generally, linear programming models in business are very complex. They involve a large number of independent variables and constraints, as does the Pilkington Libbey-Owens-Ford model. The exercises in this section are realistic but simplified examples of the product mix decision models that are routinely used by businesses. Exercise 37 is a realistic but simplified example of the production-planning decision models that are routinely used by Pilkington Libby-Owens-Ford.

Source: C. Martin, D. Dent, and J. Eckhart, "Integrated Production, Distribution, and Inventory Planning at Libbey-Owens-Ford," Interfaces, 23, no. 3, 1993, pages 68–78.

12.1 EXERCISES

In Exercises 1–6, convert the information to a linear inequality. Give the meaning of each variable used.

1. A landscape architect wants his project to use no more than 100 gallons of water per day. Each shrub requires 1 gallon of water per day, and each tree requires 3 gallons of water per day.

2. A shopper wishes to spend no more than $150. Each pair of pants costs $25 and each shirt costs $21.

3. A bookstore owner wishes to generate at least $5,000 in profit this month. Each hardback book generates $4.50 in profit, and each paperback generates $1.25 in profit.

4. Dick Rudd wants to take at least 1,000 mg of vitamin C. Each tablet of Megavite has 30 mg of vitamin C, and each tablet of Healthoboy has 45 mg of vitamin C.

5. A warehouse has 1,650 cubic feet of unused storage space. Refrigerators take up 63 cubic feet each, and dishwashers take up 41 cubic feet each.

6. A coffee shop owner has 1,000 pounds of Java beans. In addition to selling pure Java beans, the shop also sells Blend Number 202, which is 32% Java beans.

In Exercises 7–16, use the method of linear programming to solve the problem.

7. A craftswoman produces two products: floor lamps and table lamps. Production of one floor lamp requires 75 minutes of her labor and materials that cost $25. Production of one table lamp requires 50 minutes of labor, and the materials cost $20. The craftswoman wishes to work no more than 40 hours each week, and her financial resources allow her to pay no more than $900 for materials each week. If she can sell as many lamps as she can make and if her profit is $39 per floor lamp and $33 per table lamp, how many floor lamps and how many table lamps should she make each week to maximize her weekly profit? What is that maximum profit?

8. Five friends, all of whom are experienced bakers, form a company that will make bread and cakes and sell

© Jim West/The Image Works

$3.50 per pound, while profit on the Exotic Blend is $4 per pound. Each day, the shop can obtain 200 pounds of Costa Rican beans and 330 pounds of Ethiopian beans, and it uses that coffee only in the two blends. If the shop can sell all that it makes, how many pounds of Yusip Blend and of Exotic Blend should Pete's Coffees prepare each day to maximize profit? What is that maximum profit?

11. Bake-Em-Fresh sells its bread to supermarkets. Shopgood Stores needs at least 15,000 loaves each week, and Rollie's Markets needs at least 20,000 loaves each week. Bake-Em-Fresh can ship at most 45,000 loaves to these two stores each week if it wishes to satisfy its other customers' needs. If shipping costs an average of 8¢ per loaf to Shopgood Stores and 9¢ per loaf to Rollie's Markets, how many loaves should Bake-Em-Fresh allot to Shopgood and to Rollie's each week to minimize shipping costs? What shipping costs would this entail?

12. Notel Chips manufactures computer chips. Its two main customers, HAL Computers and Peach Computers, just submitted orders that must be filled immediately. HAL needs at least 130 cases of chips, and Peach needs at least 150 cases. Owing to a limited supply of silicon, Notel cannot send more than a total of 300 cases. If shipping costs $100 per case for shipments to HAL and $90 per case for shipments to Peach, how many cases should Notel send to each customer to minimize shipping costs? What shipping costs would this entail?

13. Global Air Lines has contracted with a tour group to transport a minimum of 1,600 first-class passengers and 4,800 economy-class passengers from New York to London during a 6-month time period. Global Air has two types of airplanes, the Orville 606 and the Wilbur W-1112. The Orville 606 carries 20 first-class passengers and 80 economy-class passengers and costs $12,000 to operate. The Wilbur W-1112 carries 80 first-class passengers and 120 economy-class passengers and costs $18,000 to operate. During the time period involved, Global Air can schedule no more than 52 flights on Orville 606s and no more than 30 flights on Wilbur W-1112s. How should Global Air Lines schedule its flights to minimize its costs? What operating costs would this schedule entail?

14. Compucraft sells personal computers and printers made by Peach Computers. The computers come in 12-cubic-foot boxes, and the printers come in 8-cubic-foot boxes. Compucraft's owner estimates that at least 30 computers can be sold each month and that the number of computers sold will be at least 50% more than the number of printers. The computers cost Compucraft $1,000 each and can be sold at a $1,000 profit, while the printers cost $300 each and can be sold for a $350 profit. Compucraft has 1,000 cubic feet of storage available for the Peach personal computers and printers and sufficient financing to spend $70,000 each month on computers and printers. How many computers and printers should

them to local restaurants and specialty stores. Each loaf of bread requires 50 minutes of labor and ingredients costing $0.90 and can be sold for $1.20 profit. Each cake requires 30 minutes of labor and ingredients costing $1.50 and can be sold for $4.00 profit. The partners agree that no one will work more than 8 hours a day. Their financial resources do not allow them to spend more than $190 per day on ingredients. How many loaves of bread and how many cakes should they make each day to maximize their profit? What is that maximum profit?

9. Pete's Coffees sells two blends of coffee beans: Morning Blend and South American Blend. Morning Blend is one-third Mexican beans and two-thirds Colombian beans, and South American Blend is two-thirds Mexican beans and one-third Colombian beans. Profit on the Morning Blend is $3 per pound, while profit on the South American Blend is $2.50 per pound. Each day, the shop can obtain 100 pounds of Mexican beans and 80 pounds of Colombian beans, and it uses that coffee only in the two blends. If the shop can sell all that it makes, how many pounds of Morning Blend and of South American Blend should Pete's Coffees prepare each day to maximize profit? What is that maximum profit?

10. Pete's Coffees sells two blends of coffee beans: Yusip Blend and Exotic Blend. Yusip Blend is one-half Costa Rican beans and one-half Ethiopian beans, and Exotic Blend is one-quarter Costa Rican beans and three-quarters Ethiopian beans. Profit on the Yusip Blend is

Compucraft order from Peach each month to maximize profit? What is that maximum profit?

15. The Appliance Barn has 2,400 cubic feet of storage space for refrigerators. The larger refrigerators come in 60-cubic-foot packing crates, and the smaller ones come in 40-cubic-foot crates. The larger refrigerators can be sold for a $250 profit, while the smaller ones can be sold for a $150 profit.

 a. If the manager is required to sell at least 50 refrigerators each month, how many large refrigerators and how many small refrigerators should he order each month to maximize profit?

 b. If the manager is required to sell at least 40 refrigerators each month, how many large refrigerators and how many small refrigerators should he order each month to maximize profit?

 c. Should the Appliance Barn owner require his manager to sell 40 or 50 refrigerators per month?

16. City Electronics Distributors handles two lines of televisions, the Packard and the Bell. It purchases up to $57,000 worth of television sets from the manufacturers each month and stores them in a 9,000-cubic-foot warehouse. The Packards come in 36-cubic-foot packing crates, and the Bells come in 30-cubic-foot crates. The Packards cost City Electronics $200 each and can be sold to a retailer for a $200 profit, while the Bells cost $250 each and can be sold for a $260 profit. City Electronics must stock enough sets to meet its regular customers' standing orders.

 a. If City Electronics has standing orders for 250 sets in addition to orders from other retailers, how many sets should City Electronics order each month to maximize profit?

 b. If City Electronics' standing orders increase to 260 sets, how many sets should City Electronics order each month to maximize profit?

17. How much of their available Mexican beans would be unused if Pete's Coffees of Exercise 9 maximizes its profit? How much of the available Colombian beans would be unused?

18. How much of their available time would be unused if the five friends of Exercise 8 maximize their profit? How much of their available money would be unused?

19. How much of her available time would be unused if the craftswoman of Exercise 7 maximizes her profit? How much of her available money would be unused?

20. How much of the available Costa Rican beans would be unused if Pete's Coffees of Exercise 10 maximizes its profit? How much of the available Ethiopian beans would be unused?

In Exercises 21–24, the region of possible solutions is not bounded: thus, there may not be both a maximum and a minimum. After graphing the region of possible solutions and finding each corner point, you can determine if both a maximum and a

minimum exist by choosing two arbitrary values of z and graphing the corresponding lines, as discussed in the section "Why the Corner Principle Works."

21. The objective function $z = 2x + 3y$ is subject to the constraints

$$3x + y \geq 12$$
$$x + y \geq 6$$
$$x \geq 0 \quad \text{and} \quad y \geq 0$$

 Find the following.

 a. the point at which the maximum occurs (if there is such a point)
 b. the maximum value
 c. the point at which the minimum occurs (if there is such a point)
 d. the minimum value

22. The objective function $z = 3x + 4y$ is subject to the constraints

$$2x + y \geq 10$$
$$3x + y \geq 12$$
$$x \geq 0 \quad \text{and} \quad y \geq 0$$

 Find the following.

 a. the point at which the maximum occurs (if there is such a point)
 b. the maximum value
 c. the point at which the minimum occurs (if there is such a point)
 d. the minimum value

23. U.S. Motors manufactures quarter-ton, half-ton, and three-quarter-ton panel trucks. United Delivery Service has placed an order for at least 300 quarter-ton, 450 half-ton, and 450 three-quarter-ton panel trucks. U.S. Motors builds the trucks at two plants, one in Detroit and one in Los Angeles. The Detroit plant produces 30 quarter-ton trucks, 60 half-ton trucks, and 90 three-quarter-ton trucks each week, at a total cost of $540,000. The Los Angeles plant produces 60 quarter-ton trucks, 45 half-ton trucks, and 30 three-quarter-ton trucks each week, at a total cost of $360,000. How should U.S. Motors schedule its two plants so that it can fill this order at minimum cost? What is that minimum cost?

24. Eaton's Chocolates produces semisweet chocolate chips and milk chocolate chips at its plants in Bay City and Estancia. The Bay City plant produces 3,000 pounds of semisweet chips and 2,000 pounds of milk chocolate chips each day at a cost of $1,000, while the Estancia plant produces 1,000 pounds of semisweet chips and 6,000 pounds of milk chocolate chips each day at a cost of $1,500. Eaton's has an order from SafeNShop Supermarkets for at least 30,000 pounds of semisweet chips and 60,000 pounds of milk chocolate chips. How should it schedule its production so that it can fill the order at minimum cost? What is that minimum cost?

Answer the following questions using complete sentences and your own words.

• CONCEPT QUESTIONS

25. A bicycle manufacturing firm hired a consultant who used linear programming to determine that the firm should discontinue manufacturing some of its product line. What negative ramifications might occur if this advice is followed? How could linear programming be used if the firm decided not to follow this advice?

26. Is it possible for an unbounded region to have both a maximum and a minimum? Draw a number of graphs (similar to those in Figure 12.28) to see. Justify your answer with graphs.

27. Discuss why it is that, when two corners give the same maximum (or minimum) value, the maximum (or minimum) occurs at all points on the boundary line between those corners.

28. In Example 2, Pete's Coffees mixes Colombian beans and Kona beans to make Rich Blend and Hawaiian Blend. It may be that Pete's could make slight variations in the proportions of Colombian and Kona beans in these two blends without affecting the taste of the coffee. Why would Pete's consider doing this? What effect would it have on the objective function, the constraints, the region of possible solutions, and the solution to the example?

29. In an open market, the selling price of a product is determined by the supply of the product and the demand for the product. In particular, the price falls if a larger supply of the product becomes available, and the price increases if there is suddenly a larger demand for the product. Any change in the selling price could affect the profit. However, when modeling a linear programming problem, we assume a constant profit. How could linear programming be adjusted to allow for the fact that prices change with supply and demand?

30. Make up a linear programming problem (a math problem, not a word problem) that has no maximum.

31. Make up a linear programming problem (a math problem, not a word problem) that has no minimum.

32. Make up a word problem that results in the following linear programming problem:

$$\text{maximize } z = 5x + 4y$$

subject to the constraints

$$10x + 20y \le 1,000$$
$$30x + 10y \le 2,000$$
$$x \ge 0 \quad \text{and} \quad y \ge 0$$

• HISTORY QUESTIONS

33. For whom was George Dantzig named?

34. How did Dantzig choose the topic of his Ph.D. thesis?

35. What did Dantzig do during World War II?

36. What was surprising about the awarding of a Nobel Prize to two economists for their use of linear programming?

• PROJECTS

37. Glassco, a large plate glass manufacturer, makes two products: half-inch-thick plate glass and three-eighths-inch-thick plate glass. It makes each of these two products at its two plants: Clear View and Panesville. For the first quarter in the next year, demand for half-inch-thick plate glass is forecast to be 5,000 sheets, and demand for three-eighths-inch-thick plate glass is forecast to be 6,000 sheets for the first quarter in the year. Because of the difficulties inherent in storing the product, Glassco does not make a surplus of glass above and beyond the demand. The Clear View plant is capable of producing at most 7,000 sheets per quarter, and the Panesville plant can make at most 5,000 sheets per quarter. Glass made at the Clear View plant can be sold at a profit of $820 per sheet and $760 per sheet for the half-inch and three-eighths-inch products, respectively. However, the profit on glass made at the Panesville plant is smaller by $75 per sheet, owing to higher shipping costs.

a. Translate the information on demand into two constraints. Use the following variables:

> v = amount of half-inch-thick glass made in the first quarter at Clear View
>
> w = amount of three-eighths-inch-thick glass made in the first quarter at Clear View
>
> x = amount of half-inch-thick glass made in the first quarter at Panesville
>
> y = amount of three-eighths-inch-thick glass made in the first quarter at Panesville

b. Translate the information on plant production capabilities into two constraints.

c. Translate the information on profit into an objective function.

d. Explain why this problem cannot be solved with the geometric method of linear programming.

38. You manage the Fee-Based Technical Support Division of a large computer software firm. You have sufficient resources to provide up to 1,000 hours of technical support each week. Often, this is not enough time to meet demand. As a result, you must develop a policy to allocate technical support time. Your firm's warranty specifies that customers who have purchased software in the last year can purchase technical support for $49 per incident. Software purchased more than a year ago is not under warranty. However, your firm continues to provide technical support at the same rate after the warranty has expired, to maintain customer goodwill. Each under-warranty support incident requires an average of 30 minutes of support and each out-of-warranty support incident requires an average of one hour of support. Your firm's CEO has asked that you allot between 700 and 800 hours per week to customers whose software is still under warranty to ensure that they have sufficient access to technical support. Your CEO has also asked that you allot between 100 and 250 hours per week to customers whose software is no longer under warranty to ensure that they still have some access to technical support.

a. Create a specific allocation policy that maximizes revenue.

b. One year after you instituted the allocation policy in part (a), demand has increased. As a result, you are considering increasing the fee to customers whose software is no longer under warranty from $49 to

$149 per incident. Create a new specific allocation policy that maximizes revenue.

WEB PROJECTS

39. Write an essay on the various theories about why George Dantzig did not receive a Nobel Prize for his invention of linear programming.

40. Research linear programming software packages. List at least five companies or software brands, and discuss their linear programming products.

Some useful links for this project are listed on the text companion web site. Go to **www.cengage.com/math/ johnson** to access the web site.

Sections 12.2 and 12.3 are online

12 CHAPTER REVIEW

TERMS

bounded region
constraint
corner point
Corner Principle
dependent variable

independent variable
linear equation
linear inequality
linear programming
model
objective function

optimize
region of possible solutions of a linear programming problem
region of solutions of an inequality

region of solutions of a system of linear inequalities
system of linear inequalities
unbounded region
variable

PROCEDURES

Graphing the Region of Solutions of a Linear Inequality

1. Solve the inequality for y.
2. Graph the line with a solid or dashed line.
3. Shade in one side of the line.

Linear Programming Steps:

1. List the independent variables.
2. List the constraints and translate them into linear inequalities.

Linear Programming Steps: Continued

3. Find the objective and translate it into a linear equation.
4. Graph the region of possible solutions, the region described by the constraints.
5. Find all corner points and the z-values associated with these points. Check your work.
6. Find the maximum/minimum.

REVIEW EXERCISES

In Exercises 1–4, graph the region of solutions of the given inequality.

1. $4x - 5y > 7$

2. $3x + 4y < 10$

3. $3x + 6y \leq 9$

4. $6x - 8y \geq 12$

In Exercises 5–10, graph the region of solutions of the given system of inequalities.

5. $8x - 4y < 10$
   $3x + 5y \geq 7$

6. $x - 5y < 7$
   $3x + 2y > 6$

7. $5x - y > 8$
   $x \geq -3$

8. $6x + 4y \leq 7$
   $y < 4$

9. $x - y \leq 7$
   $5x + 3y \leq 9$
   $x \geq 0$
   $y \geq 0$

10. $3x - 2y \leq 12$
   $5x + 3y \leq 15$
   $x \geq 0$
   $y \geq 0$

In Exercises 11–17, solve the given linear programming problem.

11. Minimize $z = 4x + 3y$
   subject to the constraints:
   $x + 2y \leq 20$
   $4x + y \leq 20$
   $x, y \geq 0$

12. Maximize $z = 4x + 2y$
   subject to the constraints:
   $7x + 3y \leq 21$
   $x + y \leq 5$
   $2x + y \leq 10$
   $x, y \geq 0$

13. Minimize $z = 2x + 3y$
   subject to the constraints:
   $15x + y \geq 45$
   $x + 6y \geq 40$
   $8x + 7y \geq 70$
   $x, y \geq 0$

14. Maximize $z = 8x + 6y$
   subject to the constraints:
   $x + 2y \geq 50$
   $52x + 44y \geq 1144$
   $x, y \geq 0$

15. The Mowson Audio Co. makes stereo speaker assemblies. They purchase speakers from a speaker manufacturing firm, and install them in their own cabinets. Mowson's model 110 speaker assembly, which sells for $200, has a tweeter and a midrange speaker. Their model 330 assembly, which sells for $350, has two tweeters, a midrange speaker, and a woofer. Mowson currently has in stock 90 tweeters, 60 midrange speakers, and 44 woofers. How many speaker assemblies should Mowson make to maximize their income? What is that maximum income?

16. The Stereo Guys store sells two lines of personal stereos: the Sunny and the Iwa. The Sunny comes in a 12-cubic-foot box and can be sold for a $220 profit, while the Iwa comes in an 8-cubic-foot box and can be sold for a $200 profit. The Stereo Guys marketing department estimates that at least 600 personal stereos can be sold each month and that, because of Sunny's quality reputation, the demand for the Sunny unit is at least twice that of the Iwa. If the Stereo Guys warehouse has 12,000 cubic feet of space available for personal stereos, how many Sunnys and Iwas should they stock each month to maximize profit? What is that maximum profit?

 Answer the following questions using your own words and complete sentences.

• CONCEPT AND HISTORY QUESTIONS

17. Discuss why the corner principal works.
18. What are the main elements of a linear programming model?
19. Who invented linear programming?

APPENDIX A
Using a Scientific Calculator

 WHAT KIND OF CALCULATOR DO YOU HAVE?

There are three kinds of calculators: graphing calculators, scientific calculators, and basic calculators. You probably have a:

- *graphing calculator* if your calculator has at least one button labeled "GRAPH" and its screen is about 1.5 or 2 inches tall. Read Appendix B.

- *basic calculator* if you get 20 after typing

$$2 \boxed{+} 3 \boxed{\times} 4 \boxed{=} \qquad \text{or} \qquad 2 \boxed{+} 3 \boxed{\times} 4 \boxed{\text{ENTER}}$$

  A basic calculator is insufficient for this text.

- *scientific calculator* if you get 14 after typing

$$2 \boxed{+} 3 \boxed{\times} 4 \boxed{=} \qquad \text{or} \qquad 2 \boxed{+} 3 \boxed{\times} 4 \boxed{\text{ENTER}}$$

Some scientific calculators work like graphing calculators, except that they don't graph (and therefore they don't have bigger screens). If you get:

- 1.609 . . . after typing

$$\boxed{\ln} 5 \boxed{=} \qquad \text{or} \qquad \boxed{\ln} 5 \boxed{\text{ENTER}}$$

  then your scientific calculator works more like a graphing calculator. Read Appendix B.

- 1.609 . . . after typing

$$5 \boxed{\ln}$$

  then your scientific calculator doesn't work like a graphing calculator. Read this appendix.

 USING YOUR SCIENTIFIC CALCULATOR

OBJECTIVE

- Become familiar with a scientific calculator

Read this appendix with your calculator by your side. When a calculation is discussed, do that calculation on your calculator.

Unfortunately, scientific calculators don't all work exactly the same way. Even if you do everything correctly, your answer may have a few more or less decimal places than the one given, or may differ in the last decimal place. Occasionally, the way a calculation is performed on your calculator will differ slightly from that discussed below. If so, experiment a little, or consult your instructor.

The Equals Button

Some buttons perform operations on a pair of numbers. For example, the $\boxed{+}$ button is used to add two numbers; it only makes sense to use this button in conjunction with a pair of numbers. *When using such a button, you must finish the typing with the* $\boxed{=}$ *button.* That is, to add 3 and 2, type

$$3 \boxed{+} 2 \boxed{=}$$

Some buttons perform operations on a single number. For example, the $\boxed{x^2}$ button is used to square a number; it only makes sense to use this button in conjunction with a single number. *When using such a button, you do **not** type the* $\boxed{=}$ *button.* That is, to square 5, type

$$5 \boxed{x^2}$$

The reason for this distinction is that the calculator must be told when you are done entering information and you are ready for it to compute. If you typed

$$3 \boxed{+} 2$$

the calculator would have no way of knowing whether you were ready for it to compute "3 + 2" or you were in the middle of instructing it to compute "3 + 22." You must follow "3 + 2" with the $\boxed{=}$ button to tell the calculator that you are done entering information and are ready for it to compute. On the other hand, when you type

$$5 \boxed{x^2}$$

the calculator knows that you are done entering information; if you meant to square 53, you would not have pressed the $\boxed{x^2}$ button after the "5." There is no need to follow "5 $\boxed{x^2}$" with the $\boxed{=}$ button.

The Subtraction Symbol and the Negative Symbol

If "3 − −2" were read aloud, you could say "three subtract negative two" or "three minus minus two," and you would be understood. The expression is understandable even if the distinction between the negative symbol and the subtraction symbol is not made clear. With a calculator, however, this distinction is crucial. The subtraction button is labeled $\boxed{-}$. There is no negative button; instead, there is a $\boxed{+/-}$ button or $\boxed{+\,C\,-}$ button that changes the sign of whatever number is on the display. Typing

$$5 \boxed{+/-}$$

makes the display read "−5." Typing

$$5 \boxed{+/-}\boxed{+/-}$$

makes the display read "5," because two sign changes undo each other. Typing

$$\boxed{+/-} 5$$

makes the display read "5," not "−5." *You must press the sign-change button after the number itself.*

EXAMPLE **1**

USING THE SUBTRACTION AND NEGATIVE SYMBOLS Calculate 3 − −2, both (a) by hand and (b) with a calculator.

SOLUTION

a. 3 − −2 = 3 + 2 = 5
b. To do this, we must use the subtraction button, which is labeled $\boxed{-}$, and the sign-change button, which is labeled $\boxed{+/-}$. Typing

$$3 \boxed{-} 2 \boxed{+/-} \boxed{=}$$

makes the display read "5." Remember that you must press the sign-change button *after* the number itself; you press "2" followed by $+/-$, not $+/-$ followed by "2." Also, you must finish the typing with the $=$ button, because the $=$ button performs an operation on a pair of numbers.

Order of Operations and Use of Parentheses

Scientific calculators are programmed so that they follow the standard order of operations. That is, they perform calculations in the following order:

1. Parentheses-enclosed work

2. Exponents

3. Multiplication and
Division, from left to right

4. Addition and
Subtraction, from left to right

This order can be remembered by remembering the word *PEMDAS,* which stands for

Parentheses/**E**xponents/**M**ultiplication/**D**ivision/**A**ddition/**S**ubtraction

Frequently, the fact that calculators are programmed to follow the order of operations means that you perform a calculation on your calculator in exactly the same way that it is written.

EXAMPLE **2**

PARENTHESES COME FIRST   Calculate $2(3 + 4)$, both (a) by hand and (b) with a calculator.

SOLUTION

a. $2(3 + 4) = 2(7)$ **parentheses-enclosed work comes first**
$\qquad\qquad\quad = 14$
b. Type

$\qquad$ 2 $\boxed{\times}$ $\boxed{(}$ 3 $\boxed{+}$ 4 $\boxed{)}$ $\boxed{=}$

and the display reads "14." Notice that this is typed on a calculator exactly as it is written, with two important exceptions:

- We must press the $\boxed{\times}$ button to mean multiplication; we cannot use parentheses to mean multiplication, as is done in part (a).

- We must finish the typing with the $\boxed{=}$ button.

EXAMPLE **3**

EXPONENTS BEFORE MULTIPLICATION   Calculate $2 \cdot 3^2$, both (a) by hand and (b) with a calculator.

SOLUTION

a. $2 \cdot 3^2 = 2 \cdot 9$ **exponents come before multiplication**
$\qquad\qquad\; = 18$
b. Type

$\qquad$ 2 $\boxed{\times}$ 3 $\boxed{x^2}$ $\boxed{=}$

and the display reads "18." Notice that this is typed on a calculator exactly as it is written, except that we must finish the typing with the $\boxed{=}$ button and we must press the $\boxed{\times}$ button to mean multiplication. The $\boxed{\cdot}$ button is a decimal-point button, not a multiplication-dot button.

EXAMPLE **4**

SOLUTION

MULTIPLICATION BEFORE EXPONENTS   Calculate $(2 \cdot 3)^2$, both (a) by hand and (b) with a calculator.

a. $(2 \cdot 3)^2 = 6^2$ parentheses-enclosed work comes first
$= 36$

b. Type

$\boxed{(}\ 2\ \boxed{\times}\ 3\ \boxed{)}\ \boxed{x^2}$

and the display reads "36." Notice that this is typed on a calculator exactly as it is written, except that we must use the $\boxed{\times}$ button to mean multiplication. Also, we do not use the $\boxed{=}$ button, because the $\boxed{x^2}$ button performs an operation on a single number.

EXAMPLE **5**

SOLUTION

CUBING   Calculate $4 \cdot 2^3$, both (a) by hand and (b) with a calculator.

a. $4 \cdot 2^3 = 4 \cdot 8$ exponents come before multiplication
$= 32$

b. To do this, we must use the exponent button, which is labeled either $\boxed{y^x}$ or $\boxed{x^y}$. Type

$4\ \boxed{\times}\ 2\ \boxed{y^x}\ 3\ \boxed{=}$ or $4\ \boxed{\times}\ 2\ \boxed{x^y}\ 3\ \boxed{=}$

and the display reads "32."

Sometimes, you don't perform a calculation on your calculator in the same way that it is written, even though calculators are programmed to follow the order of operations.

EXAMPLE **6**

SOLUTION

A FRACTION BAR ACTS AS PARENTHESES   Calculate $\dfrac{2}{3 \cdot 4}$ with a calculator.

Wrong   It is incorrect to type

$2\ \boxed{\div}\ 3\ \boxed{\times}\ 4\ \boxed{=}$

According to the order of operations, multiplication and division are done *from left to right,* so the above typing is algebraically equivalent to

$= \dfrac{2}{3} \cdot 4$ first dividing then multiplying, since division is on the left and multiplication is on the right

$= \dfrac{2}{3} \cdot \dfrac{4}{1} = \dfrac{2 \cdot 4}{3}$ multiplying

which is not what we want. The difficulty is that the large fraction bar in the expression $\frac{2}{3 \cdot 4}$ groups the "3 · 4" together in the denominator; in the above typing, nothing groups the "3 · 4" together, and only the 3 ends up in the denominator.

Right   The calculator needs parentheses inserted in the following manner:

$\dfrac{2}{(3 \cdot 4)}$

Thus, it is correct to type

$2\ \boxed{\div}\ \boxed{(}\ 3\ \boxed{\times}\ 4\ \boxed{)}\ \boxed{=}$

This makes the display read 0.166666667, the correct answer.

Also Right   It is correct to type

2 ÷ 3 ÷ 4 =

According to the order of operations, multiplication and division are done from left to right, so the above typing is algebraically equivalent to

$$\frac{2}{3} \div 4 \qquad \textbf{doing the left-hand division first}$$

$$= \frac{2}{3} \cdot \frac{1}{4} \qquad \textbf{inverting}$$

$$= \frac{2}{3 \cdot 4} \qquad \textbf{multiplying}$$

which is what we want. *When you're calculating something that involves only multiplication and division and you don't use parentheses, the ✕ button places a factor in the numerator, and the ÷ button places a factor in the denominator.*

EXAMPLE 7

SOLUTION

A FRACTION BAR ACTS AS PARENTHESES   Calculate $\frac{2}{3/4}$ with a calculator.

Wrong   It is incorrect to type

2 ÷ 3 ÷ 4 =

even though that matches the way the problem is written algebraically. As discussed in Example 6, this typing is algebraically equivalent to

$$\frac{2}{3 \cdot 4}$$

which is not what we want.

Right   The calculator needs parentheses inserted in the following manner:

$$\frac{2}{(3 \div 4)}$$

Thus, it is correct to type

2 ÷ (3 ÷ 4) =

since, according to the order of operations, parentheses-enclosed work is done first. This makes the display read 2.66666667, the correct answer.

EXAMPLE 8

SOLUTION

A FRACTION BAR ACTS AS PARENTHESES   Calculate $\frac{2+3}{4}$ with a calculator.

Wrong   It is incorrect to type

2 + 3 ÷ 4 =

According to the order of operations, division is done before addition, so this typing is algebraically equivalent to

$$2 + \frac{3}{4}$$

which is not what we want. The large fraction bar in the expression $\frac{2+3}{4}$ groups the "2 + 3" together in the numerator; in the above typing, nothing groups the "2 + 3" together, and only the 3 ends up in the numerator.

Right   The calculator needs parentheses inserted in the following manner:

$$\frac{(2 + 3)}{4}$$

Thus, it is correct to type

$$\boxed{(}\; 2 \;\boxed{+}\; 3 \;\boxed{)}\; \boxed{\div}\; 4 \;\boxed{=}$$

This makes the display read 1.25, the correct answer.

Also Right   It is correct to type

$$2 \;\boxed{+}\; 3 \;\boxed{=}\; \boxed{\div}\; 4 \;\boxed{=}$$

The first $\boxed{=}$ makes the calculator perform all prior calculations before continuing. This too makes the display read "1.25," the correct answer.

The 2nd Button

Many buttons have two labels and two uses. To use the label on a button, you just press that button. To use the label above a button, you press $\boxed{\text{2nd}}$ and then the button. For example, your calculator might have a button labeled "LN" on the button itself and "e^x" above the button. If so, typing

$$5 \;\boxed{\text{LN}}$$

makes the display read "1.609 . . ." because $\ln(5) = 1.609 \ldots$. Typing

$$5 \;\boxed{\text{2nd}}\; \boxed{e^x}$$

makes the display read 148.413 . . . because $e^5 = 148.143 \ldots$.

Memory

The memory is a place to store a number for later use, without having to write it down. If a number is on your display, you can place it into the memory (or **store** it) by pressing the button labeled $\boxed{\text{STO}}$ (or $\boxed{x \rightarrow \text{M}}$ or $\boxed{\text{M in}}$), and you can take it out of the memory (or **recall** it) by pressing the button labeled $\boxed{\text{RCL}}$ (or $\boxed{\text{RM}}$ or $\boxed{\text{MR}}$).

Typing

$$5 \;\boxed{\text{STO}}$$

makes the calculator store a 5 in its memory. If you do other calculations or just clear your display and later press

$$\boxed{\text{RCL}}$$

then your display will read "5."

Some calculators have more than one memory. If yours does, then pressing the button labeled $\boxed{\text{STO}}$ won't do anything; pressing $\boxed{\text{STO}}$ and then "1" will store it in memory number 1; pressing $\boxed{\text{STO}}$ and then "2" will store it in memory number 2, and so on. Pressing $\boxed{\text{RCL}}$ and then "1" will recall what has been stored in memory number 1.

EXAMPLE **9**

USING THE MEMORY   Use the quadratic formula and your calculator's memory to solve

$$2.3x^2 + 4.9x + 1.5 = 0$$

SOLUTION

The quadratic formula says that if $ax^2 + bx + c = 0$, then

$$x = \frac{-b \pm \sqrt{b^2 - 4ac}}{2a}$$

We have $2.3x^2 + 4.9x + 1.5 = 0$, so $a = 2.3$, $b = 4.9$, and $c = 1.5$. This gives

$$x = \frac{-4.9 \pm \sqrt{4.9^2 - 4 \cdot 2.3 \cdot 1.5}}{2 \cdot 2.3}$$

The quickest way to do this calculation is to calculate the radical, store it, and then calculate the two fractions.

Step 1   *Calculate the radical.* To do this, type

4.9 $\boxed{x^2}$ $\boxed{-}$ 4 $\boxed{\times}$ 2.3 $\boxed{\times}$ 1.5 $\boxed{=}$ $\boxed{\sqrt{x}}$ $\boxed{\text{STO}}$

This makes the display read "3.195309" and stores the number in the memory. Notice the use of the $\boxed{=}$ button; this makes the calculator finish the prior calculation before taking a square root. If the $\boxed{=}$ button were not used, the order of operations would require the calculator to take the square root of 1.5.

Step 2   *Calculate the first fraction.* To do this, type

4.9 $\boxed{+/-}$ $\boxed{+}$ $\boxed{\text{RCL}}$ $\boxed{=}$ $\boxed{\div}$ 2 $\boxed{\div}$ 2.3 $\boxed{=}$

This makes the display read "−0.3705849." Notice the use of the $\boxed{=}$ button.

Step 3   *Calculate the second fraction.* To do this, type

4.9 $\boxed{+/-}$ $\boxed{-}$ $\boxed{\text{RCL}}$ $\boxed{=}$ $\boxed{\div}$ 2 $\boxed{\div}$ 2.3 $\boxed{=}$

This makes the display read "−1.7598498."

The solutions to $2.3x^2 + 4.9x + 1.5 = 0$ are $x = -0.3705849$ and $x = -1.7598498$. These are approximate solutions in that they show only the first seven decimal places.

Step 4   *Check your solutions.* These solutions can be checked by seeing whether they satisfy the equation $2.3x^2 + 4.9x + 1.5 = 0$. To check the first solution, type

2.3 $\boxed{\times}$ .3705849 $\boxed{+/-}$ $\boxed{x^2}$ $\boxed{+}$ 4.9 $\boxed{\times}$ .3705849 $\boxed{+/-}$ $\boxed{+}$ 1.5 $\boxed{=}$

and the display will read either "0" or a number very close to 0.

Scientific Notation

Typing

4000000 $\boxed{\times}$ 8000000 $\boxed{=}$

makes the display read "3.2   13" rather than "32000000000000." This is because the calculator does not have enough room on its display for "32000000000000." When the display shows "3.2   13," read it as "3.2×10^{13}," which is written in scientific notation. Literally, "3.2×10^{13}" means "multiply 3.2 by 10, thirteen times," but as a shortcut, you can interpret it as "move the decimal point in the '3.2' thirteen places to the right."

Typing

.0000005 $\boxed{\times}$ .0000007 $\boxed{=}$

makes the display read "3.5   −13" rather than "0.00000000000035," because the calculator does not have enough room on its display for "0.00000000000035." Read "3.5   −13" as "3.5×10^{-13}." Literally, this means "divide 3.5 by 10, thirteen times," but as a shortcut, you can interpret it as "move the decimal point in the '3.5' thirteen places to the left."

You can type a number in scientific notation by using the button labeled $\boxed{\text{EXP}}$ (which stands for *exponent*) or $\boxed{\text{EE}}$ (which stands for *enter exponent*). For example, typing

$$5.2 \ \boxed{\text{EXP}} \ 8 \qquad \text{or} \qquad 5.2 \ \boxed{\text{EE}} \ 8$$

makes the display read "5.2   8," which means "5.2×10^8," and typing

$$3 \ \boxed{\text{EXP}} \ 17 \ \boxed{+/-} \qquad \text{or} \qquad 3 \ \boxed{\text{EE}} \ 17 \ \boxed{+/-}$$

makes the display read "3 -17," which means "3×10^{-17}." Notice that the sign-change button is used to make the exponent negative.

Be careful that you don't confuse the $\boxed{y^x}$ button with the $\boxed{\text{EXP}}$ button. The $\boxed{\text{EXP}}$ button does *not* allow you to type in an exponent; it allows you to type in scientific notation. For example, typing

$$3 \ \boxed{\text{EXP}} \ 4$$

makes the display read "3   4," which means "3×10^4," and typing

$$3 \ \boxed{y^x} \ 4$$

makes the display read "81," since $3^4 = 81$.

EXERCISES

Perform the calculations in Exercises 1–32. The correct answer is given in brackets [ ]. In your homework, write down what you type on your calculator to get that answer. Answers are not given in the back of the book.

1. $-3 - -5$ [2]
2. $-6 - 3$ [-9]
3. $4 - -9$ [13]
4. $-6 - -8$ [2]
5. $-3 - (-5 - -8)$ [-6]
6. $-(-4 - 3) - (-6 - -2)$ [11]
7. $-8 \cdot -3 \cdot -2$ [-48]
8. $-9 \cdot -3 - 2$ [25]
9. $(-3)(-8) - (-9)(-2)$ [6]
10. $2(3 - 5)$ [-4]
11. $2 \cdot 3 - 5$ [1]
12. $4 \cdot 11^2$ [484]
13. $(4 \cdot 11)^2$ [1,936]
14. $4 \cdot (-11)^2$ [484]
15. $4 \cdot (-3)^3$ [-108]

WARNING: Some calculators will not raise a negative number to a power. If yours has this characteristic, how can you use your calculator on this exercise?

16. $(4 \cdot -3)^3$ [$-1,728$]
17. $\dfrac{3 + 2}{7}$ [0.7142857]
18. $\dfrac{3 \cdot 2}{7}$ [0.8571429]
19. $\dfrac{3}{2 \cdot 7}$ [0.2142857]
20. $\dfrac{3 \cdot 2}{7 \cdot 5}$ [0.1714286]
21. $\dfrac{3 + 2}{7 \cdot 5}$ [0.1428571]
22. $\dfrac{3 \cdot -2}{7 + 5}$ [-0.5]
23. $\dfrac{3}{7/2}$ [0.8571429]
24. $\dfrac{3/7}{2}$ [0.2142857]
25. 1.8^2 [3.24]
26. $\sqrt{1.8}$ [1.3416408]
27. $47{,}000{,}000^2$ [2.209×10^{15}]
28. $\sqrt{0.0000000000027}$ [1.643168×10^{-6}]
29. $(-3.92)^7$ [$-14{,}223.368737$]
30. $(5.72 \times 10^{19})^4$ [1.070494×10^{79}]
31. $(3.76 \times 10^{-12})^{-5}$ [1.330641×10^{57}]
32. $(3.76 \times 10^{-12}) - 5$ [-5]
33. Solve $4.2x^2 + 8.3x + 1.1 = 0$ for x. Check your two answers by substituting them back into the equation.
34. Solve $5.7x^2 + 12.3x - 8.1 = 0$ for x. Check your two answers by substituting them back into the equation.
35. Which of the following buttons must be used in conjunction with the $\boxed{=}$ button, and why?

$$\boxed{+} \ \boxed{-} \ \boxed{\times} \ \boxed{\div} \ \boxed{x^2} \ \boxed{\sqrt{x}} \ \boxed{y^x} \ \boxed{1/x} \ \boxed{+/-}$$

APPENDIX B
Using a Graphing Calculator

O BJECTIVE

- Become familiar with a graphing calculator

The following discussion was written specifically for Texas Instruments (or "TI") and Casio graphing calculators, but it frequently applies to other brands as well. Texas Instruments TI-83 and TI-84 models and Casio CFX-9850, FX-9860, and CFX-9950 models are specifically addressed. Read this discussion with your calculator close at hand. When a calculation is discussed, do that calculation on your calculator.

To do any of the calculations discussed in this section on a Casio, start by pressing $\boxed{\text{MENU}}$. Then use the arrow buttons to select "RUN" on the main menu, and press $\boxed{\text{EXE}}$ (which stands for "execute").

The Enter Button

A TI graphing calculator will never perform a calculation until the $\boxed{\text{ENTER}}$ button is pressed; a Casio will never perform a calculation until the $\boxed{\text{EXE}}$ button is pressed. A Casio's $\boxed{\text{EXE}}$ button functions like a TI's $\boxed{\text{ENTER}}$ button; often, we will refer to either of these buttons as the $\boxed{\text{ENTER}}$ button.
To add 3 and 2, type

$$3 \boxed{+} 2 \boxed{\text{ENTER}}$$

and the display will read 5. To square 4, type

$$4 \boxed{x^2} \boxed{\text{ENTER}}$$

and the display will read 16. If the $\boxed{\text{ENTER}}$ button isn't pressed, the calculation will not be performed.

The 2nd and Alpha Buttons

Most calculator buttons have more than one label and more than one use; you select from these uses with the $\boxed{\text{2nd}}$ and $\boxed{\text{ALPHA}}$ buttons. A Casio's $\boxed{\text{SHIFT}}$ button functions like a TI's $\boxed{\text{2nd}}$ button; often, we will refer to either of these buttons as the $\boxed{\text{2nd}}$ button. For example, one button is labeled "x^2" on the button itself, "$\sqrt{}$" above the button, and either "I" or "K" above and to the right of the button. If it is used without the $\boxed{\text{2nd}}$ or $\boxed{\text{ALPHA}}$ buttons, it will square a number. Typing

$$4 \boxed{x^2} \boxed{\text{ENTER}}$$

makes the display read 16, since $4^2 = 16$. If it is used with the $\boxed{\text{2nd}}$ button, it will take the square root of a number. Typing

$$\boxed{\text{2nd}} \boxed{\sqrt{}} 4 \boxed{\text{ENTER}}$$

makes the display read 2, since $\sqrt{4} = 2$. If it is used with the $\boxed{\text{ALPHA}}$ button, it will display the letter I or K.

Notice that to square 4, you press the $\boxed{x^2}$ button *after* the 4, but to take the square root of 4, you press the $\boxed{\sqrt{}}$ button *before* the 4. This is because graphing calculators are designed so that the way you type something is as similar as possible to the way it is written algebraically. When you write 4^2, you write the 4 first and then the squared symbol; thus, on your graphing calculator, you press the 4 first and then the $\boxed{x^2}$ button. When you write $\sqrt{4}$, you write the square root symbol first and then the 4; thus, on your graphing calculator, you press the $\boxed{\sqrt{}}$ button first and then the 4.

Frequently, the two operations that share a button are operations that "undo" each other. For example, typing

$$3 \quad \boxed{x^2} \quad \boxed{\text{ENTER}}$$

makes the display read 9, since $3^2 = 9$, and typing

$$\boxed{\text{2nd}} \quad \boxed{\sqrt{}} \quad 9 \quad \boxed{\text{ENTER}}$$

makes the display read 3, since $\sqrt{9} = 3$. This is done as a memory device; it is easier to find the various operations on the keyboard if the two operations that share a button also share a relationship.

Two operations that *always* undo each other are called **inverses.** The x^2 and $\sqrt{x}$ operations are not inverses because $(-3)^2 = 9$, but $\sqrt{9} \neq -3$. However, there is an inverse-type relationship between the x^2 and $\sqrt{x}$ operations—they undo each other sometimes. When two operations share a button, they are inverses or they share an inverse-type relation.

Correcting Typing Errors

If you've made a typing error *and you haven't yet pressed* $\boxed{\text{ENTER}}$, you can correct that error with the $\boxed{\blacktriangleleft}$ button. For example, if you typed "$5 \times 2 + 7$" and then realized you wanted "$5 \times 3 + 7$," you can replace the incorrect 2 with a 3 by pressing the $\boxed{\blacktriangleleft}$ button until the 2 is flashing, and then press "3."

If you realize *after* you pressed $\boxed{\text{ENTER}}$ that you've made a typing error, just press $\boxed{\text{2nd}}$ $\boxed{\text{ENTRY}}$ to reproduce the previously entered line. (With a Casio, press $\boxed{\blacktriangleleft}$.) Then correct the error with the $\boxed{\blacktriangleleft}$ button, as described above.

The $\boxed{\text{INS}}$ button allows you to insert a character. For example, if you typed "5×27" and you meant to type "5×217," press the $\boxed{\blacktriangleleft}$ button until the 7 is flashing, and then insert 1 by typing

$$\boxed{\text{2nd}} \quad \boxed{\text{INS}} \quad 1$$

The $\boxed{\text{DEL}}$ button allows you to delete a character. For example, if you typed "5×217" and you meant to type "5×27," press the $\boxed{\blacktriangleleft}$ button until the 1 is flashing, and then press $\boxed{\text{DEL}}$.

If you haven't yet pressed $\boxed{\text{ENTER}}$, the $\boxed{\text{CLEAR}}$ button erases an entire line. If you have pressed $\boxed{\text{ENTER}}$, the $\boxed{\text{CLEAR}}$ button clears everything off of the screen. With a Casio, the $\boxed{\text{AC}}$ button functions in the same way. "AC" stands "all clear."

The Subtraction Symbol and the Negative Symbol

If you read "$3 - -2$" aloud, you could say, "three subtract negative two" or "three minus minus two," and you would be understood. The expression is understandable even if the distinction between the negative symbol and the subtraction symbol is not made clear. With a calculator, however, this distinction is crucial. The subtraction button is labeled "$-$," and the negative button is labeled "$(-)$."

EXAMPLE 1

THE SUBTRACTION AND NEGATIVE SYMBOLS   Calculate $3 - -2$, both (a) by hand and (b) with a calculator.

SOLUTION

a. $3 - -2 = 3 + 2 = 5$
b. Type

$$3 \boxed{-} \boxed{(-)} 2 \boxed{\text{ENTER}}$$

and the display will read 5.

   If you had typed

$$3 \boxed{-} \boxed{-} 2 \boxed{\text{ENTER}} \quad \text{or} \quad 3 \boxed{(-)} \boxed{-} 2 \boxed{\text{ENTER}}$$

the calculator would have responded with an error message.

The Multiplication Symbol

In algebra, we do not use "x" for multiplication. Instead, we use "x" as a variable, and we use "·" for multiplication. However, Texas Instruments graphing calculators use "$\times$" as the label on the multiplication button and "*" for multiplication on the display screen. (The "variable x" button is labeled "X,T,θ,n".) This is one of the few instances in which you don't type things in the same way that you write them algebraically.

Order of Operations and Use of Parentheses

Texas Instruments and Casio graphing calculators are programmed to follow the order of operations. That is, they perform calculations in the following order:

1. Parentheses-enclosed work

2. Exponents

3. Multiplication and
   Division, from left to right

4. Addition and
   Subtraction, from left to right

You can remember this order by remembering the word "PEMDAS," which stands for

Parentheses/Exponents/Multiplication/Division/Addition/Subtraction

EXAMPLE 2

PARENTHESES COME FIRST   Calculate $2(3 + 4)$, both (a) by hand and (b) with a calculator.

SOLUTION

a. $2(3 + 4) = 2(7)$ **parentheses-enclosed work comes first**
          $= 14$
b. Type

$$2 \boxed{\times} \boxed{(} 3 \boxed{+} 4 \boxed{)} \boxed{\text{ENTER}}$$

and the display will read 14.

   In the instructions to Example 2, notice that we wrote "$2(3 + 4)$" rather than "$2 \cdot (3 + 4)$"; in this case, it's not necessary to write the multiplication symbol. Similarly, it's not necessary to type the multiplication symbol. Example 2b could be computed by typing

$$2 \boxed{(} 3 \boxed{+} 4 \boxed{)} \boxed{\text{ENTER}}$$

EXAMPLE **3**

EXPONENTS BEFORE MULTIPLICATION   Calculate $2 \cdot 3^3$, both (a) by hand and (b) with a calculator.

SOLUTION

a. $2 \cdot 3^3 = 2 \cdot 27$ **exponents come before multiplication**
     $= 54$

b. To do this, we must use the exponent button, which is labeled "∧." Type

2 ⊠ 3 ∧ 3 ENTER

and the display will read 54.

EXAMPLE **4**

MULTIPLICATION BEFORE EXPONENTS   Calculate $(2 \cdot 3)^3$, both (a) by hand and (b) with a calculator.

SOLUTION

a. $(2 \cdot 3)^3 = 6^3$ **parentheses-enclosed work comes first**
     $= 216$

b. Type

(2 ⊠ 3) ∧ 3 ENTER

and the display will read 216.

   In Example 3, the exponent applies only to the 3, because the order of operations dictates that exponents come before multiplication. In Example 4, the exponent applies to the $(2 \cdot 3)$, because the order of operations dictates that parentheses-enclosed work comes before exponents. In each example, the way you type the problem matches the way it is written algebraically, because the calculator is programmed to follow the order of operations. Sometimes, however, the way you type a problem doesn't match the way it is written algebraically.

EXAMPLE **5**

A FRACTION BAR ACTS LIKE PARENTHESES   Calculate $\dfrac{2}{3 \cdot 4}$ with a calculator.

SOLUTION

Wrong   It is incorrect to type

2 ÷ 3 ⊠ 4 ENTER

even though that matches the way the problem is written algebraically. According to the order of operations, multiplication and division are done *from left to right,* so the above typing is algebraically equivalent to

$$= \frac{2}{3} \cdot 4 \quad \textbf{first dividing and then multiplying, since division is on the left and multiplication is on the right}$$

$$= \frac{2}{3} \cdot \frac{4}{1} = \frac{2 \cdot 4}{3}$$

which is not what we want. The difficulty is that the large fraction bar in the expression $\frac{2}{3 \cdot 4}$ groups the "$3 \cdot 4$" together in the denominator; in the above typing, nothing groups the "$3 \cdot 4$" together, and only the 3 ends up in the denominator.

Right   The calculator needs parentheses inserted in the following manner:

$$\frac{2}{(3 \cdot 4)}$$

Thus, it is correct to type

2 ÷ (3 × 4) ENTER

This makes the display read 0.166666667, the correct answer.

Also Right   It is correct to type

2 ÷ 3 ÷ 4 ENTER

According to the order of operations, multiplication and division are done from left to right, so the above typing is algebraically equivalent to

$$\frac{2}{3} \div 4 \quad \textbf{doing the left-hand division first}$$

$$= \frac{2}{3} \cdot \frac{1}{4} \quad \textbf{inverting and multiplying}$$

$$= \frac{2}{3 \cdot 4}$$

which is what we want. *When you're calculating something that involves only multiplication and division and you don't use parentheses, the* × *button places a factor in the numerator, and the* ÷ *button places a factor in the denominator.*

EXAMPLE 6

A FRACTION BAR ACTS AS PARENTHESES   Calculate $\dfrac{2}{3/4}$ with a calculator.

SOLUTION

Wrong   It is incorrect to type

2 ÷ 3 ÷ 4 ENTER

even though that matches the way the problem is written algebraically. As was discussed in Example 5, this typing is algebraically equivalent to

$$\frac{2}{3 \cdot 4}$$

which is not what we want.

Right   The calculator needs parentheses inserted in the following manner:

$$\frac{2}{(3/4)}$$

Thus, it is correct to type

2 ÷ (3 ÷ 4) ENTER

since, according to the order of operations, parentheses-enclosed work is done first. This makes the display read 2.6666667, the correct answer.

EXAMPLE 7

A FRACTION BAR ACTS AS PARENTHESES   Calculate $\dfrac{2+3}{4}$ with a calculator.

SOLUTION

Wrong   It is incorrect to type

2 + 3 ÷ 4 ENTER

even though that matches the way the problem is written algebraically. According to the order of operations, division is done before addition, so this typing is algebraically equivalent to

$$2 + \frac{3}{4}$$

which is not what we want. The large fraction bar in the expression $\frac{2+3}{4}$ groups the "2 + 3" together in the numerator; in the above typing, nothing groups the "2 + 3" together, and only the 3 ends up in the numerator.

Right   The calculator needs parentheses inserted in the following manner:

$$\frac{(2 + 3)}{4}$$

Thus, it is correct to type

$$\boxed{(}\ 2\ \boxed{+}\ 3\ \boxed{)}\ \boxed{\div}\ 4\ \boxed{\text{ENTER}}$$

This makes the display read 1.25, the correct answer.

Also Right   It is correct to type

$$2\ \boxed{+}\ 3\ \boxed{\text{ENTER}}\ \boxed{\div}\ 4\ \boxed{\text{ENTER}}$$

The first $\boxed{\text{ENTER}}$ makes the calculator perform all prior calculations before continuing. This too makes the display read 1.25, the correct answer.

Memory

The **memory** is a place to store a number for later use, without having to write it down. Graphing calculators have a memory for each letter of the alphabet; that is, you can store one number in memory A, a second number in memory B, and so on. Pressing

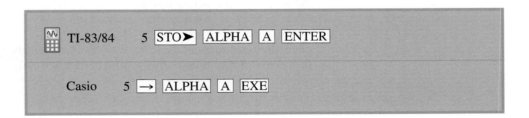

will store 5 in memory A. Similar keystrokes will store in memory B. Pressing $\boxed{\text{ALPHA}}\ \boxed{\text{A}}$ will recall what has been stored in memory A.

EXAMPLE 8

USING THE MEMORY   Calculate $\dfrac{3 + \dfrac{5 + 7}{2}}{4}$ (a) by hand and (b) by first calculating $\frac{5+7}{2}$ and storing the result.

SOLUTION

a. $\dfrac{3 + \dfrac{5 + 7}{2}}{4} = \dfrac{3 + \dfrac{12}{2}}{4} = \dfrac{3 + 6}{4} = \dfrac{9}{4} = 2.25$

b. First, calculate $\frac{5+7}{2}$ and store the result in memory A. The large fraction bar in this expression groups the "5 + 7" together in the numerator; in our typing, we must group the "5 + 7" together with parentheses. The calculator needs parentheses inserted in the following manner:

$$\frac{(5 + 7)}{2}$$

Type

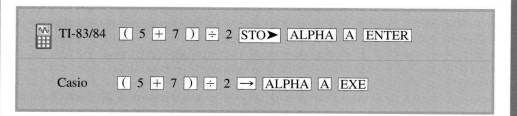

What remains is to compute $\frac{3+A}{4}$. Again, the large fraction bar groups the "3 + A" together, so the calculator needs parentheses inserted in the following manner:

$$\frac{(3 + A)}{4}$$

Type

$$\boxed{(}\ 3\ \boxed{+}\ \boxed{\text{ALPHA}}\ \boxed{A}\ \boxed{)}\ \boxed{\div}\ 4\ \boxed{\text{ENTER}}$$

and the display will read 2.25.

EXAMPLE 9

PARENTHESES WITHIN PARENTHESES   Calculate $\dfrac{3 + \dfrac{5+7}{2}}{4}$ using one line of instructions and without using the memory.

SOLUTION

There are two large fractions bars, one grouping the "5 + 7" together and one grouping the "3 + $\frac{5+7}{2}$" together. In our typing, we must group each of these together with parentheses. The calculator needs parentheses inserted in the following manner:

$$\frac{\left(3 + \dfrac{(5 + 7)}{2}\right)}{4}$$

Type

$$\boxed{(}\ 3\ \boxed{+}\ \boxed{(}\ 5\ \boxed{+}\ 7\ \boxed{)}\ \boxed{\div}\ 2\ \boxed{)}\ \boxed{\div}\ 4\ \boxed{\text{ENTER}}$$

and the display will read 2.25.

EXAMPLE 10

USING THE MEMORY

a. Use the quadratic formula and your calculator's memory to solve $2.3x^2 + 4.9x + 1.5 = 0$.

b. Check your answers.

SOLUTION

a. According to the Quadratic Formula, if $ax^2 + bx + c = 0$, then

$$x = \frac{-b \pm \sqrt{b^2 - 4ac}}{2a}$$

For our problem, $a = 2.3$, $b = 4.9$ and $c = 1.5$. This gives

$$x = \frac{-4.9 \pm \sqrt{4.9^2 - 4 \cdot 2.3 \cdot 1.5}}{2 \cdot 2.3}$$

One way to do this calculation is to calculate the radical, store it, and then calculate the two fractions.

Step 1

Calculate the radical. To do this, type

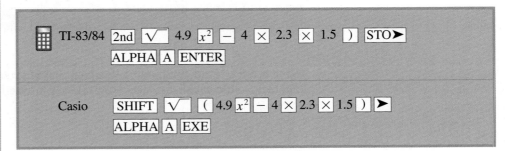

This makes the display read "3.195309062" and stores the number in the memory A. Notice the use of parentheses.

Step 2

Calculate the first fraction. The first fraction is

$$\frac{-4.9 \pm \sqrt{4.9^2 - 4 \cdot 2.3 \cdot 1.5}}{2 \cdot 2.3}$$

However, the radical has already been calculated and stored in memory A, so this is equivalent to

$$\frac{-4.9 + A}{2 \cdot 2.3}$$

Type

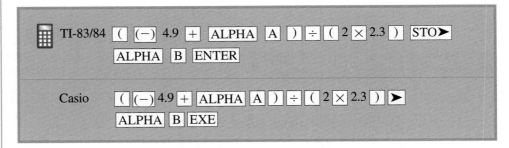

This makes the display read "−0.3705849866" and stores the number in memory B.

Step 3

Calculate the second fraction. The second fraction is

$$\frac{-4.9 - \sqrt{4.9^2 - 4 \cdot 2.3 \cdot 1.5}}{2 \cdot 2.3} = \frac{-4.9 - A}{2 \cdot 2.3}$$

Memories A and B are already in use, so we will store this in memory C. Instead of retyping the line in step 2 with the "+" changed to a "−" and the "B" to a "C," press $\boxed{\text{2nd}}$ $\boxed{\text{ENTRY}}$ ($\boxed{\leftarrow}$ with a Casio) to reproduce that line, and use the $\boxed{\blacktriangleleft}$ button to make these changes. Press $\boxed{\text{ENTER}}$ and the display will read −1.759849796, and that number will be stored in memory C. The solutions to $2.3x^2 + 4.9x + 1.5 = 0$ are $x = -0.370584966$ and $x = -1.759849796$. These are approximate solutions in that they show only the first nine decimal places.

Step 4

Check your solutions. These two solutions are stored in memories B and C; they can be checked by seeing if they satisfy the equation $2.3x^2 + 4.9x + 1.5 = 0$. To check the first solution, type

2.3 $\boxed{\times}$ $\boxed{\text{ALPHA}}$ $\boxed{\text{B}}$ $\boxed{x^2}$ $\boxed{+}$ 4.9 $\boxed{\times}$ $\boxed{\text{ALPHA}}$ $\boxed{\text{B}}$ $\boxed{+}$ 1.5 $\boxed{\text{ENTER}}$

and the display should read either "0" or a number very close to 0.

Scientific Notation

Typing

4000000 $\boxed{\times}$ 8000000 $\boxed{\text{ENTER}}$

makes the display read "3.2E13" rather than "32000000000000." This is because the calculator does not have enough room on its display for "32000000000000." When the display shows "3.2E13," read it as "3.2×10^{13}," which is written in scientific notation. Literally, "3.2×10^{13}" means "multiply 3.2 by 10, thirteen times," but as a shortcut, you can interpret it as "move the decimal point in the '3.2' thirteen places to the right."

Typing

.0000005 $\boxed{\times}$ .0000007 $\boxed{\text{ENTER}}$

makes the display read "3.5E-13" rather than "0.00000000000035," because the calculator does not have enough room on its display for "0.00000000000035." Read "3.5E-13" as "3.5×10^{-13}." Literally, this means "divide 3.5 by 10, thirteen times," but as a shortcut, you can interpret it as "move the decimal point in the '3.5' thirteen places to the left."

You can type a number in scientific notation by using the TI button labeled "EE" (which stands for "Enter Exponent") or the Casio button labeled "EXP" (which stands for "Exponent"). For example, typing

5.2 $\boxed{\text{EE}}$ 8 $\boxed{\text{ENTER}}$

makes the display read "520000000." (If the "EE" label is above the button, you will need to use the $\boxed{\text{2nd}}$ button.) On a Casio, type

5.2 $\boxed{\text{EXP}}$ 8 $\boxed{\text{EXE}}$

Be careful that you don't confuse the $\boxed{\text{EE}}$ or $\boxed{\text{EXP}}$ button with the $\boxed{\wedge}$ button. The $\boxed{\text{EE}}$ or $\boxed{\text{EXP}}$ button does *not* allow you to type in an exponent; it allows you to type in scientific notation. For example, typing

3 $\boxed{\text{EE}}$ 4 $\boxed{\text{ENTER}}$

makes the display read "30000," since $3 \times 10^4 = 30,000$. Typing

3 $\boxed{\wedge}$ 4 $\boxed{\text{ENTER}}$

makes the display read "81," since $3^4 = 81$.

EXERCISES

In Exercises 1–32, use your calculator to perform the given calculation. The correct answer is given in brackets []. In your homework, write down what you type to get that answer. Answers are not given in the back of the book.

1. $-3 - -5$ [2]
2. $-6 - 3$ [−9]
3. $4 - -9$ [13]
4. $-6 - -8$ [2]
5. $-3 - (-5 - -8)$ [−6]
6. $-(-4 - 3) - (-6 - -2)$ [11]
7. $-8 \cdot -3 \cdot -2$ [−48]
8. $-8 \cdot -3 - 2$ [22]
9. $(-3)(-8) - (-9)(-2)$ [6]
10. $2(3 - 5)$ [−4]
11. $2 \cdot 3 - 5$ [1]
12. $4 \cdot 11^2$ [484]

13. $(4 \cdot 11)^2$ [1,936]
14. $4 \cdot (-11)^2$ [484]
15. $4 \cdot (-3)^3$ [-108]
16. $(4 \cdot -3)^3$ [$-1,728$]
17. $\dfrac{3+2}{7}$ [0.7142857]
18. $\dfrac{3 \cdot 2}{7}$ [0.8571429]
19. $\dfrac{3}{2 \cdot 7}$ [0.2142857]
20. $\dfrac{3 \cdot 2}{7 \cdot 5}$ [0.1714286]
21. $\dfrac{3+5}{7 \cdot 2}$ [0.1428571]
22. $\dfrac{3 \cdot -2}{7+5}$ [-0.5]
23. $\dfrac{3}{7/2}$ [0.8571429]
24. $\dfrac{3/7}{2}$ [0.2142857]
25. 1.8^2 [3.24]
26. $\sqrt{1.8}$ [1.3416408]
27. $47,000,000^2$ [2.209×10^{15}]
28. $\sqrt{0.0000000000027}$ [1.643168×10^{-6}]
29. $(-3.92)^7$ [$-14,223.368737$]
30. $(5.72 \times 10^{19})^4$ [1.070494×10^{79}]
31. $(3.76 \times 10^{-12})^{-5}$ [1.330641×10^{57}]
32. $(3.76 \times 10^{-12}) - 5$ [-5]

In Exercises 33–36, perform the given calculation (a) by hand; (b) with a calculator, using the memory; and (c) with a calculator, using one line of instruction and without using memory, as shown in Examples 8 and 9. In your homework, for parts (b) and (c), write down what you type. Answers are not given in the back of the book.

33. $\dfrac{\frac{9-12}{5}+7}{2}$

34. $\dfrac{\frac{4-11}{6}+8}{7}$

35. $\dfrac{\frac{7+9}{5}+\frac{8-14}{3}}{3}$

36. $\dfrac{\frac{4-16}{5}+\frac{7-22}{2}}{5}$

In Exercises 37–38, use your calculator to solve the given equation for x. Check your two answers, as shown in Example 10. In your homework, write down what you type to get the answers

and what you type to check the answers. Answers are not given in the back of the book.

37. $4.2x^2 + 8.3x + 1.1 = 0$
38. $5.7x^2 + 12.3x - 8.1 = 0$
39. Discuss the use of parentheses in step 1 of Example 10a. Why are they necessary? What would happen if they were omitted?
40. Discuss the use of parentheses in step 2 of Example 10a. Why are they necessary? What would happen if they were omitted?

41. a. Calculate $\dfrac{\frac{5+7.1}{3}+\frac{2-7.1}{5}}{7}$ using one line of instruction and without using the memory. In your homework, write down what you type as well as the solution.

   b. Use the 2nd ENTRY feature to calculate $\dfrac{\frac{5+7.2}{3}+\frac{2-7.2}{5}}{7}$

   c. Use the 2nd ENTRY feature to calculate $\dfrac{\frac{5+9.3}{3}+\frac{2-4.9}{5}}{7}$

42. a. Calculate $\dfrac{3+\frac{5+\frac{6-8.3}{2}}{3}}{9}$ using one line of instruction and without using the memory. In your homework, write down what you type as well as the solution.

   b. Use the 2nd ENTRY feature to calculate $\dfrac{3+\frac{5-\frac{6-8.3}{2}}{3}}{9}$

   c. Use the 2nd ENTRY feature to calculate $\dfrac{3-\frac{5+\frac{6-8.3}{2}}{3}}{9}$

43. a. What is the result of typing "8.1 EE 4"?
   b. What is the result of typing "8.1 EE 12"?
   c. Why do the instructions in part (b) yield an answer in scientific notation, while the instructions in part (a) yield an answer that's not in scientific notation?
   d. By using the MODE button, your calculator can be reset so that all answers will appear in scientific notation. Describe how this can be done.

APPENDIX C

Graphing with a Graphing Calculator

 ## GRAPHING WITH A TEXAS INSTRUMENTS GRAPHING CALCULATOR

OBJECTIVE

● Become familiar with graphing operations

The Graphing Buttons

The graphing buttons on a TI graphing calculator are all at the top of the keypad, directly under the screen. The button labels and their uses are listed in Figure A.1.

TI-83/84	Y=	WINDOW	ZOOM	TRACE	GRAPH
use this button to tell the calculator:	what to graph	what part of the graph to draw	to zoom in or out	to give the coordinates of a highlighted point	to draw the graph

FIGURE A.1

Graphing a Line

To graph $y = 2x - 1$ on a TI graphing calculator, follow these steps.

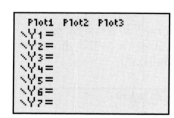

FIGURE A.2
A TI-84's "Y=" screen (other models' screens are similar).

1. *Set the calculator up for instructions on what to graph* by pressing $\boxed{Y=}$. This produces the screen similar to that shown in Figure A.2. If your screen has things written after the equals symbols, use the $\boxed{\blacktriangle}$ and $\boxed{\blacktriangledown}$ buttons along with the $\boxed{\text{CLEAR}}$ button to erase them.

2. *Tell the calculator what to graph* by typing "$2x - 1$" where the screen reads "$Y_1=$." To type the x symbol, press $\boxed{\text{X.T.θ.n}}$. After typing "$2x - 1$," press the $\boxed{\text{ENTER}}$ button.

3. *Set the calculator up for instructions on what part of the graph to draw* by pressing $\boxed{\text{WINDOW}}$.

4. *Tell the calculator what part of the graph to draw* by entering the values shown in Figure A.3. (If necessary, use the $\boxed{\blacktriangle}$ and $\boxed{\blacktriangledown}$ buttons to move from line to line).

   ● "Xmin" and "Xmax" refer to the left and right boundaries of the graph, respectively.

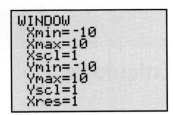

FIGURE A.3 A TI-84's "WINDOW" screen (other models' screens are similar).

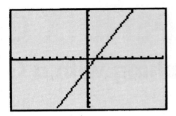

FIGURE A.4 The graph of $y = 2x - 1$.

- "Ymin" and "Ymax" refer to the lower and upper boundaries of the graph, respectively.

- "Xscl" and "Yscl" refer to the scales on the x- and y-axes (i.e., to the location of the tick marks on the axes).

5. *Tell the calculator to draw a graph* by pressing the GRAPH button. This produces the screen shown in Figure A.4.

6. *Discontinue graphing* by pressing 2nd QUIT.

GRAPHING WITH A CASIO GRAPHING CALCULATOR

To graph $y = 2x - 1$ on a Casio graphing calculator, follow these steps.

1. *Put the calculator into graphing mode* by pressing Menu, using the arrow buttons to highlight "GRAPH" on the main menu, and then pressing EXE.

2. *Tell the calculator what to graph* by typing "$2x - 1$" where the screen reads "Y1=" (see Figure A.5). (If something is already there, press F2, which now has the label "DEL" directly above it on the screen, and then press F1, which is labeled "YES.") To type the x symbol, press x,θ,T. After typing "$2x - 1$," press the EXE button.

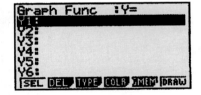

FIGURE A.5 A Casio's "Y=" screen.

3. *Tell the calculator to draw a graph* by pressing the F6 button, which now has the label "DRAW" directly above it on the screen.

4. *Set the calculator up for instructions on what part of the graph to draw* by pressing SHIFT and then the F3 button, which now has the label "V-WIN" directly above it on the screen. ("V-WIN" is short for "viewing window.")

5. *Tell the calculator what part of the graph to draw* by entering the values shown in Figure A.6. Press EXE after each entry. Then press EXIT to return to the "Y=" screen.

   - "Xmin" and "Xmax" refer to the left and right boundaries of the graph, respectively.

   - "Ymin" and "Ymax" refer to the lower and upper boundaries of the graph, respectively.

FIGURE A.6 A Casio's view window.

   - "Xscale" and "Yscale" refer to scales on the x- and y-axes, respectively (i.e., to the location of the tick marks on the axes).

6. *Tell the calculator to draw a graph* by pressing the F6 button, which now has the label "DRAW" directly above it on the screen (see Figure A.7).

7. *Discontinue graphing* by pressing MENU .

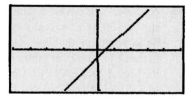

FIGURE A.7

EXERCISES

In the following exercises, you will explore some of your calculator's graphing capabilities. Answers are not given in the back of the book.

1. *Exploring the "Zoom Standard" command.* Use your calculator to graph $y = 2x - 1$, as discussed in this section. When that graph is on the screen, select the TI "Zoom Standard" command from the "Zoom menu" or the Casio "STD" command from the "V-WIN" menu by doing the following:

TI-83/84	• press ZOOM
	• select option 6 "Standard" or "ZStandard" by either:
	• using the down arrow to scroll down to that option and pressing ENTER , or
	• typing the number "6"
Casio	• press SHIFT
	• press V-WIN (i.e., F3)
	• press STD (i.e., F3)

a. What is the result?

b. How else could you accomplish the same thing without using any zoom commands?

2. *Exploring the "WINDOW" or "V-WIN" screen.* Use your calculator to graph $y = 2x - 1$ as discussed in this section. When that graph is on the screen, use the "WINDOW" or "V-WIN" screen described in this section to reset the following:

- Xmin to 1
- Xmax to 20
- Ymin to 1
- Ymax to 20

a. Why are there no axes shown?

b. Why does the graph start exactly in the lower left corner of the screen?

3. *Exploring the "WINDOW" or "V-WIN" screen.* Use your calculator to graph $y = 2x - 1$, as discussed in this

section. When that graph is on the screen, use the "WINDOW" or "V-WIN" screen described in this section to reset the following:

- Xmin to -1
- Xmax to -20
- Ymin to -1
- Ymax to -20

Why did the TI calculator respond with an error message? What odd thing did the Casio calculator do? (Casio hint: Use the TRACE button to investigate.)

4. *Exploring the "WINDOW" or "V-WIN" screen.* Use your calculator to graph $y = 2x - 1$, as discussed in this section. When that graph is on the screen, use the "WINDOW" or "V-WIN" screen described in this section to reset the following:

- Xmin to -5
- Xmax to 5
- Ymin to 10
- Ymax to 25

a. Why was only one axis shown?

b. Why was no graph shown?

5. *Exploring the "TRACE" command.* Use your calculator to graph $y = 2x - 1$, as discussed in this section. When that graph is on the screen, press TRACE (F1 on a Casio). This causes two things to happen:

- A mark appears at a point on the line. This mark can be moved with the left and right arrow buttons.
- The corresponding ordered pair is printed out at the bottom of the screen.

a. Use the "TRACE" feature to locate the line's x-intercept, the point at which the line hits the x-axis. (You may need to approximate it.)

b. Use algebra, rather than the graphing calculator, to find the x-intercept.

c. Use the "TRACE" feature to locate another ordered pair on the line.

d. Use substitution to check that the ordered pair found in part (c) is in fact a point on the line.

6. *Exploring the "ZOOM BOX" command.* Use your calculator to graph $y = 2x - 1$ as discussed in this section. When that graph is on your screen, press ZOOM (F2 on a Casio) and select option 1, "ZBOX," in the manner described in Exercise 1. This seems to have the same result as $\boxed{\text{TRACE}}$ except that the mark does not have to be a point on the line. Use the four arrow buttons to move to a point of your choice (that may be on or off the line). Press $\boxed{\text{ENTER}}$. Use the arrow buttons to move to a different point so that the resulting box encloses a part of the line. Press $\boxed{\text{ENTER}}$ again. What is the result of using the "Zoom Box" command?

7. *Exploring the "ZOOM IN" command.* Use your calculator to graph $y = 2x - 1$, as discussed in this section. When that graph is on your screen, press $\boxed{\text{ZOOM}}$ ($\boxed{\text{F2}}$ on a Casio) and select "Zoom In" in the manner described in Exercise 1. (Press $\boxed{\text{IN}}$ or $\boxed{\text{F3}}$ on a Casio.) This causes a mark to appear on the screen. Use the four arrow buttons to move the mark to a point of your choice, either on or near the line. Press $\boxed{\text{ENTER}}$.

   a. What is the result of the "Zoom In" command?

   b. How could you accomplish the same thing without using any zoom commands?

   c. The "Zoom Out" command is listed right next to the "Zoom In" command? What does it do?

8. *Zooming in on x-intercepts.* Use the "Zoom In" command described in Exercise 7, and the "Trace" command described in Exercise 5, to approximate the location of the x-intercept of $y = 2x - 1$ as accurately as possible. You might need to use these commands more than once.

   a. Describe the procedure you used to generate this answer.

   b. According to the calculator, what is the x-intercept?

   c. Is this answer the same as that of Exercise 5(b)? Why or why not?

9. *Calculating x-intercepts.* The TI-83, TI-84, and Casio will calculate the x-intercept (also called a "root" or "zero") without using the "Zoom In" and "Trace" commands. First, use your calculator to graph $y = 2x - 1$ as discussed in this section. When that graph is on the screen, do the following.

On a TI-83/84:

- Press $\boxed{\text{2nd}}$ $\boxed{\text{CALC}}$ and select option 2, "root" or "zero."

- The calculator responds by asking, "Lower Bound?" or "Left Bound?" Use the left and right arrow buttons to move the mark to a point slightly to the left of the x-intercept, and press $\boxed{\text{ENTER}}$.

- When the calculator asks, "Upper Bound?" or "Right Bound?", move the mark to a point slightly to the right of the x-intercept, and press $\boxed{\text{ENTER}}$.

- When the calculator asks, "Guess?", move the mark to a point close to the x-intercept, and press $\boxed{\text{ENTER}}$. The calculator will then display the location of the x-intercept.

On a Casio:

- Press $\boxed{\text{G-Solv}}$ (i.e., $\boxed{\text{F5}}$).

- Press $\boxed{\text{ROOT}}$. Wait and watch.

   a. According to the calculator, what is the x-intercept of the line $y = 2x - 1$?

   b. Is this answer the same as that of Exercise 5(b)? Why or why not?

APPENDIX D

Finding Points of Intersection with a Graphing Calculator

The graphs of $y = x + 3$ and $y = -x + 9$ intersect on the standard viewing screen. To find the point of intersection with a TI or Casio graphing calculator, first enter the two equations on the "Y=" screen (one as Y_1 and one as Y_2), erase any other equations, and erase the "Y= " screen by pressing 2nd QUIT. Then follow the following instructions.

On a TI-83/84

- Graph the two equations on the standard viewing screen.

- Press 2nd CALC and select option 5, "intersect."

- When the calculator responds with "First curve?" and a mark on the first equation's graph, press ENTER.

- When the calculator responds with "Second curve?" and a mark on the second equation's graph, press ENTER.

- When the calculator responds with "Guess?", use the left and right arrows to place the mark near the point of intersection, and press ENTER.

- Check your answer by substituting the ordered pair into each of the two equations.

On a Casio

- Graph the two equations on the standard viewing screen.

- Press G-Solv (i.e., F5).

- Press ISCT (i.e., F5). Watch and wait.

- After a pause, the calculator will display the location of the x-intercept.

Figure A.8 shows the results of computing the point of intersection of $y = x + 3$ and $y = -x + 9$.

The information on the screen indicates that the point of intersection is $(3, 6)$. To check this, substitute 3 for x into each of the two equations; you should get 6.

$$y = x + 3 = 3 + 3 = 6. \checkmark$$
$$y = -x + 9 = -3 + 9 = 6. \checkmark$$

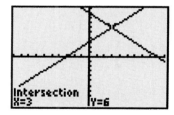

FIGURE A.8 Finding the point of intersection.

EXERCISES

In Exercises 1–6, do the following.

a. Use the graphing calculator to find the point of intersection of the given equations.

b. Check your solutions by substituting the ordered pair into each of the two equations. Answers are not given in the back of the book.

1. $y = 3x + 2$ and $y = 5x + 5$

2. $y = 2x - 6$ and $y = 3x + 4$

3. $y = 8x - 14$ and $g(x) = 11x + 23$

HINT: You will have to change Xmin, Xmax, Ymin, and Ymax to find the point.

4. $y = -7x + 12$ and $y = -12x - 71$

HINT: You will have to change Xmin, Xmax, Ymin, and Ymax to find the point.

5. $y = x^2 - 2x + 3$ and $y = -x^2 - 3x + 12$
(Find two answers.)

6. $f(x) = 8x^2 - 3x - 7$ and $y = 2x + 4$
(Find two answers.)

APPENDIX E
Dimensional Analysis

Most people know how to convert 6 feet into yards (it's $6 \div 3 = 2$ yards). Few people know how to convert 50 miles per hour into feet per minute. The former problem is so commonplace that people just remember how to do it; the latter problem is not so common, and people don't know how to do it. Dimensional analysis is an easy way of converting a quantity from one set of units to another; it can be applied to either of these two problems.

Dimensional analysis involves using a standard conversion (such as 1 yard = 3 feet) to create a fraction, including units (such as "feet" and "yards") in that fraction, and canceling units in the same way that variables are canceled. That is, in the fraction 2 feet/3 feet, we can cancel feet with feet and obtain 2/3, just as we can cancel x with x in the fraction $2x/3x$ and obtain 2/3.

EXAMPLE 1

SOLUTION

USING DIMENSIONAL ANALYSIS   Convert 6 feet into yards.

Start with the standard conversion:

$$1 \text{ yard} = 3 \text{ feet} \quad \textbf{a standard conversion}$$

Create a fraction by dividing each side by 3 feet:

$$\frac{1 \text{ yard}}{3 \text{ feet}} = \frac{3 \text{ feet}}{3 \text{ feet}}$$

$$\frac{1 \text{ yard}}{3 \text{ feet}} = 1 \qquad \textbf{a fractional version of a standard conversion}$$

To convert 6 feet into yards, multiply 6 feet by the fraction 1 yard/3 feet. This is valid because that fraction is equal to 1, and multiplying something by 1 doesn't change its value.

$$6 \text{ feet} = 6 \text{ feet} \cdot 1$$

$$= 6 \text{ feet} \cdot \frac{1 \text{ yard}}{3 \text{ feet}} \quad \textbf{substituting } \frac{1 \text{ yard}}{3 \text{ feet}} \textbf{ for 1}$$

$$= 2 \text{ yards} \qquad \textbf{canceling}$$

It is crucial to include units in this work. If at the beginning of this example, we had divided by 1 yard instead of 3 feet, we would have obtained

$$1 \text{ yard} = 3 \text{ feet} \quad \textbf{a standard conversion}$$

$$\frac{1 \text{ yard}}{1 \text{ yard}} = \frac{3 \text{ feet}}{1 \text{ yard}}$$

$$1 = \frac{3 \text{ feet}}{1 \text{ yard}} \quad \textbf{a fractional version of a standard conversion}$$

Multiplying 6 feet by 3 feet/1 yard would not allow us to cancel feet with feet and would not leave an answer in yards.

$$6 \text{ feet} = 6 \text{ feet} \cdot 1$$
$$= 6 \text{ feet} \cdot \frac{3 \text{ feet}}{1 \text{ yard}} \quad \textbf{substituting } \frac{3 \text{ feet}}{1 \text{ yard}} \textbf{ for 1}$$

It's important to include units in dimensional analysis, because it's only by looking at the units that we can tell that multiplying by 1 yard/3 feet is productive and multiplying by 3 feet/1 yard isn't.

EXAMPLE 2

USING DIMENSIONAL ANALYSIS   Convert 50 miles/hour to feet/minute.

SOLUTION

The appropriate standard conversions are

$$1 \text{ mile} = 5{,}280 \text{ feet}$$
$$1 \text{ hour} = 60 \text{ minutes} \quad \textbf{two standard conversions}$$

This problem has two parts, one for each of these two standard conversions.

Part 1   *Use the standard conversion "1 mile = 5,280 feet" to convert miles/ hour to feet/hour.* The fraction 50 miles/hour has miles in the numerator, and we are to convert it to a fraction that has feet in the numerator. To replace miles with feet, first rewrite the standard conversion as a fraction that has miles in the denominator, and then multiply by that fraction. This will allow miles to cancel.

$$1 \text{ mile} = 5{,}280 \text{ feet} \quad \textbf{a standard conversion}$$
$$\frac{\cancel{1 \text{ mile}}}{\cancel{1 \text{ mile}}} = \frac{5280 \text{ feet}}{1 \text{ mile}} \quad \textbf{placing miles in the denominator}$$
$$1 = \frac{5280 \text{ feet}}{1 \text{ mile}} \quad \textbf{a fractional version of a standard conversion}$$

Now multiply 50 miles/hour by this fraction and cancel:

$$\frac{50 \text{ miles}}{\text{hour}} = \frac{50 \text{ miles}}{\text{hour}} \cdot 1$$
$$= \frac{50 \cancel{\text{ miles}}}{\text{hour}} \cdot \frac{5280 \text{ feet}}{1 \cancel{\text{ mile}}} \quad \textbf{substituting } \frac{5280 \text{ feet}}{1 \text{ mile}} \textbf{ for 1}$$
$$= \frac{50 \cdot 5280 \text{ feet}}{\text{hour}} \quad \textbf{canceling miles}$$

Part 2   *Use the standard conversion "1 hour = 60 minutes" to convert feet/hour to feet/minute.* To replace hours with minutes, first rewrite the standard conversion as a fraction that has hours in the numerator, and then multiply by that fraction. This will allow hours to cancel.

$$1 \text{ hour} = 60 \text{ minutes} \quad \textbf{a standard conversion}$$
$$\frac{1 \text{ hour}}{60 \text{ minutes}} = \frac{\cancel{60 \text{ minutes}}}{\cancel{60 \text{ minutes}}} \quad \textbf{placing hours in the numerator}$$
$$\frac{1 \text{ hour}}{60 \text{ minutes}} = 1 \quad \textbf{a fractional version of a standard conversion}$$

Continuing where we left off, multiply by this fraction and cancel:

$$\frac{50 \text{ miles}}{\text{hour}} = \frac{50 \cdot 5280 \text{ feet}}{\text{hour}} \qquad \text{from part 1}$$

$$= \frac{50 \cdot 5280 \text{ feet}}{\text{hour}} \cdot 1$$

$$= \frac{50 \cdot 5280 \text{ feet}}{\cancel{\text{hour}}} \cdot \frac{1 \cancel{\text{hour}}}{60 \text{ minutes}} \qquad \text{substituting } \frac{1 \text{ hour}}{60 \text{ minutes}} \text{ for } 1$$

$$= \frac{4400 \text{ feet}}{1 \text{ minute}} \qquad \text{since } 50 \cdot 5280 \,/\, 60 = 4400$$

Thus, 50 miles/hour is equivalent to 4,400 feet/minute.

EXAMPLE 3

USING DIMENSIONAL ANALYSIS   How many feet will a car travel in half a minute if that car's rate is 50 miles/hour?

SOLUTION

This seems to be a standard algebra problem that uses the formula "distance = rate · time"; we're given the rate (50 miles/hour) and the time (1/2 minute), and we are to find the distance. However, the units are not consistent.

$$\text{distance} = \text{rate} \cdot \text{time}$$
$$= \frac{50 \text{ miles}}{\text{hour}} \cdot \frac{1}{2} \text{ minute}$$

None of these units cancels, and we are not left with an answer in feet. However, if the car's rate were in feet/minute rather than miles/hour, the units would cancel, and we would be left with an answer in feet:

$$\text{distance} = \text{rate} \cdot \text{time}$$
$$= \frac{50 \text{ miles}}{\text{hour}} \cdot \frac{1}{2} \text{ minute}$$
$$= \frac{4400 \text{ feet}}{1 \cancel{\text{ minute}}} \cdot \frac{1}{2} \cancel{\text{ minute}} \qquad \text{from Example 2}$$
$$= 2{,}200 \text{ feet}$$

The car would travel 2,200 feet in half a minute.

EXERCISES

In Exercises 1–6, use dimensional analysis to convert the given quantity.

1. **a.** 12 feet into yards
    b. 12 yards into feet
2. **a.** 24 feet into inches
    b. 24 inches into feet
3. **a.** 10 miles into feet
    b. 10 feet into miles (Round off to the nearest ten thousandth of a mile.)

4. **a.** 2 hours into minutes
    b. 2 minutes into hours (Round off to the nearest thousandth of an hour.)
5. 2 miles into inches

    HINT: Convert first to feet, then to inches.

6. 3 hours into seconds

    HINT: Convert first to minutes, then to seconds.

In Exercises 7–12, use the following information. The metric system is based on three different units:

- the gram (1 gram = 0.0022046 pound)
- the meter (1 meter = 39.37 inches)
- the liter (1 liter = 61.025 cubic inches)

Other units are formed by adding the following prefixes to these basic units:

- kilo-, which means one thousand (for example, 1 kilometer = 1,000 meters)
- centi-, which means one hundredth (for example, 1 centimeter = 1/100 meter)
- milli-, which means one thousandth (for example, 1 millimeter = 1/1,000 meter)

7. Use dimensional analysis to convert
   a. 50 kilometers to meters
   b. 50 meters to kilometers

8. Use dimensional analysis to convert
   a. 30 milligrams to grams
   b. 30 grams to milligrams

9. Use dimensional analysis to convert
   a. 2 centiliters to liters
   b. 2 liters to centiliters

10. Use dimensional analysis to convert
    a. 1 yard to meters (Round off to the nearest hundredth.)
    b. 20 yards to centimeters (Round off to the nearest centimeter.)

11. Use dimensional analysis to convert
    a. 1 pound to grams (Round off to the nearest gram.)
    b. 8 pounds to kilograms (Round off to the nearest tenth of a kilogram.)

12. Use dimensional analysis to convert
    a. 1 cubic inch to liters (Round off to the nearest thousandth.)
    b. 386 cubic inches to liters (Round off to the nearest tenth.)

13. a. Use dimensional analysis to convert 60 miles per hour into feet per second.
    b. Leadfoot Larry speeds through an intersection at 60 miles per hour, and he is ticketed for speeding and reckless driving. He pleads guilty to speeding but not guilty to reckless driving, telling the judge that at that speed, he would have plenty of time to react to any cross traffic. The intersection is 80 feet wide. How long would it take Larry to cross the intersection? (Round off to the nearest tenth of a second.)

14. In the United States, a typical freeway speed limit is 65 miles per hour.
    a. Use dimensional analysis to convert this to kilometers per hour. (Round off to the nearest whole number.)

    b. At this speed, how many miles can be traveled in 10 minutes? (Round off to the nearest whole number.)

15. In Germany, a typical autobahn speed limit is 130 kilometers per hour.
    a. Use dimensional analysis to convert this into miles per hour. (Round off to the nearest whole number.)
    b. How many more miles will a car traveling at 130 kilometers per hour go in 1 hour than a car traveling at 65 miles per hour? (Round off to the nearest whole number.)

16. Light travels 6×10^{12} miles per year.
    a. Convert this to miles per hour. (Round off to the nearest whole number.)
    b. How far does light travel in 1 second? (Round off to the nearest whole number.)

17. In December 2009, Massachusetts' Capital Crossing Bank offered a money market account with a 5.75% interest rate. This means that Capital Crossing Bank will pay interest at a rate of 5.75% per year.
    a. Use dimensional analysis to convert this to a percent per day.
    b. If you deposited $10,000 on September 1, how much interest would your account earn by October 1? (There are 30 days in September.) (Round off to the nearest cent.)
    c. If you deposited $10,000 on October 1, how much interest would your account earn by November 1? (There are 31 days in October.) (Round off to the nearest cent.)

18. In May 2010, Utah's Ally Bank offered a money market account with a 1.29% interest rate. This means that Republic Bank will pay interest at a rate of 1.29% per year.
    a. Use dimensional analysis to convert this to a percent per day.
    b. If you deposited $10,000 on September 1, how much interest would your account earn by October 1? (There are 30 days in September.) (Round off to the nearest cent.)
    c. If you deposited $10,000 on October 1, how much interest would your account earn by November 1? (There are 31 days in October.) (Round off to the nearest cent.)

19. You cannot determine whether a person is overweight by merely determining his or her weight; if a short person and a tall person weigh the same, the short person could be overweight and the tall person could be underweight. Body mass index (BMI) is becoming a standard way of determining if a person is overweight, since it takes both weight and height into consideration. BMI is defined as (weight in kilograms)/(height in meters)^2. According to the World Health Organization, a person is overweight if his or her BMI is 25 or greater. In October 1996, Katherine Flegal, a statistician for

the National Center for Health Statistics, said that according to this standard, one out of every two Americans is overweight. (*Source: San Francisco Chronicle*, 16 October 1996, page A6.)

a. Lenny is 6 feet tall. Convert his height to meters. (Round off to the nearest hundredth.)

b. Lenny weighs 169 pounds. Convert his weight to kilograms. (Round off to the nearest tenth.)

c. Determine Lenny's BMI. (Round off to the nearest whole number.) Is he overweight?

d. Fred weighs the same as Lenny, but he is 5′5″ tall. Determine Fred's BMI. (Round off to the nearest whole number.) Is Fred overweight?

e. Why does BMI use the metric system (kilograms and meters) rather than the English system (feet, inches, and pounds)?

20. According to the U.S. Census Bureau, in 2009, California, the most populous state, had 36,961,664 people residing in 158,706 square miles of area, while Rhode Island, the smallest state, had 1,053,209 people residing in 1,212 square miles of area. Determine which state is more crowded by computing the number of square feet per person in each state.

21. A wading pool is 4 feet wide, 6 feet long, and 11 inches deep. There are 7.48 gallons of water per cubic foot. How many gallons of water does it take to fill the pool?

22. John ate a 2,000 calorie lunch and immediately felt guilty. Jogging for one minute consumes 0.061 calorie per pound of body weight. John weighs 205 pounds. How long will he have to jog to burn off all of the calories from lunch?

APPENDIX F

Body Table for the Standard Normal Distribution

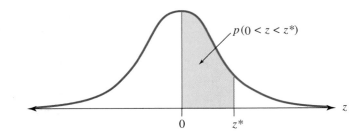

$p(0 < z < z^*)$

z*	0.00	0.01	0.02	0.03	0.04	0.05	0.06	0.07	0.08	0.09
0.0	0.0000	0.0040	0.0080	0.0120	0.0160	0.0199	0.0239	0.0279	0.0319	0.0359
0.1	0.0398	0.0438	0.0478	0.0517	0.0557	0.0596	0.0636	0.0675	0.0714	0.0753
0.2	0.0793	0.0832	0.0871	0.0910	0.0948	0.0987	0.1026	0.1064	0.1103	0.1141
0.3	0.1179	0.1217	0.1255	0.1293	0.1331	0.1368	0.1406	0.1443	0.1480	0.1517
0.4	0.1554	0.1591	0.1628	0.1664	0.1700	0.1736	0.1772	0.1808	0.1844	0.1879
0.5	0.1915	0.1950	0.1985	0.2019	0.2054	0.2088	0.2123	0.2157	0.2190	0.2224
0.6	0.2257	0.2291	0.2324	0.2357	0.2389	0.2422	0.2454	0.2486	0.2517	0.2549
0.7	0.2580	0.2611	0.2642	0.2673	0.2704	0.2734	0.2764	0.2794	0.2823	0.2852
0.8	0.2881	0.2910	0.2939	0.2967	0.2995	0.3023	0.3051	0.3078	0.3106	0.3133
0.9	0.3159	0.3186	0.3212	0.3238	0.3264	0.3289	0.3315	0.3340	0.3365	0.3389
1.0	0.3413	0.3438	0.3461	0.3485	0.3508	0.3531	0.3554	0.3577	0.3599	0.3621
1.1	0.3643	0.3665	0.3686	0.3708	0.3729	0.3749	0.3770	0.3790	0.3810	0.3830
1.2	0.3849	0.3869	0.3888	0.3907	0.3925	0.3944	0.3962	0.3980	0.3997	0.4015
1.3	0.4032	0.4049	0.4066	0.4082	0.4099	0.4115	0.4131	0.4147	0.4162	0.4177
1.4	0.4192	0.4207	0.4222	0.4236	0.4251	0.4265	0.4279	0.4292	0.4306	0.4319
1.5	0.4332	0.4345	0.4357	0.4370	0.4382	0.4394	0.4406	0.4418	0.4429	0.4441
1.6	0.4452	0.4463	0.4474	0.4484	0.4495	0.4505	0.4515	0.4525	0.4535	0.4545
1.7	0.4554	0.4564	0.4573	0.4582	0.4591	0.4599	0.4608	0.4616	0.4625	0.4633
1.8	0.4641	0.4649	0.4656	0.4664	0.4671	0.4678	0.4686	0.4692	0.4699	0.4706
1.9	0.4713	0.4719	0.4726	0.4732	0.4738	0.4744	0.4750	0.4756	0.4761	0.4767
2.0	0.4772	0.4778	0.4783	0.4788	0.4793	0.4798	0.4803	0.4808	0.4812	0.4817
2.1	0.4821	0.4826	0.4830	0.4834	0.4838	0.4842	0.4846	0.4850	0.4854	0.4857
2.2	0.4861	0.4864	0.4868	0.4871	0.4875	0.4878	0.4881	0.4884	0.4887	0.4890
2.3	0.4893	0.4896	0.4898	0.4901	0.4904	0.4906	0.4909	0.4911	0.4913	0.4916
2.4	0.4918	0.4920	0.4922	0.4925	0.4927	0.4929	0.4931	0.4932	0.4934	0.4936
2.5	0.4938	0.4940	0.4941	0.4943	0.4945	0.4946	0.4948	0.4949	0.4951	0.4952
2.6	0.4953	0.4955	0.4956	0.4957	0.4959	0.4960	0.4961	0.4962	0.4963	0.4964
2.7	0.4965	0.4966	0.4967	0.4968	0.4969	0.4970	0.4971	0.4972	0.4973	0.4974
2.8	0.4974	0.4975	0.4976	0.4977	0.4977	0.4978	0.4979	0.4979	0.4980	0.4981
2.9	0.4981	0.4982	0.4982	0.4983	0.4984	0.4984	0.4985	0.4985	0.4986	0.4986
3.0	0.4987	0.4987	0.4987	0.4988	0.4988	0.4989	0.4989	0.4989	0.4990	0.4990

Selected Answers to Odd Exercises

CHAPTER 1 **Logic**

1.1 Deductive vs. Inductive Reasoning

1. a. Valid   **b.** Invalid
3. a. Invalid   **b.** Valid
5. Invalid   **7.** Valid
9. Valid   **11.** Valid
13. Valid   **15.** Invalid
17. Invalid   **19.** Valid
21. a. Inductive
     b. Deductive
23. 23   **25.** 20
27. 25   **29.** 13
31. 5   **33.** F, S
35. T, W   **37.** r
39. f   **41.** f
43. Answers may vary.   **47.** $3 \pm \sqrt{2}$

49.

3	5	6	4	1	8	2	7	9
2	1	7	6	5	9	3	4	8
9	8	4	3	7	2	1	5	6
6	2	1	8	9	5	4	3	7
8	3	5	7	4	6	9	2	1
7	4	9	2	3	1	6	8	5
5	9	2	1	8	4	7	6	3
1	6	3	5	2	7	8	9	4
4	7	8	9	6	3	5	1	2

51.

9	5	6	3	4	1	2	8	7
8	2	3	6	5	7	1	4	9
4	7	1	9	8	2	5	3	6
6	8	7	1	3	4	9	5	2
5	9	4	2	6	8	7	1	3
3	1	2	7	9	5	4	6	8
7	6	5	8	1	9	3	2	4
2	4	8	5	7	3	6	9	1
1	3	9	4	2	6	8	7	5

53.

7	6	8	5	4	2	9	1	3
1	5	2	7	3	9	6	4	8
9	3	4	6	1	8	2	5	7
8	7	1	9	6	3	5	2	4
5	2	6	1	7	4	8	3	9
4	9	3	8	2	5	7	6	1
6	8	9	4	5	1	3	7	2
3	4	7	2	9	6	1	8	5
2	1	5	3	8	7	4	9	6

55. Deductive reasoning is the application of a general statement to a specific instance. Inductive reasoning is going from a series of specific cases to a general statement.
57. A syllogism is an argument composed of two premises followed by a conclusion.
59. Aristotle
61. He was more interested in the real world.

1.2 Symbolic Logic

1. a. Statement
     b. Statement
     c. Not a statement. (question)
     d. Not a statement. (opinion)
3. (a) and (d) are negations; (b) and (c) are negations
5. a. Her dress is red.
     b. No computer is priced under \$100. (All computers are \$100 or more.)
     c. Some dogs are not four-legged animals.
     d. Some sleeping bags are waterproof.
7. a. $p \wedge q$ **b.** $\sim p \rightarrow \sim q$ **c.** $\sim(p \vee q)$
     d. $p \wedge \sim q$ **e.** $p \rightarrow q$ **f.** $\sim q \rightarrow \sim p$
9. a. $(p \vee q) \rightarrow r$ **b.** $p \wedge q \wedge r$ **c.** $r \wedge \sim(p \vee q)$
     d. $q \rightarrow r$ **e.** $\sim r \rightarrow \sim(p \vee q)$ **f.** $r \rightarrow (q \vee p)$
11. p: A shape is a square.
     q: A shape is a rectangle.
     $p \rightarrow q$

13. p: A shape is a square.
q: A shape is a triangle.
$p \rightarrow \sim q$

15. p: A number is a whole number.
q: A number is even.
r: A number is odd.
$p \rightarrow (q \vee r)$

17. p: A number is a whole number.
q: A number is greater than 3.
r: A number is less than 4.
$p \rightarrow \sim(q \wedge r)$

19. p: A person is an orthodontist.
q: A person is a dentist.
$p \rightarrow q$

21. p: A person knows Morse code.
q: A person operates a telegraph.
$q \rightarrow p$

23. p: The animal is a monkey.
q: The animal is an ape.
$p \rightarrow \sim q$

25. p: The animal is a monkey.
q: The animal is an ape.
$q \rightarrow \sim p$

27. p: I sleep soundly.
q: I drink coffee.
r: I eat chocolate.
$(q \vee r) \rightarrow \sim p$

29. p: Your check is accepted.
q: You have a driver's license.
r: You have a credit card.
$\sim(q \vee r) \rightarrow \sim p$

31. p: You do drink.
q: You do drive.
r: You are fined.
s: You go to jail.
$(p \wedge q) \rightarrow (r \vee s)$

33. p: You get a refund.
q: You get a store credit.
r: The product is defective.
$r \rightarrow (p \vee q)$

35. a. I am an environmentalist and I recycle my aluminum cans.
b. If I am an environmentalist, then I recycle my aluminum cans.
c. If I do not recycle my aluminum cans, then I am not an environmentalist.
d. I recycle my aluminum cans or I am not an environmentalist.

37. a. If I recycle my aluminum cans or newspapers, then I am an environmentalist.
b. If I am not an environmentalist, then I do not recycle my aluminum cans or newspapers.
c. I recycle my aluminum cans and newspapers or I am not an environmentalist.
d. If I recycle my newspapers and do not recycle my aluminum cans, then I am not an environmentalist.

39. Statement #1: Cold weather is required in order to have snow.

41. Statement #2: A month with 30 days would also indicate that the month is not February.

43. A negation is the denial of a statement.

45. A disjunction is when statements are connected by "or."

47. A sufficient condition (i.e., adequate) is the premise of a conditional.

49. With inclusive or, both things may happen. With exclusive or, one statement excludes the other (so it is impossible for both to be true).

51. Law. To be able to distinguish valid arguments from invalid ones.

53. Universal character. Leibniz.

55. a. **57.** b.

1.3 Truth Tables

1.

p	q	$\sim q$	$p \vee \sim q$
T	T	F	T
T	F	T	T
F	T	F	F
F	F	T	T

3.

p	$\sim p$	$p \vee \sim p$
T	F	T
F	T	T

5.

p	q	$\sim q$	$p \rightarrow \sim q$
T	T	F	F
T	F	T	T
F	T	F	T
F	F	T	T

7.

p	q	$\sim q$	$\sim p$	$\sim q \rightarrow \sim p$
T	T	F	F	T
T	F	T	F	F
F	T	F	T	T
F	F	T	T	T

9.

p	q	$p \vee q$	$\sim p$	$(p \vee q) \rightarrow \sim p$
T	T	T	F	F
T	F	T	F	F
F	T	T	T	T
F	F	F	T	T

11.

p	q	$p \vee q$	$p \wedge q$	$(p \vee q) \rightarrow (p \wedge q)$
T	T	T	T	T
T	F	T	F	F
F	T	T	F	F
F	F	F	F	T

13.

p	q	r	$q \vee r$	$\sim(q \vee r)$	$p \wedge \sim(q \vee r)$
T	T	T	T	F	F
T	T	F	T	F	F
T	F	T	T	F	F
T	F	F	F	T	T
F	T	T	T	F	F
F	T	F	T	F	F
F	F	T	T	F	F
F	F	F	F	T	F

15.

p	q	r	$\sim q$	$\sim q \wedge r$	$p \vee (\sim q \wedge r)$
T	T	T	F	F	T
T	T	F	F	F	T
T	F	T	T	T	T
T	F	F	T	F	T
F	T	T	F	F	F
F	T	F	F	F	F
F	F	T	T	T	T
F	F	F	T	F	F

17.

p	q	r	$\sim r$	$\sim r \vee p$	$q \wedge p$	$(\sim r \vee p) \rightarrow (q \wedge p)$
T	T	T	F	T	T	T
T	T	F	T	T	T	T
T	F	T	F	T	F	F
T	F	F	T	T	F	F
F	T	T	F	F	F	T
F	T	F	T	T	F	F
F	F	T	F	F	F	T
F	F	F	T	T	F	F

19.

p	q	r	$\sim r$	$p \vee r$	$q \wedge (\sim r)$	$(p \vee r) \rightarrow (q \wedge \sim r)$
T	T	T	F	T	F	F
T	T	F	T	T	T	T
T	F	T	F	T	F	F
T	F	F	T	T	F	F
F	T	T	F	T	F	F
F	T	F	T	F	T	T
F	F	T	F	T	F	F
F	F	F	T	F	F	T

21. p: It is raining.
q: The streets are wet.
$p \rightarrow q$

p	q	$p \rightarrow q$
T	T	T
T	F	F
F	T	T
F	F	T

23. p: It rains.
q: The water supply is rationed.
$\sim p \rightarrow q$

p	q	$\sim p$	$\sim p \rightarrow q$
T	T	F	T
T	F	F	T
F	T	T	T
F	F	T	F

25. p: A shape is a square.
q: A shape is a rectangle.
$p \rightarrow q$

p	q	$p \rightarrow q$
T	T	T
T	F	F
F	T	T
F	F	T

27. p: It is a square.
q: It is a triangle.
$p \rightarrow \sim q$

p	q	$\sim q$	$p \rightarrow \sim q$
T	T	F	F
T	F	T	T
F	T	F	T
F	F	T	T

29. p: The animal is a monkey.
q: The animal is an ape.
$p \rightarrow \sim q$

p	q	$\sim q$	$p \rightarrow \sim q$
T	T	F	F
T	F	T	T
F	T	F	T
F	F	T	T

31. p: The animal is a monkey.
q: The animal is an ape.
$q \to \sim p$

p	q	$\sim p$	$q \to \sim p$
T	T	F	F
T	F	F	T
F	T	T	T
F	F	T	T

33. p: You have a driver's license.
q: You have a credit card.
r: Your check is approved.
$(p \lor q) \to r$

p	q	r	$p \lor q$	$(p \lor q) \to r$
T	T	T	T	T
T	T	F	T	F
T	F	T	T	T
T	F	F	T	F
F	T	T	T	T
F	T	F	T	F
F	F	T	F	T
F	F	F	F	T

35. p: Leaded gas is used.
q: The catalytic converter is damaged.
r: The air is polluted.
$p \to (q \land r)$

p	q	r	$q \land r$	$p \to (q \land r)$
T	T	T	T	T
T	T	F	F	F
T	F	T	F	F
T	F	F	F	F
F	T	T	T	T
F	T	F	F	T
F	F	T	F	T
F	F	F	F	T

37. p: I have a college degree.
q: I have a job.
r: I own a house.
$p \land \sim(q \lor r)$

p	q	r	$q \lor r$	$\sim(q \lor r)$	$p \land \sim(q \lor r)$
T	T	T	T	F	F
T	T	F	T	F	F
T	F	T	T	F	F
T	F	F	F	T	T
F	T	T	T	F	F
F	T	F	T	F	F
F	F	T	T	F	F
F	F	F	F	T	F

39. p: Proposition A passes.
q: Proposition B passes.
r: Jobs are lost.
s: New taxes are imposed.
$(p \land \sim q) \to (r \lor s)$

p	q	r	s	$\sim q$	$p \land \sim q$	$r \lor s$	$(p \land \sim q) \to (r \lor s)$
T	T	T	T	F	F	T	T
T	T	T	F	F	F	T	T
T	T	F	T	F	F	T	T
T	T	F	F	F	F	F	T
T	F	T	T	T	T	T	T
T	F	T	F	T	T	T	T
T	F	F	T	T	T	T	T
T	F	F	F	T	T	F	F
F	T	T	T	F	F	T	T
F	T	T	F	F	F	T	T
F	T	F	T	F	F	T	T
F	T	F	F	F	F	F	T
F	F	T	T	T	F	T	T
F	F	T	F	T	F	T	T
F	F	F	T	T	F	T	T
F	F	F	F	T	F	F	T

41. The statements are equivalent.
43. The statements are equivalent.
45. The statements are not equivalent.
47. The statements are equivalent.
49. The statements are not equivalent.
51. i. and iv. are equivalent.
ii. and iii. are equivalent.
53. i. and iv. are equivalent.
ii. and iii. are equivalent.
55. They are equivalent.
57. p: I have a college degree.
q: I am employed.
$p \land \sim q$ Negation: $\sim(p \land \sim q) \equiv \sim p \lor \sim(\sim q) \equiv \sim p \lor q$
I do not have a college degree or I am employed.
59. p: The television set is broken.
q: There is a power outage.
$p \lor q$ Negation: $\sim(p \lor q) \equiv \sim p \land \sim q$
The television set is not broken and there is not a power outage.
61. p: The building contains asbestos.
q: The original contractor is responsible.
$p \to q$ Negation: $\sim(p \to q) \equiv p \land \sim q$
The building contains asbestos and the original contractor is not responsible.
63. p: The lyrics are censored.
q: The First Amendment has been violated.
$p \to q$ Negation: $\sim(p \to q) \equiv p \land \sim q$
The lyrics are censored and the First Amendment has not been violated.
65. p: It is rainy weather.
q: I am washing my car.
$p \to \sim q$ Negation: $\sim(p \to \sim q) \equiv p \land \sim(\sim q) \equiv p \land q$
It is rainy weather and I am washing my car.

67. p: The person is talking.
q: The person is listening.
$q \rightarrow \sim p$ Negation: $\sim (q \rightarrow \sim p) \equiv q \wedge \sim (\sim p) \equiv q \wedge p$
The person is listening and talking.

69. a. A disjunction is true if either statement is true.
b. A disjunction is false if both statements are false.

71. a. A conditional is true if the hypotheses are false, or if the hypotheses are true and the conclusion is true.
b. A conditional is false if the hypotheses are true and the conclusion is false.

73. Equivalent expressions are those that have identical truth values.

75. If there are n statements, the truth table has 2^n rows.

77. It is used to determine the truth or falsity of compound statements.

1.4 More on Conditionals

1. a. If she is a police officer, then she carries a gun.
b. If she carries a gun, then she is a police officer.
c. If she is not a police officer, then she does not carry a gun.
d. If she does not carry a gun, then she is not a police officer.
e. Parts (a) and (d) are equivalent; parts (b) and (c) are equivalent. The contrapositive statement is always equivalent to the original.

3. a. If I watch television, then I do not do my homework.
b. If I do not do my homework, then I watch television.
c. If I do not watch television, then I do my homework.
d. If I do my homework, then I do not watch television.
e. Parts (a) and (d) are equivalent; parts (b) and (c) are equivalent. The contrapositive statement is always equivalent to the original.

5. a. If you do not pass this mathematics course, then you do not fulfill a graduation requirement.
b. If you fulfill a graduation requirement, then you pass this mathematics course.
c. If you do not fulfill a graduation requirement, then you do not pass this mathematics course.

7. a. If the electricity is turned on, then the television set does work.
b. If the television set does not work, then the electricity is turned off.
c. If the television set does work, then the electricity is turned on.

9. a. If you eat meat, then you are not a vegetarian.
b. If you are a vegetarian, then you do not eat meat.
c. If you are not a vegetarian, then you do eat meat.

11. a. The person not being a dentist is sufficient to not being an orthodontist.
b. Not being an orthodontist is necessary for not being a dentist.

13. a. Not knowing Morse code is sufficient to not operating a telegraph.
b. Not being able to operate a telegraph is necessary to not knowing Morse code.

15. a. *Premise*: I take public transportation. *Conclusion*: Public transportation is convenient.
b. If I take public transportation, then it is convenient.
c. The statement is false when I take public transportation and it is not convenient.

17. a. *Premise*: I buy foreign products. *Conclusion*: Domestic products are not available.
b. If I buy foreign products, then domestic products are not available.
c. The statement is false when I buy foreign products and domestic products are available.

19. a. *Premise*: You may become a U. S. senator. *Conclusion*: You are at least 30 years old and have been a citizen for nine years.
b. If you become a U. S. senator, then you are at least 30 years old and have been a citizen for nine years.
c. The statement is false when you become a U. S. senator and you are not at least 30 years old or have not been a citizen for nine years, or both.

21. If you obtain a refund, then you have a receipt and if you have a receipt, then you obtain a refund.

23. If the quadratic equation $ax^2 + bx + c = 0$ has two distinct real solutions, then $b^2 - 4ac > 0$ and if $b^2 - 4ac > 0$, then the quadratic equation $ax^2 + bx + c = 0$ has two distinct real solutions.

25. If a polygon is a triangle, then the polygon has three sides and if the polygon has three sides, then the polygon is a triangle.

27. If $a^2 + b^2 = c^2$, then the triangle has a 90° angle, and if the triangle has a 90° angle, than $a^2 + b^2 = c^2$.

29. p: I can have surgery.
q: I have health insurance.
$\sim q \rightarrow \sim p$ and $p \rightarrow q$
The statements are equivalent.

31. p: You earn less than \$12,000 per year.
q: You are eligible for assistance.
$p \rightarrow q$ and $\sim q \rightarrow \sim p$
The statements are equivalent.

33. p: I watch television.
q: The program is educational.
$p \rightarrow q$ and $\sim q \rightarrow \sim p$
The statements are equivalent.

35. p: The automobile is American-made.
q: The automobile hardware is metric.
$p \rightarrow \sim q$ and $q \rightarrow \sim p$
The statements are equivalent.

37. If I do not walk to work, then it is raining.

39. If it is not cold, then it is not snowing.

41. If you are a vegetarian, then you do not eat meat.

43. If the person does not own guns, then the person is not a policeman.

45. If the person is eligible to vote, then the person is not a convicted felon.

47. ii and iii are equivalent and i and iv are equivalent.

49. i and iii are equivalent and ii and iv are equivalent.

51. i and iii are equivalent and ii and iv are equivalent.

53. The contrapositive of "If p, then q." is "If not q, then not p."

55. The inverse of "If p, then q." is "If not p, then not q."

57. If p then q is equivalent to p only if q.

59. b.   **61.** c.

1.5 Analyzing Arguments

1. $p \rightarrow q$
$\underline{p}$
$\therefore\ q$

3. $p \rightarrow q$
$\underline{\sim q}$
$\therefore \sim p$

5. $p \rightarrow q$
$\underline{\sim p}$
$\therefore\ \sim q$

7. $q \rightarrow p$
$\underline{r \wedge p}$
$\therefore\ r \wedge q$

9. $p \rightarrow \sim q$
$\underline{r \wedge q}$
$\therefore\ r \wedge \sim p$

11. The argument is valid.

13. The argument is valid.

15. The argument is invalid if you don't exercise regularly and you are healthy.

17. The argument is invalid if (1) the person isn't Nikola Tesla, operates a telegraph, and knows Morse code; *or* (2) knows Morse code, doesn't operate a telegraph, and is Nikola Tesla; *or* (3) knows Morse code, doesn't operate a telegraph, and isn't Nikola Tesla.

19. The argument is valid.

21. p: The Democrats have a majority.
q: Smith is appointed.
r: Student loans are funded.

$$p \rightarrow (q \wedge r)$$
$$\underline{q \vee \sim r}$$
$$\sim p$$

$$\{[p \rightarrow (q \wedge r)] \wedge [q \vee \sim r]\} \rightarrow \sim p$$

The argument is invalid when the Democrats have a majority and Smith is appointed and student loans are funded.

23. p: You argue with a police officer.
q: You get a ticket.
r: You break the speed limit.

$$p \rightarrow q$$
$$\underline{\sim r \rightarrow \sim q}$$
$$r \rightarrow p$$

$$\{(p \rightarrow q) \wedge (\sim r \rightarrow \sim q)\} \rightarrow (r \rightarrow p)$$

The argument is invalid when (1) you do not argue with a police officer, you do get a ticket and you do break the speed limit, *or* (2) you do not argue with a police officer, you do not get a ticket and you do break the speed limit.

25. Rewriting the argument:
If it is a pesticide, then it is harmful to the environment.
If it is a fertilizer, then it is not a pesticide.
If it is a fertilizer, then it is not harmful to the environment.
p: It is a pesticide.
q: It is harmful to the environment.
r: It is a fertilizer.

$$p \rightarrow q$$
$$\underline{r \rightarrow \sim p}$$
$$r \rightarrow \sim q$$

$$\{(p \rightarrow q) \wedge (r \rightarrow \sim p)\} \rightarrow (r \rightarrow \sim q)$$

The argument is invalid if it is not a pesticide and if it is harmful to the environment and it is a fertilizer.

27. Rewriting the argument:
If you are a poet, then you are a loner.
If you are a loner, then you are a taxi driver.
If you are a poet, then you are a taxi driver.
p: You are a poet.
q: You are a loner.
r: You are a taxi driver.

$$p \rightarrow q$$
$$\underline{q \rightarrow r}$$
$$p \rightarrow r$$

$$\{(p \rightarrow q) \wedge (q \rightarrow r)\} \rightarrow (p \rightarrow r)$$

The argument is valid.

29. Rewriting the argument:
If you are a professor, then you are not a millionaire.
If you are a millionaire, then you are literate.
If you are a professor, then you are literate.

p: You are a professor.
q: You are a millionaire.
r: You are illiterate.

$$p \rightarrow \sim q$$
$$\underline{q \rightarrow \sim r}$$
$$p \rightarrow \sim r$$

$$\{(p \rightarrow \sim q) \wedge (q \rightarrow \sim r)\} \rightarrow (p \rightarrow \sim r)$$

The argument is invalid if you are a professor who is not a millionaire and is illiterate.

31. Rewriting the argument:
If you are a lawyer, then you study logic.
If you study logic, then you are a scholar.
You are not a scholar.
You are not a lawyer.
p: You are a lawyer.
q: You study logic
r: You are a scholar.

$$p \rightarrow q$$
$$q \rightarrow r$$
$$\underline{\sim r}$$
$$\sim p$$

$$\{(p \rightarrow q) \wedge (q \rightarrow r) \wedge \sim r\} \rightarrow \sim p$$

The argument is valid.

33. Rewriting the argument:
If you are drinking espresso, then you are not sleeping.
If you are on a diet, then you are not eating dessert.
If you are not eating dessert then you are drinking.
If you are sleeping, then you are not on a diet.
p: You are drinking espresso.
q: You are sleeping.
r: You are eating dessert.
t: You are on a diet.

$$p \rightarrow \sim q$$
$$t \rightarrow \sim r$$
$$\underline{\sim r \rightarrow p}$$
$$q \rightarrow \sim t$$

$$\{(p \rightarrow \sim q) \wedge (t \rightarrow \sim r) \wedge (\sim r \rightarrow p)\} \rightarrow (q \rightarrow \sim t)$$

The argument is valid.

35. p: The defendant is innocent.
q: The defendant goes to jail.

$$p \rightarrow \sim q$$
$$\underline{q}$$
$$\sim p$$

$$\{(p \rightarrow \sim q) \wedge q\} \rightarrow \sim p$$

The argument is valid.

37. p: You are in a hurry.
q: You eat at Lulu's Diner.
r: You eat good food.

$$\sim p \rightarrow q$$
$$p \rightarrow \sim r$$
$$\underline{q}$$
$$r$$

$$\{(\sim p \rightarrow q) \wedge (p \rightarrow \sim r) \wedge q\} \rightarrow r$$

The argument is invalid when (1) you are in a hurry and you eat at Lulu's Diner and you do not eat good food, *or* (2) you are not in a hurry and you eat at Lulu's Diner and you do not eat good food.

39. p: You listen to rock and roll.
$\quad q$: You go to heaven.
$\quad r$: You are a moral person.

$$p \to \sim q$$
$$r \to q$$
$$\overline{p \to \sim r}$$

$\{(p \to \sim q) \wedge (r \to q)\} \to (p \to \sim r)$
The argument is valid.

41. p: The water is cold.
$\quad q$: You go to swimming.
$\quad r$: You have goggles.

$$\sim p \to q$$
$$q \to r$$
$$\overline{\sim r}$$
$$p$$

$\{(\sim p \to q) \wedge (q \to r) \wedge \sim r\} \to p$
The argument is valid.

43. p: It is medicine.
$\quad q$: It is nasty.

$$p \to q$$
$$p$$
$$\overline{q}$$

$\{(p \to q) \wedge p\} \to q$
The argument is valid.

45. p: It is intelligible.
$\quad q$: It puzzles me.
$\quad r$: It is logic.

$$p \to \sim q$$
$$r \to q$$
$$\overline{r \to \sim p}$$

$\{(p \to \sim q) \wedge (r \to q)\} \to (r \to \sim p)$
The argument is valid.

47. p: A person is a Frenchman.
$\quad q$: A person likes plum pudding.
$\quad r$: A person is an Englishman.

$$p \to \sim q$$
$$r \to q$$
$$\overline{r \to \sim p}$$

$\{(p \to \sim q) \wedge (r \to q)\} \to (r \to \sim p)$
The argument is valid.

49. p: An animal is a wasp.
$\quad q$: An animal is friendly.
$\quad r$: An animal is a puppy.

$$p \to \sim q$$
$$r \to q$$
$$\overline{r \to \sim p}$$

$\{(p \to \sim q) \wedge (r \to q)\} \to (r \to \sim p)$
The argument is valid.

51. A tautology is a statement that is always true.

53. Answers may vary.

55. He wrote numerous books and essays.

57. Alice Liddell, the daughter of the dean of Christ Church, who insisted that Dodgson write down the story that he told the Liddell girls. Initially titled *Alice's Adventure Underground*, this became *Alice Adventures in Wonderland*.

Chapter 1 Review

1. a. Inductive   **b.** Deductive

3. 9   **5.** Invalid

7. Valid   **9.** Valid

11. a. Statement. It is either true or false.
b. Statement. It is either true or false.
c. Not a statement. It is a question.
d. Not a statement. It is an opinion.

13. a. His car is new.
b. No building is earthquake proof.
c. Some children do not eat candy.
d. Sometimes I cry in a movie theater.

15. a. $q \to p$ **b.** $\sim r \wedge \sim q \wedge p$
c. $r \to \sim p$ **d.** $p \wedge \sim (q \vee r)$
e. $p \to (\sim r \vee \sim q)$ **f.** $(r \wedge q) \to \sim p$

17. a. If the movie is critically acclaimed or a box office hit, then the movie is available on videotape.
b. If the movie is critically acclaimed and not a box office hit, then the movie is not available on videotape.
c. The movie is not critically acclaimed or a box office hit and it is available on DVD.
d. If the movie is not available on video tape, then the movie is not critically acclaimed and it is not a box office hit.

19.

p	q	$\sim q$	$p \wedge \sim q$
T	T	F	F
T	F	T	T
F	T	F	F
F	F	T	F

21.

p	q	$p \wedge q$	$\sim q$	$(p \wedge q) \to \sim q$
T	T	T	F	F
T	F	F	T	T
F	T	F	F	T
F	F	F	T	T

23.

p	q	r	$\sim p$	$q \vee r$	$\sim p \to (q \vee r)$
T	T	T	F	T	T
T	T	F	F	T	T
T	F	T	F	T	T
T	F	F	F	F	T
F	T	T	T	T	T
F	T	F	T	T	T
F	F	T	T	T	T
F	F	F	T	F	F

25.

p	q	r	$\sim r$	$p \vee r$	$q \wedge \sim r$	$(p \vee r) \to (q \wedge \sim r)$
T	T	T	F	T	F	F
T	T	F	T	T	T	T
T	F	T	F	T	F	F
T	F	F	T	T	F	F
F	T	T	F	T	F	F
F	T	F	T	F	T	T
F	F	T	F	T	F	F
F	F	F	T	F	F	T

27. The statements are not equivalent.

29. The statements are equivalent.

31. Jesse did not have a party or somebody came.

33. I am not the winner and you are not blind.

35. His application is ignored and the selection procedure has not been violated.

37. If a person is sleeping, then the person is not drinking espresso.

39. a. If you are an avid jogger, then you are healthy.
   b. If you are healthy, then you are an avid jogger.
   c. If you are not an avid jogger, then you are not healthy.
   d. If you are not healthy, then you are not an avid jogger.
   e. You are an avid jogger if and only if you are healthy.

41. a. Being lost is sufficient for not having a map.
   b. Not having a map is necessary for being lost.

43. a. *Premise:* The economy improves.
   Conclusion: Unemployment goes down.
   b. If the economy improves, then unemployment goes down.

45. a. *Premise:* It is a computer. *Conclusion:* It is repairable.
   b. If it is a computer, then it is repairable.

47. a. *Premise:* The fourth Thursday in November.
   Conclusion: The U. S. Post Office is closed.
   b. If it is the fourth Thursday in November, then the U. S. Post Office is closed.

49. p: You are allergic to dairy products.
   q: You can eat cheese.
   $p \to \sim q$ and $\sim q \to p$
   The statements are not equivalent.

51. i and iii are equivalent; ii and iv are equivalent.

53. p: You pay attention.
   q: You learn the new method.

$$\sim p \to \sim q$$
$$\frac{q}{p}$$

$[(\sim p \to \sim q) \wedge q] \to p$
The argument is valid.

55. p: The Republicans have a majority.
   q: Farnsworth is appointed.
   r: No new taxes are imposed.

$$p \to (q \wedge r)$$
$$\frac{\sim r}{\sim p \vee \sim q}$$

$\{[p \to (q \wedge r)] \wedge \sim r\} \to (\sim p \vee \sim q)$
The argument is valid.

57. p: You are practicing.
   q: You are making mistakes.
   r: You receive an award.

$$p \to \sim q$$
$$\sim r \to q$$
$$\frac{\sim r}{\sim p}$$

$\{[p \to \sim q] \wedge [\sim r \to q] \wedge \sim r\} \to \sim p$
The argument is valid.

59. p: I will go to the concert.
   q: You buy me a ticket.

$$p \to q$$
$$\frac{q}{p}$$

$[(p \to q) \wedge q] \to p$
The argument is invalid.

61. p: Our oil supply is cut off.
   q: Our economy collapses.
   r: We go to war.

$$p \to q$$
$$\frac{r \to \sim q}{\sim p \to \sim r}$$

$\{[p \to q] \wedge [r \to \sim q]\} \to (\sim p \to \sim r)$
The argument is invalid.

63. p: A person is a professor.
   q: A person is educated.
   r: An animal is a monkey.

$$p \to q$$
$$\frac{r \to \sim p}{r \to \sim q}$$

$\{[p \to q] \wedge [r \to \sim p]\} \to (r \to \sim q)$
The argument is invalid.

65. p: You are investing in the stock market.
   q: The invested money is to be guaranteed.
   r: You retire at an early age.

$$q \to \sim p$$
$$\frac{\sim q \to \sim r}{r \to \sim p}$$

$\{(q \to \sim p) \wedge (\sim q \to \sim r)\} \to (r \to \sim p)$
The argument is valid.

67. The argument is valid.

69. A statement is a sentence that is either true or false but not both.

71. a. A sufficient condition (i.e.,adequate) is the premise of a conditional.
   b. A necessary condition (i.e.,required) is the conclusion.

73. If the number of statements is n,the number of rows will be 2^n.

CHAPTER 2   Sets and Counting

2.1 Sets and Set Operations

1. **a.** well-defined   **b.** not well-defined
   c. well-defined   **d.** not well-defined
3. Proper: { }, {Lennon}, {McCartney}; Improper: {Lennon, McCartney}
5. Proper: { }, {yes}, {no}, {undecided}, {yes, no}, {yes, undecided}, {no, undecided}
   Improper: {yes, no, undecided}
7. **a.** {4, 5}   **b.** {1, 2, 3, 4, 5, 6, 7, 8}
   c. {0, 6, 7, 8, 9}   **d.** {0, 1, 2, 3, 9}
9. **a.** { }   **b.** {0, 1, 2, 3, 4, 5, 6, 7, 8, 9}
   c. {0, 2, 4, 6, 8}   **d.** {1, 3, 5, 7, 9}
11. {Friday}
13. {Monday, Tuesday, Wednesday, Thursday}
15. {Friday, Saturday, Sunday}
17.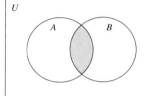

19.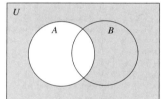

21.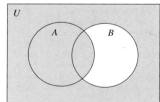

23.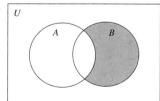

25.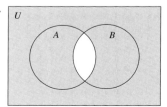

27. **a.** $n(A \cap B) = 21$
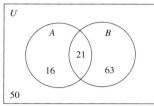
   b. $n(A \cap B) = 0$
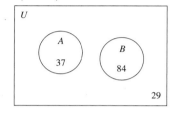

29. **a.**
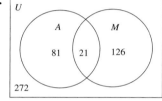
   b. 45.6%
31. **a.**

U
child   career
219 97 188
196

   b. 13.857%
33. 42   **35.** 43   **37.** 8   **39.** 5   **41.** 16
43. 32   **45.** 6   **47.** 8   **49.** 0
51. **a.** {1, 2, 3}   **b.** {1, 2, 3, 4, 5, 6}
   c. $E \subseteq F$ **d.** $E \subseteq F$
53. **a.** { }, {a} 2 subsets   **b.** { }, {a}, {b}, {a, b} 4 subsets
   c. { }, {a}, {b}, {c}, {a, b}, {a, c}, {b, c}, {a, b, c} 8 subsets
   d. { }, {a}, {b}, {c}, {d}, {a, b}, {a, c}, {a, d}, {b, c}, {b, d}, {c, d}, {a, b, c}, {a, b, d}, {a, c, d}, {b, c, d}, {a, b, c, d} 16 subsets
   e. Yes! The number of subsets of $A = 2^{n(A)}$ **f.** 64
55. The two sets are disjoint.
57. The set {0} contains one member, namely 0. The other set, $\varnothing$, contains no elements.
59. Only if $A = \{ \}$
61. The roster method is advantageous if the number of elements is small. Set-builder notation is advantageous when the number of elements is large.
63. He was a professor of moral sciences at Cambridge.
65. **a.** doesn't conform   M & O are in the same group
   b. doesn't conform   J & P are in the same group
   c. doesn't conform   N is in John's group, but P is not in Juneko's group
   d. conforms
   e. doesn't conform   M & O are in the same group
67. **c.** L, M, P   **69.** e

2.2 Applications of Venn Diagrams

1. **a.** 143 **b.** 16 **c.** 49 **d.** 57
3. **a.** 408 **b.** 1343 **c.** 664 **d.** 149
5. **a.** 106 **b.** 448 **c.** 265 **d.** 159
7. **a.** 51.351% **b.** 24.324% **c.** 16.216%
9. **a.** $x + y - z$ **b.** $x - z$
   c. $y - z$ **d.** $w - x - y + z$
11. **a.** 43.8% **b.** 10.8%
13. **a.** 66.471% **b.** 27.974%
15. **a.** 0% **b.** 54.1%
17. **a.** 44.0% **b.** 12.8%
19. **a.** 20% **b.** 58.8%
21. 16 23. {0, 4, 5} 25. {1, 2, 3, 6, 7, 8, 9}
27.
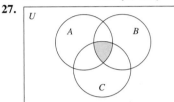
29.
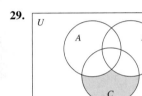
31.
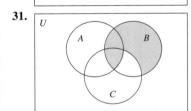
35. **a.** 84% **b.** 16% 37. **a.** 85% **b.** 15%
39. **a.** 75% **b.** 25% 41. Type B or Type O
43. Type O
45. He resigned as a protest against religious bias.
47. e 49. b 51. b
53.

If your blood type is:	You can receive:
O+	O−, O+
O−	O−
A+	O−, A−, O+, A+
A−	O−, A−
B+	O−, B−, O+, B+
B−	O−, B−
AB+	O−, A−, B−, AB−, O+, A+, B+, AB+
AB−	O−, A−, B−, AB−

2.3 Introduction to Combinatorics

1. **a.** 8
   b. {HHH, HHT, HTH, HTT, THH, THT, TTH, TTT}

3. **a.** 12
   b. {Mega-BW-BN, Mega-BW-NW, Mega-WW-BN, Mega-WW-NW, Mega-GW-BN, Mega-GW-NW, BB-BW-BN, BB-BW-NW, BB-WW-BN, BB-WW-NW, BB-GW-BN, BB-GW-NW}
5. 24 7. 720 9. 2,646 11. 216
13. $10^9 = 1,000,000,000$ 15. $10^{10} = 10,000,000,000$
17. **a.** 128 **b.** 800 **c.** The phone company needed more.
19. **a.** $1,179,360,000 = 1.17936 \times 10^9$
   b. $67,600,000,000 = 6.76 \times 10^{10}$
   c. $42,000,000,000 = 4.2 \times 10^{10}$
21. 540,000 23. 24 25. 3,628,800
27. $2,432,902,008,176,640,000 = 2.432902008 \times 10^{18}$
29. 17,280
31. **a.** 30 **b.** 360
33. 56 35. 70 37. 3,321
39. $10,461,394,944,000 = 1.046139494 \times 10^{13}$
41. 120 43. 35 45. 1
47. The Fundamental Principle of Counting is a way to find the number of possible outcomes of a series of decisions.
49. Christian Kramp. It replaced the symbol $\underline{ln}$, which was hard to print.
51. b 53. d

2.4 Permutations and Combinations

1. **a.** 210 **b.** 35
3. **a.** 120 **b.** 1
5. **a.** 14 **b.** 14
7. **a.** 970,200 **b.** 161,700
9. **a.** $x!$ **b.** x
11. **a.** $x \cdot (x - 1) = x^2 - x$ **b.** $\dfrac{x \cdot (x - 1)}{2} = \dfrac{x^2 - x}{2}$
13. **a.** 6
   b. $\{a, b\}, \{a, c\}, \{b, c\}, \{b, a\}, \{c, a\}, \{c, b\}$
15. **a.** 6
   b. $\{a, b\}, \{a, c\}, \{a, d\}, \{b, c\}, \{b, d\}, \{c, d\}$
17. **a.** 39,916,800 **b.** 1
19. 360 21. 78 23. 1,716
25. **a.** 540 **b.** 1,365 **c.** 630
27. 2,598,960
29. **a.** 4,512 **b.** 58,656
31. 123,552 33. 22,957,480
35. 376,992
37. It is easier to win a 5/36 lottery since there are fewer possible tickets.
39. **a.** 1 **b.** 2 **c.** 4
   d. 8 **e.** 16
   f. Yes. The sum of each row is twice the previous sum, or, sum of entries in n^{th} row is 2^{n-1}.
   g. 32
   h. 32. Our prediction was correct.
   i. 2^{n-1}
41. **a.** fifth row **b.** $(n + 1)^{\text{st}}$ row
   c. No. **d.** Yes.
   e. $(r + 1)^{\text{st}}$ number in the $(n + 1)^{\text{st}}$ row
43. 120 45. 3,360
47. 630 49. 831,600
51. **a.** 24 **b.** 12
53. **a.** 120 **b.** 60
55. If order matters, use $_nP_r$. If order doesn't matter, use $_nC_r$.
57. c 59. c 61. e

2.5 Infinite Sets

1. equivalent; Match each state to its capital.
3. not equivalent   5. equivalent; $3n \leftrightarrow 4n$
7. not equivalent   9. equivalent; $2n - 1 \leftrightarrow 2n + 123$
11. **a.** For n starting at 1, match the term described by n in N to the term described by $2n - 1$ in O; $n \leftrightarrow 2n - 1$
    b. $n = 918$ **c.** $n = \dfrac{x + 1}{2}$
    d. 1,563   **e.** $n \to 2n - 1$
13. **a.** For n starting at 1, match the term described by n in N to the term described by $3n$ in T; $n \leftrightarrow 3n$
    b. $n = 312$ **c.** $n = \frac{1}{3}x$
    d. 2,808   **e.** $n \to 3n$
15. **a.** -344 **b.** $248 \to 248$
    c. $n = 755$ **d.** $n(A) = \aleph_0$
17. Match any real number $x \in [0, 1]$ to $3x \in [0, 3]$.
19. From the center of the circle, match the corresponding points that lie on the same radial line.
21. From the inside circle, match the corresponding points that lie on the same radial line.
23. First, the semicircle is equivalent to $[0, 1]$. Draw a vertical line to match points. To match the semicircle and line, draw a line passing through $(\frac{1}{2}, 0)$ (not horizontal).
25. Paradoxes concerning the cardinal numbers of infinite sets, the nature of ∞ and Cantor's form of logic.
27. Paul J. Cohen. 1963.

Chapter 2 Review

1. **a.** Well defined.   **b.** Not well defined
   c. Not well defined.   **d.** Well defined.
3. **a.** $A \cup B = \{$Maria, Nobuku, Leroy, Mickey, Kelly, Rachel, Deanna$\}$
   b. $A \cap B = \{$Leroy, Mickey$\}$
5. **a.** 18
   b.

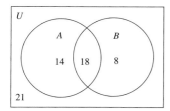

7.

8. (Venn diagram: U with circles A and B overlapping)

9. (Venn diagram: U with circles A, B, and C)

11. **a.** 31.154%   **b.** 9.038%   **c.** 17.692%
13. 27.875%
15. **a.** 85   **b.** 0   **c.** 15
17. **a.** 12
    b. {MMA-S-LC, MMA-S-L, MMA-J-LC, MMA-J-L, MMA-C-LC, MMA-C-L, NPG-S-LC, NPG-S-L, NPG-J-LC, NPG-J-L, NPG-C-LC, NPG-C-L}
19. 1,080,000
21. **a.** 165   **b.** 990
23. **a.** 660   **b.** 1,540   **c.** 880
25. 720   27. 177,100,560
29. **a.** 5,040   **b.** 2,520   **c.** 1,260
31. Order is important in permutations, whereas it is not important in combinations.
33. **a.** 3   **b.** 3   **c.** 1   **d.** 1   **e.** 8
    f. The number of subsets that set S has equals $2^{n(S)}$. $2^{n(S)} = 2^3 = 8$
35. They are equivalent. $2n + 1 \leftrightarrow 2n$
37. **a.** Match each number in N to its square in S: $n \leftrightarrow n^2$
    b. 29   **c.** $\sqrt{x}$ **d.** 20,736   **e.** $n \to n^2$
39. Match any real number $x \in [0, 1]$ to $\pi x \in [0, \pi]$.
41. {0} is a set with one element (namely 0). $\varnothing$ is a set with no elements.
43. If order of selection matters, we use permutations. If not, we use combinations.
45. e.   47. b.

CHAPTER 3 Probability

3.1 History of Probability

1. Answers will vary.
3. **a.** Answers will vary.   **b.** Answers will vary.
5. **a.** Answers will vary.   **b.** Answers will vary.
7. **a.** Answers will vary.   **b.** Answers will vary.
9. **a.** win $350   **b.** lose $10
11. **a.** win $85   **b.** win $85   **c.** lose $5
13. **a.** lose $20   **b.** win $160   **c.** lose $20
15. **a.** lose $10   **b.** win $50
17. **a.** win $50   **b.** lose $25
19. **a.** lose $50   **b.** win $50
21. **a.** lose $45   **b.** win $855   **c.** win $675
23. **a.** win $495   **b.** win $525
    c. lose $45   **d.** loss of $15

25. $3   27. $500   29. **a.** 13   **b.** $\frac{1}{4}$
31. **a.** 12   **b.** $\frac{3}{13}$
33. **a.** 4   **b.** $\frac{1}{13}$
35. Because it was connected with gambling.
37. Mendel used probability to analyze the effect of randomness in genetics.
39. Cardano was an Italian physician, mathematician and gambler who wrote the first theoretical study of probabilities and gambling.
41. dice, cards
43. During the Crusades, Europeans acquired the cards from Arabs.
45. roulette

3.2 Basic Terms of Probability

1. Choose 1 jellybean and look at its color.
3. 34%   **5.** 51%   **7.** 71%   **9.** 0%   **11.** $12:23$
13. $18:17$ **15.** Drawing or selecting a card from a deck
17. a. $\frac{1}{2}$ **b.** $1:1$
   c. According to the Law of Large Numbers, if the experiment is repeated a large number of times, we should expect to draw a black card about half the time. Also, we should expect to draw a black card one time for every time we draw a red card.
19. a. $\frac{1}{13}$ **b.** $1:12$
   c. According to the Law of Large Numbers, if the experiment is repeated a large number of times, we should expect to draw a queen about one time out of every thirteen attempts. Also, we should expect to draw a queen one time for every twelve times we do not draw a queen.
21. a. $\frac{1}{52}$ **b.** $1:51$
   c. According to the Law of Large Numbers, if the experiment is repeated a large number of times, we should expect to draw a queen of spades about one time out of every fifty-two attempts. Also, we should expect to draw a queen of spades one time for every fifty-one times we do not draw a queen of spades.
23. a. $\frac{3}{13}$ **b.** $3:10$
   c. According to the Law of Large Numbers, if the experiment is repeated a large number of times, we should expect to draw a card below a 5 about three times out of every thirteen attempts. Also, we should expect to draw a card below a 5 three times for every ten times we draw a 5 or above.
25. a. $\frac{10}{13}$ **b.** $10:3$
   c. According to the Law of Large Numbers, if the experiment is repeated a large number of times, we should expect to draw a card above a 4 about ten times out of every thirteen attempts. Also, we should expect to draw a card above a 4 ten times for every three times we draw a 4 or below.
27. a. $\frac{3}{13}$ **b.** $3:10$
   c. According to the Law of Large Numbers, if the experiment is repeated a large number of times, we should expect to draw a face card about three times out of every thirteen attempts. Also, we should expect to draw a face card three times for every ten times we do not draw a face card.
29. a. $\frac{1}{38}$ **b.** $1:37$
   c. According to the Law of Large Numbers, if the experiment is repeated a large number of times, we should expect to win a single-number bet about 1 time out of every 38 bets. Also, we should expect to win a single-number bet 1 time for every 37 times we lose.
31. a. $\frac{3}{38}$ **b.** $3:35$
   c. According to the Law of Large Numbers, if the experiment is repeated a large number of times, we should expect to win a three-number bet about 3 times out of every 38 bets. Also, we should expect to win a three-number bet 3 times for every 35 times we lose.
33. a. $\frac{5}{38}$ **b.** $5:33$
   c. According to the Law of Large Numbers, if the experiment is repeated a large number of times, we should expect to win a five-number bet about 5 times out of every 38 bets. Also, we should expect to win a five-number bet 5 times for every 33 times we lose.
35. a. $\frac{6}{19}$ **b.** $6:13$

   c. According to the Law of Large Numbers, if the experiment is repeated a large number of times, we should expect to win a twelve-number bet about 6 times out of every 19 bets. Also, we should expect to win a twelve-number bet 6 times for every 13 times we lose.
37. a. $\frac{9}{19}$ **b.** $9:10$
   c. According to the Law of Large Numbers, if the experiment is repeated a large number of times, we should expect to win an even-number bet about 9 times out of every 19 bets. Also, we should expect to win an even-number bet 9 times for every 10 times we lose.
39. a. 50.7%   **b.** 17.5%
41. a. 4%   **b.** 32%   **c.** 64%
   d. According to the Law of Large Numbers, if the experiment is repeated a large number of times, we should expect that a dart will hit the red region about one time out of every twenty-five throws, the yellow region about eight times out of every twenty-five throws, and the green region about sixteen times out of every twenty-five throws.
43. $\frac{6}{29}$ **45.** $1:4$ **47.** $\frac{3}{5}$ **49.** $a:(b-a)$
51. a. $\frac{9}{19}$ **b.** $9:10$
   c. According to the Law of Large Numbers, if the experiment is repeated a large number of times, we should expect to win an odd-number bet about nine times out of every nineteen bets. Also, we should expect to win an odd-number bet nine times for every ten times we lose.
53. a. House odds. sportsbook.com determines their own odds based on their own set of rules.
   b. New York Yankees: $p(\text{win}) = \frac{5}{7}$
      New York Mets: $p(\text{win}) = \frac{7}{8}$
   c. The New York Mets are more likely to win because $\frac{7}{8} > \frac{5}{7}$
55. a. Intentional self-harm has the highest odds ($1:9{,}085$), so it is the most likely cause of death in one year.
   b. Once again, intentional self-harm has the highest odds ($1:117$), so it is the most likely cause of death in a lifetime.
57. a. $\frac{1}{20{,}332}$.   **b.** $\frac{1}{502{,}555}$.
59. a. $\frac{1}{20{,}332}$, car transportation accident (intentional self-harm).
   b. $\frac{1}{502{,}555}$, airplane and space accidents (war).
61. a. $S = \{\text{bb, gb, bg, gg}\}$ **b.** $E = \{\text{gb, bg}\}$
   c. $F = \{\text{gb, bg, gg}\}$ **d.** $G = \{\text{gg}\}$
   e. $\frac{1}{2}$ **f.** $\frac{3}{4}$ **g.** $\frac{1}{4}$
   h. $1:1$ **i.** $3:1$ **j.** $1:3$
63. a. $S = \{\text{ggg, ggb, gbg, bgg, gbb, bgb, bbg, bbb}\}$
   b. $E = \{\text{ggb, gbg, bgg}\}$ **c.** $F = \{\text{ggb, gbg, bgg, ggg}\}$
   d. $G = \{\text{ggg}\}$
   e. $\frac{3}{8}$ **f.** $\frac{1}{2}$ **g.** $\frac{1}{8}$
   h. $3:5$ **i.** $1:1$ **j.** $1:7$
65. $S = \{\text{bb, bg, gb, gg}\}$ **a.** $\frac{1}{4}$ **b.** $\frac{1}{2}$
   c. $\frac{1}{4}$
   d. same sex: $\frac{2}{4}$, different sex: $\frac{2}{4}$. They are equally likely.
67. Same sex: $\frac{1}{4}$, Different sex: $\frac{3}{4}$ Different sex is more likely. $\frac{3}{4} > \frac{1}{4}$
69. a.

$$S = \begin{cases} (1,1),(1,2),(1,3),(1,4),(1,5),(1,6), \\ (2,1),(2,2),(2,3),(2,4),(2,5),(2,6), \\ (3,1),(3,2),(3,3),(3,4),(3,5),(3,6), \\ (4,1),(4,2),(4,3),(4,4),(4,5),(4,6), \\ (5,1),(5,2),(5,3),(5,4),(5,5),(5,6), \\ (6,1),(6,2),(6,3),(6,4),(6,5),(6,6) \end{cases}$$

   b. $E = \{(1,6),(2,5),(3,4),(4,3),(5,2),(6,1)\}$

c. $F = \{(5, 6), (6, 5)\}$
d. $G = \{(1, 1), (2, 2), (3, 3), (4, 4), (5, 5), (6, 6)\}$
e. $\frac{1}{6}$ **f.** $\frac{1}{18}$ **g.** $\frac{1}{6}$
h. $1 : 5$ **i.** $1 : 17$ **j.** $1 : 5$
71. a. $\frac{1}{4}$ **b.** $\frac{1}{4}$ **c.** $\frac{2}{4} = \frac{1}{2}$
73. a. $\frac{1}{4}$ **b.** $\frac{2}{4} = \frac{1}{2}$ **c.** $\frac{1}{4}$
75. a. $\frac{0}{4} = 0$ **b.** $\frac{2}{4} = \frac{1}{2}$ **c.** $\frac{4}{4} = 1$
77. a. $\frac{2}{4} = \frac{1}{2}$ **b.** $\frac{0}{4} = 0$ **c.** $\frac{2}{4} = \frac{1}{2}$
79. Answers will vary.
81. No. The faces are no longer equally likely.
No. The entries in the event are no longer equally likely.
83. Given the very large number of trials, this does contradict the assumption.
85. Probability is the likelihood of some event occurring. Odds is a comparison of the number of times the event occurs to the number of times it doesn't occur.
87. The gene that causes the disease has been isolated.
89. Answers will vary.   **91.** Answers will vary.

3.3 Basic Rules of Probability
1. E and F are not mutually exclusive. There are many women doctors.
3. E and F are mutually exclusive. One cannot be both single and married.
5. E and F are not mutually exclusive. A brown haired person may have some gray.
7. E and F are mutually exclusive. One can't wear both boots and sandals (excluding one on each foot!)
9. E and F are mutually exclusive. Four is not odd.
11. a. $\frac{1}{26}$ **b.** $\frac{7}{13}$ **c.** $\frac{25}{26}$
13. a. $\frac{1}{52}$ **b.** $\frac{4}{13}$ **c.** $\frac{51}{52}$
15. a. $\frac{2}{13}$ **b.** $\frac{5}{13}$ **c.** 0
d. $\frac{7}{13}$
17. a. $\frac{9}{13}$ **b.** $\frac{8}{13}$ **c.** $\frac{4}{13}$
d. 1
19. $\frac{12}{13}$ **21.** $\frac{10}{13}$ **23.** $\frac{11}{13}$
25. $\frac{9}{13}$ **27.** $9 : 5$
29. $o(E) = 2 : 5$, $o(E') = 5 : 2$
31. $b : a$ **33.** $12 : 1$
35. $10 : 3$ **37.** $10 : 3$
39. a. $\frac{71}{175}$ **b.** $\frac{104}{175}$
41. a. $\frac{151}{700}$ **b.** $\frac{97}{140}$
43. a. $\frac{8}{25}$ **b.** $\frac{17}{25}$
45. a. $\frac{9}{25}$ **b.** $\frac{16}{25}$
47. a. $\frac{1}{6}$ **b.** $\frac{1}{9}$ **c.** $\frac{1}{18}$
49. a. $\frac{2}{9}$ **b.** $\frac{7}{18}$
51. a. $\frac{1}{6}$ **b.** $\frac{3}{4}$
53. a. $\frac{1}{6}$ **b.** $\frac{1}{2}$
55. a. 0.65   **b.** 0.30   **c.** 0.35
57. Relative frequencies; data was collected from client's records and the probabilities were calculated from this data.
59. a. 76.7%   **b.** 23%   **c.** 23.3%
61. Relative frequencies; they were calculated from a poll.
63. a. $\frac{17}{25}$ **b.** $\frac{2}{25}$
65. a. $\frac{212}{1,451}$ **b.** $\frac{514}{1,451}$
67. a. $\frac{7}{58}$ **b.** $\frac{31}{174}$
69. a. $\frac{3}{4}$ **b.** $\frac{1}{2}$ **c.** 1

71. a. $\frac{1}{4}$ **b.** $\frac{1}{2}$ **c.** 0
73. a. 0.35   **b.** 0.25   **c.** 0.15
d. 0.6
75. a. 0.20   **b.** 0.95   **c.** 0.80
81. Mutually exclusive events can't both happen at the same time. Impossible events can't ever happen.
83. a. $\frac{6}{11}$ **b.** $\frac{6}{11}$
85. a. $\frac{4}{21}$ **b.** $\frac{4}{21}$
87. a. $\frac{92}{105}$ **b.** $\frac{92}{105}$
89. a. $\frac{23}{21}$ **b.** $\frac{23}{21}$
91. Don't press " $\rightarrow$ Frac"

3.4 Combinatorics and Probability
1. over 70%   **3.** 23 people
5. a. $\frac{1}{13,983,816}$ **b.** $\frac{1}{22,957,480}$
c. $\frac{1}{18,009,460}$ **d.** 64% more likely
e. Answers will vary.
7. a. $\frac{1}{575,757}$ **b.** $\frac{3}{10,000}$
9. a. $\frac{1}{324,632}$ **b.** $\frac{5}{10,000}$
11. $\frac{1}{175,711,536}$ **13.** $\frac{1}{2,718,576}$
15. 4/26 easiest;   6/54 hardest
17.

Outcome	Probability
8 winning spots	0.0000043457
7 winning spots	0.0001604552
6 winning spots	0.0023667137
5 winning spots	0.0183025856
4 winning spots	0.0815037015
Fewer than 4 winning spots	0.8976621984

19. a. 1,000
b. There are 1,000 straight play numbers.   **c.** $\frac{1}{1,000}$
21. a. 0.000495   **b.** 0.001981
c. 0.000015   **d.** 0.0019654015
23. ≈ 0.05 **25.** ≈ 0.48
27. ≈ 0.64 **29.** ≈ 0.36
31. a. ≈ 0.09 **b.** ≈ 0.42 **c.** ≈ 0.49
d. The above probabilities assume that all events are equally likely. However, the employer's task is to choose the most appropriate people for the jobs. This means that, for the employer, all events are not equally likely, and the probabilities do indicate neither presence nor absence of gender discrimination.
33. $_6C_6 = \dfrac{6!}{6!(6-6)!} = \dfrac{6!}{6!0!} = 1$
35. Answers will vary.
37. The probability of winning a lottery is ridiculously small, as shown in Exercises 5–14.
39. No. George Washington used a lottery to pay for a road through the Cumberland Moutains.

3.5 Expected Value
1. a. $-\$0.0526315789$
b. You should expect to lose about 5.3 cents for every dollar you bet, if you play a long time.

3. a. −$0.0526315789
  b. You should expect to lose about 5.3 cents for every dollar you bet, if you play a long time.
5. a. −$0.0526315789
  b. You should expect to lose about 5.3 cents for every dollar you bet, if you play a long time.
7. a. −$0.0526315789
  b. You should expect to lose about 5.3 cents for every dollar you bet, if you play a long time.
9. a. −$0.0526315789
  b. You should expect to lose about 5.3 cents for every dollar you bet, if you play a long time.
11. $549.00   **13.** 1.75 books
15. $10.05   **17.** Since $-\$1.67 < 0$, don't play.
19. No, since the expected value of 10 games, $50, is less than $100, don't play. Accept $100.

21.
$$\frac{35 + 37 \cdot (-1)}{38} = \frac{35 \cdot 1 + 37(-1)}{38} = \frac{35 \cdot 1}{38} + \frac{37(-1)}{38}$$

$$= 35 \cdot \frac{1}{38} + (-1)\frac{37}{38} = -\frac{2}{38} \approx -0.053$$

23. Decision theory indicates that the bank's savings account is the better investment.
25. The speculative investment is the better choice if $1.1p - 0.6 > 0.045$, that is, if $p > 645/1100 = 0.586\ldots \approx 0.59$.
27. a. 0   **b.** $\frac{1}{16}$ **c.** $\frac{3}{8}$
  d. If you can eliminate one or more answers, guessing is a winning strategy.
29. −$0.28; You should expect to lose about 28 cents for every dollar you bet, if you play a long time.
31. a. $p(\text{first prize}) = \dfrac{1}{14{,}950}$

  $p(\text{second prize}) = \dfrac{3}{500}$

  $p(\text{third prize}) \approx 0.09 = \dfrac{9}{100}$

  b. $p(\text{losing}) \approx 0.901 \approx \dfrac{9}{10}$

  c. $EV = 0.00147 \approx \$0.001$
33. a. $p(\text{losing}) = 0.99$
  b. $EV \approx \$0.46$
35. To make a profit, the price should be more than $1,200.
37. They should drill in the back yard.
39. a. $12.32 You should buy a ticket.
  b. −$1.34 You should not buy a ticket.
  c. −$5.89 You should not buy a ticket.
41. a. −$0.473 You should not buy a ticket.
  b. −$2.74 You should not buy a ticket.
  c. −$3.49 You should not buy a ticket.
43. $7.785 million or $7.8 million
45. a. 13,983,816   **b.** $13,983,816
  c. Since we have 100 people buying, we only need 97.109833 days.
47. You can bet and lose 6 times. The net winnings are $1.
49. Most meaningful to the casino owner, least meaningful to the occasional gambler, due to the frequency with which they play the game.
51. Answers will vary.

3.6 Conditional Probability

1. a. $p(H|Q)$; Conditional since the sample space is limited to well-qualified candidates.
  b. $p(H \cap Q)$; Not conditional.
3. a. $p(S|D)$; Conditional since the sample space is limited to users that are dropped a lot.
  b. $p(S \cap D)$; Not conditional.
  c. $p(D|S)$; Conditional since the sample space is limited to users that switch carriers.
5. a. $\frac{3}{4}$ **b.** $\frac{1}{2}$ **7. a.** $\frac{3}{8}$ **b.** $\frac{1}{8}$
9. a. ≈0.2333   About 23.3% of those surveyed said no.
  b. ≈0.5333   About 53.3% of those surveyed were women.
  c. ≈0.1406   About 14.1% of the women said no.
  d. ≈0.3214   About 32.1% of those who said no were women.
  e. ≈0.08   About 8% of the respondents said no and were women.
  f. ≈0.08   About 8% of the respondents were women and said no.
11. a. ≈0.13. In one year, 13% of those who die of a transportation accident die of a pedestrian transportation accident.
  b. ≈0.13. In a lifetime, 13% of those who die of a transportation accident die of a pedestrian transportation accident.
  c. 0. In a lifetime, none of those who die of a non-transportation accident die of a pedestrian transportation accident.
13. a. ≈0.00054. In a lifetime, approximately 0.05% of those who die of a non transportation accident die from an earthquake.
  b. ≈0.00053. In one year, approximately 0.05% of those who die of a non transportation accident die from an earthquake.
  c. ≈0.00031. In one year, approximately 0.03% of those who die from an external cause die from an earthquake.
15. a. $\frac{1}{4}$ **b.** $\frac{4}{17}$ **c.** $\frac{1}{17}$
  d.

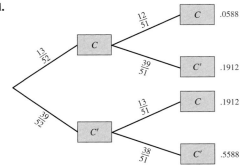

17. a. $\frac{13}{52} = \frac{1}{4}$ **b.** $\frac{13}{51}$ **c.** $\frac{13}{52} \cdot \frac{13}{51} = \frac{13}{204}$
  d.

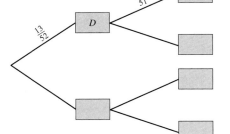

19. a. $p(B|A)$ **b.** $p(A')$ **c.** $p(C|A')$

21. a. $\frac{1}{6}$ **b.** $\frac{1}{3}$ **c.** 0   **d.** 1

23. a. $\frac{5}{36}$ **b.** $\frac{5}{18}$ **c.** 0   **d.** 1

25. a. $\frac{1}{12}$ **b.** $\frac{3}{10}$ **c.** 1

27. E_2 is most likely; E_3 is least likely

29. ≈ 0.46

About 46% of those who were happy with the service made a purchase.

31. 0.0005   **33.** 0.0020   **35.** 0.14   **37.** 0.20

39. 0.07   **41.** 0.6%   **43.** 1.3%

45. a. 0.489   **b.** 0.560

c. For those voting for Obama, a higher percentage was women.

47. a. 0.406   **b.** 0.502

c. Those over 45 years were more likely to vote for McCain as compared to those under 45 years.

49. a. $p(\text{male}) = 0.803$
$p(\text{female}) = 0.197$

b. $p(\text{injection drug use} | \text{male}) = 0.217$
$p(\text{injection drug use} | \text{female}) = 0.404$

c. $p(\text{injection drug use} \cap \text{male}) = 0.174$
$p(\text{injection drug use} \cap \text{female}) = 0.080$

d. $p(\text{heterosexual contact} | \text{male}) = 0.079$
$p(\text{heterosexual contact} | \text{female}) = 0.565$

e. $p(\text{heterosexual contact} \cap \text{male}) = 0.063$
$p(\text{heterosexual contact} \cap \text{female}) = 0.111$

f. Part (b) is the percentage of males that were exposed by injection drug use. Part (c) is the percentage of the total that are both male and exposed by injection drug use. Similarly for parts (d) and (e).

51. a. 32%   **b.** 16%   **c.** 35%

d. 18%   **e.** 34%

f. Part (c) is the percentage of adult women who are obese. Part (d) is the percentage of adults who are both female and obese.

53. a. 12.6 %   **b.** 6.7%   **c.** 0%   **d.** 5.8%

e. Answers will vary. Possible answer: When it is a mixed race case, black defendants are more likely to have the death penalty imposed than white defendants.

f. Answers will vary. Possible answer:
$p(\text{death not imposed} | \text{victim white and defendant white})$
$\approx 87.4\%$
$p(\text{death not imposed} | \text{victim white and defendant black})$
$\approx 93.3\%$
$p(\text{death not imposed} | \text{victim black and defendant white})$
$= 100\%$
$p(\text{death not imposed} | \text{victim black and defendant black})$
$\approx 94.2\%$
Also see answers to Exercise 54(c) and 54(d).

g. Answers will vary.

55. $\frac{2}{3}$ **57.** 92%   **59.** 39%   **61.** 0.05

63. a. 86%   **b.** 34%   **c.** 66%

d. $N'|W; p(N|W) = \frac{45}{320} = 0.14 = 1 - 0.86$

65. The complement is $A'|B$.   **67.** 0.14

69. Sometimes $p(A|B)$ is greater than or equal to $p(A)$, and sometimes $p(A|B)$ is less than or equal to $p(A)$, depending on the events A and B. Event B might make event A more or less likely depending on what they are.

71. Answers will vary.

73. a. $p(\text{admitted}) \approx 41\%$
$p(\text{admitted} | \text{male}) \approx 44.3\%$
$p(\text{admitted} | \text{female}) \approx 34.6\%$
The admission process does seem to be biased.

b. No.

c. $p(\text{admitted}) \approx 41.7\%$
$p(\text{admitted} | \text{male}) \approx 0.454545 = $ about 45.5%
$p(\text{admitted} | \text{female}) \approx 0.3846 = $ about 38.5%
The admission process does seem to be biased.

d. The admission probabilities for the whole school can be skewed to look bias even though the individual department probabilities are fair.

e. Maybe some departments are biased. I would obtain historical data to see if this was a trend or a one-time occurrence. I would also check the qualifications of the applicants.

3.7 Independence; Trees in Genetics

1. a. E and F are dependent. Knowing F affects E's probability.

b. E and F are not mutually exclusive. There are many women doctors.

3. a. E and F are dependent. Knowing F affects E's probability.

b. E and F are mutually exclusive. One cannot be both single and married.

5. a. E and F are independent. Knowing F does not affect E's probability.

b. E and F are not mutually exclusive. A brown haired person may have some gray.

7. a. E and F are dependent. Knowing F affects E's probability.

b. E and F are mutually exclusive. One can't wear both boots and sandals (excluding one on each foot!)

9. a. E and F are dependent.

b. E and F are mutually exclusive.

c. Knowing that you got an odd number changes the probability of getting a 4. You cannot get both a 4 and an odd number.

11. E and F are not independent.

13. a. $\frac{1}{6}$ **b.** 0

c. No. If you roll a 5, the probability of rolling an even number is zero.

d. Yes. 5 is not an even number.

e. Mutually exclusive events are always dependent.

15. a. $\frac{1}{13}$ **b.** $\frac{1}{13}$

c. Yes. Being dealt a red card doesn't change the probability of being dealt a jack.

d. No. There are red jacks.

e. Independent events are always not mutually exclusive. Exercises 14 and 15 show that events not mutually exclusive can be dependent or independent.

17. The events are dependent. Being happy with the service increases the probability of a purchase.

19. E and F are dependent. Knowing that the chip is made in Japan increases the probability that it is defective.

21. a. Since $0.616 \neq 0.673$, they are not independent.

b. Since $0.518 \neq 0$, they are not mutually exclusive.

c. Being a vegetarian increases your chances of being healthy.

23. a. Since $0.638 \neq 0.285$, they are not independent.

b. No; $p(A \cap B) \neq 0$; the circles overlap with 89 in both.

c. Living in Bishop decreases the chances of supporting proposition 3.

25. They are dependent. HAL users were more likely to quit.

27. They are independent. Smell So Good users quit at the same rate as all deodorant users.

29. a. 0.000001

    b. 4 backup systems

31. 0.010101 ≈ 1.0%

33. a. 0.47   **b.** 0.999996   **c.** 0.53   **d.** 0.000004

    e. The results from (a) and (c) would be informative because a positive test doesn't necessarily mean you are ill.

    f. (c) is a false positive because the test was positive, but he was healthy. (d) is a false negative because the test was negative, but he was ill.

    g. Answers will vary.

    h. Answers will vary.

35. Answers will vary.

37. a. Therefore, the probability of both cousins being a carrier would be 1/24. And the probability of their child having cystic fibrosis (Type A) would be 1/96 ≈ 1%.

    b. And the probability of their child having cystic fibrosis would be 1/144 ≈ 0.7%.

39. p(both test positive) $= (0.85)(0.85) = 0.7225$

    p(don't both test positive) $= 1 - 0.7225 = 0.2775$

41. $\frac{1}{3}$

43. If the first child is albino, Mrs. Jones must be a carrier. Thus, the child has $\frac{1}{2}$ chance of being albino.

45. $\frac{1}{2}$ chance of chestnut hair, $\frac{1}{2}$ chance of shiny dark brown

47. The child has a $\frac{1}{8}$ chance of the following hair colors: dark red, light brown, auburn or medium brown. The child has a $\frac{1}{4}$ chance of the following: reddish brown, chestnut.

49. The child has a $\frac{1}{8}$ chance of the following hair colors: reddish brown, chestnut, shiny dark brown, shiny black. The child has a $\frac{1}{16}$ chance of the following hair colors: dark red, light brown, auburn, medium brown, glossy dark brown, dark brown, glossy black, black.

51. a. $\frac{1}{6}$ **b.** $\frac{5}{6}$ **c.** 0.48

    d. 0.52   **e.** $0.04

53. He won because the expected value was positive. He lost because the expected value was negative. $0.04 > -$0.017

55. a. p(die) $= (0.25)(0.12)(0.9) = 0.027$

    p(live) $= 1 - 0.027 = 0.973 = 97.3\%$

    b. p(die) $= (0.25)(0.9) = 0.225$

    p(live) $= 1 - 0.225 = 0.775$

    c. p(good health returned) $= 1 - 0.02 = 0.98$

    d. Yes.

57. Both are co-dominant because it takes both to determine traits.

Chapter 3 Review

1. Experiment is pick one card from a deck of 52 cards. Sample space S = {possible outcomes} = {jack of hearts, ace of spades, . . .}; n(S) = 52.

3. $p = \frac{1}{4}$, odds 1:3 If you deal a card many times, you should expect to be dealt a club approximately 1/4 of the time, and you should expect to be dealt a club approximately one time for every 3 times you are dealt something else.

5. $p(Q \cup$ club$) = \frac{4}{13}$ odds 4:9 If you deal a card many times, you should expect to be dealt a queen or a club approximately 4/13 of the time, and you should expect to be dealt a queen or clubs approximately four times for every 9 times you are dealt something else.

7. a. Flip three coins and observe the results.

    b. S = {HHH, HHT, HTH, HTT, TTT, TTH, THT, THH}

9. F = {HTT, TTH, THT, TTT}

11. $p(F) = \frac{1}{2}$, odds 1:1. According to the Law of Large Numbers, if the experiment is repeated a large number of times, we should expect to get two or more tails about half the time. Also, we should expect to get two or more tails about 1 time for every time we get at least two heads.

13. $p = \frac{1}{6}$. According to the Law of Large Numbers, if the experiment is repeated a large number of times, we should expect to roll a 7 about one time out of every six rolls.

15. $p = \frac{7}{18}$. According to the Law of Large Numbers, if the experiment is repeated a large number of times, we should expect to roll a 7, an 11, or doubles about seven times out of every eighteen rolls.

17. p(3 or 5 or 7 or 9 or 10 or 11 or 12) $= \frac{11}{18}$. According to the Law of Large Numbers, if the experiment is repeated a large number of times, we should expect to roll a number that is either odd or greater than 8 about eleven times out of every eighteen rolls.

19. 0.013   **21.** 0.151   **23.** $\frac{12}{51}$

25. $\frac{1}{216}$ **27.** $\frac{2}{27}$ **29.** $\frac{1}{6}$

31. $\frac{1}{2}$. According to the Law of Large Numbers, if the experiment is repeated a large number of times, we should expect the offspring to be long-stemmed about half the time.

33. $\frac{1}{4}$ **35.** $\frac{1}{4}$ **37.** $\frac{1}{2}$ **39.** $\frac{1}{2}$ **41.** $\frac{1}{2}$

43. p(2 tens and 3 jacks) $\approx 9.23 \times 10^{-6}$

45. a.

Winning Spot	Probability
9	$_{20}C_9 \cdot {_{60}}C_0 / {_{80}}C_9 = 0.000000724$
8	$_{20}C_8 \cdot {_{60}}C_1 / {_{80}}C_9 = 0.000032592$
7	$_{20}C_7 \cdot {_{60}}C_2 / {_{80}}C_9 = 0.000591678$
6	$_{20}C_6 \cdot {_{60}}C_3 / {_{80}}C_9 = 0.005719558$
5	$_{20}C_5 \cdot {_{60}}C_4 / {_{80}}C_9 = 0.032601481$
4 or less	0.961053966

    b. $-$0.24; You would expect to lose $0.24.

47. $10.93

49. p(O'Neill|rural) ≈ 0.40; p(Bell|rural) ≈ 0.55

51. p(urban|Bell) ≈ 0.50; p(rural|Bell) ≈ 0.50

53. The urban residents prefer O'Neill. The rural residents prefer Bell.

55. Since the governors election depends only on who gets the most votes, O'Neill is ahead with 49.6% to Bell's $\frac{184 + 181}{800} \approx 45.6\%$.

57. Independent. Not mutually exclusive.

59. Dependent. Not mutually exclusive.

61. 14%   **63.** 8%   **65.** 43%   **67.** 0.78%

69. 1.817%   **71.** 43%

73. No, since p(AK) $= 0.39 \neq p$(AK|defective) $= 0.43$, they are not independent. Being made in Arkansas increases the probability of being defective.

75. Answers will vary.   **77.** Answers will vary.

79. The smallest chance of an event occurring is 0% (an impossible event). A certain event has the largest probability 100%.

81. Answers will vary.   **83.** Answers will vary.

85. Answers will vary.

CHAPTER 4 Statistics

4.1 Population, Sample, and Data

1. a.

Number of Visits	Frequency
1	9
2	8
3	2
4	5
5	6
	30

b. Library Habits - Number of Student Visits

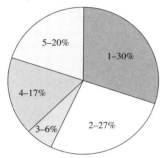

c.

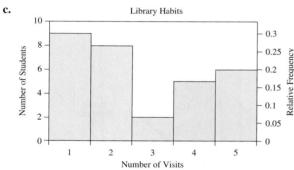

3. a.

Number of Children	Frequency
0	8
1	13
2	9
3	6
4	3
5	1
	40

b. Families in Manistee, Michigan

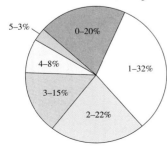
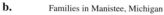

c.

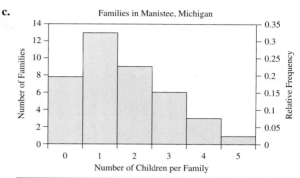

5. a.

x = Speed	Frequency	Relative Frequency
$51 \leq x < 56$	2	0.05
$56 \leq x < 61$	4	0.1
$61 \leq x < 66$	7	0.175
$66 \leq x < 71$	10	0.25
$71 \leq x < 76$	9	0.225
$76 \leq x < 81$	8	0.2
	40	

b.

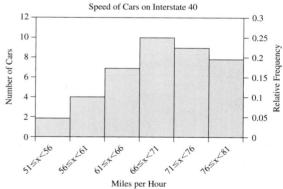

7.

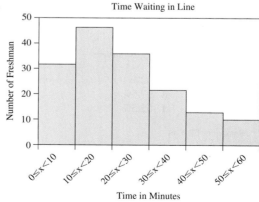

9.

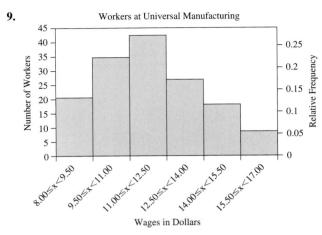

Workers at Universal Manufacturing

11.

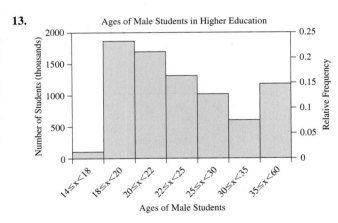

Women Giving Birth in 1997

13.

Ages of Male Students in Higher Education

15. a. 28.5%   **b.** 32%
  c. Not possible   **d.** 97.5%
  e. 50%   **f.** 52.5%

17.

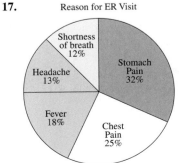

Reason for ER Visit

19. a.

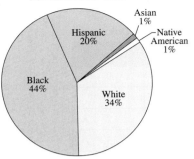

New AIDS Cases - Male

b.

New AIDS Cases - Female

c. Females have a higher percentage of blacks with new AIDS cases than males do. Males have a higher percentage of whites with new AIDS cases than females do. Both Asian and Native American males and females are less than 1%.

d.

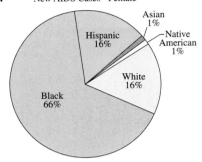

New AIDS Cases

21. a.

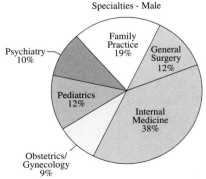

Specialties - Male

b.

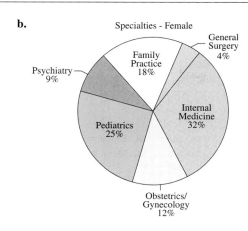

Specialties - Female

General Surgery 4%

Family Practice 18%

Psychiatry 9%

Internal Medicine 32%

Pediatrics 25%

Obstetrics/ Gynecology 12%

c. A higher percentage of males than females are general surgeons. A higher percentage of females than males are pediatricians.

d.

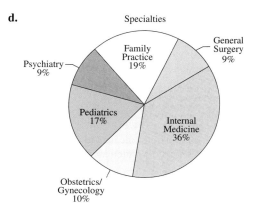

Specialties

General Surgery 9%

Family Practice 19%

Psychiatry 9%

Internal Medicine 36%

Pediatrics 17%

Obstetrics/ Gynecology 10%

23. Population–The set of all objects under study. Sample–Any subset of the population.

25. Frequency–the number of times a data point occurs. Relative Frequency–frequency expressed as a percent of the total number of data points. Relative Frequency Density–relative frequency divided by the number of data points in an interval. For group data, this allows the columns in the histogram to give a more truthful representation.

27. a. When the data consist of many nonrepeated data points, grouping is necessary.

   b. Advantages: easier.
   Disadvantages: less specific information.

29. a.

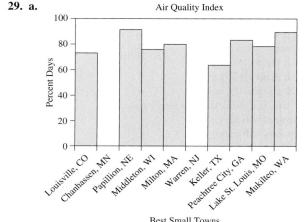

Air Quality Index

Percent Days

Best Small Towns

b. The air quality index (percent days AQI ranked good) is well over 50%, specifically from 65% to 92%.

31. a.

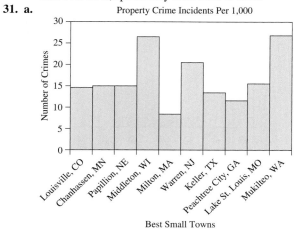

Property Crime Incidents Per 1,000

Number of Crimes

Best Small Towns

b. These small towns, considered the best places to live, have a very low rate of property crime incidents (less than 3%).

33. a. For comparison, two sets of histograms have been constructed.

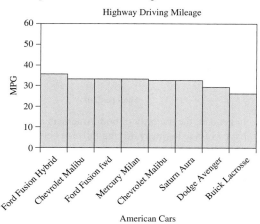

Highway Driving Mileage

MPG

American Cars

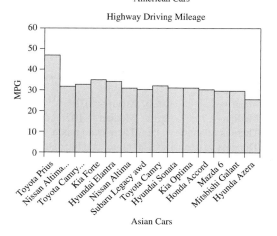

Highway Driving Mileage

MPG

Asian Cars

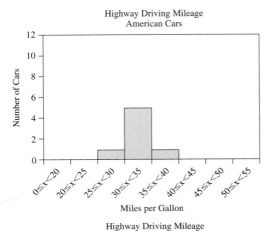

Highway Driving Mileage
American Cars

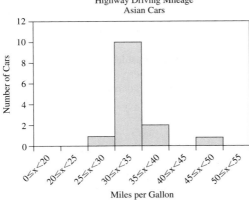

Highway Driving Mileage
Asian Cars

b. In general, the majority of both American and Asian cars have average highway driving mileage between 30 and 35 miles per gallon.

4.2 Measures of Central Tendency

1. Mean = 12.5
Median = 11.5; The mode is 9.

3. Mean = 1.45
Median = 1.5; The mode is 1.7.

5. a. Mean = 11   **b.** Mean = 25.5
Median = 10.5   Median = 10.5
The mode is 9.   The mode is 9.

  c. The mean is affected by the change. The median and mode stay the same.

7. a. Mean = 7   **b.** Mean = 107
Median = 7   Median = 107
There is no mode.   There is no mode.

  c. The data in part (b) is 100 more than the data in part (a).

  d. The mean and median are 100 more. Neither data set had a mode.

9. Mean = 3 : 14.   **11.** Mean = 9.85
Median = 3 : 14.   Median = 9.
The mode is 3 : 12.   The mode is 9 and 15.

13. Mean = 34.64
Median = 34.
Mode: 25, 33, 34, 37, 45, 46 (all have frequency = 2)

15. Mean = 7.6   **17.** Mean = 16.028 oz
Median = 8
The mode is 9 and 6.

19. a. 42   **b.** 92

  c. Impossible, it would require a score of 142.

21. Average = $51,600

23. Mean speed = 51.4 mph

25. 39 years old

27. a. mean = $40,750

  b. The median does not change: $42,000.

29. a. Envionmental Protection Agency; $10,780

  b. Department of the Army; $2,761

  c. mean = $9,627

  d. mean = $4,442

31. a. Mean = 35.7 years old   **b.** Mean = 36.5 years old

33. a. Mean = 23.0 years old   **b.** Mean = 28.5 years old

35. The mean of Group 3 is $\dfrac{S_1 + S_2}{n + m} \neq \dfrac{S_1}{n} + \dfrac{S_2}{m}$

No, the mean is not $\dfrac{A + B}{2}$.

37. Answers will vary.

4.3 Measures of Dispersion

1. a. Variance: $s^2 = 16.4$
Standard deviation: $s \approx 4.0$

  b. Variance: $s^2 = 16.4$
Standard deviation: $s \approx 4.0$

3. a. Variance: $s^2 = 0$

  b. Standard deviation: $s = 0$

5. a. Mean = 22
Standard deviation: $s \approx 7.5$

  b. Mean = 1,100
Standard deviation: $s \approx 374.2$

  c. The data in (b) are 50 times the data in (a).

  d. The mean and standard deviation are 50 times larger in (b).

7. a. Joey's mean = 168
Dee Dee's mean = 167
Joey's mean is higher.

  b. Joey: $s \approx 30.9$
Dee Dee: $s \approx 18.2$
Dee Dee's standard deviation is smaller.

  c. Dee Dee is more consistent than Joey because his standard deviation is lower.

9. $s \approx 6.74$ ounces ≈ 0.421 lb $= 0 : 07$

11. $s \approx 14.039$

13. a. $\bar{x} = 67.5$ **b.** 80%   **c.** 90%
$s \approx 7.1$

15. a. Mean = $\bar{x} = 3.175$ **b.** 58%   **c.** 100%
$s \approx 1.807$

17. a. Mean = 8   **b.** 71%   **c.** 94%   **d.** 100%
$s \approx 1.4$

19. $s \approx 0.285$ **21.** $s \approx 6.80$

23. a. If we didn't square the deviations, they would all cancel out.

  b. It makes it smaller.

  c. It makes it larger.

  d. It squares the units.

  e. Because the deviations were squared.

25. Answers will vary.

27. Answers will vary.

29. a. $\bar{x} = 80$

  b. Sample standard deviation would be more appropriate because this does not include all the cities.

  c. $s \approx 8.62$

  d. The air quality index averages 80.0% with 62.5% of all indices falling within one standard deviation of the mean air quality index.

31. a. $\bar{x} = 17.3$

   b. Sample standard deviation would be more appropriate because this does not include all the towns identified as best places to live.

   c. $s \approx 5.96$

   d. The property crime incidents averages 17.3 per 1,000 with 70% of all property crime incidents falling within one standard deviation of the mean property crime incidents.

33. a. American: $\bar{x} = \frac{261}{8} = 32.625; s \approx 2.83$

   Asian: $\bar{x} = \frac{463}{14} \approx 33.07; s \approx 4.94$

   Sample standard deviation was used as this does not include every car that could fall into this category.

   b. Asian cars have the highest mpg for city driving on the average, but it varies over 50% more than the average mpg varies for American cars.

4.4 The Normal Distribution

1. Yes, it looks like a bell curve.

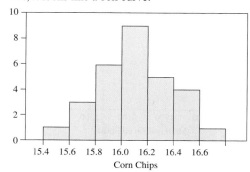

Corn Chips

3. No, it is right-tailed.

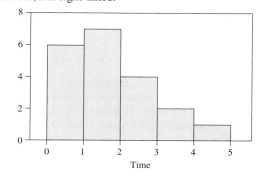

Time

5. a. 34.13%   **b.** 34.13%   **c.** 68.26%

7. a. 49.87%   **b.** 49.87%   **c.** 99.74%

9. a. One standard deviation: $[22.4, 27]$

   Two standard deviations: $[20.1, 29.3]$

   Three standard deviations: $[17.8, 31.6]$

   b. 68.26% lies in $[22.4, 27]$

   95.44% lies in $[20.1, 29.3]$

   99.74% lies in $[17.8, 31.6]$

   c.

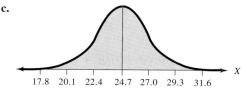

11. a. 0.4474   **b.** 0.0639   **c.** 0.5884

   d. 0.0281   **e.** 0.0960   **f.** 0.8944

13. a. 0.34   **b.** -2.08 **c.** 0.62

   d. -0.27 **e.** 1.64   **f.** -1.28

15. a. 0.42   **b.** -0.42 **c.** 2.08

   d. -1.46 **e.** 2.96   **f.** -2.21

17. a. 0.3413   **b.** 0.6272   **c.** 0.8997

   d. 0.9282   **e.** 0   **f.** 0.3300

19. a. 4.75%   **b.** 49.72%

21. a. 0.1908   **b.** 0.7745

23. a. 95.25%   **b.** 79.67%

25. a. About 87.   **b.** 62 or less.

27. 9.1 minutes or more

29. The normal distribution is a smooth, continuous, symmetric, bell-shaped curve. The frequencies of the data points nearer the center are increasingly higher than the frequencies of the data points far from the center.

31. Because area under the curve represents probability.

33. It is less than the mean.

35. No. Virtually all students are between 14 and 18 with roughly the same number of students at each age.

37. Gauss. He was trying to obtain the orbit of Ceres.

4.5 Polls and Margin of Error

1. a. $z = 0.68$ **b.** $z = 0.31$ **c.** $z = 1.39$ **d.** $z = 2.65$

3. a. $z = 1.44$ **b.** $z = 1.28$ **c.** $z = 1.15$

   d. $z = 2.575$ since it is halfway between 0.4949 and 0.4951

5. 1.75   **7.** 1.15

9. a. 1.8%   **b.** 2.6%

11. a. 75.0% ± 3.4%   **b.** 75.0% ± 4.0%

13. a. 79.0%   **b.** 21.0%   **c.** ± 1.6%

15. a. 86.0%   **b.** 70.0%

   c. For men: ± 2.8%

   For women: ± 2.5%

17. a. 71%   **b.** 29%   **c.** ± 0.3%

19. a. 45.0%   **b.** 47.0%   **c.** 8.0%   **d.** ± 0.2%

21. a. 36.0%   **b.** ± 1.6%   **c.** ± 2.3%

   d. The answer in (c) is larger than the answer in (b). In order to guarantee 98% accuracy, we must increase the error.

23. a. ± 4.7%   **b.** ± 3.5%   **c.** ± 2.8%

25. about 96.4%   **27.** about 95.3%

29. a. 1,068   **b.** 2,401   **c.** 9,604

   d. Smaller margin of errors require a larger sample size, and larger margin of errors may have a smaller sample size.

31. a. 1,509   **b.** 3,394   **c.** 13,573

   d. Smaller margin of errors require a larger sample size, and larger margin of errors may have a smaller sample size.

33. If x members in a sample of n have a certain characteristic, then sample proportion is $\frac{x}{n}$.

35. Decrease. Larger n values make the fraction in the MOE smaller.

37. Possible answer: Rather than selecting names and numbers of individuals to be called, computers are used to generate random sets of seven-digit numbers, which are then called the sample.

39. Answers will vary.   **41.** Answers will vary.

4.6 Linear Regression

1. a. $y = 1.0x + 3.5$ **b.** 14.5   **c.** 15.5

   d. 0.9529485   **e.** Yes, they are reliable since r is close to 1.

3. a. $y = -0.087x + 8.246$ **b.** 7.811   **c.** 14.322

   d. -0.095 **e.** No, they are not reliable since r is close to 0.

5. a. Yes, the ordered pairs seem to exhibit a linear trend.

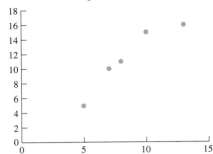

b. $y = 1.4x - 0.3$ **c.** 12.3

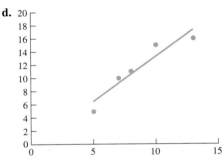

e. 0.9479459
f. Yes, the prediction is reliable since r is close to 1.

7. a. No, the ordered pairs do not seem to exhibit a linear trend.

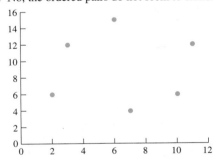

b. $y = 0.008x + 9.1$ **c.** 9.164

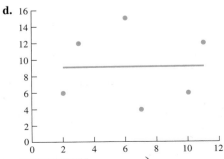

e. 0.0062782255
f. The prediction is not reliable since r is close to 0.

9. a. Yes, the data exhibit a linear trend.

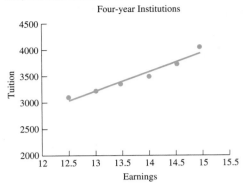

b. $y = 366.72x - 1,541.52$ **c.** \$3,867.55
d. \$15.11   **e.** 0.9732507334
f. Yes, r is close to 1.

11. a. Yes, the data exhibit a linear trend.

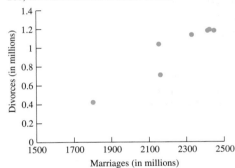

b. "y-hat" $= 1.1301x - 1.5526$ **c.** 1.555 million
d. 2.699 million   **e.** 0.93344427
f. Yes, they are reliable since r is close to 1.

13. A line of best fit is a line that best approximates the trend in a scatter of points. Given a set of points, we can calculate the slope and y-intercept by:
$$m = \frac{(\Sigma xy) - (\Sigma x)(\Sigma y)}{n(\Sigma x^2) - (\Sigma x)^2} \text{ and } b = \bar{y} - m\bar{x}, \text{ where } \bar{x} \text{ and } \bar{y} \text{ are}$$
means of the x- and y-coordinates, respectively.

15. If the value of y increases when the value of x increases, we say that there is a positive linear relation ($r > 0$). Example: $x =$ number of hours studying, $y =$ exam score.

17. Answers will vary.

19. a. $y = -24.0x + 269; r = -0.86$

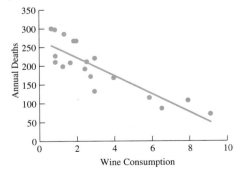

b. Yes
(2.5, 211), (3.9, 167), (2.9, 131), (2.4, 191), (2.9, 220), (0.8, 297), (9.1, 71), (0.8, 211), (0.6, 300), (7.9, 107), (1.8, 266), (1.9, 266), (0.8, 227), (6.5, 86), (1.6, 207), (5.8, 115), (1.3, 285), (1.2, 199), (2.7, 172)

Chapter 4 Review

1. **a.** 7.1   **b.** 7.5   **c.** 8   **d.** 2.1
3. **a.** 40%   **b.** 28%   **c.** 86%   **d.** 14%
   e. 50%   **f.** Cannot determine.
5. 100
7. **a.** Timo: mean = 98.8
      Henke: mean = 96.6
      Henke has a lower mean.
   b. Timo: $s \approx 6.4$
      Henke: $s \approx 10.4$
   c. Timo is more consistent because his standard deviation is lower.
9. **a.** Continuous   **b.** Neither   **c.** Discrete
   d. Neither   **e.** Discrete   **f.** Continuous
11. **a.** One standard deviation = $[71, 85]$
       Two standard deviations = $[64, 92]$
       Three standard deviations = $[57, 99]$
    b. 68.26% of the data lies in $[71, 85]$
       95.44% of the data lies in [64, 92]
       99.74% of the data lies in [57, 99]
    c.

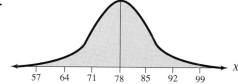

13. Cutoff at 401.
15. **a.** 66.7% ± 2.4%
    b. 66.7% ± 2.8%
17. **a.** about ± 4.1%
    b. about ± 3.1%
    c. about ± 2.5%
19. 1,414

21. **a.** Yes, the ordered pairs seem to exhibit a linear trend.

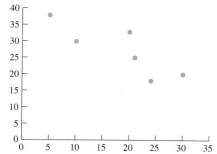

   b. $y = -0.70455x + 40.25$ **c.** 29.682
   d.

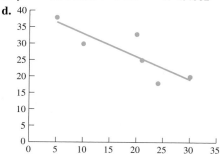

   e. −0.839839286
   f. Yes, the prediction is reliable since r is close to −1.
23. The mean is the average. The median is the middle. The mode is the most frequent.
25. A normal distribution is an ideal, bell-shaped curve. A standard normal distribution has mean 0 and standard deviation 1.
27. The strength is measured with the coefficient of linear correlation:
$$r = \frac{n(\Sigma xy) - (\Sigma x)(\Sigma y)}{\sqrt{n(\Sigma x^2) - (\Sigma x)^2}\sqrt{n(\Sigma y^2) - (\Sigma y)^2}}$$

CHAPTER 5   Finance

5.1 Simple Interest

1. **a.** 60 days   **b.** 61 days
3. **a.** 100 days   **b.** 101 days
5. $480   7. $25.24
9. **a.** $28.39   **b.** $28.87
11. $4,376.48   13. $13,750.47
15. $1,699.25   17. $6,162.51
19. $17,042.18   21. $6,692.61
23. $1,058.19   25. $1,255.76
27. 1,013 days   29. $233,027.39
31. $714.31   33. $222.86
35. **a.** $112,102.05   **b.** $2,102.05
37. Average daily balance ≈ $140.27
    Finance charge ≈ $2.08
39. Average daily balance ≈ $152.84
    Finance charge ≈ $2.73
41. **a.** $8,125.00   **b.** $146,250
    c. $8,125.00   **d.** $67.71
    e. $19,500.08   **f.** $136,500
    g. $156,000.08

43. **a.** $38,940   **b.** $311,520   **c.** $38,940
    d. $356.95   **e.** $95,013.60   **f.** $288,156
    g. $383,169.60
45. **a.** Average daily balance: $989.68
       Finance charge: $17.65
       New Balance: $997.65
    b. Average daily balance: $988.36
       Finance charge: $15.92
       New Balance: $993.57
    c. Average daily balance: $983.25
       Finance charge: $17.54
       New Balance: $991.11
    d. The minimum $20 payment is reducing your debt by about $3 each month. It will take you roughly 28 years to pay off your debt. This means you will have paid the credit card company roughly $6,720 on a $1,000 debt. (572% of your debt was total interest.)
47. **a.** Average daily balance: $979.35
       Finance charge: $17.47
       New Balance: $977.47

b. Average daily balance: $958.90
  Finance charge: $15.45
  New Balance: $952.92

c. Average daily balance: $932.27
  Finance charge: $16.63
  New Balance: $929.55

d. The change in the minimum required payment policy has had some impact, because you're reducing your debt by approximately $23 per month.

49. The Simple Interest Formula could be used for all. Future Value is Principal plus Interest. The Future Value Formula could also be used for all. Interest is Future Value minus Principal. We have both for ease of use.

51. Western Union. See 'Historical Note The History of Credit Cards,' page 338 for more information.

53. A New York lawyer.

55. Bank of America.

5.2 Compound Interest

1. a. 0.03   **b.** 0.01
   c. 0.000328767   **d.** 0.0046153846
   e. 0.005

3. a. 0.00775   **b.** 0.002583333
   c. 0.000084932   **d.** 0.001192307
   e. 0.001291667

5. a. 0.02425   **b.** 0.008083333
   c. 0.000265753   **d.** 0.003730769
   e. 0.004041667

7. a. 34 quarters   **b.** 102 months
   c. 3,102.5 days

9. a. 120 quarters   **b.** 360 months
   c. 10,950 days

11. a. $7,189.67
   b. After 15 years, the investment is worth $7,189.67.

13. a. $9,185.46
   b. After 8.5 years, the investment is worth $9,185.46.

15. a. $3,951.74
   b. After 17 years, the investment is worth $3,951.74.

17. a. 8.30%
   b. The given compound rate is equivalent to 8.30% simple interest.

19. a. 4.34%
   b. The given compound rate is equivalent to 4.34% simple interest.

21. a. 10.38%
   The given compound rate is equivalent to 10.38% simple interest.
   b. 10.47%
   The given compound rate is equivalent to 10.47% simple interest.
   c. 10.52%
   The given compound rate is equivalent to 10.52% simple interest.

23. a. $583.49
   b. One would have to invest $583.49 now to have the future value in the given time.

25. a. $470.54
   b. One would have to invest $470.54 now to have the future value in the given time.

27. a.

Month	$FV = P(1 + rt)$
1	$10,083.33
2	$10,167.36
3	$10,252.09
4	$10,337.52
5	$10,423.67
6	$10,510.53

b. $10,510.53

29. a.

Year	$FV = P(1 + rt)$
1	$15,900
2	$16,854
3	$17,865.24

b. $17,865.24

31. $4,032,299.13

33. He would have to invest at least $9,058.33.

35. a. $9,664.97
   b. The difference is the interest: $51.77
   c. $5,045.85

37. a. $19,741.51   **b.** $10,535.83

39. a. $8,108.87
   b. The difference is the interest: $690.65
   When Marlene retires, her account will not have exactly $100,000 in it, so we will not compute the monthly interest on this amount.

41. FNB: 9.55%
   CS: 9.45%
   First National Bank has the better offer.

43. a. $r \approx 3.45\%$; Does not verify.
   b. $r \approx 3.45\%$; Does not verify.   **c.** $r \approx 3.50\%$
   The annual yield for the 5-year certificate verifies if you prorate the interest over a 360-day year, and then pay interest for 365 days.

45. $r \approx 9.74\%$; Does not verify.

47. a. 5.12%   **b.** $1,025.26   **c.** $25.26   **d.** 2.53%
   e. Part (d) is for 6 months; part (a) is for 1 year.
   f. You earn interest on interest; that is what compounding means.

49. a. $387.10   **b.** $173.96   **c.** $40.71

51. $r = (1 + i)^n - 1$ **53.** 7.45%

55. a. 5.70%   **b.** 5.74%   **c.** 5.77%
   d. 5.79%   **e.** 5.78%   **f.** 5.78%

57. compound interest; $i = 5.00\%$ because
   $£1,000 \cdot (1 + 0.05)^{100} \approx £131,000$

59. 4.57%

63. It is simple interest per compound period.

65. $n = 1$ so the annual yield is equal to the interest rate.

67. Because it is simple interest. The difference reflects interest on interest for n periods in 1 year.

69. Because the future value is the principal plus a small amount of interest. Because the future value is the principal plus a large amount of interest.

71. They will be equal in the beginning and stay the same until the first interest payment to the compound interest account. After that, they will never be the same.

73. Answers will vary.

75. a. $2,000.19   **b.** $4,000.74
   c. $8,002.23   **d.** $16,005.95
   e. Every extra period of 5,061 days doubles the money, approximately.

77. a. 11.90 years   **b.** 10.24 years   **c.** 7.27 years
   d. The larger i is (the larger the interest rate), the shorter the doubling time.

79. To accumulate $25,000: 5.45 years
   To accumulate $100,000: 20.24 years

5.3 Annuities

1. $1,478.56   **3.** $5,422.51
5. a. $226.32   **b.** $226.32   **c.** $225   **d.** $1.32
7. a. $455.46   **b.** $455.46   **c.** $450   **d.** $5.46
9. a. $283,037.86   **b.** $54,600.00   **c.** $228,437.86
11. a. $251,035.03   **b.** $69,600.00   **c.** $181,435.03
13. a. $1,396.14
   b. You would have to invest a lump sum of $1,396.14 now instead of $120 per month.
15. a. $222.40
   b. You would have to invest a lump sum of $222.40 now instead of $75 per month.
17. a. $18,680.03
   b. You would have to invest a lump sum of $18,680.03 now instead of $175 per month.
19. $51.84
21. $33.93
23. $33.63
25. a. $537,986.93
   b.

	Beginning Balance	Interest	Withdrawal	Ending Balance
1	$537,986.93	$2,734.77	$650	$540,071.70
2	$540,071.70	$2,745.36	$650	$542,167.06
3	$542,167.06	$2,756.02	$650	$544,273.08
4	$544,273.08	$2,766.72	$650	$546,389.80
5	$546,389.80	$2,777.48	$650	$548,517.28

27. a. $175,384.62
   b. $111.84
29. a. $2,411.03
   b. $4,890.75
   c. $7,565.55
31. a. $1,484,909.47
   b. Shannon: $91,000
      Parents: $11,000
      Total contribution is $102,000.
   c. $1,382,909.47
   d. $979,650.96
   e. $643,631.10
33. $11,954.38 per year
35. $45.82

37. a. $595.83   **b.** $5,367.01
   c.

Period	Starting Balance	Interest	Deposit	Ending Balance
1	0	0	$5,367.01	$5,367.01
2	$5,367.01	$224.74	$5,367.01	$10,958.76
3	$10,958.76	$458.90	$5,367.01	$16,784.67
4	$16,784.67	$702.86	$5,367.01	$22,854.54
5	$22,854.54	$957.03	$5,367.01	$29,178.58
6	$29,178.58	$1,221.85	$5,367.01	$35,767.44
7	$35,767.44	$1,497.76	$5,367.01	$42,632.21
8	$42,632.21	$1,785.22	$5,367.01	$49,784.44
9	$49,784.44	$2,084.72	$5,367.01	$57,236.17
10	$57,236.17	$2,396.76	$5,367.01	$64,999.94

39. $P = pymt \cdot \dfrac{1 - (1 + i)^{-n}}{i}$ **41.** $1,396.14   **43.** $222.40

45. An annuity is based on compound interest in that each period; interest is calculated on the total balance, which includes principal payments and previous interest earned on those payments.

47. Both investments earn compound interest. The lump-sum investment has the advantage that no further payments are needed. Of course the lump-sum investment requires the investor to have a larger amount of money available for investment.

49. No. Because compound interest will make your investment grow to more than the necessary amount.

51. Answers will vary.   **53.** 447 months = 37 years 3 months

5.4 Amortized Loans

1. a. $125.62   **b.** $1,029.76
3. a. $193.91   **b.** $1,634.60
5. a. $1,303.32   **b.** $314,195.20
7. a. $378.35   **b.** $3,658.36
   c.

Payment Number	Principal Portion	Interest Portion	Total Payment	Balance
0				$14,502.44
1	$239.37	$138.98	$378.35	$14,263.07
2	$241.66	$136.69	$378.35	$14,021.41

9. a. $1,602.91   **b.** $407,047.60
   c.

Payment Number	Principal Portion	Interest Portion	Total Payment	Balance
0				$170,000.00
1	$62.28	$1,540.63	$1,602.91	$169,937.72
2	$62.85	$1,540.06	$1,602.91	$169,874.87

   d. $4,218.18 per month (assuming only home loan payment)

11. a. Dealer: $404.72
Bank: $360.29
b. Dealer: $4,597.24
Bank: $2,464.60
c. Choose the bank loan since the interest is less.
13. a. *pymt*: $354.29
Interest: $1,754.44
b. *pymt*: $278.33
Interest: $2,359.84
c. *pymt*: $233.04
Interest: $2,982.40
15. a. *pymt*: $599.55
Interest: $115,838
b. *pymt*: $665.30
Interest: $139,508
c. *pymt*: $733.76
Interest: $164,153.60
d. *pymt*: $804.62
Interest: $189,663.20
e. *pymt*: $877.57
Interest: $215,925.20
f. *pymt*: $952.32
Interest: $242,835.20
17. a. *pymt*: $877.57
Interest: $215,925.20
b. *pymt*: $404.89
Interest: $215,814.20
19. a. 30 year: $914.74
15 year: $1,066.97
Both verify, but not exactly. Their computations are actually more accurate than ours, since they are using more accurate round-off rules than we do.
b. 30 year: Total payments: $329,306.40
15 year: Total payments: $192,054.60
Savings: $137,251.80
21. a. $19,053.71
b.

Payment Number	Principal Portion	Interest Portion	Total Payment	Balance Due
0				$75,000.00
1	$18,569.33	$484.38	$19,053.71	$56,430.67
2	$18,689.26	$364.45	$19,053.71	$37,741.41
3	$18,809.96	$243.75	$19,053.71	$18,931.45
4	$18,931.45	$122.27	$19,053.72	$0

23. a. $23,693.67
b.

Payment Number	Principal Portion	Interest Portion	Total Payment	Balance Due
0				$93,000.00
1	$22,986.48	$707.19	$23,693.67	$70,013.52
2	$23,161.28	$532.39	$23,693.67	$46,852.24
3	$23,337.40	$356.27	$23,693.67	$23,514.84
4	$23,514.85	$178.81	$23,693.66	$0

25. a. $29.91
b. $797.30
c.

Payment Number	Principal Portion	Interest Portion	Total Payment	Balance
0				$6,243.00
1	$767.39	$29.91	$797.30	$5,475.61
2	$771.06	$26.24	$797.30	$4,704.55
3	$774.76	$22.54	$797.30	$3,929.79
4	$778.47	$18.83	$797.30	$3,151.32
5	$782.20	$15.10	$797.30	$2,369.12
6	$785.95	$11.35	$797.30	$1,583.17
7	$789.71	$7.59	$797.30	$793.46
8	$793.46	$3.80	$797.26	$0

d. $135.36
27. a. $89.25   **b.** $1,905.92
c.

Payment Number	Principal Portion	Interest Portion	Total Payment	Balance
0				$12,982.00
1	$1,816.67	$89.25	$1,905.92	$11,165.33
2	$1,829.16	$76.76	$1,905.92	$9,336.17
3	$1,841.73	$64.19	$1,905.92	$7,494.44
4	$1,854.40	$51.52	$1,905.92	$5,640.04
5	$1,867.14	$38.78	$1,905.92	$3,772.90
6	$1,879.98	$25.94	$1,905.92	$1,892.92
7	$1,892.92	$13.01	$1,905.93	$0

d. $359.45
29. $3,591.73   **31.** $160,234.64
33. a. $1,735.74   **b.** $151,437.74
c. $1,204.91   **d.** $472,016.40
e. $343,404.24
f. Yes, they should refinance. Why do you think so?
35. a. $1,718.80   **b.** $172,157.40
c. $240,354.60   **d.** $573,981.98
e. $1,200,543.86
f. The decision is between saving more for retirement or increasing their standard of living now. They should not prepay their loan.
If they prepay: $573,981.98
If not, 1,200,543.86 − 172,157.40 = $1,028,386.46
Even though they pay interest, they come out ahead.
37. a. $18,950   **b.** $151,600
c. $18,950   **d.** $1,501.28
e. $189.50
39. a. $271.61   **b.** $1,962.39
c. $1,501.28

41. Receiving $142,000 was not enough to pay off the loan. They did not make a profit but will have to pay $66,454.72.

43. a. $882.36
 b. The loan will be paid off sometime between the 279th and 280th payment, or 80 payments early, which is approximately 6.67 years.

45. The interest is a percentage of the principal.

47. They may need the interest figures for tax purposes. They may want to make extra payments.

49. No. The formula $pymt \dfrac{(1 + i)^n - 1}{i} = P(1 + i)^n$ shows that doubling n does not halve $pymt$.

51. Answers will vary.

53. a. $699.21   **b.** $559.97
 c. $1,670.88   **d.** $98,621.73
 e. 7.375%
    $n = 348$
 f. $687.65   **g.** $1,809.60
 h. A lower rate initially, but uncertainty later.

55. a. $474.54   **b.** $99,120.55   **c.** $602.07
 d. It is added because you owe more than you paid. This will make future payments higher.
 e. It is called negative amortization because the principal gets larger, not smaller.
 f. The fact that the interest rate can change without the monthly payment being adjusted.

57. a. $12,902.39   **b.** $768.35
 c. First year "SUM(C3:C14)"
    Last year "SUM(C171:C182)"

59. a. Now the payment is $1,529.57.   **b.** $12,165.10
 c. $724.43

61. a. $1,223.63   **b.** $13,070.29   **c.** $10,088.45
 d. The amount paid is borrowed over a longer period of time, so the total monthly payment should go down. However, the interest is not evenly spread over the loan period. The lenders collect more of their interest during the first years.

63. a. See student's spreadsheet
 b. $3,717.05   **c.** $554.10   **d.** $14,883.00

65. a. See student's spreadsheet   **b.** $13,490.99
 c. $748.06   **d.** $189,869.80

67. The answer is the same.

69. The answer is the not the same due to rounding.

71. a. $1,167.15   **b.** $1,584.02
 c. $10,769.32   **d.** $17,026.77
 e. The borrower may be able to afford a lower payment for a while and then a higher payment later. The lenders make up the difference in the potential loss of interest money by raising the rate later.

5.5 Annual Percentage Rate on a Graphing Calculator

1. 14.6%   **3.** 11.2%

5. a. $126.50
 b. $APR \approx 14.35\%$
    Verifies; this is within the tolerance.

7. a. $228.77
 b. $APR \approx 16.50\%$
    Doesn't verify; the advertised APR is incorrect.

9. Either one could be less expensive, depending on the A.P.R.

11. Really Friendly S and L will have lower payments but higher fees and/or more points.

13. $8,109.53

15. a. $819.38   **b.** $801.40
 c. $8,009.75   **d.** $9,496.12
 e. The RTC loan has a lower monthly payment but has higher fees.

17. The fees and points must be $0.

19. $246.95 \cdot \dfrac{\left(1 + \dfrac{r}{12}\right)^{24} - 1}{\dfrac{r}{12}} = 5{,}388\left(1 + \dfrac{r}{12}\right)^{24}$

You can't isolate r.
The interest rate is approximately 9.32%.

5.6 Payout Annuities

1. a. $143,465.15   **b.** $288,000
3. a. $179,055.64   **b.** $390,000
5. a. $96.26
 b. Pay in = $34,653.60
    Receive = $288,000
    Therefore, Suzanne receives $253,346.40 more than she paid.
7. a. $138.30   **b.** $331.39
9. a. $185,464.46   **b.** $14,000
 c. $14,560   **d.** $29,495.89
11. a. $36,488.32   **b.** $576,352.60
 c. $454.10   **d.** $1,079.79
13. a. $12,565.57   **b.** 10.47130674%
 c. $138,977.90   **d.** $61.48
15. a. $17,502.53   **b.** 9.2721727%
 c. $245,972.94   **d.** $372.99
17. $490,907.37
19. In a savings annuity, you make regular payments that gather interest. In a payout annuity, you make one large deposit and take monthly payments from it.
21. Answers will vary.

Chapter 5 Review

1. $8,730.15   **3.** $20,121.60
5. $4,067.71   **7.** $29,909.33
9. $23,687.72   **11.** $5,469.15
13. $6,943.26   **15.** 7.25%
17. $109,474.35   **19.** $275,327.05
21. a. $525.05   **b.** $6,503
23. $362,702.26
25. Average daily balance: $3,443.70
    $I \approx \$57.03$
27. a. $10,014.44   **b.** $1,692.71   **29.** $255,048.53
31. $37,738.26
33. a. $102,887.14
 b.

Month	Beginning	Interest Earned	Withdrawal	End
1	$102,887.14	$493.00	$1,000	$102,380.14
2	$102,380.14	$490.57	$1,000	$101,870.71
3	$101,870.71	$488.13	$1,000	$101,358.84
4	$101,358.84	$485.68	$1,000	$100,844.52
5	$100,844.52	$483.21	$1,000	$100,327.73

35. a. $1,246.53   **b.** $268,270.80
c.

Payment	Principal Portion	Interest Portion	Total Payment	Balance
0				$180,480
1	$137.33	$1,109.20	$1,246.53	$180,342.67
2	$138.17	$1,108.36	$1,246.53	$180,204.50

37. a. $268.14   **b.** $6,085.50
c.

Payment	Principal Portion	Interest Portion	Total Payment	Balance
0				$41,519.00
1	$5,817.36	$268.14	$6,085.50	$35,701.64
2	$5,854.93	$230.57	$6,085.50	$29,846.71
3	$5,892.74	$192.76	$6,085.50	$23,953.97
4	$5,930.80	$154.70	$6,085.50	$18,023.17
5	$5,969.10	$116.40	$6,085.50	$12,054.07
6	$6,007.65	$77.85	$6,085.50	$6,046.42
7	$6,046.42	$39.05	$6,085.47	$0.00

d. $1,079.50

39. a. $174.50
   b. $APR \approx 17.44\%$
      Does not verify; the advertised APR is incorrect.
41. a. $1,217.96
   b. $7,299.98
43. a. $58,409.10
   b. $881,764.23
   c. $556.55
45. Answers will vary.
47. An account that earns compound interest is earning interest on the account balance, which is the original principal and all previous interest. An annuity, earns compound interest on each periodic payment.
49. This is a law that requires lenders to disclose in writing the annual percentage rate of interest on a loan and other particular terms.
51. The first credit card, Western Union's, offered deferred payments on a customer's account and the assurance of prompt and courteous service. After World War II, credit cards provided a convenient means of paying restaurant, hotel, and airline bills.

CHAPTER 6   Voting and Apportionment

6.1 Voting Systems
1. a. 2,000   **b.** Cruz   **c.** Yes
3. a. 10,351   **b.** Edelstein   **c.** No
5. a. 30   **b.** Park   **c.** 47%   **d.** Beach wins.
   e. 53%   **f.** Beach wins.   **g.** 63 pts
   h. Beach wins.   **i.** 2 pts.
7. a. 65   **b.** Coastline wins.   **c.** 48%
   d. Coastline wins.   **e.** 63%   **f.** Coastline wins.
   g. 150 pts.   **h.** Coastline wins.   **i.** 2 pts.
9. a. 140   **b.** Shattuck wins.   **c.** 40%
   d. Nirgiotis wins.   **e.** 51%   **f.** Shattuck wins.
   g. 284 pts.   **h.** Nirgiotis wins.   **i.** 2 pts.
11. a. 1,342   **b.** Jones wins.   **c.** 44%
   d. Jones wins.   **e.** 50.4%   **f.** Jones wins.
   g. 3,960 pts.   **h.** Jones wins.   **i.** 3 pts.
13. a. 31,754   **b.** Darter wins.   **c.** 52%
   d. Darter wins   **e.** 52%   **f.** Darter wins.
   g. 138,797 pts.   **h.** Darter wins.   **i.** 4 pts.
15. 720   **17.** 15   **19. a.** 75   **b.** 25
21. a. 6   **b.** 0
23. a. A has the majority of the first-place votes and therefore should win.
   b. B wins.
   c. Yes, A received 7 votes for first choice, which is a majority of the 13 votes, but A did not win.
25. a. C wins by getting a majority of first-choice votes.
   b. Using Figure 6.38, B wins the instant runoff.
   c. Yes. C won originally and votes were changed in favor of C, but B won the instant runoff.

27. More than half.
29. A table that shows all possible rankings of the candidates and how many voters chose each ranking.
31. It is mathematically impossible to create any system of voting (involving three or more choices of candidates) that satisfies all four fairness criteria. See answer 30.

6.2 Methods of Apportionment
1. a. See table.   **b.** Standard divisor: 50.00
   c. See table.

State	A	B	C	Total
Population (thousands)	900	700	400	2,000
Std q ($d = 50.00$)	18.00	14.00	8.00	—
Lower q	18	14	8	40
Upper q	19.00	15.00	9.00	43

3. a. See table.   **b.** Standard divisor: 2.64
   c. See table.

State	NY	PA	NJ	Total
Population (millions)	18.977	12.281	8.414	39.672
Std q ($d = 2.64$)	7.19	4.65	3.19	—
Lower q ($d = 2.64$)	7	4	3	14
Upper q ($d = 2.64$)	8	5	4	17

5.

State	A	B	C	Total
Seats Hamilton	18	14	8	40

7.

State	NY	PA	NJ	Total
Seats Hamilton	7	5	3	15

9.

School	A	B	C	D	Total
Calculators Hamilton	69	53	43	35	200

11.

Region	N	S	E	W	Total
Seats Hamilton	3	8	3	10	24

13.

Country	D	F	I	N	S	Total
Seats Hamilton	5	4	0	4	7	20

15.

Country	C	E	G	H	N	P	Total
Seats Hamilton	3	4	9	4	3	2	25

17.

	State	A	B	C	Total
a.	Seats Jefferson	18	14	8	40
b.	Seats Adams	18	14	8	40
c.	Seats Webster	18	14	8	40

19.

	State	NY	PA	NJ	Total
a.	Seats Jefferson	7	5	3	15
b.	Seats Adams	7	5	3	15
c.	Seats Webster	7	5	3	15

21.

	School	A	B	C	D	Total
a.	Seats Jefferson	69	53	43	35	200
b.	Seats Adams	69	53	43	35	200
c.	Seats Webster	69	53	43	35	200

23.

	Region	N	S	E	W	Total
a.	Seats Jefferson	3	8	3	10	24
b.	Seats Adams	3	7	4	10	24
c.	Seats Webster	3	8	3	10	24

25.

	Country	D	F	I	N	S	Total
a.	Seats Jefferson	4	4	0	4	8	20
b.	Seats Adams	4	4	1	4	7	20
c.	Seats Webster	5	4	0	4	7	20

27.

	Country	C	E	G	H	N	P	Total
a.	Seats Jefferson	2	4	10	4	3	2	25
b.	Seats Adams	3	4	8	4	4	2	25
c.	Seats Webster	3	4	9	4	3	2	25

29.

State	A	B	C	Total
Seats Hill-Huntington	18	14	8	40

31.

State	NY	PA	NJ	Total
Seats Hill-Huntington	7	5	3	15

33.

School	A	B	C	D	Total
Calculators Hill-Huntington	69	53	43	35	200

35.

Region	N	S	E	W	Total
Seats Hill-Huntington	3	8	3	10	24

37.

Country	D	F	I	N	S	Total
Seats Hill-Huntington	4	4	1	4	7	20

39.

Country	C	E	G	H	N	P	Total
Seats Hill-Huntington	3	4	9	4	3	2	25

41. The satellite campus should receive the new instructor.

43. Banach school should receive the new instructor.

45. The hypothetical Delaware has one more and Virginia has one fewer than the actual. See Table below.

State	Seats Actual	45. Seats Hamilton	46. Seats Adams	47. Seats Webster	48. Seats Hill-Huntington
VA	19	18	18	18	18
MA	14	14	14	14	14
PA	13	13	12	13	12
N.C.	10	10	10	10	10
N.Y.	10	10	10	10	10
MD	8	8	8	8	8
CT	7	7	7	7	7
S.C.	6	6	6	6	6
N.J.	5	5	5	5	5
N.H.	4	4	4	4	4
VT	2	2	3	2	3
GA	2	2	2	2	2
KY	2	2	2	2	2
RI	2	2	2	2	2
DE	1	2	2	2	2
Total	105	105	105	105	105

47. Virginia has one fewer and Delaware has one more than the actual. See table in answer 45.

49. A method of dividing and distributing

51. A divisor close to the standard divisor

53. Quota close to standard quota

55. Hamilton's method

57. President George Washington vetoed Congress and the Hamilton method.

59. Method of Equal Proportions Hill method

6.3 Flaws of Apportionment

1. a. See table. **b.** Standard divisor: 0.77 **c.** See table.

d.

State	A	B	C	Total	Comments
Population (millions)	3.5	4.2	16.8	24.5	Part (a)
Std q ($d = 0.77$)	4.55	5.45	21.82	—	Part (c)
Lower q ($d = 0.77$)	4	5	21	30	Part (c)
Upper q ($d = 0.77$)	5	6	22	33	Part (c)
Modified lower q ($d = 0.73$)	4	5	23	32	d. Jefferson

e. Yes. C has 23 seats, which is neither upper nor lower quota.

3. a. See table. **b.** Standard divisor: 0.31 **c.** See table.

d.

State	A	B	C	Total	Comments
Population (millions)	3.5	4.2	16.8	24.5	Part (a)
Std q ($d = 0.31$)	11.29	13.55	54.19	—	Part (c)
Lower q ($d = 0.31$)	11	13	54	78	Part (c)
Upper q ($d = 0.31$)	12	14	55	81	Part (c)
Modified upper q ($d = 0.317$)	12	14	53	79	d. Adams

e. Yes. C has 53 seats, which is neither upper nor lower quota.

5. a. See table.   **b.** Standard divisor: 0.21   **c.** See table.   **d.** (Note: For answer, delete line 6, "Rounded q . . .".)

State	A	B	C	D	Total	Comments
Population (millions)	1.2	3.4	17.5	19.4	41.5	Part (a)
Std q ($d = 0.21$)	5.71	16.19	83.33	92.38	—	Part (c)
Lower q ($d = 0.21$)	5	16	83	92	196	Part (c)
Upper q ($d = 0.21$)	6	17	84	93	200	Part (c)
Rounded q ($d = 0.21$)	6	16	83	92	197	
Modified rounded q ($d = 0.20715$)	6	16	84	94	200	d. Webster

e. Yes. D has 94 seats, which is neither upper nor lower quota.

7. a. See table.   **b.** Standard divisor: 0.21   **c.** See table.
d. (Note: For answer, delete line 6 and line 7, "Geometric mean . . ." and "Rounded q . . .")

State	A	B	C	D	Total	Comments
Population (millions)	1.2	3.4	17.5	19.4	41.5	Part (a)
Std q ($d = 0.21$)	5.71	16.19	83.33	92.38	—	Part (c)
Lower q ($d = 0.21$)	5	16	83	92	196	Part (c)
Upper q ($d = 0.21$)	6	17	84	93	200	Part (c)
Geometric mean	5.48	16.49	83.50	92.50	—	
Rounded q ($d = 0.21$)	6	16	83	92	197	
Modified rounded q ($d_m = 0.2085$)	6	16	84	94	200	d. H–H

e. Yes. D has 94 seats, which is neither upper nor lower quota.

9. a. See table.   **b.** Standard divisor: 106.67   **c.** See table.   **d.** New standard divisor: 105.79

	State	A	B	C	Total
a.	Population (thousands)	690	5700	6410	12800
c.	Seats Hamilton ($d = 106.67$)	7	53	60	120
d.	Seats Hamilton (d = 105.79)	6	54	61	121

e. Yes. State A has lost a seat at the expense of the two larger states even though its population didn't change.

11. a. See table.   **b.** Standard divisor: 45.45   **c.** See table.

	State	A	B	C	Total
a.	Population (thousands)	1056	1844	2100	5000
c.	Seats Hamilton ($d = 45.45$)	23	41	46	110

d. Add 53 new seats.
e. New standard divisor: 45.64

	State	A	B	C	D	Total
a.	Population (thousands)	1056	1844	2100	2440	7440
e.	Seats Hamilton ($d = 45.64$)	23	40	46	54	163

f. Yes, apportionment changes. State B is altered.

13. a.

	Campus	A	B	C	Total
a.	2005 Specialists Hamilton	1	3	7	11
b.	2006 Specialists Hamilton	2	3	6	11
c.	% increase	5.1	10.6	6.25	
	Change in seats	+1	0	−1	

d. Yes. C lost a seat to A, but C grew at a faster rate.

15. The apportionment of a group should equal either the lower or upper quota of the group.

17. Adding a new state and the corresponding number of seats on the basis of its population alters the apportionment of the other states.

19. Satisfying the quota rule implies the existence of paradoxes. See answers 15–18.

Chapter 6 Review

1. a. 74   **b.** Beethoven wins.   **c.** 34%   **d.** Vivaldi wins.   **e.** 59%   **f.** Beethoven wins.
g. Beethoven received 197 points.   **h.** There is a tie between Beethoven and Vivaldi.   **i.** They each received 2 points.
3. See Table.

	State	A	B	C	Total
3.	Seats Hamilton	15	29	32	76
4.	Seats Jefferson	15	29	32	76
5.	Seats Adams	15	29	32	76
6.	Seats Webster	15	29	32	76
7.	Seats Hill-Huntington	15	29	32	76

5. See table above.   **7.** See table above.

	Country	A	B	C	P	U	Total
8.	Seats Hamilton	13	3	6	2	1	25
9.	Seats Jefferson	14	3	5	2	1	25
10.	Seats Adams	13	3	5	2	2	25
11.	Seats Webster	14	3	5	2	1	25
12.	Seats Hill-Huntington	14	3	5	2	1	25

9. See table above.   **11.** See table above.   **13.** Napier school should get the new instructor.
15. a. See table.   **b.** Standard divisor: 0.28   **c.** See table.
d. See the apportionment shown in the table. (Note: For answer, delete top part only of line 6, "Modified std....")

State	A	B	C	Total	Comments
Population (millions)	1.6	3.5	15.3	20.4	Part (a)
Std q ($d = 0.28$)	5.71	12.5	54.64	—	Part (c)
Lower q ($d = 0.28$)	5	12	54	71	Part (c)
Upper q ($d = 0.28$)	6	13	55	74	Part (c)
Modified std q ($d_m = 0.29$)	5.52	12.07	52.76	—	Use $d_m > d$
Upper q	6	13	53	72	**Adams**

e. Yes. C has 53 seats, which is neither an upper nor a lower quota.

17. a. See table.   **b.** Standard divisor: 0.21   **c.** See table.
   d. See the apportionment shown in the table. (Note: For answer, delete line 6 and line 7, "Geometric mean..." and "Rounded q...".
   Also, delete top part <u>only</u> of line 8, "Modified std q...")

State	A	B	C	D	Total	Comments
Population (millions)	1.100	3.500	17.600	19.400	41.6	Part (a)
Std q ($d = 0.21$)	5.24	16.67	83.81	92.38	—	Part (c)
Lower q ($d = 0.21$)	5	16	83	92	196	Part (c)
Upper q ($d = 0.21$)	6	17	84	93	200	Part (c)
Geometric mean ($d = 0.21$)	5.48	16.49	83.50	92.50	—	
Rounded q ($d = 0.21$)	5	17	84	92	198	Use $d_m < d$
Modified std q ($d_m = 0.208$)	5.29	16.83	84.62	93.27		Part (d)
Rounded q	5	17	85	94	201	H–H

   e. Yes. Neither C nor D has a number of seats that is either an upper or lower quota.

19. a. See table.   **b.** Standard divisor: 45.45   **c.** See table.

	State	A	B	C	Total
a.	Population (thousands)	1057	1942	2001	5000
c.	Seats Hamilton ($d = 45.45$)	23	43	44	110

   d. Add 53 seats.
   e. New standard divisor: 45.71

	State	A	B	C	D	Total
a.	Population (thousands)	1057	1942	2001	2450	7450
e.	Seats Hamilton ($d = 45.71$)	23	42	44	54	163

   f. Yes. The apportionment changed.
21. More than half.
23. A table that shows all possible rankings of the candidates and how many voters chose each ranking.
25. States population divided by standard divisor
27. The apportionment of a group should equal either the lower or upper quota of the group.
29. Adding a new state and the corresponding number of seats on the basis of its population alters the apportionment of the other states.
31. Satisfying the quota rule implies the existence of paradoxes. See answers 27–30.
33. It is mathematically impossible to create any system of voting (involving three or more choices of candidates) that satisfies all four
   fairness criteria. See answer 32.
35. Method of Equal Proportions Hill method

CHAPTER 7   Number Systems and Number Theory

7.1 Place Systems
1. $891 = 8 \cdot 10^2 + 9 \cdot 10^1 + 1 \cdot 10^0$ or $8 \cdot 10^2 + 9 \cdot 10 + 1$
3. $3{,}258 = 3 \cdot 10^3 + 2 \cdot 10^2 + 5 \cdot 10^1 + 8 \cdot 10^0$
   or $3 \cdot 10^3 + 2 \cdot 10^2 + 5 \cdot 10 + 8$
5. $372_8 = 3 \cdot 8^2 + 7 \cdot 8^1 + 2 \cdot 8^0$ or $3 \cdot 8^2 + 7 \cdot 8 + 2$
7. $3592_{16} = 3 \cdot 16^3 + 5 \cdot 16^2 + 9 \cdot 16^1 + 2 \cdot 16^0$
   or $3 \cdot 16^3 + 5 \cdot 16^2 + 9 \cdot 16 + 2$
9. $ABCDE0_{16} = 10 \cdot 16^5 + 11 \cdot 16^4 + 12 \cdot 16^3 + 13 \cdot 16^2 +$
   $14 \cdot 16^1 + 0 \cdot 16^0$ or $10 \cdot 16^5 + 11 \cdot 16^4 + 12 \cdot 16^3 +$
   $13 \cdot 16^2 + 14 \cdot 16$
11. $1011001_2 = 1 \cdot 2^6 + 0 \cdot 2^5 + 1 \cdot 2^4 + 1 \cdot 2^3 + 0 \cdot 2^2 +$
   $0 \cdot 2^1 + 1 \cdot 2^0$ or $2^6 + 2^4 + 2^3 + 1$
13. 1324_2 does not exist because base two only uses numerals
   0 and 1.

15. $5,32,85_{60}$ does not exist because base 60 only uses numerals
   0 through 59.
17. $4312_5 = 4 \cdot 5^3 + 3 \cdot 5^2 + 1 \cdot 5^1 + 2 \cdot 5^0$
   or $4 \cdot 5^3 + 3 \cdot 5^2 + 1 \cdot 5 + 2$
19. $123_4 = 1 \cdot 4^2 + 2 \cdot 4^1 + 3 \cdot 4^0$ or $1 \cdot 4^2 + 2 \cdot 4 + 3$
21. 250   **23.** 13,714   **25.** 11,259,360
27. 89   **29.** $5,32,85_{60}$ does not exist.
31. 582   **33.** 27   **35.** 965
37. 219   **39.** 11,982,839   **41.** 22,424
43. 3,732   **45.** 59,714   **47.** 136
49. 100101010_2 **51.** 1010101100000_2
53. 12_8 **55.** 266_8 **57.** 101001110100010_2
59. 11101010110000_2 **61.** 14_{16}
63. BA_{16} **65. a.** 1010111010_2 **b.** 1272_8

67. 22_{16} **69.** 160_{16} **71.** 5270_8
73. 100035_8 **75.** 1,332   **77.** 119,665
79. $6,4_{60}$ **81.** $36,4,5_{60}$ **83.** 1162_7
85. 221_5 **87. a.** The base is 3.   **b.** The base is 5.
89. a. The base is 8.   **b.** The base is 9.
91. a. 0, 1, 2, 3, 4   **b.** 5^0 5^1 5^2 5^3
   c. 0_5 1_5 2_5 3_5 4_5
   10_5 11_5 12_5 13_5 14_5
   20_5 21_5 22_5 23_5 24_5
   30_5 31_5 32_5 33_5 34_5
   40_5 41_5 42_5 43_5 44_5
93. a. 0, 1, 2, 3, 4, 5, 6, 7, 8, 9, A   **b.** 11^0 11^1 11^2 11^3
   c. 0_{11} 1_{11} 2_{11} 3_{11} 4_{11} 5_{11} 6_{11} 7_{11} 8_{11} 9_{11} A_{11}
   10_{11} 11_{11} 12_{11} 13_{11} 14_{11} 15_{11} 16_{11} 17_{11} 18_{11} 19_{11} $1A_{11}$
   20_{11} 21_{11} 22_{11}
95. 16-bit color has $2^{16} = 65,536$ shades, and 24-bit color has $2^{24} = 16,777,216$ shades.
97. Answers may vary. Possible answer:
   Base 10 (decimal): The Indian culture developed the decimal system. The use of 10 digits for a numbering system may be seen to arise from counting on our 10 fingers. Count on your fingers up to ten, put a mark in the sand and continue counting on fingers.
   Base 20: The use of 20 as a grouping number was used by many cultures throughout our earth's history. This was because people have twenty digits (fingers and toes). The Mayan culture developed a rigorous numbering system that was consistent with place values being multiples of 20.
99. Convert 531 to base seven.
   $531 = 1 \cdot 7^3 + 3 \cdot 7^2 + 5 \cdot 7^1 + 6 \cdot 7^0$
   So, it would appear with 6 beads raised on the rightmost wire, then 5 beads raised on the next wire, then 3 beads on the next wire and finally 1 bead raised on the fourth wire.
   http://mathforum.org/library/drmath/view/57564.html
101. Answers may vary. Possible answer: In a place system the value of a digit is determined by its place. This system is used on an abacus because the value of the number of beads raised on the wire is determined by which wire it is.
103. Answers may vary. Possible answer: Arab scholars translated many Greek and Hindu works in mathematics and the sciences. The Arab mathematician Mohammed ibn Musa al-Khowarizmi wrote two important books. See Exercises 104 and 105.
105. The title of the first book is the origin of the word algorithm. The title of the second book is the origin of the word algebra.

7.2 Addition and Subtraction in Different Bases
1. 12_8 **3.** 11_8 **5.** 10_{16}
7. 14_{16} **9.** 10_2 **11.** 110_2
13. a. 105_8 **15. a.** 751_8 **17. a.** 124_{16}
19. a. 1889_{16} **21. a.** 1011_2 **23. a.** 11010011_2
25. 4_8 **27.** 142_{16} **29.** 0011_2 or 11_2
31. a. 7_8 **33. a.** AE_{16} **35. a.** 00101_2 or 101_2
37. a. 43_5
39. Answers may vary. Possible answer:
   When using an abacus to solve problems of addition and subtraction, the process can often be quite straightforward and easy to understand. In some cases, the beads are either added or subtracted as needed. But what happens when an operator is presented with a situation where rods don't contain enough

beads to complete addition or subtraction problems in a simple, straightforward manner? This is where one technique employed by the operator is the use of complementary numbers with to 5 and 10.
   - In the case of 5; the operator uses two groups of complementary numbers: 4 & 1 and 3 & 2
   - In the case of 10, the operator uses five groups of complementary numbers: 9 & 1, 8 & 2, 7 & 3, 6 & 4, 5 & 5
   With time and practice using complementary numbers becomes effortless and mechanical. Once these techniques are learned, a good operator has little difficulty in keeping up with (even surpassing) someone doing the same addition and subtraction work on an electronic calculator.
   http://webhome.idirect.com/~totton/abacus/pages.htm
   Example: 38 + 99
   Step 1: Set 38 on the first (ones) and second (tens) rods.
   Step 2: To add 99, first consider adding the 9 in the tens place. Since there are not 9 beads to add to the tens rod, subtract 9's complement (1) from the tens rod and add 1 to the third rod (hundreds).
   Step 3: Now add the 9 in the ones place. Once again, since there are not 9 beads to add to the ones rod, subtract 9's complement (1) from the ones rod and add 1 to the second rod (tens).
   Step 4: Read the result, 137.
41. Answers may vary. In subtraction, always add the complement.
   Example: 251 − 89:
   Step 1: Set 251 on the first (ones) and second (tens) and third (hundreds) rods.
   Step 2: To subtract 89, first consider subtracting the 8 in the tens place. Subtract 1 bead from the hundreds rod and add 8's complementary (2) to the tens rod.
   Step 3: Now subtract the 9 in the ones place. Once again, subtract 1 bead from the tens rod and add 9's complement (1) to the ones rod.
   Step 4: Read the result, 162.

7.3 Multiplication and Division in Different Bases
1. 74_8 **3.** $3C5_{16}$
5. 1111_2 **7.** 2122_4
9. a. 151_8 **11. a.** 2442_{16}
13. a. 100011110_2 **15. a.** 1313_5
17. 23_8 **19. a.** $6F_{16}$
21. a. 4_8 **23. a.** 10_8 Rem 3_8
25. a. 1_8 Rem 125_8 **27. a.** $22C_{16}$ Rem 11_{16}
29. a. 39_{16} Rem $1B3_{16}$ **31. a.** 10001_2 Rem 1_2
33. a. 1001_2 Rem 11_2
35.

	1_5	2_5	3_5	4_5	10_5	11_5
1_5	1_5	2_5	3_5	4_5	10_5	11_5
2_5	2_5	4_5	11_5	13_5	20_5	22_5
3_5	3_5	11_5	14_5	22_5	30_5	33_5
4_5	4_5	13_5	22_5	31_5	40_5	44_5
10_5	10_5	20_5	30_5	40_5	100_5	110_5
11_5	11_5	22_5	33_5	44_5	110_5	121_5

37.

	1_{14}	2_{14}	3_{14}	4_{14}	5_{14}	6_{14}	7_{14}	8_{14}	9_{14}	A_{14}	B_{14}	C_{14}	D_{14}	10_{14}	11_{14}
1_{14}	1_{14}	2_{14}	3_{14}	4_{14}	5_{14}	6_{14}	7_{14}	8_{14}	9_{14}	A_{14}	B_{14}	C_{14}	D_{14}	10_{14}	11_{14}
2_{14}	2_{14}	4_{14}	6_{14}	8_{14}	A_{14}	C_{14}	10_{14}	12_{14}	14_{14}	16_{14}	18_{14}	$1A_{14}$	$1C_{14}$	20_{14}	22_{14}
3_{14}	3_{14}	6_{14}	9_{14}	C_{14}	11_{14}	14_{14}	17_{14}	$1A_{14}$	$1D_{14}$	22_{14}	25_{14}	28_{14}	$2B_{14}$	30_{14}	33_{14}
4_{14}	4_{14}	8_{14}	C_{14}	12_{14}	16_{14}	$1A_{14}$	20_{14}	24_{14}	28_{14}	$2C_{14}$	32_{14}	36_{14}	$3A_{14}$	40_{14}	44_{14}
5_{14}	5_{14}	A_{14}	11_{14}	16_{14}	$1B_{14}$	22_{14}	27_{14}	$2C_{14}$	33_{14}	38_{14}	$3D_{14}$	44_{14}	49_{14}	50_{14}	55_{14}
6_{14}	6_{14}	C_{14}	14_{14}	$1A_{14}$	22_{14}	28_{14}	30_{14}	36_{14}	$3C_{14}$	44_{14}	$4A_{14}$	52_{14}	58_{14}	60_{14}	66_{14}
7_{14}	7_{14}	10_{14}	17_{14}	20_{14}	27_{14}	30_{14}	37_{14}	40_{14}	47_{14}	50_{14}	57_{14}	60_{14}	67_{14}	70_{14}	77_{14}
8_{14}	8_{14}	12_{14}	$1A_{14}$	24_{14}	$2C_{14}$	36_{14}	40_{14}	48_{14}	52_{14}	$5A_{14}$	64_{14}	$6C_{14}$	76_{14}	80_{14}	88_{14}
9_{14}	9_{14}	14_{14}	$1D_{14}$	28_{14}	33_{14}	$3C_{14}$	47_{14}	52_{14}	$5B_{14}$	66_{14}	71_{14}	$7A_{14}$	85_{14}	90_{14}	99_{14}
A_{14}	A_{14}	16_{14}	22_{14}	$2C_{14}$	38_{14}	44_{14}	50_{14}	$5A_{14}$	66_{14}	72_{14}	$7C_{14}$	88_{14}	94_{14}	$A0_{14}$	AA_{14}
B_{14}	B_{14}	18_{14}	25_{14}	32_{14}	$3D_{14}$	$4A_{14}$	57_{14}	64_{14}	71_{14}	$7C_{14}$	89_{14}	96_{14}	$A3_{14}$	$B0_{14}$	BB_{14}
C_{14}	C_{14}	$1A_{14}$	28_{14}	36_{14}	44_{14}	52_{14}	60_{14}	$6C_{14}$	$7A_{14}$	88_{14}	96_{14}	$A4_{14}$	$B2_{14}$	$C0_{14}$	CC_{14}
D_{14}	D_{14}	$1C_{14}$	$2B_{14}$	$3A_{14}$	49_{14}	58_{14}	67_{14}	76_{14}	85_{14}	94_{14}	$A3_{14}$	$B2_{14}$	$C1_{14}$	$D0_{14}$	DD_{14}
10_{14}	10_{14}	20_{14}	30_{14}	40_{14}	50_{14}	60_{14}	70_{14}	80_{14}	90_{14}	$A0_{14}$	$B0_{14}$	$C0_{14}$	$D0_{14}$	100_{14}	110_{14}
11_{14}	11_{14}	22_{14}	33_{14}	44_{14}	55_{14}	66_{14}	77_{14}	88_{14}	99_{14}	AA_{14}	BB_{14}	CC_{14}	DD_{14}	110_{14}	121_{14}

39. 133_5 Rem 11_5
41. $B9_{14}$ Rem 48_{14}

7.4 Prime Numbers and Perfect Numbers

1. a. $2 \cdot 3 \cdot 7$ **b.** composite
3. a. 23   **b.** prime
5. a. $2 \cdot 3^3$ **b.** composite
7. composite
9. composite
11. prime
13. The prime numbers are 29, 31, 37, 41, 43, 47.
15. Since $17 = 17 \cdot 1$, it is prime.
17. Since $65 = 5(13)$, it is composite.
19. Since $\frac{15}{3} = 5$, it is composite.
21. Since $\frac{511}{7} = 73$, it is composite.
23. The proper factors are 1, 2, 3, 6, 7, 14, 21.
25. The proper factors are 1, 2, 3, 6, 9, 18, 27.
27. Since $1 + 2 + 3 + 6 + 7 + 14 + 21 = 54 > 42$, it is abundant.
29. Since $1 + 2 + 3 + 6 + 9 + 18 + 27 = 66 > 54$, it is abundant.
31. The proper factor is 1.
Since $1 < 61$, it is deficient.
33. The proper factors are 1, 2, 31.
Since $1 + 2 + 31 = 34 < 62$, it is deficient.
35. a. 13   **b.** 4,096
    c. 8,191   **d.** 33,550,336
37. a. 19   **b.** 262,144
    c. 524,287   **d.** $1.374386913 \times 10^{11}$
39. a. $(2^{127-1})(2^{127} - 1)$ **b.** 1.45×10^{76}
41. The private key is 5, 7.
43. The private key is 11, 13.

45. a. $6 = 110_2$
    $28 = 11100_2$
    $496 = 111110000_2$
    $8128 = 1111111000000_2$
    b. The nth perfect number is a sequence of m ones, where m is the nth prime, followed by $2(n - 1)$ zeros, $n \neq 1$. (When $n = 1$, there is 1 zero.)
47. See answer 46.
49. Answers may vary. Possible answer: The Pythagoreans saw a mystical significance in numbers. See *Historical Note Pythagoras (approximately 560–480 B.C.)* in Section 7.4.

7.5 Fibonacci Numbers and the Golden Ratio

1. After six years, there are 13 total, and 5 newly born.
3. 21, 34, 55, 89, 144, 233, 377, 610, 987, 1597
5. 8 great-great-great grandparents
7. Answers will vary.
9.

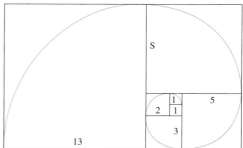

11. a. $n = 30,\ 832,040$
    $n = 40,\ 102,334,155$
    b. Possible answer: It allows you to find a Fibonacci number without listing all the preceding numbers.

c. Possible answer: It is an unnecessarily lengthy formula when dealing with the first numbers, say the first 24 or so numbers.

13. a. $\dfrac{a + b}{a} = \dfrac{a}{b}$

b. $ab \cdot \dfrac{a + b}{a} = ab \cdot \dfrac{a}{b}$

$b(a + b) = a^2$

c. $ab + b^2 = a^2$

d. $\dfrac{ab + b^2}{b^2} = \dfrac{a^2}{b^2}$

$\dfrac{ab}{b^2} + \dfrac{b^2}{b^2} = \dfrac{a^2}{b^2}$

$\dfrac{ab}{b^2} + 1 = \dfrac{a^2}{b^2}$

e. $\dfrac{a}{b} + 1 = \left(\dfrac{a}{b}\right)^2$

f. Let $\dfrac{a}{b} = x$.

$x + 1 = x^2$

g. $x^2 - x - 1 = 0$

$x = \dfrac{1 + \sqrt{5}}{2}$ is the golden ratio.

h. x is a ratio of lengths.

15. Answers may vary.   **17.** Answers may vary.

19. 1.61803398896 is an accurate approximation to the golden ratio up to nine decimal places.

21. Answers may vary.

23. Answers may vary. Possible answer: the heads in the bottom right exhibit a golden ratio.

25. Answers will vary.

Chapter 7   Review

1. $5372 = 5 \cdot 10^3 + 3 \cdot 10^2 + 7 \cdot 10^1 + 2 \cdot 10^0$ or $5 \cdot 10^3 + 3 \cdot 10^2 + 7 \cdot 10 + 2$

3. $325_8 = 3 \cdot 8^2 + 2 \cdot 8^1 + 5 \cdot 8^0$ or $3 \cdot 8^2 + 2 \cdot 8 + 5$

5. 390_8; does not exist because base eight only uses numerals 0 through 7.

7. $905_{16} = 9 \cdot 16^2 + 0 \cdot 16^1 + 5 \cdot 16^0$ or $9 \cdot 16^2 + 5$

9. $ABC_{16} = 10 \cdot 16^2 + 11 \cdot 16^1 + 12 \cdot 16^0$ or $10 \cdot 16^2 + 11 \cdot 16 + 12$

11. $11011001_2 = 1 \cdot 2^7 + 1 \cdot 2^6 + 0 \cdot 2^5 + 1 \cdot 2^4 + 1 \cdot 2^3 + 0 \cdot 2^2 + 0 \cdot 2^1 + 1 \cdot 2^0$ or $2^7 + 2^6 + 2^4 + 2^3 + 1$

13. 101112_2 does not exist because base two only uses numerals 0 and 1.

15. $39,22,54_{60} = 39 \cdot 60^2 + 22 \cdot 60^1 + 54 \cdot 60^0$ or $39 \cdot 60^2 + 22 \cdot 60 + 54$

17. $25,44,34_{60} = 25 \cdot 60^2 + 44 \cdot 60^1 + 34 \cdot 60^0$ or $25 \cdot 60^2 + 44 \cdot 60 + 34$

19. 332   **21.** 53   **23.** 48,282   **25.** 56

27. 940   **29.** 1002_8 **31.** $20EE_{16}$ **33.** $10,54,47_{60}$

35. 10111011100011_2 **37.** 335051_7 **39.** The base is 6.

41. The base is 9.

43. a. 134_8 **45. a.** 1011_2

47. a. 144_{16} **49. a.** $1,18,18,17_{60}$

51. a. $6\ 7_8$ **53. a.** 0011_2 or 11_2

55. a. $98D_{16}$ **57. a.** $11,49,11_{60}$

59. a. 2024_8 **61. a.** 11110_2

63. a. 6616_{16} **65. a.** 3044_5

67. a. 23_8 Rem 3_8 **69. a.** 10_2 Rem 11_2

71. a. $5F_{16}$ Rem 3_{16} **73. a.** 11011_2

75. a. $3 \cdot 11$ **b.** composite

77. a. 41   **b.** prime

79. The prime numbers are 53 and 59.

81. $n = 2, 2^n + 1 = 2^2 + 1 = 4 + 1 = 5$; prime
$n = 3, 2^n + 1 = 2^3 + 1 = 8 + 1 = 9$; composite, $9 = 3 \cdot 3$

83. $2^n - 1 = 2^3 - 1 = 8 - 1 = 7$; prime because $7 = 1 \cdot 7$.

85. The proper factors are 1, 2, 5, 7, 10, 14, 35.

87. The proper factors are 1, 2, 4, 5, 10.
Since $1 + 2 + 4 + 5 + 10 = 22 > 20$, it is abundant.

89. The proper factors are 1 and 7.
Since $1 + 7 = 8 < 49$, it is deficient.

91. a. $(2^{17-1})(2^{17} - 1)$ **b.** 8.59×10^9

93. 1, 1, 2, 3, 5, 8, 13, 21, 34, 55, 89, 144, 233, 377, 610, 987, 1597, 2584, 4181, 6765

95. 5 great-great grandparents

97. After 4 years, there are 5 total, and 2 newly born.

99. a. $n = 25$: 75,025
$n = 26$: 121,393
b. 196,418

101. The rectangle's length is about 1.6 times its width.

103. Answers may vary.

105. Answers will vary.

107. Answers may vary. Possible answer: 0 represents none in that place value.

109. Answers will vary.

111. Answers may vary. Possible description: $321 - 74$
Step 1: Set 321 on the first (ones) and second (tens) and third (hundreds) rods.
Step 2: To subtract 74, first consider subtracting the 7 in the tens place. Subtract 1 bead from the hundreds rod and add 7's complement (3) to the tens rod.
Step 3: Now subtract the 4 in the ones place. Once again, subtract 1 bead from the tens rod and add 4's complement (6) to the ones rod.
Step 4: Read the result, 247.

113. His first book was on Hindu numerals. It introduced Europe to the simpler calculation techniques of the Hindu system, including the multiplication algorithm and the division algorithm that we use today. His second book was on algebra. It discusses linear and quadratic equations.

115. Answers may vary. Possible answer: They believed that numbers had mystical properties.

117. Answers may vary. Possible answer: He was trying to answer questions in nature.

CHAPTER 8   Geometry

8.1 Perimeter and Area

1. 16.1 square centimeters   **3.** 12.5 square inches

5. 21.7 square feet   **7.** 176 square meters

9. a. 33.2 square inches   **b.** 20.4 inches

11. a. 30 square meters   **b.** 30 meters

13. a. 27.7 square meters   **b.** 24 meters

15. a. 80 square feet   **b.** 44 feet

17. a. 3927.0 square yards   **b.** 257.1 yards

19. a. 54 square yards   **b.** 30 yards
21. a. 49.8 square feet   **b.** 28.9 feet
23. a. 1963.5 square feet   **b.** 863.9 square feet
   c. 3,572.6 square feet
25. a. 3,136 square inches   **b.** 21.8 square feet
27. 8.0 square feet   **29.** 3.9 miles
31. 8 feet
33. a. 5,256.6 square yards   **b.** 325.7 yards
35. You need 7 cans.
37. The large pizza is the best deal.
39. 0.25 A   **41.** 640 A
43. 525.03 square miles
45. a. 36 million miles   **b.** 0.39 AU
   c. The distance between Mercury and the earth is 0.39 times
   the distance between the earth and sun.
47. a. 27,200,000 AU   **b.** 431 ly   **c.** 132 pc
49. Area of shaded region $= 16 - 4\sqrt{3}$, which is choice **c.**
51. The correct choice is **f.** none of these is correct.
53. Circumference of a circle $= 12\pi$, which is choice **e.**
55. area of a circle $= \pi r^2 = \pi(\sqrt{2})^2 = 2\pi$, which is choice **b.**

8.2 Volume and Surface Area

1. a. 38.22 cubic meters   **b.** 72.94 square meters
3. a. 785.40 cubic inches   **b.** 471.24 square inches
5. a. 2.81 cubic inches   **b.** 9.62 square inches
7. 16.76 cubic feet   **9.** 21.33 cubic feet
11. 36 cubic feet   **13.** 47.12 cubic feet
15. a. 136.5 cubic inches   **b.** 151 square inches
17. 17,802.36 cubic feet
19. a. Hardball: 12.31 cubic inches
   Softball: 29.18 cubic inches
   Volume of softball is 137% more than that of hardball.
   b. Hardball: 25.78 square inches
   Softball: 45.83 square inches
   Surface area of softball is 78% more than that of hardball.
21. about 49 moons   **23.** about 147 Plutos
25. You did not get an honest deal. You should have paid $118.75.
27. 36.82 cubic inches   **29.** $175.20
31. 91,445,760 cubic feet   **33.** 301.59 cubic feet
35. 1,728 cubic inches, which is choice **e.**
37. $\frac{1}{27}$, which is choice **a.**
39. 216 square units, which is choice **c.**
41. $\frac{C^3}{2\pi}$, which is choice **b.**
43. Answers will vary.

8.3 Egyptian Geometry

1. This is a right triangle.   **3.** This is not a right triangle.
5. The regular pyramid (part **b**) holds more.
7. 36; 18; 9; 36; 18; 9; 63; 9; 3; 63; 189; 189
9. square of side 21   **11.** $\pi = \frac{49}{16} = 3.0625$
13. a. 28.4444 square palms   **b.** 28.2743 square palms
   c. $0.00623 \approx 0.6\%$
15. a. 113.7778 cubic palms   **b.** 113.0973 cubic palms
   c. $0.00602 \approx 0.6\%$
17. 11 cubits
19. 1 khar is larger.   **21.** 1.5 setats
23. a. 2 khet   **b.** 200 cubits
25. a. 960 khar   **b.** 954.2588 khar
   c. $0.00602 \approx 0.6\%$

27. A rope with 12 equally spaced knots is stretched to form a
   3-4-5 right triangle.
29. The area of a circle of diameter 9 is the same as the area of a
   square of side 8.

8.4 The Greeks

1. $x = 5$ **3.** $x = 64.6$
   $y = 60$ $y = 2.5$
5. 36 feet
9. 5.3 feet, (rounded down so the object will fit)
11. 7.5 inches
13.
1.	$AD = CD$	Given
2.	$AB = CB$	Given
3.	$DB = DB$	Anything is equal to itself.
4.	$\triangle DBA \cong \triangle DBC$	SSS
5.	$\angle DBA = \angle DBC$	Corresponding parts of congruent triangles are equal.

15.
1.	$AD = BD$	Given
2.	$\angle ADC = \angle BDC$	Given
3.	$DC = DC$	Anything is equal to itself.
4.	$\triangle ACD \cong \triangle BCD$	SAS
5.	$AC = BC$	Corresponding parts of congruent triangles are equal.

17.
1.	$AE = CE$	Given
2.	$AB = CB$	Given
3.	$BE = BE$	Anything is equal to itself.
4.	$\triangle ABE \cong \triangle CBE$	SSS
5.	$\angle ABE = \angle CBE$	Corresponding parts of congruent triangles are equal.
6.	$BD = BD$	Anything is equal to itself.
7.	$\triangle ADB \cong \triangle CDB$	SAS
8.	$\angle ADB = \angle CDB$	Corresponding parts of congruent triangles are equal.

19. a. $\pi \approx 4\sqrt{2 - \sqrt{2}} = 3.061467459$
   b. $\pi \approx 8(\sqrt{2} - 1) = 3.313708499$
21. a. $\pi \approx 8\sqrt{2 - \sqrt{2 + \sqrt{2}}} = 3.121445152$
   b. $\pi \approx 8\left(\dfrac{2\sqrt{2 - \sqrt{2}}}{2 + \sqrt{2 + \sqrt{2}}}\right) = 3.182597878$
23. $\dfrac{V_{\text{sphere}}}{V_{\text{cylinder}}} = \dfrac{\frac{4}{3}\pi^3}{\pi r^2(2r)} = \dfrac{\frac{4}{3}}{2} = \dfrac{2}{3}$
25. All is number.
27. Given a line and a point not on that line, there is one and only
   one line through the point parallel to the original line.

8.5 Right Triangle Trigonometry

1. $x = 3$ **3.** $x = 14$
   $y = 3\sqrt{3}$ $y = 7\sqrt{3}$
   $\theta = 60°$ $\theta = 30°$
5. $y = 3\sqrt{2}$ **7.** $x = \frac{7\sqrt{2}}{2}$
   $x = 3$ $y = \frac{7\sqrt{2}}{2}$
   $\theta = 45°$ $\theta = 45°$
9. $B = 53°$ **11.** $B = 35.7°$
   $c = 19.9$ $c = 9.6$
   $b = 15.9$ $a = 7.8$
13. $A = 40.1°$ **15.** $A = 80.85°$
   $b = 0.70$ $c = 1,565.9$
   $a = 0.59$ $b = 249$

17. $B = 36.875°$
  $a = 43.52$
  $b = 32.64$
19. $A = 85.00°$
  $c = 11.497$
  $a = 11.453$
21. $b = 17.4$
  $A = 40.7°$
  $B = 49.3°$
23. $c = 9.2$
  $A = 40.6°$
  $B = 49.4°$
25. $a = 0.439$
  $A = 74.4°$
  $B = 15.6°$
27. a. 64.0 feet   **b.** 92.3 feet
29. 150 feet   **31.** 18.4°
33. a. 176.1 feet   **b.** 26.7 feet
35. 26.2 feet   **37.** 167 feet
39. 630 feet   **41.** 1,454 feet
43. 24.6 feet   **45.** 0.5 pc
47. a. 185,546 AU   **b.** 1.11 seconds
49. $32 + 8\sqrt{3}$, which is choice **e**.
51. 48°, which is choice **d**.
53. Answers will vary.

8.6 Linear Perspective

1. a. one-point perspective   **b.** above
  c. approximately (17, 14)   **d.** $y = 14$
3. a. two-point perspective   **b.** below
  c. approximately (1, 2) and (18, 2)   **d.** $y = 2$
5. Answers will vary.
7. Anwers will vary.   **9.** Answers will vary.
11. Answers may vary.
  a. One-point perspective; there is one vanishing point and the wall faces are parallel to the surface of the painting.
  b. Between the two figures at the center of the last arch (Plato and Aristotle). Raphael might have chosen that point to put the focus on Plato and Aristotle.
  c. In the front, a man is leaning against a small table. Two-point perspective was used.
  d. The arch in the very front has its own vanishing point, directly below Plato and Aristotle. Raphael might have done this to make the front arch seem separate from the rest of the painting, thus giving the work more three-dimensionality
  e. Yes, the tiles in the front.
  f. Yes, the tiles.
  g. Albertian grid is used on the tiles.
  h. Yes, along the row of people, which includes Plato and Aristotle.
13. Answers may vary.
  a. Both one-point and two-point perspective. The work has two parts. The left side includes the big arch and uses one-point perspective. The right side includes the two buildings, and uses two-point perspective.
  On the floor at the very bottom of the painting, in the middle (horizontally). It is between the two parts, but not on either part. This unites the two separate parts.
  c. No
  d. The left side's one vanishing point is on the floor at the very bottom of the painting, in the middle (horizontally). The right part has two vanishing points, one of which is off the work. Their visible vanishing point is the same as the large arch's vanishing point. All vanishing points are on the same horizon.
  e. Yes, the tiles on the ceiling.
  f. The tiles on the ceiling are somewhat like pavement

  g. On the tiles in the ceiling.
  h. Yes, it is the floor.
15. Answers may vary.
  a. One-point perspective; there is one vanishing point
  b. The vanishing point appears to be at the small grove of trees directly above the walking people. Pissaro might have chosen that point to emphasize the distance between the travelers and the vanishing point, thus giving the work more three-dimensionality.
  c. No   **d.** No
  e. Yes, the houses get smaller and smaller.
  f. No   **g.** No
  h. Yes, horizon is along the tops of trees and buildings in the distance.
17. Answers may vary.
  a. One-point perspective; there is one vanishing point and the walls are parallel to the surface of the painting.
  b. The vanishing point is on the column that separates the painting's two halves. This serves to emphasize this separation.
  c. No
  d. No, There are two separate parts, the left side of the column and the right side. They share the same vanishing point.
  e. Yes, the tiles in the ceiling and on the floor.
  f. Yes, both sets of tiles.
  g. Albertian grid is used on both sets of tiles.
  h. No
19. Answers may vary.
  a. One-point perspective; there is one vanishing point.
  b. The vanishing point is at the left end of the line of boys. Homer might have chosen that point to give the sense of movement of the 'whip.'
  c. No   **d.** No
  e. Yes, the boys get smaller
  f. No   **g.** No
  h. Yes, the horizon is at the edge of the level ground.
21. Eakins drew center lines on the horizontal and vertical. Directly below that intersection, he placed the vanishing point. An Albertian grid was used to draw the boat. Eakins was so accurate that scholars have been able to determine the precise length of the boat and to pinpoint the exact time of day to 7:20 p.m. http://www.philamuseum.org/micro_sites/exhibitions/eakins/1872/main_frameset.html
23. Answers may vary.
  a. One-point perspective; there is one vanishing point.
  b. The vanishing point appears to be off center at the height of the people's heads. Sargent might have chosen that point to give a sense of the length of the street that the woman walked.
  c. No
  d. No, there is a central vanishing point.
  e. Yes, on the street pavement and the windows.
  f. Yes, on the street.
  g. Albertian grid is used on the street pavements.
  h. No
25. Answers may vary.
  a. Three-point perspective portrays three dimensions.
  b. Near the left roof, near the right tower, and below the etching.
  c. No   **d.** See (b).
  e. All of the windows, and the marching people
  f. No   **g.** No.   **h.** No.

8.7 Conic Sections and Analytic Geometry

1. Center $(0, 0)$
$r = 1$

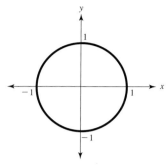

3. Center $(2, 0)$
$r = 3$

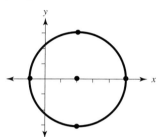

5. Center $(5, -2)$
$r = 4$

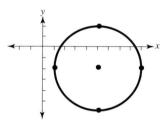

7. Center $(5, 5)$
$r = 5$

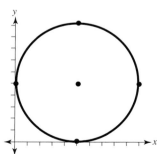

9. Focus $(0, \frac{1}{4})$

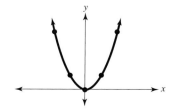

11. Focus $(0, \frac{1}{2})$

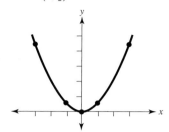

13. The water container should be placed 2.9 ft above the bottom of the dish.

15. The light bulb should be located $\frac{9}{16}$ inch above the bottom of the reflector.

17. Foci $(-\sqrt{5}, 0)$ and $(\sqrt{5}, 0)$

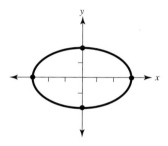

19. Foci $(0, -\sqrt{21})$ and $(0, \sqrt{21})$

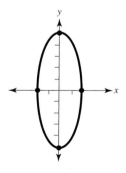

21. Foci $(0, -\sqrt{7})$ and $(0, \sqrt{7})$

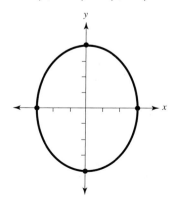

23. Foci $(1, 2 - \sqrt{3})$ and $(1, 2 + \sqrt{3})$

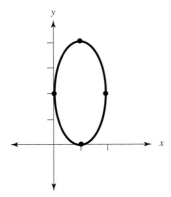

25. They should stand at the foci, which are located 9 ft from the center, in the long direction.

27. $\dfrac{x^2}{92.955^2} + \dfrac{y^2}{92.942^2} = 1$

29. a. Foci $(-\sqrt{2}, 0)$ and $(\sqrt{2}, 0)$

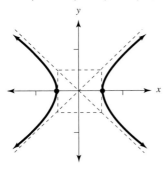

b. Foci $(0, -\sqrt{2})$ and $(0, \sqrt{2})$

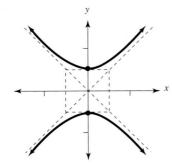

31. Foci $(-\sqrt{13}, 0)$ and $(\sqrt{13}, 0)$

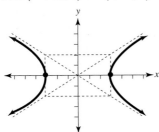

33. Foci $(0, -\sqrt{29})$ and $(0, \sqrt{29})$

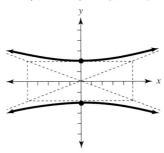

35. Foci $(3 - \sqrt{5}, 0)$ and $(3 + \sqrt{5}, 0)$

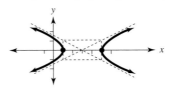

37. Hypatia. Bishop Cyril incited a Christian mob to kill her since he associated her with paganism and viewed her as a threat to his quest for power.

8.8 Non-Euclidean Geometry

1. One   **3.** Zero or one
5. Zero or one   **7.** None
9. Zero, one, two, or three   **11.** Two
13. Infinitely many
15. Zero or one
17. Zero or one
19. Given a line and a point not on that line, there is one and only one line through the point parallel to the original line.
21. Girolamo Sacchein. He couldn't find a contradiction. He was distressed so he manufactured a spurious contradiction so he could support the Parallel Postulate.
23. Janos Bolyai. He was worried that Gauss was stealing his ideas and there was a lack of interest by other mathematicians.
25. Bernhard Riemann

8.9 Fractal Geometry

1. a.

b.

3.

step 1

step 2

step 3

5. $d = 2$

7. $d = \frac{1}{3}$

9. $d \approx 1.6$ The dimension of the Sierpinski gasket is larger than the dimension of a circle, but smaller than the dimension of a square. It's not a two-dimensional as a regular triangle is.

11. $d \approx 1.6$ The dimension of the Mitsubishi gasket is larger than the dimension of a circle, but smaller than the dimension of a square. It's not a two-dimensional as a regular triangle is.

13. $d = 1.5$ The dimension of the square snowflake is larger than the dimension of a circle, but smaller than the dimension of a square. It's not two-dimensional.

15. $d = 3$, the same dimension that we observed before.

17. $d = 2$. Also, using the formula $s^d = n$: solve $3^d = 9$. Again, $d = 2$.

19. Because you always move to a point halfway between two points, the worst that can happen is you end up on the boundary of the hole. You can never end up inside it. This is important because the hole must remain untouched otherwise we wouldn't generate the Sierpinski gasket.

21. It is the 6th step in forming the Sierpinski gasket. No finite number of rows could illustrate the Sierpinski gasket.

23. *Sierpinski carpet*—Looking at any of the squares that are not removed reveals a pattern like the entire carpet. This is exact similarity.
Mitsubishi gasket—Looking at any of the triangles that are not removed reveals a pattern like the entire gasket. This is exact similarity.
Square snowflake—Looking at any length of the perimeter reveals a pattern like the entire snowflake. This is exact similarity.

25. *Sierpinski carpet*—For any square that is not removed, the recursive rule is to remove the center square.
Mitsubishi gasket—For any triangle that is not removed, the recursive rule is to remove three triangles from the center.
Square snowflake—For every line segment, the recursive rule is to replace the segment with:

27. In nature, approximate self-similarity. In geometry, exact self-similarity.

29. No. In nature there is a lower limit to the scale.

31. Benoit Mandelbrot

33. Answers will vary.

8.10 The Perimeter and Area of a Fractal

1. a. In step 1, there is only one triangle. If each side is 1 ft. in length, then the perimeter is 3 ft. In step 2, the original triangle is modified by removing a center triangle. This results in three smaller triangles. Each side of the original triangle now has two triangles with sides of length $\frac{1}{2}$ ft. Thus, the perimeter of one of these smaller triangles is $\frac{1}{2} + \frac{1}{2} + \frac{1}{2} = 3 \cdot \frac{1}{2} = \frac{3}{2}$ ft. Since there are three such triangles, the total perimeter is $3 \cdot \frac{3}{2} = \frac{9}{2}$ ft.

b.

Step	Number of Triangles	Length of Each Side (feet)	Perimeter of One Triangle (feet)	Total Perimeter of All Triangles (feet)
1	1	1	3	3
2	3	$\frac{1}{2}$	$3 \cdot \frac{1}{2} = \frac{3}{2}$	$3 \cdot \frac{3}{2} = \frac{9}{2}$
3	9	$\frac{1}{4}$	$3 \cdot \frac{1}{4} = \frac{3}{4}$	$9 \cdot \frac{3}{4} = \frac{27}{4}$
4	27	$\frac{1}{8}$	$3 \cdot \frac{1}{8} = \frac{3}{8}$	$27 \cdot \frac{3}{8} = \frac{81}{8}$
5	81	$\frac{1}{16}$	$3 \cdot \frac{1}{16} = \frac{3}{16}$	$81 \cdot \frac{3}{16} = \frac{243}{16}$
6	243	$\frac{1}{32}$	$3 \cdot \frac{1}{32} = \frac{3}{32}$	$243 \cdot \frac{3}{32} = \frac{729}{32}$

c. 3. Each triangle is divided into four smaller triangles, one of which is deleted.

d. $\frac{1}{2}$. When the middle triangle is taken out, it creates two triangles along each side.

e. $\frac{1}{2}$. Since each side is half as long, the perimeter is half as much. The perimeter of one triangle is decreasing.

f. $\frac{3}{2}$. Three times the number of triangles, but each one has half the perimeter. The total perimeter of all triangles is increasing.

g. $3\left(\frac{3}{2}\right)^{n-1}$

h. The perimeter is infinite. The numbers get larger and larger without bound.

3. a. In step 1, there is only one square. If each side is 1 ft. in length, then the perimeter is 4 ft. In step 2, the original square is divided into nine smaller squares and modified by removing the center square. The removed square contributes to the perimeter. It has sides of length $\frac{1}{3}$ ft. Thus, the perimeter of this smaller square is $4 \cdot \frac{1}{3} = \frac{4}{3}$ ft. The total perimeter is the sum of the previous perimeter and the new contribution: $4 + \frac{4}{3}$ ft.

b.

Step	Number of New Squares	Length of Each Side (feet)	Perimeter of One New Square (feet)	Total Perimeter of All New Squares (feet)	Total Perimeter of All Squares (feet)
1	1	1	$4 \cdot 1 = 4$	$1 \cdot 4 = 4$	4
2	1	$\frac{1}{3}$	$4 \cdot \frac{1}{3} = \frac{4}{3}$	$1 \cdot \frac{4}{3} = \frac{4}{3}$	$4 + \frac{4}{3}$
3	8	$\frac{1}{9}$	$4 \cdot \frac{4}{9} = \frac{4}{9}$	$8 \cdot \frac{4}{9} = \frac{32}{9}$	$4 + \frac{4}{3} + \frac{32}{9}$
4	64	$\frac{1}{27}$	$4 \cdot \frac{1}{27} = \frac{4}{27}$	$64 \cdot \frac{4}{27} = \frac{256}{27}$	$4 + \frac{4}{3} + \frac{32}{9} + \frac{256}{27}$
5	512	$\frac{1}{81}$	$4 \cdot \frac{1}{81} = \frac{4}{81}$	$512 \cdot \frac{4}{81} = \frac{2,048}{81}$	$4 + \frac{4}{3} + \frac{32}{9} + \frac{256}{27} + \frac{2048}{81}$
6	4,096	$\frac{1}{243}$	$4 \cdot \frac{1}{243} = \frac{4}{243}$	$4,096 \cdot \frac{4}{243} = \frac{16,384}{243}$	$4 + \frac{4}{3} + \frac{32}{9} + \frac{256}{27} + \frac{2048}{81} + \frac{16384}{243}$

c. 8. Each step creates nine new squares, one of which is deleted.

d. $\frac{1}{3}$. Each original side is divided into three new squares.

e. $\frac{1}{3}$. Since each side is one third as long, the perimeter is one third as long.

f. $\frac{8}{3}$. Eight new squares each of which has perimeter one third as much as before. The total perimeter of all new squares is increasing.

g. $\dfrac{4 \cdot 8^{n-2}}{3^{n-1}}$ or $\dfrac{4}{3} \cdot \left(\dfrac{8}{3}\right)^{n-2}$, valid for $n > 1$

h. The total perimeter is infinite. The numbers increase without bound.

i. The total perimeter is infinite. It increases without bound.

5. a. $3 \cdot 2^{n-1}$ ft. (If the original triangle has sides of length 1 ft.)

b. Infinite ∞

c. $\left(\dfrac{2}{3}\right)^{n-1} \cdot \dfrac{\sqrt{3}}{4}$ sq. ft. (If the original triangle has sides of length 1 ft.)

d. 0

7. Inductive: noticing the pattern that the number of sides is increasing by a factor of 4, that the length of each side is decreasing by a factor of $\frac{1}{3}$, and that the perimeter is increasing by a factor of $\frac{4}{3}$.

Deductive: applying the general rule that the perimeter of a square is four times the length of a side, applying the general rule that the perimeter of a shape is the sum of all lengths of sides.

9. The kidney has an almost infinite surface area, yet occupies a small volume. Understanding fractals may help us understand the human body better.

11. a.

Step	Number of New Squares	Base of New Triangle (feet)	Height of New Triangle (feet)	Area of Each New Triangle (square feet)	Total Area of All New Triangles (square feet)	Total Area (square feet)
1	1	1	$\frac{\sqrt{3}}{2}$	$\frac{\sqrt{3}}{4}$	$\frac{\sqrt{3}}{4}$	$\frac{\sqrt{3}}{4}$
2	3	$\frac{1}{3}$	$\frac{\sqrt{3}}{6}$	$\frac{\sqrt{3}}{36}$	$\frac{\sqrt{3}}{12}$	$\frac{\sqrt{3}}{3}$
3	12	$\frac{1}{9}$	$\frac{\sqrt{3}}{18}$	$\frac{\sqrt{3}}{324}$	$\frac{\sqrt{3}}{27}$	$\frac{10\sqrt{3}}{27}$

b.

Step	Number of New Squares	Base of New Triangle (feet)	Height of New Triangle (feet)	Area of Each New Triangle (square feet)	Total Area of All New Triangles (square feet)	Total Area (square feet)
1	1	1	$\frac{\sqrt{3}}{2}$	$\frac{\sqrt{3}}{4}$	$\frac{\sqrt{3}}{4}$	$\frac{\sqrt{3}}{4}$
2	3	$\frac{1}{3}$	$\frac{\sqrt{3}}{6}$	$\frac{\sqrt{3}}{36}$	$\frac{\sqrt{3}}{12}$	$\frac{\sqrt{3}}{3}$
3	12	$\frac{1}{9}$	$\frac{\sqrt{3}}{18}$	$\frac{\sqrt{3}}{324}$	$\frac{\sqrt{3}}{27}$	$\frac{10\sqrt{3}}{27}$
4	48	$\frac{1}{27}$	$\frac{\sqrt{3}}{54}$	$\frac{\sqrt{3}}{2,916}$	$\frac{4\sqrt{3}}{243}$	$\frac{94\sqrt{3}}{243}$
5	192	$\frac{1}{81}$	$\frac{\sqrt{3}}{162}$	$\frac{\sqrt{3}}{26,244}$	$\frac{16\sqrt{3}}{2,187}$	$\frac{862\sqrt{3}}{2,187}$
6	768	$\frac{1}{243}$	$\frac{\sqrt{3}}{486}$	$\frac{\sqrt{3}}{236,196}$	$\frac{64\sqrt{3}}{19,683}$	$\frac{7,822\sqrt{3}}{19,683}$
7	3,072	$\frac{1}{729}$	$\frac{\sqrt{3}}{1,458}$	$\frac{\sqrt{3}}{2,125,764}$	$\frac{256\sqrt{3}}{177,147}$	$\frac{70,654\sqrt{3}}{177,147}$
8	12,288	$\frac{1}{2,187}$	$\frac{\sqrt{3}}{4,374}$	$\frac{\sqrt{3}}{19,131,876}$	$\frac{1,024\sqrt{3}}{1,594,323}$	$\frac{636,910\sqrt{3}}{1,594,323}$

c. $A_1 = 0.4430$, $A_2 = 0.5774$, $A_3 = 0.6415$, $A_4 = 0.6700$,
$A_5 = 0.6827$, $A_6 = 0.6883$, $A_7 = 0.6908$, $A_8 = 0.6919$

d. The total area must be larger because we are adding more area at each step. Estimate: Answers will vary.

Chapter 8 Review

1. Area = $(x^2 - 2x)$ square feet
Perimeter = $4x - 4$ feet

3. $A = 73.5$ square inches
$P = 41$ inches

5. $A = 69$ square yards
$P \approx 37.4$ yards

7. a. 1,598,000 AU   **b.** 25.3 ly   **c.** 7.8 pc

9. $V = 72.8$ cubic centimeters
$SA = 153.9$ square centimeters

11. $V = 1,056$ cubic inches
$A = 664$ square inches

13. He will need 9 bags.

15. a. 224 cubic inches   **b.** 200 square inches

17. a. 450.7 cubic feet   **b.** 268.0 square feet

19. a. 4.2140 cubic cubits   **b.** 4.1888 cubic cubits
c. $0.00602 \approx 0.6\%$

21. square with side 16

23. 6.5 ft (rounded down so the object will fit)

25. The first square has area c^2.
Rearranging the area creates two squares: one with area a^2 and another with area b^2. Thus, $a^2 + b^2 = c^2$.

27. 1. $\angle CBA = \angle DAB$ Given.
 2. $BC = AD$ Given.
 3. $AB = AB$ Anything is equal to itself.
 4. $\triangle CAB \cong \triangle DBA$ SAS
 5. $\angle CAB = \angle DBA$ Corresponding parts of congruent triangles are equal.
 6. $\angle CAD = \angle DBC$ Since $\angle CAD = \angle CAB - \angle DAB$
 $= \angle DBA - \angle CBA$
 $= \angle DBC$

29. $\pi \approx 6[(\sqrt{6} - \sqrt{2})/2] = 3.105828541$

31. a. Foci $(-\sqrt{13}, 0)$ and $(\sqrt{13}, 0)$

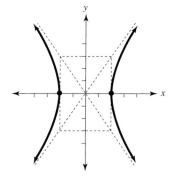

b. Foci $(0, -\sqrt{13})$ and $(0, \sqrt{13})$

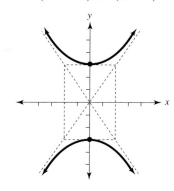

33. Foci $(0, -\sqrt{5})$ and $(0, \sqrt{5})$

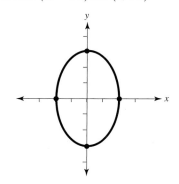

35. $B = 64.6°$
$b = 50.7$ feet
$a = 24.1$ feet

37. 213.2 feet

39. 0.33 pc

41. a. 175,238 AU
b. 1.18 seconds

43. a. One
b. More than one
c. None

45. a. Step 1

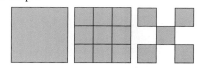

b.

Step	Number of Squares	Length of Side	Perimeter of Square	Total Perimeter
1	1	1	4	4
2	5	$\frac{1}{3}$	$\frac{4}{3}$	$\frac{20}{3}$
3	25	$\frac{1}{9}$	$\frac{4}{9}$	$\frac{100}{9}$
4	125	$\frac{1}{27}$	$\frac{4}{27}$	$\frac{500}{27}$
5	625	$\frac{1}{81}$	$\frac{4}{81}$	$\frac{2500}{81}$
n	5^{n-1}	$\frac{1}{3^{n-1}}$	$\frac{4}{3^{n-1}}$	$4 \cdot \frac{5^{n-1}}{3^{n-1}} = 4\left(\frac{5}{3}\right)^{n-1}$

c. $4\left(\dfrac{5}{3}\right)^{n-1}$

d. The perimeter is infinite.

e.

Step	Number of Squares	Length of Side	Area of Square	Total Area
1	1	1	1	1
2	5	$\frac{1}{3}$	$\frac{1}{9}$	$\frac{5}{9}$
3	25	$\frac{1}{9}$	$\frac{1}{81}$	$\frac{25}{81}$
4	125	$\frac{1}{27}$	$\frac{1}{729}$	$\frac{125}{729}$
5	625	$\frac{1}{81}$	$\frac{1}{6561}$	$\frac{625}{6561}$
n	5^{n-1}	$\dfrac{1}{3^{n-1}}$	$\left(\dfrac{1}{3^{n-1}}\right)^2$	$\left(\dfrac{5}{9}\right)^{n-1}$

f. 0   **g.** $d = 1.465$.

h. If we focus in on any smaller square it will look like the entire box fractal.

i. For any square that is not removed, the recursive rule is to remove the same four squares in the middle of the edges.

47. a. one-point perspective

b. below   **c.** approximately $(1, 2)$ **d.** $y = 2$

49. Answers will vary. Possible drawing:

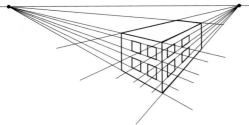

51. Answers may vary.

a. Both. The floor tiles use two-point perspective while the rest uses one-point perspective.

b. The vanishing point is on the girl playing. Vermeer might have chosen that point to draw the viewer's eye to the girl at the piano.

c. No

d. Yes, the floor's two vanishing points are off of the painting.

e. Yes, on the tiles in the floor and on the windows.

f. Yes, there are tiles on the floor.

g. Albertian grid is used on the tiles in the floor. See diagonal lines on painting.

h. Yes, horizon follows the line of the windows and piano.

53. <u>Archimedes</u> was the father of physics and occupied himself with the calculation of π and with surface area and volume of spheres, cylinders, and cones. <u>Euclid</u> wrote *Elements* and brought a method of organization to mathematics. <u>Hypatia</u> was one of the first women recognized for her mathematical accomplishments. She wrote commentaries on the geometry of Euclid. <u>Kepler</u> discovered the elliptical orbits of the planets. <u>Pythagoras</u> used rearrangement of areas as a tool to develop formulas. <u>Saccheri</u> first attempted to prove the Parallel Postulate by contradiction. <u>Bolyai</u> realized that geometries in which the Parallel Postulate does not hold could exist. <u>Gauss</u> realized that geometries based on axioms different from Euclid's could exist. <u>Lobachevsky</u> published the first book on non-Euclidean geometry. <u>Riemann</u> was the first person to really convince the academic world of the merits of non-Euclidean geometry. <u>Thales</u> insisted that geometric statements be established by deductive reasoning. <u>Mandelbrot</u> is known as the 'father of fractal geometry.'

CHAPTER 9   Graph Theory

9.1 A Walk Through Konigsberg

1. Yes. There are four vertices. If you eliminate one bridge, then two of the vertices have an odd number of edges, and the rest have an even number of edges. Start at one of the odd vertices and end at the other.

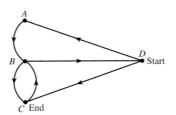

3. No; the closure of any one bridge is sufficient to create a bridge walk.

5. Yes. Start at one odd vertex and end at the other.

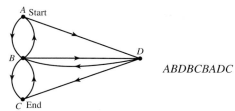

ABDBCBADC

7. 4 vertices, 4 edges, 0 loops

9. 5 vertices, 5 edges (one of which is a loop), 1 loop

11. a. 7 vertices, 6 edges, 0 loops

   b. A family member is not its own child.

13.

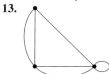

15. a. They are all graphs with 4 edges, 4 vertices, and 0 loops, and the edges in both graphs connect the same points.

   b.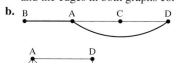

17.

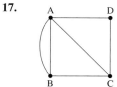

19. Answers will vary.

9.2 Graphs and Euler Trials

1. a. AB and AC   **b.** A and B

   c. A-3, B-1, C-2, D-2

   d. Yes; every pair of vertices is connected by a trail.

3. a. AB and BD   **b.** A and B

   c. A-3, B-4, C-2, D-3

   d. Yes; every pair of vertices is connected by a trail.

5. a. AB and AC   **b.** A and B

   c. A-4, B-3, C-3, D-2

   d. Yes; every pair of vertices is connected by a trail.

7. a.   **b.**   **c.**

9. a.   **b.**

11. a.

   b.

   c.   **d.**

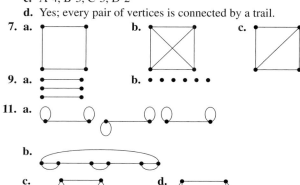

13. a. Euler trail (2 odd)   **b.** BACDA

15. a. There is an Euler trail because there are exactly two odd vertices.

   b. ACABDBD (Label the vertices AB from L to R in the top row and CD from L to R in the bottom row.)

17. a. There is an Euler trail because there are exactly two odd vertices.

   b. Start at B; end at C. BADCABC

19. Answers will vary. Possible answer:

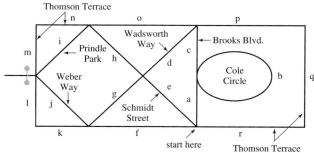

21. a.

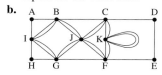

Start in the upper left-hand corner, go clockwise one block, and detour to take Prindle, Weber, Wadsworth, Schmidt, Brooks Road, Brooks Circle, Brooks Road, and Wadsworth and Schmidt back to where you detoured. Continue along the outer border to where you started, and repeat for the other side.

   b.

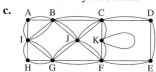

Start in the upper left-hand corner and go one block clockwise. Make a detour around the inside. (See part (a).) When you come back, continue along the border clockwise to your return.

   c.

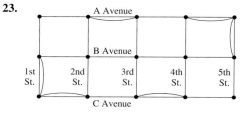

Everything is the same in part (a) except that when you repeat the inside circuit, you skip Cole Circle the second time.

23.

25.

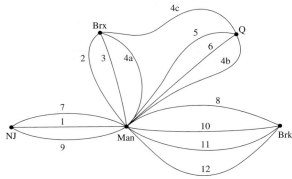

a. Manhattan has degree 13, NJ has degree 3, Brooklyn, Bronx, and Queens all have degree 4. Label the three edges of the Triborough Bridge as 4a Manhattan to Bronx, 4b Manhattan to Queens, and 4c Bronx to Queens.

b. No, there are two odds. To Eulerize, revisit any of the three bridges between NJ and Manhattan. Start in Manhattan: 7, 1, 9, 1(or 7 or 9), 2, 3, 4a, 4c, 5, 6, 4b, 8, 10, 11, 12, ending in Manhattan.

c. There are exactly two vertices with odd degrees. Start in NJ and end in M. Here is the route NJ, 1, 2, 3, 4a, 4c, 5, 6, 4b, 8, 10, 11, 12, 9, 7, M.

27. a.

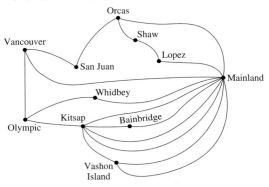

b. No, there are four odd vertices. To Eulerize, revisit a ferry from Vancouver to Orcas, and a ferry from Vashon Island to Olympic Peninsula. Now all the vertices could be considered even, and you can start and end at the same point.

c. No, there are four odd vertices. To Eulerize, revisit a ferry connecting two of the odd vertices. The other pair of odd vertices are the starting and stopping points.

29. a.

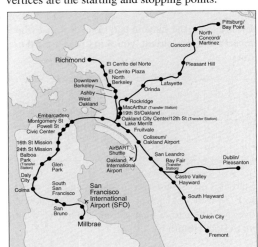

b. No, because there are only two stations with odd-degree: Pittsburg and Dublin, both with 1. Millbrae has 2 (the red and the yellow); Daly City has 6 (2 reds, 2 yellows, 1 blue, and 1 green). To Eulerize, take the yellow from Pittsburg to Millbrae, then the red from Millbrae to Richmond, the orange from Richmond to Fremont, the green from Fremont to Daly City, the blue from Daly City to Dublin, the blue from Dublin to West Oakland and the yellow back to Pittsburg.

c. Yes, it is possible to start and end at different points since there are only two odd-degrees stations. Therefore, follow the route described in part (b) except end at Dublin. Do not proceed from Dublin back to Pittsburg.

d. They can patrol all stations efficiently.

e. Part (b) is more useful because they can start and end their day at the same place.

31. a. Figure 9.15:
Number of edges in the graph: 11
Sum of the degrees of all of the vertices: 22
Figure 9.17:
Number of edges in the graph: 7
Sum of the degrees of all of the vertices: 14
Figure 9.20:
Number of edges in the graph: 18
Sum of the degrees of all of the vertices: 36

b. Sum of degrees of vertices = 2 × number of edges or

$$\text{number of edges} = \frac{\text{sum of degrees of vertices}}{2}$$

c. 2

d. 1 to each

e. An edge connects two vertices.

9.3 Hamilton Circuits

1. a. 6

b. $A \rightarrow B \rightarrow D \rightarrow P \rightarrow A$: \$927

c. $A \rightarrow B \rightarrow P \rightarrow D \rightarrow A$: \$848

d. $A \rightarrow B \rightarrow P \rightarrow D \rightarrow A$: \$848

e. $A \rightarrow B \rightarrow P \rightarrow D \rightarrow A$: \$848

3. a. 24

b. $A \rightarrow B \rightarrow Po \rightarrow Px \rightarrow D \rightarrow A$: \$685

c. $A \rightarrow B \rightarrow Po \rightarrow Px \rightarrow D \rightarrow A$: \$685

d. $A \rightarrow B \rightarrow Px \rightarrow Po \rightarrow D \rightarrow A$: \$918

e. There are too many possibilities.

5. $F \rightarrow I \rightarrow K \rightarrow B \rightarrow C \rightarrow L \rightarrow F$: 123 min

7. $F \rightarrow K \rightarrow I \rightarrow C \rightarrow B \rightarrow L \rightarrow F$: 108 min

9. $F \rightarrow I \rightarrow L \rightarrow B \rightarrow C \rightarrow R \rightarrow F$: 84 min

11. $F \rightarrow I \rightarrow L \rightarrow B \rightarrow C \rightarrow R \rightarrow F$: 84 min

13. a. 5 mm   **b.** 7 mm   **c.** 10 mm   **d.** 22 mm

15. $(0, 0) \rightarrow (2, 2) \rightarrow (3, 1) \rightarrow (4, 5) \rightarrow (6, 4) \rightarrow (7, 5) \rightarrow (0, 0)$: 28 mm
Or
$(0, 0) \rightarrow (3, 1) \rightarrow (2, 2) \rightarrow (4, 5) \rightarrow (6, 4) \rightarrow (7, 5) \rightarrow (0, 0)$: 28 mm

17. $(0, 0) \rightarrow (2, 2) \rightarrow (3, 1) \rightarrow (4, 5) \rightarrow (6, 4) \rightarrow (7, 5) \rightarrow (0, 0)$: 28 mm

19. $(0, 0) \rightarrow (5, 1) \rightarrow (4, 3) \rightarrow (3, 4) \rightarrow (1, 8) \rightarrow (2, 7) \rightarrow (0, 0)$: 28 mm

21. $(0, 0) \rightarrow (1, 8) \rightarrow (2, 7) \rightarrow (3, 4) \rightarrow (4, 3) \rightarrow (5, 1) \rightarrow (0, 0)$: 26 mm

23. In some cases answers may differ because of a tie.
$ATL \rightarrow WASH \rightarrow BOS \rightarrow SFO \rightarrow PORT \rightarrow PHX \rightarrow DEN \rightarrow ATL$: \$748 The cheapest route is \$748.

25. In some cases, answers may differ because of a tie.
ATL → WASH → BOS → SFO → PORT → PHX →
DEN → ATL: $748
Or
ATL → DEN → PHX → PORT → SFO → BOS →
WASH → ATL: $748

27. In some cases, answers may differ because of a tie.
ATL → DEN → SFO → PORT → PHX → BOS →
WASH → ATL: $1,096

29. Exercises 23 and 25 both yielded $748.

31. In some cases, answers may differ because of a tie.
NYC → SEA → CHI → MIA → HOU → LAX →
NYC: $1,132

9.4 Networks

1. Yes; it is a connected graph that has no circuits.
3. No; it is not connected.
5. No; it has a circuit.
7. No; it has a circuit.
9. Yes; it is a connected graph that has no circuits.
11. No; it is not connected.
19. 65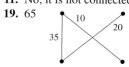
21. 115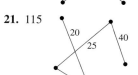
23. 531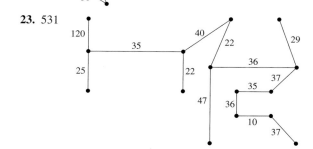

25. N–B–C–K–F: 2808 miles
27. N → Ph → A → D → S → Po: 3,442 miles
29. S → Po → K → C → A → Ph → N → B: 3,954 miles
31. 131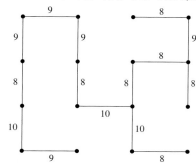

33. B because the sum of the distances is the shortest and it is given that one of them is a Steiner point.
35. Answers will vary. Possible answer: 10.5 km
37. Answers will vary. Possible answer: 14 mi
39. Answers may vary. Possible answer: 543 mi
41. Answers will vary. Possible answer: 263 mi
43. **a.** 100 tan 40°   **b.** 100 tan 20°
   c. 100(tan 40° tan 20°)
45. **a.** 50°   **b.** 70°   **c.** 110°
47. **a.** 100   **b.** 100 tan 25°   **c.** 100 − 100 tan 25°
49. **a.** 45°   **b.** 65°   **c.** 115°
51. Since the three points form a triangle with angles that are less than 120°, find a Steiner point inside the triangle. Let y = perpendicular edge length and let h = length of each other edge.
$y + 2h = 50(\tan 40° + \sqrt{3})$
53. Since the three points form a triangle with an angle that is 120° or more (Durham, 130°), then the shortest network consists of the sum of the two shortest sides of the triangle. Let h = length of one short side.
$2h = \frac{560}{\sin 65°}$
55. Therefore, **c**, $500\sqrt{3} + 600$, has the shortest network.
57. Use two Steiner points to form triangles with the shorter sides: $600\sqrt{3} + 700$
59. Therefore, **c**, $500\sqrt{3} + 500$, has the shortest network.
61. Answers will vary. Possible answer: Financial considerations

9.5 Scheduling

1.

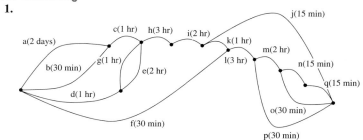

The critical path is a, c, h, i, k, l, m, n, q.
3. 4 workers

5.

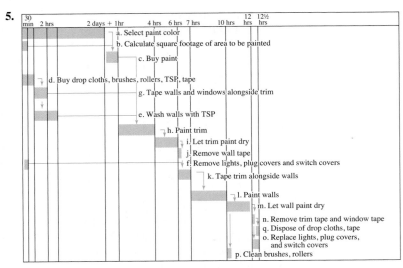

7. 2 days 12 hours. **9.** $8\frac{3}{4}$ hours **11.** $12\frac{1}{4}$ hours

13. Start: $4\frac{1}{2}$ hours
Finish: $5\frac{1}{2}$ hours

15.

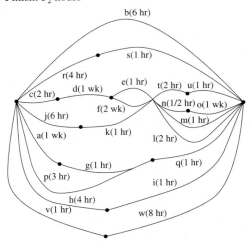

The critical path is c, d, f, n, o.

17. 3 workers

19.

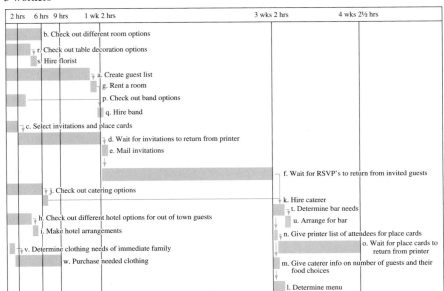

21. 3 weeks, 5 hours

23. 37 wk

25. Start: 24 weeks
Finish: 28 weeks

27. The inspection takes place at p. So start right before p, 35 weeks, and finish right after p, 37 weeks.

29. It was dependent on running closed ESS in the lab. The task's approximate length is 1 year. The task's completion date is end of the 1999.

31. It was dependent on install solar array, perform targeted lightweighting, improve reliability of motors, complete environmental control system installation, upgrade PMRF facilities, solar cell procurement, and Helios prototype functional test. The task's approximate length is 1 year. The task's completion date is the end of 2001.

33. 9 years
The approximate completion date is the end of 2003.

Chapter 9 Review

1. a. 3 vertices, 6 edges, 0 loops   **b.** Answers will vary.
  c. AB, BC   **d.** A and B   **e.** A-4, B-6, C-2
  f. Yes; every pair of vertices is connected by a trail.

3. a. 5 vertices, 5 edges, 1 loop   **b.** Answers will vary.
  c. AB, BE   **d.** A and B   **e.** A-2, B-3, C-2, D-2, E-3
  f. No; there is no trail connecting any of the 4 vertices on the left to vertex C.

5. a. 6 vertices, 12 edges, 0 loops   **b.** Answers will vary.
  c. AB, BC   **d.** A and B
  e. A-4, B-4, C-5, D-3, E-4, F-4
  f. Yes; every pair of vertices is connected by a trail.

7. **9.**

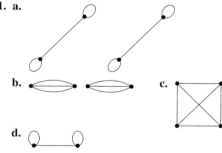

11. a.

b.   **c.**

d.

13. a. All even vertices means that there is an Euler circuit.
  b. BABABCB

15. a. None because the graph is not connected.
  b.

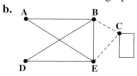

  ABCCEBDEA

17. a. There are exactly two odd vertices. There is an Euler trail starting at C and ending at D.
  b. CAFCEBFEDCBAD

19. a. There are more than two odd vertices. There is no Euler circuit. Eulerize.
  b.

Add two edges, .one connecting the two top vertices and the other connecting the two bottom vertices.

21. a. No. There are odd vertices.
  b.

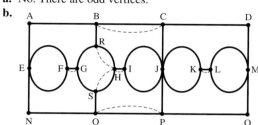

The route is E-F-G-R-B-C-J-P-O-S-H-I-J-K-L-M-L-K-J-I-H-R-H-S-G-F-E-A-B-C-D-M-Q-P-O-N-E.

23.

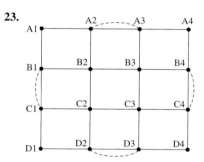

Start on A and 1st, go right on A one block, down to D, over to 3rd, back to A, loop to 2nd and back, over to 4th, down to D, over to 1st, up to C, over to 4th, up to B over to 1st, loop to C and back to A1.

25. a. 6   **b.** CMNLC: $571   **c.** CNLMC: $560
  d. CMNLC:$571   **e.** CNLMC: $560

27. a. 24   **b.** CMNSLC: $650   **c.** CSLNMC: $490
  d. CSLNMC: $490   **e.** Too many possibilities.

29. CMNSLHC: $638

31. CMNSLHC: $638

33. $(0, 0) \rightarrow (2, 4) \rightarrow (5, 7) \rightarrow (6, 3) \rightarrow (7, 2) \rightarrow (1, 1) \rightarrow (0, 0)$: 28 mm

35. Answers will vary. Possible answer:

37. Answers will vary. Possible answer:

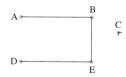

39. Answers will vary. Possible answer:

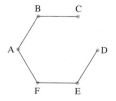

41. Answers will vary. Possible answer:

43. 60

45. 200

47. 1,000

49. BNCKD: 1,866 mi   **51.** BNPhCKDS: 2,849 mi

53. Since the three points form a triangle with angles that are less than 120°, find a Steiner point inside the triangle. Let $y =$ perpendicular edge length and let $h =$ length of each other edge. $y + 2h = 125(\tan 35° + \sqrt{3})$

55. Since the three points form a triangle with an angle that is 120° or more (Pleasant Hill, 124°), then the shortest network consists of the sum of the two shortest sides of the triangle. Let $h =$ length of one short side.
$$2h = \frac{300}{\sin 62°} \text{ or } \frac{300}{\cos 28°}$$

57. Place two Steiner points, so the shorter sides are bases of isosceles triangles. $700\sqrt{3} + 900$

59.

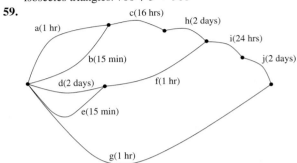

The critical path is a, c, h, i, j.

61. 2 workers

63. Create the Gantt chart from the PERT chart.

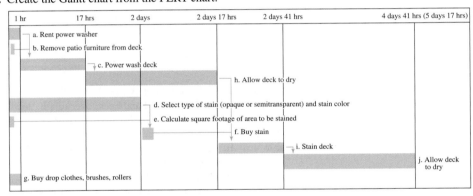

CHAPTER 10   Exponential and Logarithmic Functions

10.0A Review of Exponentials and Logarithms

1. $v = 2$ **3.** $v = -4$ **5.** $25 = u$

7. $1 = u$ **9.** $b = 4$ **11.** $b = \frac{1}{2}$

13. $b^P = Q$ **15.** $b^M = N + T$ **17.** $b^{M+R} = N + T$

19. $\log_b G = F$ **21.** $\log_b G = F + 2$

23. $\log_b (E - F) = CD$ **25.** $\log_b (Z + 3) = 2 - H$

27. a. 4.05519996684   **b.** 25.1188643151

29. a. 2.14501636251   **b.** 2.34979510988

31. a. 0.301194211912   **b.** 0.063095734448

33. a. 3.93313930533   **b.** 2.87719968669

35. a. 4.51250674972   **b.** 79.2446596231

37. a. 18.0500269989   **b.** 316.978638492

39. a. 5.67969701231   **b.** 0.176065729885

41. a. 0.982078472412   **b.** 0.426511261365

43. a. −0.162518929498   **b.** −0.070581074286

45. a. 4.1   **b.** 4.1

47. a. 4.79314718056   **b.** 4.40102999566

49. a. 2.3   **b.** 2.3

51. a. 8   **b.** 8

53. a. 2.30258509299   **b.** 4.60517018599
c. 6.90775527898

55. a. 0.434294481903   **b.** 0.868588963807
c. 1.30288344571

57. $\log 2.9$, $\ln 2.9$, $e^{2.9}$, $10^{2.9}$

59. An exponential function has the form $y = b^x$, where b is a positive constant.

61. 10^x. Exponential functions grow faster than logarithmic functions.

63. $\ln x$. The base of the natural log function, e, must be raised to a higher exponent than the base of the common log function, 10. In other words, you must raise e to a power higher than the power to which 10 is raised.

65. 10^x. The larger base makes the values larger.

67. Euler. He also introduced the symbols: $f(x)$, π, i, e

10.0B Review of Properties of Logarithms

1. $6x$ **3.** $-0.036x$ **5.** $2x + 5$

7. $1 - x$ **9. a.** x^2 **b.** x^2

11. a. $9x^2$ **b.** $9x^2$

13. $\log x - \log 4$ **15.** $\ln 1.8 + \ln x$

17. $x \log 1.225$ **19.** $\ln 3 + 4 \ln x$

21. $\log 5 + 2 \log x - \log 7$

23. $\ln 20x$ **25.** $\log 3x$

27. $\ln 2x$ **29.** $\log x^3$

31. 0

(Note: The <u>solution</u> of <u>every other odd</u>-numbered exercise is given for Exercises 33–53; <u>answers</u> to the remaining exercises are given.)

33. a. $e^x = 0.35$
$\quad \ln e^x = \ln 0.35$
$\quad\quad x = \ln 0.35$

b. $10^x = 0.35$
$\quad \log 10^x = \log 0.35$
$\quad\quad x = \log 0.35$

35. a. $\dfrac{\ln 2}{0.024}$ **b.** $\dfrac{\log 2}{0.024}$

37. a. $2000e^{0.004x} = 8500$
$\quad e^{0.004x} = \frac{8500}{2000}$
$\quad e^{0.004x} = \frac{17}{4}$
$\quad \ln e^{0.004x} = \ln \frac{17}{4}$
$\quad 0.004x = \ln \frac{17}{4}$
$\quad x = \dfrac{1}{0.004} \ln \frac{17}{4}$

b. $2000(10)^{0.004x} = 8500$
$\quad 10^{0.004x} = \frac{8500}{2000}$
$\quad 10^{0.004x} = \frac{17}{4}$
$\quad \log 10^{0.004x} = \log \frac{17}{4}$
$\quad 0.004x = \log \frac{17}{4}$
$\quad x = \dfrac{1}{0.004} \log \frac{17}{4}$

39. a. $\dfrac{1}{0.035} \ln 2$ **b.** $\dfrac{1}{0.035} \log 2$

41. a. $80e^{-0.0073x} = 65$
$\quad e^{-0.0073x} = \frac{65}{80}$
$\quad e^{-0.0073x} = \frac{13}{16}$
$\quad e^{0.0073x} = \frac{16}{13}$
$\quad \ln e^{0.0073x} = \ln \frac{16}{13}$
$\quad 0.0073x = \ln \frac{16}{13}$
$\quad x = \dfrac{1}{0.0073} \ln \frac{16}{13}$

b. $80(10)^{-0.0073x} = 65$
$\quad 10^{-0.0073x} = \frac{65}{80}$
$\quad 10^{-0.0073x} = \frac{13}{16}$
$\quad 10^{0.0073x} = \frac{16}{13}$
$\quad \log 10^{0.0073x} = \log \frac{16}{13}$
$\quad 0.0073x = \log \frac{16}{13}$
$\quad x = \dfrac{1}{0.0073} \log \frac{16}{13}$

43. a. $x = e^{0.66}$ **b.** $x = 10^{0.66}$

45. a. $\ln x = 3.66$
$\quad e^{\ln x} = e^{3.66}$
$\quad x = e^{3.66}$

b. $\log x = 3.66$
$\quad 10^{\log x} = 10^{3.66}$
$\quad x = 10^{3.66}$

47. a. $x = \dfrac{e^2}{6}$ **b.** $x = \frac{50}{3}$

49. a. $\ln x - \ln 6 = 2$
$\quad \ln \dfrac{x}{6} = 2$
$\quad e^{\ln \frac{x}{6}} = e^2$
$\quad \dfrac{x}{6} = e^2$
$\quad x = 6e^2$

b. $\log x - \log 6 = 2$
$\quad \log \dfrac{x}{6} = 2$
$\quad 10^{\log \frac{x}{6}} = 10^2$
$\quad \frac{x}{6} = 100$
$\quad x = 6 \cdot 100$
$\quad x = 600$

51. a. $x = 6.9e^{4.8}$ **b.** $x = \dfrac{10^{4.8}}{6.9}$

53. a. $\ln 0.9 = 3.1 - \ln(4x)$
$\quad \ln 0.9 + \ln(4x) = 3.1$
$\quad \ln (0.9 \cdot 4x) = 3.1$
$\quad \ln 3.6x = 3.1$
$\quad e^{\ln 3.6x} = e^{3.1}$
$\quad 3.6x = e^{3.1}$
$\quad x = \dfrac{1}{3.6} \cdot e^{3.1}$

b. $\log 0.9 = 3.1 - \log (4x)$
$\quad \log 0.9 + \log (4x) = 3.1$
$\quad \log (0.9 \cdot 4x) = 3.1$
$\quad \log 3.6x = 3.1$
$\quad 10^{\log 3.6} = 10^{3.1}$
$\quad 3.6x = 10^{3.1}$
$\quad x = \dfrac{1}{3.6} \cdot 10^{3.1}$

55. Let $a = \ln A$, $b = \ln B$
Then $A = e^a$, $B = e^b$
$\ln \dfrac{A}{B} = \ln \dfrac{e^a}{e^b} = \ln e^{(a-b)} = a - b = \ln A - \ln B$

57. Let $a = \log A$, $b = \log B$
Then $A = 10^a$, $B = 10^b$
$\log(A \cdot B) = \log (10^a \cdot 10^b) = \log (10^{a+b})$
$\quad\quad\quad\quad = a + b = \log A + \log B$

59. pH $\approx 3.5 < 7$ acid

61. pH $= 7.4 > 7$ base

63. pH $= -\log(1.3 \times 10^{-5}) = 4.9 < 7$ acid

65. $x = 10^{-7}$ mole per liter

67. a. Yes; plant because pH $= 6.5$ is acceptable.
b. No; do not plant because pH $= 3.5$ is not acceptable.

69. Paprika prefers soil that has a hydrogen ion concentration from $3.16 \times 10^{-8.5}$ to 10^{-7} mole per liter.

71. a. $t = 14.21$ years; $n = 14.21$ periods
b. $t = 13.95$ years
c. $t = 13.89$ years
d. $t = 13.86$ years
e. The larger n is (the shorter the compounding period), the shorter the doubling time.

73. To accumulate \$15,000:
$t = 4.99$ years
To accumulate \$100,000:
$t = 28.34$ years

75. To accumulate \$30,000: $t = 6.49$ years
To accumulate \$100,000: $t = 25.75$ years

77. a. $d = \dfrac{\log 3}{\log 2}$
b. (Note: The answers for part b for exercises 77–81 are not given per text reference.)

79. a. $d = \dfrac{\log 6}{\log 3}$

81. a. $d = \dfrac{\log 8}{\log 4}$

83. Easy calculation of products, quotients, powers and roots.

85. Napier was opposed. He thought James VI was arranging an invasion of Scotland.

10.1 Exponential Growth

1. Population is 32,473.

3. July 2012

5. a. (0,9392) and (4,9786)
   b. 4 years
   c. 394 thousand people
   d. 98.5 thousand people/year
   e. 1.049% per year

7. a. (0,18731) and (4,19007)
   b. 4 years
   c. 276 thousand people
   d. 69 thousand people/year
   e. 0.368%/year

9. a. $p = 9392e^{0.0102736324r}$
   b. 10,197 thousand
   c. 10,734 thousand
   d. In 2071

11. a. $P(t) = 18731e^{0.0036568568t}$
   b. 19,147 thousand
   c. 19,643 thousand
   d. In 2115

13. a. $p(t) = 2,510e^{0.2541352t}$
   b. 14,868
   c. 2.7 days

15. a. $p(t) = 230,000e^{0.0497488314t}$
   b. October 2010
   c. $417,826
   d. $15,652 per month

17. a. $p(t) = 1.6e^{0.1899057139t}$
   b. $t = 9.65$ months from April 2009 February 2010

19. a. $p(t) = 109,478e^{0.1256876607t}$
   b. 180,996 thousand
   c. 232,723 thousand

21. a. $p(t) = 3,929,214e^{0.0300866701t}$
   b. 1810. The exponential growth can't continue forever.
   c. 1810: 7,171,916
       2000: 2,179,040,956

23. Africa: 32.9 years
North America: 55.3 years
Europe: 19.4 years

25. 49 years

27. a. Each year the house is worth 10% more than its current value (not the original value)
   b. 4.25 years

29. $P(t) = 18,731(1.0036837328)^t$

31. $P(t) = 9392(1.0104876491)^t$

35. 5060.3 days or 13.86 years

37. 1989 days

39. Most accurate is population of the world over a short time because no immigration, no emigration, little time for significant change in birth/death rate); least accurate is population of a specific country over a long time period.

10.2 Exponential Decay

1. 3.2 g
3. 7.4

5. a. $Q(t) = 50e^{-0.266595069t}$
   b. 38.3 mg
   c. 0.08 mg
   d. -11.7 mg per hour
   e. -0.234 per hour or -23.4% per hour.
   f. -2.1 mg per hour
   g. -0.042 per hour or -4.2% per hour.
   h. Radioactive substances decay faster when there is more substance present. The rate of decay is proportional to the amount present.

7. a. 21.0 hours   **b.** 42.0 hours   **c.** 63.0 hours

9. 30.2 years
11. 50.3 years
13. 81,055 years
15. 99.7 seconds

17. 53% of the original amount expected in a living organism

19. 63.5 days
21. 1,441 years old

23. 14,648 years old
25. 2,949 years old

27. Shroud made in 1350 A.D.: 92.6% of the original amount expected in a living organism
Shroud made in 33 A.D.: 78.9% of the original amount expected in a living organism

29. 50% of the original amount expected in a living organism

31. A 5,000 year old mummy should have about 55% of the original amount.
62% remaining carbon-14 would indicate approximately 3,950 years.
The museum's claim is not justified.

33. 8,300 years old

35.

t Hours After Injection (hr)	1	2	3	4	5
Q Portion Remaining (%)	0.891	0.794	0.708	0.631	0.562

37. 46.0 hours

39. a. From the article on pg. 777, there are 3.7×10^7 atomic disintegrations/second.
   b. 3.7×10^8 atomic disintegrations/second
   c. 7.4×10^8 atomic disintegrations/second

41. The amount of time it takes for half of the initial amount to disintegrate.

43. The units for average decay rate are for example grams per year. The units for relative decay rate are percent per year.

45. The assumption that the current ratio of C-14 to C-12 in the biosphere remains constant over time.

47. Becquerel, one radioactive disintegration per second.

49. Marie Curie, Pierre Curie and Antoine Henri Becquerel won the prize in 1903.

10.3 Logarithmic Scales

1. 7.6

3. Each recording yields a magnitude of 5.2

5. a. The 1906 earthquake's amplitude was almost 16 times that of the 1989 quake.
   b. The 1906 quake released about 55 times as much energy as the 1989 quake.

7. a. The San Francisco earthquake's amplitude was about 32 times that of the LA quake.
   b. The San Francisco quake released about 150 times as much energy as the LA quake.

9. a. The Indian Ocean earthquake's amplitude was about 126 times that of the 1989 San Francisco quake.
   b. The Indian Ocean quake released about 1,109 times as much energy as the 1989 San Francisco quake.

11. a. The New Madrid earthquake's amplitude was more than 158 times that of the Coalinga quake.
   b. The New Madrid quake released about 1,549 times as much energy as the Coalinga quake.

13. a. About a 26% increase.
   b. About a 40% increase.

15. Energy released is magnified by a factor of 28.

17. 70 dB
19. 74 dB
21. 10 dB gain

23. 9.1 dB gain

25. It requires 5 singers to reach the higher dB level. Thus, 4 singers joined him.

27. It requires 2 singers to reach the higher dB level. Thus, 1 singer joined her.

29. It requires 6 players to reach the higher dB level. Thus, 5 players joined in.

31. 3 dB gain

33. a. The American iPod's impact is about 1,000 times that of the European iPod.
b. American: less than 2 minutes, European: 2 hours

35. $I_1 = 10^{-3.5} I_2$

37. The two scales are very similar. They both use ratios of intensities. They both use logarithms to compress large variations into more manageable ranges.

39. California Institute of Technology

Chapter 10 Review

1. $x = 4$ **3.** $x = 5^5 = 3{,}125$

5. $\ln(e^x) = x$ and $e^{\ln x} = x$

7. $\ln\dfrac{A}{B} = \ln A - \ln B$

9. $\ln(A^n) = n \cdot \ln A$

11. $\ln(A \cdot B) = \ln A + \ln B$

13. $\log(x + 2)$

15. $x = 11.30719075$

17. $x = 447{,}213.6$

19. a. $p(t) = 300e^{-0.130782486t}$ **b.** They lose about 1.5 g
c. 263.2 g   **d.** 17.6 years

21. 3.1 on the Richter scale

23. 102 dB

25. It requires 4 players to reach the highest dB level. Thus, 3 trumpet players have joined in.

27. A logarithmic function has the form $y = \log x$, where b is a positive constant. Also, $\log_b u = v$ means $b^v = u$, $b \neq 1$.

29. Scientists measure the amount of C-14 and C-12 in an item and use this to estimate age. C-14 is radioactive so the amount decreases as the item ages.

31. He invented wax recorders for phonographs, the photophone and the audiometer.

33. John Napier. He wanted a way to easily calculate products, quotients, powers and roots.

35. Marie Curie. The first prize was for the discovery of radioactivity.

CHAPTER 11   Matrices and Markov Chains

11.0 Review of Matrices

1. a. 3×2 **b.** none of these

3. a. 2×1 **b.** column matrix

5. a. 1×2 **b.** row matrix

7. a. 3×3 **b.** square matrix

9. a. 3×1 **b.** column matrix

11. $a_{21} = 22$ **13.** $c_{21} = 41$

15. $e_{11} = 3$ **17.** $g_{12} = -11$

19. $j_{21} = -3$

21. a. $AC = \begin{bmatrix} 5 \cdot 23 + 0 \cdot 41 \\ 22 \cdot 23 - 3 \cdot 41 \\ 18 \cdot 23 + 9 \cdot 41 \end{bmatrix} = \begin{bmatrix} 115 \\ 383 \\ 783 \end{bmatrix}$

b. CA does not exist.

23. a. AD does not exist.   **b.** DA does not exist.

25. a. CG does not exist.   **b.** GC does not exist.

27. a. JB does not exist.

b. $BJ = \begin{bmatrix} 2243 \\ 1056 \\ 52 \end{bmatrix}$

29. a. $AF = \begin{bmatrix} -10 & 50 \\ -56 & 229 \\ 0 & 153 \end{bmatrix}$

b. FA does not exist.

31. Sale: $130, Regular: $170

33. Blondie: $3.45, Slice Man: $3.70

35. Jim: 3.07, Eloise: 3.00, Sylvie: 3.50

37. a. $\begin{bmatrix} -0.3 \\ 4.4 \end{bmatrix}$

b. This represents the change in sales from 2002 to 2003 for hotels and restaurants.

39. a. $\begin{bmatrix} 6.25 & 8.97 & 4.97 & 24.85 & 6.98 & 3.88 \\ 6.10 & 8.75 & 5.25 & 22.12 & 6.98 & 3.75 \end{bmatrix} \begin{bmatrix} 2 \\ 1 \\ 2 \\ 1 \\ 9 \\ 4 \end{bmatrix}$

b. The cost is $134.60 at Piedmont Lumber and $131.39 at Truitt and White.

41. a. $[53{,}594 \quad 64{,}393 \quad 100{,}237 \quad 63{,}198]$

b. $\begin{bmatrix} 0.9885 & 0.0015 & 0.0076 & 0.0024 \\ 0.0011 & 0.9901 & 0.0057 & 0.0032 \\ 0.0018 & 0.0042 & 0.9897 & 0.0043 \\ 0.0017 & 0.0035 & 0.0077 & 0.9870 \end{bmatrix}$

c. $[53{,}336 \quad 64{,}478 \quad 100{,}466 \quad 63{,}142]$

This is the new population in 2001.

43. $BC = \begin{bmatrix} 1 \\ 22 \end{bmatrix}$, so $A(BC) = \begin{bmatrix} 106 \\ 68 \end{bmatrix}$

$AB = \begin{bmatrix} 17 & 20 & -6 \\ -3 & 12 & -8 \end{bmatrix}$, so $(AB)C = \begin{bmatrix} 106 \\ 68 \end{bmatrix}$

45. $\begin{bmatrix} 3 & -2 \\ 4 & 0 \end{bmatrix}$

47. Does not exist

49. $\begin{bmatrix} 19 & 7 & 34 \\ 74 & 0 & -11 \\ 13 & -2 & 44 \end{bmatrix}$

51. Examples may vary. Possible answer:

$\begin{bmatrix} 2 & 3 \\ -5 & 4 \end{bmatrix} \cdot \begin{bmatrix} -6 & 1 \\ 5 & -4 \\ 1 & 3 \end{bmatrix}$

53. Because multiplying a matrix by the identity doesn't change any values.

55. Examples may vary. Possible answer: Let

$$A = \begin{bmatrix} 1 & 0 \\ 0 & 1 \end{bmatrix} \text{ and } B = \begin{bmatrix} 1 & 2 & 6 \\ 0 & 4 & 8 \end{bmatrix}.$$

$$AB = \begin{bmatrix} 1 & 0 \\ 0 & 1 \end{bmatrix}\begin{bmatrix} 1 & 2 & 6 \\ 0 & 4 & 8 \end{bmatrix} = \begin{bmatrix} 1 & 2 & 6 \\ 0 & 4 & 8 \end{bmatrix}$$

$$BA = \begin{bmatrix} 1 & 2 & 6 \\ 0 & 4 & 8 \end{bmatrix}\begin{bmatrix} 1 & 0 \\ 0 & 1 \end{bmatrix} = \text{ does not exist}$$

57. Answers will vary.

59. Because Sylvester was Jewish.

61. Sylvester taught mathematics at the University of Virginia.

63. $AB = \begin{bmatrix} 62 & 32 \\ -40 & 56 \end{bmatrix}$

65. a. $EF = \begin{bmatrix} -1178 & -2101 & -2378 \\ 5970 & -3091 & 498 \\ 5580 & -660 & 2340 \end{bmatrix}$

b. $FE = \begin{bmatrix} 1892 & -2520 \\ 10{,}723 & -3821 \end{bmatrix}$

67. a. $(BF)C = \begin{bmatrix} 109{,}348 & 23{,}813 & -16{,}663 \\ -23{,}840 & 1688 & -7720 \end{bmatrix}$

b. $B(FC) = \begin{bmatrix} 109{,}348 & 23{,}813 & -16{,}663 \\ -23{,}840 & 1688 & -7720 \end{bmatrix}$

69. a. $C^2 = \begin{bmatrix} 376 & 64 & 281 \\ -932 & -1203 & 614 \\ 952 & -808 & 293 \end{bmatrix}$

b. $C^5 = \begin{bmatrix} 4{,}599{,}688 & -2{,}586{,}492 & 9{,}810{,}101 \\ -1{,}887{,}820 & -24{,}086{,}567 & 2{,}293{,}357 \\ 42{,}244{,}296 & -7{,}140{,}820 & -4{,}471{,}911 \end{bmatrix}$

11.1 Markov Chains

1. a. 0.9
b. 0.2 = 20%
c. [0.22 0.78]
d. [0.568 0.432]
e.

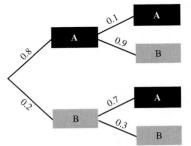

3. a. 0.8
b. 0.4 = 40%
c. [0.12 0.88]
d. [0.024 0.976]
e.

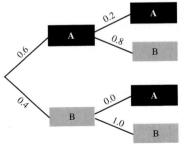

5. a. p(a cola drinker chooses KickKola) = 0.14
p(a cola drinker doesn't choose KickKola) = 0.86
b. [0.14 0.86]

7. a. p(health club user in Metropolis uses Silver's Gym) = 0.48
p(health club user in Metropolis uses Fitness Lab) = 0.37
p(health club user in Metropolis uses ThinNFit) = 0.15
b. [0.48 0.37 0.15]

9. a. p(next purchases is KickKola | current purchase is not KickKola) = 0.12
p(next purchase is not KickKola | current purchase is not KickKola) = 0.88
p(next purchase is KickKola | current purchase is KickKola) = 0.63
p(next purchase is not KickKola | current purchase is KickKola) = 0.37
b. The transition matrix is given by:
$$\begin{matrix} K & K' \\ \begin{bmatrix} 0.63 & 0.37 \\ 0.12 & 0.88 \end{bmatrix} & \begin{matrix} K \\ K' \end{matrix} \end{matrix}$$

11. a. p(go next to Silvers | go now to Silvers) = 0.71
p(go next to Fitness Lab | go now to Silvers) = 0.12
p(go next to ThinNFit | go now to Silvers) = 1 − 0.71 − 0.12 = 0.17
p(go next to Silvers | go now to Fitness Lab) = 0.32
p(go next to ThinNFit | go now to Fitness Lab) = 0.34
p(go next to Fitness lab | go now to Fitness Lab) = 1 − 0.32 − 0.34 = 0.34
p(go next to ThinNFit | go now to ThinNFit) = 0.96
0 p (go next to Silvers | go now to ThinNFit) = 0.02
p(go next to Fitness Lab | go now to ThinNFit) = 0.02
b. The transition matrix is given by:
$$\begin{matrix} S & F & T \\ \begin{bmatrix} 0.71 & 0.12 & 0.17 \\ 0.32 & 0.34 & 0.34 \\ 0.02 & 0.02 & 0.96 \end{bmatrix} & & \begin{matrix} S \\ F \\ T \end{matrix} \end{matrix}$$

13. a. Market share: 0.0882 + 0.1032 = 0.1914 ≈ 19%
b. Market share: 0.217614 ≈ 22%
c. Market share: 0.1914 ≈ 19%
d. Market share: 0.217614 ≈ 22%
e. Market share: 0.2431 ≈ 24%
f. Trees don't require any knowledge of matrices and are very visual. The matrices are quicker and don't require drawings.

15. a. Market share: Silver's: 0.4622 ≈ 46%
Fitness Lab: 0.1864 ≈ 19%
ThinNFit: 0.3514 ≈ 35%
b. Market share: Silver's = 0.4622 ≈ 46%, Fitness Lab = 0.1864 ≈ 19%, ThinNFit = 0.3514 ≈ 35%
c. Market share: Silver's = 0.3948 ≈ 39%, Fitness Lab = 0.1259 ≈ 13%, ThinNFit = 0.4793 ≈ 48%
d. Market share: Silver's = 0.3302 ≈ 33%, Fitness Lab = 0.0998 ≈ 10%, ThinNFit = 0.5700 ≈ 57%
e. Market share: Silver's = 0.2371 ≈ 24%, Fitness Lab = 0.0750 ≈ 8%, ThinNFit = 0.6879 ≈ 69% (More than 100% due to rounding.)

17. a. Homeowners: 39%; Renters: 61%
b. Homeowners: 45%; Renters: 55%
c. Homeowners: 51%; Renters: 49%

19. Market share: 0.2659 ≈ 27%

21. a. central cities: 0.313
   suburb: 0.462
   nonmetropolitan: 0.225
   $[0.313 \quad 0.462 \quad 0.225]$

b. $\begin{bmatrix} 0.927 & 0.064 & 0.009 \\ 0.022 & 0.970 & 0.008 \\ 0.013 & 0.019 & 0.968 \end{bmatrix}$

   c. City: 30.3%, Suburb: 47.2%, Nonmetropolitan: 22.4%
   d. City: 29.4%, Suburb: 48.2%, Nonmetropolitan: 22.4%
   e. Part (c). The percentages may change over time.
23. a. Cropland = 0.296
   High Potential = 0.091
   Low Potential = 0.613
   $[0.296 \quad 0.091 \quad 0.613]$

b. $\begin{bmatrix} 0.874 & 0.041 & 0.085 \\ 0.268 & 0.732 & 0 \\ 0 & 0 & 1 \end{bmatrix}$

   c. Cropland: 395 million acres, Potential: 110 million acres,
   No Potential: 891 million acres
   d. Cropland: 375 million acres, Potential: 97 million acres,
   No Potential: 925 million acres
25. No. Although the number of observations is finite, the probability at one time does not depend on the probability of another time.
27. Transition matrices. We can tell by the dimensions.
29. They would need the current market share and the probabilities that people will switch to a new product or stay with their current product. These could be obtained through sales records and surveys.
31. The tree method is visual and easy to implement. The matrix method is efficient and can easily be programmed into a computer.
33. The current state gives the current percentages. The following states give the future percentages.
35. Markov was motivated purely by theoretical concerns.
37. Markov organized a celebration of the 200th anniversary of the publishing of Jacob Bernoulli's book on probabilities.

11.2 Systems of Linear Equations
1. Solution   **3.** Not a solution
5. Not a solution
7. a. $2y = -5x + 4$:   slope $= -\frac{5}{2}$ y − int $= 2$
   $-19y = -6x + 72$:   slope $= \frac{6}{19}$ y − int $= -\frac{72}{19}$
b. One solution
9. a. $3y = -4x + 12$:   slope $= -\frac{4}{3}$ y − int $= 4$
   $6y = -8x + 24$:   slope $= -\frac{4}{3}$ y − int $= 4$
b. Infinite number of solutions
11. a. $y = -x + 7$:   slope $= -1$ y − int $= 7$
   $2y = -3x + 8$:   slope $= -\frac{3}{2}$ y − int $= 4$
   $2y = -2x + 14$:   slope $= -1$ y − int $= 7$
b. One solution
13. This system could have a single solution since it has 3 equations and 3 unknowns.
15. This system could not have a single solution because the first and second equations are equivalent.
17. This system could not have a single solution because there are less equations than unknowns.
25. The system could have no solution, one solution or an infinite number of solutions. If the equations are unique, it could have no solution or one solution.

27. Graphs aren't accurate.
29. They are the same line. The same sets of ordered pairs will solve each.
37. The graph would be two planes. TI-83, TI-84, or TI-86 cannot graph planes. The TI-89 can only graph one plane at a time on the screen so it would be difficult to discern the relationship between the two planes.

11.3 Long-Range Predictions with Markov Chains
5. Market share: 24%
7. 20% will rent, 80% will own
   We assume that the trend won't change and that the residents' moving plans are realized.
9. Sierra Cruiser will eventually control 26% of the market.

11.4 Solving Larger Systems of Equations
1. $(4, -2, 1)$ **3.** $(2, 5, 7)$
5. $(5, 10, -2)$ **7.** $(3, 5, 7)$
9. $(4, 0, -3)$ **11.** $(5, -6, 0)$
13. $(4, -2, 1)$ **15.** $(2, 5, 7)$
17. $(5, 10, -2)$ **19.** $(3, 5, 7)$
21. $(4, 0, -3)$ **23.** $(5, -6, 0)$
25. $(5, -4)$ **27.** $(2, 3)$

11.5 More on Markov Chains
1. a. $L = \begin{bmatrix} \frac{11}{52} & \frac{29}{52} & \frac{3}{13} \end{bmatrix} \approx [0.2115 \quad 0.5577 \quad 0.2308]$
3. a. $L = [0\ 0\ 1]$
5. Silver's 10.8%, Fitness Lab 4.5%, ThinNFit 84.6%
7. Safe Shop 48.6%, PayNEat 37.6%, other markets 13.8%
9. a. The transition matrix, with columns: CS, S, NM:

$\begin{bmatrix} 0.927 & 0.064 & 0.009 \\ 0.022 & 0.970 & 0.008 \\ 0.013 & 0.019 & 0.968 \end{bmatrix}$

   b. Central City: 21.2%, Suburb: 58.3%, Non-metropolitan: 20.5%
11. a. bottom 10%: 8.4%
   10% to 50%: 41.6%
   50% to 90%: 41.6%
   top 10%: 8.4%
   b. No. Sons may do better or worse than their fathers.
   c. It assumes that each father has only 1 son. It ignores the effects of mothers' salaries.

Chapter 11 Review
1. Neither; the dimensions are 3×2.
3. Column; the dimensions are 4×1.
5. $\begin{bmatrix} -3 & -24 \\ -24 & 28 \end{bmatrix}$
7. Does not exist
9. $\begin{bmatrix} 35 & -19 & -72 & 24 \\ 12 & 34 & 51 & 24 \end{bmatrix}$
11. NYC: $43,000, DC: $41,300
13. a. $(2, -4)$
   b. Substituting $x = 2$ and $y = -4$ into the original equations yields true statements.
15. a. $(-3, 2)$
   b. Substituting $x = -3$ and $y = 2$ into the original equations yields true statements.

17. $(1, 7, 2)$

   b. Substituting $x = 1$, $y = 7$, and $z = 2$ into the original equations yields true statements.

19. a. 0.7   **b.** 0.37

   c. [0.226 0.774]   **d.** [0.1452 0.8548]

   e.

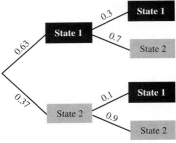

21. a. $L = [0.4 \quad 0.6]$

   b. $LT = [0.4 \quad 0.6]\begin{bmatrix} 0.7 & 0.3 \\ 0.2 & 0.8 \end{bmatrix} = [0.4 \quad 0.6]$

23. a. $L = \begin{bmatrix} \frac{1}{7} & \frac{6}{7} \end{bmatrix}$

   b. $LT = \begin{bmatrix} \frac{1}{7} & \frac{6}{7} \end{bmatrix}\begin{bmatrix} 0.4 & 0.6 \\ 0.1 & 0.9 \end{bmatrix} = \begin{bmatrix} \frac{1}{7} & \frac{6}{7} \end{bmatrix}$

25. a. $L = \begin{bmatrix} \frac{2}{5} & \frac{1}{5} & \frac{2}{5} \end{bmatrix}$

   b. $LT = \begin{bmatrix} \frac{2}{5} & \frac{1}{5} & \frac{2}{5} \end{bmatrix}\begin{bmatrix} 0.5 & 0.1 & 0.4 \\ 0.2 & 0.2 & 0.6 \\ 0.4 & 0.3 & 0.3 \end{bmatrix} = \begin{bmatrix} \frac{2}{5} & \frac{1}{5} & \frac{2}{5} \end{bmatrix}$

27. Market share: 21.5%

29. Long term prediction 29.6%.

31. Apartment: 22.1%, Condo/Townhouse: 19.6%, House: 58.3%

   Thus, more condominiums and townhouses will be needed.

33. Emigration and immigration, also affordability based or economy. We don't know if the prospective buyers can afford houses. It ignores economic conditions.

CHAPTER 12   Linear Programming

12.0   Review of Linear Inequalities

1.

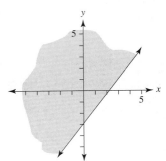

3.

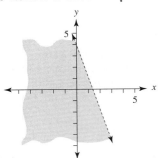

5.

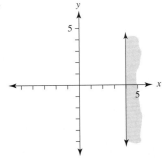

7.

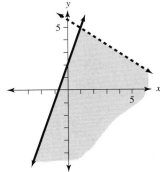

9. a.

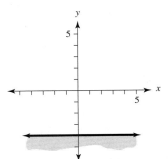

   b. Unbounded   **c.** (1, 3)

11. a.

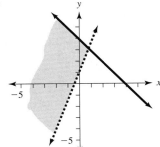

b. Unbounded

c. $(1, 5)$

13. a.

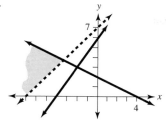

b. Unbounded

c. $\left(-3\frac{1}{3}, 3\frac{2}{3}\right)$

15. a.

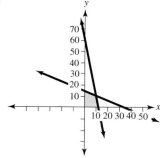

b. Bounded

c. Corner Points: $(0, 14), (0, 0), (12, 0), (10, 10)$

17. a.

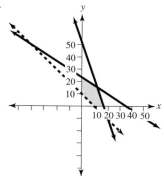

b. Bounded

c. Corner Points:

$(0, 10), \left(0, 23\frac{2}{11}\right), \left(17\frac{1}{7}, 0\right), (10, 0), (12, 15)$

19. a.

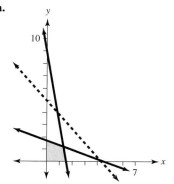

b. Bounded

c. Corner Points: $(0, 0), (0, 1.7), (1.5, 0), (1.3, 1.2)$

21. a.

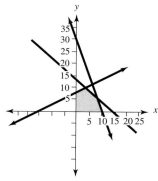

b. Bounded

c. Corner Points: $(0, 0), (0, 8), (10, 0), (4, 10), (8, 6)$

23. Answers will vary.   **25.** Answers will vary.

27. The petroleum industry uses it to blend gasoline and decide what crude oil to buy. The steel industry uses it to evaluate ores and when to build new furnaces. Airlines use it to minimize costs related to scheduling flights.

29–49: See Exercises 1-21.

12.1 The Geometry of Linear Programming

1. Let x = number of shrubs

y = number of trees

$1x + 3y \leq 100$

3. Let x = number of hardbacks

y = number of paperbacks

$4.50x + 1.25y \geq 5,000$

5. Let x = number of refrigerators

y = number of dishwashers

$63x + 41y \leq 1,650$

7. She should make 45 table lamps and no floor lamps. The profit is $1,485.

9. They should make 60 pounds of Morning Blend and 120 pounds of South American Blend. The profit is $480.

11. They should send 15,000 loaves to Shopgood and 20,000 loaves to Rollie's. The cost is $3,000.

13. Global has many choices, including the following:

15 Orvilles and 30 Wilbers

48 Orvilles and 8 Wilburs

Each of these generates a cost of at least $720,000. Another way to express the answer would be any point on the line $y = -\frac{2}{3}x + 40$ produces a minimum of $720,000.

15. a. He should order 20 large refrigerators and 30 smaller refrigerators. The profit is $9,500.

b. He should order 40 large refrigerators and no small refrigerators. The profit is $10,000.

c. 40

17. 0 Mexican; 0 Columbian

19. 150 minutes unused; $0

21. a. There is no maximum.

b. No maximum

c. $(6, 0)$

d. $z = 12$

23. They should use 3 weeks of production in Detroit and 6 weeks of production in Los Angeles. The cost is $3,780,000.

25. Customers might be upset about the loss of their favorite model. The firm could set small maximum production constraints for the low selling models.

27. If two points satisfy an objective function $C = Ax + By$, then $C = Ax + By$ is the equation of the line connecting the two points.

29. The prices could be made variable. This means that the coefficients of the constraint equations and the objective function would be variable instead of constant.

31. Answers will vary.   **33.** George Bernard Shaw.

35. Dantzig was hired by the Air Force to find ways to distribute personnel, weapons and supplies.

37. a. The demand for half-inch plate glass made at both plants (v = half-inch glass made at Clear View and x = half-inch glass made at Panesville) is 5,000 sheets.
$v + x \leq 5,000$
The demand for three-eighths-inch-thick plate glass made at both plants (w = three-eighths-inch glass at Clear View and y = three-eighths-inch glass at Panesville) is 6,000 sheets.
$w + y \leq 6,000$

b. The Clear View plant (v and w) is capable of making at most 7,000 sheets.
$v + w \leq 7,000$
The Panesville plant (x and y) is capable of making at most 5,000 sheets.
$x + y \leq 5,000$

c. Clear View: $820 per sheet for half-inch glass (v)
$760 per sheet for three-eighths-inch glass (w)
Panesville: $745 per sheet for half-inch glass (x)
$685 per sheet for three-eighths-inch glass (y)
$z = 820v + 760w + 745x + 685y$

d. There are four variables.

Chapter 12 Review

1.

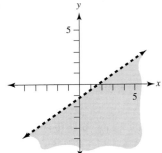

3.

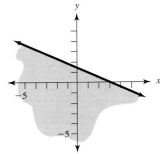

5. Corner Point: $(\frac{3}{2}, \frac{1}{2})$

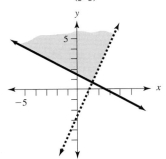

7. Corner Point: $(-3, -23)$

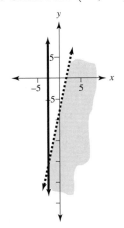

9. Corner Points: $(\frac{9}{5}, 0), (0, 3)$

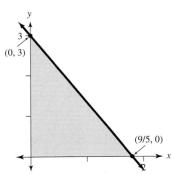

11. The minimum value of z is 0 at $(0, 0)$.

13. The minimum value of z is 25.12 at $(\frac{140}{41}, \frac{250}{41})$.

15. They should make 30 of each assembly for a maximum income of $16,500.

17. Answers will vary. Possible answer: Think of the objective function as part of a 'profit line.' For instance, if the profit was to be 100, all points in the feasible region that satisfy the equation, objective function = 100, would make a profit line. Increasing the profit, would produce another profit line. The farther the profit lines move away from the origin, the higher the profit represented by the line. The highest profit line would be located at a corner point or along a line segment joining two corners as that is the farthest the line could go and still have points in the feasible region.

19. George Dantzig

INDEX

Networks, 696–713
  adding a vertex, 699
  installing, 696–697
  Kruskal's algorithm for finding a
    minimum spanning tree, 698–699, 726
  right triangle trigonometry, 703–704
  shortest edges first, 697
  Steiner points, 700–703
  trees, 697–698
  where to add a vertex, 699–700
Nevada, map of, 709
*New Elements of Geometry, with a
  Complete Theory of Parallels*
  (Lobachevsky), 621
New states paradox, 462–463
*New Technique for Objective Methods for
  Measuring Reader Interest in
  Newspapers, A* (Gallup), 298
Newton, Isaac, 13-38–13-39
  antiderivative and areas, 13-72–13-75
  area of any shape, 13-68–13-76
  derivative, 13-46–13-52
  falling objects, 13-42–13-43
  gravity, 13-43–13-46
  method for finding slope of tangent line,
    13-35–13-37
  tangent lines and, 13-35–13-42
New York, map of, 710
90 degrees, 580
Noise-induced hearing loss, 800
Nominal rate, 347
Non-Euclidean geometries
  in art, 624–626
  Bolyai, Janos and, 619–620
  comparison of triangles and, 624
  Gauss, Carl Friedrich and, 619
  geometric models, 622–623
  Lobachevsky, Nikolai, and, 620–621
  parallel postulate, 618
  Riemann, Bernhard and, 621–622
  Saccheri, Girolamo, and, 618–619
Normal distribution
  bell-shaped, 277–278
  body table, 283
  converting to standard normal from,
    288–292
  defined, 277–278, 279–280
  discrete versus continuous variables, 278
  probability and area in, 280–281
  standard, 282–288
  tail, end of bell-shaped curve, 283,
    284–285
  z-distribution, 283
Notation
  capital letters for money, 331
  cardinal number of a set, 69
  element of a set, 69
  empty set, 71
  equal set, 70
  roster, 69
  set-builder, 70
  summation, 248

Note. *See* Loan
Nuclear medicine, radioactivity and, 784
Null sets and probability, 158
Number of days, 337
Number Systems and Number Theory,
  473–525
  Arabic mathematics, 477–478
  arithmetic in different bases, 490–500
  bases, 478–488
  composite numbers, 501–504
  converting between computer bases,
    484–486
  converting number bases (*see* Converting
    number bases)
  Fibonacci numbers and the golden ratio,
    513–523
  Hindu number system, 476
  place systems, 475–490
  prime numbers and perfect numbers
    Prime numbers); (*see* Perfect numbers)
  reading numbers in different bases, 478

O

Obama family tree, 667–668
Objective function, 872
Octagon, 529
Octal system, 475, 478
  addition, 492
  converting to/from base sixteen, 488
  converting to/from base two, 484–486
  division, 498–500
  multiplication, 496–498
  subtraction, 494
Odds and probabilities, 141–143
  odds dying due to accident or injury,
    table of, 154
Odd vertex, 671
O'Keefe, Georgia, 604
One-bit image, 487
One-point perspective, 600
One-to-one correspondence between sets,
  118–120
"Only if" connective, 46–47
On Poetry. A Rhapsody (Swift), 630–631
On the Foundations of Geometry
  (Lobachevsky), 621
Optical and Geometrical Lectures
  (Barrow), 13-28
Optimize, 860
Ordinary annuity, 358–359
Oresme, Nicole, 13-17
  distance traveled by falling object,
    13-24–13-26
  Oresme's triangle, 13-25
Organon (Aristotle), 7
Outcomes in probability, 140
Outlier, 252
Outstanding principal, 371

P

Pacioli, Luca, 518
Pair-Oared Shell, The (Eakins), 605

Pairwise comparison voting system,
  418–421
Palm, 558
Parabola
  in analytic geometry, 608–610
  in calculus, 13-6–13-7
  as conic section, 606
  defined, 608–609
  focus and, 609
  line of symmetry and, 13-6
  vertex of, 13-6
Parallel lines, 622
Parallelogram, 529
  area of, 531
Parallel Postulate, 618
Parallel tasks, 714, 715
Parsec (pc), 539, 590
Parthenon and the golden rectangle, 520
Pascal, Blaise, 24, 110, 112, 135
Pascal's triangle, 110, 112
  Sierpinski Gasket and, 634–635
Pasquale, Don, 133
Pavements, 598
Payment period of annuity, 357
Payout annuities, 394–402
  formula, 397
  formula for annual payout with COLA
    formula, 398–400
  long-term, 396–398
  savings annuities versus, 395–396
  short-term, 394–395
Percentage, writing probability as, 173
Perfect numbers, 508, 509
  Mersenne primes and, 507
Perimeter
  defined, 530
  of fractals, 648–657
  of Koch snowflake, 649–651
  of polygons, 530
Periodic rate, 343
Permutations
  counting technique, 105–106
  defined, 103
  distinguishable, 114
  formula, 104
  identical items and, 113–114
  with versus without replacement,
    102–104
PERT chart, 714, 716–718
  and NASA, 720–721
Perugino, 596, 600
Philosophical Principles (Descartes), 13-20
pi, 536, 561, 575, 733–734
  area of a circle, 563–565
Picture cards, 137
Pie charts, 233–234
  categorical data for, 233
  computerized spreadsheet for, 241–247
  data preparation for, 242–243
  drawing a, 244–245
Pilkington Libbey-Owens-Ford, 882
Pink Floyd, 803